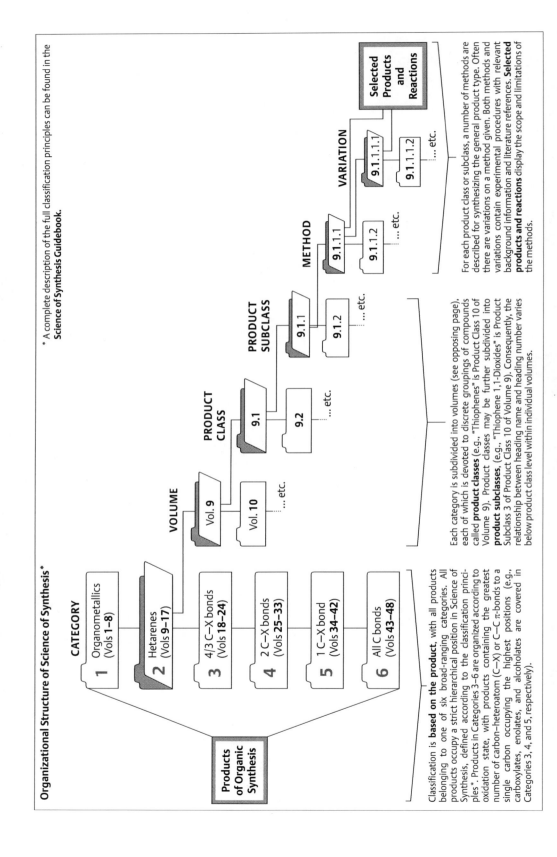

Science of Synthesis Reference Library

The **Science of Synthesis Reference Library** comprises volumes covering special topics of organic chemistry in a modular fashion, with six main classifications: (1) Classical, (2) Advances, (3) Transformations, (4) Applications, (5) Structures, and (6) Techniques. Volumes in the **Science of Synthesis Reference Library** focus on subjects of particular current interest with content that is evaluated by experts in their field. **Science of Synthesis**, including the **Knowledge Updates** and the **Reference Library**, is the complete information source for the modern synthetic chemist.

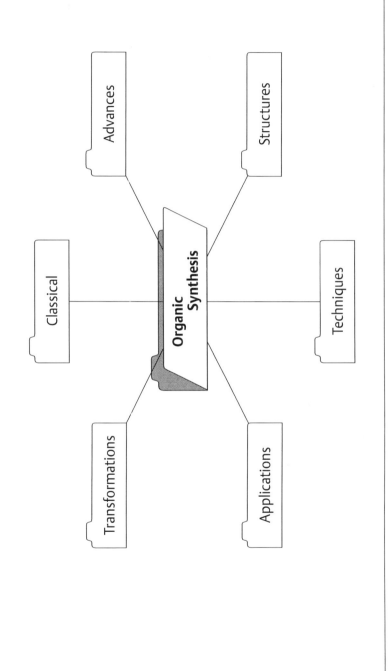

Science of Synthesis

Science of Synthesis is the authoritative and comprehensive reference work for the entire field of organic and organometallic synthesis.

Science of Synthesis presents the important synthetic methods for all classes of compounds and includes:
- Methods critically evaluated by leading scientists
- Background information and detailed experimental procedures
- Schemes and tables which illustrate the reaction scope

 Science of Synthesis

Editorial Board	E. M. Carreira	E. Schaumann
	C. P. Decicco	M. Shibasaki
	A. Fuerstner	E. J. Thomas
	G. Koch	B. M. Trost
	G. A. Molander	
Managing Editor	M. F. Shortt de Hernandez	
Senior Scientific Editors	K. M. Muirhead-Hofmann	
	T. B. Reeve	
	A. G. Russell	
Scientific Editors	E. L. Hughes	M. J. White
	J. S. O'Donnell	F. Wuggenig
	E. Smeaton	

Georg Thieme Verlag KG
Stuttgart · New York

 # Science of Synthesis

Applications of Domino Transformations in Organic Synthesis 1

Volume Editor	S. A. Snyder
Responsible Member of the Editorial Board	E. Schaumann

Authors:

D. Adu-Ampratwum
E. A. Anderson
K. W. Armbrust
J. J. Devery, III
J. J. Douglas
M. P. Doyle
K. M. Engle
C. J. Forsyth
F. Gille
T. Halkina
X. Hu
T. F. Jamison

E. H. Kelley
A. Kirschning
D. Lee
T. J. Maimone
E. Merino
C. Nevado
M. O'Connor
T. Ohshima
K. A. Parker
H. Renata
A. Salvador
R. A. Shenvi

L. Shi
S. Sittihan
C. R. J. Stephenson
M. Tang
P. Truong
Y.-Q. Tu
K. K. Wan
S.-H. Wang
M. Wolling
X. Xu
Z. Yang

2016
Georg Thieme Verlag KG
Stuttgart · New York

© 2016 Georg Thieme Verlag KG
Rüdigerstrasse 14
D-70469 Stuttgart

Printed in Germany

Typesetting: Ziegler + Müller, Kirchentellinsfurt
Printing and Binding: AZ Druck und Datentechnik
GmbH, Kempten

Bibliographic Information published by
Die Deutsche Bibliothek

Die Deutsche Bibliothek lists this publication in the
Deutsche Nationalbibliografie; detailed bibliographic
data is available on the internet at <http://dnb.ddb.de>

Library of Congress Card No.: applied for

British Library Cataloguing in Publication Data

A catalogue record for this book is available from the
British Library

Date of publication: May 11, 2016

Copyright and all related rights reserved, especially
the right of copying and distribution, multiplication
and reproduction, as well as of translation. No part of
this publication may be reproduced by any process,
whether by photostat or microfilm or any other procedure, without previous written consent by the pub-
lisher. This also includes the use of electronic media
of data processing or reproduction of any kind.

This reference work mentions numerous commercial
and proprietary trade names, registered trademarks
and the like (not necessarily marked as such), patents,
production and manufacturing procedures, registered
designs, and designations. The editors and publishers
wish to point out very clearly that the present legal situation in respect of these names or designations or
trademarks must be carefully examined before making any commercial use of the same. Industrially produced apparatus and equipment are included to a necessarily restricted extent only and any exclusion of
products not mentioned in this reference work does
not imply that any such selection of exclusion has
been based on quality criteria or quality considerations.

Warning! Read carefully the following: Although
this reference work has been written by experts, the
user must be advised that the handling of chemicals,
microorganisms, and chemical apparatus carries potentially life-threatening risks. For example, serious
dangers could occur through quantities being incor-
rectly given. The authors took the utmost care that
the quantities and experimental details described
herein reflected the current state of the art of science
when the work was published. However, the authors,
editors, and publishers take no responsibility as to the
correctness of the content. Further, scientific knowledge is constantly changing. As new information becomes available, the user must consult it. Although
the authors, publishers, and editors took great care in
publishing this work, it is possible that typographical
errors exist, including errors in the formulas given
herein. Therefore, **it is imperative that and the responsibility of every user to carefully check
whether quantities, experimental details, or other information given herein are correct based on
the user's own understanding as a scientist.** Scale-
up of experimental procedures published in **Science
of Synthesis** carries additional risks. In cases of doubt,
the user is strongly advised to seek the opinion of an
expert in the field, the publishers, the editors, or the
authors. When using the information described herein, the user is ultimately responsible for his or her
own actions, as well as the actions of subordinates
and assistants, and the consequences arising there-
from.

ISBN 978-3-13-173141-8

Preface

As the pace and breadth of research intensifies, organic synthesis is playing an increasingly central role in the discovery process within all imaginable areas of science: from pharmaceuticals, agrochemicals, and materials science to areas of biology and physics, the most impactful investigations are becoming more and more molecular. As an enabling science, synthetic organic chemistry is uniquely poised to provide access to compounds with exciting and valuable new properties. Organic molecules of extreme complexity can, given expert knowledge, be prepared with exquisite efficiency and selectivity, allowing virtually any phenomenon to be probed at levels never before imagined. With ready access to materials of remarkable structural diversity, critical studies can be conducted that reveal the intimate workings of chemical, biological, or physical processes with stunning detail.

The sheer variety of chemical structural space required for these investigations and the design elements necessary to assemble molecular targets of increasing intricacy place extraordinary demands on the individual synthetic methods used. They must be robust and provide reliably high yields on both small and large scales, have broad applicability, and exhibit high selectivity. Increasingly, synthetic approaches to organic molecules must take into account environmental sustainability. Thus, atom economy and the overall environmental impact of the transformations are taking on increased importance.

The need to provide a dependable source of information on evaluated synthetic methods in organic chemistry embracing these characteristics was first acknowledged over 100 years ago, when the highly regarded reference source **Houben–Weyl Methoden der Organischen Chemie** was first introduced. Recognizing the necessity to provide a modernized, comprehensive, and critical assessment of synthetic organic chemistry, in 2000 Thieme launched **Science of Synthesis, Houben–Weyl Methods of Molecular Transformations**. This effort, assembled by almost 1000 leading experts from both industry and academia, provides a balanced and critical analysis of the entire literature from the early 1800s until the year of publication. The accompanying online version of **Science of Synthesis** provides text, structure, substructure, and reaction searching capabilities by a powerful, yet easy-to-use, intuitive interface.

From 2010 onward, **Science of Synthesis** is being updated quarterly with high-quality content via **Science of Synthesis Knowledge Updates**. The goal of the **Science of Synthesis Knowledge Updates** is to provide a continuous review of the field of synthetic organic chemistry, with an eye toward evaluating and analyzing significant new developments in synthetic methods. A list of stringent criteria for inclusion of each synthetic transformation ensures that only the best and most reliable synthetic methods are incorporated. These efforts guarantee that **Science of Synthesis** will continue to be the most up-to-date electronic database available for the documentation of validated synthetic methods.

Also from 2010, **Science of Synthesis** includes the **Science of Synthesis Reference Library**, comprising volumes covering special topics of organic chemistry in a modular fashion, with six main classifications: (1) Classical, (2) Advances, (3) Transformations, (4) Applications, (5) Structures, and (6) Techniques. Titles will include *Stereoselective Synthesis*, *Water in Organic Synthesis*, and *Asymmetric Organocatalysis*, among others. With expert-evaluated content focusing on subjects of particular current interest, the **Science of Synthesis Reference Library** complements the **Science of Synthesis Knowledge Updates**, to make **Science of Synthesis** the complete information source for the modern synthetic chemist.

The overarching goal of the **Science of Synthesis** Editorial Board is to make the suite of **Science of Synthesis** resources the first and foremost focal point for critically evaluated information on chemical transformations for those individuals involved in the design and construction of organic molecules.

Throughout the years, the chemical community has benefited tremendously from the outstanding contribution of hundreds of highly dedicated expert authors who have devoted their energies and intellectual capital to these projects. We thank all of these individuals for the heroic efforts they have made throughout the entire publication process to make **Science of Synthesis** a reference work of the highest integrity and quality.

The Editorial Board   July 2010

E. M. Carreira (Zurich, Switzerland)   E. Schaumann (Clausthal-Zellerfeld, Germany)
C. P. Decicco (Princeton, USA)   M. Shibasaki (Tokyo, Japan)
A. Fuerstner (Muelheim, Germany)   E. J. Thomas (Manchester, UK)
G. A. Molander (Philadelphia, USA)   B. M. Trost (Stanford, USA)
P. J. Reider (Princeton, USA)

Science of Synthesis Reference Library

Applications of Domino Transformations in Organic Synthesis (2 Vols.)
Catalytic Transformations via C—H Activation (2 Vols.)
Biocatalysis in Organic Synthesis (3 Vols.)
C-1 Building Blocks in Organic Synthesis (2 Vols.)
Multicomponent Reactions (2 Vols.)
Cross Coupling and Heck-Type Reactions (3 Vols.)
Water in Organic Synthesis
Asymmetric Organocatalysis (2 Vols.)
Stereoselective Synthesis (3 Vols.)

Volume Editor's Preface

Domino reactions have been a mainstay of synthetic chemistry for much of its history. Domino chemistry's roots trace to achievements such as the one-pot synthesis of tropinone in 1917 by Robinson and the generation of steroidal frameworks through polyene cyclizations, as originally predicted by the Stork–Eschenmoser hypothesis. In the ensuing decades, chemists have used these, and other inspiring precedents, to develop even more complicated domino sequences that rapidly and efficiently build molecular complexity, whether in the form of natural products, novel pharmaceuticals, or materials such as buckminsterfullerene.

Despite this body of achievements, however, the development of such processes remains a deeply challenging endeavor. Indeed, effective domino chemistry at the highest levels requires not only creativity and mechanistic acumen, but also careful planning at all stages of a typical experiment, from substrate design, to reagent and solvent choice, to timing of additions, and even the quench. Thus, if the frontiers are to be pushed even further, there is certainly much to master.

It was with these parameters in mind that the Editorial Board of **Science of Synthesis** decided to focus one of its Reference Library works on domino chemistry, covering the myriad ways that these sequences can be achieved with the full array of reactivity available, whether in the form of pericyclic reactions, radical transformations, anionic and cationic chemistry, metal-based cross couplings, and combinations thereof. In an effort to provide a unique approach in organizing and presenting such transformations relative to other texts and reviews on the subject, the sections within this book have been organized principally by the type of reaction that initiates the sequence. Importantly, only key and representative examples have been provided to highlight the best practices and procedures that have broad applicability. The hope is that this structure will afford a clear sense of current capabilities as well as highlight areas for future development and research.

A work on such a vibrant area of science would not have been possible, first and foremost, without a talented and distinguished author team. Each is mentioned in the introductory chapter, and I wish to thank all of them for their professionalism, dedication, and expertise. I am also grateful to all of the coaching, advice, and assistance provided by Ernst Schaumann, member of the Editorial Board of **Science of Synthesis**. Deep thanks also go, of course, to the entire editorial team at Thieme, particularly to Robin Padilla and Karen Muirhead-Hofmann who served as the scientific editors in charge of coordinating this reference work; Robin started the project, and Karen saw it through to the end. Their attention to detail and passion to produce an excellent final product made this project a true pleasure. Last, but not least, I also wish to thank my wife Cathy and my son Sebastian for their support of this project over the past two years.

Finally, I wish to dedicate this work, on behalf of the chapter authors and myself, to our scientific mentors. It was through their training that we learned how to better understand reactivity, propose novel chemistry, and identify the means to actually bring those ideas to fruition. Hopefully this text will serve the same role to those who study its contents, with even greater wisdom achieved as a result.

Scott A. Snyder   Chicago, October 2015

Abstracts

--- p 13 ---

1.1 Polyene Cyclizations
R. A. Shenvi and K. K. Wan

A domino transformation consists of a first chemical reaction enabling a second reaction, which can then effect a third reaction, and so on, all under the same reaction conditions. A polyene cyclization is defined as a reaction between two or more double bonds contained within the same molecule to form one or more rings via one or more C–C bond-forming events. Herein, domino polyene cyclizations are discussed, with an emphasis on operationally simple methods of broad utility. From the perspective of synthesis theory, polyene cyclizations are a powerful approach for the efficient generation of both complexity and diversity, with the potential for a single synthetic route to generate a series of both constitutional and stereochemical isomers. However, with some noteworthy exceptions, the ability to controllably cyclize a linear chain to multiple products with high selectivity still generally eludes synthetic chemists and represents a significant chemical frontier for further development.

Keywords: polyenes · cyclization · carbocations · radicals · polycycles

--- p 43 ---

1.2 Cation–π Cyclizations of Epoxides and Polyepoxides
K. W. Armbrust, T. Halkina, E. H. Kelley, S. Sittihan, and T. F. Jamison

This chapter describes the formation of complex polycyclic fragments from linear epoxide and polyepoxide precursors via domino reactions. Depending on the reaction conditions employed, either *exo* or *endo* epoxide opening can be selectively achieved. Applications of these domino reactions toward the synthesis of complex natural products are discussed.

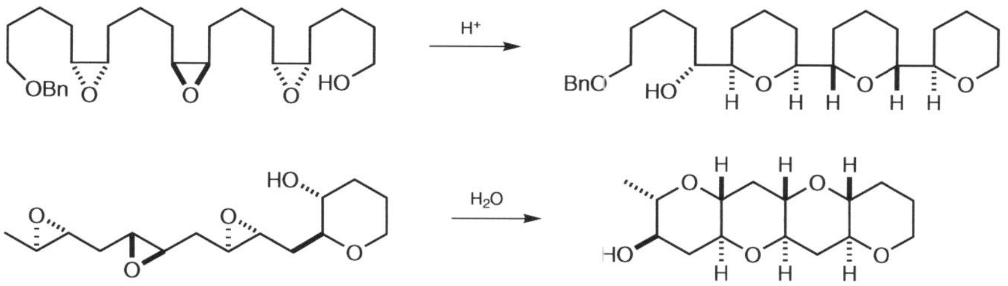

Keywords: oxiranes · cascades · natural products · marine ladder polyethers · ionophores · ethers · oxygen heterocycles · tetrahydrofurans · tetrahydropyrans · oxepanes

XII Abstracts

— p 67 —

1.3.1 **Enyne-Metathesis-Based Domino Reactions in Natural Product Synthesis**
D. Lee and M. O'Connor

Enyne-metathesis-based domino processes are highlighted in the context of natural product synthesis; these include domino double ring-closing metathesis, enyne metathesis/metallotropic [1,3]-shifts, enyne metathesis/Diels–Alder reaction, and other variations of their domino combinations. Issues regarding selectivity and mechanism are also discussed.

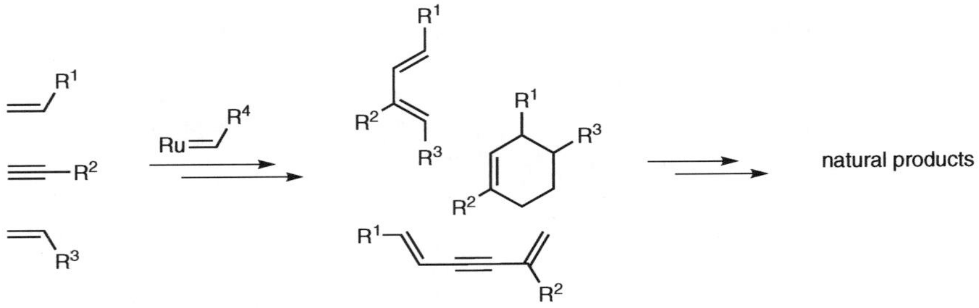

Keywords: enyne metathesis · π-bond exchange · domino transformations · natural products · total synthesis

— p 135 —

1.3.2 **Domino Metathesis Reactions Involving Carbonyls**
H. Renata and K. M. Engle

This review describes different methods to perform net carbonyl–alkene metathesis. Reactions of this type generally involve domino transformations employing organometallic reagents. Different conditions and procedures are surveyed and strategic applications of carbonyl–alkene metathesis in the synthesis of natural products are highlighted.

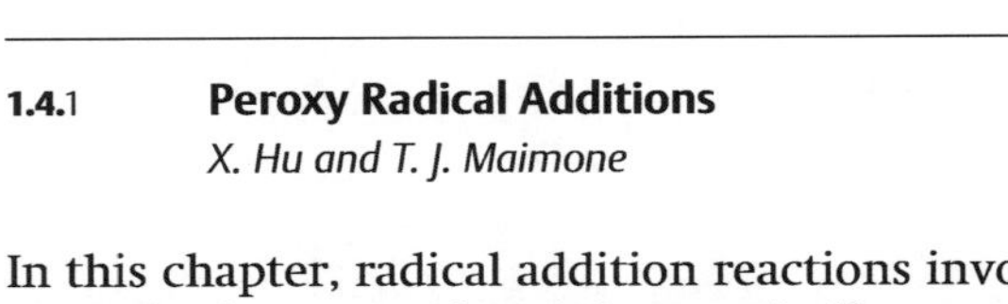

Keywords: metathesis · alkenylation · carbonyl compounds · alkenes · ring closure · transition metals · titanium complexes · organometallic reagents · organocatalysts

— p 157 —

1.4.1 **Peroxy Radical Additions**
X. Hu and T. J. Maimone

In this chapter, radical addition reactions involving peroxy radical intermediates are reviewed. These transformations typically generate a carbon radical intermediate which then reacts with molecular oxygen forming a peroxy radical species. Following peroxy radical cyclization, various endoperoxide rings are constructed. Two major classes of reactions are discussed: (1) radical additions to alkenes and quenching with molecular oxygen, and (2) radical formation from the opening of cyclopropanes and incorporation of molecular oxygen. Various methods for radical initiation that are compatible with the presence of molecular oxygen are described.

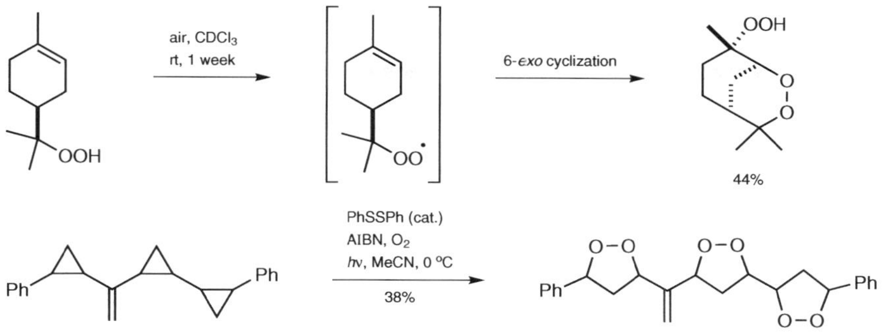

Abstracts

Keywords: peroxide synthesis · endoperoxides · cyclic peroxides · radical addition · peroxy radicals · thiyl radicals · hydroperoxidation · cyclopropane cleavage · 1,2-dioxolanes · 1,2-dioxanes · 1,2-dioxepanes

— p 187 —

1.4.2 **Radical Cyclizations**
J. J. Devery, III, J. J. Douglas, and C. R. J. Stephenson

This chapter details recent examples of domino radical reactions that are initiated via an intramolecular radical cyclization.

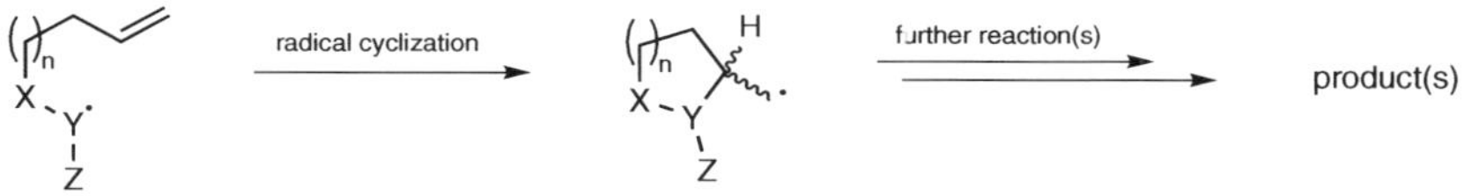

Keywords: radicals · domino reactions · cyclization · tin · samarium · organo-SOMO · ammonium cerium(IV) nitrate (CAN) · visible light

— p 217 —

1.4.3 **Tandem Radical Processes**
K. A. Parker

This review presents selected examples of regio- and stereospecific domino radical reactions developed in the context of total synthesis studies. The underlying strategies demonstrate the variety of connectivity patterns that can be generated by cascades of intra- and intermolecular bond-forming steps.

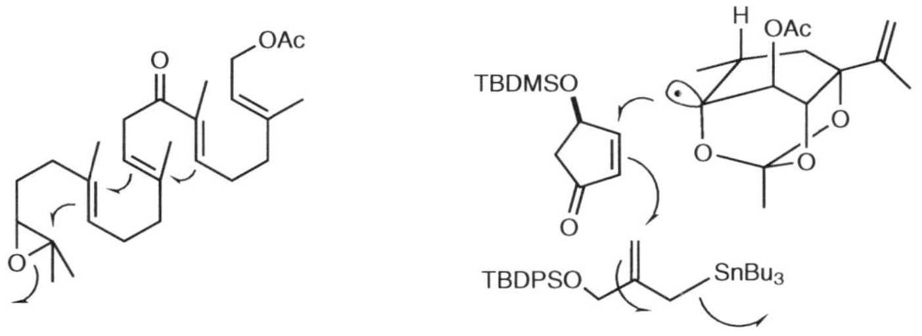

Keywords: tandem radical cyclization · radical domino cyclization · radical cascade cyclization · intermolecular reactions · radical trapping · manganese(III) acetate · titanocene dichloride · tris(trimethylsilyl)silane · triethylborane · tri-sec-butylborane · TEMPO · 1,1,3,3-tetramethylguanidine · samarium(II) iodide · cobaloxime

XIV Abstracts

— p 243 —

1.5.1 **Protic Acid/Base Induced Reactions**
D. Adu-Ampratwum and C. J. Forsyth

This chapter covers synthetic domino processes that are induced by protic acid or base. They are broadly classified into those that capitalize upon the release of oxirane ring strain under acidic or basic conditions, and carbocyclic ring expansions and contractions under protic acid or basic conditions. The focus here is upon single substrate, monocomponent domino processes, rather than multicomponent variants.

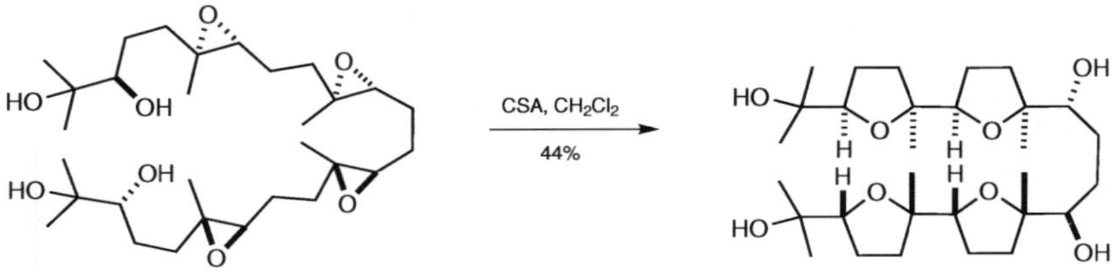

Keywords: carbocyclic compounds · cyclization · epoxy compounds · ethers · Favorskii rearrangement · intramolecular reactions · Nazarov cyclization · pinacol rearrangement · ring contraction · ring expansion · tandem reactions · Wagner–Meerwein rearrangement

— p 269 —

1.5.2 **Lewis Acid/Base Induced Reactions**
S.-H. Wang, Y.-Q. Tu, and M. Tang

The efficient construction of complex molecular skeletons is always a hot topic in organic synthesis, especially in the field of natural product synthesis, where many cyclic structural motifs can be found. Under the assiduous efforts of synthetic chemists, more and more methodologies are being developed to achieve the construction of cyclic skeletons. In particular, the beauty and high efficiency of organic synthesis are expressed vividly among those transformations realized through a domino strategy. Based on these important methodologies, selected Lewis acid/base induced domino reactions leading to ring expansions, contractions, and closures are presented in this chapter.

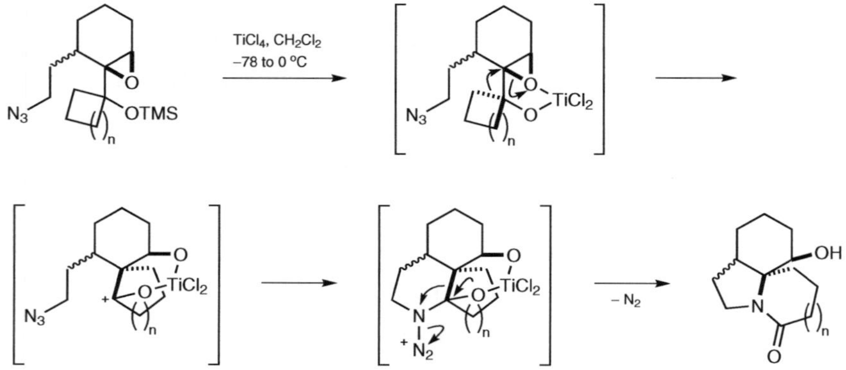

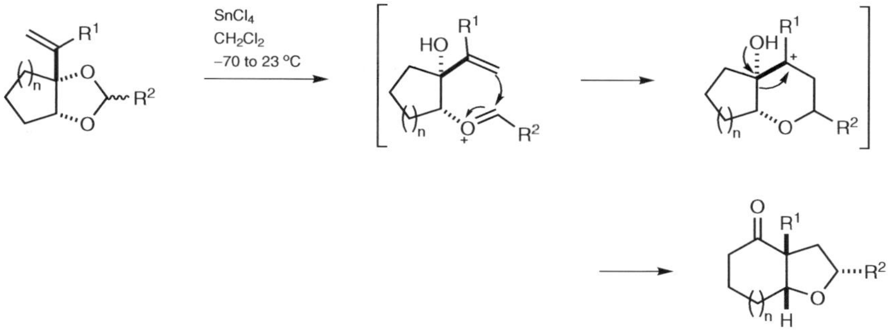

Keywords: tandem reactions · Lewis acid · Lewis base · ring expansion · ring contraction · ring closure

—————————————————————————————————————— p 355 ——

1.5.3   Brook Rearrangement as the Key Step in Domino Reactions
A. Kirschning, F. Gille, and M. Wolling

The Brook rearrangement has lost its Cinderella status over the past twenty years since being embedded into cascade reaction sequences. The powerful formation of carbanions through silyl migration has been exploited for the development of many new methodologies and has been used as a key transformation in complex natural product syntheses. Now, the Brook rearrangement belongs to the common repertoire of synthetic organic chemists.

Keywords: Brook rearrangement · domino reactions · migration · organosilicon chemistry · total synthesis

—————————————————————————————————————— p 449 ——

1.6.1   Palladium-Mediated Domino Reactions
E. A. Anderson

Palladium catalysis offers excellent opportunities to engineer domino reactions, due to the ability of this transition metal to engage with a variety of electrophiles and to effect stereocontrolled bond formations in complex settings. This review covers palladium-catalyzed domino processes, categorized according to the initiating species (alkenyl-, aryl-, allyl-, allenyl-, or alkylpalladium complexes), with a particular focus on applications in natural product synthesis that exemplify more general methodology.

Keywords: palladium · domino · cascade · total synthesis

―――――――――――――――――――――――――――――――――――― p 511 ――

1.6.2 Dirhodium-Catalyzed Domino Reactions
X. Xu, P. Truong, and M. P. Doyle

With dirhodium carbenes generated from diazocarbonyl compounds, 1-sulfonyl-1,2,3-triazoles, or cyclopropenes, a subsequent intramolecular cyclization forms a reactive intermediate that undergoes a further transformation that usually terminates the reaction process. Commonly, the electrophilic dirhodium carbene adds intramolecularly to a C≡C bond to provide a second rhodium carbene. Catalytically generated dirhodium-bound nitrenes initiate domino reactions analogously, and recent examples (nitrene to carbene to product) have also been documented.

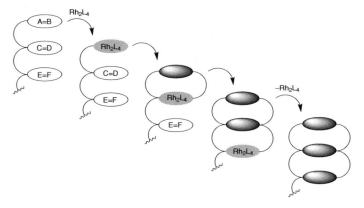

A=B = C=N$_2$, triazolyl, cyclopropenyl; C=D = C=C, C≡C; E=F = C—H

= newly formed bond

Keywords: α-carbonyl carbenes · (azavinyl)carbenes · cyclopropenes · [3 + 2] annulation · cyclopropenation · carbene/alkyne metathesis · carbonyl ylide reactions · Claisen/Cope rearrangement · C—H insertion · oxonium ylides · dipolar cycloaddition · aromatic substitution

—————— p 535 ——————

1.6.3   Gold-Mediated Reactions
E. Merino, A. Salvador, and C. Nevado

In this review, a selection of the most relevant examples featuring gold-catalyzed domino transformations are presented. Processes catalyzed by both gold(I) and gold(III) complexes are described, including multicomponent reactions, annulations, cycloisomerizations, and cycloadditions. The scope, limitations, and mechanistic rationalization of these transformations are also provided.

Keywords: domino transformations · multicomponent reactions · cycloisomerizations · cycloadditions · rearrangements · gold

—————— p 577 ——————

1.6.4   Rare Earth Metal Mediated Domino Reactions
T. Ohshima

Rare earth metals, comprising 17 chemical elements in the periodic table, are relatively abundant in the Earth's crust despite their name. In the series of lanthanides, a systematic contraction of the ionic radii is observed when going from lanthanum to lutetium (often referred to as the lanthanide contraction), but this variation is so smooth and limited, with ca. 1% contraction between two successive lanthanides, that it is possible to fine-tune the ionic radii, Lewis acidity, and Brønsted basicity of rare earth complexes. As a result of the large size of the lanthanide ions compared to other metal ions, lanthanide ions have high coordination numbers, varying from 6 to 12. Due to the strong oxophilicity of rare earth elements, their metal ions have a hard Lewis acidic nature. Most particularly, rare earth metal trifluoromethanesulfonates [M(OTf)$_3$] have been regarded as new types of Lewis acids and are stable and active in the presence of many Lewis bases. Another important type of rare earth metal species, the rare earth metal alkoxides [M(OR)$_3$], exhibit both Lewis acidity and Brønsted basicity. These collated characteristic features of rare earth based complexes, such as high coordination numbers, a hard Lewis acidic nature, high compatibility with various functional groups, ease of fine-tuning, and multifunctionality, have led to the development of a variety of domino reactions catalyzed largely by rare earth metal trifluoromethanesulfonates and alkoxides.

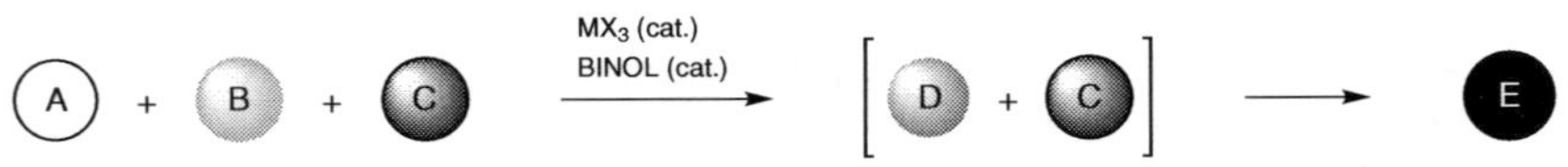

XVIII   Abstracts

M = rare earth metal; X = OTf, OR¹

Keywords: lanthanide contraction · rare earth metal trifluoromethanesulfonates · rare earth metal alkoxides · Lewis acidity · Brønsted basicity · high coordination numbers · multifunctionality

——————————————————————————————————— p 601 ——

1.6.5   Cobalt and Other Metal Mediated Domino Reactions: The Pauson–Khand Reaction and Its Use in Natural Product Total Synthesis
L. Shi and Z. Yang

The Pauson–Khand reaction constitutes one of the most formidable additions to the repertoire of synthetically useful reactions. It rapidly affords a cyclopentenone skeleton from an alkene, an alkyne, and carbon monoxide, based on a domino sequence of bond constructions. In this chapter, the prowess of the Pauson–Khand reaction is illustrated by judicious selection of complex target molecules, the total syntheses of which are cleverly orchestrated by the key Pauson–Khand reaction sequence. Emphasis is placed on cobalt-mediated processes to exemplify the applicability of this classical reaction.

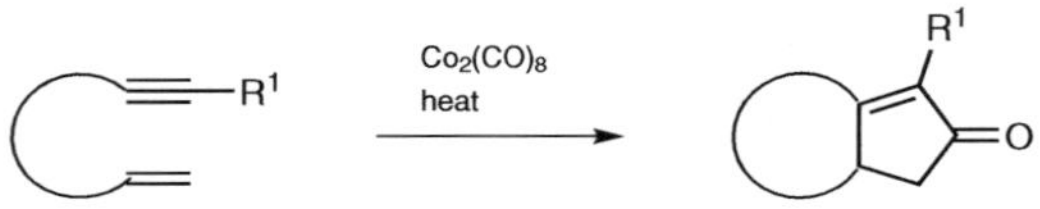

Keywords: Pauson–Khand reaction · alkenes · carbon monoxide · alkynes · cyclopentenones · natural product synthesis · octacarbonyldicobalt · thioureas · allenic alkynes · asymmetric synthesis

Applications of Domino Transformations in Organic Synthesis 1

Preface	V
Volume Editor's Preface	IX
Abstracts	XI
Table of Contents	XXI

Introduction
S. A. Snyder · 1

1.1 Polyene Cyclizations
R. A. Shenvi and K. K. Wan · 13

1.2 Cation–π Cyclizations of Epoxides and Polyepoxides
K. W. Armbrust, T. Halkina, E. H. Kelley, S. Sittihan, and T. F. Jamison · 43

1.3 Metathesis Reactions · 67

1.3.1 Enyne-Metathesis-Based Domino Reactions in Natural Product Synthesis
D. Lee and M. O'Connor · 67

1.3.2 Domino Metathesis Reactions Involving Carbonyls
H. Renata and K. M. Engle · 135

1.4 Radical Reactions · 157

1.4.1 Peroxy Radical Additions
X. Hu and T. J. Maimone · 157

1.4.2 Radical Cyclizations
J. J. Devery, III, J. J. Douglas, and C. R. J. Stephenson · 187

1.4.3 Tandem Radical Processes
K. A. Parker · 217

1.5 Non-Radical Skeletal Rearrangements · 243

1.5.1 Protic Acid/Base Induced Reactions
D. Adu-Ampratwum and C. J. Forsyth · 243

1.5.2 Lewis Acid/Base Induced Reactions
S.-H. Wang, Y.-Q. Tu, and M. Tang · 269

1.5.3	**Brook Rearrangement as the Key Step in Domino Reactions**	
	A. Kirschning, F. Gille, and M. Wolling	355
1.6	**Metal-Mediated Reactions**	449
1.6.1	**Palladium-Mediated Domino Reactions**	
	E. A. Anderson	449
1.6.2	**Dirhodium-Catalyzed Domino Reactions**	
	X. Xu, P. Truong, and M. P. Doyle	511
1.6.3	**Gold-Mediated Reactions**	
	E. Merino, A. Salvador, and C. Nevado	535
1.6.4	**Rare Earth Metal Mediated Domino Reactions**	
	T. Ohshima	577
1.6.5	**Cobalt and Other Metal Mediated Domino Reactions: The Pauson–Khand Reaction and Its Use in Natural Product Total Synthesis**	
	L. Shi and Z. Yang	601
	Keyword Index	633
	Author Index	669
	Abbreviations	693

Table of Contents

Introduction
S. A. Snyder

Introduction ⋯⋯⋯⋯⋯⋯⋯⋯⋯⋯⋯⋯⋯⋯⋯⋯⋯⋯⋯⋯⋯⋯⋯⋯⋯⋯⋯⋯⋯⋯⋯ 1

1.1	**Polyene Cyclizations** R. A. Shenvi and K. K. Wan	

1.1	**Polyene Cyclizations** ⋯⋯⋯⋯⋯⋯⋯⋯⋯⋯⋯⋯⋯⋯⋯⋯⋯⋯⋯⋯⋯⋯⋯⋯	13
1.1.1	Cationic Polyene Cyclizations Mediated by Brønsted or Lewis Acids ⋯⋯⋯⋯	14
1.1.1.1	Most Used Cationic Polyene Cyclization Methods ⋯⋯⋯⋯⋯⋯⋯⋯⋯⋯⋯	15
1.1.1.1.1	Polyene Cyclization via Biomimetic Heterolytic Opening of Epoxides by Alkylaluminum Lewis Acids ⋯⋯⋯⋯⋯⋯⋯⋯⋯⋯⋯⋯⋯	15
1.1.1.1.2	Polyene Cyclization Mediated by Carbophilic Lewis Acids ⋯⋯⋯⋯⋯⋯⋯	17
1.1.1.2	Recent Advances in Cationic Polycyclization: Halonium-Initiated Polycyclization	17
1.1.1.3	Other Common Cationic Polyene Cyclization Methods ⋯⋯⋯⋯⋯⋯⋯⋯	19
1.1.1.3.1	Catalytic, Enantioselective, Protonative Polycyclization ⋯⋯⋯⋯⋯⋯⋯	19
1.1.1.3.1.1	Chiral Transfer from a Brønsted Acid ⋯⋯⋯⋯⋯⋯⋯⋯⋯⋯⋯⋯⋯⋯	19
1.1.1.3.1.2	Chiral Transfer from (*R*)-2,2′-Dichloro-1,1′-bi-2-naphthol–Antimony(V) Chloride Complex ⋯⋯⋯⋯⋯⋯⋯⋯⋯⋯⋯⋯⋯⋯⋯⋯⋯⋯⋯⋯⋯⋯	19
1.1.1.3.1.3	Chiral Transfer via Nucleophilic Phosphoramidites ⋯⋯⋯⋯⋯⋯⋯⋯⋯	20
1.1.1.3.2	Polyene Cyclization Initiated by Unsaturated Ketones and Mediated by Aluminum Lewis Acids ⋯⋯⋯⋯⋯⋯⋯⋯⋯⋯⋯⋯⋯⋯	21
1.1.1.3.3	Gold-Mediated Enantioselective Polycyclization ⋯⋯⋯⋯⋯⋯⋯⋯⋯⋯	22
1.1.1.3.4	Polycyclization Initiated by an Episulfonium Ion ⋯⋯⋯⋯⋯⋯⋯⋯⋯⋯	23
1.1.1.3.5	Polycyclization Initiated by a π-Lewis Acidic Metal ⋯⋯⋯⋯⋯⋯⋯⋯⋯	23
1.1.1.3.6	Enantioselective Polyene Cyclization Mediated by Chiral Scalemic Iridium Complexes ⋯⋯⋯⋯⋯⋯⋯⋯⋯⋯⋯⋯⋯⋯⋯⋯⋯	24
1.1.1.3.7	Acyliminium-Initiated Polyene Cyclization Mediated by Thioureas ⋯⋯⋯⋯	24
1.1.1.3.8	Tail-to-Head Polycyclization ⋯⋯⋯⋯⋯⋯⋯⋯⋯⋯⋯⋯⋯⋯⋯⋯⋯⋯	25
1.1.2	Radical Polyene Cyclizations ⋯⋯⋯⋯⋯⋯⋯⋯⋯⋯⋯⋯⋯⋯⋯⋯⋯⋯⋯	26
1.1.2.1	Most Used Radical Polycyclization Methods ⋯⋯⋯⋯⋯⋯⋯⋯⋯⋯⋯⋯	26
1.1.2.1.1	Cyclization of Mono-and Polyunsaturated β-Oxo Esters Mediated by Manganese(III) Acetate ⋯⋯⋯⋯⋯⋯⋯⋯⋯⋯⋯⋯⋯⋯⋯⋯⋯⋯	26
1.1.2.1.2	Titanocene-Catalyzed Polycyclization ⋯⋯⋯⋯⋯⋯⋯⋯⋯⋯⋯⋯⋯⋯	27
1.1.2.2	Recent Advances in Radical Polycyclization ⋯⋯⋯⋯⋯⋯⋯⋯⋯⋯⋯⋯	28
1.1.2.2.1	Manganese- and Cobalt-Catalyzed Cyclization ⋯⋯⋯⋯⋯⋯⋯⋯⋯⋯⋯	28

1.1.2.2.2	Manganese-Catalyzed Hydrogenative Polycyclization	29
1.1.2.2.3	Radical Isomerization, Cycloisomerization, and Retrocycloisomerization with a Cobalt–salen Catalyst	30
1.1.2.3	Other Examples of Radical Polycyclization	31
1.1.2.3.1	Radical Polycyclization via Photoinduced Electron Transfer	31
1.1.2.3.2	Polycyclization via Organo-SOMO Catalysis	31
1.1.2.3.3	Polyene Radical Cascades in Complex Molecule Synthesis	31
1.1.3	Polyene Cyclization via Reductive Elimination from a Metal Center or Metathesis	32
1.1.3.1	Most Used Polycyclization Methods via Reductive Elimination	32
1.1.3.1.1	Palladium Zipper Cyclization Cascades	32
1.1.3.1.2	Polycyclization Cascades via Metathesis	33
1.1.3.2	Other Polycyclization Methods via Reductive Elimination	34
1.1.3.2.1	Cyclization via π-Allylpalladium Complexes	34
1.1.3.2.2	Palladium-Catalyzed Ene–Yne Cycloisomerization	35
1.1.3.2.3	Cycloisomerization in Complex Molecule Synthesis	35
1.1.4	Anionic Polyene Cyclizations	36
1.1.4.1	Examples of Anionic Polyene Cyclizations	37
1.1.4.1.1	Stereoselective Polycyclization via Intramolecular Diels–Alder Cycloaddition Followed by Aldol Condensation	37
1.1.4.1.2	Transannular Double Michael Cyclization Cascades	38
1.2	**Cation–π Cyclizations of Epoxides and Polyepoxides** K. W. Armbrust, T. Halkina, E. H. Kelley, S. Sittihan, and T. F. Jamison	
1.2	**Cation–π Cyclizations of Epoxides and Polyepoxides**	43
1.2.1	*exo*-Selective Polyepoxide Cascades	43
1.2.1.1	Brønsted Acid Promoted Cascades	43
1.2.1.2	Brønsted Base Promoted Cascades	45
1.2.1.3	Oxocarbenium-Initiated Cascades via Photooxidative Cleavage	46
1.2.2	*endo*-Selective Polyepoxide Cascades	47
1.2.2.1	Lewis Acid Activation	47
1.2.2.1.1	Alkyl-Directed Cascades	47
1.2.2.2	Brønsted Base Activation	51
1.2.2.2.1	Trimethylsilyl-Directed Cascades	51
1.2.2.3	Distal Electrophilic Activation	52
1.2.2.3.1	Bromonium-Initiated Cascades	52
1.2.2.3.2	Oxocarbenium-Initiated Cascades via Photooxidative Cleavage	54

1.2.2.3.3	Carbocation-Initiated Cascades via Halide Abstraction	55
1.2.2.4	Water-Promoted Cascades	56
1.2.2.4.1	Cascades of Diepoxides Templated by a Tetrahydropyran	56
1.2.2.4.2	Cascades of Diepoxides Templated by a Dioxane	57
1.2.2.4.3	Cascades of Triepoxides Templated by a Tetrahydropyran	57
1.2.3	Epoxide Cascades with C—C π Bonds	58
1.2.3.1	Cascades Terminated by Alkenes and Alkynes	58
1.2.3.2	Cascades Terminated by Arenes	60
1.2.3.3	Cascades Terminated by Protected Phenols	63

1.3	**Metathesis Reactions**	

1.3.1	**Enyne-Metathesis-Based Domino Reactions in Natural Product Synthesis** D. Lee and M. O'Connor	

1.3.1	**Enyne-Metathesis-Based Domino Reactions in Natural Product Synthesis**	67
1.3.1.1	Mechanism	71
1.3.1.1.1	Alkylidene Carbene Catalyzed Reactions	71
1.3.1.1.2	π-Lewis Acid Catalyzed Reactions	75
1.3.1.1.3	Metallotropic [1,3]-Shift	75
1.3.1.2	Selectivity	76
1.3.1.2.1	Regioselectivity in Cross Metathesis	76
1.3.1.2.2	*exo/endo*-Mode Selectivity in Ring-Closing Metathesis	78
1.3.1.2.3	Stereoselectivity in Cross Metathesis and Ring-Closing Metathesis	79
1.3.1.2.4	Regioselectivity in Metallotropic [1,3]-Shift	80
1.3.1.3	Applications in Natural Product Synthesis	86
1.3.1.3.1	Simple Enyne Metathesis	88
1.3.1.3.1.1	Enyne Cross Metathesis	88
1.3.1.3.1.2	Enyne Ring-Closing Metathesis for Small Rings	92
1.3.1.3.2	Domino Enyne Metathesis	99
1.3.1.3.2.1	Double Ring-Closing Metathesis with Dienynes	99
1.3.1.3.2.2	Domino Ring-Closing Metathesis/Cross Metathesis, Cross Metathesis/ Ring-Closing Metathesis, and Cross Metathesis/Cross Metathesis Sequences	112
1.3.1.3.3	Enyne Metathesis/Metallotropic [1,3]-Shift Sequences	120
1.3.1.3.4	Enyne Metathesis/Diels–Alder Reaction Sequences	123
1.3.1.3.5	Enyne Metathesis with π-Lewis Acids	128
1.3.1.4	Conclusions	130

1.3.2 **Domino Metathesis Reactions Involving Carbonyls**
H. Renata and K. M. Engle

1.3.2	**Domino Metathesis Reactions Involving Carbonyls**	135
1.3.2.1	Two-Pot Reactions	136
1.3.2.1.1	Reaction with In Situ Generated Titanium–Alkylidene Complexes Followed by Metathesis	137
1.3.2.2	One-Pot Reactions	139
1.3.2.2.1	Reaction with Bis(η^5-cyclopentadienyl)methylenetitanium(IV)-Type Complexes	139
1.3.2.2.2	Reaction with In Situ Generated Titanium–Alkylidene Complexes	143
1.3.2.2.3	Reaction with Stoichiometric Molybdenum or Tungsten Complexes	148
1.3.2.2.4	Organocatalytic Reactions	150

1.4 **Radical Reactions**

1.4.1 **Peroxy Radical Additions**
X. Hu and T. J. Maimone

1.4.1	**Peroxy Radical Additions**	157
1.4.1.1	Initiation from a Preexisting Hydroperoxide	157
1.4.1.1.1	Using Peroxide Initiators	158
1.4.1.1.2	Using Copper(II) Trifluoromethanesulfonate/Oxygen	161
1.4.1.1.3	Using Samarium(II) Iodide/Oxygen	162
1.4.1.2	Initiation by Metal-Catalyzed Hydroperoxidation	163
1.4.1.2.1	The Mukaiyama Hydration/Hydroperoxidation	163
1.4.1.2.2	Hydroperoxidation-Initiated Domino Transformations	164
1.4.1.2.3	Manganese-Catalyzed Domino Hydroperoxidation	166
1.4.1.3	Initiation by Radical Addition/Oxygen Quenching	168
1.4.1.3.1	Thiyl Radical Initiation	168
1.4.1.3.1.1	Thiol–Alkene Co-oxygenation Reactions	168
1.4.1.3.1.2	Domino Transformations of Vinylcyclopropanes	171
1.4.1.3.2	Carbon-Centered Radical Additions	173
1.4.1.3.2.1	By C—H Abstraction	174
1.4.1.3.2.2	Manganese(III)-Mediated Oxidation of 1,3-Dicarbonyls	175
1.4.1.4	Heteroatom Oxidation/Cyclopropane Cleavage Pathways	177
1.4.1.5	Radical Cation Intermediates	178
1.4.1.5.1	1,2-Diarylcyclopropane Photooxygenation	179
1.4.1.5.2	Alkene/Oxygen [2+2+2] Cycloaddition	180

1.4.2	**Radical Cyclizations**	
	J. J. Devery, III, J. J. Douglas, and C. R. J. Stephenson	
1.4.2	**Radical Cyclizations**	187
1.4.2.1	Tin-Mediated Radical Cyclizations	188
1.4.2.1.1	Tin-Mediated Synthesis of Hexahydrofuropyrans	188
1.4.2.1.2	Tin-Mediated Radical [3+2] Annulation	191
1.4.2.2	Reductive Radical Domino Cyclizations	193
1.4.2.2.1	Samarium(II) Iodide Mediated Radical Cyclizations	193
1.4.2.2.2	Samarium(II) Iodide Mediated Radical–Anionic Cyclizations	196
1.4.2.3	Oxidative Radical Cyclizations	198
1.4.2.3.1	Organo-SOMO-Activated Polyene Cyclization	198
1.4.2.3.2	Oxidative Rearrangement of Silyl Bis(enol ethers)	201
1.4.2.3.3	Diastereoselective Oxidative Rearrangement of Silyl Bis(enol ethers)	204
1.4.2.4	Visible-Light-Mediated Reactions	207
1.4.2.4.1	Light-Mediated Radical Cyclizatior/Divinylcyclopropane Rearrangement	207
1.4.2.4.2	Visible-Light-Mediated Radical Fragmentation and Bicyclization	211
1.4.3	**Tandem Radical Processes**	
	K. A. Parker	
1.4.3	**Tandem Radical Processes**	217
1.4.3.1	General and Specialized Reviews on Radical Cyclization Reactions	217
1.4.3.2	A Brief History of Tandem Radical Cyclization Chemistry	218
1.4.3.2.1	The Tandem Radical Cyclization Concept: Fused Rings	218
1.4.3.2.2	The Biomimetic Tandem Radical Cyclization Postulate	219
1.4.3.2.3	A Vinyl Radical Tandem Radical Cyclization: A Product with Linked Rings	219
1.4.3.2.4	Introduction to Selectivity: A Bridged Ring System	220
1.4.3.3	Alternative Reagents for Cascade Initiation: Getting Away from Tin, 2,2′-Azobisisobutyronitrile, and Peroxides	221
1.4.3.3.1	The Manganese(III) System	221
1.4.3.3.2	The Titanium(III) System	222
1.4.3.3.3	Using Silanes Rather than Stannanes	222
1.4.3.3.3.1	Carboxyarylation	222
1.4.3.3.3.2	Reactions Terminated by Azide	223
1.4.3.3.4	Reactions with Borane Initiators	225
1.4.3.3.4.1	Tin Hydrides with Triethylborane for Initiation and Fragmentation with Samarium(II) Iodide	225

XXVI Table of Contents

1.4.3.3.4.2 Triethylborane-Mediated Atom Transfer and Cobaloxime-Initiated Reductive Tandem Cyclization 227

1.4.3.3.4.3 Tri-*sec*-butylborane/Oxygen/Tris(trimethylsilyl)silane Induced Reductive Cyclization 228

1.4.3.4 Nitrogen- and Oxygen-Centered Radicals 229

1.4.3.5 Intramolecular Plus Intermolecular Pathways 231

1.4.3.5.1 Cyclization/Trapping 231

1.4.3.5.2 Trapping/Cyclization 236

1.4.3.6 Intermolecular Trapping/Trapping Pathways 237

1.4.3.7 Conclusions 238

1.5 **Non-Radical Skeletal Rearrangements**

1.5.1 **Protic Acid/Base Induced Reactions**
D. Adu-Ampratwum and C. J. Forsyth

1.5.1 **Protic Acid/Base Induced Reactions** 243

1.5.1.1 Intramolecular Epoxide-Opening Cyclizations 244

1.5.1.1.1 Protic Acid Induced Intramolecular Epoxide Openings 245

1.5.1.1.1.1 *exo* Epoxide Ring Expansions 246

1.5.1.1.1.2 *endo* Epoxide Ring Expansions 248

1.5.1.1.2 Base-Induced Intramolecular Epoxide Openings 250

1.5.1.2 Carbocyclic Ring Expansions/Ring Contractions 252

1.5.1.2.1 Acid-Induced Carbocyclic Ring Expansions/Ring Contractions 252

1.5.1.2.1.1 Wagner–Meerwein Rearrangements 254

1.5.1.2.1.1.1 Ring-Expansion Rearrangements 254

1.5.1.2.1.1.2 Ring-Contraction Rearrangements 255

1.5.1.2.1.2 Pinacol Rearrangements 256

1.5.1.2.1.3 Semipinacol Rearrangements 258

1.5.1.2.2 Base-Induced Carbocyclic Ring Expansions/Ring Contractions 259

1.5.1.2.2.1 Benzilic Acid Rearrangements 259

1.5.1.2.2.2 Retro-Benzilic Acid Rearrangements 260

1.5.1.2.2.3 Favorskii Rearrangements 261

1.5.1.2.2.3.1 Homo-Favorskii Rearrangements 263

1.5.1.2.2.4 α-Hydroxy Ketone Rearrangements 263

| 1.5.2 | **Lewis Acid/Base Induced Reactions** | |
| | S.-H. Wang, Y.-Q. Tu, and M. Tang | |

1.5.2	**Lewis Acid/Base Induced Reactions**	269
1.5.2.1	Ring Expansions	269
1.5.2.1.1	Semipinacol Rearrangement of 2,3-Epoxy Alcohols and Their Derivatives	269
1.5.2.1.2	Reductive Rearrangement of 2,3-Epoxy Alcohols with Aluminum Triisopropoxide	272
1.5.2.1.3	Tandem Semipinacol/Schmidt Reaction of α-Siloxy Epoxy Azides	273
1.5.2.1.4	Prins–Pinacol Rearrangement	277
1.5.2.2	Ring Contractions	282
1.5.2.2.1	Rearrangement of Epoxides	282
1.5.2.2.2	Favorskii Rearrangement and Quasi-Favorskii Rearrangement	286
1.5.2.3	Ring Closures	289
1.5.2.3.1	Induction by an Electrophilic Step	289
1.5.2.3.1.1	Initiation by Epoxide Ring Opening	290
1.5.2.3.1.1.1	Termination with a Carbon Nucleophile	290
1.5.2.3.1.1.2	Termination with an Oxygen Nucleophile	297
1.5.2.3.1.1.3	Termination with a Rearrangement	300
1.5.2.3.1.1.4	Termination with a Pericyclic Reaction	305
1.5.2.3.1.2	Initiation with a Carbonyl and Its Derivatives	306
1.5.2.3.1.2.1	Termination with a Nucleophile	307
1.5.2.3.1.2.2	Termination with a Rearrangement	316
1.5.2.3.1.3	Initiation by Activation of a π-Bond	318
1.5.2.3.1.3.1	Initiation by a Lewis Acid	318
1.5.2.3.1.3.2	Initiation by a π-Acid	321
1.5.2.3.1.3.2.1	Activation of Alkenes	321
1.5.2.3.1.3.2.2	Activation of Alkynes	328
1.5.2.3.2	Induction by a Pericyclic Reaction	338
1.5.2.3.3	Induction by a Nucleophilic Step	344

| 1.5.3 | **Brook Rearrangement as the Key Step in Domino Reactions** | |
| | A. Kirschning, F. Gille, and M. Wolling | |

1.5.3	**Brook Rearrangement as the Key Step in Domino Reactions**	355
1.5.3.1	1,2-Brook Rearrangement	356
1.5.3.1.1	1,2-Brook Rearrangement with Aldehydes, Ketones, or Acyl Chlorides	356

1.5.3.1.2	1,2-Brook Rearrangement with Acylsilanes	358
1.5.3.1.2.1	Domino Reactions of Acylsilanes by Addition of Nucleophiles	358
1.5.3.1.2.2	Domino Reactions of Acylsilanes Initiated by Nucleophiles Acting as Catalysts	377
1.5.3.1.2.3	Domino Reactions of Acylsilanes Initiated by Enolization	380
1.5.3.1.3	1,2-Brook Rearrangement with α-Silyl Carbinols	381
1.5.3.1.4	1,2-Brook Rearrangement with Epoxy Silanes	382
1.5.3.1.5	Miscellaneous Examples of 1,2-Brook Rearrangement	386
1.5.3.1.6	Retro-1,2-Brook Rearrangement	387
1.5.3.2	1,3-Brook Rearrangement	392
1.5.3.2.1	Addition of Silyl-Substituted Stabilized Organolithium Agents to Carbonyl Groups	393
1.5.3.2.2	1,3-Brook Rearrangement at sp^2-Hybridized Carbon Atoms	399
1.5.3.2.3	1,3-Brook Rearrangement Accompanied by β-Elimination	399
1.5.3.2.4	Carbon to Nitrogen Rearrangement	400
1.5.3.2.5	Carbon to Sulfur Rearrangement	401
1.5.3.2.6	Retro-1,3-Brook Rearrangement	402
1.5.3.3	1,4-Brook Rearrangement	403
1.5.3.3.1	1,4-Brook Rearrangement of Silyl-Substituted Carbanions with Epoxides	404
1.5.3.3.2	1,4-Brook Rearrangement with Dihalosilyl-Substituted Methyllithium	416
1.5.3.3.3	1,4-Brook Rearrangement with Allylsilanes	417
1.5.3.3.4	1,4-Brook Rearrangement with Silylated Benzaldehydes	422
1.5.3.3.5	1,4-Brook Rearrangement with Vinylsilanes	426
1.5.3.3.6	Sulfur to Oxygen Rearrangement	430
1.5.3.3.7	Retro-1,4-Brook Rearrangement	431
1.5.3.4	Applications in the Total Synthesis of Natural Products	432
1.5.3.4.1	The 1,2-Brook Rearrangement in Natural Product Synthesis	432
1.5.3.4.2	The 1,4-Brook Rearrangement in Natural Product Synthesis	433
1.5.3.4.2.1	Synthesis of Polyketides	433
1.5.3.4.2.2	Synthesis of Terpenes	442
1.5.3.4.2.3	Synthesis of Alkaloids	442
1.5.3.5	Conclusions	444

1.6	**Metal-Mediated Reactions**	

1.6.1	**Palladium-Mediated Domino Reactions**	
	E. A. Anderson	

1.6.1	**Palladium-Mediated Domino Reactions** ·······················	449
1.6.1.1	Reactions Initiating with Alkenylpalladium Intermediates ·················	449
1.6.1.2	Reactions Initiating with Arylpalladium Species ·························	473
1.6.1.3	Reactions Initiating with Allylpalladium Intermediates ·················	491
1.6.1.4	Reactions Initiating with Allenylpalladium Intermediates ··············	496
1.6.1.5	Reactions Initiating with Alkylpalladium Intermediates ·················	498
1.6.1.6	Conclusions ···	505

1.6.2	**Dirhodium-Catalyzed Domino Reactions**	
	X. Xu, P. Truong, and M. P. Doyle	

1.6.2	**Dirhodium-Catalyzed Domino Reactions** ·······················	511
1.6.2.1	1-Sulfonyl-1,2,3-triazoles as (Azavinyl)carbene Precursors in Domino Reactions	512
1.6.2.2	Dirhodium(II)-Catalyzed Generation of Rhodium–Carbenes from Cyclopropenes and Their Subsequent Reactions ·························	516
1.6.2.3	Dirhodium(II)-Catalyzed Carbene/Alkyne Metathesis ·················	521
1.6.2.4	Nitrene Cascade Reactions Catalyzed by a Dirhodium Complex ··········	527
1.6.2.5	Conclusions ···	532

1.6.3	**Gold-Mediated Reactions**	
	E. Merino, A. Salvador, and C. Nevado	

1.6.3	**Gold-Mediated Reactions** ·····································	535
1.6.3.1	Gold-Catalyzed Annulations ···	535
1.6.3.1.1	Using *ortho*-Alkynylbenzaldehydes ·································	535
1.6.3.1.2	Using Arylimines and Alkynes ··	538
1.6.3.1.3	Using Alcohols and Dienes ··	538
1.6.3.1.4	Using Carbonyl Compounds, Alkynes, and Nitrogen-Containing Compounds ·	540
1.6.3.2	Gold-Catalyzed Domino Reactions via Addition of Carbon Nucleophiles to π-Electrophiles ··································	542
1.6.3.2.1	1,n-Enynes ···	542
1.6.3.2.2	1,n-Diynes ··	545
1.6.3.2.3	1,n-Allenenes ··	549
1.6.3.2.4	1,n-Allenynes ··	550

1.6.3.3	Gold-Catalyzed Domino Reactions via Addition of Heteroatom Nucleophiles to π-Electrophiles	551
1.6.3.3.1	Addition of Nitrogen Nucleophiles to Alkynes	551
1.6.3.3.2	Addition of Oxygen Nucleophiles to Alkynes and Allenes	552
1.6.3.3.2.1	Alcohols as Nucleophiles	552
1.6.3.3.2.2	Epoxides as Nucleophiles	553
1.6.3.3.3	Addition of Heteroatom Nucleophiles to Alkenes	555
1.6.3.4	Gold-Catalyzed Domino Reactions Involving the Rearrangement of Propargyl Esters	557
1.6.3.4.1	Synthesis of α-Ylidene β-Diketones	557
1.6.3.4.2	Synthesis of Dienes	558
1.6.3.4.3	Synthesis of α-Substituted Enones	561
1.6.3.4.3.1	Synthesis of α-Halo-Substituted Enones	561
1.6.3.4.3.2	Synthesis of α-Aryl-Substituted Enones	562
1.6.3.4.4	Synthesis of Cyclopentenones	563
1.6.3.4.5	Acetate Migration and Reaction with π-Electrophiles	564
1.6.3.4.5.1	Acetate Migration and Reaction with Alkynes	564
1.6.3.4.5.2	Acetate Migration and Reaction with Alkenes	565
1.6.3.4.6	Acetate Migration and Ring-Opening Reactions	566
1.6.3.4.6.1	Cyclopentannulations	566
1.6.3.4.6.2	Cyclohexannulations	568
1.6.3.4.6.3	Cycloheptannulations	569

1.6.4	**Rare Earth Metal Mediated Domino Reactions** T. Ohshima	
1.6.4	**Rare Earth Metal Mediated Domino Reactions**	577
1.6.4.1	Addition to C=O or C=C—C=O as a Primary Step	578
1.6.4.1.1	Aldol-Type Reactions	578
1.6.4.1.2	1,4-Addition Reactions	582
1.6.4.2	Addition to C=N or C=C—C=N as a Primary Step	583
1.6.4.2.1	Strecker-Type Reactions	583
1.6.4.2.2	Other Reactions Initiated by Imine Formation	584
1.6.4.3	Enamine Formation as a Primary Step	589
1.6.4.3.1	Enamines from β-Keto Esters	589
1.6.4.3.2	Enamines from Alkynes	590
1.6.4.4	Ring-Opening or Ring-Closing Reactions as a Primary Step	592
1.6.4.4.1	Ring-Opening Reactions	592

1.6.4.4.2	Ring-Closing Reactions	594
1.6.4.5	Rearrangement Reactions	596
1.6.4.6	Miscellaneous Reactions	598
1.6.4.6.1	Domino Reactions with Transition-Metal Catalysts	598

1.6.5	**Cobalt and Other Metal Mediated Domino Reactions:** **The Pauson–Khand Reaction and Its Use in Natural Product Total Synthesis** L. Shi and Z. Yang	
1.6.5	**Cobalt and Other Metal Mediated Domino Reactions:** **The Pauson–Khand Reaction and Its Use in Natural Product Total Synthesis**	601
1.6.5.1	Enyne-Based Pauson–Khand Reactions	606
1.6.5.1.1	A Short Synthesis of Racemic 13-Deoxyserratine	606
1.6.5.1.2	Total Synthesis of Paecilomycine A	606
1.6.5.1.3	Stereoselective Total Syntheses of (−)-Magellanine, (+)-Magellaninone, and (+)-Paniculatine	607
1.6.5.1.4	Concise, Enantioselective Total Synthesis of (−)-Alstonerine	608
1.6.5.1.5	Pauson–Khand Approach to the Hamigerans	609
1.6.5.1.6	Enantioselective Synthesis of (−)-Pentalenene	610
1.6.5.1.7	Formal Synthesis of (+)-Nakadomarin A	611
1.6.5.1.8	Diastereoselective Total Synthesis of Racemic Schindilactone A	612
1.6.5.1.9	Asymmetric Total Synthesis of (−)-Huperzine Q	614
1.6.5.1.10	Total Synthesis of (−)-Jiadifenin	615
1.6.5.1.11	Total Synthesis of Penostatin B	617
1.6.5.1.12	Total Synthesis of Racemic Pentalenolactone A Methyl Ester	617
1.6.5.1.13	Asymmetric Total Synthesis of (+)-Fusarisetin A	618
1.6.5.2	Heteroatom-Based Pauson–Khand Reaction	619
1.6.5.2.1	Total Synthesis of Physostigmine	620
1.6.5.2.2	Asymmetric Total Synthesis of Racemic Merrilactone A	621
1.6.5.3	Allenic Pauson–Khand Reaction	622
1.6.5.3.1	Total Synthesis of (+)-Achalensolide	623
1.6.5.3.2	Synthesis of 6,12-Guaianolide	624
1.6.5.3.3	Stereoselective Total Syntheses of Uncommon Sesquiterpenoids	625
1.6.5.3.4	14-Step Synthesis of (+)-Ingenol from (+)-3-Carene	626
1.6.5.4	Conclusions	627

Keyword Index 633

Author Index 669

Abbreviations 693

Introduction

S. A. Snyder

There is something inherently satisfying about setting up a number of dominoes in series, pushing one down at the end of the chain, and watching the rest tumble as if in slow motion until all have been knocked over; I remember spending hours doing this as a child, once my father showed me the principle and I had enough dexterity to not knock anything down before the right moment. I felt the same way about playing with a Newton's cradle, where, although the movement does not cease in the same way as a domino chain, the idea of setting something into motion and causing a series of subsequent reactions was certainly pleasing and mesmerizing to watch.

My first experience with a "domino", or "cascade", sequence in the chemistry world came right at the start of my graduate career in K. C. Nicolaou's laboratory at The Scripps Research Institute, in the summer of 1999. For weeks I watched my talented neighbors in the lab strive to achieve a biomimetic synthesis of trichodimerol (**2**) (Scheme 1) from protected forms of the natural product sorbicillin, e.g. **1**. Although they registered a number of successes in merging **1** with itself to create other members of the bisorbicillinoid family upon exposure to different reaction conditions, trichodimerol itself seemed elusive.[1] Each day, several different products were generated from individual experiments, and I remember my colleagues picking out which ones they would try and characterize, and then thinking about what conditions to change in the hope of achieving success based on the structures they deduced that they had produced. Ultimately, in near-concomitant reports, E. J. Corey (my future postdoctoral supervisor)[2] and the Nicolaou team[3] determined that low water content in the presence of a soluble base was the key to the domino series of Michael reactions and ketalizations needed to form the unique caged structure of **2**; this sequence of events was impressive, even though only relatively modest yields could be achieved.

for references see p 11

Domino Transformations   Introduction

Scheme 1   Domino Construction of Trichodimerol through Sequential Michael Reactions and Ketalizations[2,3]

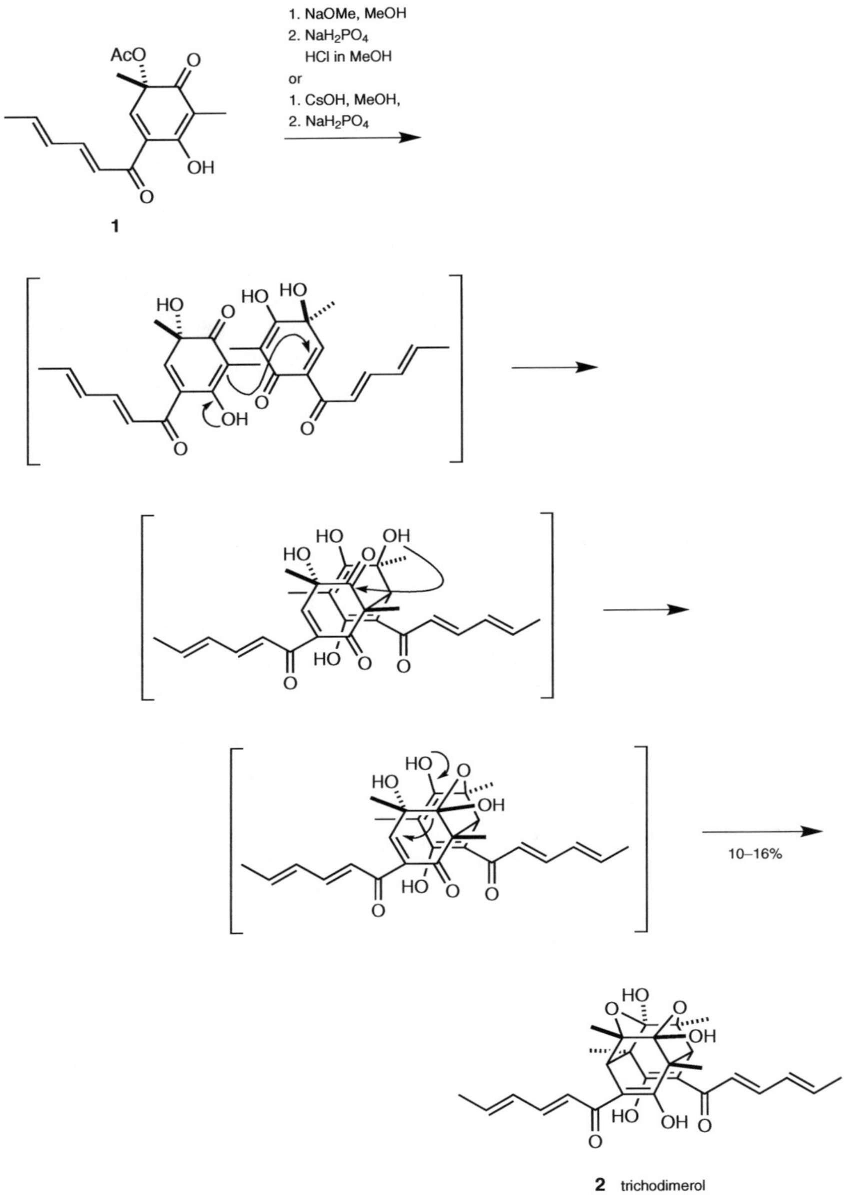

1. NaOMe, MeOH
2. NaH$_2$PO$_4$
   HCl in MeOH
or
1. CsOH, MeOH,
2. NaH$_2$PO$_4$

10–16%

2   trichodimerol

Watching this experience was transformative for me as a chemist. First, it immediately highlighted the artistry, skill, and talent needed by a practicing synthetic chemist to achieve success. Indeed, although individual reaction parameters certainly have a dramatic effect on the outcome, it is deep thinking based on first principles coupled with proper technique that can change those outcomes, ultimately putting everything into place so that the domino process can properly orchestrate itself. Second, it opened up that same joy I felt as a child with my toy dominoes, indicating that I would want to devote a significant portion of my research career into pursuing strategies, reactions, and tactics that could similarly bring molecules together and forge bonds with ballet-like efficiency and

Introduction   **3**

precision. Finally, it made me realize how much I still needed to learn and master if I was ever to reach that goal.

The objective of these two volumes for the *Science of Synthesis Reference Library* is to communicate that same sense of inherent pleasure in the power, artistry, and challenge of domino chemistry, highlighting the current state of the art by teaching the skills and providing the tools needed for success, while also demarcating frontiers for future developments. With the efforts of a spectacular group of world-class authors who have contributed their time and talent to this work, I am confident that this goal has been achieved. Unlike other excellent reviews and monographs that exist on the subject (and there are many, for which no attempt at referencing will be made here; the leading and influential works are cited in the subsequent individual chapters), a different organization framework for domino chemistry has been sought for these volumes based on the reaction type that initiates the sequence. In some cases, this is a specific process such as the Diels–Alder reaction, while in other chapters it is a slightly broader presentation, for example reaction sequences initiated by gold. What has resulted is an effective way to compare and contrast approaches. In addition, because the contributors have provided only the most representative examples along with experimental procedures for processes that have high generality, rather than attempt to be comprehensive, key lessons can be imparted effectively for developing even more powerful approaches.

The first volume opens with polyene cyclizations, a classic domino sequence dating back over half a century to the pioneering ideas inherent in the Stork–Eschenmoser hypothesis, which have been explored non-stop in the ensuing decades. In a welcome contribution by Shenvi and Wan (Section 1.1), not only is a clear sense of the history of the reaction provided with several classic transformations, but the vitality of research in this arena is also evident by the inclusion of newer advances, such as the Corey group's use of indium(III) bromide to activate terminal alkynes for highly efficient domino bond constructions, as shown in the conversion of **3** into **4** (Scheme 2).[4]

Scheme 2   Indium(III) Bromide Catalyzed Polyene Cyclization Involving an Alkyne[4]

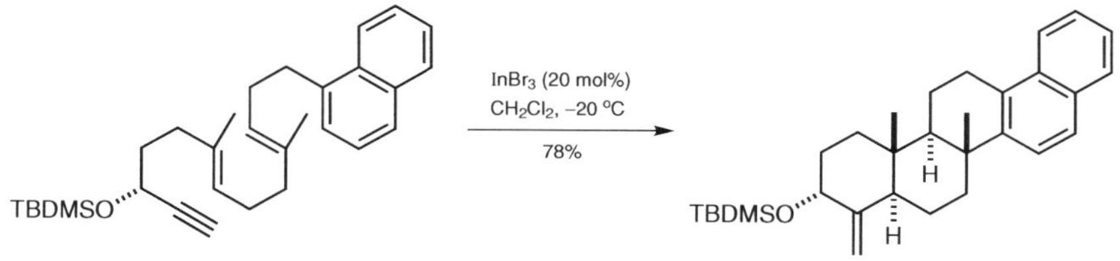

In the very next chapter, an excellent contribution by Jamison, Halkina, Kelley, Sittihan, and Armbrust (Section 1.2) shows how after those polyenes are converted into polyepoxides, appropriate activation can afford polyether natural products along the lines of the original Nakanishi hypothesis for the formation of brevetoxins and related targets. One example of such a powerful sequence is shown in Scheme 3.[5] Specific reaction conditions, careful substrate design, and knowledge of key physical organic principles prove essential to the successful examples that have been achieved to date, as this chapter effectively illustrates.

for references see p 11

Scheme 3 An Epoxide-Opening Domino Sequence To Form an Oxepane[5]

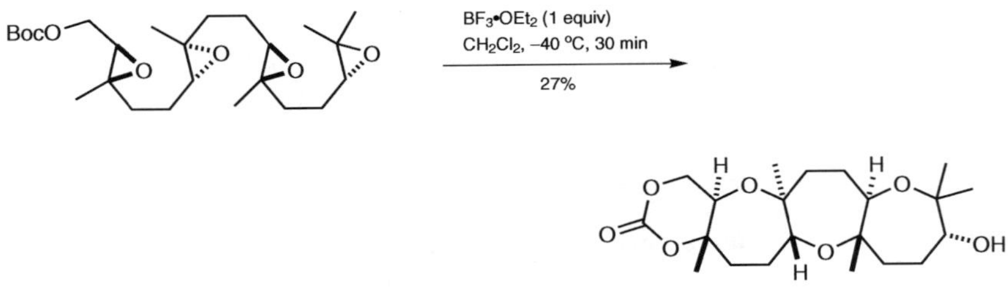

Next, transformations initiated by metathesis events are presented, with an excellent chapter by Lee and O'Connor highlighting various ways that alkenes and alkynes can be manipulated to forge a variety of complex materials, particularly natural products (Section 1.3.1). Their coverage includes detailed mechanistic analyses of several examples, emphasizing both current knowledge and areas where further studies are needed to advance such domino sequences. In the following contribution by Engle and Renata (Section 1.3.2), the extension of metathesis processes to carbonyl substrates highlights some additional directions for the prosecution of domino sequences. One particularly instructive and classic example using the Tebbe reagent is highlighted in Scheme 4.[6] Here, ring-opening metathesis of a strained alkene within bicycle **5** sets the stage for cyclization onto a neighboring carbonyl that affords a highly strained, four-membered ring en route to capnellene.

Scheme 4 A Domino Ring-Opening Metathesis/Carbonyl Cyclization Reaction[6]

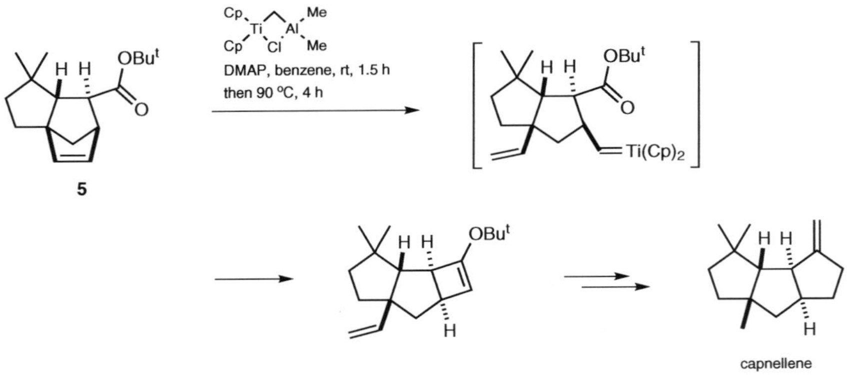

We then move on to domino events that begin with a radical reaction. An insightful chapter from Maimone and Hu (Section 1.4.1) opens this section of the volume by focusing on peroxide-initiated additions to functionalized molecules. Given their recent elegant synthesis of cardamom peroxide (Scheme 5) through a key step that features two such additions catalyzed by manganese (i.e., from **6** to **7**),[7] they are able to provide an excellent sense of the state of the art in this important area.

Introduction

Scheme 5 Domino Hydroperoxidation Reactions Catalyzed by Manganese[7]

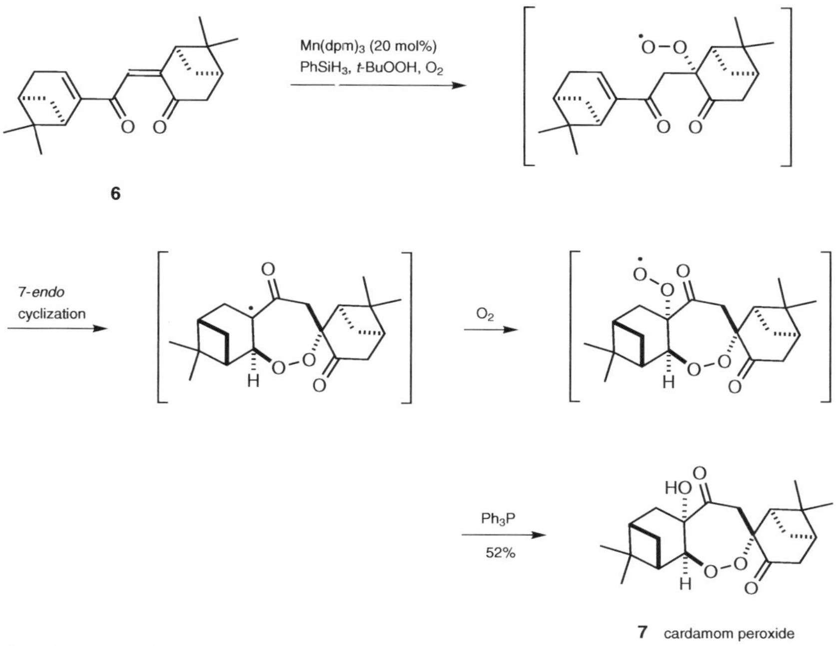

dpm = 2,2,6,6-tetramethyl-3,5-heptanedionato

Two chapters on more traditional radical chemistry (particularly involving carbon-centered radicals) follow, with the first from Stephenson, Devery, and Douglas (Section 1.4.2) and the second from Parker (Section 1.4.3). These contributions collectively cover historical cases as well as more modern applications such as photoredox-based radical sequences. They also cover the full gamut of initiators and quenching sources one might wish to consider for both inter- and intramolecular variants of radical-based domino chemistry, whether the goal is cyclizations, rearrangements, and/or fragmentations.

Volume 1 then continues with the presentation of some other, non-radical-based, domino transformations, starting with skeletal rearrangements. Such processes comprise a panoply of chemical reactions, be they the result of ring contractions, ring expansions, ring closures, or fragmentations, noting that all of these can, of course, be effected by many different reagents. Here, Forsyth and Adu-Ampratwum beautifully open the discussions of this body of domino chemistry by focusing on events initiated by protic acids and bases (Section 1.5.1). The following chapter by Tu, Wang, and Tang (Section 1.5.2) then outlines occasions when Lewis acids and bases can provide complementary and/or unique opportunities for related chemistry. The range of reactions covered in these two chapters is quite impressive, and the selections found within should hopefully inspire a number of future studies, be it in the area of transformations such as semipinacol rearrangements or Favorskii chemistry, or the myriad other processes that are touched upon for these types of processes. A final and comprehensive contribution from Kirschning, Gille, and Wolling (Section 1.5.3) then affords a sense of the power of silicon to induce such chemistry, particularly under Brook-rearrangement-type manifolds. One example from their chapter, part of the Smith total synthesis of mycoticin A,[8] is shown in Scheme 6. Here, carefully orchestrated 1,4-Brook rearrangements (i.e., anion relay chemistry) prove capable, in short order, of setting the key chiral centers in structure **8** that constitutes the upper half of the target molecule.

for references see p 11

6　Domino Transformations　Introduction

Scheme 6　1,4-Brook Rearrangements in a Total Synthesis of Mycoticin A[8]

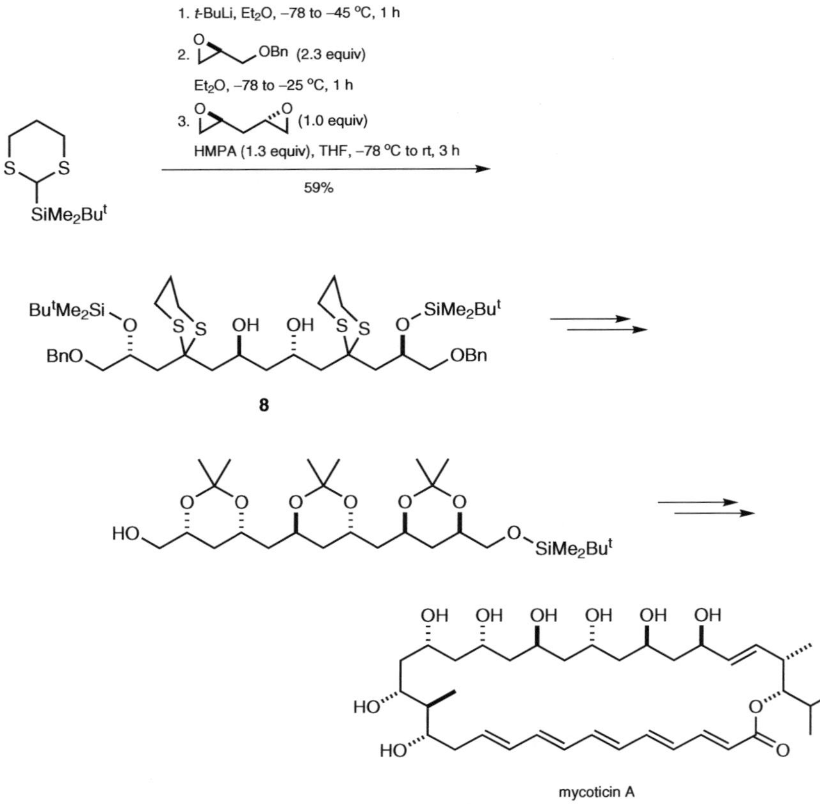

mycoticin A

Volume 1 then concludes with five chapters devoted to metal-mediated domino reactions (excluding those cases where the metals are behaving simply as Lewis acids, as those cases are covered earlier in the volume, as already mentioned). Here, the reader is treated to a world-class series of presentations from leading practitioners. We open with palladium-initiated chemistry, where Anderson has provided an excellent collection of domino processes that are possible with the right tools and properly designed substrates (Section 1.6.1). One example found in his chapter is shown in Scheme 7;[9] this work, from the Curran group, highlights beautifully the range of bonds, rings, and materials that are accessible when everything is carefully set up to be orchestrated into greater complexity.

Introduction

Scheme 7 Palladium-Initiated Domino Sequence En Route to DB-67[9]

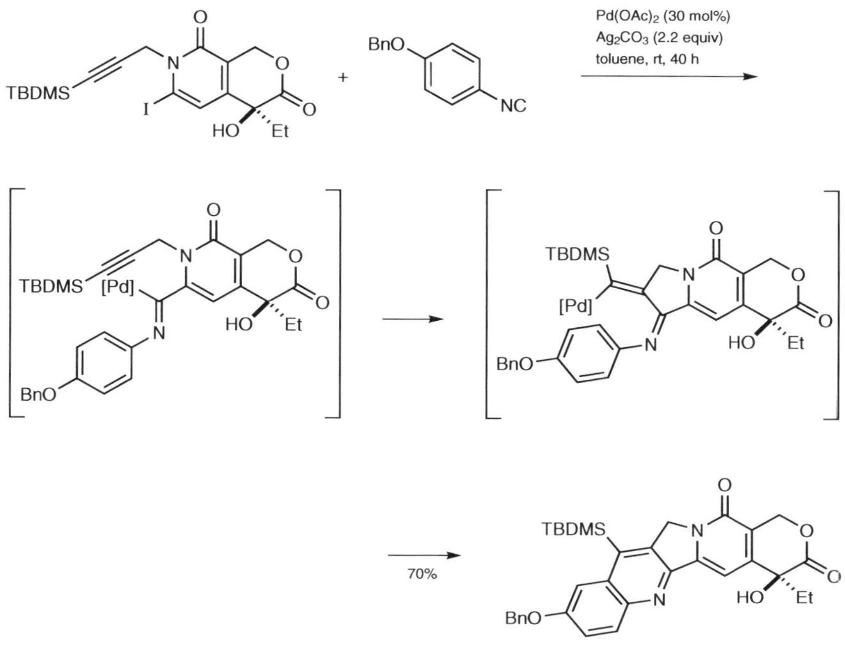

Next is a chapter by Doyle, Truong, and Xu on rhodium-mediated domino sequences (Section 1.6.2). A transformation from one of their own reports is presented here in Scheme 8,[10] which highlights, in particular, the value of rhodium in forming reactive carbenes that can effect a range of reactions, in this case the synthesis of fully substituted furan **9**. Well-written, thorough, and interesting chapters then follow on gold chemistry (Nevado, Merino, and Salvador; Section 1.6.3), rare earth metal chemistry (Ohshima; Section 1.6.4), and cobalt chemistry (Yang and Shi; Section 1.6.5). These contributions collectively highlight how diverse and important processes such as Conia-ene reactions and Pauson–Khand chemistry can be effectively orchestrated by many different metals to rapidly build molecular complexity in highly challenging settings.

Scheme 8 Rhodium-Initiated Domino Sequence Involving Carbene/Alkyne Metathesis and Carbonyl Ylide Formation[10]

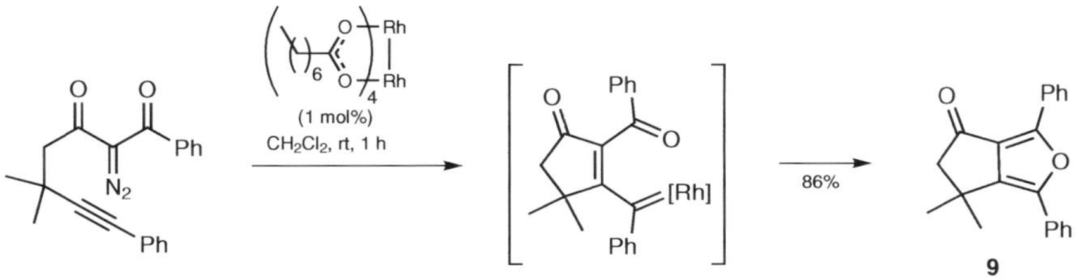

Volume 2 begins with coverage of domino processes that are started by pericyclic reactions. Of the possible ways to achieve domino sequences, pericyclic reactions in all their forms, be it Diels–Alder reactions, Cope rearrangements, electrocyclizations, ene reactions, etc., have perhaps provided some of the most fertile ground for creative approaches to rapidly build molecular complexity; often these domino sequences may well be biomi-

for references see p 11

Domino Transformations   Introduction

metic bond constructions. The five chapters in this section are, like those in Volume 1, arranged based on the initiating event, with the opening contribution by Sorensen and West (Section 2.1.1) describing domino events that commence with the Diels–Alder reaction. One example from their scholarly presentation is shown below in Scheme 9,[11] highlighting how a tandem Diels–Alder reaction/retro-Diels–Alder sequence involving an alkyne dienophile and a pyrone diene in the form of **10** could lead to the "bent" aromatic ring needed to complete a total synthesis of haouamine A.

Scheme 9   Diels–Alder/Retro-Diels–Alder Domino Sequence for Haouamine A[11]

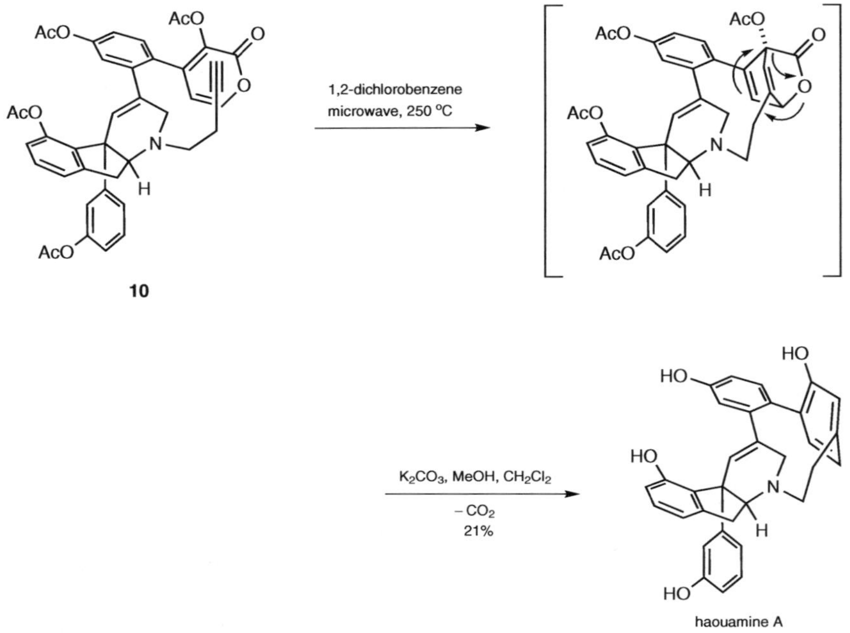

The subsequent chapter from Coldham and Sheikh (Section 2.1.2) beautifully illustrates the power of domino constructions started through [2+2], [3+2], and [5+2] cycloadditions, and other related manifolds that are not of a Diels–Alder-type. We then move onto electrocyclic ring constructions and ring openings in a well-fleshed-out chapter from Suffert, Gulea, Blond, and Donnard (Section 2.1.3). One example discussed there, a tandem 8π-electrocyclization/Diels–Alder sequence that follows a Stille coupling between **11** and **12**, is shown in Scheme 10;[12] this event produces the fused-ring system of **13** via a process that might be similar to Nature's construction of the target PF-1018; it is notable that only a single diastereomer results from this laboratory domino sequence. This example could easily have been put into the palladium chapter, but the centrality of the electrocyclization process in light of other examples in the chapter suggested this as the best location from a didactic perspective. Finally, well-framed and thoroughly presented chapters from Zakarian and Novikov on sigmatropic shifts and ene reactions (Section 2.1.4), and from Guerrero on [3,3]-rearrangements (i.e., Cope and Claisen chemistry; Section 2.1.5) round out this impressive collection of domino chemistry examples.

Introduction

Scheme 10   Stille Coupling/Electrocyclization/Diels–Alder Domino Sequence toward (−)-PF-1018[12]

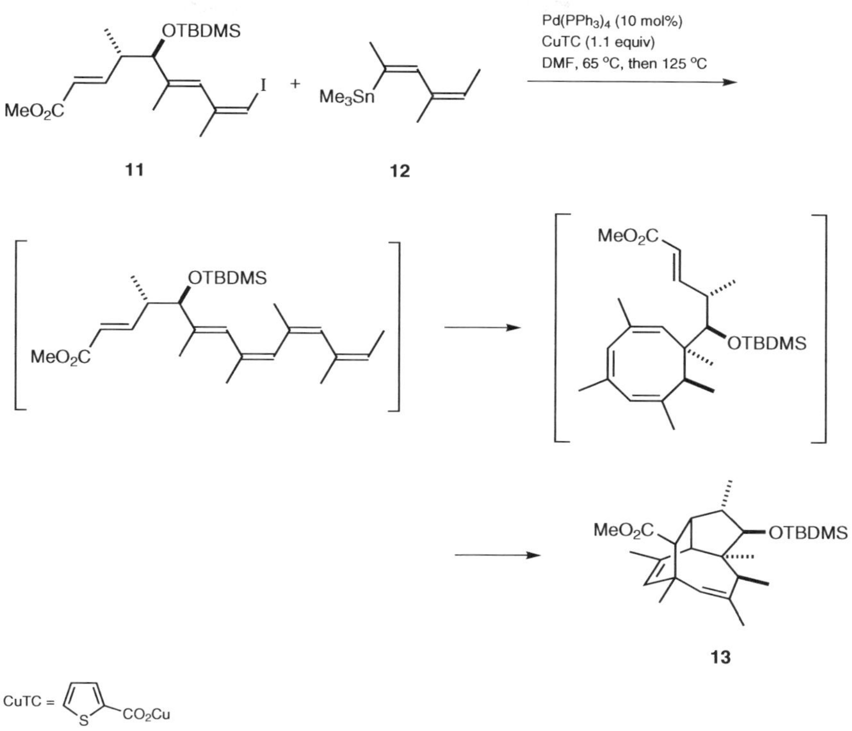

The next chapter comes from Porco and Boyce (Section 2.2) and covers attempts to use dearomatization events as part of domino sequences, an area in which the lead author's group has long been engaged. An example of their work is shown in Scheme 11, with a series of alkylative processes effecting subsequent dearomatizations leading to a functionalized cage structure (**14**) in short order if the right conditions are used. Thorough tables in this chapter help the reader identify optimal conditions for each desired transformation.

Scheme 11   Alkylative Dearomatization via a Domino Sequence[13]

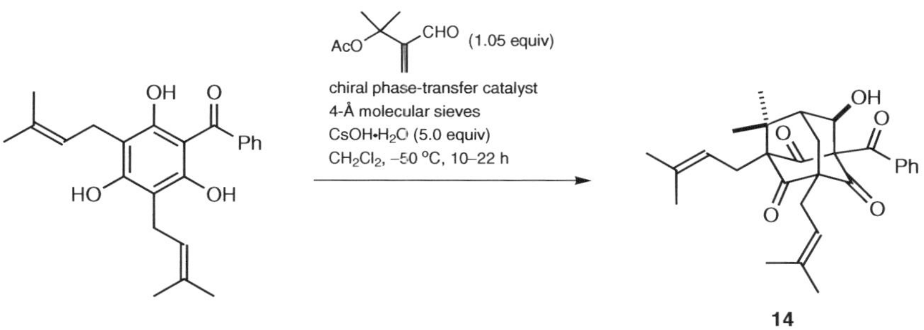

Finally, the volume rounds out with domino sequences that commence with various addition reactions. Yeung and Yu begin these presentations with additions to nonactivated alkenes (Section 2.3.1), covering hydroaminations, hydroetherifications, haloetherifications, and halocyclizations in particular, drawing in many cases on their own expertise

for references see p 11

and scholarship in the area. The following chapter, by Bella, Moliterno, Renzi, and Salvio (Section 2.3.2), focuses on additions to activated double bonds, such as Michael reactions, enamine chemistry, and the reactions of enol ethers to achieve domino transformations. Much of the discussion here focuses on the use of organic catalysts to effect such processes, with a particularly informative and timely discussion on the large-scale industrial synthesis of molecules such as **15A** and/or **15B** (Scheme 12), which can subsequently be converted into chiral diene ligands.[14–16] Other high-value targets discussed at some length through varied approaches include (−)-oseltamivir (Tamiflu).

Scheme 12   A Michael/Aldol Domino Sequence[14]

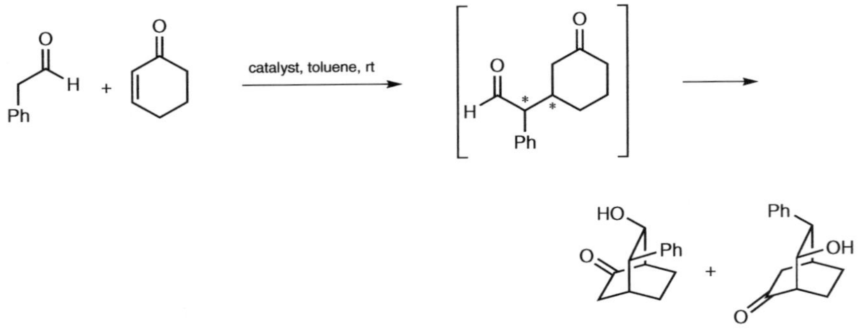

The next chapter, by Wang and Song, on additions to carbonyls (Section 2.3.3) includes continued discussion of the importance of small molecule organic catalysts, for example to achieve aldol reactions, among other types of additions such as Grignard chemistry. The final chapter, from Dömling, Zarganes Tzitzikas, Neochoritis, and Kroon (Section 2.3.4), moves on to additions to carbon–nitrogen bonds in the form of imines, nitriles, and related functional groups. These final two chapters certainly highlight that there are many diverse targets that can be made through appropriate substrate design coupled with mechanistic creativity and high experimental acumen.

In conclusion, I hope you will find these volumes to be as informative and instructive as I have while editing their pages. They definitely provide a tool-box that can hopefully propel domino chemistry to the next stages of complexity generation and enhanced synthetic efficiency, and I hope you will find inspiration within their pages, whether your goal is new synthetic methodology, target construction, or the development of commercially viable routes to high-value materials. The team of authors has certainly set up several dominoes in place; it is for you, the reader, to add to them (whether at the front or the back of the chain) and/or rearrange them to make even more intricate designs and impressive combinations.

References

[1] Nicolaou, K. C.; Jautelat, R.; Vassilikogiannakis, G.; Baran, P. S.; Simonsen, K. B., *Chem. Eur. J.*, (1999) **5**, 3651.
[2] Barnes-Seeman, D.; Corey, E. J., *Org. Lett.*, (1999) **1**, 1503.
[3] Nicolaou, K. C.; Simonsen, K. B.; Vassilikogiannakis, G.; Baran, P. S.; Vidali, V. P.; Pitsinos, E. N.; Couladouros, E. A., *Angew. Chem. Int. Ed.*, (1999) **38**, 3555.
[4] Surendra, K.; Qui, W.; Corey, E. J., *J. Am. Chem. Soc.*, (2011) **133**, 9724.
[5] McDonald, F. E.; Bravo, F.; Wang, X.; Wei, X.; Toganoh, M.; Rodríguez, J. R.; Do, B.; Neiwert, W. A.; Hardcastle, K. I., *J. Org. Chem.*, (2002) **67**, 2515.
[6] Stille, J. R.; Grubbs, R. H., *J. Am. Chem. Soc.*, (1986) **108**, 855.
[7] Hu, X.; Maimone, T. J., *J. Am. Chem. Soc.*, (2014) **136**, 5287.
[8] Smith, A. B., III; Pitram, S. M., *Org. Lett.*, (1999) **1**, 2001.
[9] Curran, D. P.; Du, W., *Org. Lett.*, (2002) **4**, 3215.
[10] Qian, Y.; Shanahan, C. S.; Doyle, M. P., *Eur. J. Org. Chem.*, (2013), 6032.
[11] Baran, P. S.; Burns, N. Z., *J. Am. Chem. Soc.*, (2006) **128**, 3908.
[12] Webster, R.; Gaspar, B.; Mayer, P.; Trauner, D., *Org. Lett.* (2013) **15**, 1866.
[13] Qi, J.; Beeler, A. B.; Zhang, Q.; Porco, J. A., Jr., *J. Am. Chem. Soc.*, (2010) **132**, 13642.
[14] Bella, M.; Scarpino Schietroma, D. M.; Cusella, P. P.; Gasperi, T.; Visca, V., *Chem. Commun. (Cambridge)*, (2009), 597.
[15] Abele, S.; Inauen, R.; Spielvogel, D.; Moessner, C., *J. Org. Chem.*, (2012) **77**, 4765.
[16] Brönnimann, R.; Chun, S.; Marti, R.; Abele, S., *Helv. Chim. Acta*, (2012) **95**, 1809.

1.1 Polyene Cyclizations

R. A. Shenvi and K. K. Wan

General Introduction

Saturated and partially unsaturated (nonaromatic) carbocycles find broad distribution across the planet as endogenous compounds in all organisms and also as products of the chemical industry. Their isolation from natural sources, in combination with their laboratory chemical synthesis, enables large-scale access to these carbocycles for human use as fragrances, flavors, pharmaceuticals, and polymers.[1] Dissection of any given carbocycle into one or more alkenes benefits from a decrease in complexity, sometimes dramatically, to simple unsaturated building blocks.[2] Steroid biosynthesis is a classic illustration of this abstract principle, whereby (S)-2,3-oxidosqualene (**1**) is converted in one step into lanosterol (**2**) (Scheme 1).[3]

Scheme 1 Biosynthesis of Lanosterol from Oxidosqualene in One Step[3]

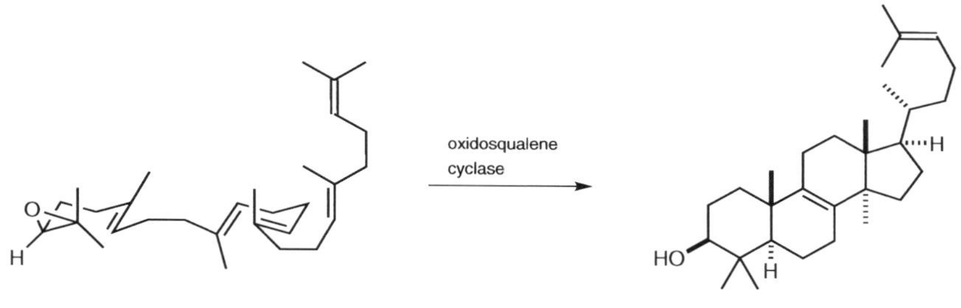

The Diels–Alder reaction is a more familiar example, and generates two C—C bonds, one or more rings, and as many as four stereocenters. In this chapter, however, the Diels–Alder reaction (Section 2.1.1) and other pericyclic ring-forming reactions (Section 2.1) will not be discussed because these are cycloadditions not cyclizations, and because each pericyclic reaction could constitute its own chapter given the diversity of application {e.g., see *Science of Synthesis*, Vol. 47 [Alkenes (Section 47.1.3.1)]}. Instead, the focus will be on domino (cascade) polyene cyclizations, with an emphasis on those processes with practicality and broad utility rather than idiosyncratic examples with narrower applications. To begin, some terms must be defined.

Domino transformation: A first reaction sets the stage for a second reaction, which can then effect a third reaction, and so on.

Polyene cyclization: A reaction between two or more double bonds contained within the same molecule (a polyene) to form one or more rings via one or more C—C bond-forming events.

In this chapter, we spend very little time on the discussion of intramolecular Michael reactions. Although these reactions fulfill the criteria presented above, the actual building blocks are not alkenes, but rather carbonyls, which are converted into nucleophilic enols, enolates, or enol ethers. Therefore, we will discuss only the most dramatic examples of successful anionic polycyclizations using carbonyl chemistry; more rigorous dis-

for references see p 39

cussions can be found in chapters devoted to domino Michael reactions [see Section 2.3.2 and *Science of Synthesis: Asymmetric Organocatalysis*, Vol. 1 (Section 1.1.4.2.5)]. Our focus here will be primarily on electron-neutral alkenes, which for reasons discussed below, must be evaluated using very different criteria.

Polyene cyclizations can be organized according to substrate structure and/or the mechanism of cyclization. This chapter will be organized according to reaction mechanism. It is important to point out that the substrate scope of polyene cyclizations is, in general, rather narrow and has mostly been restricted to naturally occurring polyisoprenoids and their derivatives. The reasons for this restriction are primarily two-fold: (1) the inexpensive, large-scale availability of geometrically pure geranyl, neryl, and farnesyl derivatives, and (2) the biological utility of cyclized polyisoprenoids, including steroids. It should also be noted that the cationic cyclizations of polyisoprenoids to cyclic terpenes are predictable and reliable, meaning that the all-Markovnikov patterning of cyclization sites in all-head-to-tail isoprenyls predispose the reactions to succeed. In contrast, it is challenging to design polyene cyclizations outside this class that proceed in general, or with high degrees of stereocontrol through appropriate transition states, so cyclizations of other polyene systems are infrequent and occasionally idiosyncratic. Some examples will be provided in this chapter, but it is worth noting at the outset that these events are currently hard to generalize.

From the perspective of synthesis theory, polyene cyclizations are a powerful approach for the efficient generation of both complexity and diversity, because each prochiral alkene within the substrate can generate two new stereocenters in the product. A notable feature of polyene cyclizations is the potential for a single synthetic route to a single target structure to be explosively elaborated to a series of both constitutional and stereochemical isomers. However, the general ability to controllably cyclize linear chains to multiple products, each with high selectivity, is unknown outside of enzymatic mediation. There are some noteworthy exceptions in the synthesis literature (see Section 1.1.3.2.3), but the control exerted by enzymes still generally eludes synthetic chemistry and represents a significant chemical frontier for further development.

1.1.1 Cationic Polyene Cyclizations Mediated by Brønsted or Lewis Acids

The steroid biosynthesis depicted in Scheme 1 is one fragment of an immense area of study in chemistry that can be collectively referred to as cationic polyene (or polyolefin) cyclization (or polycyclization), and also as cation–π cyclizations.[4,5] Formation of carbocations that are unstabilized by conjugation or attached heteroatoms has attracted immense interest over the last 90 years[6] due to the extremely high energy of these electron-deficient species.[7,8] The instability of these unstabilized carbocations can be observed in their short lifetimes, estimated to be between 100 femtoseconds to 1 picosecond[9] in water (close to the time required for a single bond vibration, i.e. 10^{-13} s),[10] and the consequent low activation energies associated with their reactions. The free energy difference between a carbocation and the transition state of its electrophilic reaction with an alkene is nearly "barrier-less".[11,12] Therefore, the ability of cyclase enzymes arrayed with multiple Lewis basic sites to stabilize and direct the reactions of cationic isoprenoids is nothing short of astonishing. It is not surprising that translating the enzyme-mediated reactions into bulk solvent for large-scale chemical preparation meets with numerous obstacles, some of which will be mentioned later in this chapter. However, despite these inherent challenges, polyisoprenyl chains undergo predictable, stereoselective, and occasionally high-yielding cyclizations (head-to-tail) to form multiple rings in bulk organic solvent. Methods to effect this transformation vary in the initiation and termination step, but propagation is generally carried through a single or repeating isoprenyl unit. The benefit of this conserved propagation, and the reason it has seen little variation, is that (1) many common natural product scaffolds can be accessed from these poly-

1.1.1.1 Most Used Cationic Polyene Cyclization Methods

1.1.1.1.1 Polyene Cyclization via Biomimetic Heterolytic Opening of Epoxides by Alkylaluminum Lewis Acids

Early work on the reactions of alkenes with Brønsted acid mediation originally did not recognize the intermediacy of carbocations (or carbenium ions) because the theoretical instability of such structures appeared to preclude their existence.[14] However, once stabilized[15–17] and unstabilized carbocations[6] had been established as viable intermediates, it was not long before they were recognized as possible intermediates in sesquiterpene biosynthesis[18,19] and in steroid biosynthesis.[20,21] Extensive work proved both the possibility and strategic benefit of utilizing cationic polyene cyclization in the synthesis of steroids and other molecules,[22] as well as many rules for carrying out the reactions in bulk solvent, all of which has been reviewed previously.[4,23,24]

The most consistently utilized method for carrying out polyene cyclization in solution relies on the biomimetic heterolytic opening of epoxides[25–27] by alkylaluminum Lewis acids.[28] There are several advantages to this strategy. First, the strain of the oxirane ring allows reactions at low temperature, which is important for the highly organized transition states of cyclizations because it minimizes the contribution of the entropic term to the transition state free energy (Scheme 2 und Scheme 3).[29,30] Second, alkylaluminum Lewis acids are "self-quenching", so any strong acid generated by adventitious water contamination will react with the strongly basic C—Al bond. Third, the Lewis acids are completely soluble in nonbasic aprotic solvents such as dichloromethane, which does not itself cause the elimination of carbocations ($H_0 \sim -17$, Hammett acidity), whereas basic solvents such as tetrahydrofuran (pK_a −2.05) or diethyl ether (pK_a −3.05) are problematic. Finally, several methods exist for the enantioselective synthesis of epoxyisoprenoids with high levels of absolute stereocontrol.[31–35] Another benefit of this method has been proposed to be a ligand exchange that can occur between oxirane-ligated and unligated molecules of the alkylaluminum Lewis acids,[28] because similar exchange is observed with pyridine[36] or carbonyl ligands,[37] but whether this exchange is kinetically relevant on the time scale of polycyclization is unknown. It should also be noted that the first ring formation in Brønsted acid mediated epoxide–alkene cyclizations is concerted with epoxide scission rather than a stepwise process,[38] and some evidence suggests that subsequent cyclizations may also be concerted with the initial ionization,[39] but the relevance of this latter data to the methods discussed herein is unclear. An example of a polyene cyclization mediated by ethylaluminum dichloride is shown in Scheme 3 for the formation of decalins **4** and **5** from diene epoxide **3**.

for references see p 39

Domino Transformations 1.1 Polyene Cyclizations

Scheme 2 Corey's Cationic Polyene Cyclization[29]

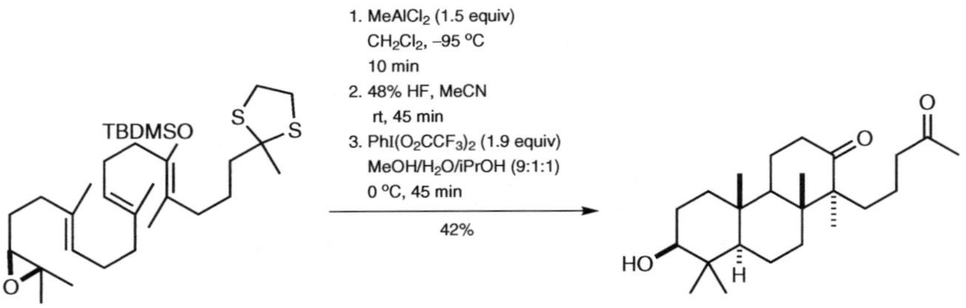

1. MeAlCl$_2$ (1.5 equiv)
   CH$_2$Cl$_2$, –95 °C
   10 min
2. 48% HF, MeCN
   rt, 45 min
3. PhI(O$_2$CCF$_3$)$_2$ (1.9 equiv)
   MeOH/H$_2$O/iPrOH (9:1:1)
   0 °C, 45 min

42%

Scheme 3 Corey's Cationic Polyene Cyclization with Ethylaluminum Dichloride as Lewis Acid[30]

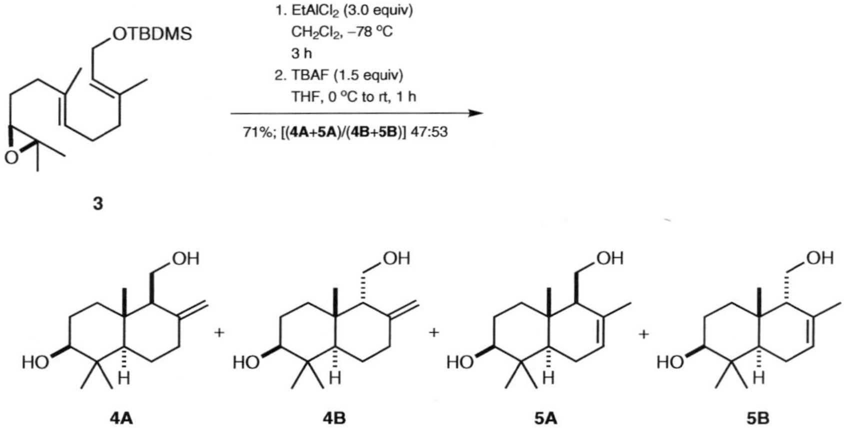

1. EtAlCl$_2$ (3.0 equiv)
   CH$_2$Cl$_2$, –78 °C
   3 h
2. TBAF (1.5 equiv)
   THF, 0 °C to rt, 1 h

71%; [(**4A+5A**)/(**4B+5B**)] 47:53

4A   **4B**   **5A**   **5B**

Decalins 4 and 5; Typical Procedure:[30]

> **CAUTION:** *Ethylaluminum dichloride is highly pyrophoric and corrosive, and is known to ignite spontaneously at room temperature when exposed to air. Extreme caution should be taken during its synthesis, storage, and handling in an effort to avoid exposure to air.*

Polyene epoxide **3** (50 mg, 0.142 mmol, 1.0 equiv) was dried azeotropically with toluene (2 × 0.5 mL) under high vacuum for 30 min. The vacuum was then filled with dry N$_2$ and the flask was fitted with a septum and an argon balloon. Anhyd CH$_2$Cl$_2$ (14 mL) was then added and the resulting soln was cooled to –78 °C. A 1 M soln of EtAlCl$_2$ (425 µL, 0.425 mmol, 3.0 equiv) diluted with anhyd CH$_2$Cl$_2$ (4.5 mL) was then added down the inner wall of the reaction flask (so it can cool before hitting the reaction mixture) via a 5-mL glass syringe (which was well greased and oven dried) over the course of 3 h while the mixture was stirred rapidly at –78 °C. After the addition was complete, the bright yellow mixture was stirred for an additional 30 min at –78 °C. Et$_3$N (425 µL, 3.05 mmol, 21.5 equiv) was added slowly over 1 min down the inner wall of the reaction flask. The flask was then taken out of the cooling bath and the septum was removed. Sat. aq potassium sodium tartrate (3 mL) was added while the mixture was stirred rapidly. The mixture was then allowed to warm to ambient temperature, and the volatile components were removed under reduced pressure at 30 °C on a rotary evaporator. The residue was then extracted with EtOAc (3 × 20 mL), and the combined organic layer was dried (Na$_2$SO$_4$), filtered, and concentrated. The crude mixture was taken up in EtOAc/hexanes (1:1) and filtered through silica gel to remove any residual amine to afford the crude cyclized product,

which was reacted further without characterization. To a 0.5 M soln of the intermediate in anhyd THF (0.3 mL) at 0 °C was added a 1.0 M soln of TBAF in THF (213 µL, 0.21 mmol, 1.5 equiv). The mixture was then warmed to ambient temperature and TLC analysis indicated complete consumption of the starting materials after 1 h. The mixture was treated with sat. aq NaHCO$_3$ (1 mL) and diluted with EtOAc (10 mL). The organic layer was removed and the aqueous layer was extracted with EtOAc (2 × 5 mL). The combined organic layers were then dried (Na$_2$SO$_4$), filtered, and concentrated. The residue was purified by flash chromatography (silica gel, EtOAc/hexanes 65:35) to give a mixture of the products **4** and **5**; yield: 24 mg (71%); ratio [(**4A+5A**)/(**4B+5B**)] 47:53.

1.1.1.1.2   **Polyene Cyclization Mediated by Carbophilic Lewis Acids**

A disadvantage of the method described in Section 1.1.1.1.1 is the intolerance of Lewis basic functional groups, which compete with the epoxide for coordination to the aluminum-based Lewis acid. Whereas silyl ethers are generally tolerated due to only weak coordination ability,[30] alcohols, ethers, esters, and carbonates inhibit the initiation step. However, carbophilic Lewis acids such as indium(III) bromide will initiate polycyclization[40] via selective coordination to an appropriately placed alkyne [cationic gold(I) behaves similarly; see Section 1.1.1.3.3]. Notably, free phenols, phenyl ethers, and carbonates are tolerated and serve as efficient terminators via carbocation capture. Furthermore, the ability to initiate the reaction from a chiral nonracemic propargylic ether or alcohol relays this absolute stereochemistry into the rest of the polycyclic core (Scheme 4).

Scheme 4   Indium(III) Bromide Catalyzed Polyene Cyclization[40]

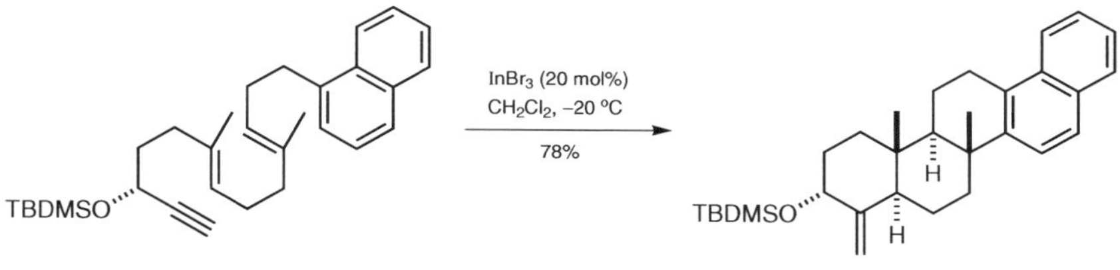

1.1.1.2   **Recent Advances in Cationic Polycyclization: Halonium-Initiated Polycyclization**

As noted above, the pK_a of protons adjacent to carbocations is extremely low,[8] and preemptive termination of cyclizations through E1 elimination is a general problem. The Snyder group therefore initiated a program[41] to improve the efficiency of halonium-initiated polyene cyclizations[42,43] leading to marine-derived haloterpenes,[44,45] a route which in the past suffered from generally poor yields. Key to the improvement in efficiency of these reactions is the development of electrophilic halogen sources that generate nonbasic anions upon reaction with nucleophiles, in contrast to widely used halosuccinimide or dihalogen reagents. Thus, the Snyder group developed a suite of chloronium, bromonium (e.g., **6**), and iodonium reagents that generate nonbasic, nonnucleophilic antimonate anions and thereby promote polyene cyclizations in good yield and generally excellent diastereoselectivity. A notable advantage of these reagents is that Lewis basic substituents and terminating groups are tolerated, whereas such groups are generally problematic for the aforementioned aluminum Lewis acid reagents, which are competitively coordinated and deactivated. As an example, Scheme 5 shows the conversion of 1,5-diene **7** into tricycle **8** using bromodiethylsulfonium bromopentachloroantimonate(V) (**6**).[42]

for references see p 39

Scheme 5 Bromonium-Initiated Cation–π Cyclization En Route to 4-Isocymobarbatol[42]

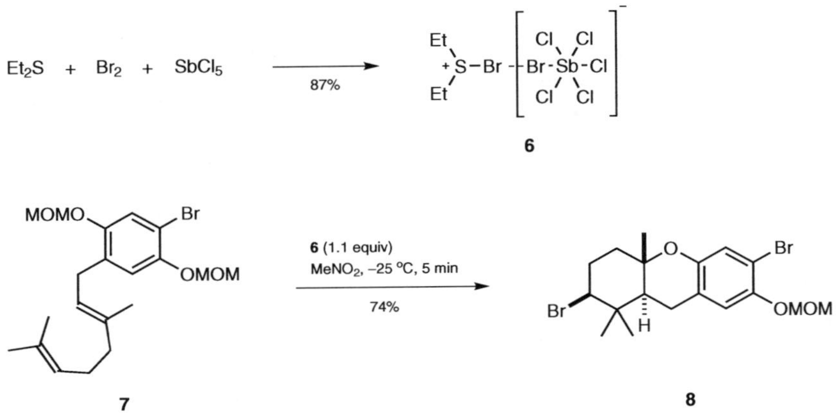

Bromodiethylsulfonium Bromopentachloroantimonate(V) (6); Typical Procedure:[42]

> **CAUTION:** *Bromine, diethyl sulfide, and antimony(V) chloride solution are highly toxic, caustic liquids that may be fatal if inhaled, swallowed, or absorbed through skin. All manipulations should be carefully carried out in a well-ventilated fume hood.*

Diethyl sulfide (2.97 mL, 27.5 mmol, 1.1 equiv) and a 1.0 M soln of $SbCl_5$ in CH_2Cl_2 (30.0 mL, 30.0 mmol, 1.2 equiv) were added slowly and sequentially to a soln of Br_2 (1.28 mL, 25.0 mmol, 1.0 equiv) in 1,2-dichloroethane (60 mL) at −30 °C. The dark red heterogeneous mixture was stirred at −30 °C for 20 min, then warmed slowly using a water bath until the soln became homogeneous (~30 °C). At this time, the reaction flask was allowed to cool slowly to 0 °C (4 h), then to −20 °C (12 h), and large orange plates crystallized from the soln. The solvent was decanted and the crystals were rinsed with cold CH_2Cl_2 (2 × 5 mL), then dried under reduced pressure; yield: 11.9 g (87%); mp 104 °C (with decomposition). A checked procedure can be found in *Organic Syntheses*.[46]

(2R*,4aR*,9aR*)-2,6-Dibromo-7-(methoxymethoxy)-1,1,4a-trimethyl-2,3,4,4a,9,9a-hexahydro-1H-xanthene (8); Typical Procedure:[42]

> **CAUTION:** *Nitromethane is flammable, a shock- and heat-sensitive explosive, and an eye, skin, and respiratory tract irritant.*

A soln of bromodiethylsulfonium bromopentachloroantimonate(V) (**6**; 55 mg, 0.100 mmol, 1.0 equiv) in $MeNO_2$ (0.5 mL) was added quickly via syringe to a soln of 1,5-diene **7** (0.100 mmol, 1.0 equiv) in $MeNO_2$ (1.5 mL) at −25 °C. After the mixture had been stirred at that temperature for 5 min, the reaction was quenched with 5% aq $NaHCO_3$/5% aq Na_2SO_3 (1:1; 5 mL). The mixture was stirred for 15 min, poured into H_2O (5 mL), and extracted with CH_2Cl_2 (3 × 10 mL). The combined organic layers were dried ($MgSO_4$), concentrated, and purified by flash column chromatography (silica gel); yield: 74%.

1.1.1.3 Other Common Cationic Polyene Cyclization Methods

1.1.1.3.1 Catalytic, Enantioselective, Protonative Polycyclization

1.1.1.3.1.1 Chiral Transfer from a Brønsted Acid

Enzymes can initiate polyene cyclization through direct protonation of polyisoprenoids to give single enantiomers of a polycycle,[3] a process which is challenging to replicate in bulk solvent because the corresponding unfolded, solvated isoprenyl carbocations are achiral. Therefore, the catalytic, enantioselective protonative polycyclization is truly remarkable in its transmission of chiral information from a Brønsted acid to a carbocationic intermediate in bulk solvent and onto a polycyclic scaffold that possesses multiple points of asymmetry.[47] The difficulty of the reaction is compounded by the facile reversibility of alkene protonation by E1 elimination. However, the Yamamoto group has shown that coordination of 1,1′-bi-2-naphthol derivatives with a strong Lewis acid ($SnCl_4$), e.g. to give **9**, sufficiently lowers the pK_a of these phenols to promote alkene protonation. This initial reaction is highly site-selective for the terminal alkene and provides a sufficiently long-lived cation to allow cyclization (or the cyclization is concerted) and a remarkable induction of asymmetry (Scheme 6).

Scheme 6 Combined Lewis Acid and Chiral Brønsted Acid Catalyzed Cyclization[47]

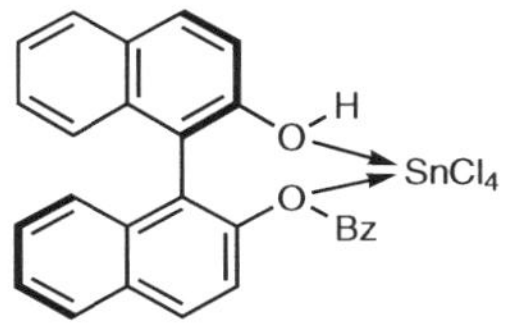

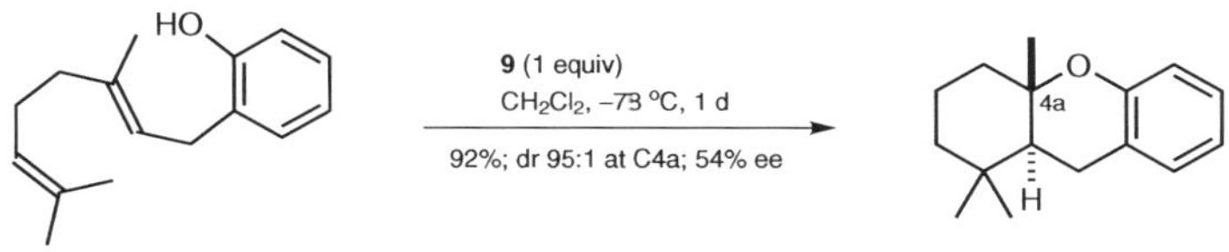

1.1.1.3.1.2 Chiral Transfer from (*R*)-2,2′-Dichloro-1,1′-bi-2-naphthol–Antimony(V) Chloride Complex

An additional, and arguably more convenient, method for site-selective and enantioselective protonation/polyene cyclization developed recently by the Corey group uses simple 2,2′-dichloro-1,1′-bi-2-naphthol complexes with a strong antimony Lewis acid ($SbCl_5$) such as **10**, and causes sufficient alkene protonation at −78 °C and transmission of absolute stereochemistry at 89% enantiomeric excess, on average (Scheme 7).[48] The efficiency of the reaction is attributed to both the electronegativity of the *ortho* chlorides and the increased Lewis acidity of antimony(V) chloride over tin(IV) chloride, rendering the phenols more acidic and the reaction therefore more rapid. The steric bulk of the antimonate complex is proposed to provide enhanced stereoselection, given the only modest steric size of the *ortho* chlorides.

for references see p 39

Scheme 7 Polycyclization with (*R*)-2,2′-Dichloro-1,1′-bi-2-naphthol–Antimony(V) Chloride Complex[48]

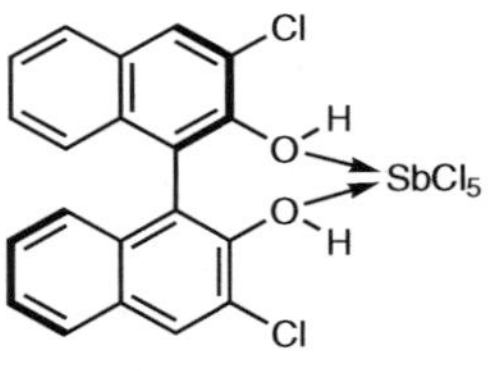

10

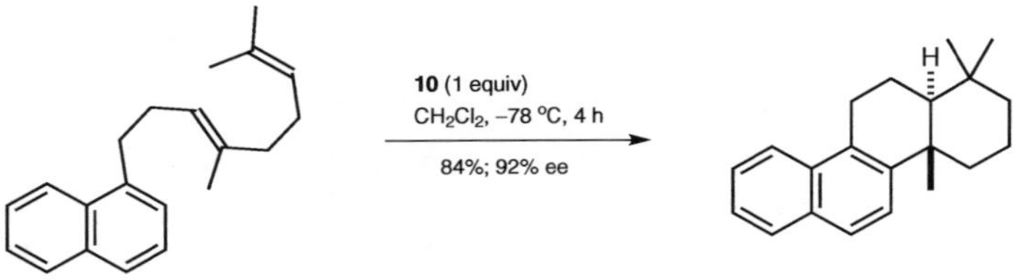

10 (1 equiv)
CH₂Cl₂, −78 °C, 4 h

84%; 92% ee

Remarkably, this reactivity is generalizable and can be extended beyond the standard polyisoprenyl substrates to more unusual systems such as triene **11**, which cyclizes to enantiomerically enriched tricycle **12**. It is hard to imagine other methods to synthesize **12** easily as a single enantiomer (Scheme 8).[49]

Scheme 8 Polycyclization En Route to Dehydroabietic Acid[49]

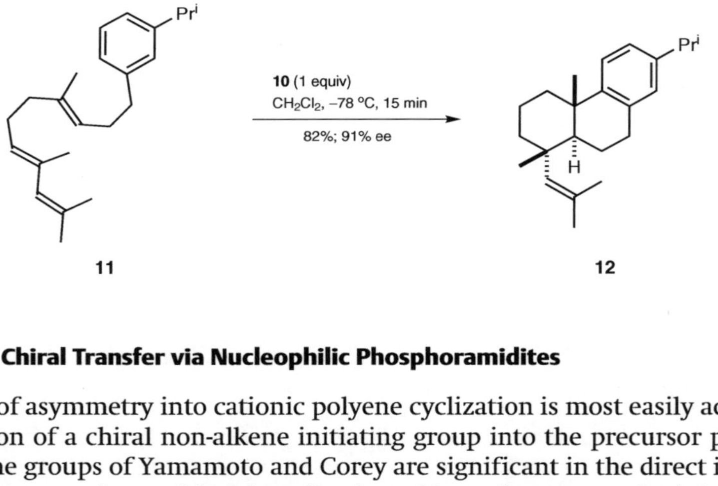

10 (1 equiv)
CH₂Cl₂, −78 °C, 15 min

82%; 91% ee

11

12

1.1.1.3.1.3 Chiral Transfer via Nucleophilic Phosphoramidites

Introduction of asymmetry into cationic polyene cyclization is most easily accomplished by introduction of a chiral non-alkene initiating group into the precursor polyene. The methods of the groups of Yamamoto and Corey are significant in the direct introduction of asymmetry concomitant with polycyclization, akin to the protonative initiation of isoprenes by cyclase enzymes. In a similar vein, the Ishihara group demonstrated that stoichiometric 1,1′-bi-2-naphthol-based phosphoramidites (e.g., **13**) can mediate the enantioselective transfer of iodonium to a terminal alkene to initiate polyene cyclization as long as electron-rich alkenes and terminating groups are used (Scheme 9).[50]

Scheme 9 Enantioselective Halocyclization Induced by Nucleophilic Phosphoramidites[50]

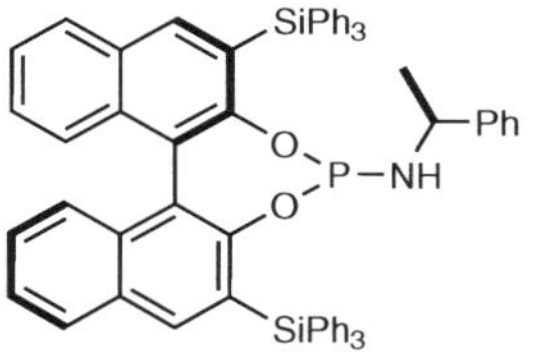

13

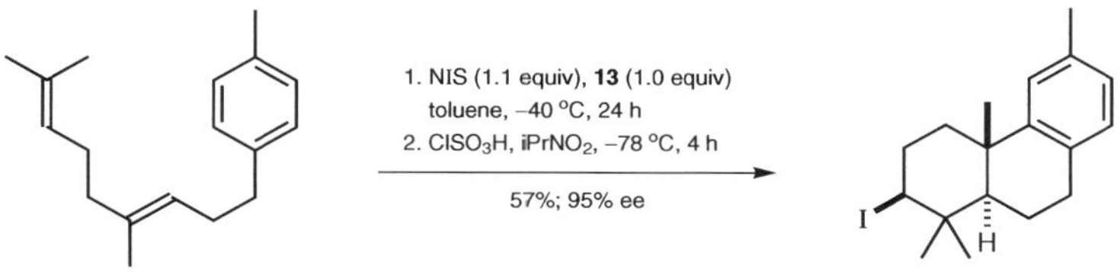

1. NIS (1.1 equiv), **13** (1.0 equiv)
   toluene, −40 °C, 24 h
2. ClSO₃H, iPrNO₂, −78 °C, 4 h

57%; 95% ee

1.1.1.3.2 Polyene Cyclization Initiated by Unsaturated Ketones and Mediated by Aluminum Lewis Acids

High energy carbocations have a tendency to undergo Wagner–Meerwein shifts of adjacent hydrogen and carbon atoms to produce rearranged structures.[6] Within cyclase enzymes, these reactions are not restricted to the energy surface of the carbon skeleton itself, but instead are highly controlled by the entire cyclase–substrate complex,[3] although promiscuity abounds in some cases.[51] In bulk solvent, carbocations slip across a very high energy landscape[52] and therefore the ability to mimic and direct the postcyclization rearrangements found in terpene biosynthesis has evaded chemists for decades. The Snider group has developed aluminum Lewis acid mediated reactions that predictably lead to rearranged products (e.g., **15**) in certain substrates by using unsaturated ketones (e.g., **14**) as the initiating moiety (Scheme 10).[53] The intermediate enolate acts as a cation sink, i.e. bond migration places the cationic charge allylic to the enolate as a highly stabilized vinylogous oxonium. The predictability of these Wagner–Meerwein shifts make this method a very useful tool for relaying oxidation state and stereochemistry.

Scheme 10 Lewis Acid Induced Conjugate Addition[53]

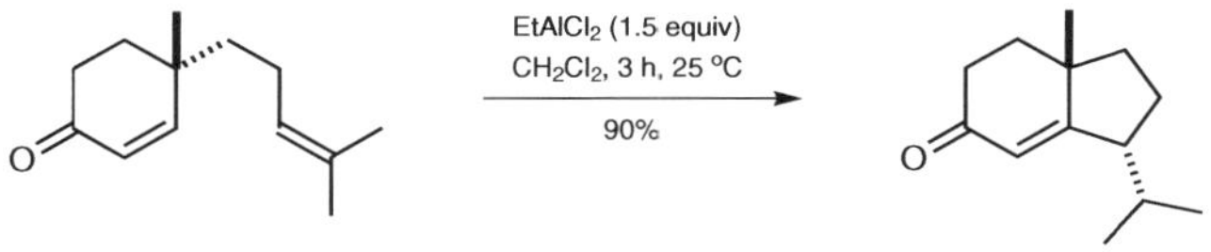

EtAlCl₂ (1.5 equiv)
CH₂Cl₂, 3 h, 25 °C

90%

14　　　　　　　　　　**15**

(3R,7aS)-3-Isopropyl-7a-methyl-1,2,3,6,7,7a-hexahydro-5H-inden-5-one (15); Typical Procedure:[53]

> **CAUTION:** *Ethylaluminum dichloride is highly pyrophoric and corrosive, is known to ignite spontaneously at room temperature when exposed to air. Extreme caution should be taken during its synthesis, storage, and handling in order to avoid exposure to air.*

A soln of (S)-4-methyl-4-(4-methylpent-3-enyl)cyclohex-2-en-1-one (**14**; 200 mg, 1.05 mmol, 1.0 equiv) in CH₂Cl₂ (5 mL) was treated with a 1.57 M soln of EtAlCl₂ in heptane (1.00 mL, 1.57 mmol, 1.5 equiv). After the mixture had been stirred at 25 °C for 3 h, sat. aq NH₄Cl

for references see p 39

(10 mL) was slowly added followed by just enough 10% aq HCl to dissolve the precipitated alumina. The organic layer was separated and the aqueous layer was extracted with Et_2O (3 ×). The combined organic layers were dried ($MgSO_4$) and concentrated to give 200 mg of the cyclized product which was >95% pure. The resulting crude material was purified by flash chromatography (silica gel, EtOAc/hexanes 1:2) to give the pure product; yield: 180 mg (90%).

1.1.1.3.3 Gold-Mediated Enantioselective Polycyclization

Gold(I) catalysts supported by a strong, neutral ligand but unbound by a strong anionic ligand (referred to as "cationic gold") exhibit strong carbophilicity/π-Lewis acidity, weak oxophilicity, and will activate alkynes toward nucleophilic attack. The groups of Fürstner, Kozmin, and others have illustrated the cationic nature of intermediates in gold-catalyzed ene–yne cyclizations[54] by trapping the nascent carbocations[55] with pendent nucleophiles such as sulfonamides[56] and carboxylic acids. Although this early work focused on ene–yne cyclizations, it laid the mechanistic groundwork to prompt investigations into polyisoprenoid cyclizations. A major challenge to π-Lewis acidic gold catalysis is the induction of enantioselectivity, because the alkyne–gold(I)–ligand dihedral angle is nearly linear[57] and therefore the chiral environment of most ligands is difficult to transmit. The Toste group identified[58] extremely sterically encumbered 2,2′-bis(diphenylphosphino)-1,1′-binaphthyl derivatives capable of initiating asymmetric polyene cyclization via coordination and electrophilic activation of an initiating alkyne.[59] Due to the high carbophilicity of gold(I), multiple Lewis basic sites such as esters, phenols, phenyl ethers, and sulfonamides are tolerated in the reaction process. An example of this enantioselective polycyclization is shown in Scheme 11 for the synthesis of tricycle 18 from enyne 17 catalyzed by the digold(I) chloride complex 16.[60]

Scheme 11 Gold-Catalyzed Enantioselective Polycyclization Reaction[60]

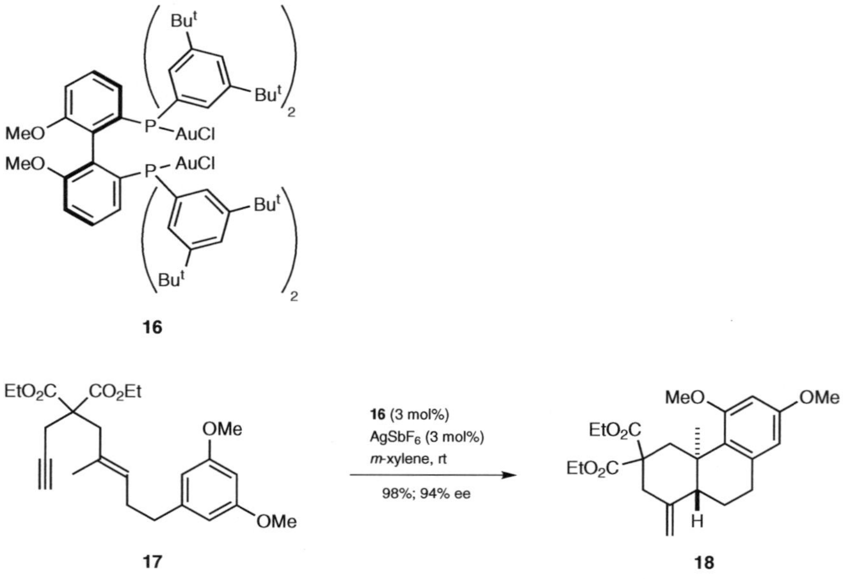

Diethyl (4aR,10aR)-5,7-Dimethoxy-4a-methyl-1-methylene-1,4,4a,9,10,10a-hexahydro-phenanthrene-3,3(2H)-dicarboxylate (18); Typical Procedure:[60]
A mixture of $AgSbF_6$ (0.8 mg, 2.2 µmol, 0.05 equiv) and the bisphosphine digold(I) chloride complex 16 (3.32 mg, 2.22 µmol, 0.05 equiv) was suspended in *m*-xylene (300 µL) in a

sealed vial, and sonicated or stirred magnetically at rt for 15 min. The resulting suspension was filtered through a glass microfiber plug directly into a soln of diethyl (E)-2-[5-(3,5-dimethoxyphenyl)-2-methylpent-2-enyl]-2-(prop-2-ynyl)malonate (**17**; 18 mg, 0.044 mmol, 1 equiv) in *m*-xylene (600 µL). Thorough mixing was ensured and the resulting homogenous soln was allowed to stand until the substrate had been fully consumed as judged by TLC or ^{1}H NMR analysis. Upon consumption of the starting material, the mixture was concentrated to a volume of ca. 100 µL, which was then eluted through a short silica gel column to obtain the pure cyclized product; yield: 98%; 94% ee.

1.1.1.3.4 Polycyclization Initiated by an Episulfonium Ion

Whereas enzymes utilize a litany of electrophilic agents to initiate polycyclization, electrophilic sulfur, which is present in many biological systems, has not been observed in a cyclase catalytic active site. In contrast, the Livinghouse group demonstrated that combining sulfenic esters and amides with strong Lewis acids will initiate polyene cyclizations that can be terminated with arenes and vinyl esters to provide versatile sulfurated polycycles in good yield (Scheme 12).[61]

Scheme 12 Episulfonium Ion Initiated Polyene Cyclization[61]

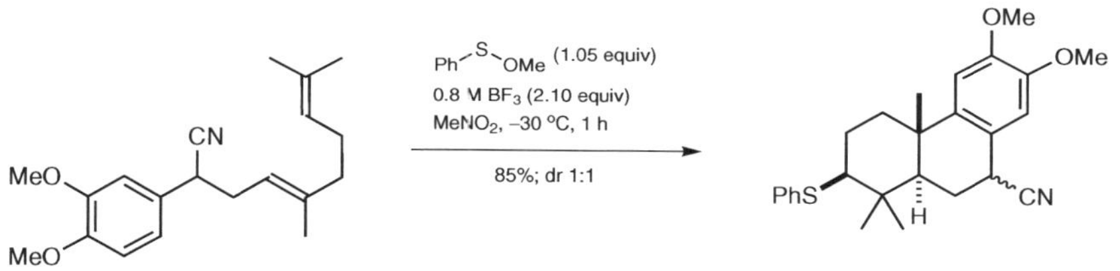

1.1.1.3.5 Polycyclization Initiated by a π-Lewis Acidic Metal

Cyclase enzymes catalyze the polyene cyclization of oxidosqualene and other epoxy-isoprenoids using alkali metal co-factors (such as Mg^{2+}) that serve as Lewis acids to initiate heterolytic C—O bond cleavage.[3] To circumvent the intermediacy of an epoxide initiating group, a π-Lewis acidic metal is an ideal candidate to coordinate and polarize the precursor alkene. Because numerous chiral ligands are available for screening, the possibility of obtaining enantioselectivity in the reaction is very good. Unlike Yamamoto's method for asymmetric protonative initiation (see Section 1.1.1.3.1.1), an initiation by a metal–alkene complex does not require multiply substituted alkenes as substrates, and in fact current methods require minimal steric encumbrance on the initiating alkene, as demonstrated by the Gagné group (Scheme 13).[62,63] A disadvantage of this method is that naturally occurring substrates, such as geranyl or farnesyl chains, are not directly amenable to these reactions; an advantage is orthogonality in substrate requirements as compared to other methods. It is noteworthy that bulky chiral bidentate phosphine ligands induce enantioselectivity even using 10 mol% platinum loading, although the selectivity between enantiomers only reaches 79% enantiomeric excess.

for references see p 39

Scheme 13 Palladium-Mediated Polycyclization[62]

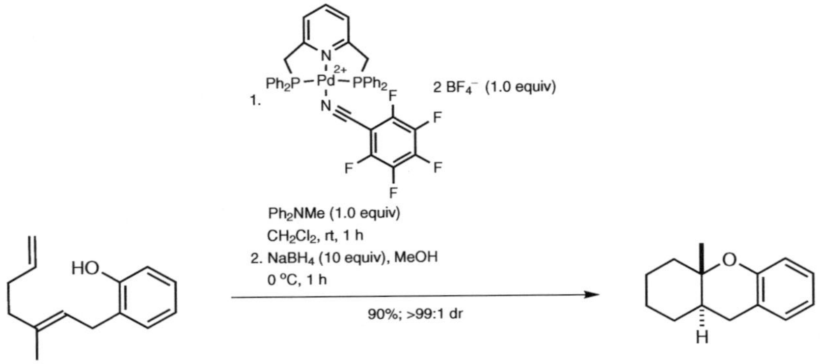

1.1.1.3.6 Enantioselective Polyene Cyclization Mediated by Chiral Scalemic Iridium Complexes

Occasionally, use of an unnatural initiating group holds strategic advantages for solving problems in cationic polyene cyclization. Whereas allylic alcohols are not normally electrophilic substrates for head-to-tail polycyclization, the Carreira group has utilized these initiators to good effect in devising a highly enantioselective cascade reaction (>99% ee) that is also high yielding and experimentally simple (Scheme 14).[64] The reaction relies on the ability of chiral iridium complexes, for example with dioxaphosphepin ligand **19**, to intercept racemic allylic alcohols (e.g., **20**) in the presence of an additional Lewis acid and initiate highly enantioselective C—C bond formation. This procedure can lead to short, enantioselective syntheses of challenging terpenes such as asperolide C.[65]

Scheme 14 Iridium-Catalyzed Polyene Cyclization[64]

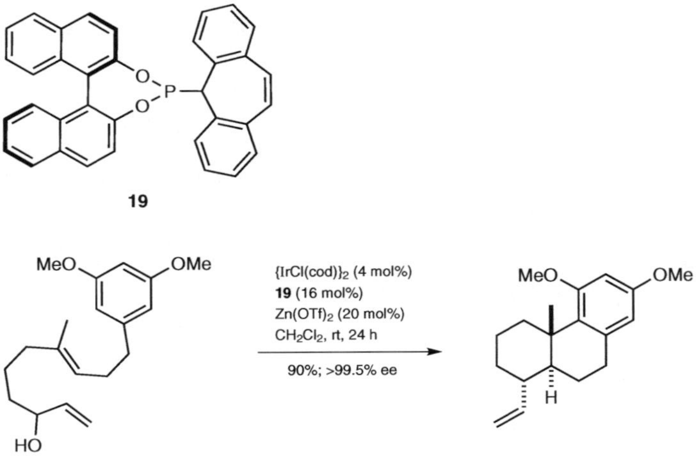

1.1.1.3.7 Acyliminium-Initiated Polyene Cyclization Mediated by Thioureas

It is believed that aromatic amino acid residues embedded in cyclase active sites undergo stabilizing cation–π interactions with their terpenoid substrates to direct product formation in cationic polyene cyclizations and Wagner–Meerwein rearrangements.[3,66–69] Appli-

cation of these ideas to biomimicry in bulk solvent has been limited by the difficulty of designing systems that capitalize on these interactions. However, Jacobsen, Lin, and Knowles determined that the arene surface area can be correlated to yield and absolute stereoselectivity in an acyliminium-initiated polyene cyclization mediated by a thiourea catalyst (e.g., **21**) (Scheme 15).[70] The authors hypothesize that the cation–π interaction leads to a lower energy pathway than the uncatalyzed, unorganized background reaction. Although the system is idiosyncratic in its initiation and therefore unlikely to see widespread use, the observation of productive and synthetically relevant carbocation–π interactions may lead to widespread redesign of catalysts to control cationic reaction pathways in other manifolds.

Scheme 15   Enantioselective Thiourea-Catalyzed Polycyclizations[70]

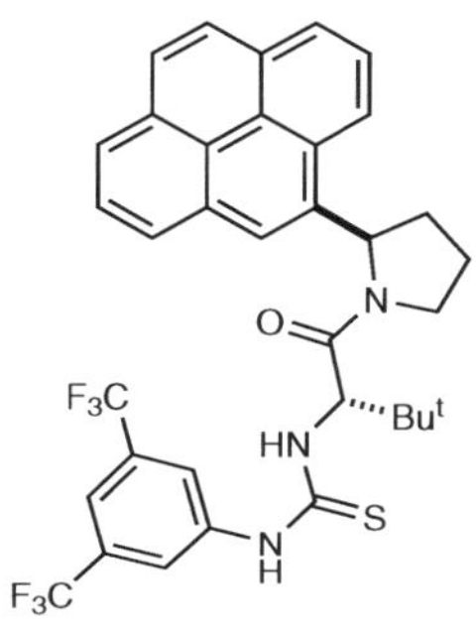

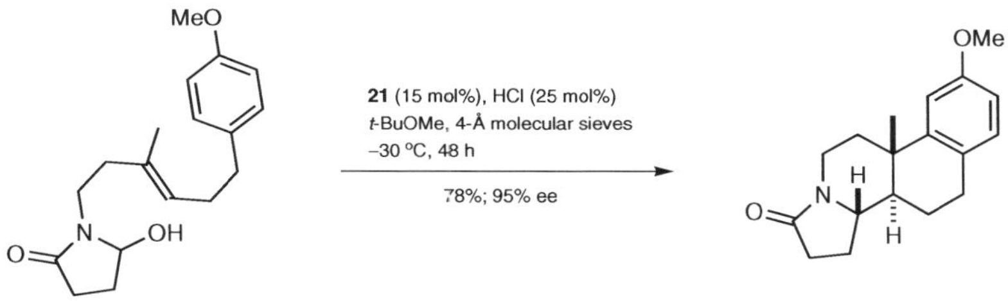

1.1.1.3.8   Tail-to-Head Polycyclization

The examples of biomimicry discussed above primarily explore cationic polyene cyclization initiated at the head terminus of a polyisoprenyl chain, as often occurs in nature. An additional, major branch of terpene biosynthesis instead relies on initiation at the tail terminus (tail-to-head polycyclization). This mode of polyene cyclization has seen relatively little exploration, because multiple ring formations are not kinetically competent to outcompete early termination through elimination, unlike the head-to-tail cyclizations discussed above. The Shenvi group demonstrated that a key obstacle to these cascades is the position of the counteranion relative to the carbocation.[71] If the counteranion is allowed to travel with the carbocation, then elimination occurs due to the highly acidic nature of protons adjacent to carbocations. If instead the counteranion is bound distal to the carbocation, then Wagner–Meerwein shifts become kinetically competent and polycyclic products (e.g., **22**) can be formed (Scheme 16). A major barrier to the broad application of this strategy is the lack of stereocontrol encountered in these cascades, which stands in stark contrast to the head-to-tail cascades covered in Sections 1.1.1.1.1–1.1.1.3.7.

for references see p 39

26   Domino Transformations   **1.1** Polyene Cyclizations

Scheme 16   Tail-to-Head Polycyclization[71]

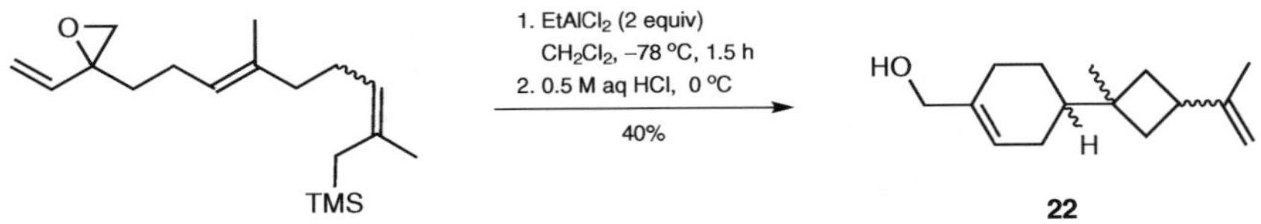

1. EtAlCl₂ (2 equiv)
   CH₂Cl₂, –78 °C, 1.5 h
2. 0.5 M aq HCl, 0 °C

40%

22

1.1.2   Radical Polyene Cyclizations

The inherent high energy of carbocations and the difficulty in controlling their behavior has led to the exploration of alternative modes for effecting polycyclization, especially through carbon-centered radicals. Radicals are no panacea for polyene reactions, but there are different pros and cons associated with their use. Whereas carbon radicals are more stable than carbocations due to their lower relative free energy, they are not necessarily more persistent.[72] Persistence, or lifetime, depends on the reaction environment, and carbocations can be rendered remarkably persistent in stabilizing solvent and when the counteranion is the conjugate base of a superacid.[7] Radicals are generally more persistent, but must be formed at low concentrations to prevent homocoupling and in the absence of radicalophiles, especially triplet oxygen, to prevent preemptive capture. However, many C—C bond-forming reactions can use radical intermediates to unique effect due to their geometry, predictability, ease of generation, and the efficiency of intramolecular reactions, especially with alkenes. In the sections that follow, we document the most efficient methods for effecting polyene cyclizations via radical intermediates, some of which can mimic the reaction outcomes of biosynthetic cationic cascades.

1.1.2.1   Most Used Radical Polycyclization Methods

Manganese(III) acetate has been recognized for several decades as a useful reagent to mediate the single-electron oxidation of carbonyls to their α-radical equivalents. Initially, manganese(III) acetate was used for the addition[73,74] and annulation[75] of carboxylic acids to alkenes. However, it has since been recognized as one of the most powerful reagents for initiating radical polyene cyclization.

1.1.2.1.1   Cyclization of Mono-and Polyunsaturated β-Oxo Esters Mediated by Manganese(III) Acetate

The Snider group has extensively studied the initiation, propagation, and termination of the cyclization of mono- and polyunsaturated β-oxo esters as mediated by manganese(III) acetate. This strategy has several advantages: (1) ease of preparation of substrate, because the polyene can be alkylated efficiently onto the initiating β-oxo ester under mild conditions; (2) the initiating group possesses very different reactivity to the polyene, and therefore chemoselectivity of initiation is not a problem; and (3) the β-oxo ester is retained in the product and can be further functionalized to many different moieties. The biggest disadvantages are (1) early cascade termination, which depends on the relative rate of cyclization versus oxidation to the carbocation/elimination and can only be determined empirically; and (2) that the reaction must be run in acetic acid (only a minor drawback). As long as the relative rates of radical alkene cyclizations are understood, the products are simple to predict and therefore the domino reactions in this manifold are easy to design. A typical example of a manganese(III) acetate mediated cyclization is the reaction of methyl 2-allyl-6-methyl-3-oxohept-6-enoate (**23**) to give **24** (Scheme 17).[76]

Scheme 17 Manganese-Based Oxidative Cyclization[76]

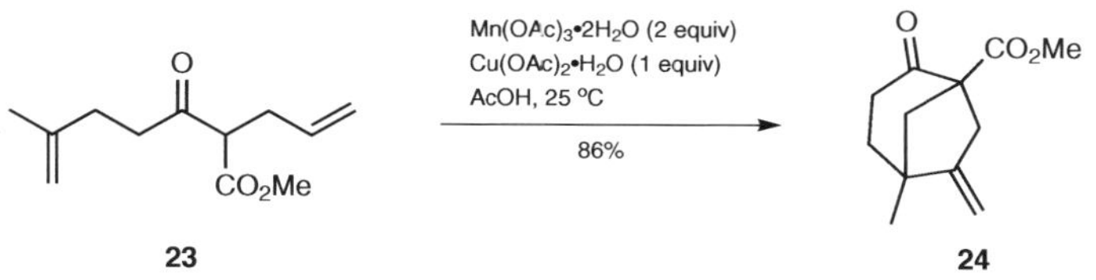

**Methyl 5-Methyl-6-methylene-2-oxobicyclo[3.2.1]octane-1-carboxylate (24);
Typical Procedure:**[76]

CAUTION: *Glacial acetic acid is a caustic liquid that may be an irritant if inhaled, swallowed, or absorbed through skin. All manipulations should be carefully carried out in a well-ventilated fume hood.*

To a stirred soln of $Mn(OAc)_3 \cdot 2H_2O$ (0.804 g, 3.0 mmol, 2.0 equiv) and $Cu(OAc)_2 \cdot H_2O$ (0.300 g, 1.5 mmol, 1.0 equiv) in glacial AcOH (13.5 mL) was added methyl 2-allyl-6-methyl-3-oxohept-6-enoate (**23**; 0.307 g, 1.5 mmol, 1.0 equiv) in glacial AcOH (4 mL). The mixture was stirred at rt for 26 h. After H_2O (100 mL) had been added, 10% aq $NaHSO_3$ was added dropwise to the mixture to decompose any residual $Mn(OAc)_3$. The resulting soln was extracted with CH_2Cl_2 (3 × 30 mL). The combined organic extracts were washed with sat. aq $NaHCO_3$, dried (Na_2SO_4), and concentrated to give a yellow solid, which was recrystallized (pentane) to give the pure product; yield: 270 mg (86%).

1.1.2.1.2 Titanocene-Catalyzed Polycyclization

A major benefit of using epoxyisoprenyl substrates in polyene cyclizations is that they map onto numerous bioactive terpene scaffolds. As discussed above, Lewis acids mediate cationic polyene cyclization of these substrates, but there are numerous challenges to this approach, especially early termination of the cascade and incompatibility with strong Lewis bases in the substrate. In the 1980s, RajanBabu and Nugent pioneered the use of low-valent titanium complexes to reductively cleave epoxides via carbon-centered radicals, which could also intercept alkenes intramolecularly.[77,78] This reaction was subsequently rendered catalytic.[79] The Barrero group explored the application of this reactivity to the polycyclization of epoxyisoprenyls in the anticipation that radical intermediates may offer a strategic advantage over carbocations (Scheme 18).[80] In general, these reactions are easy to carry out, and do allow the inclusion of Lewis basic groups such as acetates, ketals, and alcohols within the substrates (e.g., **25**). However, the yields of the products (e.g., **26**) are not appreciably better than the ones for the best cationic polyene cyclizations achieved to date. The most important feature of this reaction is the ability to terminate the cascade with an enone function, which is resistant to further radical cyclization and instead forms a stable metalloenolate.[81]

Scheme 18 Titanocene-Catalyzed Polyene Cyclization[80]

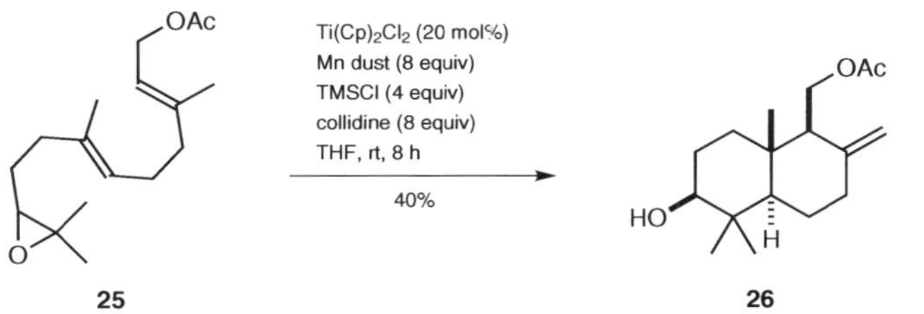

for references see p 39

[(1R*,4aS*,6R*,8aR*)-6-Hydroxy-5,5,8a-trimethyl-2-methylenedecahydronaphthalen-1-yl]methyl Acetate (26); Typical Procedure:[80]

Strictly deoxygenated THF (20 mL) was added to a mixture of Ti(Cp)$_2$Cl$_2$ (0.5 mmol, 0.2 equiv) and Mn dust (20 mmol, 8.0 equiv) under argon and the suspension was stirred at rt until it turned lime green (after about 15 min). Then, a soln of epoxide **25** (2.5 mmol, 1.0 equiv) and 2,4,6-collidine (20 mmol, 8.0 equiv) in THF (2 mL) and TMSCl (10 mmol, 4 equiv) were added and the soln was stirred for 8 h. The reaction was then quenched with 2 M aq HCl and the mixture was extracted with *t*-BuOMe. The organic layer was washed with brine, dried (Na$_2$SO$_4$), and concentrated. The residue was dissolved in THF (20 mL) and stirred with TBAF (10 mmol, 4.0 equiv) for 2 h. The mixture was then diluted with *t*-BuOMe, washed with brine, dried (Na$_2$SO$_4$), and concentrated. The residue was purified by flash chromatography (silica gel, hexanes/*t*-BuOMe); yield: 40%.

1.1.2.2 Recent Advances in Radical Polycyclization

1.1.2.2.1 Manganese- and Cobalt-Catalyzed Cyclization

Almost thirty years ago, Mukaiyama introduced an array of manganese, iron, and cobalt catalysts that promote the Markovnikov hydrofunctionalization of alkenes via carbon-centered radical intermediates.[82] Hydration and hydroperoxidation of electron-neutral alkenes were demonstrated by Mukaiyama, as was hydronitrosation of electron-deficient alkenes (unsaturated amides and esters). The detailed mechanisms of these reactions have not been elucidated, but they were proposed by the Shenvi group[83] to result from metal hydride hydrogen-atom transfer, in analogy to the reactions studied by the groups of Halpern,[84,85] Bullock,[86,87] Norton,[88–90] and Eisenberg.[91] This proposal contrasts with the hydrometalation mechanisms proposed extensively in the hydrofunctionalization literature originating with Mukaiyama. These mechanisms instead invoke metal hydride addition[92] across the alkene[93] to form an alkylcobalt species,[82,94] similar to hydroboration but with reversed polarity preferences. However, the regioselectivity (Markovnikov) is opposite to that normally observed for hydrometalation, and the porphyrin–metal complexes that can catalyze these reactions[95,96] do not possess a binding site adjacent to the hydride to allow this concerted addition to occur.[97] Additional related reactions in this area were reported by the Magnus group who demonstrated enone reduction[98] and hydration,[99] the Krische group who demonstrated an enone [2+2] cycloaddition from a low-valent cobalt catalyst,[100] and the Carreira group who greatly expanded this area by contributing hydrazidation,[93,101] hydrohydrazidation,[102–105] hydrocyanation,[106] hydrochlorination,[107,108] and hydrooximation.[109] More recent contributions have been made by the Boger group using a combination of iron salts and borohydrides, which can achieve, notably, Markovnikov hydrofluorination[110] in addition to many other transformations,[111] by the Baran group, who have demonstrated conjugate addition to enones,[112,113] and by the Shenvi group, who demonstrated hydrogenation of electron-neutral alkenes with thermodynamic stereocontrol and alkene isomerization.[83] A related reduction of haloalkenes with a cobalt catalyst was reported subsequently by the Herzon group,[114] who also proposed hydrogen-atom transfer to be operative in these reactions.

A significant benefit of this reactivity is that it allows for direct generation of a carbon-centered radical from any alkene, whether electron-neutral, electron-rich, or electron-deficient. There is excellent proof of principle for initiating radical polyene cyclizations using these methods, although there has been no rigorous study of the breadth of this strategy. For example, the Carreira group has used their hydrazidation reaction to initiate five-membered ring cyclization/azidation of diene **27** to give cyclopentane **28** catalyzed by tris(dipivaloylmethanato)manganese(III) [tris(2,2,6,6-tetramethylheptane-3,5-dionato)manganese(III)] (Scheme 19).[93]

1.1.2 Radical Polyene Cyclizations **29**

Scheme 19 Tris(dipivaloylmethanato)manganese(III)-Catalyzed Hydrazidation of Alkenes[93]

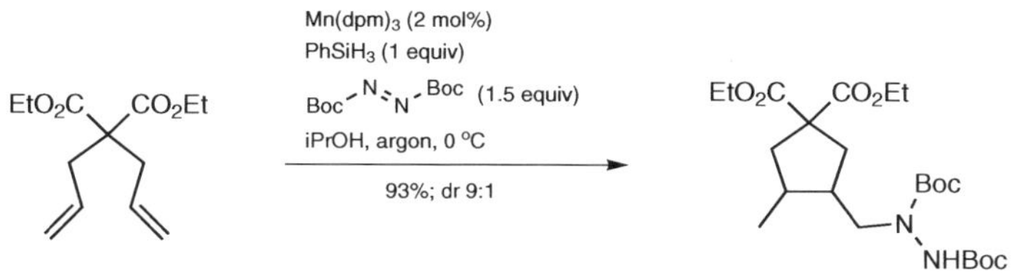

dpm = 2,2,6,6-tetramethylheptane-3,5-dionato

Diethyl 3-[*N*,*N*′-Di-(*tert*-butoxycarbonyl)hydrazinomethyl]-4-methylcyclopentane-1,1-dicarboxylate (28); Typical Procedure:[93]

Mn(dpm)$_3$ (6 mg, 0.01 mmol, 0.02 equiv) was dissolved in iPrOH (2.5 mL) at 23 °C under argon and the dark brown-green soln was cooled to 0 °C. Diethyl 2,2-diallylmalonate (**27**; 121 mg, 0.50 mmol, 1.0 equiv) and PhSiH$_3$ (65 µL, 0.52 mmol, 1.0 equiv) were added, followed by di-*tert*-butyl azodicarboxylate (0.17 g, 0.75 mmol, 1.5 equiv) in one portion. The resulting suspension was stirred at 0 °C and the reaction was monitored by TLC (EtOAc/hexane 1:5). After 4 h (color change to yellow), the reaction was quenched with H$_2$O (1 mL). Brine (5 mL) was added and the mixture was extracted with EtOAc (3 × 10 mL). The combined organic layers were dried (Na$_2$SO$_4$), filtered, and concentrated. The residue was purified by chromatography (silica gel, EtOAc/hexanes 1:10 to 1:5); yield: 221 mg (93%); dr 9:1.

1.1.2.2.2 Manganese-Catalyzed Hydrogenative Polycyclization

If the introduction of further functionality is unnecessary, a simple hydrogenative cyclization procedure is available. It is noteworthy that radical alkene cyclizations usually possess very early transition states and therefore tolerate severe steric clash in the product. Unlike the hydrogenative hydrogen-atom transfer cyclizations previously reported,[89,115] these reactions work well on electron-neutral alkenes. Furthermore, compared to most other metal-mediated cyclizations of polyunsaturated carbon chains, this reaction does not require one more alkyne in the substrate for initial metal coordination, substituted alkenes are tolerated, and the catalysts are based on abundant, non-precious metals with low-cost ligands. As an example, Scheme 20 shows the formation of cyclic diesters **30** and **31** from diene **29** catalyzed by tris(dipivaloylmethanato)manganese(III) [tris(2,2,6,6-tetramethylheptane-3,5-dionato)manganese(III)].[83]

Scheme 20 Reductive Manganese-Catalyzed Radical Cyclization[83]

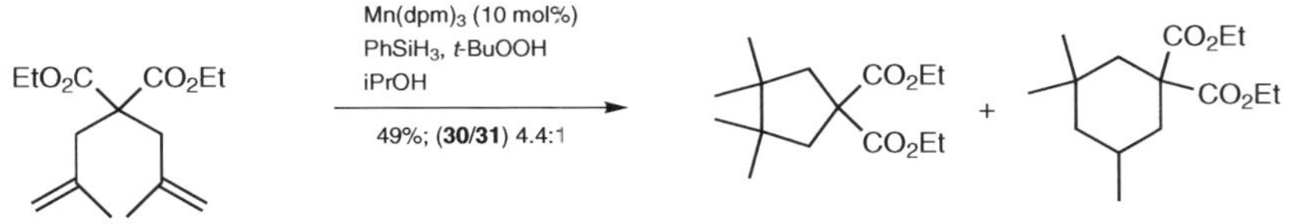

dpm = 2,2,6,6-tetramethylheptane-3,5-dionato

for references see p 39

Diethyl 3,3,4,4-Tetramethylcyclopentane-1,1-dicarboxylate (30) and Diethyl 3,3,5-Tri-methylcyclohexane-1,1-dicarboxylate (31); Typical Procedure:[83]

Diethyl 2,2-bis(2-methylallyl)malonate (**29**; 53.6 mg, 0.20 mmol, 1.0 equiv) was dissolved in anhyd iPrOH (1.0 mL, 0.2 M) under argon. To the stirred soln were added PhSiH$_3$ (24.4 µL, 0.20 mmol, 1.0 equiv) and an 8.0 M soln of *t*-BuOOH in hexanes (37.5 µL, 0.30 mmol, 1.5 equiv) and the resulting mixture was degassed by bubbling argon through the soln for 10 min. Mn(dpm)$_3$ (12 mg, 0.02 mmol, 10 mol%) was added in one portion and the mixture was degassed for an additional 30 s. After the reaction was complete (GC/MS), iPrOH was evaporated and the crude mixture [yield: 85%; ratio (**30/31**) 2.4:1.0] was subjected to flash chromatography (silica gel, Et$_2$O/hexanes 1:99) to obtain an enriched mixture of the products **30** and **31**; yield: 27 mg (49%); ratio (**30/31**) 4.4:1.

1.1.2.2.3 Radical Isomerization, Cycloisomerization, and Retrocycloisomerization with a Cobalt–salen Catalyst

Within the mechanistic framework of hydrogen-atom transfer proposed by the Shenvi group[83] to be operative within Mukaiyama hydrofunctionalizations, a broadly applicable radical cycloisomerization of linear polyenes was developed by the Shenvi group,[116] based on observations in related systems by the Norton group.[117] The initiating alkene must be terminal due to the steric constraints of the catalyst, but this allows the cyclo-isomerization to terminate without further isomerization and loss of stereochemistry. This chemoselective and controlled reorganization of simple alkenes has potential for general use in complex molecule synthesis. In addition to cycloisomerization, the linear isomerization and retrocycloisomerization (formal retro-ene reaction) of terminal alkenes can be effected with the same chemistry. Because carbon radical formation does not require reaction with a Lewis acidic metal, strongly Lewis basic functional groups such as tertiary amines are tolerated. As an example, Scheme 21 shows the formation of 1-benzyl-3,3-dimethyl-4-(prop-1-en-2-yl)pyrrolidine (**34**) from diene **33** catalyzed by co-balt–salen complex **32**.[116]

Scheme 21 Cobalt-Catalyzed Radical Retrocycloisomerization[116]

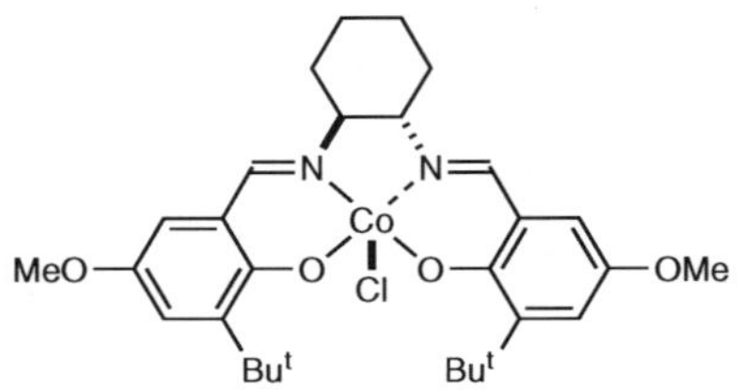

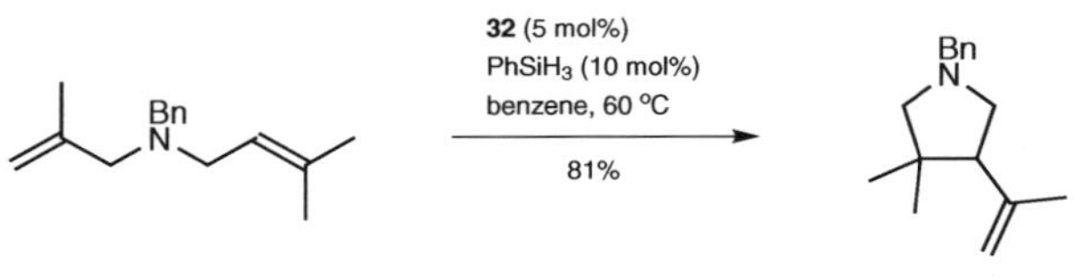

1-Benzyl-3,3-dimethyl-4-(prop-1-en-2-yl)pyrrolidine (34); Typical Procedure:[116]

Cobalt catalyst **32** (4 mg, 7 µmol, 0.03 equiv) was added to a flame-dried small vial under argon, then dissolved in benzene (previously degassed with argon; 2.2 mL) (**CAUTION:** *carcinogen*). This dark green soln was then added to a flame-dried flask containing *N*-ben-

zyl-3-methyl-N-(2-methylallyl)but-2-en-1-amine (**33**; 50 mg, 0.22 mmol, 1.0 equiv) under argon. PhSiH$_3$ (1.6 µL, 0.01 mmol, 0.06 equiv) was then added to the stirred soln and the resulting red-orange soln was heated at 60 °C. The reaction was monitored by GC/MS, and after the mixture had been stirred for 12 h, more cobalt catalyst **32** (4 mg, 7 µmol, 0.03 equiv) and phenylsilane (1.6 µL, 0.01 mmol, 0.06 equiv) were added. The mixture was heated at 60 °C until the reaction was complete and concentrated directly under reduced pressure. The residue was purified by flash chromatography (silica gel, Et$_2$O/hexanes 5:95) to obtain the product as a clear yellow oil; yield: 41 mg (81%).

1.1.2.3 Other Examples of Radical Polycyclization

There are numerous other reports of radical polyene cyclizations, but many utilize unique/nongeneralized initiating groups, engineered propagating groups, or reagents that are inconvenient due to instability or toxicity. Three examples worth mentioning that have not been extensively applied, but showcase important methods for carrying out radical polyene cyclizations are shown in Sections 1.1.2.3.1–1.1.2.3.3.

1.1.2.3.1 Radical Polycyclization via Photoinduced Electron Transfer

Although radical anions and radical cations derived from electron-neutral alkenes are higher in energy than the corresponding neutral carbon-centered radicals, their formation can occasionally be realized with some efficiency. Demuth and co-workers have demonstrated[118] that photoinduced electron transfer from 1,4-dicyano-2,3,5,6-tetramethylbenzene/biphenyl will oxidize the terminal head unit of a polyisoprenyl chain with surprising site selectivity and, after capture with water, radical polyene cyclization ensues. Although this method for polyene cyclization has not been applied extensively outside the Demuth laboratory, the orthogonality of its reactivity compared to most methods make it an enticing prospect for further exploration.

1.1.2.3.2 Polycyclization via Organo-SOMO Catalysis

The most closely related reaction to parallel Demuth's method is a recent report of polyene cyclization initiated by single-electron oxidation of a catalytically generated enamine (see also Section 1.4.2.3.1). A great benefit of this strategy is the ability to induce chirality catalytically using a chiral amine and along a radical reaction pathway, which means that electron-poor arenes can be efficiently utilized as terminating groups. A deficiency of this reaction is that electron-deficient alkenes must be used to propagate the cascade beyond an initial bicyclization. However, as a general route to some specific terpene skeletons in asymmetric fashion, the method is excellent.[119]

1.1.2.3.3 Polyene Radical Cascades in Complex Molecule Synthesis

There are numerous, arguably esoteric, examples of radical polyene cyclization in synthesis, and so it is challenging to select the "best" example. However, certainly some of the best benefits of using a polyene radical cascade to construct a complex molecule (e.g., **36** from polyene **35**) are encapsulated by the Parker group in the synthesis of morphine using simple building blocks (Scheme 22).[120] First, carbon-centered radicals are not sensitive to acidic hydrogens, unlike their organometallic equivalents, so unprotected functional groups (e.g., hydroxy groups) can be used. Second, quaternary centers can be formed efficiently. Third, termination need not arise from hydrogen-atom abstraction from initiator or solvent, but rather can occur by elimination of, in this case, a thiol radical.

for references see p 39

Scheme 22 A Radical Cyclization Cascade En Route to Morphine[120]

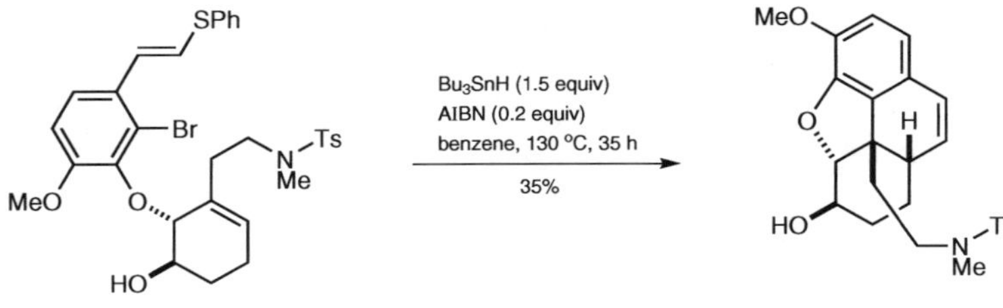

N-[2-{(3*R*,3a*R*,3a¹*S*,9a*S*)-3-Hydroxy-5-methoxy-1,3,3a,9a-tetrahydrophenanthro[4,5-*bcd*]furan-3a¹(2*H*)-yl}ethyl]-*N*,4-dimethylbenzenesulfonamide (36); Typical Procedure:[120]
A soln of alcohol **35** (120 mg, 0.187 mmol, 1.0 equiv), Bu₃SnH (75 μL, 0.280 mmol, 1.5 equiv), and a catalytic amount of AIBN (0.1–0.2 equiv) in benzene (8 mL) (**CAUTION:** *carcinogen*) was heated in a sealed tube at 130 °C. A small amount of AIBN was added every 8 h to maintain the radical chain. After 35 h, the mixture was concentrated, and the residue was dissolved in Et₂O and then stirred vigorously with 10% aq KF for 2 h. The ether phase was separated, washed with brine, dried (Na₂SO₄), and concentrated. Purification by preparative TLC (silica gel, EtOAc/hexanes/acetone 1:3:1) afforded the pure product; yield: 30 mg (35%).

1.1.3 Polyene Cyclization via Reductive Elimination from a Metal Center or Metathesis

The advent of organometallic chemistry, and especially metal catalysis, in complex organic molecule synthesis over the past four decades has opened many avenues to explore in polyene cyclization. In some sense, these reactions are intramolecular variants of metal-catalyzed polymerization reactions.

1.1.3.1 Most Used Polycyclization Methods via Reductive Elimination

1.1.3.1.1 Palladium Zipper Cyclization Cascades

Unlike cationic and radical methods for polyene cyclization, where reactive intermediates are centered at the more-substituted position of an alkene, reactions that proceed via the intermediacy of carbon—metal bonds tend to propagate at the less substituted position. The groups of Overman,[121–123] Trost,[124] Oppolzer,[125] and Negishi[126,127] were among the first to recognize the power of palladium catalysts to mediate polyene cyclization along these less accessible reaction pathways. In particular, the Overman group demonstrated that quaternary centers (e.g., in compounds **37**, **39**, and **40**, Scheme 23) are easily accessible via intramolecular Heck polyene cyclizations, and that these reactions can be rendered asymmetric using chiral ligands on palladium.[121–123] From the perspective of retrosynthetic analysis, the ability to initiate the cascades using insertion into enol trifluoromethanesulfonates (e.g., **38**) is a powerful asset, because substrates are then easily accessible from functionalized ketones, and in turn from simple unfunctionalized ketones.

1.1.3 Polyene Cyclization via Reductive Elimination from a Metal Center/Metathesis 33

Scheme 23 Palladium-Catalyzed Heck Polyene Cyclizations[121–123]

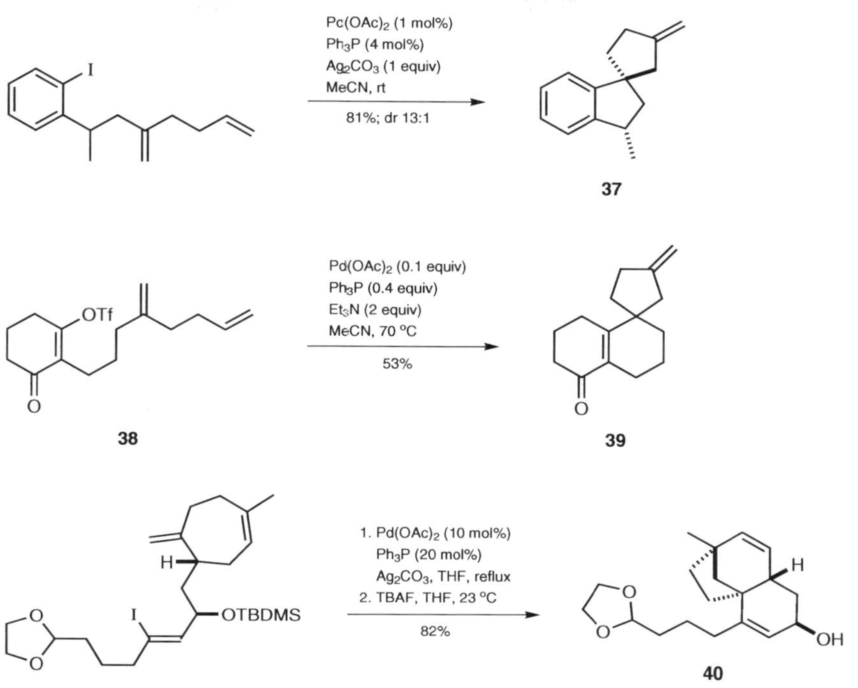

3-Methylene-2′,3′,4′,6′,7′,8′-hexahydro-5′H-spiro[cyclopentane-1,1′-naphthalen]-5′-one (39); Typical Procedure:[122]

Et₃N (0.31 mL, 2.2 mmol, 2.0 equiv), Ph₃P (60 mg, 0.23 mmol, 0.2 equiv), and Pd(OAc)₂ (13 mg, 0.057 mmol, 0.05 equiv) were sequentially added to a soln of vinyl trifluoromethanesulfonate **38** (407 mg, 1.11 mmol, 1.0 equiv) in anhyd MeCN (37 mL). The resulting green soln was heated to 70 °C. After 22 h, the reaction was ~30% complete. Additional Ph₃P (60 mg, 0.23 mmol, 0.2 equiv) and Pd(OAc)₂ (13 mg, 0.057 mmol. 0.05 equiv) were added and heating was continued for an additional 19 h, at which time GLC analysis showed complete consumption of the starting material. The mixture was adsorbed onto Florisil (with a rotary evaporator) and the Florisil was extracted by filtration with Et₂O (ca. 20 mL). Concentration and chromatography of the residue (silica gel, hexanes/EtOAc 20:1 to 10:1) gave the pure product; yield: 0.13 g (53%).

1.1.3.1.2 Polycyclization Cascades via Metathesis

The polyene cyclizations discussed in prior sections differ from those covered in this section in that C—C bonds are only formed, not broken. However, implicit in alkene metathesis based approaches to complex molecule synthesis[128] is the fact that a carbon must be lost, if only temporarily, for the reaction to occur. This loss can involve the fragmentation of a single methylene in the case of a terminal vinyl group, the fragmentation of a longer chain, or, in some cases, the fragmentation of a ring. Although counterintuitive, ring fragmentation can be used to advantage if gross structure assembly of the cycle is facile and strategically important for generating overall complexity more quickly.[2] In this vein, the approach of the Phillips group to the cyanthiwigins is instructive. The polyene cyclization precursor can be synthesized in short order via a chiral auxiliary controlled Diels–Alder reaction followed by some functional group interconversions. Subjection of intermediate **42** to Grubbs' second-generation initiator **41** fragments the cyclohexene ring and engages

for references see p 39

34   Domino Transformations   **1.1** Polyene Cyclizations

the proximal enones in two ring-closing metathesis reactions to yield the cyanthiwigin core **43** (Scheme 24).[129] So, although one element of complexity – the bridging cyclohexene – is sacrificed, its destruction leads to a rapid assembly of the more challenging 7/6/5 core. In addition, indenol **45** can also be formed with the same strategy from dienol **44**.[130] In a general sense, this is an important maneuver to remember in synthesis, the equivalent of baseball's sacrifice fly.[2]

Scheme 24   Metathesis Cyclization[129,130]

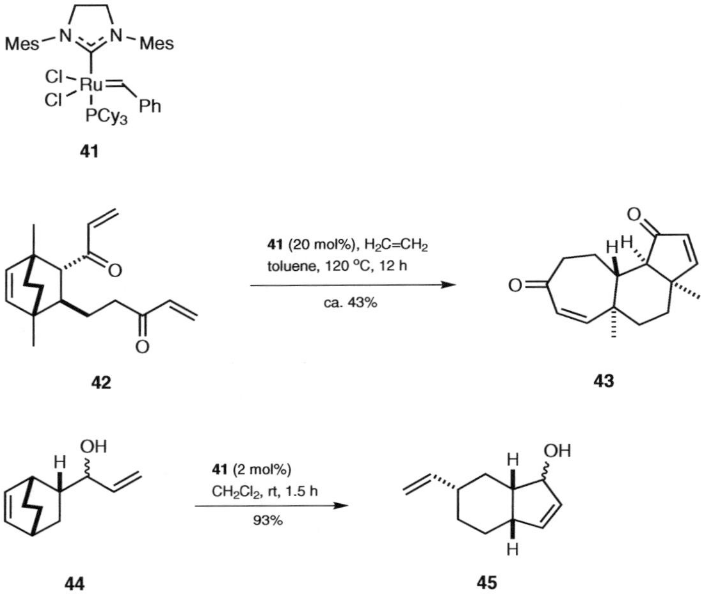

(3aS,6S,7aS)-6-Vinyl-3a,4,5,6,7,7a-hexahydro-1H-inden-1-ol (45); Typical Procedure:[130]
Grubbs' catalyst **41** (4.8 mg, 5.8 µmol, 2 mol%) was dissolved in CH_2Cl_2 (2 mL) and added by syringe to a soln of dienol **44** (48 mg, 0.29 mmol, 1.0 equiv) in CH_2Cl_2 (28 mL, 0.01 M). This soln was then sparged with ethene three times over a 30-min period for 60 s each time. At 30 min, the remaining ethene was purged from the soln with N_2 and the mixture was stirred at rt under N_2 for 1 h. The solvent was then removed by rotary evaporation and the residue was purified by flash chromatography (silica gel, hexanes/EtOAc 4:1); yield: 45 mg (93%).

1.1.3.2   Other Polycyclization Methods via Reductive Elimination

1.1.3.2.1   Cyclization via π-Allylpalladium Complexes

Palladium catalysis is a powerful approach to polyene cyclization because of the myriad of C—C bond-forming reactions available to this privileged metal. Of the numerous reaction "silos" that categorize palladium chemistry,[131] among the largest are Heck reactivity (see Section 1.1.3.1.1) and π-allylpalladium reactivity.[132,133] Polyene cyclizations using this latter reaction manifold are difficult to achieve because alkenes are not usually strong enough nucleophiles to engage in outer-sphere attack of the electrophilic η^2 π-allylpalladium complex. However, given the correct geometry, the palladium complex can engage in an alternative mode of reactivity and instead react with proximal alkenes through a metallo-ene reaction. These reactions were described by the Oppolzer group to form

1.1.3 Polyene Cyclization via Reductive Elimination from a Metal Center/Metathesis **35**

rings with high levels of stereocontrol for point chirality and alkene geometry (Scheme 25).[125,134]

Scheme 25   Palladium-Catalyzed Cyclization[125,134]

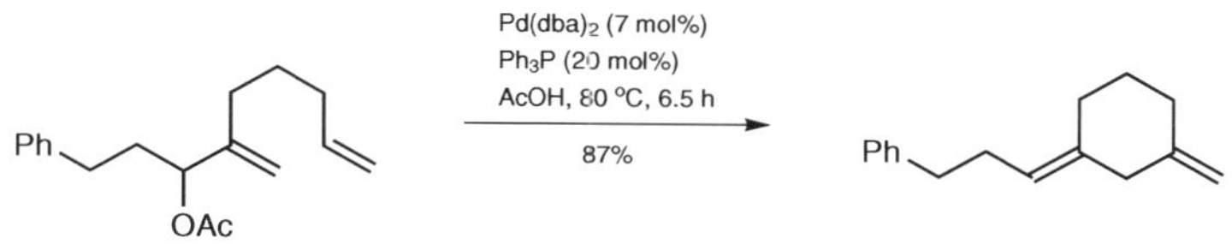

1.1.3.2.2 **Palladium-Catalyzed Ene–Yne Cycloisomerization**

Cycloisomerization is a powerful subset of ring-forming cascade reactions that, as the name suggests, generate cyclic molecules that are isomeric with their acyclic precursors [for an extended discussion see *Science of Synthesis: Stereoselective Synthesis*, Vol. 3 (Section 3.5)]. The discovery of palladium-catalyzed ene–yne cycloisomerizations by Trost opened new vistas of reactivity available to these simple and readily accessed unsaturated hydrocarbon chains. Importantly, these metal-catalyzed reactions extended cycloisomerization chemistry from a high-temperature and little-applied niche reaction into the synthesis mainstream by virtue of their predictability, ease of execution, and interface with the broad field of palladium catalysis. Most early examples of these reactions require enyne substrates, because electron-neutral alkenes are poor ligands for carbophilic metals relative to alkynes and allenes. However, Trost demonstrated that alkynes could initiate a cascade that would be propagated by a polyene chain and terminated by β-hydride elimination. In a dramatic example of this cascade, the polycyclization of pentaenyne **46** can be initiated with catalytic palladium(0) and a Brønsted acid via alkyne hydropalladation, followed by iterative alkene insertions, and hydride elimination termination. Although the reaction yields a mixture of two diastereomers, the reaction mechanism and substrate design constitutes an important proof of principle for designing related cascades (Scheme 26).[124]

Scheme 26   Palladium-Catalyzed Zipper Reaction[124]

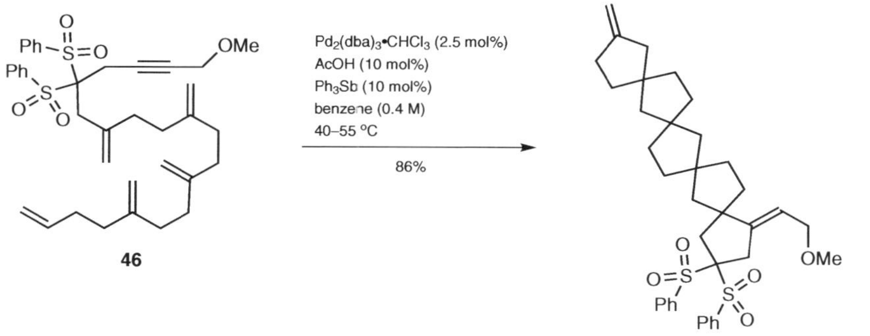

1.1.3.2.3 **Cycloisomerization in Complex Molecule Synthesis**

The value of cycloisomerization to the synthesis of complex carbocycles is nicely illustrated by work from the Winssinger group, who took inspiration from the polyisoprene polyene cyclization utilized by cells to rapidly build molecular complexity (Scheme 27).[135] Sesquiterpenes in particular display a dazzling array of diversity of carbon skeletons,[136] but their divergent syntheses from farnesyl pyrophosphate is challenging to reproduce

for references see p 39

36 Domino Transformations **1.1** Polyene Cyclizations

in the laboratory.[71] However, metal-catalyzed polyene cycloisomerization reactions can circumvent this biomimicry and yield divergent routes to known terpene skeletons. A benefit of this approach is relatively short access to different structural motifs from a single, keystone intermediate. The sole deficits are that this intermediate is not as inexpensive or as easily obtained as farnesol, and that the diversity available still does not match that found in nature, both of which are extraordinarily high benchmarks to reach. This work by the Winssinger group does, however, allow numerous scaffolds to be obtained selectively and in good yield using the now vast arsenal of cycloisomerizations available from transition metals.

Scheme 27 Metal-Mediated Cyclizations[135]

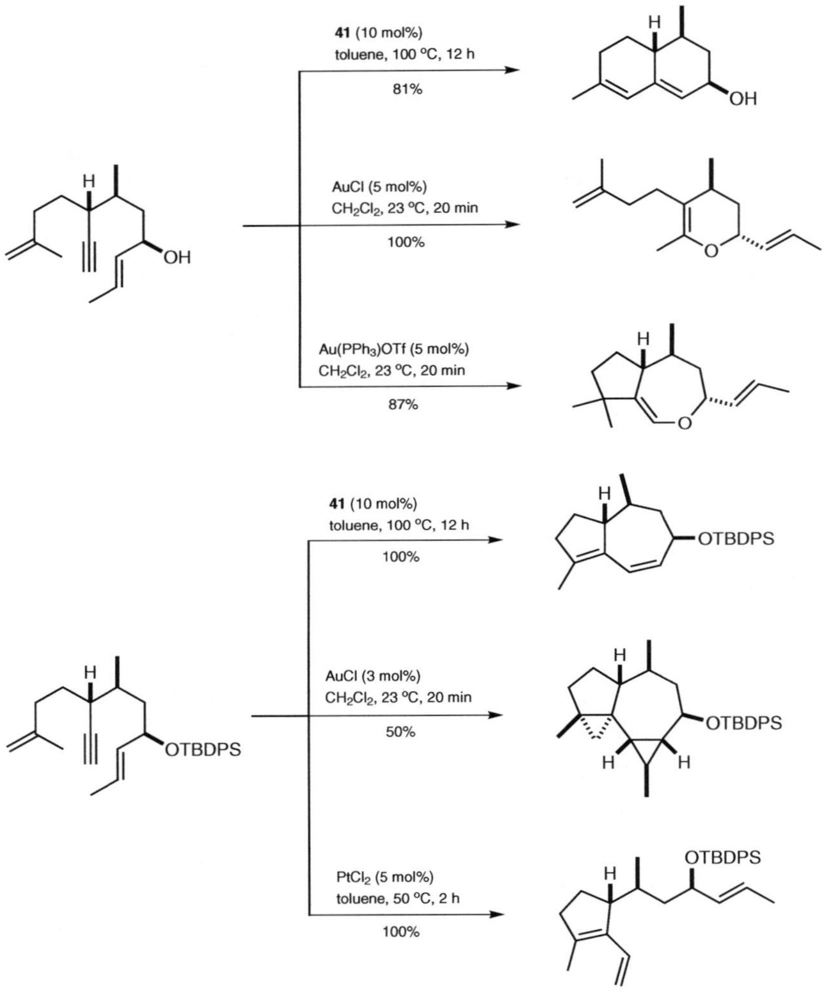

1.1.4 **Anionic Polyene Cyclizations**

There are two problems with anionic polyene cyclizations. First, there is a problem of definition, because most "carbanions" actually exist as metal-bound organometallics instead of free ionic species and most species involved in polyene cyclization are not carbanions per se, but rather resonance-stabilized anionic functional groups such as enolates, which are similarly metal bound, not free ions. Thus, this section will discuss polyene cycliza-

1.1.4 Anionic Polyene Cyclizations

tions that are initiated by a deprotonation event – a relatively weak definition. Second, there are few examples of polyene cyclizations using electron-neutral alkenes. Instead, the two examples shown below involve Michael reaction cascades of enolates and alkenes conjugated to carbonyls. The obvious consequence is that functionality must be installed and/or excised to achieve reactivity, which can lead to inefficient syntheses. On the other hand, if the required functionality is present in the targeted molecule, it can be used not only to effect polycyclization, but also to efficiently construct the substrate polyene using its inherent reactivity.

1.1.4.1   **Examples of Anionic Polyene Cyclizations**

1.1.4.1.1   **Stereoselective Polycyclization via Intramolecular Diels–Alder Cycloaddition Followed by Aldol Condensation**

The first example demonstrates a powerful transformation; a stereocontrolled anionic polycyclization reaction that establishes a steroid core with six contiguous chiral centers in a single step. Deslongchamps and Lavallée first reported this transformation in 1988.[137] The intermediate cesium enolate, formed by the deprotonation of the cyclohexenone substrate by cesium carbonate, undergoes an intramolecular Diels–Alder cycloaddition to set the first three stereocenters. This event is followed by a highly stereoselective aldol reaction, which traps the resulting enolate in situ with one of the carbonyl groups of the cyclopentadione to form the remaining three stereocenters of the product in one pot. In 2002, Deslongchamps and Rouillard applied this reaction to the synthesis of a pentacyclic lactone via decalin **49**, formed from methyl 6-oxocyclohex-1-ene-1-carboxylate (**47**) and diketal **48** (Scheme 28).[138]

Scheme 28   Anionic Polycyclization[137,138]

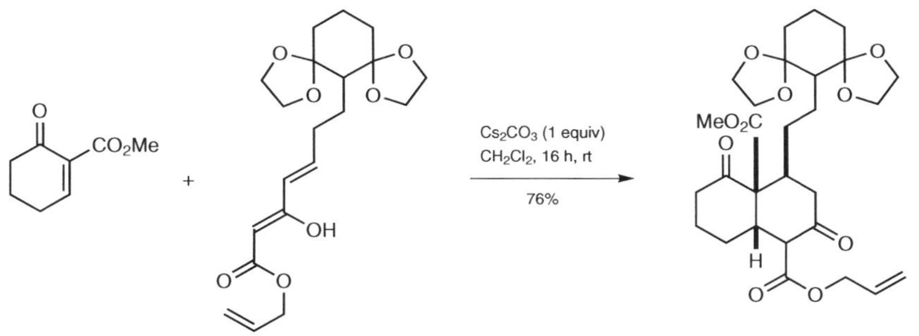

1-Allyl 4a-Methyl (4*S*,4a*S*,8a*S*)-4-{2-(1,4,8,11-Tetraoxadispiro[4.1.4⁷.3⁵]tetradecan-6-yl)ethyl}-2,5-dioxooctahydronaphthalene-1,4a(2*H*)-dicarboxylate (49); Typical Procedure:[138]
A freshly prepared soln of methyl 6-oxocyclohex-1-ene-1-carboxylate (**47**; 1.36 g, 8.85 mmol, 1.5 equiv) in CH_2Cl_2 (160 mL) was added to a soln of diketal **48** (2.34 g, 5.90 mmol) and Cs_2CO_3 (2.89 g, 8.85 mmol) in CH_2Cl_2 (320 mL). The mixture was stirred for 16 h, filtered through silica gel, and concentrated. The crude residue was purified by flash chromatography (silica gel, hexanes/Et_2O 1:3) to give the pure product; yield: 2.36 g (76%).

for references see p 39

38 Domino Transformations **1.1** Polyene Cyclizations

1.1.4.1.2 **Transannular Double Michael Cyclization Cascades**

The second example features a transannular double Michael reaction cascade of a bisenone macrocycle to form three stereocenters in one step. The Evans group reported this transformation as part of their synthesis of salvinorin A (Scheme 29).[139] Treatment of the β-oxo lactone **50** with tetrabutylammonium fluoride induces a transannular reaction cascade to give the tricycle **51** as a single diastereomer. While it is assumed that the reaction proceeds via a stepwise process, the authors note that a concerted *exo*-selective Diels–Alder cycloaddition is also possible via the enolate of the β-oxo lactone **50** instead.

Scheme 29 Anionic Polycyclization En Route to Salvinorin A[139]

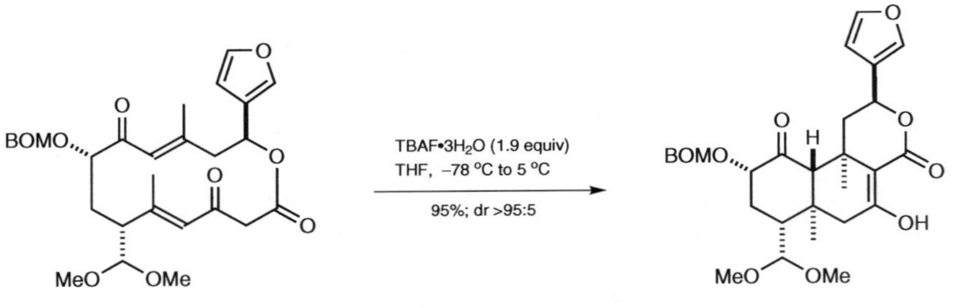

(2S,6aS,7R,9S,10aS,10bR)-9-[(Benzyloxy)methoxy]-7-(dimethoxymethyl)-2-(furan-3-yl)-5-hydroxy-6a,10b-dimethyl-1,2,6a,7,8,9,10a,10b-octahydronaphtho[2,1-c]pyran-4,10(2H)-dione (51); Typical Procedure:[139]

To a freshly prepared soln of TBAF•3H$_2$O (172 mg, 0.578 mmol, 1.9 equiv) in THF (13 mL) at −78 °C was added a soln of dioxo macrocycle **50** (162 mg, 0.300 mmol, 1.0 equiv) in THF (10 mL) via cannula over 5 min. The cannula wire was rinsed with DMF (2 × 10 mL). The reaction vessel was transferred to a cold water bath at 5 °C. The mixture was stirred for 2 h, diluted with sat. aq NH$_4$Cl (10 mL), and extracted with Et$_2$O (3 × 100 mL). The combined organic extracts were washed with half-sat. brine (2 × 75 mL), dried (Na$_2$SO$_4$), and concentrated; yield: 160 mg (95%).

References

[1] Breitmaier, E., *Terpenes: Flavors, Fragrances, Pharmaca, Pheromones*, Wiley-VCH: Weinheim, Germany, (2006).
[2] Corey, E. J.; Cheng, X.-M., *The Logic of Chemical Synthesis*, Wiley: New York, (1995); pp 47–57.
[3] Wendt, K. U.; Schultz, G. E.; Corey, E. J.; Liu, D. R., *Angew. Chem. Int. Ed.*, (2000) **39**, 2812.
[4] Johnson, W. S., *Bioorg. Chem.*, (1976) **5**, 51.
[5] Yoder, R. A.; Johnston, J. N., *Chem. Rev.*, (2005) **105**, 4730.
[6] Meerwein, H.; van Emster, K., *Ber. Dtsch. Chem. Ges. B*, (1922) **55**, 2500.
[7] Olah, G. A., *J. Org. Chem.*, (2001) **66**, 5943.
[8] Reed, C. A., *Chem. Commun. (Cambridge)*, (2005), 1669.
[9] Bentley, T. W.; Llewellyn, G.; Ryu, Z. H., *J. Org. Chem.*, (1998) **63**, 4654.
[10] Toteva, M. M.; Richard, J. P., *J. Am. Chem. Soc.*, (1996) **118**, 11434.
[11] Jenson, C.; Jorgensen, W. L., *J. Am. Chem. Soc.*, (1997) **119**, 10846.
[12] Tantillo, D. J., *J. Phys. Org. Chem.*, (2008) **21**, 561.
[13] Corey, E. J.; Wood, H. B., Jr., *J. Am. Chem. Soc.*, (1996) **118**, 11982.
[14] Kohler, R. E., Ph.D. Thesis, Harvard University, (1965).
[15] Stieglitz, J., *Am. Chem. J.*, (1899) **21**, 110.
[16] Baeyer, A.; Villiger, V., *Ber. Dtsch. Chem. Ges.*, (1902) **35**, 1189.
[17] Baeyer, A.; Villiger, V., *Ber. Dtsch. Chem. Ges.*, (1902) **35**, 3013.
[18] Eschenmoser, A., Ph.D. Thesis, ETH Zürich, (1952).
[19] Ruzicka, L.; Eschenmoser, A.; Heusser, H., *Experientia*, (1953) **9**, 357.
[20] Eschenmoser, A.; Ruzicka, L.; Jeger, O.; Arigoni, D., *Helv. Chim. Acta*, (1955) **38**, 1890.
[21] Stork, G.; Burgstahler, A. W., *J. Am. Chem. Soc.*, (1955) **77**, 5068.
[22] Volkmann, R. A.; Andrews, G. C.; Johnson, W. S., *J. Am. Chem. Soc.*, (1975) **97**, 4777.
[23] van Tamelen, E. E., *Acc. Chem. Res.*, (1968) **1**, 111.
[24] van Tamelen, E. E., *Acc. Chem. Res.*, (1975) **8**, 152.
[25] Corey, E. J.; Russey, W. E.; Ortiz de Montellano, P. R., *J. Am. Chem. Soc.*, (1966) **88**, 4750.
[26] Corey, E. J.; Russey, W. E., *J. Am. Chem. Soc.*, (1966) **88**, 4751.
[27] van Tamelen, E. E.; Willett, J. D.; Clayton, R. B.; Lord, K. E., *J. Am. Chem. Soc.*, (1966) **88**, 4752.
[28] Corey, E. J.; Sodeoka, M., *Tetrahedron Lett.*, (1991) **32**, 7005.
[29] Lin, S.; Corey, E. J., *J. Am. Chem. Soc.*, (1996) **118**, 8765.
[30] Shenvi, R. A.; Corey, E. J., *Org. Lett.*, (2010) **12**, 3548.
[31] Sharpless, K. B.; Katsuki, T., *J. Am. Chem. Soc.*, (1980) **102**, 5974.
[32] Corey, E. J.; Noe, M. C.; Lin, S., *Tetrahedron Lett.*, (1995) **36**, 8741.
[33] Corey, E. J.; Zhang, J., *Org. Lett.*, (2001) **3**, 3211.
[34] Corey, E. J.; Huang, J., *Org. Lett.*, (2003) **5**, 3455.
[35] Lichtor, P. A.; Miller, S. J., *Nat. Chem.*, (2012) **4**, 990.
[36] Lehmkuhl, H.; Kobs, H.-D., *Justus Liebigs Ann. Chem.*, (1968) **729**, 11.
[37] Evans, D. A.; Allison, B. D.; Yang, M. G., *Tetrahedron Lett.*, (1999) **40**, 4457.
[38] Corey, E. J.; Staas, D. D., *J. Am. Chem. Soc.*, (1998) **120**, 3526.
[39] Kronja, O.; Orlović, M.; Humski, K.; Borčić, S., *J. Am. Chem. Soc.*, (1991) **113**, 2306.
[40] Surendra, K.; Qui, W.; Corey, E. J., *J. Am. Chem. Soc.*, (2011) **133**, 9724.
[41] Snyder, S. A.; Tang, Z.-Y.; Gupta, R., *J. Am. Chem. Soc.*, (2009) **131**, 5744.
[42] Snyder, S. A.; Treitler, D. S., *Angew. Chem. Int. Ed.*, (2009) **48**, 7899.
[43] Snyder, S. A.; Treitler, D. S.; Brucks, A. P., *J. Am. Chem. Soc.*, (2010) **132**, 14303.
[44] Snyder, S. A.; Treitler, D. S.; Brucks, A. P.; Sattler, W. I., *J. Am. Chem. Soc.*, (2011) **133**, 15898.
[45] Snydt, S. A.; Brucks, A. P.; Treitler, D. S.; Moga, I., *J. Am. Chem. Soc.*, (2012) **134**, 17714.
[46] Snyder, S. A.; Treitler, D. S., *Org. Synth.*, (2011) **88**, 54.
[47] Ishihara, K.; Nakamura, S.; Yamamoto, H., *J. Am. Chem. Soc.*, (1999) **121**, 4906.
[48] Surendra, K.; Corey, E. J., *J. Am. Chem. Soc.*, (2012) **134**, 11992.
[49] Surendra, K.; Rajendar, G.; Corey, E. J., *J. Am. Chem. Soc.*, (2014) **136**, 642.
[50] Sakakura, A.; Ukai, A.; Ishihara, K., *Nature (London)*, (2007) **445**, 900.
[51] Lodeiro, S.; Xiong, Q. B.; Wilson, W. K.; Kolesnikova, M. D.; Onak, C. S.; Matsuda, S. P. T., *J. Am. Chem. Soc.*, (2007) **129**, 11213.
[52] Corey, E. J.; Roberts, B. E., *Tetrahedron Lett.*, (1997) **38**, 8921.
[53] Snider, B. B.; Rodini, D. J.; Straten, J. V., *J. Am. Chem. Soc.*, (1980) **102**, 5872.
[54] Fürstner, A.; Szillat, H.; Gabor, B.; Mynott, R., *J. Am. Chem. Soc.*, (1998) **120**, 8305.

[55] Fürstner, A.; Morency, L., *Angew. Chem. Int. Ed.*, (2008) **47**, 4030.

[56] Zhang, L.; Kozmin, S. A., *J. Am. Chem. Soc.*, (2005) **127**, 6962.

[57] Brooner, R. E. M.; Widenhoefer, R. A., *Angew. Chem. Int. Ed.*, (2013) **52**, 11714.

[58] Johannson, M. J.; Gorin, D. J.; Staben, S. T.; Toste, F. D., *J. Am. Chem. Soc.*, (2005) **127**, 18002.

[59] Muñoz, M. P.; Adrio, J.; Carretero, J. C.; Echavarren, A. M., *Organometallics*, (2005) **24**, 1293.

[60] Sethofer, S. G.; Mayer, T.; Toste, F. D., *J. Am. Chem. Soc.*, (2010) **132**, 8276.

[61] Edstrom, E. D.; Livinghouse, T., *J. Org. Chem.*, (1987) **52**, 949.

[62] Koh, J. H.; Gagné, M. R., *Angew. Chem. Int. Ed.*, (2004) **43**, 3459.

[63] Sokol, J. G.; Cochrane, N. A.; Becker, J. J.; Gagné, M. R., *Chem. Commun. (Cambridge)*, (2013) **49**, 5046.

[64] Schafroth, M. A.; Sarlah, D.; Krautwald, S.; Carreira, E. M., *J. Am. Chem. Soc.*, (2012) **134**, 20276.

[65] Jeker, O. F.; Kravina, A. G.; Carreira, E. M., *Angew. Chem. Int. Ed.*, (2013) **52**, 12166.

[66] Hoshino, T.; Sato, T., *Chem. Commun. (Cambridge)*, (2002), 291.

[67] Johnson, W. S.; Telfer, S. J.; Cheng, S.; Schubert, U., *J. Am. Chem. Soc.*, (1987) **109**, 2517.

[68] Abe, I.; Rohmer, M.; Prestwich, G. D., *Chem. Rev.*, (1993) **93**, 2189.

[69] Paschall, C. M.; Hasserodt, J.; Jones, T.; Lerner, R. A.; Janda, K. D.; Christianson, D. W., *Angew. Chem. Int. Ed.*, (1999) **38**, 1743.

[70] Knowles, R. R.; Lin, S.; Jacobsen, E. N., *J. Am. Chem. Soc.*, (2010) **132**, 5030.

[71] Pronin, S. V.; Shenvi, R. A., *Nat. Chem.*, (2012) **4**, 915.

[72] Anslyn, E. V.; Dougherty, D. A., *Modern Physical Organic Chemistry*, University Science Books: Sausalito, CA, (2006); pp 82–84.

[73] Bush, J. B., Jr.; Finkbeiner, H., *J. Am. Chem. Soc.*, (1968) **90**, 5903.

[74] Heiba, E. I.; Dessau, R. M.; Koehl, W. J., Jr., *J. Am. Chem. Soc.*, (1968) **90**, 5905.

[75] Corey, E. J.; Kang, M., *J. Am. Chem. Soc.*, (1984) **106**, 5384.

[76] Dombroski, M. A.; Kates, S. A.; Snider, B. B., *J. Am. Chem. Soc.*, (1990) **112**, 2759.

[77] Nugent, W. A.; RajanBabu, T. V., *J. Am. Chem. Soc.*, (1988) **110**, 8561.

[78] RajanBabu, T. V.; Nugent, W. A., *J. Am. Chem. Soc.*, (1994) **116**, 986.

[79] Gansäuer, A.; Pierobon, M.; Bluhm, H., *Angew. Chem. Int. Ed.*, (1998) **37**, 101.

[80] Justicia, J.; Rosales, A.; Buñuel, E.; Oller-López, J. L.; Valdivia, M.; Haïdour, A.; Oltra, J. E.; Barrero, A. F.; Cárdenas, D. J.; Cuerva, J. M., *Chem.–Eur. J.*, (2004) **10**, 1778.

[81] Morcillo, S. P.; Miguel, D.; Resa, S.; Martín-Lasanta, A.; Millán, A.; Choquesillo-Lazarte, D.; García-Ruiz, J. M.; Mota, A. J.; Justicia, J.; Cuerva, J. M., *J. Am. Chem. Soc.*, (2014) **136**, 6943.

[82] Mukaiyama, T.; Yamada, T., *Bull. Chem. Soc. Jpn.*, (1995) **68**, 17.

[83] Iwasaki, K.; Wan, K. K.; Oppedisano, A.; Crossley, S. W. M.; Shenvi, R. A., *J. Am. Chem. Soc.*, (2014) **136**, 1300.

[84] Sweany, R. L.; Halpern, J., *J. Am. Chem. Soc.*, (1977) **99**, 8335.

[85] Halpern, J., *Pure Appl. Chem.*, (1986) **58**, 575.

[86] Bullock, R. M.; Samsel, E. G., *J. Am. Chem. Soc.*, (1987) **109**, 6542.

[87] Bullock, R. M.; Samsel, E. G., *J. Am. Chem. Soc.*, (1990) **112**, 6886.

[88] Choi, J.; Tang, L.; Norton, J. R., *J. Am. Chem. Soc.*, (2007) **129**, 234.

[89] Li, G.; Pulling, M. E.; Estes, D. P.; Norton, J. R., *J. Am. Chem. Soc.*, (2012) **134**, 14662.

[90] Tang, L.; Papish, E. T.; Abramo, G. P.; Norton, J. R.; Baik, M.-H.; Friesner, R. A.; Rappé, A., *J. Am. Chem. Soc.*, (2003) **125**, 10093.

[91] Eisenberg, D. C.; Norton, J. R., *Isr. J. Chem.*, (1991) **31**, 55.

[92] Hegedus, L. S., *Transition Metals in the Synthesis of Complex Organic Molecules*, University Science Books: Sausalito, CA, (1999); p 39.

[93] Waser, J.; Gaspar, B.; Nambu, H.; Carreira, E. M., *J. Am. Chem. Soc.*, (2006) **128**, 11693.

[94] Okamoto, T.; Oka, S., *Tetrahedron Lett.*, (1981) **22**, 2191.

[95] Okamoto, T.; Oka, S., *J. Org. Chem.*, (1984) **49**, 1589.

[96] Inoue, S.; Ohkatsu, Y.; Ohno, M.; Ooi, T., *Nippon Kagaku Kaishi*, (1985), 387.

[97] de Bruin, B.; Dzik, W. I.; Li, S.; Wayland, B. B., *Chem.–Eur. J.*, (2009) **15**, 4312.

[98] Magnus, P.; Waring, M. J.; Scott, D. A., *Tetrahedron Lett.*, (2000) **41**, 9731.

[99] Magnus, P.; Payne, A. H.; Waring, M. J.; Scott, D. A.; Lynch, V., *Tetrahedron Lett.*, (2000) **41**, 9725.

[100] Wang, L.-C.; Jang, H.-Y.; Roh, Y.; Lynch, V.; Schultz, A. J.; Wang, X.; Krische, M. J., *J. Am. Chem. Soc.*, (2002) **124**, 9448.

[101] Gaspar, B.; Waser, J.; Carreira, E. M., *Synthesis*, (2007), 3839.

[102] Waser, J.; Carreira, E. M., *J. Am. Chem. Soc.*, (2004) **126**, 5676.

[103] Waser, J.; Carreira, E. M., *Angew. Chem. Int. Ed.*, (2004) **43**, 4099.

References

[104] Waser, J.; González-Gómez, J. C.; Nambu, H.; Huber, P.; Carreira, E. M., *Org. Lett.*, (2005) **7**, 4249.

[105] Waser, J.; Nambu, H.; Carreira, E. M., *J. Am. Chem. Soc.*, (2005) **127**, 8294.

[106] Gaspar, B.; Carreira, E. M., *Angew. Chem. Int. Ed.*, (2007) **46**, 4519.

[107] Gaspar, B.; Carreira, E. M., *Angew. Chem. Int. Ed.*, (2008) **47**, 5758.

[108] Gaspar, B.; Waser, J.; Carreira, E. M., *Org. Synth.*, (2010) **87**, 88.

[109] Gaspar, B.; Carreira, E. M., *J. Am. Chem. Soc.*, (2009) **131**, 13214.

[110] Barker, T. J.; Boger, D. L., *J. Am. Chem. Soc.*, (2012) **134**, 13588.

[111] Leggans, E. K.; Barker, T. J.; Duncan, K. K.; Boger, D. L., *Org. Lett.*, (2012) **14**, 1428.

[112] Lo, J. C.; Yabe, Y.; Baran, P. S., *J. Am. Chem. Soc.*, (2014) **136**, 1304.

[113] Lo, J. C.; Gui, J.; Yabe, Y.; Pan, C.-M.; Baran, P. S., *Nature (London)*, (2014) **516**, 343.

[114] King, S. M.; Ma, X.; Herzon, S. B., *J. Am. Chem. Soc.*, (2014) **136**, 6884.

[115] Hartung, J.; Pulling, M. E.; Smith, D. M.; Yang, D. X.; Norton, J. R., *Tetrahedron*, (2008) **64**, 11822.

[116] Crossley, S. W. M.; Barabé, F.; Shenvi, R. A., *J. Am. Chem. Soc.*, (2014) **136**, 16788.

[117] Han, A.; Spataru, T.; Hartung, J.; Li, G.; Norton, J. R., *J. Org. Chem.*, (2014) **79**, 1938.

[118] Heinemann, C.; Demuth, M., *J. Am. Chem. Soc.*, (1997) **119**, 1129.

[119] Rendler, S.; MacMillan, D. W. C., *J. Am. Chem. Soc.*, (2010) **132**, 5027.

[120] Parker, K. A.; Fokas, D., *J. Org. Chem.*, (2006) **71**, 449.

[121] Abelman, M. M.; Overman, L. E., *J. Am. Chem. Soc.*, (1988) **110**, 2328.

[122] Carpenter, N. E.; Kucera, D. J.; Overman, L. E., *J. Org. Chem.*, (1989) **54**, 5846.

[123] Kucera, D. J.; O'Connor, S. J.; Overman, L. E., *J. Org. Chem.*, (1993) **58**, 5304.

[124] Trost, B. M.; Shi, Y., *J. Am. Chem. Soc.*, (1991) **113**, 701.

[125] Oppolzer, W.; Swenson, R. E.; Gaudin, J., *Tetrahedron Lett.*, (1988) **29**, 5529.

[126] Zhang, Y.; Negishi, E.-i., *J. Am. Chem. Soc.*, (1989) **111**, 3454.

[127] Wu, G.-z.; Lamaty, F.; Negishi, E.-i., *J. Org. Chem.*, (1989) **54**, 2507.

[128] Nicolaou, K. C.; Bulger, P. G.; Sarlah, D., *Angew. Chem. Int. Ed.*, (2005) **44**, 4490.

[129] Pfeiffer, M. W. B.; Phillips, A. J., *J. Am. Chem. Soc.*, (2005) **127**, 5334.

[130] Minger, T. L.; Phillips, A. J., *Tetrahedron Lett.*, (2002) **43**, 5357.

[131] Nicolaou, K. C.; Bulger, P. G.; Sarlah, D., *Angew. Chem. Int. Ed.*, (2005) **44**, 4442.

[132] Trost, B. M.; Van Vranken, D. L., *Chem. Rev.*, (1996) **96**, 395.

[133] Trost, B. M.; Crawley, M. L., *Chem. Rev.*, (2003) **103**, 2921.

[134] Oppolzer, W., *Angew. Chem. Int. Ed. Engl.*, (1989) **28**, 38.

[135] Valot, G.; Garcia, J.; Duplan, V.; Serba, C.; Barluenga, S.; Winssinger, N., *Angew. Chem. Int. Ed.*, (2012) **51**, 5391.

[136] Tantillo, D. J., *Nat. Prod. Rep.*, (2011) **28**, 1035.

[137] Lavallée, J.; Deslongchamps, P., *Tetrahedron Lett.*, (1988) **29**, 6033.

[138] Rouillard, A.; Deslongchamps, P., *Tetrahedron*, (2002) **58**, 6555.

[139] Scheerer, J. R.; Lawrence, J. F.; Wang, G. C.; Evans, D. A., *J. Am. Chem. Soc.*, (2007) **129**, 8968.

1.2 Cation–π Cyclizations of Epoxides and Polyepoxides

K. W. Armbrust, T. Halkina, E. H. Kelley, S. Sittihan, and T. F. Jamison

General Introduction

Domino reactions of polyepoxides are of general interest for the synthesis of a variety of polycyclic ethers, including polyether ionophores, squalene-derived polyethers, and marine ladder polyethers.[1] A number of biosynthetic hypotheses invoke polyepoxide cascades to rapidly generate complex polyethers; the two most cited are the Cane–Celmer–Westley and Nakanishi proposals. Respectively, these hypotheses account for the observed formation of all-*exo* and all-*endo* cascade products, borrowing terminology from Baldwin's classification of ring-closing processes.[2–4] In general, smaller-ring products (formed by what is generally termed an *exo* pathway) are kinetically favored over larger-ring products (*endo*). That is, five-membered tetrahydrofuran rings form preferentially to their six-membered tetrahydropyran counterparts. Similarly, six-membered tetrahydropyrans form preferentially to seven-membered oxepane rings. A number of total syntheses have been inspired by these proposed cascades, and many creative methods have been developed to generate either *exo* or *endo* products in a selective fashion. The polyepoxide starting materials are often prepared via the powerful Sharpless and Shi asymmetric epoxidations of alkenes.[5,6] Complete transfer of stereochemical information is generally observed, as most epoxide openings proceed with stereospecific inversion of configuration.

For the large majority of cascades discussed in this review, mechanistic details remain unresolved, but most can nonetheless be categorized on the basis of how they are designed. In some cases, the epoxide undergoing ring opening is activated to increase its electrophilicity (which we will thus classify as an "electrophilic cascade"), and in others, an attacking nucleophile is activated to increase its nucleophilicity ("nucleophilic cascade").

1.2.1 *exo*-Selective Polyepoxide Cascades

exo-Selective polyepoxide cascades have found considerable use in the synthesis of ionophore and squalene-derived polycyclic ethers, capitalizing on the kinetic preference for the formation of the smaller-ring product in each epoxide-opening event.[1] Drawing inspiration from the Cane–Celmer–Westley proposal, a variety of syntheses have been completed utilizing Brønsted acid or base activation to rapidly assemble complex fragments from polyepoxides.[2]

1.2.1.1 Brønsted Acid Promoted Cascades

Activation of polyepoxide cascades can be accomplished with a variety of Brønsted acids, but the most common activator is 10-camphorsulfonic acid, generally used in non-polar solvents such as dichloromethane and toluene. For the cascade of diepoxide **1** to form tris-tetrahydrofuran **2**, treatment with camphorsulfonic acid rapidly forms the two new rings with high *exo* selectivity under mild conditions (Scheme 1).[7] Bidirectional cascades are also possible, as demonstrated by the reaction of epoxide **3** to provide tetrakis-tetrahydro-

for references see p 65

44 Domino Transformations **1.2** Cation–π Cyclizations of Epoxides and Polyepoxides

furan **4**.[8] Low temperatures are used to prevent the undesired unidirectional cascade. Polytetrahydropyran systems are also possible, shown by the synthesis of tricycle **6** from triepoxide **5**, although detailed experimental procedures were not provided.[9]

Scheme 1 *exo*-Selective Epoxide-Opening Cascades Promoted by Brønsted Acid[7–9]

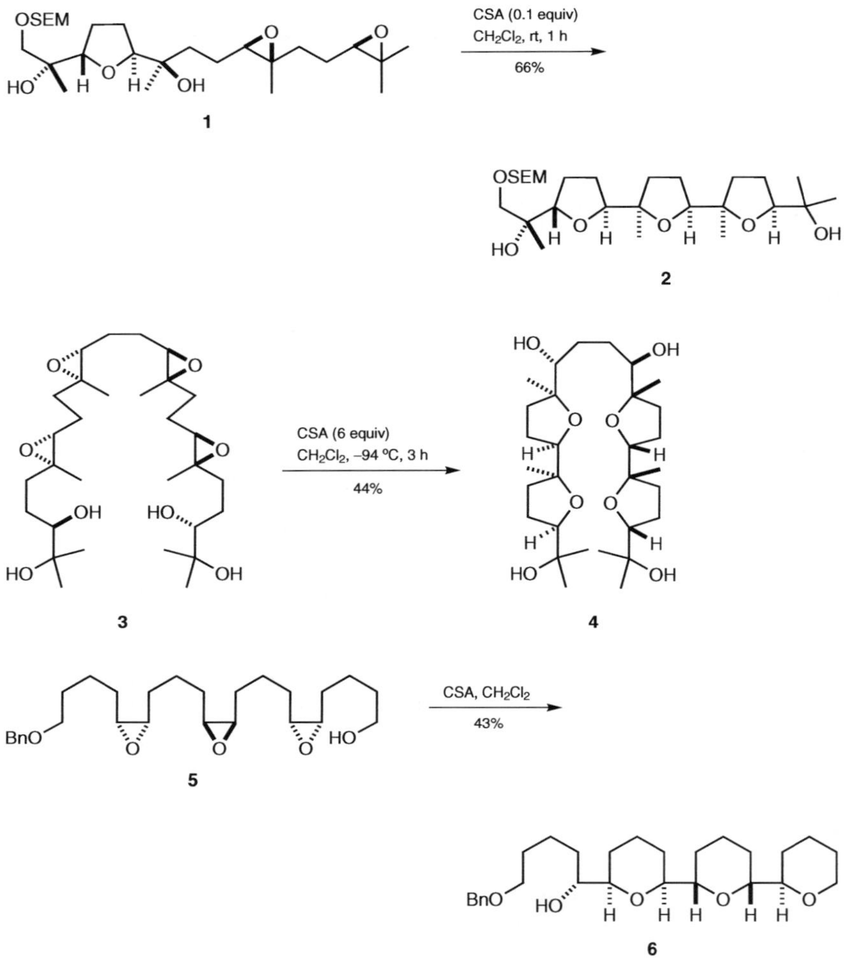

(R)-2-{(2S,2′R,2″R,5S,5′S,5″S)-5″-(2-Hydroxypropan-2-yl)-2′,2″-dimethyldodecahydro-[2,2′:5′,2″-terfuran]-5-yl}-1-{[2-(trimethylsilyl)ethoxy]methoxy}propan-2-ol (2):[7]
To a soln of diepoxide **1** (1.0 g, 2.0 mmol) in CH$_2$Cl$_2$ (70 mL) at rt under N$_2$ was added CSA (46 mg, 0.20 mmol). The soln was kept at rt for 1 h. Sat. aq NaHCO$_3$ (50 mL) was added, and the aqueous layer was extracted with CH$_2$Cl$_2$ (3 × 30 mL). The combined organic phase was washed with brine, dried (MgSO$_4$), and concentrated. The residue was purified by column chromatography (EtOAc/hexane 1:1) to afford **2** as a colorless oil; yield: 0.66 g (66%).

(1R,4R)-1,4-Bis{(2R,2′S,5S,5′R)-5′-(2-hydroxypropan-2-yl)-2′,5-dimethyloctahydro-[2,2′-bi-furan]-5-yl}butane-1,4-diol (4):[8]
To a soln of tetraepoxide **3** (23 mg, 43 μmol) in CH$_2$Cl$_2$ (15 mL) at −94 °C under N$_2$ was added a mixture of CSA (60 mg, 0.26 mmol) in CH$_2$Cl$_2$ via cannula. The mixture was stirred at −94 °C for 3 h. Et$_3$N (0.3 mL) was then added. The mixture was allowed to warm up to rt

1.2.1 *exo*-Selective Polyepoxide Cascades   **45**

and concentrated. The residue was purified by column chromatography (EtOAc/hexanes 2:1) to afford **4** as a colorless solid; yield: 10.1 mg (44%).

1.2.1.2 **Brønsted Base Promoted Cascades**

An alternative means of promoting *exo*-selective epoxide cascades is to make the pendent alcohol more nucleophilic via deprotonation with base, with each successive epoxide-opening event generating the alkoxide for the next ring-opening event. The effectiveness of this reaction manifold is demonstrated in the cascade of diepoxide **7**, which generates the C_s tetrahydrofuran tetraol **8**, upon heating in aqueous sodium hydroxide (Scheme 2).[10,11] The C_2 tetrahydrofuran tetraol **10** can be synthesized in a similar manner from the monoprotected diepoxyalcohol **9**, with a single alcohol functional group controlling the directionality of the cascade, and ultimately the stereochemistry of the final product.[11] A similar cascade generates the more elaborated tetrahydropyran fragment **12** from diepoxide **11**, in a synthesis of (+)-omaezakianol.[12] In addition to unidirectional cascades, bidirectional cascades are also known, exemplified by the domino reaction of diol **13** to synthesize (−)-glabrescol (**14**).[13] An acid-promoted cascade of diol **13** is also possible but gives lower yields, attributed to the nondiscriminating nature of epoxide activation by acid. In contrast, the initiation of the cascade by base helps to control the directionality of the cascade.

Scheme 2   *exo*-Selective Epoxide-Opening Cascades Promoted by Brønsted Base[10–13]

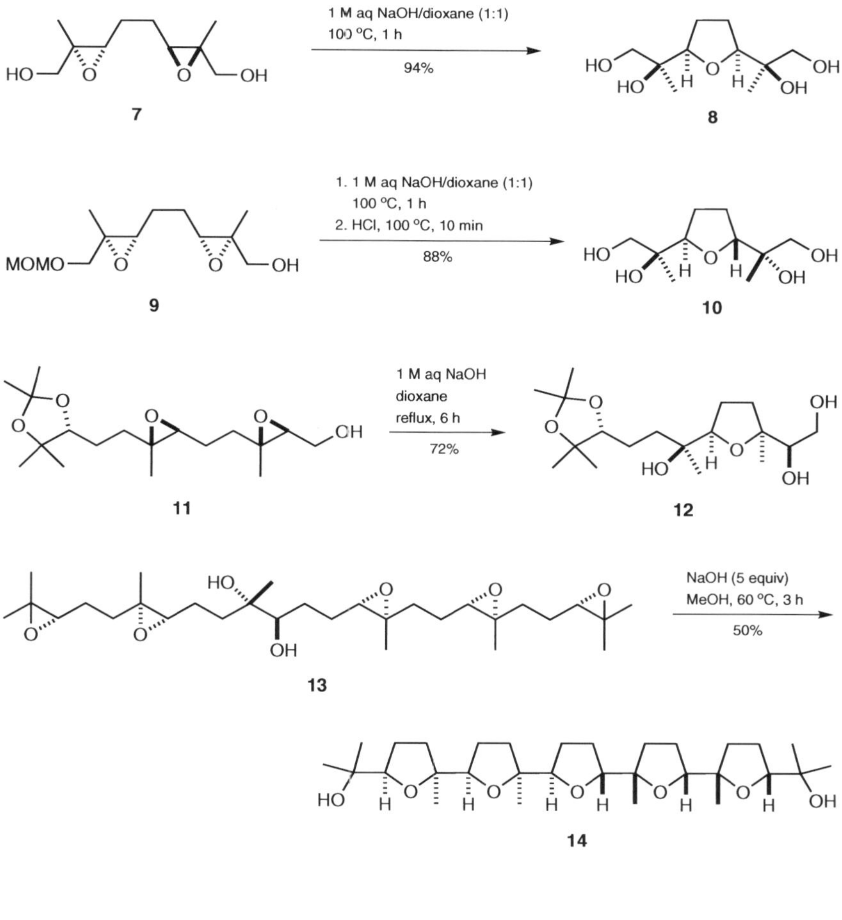

for references see p 65

(S)-2-{(2R,5S)-5-[(R)-1,2-Dihydroxypropan-2-yl]tetrahydrofuran-2-yl}propane-1,2-diol (8):[11]
A soln of diol **7** (4.06 g, 20.1 mmol) in dioxane (60 mL) and 1 M aq NaOH (60 mL) was stirred at 100 °C for 1 h. The mixture was cooled to rt and neutralized with concd HCl. The organic and aqueous layers were concentrated under reduced pressure, and EtOH was added to the residue to precipitate NaCl. After filtration through a pad of Celite, the filtrate was concentrated. The residue was purified by column chromatography (CHCl$_3$/MeOH 85:15) to afford **8** as a colorless oil; yield: 4.14 g (94%).

(2S,2'S)-2,2'-[(2R,5R)-Tetrahydrofuran-2,5-diyl]bis(propane-1,2-diol) (10):[11]
A soln of alcohol **9** (3.55 g 14.4 mmol) in dioxane (45 mL) and 1 M aq NaOH (45 mL) was stirred at 100 °C for 1 h. The mixture was cooled to rt, adjusted to pH 2 with concd HCl, and stirred at 100 °C for 10 min. The mixture was cooled to rt and neutralized with 1 M aq NaOH. The organic and aqueous layers were concentrated under reduced pressure, and EtOH was added to the residue to precipitate NaCl. After filtration through a pad of Celite, the filtrate was concentrated. The residue was purified by column chromatography (EtOAc/MeOH 4:1) to afford **10** as a colorless oil; yield: 2.80 g (88%).

(R)-1-[(2S,5R)-5-{(S)-2-Hydroxy-4-[(R)-2,2,5,5-tetramethyl-1,3-dioxolan-4-yl]butan-2-yl}-2-methyltetrahydrofuran-2-yl]ethane-1,2-diol (12):[12]
A soln of alcohol **11** (670 mg, 2.04 mmol) in dioxane (6.1 mL) and 1 M aq NaOH (6.10 mL) was stirred at 100 °C for 6 h. The mixture was cooled to rt, and 1 M aq HCl (6.1 mL) was added to the mixture. The organic and aqueous layers were concentrated under reduced pressure, and EtOH was added to the residue to precipitate NaCl. After filtration through a pad of Celite, the filtrate was concentrated. The residue was purified by column chromatography (hexanes/EtOAc 1:1 to EtOAc/MeOH 95:5) to provide **12** as a colorless oil; yield: 510 mg (72%).

(–)-Glabrescol (14):[13]
To a soln of diol **13** (126 mg, 0.24 mmol) in MeOH (2.4 mL) was added NaOH (48 mg, 1.2 mmol) and the mixture was heated to 60 °C. After 3 h, sat. aq NH$_4$Cl (1 mL) was added, and the mixture was extracted with EtOAc (3 ×). The combined extracts were washed with brine, dried (Na$_2$SO$_4$), and concentrated. The residue was purified by column chromatography (t-BuOMe/hexanes 2:3) to provide (–)-glabrescol (**14**) as a colorless oil; yield: 62 mg (50%).

1.2.1.3 Oxocarbenium-Initiated Cascades via Photooxidative Cleavage

In situ generation of an oxocarbenium ion via photooxidative cleavage of C—C bonds represents a mild method for initiating polyepoxide cascades capable of leading to *exo*-type products.[14] The mild conditions allow for the inclusion of acid- or base-sensitive functional groups, while still proceeding through an electrophilic cascade. Reaction of diastereomeric diepoxides **15** and **17** leads cleanly to bis-tetrahydrofuran products **16** and **18** (Scheme 3), respectively.

Scheme 3 Oxocarbenium-Initiated Epoxide-Opening Cascades via Photooxidative Cleavage[14]

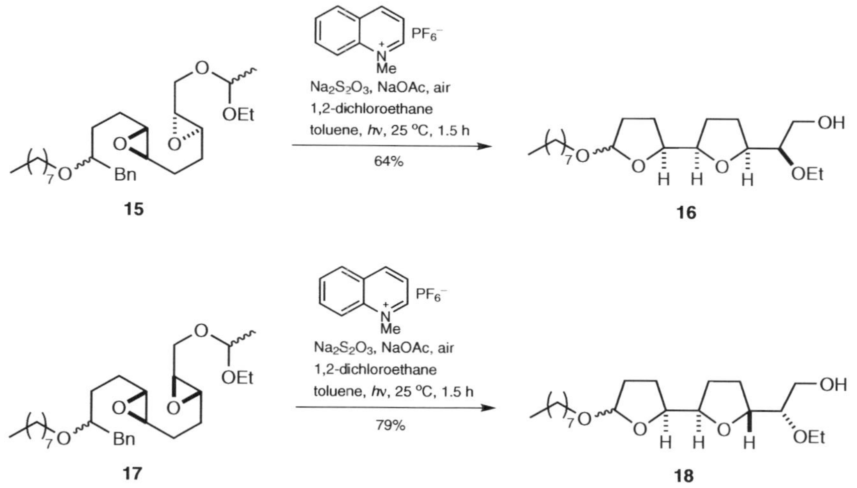

**(2R)-2-Ethoxy-2-{(2R,2′S,5S)-5′-octyloxyoctahydro-[2,2′-bifuran]-5-yl}ethanol (16);
Typical Procedure:**[14]
To a borosilicate flask containing diepoxide **15** (50 mg, 0.10 mmol) in 1,2-dichloroethane (6 mL) was added *N*-methylquinolinium hexafluorophosphate (0.7 mg, 2.6 μmol), NaOAc (100 mg, 1.21 mmol), anhyd Na$_2$S$_2$O$_3$ (10 mg, 0.63 mmol), and toluene (1 mL). The mixture was photoirradiated with gentle aeration for 1.5 h while stirring at rt. The mixture was filtered through a small plug of silica gel, and the filtrate was concentrated. The resulting residue was purified via column chromatography (EtOAc/hexanes 1:9) to provide **16** as a mixture of two diastereomers; yield: 28 mg (64%).

1.2.2   *endo*-Selective Polyepoxide Cascades

endo-Selective polyepoxide cascades have found considerable use in the synthesis of natural products containing fused tetrahydropyran and oxepane ring systems, especially in the marine ladder polyether family.[1] Following the Nakanishi proposal for the biosynthesis of these ladder polyethers via *endo*-selective cascades of polyepoxide precursors, a variety of synthetic studies have been reported to rapidly assemble the polycyclic ether framework from polyepoxides.[3] In most cases, substitution patterns of the epoxides play a significant role in the success of the *endo*-selective cascades. More highly substituted epoxide carbons can better stabilize the incipient positive charge buildup during the epoxide-opening event. Due to their ability to bias epoxide opening, these additional substituents are generally described as "directing groups", as they govern the attack of the incoming nucleophile toward the desired epoxide carbon.

1.2.2.1   Lewis Acid Activation

1.2.2.1.1   Alkyl-Directed Cascades

Fused polytetrahydropyrans and polyoxepanes can be accessed via Lewis acid catalyzed *endo*-selective domino reactions of electronically biased polyepoxide precursors. Lewis acids not only activate the epoxides but also enhance the effect of the directing groups, if present. The most widely used Lewis acid is boron trifluoride–diethyl ether complex in non-polar solvents such as dichloromethane. In the case of tetrahydropyran formation,

for references see p 65

the identity of the terminating nucleophile determines the product outcome. For instance, *N,N*-dimethylcarbamates, such as **19** and **21**, afford *trans*-fused products **20** and **22**, respectively, with complete inversion of the stereochemistry of the epoxides (Scheme 4).[15] In contrast, *tert*-butyl carbonate derivative **23**, under the same conditions though at a higher temperature, results in significant formation of *cis*-fused product **24** presumably via the intermediacy of a tertiary carbocation. This method requires a directing group at every epoxide. In addition, yields generally drop precipitously as the number of epoxides increases.

Scheme 4 Synthesis of Tetrahydropyrans via Epoxide-Opening Cascades[15]

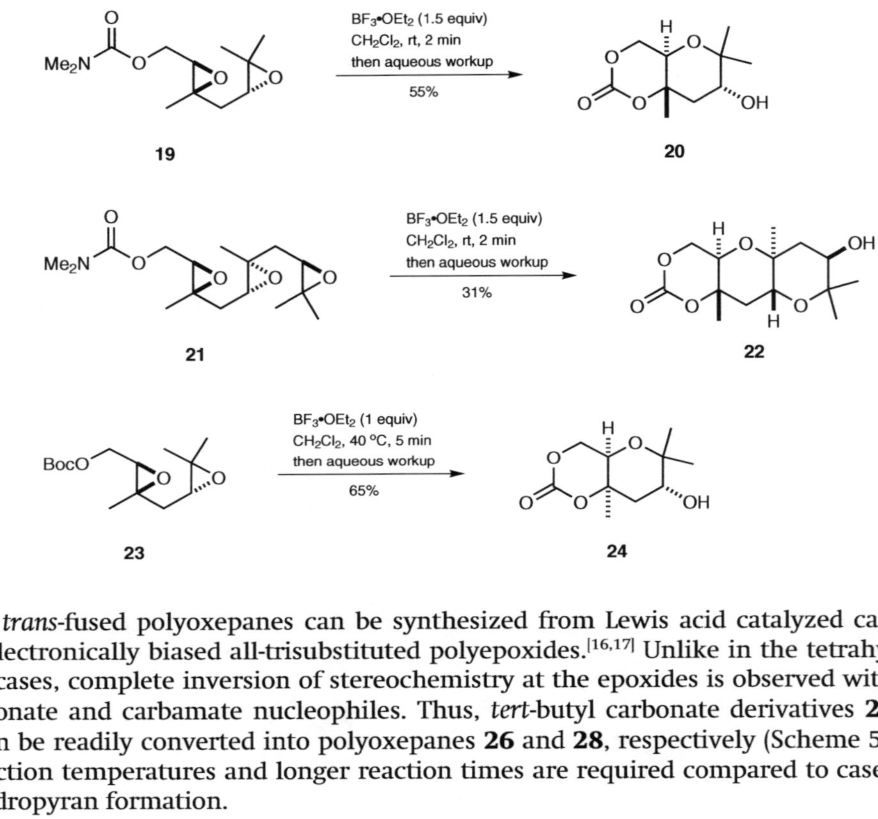

Similarly, *trans*-fused polyoxepanes can be synthesized from Lewis acid catalyzed cascades of electronically biased all-trisubstituted polyepoxides.[16,17] Unlike in the tetrahydropyran cases, complete inversion of stereochemistry at the epoxides is observed with both carbonate and carbamate nucleophiles. Thus, *tert*-butyl carbonate derivatives **25** and **27** can be readily converted into polyoxepanes **26** and **28**, respectively (Scheme 5). Lower reaction temperatures and longer reaction times are required compared to cases of tetrahydropyran formation.

Scheme 5 Synthesis of Oxepanes via Epoxide-Opening Cascades with Carbonate Trapping Nucleophiles[16,17]

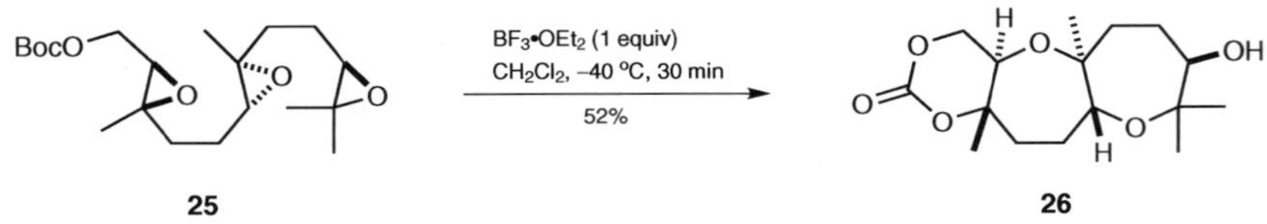

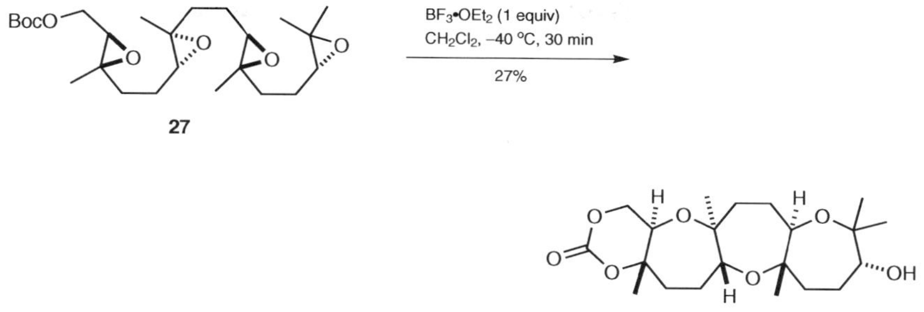

1.2.2 *endo*-Selective Polyepoxide Cascades

This method has been applied in synthetic studies toward *ent*-dioxepandehydro-thyrsiferol (**29**, Scheme 6) with a *tert*-butyl ester as a trapping nucleophile.[18] Upon Lewis acid catalysis, *tert*-butyl ester **30** is converted into tricycle **31** after silylation. Epoxides with opposite absolute configurations are used to access the desired *trans–anti* topology present in **29**.

Scheme 6   Synthesis of Oxepanes via Epoxide-Opening Cascades with Ester Trapping Nucleophiles[18]

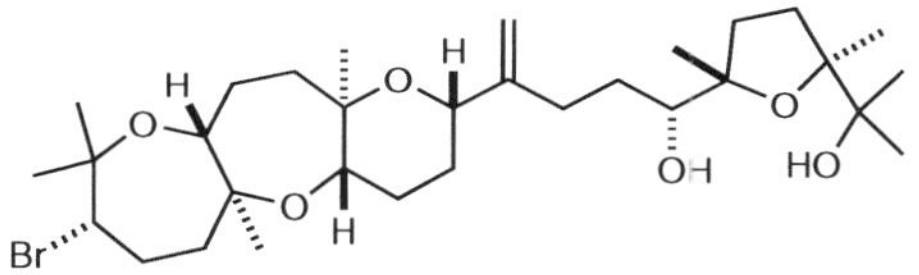

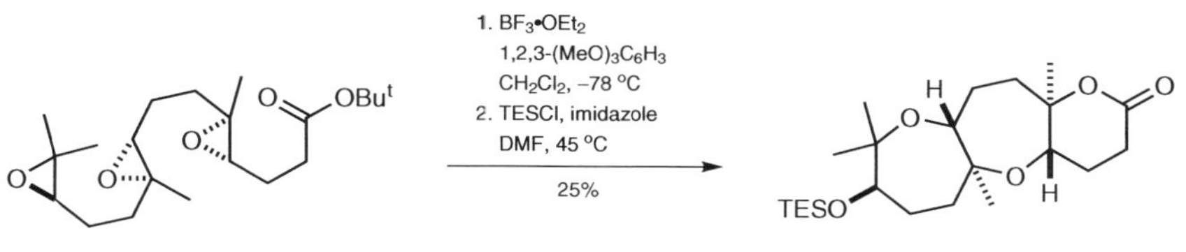

This method tolerates unbiased epoxides, provided that the first and last epoxides are electronically biased by additional alkyl groups.[19] More nucleophilic *N,N*-dimethylcarbamate trapping nucleophiles afford slightly higher yields than their *tert*-butyl carbonate counterparts (Scheme 7). Thus, after acylation, tricycle **33** and tetracycle **35** can be obtained from Lewis acid catalyzed cascades of polyepoxides **32** and **34**, respectively.

for references see p 65

Domino Transformations 1.2 Cation-π Cyclizations of Epoxides and Polyepoxides

Scheme 7 Cascades of Polyepoxides Bearing Internal Disubstituted Epoxides[19]

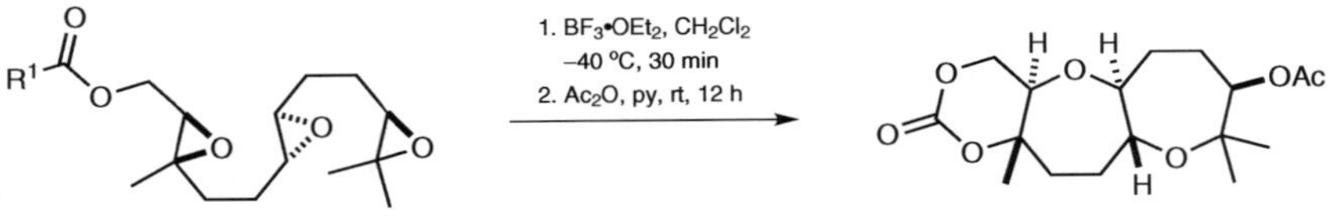

R¹	Yield (%)	Ref
O*t*-Bu	30	[19]
NMe₂	35	[19]

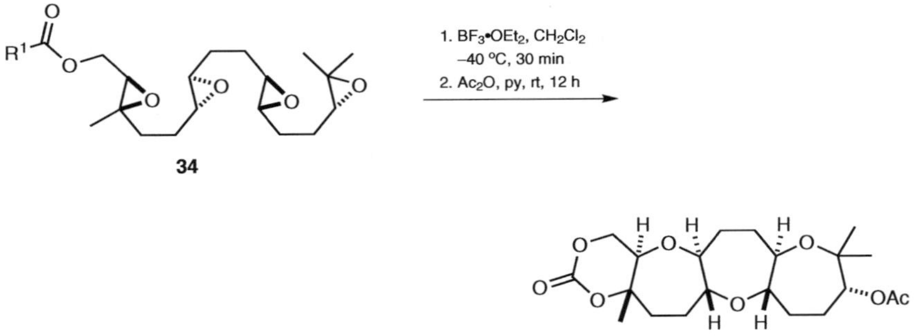

R¹	Yield (%)	Ref
O*t*-Bu	20	[19]
NMe₂	25	[19]

(4a*R*,5a*S*,7*R*,9a*R*,10a*S*)-7-Hydroxy-5a,8,8,10a-tetramethyloctahydro-4*H*-pyrano-[2′,3′:5,6]pyrano[3,2-*d*][1,3]dioxin-2-one (22); Typical Procedure:[15]

To a soln of triepoxide **21** (155 mg, 0.49 mmol) in CH_2Cl_2 (50 mL) at rt under N_2 was added 0.2 M BF_3•OEt_2 in CH_2Cl_2 (3.7 mL, 0.74 mmol). The soln was stirred at rt for 2 min. A 1:1 mixture of H_2O and sat. aq $NaHCO_3$ (10 mL) was added and stirred overnight. The aqueous layer was extracted with CH_2Cl_2. The combined organic phase was dried ($MgSO_4$), filtered, and concentrated. The residue was purified by column chromatography (EtOAc/hexanes 1:3 to 5:1) to afford **22**; yield: 44 mg (31%).

(4a*R*,7*R*,8a*R*)-7-Hydroxy-6,6,8a-trimethylhexahydropyrano[3,2-*d*][1,3]dioxin-2-one (24):[15]

A soln of diepoxide **23** (138 mg, 0.51 mmol) in CH_2Cl_2 (10.1 mL) under N_2 was heated to reflux, and 0.2 M BF_3•OEt_2 in CH_2Cl_2 (2.6 mL, 0.52 mmol) was added. After 5 min, the soln was quenched with sat. aq $NaHCO_3$ (3 mL) and cooled to rt. The aqueous layer was extracted with CH_2Cl_2. The combined organic phase was dried ($MgSO_4$), filtered, and concentrated. The residue was purified by column chromatography (EtOAc/hexanes 1:3 to 5:1) to afford **24**; yield: 71 mg (65%).

(4a*R*,5a*S*,8*R*,10a*R*,12a*S*)-8-Hydroxy-5a,9,9,12a-tetramethyldecahydro-4*H*-[1,3]dioxino[5,4-*b*]oxepino[2,3-*f*]oxepin-2-one (26); Typical Procedure:[16,17]

To a soln of triepoxide **25** (111 mg, 0.3 mmol) in CH_2Cl_2 (50 mL) at −40 °C under N_2 was added dropwise 0.3 M BF_3•OEt_2 in CH_2Cl_2 (precooled to −40 °C; 1 mL, 0.3 mmol). The mix-

ture was stirred at −40 °C for 30 min, quenched with H_2O (0.2 mL), and allowed to warm to rt. The aqueous layer was extracted with CH_2Cl_2 (3 × 15 mL). The combined organic phase was dried ($MgSO_4$), filtered, and concentrated. The residue was purified by column chromatography (EtOAc/hexanes 1:9 to 4:1) to afford **26**; yield: 49 mg (52%).

(4a*S*,5a*S*,8*R*,10a*R*,12a*R*)-5a,9,9,12a-Tetramethyl-8-(triethylsilyloxy)dodecahydro-2*H*-oxepino[3,2-*b*]pyrano[2,3-*f*]oxepin-2-one (31):[18]

To a soln of triepoxide **30** (5.15 g, 13.98 mmol) and 1,2,3-trimethoxybenzene (4.7 g, 26.95 mmol) in CH_2Cl_2 (280 mL) at −78 °C under N_2 was added $BF_3 \cdot OEt_2$ (1.77 mL, 13.98 mmol). The soln was stirred at −78 °C for 1 h, quenched with sat. aq $NaHCO_3$ (50 mL), and allowed to warm to rt. The aqueous layer was extracted with CH_2Cl_2 (2 × 100 mL). The combined organic phase was dried ($MgSO_4$), filtered, and concentrated. The residue was dissolved in DMF (20 mL). To this soln was added imidazole (3 g, 44.07 mmol), and the reaction vessel was fitted with a condenser and purged with argon. TESCl (3.0 mL, 17.87 mmol) was added to the soln at rt. The soln was heated to 45 °C for 16 h and quenched with MeOH (3 mL). The soln was stirred at 45 °C for 45 min and allowed to cool to rt. The soln was diluted with Et_2O, washed with sat. aq NH_4Cl, and dried ($MgSO_4$). The residue was purified by column chromatography to afford **31**; yield: 1.49 g (25%).

(4a*R*,5a*S*,8*R*,10a*R*,12a*S*)-9,9,12a-Trimethyl-2-oxodecahydro-4*H*-[1,3]dioxino[5,4-*b*]oxepino[2,3-*f*]oxepin-8-yl Acetate (33); Typical Procedure:[19]

To a soln of triepoxide **32** (R^1 = O*t*-Bu; 36 mg, 0.10 mmol) in CH_2Cl_2 (2 mL) at −40 °C under N_2 was added dropwise 0.1925 M $BF_3 \cdot OEt_2$ in CH_2Cl_2 (0.52 mL, 0.10 mmol) over 5 min. The soln was stirred at −40 °C for 30 min and quenched with sat. aq $NaHCO_3$ (2 mL). The aqueous layer was extracted with CH_2Cl_2 (3 × 15 mL). The combined organic phase was washed with brine (10 mL), dried ($MgSO_4$), filtered, and concentrated. The residue was dissolved in pyridine (1 mL) and Ac_2O (50 µL) was added. After 12 h at rt, the soln was concentrated and purified by column chromatography (EtOAc/hexanes 1:10 to 2:1) to afford **33** as a white crystalline solid; yield: 10 mg (30%).

1.2.2.2 Brønsted Base Activation

1.2.2.2.1 Trimethylsilyl-Directed Cascades

Under basic conditions, epoxide-opening reactions generally have a strong preference for *exo* selectivity. Methods have been developed to overcome this intrinsic bias by utilizing the directing effect of trimethylsilyl groups to construct up to three tetrahydropyrans.[20] The trimethylsilyl groups at each epoxide are conveniently cleaved in situ by hydroxy-assisted protodesilylation, which results in the disappearance of the trimethylsilyl groups at ring junctions. Construction of one tetrahydropyran as a template prior to the cascade is crucial for successful cascades. A combination of a Brønsted base, a fluoride source, and a hydroxylic solvent affords satisfactory yields. For instance, after acylation, tetrahydro-pyran triad **37** and tetrad **39** can be readily derived from diepoxide **36** and triepoxide **38**, respectively (Scheme 8). Although effective, it is worth noting that the transformation still suffers from a significant decrease in yields with an increasing number of epoxides.

for references see p 65

52 Domino Transformations 1.2 Cation–π Cyclizations of Epoxides and Polyepoxides

Scheme 8 Synthesis of Tetrahydropyrans via Epoxide-Opening Cascades with Disappearing Trimethylsilyl Directing Groups[20]

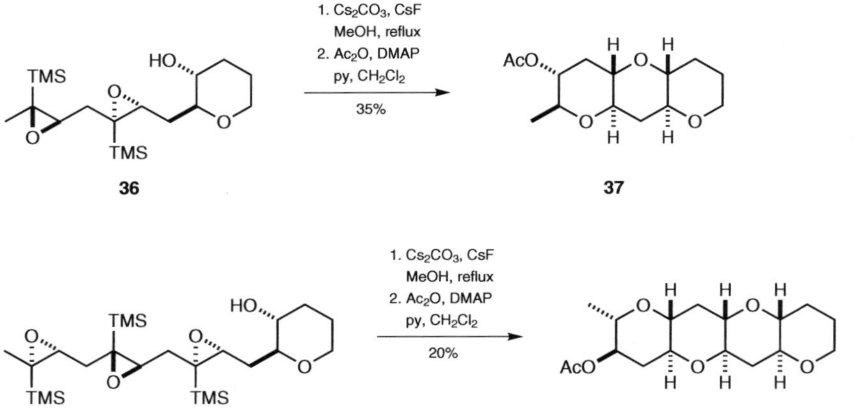

(2S,3R,4aS,5aR,9aS,10aR)-2-Methyldecahydro-2H-dipyrano[3,2-b:2′,3′-e]pyran-3-yl Acetate (37); Typical Procedure:[20]

Cs_2CO_3 (300 mg, 0.91 mmol) and CsF (140 mg, 0.91 mmol) were added to a flame-dried Schlenk tube in a glovebox under N_2. A soln of diepoxide **36** (17 mg, 46 µmol) in MeOH (480 mL) was added to the tube. The tube was sealed, and the slurry was heated to 65 °C for 3 d. The soln was concentrated and quenched with sat. aq NH_4Cl (10 mL). The aqueous layer was extracted with EtOAc (4 × 10 mL). The combined organic phase was washed with brine (10 mL), dried ($MgSO_4$), filtered, and concentrated. The residue was dissolved in CH_2Cl_2 (810 µL). DMAP (14 mg, 0.11 mmol), pyridine (16 µL, 0.11 mmol), and Ac_2O (11 µL, 0.11 mmol) were added at rt. After 3 h at rt, the soln was quenched with sat. aq NH_4Cl (5 mL). The aqueous layer was extracted with CH_2Cl_2 (4 × 5 mL). The combined organic phase was washed with brine (5 mL), dried ($MgSO_4$), filtered, and concentrated. The residue was purified by column chromatography (EtOAc/hexanes 1:1) to afford **37**; yield: 4.3 mg (35%).

1.2.2.3 Distal Electrophilic Activation

1.2.2.3.1 Bromonium-Initiated Cascades

Rapid access to fused polyether ring systems can be achieved through bromonium-initiated cascades. The activation of a distal alkene of a cascade precursor proceeds smoothly with N-bromosuccinimide to furnish desired tricycles **41**, **43**, and **45** with complete regioselectivity for the desired *endo* cyclization product in moderate to high yields (Scheme 9).[21] Bromonium formation is nonselective, however, resulting in the isolation of products as a 1:1 mixture of diastereomers. The identity of the internal trapping nucleophile affects the efficiency of the transformation, with higher yields being reported for carbonate and ester nucleophiles **40** and **42** than for a pendent alcohol nucleophile **44**. The polar, non-nucleophilic solvent 1,1,1,3,3,3-hexafluoropropan-2-ol (HFIP) was found to be crucial to the success of this transformation.

1.2.2 *endo*-Selective Polyepoxide Cascades **53**

Scheme 9 Bromonium-Initiated Diepoxide Cascades with Internal Nucleophiles[21]

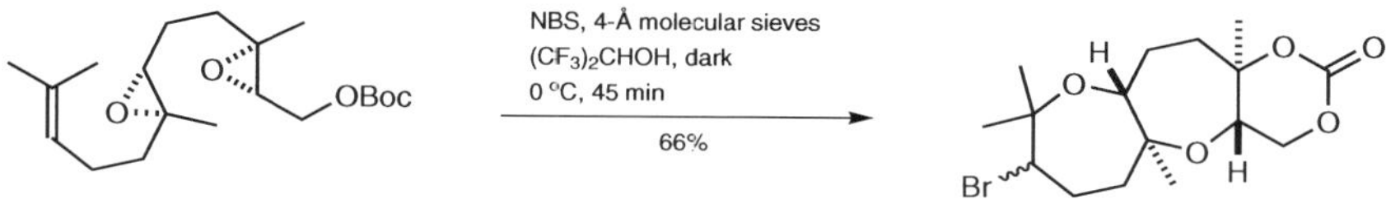

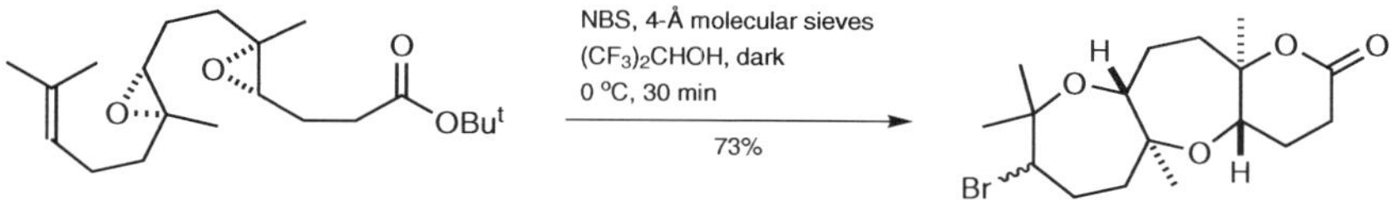

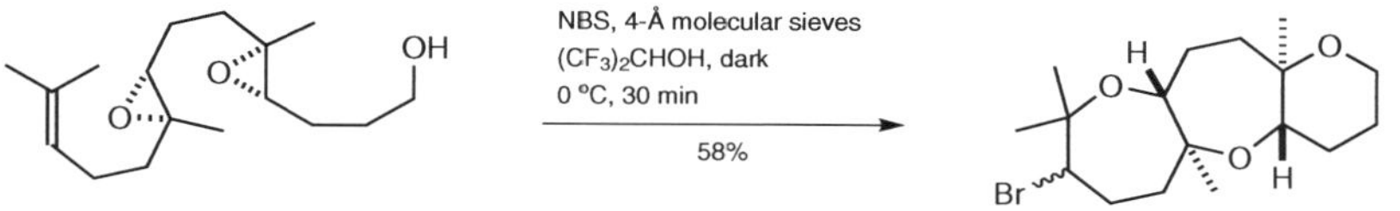

This efficient bromonium-triggered epoxide-opening cascade has been utilized in the synthesis of *ent*-dioxepandehydrothyrsiferol (**29**, Section 1.2.2.1.1, Scheme 6). Synthesis of advanced intermediate **47** was achieved in 36% yield under mild conditions in the presence of *N*-bromosuccinimide (Scheme 10).

Scheme 10 Bromonium-Initiated Cascade in the Synthesis of *ent*-Dioxepandehydrothyrsiferol[21]

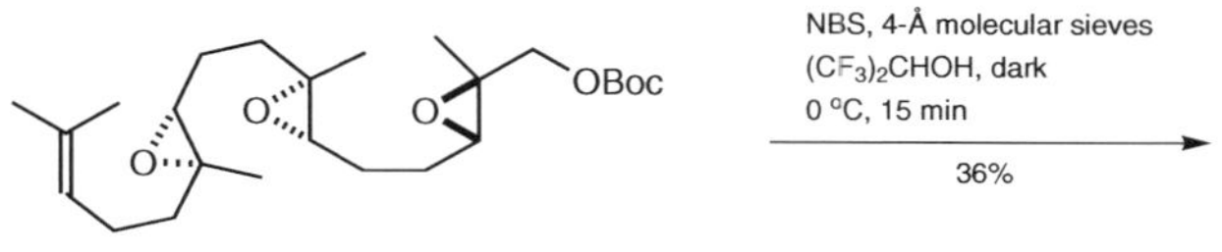

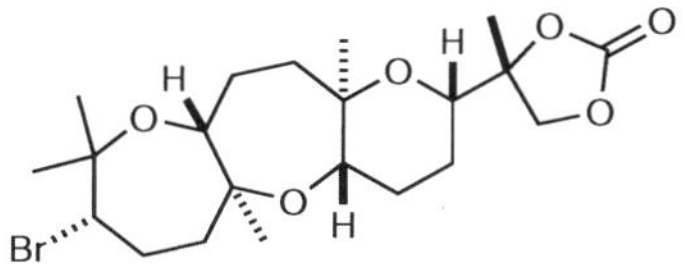

Termination of bromonium-initiated cascades can also be achieved intermolecularly with external oxygen nucleophiles (Scheme 11).[18] The cascade can be successfully initiated with either *N*-bromosuccinimide or the collidine-based electrophilic bromine source bis(2,4,6-collidine)bromonium perchlorate [Br(coll)$_2$ClO$_4$].[22] Again, nonselective bromonium formation leads to a 1:1 mixture of diastereomers in all cases. Slightly higher yields are observed for sp^2 (Bu$_4$NOAc) versus sp^3 oxygen nucleophiles (EtOH, H$_2$O, and 1,1,1,3,3,3-hexafluoropropan-2-ol). Interestingly, even a solvent molecule of 1,1,1,3,3,3-hexafluoropropan-2-ol (HFIP) can act as a nucleophile in the absence of an added exogenous nucleophile.

for references see p 65

Scheme 11 Bromonium-Initiated Diepoxide Cascades with External Nucleophiles[18]

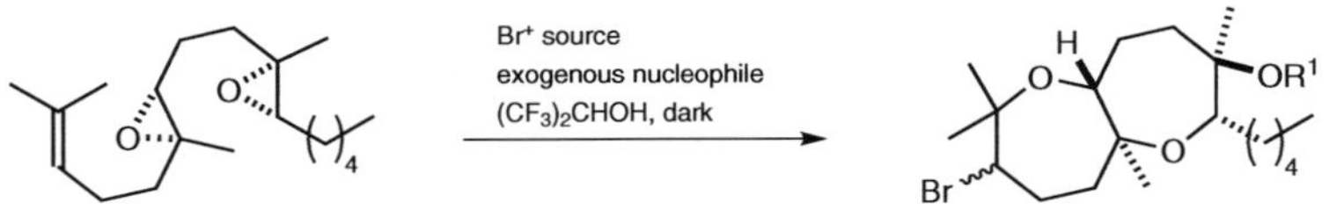

Nucleophile	R¹	Br⁺ source	Yield[a] (%)	Ref
Bu₄NOAc	Ac	NBS	74	[18]
EtOH	Et	Br(coll)₂ClO₄	56	[18]
H₂O	H	NBS	41	[18]
–[b]	CH(CF₃)₂	Br(coll)₂ClO₄	33	[18]

[a] Isolated product was a 1:1 mixture of diastereomers in all cases.
[b] Solvent acts as nucleophile.

(R)-4-{(2S,4aS,5aS,8S,10aR,12aR)-8-Bromo-5a,9,9,12a-tetramethyldodecahydro-2H-oxepino[3,2-b]pyrano[2,3-f]oxepin-2-yl}-4-methyl-1,3-dioxolan-2-one (47); Typical Procedure:[21]

To activated 4-Å molecular sieves (760 mg) in 1,1,1,3,3,3-hexafluoropropan-2-ol (7.60 mL) under an atmosphere of argon was added polyepoxide **46** (91.9 mg, 0.2095 mmol) and the soln was cooled to 0 °C. NBS (122 mg, 0.684 mmol) was added and the mixture was stirred vigorously for 15 min in the absence of light. Following dilution with Et₂O, filtration through Celite, and concentration, the residue was redissolved in Et₂O and washed with sat. aq Na₂S₂O₃/brine (1:1). The aqueous layer was extracted with Et₂O, dried (MgSO₄), filtered, and concentrated. The residue was purified by column chromatography (EtOAc/hexanes 1:1) to provide **47**; yield: 31.6 mg (36%).

(5aS,7S,8R,10aR)-3-Bromo-8-ethoxy-2,2,5a,8-tetramethyl-7-pentyldecahydrooxepino[3,2-b]oxepin (49, R¹ = Et); Typical Procedure:[18]

To a cooled (0 °C) soln of diepoxide **48** (25.2 mg, 0.0856 mmol) and EtOH (0.28 mL) in 1,1,1,3,3,3-hexafluoropropan-2-ol (1.4 mL) under argon was added Br(coll)₂ClO₄ (72.2 mg, 0.171 mmol). The mixture was stirred in the absence of light for 17 h at 0 °C. Quenching with sat. aq Na₂S₂O₃ was followed by filtration through Celite, eluting with Et₂O, to provide a soln, which was concentrated to dryness. The crude residue was repartitioned between Et₂O and sat. aq Na₂S₂O₃/brine (1:1). Extraction with Et₂O, drying (MgSO₄), and concentration provided the crude product, which was purified by column chromatography (Et₂O/hexanes 1:19 to 1:9) to afford **49** (R¹ = Et); yield: 10.0 mg (56%).

1.2.2.3.2 Oxocarbenium-Initiated Cascades via Photooxidative Cleavage

Diepoxide cascades can be initiated by in situ formation of an oxocarbenium ion via triplet sensitizer N-methylquinolinium hexafluorophosphate.[23] As with the bromonium-initiated cascades, clean inversion occurs at each epoxide undergoing attack. However, a 1:1 mixture of diastereomers at the carbocationic cascade-initiation center is observed. Moderate to good yields of all-*endo* cyclization products are observed with methyl directing groups providing the highest yields (e.g., **50** to **51** and **52** to **53** vs **54** to **55**, Scheme 12).

1.2.2 *endo*-Selective Polyepoxide Cascades

55

Scheme 12   Oxocarbenium-Initiated Epoxide-Opening Cascades via Photooxidative Cleavage[23]

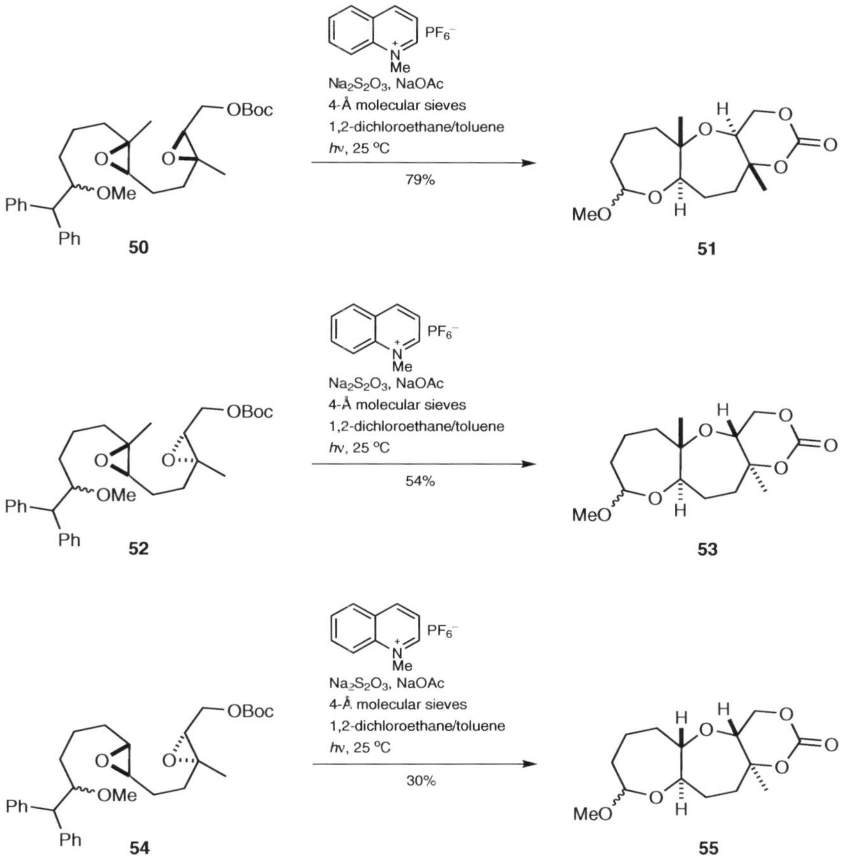

(4a*S*,5a*S*,10a*R*,12a*R*)-9-Methoxy-5a,12a-dimethyldecahydro-4*H*-[1,3]dioxino[5,4-*b*]oxepino[2,3-*f*]oxepin-2-one (51); Typical Procedure:[23]
To a borosilicate flask containing diepoxide **50** (48.2 mg, 91.9 µmol) in 1,2-dichloroethane/toluene (5:1; 3.5 mL) were added activated 4-Å molecular sieves (96 mg), anhyd Na₂S₂O₃ (96 mg, 0.608 mmol), NaOAc (96 mg, 1.17 mmol), and *N*-methylquinolinium hexafluorophosphate (2.6 mg, 9.2 µmol). The mixture was photoirradiated while being stirred with gentle aeration at rt for 3 h. Following filtration through silica gel, and rinsing of the residue with EtOAc, the crude organic material was concentrated and purified via column chromatography (EtOAc/hexanes 25:75 to 35:65) to furnish **51**; yield: 22 mg (79%).

1.2.2.3.3   **Carbocation-Initiated Cascades via Halide Abstraction**

One-pot conversion of bromo diepoxides into fused bicyclic tetrahydropyran ring systems has been reported.[24,25] Silver(I) trifluoromethanesulfonate acts as a halide-abstraction agent, initiating the electrophilic cascade, which is eventually terminated by the trifluoromethanesulfonate anion to provide the *cis*-fused 6,6 tetrahydropyran ring system as in **57** (Scheme 13). Successful formation of the *endo* product is highly sensitive to the solvent choice and the stereochemistry of the diepoxide starting materials. Reactions conducted

for references see p 65

56 Domino Transformations **1.2** Cation–π Cyclizations of Epoxides and Polyepoxides

using mixtures of tetrahydrofuran and water, or with diastereomeric epoxide starting materials, led to undesired domino reaction products such as tetrahydropyran–epoxide containing or 6,5-fused ring systems.

Scheme 13 A Carbocation-Initiated Epoxide-Opening Cascade via Halide Abstraction[24,25]

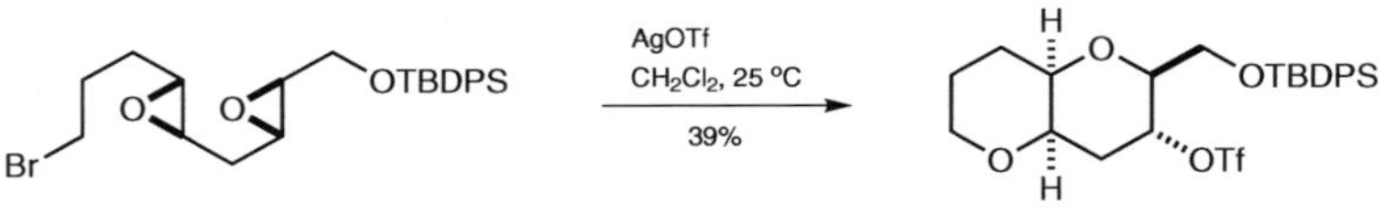

56 **57**

(2S,3R,4aR,8aR)-2-[(tert-Butyldiphenylsiloxy)methyl]octahydropyrano[3,2-b]pyran-3-yl Trifluoromethanesulfonate (57); Typical Procedure:[24]

AgOTf (22.0 mg, 0.0858 mmol) was added to a soln of bromo diepoxide **56** (35.0 mg, 0.0715 mmol) in CH$_2$Cl$_2$ at 25 °C. After being stirred at rt for 30 min, the mixture was poured into sat. aq NaHCO$_3$ at 0 °C, and extracted with Et$_2$O. The resulting organic layer was washed with brine, dried (MgSO$_4$), and concentrated. Purification by column chromatography (EtOAc/hexanes 1:10) afforded **57**; yield: 15.6 mg (39%).

1.2.2.4 **Water-Promoted Cascades**

Water-promoted epoxide-opening cascades of di- and triepoxides efficiently construct fused *trans–syn–trans* polytetrahydropyran motifs characteristic of the ladder polyether natural products.[26,27] A synergistic relationship between a tetrahydropyran or dioxane template ring and neutral water as the reaction solvent overcomes the intrinsic preference for the formation of smaller tetrahydrofuran rings even in the presence of *exo*-directing methyl groups.[28–30] Cascades proceed in a stepwise fashion, one epoxide at a time, in the direction from the template to the terminal epoxide.[31]

1.2.2.4.1 **Cascades of Diepoxides Templated by a Tetrahydropyran**

Water-promoted cascades of diepoxides templated by a tetrahydropyran ring proceed in moderate to good yields depending on the methyl substitution pattern (Scheme 14).[30] Reaction of unsubstituted diepoxide **58** (R^1 = R^2 = R^3 = H) in water provides the desired fused tricycle **59** in 74% yield (86% per epoxide). Incorporation of an *endo*-directing methyl group on one (R^1 = H; R^2 = Me) or both (R^1 = R^2 = Me) epoxides affords product in slightly decreased yields (82 and 73% yield per epoxide, respectively). A more challenging *exo*-directing methyl group at the internal epoxide (R^3 = Me) results in 32% yield (57% per epoxide), which represents the first example of an *endo*-selective epoxide-opening cascade to accommodate a methyl group at the *exo* site of the epoxide. In all four cases, neutral water is a superior reaction promoter compared to acids or bases in organic solvents.

Scheme 14 Tetrahydropyran-Templated Cascades of Diepoxides with Varying Methyl Substitution Patterns[30]

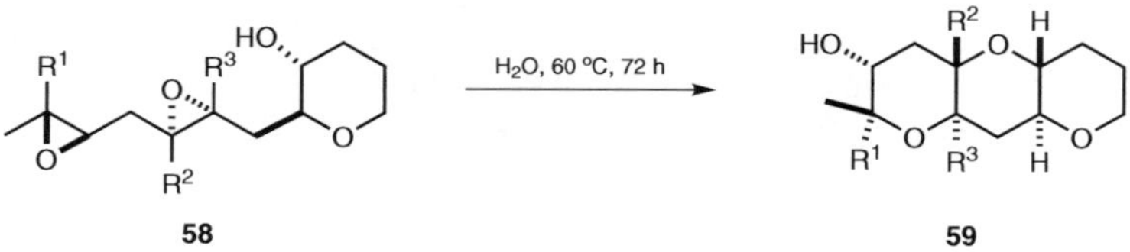

58 **59**

1.2.2 *endo*-Selective Polyepoxide Cascades

R¹	R²	R³	Yield[a] (%)	Ref
H	H	H	74	[30]
H	Me	H	67	[30]
Me	Me	H	54	[30]
H	H	Me	32	[30]

[a] Corrected for dr of starting material.

Decahydro-2*H*-dipyrano[3,2-*b*:2′,3′-*e*]pyrans 59; General Procedure:[30]
A glass vial was charged with a diepoxy alcohol **58** in deionized H_2O (0.02 M). The threads of the vial were lined with Teflon tape, the cap was sealed and covered with Parafilm, and the soln was heated to 60 °C under air for 3 d. The soln was then cooled to rt and concentrated under reduced pressure (10 Torr, 40 °C). The crude product mixture was purified by column chromatography (EtOAc/hexanes 3:7 or 1:1, depending on product polarity) to separate the desired tris-tetrahydropyran product (always the least polar product) from undesired cyclization products.

1.2.2.4.2 Cascades of Diepoxides Templated by a Dioxane

Substituted 1,3-dioxanes are competent templates for water-promoted, *endo*-selective epoxide-opening cascades. Such a framework provides approximately a tenfold increase in the *endo/exo* selectivity for the opening of a single epoxide compared to the tetrahydropyran template.[29] However, a decreased cyclization rate results in a larger amount of epoxide hydrolysis side products in the context of polyepoxide cascades. When diepoxide **60** is heated in buffered neutral water for 5 days, a 50% yield of the desired tricycle **61** is obtained (Scheme 15).

Scheme 15 A Dioxane-Templated Diepoxide Cascade[29]

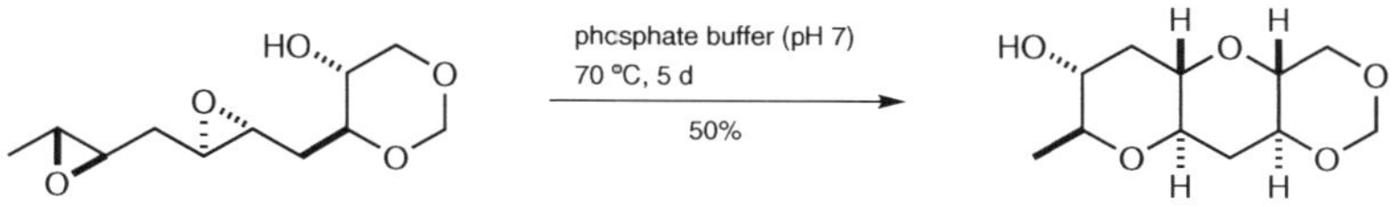

(4a*R*,5a*S*,7*R*,8*S*,9a*R*,10a*S*)-8-Methyloctahydro-4*H*-pyrano[2′,3′:5,6]pyrano[3,2-*d*]-[1,3]dioxin-7-ol (61):[29]
A glass vial was charged with diepoxy alcohol **60** (15 mg, 0.065 mmol) in 0.1 M phosphate buffer (pH 7; 3.25 mL) in a glass vial. The threads of the vial were lined with Teflon tape, the cap was sealed and covered with Parafilm, and the soln was stirred at 70 °C for 5 d. The soln was then washed out of the reaction vial with MeOH and concentrated under reduced pressure (2 Torr, 40 °C). Purification by column chromatography provided the desired tricycle **61**; yield: 7.5 mg (50%).

1.2.2.4.3 Cascades of Triepoxides Templated by a Tetrahydropyran

Triepoxide cascades have been demonstrated both in a simple model system and in the context of a natural product synthesis (Scheme 16). Triepoxide **62** cyclizes in water with remarkable efficiency to provide tetracycle **63** in 71% yield (89% per epoxide opening).[26] Triepoxide **64** has been prepared as part of the synthesis of the HIJK ring fragment of gymnocin A (**65**).[32] Exposure to water, followed by acetylation of the secondary alcohol, af-

for references see p 65

fords the desired tetracycle **65** in 38% yield (72% per epoxide). Importantly, previously unexplored substitution of the template tetrahydropyran ring, as well as the (benzyloxy)methyl substitution of the terminal epoxide, is tolerated.

Scheme 16 Water-Promoted Cascades of Triepoxides[26,32]

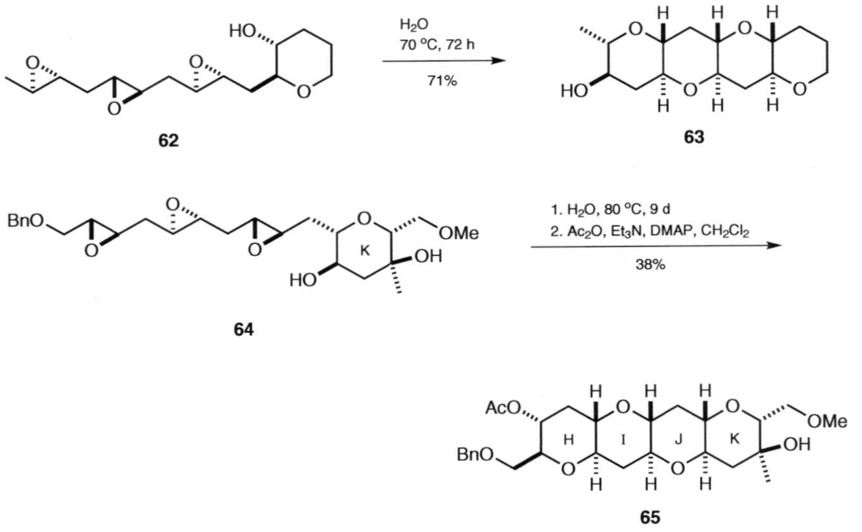

(2S,3R,4aS,5aR,6aS,10aR,11aS,12aR)-2-Methyltetradecahydropyrano[3,2-b]pyrano-[2′,3′:5,6]pyrano[2,3-e]pyran-3-ol (63):[26]
Triepoxide **62** (dr ~3:1; 32 mg, 0.11 mmol) was dissolved in H_2O (32 mL) and the resulting soln was stirred for 3 d at 70 °C in a sealed tube. H_2O was removed under reduced pressure and the residue was purified by column chromatography (EtOAc/hexane 1:1) to afford tetracycle **63**; yield: 17 mg (53%); yield corrected for dr of starting material: 71%.

1.2.3 Epoxide Cascades with C–C π-Bonds

Cascades involving one or more epoxides and C–C π-bonds are enabled by electrophilic activation of the terminal epoxide. *endo* Selectivity is usually directed by methyl groups, and trapping nucleophiles include alkenes, alkynes, arenes, and protected phenolic oxygens. Formation of up to three new rings has been reported.

1.2.3.1 Cascades Terminated by Alkenes and Alkynes

endo-Regioselective epoxide-opening cascades terminated by alkenes and alkynes have been successfully employed in biomimetic syntheses of terpene-derived polyether natural products. In the first-generation synthesis of marine triterpenoid *ent*-abudinol B (**70**), subunits **67** and **69** are accessed quickly from cascade cyclizations of their corresponding diepoxy alkene precursors **66** and **68** under activation with trimethylsilyl or *tert*-butyldimethylsilyl trifluoromethanesulfonate, in the presence of 2,6-di-*tert*-butyl-4-methylpyridine (DTBMP) (Scheme 17).[33] The desired *endo* selectivity is achieved by the directing effect of the methyl substituents on epoxides and alkenes. In the second-generation approach, *ent*-abudinol B is prepared from the linear precursor **71** in two sequential cascades similar to the synthesis of the individual fragments above (Scheme 17).[34,35] First, diepoxide enol ether **71** cyclizes upon treatment with trimethylsilyl trifluoromethanesulfonate to provide tricycle diene **72** in 50% yield. This intermediate is then elaborated in

1.2.3 Epoxide Cascades with C–C π Bonds   **59**

two steps to diepoxide alkene **73**, which affords a 15% yield of *ent*-abudinol B (**70**) after cascade cyclization.

Scheme 17   Synthesis of *ent*-Abudinol B via Epoxide-Opening Cascades Terminated by Alkenes and Alkynes[33–35]

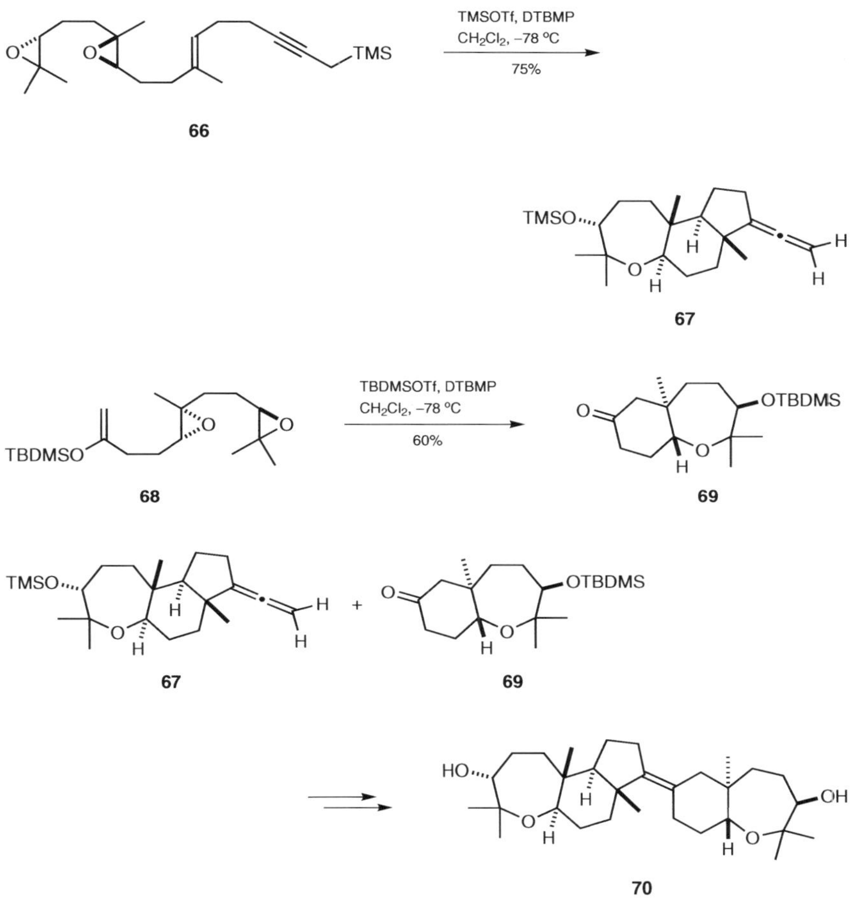

for references see p 65

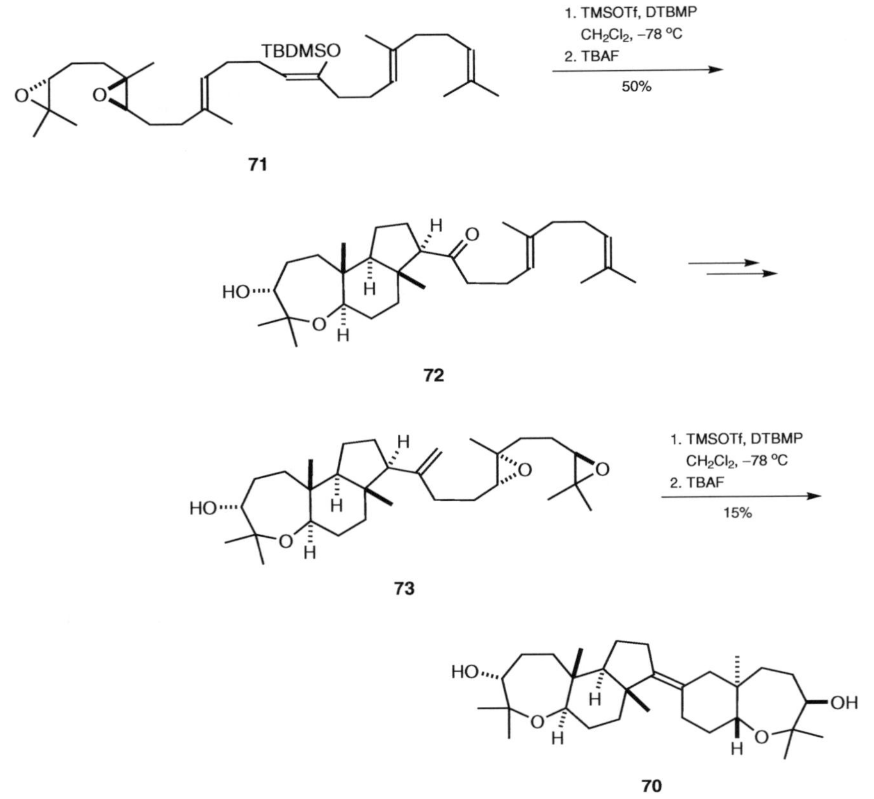

The reaction schemes with structures **71**, **72**, **73**, and **70** are shown, with reagents:

For **71** → **72**:
1. TMSOTf, DTBMP
   CH$_2$Cl$_2$, −78 °C
2. TBAF
50%

For **73** → **70**:
1. TMSOTf, DTBMP
   CH$_2$Cl$_2$, −78 °C
2. TBAF
15%

DTBMP = 2,6-di-*tert*-butyl-4-methylpyridine

(3*R*,5a*R*,7a*R*,10a*S*,10b*R*)-4,4,7a,10b-Tetramethyl-3-(trimethylsilyloxy)-8-vinylidenedodecahydro-1*H*-indeno[5,4-*b*]oxepin (67):[33]

A flame-dried 250-mL flask was charged with diepoxy enyne **66** (0.60 g, 1.70 mmol), CH$_2$Cl$_2$ (100 mL), and DTBMP (68.0 mg, 0.33 mmol) and then cooled to −78 °C. TMSOTf (0.06 mL, 0.33 mmol) was added dropwise with vigorous stirring at −78 °C. The reaction was usually complete within 1 h. If not complete, additional TMSOTf (0.03 mL, 0.16 mmol) was added to consume all starting material. Et$_3$N (1.0 mL) was added to the mixture, which was stirred for another 10 min before H$_2$O (10 mL) was added to quench the reaction at −78 °C. The organic fraction was separated by decanting the organic layer from the frozen aqueous component while the mixture was still cold. After the ice had melted, the aqueous layer was extracted with CH$_2$Cl$_2$ (2 × 50 mL). The combined organic fractions were washed with brine (50 mL), dried (MgSO$_4$), and concentrated. Purification by column chromatography (hexanes/EtOAc 30:1) afforded allene **67**; yield: 0.45 g (75%).

1.2.3.2 Cascades Terminated by Arenes

Indium(III) bromide efficiently promotes epoxy alkene cascades terminated by an arene.[36] A variety of tri- and tetracyclic products bearing arene substitution can be obtained in up to 58% yield (up to 76% per ring; selected examples in Table 1). A furan ring is also tolerated as a terminating nucleophile. This method has been applied to the total synthesis of totaradiol (**76**) and two related diterpene natural products, totarolone and totarol (Scheme 18).[37]

1.2.3 Epoxide Cascades with C–C π Bonds 61

Table 1 Indium(III) Bromide Promoted Epoxy Alkene Cascades Terminated by Arenes[36]

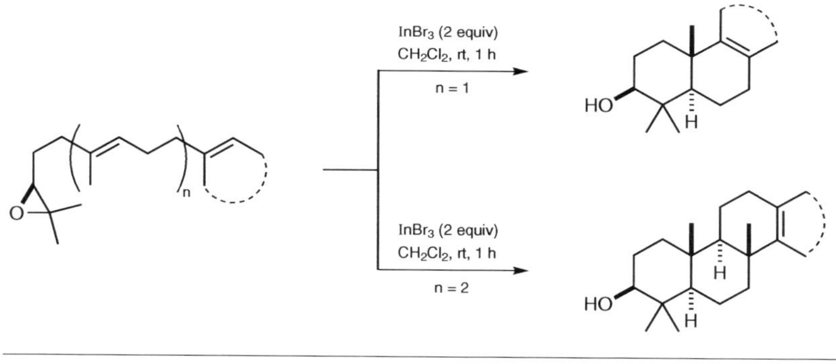

Starting Material	Product	Yield (%)	Ref
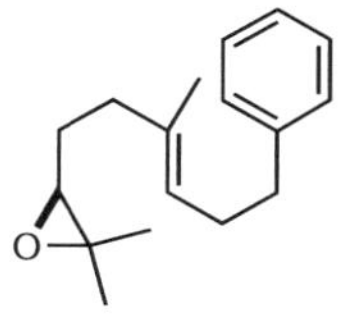	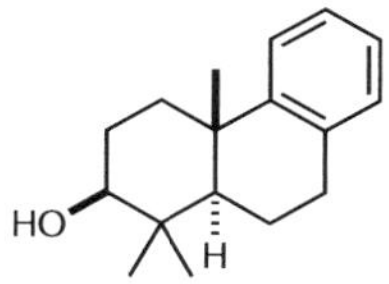	57	[36]
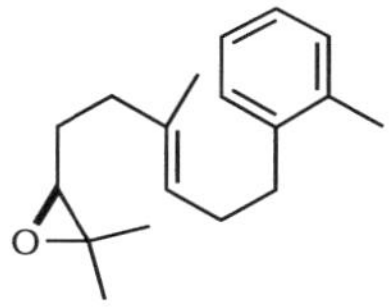	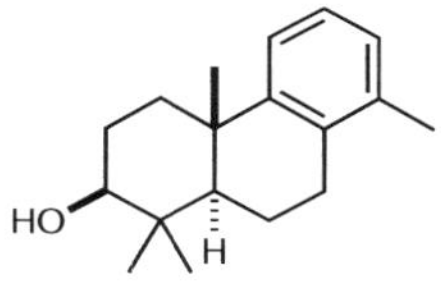	54	[36]
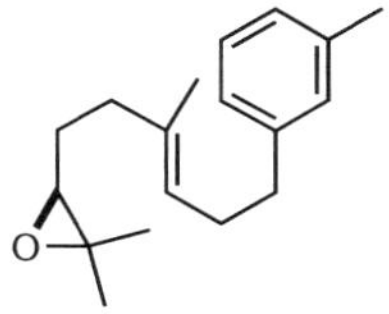	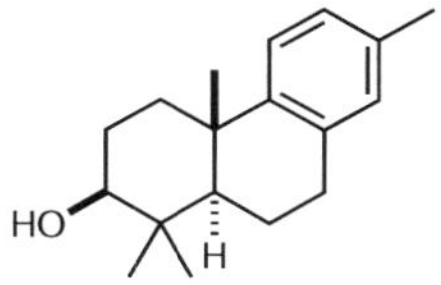	51	[36]
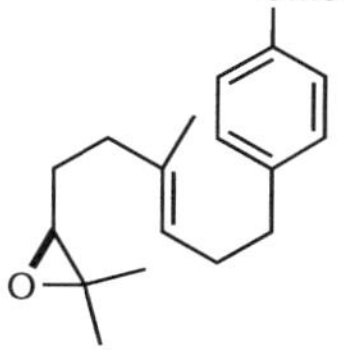	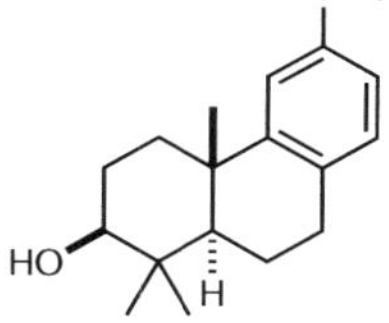	58	[36]

for references see p 65

Domino Transformations 1.2 Cation–π Cyclizations of Epoxides and Polyepoxides

Table 1 (cont.)

Starting Material	Product	Yield (%)	Ref
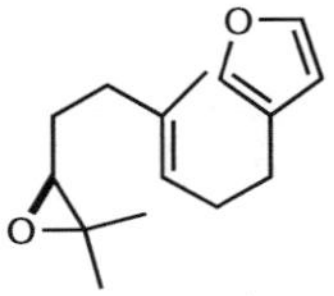	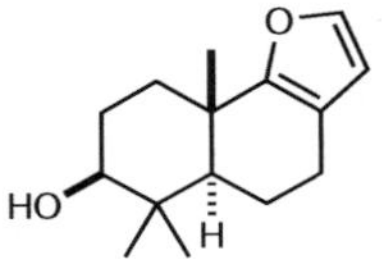	52[a]	[36]
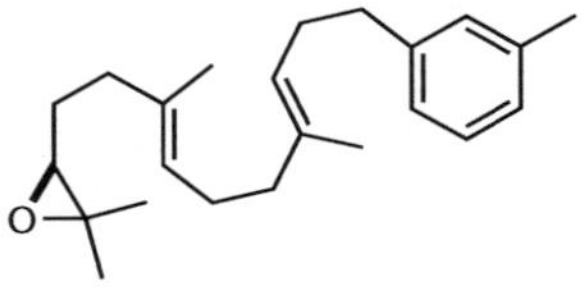	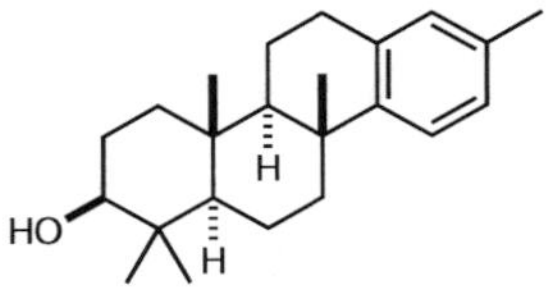	31	[36]
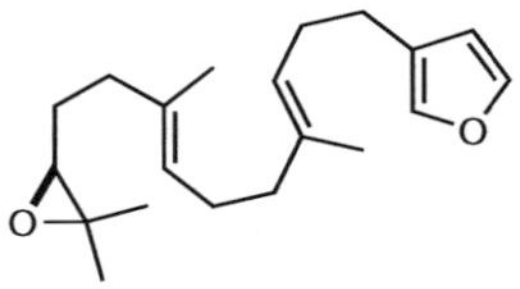	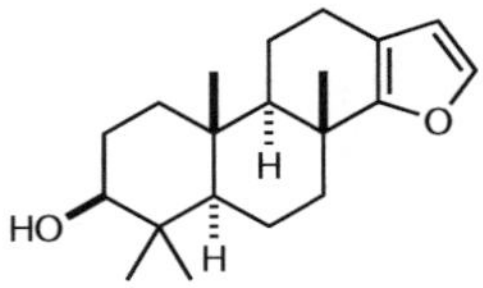	37[a]	[36]

[a] 0.2 Equiv InBr₃ was used.

Scheme 18 An Epoxy Alkene Cascade in the Synthesis of Totaradiol[37]

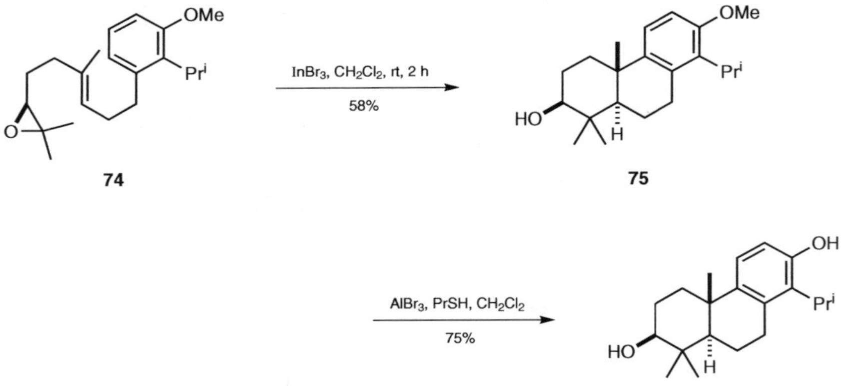

(2S,4aS,10aR)-8-Isopropyl-7-methoxy-1,1,4a-trimethyl-1,2,3,4,4a,9,10,10a-octahydro-phenanthren-2-ol (75); Typical Procedure:[37]

Epoxy alkene **74** (1.4 g, 2.5 mmol) was dissolved in dry CH_2Cl_2 (50 mL) in a flame-dried flask. InBr₃ (3.2 g, 8.9 mmol) was added. The soln was stirred for 2 h and then quenched with sat. aq NH₄Cl (20 mL). The two layers were separated and the aqueous layer was extracted with CH_2Cl_2 (3 × 40 mL). The combined organic layers were washed with brine and dried (MgSO₄). The solvent was removed under reduced pressure and the crude material was purified by column chromatography (EtOAc/hexane 1:4) to provide tricycle **75** as a white solid; yield: 0.82 g (58%).

1.2.3.3 Cascades Terminated by Protected Phenols

Epoxy alkene cascades terminated by a protected phenolic oxygen have been developed. Of the Brønsted and Lewis acids tested, boron trifluoride–diethyl ether complex best promotes the cyclization to provide good yields of the requisite hexahydroxanthene cores **78** (Scheme 19). The only byproduct is the methoxymethyl-protected secondary alcohol **79**, which can be separately converted into the free alcohol **78** to afford a higher combined yield.

Scheme 19 Epoxy Alkene Cascades Terminated by a Phenolic Oxygen[38,39]

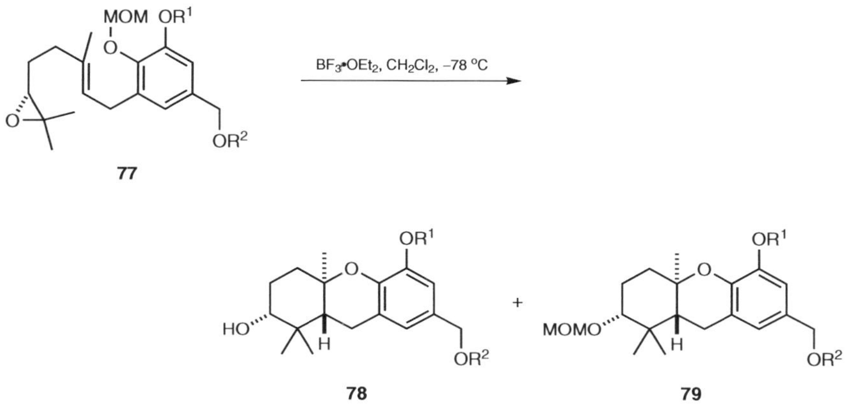

R[1]	R[2]	Yield (%) of **78**	Yield (%) of **79**	Ref
MOM	TBDMS	52	30	[38]
Me	Me	69	17	[39]

A variation of this cascade reported recently includes a tandem electrophilic aromatic substitution, which is enabled when the protecting group on phenolic oxygen is lost as a stabilized electrophile and is subsequently attacked by the arene (Scheme 20).[40] The success and regioselectivity of the electrophilic aromatic substitution depend on the identity of the protecting group in substrate **80**, whereby the groups most able to stabilize positive charge provide the highest yields of the desired product **81**. Byproducts **82–84** become more abundant when less-electron-rich protecting groups are used. This methodology was applied in the first total synthesis of schweinfurthin A.[40]

for references see p 65

Scheme 20 Tandem Epoxy Alkene/Electrophilic Substitution Cascades[40]

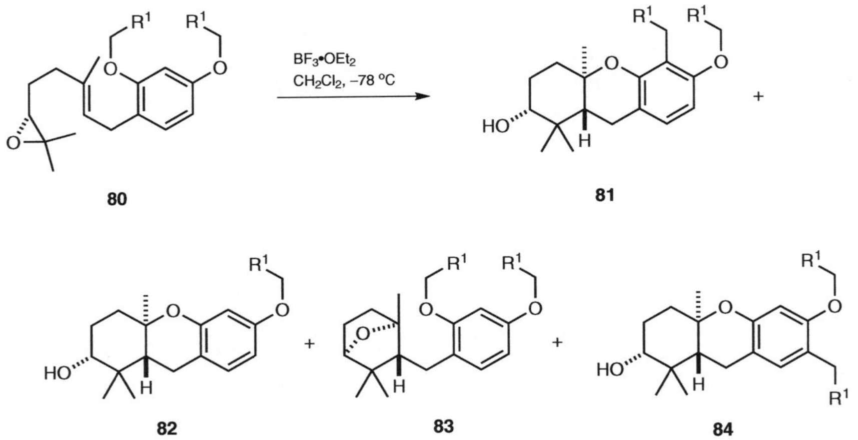

R^1	Yield (%) of **81**	Yield (%) of **82**	Yield (%) of **83**	Yield (%) of **84**	Ref
OBn	62	–	–	–	[40]
O(CH$_2$)$_2$TMS	57	–	–	–	[40]
OMe	52	30	–	–	[40]
3-furyl	49	–	–	–	[40]
Ph	43	18	–	2	[40]
4-MeOC$_6$H$_4$	33	12[a]	–	19	[40]
4-ClC$_6$H$_4$O	–	56	37	–	[40]
H	–	–	72	–	[40]
4-O$_2$NC$_6$H$_4$	–	–	20	–	[40]
OCOt-Bu	–	–	47	–	[40]

[a] Isolated as the 4-methoxybenzyl ether of the secondary alcohol.

(2R,4aR,9aR)-5-Methoxy-7-(methoxymethyl)-1,1,4a-trimethyl-2,3,4,4a,9,9a-hexahydro-1H-xanthen-2-ol (78, R^1 = R^2 = Me); Typical Procedure:[39]

To a soln of epoxy alkene **77** (R^1 = R^2 = Me; 958 mg, 2.6 mmol) in CH$_2$Cl$_2$ (350 mL) at −78 °C was added BF$_3$•OEt$_2$ (2.0 mL, 16 mmol). After 7 min, the reaction was quenched by addition of Et$_3$N (4.1 mL, 29 mmol). The resulting soln was concentrated under reduced pressure and the residue was dissolved in CH$_2$Cl$_2$ and washed with H$_2$O and brine. The organic phase was dried (MgSO$_4$) and concentrated under reduced pressure. Purification by column chromatography (hexanes/EtOAc 1:1) afforded desired tricyclic ether **78** (R^1 = R^2 = Me) as a yellow oil; yield: 583 mg (69%).

References

[1] Vilotijevic, I.; Jamison, T. F., *Angew. Chem. Int. Ed.*, (2009) **48**, 5250.

[2] Cane, D. E.; Celmer, W. D.; Westley, J. W., *J. Am. Chem. Soc.*, (1983) **105**, 3594.

[3] Nakanishi, K., *Toxicon*, (1985) **23**, 473.

[4] Baldwin, J. E., *J. Chem. Soc., Chem Commun.*, (1976), 734.

[5] Katsuki, T.; Sharpless, K. B., *J. Am. Chem. Soc.*, (1980) **102**, 5974.

[6] Wong, O. A.; Shi, Y., *Chem. Rev.*, (2008) **108**, 3958.

[7] Morimoto, Y.; Yata, H.; Nishikawa, Y., *Angew. Chem. Int. Ed.*, (2007) **46**, 6481.

[8] Xiong, Z.; Corey, E. J., *J. Am. Chem. Soc.*, (2000) **122**, 9328.

[9] Iimori, T.; Still, W. C.; Rheingold, A. L.; Staley, D. L., *J. Am. Chem. Soc.*, (1989) **111**, 3439.

[10] Hoye, T. R.; Jenkins, S. A., *J. Am. Chem. Soc.*, (1987) **109**, 6196.

[11] Morimoto, Y.; Iwai, T.; Nishikawa, Y.; Kinoshita, T., *Tetrahedron: Asymmetry*, (2002) **13**, 2641.

[12] Morimoto, Y.; Okita, T.; Kambara, H., *Angew. Chem. Int. Ed.*, (2009) **48**, 2538.

[13] Yang, P.; Li, P.-F.; Qu, J.; Tang, L.-F., *Org. Lett.*, (2012) **14**, 3932.

[14] Kumar, V. S.; Aubele, D. L.; Floreancig, P. E., *Org. Lett.*, (2002) **4**, 2489.

[15] Bravo, F.; McDonald, F. E.; Neiwert, W. A.; Do, B.; Hardcastle, K. I., *Org. Lett.*, (2003) **5**, 2123.

[16] McDonald, F. E.; Wang, X.; Do, B.; Hardcastle, K. I., *Org. Lett.*, (2000) **2**, 2917.

[17] McDonald, F. E.; Bravo, F.; Wang, X.; Wei, X.; Toganoh, M.; Rodríguez, J. R.; Do, B.; Neiwert, W. A.; Hardcastle, K. I., *J. Org. Chem.*, (2002) **67**, 2515.

[18] Underwood, B. S.; Tanuwidjaja, J.; Ng, S.-S.; Jamison, T. F., *Tetrahedron*, (2013) **69**, 5205.

[19] Valentine, J. C.; McDonald, F. E.; Neiwert, W. A.; Hardcastle, K. I., *J. Am. Chem. Soc.*, (2005) **127**, 4586.

[20] Simpson, G. L.; Heffron, T. P.; Merino, E.; Jamison, T. F., *J. Am. Chem. Soc.*, (2006) **128**, 1056.

[21] Tanuwidjaja, J.; Ng, S.-S.; Jamison, T. F., *J. Am. Chem. Soc.*, (2009) **131**, 12084.

[22] Neverov, A. A.; Feng, H. X.; Hamilton, K.; Brown, R. S., *J. Org. Chem.*, (2003) **68**, 3802.

[23] Wan, S.; Gunaydin, H.; Houk, K. N.; Floreancig, P. E., *J. Am. Chem. Soc.*, (2007) **129**, 7915.

[24] Hayashi, N.; Fujiwara, K.; Murai, A., *Tetrahedron*, (1997) **53**, 12425.

[25] Hayashi, N.; Fujiwara, K.; Murai, A., *Tetrahedron Lett.*, (1996) **37**, 6173.

[26] Vilotijevic, I.; Jamison, T. F., *Science (Washington, D. C.)*, (2007) **317**, 1189.

[27] Morten, C. J.; Byers, J. A.; Van Dyke, A. R.; Vilotijevic, I.; Jamison, T. F., *Chem. Soc. Rev.*, (2009) **38**, 3175.

[28] Byers, J. A.; Jamison, T. F., *J. Am. Chem. Soc.*, (2009) **131**, 6383.

[29] Mousseau, J. J.; Morten, C. J.; Jamison, T. F., *Chem.–Eur. J.*, (2013) **19**, 10004.

[30] Morten, C. J.; Jamison, T. F., *J. Am. Chem. Soc.*, (2009) **131**, 6678.

[31] Morten, C. J.; Byers, J. A.; Jamison, T. F., *J. Am. Chem. Soc.*, (2011) **133**, 1902.

[32] Van Dyke, A. R.; Jamison, T. F., *Angew. Chem. Int. Ed.*, (2009) **48**, 4430.

[33] Tong, R.; Valentine, J. C.; McDonald, F. E.; Cao, R.; Fang, X.; Hardcastle, K. I., *J. Am. Chem. Soc.*, (2007) **129**, 1050.

[34] Tong, R.; McDonald, F. E., *Angew. Chem. Int. Ed.*, (2008) **47**, 4377.

[35] Boone, M. A.; Tong, R.; McDonald, F. E.; Lense, S.; Cao, R.; Hardcastle, K. I., *J. Am. Chem. Soc.*, (2010) **132**, 5300.

[36] Zhao, J.-F.; Zhao, Y.-J.; Loh, T.-P., *Chem. Commun. (Cambridge)*, (2008), 1353.

[37] Kim, M. B.; Shaw, J. T., *Org. Lett.*, (2010) **12**, 3324.

[38] Mente, N. R.; Neighbors, J. D.; Wiemer, D. F., *J. Org. Chem.*, (2008) **73**, 7963.

[39] Topczewski, J. J.; Neighbors, J. D.; Wiemer, D. F., *J. Org. Chem.*, (2009) **74**, 6965.

[40] Topczewski, J. J.; Kodet, J. G.; Wiemer, D. F., *J. Org. Chem.*, (2011) **76**, 909.

1.3 Metathesis Reactions

1.3.1 Enyne-Metathesis-Based Domino Reactions in Natural Product Synthesis

D. Lee and M. O'Connor

General Introduction

Since the synthesis of urea by Friedrich Wöhler in 1828, synthetic organic chemistry has blossomed into an intricate science where chemo-, regio-, and stereoselectivity can increasingly be controlled at will in complex natural product syntheses. With many significant accomplishments made over the past century, the question posed is arguably no longer "what can we synthesize?". New challenges involve compound syntheses that have robustness, scalability, and appropriate safety/minimal environmental impact.

To overcome these obstacles, significant efforts have been made to improve the power of existing bond-forming processes by inventing new approaches that enable the formation or reorganization of many bonds in a single operation.[1–12] One of the measures indicating the effectiveness of a given synthetic transformation is defined by step economy, which provides a theoretical framework for synthetic chemists considering the efficiency of synthetic operations. Central to this idea of step economy is how to attain maximum complexity with a minimum number of synthetic operations, which has co-evolved with several related concepts such as domino, tandem, cascade, and sequential reactions, and so forth. Although these terms are similar and conceptually related, the definition of a domino reaction should include the most crucial element stipulating, as initially introduced by Tietze,[1] that in a multibond-forming event, the latter bond forming event should result via ramification of the bond formed in the former event.

Following this definition, we can consider the four processes shown in Scheme 1 to see what combination of σ- and π-bond-exchange processes has the potential to become a domino reaction. From a simple analysis, we can readily recognize that the exchange processes between a total odd number of π-bonds, such as in the first and third transformations, will generate temporary products that still contain a reactive functionality; thus, they can initiate the next round of exchange processes. On the other hand, the corresponding exchange processes with a total even number of π-bonds, such as in the second and fourth transformations, will lead to the formation of end products. Under a conventional definition of metathesis, the π-bond-exchange represented in the third process is termed as enyne metathesis, one which should have an inherent potential to be accommodated in the development of various domino reactions. In addition, succeeding an enyne-metathesis sequence is the formation of a 1,3-diene product that can serve as an ideal substrate for additional metathesis or Diels–Alder reactions, or other non-metathetic addition reactions, such as cationic, anionic, radical, carbenoid, photochemical, or further transition-metal-induced transformations.[1]

for references see p 131

Scheme 1 π-Bond Exchanges for Defining Potential Domino Reactions

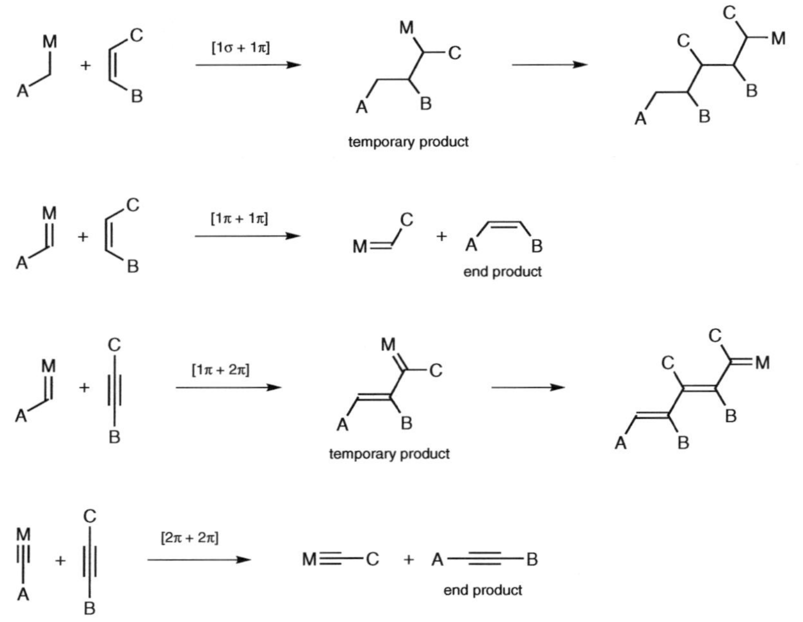

Enyne metathesis dates back to 1985 when Katz and Sivavec demonstrated that tungsten Fischer carbene complexes could facilitate ring-closing metathesis (RCM) when used in stoichiometric quantities.[13,14] Soon after, the Hoye and Mori groups independently demonstrated that the chromium variants of these complexes were also effective.[15–20] Although groundbreaking, these seminal reports were limited by functional-group tolerance, selectivity, and the need for ample catalyst/reagent loading. A key breakthrough came about a decade later when Mori and Konoshita showed that enyne **1** is effectively converted into 1,3-diene **3** via metathesis in the presence of a catalytic amount of Grubbs ruthenium alkylidene complex **2** (Scheme 2).[21]

Scheme 2 First Catalytic Ruthenium-Based Enyne Metathesis[21]

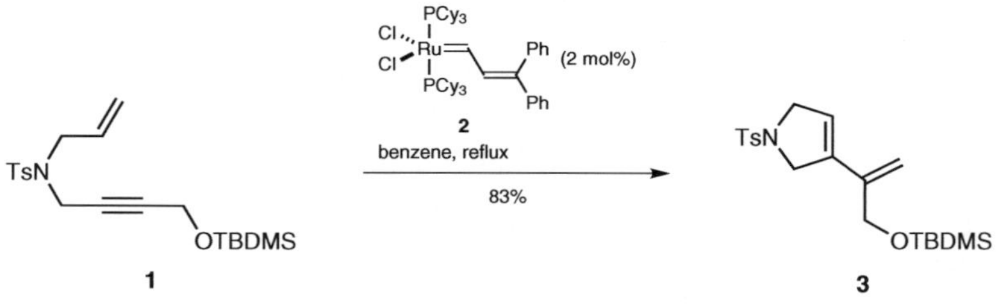

This report largely instigated the exploration of enyne metathesis among synthetic chemists, with Grubbs and co-workers describing several examples of domino enyne metathesis catalyzed by similar ruthenium carbenes. This study provided the initial framework for further development of enyne metathesis in various domino processes.[22]

Because of their user-friendly nature, functional-group tolerance, and ready availability, Grubbs-type ruthenium complexes **2** and **4–7** are the most common initiators and have been extensively studied and utilized in enyne-metathesis-based synthetic applications (Scheme 3).[23] The molybdenum catalyst **8**, developed by Schrock, was the first cata-

lyst to show a good reactivity profile for enyne metathesis, since the initial tungsten catalysts originally used by Katz had sensitivity to many functional groups as well as air and moisture.[24] Some of these limitations were overcome by molybdenum-based alkylidene complexes containing pyrrolide ligands, such as 9. These catalysts displayed good reaction profiles, but an opposite ring-closure mode is observed compared to that of ruthenium-based complexes, a finding that may have important implications for the reaction mechanism.[25,26]

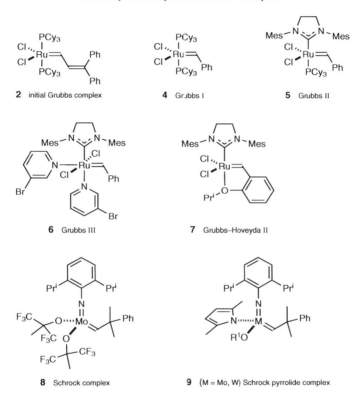

Scheme 3 Commonly Used Enyne Metathesis Catalysts[23–26]

Among the three general classes of metathesis (i.e., alkene, alkyne, and enyne metathesis), only enyne metathesis can generate new functionality compared to the starting material. For example, a 1,3-diene can arise from the metathesis reaction of an alkene and an alkyne, while only a new alkene is generated from two different alkenes in alkene metathesis and a new alkyne is formed from two alkynes in alkyne metathesis. Furthermore, distinct from its diene and diyne metathesis cousins, which are overall substitution reactions of the component alkylidenes generating gaseous byproduct (typically ethene and but-2-yne, respectively), enyne metathesis is an addition reaction of two alkylidenes across a triple bond. Thus, no alkene byproduct is formed during the metathesis sequence. Depending on the disposition of the reacting alkene and alkyne moieties, enyne metathesis can be classified into three categories: cross metathesis (CM), ring-closing metathesis (RCM), and metallotropic [1,3]-shifts (MS), as shown in Scheme 4. In cross metathesis, the reacting alkene and alkyne functionality reside in separate molecules while in ring-closing metathesis both functional groups are on the same molecule connected by a suitable tether.

for references see p 131

Scheme 4 Enyne Metathesis and Metallotropic [1,3]-Shift

enyne cross metathesis

enyne ring-closing metathesis

metallotropic [1,3]-shift

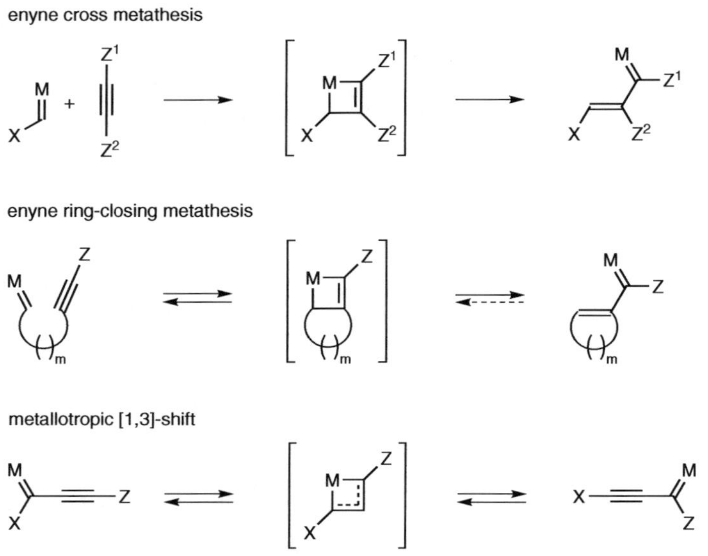

The metallotropic [1,3]-shift indicates a process that involves a propargylic metal alkylidene shift across a triple bond to form a new propargylic alkylidene species. From a mechanistic viewpoint, enyne ring-closing metathesis and metallotropic [1,3]-shift belong to the same reaction class because these two processes involve identical bond-forming and bond-breaking steps (except for the difference in the length of the tether connecting the alkene and alkyne moieties). Therefore, metallotropic [1,3]-shifts can be thought of as an enyne ring-closing metathesis with no tether (m = 0) between the ene and yne partners (Scheme 4).[27–29] On a more accurate level of analysis of each mechanistic step, however, the reversible nature of metallotropic [1,3]-shifts may be a unique feature not shared by enyne ring-closing metathesis because density functional theory (DFT) based theoretical investigations indicate that the steps to form a ruthenacyclobutene followed by its electrocyclic ring opening to a vinyl allylidene has a high barrier (~42 kcal·mol^{-1}); thus, these steps most likely are not reversible.[29] Yet, the overall mechanistic similarity shared by enyne ring-closing metathesis and metallotropic [1,3]-shifts mediated by metal alkylidene species would provide a possibility for their combination in domino bond-forming events. If realized, this enyne ring-closing metathesis/metallotropic [1,3]-shift (RCM–MS) domino process would remarkably expand the scope of enyne metathesis because their combined domino reactions can install an alkynyl moiety into the product, which is not possible in conventional enyne metathesis (Scheme 5).

Scheme 5 Domino Enyne Ring-Closing Metathesis versus Enyne Ring-Closing Metathesis/Metallotropic [1,3]-Shift

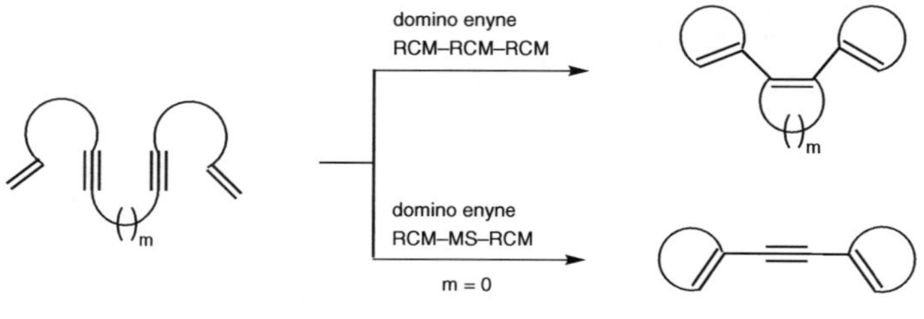

1.3.1.1 Mechanism

1.3.1.1.1 Alkylidene Carbene Catalyzed Reactions

One of the major limitations in utilizing domino enyne metathesis in complex molecule synthesis is the inherent problem of chemo-, regio-, and stereoselectivity. This issue is further complicated by the possibility of two different reaction pathways involving different propagating alkylidene species resulting either from the "yne-first" and "ene-first" mechanisms, a facet that is unique to enyne metathesis when compared to standard diene and diyne metathesis. Indeed, the metathesis event can commence at either the alkyne or alkene moiety leading to two unique reaction modes (Scheme 6). This initiation event by different metal alkylidene species, as well as the possibility of forming two regioisomeric vinyl alkylidenes from each site of initiation, is what determines the identity of the propagating alkylidene. In the yne-first mechanism, the propagating species is a ruthenium methylidene (M = Ru), a species which then adds to an alkyne counterpart to form two isomeric vinyl ruthenium alkylidene intermediates. After cycloreversion and alkylidene transfer, two isomeric products (i.e., 1,2-substituted and 1,3-substituted dienes) result from these intermediates. On the other hand, in the ene-first mechanism, the propagating species is a substituted ruthenium alkylidene that regioisomerically adds to an alkyne to produce two completely different end products.[29–34] Traditionally, when catalysts derived from group VI (W, Cr, Mo) metals are used in enyne metathesis, the reaction was believed to initiate from the alkyne counterpart.[13–16] Therefore, without any solid experimental evidence, enyne metathesis catalyzed by ruthenium alkylidene complexes was also proposed to proceed via the yne-first pathway in the early days of its exploration, though as will be discussed, this did not prove to be accurate.

Scheme 6 Yne-First and Ene-First Mechanisms for Enyne Cross Metathesis

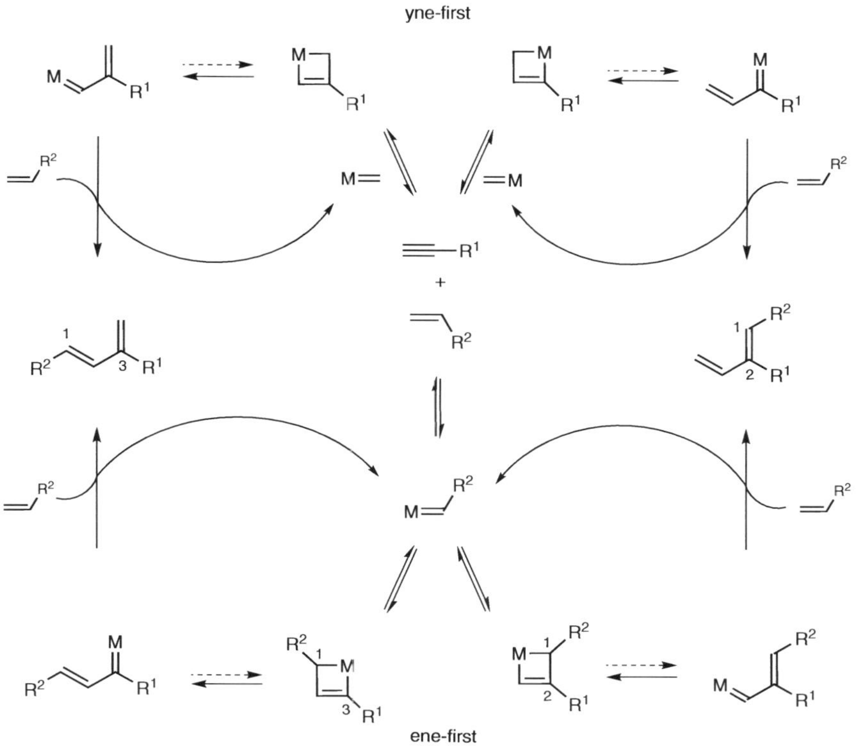

for references see p 131

In an initial effort to elucidate the initiation mode and mechanism of enyne metathesis, Mori and co-workers performed an intramolecular metathesis experiment with enyne **10**, a reaction that ultimately afforded products **11–13** in low yields (Scheme 7).[21] Triene **11** (19%) should be the result of enyne metathesis, while cyclopentene derivative **12** (5%) and cyclopropanated product **13** (3%) are the result of diene metathesis and cyclopropanation, respectively. Based on this product distribution, it was proposed that **12** is derived from an initiation event occurring from one of the alkenes, while **13** arises from initiation at the alkyne. Unfortunately, however, product **11** does not provide conclusive evidence to the initiation that is responsible for its formation.

Scheme 7   Enyne Metathesis Competition Experiment[21]

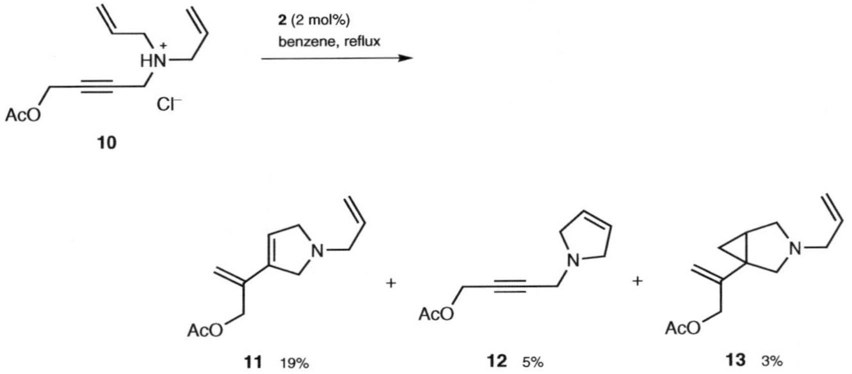

The first compelling evidence supporting the ene-first mechanism was provided by Grubbs, who reported that kinetically favored alkylidene products formed from terminal alkenes.[35] Later, NMR studies by Hoye and Kozmin also supported initial reaction at the alkene reaction partner, though it must be noted that the probe enyne substrates in these cases are electronically biased such that fair competition between the alkene and the alkyne counterparts was not established.[36,37] Nevertheless, the body of evidence grew. In addition, Nolan and co-workers demonstrated that the ene-first mechanism was favored with both phosphine and N-heterocyclic carbenes (NHC) containing ruthenium indenylidene complexes,[38] while a report using Hammett studies showed that cross metathesis of butynyl benzoate with styrenes in the presence of Grubbs second-generation catalyst **5** also occurs via the ene-first pathway.[31]

In an alternative investigation, Lloyd-Jones and co-workers employed deuterium-labeled substrates and examined the stereochemical outcome in the product formation as a reporter for the mechanism (Scheme 8).[32] If the yne-first mechanism is active, then the addition of deuterium-labeled methylidene **15** to the alkyne moiety of **14** would form ruthenacyclobutene intermediate **16**; this event, when followed by its ring-opening to form product (E/Z)-**17**, would be nonstereoselective, forming a 1:1 mixture of E/Z-isomers. By contrast, if the ene-first mechanism is active, then the addition of vinyl alkylidene **18** to an alkene moiety of **14** should form *trans*-**19** selectively over *cis*-**19** due to the unfavorable steric interaction; thus, this mechanism should generate (Z)-**17** as the major or exclusive product. As predicted, the ring-closing metathesis of deuterium-labeled enynes (E)-**20** and (Z)-**22** provides a high level of enrichment of one of the E/Z-stereoisomers (Z)-**21** and (E)-**23**, respectively. These results also strongly support the ene-first mechanism, at least for enyne ring-closing metathesis with simple substrates containing both a terminal alkene and an alkyne.

1.3.1 Enyne-Metathesis-Based Domino Reactions in Natural Product Synthesis   **73**

Scheme 8   Mechanistic Support for the Ene-First Mechanism[32]

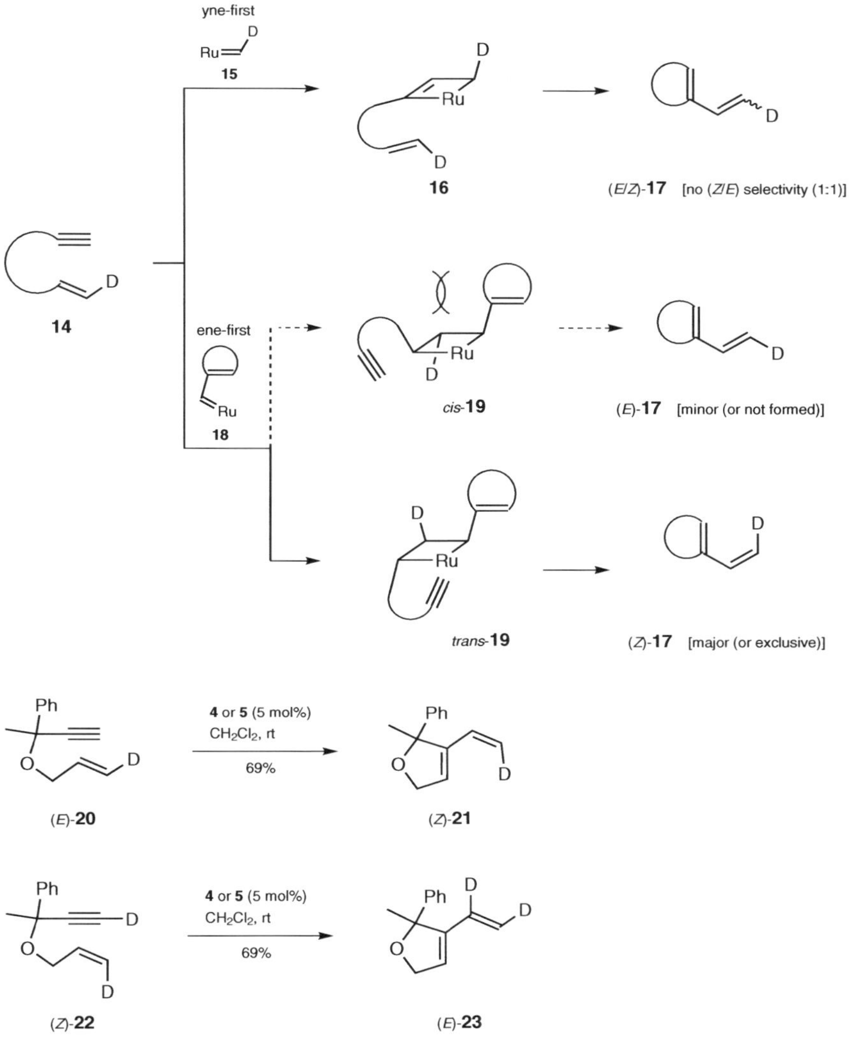

It is notable that electronic and steric factors can affect the enyne metathesis mechanism; small changes in the substrate can lead to a switch from an ene-first to an yne-first reaction pathway, or vice versa. For example, when 1,6-enyne **24** is exposed to initiator **5**, it undergoes enyne ring-closing metathesis to generate products **25** and **26** via *exo-* and *endo-*mode ring closure, respectively (Scheme 9). This outcome is believed to be a direct consequence of the retarded initiation due to steric hindrance of the 1,1-disubstituted alkene moiety present in the substrate.[39,40] Similarly, aminophosphonate-containing 1,6-enyne **27** affords dihydropyrrole **28**, which should be the result of the yne-first mechanism because the initiation from the trisubstituted alkene is difficult; this pathway was further confirmed by NMR studies and careful analysis of byproducts.[33]

for references see p 131

Scheme 9 Enyne Ring-Closing Metathesis Supporting the Yne-First Mechanism[33]

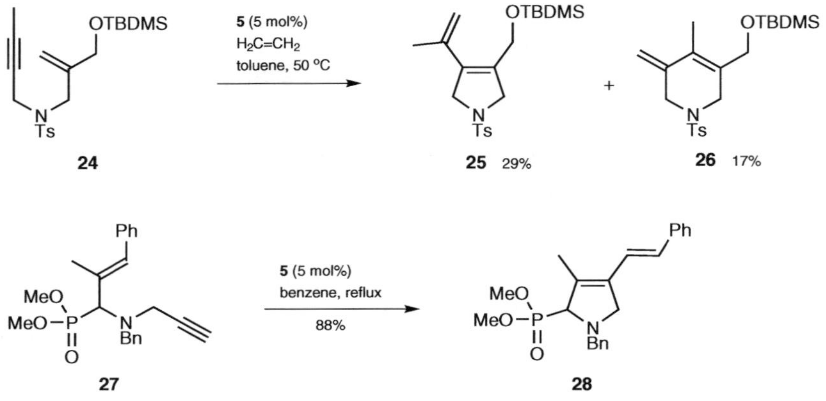

It is worth noting that the formation of a metallacyclobutene intermediate is often taken for granted in the vast majority of enyne metathesis mechanisms (in line with that proposed for other metathesis reactions) although there is actually no direct evidence for the existence of this species. Calculations by Lippstreu and Straub have shown that vinyl carbene complexes are formed directly from the association of a propagating alkylidene carbene and an alkyne without forming the putative metallacyclobutene intermediate. Thus, it was proposed that the metallacyclobutene is a transition state rather than an intermediate corresponding to a local minimum.[29] Furthermore, alkynes bind more strongly than alkenes to ruthenium alkylidenes, but their insertion is 5–7 $kcal \cdot mol^{-1}$ higher than that of alkenes; this fact makes insertion disfavored kinetically although thermodynamically alkyne insertion is favored by 33 $kcal \cdot mol^{-1}$, an operation which thus becomes the sole irreversible step in the enyne metathesis catalytic cycle. Likewise, this step determines the regioselectivity and governs the formation of 1,3-disubstituted 1,3-dienes in enyne cross metathesis, which is mainly affected by the steric interaction between the substituent on the ruthenium alkylidene and the interacting alkyne. For enyne cross metathesis, the rate-limiting step is either the formation of a vinyl alkylidene or alkyne insertion, which is in agreement with kinetic studies by Diver.[30] Interestingly, other DFT studies on the enyne ring-closing metathesis mechanism by the Solans-Monfort group showed no clear energetic bias for the ene-first or yne-first mechanism.[41]

Finally, Sohn and co-workers investigated the initiation event for enyne metathesis by using a FRET (fluorescence resonance energy transfer) based direct monitoring of the interaction between Grubbs first-generation catalyst and alkenes and alkynes containing a fluorophore.[42] The data obtained from this study strongly support the favorable initial association of the catalyst with an alkene over an alkyne moiety. Other ruthenium and molybdenum catalysts, when probed under the same conditions, showed that the association preference of unsaturated functionality depends on the catalysts as well as the structure of alkenes and alkynes.

Based on these experimental and theoretical studies, one conclusion is evident: the mechanism of enyne metathesis is affected by many factors and can be easily changed. Therefore, to clearly understand the mechanism of any given enyne metathesis, various parameters including the type of catalyst and substrate structures should be considered if predictive power is to be achieved.

1.3.1.1.2 π-Lewis Acid Catalyzed Reactions

It is also important to realize that metal species other than metal alkylidenes can trigger an enyne metathesis event. The most common metals that can achieve such a skeletal reorganization are later d- or p-block metals (e.g., Ru, Rh, Ir, Pd, Pt, Au, Ga, and In). Catalyzed by these π-Lewis acidic metals, 1,n-enynes can take one of two distinct skeletal-reorganization pathways (Scheme 10). In the traditional organometallic pathway, a 1,6-enyne and the catalyst will form metallacycle **29**, reductive elimination of which then generates a cyclobutene derivative **30**. Subsequent thermal electrocyclic ring opening will lead to an *exo*-type-I product. On the other hand, in the Lewis acid pathway, η^2-activation of the alkyne by π-philic Lewis acids forms a metal alkylidene **31**, a species which rearranges to carbocations **32** and **33** and ultimately results in *exo*-products.[43–45] Although only sporadically reported, metal alkylidenes of type **34**, formed from an endocyclic mode of ring closure, can transform into nonclassical carbocation **35**, ultimately giving rise to an *endo*-product. The outcome of these bond reorganization reactions is governed largely by the substituents (steric, electronic, and substitution) on enyne substrates and the catalyst properties.[46,47]

Scheme 10   Metal-Catalyzed Skeletal Reorganization of Enynes

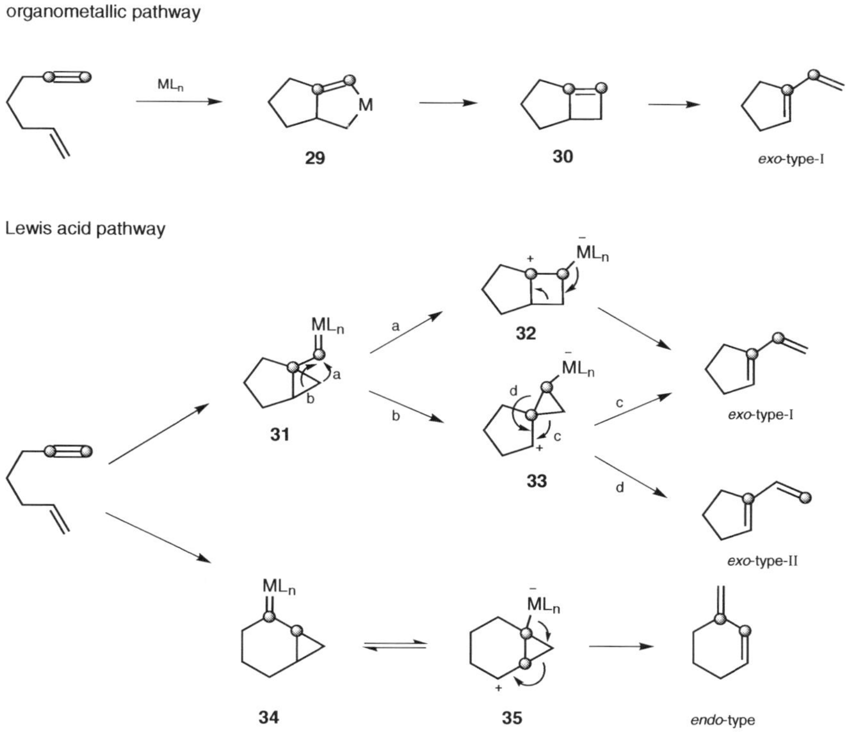

organometallic pathway

Lewis acid pathway

1.3.1.1.3 Metallotropic [1,3]-Shift

The [1,3]-bond shift of alkynyl carbenes and their metal complexes has been reviewed.[48] The bond shift (rearrangement) of acetylenic carbenes was first observed by Skell and Klebe when they generated propynylidene via photochemical methods.[49,50] Bond shifts of these free carbenes were only later deduced by trapping experiments and dimerizations,[51–56] and the necessity for more controllable metal–carbenes is apparent in the

for references see p 131

76 Domino Transformations **1.3** Metathesis Reactions

fields of theoretical and materials science.[57–59] Although there are many examples of metallotropic [1,3]-shifts of metal–carbenoids, which is effectively an equivalent of homotropic shifts of acetylenic carbenes (Scheme 11), an understanding of their mechanisms is only in its infancy. Similarly, despite the mechanistic similarity between enyne ring-closing metathesis and metallotropic [1,3]-shifts in terms of their bond-making and bond-breaking steps, their actual mechanisms cannot be directly translated from each other without any solid theoretical basis. What has been deduced from many experiments is that the metallotropic [1,3]-shift of ruthenium alkylidenes has a low activation barrier, one which is yet to be quantified by theoretical studies.[60]

Scheme 11 Homotropic versus Metallotropic [1,3]-Shift

free carbene homotropic shift

metal–carbenoid metallotropic shift

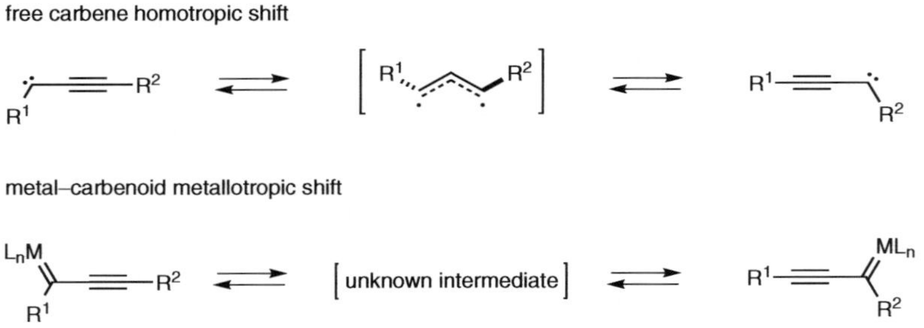

1.3.1.2 **Selectivity**

1.3.1.2.1 **Regioselectivity in Cross Metathesis**

In contrast to the cross metathesis of alkenes, the cross metathesis between an alkene and an alkyne renders a significant regioselectivity problem that limits the structural space of alkyne substrates that can be used (Scheme 12). Thus, to circumvent the typically inseparable mixture of isomers that can result, enyne cross metathesis has been confined to pairings of internal alkynes and ethene[6] or a terminal alkyne with a terminal alkene.[7] Critically, it has been shown that the cross metathesis of unsymmetrical internal alkynes containing two alkyl groups provides a mixture of regioisomers regardless of the structure of the reacting alkene counterpart (i.e., whether terminal or internal), whereas that of a terminal alkyne with a terminal alkene is highly regioselective and can generate a 1,3-disubstituted 1,3-diene without forming any 1,2-disubstituted 1,3-diene products.

Scheme 12 Regioselectivity in Enyne Cross Metathesis

internal alkene and alkyne

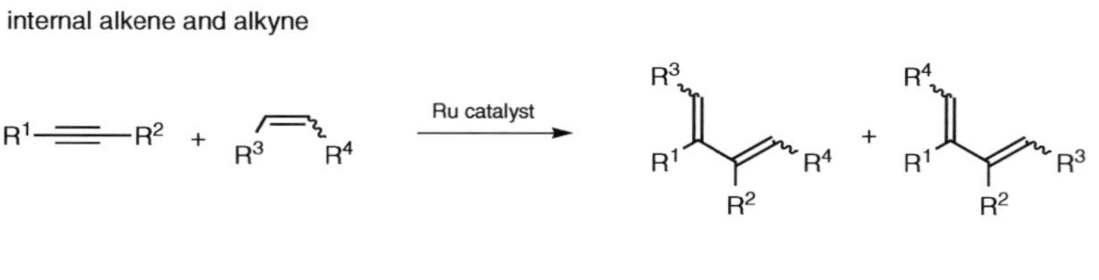

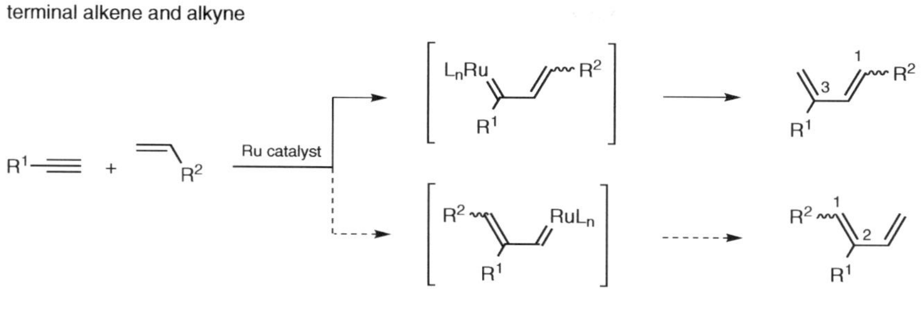

terminal alkene and alkyne

Of note, the observed regiochemical preference for the cross metathesis of terminal alkyne–terminal alkene combinations is consistent with DFT calculations, but only when the propagating species is a monosubstituted alkylidene in the "ene-first" mechanism (Scheme 13).[29] The opposite outcome was predicted when calculated with methylidene as the propagating species in an "yne-first" mechanism. This finding may imply that the ene-first mechanism is the operating mechanism in enyne cross metathesis in general, although the yne-first mechanism cannot always be excluded as noted in the previous sections.

Scheme 13 Regioselectivity of Enyne Cross Metathesis[29]

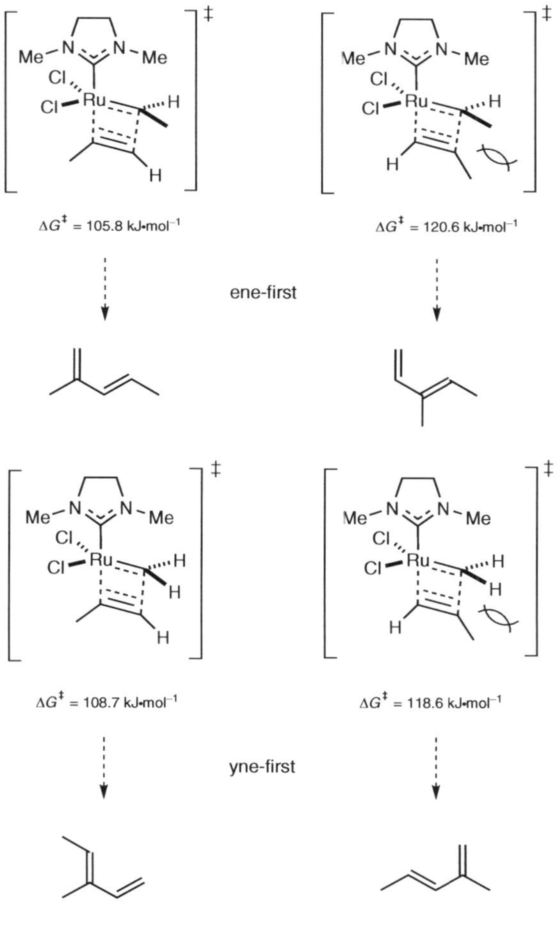

for references see p 131

The regioselectivity problem associated with unsymmetrical internal alkynes can be improved by using intricately substituted alkynes with silyl, boryl, or alkynyl groups as first demonstrated by Lee and co-workers (Scheme 14).[60–63] These biasing elements (mostly electronic rather than steric), effectively direct the cross metathesis to generate a single regioisomer of the corresponding 1,3-diene in all cases, though the origin of this observed regioselectivity is yet to be supported by theoretical investigations. Interestingly, the cross metathesis of terminal 1,3-diynes under identical conditions shows the opposite regioselectivity in the formation of the propagating alkylidene, such that a metallotropic [1,3]-shift became viable and thus two different products resulted. Interestingly, the products involving a metallotropic shift contain only a Z double bond while the products arising from pathways devoid of such shifts contain both E/Z-isomers with varying ratios dependent on the nature of the group labeled as R^1.[56]

Scheme 14 Regioselectivity in Enyne Cross Metathesis[60–63]

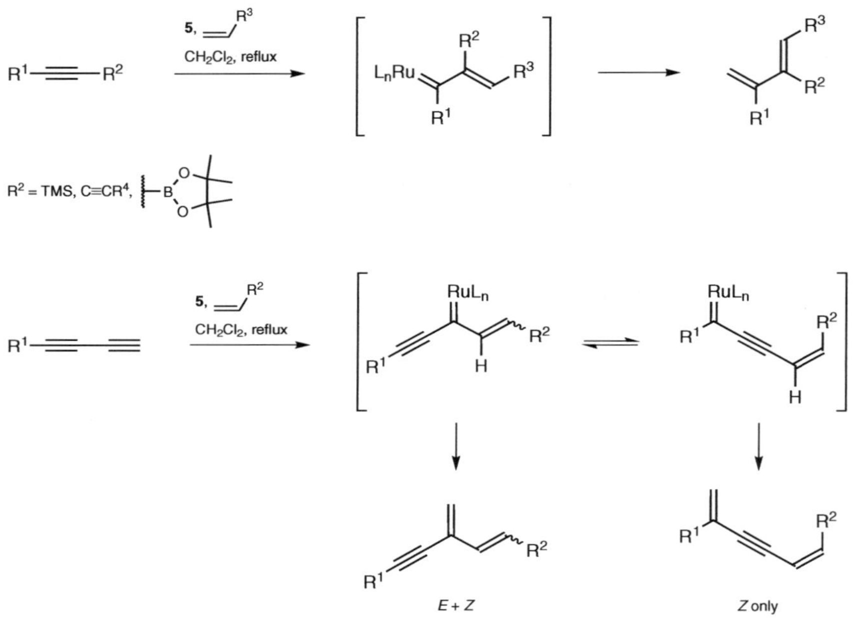

1.3.1.2.2 *exo/endo*-Mode Selectivity in Ring-Closing Metathesis

In contrast to the selectivity associated with enyne cross metathesis, enyne ring-closing metathesis (RCM) has a different type of selectivity that is defined as either *exo* or *endo*, with the outcome dictated largely by whether the "ene-first" or "yne-first" mechanism (see Section 1.3.1.1.1) is operative (Scheme 15). For example, when Grubbs-type ruthenium alkylidenes are used as promotor with enynes possessing a terminal alkyne and a terminal alkene, enyne ring-closing metathesis leading to small-membered rings (5–11) is generally *exo* selective, while for large-membered rings (≥12), it is *endo* selective.[28,64,65] This trend can be easily switched to form *endo*-products if enynes contain a trisubstituted alkene (Scheme 9) or molybdenum and tungsten complexes are used instead of ruthenium complexes.[25,26]

1.3.1 Enyne-Metathesis-Based Domino Reactions in Natural Product Synthesis **79**

Scheme 15 *exo/endo*-Mode Selectivity in Ring-Closing Metathesis

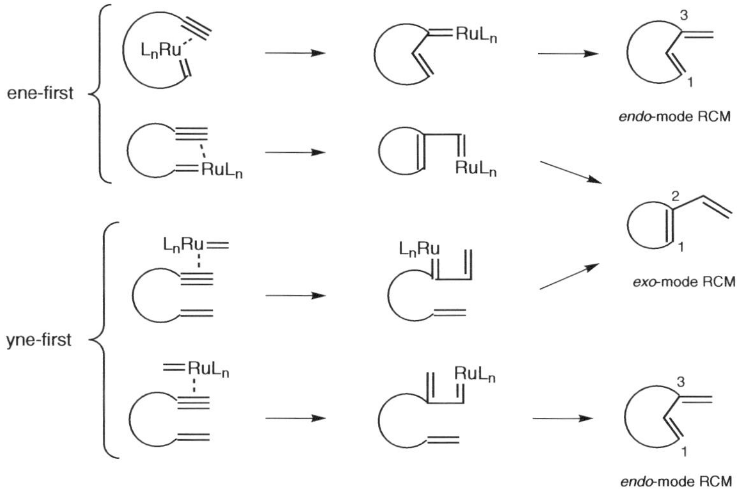

1.3.1.2.3 **Stereoselectivity in Cross Metathesis and Ring-Closing Metathesis**

The cross metathesis of internal alkynes with either terminal or internal alkenes is non-stereoselective. Similarly, cross metathesis of terminal alkynes with terminal alkenes is also of low selectivity unless the reactions are run under an ethene atmosphere to promote degenerative metathesis that can equilibrate the E/Z mixture of products (Scheme 16).[66,67] For the cross metathesis of internal alkynes, however, the outcome can be significantly influenced by, or more appropriately coupled with, its regioselectivity for the formation of propagating alkylidenes as shown in Scheme 14 (Section 1.3.1.2.1).[68]

The stereoselectivity of enyne ring-closing metathesis is relatively straightforward to predict. For an *exo*-mode ring-closing metathesis, favored for small- to medium-sized rings (5–11), the closure will lead to a (Z)-endocyclic alkene, whereas for an *endo*-mode ring-closing metathesis as favored for macrocycles (>12), the alkene selectivity trend will follow that of enyne cross metathesis because the ring strain is no longer a constraint.

Scheme 16 Stereoselectivity in Enyne Cross Metathesis and Ring-Closing Metathesis[66,67]

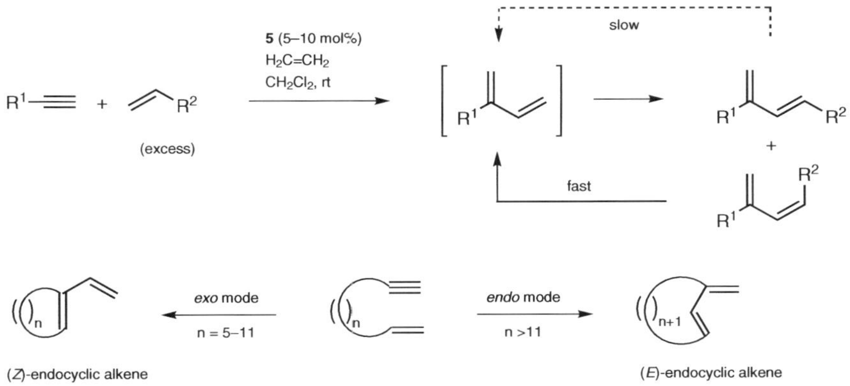

for references see p 131

1.3.1.2.4 **Regioselectivity in Metallotropic [1,3]-Shift**

For the development of domino reactions that involve both enyne metathesis and a metallotropic [1,3]-shift, fine control over the issues of chemo- and regioselective initiation is crucial. Especially for pathways involving a metallotropic [1,3]-shift, the formation of a propargylic alkylidene intermediate is a prerequisite, which can be achieved in three different ways: cross metathesis, ring-closing metathesis, and relay metathesis (Scheme 17). Initiation via cross metathesis is the most straightforward of these, requiring the least effort for the preparation of substrates; the price, however, is that the desired initiation regiochemistry is possible only with terminal diynes. Initiation with internal diynes forms the alternative non-propargylic alkylidene with a high degree of preference. This initiation regioselectivity can be easily reversed by tethering the alkene and alkyne moieties, such that the favorable *exo*-mode ring-closing metathesis generates only propargylic alkylidene species when the incipient rings are relatively small. The *endo*-mode ring closure, however, can become more favorable if the ring size becomes large (>11-membered ring). Finally, the relay-metathesis-based initiation is also quite reliable for the formation of both the mono- and disubstituted propargylic alkylidenes.

Scheme 17 Regioselectivity in Propargylic Alkylidene Formation

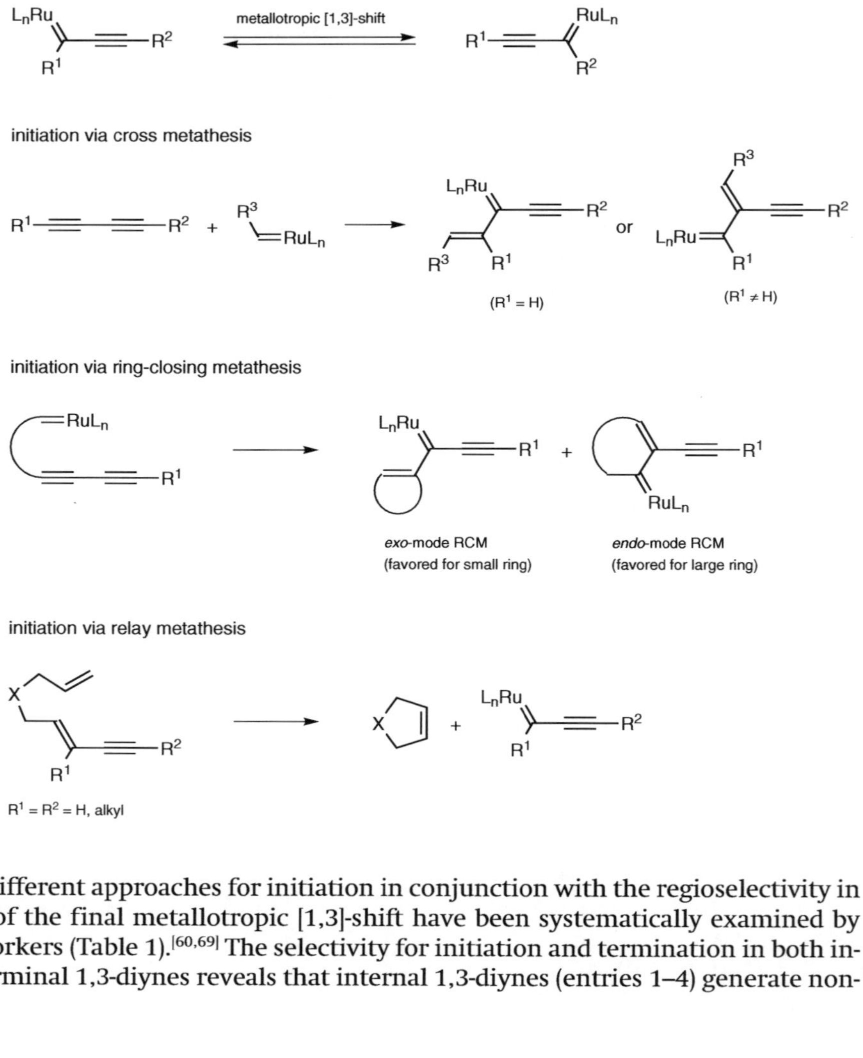

These three different approaches for initiation in conjunction with the regioselectivity in termination of the final metallotropic [1,3]-shift have been systematically examined by Lee and co-workers (Table 1).[60,69] The selectivity for initiation and termination in both internal and terminal 1,3-diynes reveals that internal 1,3-diynes (entries 1–4) generate non-

1.3.1 Enyne-Metathesis-Based Domino Reactions in Natural Product Synthesis **81**

propargylic alkylidenes with high selectivity, species that are irrelevant to further exploration of their metallotropic [1,3]-shift behavior, leading to products where only one alkynyl moiety of the 1,3-diyne is metathesized. On the other hand, the corresponding terminal 1,3-diynes (entries 6–9) undergo regioselective initiation to generate propargylic alkylidenes, which, depending on the structural characteristics of the substituent on the other end of the diynyl moiety, show a predictable metallotropic-shift-coupled termination.

Table 1 Initiation of Metallotropic [1,3]-Shift via Cross Metathesis[60,69]

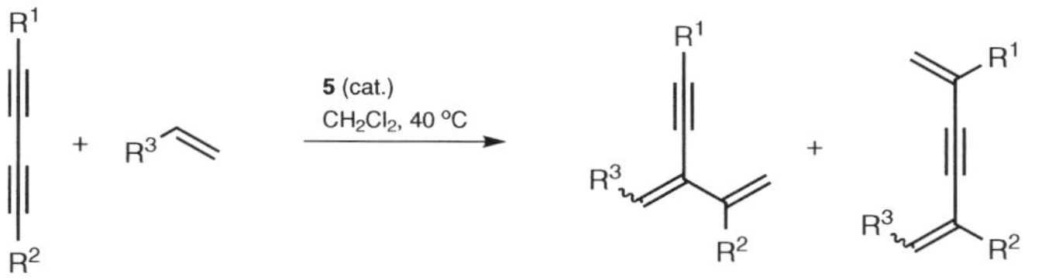

Entry	Starting Materials		Products	Yield (%)	Ref
	Diyne	Alkene			
1	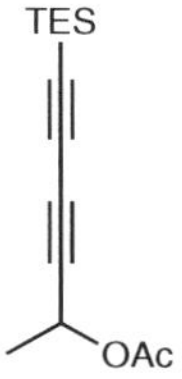	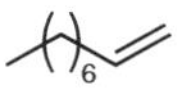	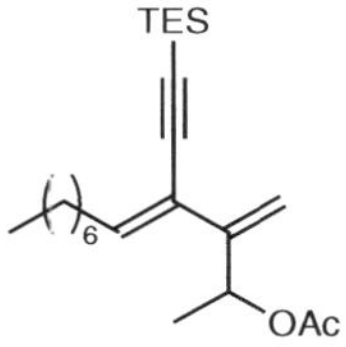	76	[60]
2	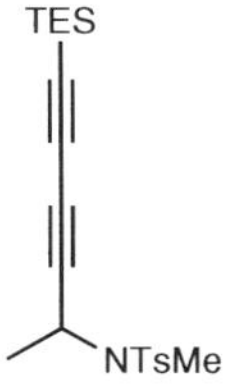	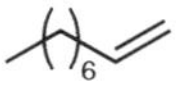	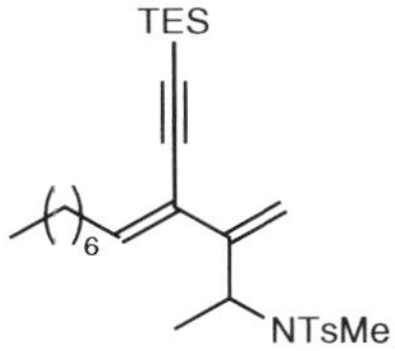	91	[60]
3		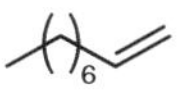	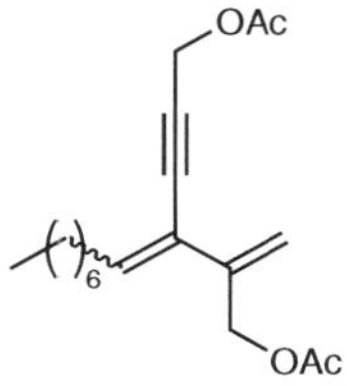 (*E/Z*) 64:36	72	[60]
4		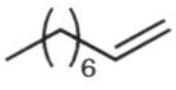	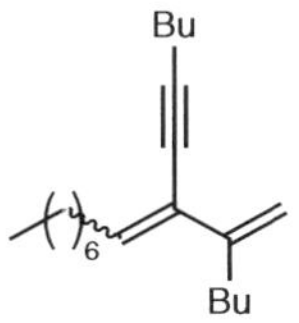 (*E/Z*) 1:1	62	[60]

for references see p 131

Table 1 (cont.)

Entry	Starting Materials		Products	Yield (%)	Ref
	Diyne	Alkene			
5	 TMS ... TMS	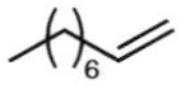 ⟨⟩₆	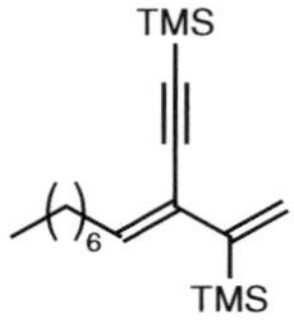 TMS ... TMS	0	[60]
6	 TIPS	 AcO	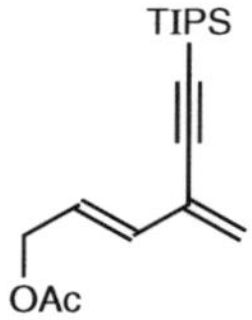 TIPS, OAc	47	[69]
7	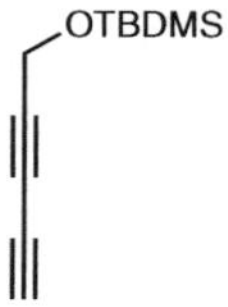 OTBDMS	 AcO	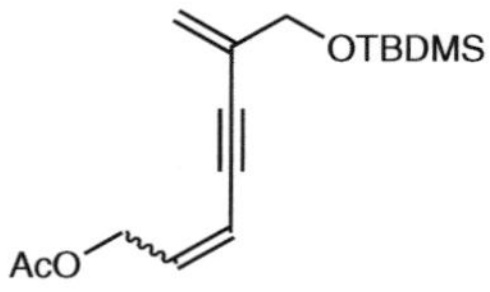 OTBDMS, AcO (E/Z) 1.2:1	77	[69]
8	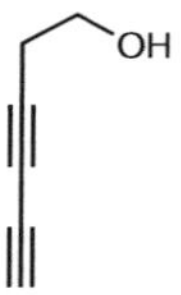 OH	 AcO	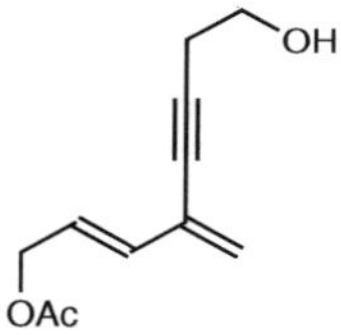 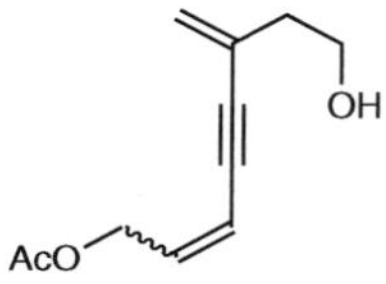 OH, OAc + OH, AcO 1:4 (E/Z) 3:1	73	[69]
9	 Cy	 AcO	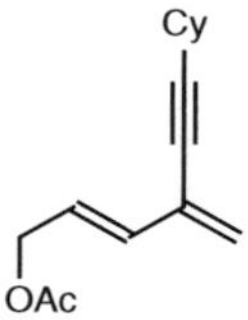 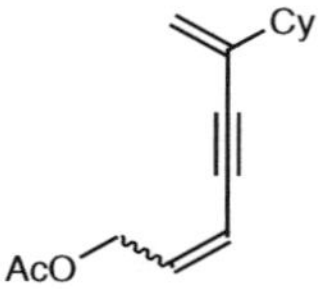 Cy, OAc + Cy, AcO 1:1 (E/Z) 4:1	73	[69]

The ring-closing metathesis of alkene-tethered 1,3-diynes **36** and **40** has been explored (Scheme 18 and Scheme 19).[60] As expected, the *exo*-mode selective ring-closing metathesis of **36** and **40** affords products **39** and **42** where the latter is the consequence of the direct termination of alkylidene **41** without a metallotropic [1,3]-shift, while **39** is derived from a sequence of events involving a metallotropic [1,3]-shift of **37** to **38** followed by regioselective trapping. Clearly, the termination is affected by the substituent on the alkyne, one that ultimately controls the equilibrium of alkylidenes **37** and **38**. The ring-closing metathesis and termination of other substrates having different terminal alkyne substituents and tethers clearly reveals that the formation of fully conjugated products derived from a sequence involving a metallotropic [1,3]-shift is more favorable except in the case of the substrate containing a silyl group attached directly to the 1,3-diyne, e.g. **40**.

1.3.1 Enyne-Metathesis-Based Domino Reactions in Natural Product Synthesis **83**

Scheme 18 Initiation of Metallotropic [1,3]-Shift via Ring-Closing Metathesis[60]

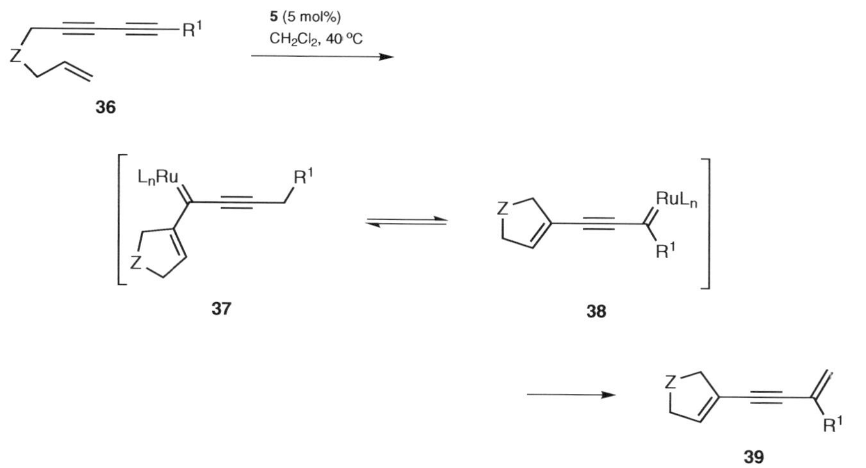

Z	R[1]	Yield (%) of **39**	Ref
$(CH_2)_2$	Bu	96	[60]
O	CH(OAc)Me	89	[60]
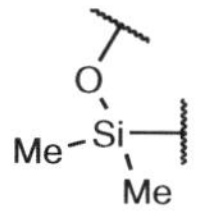 (structure)	CH_2OMe	quant	[60]
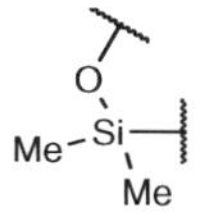 (structure)	H	69	[60]

Scheme 19 Ring-Closing Metathesis without a Metallotropic [1,3]-Shift[60]

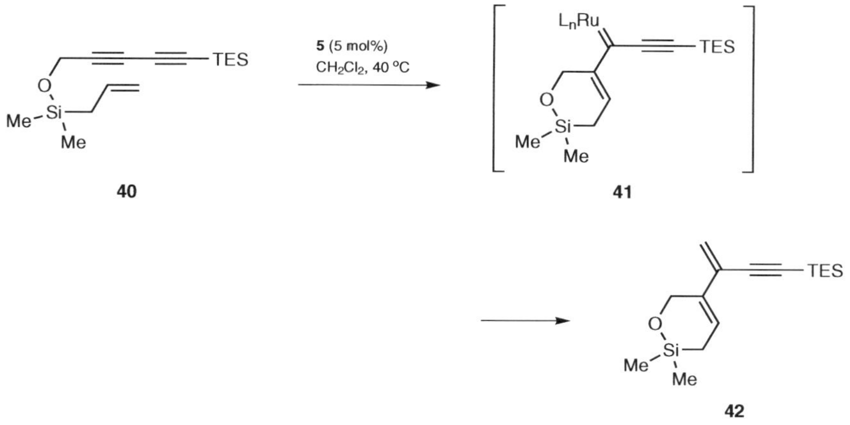

Even a subtle change in steric and electronic factors on the alkyne moiety of multiynes significantly influences the metallotropic [1,3]-shift behavior of the equilibrating alkyli-

for references see p 131

dene species (Scheme 20).[70] Using triyne **43**, which is terminally unsubstituted on the alkyne, catalytic turnover is completely inhibited and complex **45** is isolated when a stoichiometric amount of ruthenium complex **5** is used. On the other hand, in the reaction with **46**, the triethylsilyl group blocks the [1,3]-shift, and diyne **48** is isolated exclusively. These dichotomic behaviors can be explained by the existence of propagating alkylidenes **44** and **47**, where the former containing a terminal alkyne can undergo rapid double [1,3]-shifts to form **45**, a species stable enough to be isolated. On the other hand, the triethylsilyl group at the terminus of the diyne within alkylidene **47** prohibits the [1,3]-shift (similar to the behavior of **41**); thus, termination occurs directly after ring-closing metathesis. To further investigate these unexpected behaviors of certain putative alkylidene species, relay metathesis was employed.

This metathesis technology allows for the formation of a metal alkylidene at a specific location in the substrate when it contains more than one metathesis-active functionality. This tool is especially valuable for the exploration of metal alkylidene species generated in different electronic and steric environments so that the existence or lack of metallotropic [1,3]-shift of these alkylidenes can be accurately probed. Under typical conditions, substrates **49** and **51** (differing by the substituent on the alkene, i.e. H versus Me) engage in completely different reaction courses. The former generates a stable complex **50** in the presence of a stoichiometric amount of **5**, whereas **51** provides only **53** with a catalytic amount of **5**, likely via the initially formed alkylidene **52**. This relay-metathesis-based approach led to the conclusion that alkylidene complex **45**, once formed, is not capable of undergoing a [1,3]-shift in analogy to the lack of the [1,3]-shift in complex **50**, whereas the formation of **53** confirms the existence and effective participation in a [1,3]-shift.

Scheme 20   Initiation of Metallotropic [1,3]-Shift via Relay Metathesis[70]

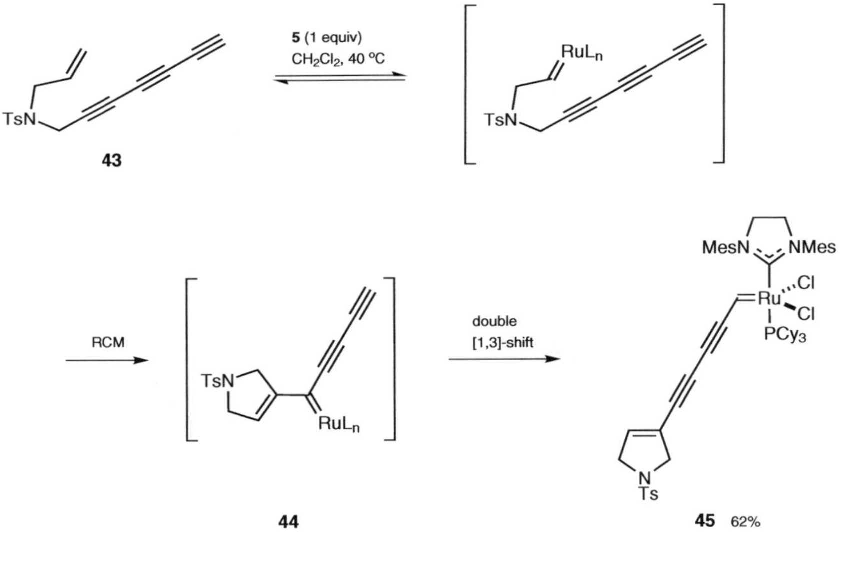

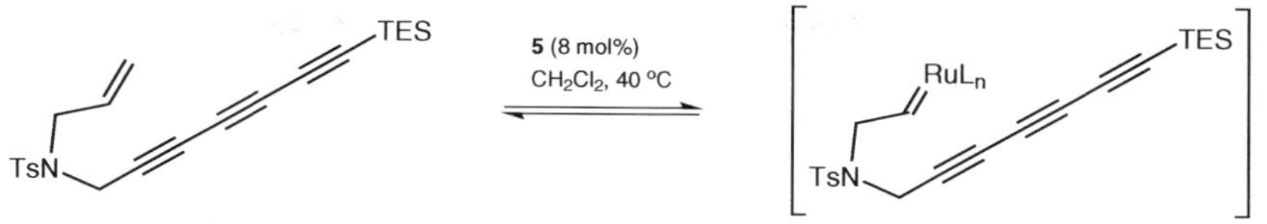

1.3.1 Enyne-Metathesis-Based Domino Reactions in Natural Product Synthesis **85**

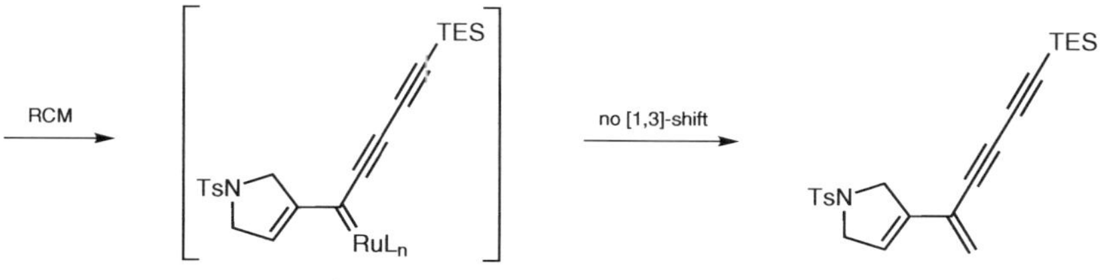

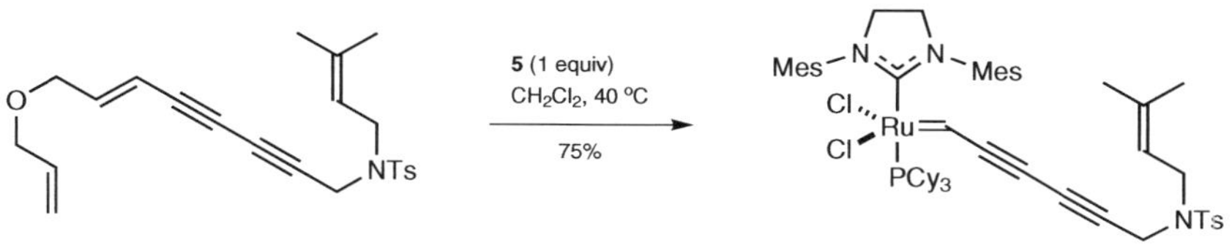

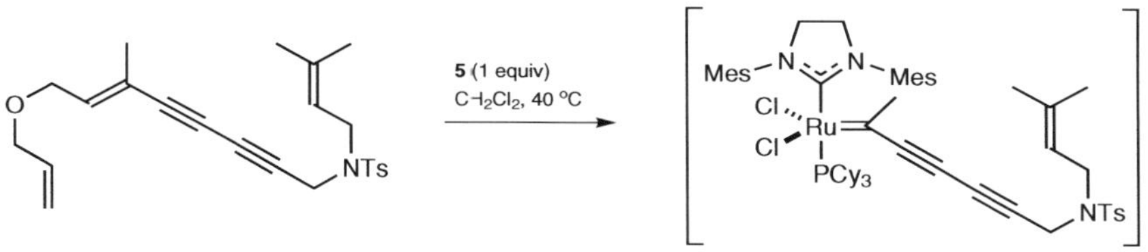

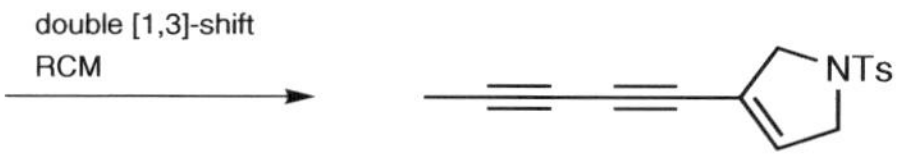

for references see p 131

1.3.1.3 **Applications in Natural Product Synthesis**

Enyne metathesis has become an important synthetic tool in natural product synthesis for the construction of 1,3-dienes as well as more conjugated unsaturated linear and cyclic carbon frameworks because of its powerful and atom-economical nature. The ability to generate new cyclic compounds while tolerating a broad range of functional groups from readily accessed enyne building blocks is just one of its merits. Not surprisingly, the utility of this method has been showcased in a variety of complex molecule syntheses via the use of simple enyne metathesis as well as more sophisticated domino transformations in the form of cross metathesis, ring-closing metathesis, metallotropic [1,3]-shifts, and combinations thereof.

Although uniquely effective in generating rings and conjugated unsaturation in molecules of interest, it must be noted that enyne metathesis has issues in regioselectivity that do not exist in other classes of metathesis. This problem ultimately narrows the substrate scope and the possible combinations of alkene and alkyne counterparts (Table 2); however, if past precedent is studied carefully, it is possible to predict when the process can be deployed successfully.

For example, terminal alkynes are good for their metathetic merger with both ethene and terminal alkenes (entries 1 and 2). For the metathesis of internal alkynes, only reaction with ethene can avoid regio- and stereochemistry problems (entry 3). Terminal alkenes can be used, however, with borylated internal alkynes to achieve high regio- and stereoselectivities (entry 4). The cross metathesis of internal alkynes with internal alkenes has rarely been engaged because their metathesis tends to generate a statistical mixture of 4 possible isomers (entry 5). Compared to cross metathesis of enynes, the corresponding ring-closing metathesis processes have less severe regio- and stereochemical uncertainty due to the constraint imposed by a tether between the reacting alkene and alkyne counterparts. The ring-closing metathesis of enynes to form small rings (5–11 members) is *exo*-mode selective (entry 6), while that of large rings (≥12 members) are generally *endo*-mode selective (entry 7). In the later ring-closing metathesis, however, the stereoselectivity of the incipient endocyclic double bond is highly ring-size and substrate dependent, and thus is not predictable a priori. The domino double-ring-closing metathesis of dienynes is known to be quite efficient for the formation of fused small bicycles (entry 8), yet the closely related domino triple-ring-closing metathesis is yet to be demonstrated (entry 9). The domino ring-closing metathesis/metallotropic [1,3]-shift sequence has been demonstrated with a broad spectrum of substrates, generating fully conjugated products rather than cross-conjugated ones except for the cases where R^1 is a silyl group. Most of these types of enyne ring-closing metathesis have been employed as a key step in the successful synthesis of natural products ranging from small terpenes to macrocyclic polyketides; specific examples of these synthesis applications are described in the following sections to illustrate their power in the most complex contexts yet explored.

1.3.1 Enyne-Metathesis-Based Domino Reactions in Natural Product Synthesis **87**

Table 2 Substrate-Based Types of Enyne Cross Metathesis and Ring-Closing Metathesis

Entry	Alkyne	Alkene	Metathesis Type[a]	Product Alkene	Selectivity	Applied to Natural Product Synthesis
1	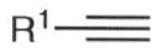		CM	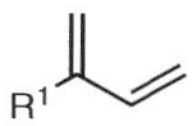	highly selective	yes
2		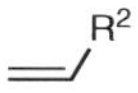	CM	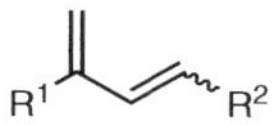	substrate-dependent stereoselective	no
3			CM	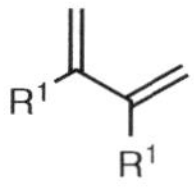	highly selective	yes
4	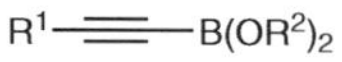	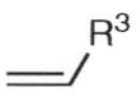	CM	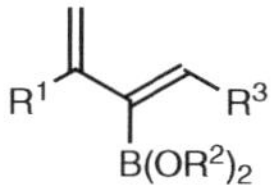	highly selective	yes
5		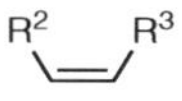	CM	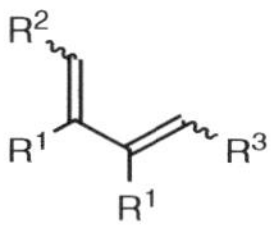	nonregio- and nonstereoselective	no
6	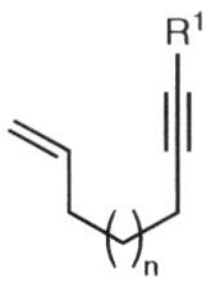		exo-RCM	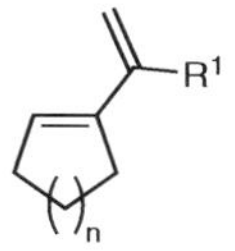	highly selective	yes
7	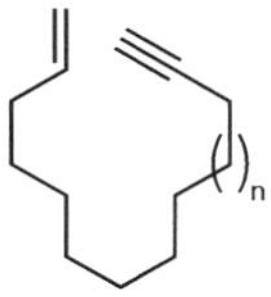		endo-RCM	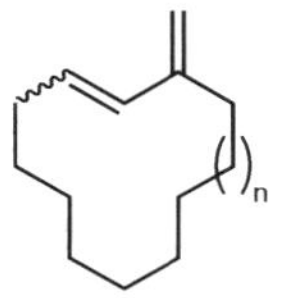	substrate-dependent stereoselective	yes
8	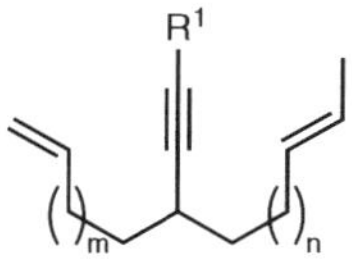		double RCM	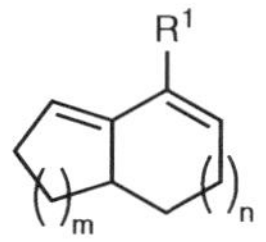	highly selective	yes
9	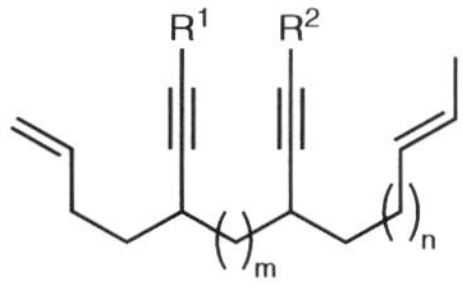		triple RCM	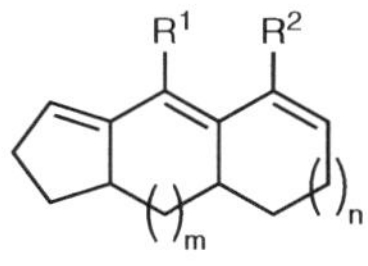	highly selective[b]	no
10	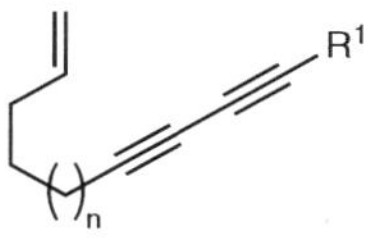		RCM–MS	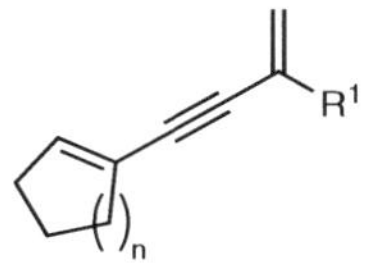	substrate-dependent regioselective	yes

[a] CM = cross metathesis; RCM = ring-closing metathesis; MS = metallotropic [1,3]-shift.
[b] Experimental data is required to confirm the selectivity.

for references see p 131

1.3.1.3.1 Simple Enyne Metathesis

1.3.1.3.1.1 Enyne Cross Metathesis

One of the most convenient ways to use enyne metathesis to install a 1,3-diene substructure without having to consider issues of regio- and stereoselectivity is to combine ethene and both terminal and internal alkynes (see Table 2, entries 1 and 3). One creative use of this effective 1,3-diene synthesis is in a strategy whereby the internal alkyne within a macrocyclic scaffold of a precursor of amphidinolide V is installed using alkyne ring-closing metathesis, and this internal alkyne is then metathesized with ethene. To install the prerequisite macrocyclic alkyne, ester-tethered 1,15-diyne **54** is treated with tris[aryl(*tert*-butyl)amido]molybdenum complex **55** in dichloromethane/toluene at 85 °C. The choice of solvent is key, as dichloromethane activates the metathesis-active molybdenum/alkylidyne catalyst[71,72] to afford a 14-membered macrocyclic alkyne **56** (Scheme 21).[73] The macrocyclic alkyne **56** is subsequently treated with Grubbs second-generation complex **5** under ethene to generate **57**, which is successfully elaborated to *ent*-amphidinolide V via the installation of the triene side chain with sulfone **58** followed by final removal of the silyl protecting group.

1.3.1 Enyne-Metathesis-Based Domino Reactions in Natural Product Synthesis **89**

Scheme 21 Fürstner's Synthesis of *ent*-Amphidinolide V[73]

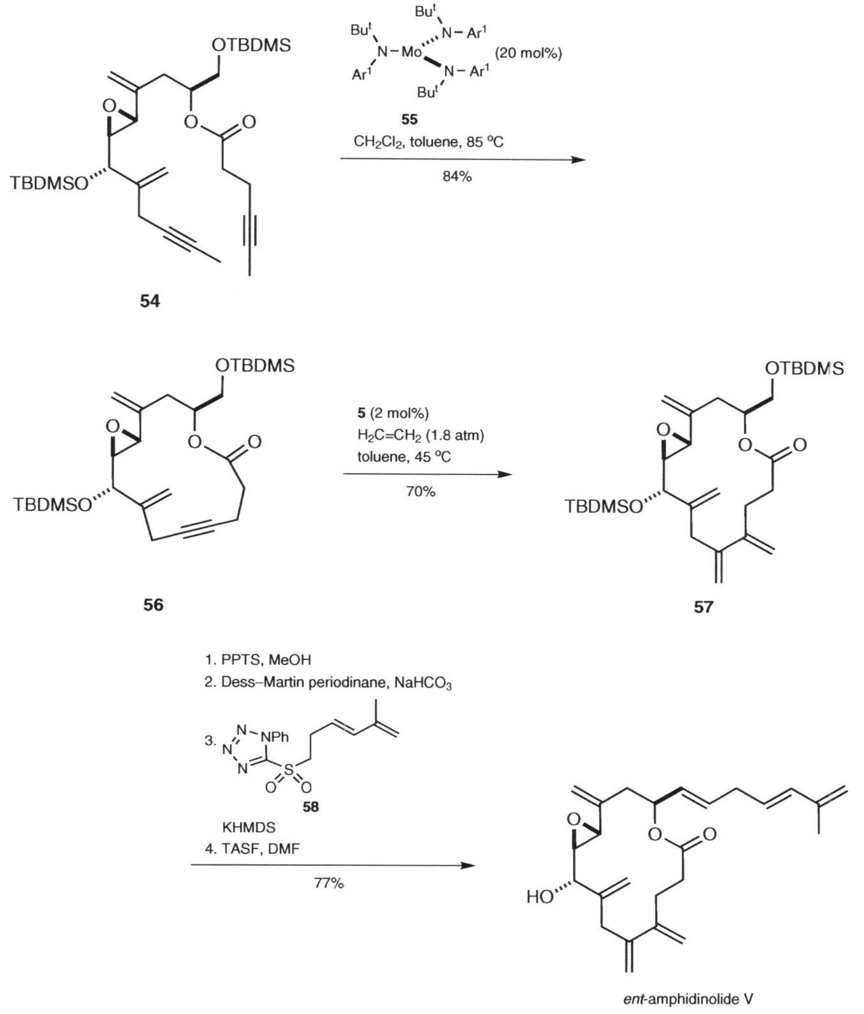

Ar¹ = 3,5-Me₂C₆H₃; TASF = tris(diethylamino)sulfonium difluorotrimethylsilicate

for references see p 131

90 Domino Transformations **1.3** Metathesis Reactions

In another unique approach, enyne metathesis is utilized to construct both 1,3-diene subunits of different connectivity of the *ent*-amphidinolide V structure (Scheme 22). Similarly to the previous approach, the 1,2-dimethylene moiety of the target is constructed by cross metathesis between internal alkyne **59** and ethene to give primary alcohol **60**, which is oxidized to aldehyde **61**. The other 1,3-diene-containing building block **64** is prepared starting from an enyne ring-closing metathesis of alkynylsilyl ether **62** to give **63** followed by its elaboration to **64**. Subsequent proline-mediated aldol condensation of aldehydes **61** and **64** affords enal **65** with good E/Z selectivity; this material is elaborated to macrocyclic allylic alcohol **66** in four steps. To complete the synthesis, 1,3-transposition of the allylic alcohol is required, and that goal is accomplished via a three-step protocol employing: (1) the Sharpless asymmetric epoxidation; (2) iodination of the primary alcohol; and (3) reductive opening of the iodo epoxide with *tert*-butyllithium. Finally, removal of the trimethylsilyl group completes the total synthesis of *ent*-amphidinolide V.[74]

Scheme 22 Lee's Synthesis of *ent*-Amphidinolide V[74]

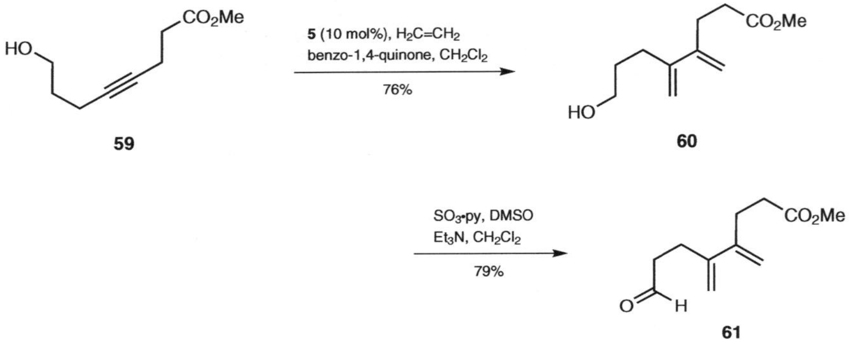

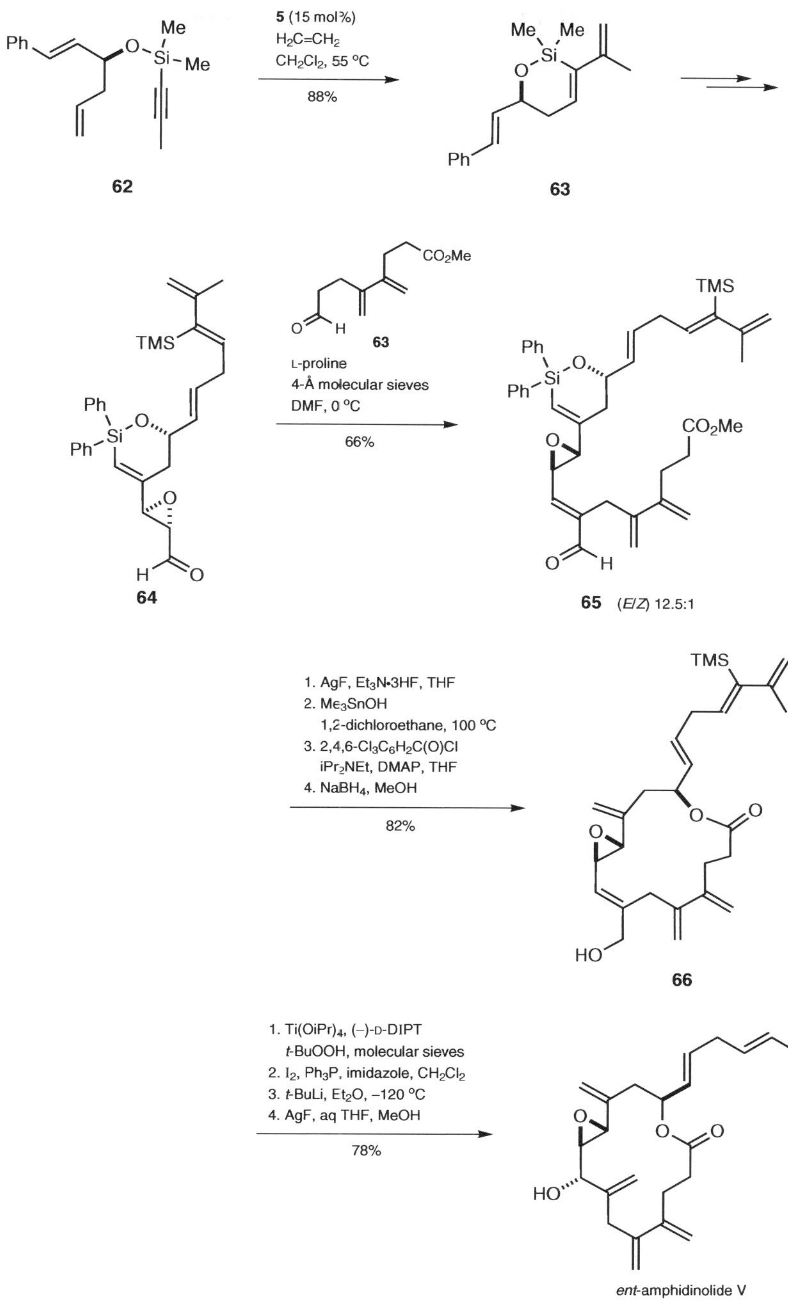

The enyne cross metathesis between borylated internal alkynes and terminal alkenes generates 2-boryl-1,3-dialkyl-1,3-dienes with high regio- and stereoselectivity (Table 2, entry 4). The sterically bulky and electronically biased boryl group is assumed to serve as an effective directing group, resulting in the observed excellent regioselectivity and E selectivity.[75] This enyne cross-metathesis technology is employed in the stereoselective construction of an advanced intermediate for the total synthesis of (−)-amphidinolide K (Scheme 23). Here, cross metathesis between borylated alkyne **67**, generated from the corresponding terminal alkyne, and alkene **68** provides trisubstituted 1,3-diene **69**. Palladium-cata-

for references see p 131

lyzed cross coupling of the vinylboronate moiety within **69** with iodomethane affords a trisubstituted-1,3-diene-containing advanced intermediate **70**, which can be successfully elaborated to the target natural product.

Scheme 23 Synthesis of Amphidinolide K[75]

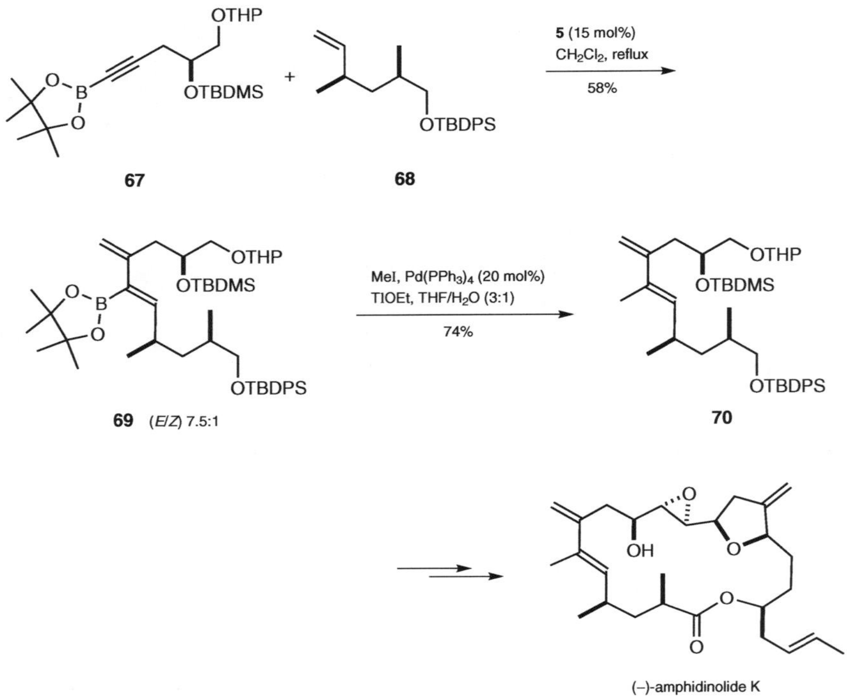

(–)-amphidinolide K

(S,E)-2,2-Dimethyl-6-(2-phenylvinyl)-3-(prop-1-en-2-yl)-5,6-dihydro-2H-1,2-oxasilin (63); Typical Procedure for Enyne Cross Metathesis:[74]

To a soln of alkyne **62** (2.23 g, 8.25 mmol) in CH_2Cl_2 (825 mL) was added Grubbs second-generation initiator **5** (1.02 g, 1.23 mmol, 15 mol%). Ethene was then passed through the soln for 25 min followed by argon for 30 min to remove excess ethene. The mixture was refluxed at 55 °C for 12 h and the solvent was evaporated under reduced pressure. The residue was quickly purified by flash column chromatography (hexanes/EtOAc 50:1) to afford a red-brown liquid. In order to remove ruthenium byproducts generated during the reaction, the crude material was dissolved in CH_2Cl_2 (10 mL) and DMSO (10 mL) was added. The resultant soln was stirred at rt for 12 h and concentrated under reduced pressure. The residue was purified by gravity column chromatography (hexanes/EtOAc 50:1 to 30:1) to afford the product as a yellow liquid; yield: 1.97 g (88%).

1.3.1.3.1.2   Enyne Ring-Closing Metathesis for Small Rings

Due to their strained nature, small-sized rings containing unsaturation usually undergo ring-opening metathesis; therefore, their formation by ring-closing metathesis is quite challenging under inherently reversible metathesis conditions. Contrary to this general notion, it has been demonstrated that cyclobutenes can be generated from 1,5-enynes via enyne ring-closing metathesis.[76] This unusual ring-closing metathesis subsequently inspired the ring-closing metathesis of 1,5-enyne **71** using the Hoveyda–Grubbs second-generation ruthenium alkylidene complex **7** under microwave irradiation to generate

72 in high yield [83%; an alternative procedure not employing microwave irradiation affords the product in lower yield (66%)]. This product is elaborated to (±)-grandisol via chemo- and stereoselective hydrogenation followed by deprotection of the silyl group (Scheme 24).[77]

Scheme 24 Synthesis of Grandisol[77]

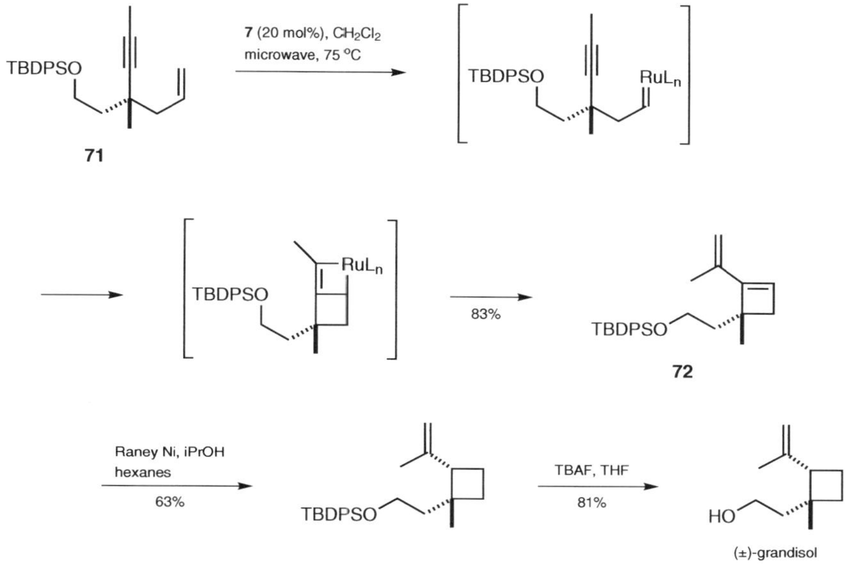

An enyne substrate containing a silyl ynol ether moiety has been employed to directly install a ketone functional group equivalent in the ring-closing metathesis product (Scheme 25).[78] Although conceptually straightforward, in practice, the ring-closing metathesis of enyne **73** could have a potential problem because the initially formed propagating alkylidene **74** is a Fischer-type ruthenium alkylidene complex, which might be so stable that it cannot undergo catalytic turnover. When treated under typical ring-closing metathesis conditions with Grubbs second-generation complex **5**, however, the desired 1,3-diene **75** (a form of a silyl enol ether) is obtained efficiently, and can then be converted into α,β-unsaturated enone **76** by treatment with hydrogen fluoride. After stereoselective hydrogenation, the resultant methyl ketone **77** can be further transformed into both α- and β-eremophilane in a straightforward manner.

for references see p 131

Scheme 25 Synthesis of β-Eremophilane[78]

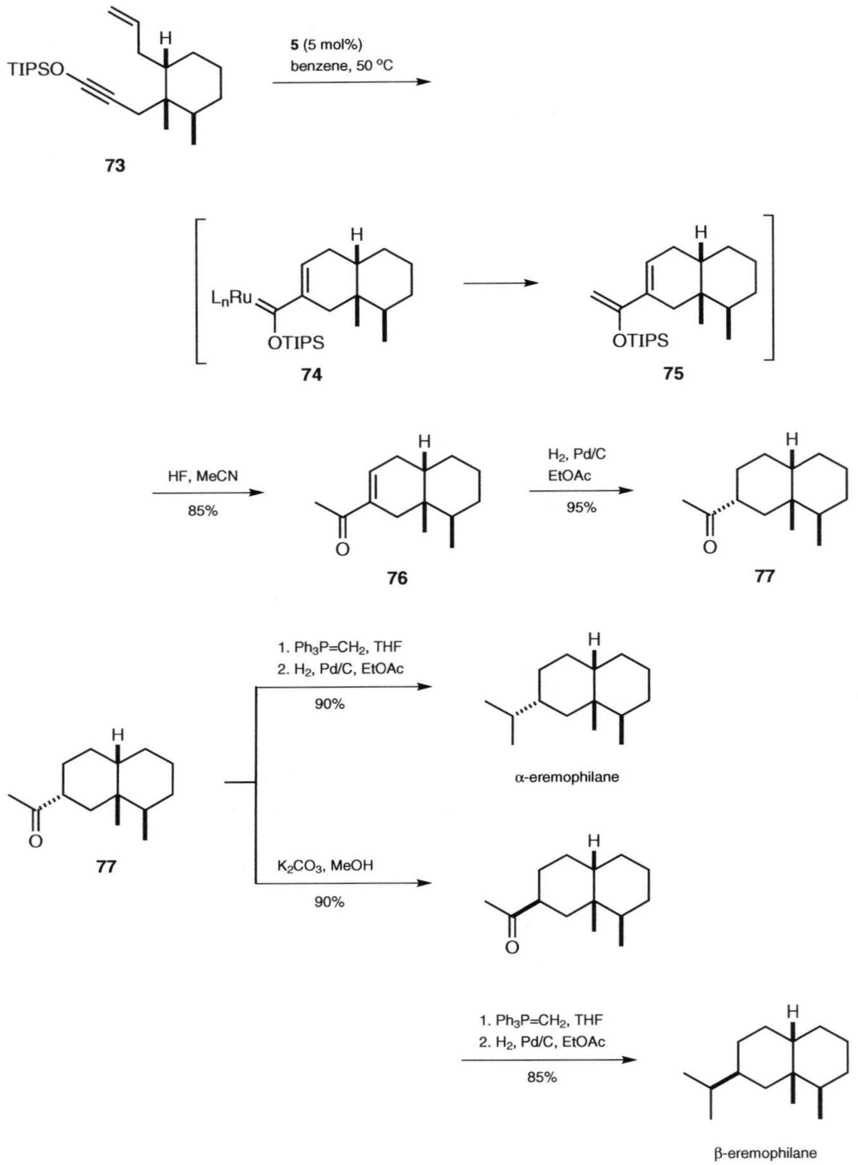

The rate-accelerating effect of an allylic hydroxy substituent in metathesis reactions has been frequently reported, and it has found great use in natural product synthesis.[79–81] For example, endeavors for efficient syntheses of nitrogen-containing natural products employ a ring-closing metathesis based strategy (Scheme 26)[82] where it is observed that among various 1,7-enynes with an allylic substituent, those enynes containing a free hydroxy group show a significantly more favorable reaction profile, generating product **78** (R^1 = OH) quantitatively. On the other hand, benzyl- and silyl-protected enynes, as well as the non-oxygenated enyne, provide only low to moderate yields of the corresponding dehydropiperidine derivative **78**. To complete the synthesis of a potent glycosidase inhibitor isofagomine,[83] the hydroxy group of **78** (R^1 = OH) is protected as its *tert*-butyldiphenylsilyl ether **78** (R^1 = OTBDMS), the vinyl group of which is oxidatively cleaved and the resulting

aldehyde is reduced to generate allylic primary alcohol **79**. Hydroboration–oxidation of the trisubstituted double bond affords diol **80** with 9:1 diastereoselectivity, and subsequent global deprotection of the silyl and *tert*-butoxycarbonyl groups under acidic conditions provides isofagomine. The route as a whole proceeded in 34% total yield from commercial vinyloxirane.

Scheme 26 Synthesis of Isofagomine[82]

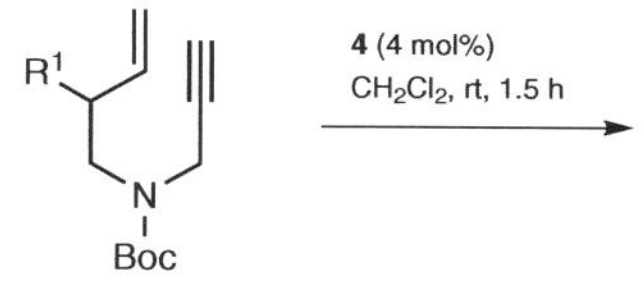

R¹	Yield (%) of **78**	Ref
OH	>98	[82]
OBn	>44	[82]
OTBDPS	>7	[82]
H	>32	[82]

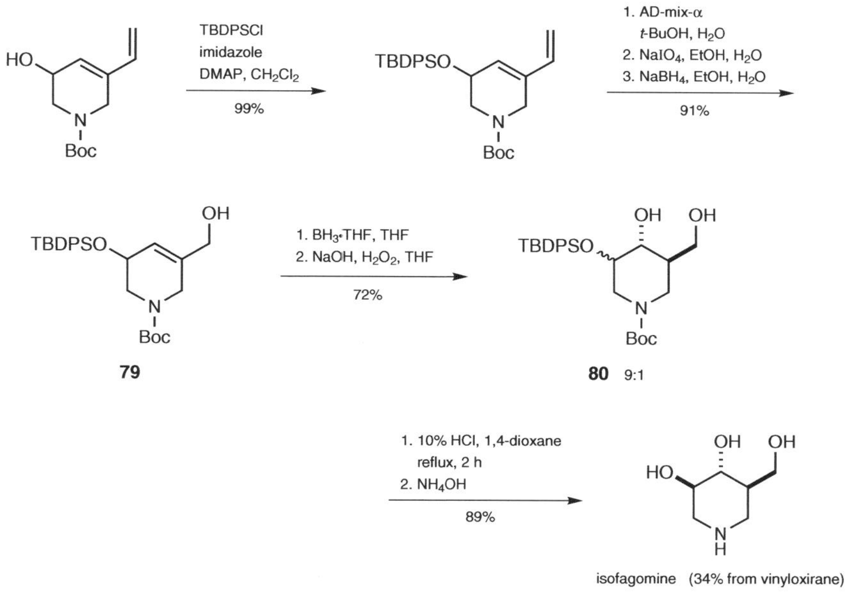

isofagomine (34% from vinyloxirane)

An enyne ring-closing metathesis based approach has been employed in the synthesis of (+)-pericosine C (Scheme 27).[84] Ring-closing metathesis of the ribose-derived enyne **81** affords highly functionalized cyclohexene derivative **84** in good yield. For this enyne substrate, it is not clear whether the ring-closing metathesis occurs via propagating alkylidene **82** or **83** due to the similarity of the steric and electronic environment of the terminal alkene and alkyne moieties; however, regardless of the structure of the propagating species, the same product **84** is formed. Oxidative cleavage of the vinyl group to generate aldehyde **85**, followed by its conversion into a methyl ester and removal of the acetonide protecting group, results in (+)-pericosine C.

for references see p 131

Scheme 27 Synthesis of (+)-Pericosine C[84]

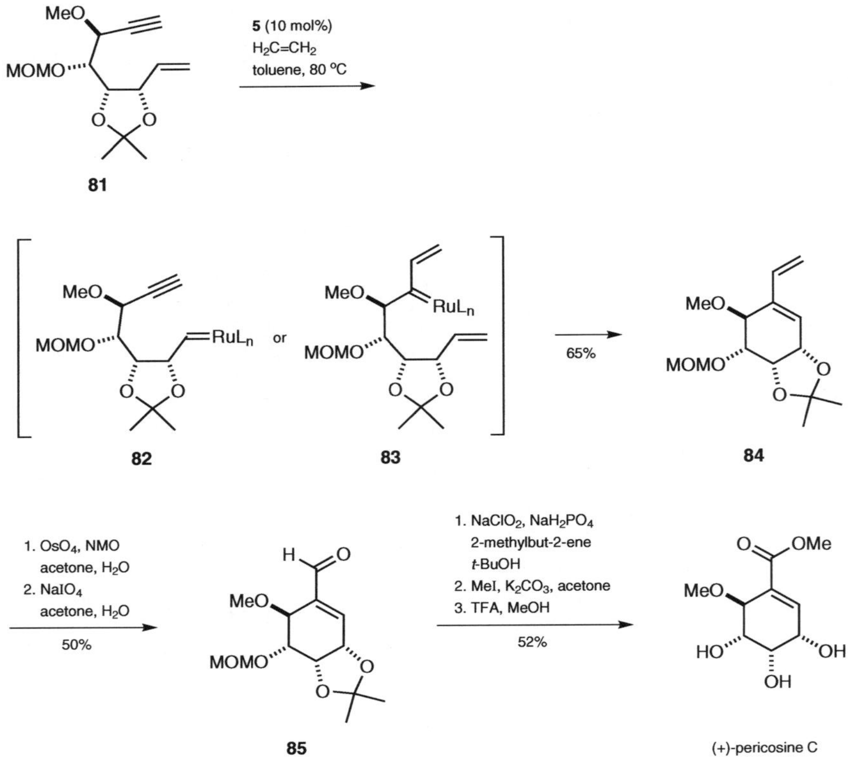

Conceptually, a similar enyne ring-closing metathesis approach is also employed as the key step in a synthesis of (+)-valienamine (Scheme 28).[85] Here, highly functionalized 1,7-enyne **86**, readily prepared from serine, is exposed to a catalytic amount of the Grubbs second-generation complex **5** in the presence of ethene, generating cyclohexene derivative **87** in excellent yield. Oxidative cleavage of the vinyl group and reduction of the resulting aldehyde then affords allylic alcohol **88**. Both the methoxymethyl and *tert*-butoxycarbonyl protecting groups are removed under acidic conditions, and the resulting compound is treated with acetic anhydride to form peracetylated valienamine **89**, thereby completing a formal synthesis of (+)-valienamine.

1.3.1 Enyne-Metathesis-Based Domino Reactions in Natural Product Synthesis **97**

Scheme 28 Synthesis of Valienamine[85]

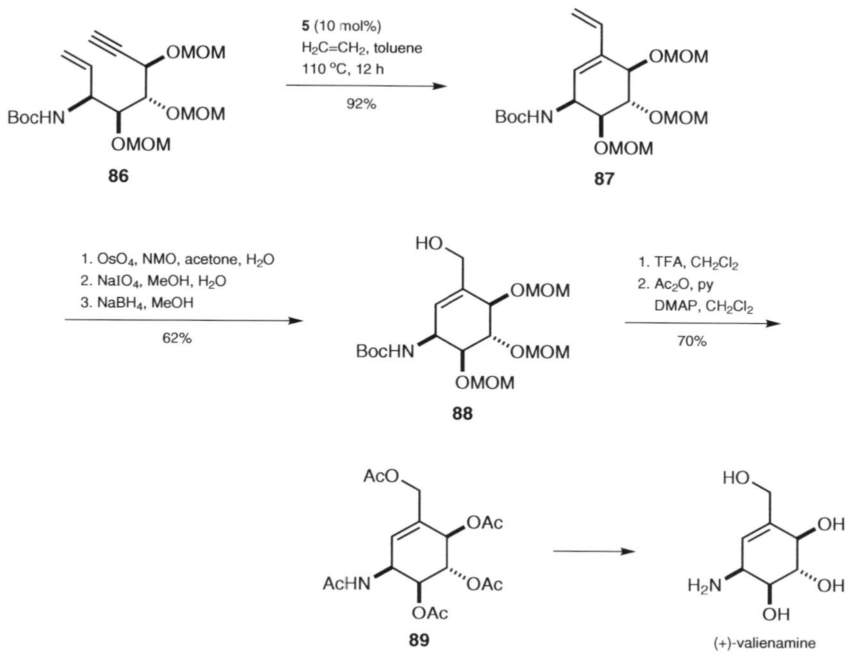

Enyne ring-closing metathesis also serves as a convenient method for the synthesis of tropane-type alkaloids (Scheme 29). For example, a concise synthesis of (+)-ferruginine employs ring-closing metathesis of 1,7-enyne **90** as the key step to construct bridged bicycle **91**. This product is converted into the target molecule through the Wacker oxidation, *tert*-butoxycarbonyl deprotection, and reductive amination for N-methylation.[86] Likewise, ring-closing metathesis of 1,8-enyne **92** forms cyclic diene **93** with simultaneous installation of the 9-azabicyclo[4.2.1]nonane skeleton and the necessary functionality for its conversion into anatoxin-a. Thus, oxidative cleavage of the propenyl group in **93** followed by benzyloxycarbonyl deprotection results in (+)-anatoxin-a.[87] Finally, bridged azabicycle **95** is generated from the ring-closing metathesis of 1,8-enyne **94**. During this ring-closing metathesis, the silyl group is removed; the vinyl group within **95** is elaborated into the corresponding acetyl group in *N*-tosylanatoxin-a.[88,89]

for references see p 131

Domino Transformations   1.3 Metathesis Reactions

Scheme 29   Synthesis of Tropane-Type Alkaloids[86–89]

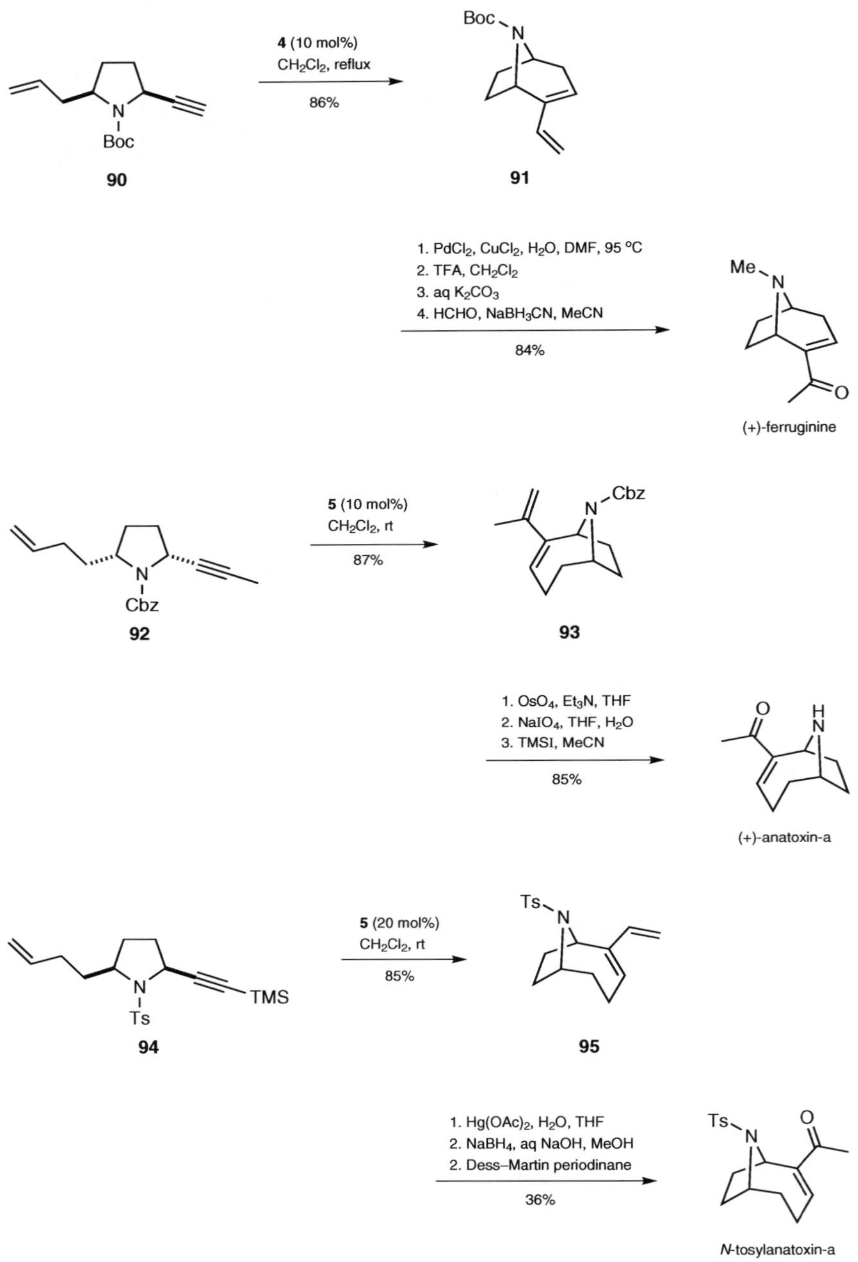

(R)-4-[2-(tert-Butyldiphenylsiloxy)ethyl]-4-methyl-1-(prop-1-en-2-yl)cyclobuta-1-ene (72);
Typical Procedure for Thermal Enyne Ring-Closing Metathesis for Small Rings:[77]
To a soln of enyne **71** (467 mg, 1.19 mmol) in anhyd CH_2Cl_2 (31 mL) was added Hoveyda–
Grubbs second-generation initiator **7** (150 mg, 20 mol%). The flask was then briefly purged
with argon and sealed, and was then submerged in a preheated oil bath at 75 °C and
stirred for 35 min. The mixture was cooled in an ice bath, and concentrated to give a
green residue, which was purified by column chromatography (isocratic, Et_2O/pentane
1:49) to afford the product as a colorless syrup; yield: 309 mg (66%).

(R)-4-[2-(*tert*-Butyldiphenylsiloxy)ethyl]-4-methyl-1-(prop-1-en-2-yl)cyclobuta-1-ene (72); Typical Procedure for Microwave Enyne Ring-Closing Metathesis for Small Rings:[77]

To a soln of **71** (73.0 mg, 0.187 mmol) in anhyd CH_2Cl_2 (5 mL) was added Hoveyda–Grubbs second-generation catalyst **7** (23 mg, 20 mol%). The vial was briefly purged with dry argon and then sealed. The vial was microwave irradiated (75 °C) for 30 min, then concentrated and purified via column chromatography (isocratic, Et_2O/pentane 1:49) to give the product as a colorless syrup; yield: 60 mg (83%).

1.3.1.3.2 Domino Enyne Metathesis

1.3.1.3.2.1 Double Ring-Closing Metathesis with Dienynes

The prowess of enyne metathesis for the construction of multiply fused cyclic frameworks containing conjugated unsaturation is most clearly illustrated in the domino double ring-closing metathesis of dienynes containing appropriate tethers between the alkyne and alkenes (Table 2, entry 8).

In an approach to guanacastepene A, domino enyne ring-closing metathesis of dienyne **96** is employed to generate the complete guanacastepene carbon skeleton **98**, containing a conjugated triene moiety, in good yield (Scheme 30).[90,91] In the domino double ring-closing metathesis, the sterically and electronically differentiated alkenes and alkynes allows the initiation selectively at the least-hindered vinyl group attached to the cyclopentene moiety to form a propagating alkylidene **97**. This adduct then undergoes subsequent ring-closing metathesis to generate the observed product **98**. The electronic and ring strain differences of each double bond within **98** also allows regioselective epoxidation followed by its opening to generate allyl ether **99A** and its diastereomer **99B**. *tert*-Butyldimethylsilyl protection of **99A** was followed by nickel-catalyzed reductive cleavage employing excess diisobutylaluminum hydride, which removed the allyl ether and reduced the neighboring ester group to afford **100** following *tert*-butyldimethylsilyl deprotection. Acetonide formation from the 1,3-diol moiety of **100** followed by Dess–Martin oxidation of the remaining secondary alcohol then delivered an advanced intermediate **101** from which Danishefsky completed a total synthesis of guanacastepene A.[92]

for references see p 131

Scheme 30 Synthesis of Guanacastepene A[90,91]

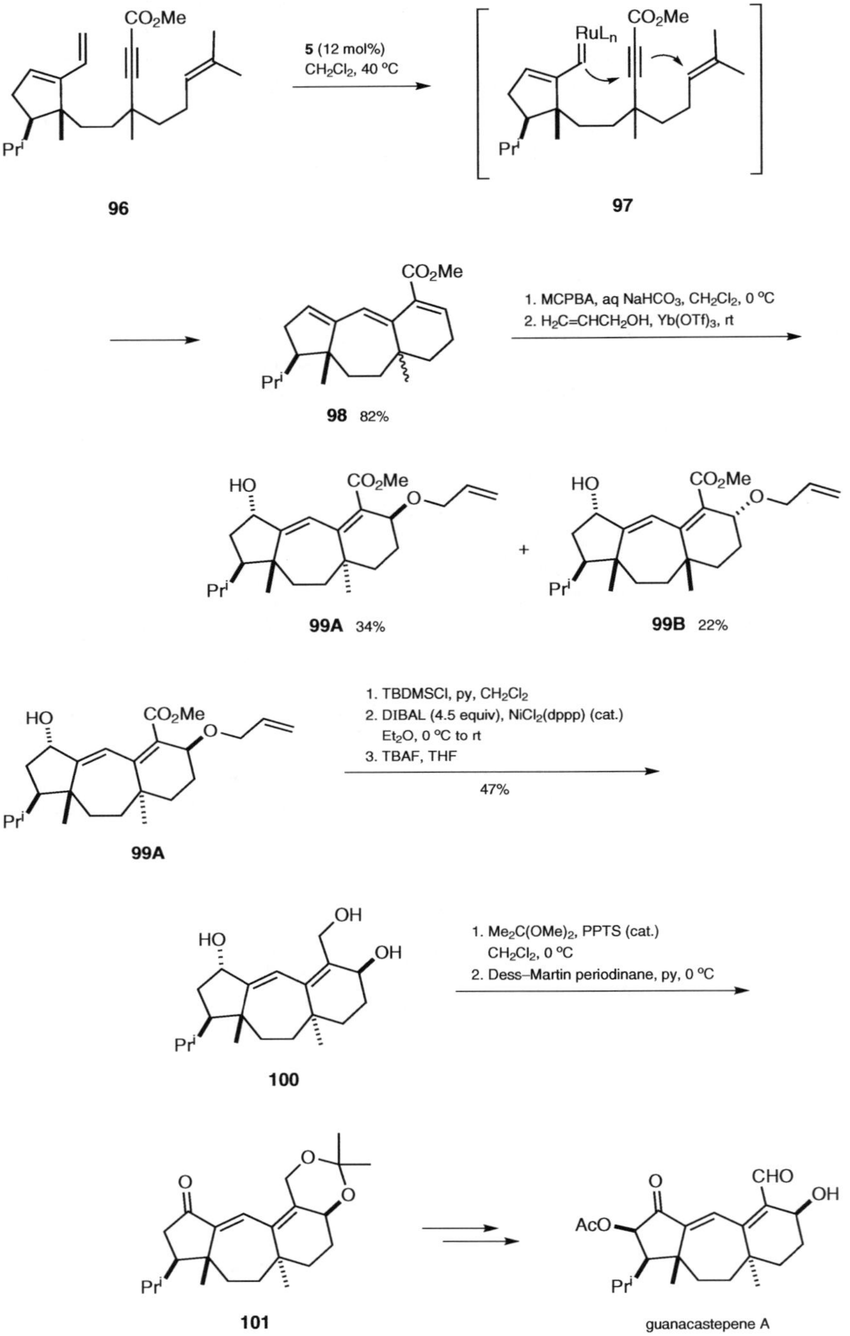

A formal synthesis of englerin A utilizes a domino double enyne ring-closing metathesis reaction as a platform to construct the complete guaiane core (Scheme 31).[93] When enyne **103** is treated with Grubbs–Stewart initiator **102**, a mixture of diastereomeric $\Delta^{4,6}$-guaia-diene-9,10-diol carbonates **105A/105B** is isolated in good yield. To initiate the domino process effectively in the correct order, a relay-metathesis strategy is employed to gener-

ate the required propagating alkylidene **104**. This species then engages the nearby alkyne first and subsequently the 1,1-disubstituted alkene to complete the catalytic cycle, forming the observed products. Once formed, the desired diastereomer **105B** is converted into allylic transposed *tert*-alcohols **108A** and **108B** via **106** and oxygen trapping of organomercury intermediate **107**. The elaboration of diastereomer **108A** to englerin A had already been reported by Echavarren and co-workers,[94] and thus a formal synthesis is achieved by relying on this powerful domino double enyne ring-closing metathesis strategy.

Scheme 31   Synthesis of Englerin A[93]

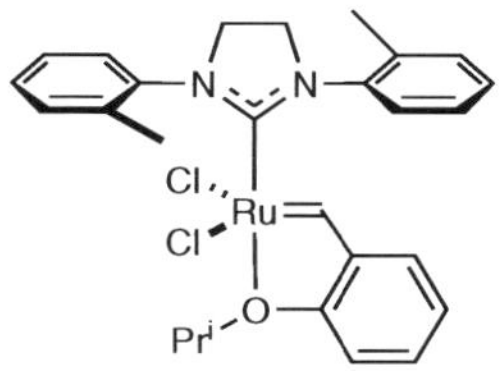

102   Grubbs–Stewart catalyst

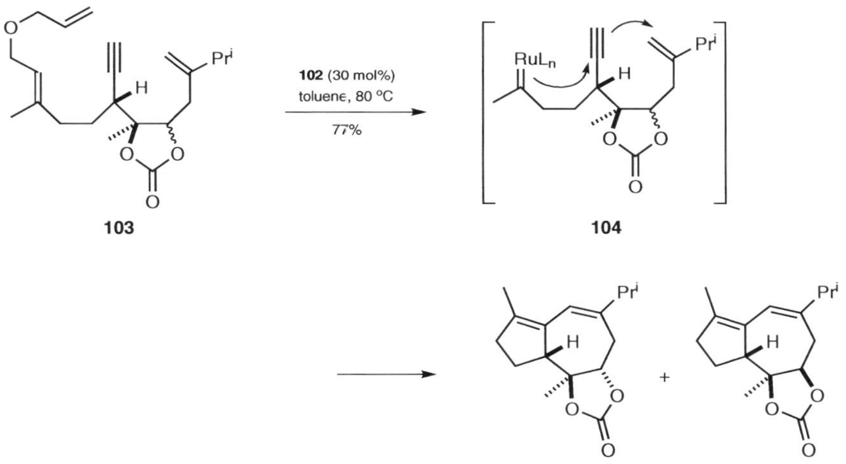

103   **102** (30 mol%)
toluene, 80 °C

77%

104

105A 32%   +   **105B** 45%

for references see p 131

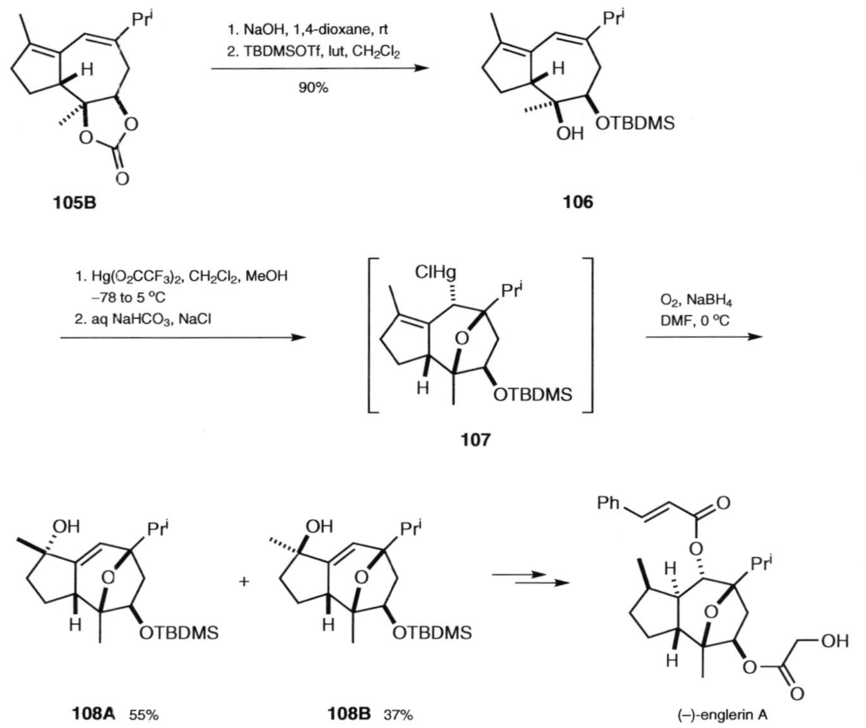

Dienyne ring-closing metathesis is used in the synthesis of (–)-acylfulvene and (–)-irofulven (Scheme 32).[95,96] In this work, the carbocyclic core of these compounds is constructed via consecutive double enyne ring-closing metathesis and diene ring-closing metathesis. In the key step seeking to install the highly functionalized cyclohexene moiety, dienyne **109** is treated with initiator **5** to form dihydrodioxasilepin **112** through the initially formed alkylidene **110**. Ring closure of **110** to give **111** is followed by its termination to generate the second ring. For the planned diene ring-closing metathesis projected to install a fused five-membered ring, **112** is treated with tetrabutylammonium fluoride to afford triene **113**, which after installing a cyclic carbonate, is subjected to a Myers' reductive allylic transposition conditions to form triene **114**. Ring-closing metathesis of **114** generates **115** and this event is followed by in situ removal of the carbonate to form **116**. Sequential 2,3-dichloro-5,6-dicyanobenzo-1,4-quinone oxidation then affords acylfulvene, which is further converted into irofulven by introducing a hydroxymethyl group with paraformaldehyde and sulfuric acid.

1.3.1 Enyne-Metathesis-Based Domino Reactions in Natural Product Synthesis **103**

Scheme 32 Synthesis of (−)-Acylfulvene and (−)-Irofulven[95,96]

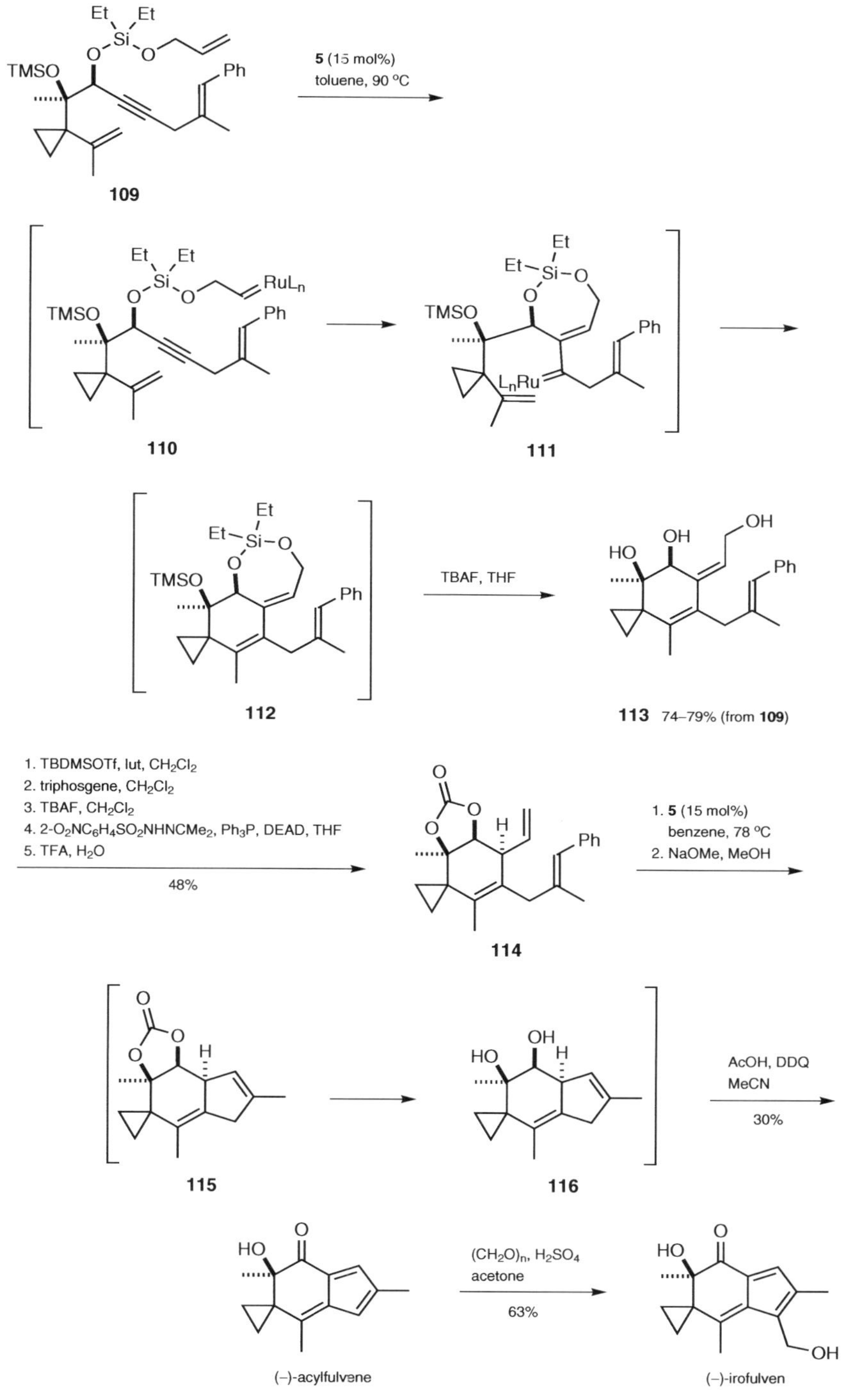

The core of the kempene diterpenes has been synthesized employing domino enyne ring-closing metathesis as the central bond-forming event to directly install a hydroazulene moiety fused to a decalin core (Scheme 33).[97] Here, treating dienyne **117** with initiator

for references see p 131

104 Domino Transformations **1.3** Metathesis Reactions

5 affords tetracyclic compound **119** via the formation of an initial propagating ruthenium alkylidene **118** followed by an uninterrupted second ring-closing metathesis with the prenyl group. Removal of the *tert*-butyldimethylsilyl group from **119** generates free alcohol **120**, a compound that serves as a common precursor for three kempene diterpenes secreted by termite soldiers of the species *Nasutitermes kempae*. Indeed, while the direct acetylation of **120** affords (+)-kempene-2, reduction of the ketone functionality followed by diacetylation provides a mixture of (+)-kempene-1 and (+)-3-*epi*-kempene-1 in high yield and 2.3:1 ratio.

Scheme 33 Synthesis of Kempene Diterpenes[97]

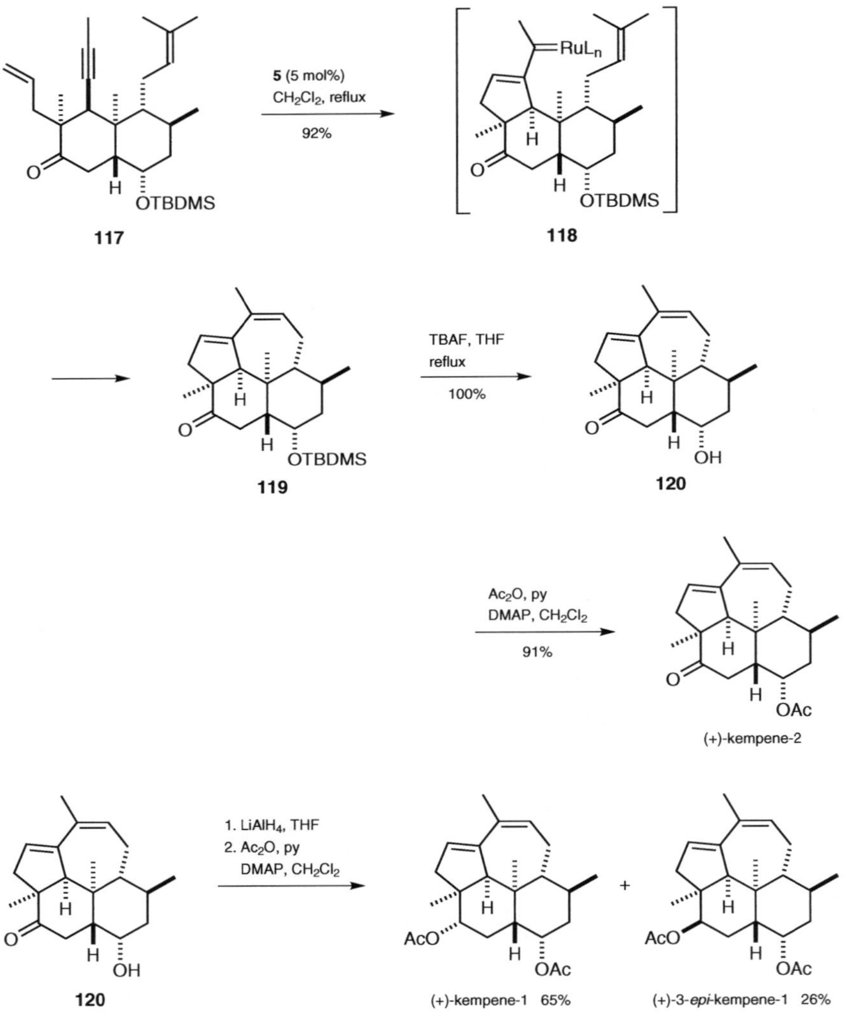

A silaketal-based temporary tether strategy has been developed for domino double enyne ring-closing metathesis for the construction of E/Z-1,3-dienes with high E/Z selectivity and efficiency. This strategy is deployed in a synthetic approach toward tartrolon B, a target that contains an E/Z-1,3-diene moiety, as well as a synthesis of cochleamycin A where a pivotal intermediate contains an E/Z-1,3-diene suitable for a Diels–Alder reaction.[98] For the synthesis of cochleamycin A, readily available alcohol building blocks and cyclopentyltriethynyl silane are easily converted into silaketal **121** in two consecutive sodium hydride catalyzed alcoholysis steps. Subsequent treatment of **121** under typical ring-closing metathesis conditions with Grubbs' second-generation complex **5** followed by disilylation delivers bicyclic 1,3-diene **122** in good yield, which is further transformed into an advanced intermediate **123**.[99] Compound **123** is identical to the Roush intermediate[100] used for the same target, and thus, a formal synthesis of (−)-cochleamycin A is achieved (Scheme 34).

Scheme 34 Formal Synthesis of Cochleamycin A[99,100]

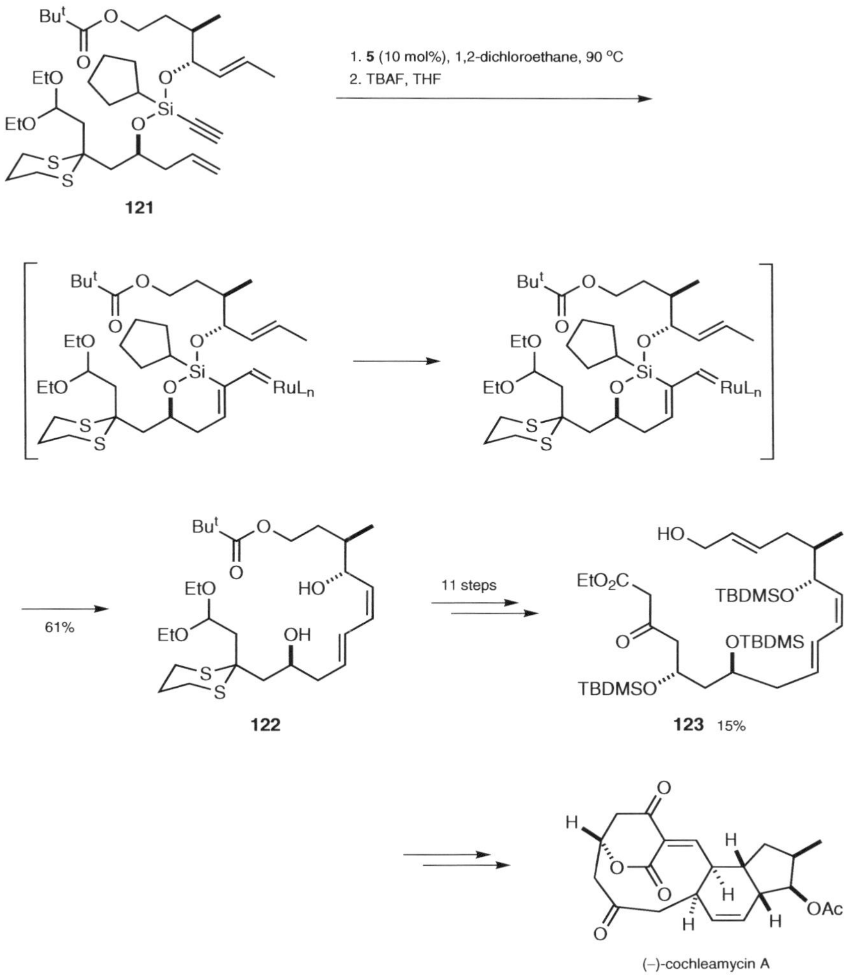

for references see p 131

The lycopodium alkaloid lycoflexine is synthesized using dienyne ring-closing metathesis as the pivotal step.[101] As shown in Scheme 35, dienyne **124** undergoes double ring-closing metathesis, most likely initiated from the least-hindered terminal alkene, to generate a propagating vinyl alkylidene **125**, which then closes up the second ring to provide diene **126** containing a Z-disubstituted double bond within the newly formed nine-membered ring. Diene **126** is directly hydrogenated selectively on the Z-disubstituted double bond to afford monoalkene **127**. Subsequent conversion of the trisubstituted alkene into a ketone via hydroboration followed by oxidation with 2-iodoxybenzoic acid, followed by a Mannich reaction under acidic conditions with formaldehyde, completes the synthesis of lycoflexine.

Scheme 35 Synthesis of Lycoflexine[101]

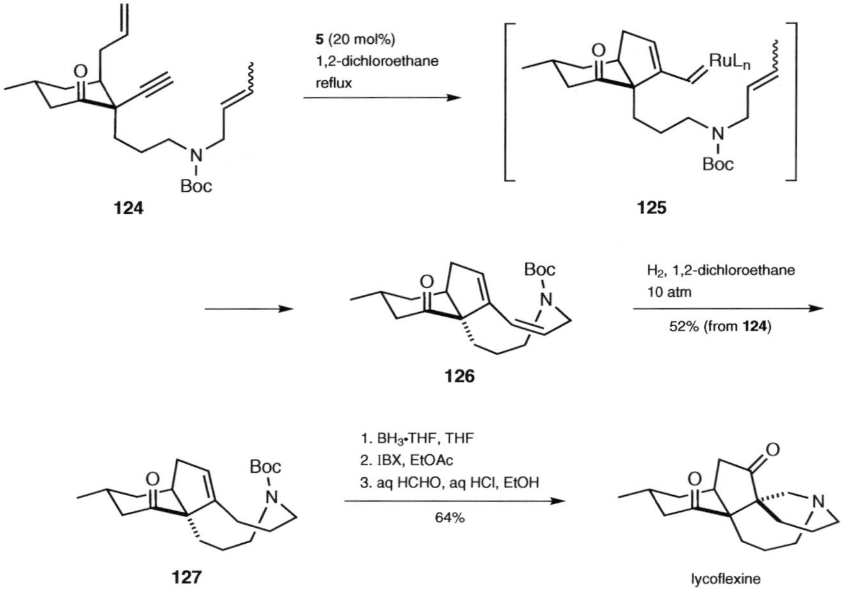

IBX = 2-iodoxybenzoic acid

Good functional-group tolerance is also shown when enyne **128** is subjected to a group-selective ring-closing metathesis (Scheme 36). The free hydroxy group in **128** is not only well tolerated, but it also activates the initiation at the alkene next to it, thus selectively generating a propagating species **129**. Domino double ring-closing metathesis with a putative intermediate **129** followed by protection of the hydroxy group as the *tert*-butyldimethylsilyl ether yields fused bicyclic 1,3-diene **130**; this material is elaborated to saturated system **133** in a two-step sequence: (1) oxidation of the primary alcohol in **131**, and (2) Julia–Kocienski alkenation with sulfone **132** to afford E-alkene **133** selectively. This compound is further elaborated into **134** by removal of the newly formed double bond and inversion of the secondary hydroxy group by oxidation and stereoselective reduction. Final acylation of **134** with (E)-oct-2-enoic acid or (E,E)-octa-2,4-dienoic acid followed by *tert*-butoxycarbonyl deprotection provides *ent*-lepadin F and G, respectively.[102]

1.3.1 Enyne-Metathesis-Based Domino Reactions in Natural Product Synthesis **107**

Scheme 36 Synthesis of *ent*-Lepadin F and G[102]

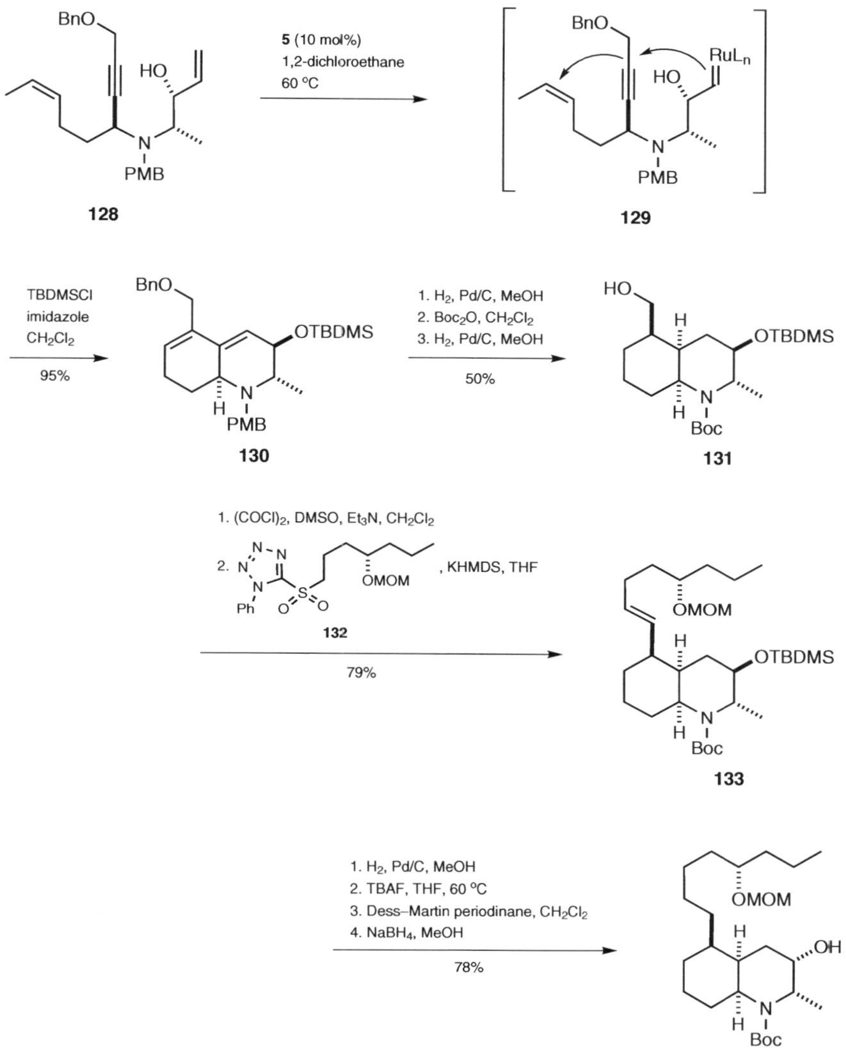

for references see p 131

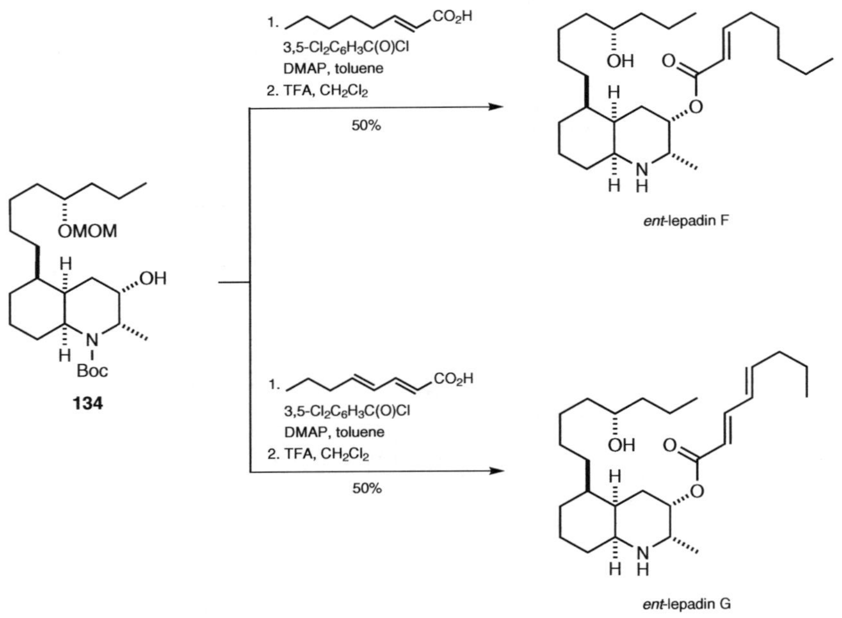

108 Domino Transformations 1.3 Metathesis Reactions

A synthesis of (–)-securinine relies on selective initiation of enyne ring-closing metathesis from the sterically least-hindered double bond of dienyne **136** (Scheme 37).[103] Upon treating **136** with the more reactive Hoveyda–Grubbs-type ruthenium complex **135**, a putative propagating alkylidene, i.e. **137**, forms that can then undergo a domino double ring-closing metathesis to deliver **138**. Chromium-mediated allylic oxidation of the activated methylene on **138** to generate the lactone moiety in **139**, followed by allylic bromination and subsequent ring closure upon removal of the *tert*-butoxycarbonyl group, affords (–)-securinine.

1.3.1 Enyne-Metathesis-Based Domino Reactions in Natural Product Synthesis **109**

Scheme 37 Synthesis of (−)-Securinine[103]

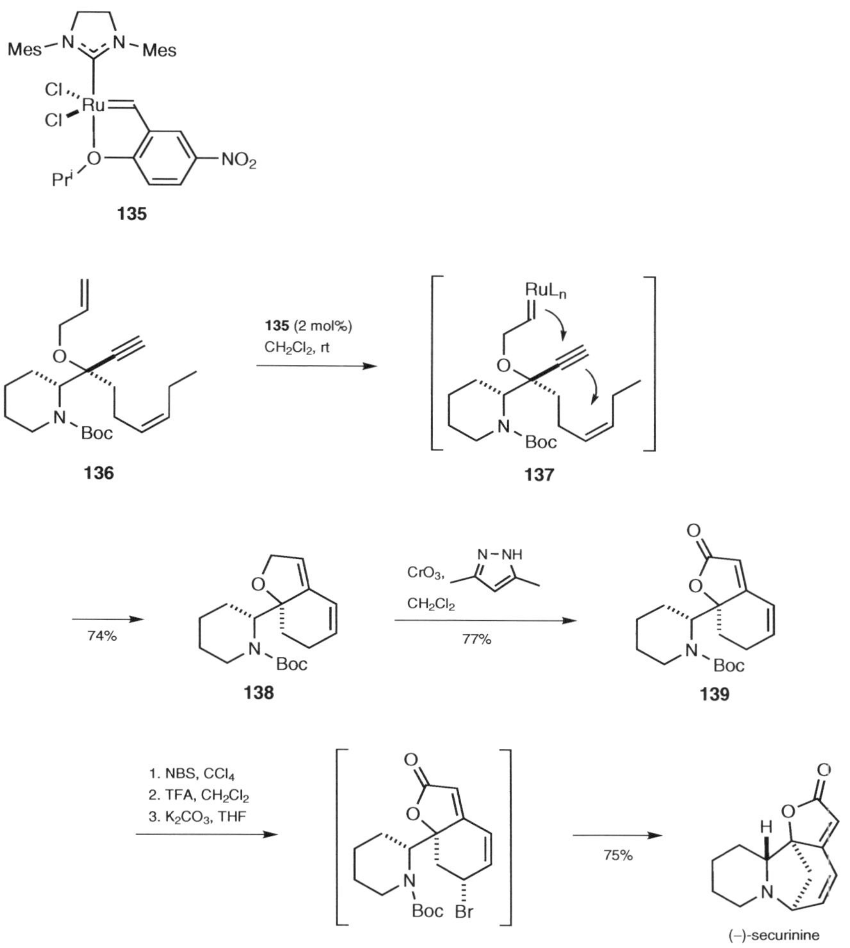

A domino double ring-closing metathesis strategy is also employed in the syntheses of (±)-erythrocarine[104] (Scheme 38) and (±)-erythravine.[105] The pentacyclic core of these natural products is obtained upon treating dienyne hydrochloride **140** with Grubbs first-generation complex **4**. Although there are two terminal alkenes within **140**, the allylic acetate significantly slows down initiation on the alkene next to it, thus the propagating alkylidene **141** should form selectively; this outcome would ultimately provide only the desired connectivity in the product **142** after double ring-closing metathesis. Upon removal of the acetate from **142**, (±)-erythrocarine (**143A**) and its epimer **143B** are obtained in good yield.

for references see p 131

Scheme 38 Synthesis of Erythrocarine[104]

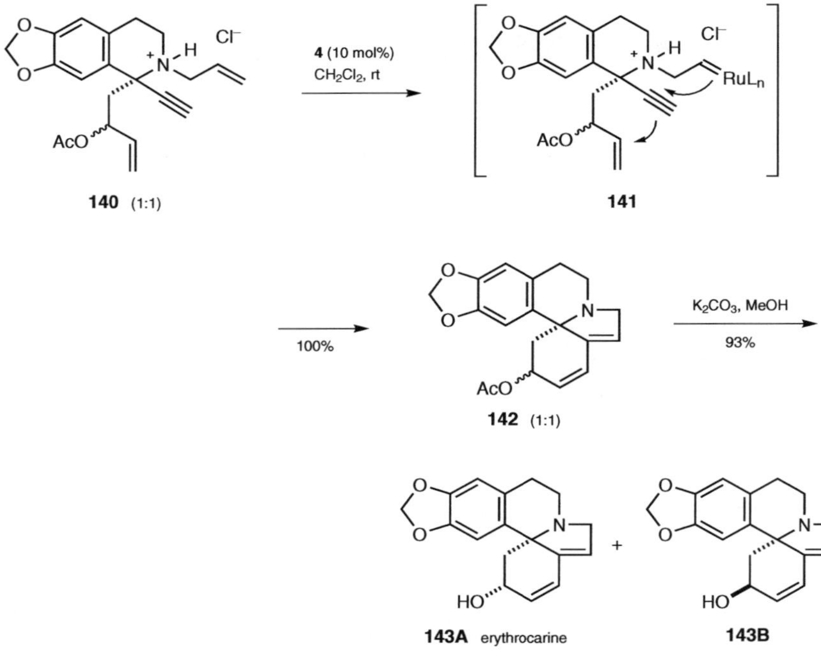

140 (1:1) → **141**

142 (1:1)

143A erythrocarine + **143B**

Using a similar concept, total syntheses of some members of the family of securinine alkaloids use enyne ring-closing metathesis as a pivotal step to construct the dihydrobenzofuranone framework (Scheme 39).[106] To initiate the domino double ring-closing metathesis from an electronically deactivated acrylate moiety, a relay ring-closing metathesis strategy is incorporated to effectively generate highly unstable propagating alkylidene **146A** or **146B**. Thus, treating individual epimers **145A** and **145B** with a more reactive version of the Hoveyda–Grubbs second-generation complex **144** at an elevated temperature provides epimers **147A** and **147B**, respectively, in similar yields. Both epimers are separately elaborated to (−)-norsecurinine (**148A**) and (+)-allonorsecurinine (**148B**), respectively, via allylic bromination followed by a deprotection-induced N-alkylation sequence. Subsequently, (+)-allonorsecurinine and (−)-norsecurinine are converted into (+)-virosaine B and flueggine A, respectively, by the proposed biosynthetic pathway wherein 3-chloroperoxybenzoic acid mediated oxidation of each compound leads to respective intermediate **149B** and **149A**, with another 3-chloroperoxybenzoic acid oxidation ultimately providing (+)-virosaine B (**151**) and flueggine A (**150**) efficiently.

Scheme 39 Synthesis of the Securinine Alkaloids[106]

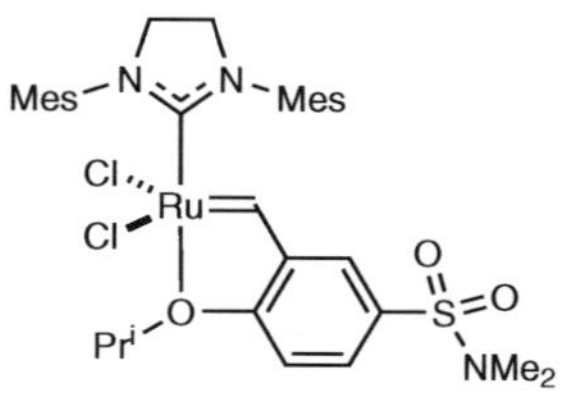

144

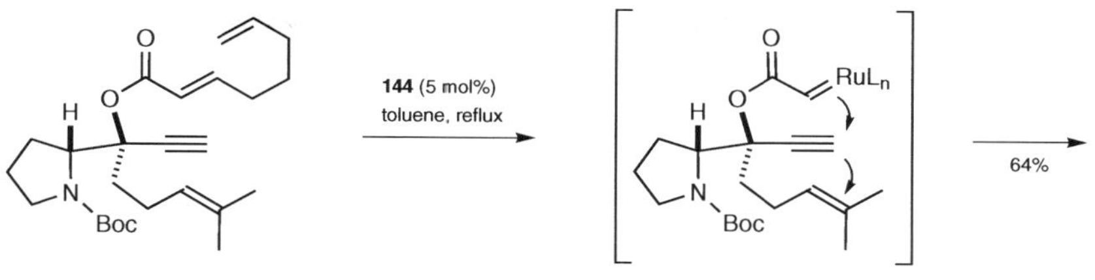

1.3.1 Enyne-Metathesis-Based Domino Reactions in Natural Product Synthesis **111**

145A
146A
64%

147A

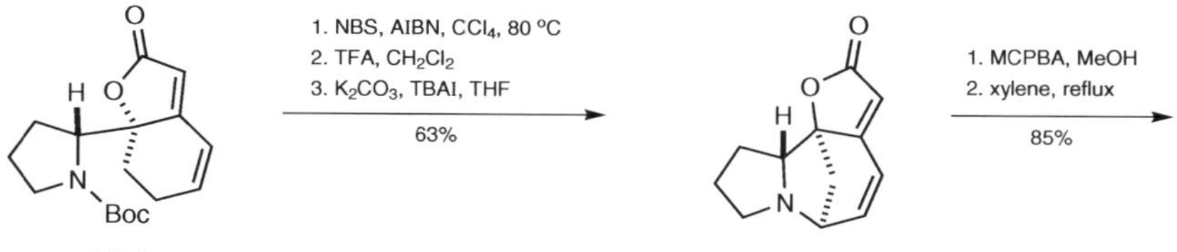

1. NBS, AIBN, CCl₄, 80 °C
2. TFA, CH₂Cl₂
3. K₂CO₃, TBAI, THF
63%

148A (−)-norsecurinine

1. MCPBA, MeOH
2. xylene, reflux
85%

149A

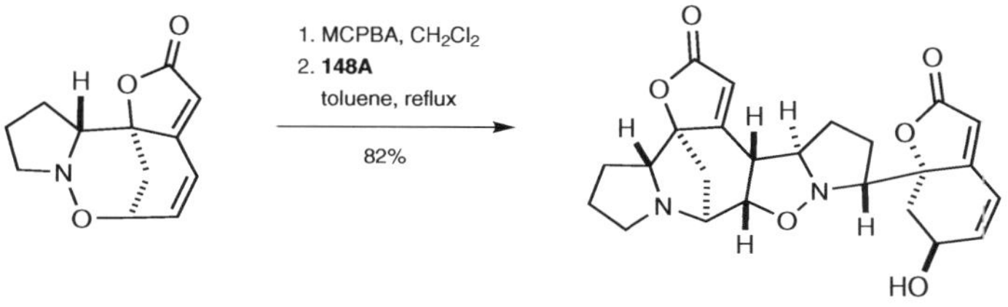

1. MCPBA, CH₂Cl₂
2. **148A**
toluene, reflux
82%

150 (−)-flueggine A

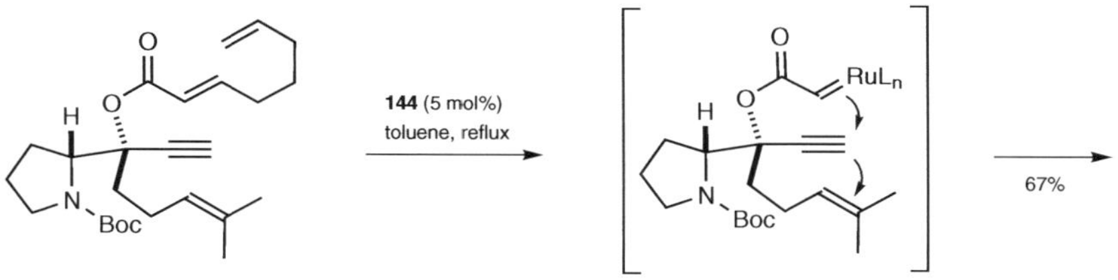

145B
146B
67%

147B

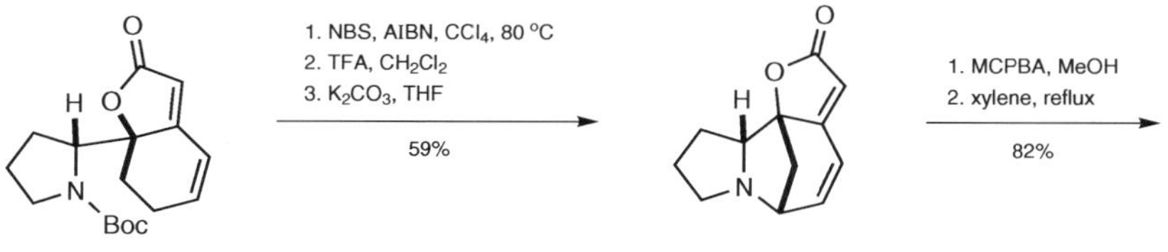

1. NBS, AIBN, CCl₄, 80 °C
2. TFA, CH₂Cl₂
3. K₂CO₃, THF
59%

148B (+)-allonorsecurinine

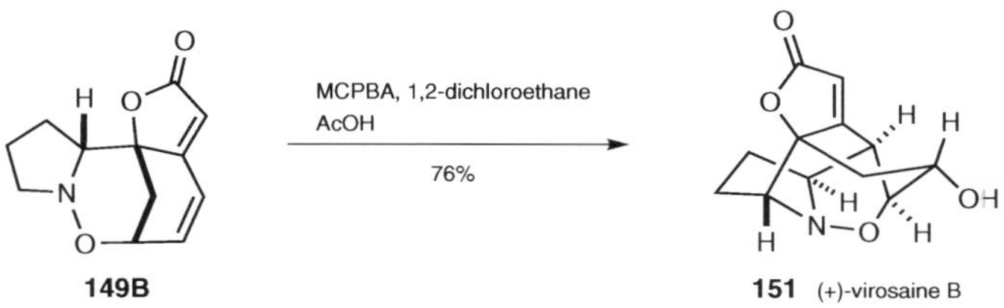

1. MCPBA, MeOH
2. xylene, reflux
82%

149B

MCPBA, 1,2-dichloroethane
AcOH
76%

151 (+)-virosaine B

for references see p 131

(3R,4R,5Z,7E,10S)-11-[2-(2,2-Diethoxyethyl)-1,3-dithian-2-yl]-4,10-dihydroxy-3-methylundeca-5,7-dienyl Pivalate (122); Typical Procedure for Double Ring-Closing Metathesis with Dienynes:[99]

A soln of dienyne **121** (70 mg, 0.11 mmol) in dry 1,2-dichloroethane (6 mL) in a Schlenk tube was degassed under vacuum by two freeze–pump–thaw cycles, and treated with Grubbs second-generation initiator **5** (9.3 mg, 0.01 mmol). The mixture was then heated at 80–90 °C for 3 h, followed by cooling to rt and concentration under reduced pressure. The residue was filtered through a small plug of silica gel (EtOAc/hexanes 1:4) to obtain a mixture of the bicyclic siloxane, monocyclic siloxane, and partially desilylated product. This mixture in THF (5 mL) was subjected to a soln of 1 M TBAF in THF (0.3 mL, 0.3 mmol) and refluxed for 4–5 h. The mixture was cooled to rt and then concentrated, and the residue was subjected to column chromatography (EtOAc/hexanes 20:80 to 35:65) to obtain the product as a yellow oil; yield: 34.2 mg (61%).

1.3.1.3.2.2   **Domino Ring-Closing Metathesis/Cross Metathesis, Cross Metathesis/Ring-Closing Metathesis, and Cross Metathesis/Cross Metathesis Sequences**

Various modes of enyne and diene metatheses can be merged by strategically tethering alkene and alkyne moieties with appropriate tethers, and then performing a ring-closing metathesis in a controlled manner in the presence of an external alkene component (Scheme 40).[107,108] For short-tethered enyne ring-closing metathesis, the rate of ring-closing metathesis is generally higher than that of diene cross metathesis or enyne cross metathesis; thus, a domino enyne ring-closing metathesis/diene cross metathesis can be achieved with good efficiency, the synthetic applications of which are illustrated in Schemes 41 and 42. On the other hand, for enynes with a long tether, the enyne ring-closing metathesis to form macrocycles is typically slower than the enyne cross metathesis with ethene, and in this case, an enyne cross metathesis/diene ring-closing metathesis domino sequence generates a formal *endo*-mode enyne ring-closing metathesis product. An excellent synthetic application is illustrated with the synthesis of longithorone A in Scheme 43. If the alkene components are not tethered to the reacting alkyne component, then two consecutive domino diene cross metathesis/diene cross metathesis will be a typical mode of domino process; an application of this process is illustrated in Scheme 44.

Scheme 40   Different Modes of Domino Metathesis[107,108]

ring-closing metathesis/cross metathesis

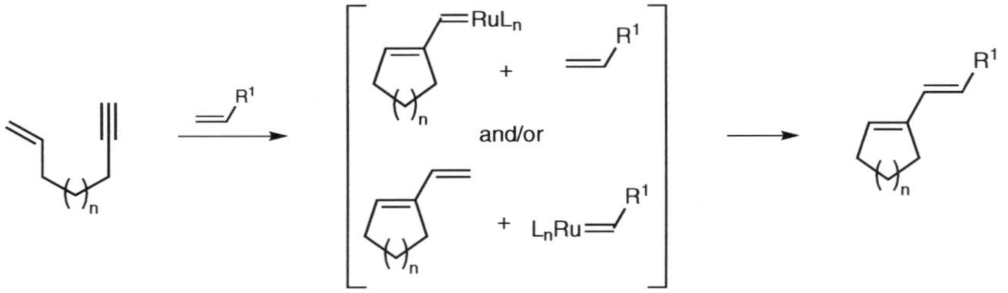

1.3.1 Enyne-Metathesis-Based Domino Reactions in Natural Product Synthesis **113**

cross metathesis/ring-closing metathesis

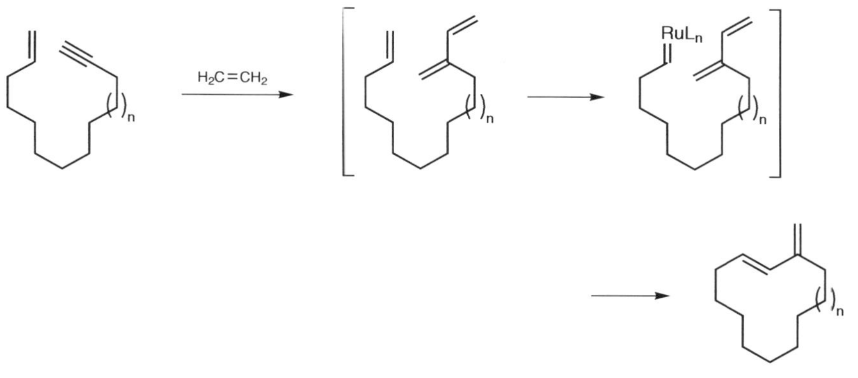

cross metathesis/cross metathesis

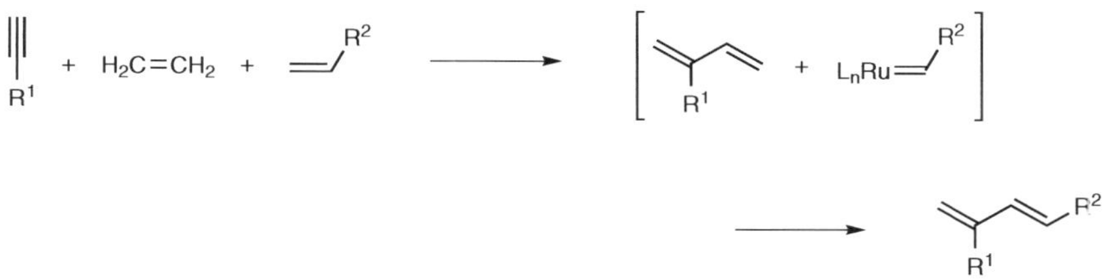

A concise total synthesis of (+)-8-*epi*-xanthatin relies on a domino ring-closing metathesis/ cross metathesis sequence (Scheme 41).[109] Under this metathesis regime, 1,8-enyne **152** undergoes ring-closing metathesis to form a propagating alkylidene **154**, a species which most likely is in equilibrium with ring-closing metathesis product **155**, and slowly participates in a cross metathesis with methyl vinyl ketone to generate (+)-8-*epi*-xanthatin. Probably, some portion of the initially formed **154** from **152** may react with methyl vinyl ketone (**153**) to generate the final product directly without the involvement of **155** in the domino metathesis catalytic cycle. A related strategy for the synthesis of (−)-11α,13-dihydroxanthatin showcases the use of enyne metathesis starting from enyne **156**, but here the ring-closing metathesis and cross metathesis reactions are interrupted by another synthetic manipulation after the formation of ring-closing metathesis product **157**.[110]

Scheme 41 Synthesis of 8-*epi*-Xanthatin and (−)-11α,13-Dihydroxanthatin[109,110]

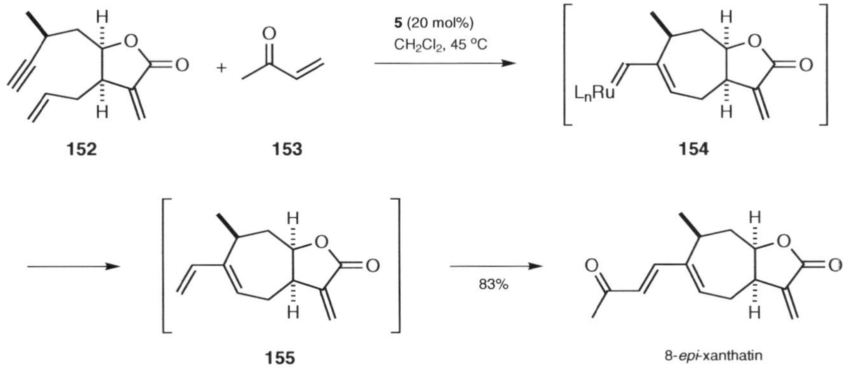

for references see p 131

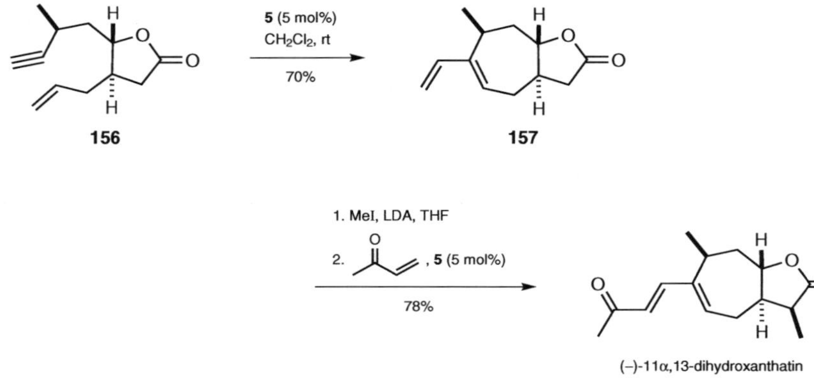

156 → **157**

5 (5 mol%)
CH₂Cl₂, rt
70%

1. MeI, LDA, THF

2. (acetyl vinyl), **5** (5 mol%)
78%

(–)-11α,13-dihydroxanthatin

A total synthesis of (+)-panepophenanthrin is accomplished employing enyne metathesis combined with other types of metathesis processes, such as diene ring-closing metathesis and cross metathesis (Scheme 42).[111] To improve the efficiency of metathesis in this case, a relay-metathesis strategy was incorporated in the design of enyne substrate **158**. It was expected that propagating alkylidene **159** should be formed in the first step and subsequently **160** would be generated after extrusion of dihydrofuran. This species then participates in cross metathesis with but-3-en-2-ol, providing **161**. Without the relay device, the metathesis of simpler 1,7-enyne **162** in the presence of but-3-en-2-ol would first form alkylidene **163**, due to the activating role of an allylic hydroxy group, and this would then undergo cross metathesis with **162** to generate **164**, and finally the ring-closing metathesis of which would deliver product **161**. Elaboration of **161** through simple functional-group and protecting-group manipulation leads to monomer **165**, which spontaneously dimerizes to form (+)-panepophenanthrin.[112,113]

Scheme 42   Synthesis of Panepophenanthrin[111–113]

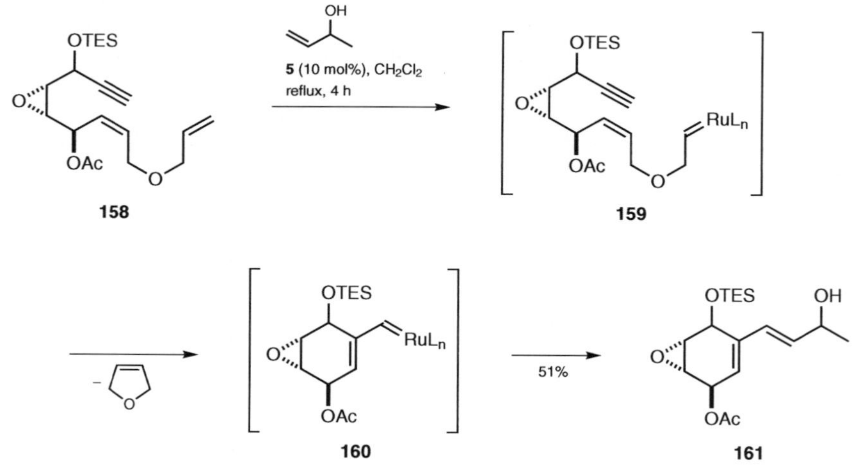

158 → **159**

OH

5 (10 mol%), CH₂Cl₂
reflux, 4 h

160 → **161**

51%

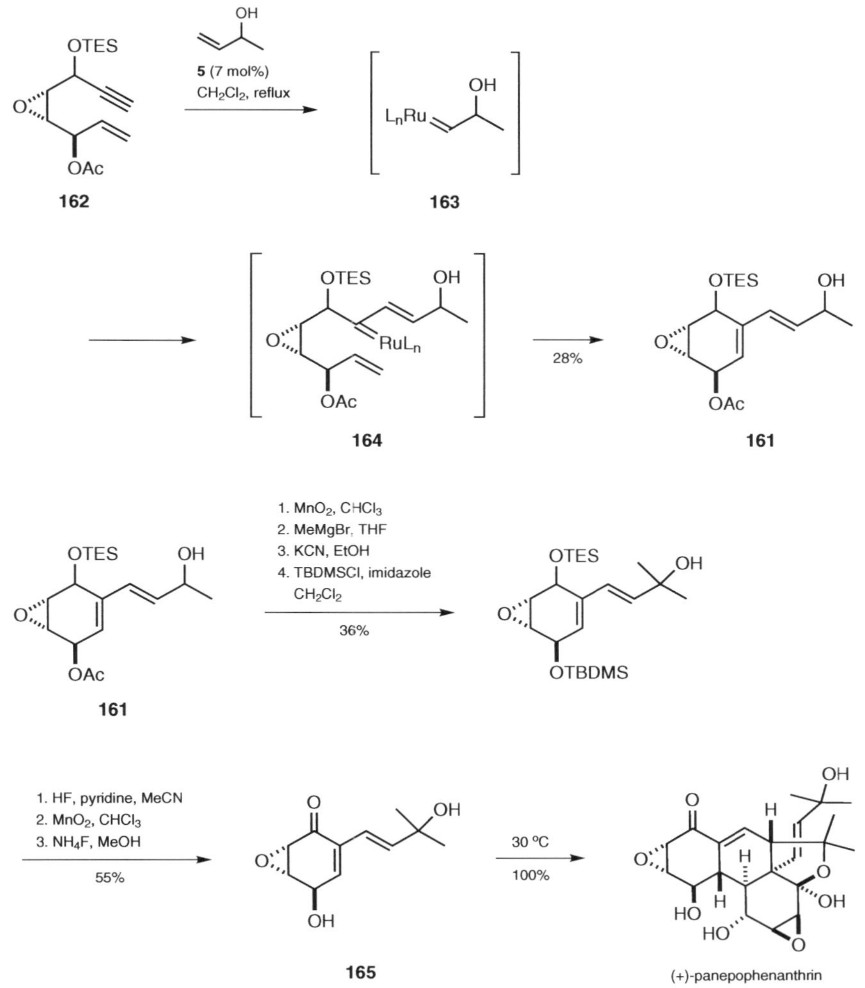

A formal *endo*-selective macrocyclic enyne ring-closing metathesis was employed as a pivotal strategy in the total synthesis of (−)-longithorone A (Scheme 43).[114] Enyne ring-closing metathesis of **166** under ethene is known to proceed through fast cross metathesis of the terminal alkyne with ethene followed by the formation of alkylidene **167**; this species should undergo slow ring-closing metathesis engaging the less sterically hindered monosubstituted double bond of the initially formed 1,3-diene moiety, providing a formal *endo*-mode ring-closing metathesis product in the form of **168**.[115] Similarly, macrocyclic enyne ring-closing metathesis of **170** under identical conditions generates macrocycle **171**, albeit in marginal yield even at high catalyst loading. For the Diels–Alder heterodimerization, cyclophanes **168** and **171** are converted into penultimate Diels–Alder precursors **169** and **172** via a three-step sequence for each. The Lewis acid promoted initial Diels–Alder reaction between diene **169** and enal **172** affords the desired product as a 1:1.4 mixture of diastereomers, and is followed by removal of both silyl groups and oxidation of the resulting phenolic intermediate to the corresponding quinones. The second Diels–Alder reaction ensues directly, providing (−)-longithorone A in excellent yield.

for references see p 131

Scheme 43 Synthesis of (−)-Longithorone A[114]

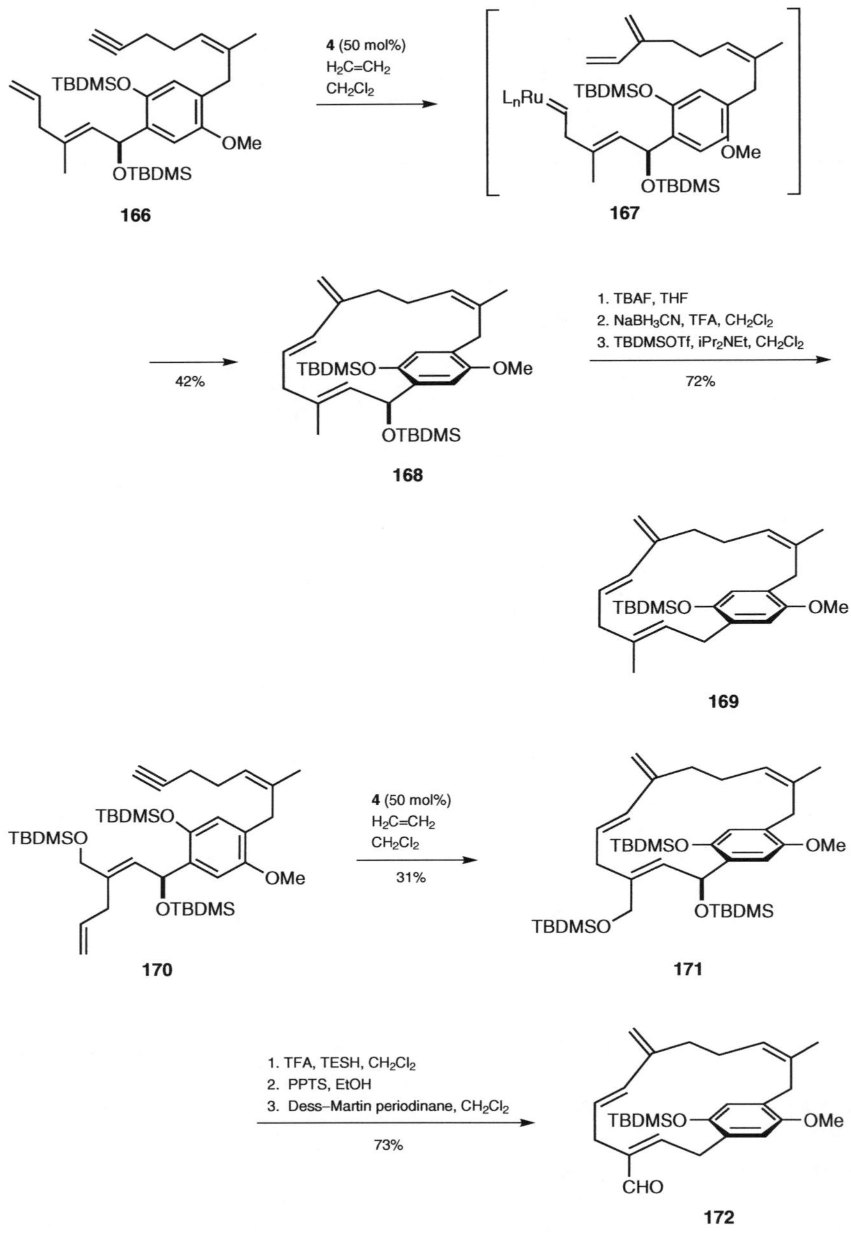

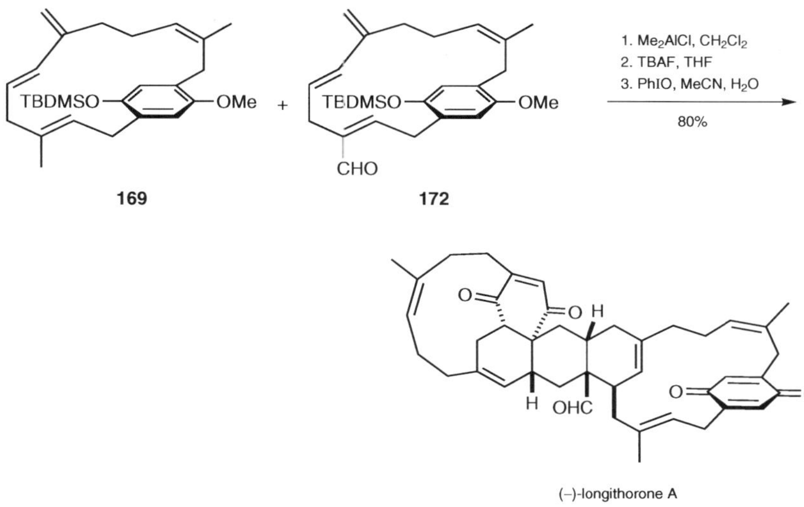

(−)-longithorone A

A domino enyne cross metathesis and diene cross metathesis sequence is employed in the synthesis of (−)-amphidinolide E for the installation of a 1,3-diene moiety in the side chain. Treating alkyne **173** with Grubbs second-generation complex **5** under ethene followed by subsequent cross metathesis of the resulting intermediate 1,3-diene **175** with another alkene (2-methylpenta-1,4-diene) affords 1,3-diene **176** and residual **175** (Scheme 44).[116,117] Mechanistically, there are two plausible metathesis pathways that involve different propagating species **174** and **178**; however, due to the well-known preference for the formation of an internal alkylidene **174** over **178**, the initial cross metathesis should generate 1,3-diene **175** as the sole product instead of directly forming triene **176** by the merger of **178** and 2-methylpenta-1,4-diene. In the second round of cross metathesis, the initially formed 1,3-diene **175** reacts with a major propagating species **177** that is generated from 2-methylpenta-1,4-diene to yield triene product **176**, rather than generating vinyl alkylidene **178** followed by its reaction with 2-methylpenta-1,4-diene. Once generated, triene **176** is elaborated to sulfone **179** via a four-step sequence. This material is then coupled with aldehyde **180** to install the required *E*-double bond in **181**. Compound **181** is converted in 4 steps into advanced intermediate **182**, and then macrolactonization and final deprotection of the acetonide completes the total synthesis of (−)-amphidinolide E.

for references see p 131

118 Domino Transformations **1.3** Metathesis Reactions

Scheme 44 Synthesis of Amphidinolide E[116,117]

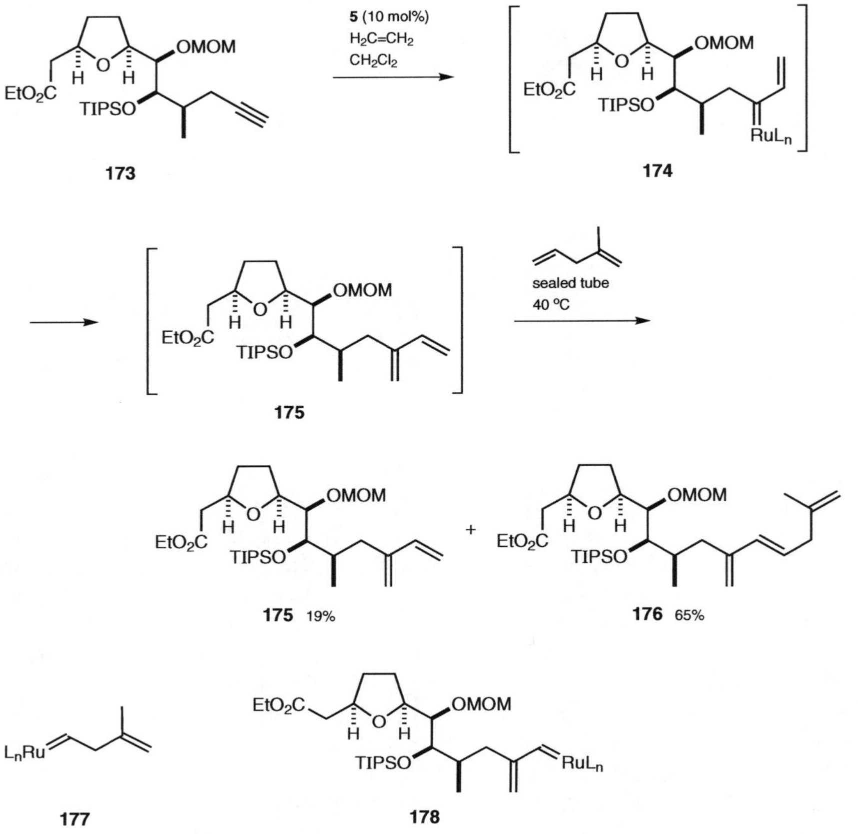

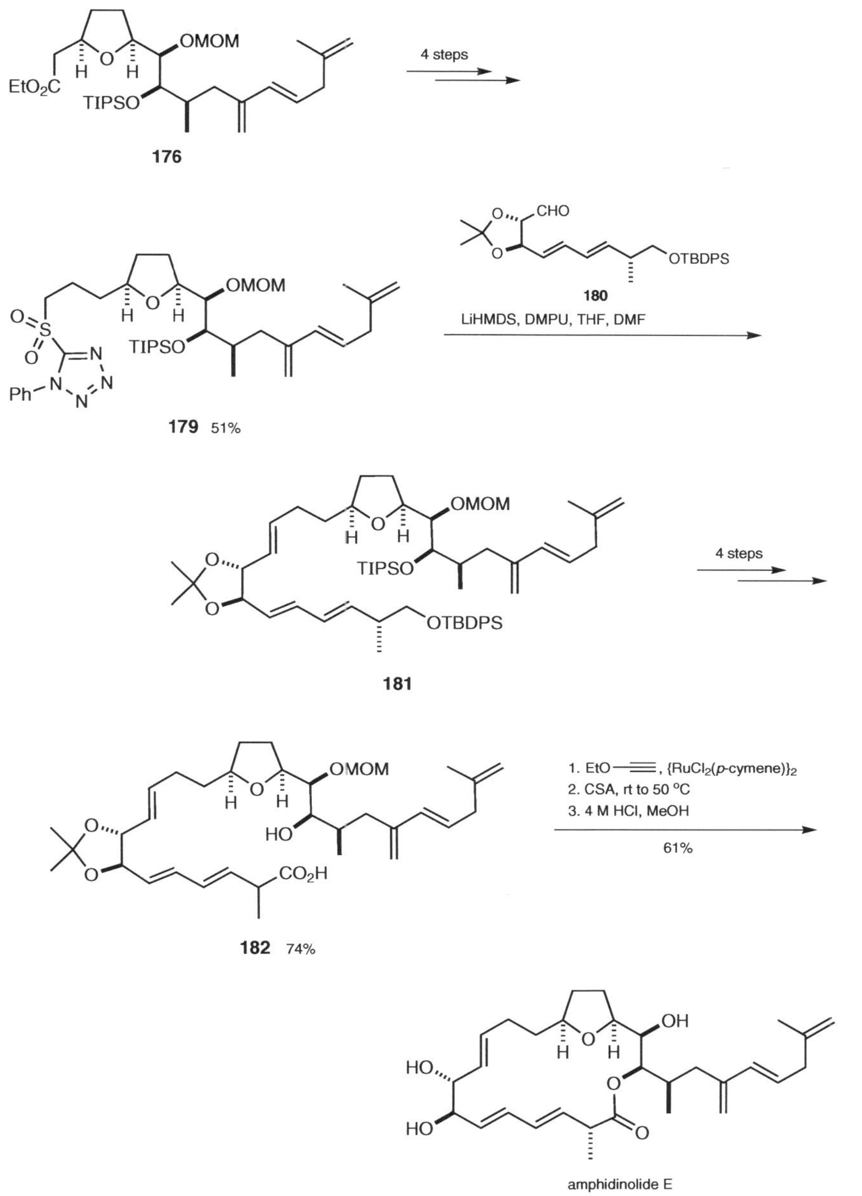

1.3.1 Enyne-Metathesis-Based Domino Reactions in Natural Product Synthesis **119**

176

4 steps

179 51%

180

LiHMDS, DMPU, THF, DMF

181

4 steps

182 74%

1. EtO—≡, {RuCl$_2$(*p*-cymene)}$_2$
2. CSA, rt to 50 °C
3. 4 M HCl, MeOH

61%

amphidinolide E

**(*E*)-4-[(3*R*,4*S*,5*S*)-3-Acetoxy-4,5-epoxy-6-(triethylsiloxy)cyclohex-1-enyl]but-3-en-2-ol (161);
Typical Procedure:**[111]

To a soln of enyne **158** (80 mg, 0.21 mmol) and but-3-en-2-ol (38 mg, 0.52 mmol) in CH$_2$Cl$_2$
(3 mL) was added a soln of the Grubbs second-generation initiator **5** (18 mg, 0.02 mmol) in
CH$_2$Cl$_2$ (2 mL) under N$_2$. The Schlenk tube was then placed in an oil bath at 50 °C and the
reaction was run under reflux for 4 h. After concentration under reduced pressure, the
residue was purified by chromatography (silica gel) to give the cyclized products as a colorless oil; yield: 38 mg (51%); dr 2:1.

for references see p 131

1.3.1.3.3 Enyne Metathesis/Metallotropic [1,3]-Shift Sequences

A domino bond-forming strategy involving enyne ring-closing metathesis initiated metallotropic [1,3]-shift is considered to be ideal for the construction of the 1,5-dien-3-yne moiety contained in many epoxyquinoids such as (+)-asperpentyn, (−)-harveynone, and (−)-tricholomenyn A (Scheme 45).[118] The prowess of this enyne ring-closing metathesis/metallotropic [1,3]-shift domino sequence is illustrated by the concise synthesis of these natural products. Compared to other previously reported synthetic approaches that accommodate a late-stage Sonogashira-type coupling reaction between a preformed cyclohexene derivative and a suitable alkyne moiety, the metathesis-based approach is unique because it directly installs the alkyne moiety on the cyclohexene core while the cyclohexene is constructed. Here, to construct a fully conjugated 1,5-dien-3-yne moiety, relay-metathesis-based initiation of a domino sequence with enediyne **183** provides separable epimeric mixtures of **187A** and **187B** through the formation of putative propagating species **184–186**. Subsequent coordinated protecting-group manipulations and appropriate oxidation level adjustment of these metathesis products, in operations dependent on their stereochemistry and the structural characteristics, afforded total syntheses of epoxyquinoid natural products (+)-asperpentyn and (−)-tricholomenyn A in just one- and two-step manipulations. On the other hand, due to the difficulty in removing the acetate at the keto-epoxide stage, the acetate was replaced with triisopropylsilyl to generate **188** and elaborated to (−)-harveynone via a three-step sequence.

Scheme 45 Synthesis of (+)-Asperpentyn, (−)-Harveynone, and (−)-Tricholomenyn A[118]

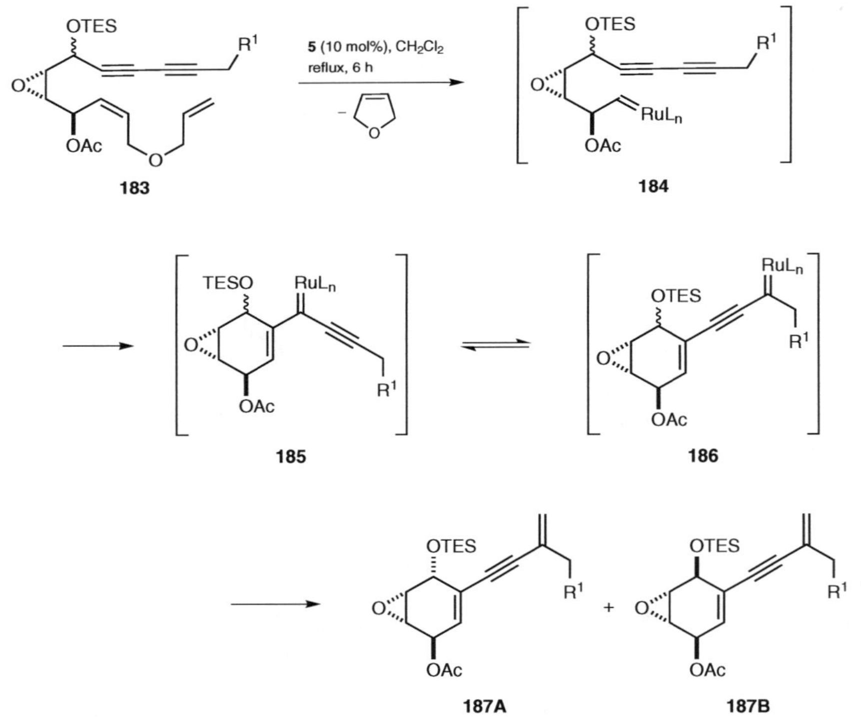

R^1	Yield (%) of **187A/187B**	Ref
H	62	[118]
CH$_2$CH=CMe$_2$	58	[118]

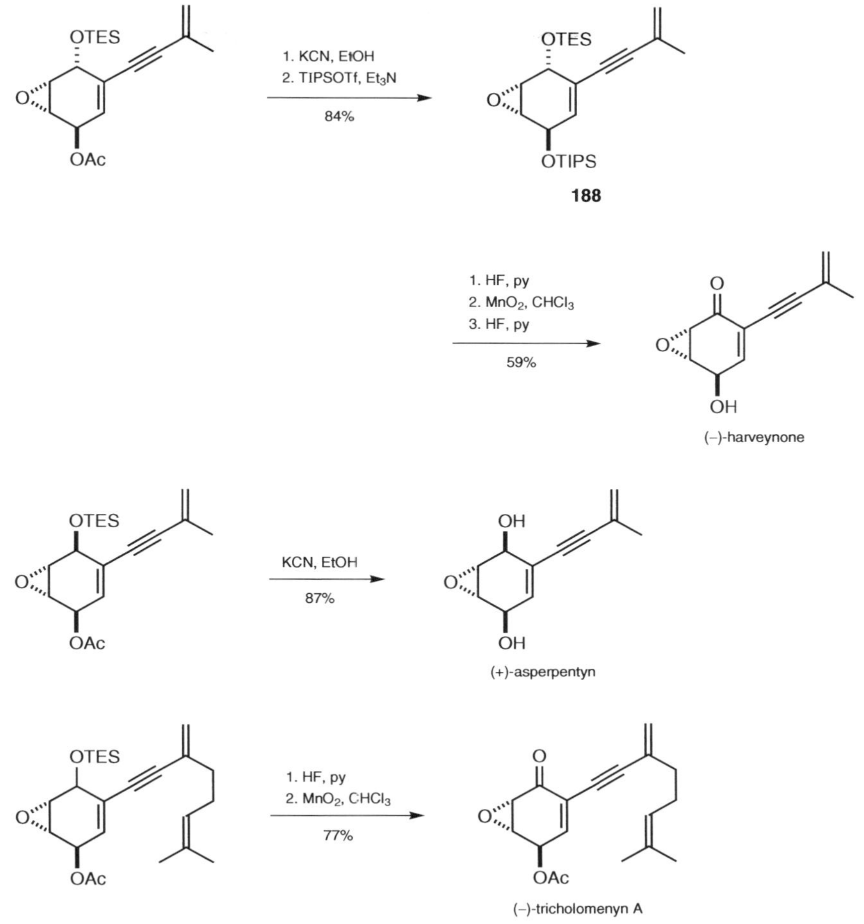

1.3.1 Enyne-Metathesis-Based Domino Reactions in Natural Product Synthesis **121**

The utility of the metallotropic-[1,3]-shift-based domino bond-forming strategy has been further employed in the novel synthesis of (3*R*,9*R*,10*R*)-panaxytriol (Scheme 46).[119] In this approach, a domino relay metathesis, double metallotropic [1,3]-shift, and cross metathesis sequence starting from terminal 1,3-diyne **189** provides enediyne **193** with *Z* selectivity when promoted by Grubbs second-generation complex **5** in the presence of excess external alkene **190**. The sequence proceeds through the formation of propagating alkylidene species **191** and then its double-metallotropic shifted regioisomer **192** in the penultimate stage. Although metallotropic [1,3]-shifts of ruthenium alkylidenes are reversible, in general, it is found that terminal propargylic alkylidenes, such as **192**, once formed tend not to undergo metallotropic shift in the reverse direction. Finally, the alkene moiety of **193** is elaborated through a formal [1,3]-allylic alcohol transposition via **194** to complete a total synthesis of (3*R*,9*R*,10*R*)-panaxytriol.

for references see p 131

Scheme 46 Synthesis of Panaxytriol[119]

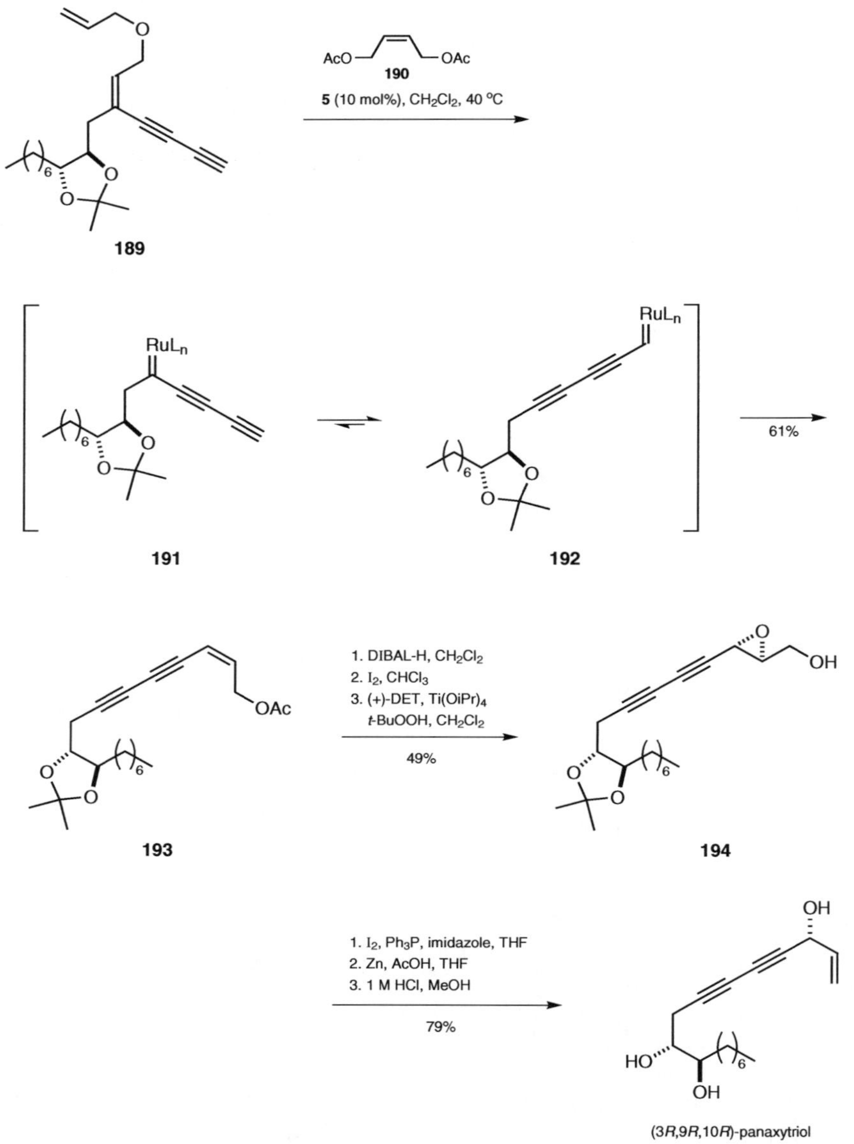

An elegant display of domino relay cross metathesis is shown in the synthesis of (Z)-laureatin (Scheme 47).[120] Although a direct enyne metathesis is not involved in the sequence, the formation of an alkynyl alkylidene from an enyne precursor is quite similar to the domino process in Scheme 50. Due to the preferred Z-alkene formation in the cross metathesis involving an alkynyl alkylidene, cross metathesis between **195** and **196**, as promoted by the Grubbs third-generation complex **6**, provides Z-enyne **197** as the major isomer. Because of the preferred ring-closing metathesis of **198** to generate penultimate propagating alkylidene **199** via dihydrofuran extrusion, terminal alkene **195** does not compete by initiating alternative cross-metathesis events other than putative slow alkylidene exchanges with itself or **196**.

1.3.1 Enyne-Metathesis-Based Domino Reactions in Natural Product Synthesis **123**

Scheme 47 Domino Relay Cross Metathesis in the Synthesis of (Z)-Laureatin[120]

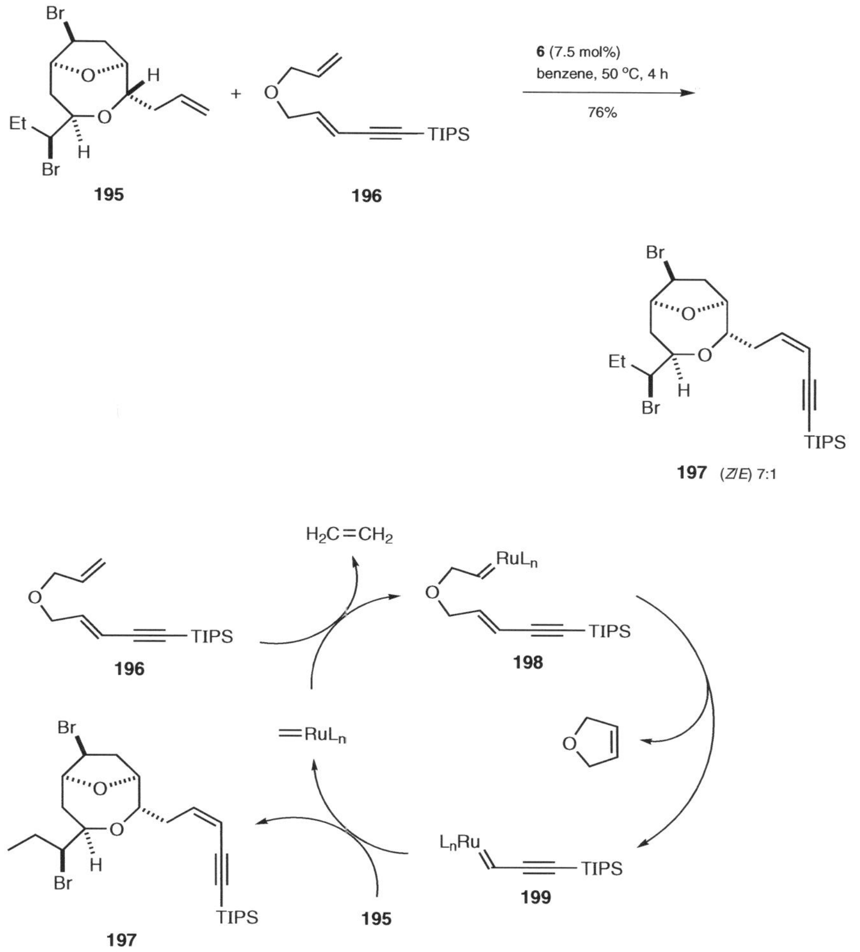

4-[(3R,4S,5S,6R/S)-3-Acetoxy-4,5-epoxy-6-(triethylsiloxy)cyclohex-1-enyl]-2-methylbut-1-en-3-yne (187, R¹ = H); General Procedure for Enyne Metathesis/Metallotropic [1,3]-Shift Sequence:[118]

A soln of diyne **183** (R¹ = H; 82 mg, 0.19 mmol) in CH₂Cl₂ (50 mL) was added dropwise to a soln of Grubbs second-generation initiator **5** (15 mg, 0.02 mmol) under N₂. During the addition, the mixture was stirred under reflux for 6 h. After removal of the solvent under reduced pressure, the residue was purified by chromatography (silica gel, pentane/Et₂O 10:1) to afford the product as a colorless oil; yield: 41 mg (62%).

1.3.1.3.4 Enyne Metathesis/Diels–Alder Reaction Sequences

A bond-making process that involves in situ generation of a 1,3-diene and its subsequent trapping with a dienophile should constitute an elegant domino reaction strategy, one which would have a significant potential in complex molecule synthesis. There are many examples demonstrating this concept with simple substrates, but only a small number of natural product total syntheses have been reported that rely on this domino transformation.

for references see p 131

One of the simplest and earliest total syntheses of a natural product using domino enyne ring-closing metathesis/Diels–Alder reaction is shown in Scheme 48.[36] The enyne ring-closing metathesis of an ester-linked 1,6-enyne **200** provides vinylbutenolide **202** in 40% yield. This material spontaneously undergoes Diels–Alder dimerization at 50 °C to generate (±)-differolide in 91% yield. The existence of the initially formed ruthenium alkylidene **201** has been confirmed by NMR monitoring of the reaction, an outcome which implies that the ring-closure rate of this intermediate is relatively slow, probably because of the difficulty in attaining the necessary, but unfavorable, *s-cis* conformation of the ester moiety to reach a ring-closing metathesis transition state. Of note, although the existence of alkylidene **201** does not exclude the "yne-first" pathway (see Section 1.3.1.1.1), it is more likely that the "ene-first" pathway is operating in this ring-closing metathesis.

Scheme 48   Synthesis of (±)-Differolide[36]

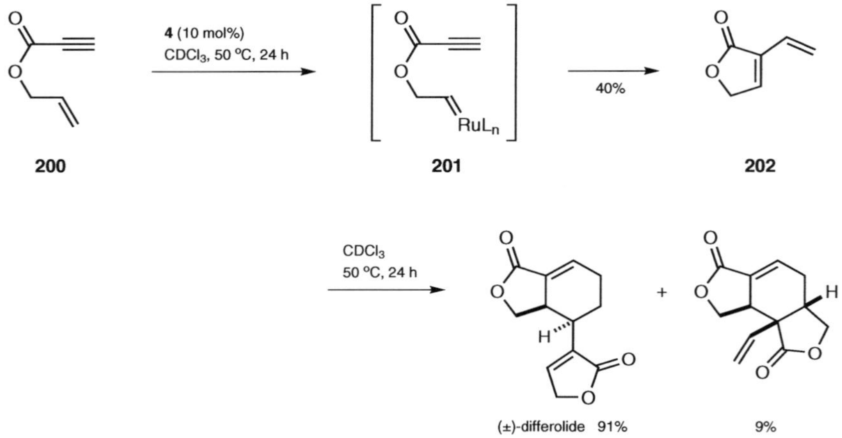

An enyne ring-closing metathesis/Diels–Alder sequence is also used in the synthesis of the GABA$_A$ receptor modulator valerenic acid (Scheme 49).[121] The initial enyne ring-closing metathesis of **203** yields vinylcyclopentene **204** with Grubbs first-generation complex **4** under an ethene atmosphere. To ensure the regio- and stereoselectivity, intermolecular Diels–Alder cycloaddition is carried out after removing the *tert*-butyldimethylsilyl group from **204** to give alcohol **205**. Reaction of **205** with methyl acrylate in the presence of magnesium bromide gives the Diels–Alder adduct, which is treated with diisobutylaluminum hydride and subsequently with a methylphosphorane to generate homologated ester **206**. Directed hydrogenation to form a *trans*-hydrindane skeleton, followed by oxidation of the secondary alcohol, provides ketone **207**. Formation of a trifluoromethanesulfonate from the ketone followed by its coupling with dimethylzinc yields **208**, from which valerenic acid was obtained via ethyl ester hydrolysis.

1.3.1 Enyne-Metathesis-Based Domino Reactions in Natural Product Synthesis

Scheme 49 Synthesis of Valerenic Acid[121]

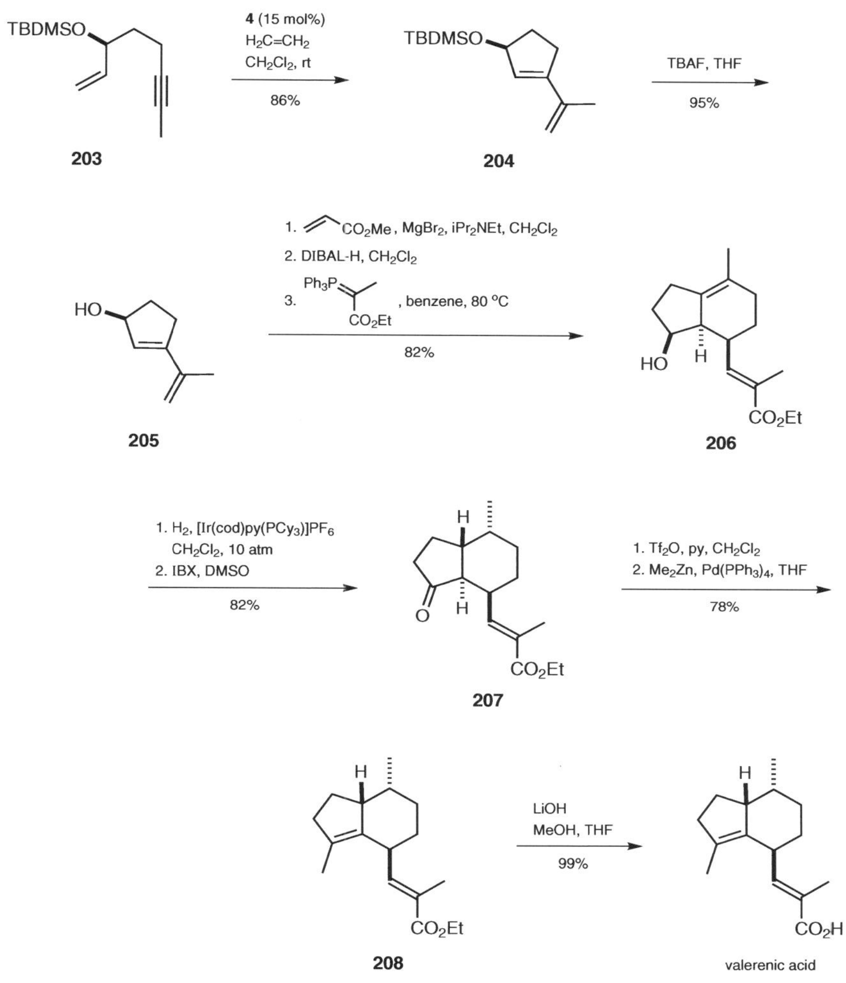

IBX = 2-iodoxybenzoic acid

Benz[*a*]anthraquinone derivatives can be effectively accessed via a domino sequence of enyne metatheses to generate 1,3-diene substructures followed by their direct engagement in Diels–Alder cycloaddition. This strategy has been exploited in the synthesis of several angucyclinone-type natural products (Scheme 50).[122,123] Treatment of the readily available chiral 1,7-enyne **209** with the Grubbs second-generation complex **5** affords cyclohexene-based 1,3-diene **210**. This material is subjected in situ to a Diels–Alder cycloaddition with 2-bromonaphtho-1,4-quinone **211**, generating cycloadduct **212**. Base-induced elimination of hydrogen bromide effects spontaneous aromatization to provide a tetracyclic anthraquinone derivative **213** as an inseparable 9:1 mixture of regioisomers.[124] Singlet-oxygen-based benzylic oxidation, followed by acetate removal, provides the natural product YM-181741. Following the same protocol but with slightly different reactants, total syntheses of other members of angucyclinone-type natural products have been realized including (+)-rubiginone B2 and (+)-tetrangomycin, thereby highlighting the power of the overall approach.

for references see p 131

Scheme 50 Synthesis of Angucyclinone-Type Natural Products[122–124]

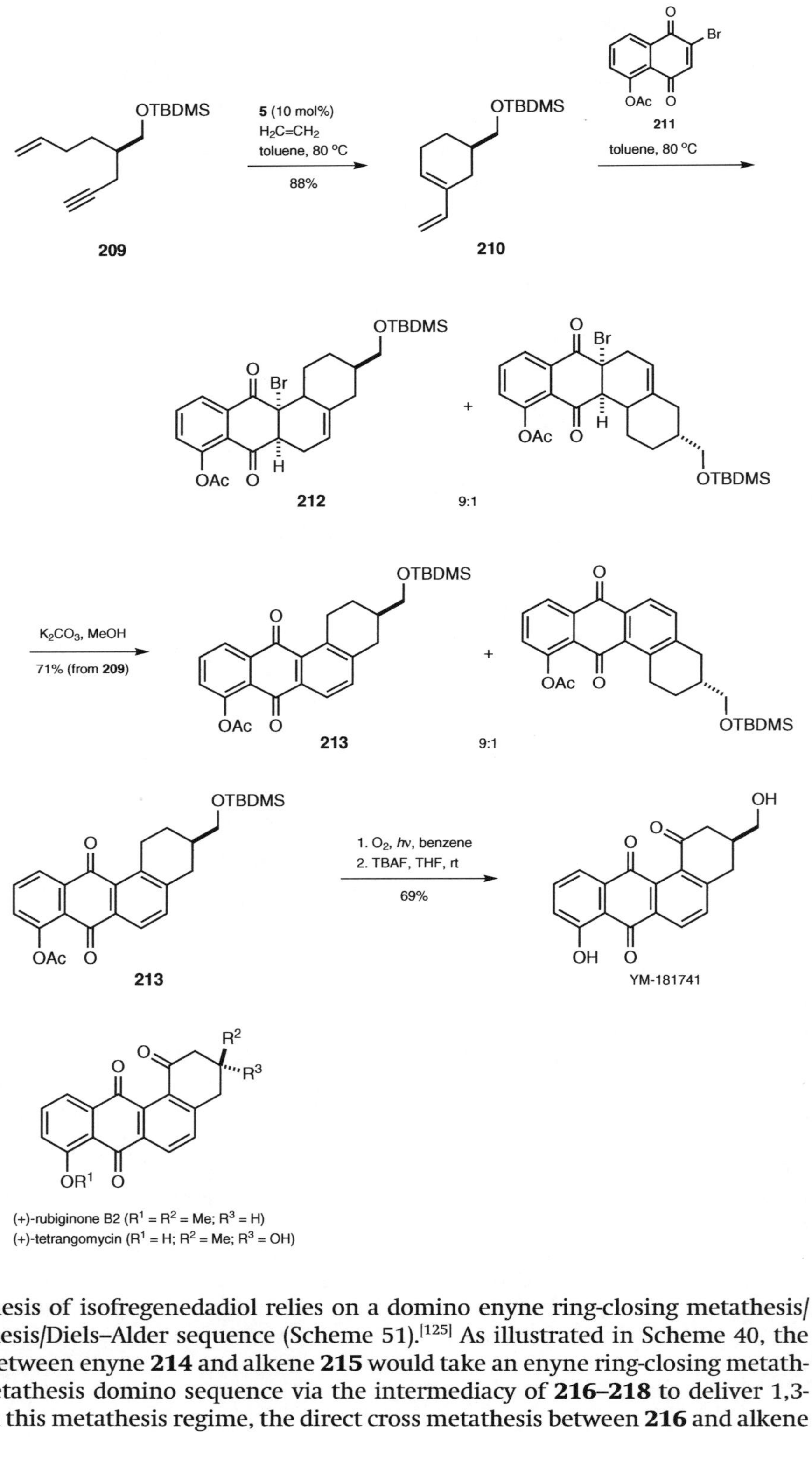

A total synthesis of isofregenedadiol relies on a domino enyne ring-closing metathesis/ cross metathesis/Diels–Alder sequence (Scheme 51).[125] As illustrated in Scheme 40, the metathesis between enyne **214** and alkene **215** would take an enyne ring-closing metathesis/cross metathesis domino sequence via the intermediacy of **216–218** to deliver 1,3-diene **219**. In this metathesis regime, the direct cross metathesis between **216** and alkene

1.3.1 Enyne-Metathesis-Based Domino Reactions in Natural Product Synthesis **127**

217 is less likely to be operating. Once 1,3-diene **219** is generated, its Diels–Alder reaction readily takes place to deliver the final product **220**, following in situ oxidation of the Diels–Alder adduct with 2,3-dichloro-5,6-dicyanobenzo-1,4-quinone. The conversion of the esters into benzylic bromides, hydrogenolytic cleavage of those benzylic bromides, removal of the *tert*-butyldimethylsilyl group, and a final acetylation of the primary alcohol delivers isofregenedadiol in good yield.

Scheme 51 Synthesis of Isofregenedadiol[125]

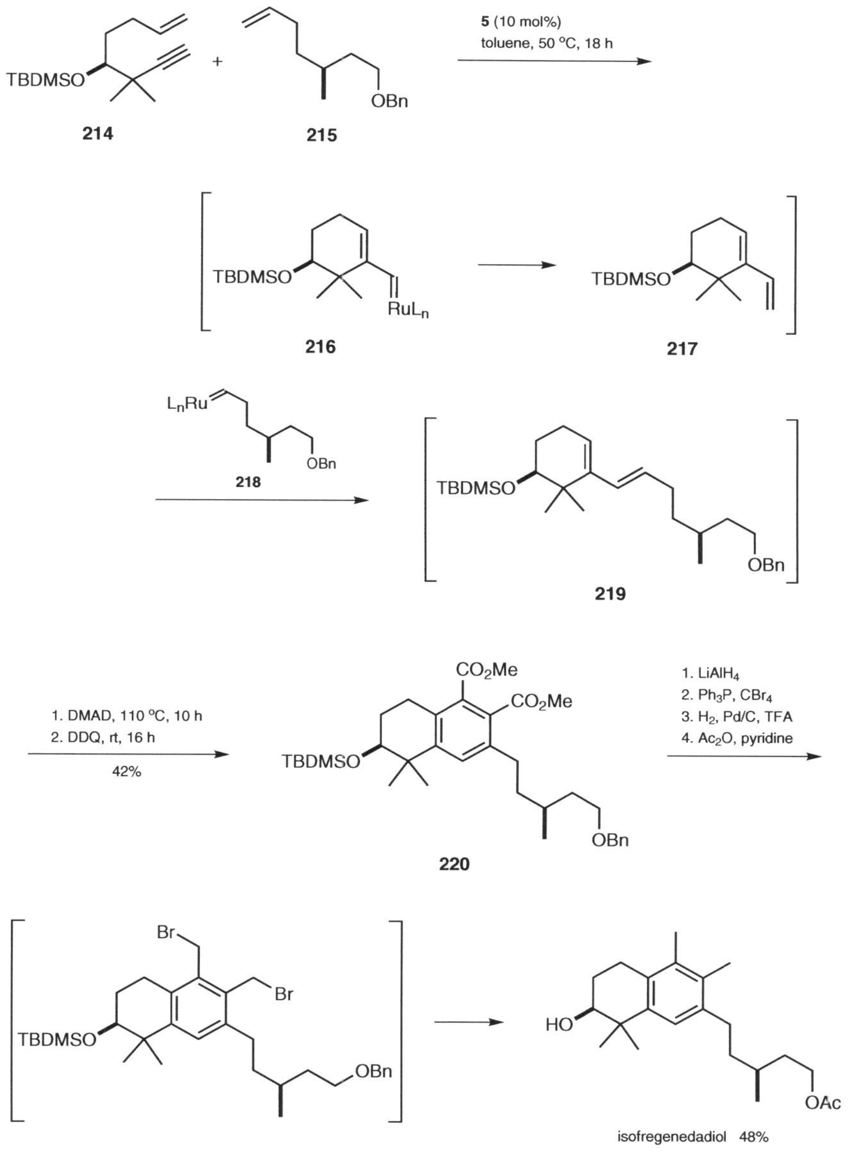

for references see p 131

(S)-3-(tert-Butyldimethylsiloxy)-1-(prop-1-en-2-yl)cyclopentene (204); Typical Procedure for Enyne Metathesis/Diels–Alder Reaction Sequence:[121]

A soln of enyne **203** (263.8 mg, 1.106 mmol) in degassed CH_2Cl_2 (37 mL) was cooled to −78 °C. Ethene gas was bubbled through the soln for 10 min. Grubbs first-generation initiator **4** (136.4 mg, 0.1657 mmol) was added and once again ethene gas was bubbled through the resulting mixture for 1 min. The ice bath was removed and the mixture was stirred for 24 h. The flask was flushed with argon and the solvent was evaporated. Chromatography of the residue (pentane/Et_2O 200:1) furnished the product as a clear, colorless oil; yield: 226.6 mg (86%).

1.3.1.3.5 Enyne Metathesis with π-Lewis Acids

Metathesis of enynes is known to be catalyzed by π-acidic transition-metal complexes that do not contain metal alkylidene functionality. Trost and Fürstner are pioneers in this field and significant advances have been made by their research. For example, Trost and co-workers have developed an enyne-metathesis-based synthetic method to construct macrocyclic bridged bicycles. Relying on this methodology, an efficient synthetic strategy was developed for roseophilin (Scheme 52).[126] For macrocycle construction in this case, monocyclic enyne **221** is treated with platinum(II) chloride to generate *exo*-type-I product **222** in quantitative yield. This product is converted into a diastereomeric mixture of epoxides **223** via bromohydrin formation with opening of the epoxide via S_N2' addition of isopropylmagnesium bromide as catalyzed by copper(I) bromide affording allylic alcohols **224**. Oxidation and double-bond isomerization results in enone **225**, which is subjected to hydroboration–oxidation followed by tetrapropylammonium perruthenate oxidation to provide 1,4-dione **226**. N-Protected pyrrole formation from the 1,4-diketone in **226**, followed by desilylation and oxidation results in Fürstner's advanced intermediate **227**,[127] thereby completing a formal synthesis of roseophilin.

1.3.1 Enyne-Metathesis-Based Domino Reactions in Natural Product Synthesis **129**

Scheme 52 Synthesis of Roseophilin[126]

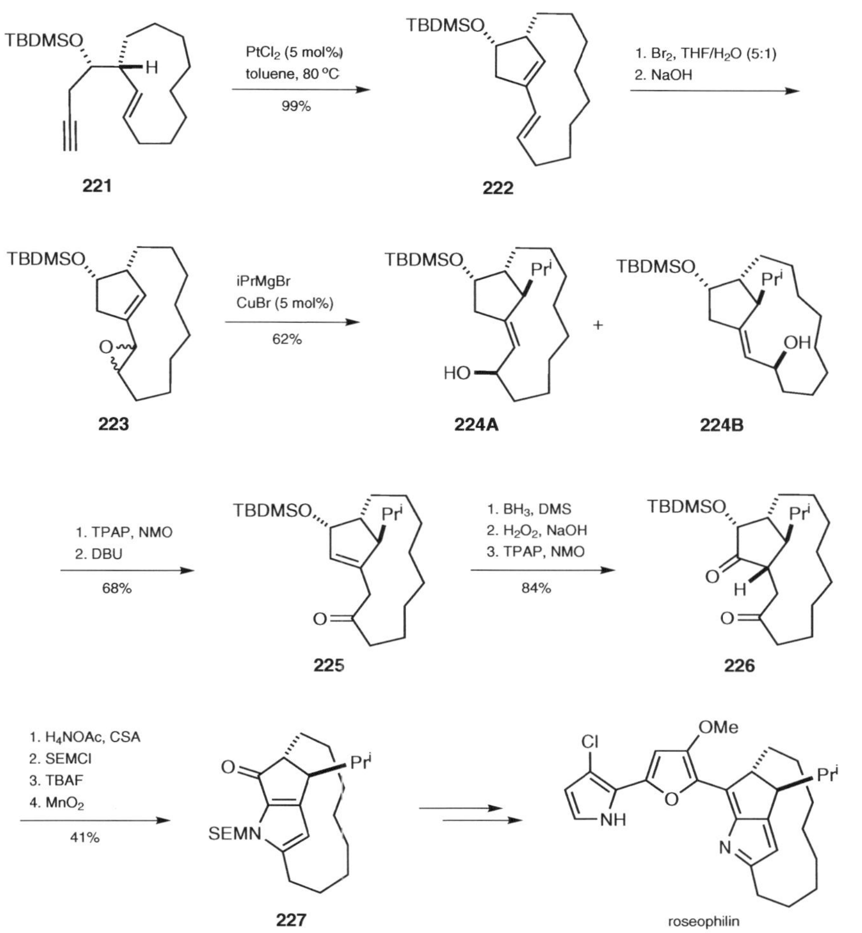

TPAP = tetrapropylammonium perruthenate

On the basis of π-acidic transition-metal-catalyzed formation of *exo*-type-I product from enynes, Fürstner and co-workers converted nitrogen-tethered 1,6-enyne **228** into five-membered ring-bridged macrocycle **229**.[128] Reduction of the conjugated enone to the corresponding secondary alcohol **230** via a two-step sequence, followed by the Barton–McCombie deoxygenation of the secondary alcohol, delivers an ethyl-substituted macrocycle **231**. Pyrrole formation is achieved by treating **231** with excess potassium 3-amino-propylamide (KAPA), thereby intercepting a known advanced synthetic intermediate **232**, and thus achieving a formal synthesis of metacycloprodigiosin (Scheme 53).

for references see p 131

Scheme 53 Synthesis of Metacycloprodigiosin[128]

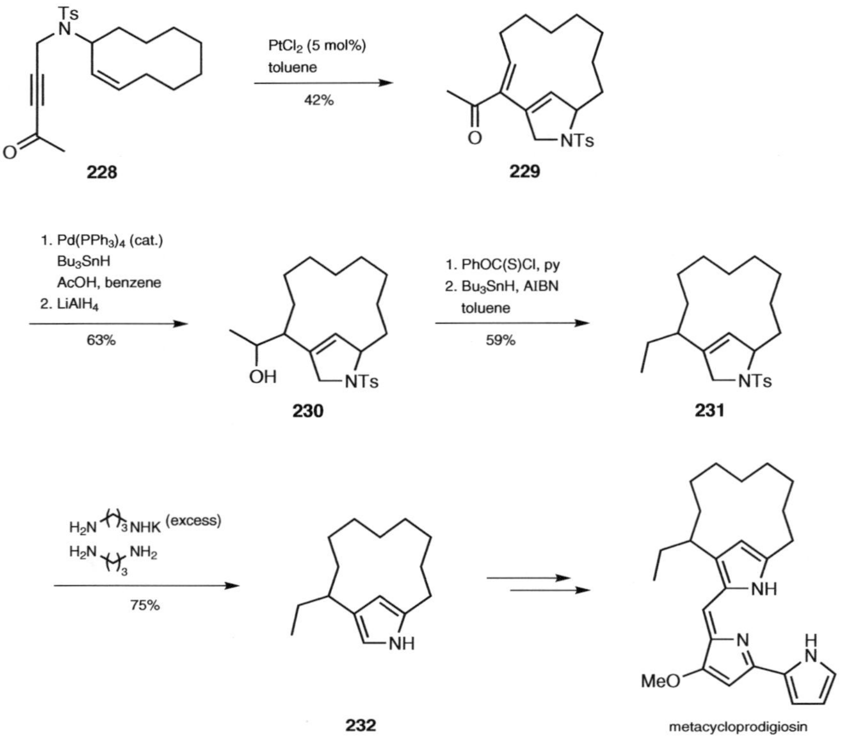

1-{(1Z,2Z)-12-Tosyl-12-azabicyclo[9.2.1]tetradeca-1(14),2-dien-2-yl}ethan-1-one (229); Typical Procedure for Enyne Metathesis with a π-Lewis Acid:[128]

A suspension of substrate **228** (300 mg, 0.774 mmol) and PtCl₂ (21 mg, 0.077 mmol, 10 mol%) in toluene (80 mL) was stirred for 21 h at 100 °C. Removal of the solvent under reduced pressure and flash chromatography (hexanes/EtOAc 6:1 to 4:1) provided the product as a bright yellow oil; yield: 126 mg (42%).

1.3.1.4 Conclusions

Following a brief survey of the general aspects of enyne metathesis, including both mechanism and selectivity, selected sets of completed formal and total syntheses of various natural products have been discussed to illustrate the prowess of domino transformations using enyne metathesis as a component step. These examples clearly show that enyne metathesis offers a unique platform for the development of domino reaction strategies for natural product synthesis, though it is hoped that even more sophisticated applications are to come. In theory, domino transformations mediated by enyne metathesis, such as domino double ring-closing metathesis, enyne metathesis/metallotropic [1,3]-shifts, enyne metathesis/Diels–Alder, and other variations of their domino combinations would be potentially more atom- and step-economical compared to the corresponding stepwise processes if the associated chemo-, regio-, stereoselectivity constraints of the reaction in the current form are relieved. The availability of various catalysts entrusted with the desired reactivity and selectivity profiles toward a broad range of substrates is the key to further development of this field, as is the creative strategies which chemists can devise as a result.

References

[1] Tietze, L. F.; Beifuss, U., *Angew. Chem.*, (1993) **105**, 137; *Angew. Chem. Int. Ed. Engl.*, (1993) **32**, 131.
[2] Tietze, L. F., *Chem. Ind. (London)*, (1995), 453.
[3] Waldmann, H., *Organic Synthesis Highlights II*, Waldmann, H., Ed.; VCH: Weinheim, Germany, (1995); p 193.
[4] Hall, N., *Science (Washington, D. C.)*, (1994) **266**, 32.
[5] *Domino Reactions in Organic Synthesis*, Tietze, L. F.; Brasche, G.; Gericke, K. M., Eds.; Wiley-VCH: Weinheim, Germany, (2006).
[6] Wasilke, J.-C.; Obrey, S. J.; Baker, R. T.; Bazan, G. C., *Chem. Rev.*, (2005) **105**, 1001.
[7] Parsons, P. J.; Penkett, C. S.; Shell, A. J., *Chem. Rev.*, (1996) **96**, 195.
[8] Tietze, L. F.; Rackelmann, N., *Pure Appl. Chem.*, (2004) **76**, 1976.
[9] Trost, B. M., *Angew. Chem.*, (1995) **107**, 285; *Angew. Chem. Int. Ed. Engl.*, (1995) **34**, 259.
[10] Sheldon, R. A., *Pure Appl. Chem.*, (2000) **72**, 1233.
[11] Trost, B. M., *Acc. Chem. Res.*, (2002) **35**, 695.
[12] Wender, P. A.; Verma, V. A.; Paxton, T. J.; Pillow, T. H., *Acc. Chem. Res.*, (2008) **41**, 40.
[13] Sivavec, T. M.; Katz, T. J., *Tetrahedron Lett.*, (1985) **26**, 2159.
[14] Katz, T. J.; Sivavec, T. M., *J. Am. Chem. Soc.*, (1985) **107**, 737.
[15] Korkowski, P. F.; Hoye, T. R.; Rydberg, D. B., *J. Am. Chem. Soc.*, (1988) **110**, 2676.
[16] Hoye, T. R.; Rehberg, G. M., *Organometallics*, (1989) **8**, 2070.
[17] Hoye, T. R.; Rehberg, G. M., *J. Am. Chem. Soc.*, (1990) **112**, 2841.
[18] Watanuki, S.; Ochifuji, N.; Mori, M., *Organometallics*, (1994) **13**, 4129.
[19] Watanuki, S.; Mori, M., *Organometallics*, (1995) **14**, 5054.
[20] Watanuki, S.; Ochifuji, N.; Mori, M., *Organometallics*, (1995) **14**, 5062.
[21] Kinoshita, A.; Mori, M., *Synlett*, (1994), 1020.
[22] Kim, S.-H.; Bowden, N.; Grubbs, R. H., *J. Am. Chem. Soc.*, (1994) **116**, 10801.
[23] Mori, M., In *Handbook of Metathesis*, Grubbs, R. H., Ed.; Wiley-VCH: Weinheim, Germany, (2003); Vol. 2, p 176.
[24] Schrock, R. R.; Murdzek, J. S.; Bazan, G. C.; Robbins, J.; DiMare, M.; O'Regan, M., *J. Am. Chem. Soc.*, (1990) **112**, 3875.
[25] Lee, Y.-J.; Schrock, R. R.; Hoveyda, A. H., *J. Am. Chem. Soc.*, (2009) **131**, 10652.
[26] Zhao, Y.; Hoveyda, A. H.; Schrock, R. R., *Org. Lett.*, (2011) **13**, 784.
[27] Casey, C. P.; Kraft, S.; Powell, D. R., *J. Am. Chem. Soc.*, (2002) **124**, 2584.
[28] Hansen, E. C.; Lee, D., *J. Am. Chem. Soc.*, (2004) **126**, 15074.
[29] Lippstreu, J. J.; Straub, B. F., *J. Am. Chem. Soc.*, (2005) **127**, 7444.
[30] Galan, B. R.; Giessert, A. J.; Keister, J. B.; Diver, S. T., *J. Am. Chem. Soc.*, (2005) **127**, 5762.
[31] Giessert, A. J.; Diver, S. T., *Org. Lett.*, (2005) **7**, 351.
[32] Lloyd-Jones, G. C.; Margue, R. G.; de Vries, J. G., *Angew. Chem.*, (2005) **117**, 7608; *Angew. Chem. Int. Ed.*, (2005) **44**, 7442.
[33] Dieltiens, N.; Moonen, K.; Stevens, C. V., *Chem.–Eur. J.*, (2007) **13**, 203.
[34] Marshall, J. E.; Keister, J. B.; Diver, S. T., *Organometallics*, (2011) **30**, 1319.
[35] Ulman, M.; Grubbs, R. H., *Organometallics*, (1998) **17**, 2484.
[36] Hoye, T. R.; Donaldson, S. M.; Vos, T. J., *Org. Lett.*, (1999) **1**, 277.
[37] Schramm, M. P.; Reddy, D. S.; Kozmin, S. A., *Angew. Chem.*, (2001) **113**, 4404; *Angew. Chem. Int. Ed.*, (2001) **40**, 4274.
[38] Clavier, H.; Correa, A.; Escudero-Adán, E. C.; Benet-Buchholz, J.; Cavallo, L.; Nolan, S. P., *Chem.–Eur. J.*, (2009) **15**, 10244.
[39] Kitamura, T.; Sato, Y.; Mori, M., *Chem. Commun. (Cambridge)*, (2001), 1258.
[40] Kitamura, T.; Sato, Y.; Mori, M., *Adv. Synth. Catal.*, (2002) **344**, 678.
[41] Nuñez-Zarur, F.; Solans-Monfort, X.; Rodríguez-Santiago, L.; Pleixats, R.; Sodupe, M., *Chem.–Eur. J.*, (2011) **17**, 7506.
[42] Sohn, J.-H.; Kim, K. H.; Lee, H.-Y.; No, Z. S.; Ihee, H., *J. Am. Chem. Soc.*, (2008) **130**, 16506.
[43] Miyanohana, Y.; Inoue, H.; Chatani, N., *J. Org. Chem.*, (2004) **69**, 8541.
[44] Bajracharya, G. B.; Nakamura, I.; Yamamoto, Y., *J. Org. Chem.*, (2005) **70**, 892.
[45] Nieto-Oberhuber, C.; Muñoz, M. P.; Buñuel, E.; Nevado, C.; Cárdenas, D. J.; Echavarren, A. M., *Angew. Chem.*, (2004) **116**, 2456; *Angew. Chem. Int. Ed.*, (2004) **43**, 2402.
[46] Nieto-Oberhuber, C.; Muñoz, M. P.; López, S.; Jiménez-Núñez, E.; Nevado, C.; Herrero-Gómez, E.; Raducan, M.; Echavarren, A. M., *Chem.–Eur. J.*, (2006) **12**, 1677.

[47] Ferrer, C.; Raducan, M.; Nevado, C.; Claverie, C. K.; Echavarren, A. M., *Tetrahedron*, (2007) **63**, 6306.

[48] Lee, D.; Kim, M., *Org. Biomol. Chem.*, (2007) **5**, 3418.

[49] Skell, P. S.; Klebe, J., *J. Am. Chem. Soc.*, (1960) **82**, 247.

[50] Bergman, R. G.; Rajadhyaksha, V. J., *J. Am. Chem. Soc.*, (1970) **92**, 2163.

[51] Franck-Neumann, M.; Geoffroy, P.; Lohmann, J. J., *Tetrahedron Lett.*, (1983) **24**, 1775.

[52] Franck-Neumann, M.; Geoffroy, P., *Tetrahedron Lett.*, (1983) **24**, 1779.

[53] Padwa, A.; Gareau, Y.; Xu, S. L., *Tetrahedron Lett.*, (1991) **32**, 983.

[54] Noro, M.; Masuda, T.; Ichimura, A. S.; Koga, N.; Iwamura, H., *J. Am. Chem. Soc.*, (1994) **116**, 6179.

[55] Hauptmann, H., *Tetrahedron*, (1976) **32**, 1293.

[56] Hori, Y.; Noda, K.; Kobayashi, S.; Taniguchi, H., *Tetrahedron Lett.*, (1969), 3563.

[57] Okada, S.; Peng, S.; Spevak, W.; Charych, D., *Acc. Chem. Res.*, (1998) **31**, 229.

[58] Song, J.; Cheng, Q.; Zhu, S.; Stevens, R. C., *Biomed. Microdevices*, (2002) **4**, 213.

[59] Koyanagi, T.; Muratsubaki, M.; Hosoi, Y.; Shibata, T.; Tsutsui, K.; Wada, Y.; Furukawa, Y., *Chem. Lett.*, (2006) **35**, 20.

[60] Kim, M.; Miller, R. L.; Lee, D., *J. Am. Chem. Soc.*, (2005) **127**, 12818.

[61] Kim, M.; Park, S.; Maifeld, S. V.; Lee, D., *J. Am. Chem. Soc.*, (2004) **126**, 10242.

[62] Park, S.; Kim, M.; Lee, D., *J. Am. Chem. Soc.*, (2005) **127**, 9410.

[63] Kim, M.; Lee, D., *Org. Lett.*, (2005) **7**, 1865.

[64] Hansen, E. C.; Lee, D., *J. Am. Chem. Soc.*, (2003) **125**, 9582.

[65] Miller, R. L.; Maifeld, S. V.; Lee, D., *Org. Lett.*, (2004) **6**, 2773.

[66] Lee, H.-Y.; Kim, B. G.; Snapper, M. L., *Org. Lett.*, (2003) **5**, 1855.

[67] Giessert, A. J.; Diver, S. T., *J. Org. Chem.*, (2005) **70**, 1046.

[68] Watanabe, K.; Minato, H.; Murata, M.; Oishi, T., *Heterocycles*, (2007) **72**, 207.

[69] Wang, K.-P.; Cho, E. J.; Yun, S. Y.; Rhee, J. Y.; Lee, D., *Tetrahedron*, (2013) **69**, 9105.

[70] Yun, S. Y.; Kim, M.; Lee, D.; Wink, D. J., *J. Am. Chem. Soc.*, (2009) **131**, 24.

[71] Laplaza, C. E.; Odom, A. L.; Davis, W. M.; Cummins, C. C.; Protasiewicz, J. D., *J. Am. Chem. Soc.*, (1995) **117**, 4999.

[72] Cummins, C. C., *Chem. Commun. (Cambridge)*, (1998), 1777.

[73] Fürstner, A.; Flügge, S.; Larionov, O.; Takahashi, Y.; Kubota, T.; Kobayashi, J., *Chem.–Eur. J.*, (2009) **15**, 4011.

[74] Volchkov, I.; Lee, D., *J. Am. Chem. Soc.*, (2013) **135**, 5324.

[75] Ko, H. M.; Lee, C. W.; Kwon, H. K.; Chung, H. S.; Choi, S. Y.; Chung, Y. K.; Lee, E., *Angew. Chem.*, (2009) **121**, 2400; *Angew. Chem. Int. Ed.*, (2009) **48**, 2364.

[76] Debleds, O.; Campagne, J.-M., *J. Am. Chem. Soc.*, (2008) **130**, 1562.

[77] Graham, T. J. A.; Gray, E. E.; Burgess, J. M.; Goess, B. C., *J. Org. Chem.*, (2010) **75**, 226.

[78] Reddy, D. S.; Kozmin, S. A., *J. Org. Chem.*, (2004) **69**, 4860.

[79] Hoye, T. R.; Zhao, H., *Org. Lett.*, (1999) **1**, 1123.

[80] Hoye, T. R.; Promo, M. A., *Tetrahedron Lett.*, (1999) **40**, 1429.

[81] Imahori, T.; Ojima, H.; Yoshimura, Y.; Takahata, H., *Chem.–Eur. J.*, (2008) **14**, 10762.

[82] Imahori, T.; Ojima, H.; Tateyama, H.; Mihara, Y.; Takahata, H., *Tetrahedron Lett.*, (2008) **49**, 265.

[83] Zhu, X.; Sheth, K. A.; Li, S.; Chang, H.-H.; Fan, J.-Q., *Angew. Chem.*, (2005) **117**, 7616; *Angew. Chem. Int. Ed.*, (2005) **44**, 7450.

[84] MuniRaju, C.; Rao, J. P.; Rao, B. V., *Tetrahedron: Asymmetry*, (2012) **23**, 86.

[85] Krishna, P. R.; Reddy, P. S., *Synlett*, (2009), 209.

[86] Aggarwal, V. K.; Astle, C. J.; Rogers-Evans, M., *Org. Lett.*, (2004) **6**, 1469.

[87] Brenneman, J. B.; Martin, S. F., *Org. Lett.*, (2004) **6**, 1329.

[88] Mori, M.; Tomita, T.; Kita, Y.; Kitamura, T., *Tetrahedron Lett.*, (2004) **45**, 4397.

[89] Tomita, T.; Kita, Y.; Kitamura, T.; Sato, Y.; Mori, M., *Tetrahedron*, (2006) **62**, 10518.

[90] Boyer, F.-D.; Hanna, I.; Ricard, L., *Org. Lett.*, (2004) **6**, 1817.

[91] Boyer, F.-D.; Hanna, I., *Eur. J. Org. Chem.*, (2006), 471.

[92] Mandal, M.; Yun, H.; Dudley, G. B.; Lin, S.; Tan, D. S.; Danishefsky, S. J., *J. Org. Chem.*, (2005) **70**, 10619.

[93] Lee, J.; Parker, K. A., *Org. Lett.*, (2012) **14**, 2682.

[94] Molawi, K.; Delpont, N.; Echavarren, A. M., *Angew. Chem.*, (2010) **122**, 3595; *Angew. Chem. Int. Ed.*, (2010) **49**, 3517.

[95] Movassaghi, M.; Piizzi, G.; Siegel, D. S.; Piersanti, G., *Angew. Chem.*, (2006) **118**, 5991; *Angew. Chem. Int. Ed.*, (2006) **45**, 5859.

References

[96] Siegel, D. S.; Piizzi, G.; Piersanti, G.; Movassaghi, M., *J. Org. Chem.*, (2009) **74**, 9292.

[97] Schubert, M.; Metz, P., *Angew. Chem.*, (2011) **123**, 3011; *Angew. Chem. Int. Ed.*, (2011) **50**, 2954.

[98] Kim, Y. J.; Lee, D., *Org. Lett.*, (2006) **8**, 5219.

[99] Mukherjee, S.; Lee, D., *Org. Lett.*, (2009) **11**, 2916.

[100] Dineen, T. A.; Roush, W. R., *Org. Lett.*, (2004) **6**, 2043.

[101] Ramharter, J.; Weinstabl, H.; Mulzer, J., *J. Am. Chem. Soc.*, (2010) **132**, 14 338.

[102] Niethe, A.; Fischer, D.; Blechert, S., *J. Org. Chem.*, (2008) **73**, 3088.

[103] Honda, T.; Namiki, H.; Kaneda, K.; Mizutani, H., *Org. Lett.*, (2004) **6**, 87.

[104] Shimizu, K.; Takimoto, M.; Mori, M., *Org. Lett.*, (2003) **5**, 2323.

[105] Fukumoto, H.; Esumi, T.; Ishihara, J.; Hatakeyama, S., *Tetrahedron Lett.*, (2003) **44**, 8047.

[106] Wei, H.; Qiao, C.; Liu, G.; Yang, Z.; Li, C.-C., *Angew. Chem.*, (2013) **125**, 648; *Angew. Chem. Int. Ed.*, (2013) **52**, 620.

[107] Hansen, E. C.; Lee, D., *Acc. Chem. Res.*, (2006) **39**, 509.

[108] Wallace, D. J., *Angew. Chem.*, (2005) **117**, 1946; *Angew. Chem. Int. Ed.*, (2005) **44**, 1912.

[109] Kummer, D. A.; Brenneman, J. B.; Martin, S. F., *Tetrahedron*, (2006) **62**, 11 437.

[110] Evans, M. A.; Morken, J. P., *Org. Lett.*, (2005) **7**, 3371.

[111] Li, J.; Lee, D., *Chem.–Asian J.*, (2010) **5**, 1298.

[112] Lei, X.; Johnson, R. P.; Porco, J. A., Jr., *Angew. Chem.*, (2003) **115**, 4043; *Angew. Chem. Int. Ed.*, (2003) **42**, 3913.

[113] Moses, J. E.; Commeiras, L.; Baldwin, J. E.; Adlington, R. M., *Org. Lett.*, (2003) **5**, 2987.

[114] Layton, M. E.; Morales, C. A.; Shair, M. D., *J. Am. Chem. Soc.*, (2002) **124**, 773.

[115] Morales, C. A.; Layton, M. E.; Shair, M. D., *Proc. Natl. Acad. Sci. U. S. A.*, (2004) **101**, 12 036.

[116] Kim, C. H.; An, H. J.; Shin, W. K.; Yu, W.; Woo, S. K.; Jung, S. K.; Lee, E., *Chem.–Asian J.*, (2008) **3**, 1523.

[117] Kim, C. H.; An, H. J.; Shin, W. K.; Yu, W.; Woo, S. K.; Jung, S. K.; Lee, E., *Angew. Chem.*, (2006) **118**, 8187; *Angew. Chem. Int. Ed.*, (2006) **45**, 8019.

[118] Li, J.; Park, S.; Miller, R. L.; Lee, D., *Org. Lett.*, (2009) **11**, 571.

[119] Cho, E. J.; Lee, D., *Org. Lett.*, (2008) **10**, 257.

[120] Kim, H.; Lee, H.; Lee, D.; Kim, S.; Kim, D., *J. Am. Chem. Soc.*, (2007) **129**, 2269.

[121] Ramharter, J.; Mulzer, J., *Org. Lett.*, (2009) **11**, 1151.

[122] Kaliappan, K. P.; Subrahmanyam, A. V., *Org. Lett.*, (2007) **9**, 1121.

[123] Kaliappan, K. P.; Ravikumar, V., *Synlett*, (2007), 977.

[124] Kaliappan, K. P.; Ravikumar, V., *J. Org. Chem.*, (2007) **72**, 6116.

[125] Kurhade, S. E.; Sanchawala, A. I.; Ravikumar, V.; Bhuniya, D.; Reddy, D. S., *Org. Lett.*, (2011) **13**, 3690.

[126] Trost, B. M.; Doherty, G. A., *J. Am. Chem. Soc.*, (2000) **122**, 3801.

[127] Fürstner, A.; Weintritt, H., *J. Am. Chem. Soc.*, (1998) **120**, 2817.

[128] Fürstner, A.; Szillat, H.; Gabor, B.; Mynott, R., *J. Am. Chem. Soc.*, (1998) **120**, 8305.

1.3.2   **Domino Metathesis Reactions Involving Carbonyls**

H. Renata and K. M. Engle

General Introduction

Carbonyl–alkene metathesis is a powerful means to form highly substituted alkenes (Scheme 1). Tandem, or domino, reactions are an established approach to achieve this type of transformation, and they have thus been widely used in organic synthesis.[1–3] In particular, carbonyl–alkene metathesis enables straightforward access to vinyl ether motifs that would otherwise be challenging to prepare. By using carbonyl–alkene metathesis, vinyl ethers can be synthesized through the coupling of comparatively simple esters and alkenes, a retrosynthetic disconnection that has proven to be indispensable in the synthesis of glycal motifs, including precursors to ladder polyether natural products.[4–23]

Scheme 1   Carbonyl–Alkene Metathesis: General Depiction

Although carbonyl–alkene metathesis could, in principle, be effected directly by the action of a single catalyst and/or stoichiometric reagent, the vast majority of examples in the literature involve a sequence of carefully orchestrated discrete steps that take place in one or two pots. There are two general mechanistic schemes to perform this overall transformation: one, reaction of a [M=CH$_2$] reagent with an alkene to generate a new metal alkylidene, which then couples with a carbonyl group to form the desired substituted alkene and an inactive [M=O] species (Scheme 2, type A); two, conversion of the carbonyl moiety into an alkene through Wittig-type alkenation, followed by metathesis between this newly formed alkene and a second alkene (Scheme 2, type B). Many different protocols for carbonyl–alkene metathesis have been reported using transition-metal complexes **1–7** (Scheme 3). In some cases, the mechanism has been elucidated. In other cases, the mechanism remains unclear, or both types of mechanism could be operative simultaneously. Both of the above reaction types have in common the fact that they generate a stoichiometric [M=O] species, which serves as a thermodynamic driving force for the process. As a consequence, transition-metal species that promote these reactions are generally used stoichiometrically. The development of catalytic carbonyl–alkene metathesis mediated by transition metals remains a tremendous challenge. Recently, alternative approaches involving the use of small-molecule organocatalysts to effect carbonyl–alkene metathesis have been reported, and this area appears to be a promising direction for further development.

for references see p 154

Scheme 2 Carbonyl–Alkene Metathesis: Potential Types of Reaction

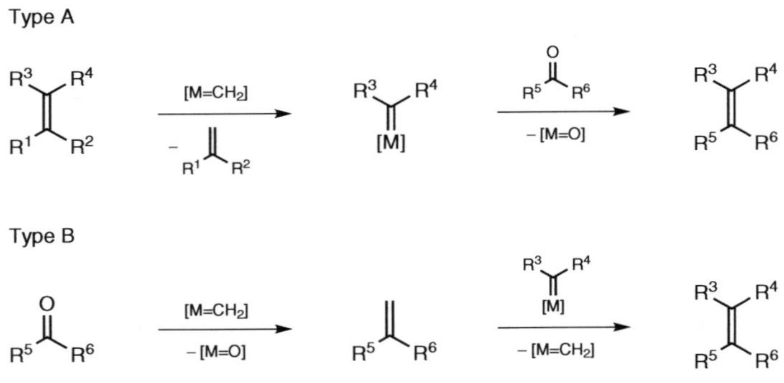

Type A

Type B

This chapter reviews reactions to effect net carbonyl–alkene metathesis and highlights applications of these methods in the synthesis of structurally complex targets. Each section covers one thematically similar collection of transformations. First, two-pot protocols consisting of Wittig or Tebbe alkenation of a carbonyl group followed by alkene metathesis using a Schrock (**5**)[24,25] or Grubbs (**6** or **7**)[26–28] initiator are described. Second, one-pot reactions that employ multiple equivalents of stoichiometric titanium complexes, as exemplified by bis(η^5-cyclopentadienyl)methylenetitanium(IV) (**1**, titanocene methylidene),[29–31] which is generated in situ from the Tebbe (**2**),[32,33] Grubbs (**3**),[34,35] and Petasis (**4**)[36] reagents, are presented. Third, reactions involving a combination of titanium(IV) chloride, zinc, lead(II) chloride, and either dibromomethane or 1,1-dibromoethane (Takai–Utimoto[37] or Rainier[17] conditions), which all together generate a titanium–alkylidene in situ, are discussed. Fourth, transformations involving stoichiometric molybdenum- or tungsten-based complexes are summarized. The fifth and final section highlights recent approaches to carbonyl–alkene metathesis based on the use of organic catalysts.

Scheme 3 Reagents and Catalysts Relevant to Carbonyl–Alkene Metathesis

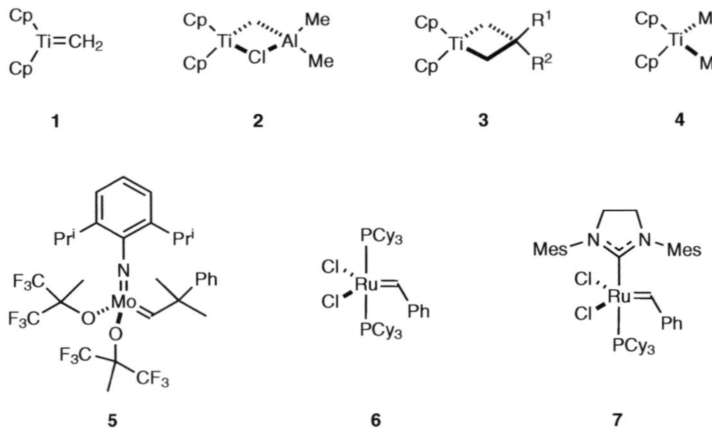

1.3.2.1 Two-Pot Reactions

Arguably the most straightforward way to carry out a type B carbonyl–alkene metathesis reaction is to perform a two-step, two-pot reaction sequence in which the carbonyl moiety is first converted into an alkene. Next, treatment with a Grubbs-[26–28] or Schrock-type[24,25] initiator promotes alkene metathesis. The advantages of this approach are that there are

1.3.2 Domino Metathesis Reactions Involving Carbonyls **137**

no potential issues in terms of the incompatibility of the reaction conditions, reagents, and/or byproducts of each of the two transformations and that the two steps can be optimized independently. The disadvantages are the comparatively poor atom and step economy and the process is operationally complex, as two distinct reactions and purifications are required.

1.3.2.1.1 Reaction with In Situ Generated Titanium–Alkylidene Complexes Followed by Metathesis

One of the first examples of this two-pot approach was the catalytic ring-closing alkene metathesis of alkene esters to give cyclic enol ethers, as reported by Grubbs and co-workers.[38] First, carbonyl alkenation using titanium-based reagents, such as the Tebbe reagent (**2**)[32,33] or the Takai–Utimoto protocol,[37] generates an acyclic enol ether. Next, ring-closing alkene metathesis using Schrock molybdenum catalyst **5** generates the cyclic products.[24,25]

The utility of this transformation has been demonstrated in the synthesis of naturally occurring phytoalexin **12**.[38] The ester precursor is synthesized by first converting acid **8** into the corresponding acid chloride, which is followed by coupling with phenol **9** to afford intermediate **10**. After Takai–Utimoto alkenation, treatment of resulting enol ether **11** with Schrock molybdenum catalyst **5** furnishes the benzofuran product, which is hydrogenated to effect global benzyl deprotection and give the desired synthetic target **12** (Scheme 4). A similar strategy has been employed by Rainier and co-workers in the total synthesis of gambierol.[15,16]

Scheme 4 Two-Pot Takai–Utimoto Alkenation/Alkene Metathesis in the Synthesis of a Phytoalexin[38]

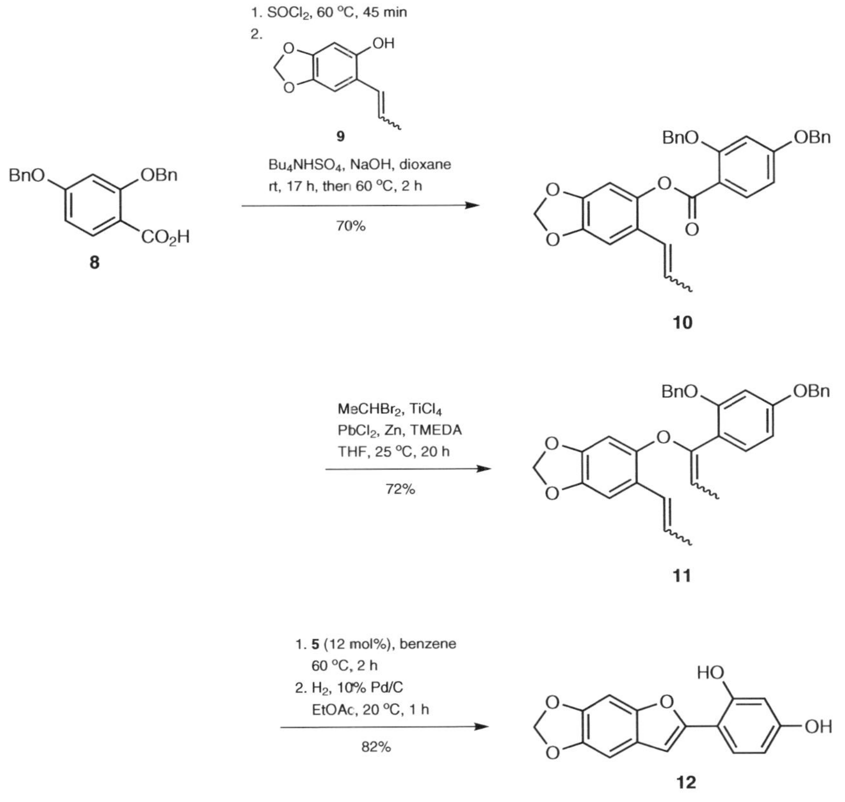

for references see p 154

Postema has employed a two-step, two-pot sequence consisting of Wittig/Takai–Utimoto alkenation followed by alkene metathesis to synthesize C1 glycals.[6,7] This sequence typically commences with dehydrative coupling of an alkene alcohol with the appropriate carboxylic acid. The resulting ester (e.g., **13**) is then alkenylated and the corresponding enol ether (e.g., **14**) is cyclized with Schrock molybdenum initiator **5** to give the final product (e.g., **15**) (Scheme 5). Similarly, Grubbs first- and second-generation initiators (**6** and **7**) are capable of cyclizing a range of vinyl ethers[39] and silyl enol ethers[40,41] to furnish the corresponding dihydropyrans and carbocycles, respectively.

Scheme 5 Two-Pot Takai–Utimoto Alkenation/Alkene Metathesis in the Synthesis of a C1 Glycal[6]

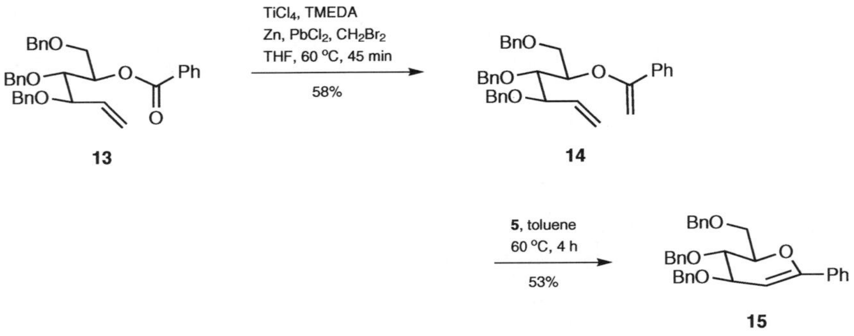

In a similar vein, amides have been found to be a suitable class of substrate for the same cyclization sequence. Bennasar has demonstrated that sequential Tebbe alkenation of amides to generate the corresponding enamines, followed by alkene metathesis mediated by second-generation Grubbs catalyst **7**, can be used to synthesize N-heterocycles.[42,43]

(3S,4R)-1-{[1,3,4-Tris(benzyloxy)hex-5-en-2-yl]oxy}ethenylbenzene (14); Typical Procedure:[6]

A 1 M soln of $TiCl_4$ in CH_2Cl_2 (5.1 mL, 5.10 mmol, 17.0 equiv) was added to THF (4 mL) precooled to 0 °C. The resulting mixture was stirred for 5 min, at which point TMEDA (1.5 mL, 9.84 mmol, 32.8 equiv) was added in one portion. The resulting suspension was allowed to warm to ambient temperature and was stirred for 15 min. Then, Zn dust (725 mg, 11.1 mmol, 37.0 equiv) and $PbCl_2$ (9 mg, 0.03 mmol, 0.10 equiv) were added in one portion, and the soln was stirred at ambient temperature for 15 min. A soln of ester **13** (154 mg, 0.30 mmol, 1.00 equiv) and CH_2Br_2 (0.2 mL, 2.79 mmol, 9.30 equiv) in THF (3 mL) was added by cannula to the reaction flask in one portion. The mixture was stirred at 60 °C for 45 min and then cooled to 0 °C. The reaction was then quenched by the addition of sat. aq K_2CO_3 (3 mL) and the resulting mixture was stirred for 30 min (while warming to ambient temperature), diluted with Et_2O (30 mL), and vigorously stirred for 15 min. The resulting mixture was filtered through neutral alumina (Et_2O, 20 mL) and the greenish-blue precipitate that resulted was crushed (mortar and pestle) and thoroughly extracted with Et_2O (vigorous stirring, 15–20 mL). The combined ethereal extracts were concentrated under reduced pressure, and flash chromatography of the residue (silica gel, Et_2O/hexanes 1:9 with 1% Et_3N) gave the title compound as an oil; yield: 91 mg (58%).

(2R,3S)-3,4-Bis(benzyloxy)-2-[(benzyloxy)methyl]-6-phenyl-3,4-dihydro-2H-pyran (15); Typical Procedure:[6]

In a glovebox, catalyst **5** (17 mg, 22.5 μmol, 0.47 equiv) was added in one portion to a soln of alkene **14** (25 mg, 0.048 mmol, 1.00 equiv) in toluene (1.4 mL), and the resulting mixture was heated to 60 °C and stirred for 4 h. The mixture was removed from the glovebox

1.3.2 Domino Metathesis Reactions Involving Carbonyls

and concentrated under reduced pressure. Flash chromatography of the residue (silica gel, Et$_2$O/hexanes 7:93 with 1% Et$_3$N) afforded the title compound as an oil; yield: 20 mg (53%, as reported).

1.3.2.2   **One-Pot Reactions**

1.3.2.2.1   **Reaction with Bis(η^5-cyclopentadienyl)methylenetitanium(IV)-Type Complexes**

Bis(η^5-cyclopentadienyl)methylenetitanium(IV)-type complexes (e.g., **1**), including the Tebbe (**2**),[32,33] Grubbs (**3**),[34,35] and Petasis (**4**)[36] reagents, are capable of reacting with alkenes to effect cycloaddition/metathesis reactions or with carbonyl groups to perform Wittig-type alkenations.[29–31] This dual mode of reactivity enables this family of organometallic compounds to mediate tandem carbonyl–alkene reactions by both mechanisms shown in Scheme 2. Given the high oxophilicity of titanium, these reagents typically react faster with carbonyl moieties than with alkenes. Thus, if used in excess, titanium–methylene complexes can mediate one-pot type B carbonyl–alkene metatheses by initial Wittig-type alkenation followed by alkene metathesis. Type A carbonyl–alkene metatheses, however, are also possible with highly reactive alkenes in strained ring systems.

A series of seminal papers by Grubbs demonstrates that the Tebbe reagent (**2**) is able to engage alkenes to generate titanacyclobutanes (Scheme 6).[29–31,34,35] Although, in general, this process is reversible, cyclic substrates with inherent ring strain can undergo ring-opening retro-cycloaddition to generate the corresponding titanium–alkylidenes. In fact, treatment of norbornene (**16**) with the Tebbe reagent (**2**) generates titanacycle **17**, which is unusually stable up to 55 °C.[44] In the presence of benzophenone at 65 °C, this titanacycle undergoes a ring-opening retro-cycloaddition/carbonyl alkenation sequence to give product **18** in good yield.[45] This sequence has been employed by Grubbs and co-workers in the total synthesis of capnellene.[44,46] Owing to the ring strain of the norbornene system in **19**, active titanium–methylene species **2** selectively reacts with the alkene first, rather than engaging with the ester. Ring opening of titanacycle **20** upon heating subsequently generates the titanium–alkylidene species, which is then trapped by the ester moiety in an intramolecular fashion to generate cyclobutenol ether product **21**. Following acid-mediated hydrolysis of the enol ether to the corresponding ketone and in situ protection with ethylene glycol, 1,3-dioxolane **22** is obtained. Finally, a sequence of deprotection, ring expansion, and Tebbe alkenation leads to capnellene. The fact that reagent **2** undergoes a chemoselective reaction with the alkene moiety is a feature that is conceptually shared with the one-pot conditions developed by Rainier and with the conditions involving the employment of a stoichiometric amount of molybdenum or tungsten (see Section 1.3.2.2.3).

for references see p 154

Scheme 6 Ring-Opening Alkene Metathesis/Carbonyl Cyclization Reactions with the Tebbe Reagent[44–46]

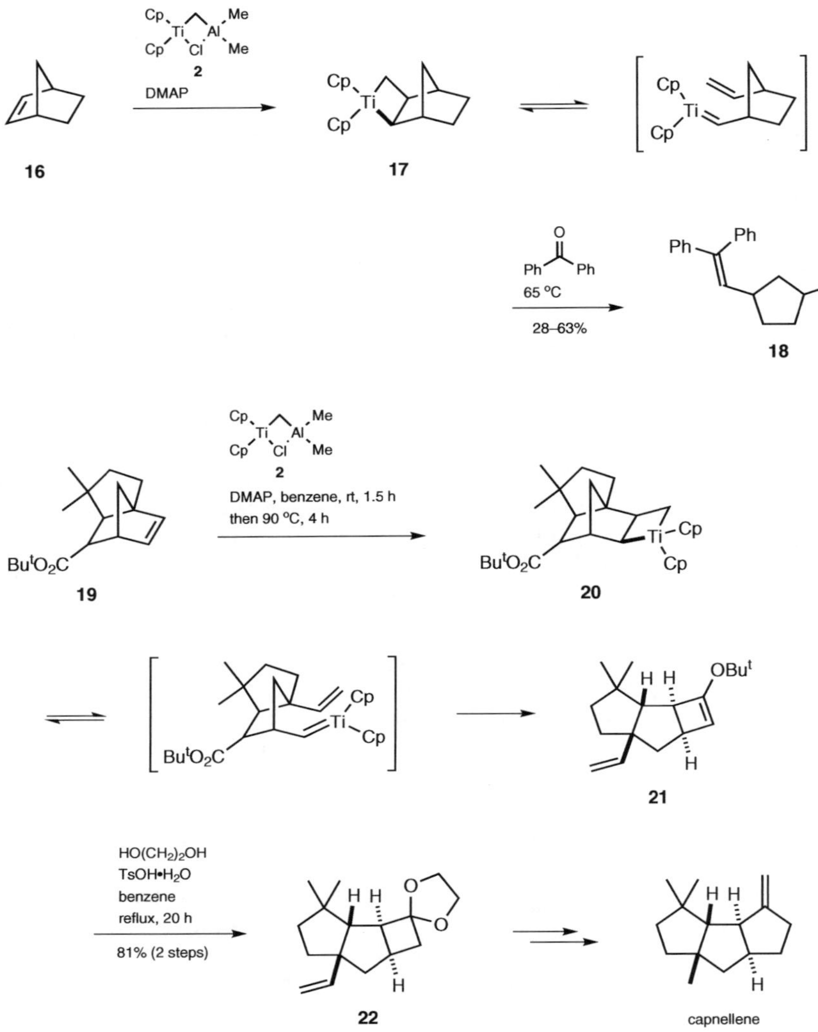

Most synthetic applications of carbonyl–alkene metathesis reactions relate to the preparation of ladder polyether natural products. For example, pioneering work by Nicolaou and co-workers demonstrates the general applicability of titanium-based reagents, such as the Tebbe (**2**) and Petasis (**4**) reagents, to convert alkene esters directly into the corresponding cyclic enol ethers (e.g., **23**) in a cascade process if used in stoichiometric excess.[4] These reactions are thought to be type B, involving an enol intermediate. This methodology has found applications in the synthesis of fragments of maitotoxin, and Table 1 illustrates the scope of this method.[5]

In general, the reaction can furnish a series of six- and seven-membered cyclic products in good yields. It is noted, however (Table 2), that open-chain products (e.g., **24**) are observed with some substrates if the Tebbe reagent (**2**) is used (Table 2, entries 1, 3, 5 and 6). These products are presumably formed by initial generation of the cyclic enol ethers, followed by sequential hydrolysis and another alkenation event at the resulting ketone.

1.3.2 Domino Metathesis Reactions Involving Carbonyls

In some cases, the use of the Petasis reagent (**4**) circumvents this issue and allows the isolation of the desired cyclic enol ethers (Table 2, entries 2 and 4).

Table 1 One-Pot Carbonyl–Alkene Metathesis with an Excess of the Tebbe Reagent[4]

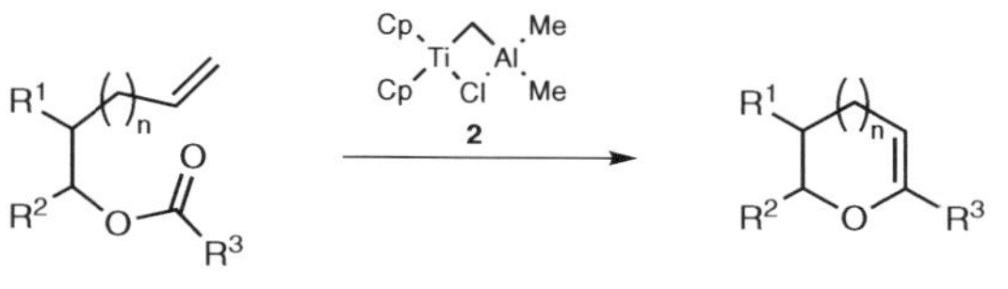

Entry	Starting Material	Product	Yield (%)	Ref
1	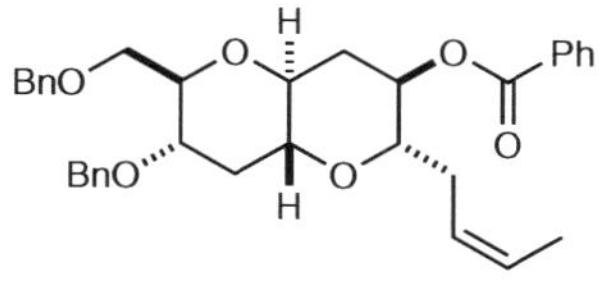	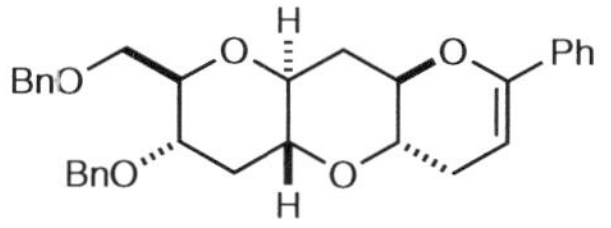	70	[4]
2	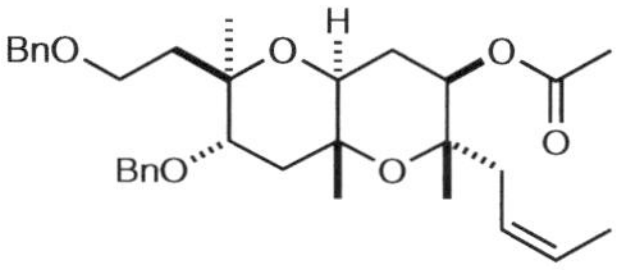	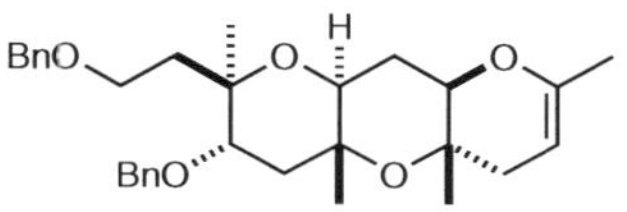	64	[4]
3	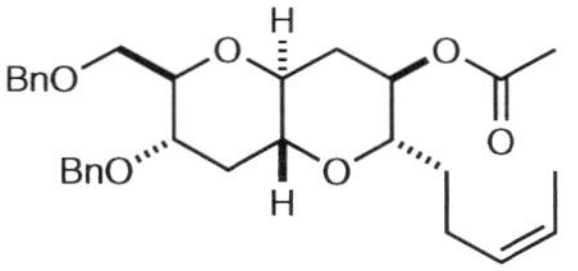	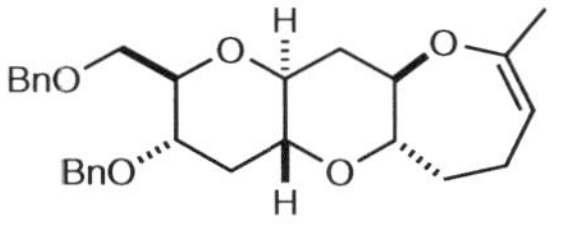	45	[4]
4	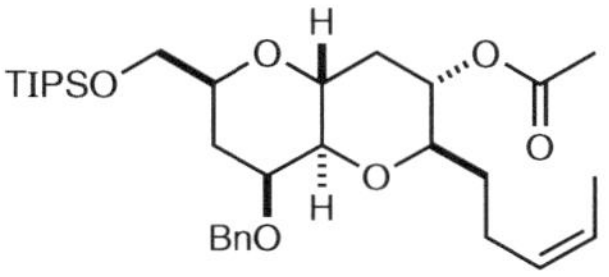	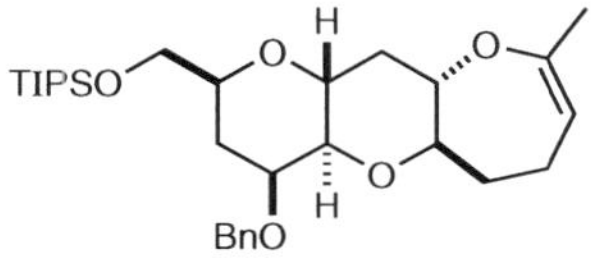	30	[4]

for references see p 154

Table 2 Comparison of the Reactivity of the Tebbe and Petasis Reagents in One-Pot Carbonyl–Alkene Metathesis[4]

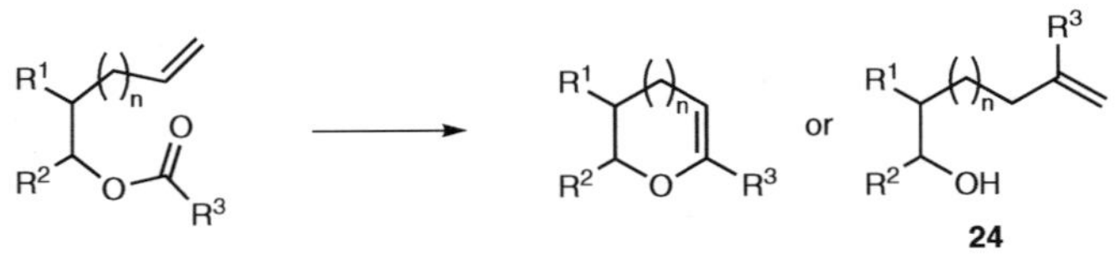

Entry	Starting Material	Reagent	Product	Yield (%)	Ref
1	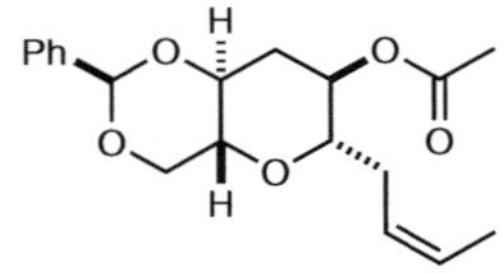 (substrate)	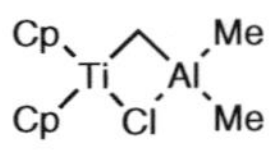 Cp$_2$Ti(Cl)AlMe$_2$ **2**	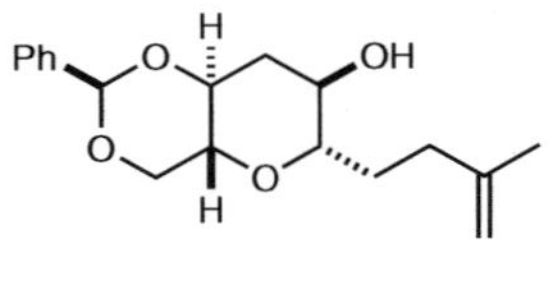 (product)	41	[4]
2	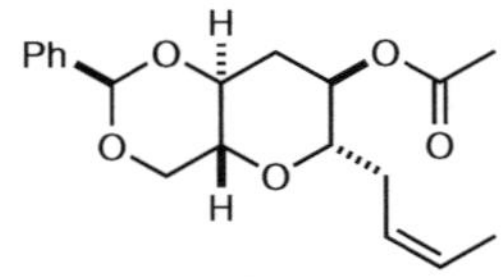 (substrate)	TiMe$_2$(Cp)$_2$ **4**	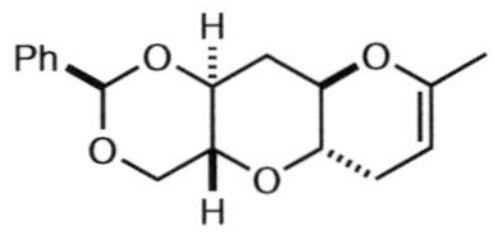 (product)	60	[4]
3	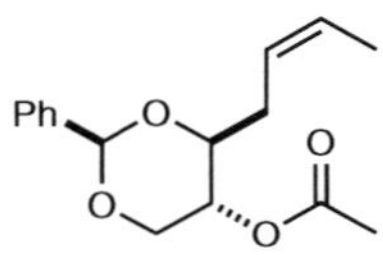 (substrate)	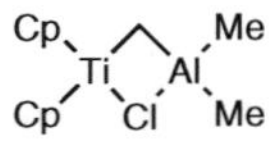 Cp$_2$Ti(Cl)AlMe$_2$ **2**	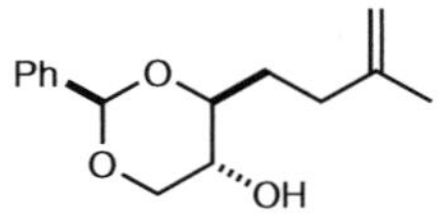 (product)	41	[4]
4	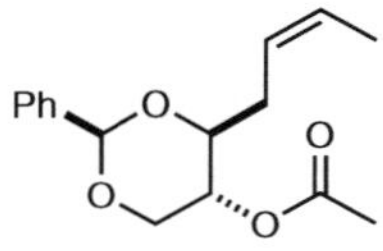 (substrate)	TiMe$_2$(Cp)$_2$ **4**	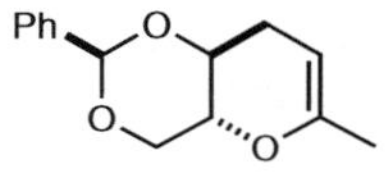 (product)	30	[4]
5	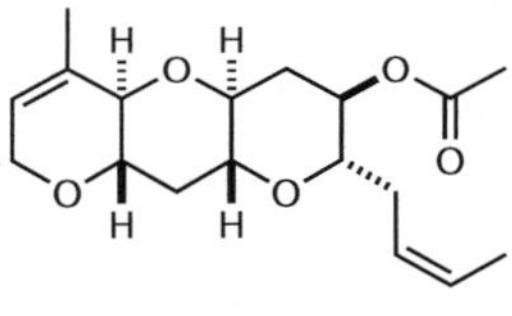 (substrate)	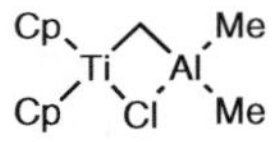 Cp$_2$Ti(Cl)AlMe$_2$ **2**	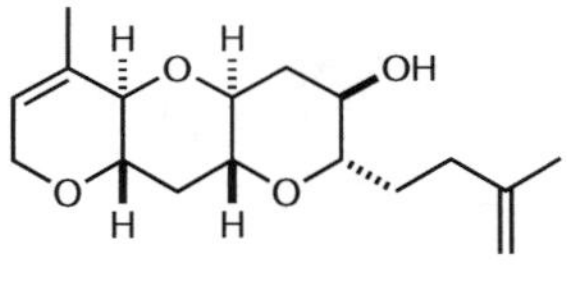 (product)	77	[4]
6	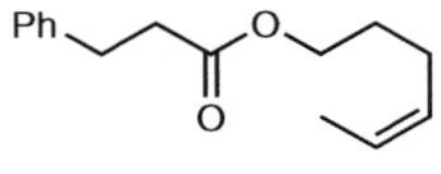 (substrate)	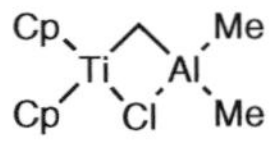 Cp$_2$Ti(Cl)AlMe$_2$ **2**	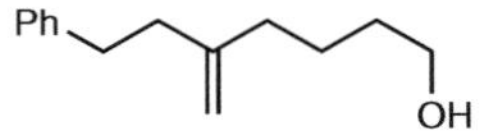 (product)	71	[4]

Subsequent studies have revealed that the success or failure of this approach to carbonyl–alkene metathesis using an excess amount of the Tebbe (**2**) or Petasis (**4**) reagent is highly context dependent.[8,10,11,13,20,43,47] For cases in which carbonyl–alkene metathesis is not seen, the enol ether intermediate and/or decomposition are typically observed.

Takeda and co-workers have developed another variant of the above procedure by employing thioacetals and carbonyl compounds as coupling partners using bis(cyclopentadienyl)bis(triethylphosphite)titanium(II) as the stoichiometric reagent to effect the process.[48] It is hypothesized that, under these conditions, thioacetals react with the titanium reagent to generate either a *gem*-dititanium or a titanium–alkylidene species that proceeds to add to the carbonyl compound. Elimination of a titanium oxide species then gen-

1.3.2 Domino Metathesis Reactions Involving Carbonyls

erates the requisite alkene product. When the carbonyl and thioacetal groups are tethered intramolecularly, the reaction generates cyclic alkene products that can be of high synthetic value. For example, Hirama and co-workers have successfully applied the Takeda method in the synthesis of complex fragments of ciguatoxin CTX3C,[8,11,12] a case in which the Tebbe (**2**) and Petasis (**4**) reagents prove problematic.

Spiro-(1*S**,2*S**,5*R**,7*S**)-10,10-dimethyl-7-vinyltricyclo[5.3.0.0^{2,5}]decane-3,2′-[1,3]-dioxolane (22); Typical Procedure:[44]

A soln of **2** (2.73 g, 9.6 mmol, 1.50 equiv) in benzene (6 mL) (**CAUTION:** *carcinogen*) was added slowly to a soln of DMAP (1.56 g, 12.8 mmol, 2.00 equiv) and alkene **19** (1.68 g, 6.4 mmol, 1.00 equiv) in benzene (13 mL). The reaction temperature was maintained below 30 °C through the intermittent use of an ice bath. Once the addition was complete, the mixture was stirred at rt for 1.5 h. The solvent volume was reduced by 1 mL under reduced pressure, and the reaction vessel was sealed and then placed in a 90 °C oil bath for 4 h. After the mixture had cooled to rt, it was added slowly to a flask containing petroleum ether (1 L) and allowed to stir at rt for 72 h open to the atmosphere. The mixture was then cooled to −30 °C, the precipitate was removed by filtration, and the soln was concentrated to a slushy solid. The precipitate that was removed from the soln was further extracted by suspension in benzene (20 mL) with stirring for 3 h. This mixture was subsequently diluted with petroleum ether (1 L), cooled to −30 °C, and filtered, and the resulting soln was combined with the previously obtained slushy solid. After concentration of the mixture, the solid obtained was dissolved in petroleum ether (800 mL), and the mixture was cooled to −30 °C and filtered. The resulting soln was concentrated under reduced pressure to give an oil, which was dissolved in a mixture of benzene (150 mL) and ethylene glycol (20 mL). TsOH•H$_2$O (500 mg) was added to this mixture. The reaction flask was equipped with a Dean–Stark trap containing 4-Å molecular sieves, and the contents were heated to reflux for 20 h. Once the mixture had cooled to rt, it was poured into a sat. aq soln of NaHCO$_3$ (100 mL), diluted with benzene (150 mL), and separated. The aqueous layer was extracted with benzene (2 × 100 mL) and the organic extracts were combined, washed with sat. aq NaHCO$_3$, and then dried (MgSO$_4$). Filtration through a pad of silica gel, followed by washing thoroughly with benzene and concentration afforded the title compound as an oil; yield: 1.30 g (81%).

1.3.2.2.2 Reaction with In Situ Generated Titanium–Alkylidene Complexes

A major advance in titanium-based carbonyl–alkene metathesis methodology was made by Rainier and co-workers in the early to mid 2000s. Key to this development was the observation that Takai–Utimoto conditions, involving the use of a mixture of titanium(IV) chloride, zinc, lead(II) chloride, and dibromomethane, could also effect the same carbonyl alkenation reaction as the Tebbe (**2**) and Petasis (**4**) reagents.[9] In addition, this set of conditions has the added advantages of in situ preparation of the reactive organometallic species from widely available starting materials, improved reactivity relative to the Petasis reagent (**4**), and decreased Lewis acidity relative to the Tebbe reagent (**2**). In a full study of the process, it was serendipitously found that increasing the amount of lead(II) chloride and decreasing the concentration of starting material **25** led to moderate formation of cyclic enol ether **27** along with expected acyclic enol ether **26** (Scheme 7).[10]

for references see p 154

Scheme 7 Carbonyl–Alkene Metathesis with an In Situ Generated Titanium Alkylidene[10]

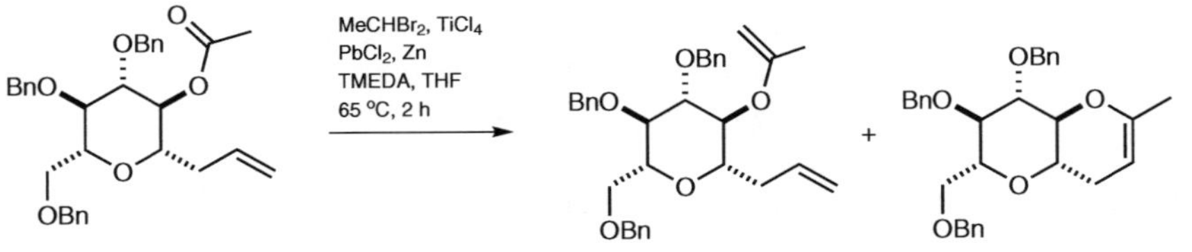

TiCl₄ (Equiv)	Zn (Equiv)	PbCl₂ (Equiv)	CH₂Br₂ (Equiv)	[25] (M)	Ratio (CH₂Cl₂/THF/TMEDA)	Yield (%) of **26**	Yield (%) of **27**	Ref
4	9	0.045	2.2	0.098	1:3.5:0.6	0	0	[10]
16	36	0.18	8.8	0.020	1:6:0.3	65	15	[10]
6	13	0.72	6	0.006	16:1:1	30	50	[10]

In an effort to rationalize this observation, acyclic enol ether **26** was prepared and isolated. Critically, upon its exposure to the same Takai–Utimoto conditions, corresponding cyclic enol ether **27** was not observed.[9,13] The authors conclude that this result is inconsistent with cyclic enol ether formation proceeding through a type B mechanism but is consistent with it resulting from a type A mechanism. Consequently, the reaction proceeds through a domino process involving metathesis between a titanium–methylene and the alkene followed by intramolecular carbonyl alkenation. The authors further reason that the cyclic and acyclic products likely arise from divergent reaction pathways in which the active titanium–methylene species reacts unselectively with either the alkene or the ester.

In a follow-up publication, Roberts and Rainier observed that the use of 1,1-dibromo-ethane instead of dibromomethane as the titanium alkylidene precursor led to the sole formation of the cyclic enol ether product (e.g., **28**) with no observable acyclic enol ether (e.g., **29**).[18] This result stands in contrast to the outcome of the reaction with dibromomethane, for which a mixture of cyclic and acyclic enol ethers is commonly observed. The authors have attributed this difference in reactivity to the fact that the titanium–ethylidene species is more sterically encumbered than the titanium–methylene species and thus less reactive toward the carbonyl group. It is also plausible that the titanium–ethylidene species is less Lewis acidic and hence less oxophilic. The titanium–ethylidene is thus proposed to preferentially react with the alkene moiety by metathesis as opposed to directly engaging the carbonyl group. The scope of this improved one-pot cyclization procedure employing Takai–Utimoto conditions has been investigated (Table 3).[17] In addition, the same conditions have been found to effect ring-closing metathesis of unstrained dienes.[17]

1.3.2 Domino Metathesis Reactions Involving Carbonyls **145**

Table 3 Reactivity Differences in Carbonyl–Alkene Metathesis with In Situ Generated Titanium–Alkylidenes Derived from Dibromomethane and 1,1-Dibromoethane[17]

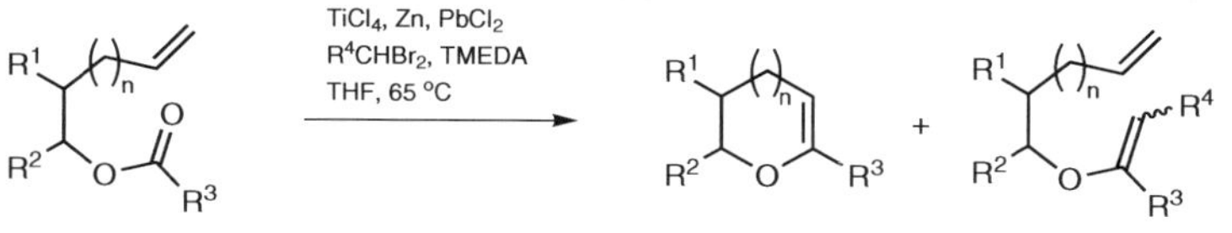

En-try	Starting Materials		Cyclic Product	Ratio (28/29)	Yield (%) of 28 + 29	Ref
	Ester	Dibromoalkane				
1	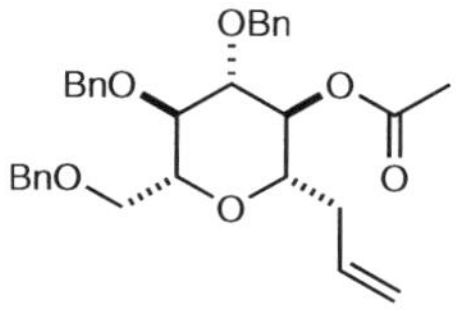	CH_2Br_2	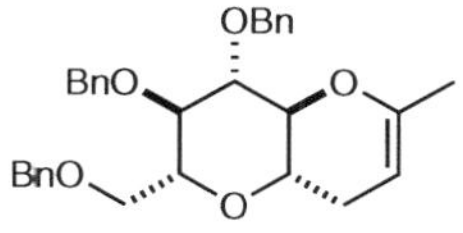	5:3	80	[17]
2	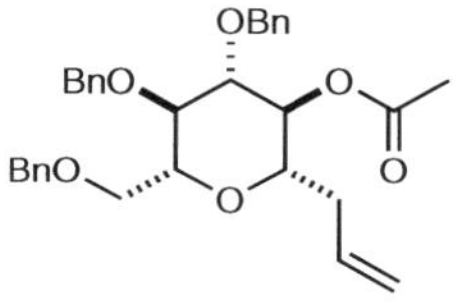	$MeCHBr_2$	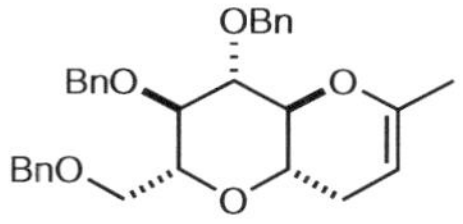	>95:5	75	[17]
3	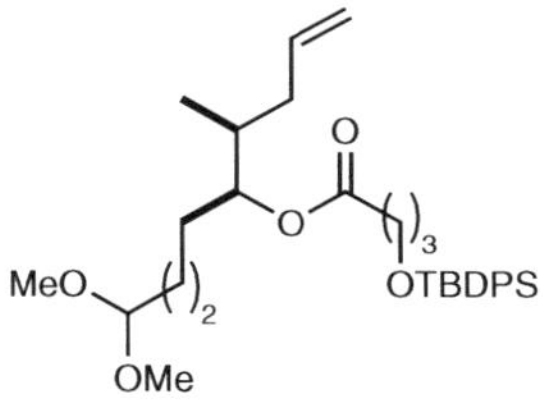	CH_2Br_2	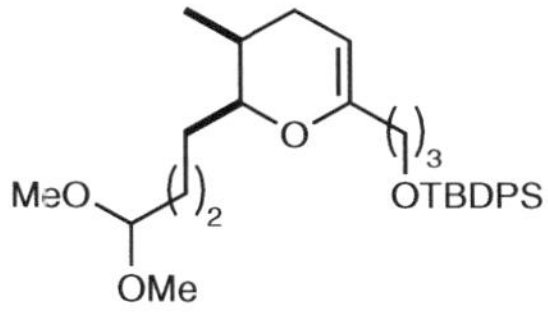	1:1	70	[17]

for references see p 154

146 Domino Transformations **1.3** Metathesis Reactions

Table 3 (cont.)

En-try	Starting Materials		Cyclic Product	Ratio (28/29)	Yield (%) of 28 + 29	Ref
	Ester	Dibromoalkane				
4	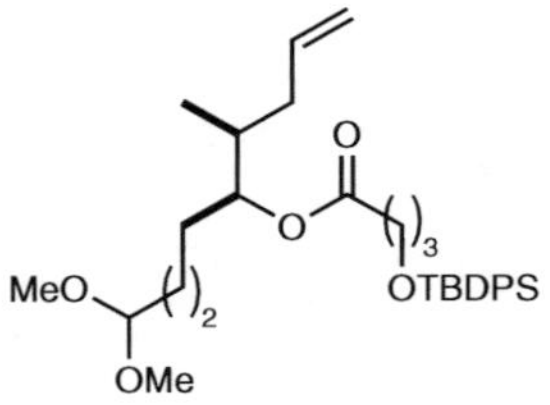	MeCHBr₂	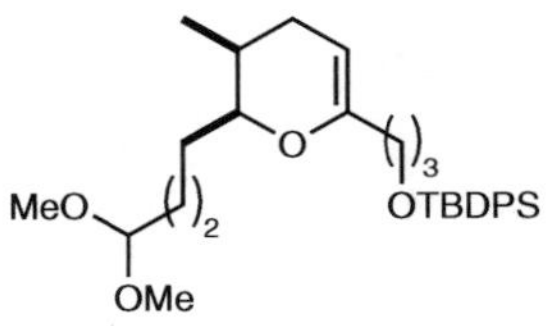	>95:5	70	[17]
5	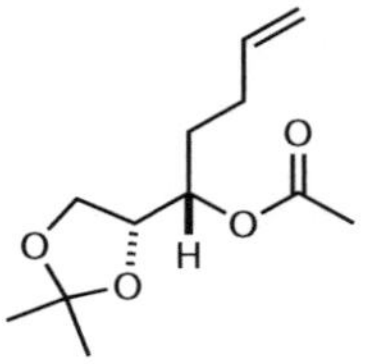	MeCHBr₂	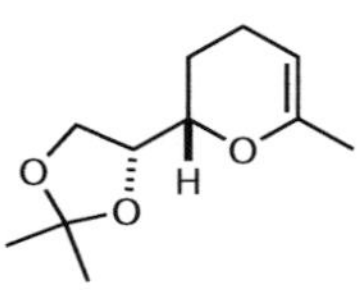	>95:5	78	[17]
6	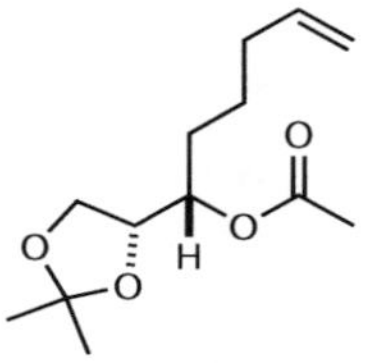	MeCHBr₂	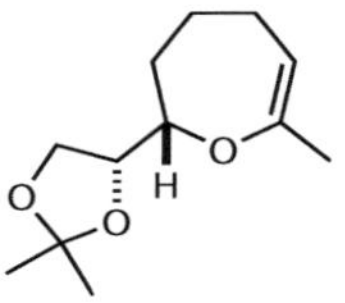	>95:5	82	[17]

This protocol has been applied in a range of different contexts, including the formation of macrocyclic enol ethers,[49] the cyclization of amides and lactams to the corresponding enamides,[50] and bidirectional cyclization (Scheme 8).[19]

Scheme 8 Synthetic Applications of Carbonyl–Alkene Metathesis with an In Situ Generated Titanium–Ethylidene[19,49,50]

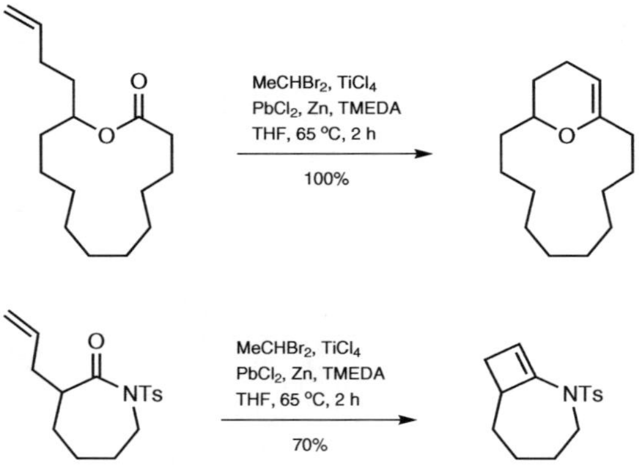

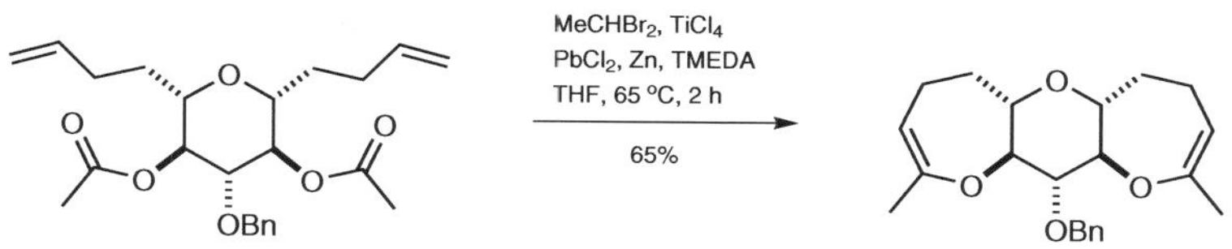

1.3.2 Domino Metathesis Reactions Involving Carbonyls

147

This methodology has recently been applied by Keck and co-workers in the preparation of the C-ring fragment of bryostatin 1, a polyketide natural product that has shown promising activity against a range of cancers (Scheme 9).[22] Treatment of ester **30** under Rainier alkenation conditions yields desired glycal product **31**, which is then further elaborated to product **32**, the complete C-ring fragment of the target molecule. In addition, Rainier and co-workers have also employed this cyclization procedure as one of the key steps in the total syntheses of brevenal[23] and gambierol[15,16] and to a subunit of gambieric acid.[18] Similarly, Neumaier and Maier have utilized this protocol in the synthesis of the core structure of spirofungin A.[51]

Scheme 9 Rainier-Type Carbonyl–Alkene Metathesis in Keck's Total Synthesis of Bryostatin 1[22]

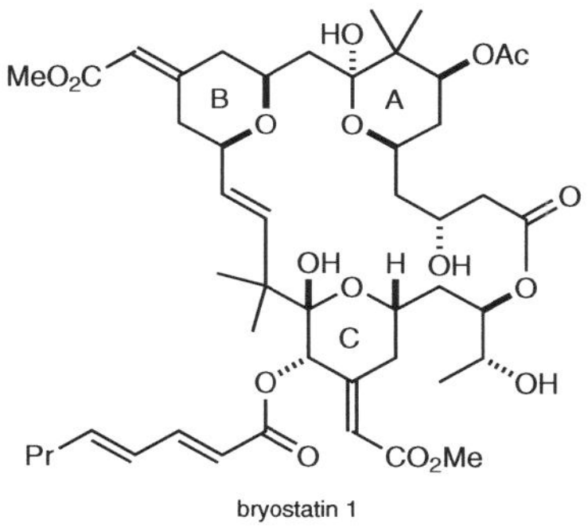

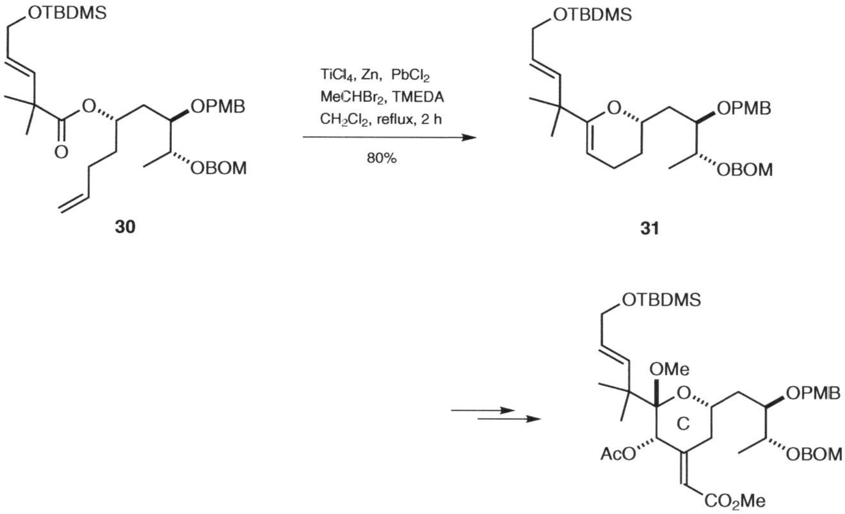

bryostatin 1

for references see p 154

(2R,3R,4S,4aS,8aS)-3,4-Dibenzyloxy-2-(benzyloxymethyl)-6-methyl-2,3,4,4a,8,8a-hexahydropyrano[3,2-b]pyran (Table 3, Entry 2); Typical Procedure:[17]

An oven-dried, two-necked flask fitted with a condenser was cooled to 0 °C and charged with CH_2Cl_2 (3.8 mL) followed by $TiCl_4$ (0.1 mL, 0.9 mmol, 31.0 equiv). THF (0.49 mL, 5.58 mmol, 192 equiv) was added dropwise, during which time the soln turned yellow. The addition of THF was followed by the dropwise addition of TMEDA (0.84 mL, 5.6 mmol, 193 equiv), which resulted in the formation of a brown soln. The ice bath was removed, and the mixture was stirred for 20 min. Activated Zn dust (0.14 g, 2.1 mmol, 72.4 equiv) and $PbCl_2$ (0.030 g, 0.11 mmol, 3.79 equiv) were then added sequentially. The resulting mixture went through a series of color changes from brown to green to purple and finally to blue-green over the course of 3–5 min. A soln of the ester (0.015 g, 0.029 mmol, 1.00 equiv) and $MeCHBr_2$ (0.085 mL, 0.93 mmol, 32.1 equiv) in CH_2Cl_2 (0.4 mL + 0.4 mL rinse) was transferred to this slurry by cannula. The mixture was then heated to reflux for 2 h. Following this time period, the mixture was cooled to 0 °C and the reaction was quenched with sat. aq K_2CO_3 (0.5 mL). After stirring at 0 °C for 30 min, the resulting mixture was filtered through a cotton plug, and the filtrate was concentrated. The residue was taken up in hexanes/EtOAc (100:1) and filtered through a plug of silica gel (hexanes/EtOAc 100:1) to give a yellow oil. Flash chromatography (hexanes/EtOAc 100:1 to 20:1) gave the title compound as a colorless oil; yield: 0.0105 g (75%).

1.3.2.2.3 Reaction with Stoichiometric Molybdenum or Tungsten Complexes

Carbonyl–alkene metathesis reactions mediated by stoichiometric quantities of molybdenum or tungsten complexes are a category of type A transformations that are mechanistically analogous to the Rainier reactions described in Section 1.3.2.2.2.

Fu and Grubbs were the first to report ring-closing carbonyl–alkene metathesis using stoichiometric quantities of Schrock initiator **5**.[52] Key to the success of this reaction was the recognition that Schrock catalyst **5** metathesizes alkenes at a faster rate than carbonyls. Thus, compound **5** reacts with alkene ketones preferentially at the alkene to generate a molybdenum–alkylidene, which then reacts intramolecularly with the ketone to close the ring with concomitant formation of an unreactive [Mo=O] species.

Tungsten-based complexes have also proven effective for this transformation. Fu and Grubbs reported that the more reactive tungsten analogue of Schrock catalyst **5** is able to effect intramolecular ester–alkene cyclization.[38,52] Similarly, Lokare and Odom report the preparation of a tungsten metallacycle that promotes a stoichiometric ring-closing carbonyl–alkene metathesis sequence in the presence of sodium tetrakis(pentafluorophenyl)borate salts.[53]

More recently, Lei and co-workers have utilized the carbonyl–alkene metathesis protocol developed by Fu and Grubbs as one of the key ring-forming steps in the total synthesis of huperzine Q and related natural products starting from cyclohexenone **33**, allylmagnesium bromide, and aldehyde **34** (Scheme 10).[54] In this work, spiro compound **35** is treated with Schrock initiator **5** to furnish tricycle **36**. Even though two different carbonyl groups are present in the starting material, only product **36** is observed, likely as a result of the high strain energy associated with the *anti*-Bredt alkene in alternative product **37**.

1.3.2 Domino Metathesis Reactions Involving Carbonyls **149**

Scheme 10 Stoichiometric Molybdenum-Mediated Carbonyl–Alkene Metathesis in Lei's Total Synthesis of Huperzine Q[54]

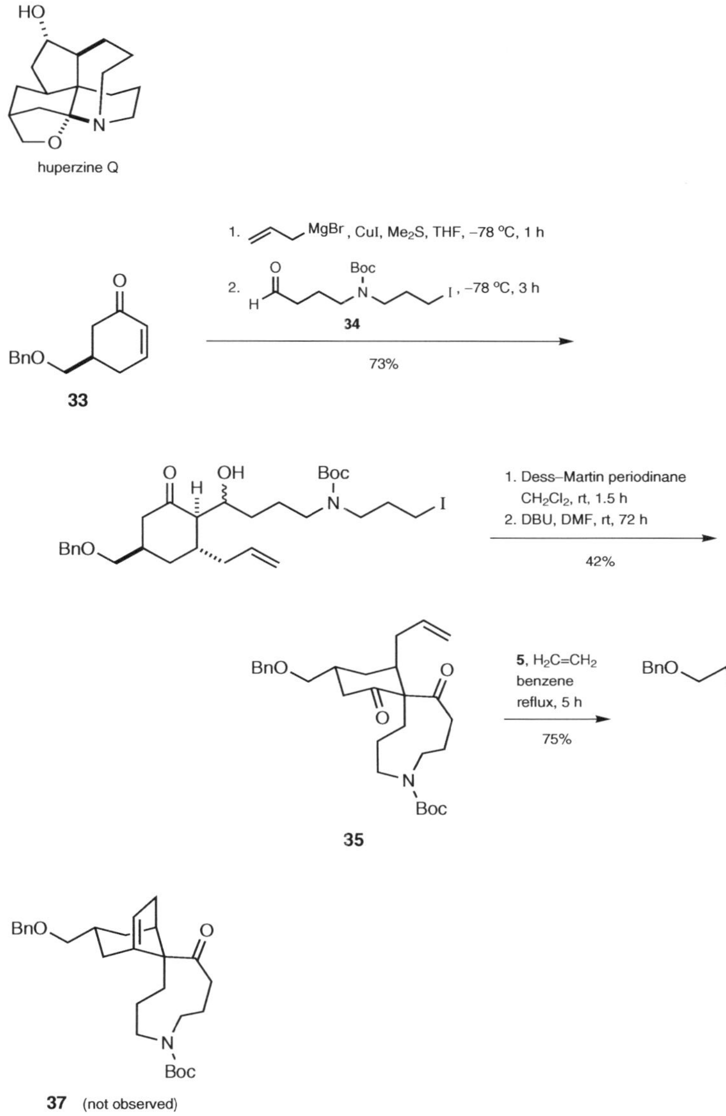

tert-Butoxycarbonyl 11-Benzyloxy-13-oxo-2,3,6,7,9,9a,10,11,12,13-decahydro-1H-indeno-[1,7a-e]azonine-4(5H)-carboxylate (36); Typical Procedure:[54]

In a glovebox, a round-bottomed flask equipped with a magnetic stirrer bar and a reflux condenser was charged with Schrock initiator **5** (678 mg, 0.868 mmol, 1.00 equiv) and diketone **35** (420 mg, 0.868 mmol, 1.00 equiv). This suspension was degassed and backfilled with ethene (3×). Then, anhyd benzene (87 mL) (**CAUTION:** *carcinogen*) was added by syringe. The resulting mixture was heated at reflux for 5 h with an attached ethene-filled balloon. The mixture was then quenched by exposure to air, concentrated, and purified (silica gel, petroleum ether/EtOAc 15:1) to afford the title compound as a light yellow oil; yield: 189 mg (48%, 75% based on recovered starting material); and recovered diketone **35**; yield: 151 mg (36%).

for references see p 154

150 Domino Transformations **1.3** Metathesis Reactions

1.3.2.2.4 **Organocatalytic Reactions**

Sections 1.3.2.1.1 and 1.3.2.2.1 to 1.3.2.2.3 illustrate the unique synthetic utility of carbonyl–alkene metathesis but also demonstrate some of the limitations of the existing methods. In particular, many of these reactions are operationally complex, utilize air- and moisture-sensitive reagents, possess limited scope, and exhibit poor atom economy. Owing to the fact that these protocols inherently generate one equivalent of [M=O] per reaction at a carbonyl site, stoichiometric equivalents of the transition-metal complexes, which can require multiple steps to synthesize and/or be expensive to purchase from commercial suppliers, are required. These considerations underscore a significant challenge in using transition-metal mediators for carbonyl–alkene metathesis and highlight the need for novel catalytic approaches that are efficient, inexpensive, and operationally convenient.

To this end, Lambert and co-workers have developed hydrazine organocatalyst **38** that promotes ring-opening cross-metathesis between strained cyclopropenes **40** and substituted aldehydes **39** (Scheme 11).[55] In this scenario, the organocatalyst condenses with the aldehyde to form the corresponding azomethine imine, which then engages the alkene in a [3+2] fashion. A strain-promoted cycloreversion forms desired metathesis product **41** and regenerates the catalyst after hydrolysis.

Scheme 11 Organocatalytic Carbonyl–Alkene Metathesis of Cyclopropenes and Aldehydes[55]

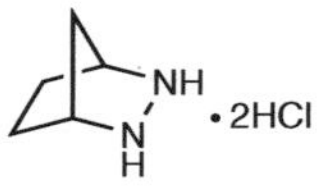

Ar[1]	R[1]	Yield (%)	Ref
Ph	OBn	80[a]	[55]
2-Tol	OBn	85	[55]
2,4-Me$_2$C$_6$H$_3$	OBn	66	[55]
4-MeOC$_6$H$_4$	OBn	50	[55]
4-O$_2$NC$_6$H$_4$	OBn	60	[55]
3-BrC$_6$H$_4$	OBn	60[a]	[55]
1-naphthyl	OBn	67	[55]
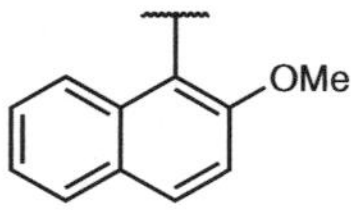 (naphthyl-OMe)	OBn	68	[55]
2-furyl	OBn	80	[55]

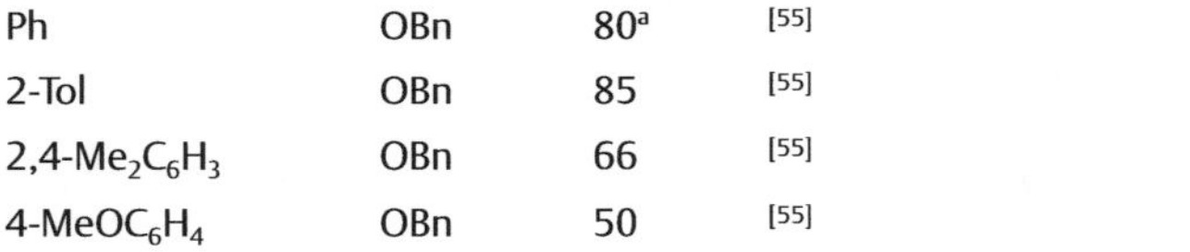

Ar[1]	R[1]	Yield (%)	Ref
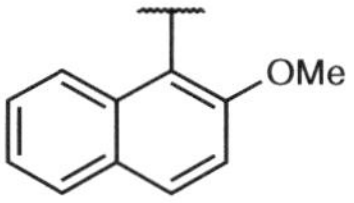 (1-methoxynaphthalen-2-yl, OMe)	OAc	68	[55]
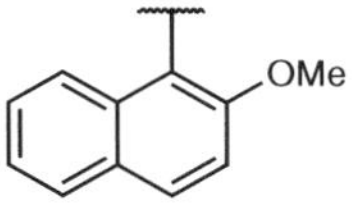 (1-methoxynaphthalen-2-yl, OMe)	OTBDPS	79	[55]
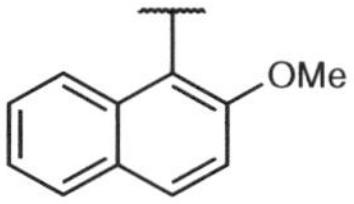 (1-methoxynaphthalen-2-yl, OMe)	4-BrC$_6$H$_4$S	50	[55]

[a] Yield determined for the reduced (alcohol) form of the product.

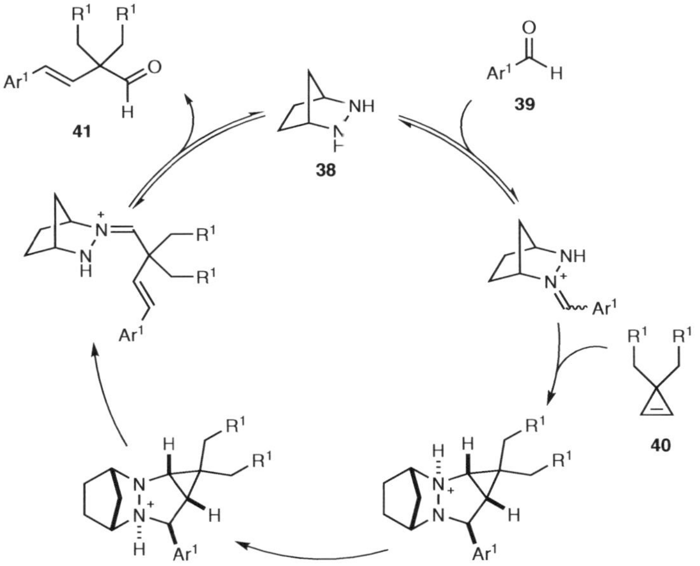

In an alternative approach, Franzén and co-workers have found that the trityl cation catalyzes cross-metathesis between trisubstituted alkenes (e.g., **43**) and arenecarbaldehydes **42** to give β-alkylstyrenes **44** and acetone (Scheme 12).[56] This reaction is thought to involve initial coordination of the trityl cation to the carbonyl group as a Lewis acid. Nucleophilic attack of the alkene on the oxonium ion generates a carbocation that then cyclizes to give an oxetane. Opening of the oxetane, fragmentation, and catalyst dissociation leads to the alkene product and closes the catalytic cycle.

for references see p 154

Domino Transformations — 1.3 Metathesis Reactions

Scheme 12 Trityl Cation Catalyzed Carbonyl–Alkene Metathesis[56]

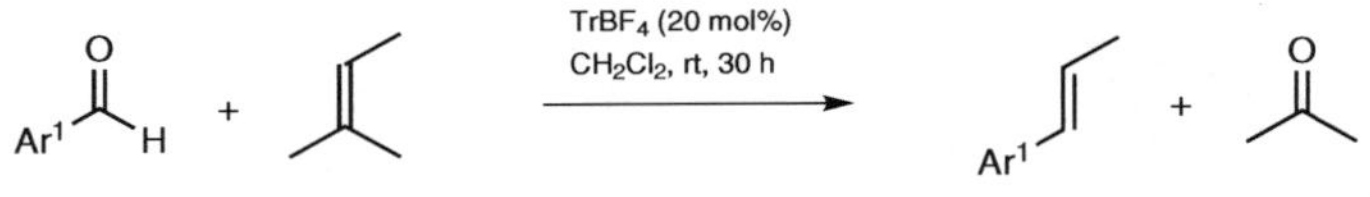

TrBF$_4$ (20 mol%)
CH$_2$Cl$_2$, rt, 30 h

42 (5 equiv)   **43** (1 equiv)   **44**

Ar^1	Yield (%)	Ref
Ph	60	[56]
2-naphthyl	54	[56]
4-Tol	52	[56]
4-BrC$_6$H$_4$	52	[56]
4-FC$_6$H$_4$	49	[56]

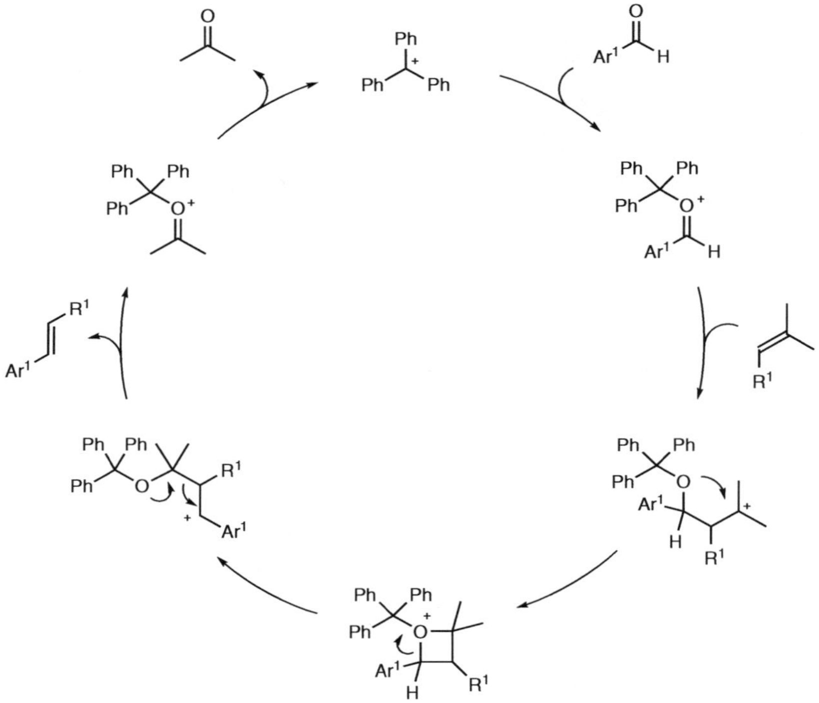

(E)-2,2-Bis(benzyloxymethyl)-4-phenylbut-3-enal (41, Ar^1 = Ph; R^1 = OBn); Typical Procedure:[55]

A soln of benzaldehyde (**39**, Ar^1 = Ph; 0.3 mmol, 2.00 equiv) in 1,2-dichloroethane (0.5 mL) was added to hydrazine **38** (2.6 mg, 0.015 mmol, 0.10 equiv) followed by a soln of cyclopropene **40** (R^1 = OBn; 42 mg, 0.15 mmol, 1.00 equiv) in 1,2-dichloroethane (0.25 mL). The resulting mixture was heated to 90 °C for 24 h and, upon cooling to rt, was diluted with CH$_2$Cl$_2$, washed with 1 M NaOH, extracted into CH$_2$Cl$_2$, dried (Na$_2$SO$_4$), and concentrated. The excess amount of benzaldehyde was removed by leaving the flask under vacuum (several hours), and the resulting crude material was dissolved in MeOH. NaBH$_4$ (11.3 mg, 0.30 mmol, 2.00 equiv) was added to reduce the aldehyde moiety in **41** to facilitate purification. The mixture was stirred at rt for 30 min, diluted with H$_2$O, extracted into EtOAc, dried (Na$_2$SO$_4$), and concentrated. The resulting residue was purified by column chroma-

tography (silica gel, EtOAc/hexanes 1:19) to yield the reduced form of the title compound as a clear oil; yield: 46.6 mg (80%).

(*E*)-Prop-1-enylbenzene (44, Ar¹ = Ph); Typical Procedure:[56]

Benzaldehyde (**42**, Ar¹ = Ph; 636 mg, 6.0 mmol, 5.00 equiv) and 2-methylbut-2-ene (**43**; 0.127 mL, 1.2 mmol, 1.00 equiv) were added to a soln of TrBF$_4$ (79 mg, 0.24 mmol, 0.20 equiv) in CH$_2$Cl$_2$ (4.0 mL) containing 1-methylnaphthalene (171 mg, 1.2 mmol, 1.00 equiv) as an internal standard. After stirring at rt for 30 h, the reaction was quenched with sat. aq NaHCO$_3$, and the resulting mixture was extracted with CH$_2$Cl$_2$ (3 ×). Analysis of the crude material by ¹H NMR spectroscopy revealed the yield to be 68% according to the internal standard. The combined organic phases were dried (MgSO$_4$), and the solvent was removed by distillation at ambient pressure. The residue was purified by column chromatography (silica gel, pentane) to give the title compound as a colorless oil; yield: 85 mg (60%).

for references see p 154

References

[1] Hartley, R. C.; McKiernan, G. J., *J. Chem. Soc., Perkin Trans. 1*, (2002), 2763.

[2] Van de Weghe, P.; Bisseret, P.; Blanchard, N.; Eustache, J., *J. Organomet. Chem.*, (2006) **691**, 5078.

[3] Hartley, R. C.; Li, J.; Main, C. A.; McKiernan, G. J., *Tetrahedron*, (2007) **63**, 4825.

[4] Nicolaou, K. C.; Postema, M. H. D.; Claiborne, C. F., *J. Am. Chem. Soc.*, (1996) **118**, 1565.

[5] Nicolaou, K. C.; Postema, M. H. D.; Yue, E. W.; Nadin, A., *J. Am. Chem. Soc.*, (1996) **118**, 10335.

[6] Calimente, D.; Postema, M. H. D., *J. Org. Chem.*, (1999) **64**, 1770.

[7] Postema, M. H. D.; Calimente, D.; Liu, L.; Behrmann, T. L., *J. Org. Chem.*, (2000) **65**, 6061.

[8] Oishi, T.; Uehara, H.; Nagumo, Y.; Shoji, M.; Le Brazidec, J.-Y.; Kosaka, M.; Hirama, M., *Chem. Commun. (Cambridge)*, (2001), 381.

[9] Rainier, J. D.; Allwein, S. P.; Cox, J. M., *J. Org. Chem.*, (2001) **66**, 1380.

[10] Allwein, S. P.; Cox, J. M.; Howard, B. E.; Johnson, H. W. B.; Rainier, J. D., *Tetrahedron*, (2002) **58**, 1997.

[11] Uehara, H.; Oishi, T.; Inoue, M.; Shoji, M.; Nagumo, Y.; Kosaka, M.; Le Brazidec, J.-Y.; Hirama, M., *Tetrahedron*, (2002) **58**, 6493.

[12] Inoue, M.; Yamashita, S.; Tatami, A.; Miyazaki, K.; Hirama, M., *J. Org. Chem.*, (2004) **69**, 2797.

[13] Majumder, U.; Rainier, J. D., *Tetrahedron Lett.*, (2005) **46**, 7209.

[14] Clark, J. S., *Chem. Commun. (Cambridge)*, (2006), 3571.

[15] Majumder, U.; Cox, J. M.; Johnson, H. W. B.; Rainier, J. D., *Chem.–Eur. J.*, (2006) **12**, 1736.

[16] Johnson, H. W. B.; Majumder, U.; Rainier, J. D., *Chem.–Eur. J.*, (2006) **12**, 1747.

[17] Iyer, K.; Rainier, J. D., *J. Am. Chem. Soc.*, (2007) **129**, 12604.

[18] Roberts, S. W.; Rainier, J. D., *Org. Lett.*, (2007) **9**, 2227.

[19] Zhang, Y.; Rainier, J. D., *Org. Lett.*, (2009) **11**, 237.

[20] Nicolaou, K. C.; Gelin, C. F.; Seo, J. H.; Huang, Z.; Umezawa, T., *J. Am. Chem. Soc.*, (2010) **132**, 9900.

[21] Nicolaou, K. C.; Baker, T. M.; Nakamura, T., *J. Am. Chem. Soc.*, (2011) **133**, 220.

[22] Keck, G. E.; Poudel, Y. B.; Cummins, T. J.; Rudra, A.; Covel, J. A., *J. Am. Chem. Soc.*, (2011) **133**, 744.

[23] Zhang, Y.; Rohanna, J.; Zhou, J.; Iyer, K.; Rainier, J. D., *J. Am. Chem. Soc.*, (2011) **133**, 3208.

[24] Schrock, R. R.; Murdzek, J. S.; Bazan, G. C.; Robbins, J.; DiMare, M.; O'Regan, M., *J. Am. Chem. Soc.*, (1990) **112**, 3875.

[25] Bazan, G. C.; Oskam, J. H.; Cho, H.-N.; Park, L. Y.; Schrock, R. R., *J. Am. Chem. Soc.*, (1991) **113**, 6899.

[26] Nguyen, S. T.; Grubbs, R. H.; Ziller, J. W., *J. Am. Chem. Soc.*, (1993) **115**, 9858.

[27] Schwab, P.; France, M. B.; Ziller, J. W.; Grubbs, R. H., *Angew. Chem.*, (1995) **107**, 2179; *Angew. Chem. Int. Ed.*, (1995) **34**, 2039.

[28] Scholl, M.; Ding, S.; Lee, C. W.; Grubbs, R. H., *Org. Lett.*, (1999) **1**, 953.

[29] Brown-Wensley, K. A.; Buchwald, S. L.; Cannizzo, L.; Clawson, L.; Ho, S.; Meinhardt, D.; Stille, J. R.; Straus, D.; Grubbs, R. H., *Pure Appl. Chem.*, (1983) **55**, 1733.

[30] Grubbs, R. H.; Pine, S. H., In *Comprehensive Organic Synthesis*, Trost, B. M.; Fleming, I., Eds.; Pergamon: Oxford, (1991); Vol. 5, p 1115.

[31] Grubbs, R. H.; Pine, S. H., In *Comprehensive Organic Synthesis*, Trost, B. M.; Fleming, I., Eds.; Pergamon: Oxford, (1991); Vol. 5, p 1127.

[32] Tebbe, F. N.; Parshall, G. W.; Reddy, G. S., *J. Am. Chem. Soc.*, (1978) **100**, 3611.

[33] Pine, S. H.; Zahler, R.; Evans, D. A.; Grubbs, R. H., *J. Am. Chem. Soc.*, (1980) **102**, 3270.

[34] Howard, T. R.; Lee, J. B.; Grubbs, R. H., *J. Am. Chem. Soc.*, (1980) **102**, 6876.

[35] Ott, K. C.; Grubbs, R. H., *J. Am. Chem. Soc.*, (1981) **103**, 5922.

[36] Petasis, N. A.; Bzowej, E. I., *J. Am. Chem. Soc.*, (1990) **112**, 6392.

[37] Takai, K.; Kakiuchi, T.; Kataoka, Y.; Utimoto, K., *J. Org. Chem.*, (1994) **59**, 2668.

[38] Fujimura, O.; Fu, G. C.; Grubbs, R. H., *J. Org. Chem.*, (1994) **59**, 4029.

[39] Sturino, C. F.; Wong, J. C. Y., *Tetrahedron Lett.*, (1998) **39**, 9623.

[40] Okada, A.; Ohshima, T.; Shibasaki, M., *Tetrahedron Lett.*, (2001) **42**, 8023.

[41] Aggarwal, V. K.; Daly, A. M., *Chem. Commun. (Cambridge)*, (2002), 2490.

[42] Bennasar, M. L.; Roca, T.; Monerris, M.; García-Díaz, D., *Tetrahedron Lett.*, (2005) **46**, 4035.

[43] Bennasar, M. L.; Roca, T.; Monerris, M.; García-Díaz, D., *J. Org. Chem.*, (2006) **71**, 7028.

[44] Stille, J. R.; Santarsiero, B. D.; Grubbs, R. H., *J. Org. Chem.*, (1990) **55**, 843.

[45] Gilliom, L. R.; Grubbs, R. H., *J. Am. Chem. Soc.*, (1986) **108**, 733.

[46] Stille, J. R.; Grubbs, R. H., *J. Am. Chem. Soc.*, (1986) **108**, 855.

[47] Kadota, I.; Kadowaki, C.; Park, C.-H.; Takamura, H.; Sato, K.; Chan, P. W. H.; Thorand, S.; Yamamoto, Y., *Tetrahedron*, (2002) **58**, 1799.

[48] Horikawa, Y.; Watanabe, M.; Fujiwara, T.; Takeda, T., *J. Am. Chem. Soc.*, (1997) **119**, 1127.
[49] Rohanna, J. C.; Rainier, J. D., *Org. Lett.*, (2009) **11**, 493.
[50] Zhou, J.; Rainier, J. D., *Org. Lett.*, (2009) **11**, 3774.
[51] Neumaier, J.; Maier, M. E., *Synlett*, (2011), 187.
[52] Fu, G. C.; Grubbs, R. H., *J. Am. Chem. Soc.*, (1993) **115**, 3800.
[53] Lokare, K. S.; Odom, A. L., *Inorg. Chem.*, (2008) **47**, 11191.
[54] Hong, B.; Li, H.; Wu, J.; Zhang, J.; Lei, X., *Angew. Chem.*, (2015) **127**, 1025; *Angew. Chem. Int. Ed.*, (2015) **54**, 1011.
[55] Griffith, A. K.; Vanos, C. M.; Lambert, T. H., *J. Am. Chem. Soc.*, (2012) **134**, 18581.
[56] Naidu, V. R.; Bah, J.; Franzén, J., *Eur. J. Org. Chem.*, (2015), 1834.

1.4   **Radical Reactions**

1.4.1   **Peroxy Radical Additions**

X. Hu and T. J. Maimone

General Introduction

The cyclization of peroxy radicals has captivated practitioners of synthetic, mechanistic, and bioorganic chemistry for decades. Owing to the relevance of such processes to various key biochemical and biosynthetic transformations,[1] in addition to the discovery of numerous biologically active peroxide-containing natural products,[2] a substantial body of work exists on the topic. Because of their perceived unpredictable nature, coupled with the potential dangers associated with handling peroxides in general, peroxy radical reactions are not typically a first choice synthetic method for the construction of C—O bonds. Nevertheless, if used in domino transformations, these intermediates can be successfully deployed to assemble complex oxygenated architectures rapidly, with the advantage that these events can be performed by using inexpensive reagents and sometimes with just air itself. Critically important for synthetic applications are the methods by which the peroxy radicals are generated and their common cyclization and termination modes. Though a growing number of methods exist for the synthesis of peroxides, this chapter focuses on domino transformations involving (or believed to involve) peroxy radical intermediates. The interested reader can find several informative reviews on the synthesis of organic peroxides as a whole.[3,4]

SAFETY: Organic peroxides can undergo spontaneous and exothermic decomposition; great care must be taken upon handling them in pure form as well as during reactions in which they are generated. It is highly recommended that synthetic reactions involving peroxide-containing intermediates be conducted behind a blast shield in a well-ventilated fume hood, especially if elevated temperatures are employed. Caution must also be exercised upon handling oxygen gas in the presence of organic solvent vapors. The stability of organic peroxides varies greatly, and it should not be assumed that a given procedure is safe on a new or untested substrate or on scales that deviate from literature reports. Thus, all synthetic reactions should be initially performed on small scales and the potential hazards carefully evaluated before scaling up. Experimentalists unfamiliar with peroxides are encouraged to consult one of several useful references pertaining to their properties and safety.[5,6]

1.4.1.1   **Initiation from a Preexisting Hydroperoxide**

The simplest method to generate a peroxy radical for use in a domino transformation is through O—H abstraction from a preexisting organic hydroperoxide. Owing to the weak R^1OO—H bond, a wide range of reagents can selectively initiate this process over potentially competitive C—H bond abstraction pathways, including those involving allylic hydrogen atoms.[7,8] The multitude of synthetic methods available for hydroperoxide synthesis also bodes well for implementing such strategies.[3,4] Although even air itself can initiate

for references see p 184

158 Domino Transformations **1.4** Radical Reactions

this radical process [Scheme 1; note that the same reaction in the presence of 2,6-di-*tert*-butyl-4-methylphenol (BHT) gives 0% yield],[9] the majority of synthetically useful transformations of this type involve reacting a hydroperoxide in the presence of a radical initiator, which is often either itself a peroxide or a metal salt. Notably, the low to moderate yields often observed in peroxy radical cyclizations can, in certain cases, be attributed to the instability of the endoperoxide products themselves and not necessarily to the efficiency of the parent transformation.

Scheme 1 Air-Initiated Peroxy Radical Cyclization[9]

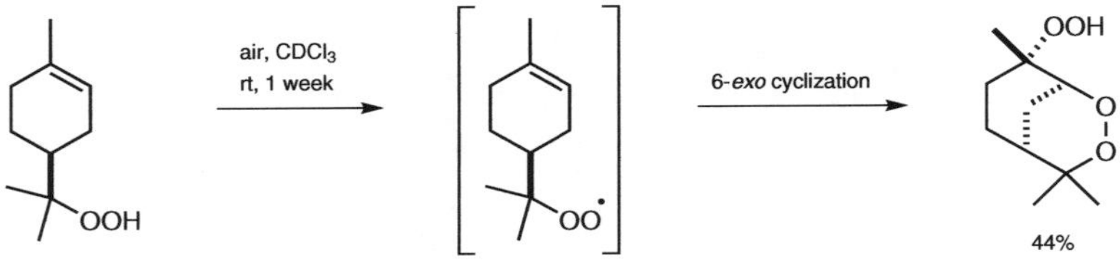

1.4.1.1.1 **Using Peroxide Initiators**

A number of radical initiators have been used to induce peroxy radical cyclizations through O—H abstraction, most notably the reagents di-*tert*-butyl peroxyoxalate[10] and di-*tert*-butyl hyponitrite as shown in Scheme 2.[11] With a half-life of 6.8 min at 60 °C, di-*tert*-butyl peroxyoxalate is a convenient source of *tert*-butoxy radicals under mild conditions. It is typically employed in an inert solvent such as benzene or carbon tetrachloride. Peroxy radicals, once formed, add easily to electron-rich alkenes. The cyclization mode and stereochemistry of peroxy radical addition to alkenes have been studied,[7,8,12] and in accordance with the Houk–Beckwith guidelines for radical cyclizations, 5-*exo*- and 6-*exo*-cyclization pathways dominate.[13,14] Representative di-*tert*-butyl peroxyoxalate initiated cyclization processes are shown in Scheme 2. After initial cyclization of the peroxy radical, the newly formed carbon-centered radical (e.g., **1**) is trapped with oxygen and typically subjected to in situ hydroperoxide reduction. This reduction is chemoselective, as the endoperoxide unit remains intact.[15] For cases in which competing 5-*exo* and 6-*exo* pathways are possible, the 5-*exo* cyclization typically dominates.[16] Following cyclization, hydroperoxides bearing α-hydrogen atoms can also be converted directly into ketones through acetylation and peroxide fragmentation (e.g., **4** to **5**). This particular example by Mayrargue and co-workers also highlights the difficulty encountered in forming larger, seven-membered endoperoxides through peroxy radical cyclizations.[17]

Scheme 2 Structure of Di-*tert*-butyl Peroxyoxalate and Di-*tert*-butyl Hyponitrite and Representative Di-*tert*-butyl Peroxyoxalate Initiated Peroxy Radical Cyclizations[7,15,17]

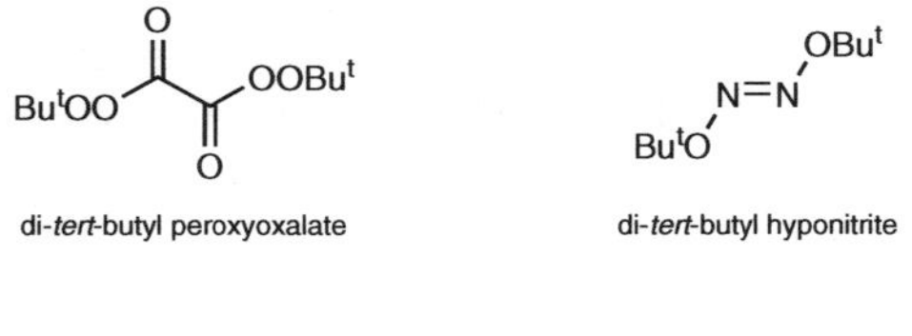

di-*tert*-butyl peroxyoxalate

di-*tert*-butyl hyponitrite

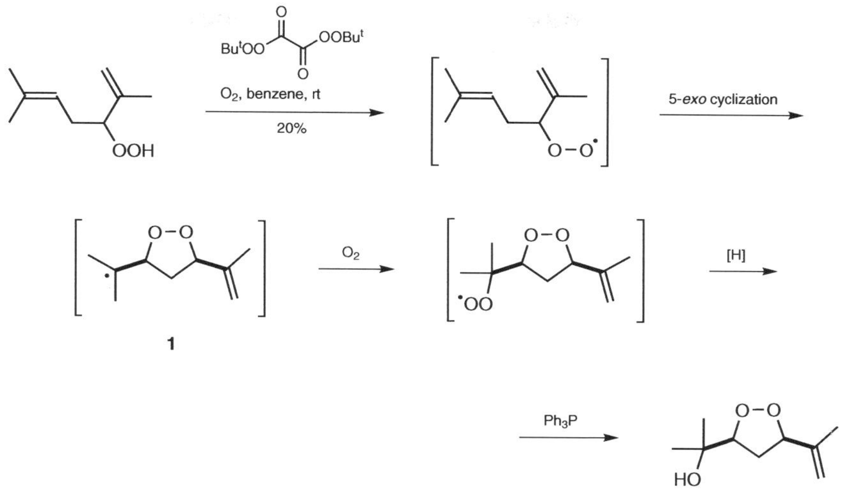

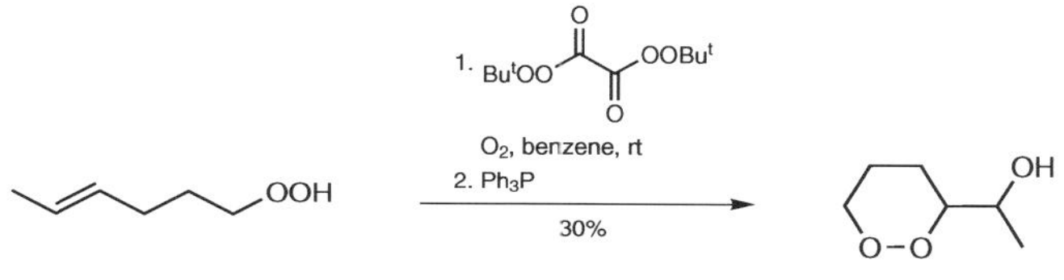

1.4.1 Peroxy Radical Additions

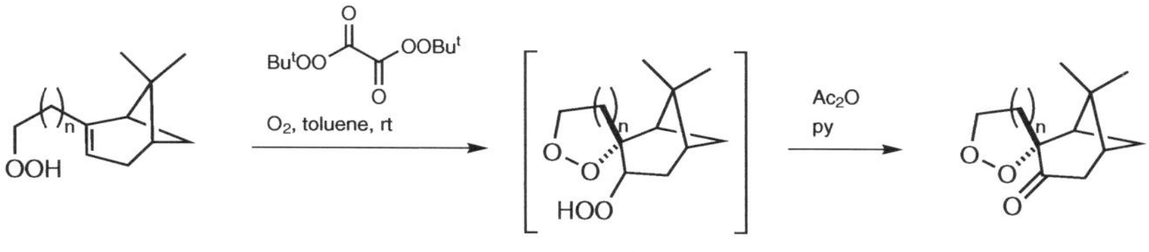

159

If multiple alkenes are positioned accordingly, polyene substrates can participate in a domino oxygenation process to afford multiple endoperoxide rings (Scheme 3).[18,19] For example, Porter and co-workers have reported the formation of **9** (as a mixture of four isomers) and epoxide **11** from hydroperoxide **6** through a di-*tert*-butyl peroxyoxalate induced peroxy radical cascade cyclization (Scheme 3).[18] Presumably, two pathways are available to the initially formed carbon-centered radical **7**. In the first route, a second oxygen-trapping/cyclization/oxygen-trapping sequence leads to peroxy radical **8**, which is ultimately reduced to **9**. Intermediate **7**, however, can also presumably undergo a $S_{H}i$ reaction (intramolecular carbon radical substitution) to afford oxy radical **10** and, ultimately, epoxide **11**.

for references see p 184

Scheme 3 Di-*tert*-butyl Peroxyoxalate Mediated Domino Cyclization of a Polyene[18]

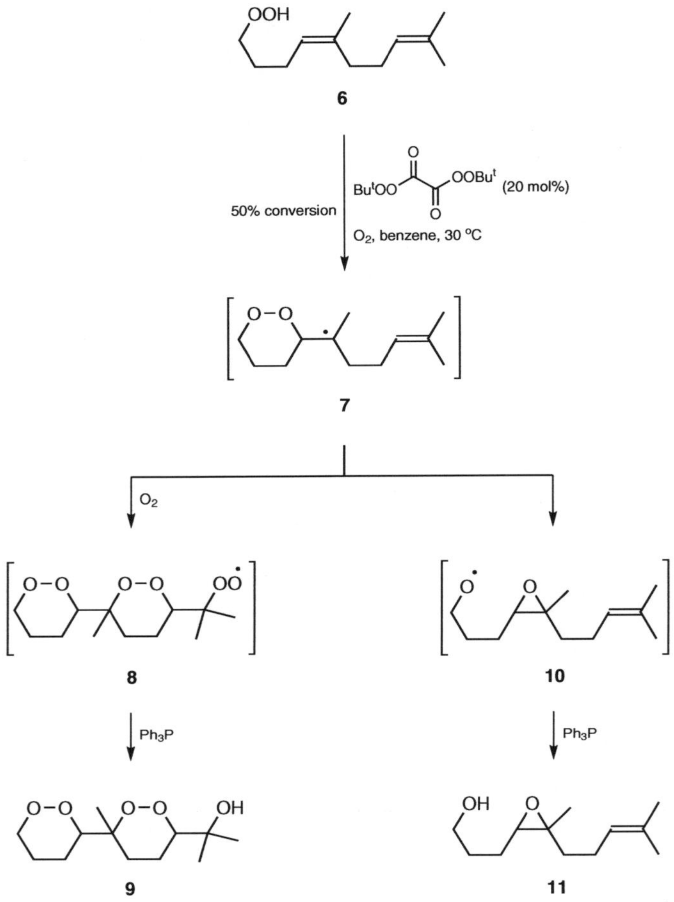

1-(1,2-Dioxan-3-yl)ethanol (3); Typical Procedure:[7]

> **CAUTION:** *Owing to the potential risk of explosion, it is strongly recommended that all operations involving the use or preparation of di-tert-butyl peroxyoxalate or di-tert-butyl hyponitrite be performed behind a safety shield.*[5,6,10,11] *Di-tert-butyl peroxyoxalate is shock and scratch sensitive and solid di-tert-butyl hyponitrite can detonate if struck. These reagents can be stored for extended periods at −10 °C or below, but their preparation and storage on large scales is not recommended.*

Hydroperoxide **2** (605 mg, 5.2 mmol) was dissolved in O_2-saturated benzene (500 mL) (**CAUTION:** *carcinogen*) and di-*tert*-butyl peroxyoxalate (475 mg, 2.0 mmol) was added. The mixture was stirred at rt for 48 h under an O_2 atmosphere and then cooled to 5 °C. Ph$_3$P (1.01 g, 3.8 mmol) was added. Upon completion of the hydroperoxide reduction, the mixture was concentrated and purified by chromatography (silica gel) to afford the title compound; yield: 201 mg (30%).

1.4.1.1.2 Using Copper(II) Trifluoromethanesulfonate/Oxygen

Catalytic quantities of copper(II) trifluoromethanesulfonate in the presence of oxygen can serve as a safer alternative to di-*tert*-butyl peroxyoxalate for the generation of peroxy radicals from hydroperoxides.[20] These conditions are also notable in that they can efficiently form such species at low temperatures. During model studies directed toward the synthesis of prostaglandins, Haynes and co-workers used copper(II) trifluoromethanesulfonate/oxygen to initiate the peroxy radical cyclization of unsaturated hydroperoxide **12** (Scheme 4).[20] Following cyclization, carbon-centered radical **13** engages the pendent diene in a 5-*exo* radical-cyclization/oxygen-recombination process to afford **14A–14C** [ratio (**14A/14B/14C**) 14:77:9], compounds that feature the connectivity (but incorrect stereochemistry) found in natural prostanoids. A certain amount of intermediate **13** is also partitioned down an oxygen-recombination pathway, which results in the formation of **15** as a mixture of isomers. Interestingly, **15** is the sole product generated if the reaction is performed at −20 °C.

Scheme 4 Copper(II) Trifluoromethanesulfonate Induced Domino Transformation[20]

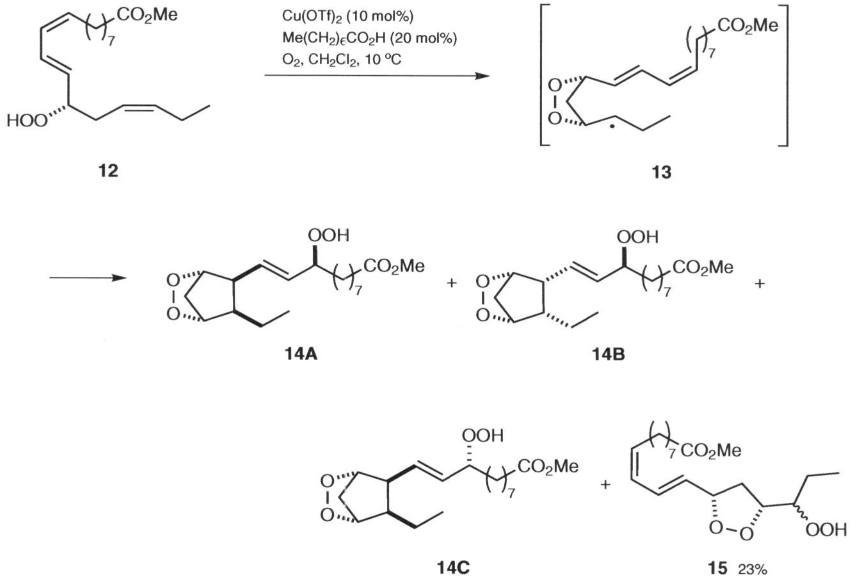

Methyl (E)-11-{(1R,4S)-6-Ethyl-2,3-dioxabicyclo[2.2.1]heptan-5-yl}-9-hydroperoxyundec-10-enoates 14A–14C and Methyl (9Z,11E)-12-[(3S,5R)-5-(1-Hydroperoxypropyl)-1,2-dioxolan-3-yl]dodeca-9,11-dienoate (15); Typical Procedure:[20]
$Me(CH_2)_6CO_2H$ (0.2 equiv) and a 0.1 M soln of $Cu(OTf)_2$ (0.1 equiv) in MeCN were added to a soln of hydroperoxide **12** (78 mg, 0.24 mmol) in CH_2Cl_2 (4 mL) at 10 °C under an atmosphere of O_2. After 40 min, the reaction was judged complete. The solvent was removed, and the material was purified by chromatography (silica gel) to afford **15** as a mixture of isomers; yield: 23%; further purification by HPLC (EtOAc/petroleum ether 1:3) afforded **14A–14C**.

for references see p 184

162 Domino Transformations **1.4** Radical Reactions

1.4.1.1.3 **Using Samarium(II) Iodide/Oxygen**

The combination of samarium(II) iodide and oxygen, originally reported by Corey and Wang,[21] has found general use in the generation of peroxy radicals from hydroperoxides. Presumably generating a species of the form "$I_2SmOOSmI_2$", this reagent combination has proven to be a convenient initiator of peroxy radicals under mild catalytic conditions. Representative cyclizations are shown in Scheme 5, e.g. **16** to **17**; the reported yields are typically higher than those seen in di-*tert*-butyl peroxyoxalate mediated cyclization reactions. Sherburn and co-workers have utilized this reagent combination to cyclize sensitive hydroperoxide **18** into endoperoxide **19** in an impressive 82% overall yield.[22] These conditions also served to initiate peroxy radical cyclization of hydroperoxide **20** during Corey's investigations into the biosynthesis of prostaglandin G_2 (PGG_2) (Scheme 5).[21]

Scheme 5 Initiation by Samarium(II) Iodide/Oxygen[21,22]

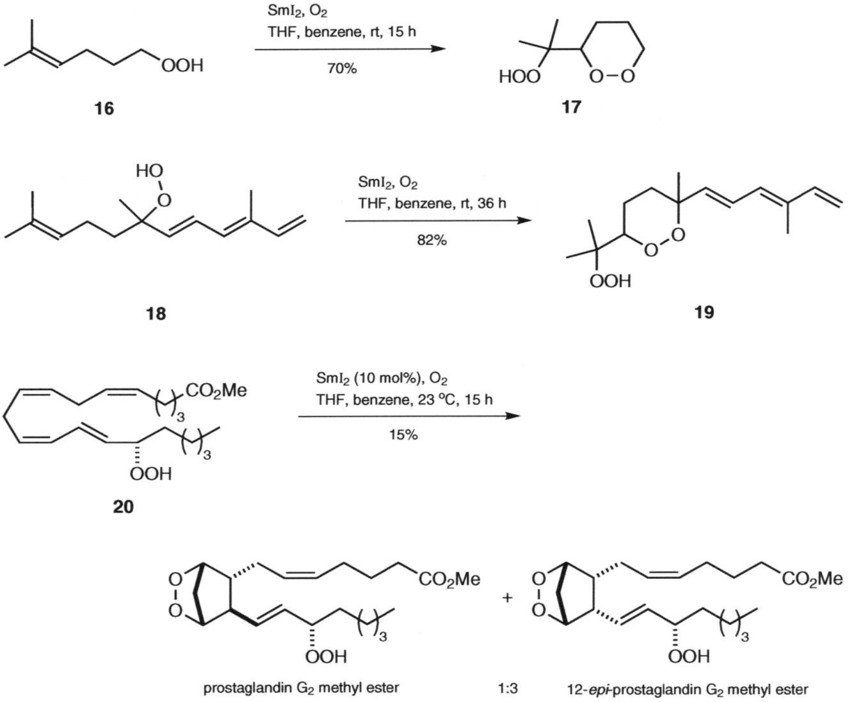

prostaglandin G_2 methyl ester   1:3   12-*epi*-prostaglandin G_2 methyl ester

Boukouvalas and co-workers have employed a 6-*exo* peroxy radical cyclization of hydroperoxide **21** as a key step during an exceedingly short biomimetic total synthesis of ying-zhaosu C and *epi*-yingzhaosu C (Scheme 6).[23] Of particular note is the comparison of previous peroxidation methods for the conversion of **21** into **22A** and **22B** (Scheme 6); conditions involving the employment of copper(II) trifluoromethanesulfonate/oxygen are completely ineffectual at eliciting this transformation, whereas di-*tert*-butyl peroxyoxalate/oxygen and samarium(II) iodide/oxygen give somewhat similar results. Interestingly, the addition of 10 equivalents of *tert*-butyl hydroperoxide to a typical di-*tert*-butyl peroxyoxalate/oxygen initiated cyclization has been found to be the most effective combination, affording **22A** and **22B** as a 1.7:1 mixture of isomers in an outstanding 89% yield.

1.4.1 Peroxy Radical Additions **163**

Scheme 6 Total Synthesis of Yingzhaosu C[23]

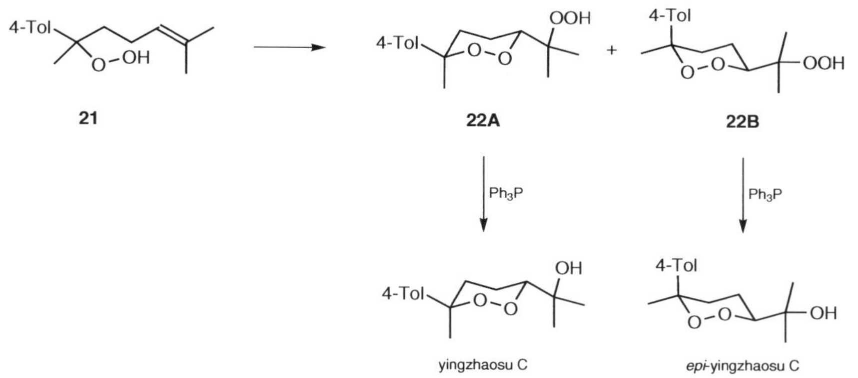

Conditions	Yield (%)	Ref
Cu(OTf)$_2$ (10 mol%), Me(CH$_2$)$_6$CO$_2$H (20 mol%), O$_2$, MeCN, −25 °C	0	[23]
SmI$_2$ (10 mol%), O$_2$, benzene, rt, 46 h	79	[23]
di-*tert*-butyl peroxyoxalate (50 mol%), O$_2$, benzene, rt, 7 h	66	[23]
di-*tert*-butyl peroxyoxalate (50 mol%), O$_2$, *t*-BuOOH (10 equiv), benzene, rt, 4.5 h	89^a	[23]

^{a} Obtained as a 1.7:1 mixture of **22A/22B**.

2-{6-Methyl-6-[(1*E*,3*E*)-4-methylhexa-1,3,5-trienyl]-1,2-dioxan-3-yl}propan-2-yl Hydroperoxide (19); Typical Procedure:[22]
An aged, yellow-colored 0.1 M soln of SmI$_2$ in THF (0.125 mL) was diluted with benzene (1 mL) (**CAUTION:** *carcinogen*) and added dropwise over the course of 2 h to a soln of hydroperoxide **18** (30 mg) in benzene (12 mL) under an atmosphere of dry O$_2$ at rt. After 36 h, the solvent was evaporated, and the residue was purified by preparative centrifugal chromatography [cold (0 °C) hexane/EtOAc 4:1] under an argon atmosphere to afford the title compound as a clear oil; yield: 28.2 mg (82%).

1.4.1.2   Initiation by Metal-Catalyzed Hydroperoxidation

The addition of a metal hydride across an alkene produces a metal alkyl complex capable of producing a carbon-centered radical through metal–carbon bond homolysis. In the presence of oxygen, a peroxy radical (or metal peroxo species) is formed, which can undergo domino cyclization processes.[24] Catalysts and reagents based on cobalt,[25–27] manganese,[28–30] and iron[31,32] are frequently utilized to initiate this chemistry.

1.4.1.2.1   The Mukaiyama Hydration/Hydroperoxidation

Mukaiyama hydroperoxidation is one of the mildest methods available to synthesize silylated peroxides from alkenes (see compound **27**, Scheme 7).[25–27] This transformation is also notable in that both an oxidant (oxygen) and reductant (triethylsilane) are present in the reaction simultaneously. The proposed mechanism of this transformation, and several common metal precatalysts that can be employed, including bis[4,4-dimethyl-1-(morpholinocarbonyl)pentane-1,3-dionato]cobalt(II) {[Co(modp)$_2$]; modp = 4,4-dimethyl-1-(morpholinocarbonyl)pentane-1,3-dionato} and tris(2,2,6,6-tetramethyl-3,5-heptanedionato)manganese(III) {[Mn(dpm)$_3$]; dpm = 2,2,6,6-tetramethyl-3,5-heptanedionato}, are shown in Scheme 7.[24] It is proposed that a cobalt(III) hydride **23** enters the catalytic

for references see p 184

164   Domino Transformations   **1.4** Radical Reactions

cycle and engages the alkene in a hydrocobaltation process to afford intermediate cobalt(III)–alkyl complex **24**. The weak C—Co bond is then cleaved to produce radical **25**, which combines with oxygen to form complex **26**. Complex **26** is converted into product **27** by way of a σ-bond metathesis reaction with triethylsilane. Intermediates **25** and **26**, which both display radical character, serve as entry points into domino transformations involving peroxy radicals.

Scheme 7   Mukaiyama Hydration: Precatalysts and Mechanism[9,24]

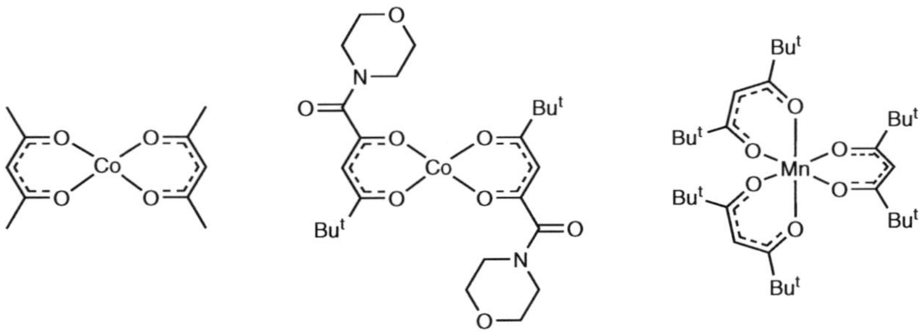

Co(acac)₂   Co(modp)₂   Mn(dpm)₃

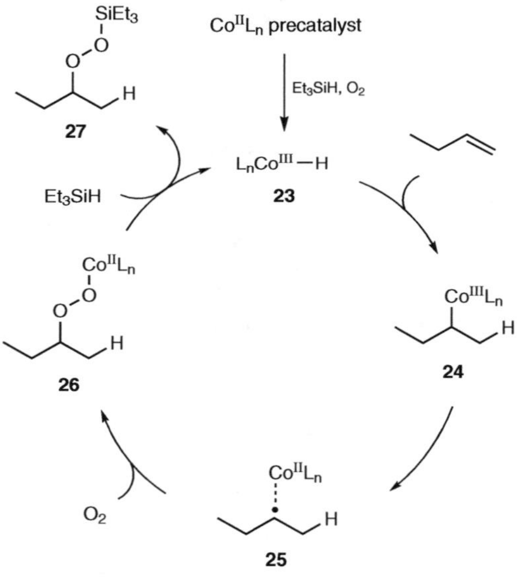

1.4.1.2.2   **Hydroperoxidation-Initiated Domino Transformations**

The growing number of endoperoxides isolated from nature with antimalarial activity has fueled interest in synthetic strategies toward both complex endoperoxide natural products and simplified analogues.[33–35] Domino peroxidation reactions feature prominently in this arena. For example, Nojima and co-workers have subjected (S)-limonene [(S)-**28**] to Mukaiyama conditions (Scheme 8); this process results in the chemoselective formation of presumed peroxy radical **29** at the exocyclic alkene position. This species undergoes 6-exo cyclization and reaction with oxygen to afford bicyclic endoperoxides **30** and **31**.[9] A small amount of standard Mukaiyama product **32** is also formed. The synthetic utility of such processes is largely dictated by the efficiency of the final peroxy radical cyclization; if this step is slow, the peroxy radical is either rapidly reduced to a hydro-

1.4.1 Peroxy Radical Additions **165**

peroxide or is silylated. Cyclization of diene **33** is illustrative of this point, as less favorable addition to the conjugated ester results mainly in the formation of **34** and **35**, with a very small amount of domino cyclization product **36**.[9] Increasing dilution, however, can be beneficial in promoting the domino cyclization process.

Scheme 8 Cobalt-Catalyzed Domino Hydroperoxidation[9]

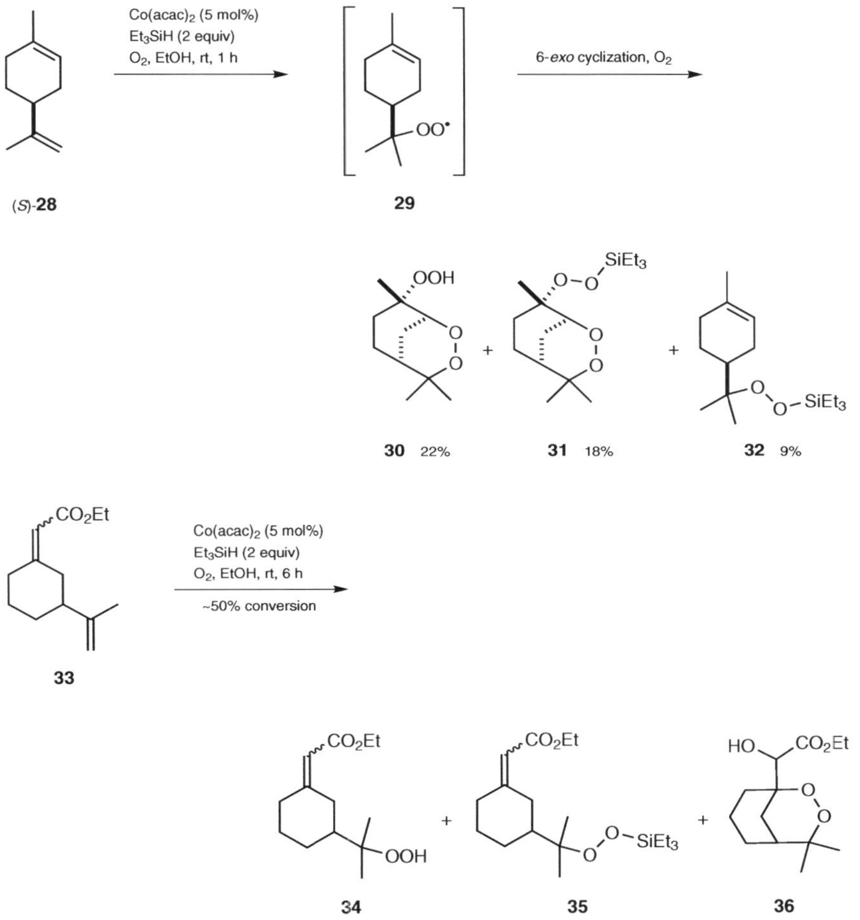

Concentration (M) of **33**	Yield (%) of **34**	Yield (%) of **35**	Yield (%) of **36**	Ref
0.4	18	34	0	[9]
0.1	47	17	20	[9]

Cyclopropane cleavage reactions of the putative Mukaiyama carbon-centered radical intermediate can also trigger domino transformations (Scheme 9). During mechanistic investigations into the cobalt-catalyzed hydroperoxidation reaction, Nojima and co-workers found that if vinylcyclopropane **37** is exposed to a catalytic amount of bis[4,4-dimethyl-1-(morpholinocarbonyl)pentane-1,3-dionato]cobalt(II)/triethylsilane/oxygen, endoperoxide **39** is formed in 38% yield.[24] Presumably, carbon-centered radical **38** is formed following C—C bond cleavage. This species combines with oxygen to undergo 5-*exo* peroxy

for references see p 184

radical cyclization and oxygen trapping, which affords **39** after reaction with triethylsilane. Formally, this represents an insertion of oxygen into a cyclopropane C—C bond, a process that will be discussed further in Section 1.4.1.5.1.

Scheme 9   Domino Hydroperoxidation of a Vinylcyclopropane[24]

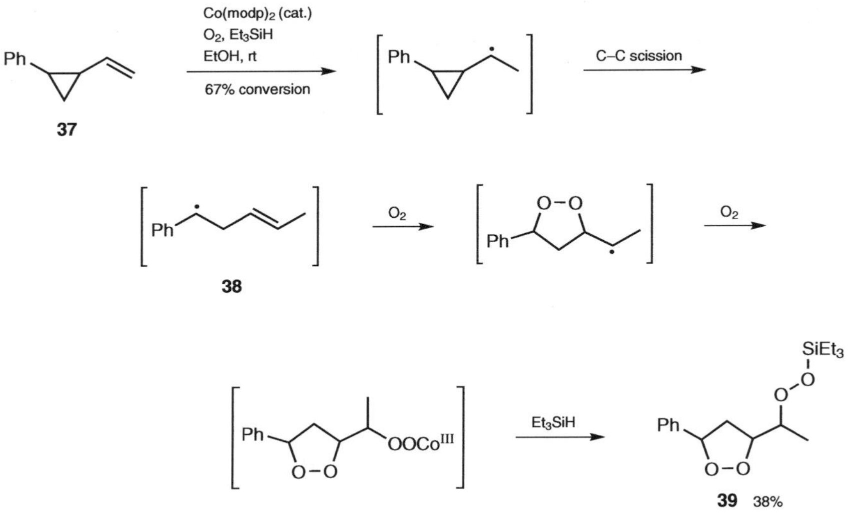

Ethyl [3-(2-Hydroperoxypropan-2-yl)cyclohexylidene]acetate (34), Ethyl (3-{2-[(Triethylsilyl)peroxy]propan-2-yl}cyclohexylidene)acetate (35), and Ethyl (4,4-Dimethyl-2,3-dioxabicyclo[3.3.1]nonan-1-yl)(hydroxy)acetate (36); Typical Procedure:[9]
A two-necked, 50-mL flask was charged with diene **33** (210 mg, 1.0 mmol), Co(acac)$_2$ (13 mg, 0.05 mmol), EtOH (10 mL), and O$_2$. Et$_3$SiH (230 mg, 2.0 mmol) was added by syringe, and the mixture was stirred vigorously under an atmosphere of O$_2$ at rt for 6 h. The solvent was removed under reduced pressure, and the residue was purified by chromatography (silica gel, Et$_2$O/hexane 10:90 to 15:85) to afford hydroperoxide **34**; yield: 47% (based on consumed **33**); bicycle **36**; yield: 20% (based on consumed **33**); and a mixture of recovered **33** and **35**. The **33/35** mixture was treated with a drop of concd HCl in MeOH (1 mL) for 1 min, and then solid NaHCO$_3$ and anhyd MgSO$_4$ were added. The mixture was stirred for an additional 5 min, and the solid materials were then removed by filtration through Celite. The solvent was removed under reduced pressure, and the residue was purified by column chromatography (silica gel, Et$_2$O/hexane 10:90 to 25:75) to afford recovered **33**; conversion: 49%; and **34**; yield: 17% (based on consumed **33**).

1.4.1.2.3   Manganese-Catalyzed Domino Hydroperoxidation

Tris(2,2,6,6-tetramethyl-3,5-heptanedionato)manganese(III) [Mn(dpm)$_3$] has found widespread use as a catalyst in hydrofunctionalization chemistry.[36–38] Initial work by Magnus and co-workers details its particular usefulness in the hydration and reduction of conjugated alkenes, wherein a manganese hydride is proposed to initiate conjugate reduction.[28–30] Maimone and Hu have recently reported a domino peroxy radical cyclization proceeding in an unusual 7-*endo*-cyclization mode en route to a four-step total synthesis of the antimalarial natural product cardamom peroxide (**43**) (Scheme 10).[39] Dienedione **40** is subjected to modified Magnus hydration conditions [catalytic tris(2,2,6,6-tetramethyl-3,5-heptanedionato)manganese(III), phenylsilane, propan-2-ol, oxygen], which presumably affords peroxy radical **41** in a chemo- and regioselective manner. This intermediate

1.4.1 Peroxy Radical Additions

167

then undergoes diastereoselective peroxy radical cyclization to form a carbon-centered radical, which is stereoselectively quenched with oxygen to afford diperoxide **42** after reduction. In situ hydroperoxide reduction with triphenylphosphine yields cardamom peroxide (**43**) in 52% yield. Although manganese catalysis is superior to cobalt catalysis for this transformation, several additional experimental parameters must be tuned to obtain acceptable yields of the final product. As in the case of diene **33**, products (not shown) arising from premature reduction of **41** negatively impact the yield of cardamom peroxide; slow addition of phenylsilane is beneficial in this regard. This finding, combined with the incorporation of an added external oxidant (*tert*-butyl hydroperoxide), which presumably facilitates catalyst turnover, leads to optimal results.

Scheme 10 Manganese-Catalyzed Tandem Hydroperoxidation[39]

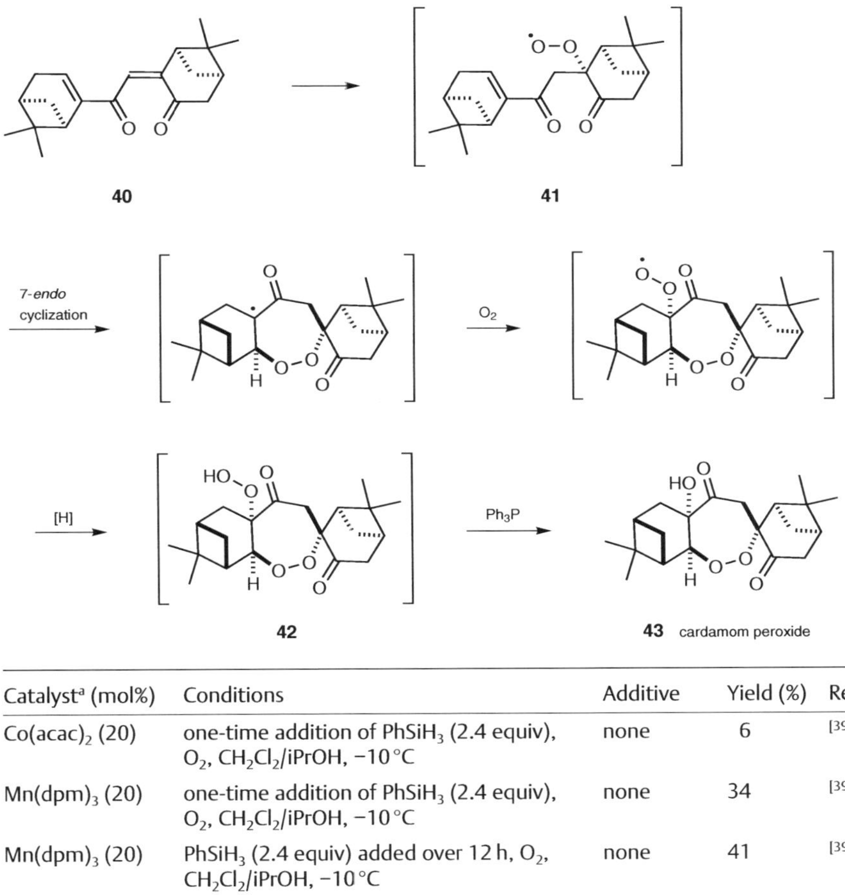

Catalyst[a] (mol%)	Conditions	Additive	Yield (%)	Ref
Co(acac)$_2$ (20)	one-time addition of PhSiH$_3$ (2.4 equiv), O$_2$, CH$_2$Cl$_2$/iPrOH, −10 °C	none	6	[39]
Mn(dpm)$_3$ (20)	one-time addition of PhSiH$_3$ (2.4 equiv), O$_2$, CH$_2$Cl$_2$/iPrOH, −10 °C	none	34	[39]
Mn(dpm)$_3$ (20)	PhSiH$_3$ (2.4 equiv) added over 12 h, O$_2$, CH$_2$Cl$_2$/iPrOH, −10 °C	none	41	[39]
Mn(dpm)$_3$ (20)	PhSiH$_3$ (2.4 equiv) added over 12 h, O$_2$, CH$_2$Cl$_2$/iPrOH, −10 °C	*t*-BuOOH (1.5 equiv)	52	[39]

[a] dpm = 2,2,6,6-tetramethyl-3,5-heptanedionato.

Cardamom Peroxide (43); Typical Procedure:[39]

A flame-dried, round-bottomed flask was charged with dienedione **40** (30 mg, 0.1 mmol, 1.0 equiv), Mn(dpm)$_3$ (12 mg, 0.02 mmol, 20 mol%), CH$_2$Cl$_2$ (1.6 mL), and iPrOH (0.4 mL). O$_2$ was vigorously bubbled through the soln for 5 min, which was followed by the addition of a soln of 5 M *t*-BuOOH in decane (30 μL, 0.15 mmol, 1.5 equiv). The soln was cooled to −10 °C under an atmosphere of O$_2$, and PhSiH$_3$ (30 μL, 0.24 mmol, 2.4 equiv) in CH$_2$Cl$_2$

for references see p 184

168 Domino Transformations **1.4** Radical Reactions

(1 mL) was added dropwise over 12 h using a syringe pump. After the addition was complete, a soln of Ph_3P (56 mg, 0.21 mmol) in CH_2Cl_2 was added dropwise at −10 °C to quench intermediate diperoxide **42**. The mixture was diluted with H_2O (5 mL) and extracted with CH_2Cl_2 (3 × 10 mL). The combined organic layer was washed with brine, dried (Na_2SO_4), filtered through Celite, and concentrated under reduced pressure. The crude mixture was purified by flash column chromatography (CH_2Cl_2) to afford the title compound as a white solid; yield: 18.2 mg (52%).

1.4.1.3 Initiation by Radical Addition/Oxygen Quenching

The addition of a free radical to an alkene and its subsequent reaction with triplet oxygen offers convenient access to peroxy radicals for use in subsequent transformations. Key to the success of this process is the identification of radical initiators that are compatible with the aerobic conditions required. Although the aforementioned metal-catalyzed hydroperoxidation reaction formally accomplishes the addition of a hydrogen radical across an alkene, it is in some cases desirable to be able to add more diverse atoms across alkenes, as this would allow for further synthetic manipulations downstream.

1.4.1.3.1 Thiyl Radical Initiation

The addition of sulfur-based radicals (thiyl radicals) to alkenes,[40] its reversibility,[41] and the outcome of such processes in the presence of oxygen[42] have been studied for nearly a century. Although largely focused on polymer applications, this chemistry has been developed into useful methodology for the synthesis of endoperoxides. The combination of a thiol, oxygen, and a radical initiator is typically employed to form the requisite thiyl radicals. Chalcogen-centered radicals (thiyl and selenyl) can also be efficiently generated from disulfide or diselenide precursors.

1.4.1.3.1.1 Thiol–Alkene Co-oxygenation Reactions

Beckwith first examined the classic thiol–alkene co-oxygenation (thiol–olefin co-oxygenation; TOCO) reaction from the vantage point of endoperoxide construction (Scheme 11).[43] The thiyl radical derived from benzenethiol smoothly adds to 1,4-diene **44**. The resulting radical (not shown) is trapped by oxygen and undergoes a 5-*exo*-cyclization/oxygen-trapping sequence leading to endoperoxides **45A** and **45B** after treatment with triphenylphosphine. The preferential formation of the *cis*-isomer **45A** parallels carbon-based radical cyclizations, and the addition of the thiyl radical to the less-hindered, terminal position of the alkene is general in these reactions. The nature of the substituents bound to the alkene that accepts the peroxy radical can play a role in determining product outcome. In the TOCO reaction of diene **46**, in addition to expected products **47A** and **47B**, some uncyclized products (compounds **48** and **49**), presumably formed by premature peroxy radical reduction, are observed if the pendent alkene does not carry full substitution.

1.4.1 Peroxy Radical Additions

Scheme 11 Representative Thiol–Alkene Co-oxygenation Reactions[43]

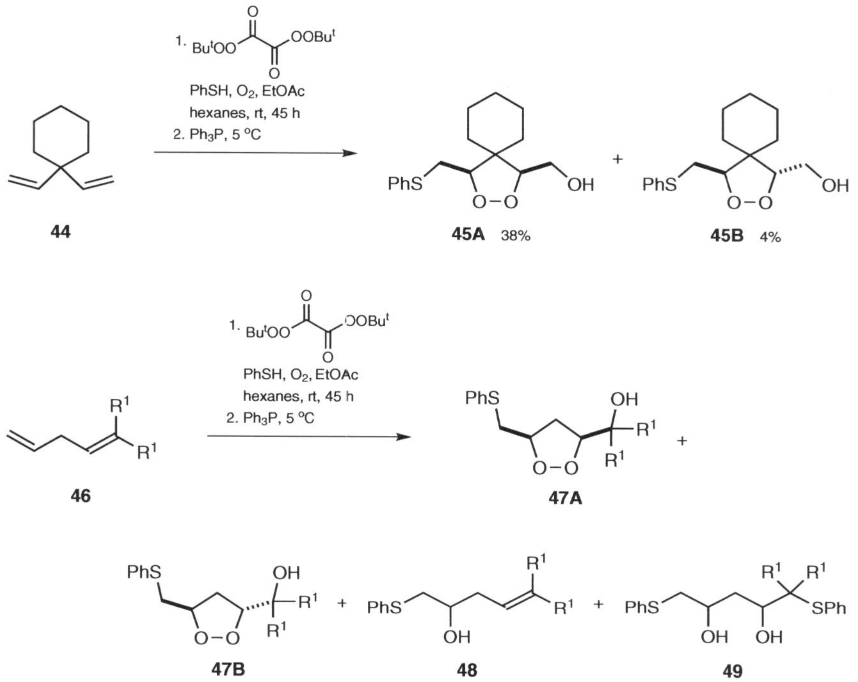

R¹	Yield (%) of **47A**	Yield (%) of **47B**	Yield (%) of **48**	Yield (%) of **49**	Ref
H	32	8	23	12	[43]
Me	20	5	_a	_a	[43]

ᵃ Not observed.

Bachi and co-workers have demonstrated the power of TOCO chemistry en route to the total synthesis of the antimalarial natural product yingzhaosu A (**53**) (Scheme 12).[44,45] In a model study,[45] utilizing the combination of di-*tert*-butyl peroxyoxalate/oxygen/benzenethiol to initiate thiyl radical formation, (R)-limonene [(R)-**28**] can be successfully engaged in a domino cyclization to afford bicyclic endoperoxides **50A** und **50B**, uncyclized compound **51**, and diphenyl disulfide (**52**) following in situ reduction with triphenylphosphine. The endoperoxides **50C** and **50D** derived from (S)-limonene [(S)-**28**] can be carried on to the natural product in just seven additional operations.[44] The multigram scales that this reaction can be conducted on are notable for preparative peroxy radical chemistry. Similarly to previous cases, the amount of limonene and the manner in which the thiol is added greatly affect the outcome of this transformation (Scheme 12). Optimal results are obtained in the presence of an excess amount of limonene and with the slow addition of benzenethiol. These conditions minimize both the dimerization of the thiyl radicals and the premature reduction of the initially formed peroxy radical leading to **51**.

for references see p 184

Scheme 12 Thiol–Alkene Co-oxygenation Reactions of Limonene[44,45]

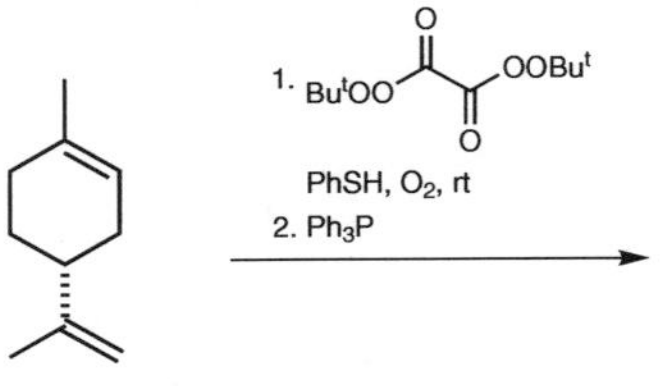

(R)-28

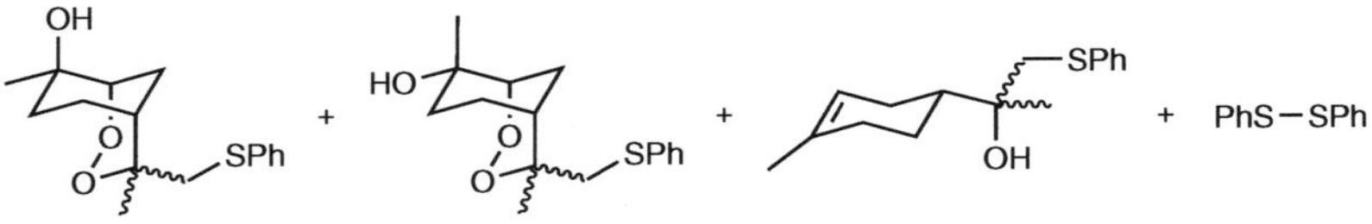

50A   50B   51   52

(R)-28 (Equiv)	Conditions	Solvent	Yield[a,b] (%) of 50A/50B	Yield (%) of 51	Yield (%) of 52	Ref
1	one-time addition of PhSH	benzene	14	32	25	[45]
3	one-time addition of PhSH	heptane/benzene	29	46	10	[45]
3	slow addition of PhSH over 10 h	heptane/benzene	54[c]	25	10	[45]

[a] Yield of isolated products based on PhSH.
[b] Ratio (**50A/50B**) was about 1:1.
[c] Use of the enantiomeric substrate (S)-28 also gave the corresponding products **50C/50D** in 54% yield.

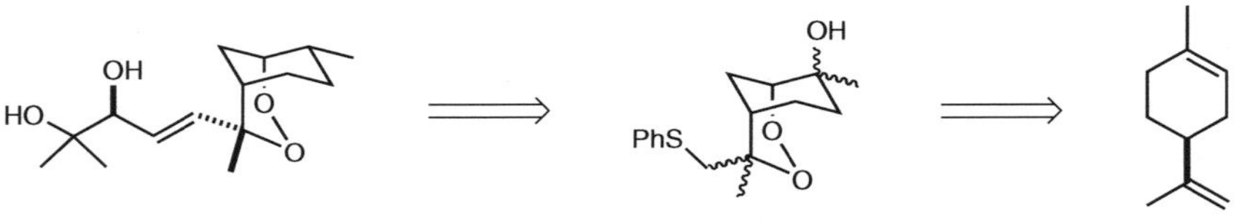

53 yingzhaosu A   **50C/50D**   (S)-**28**

1,3-Dienes readily react with thiyl radicals to give stabilized allylic radicals capable of undergoing further peroxy radical cyclizations. Beckwith has examined the TOCO reaction of 1,3,6-triene **54** using catalytic di-*tert*-butyl peroxyoxalate/4-methylbenzenethiol/oxygen (Scheme 13).[46,47] Several products **56–58** are formed owing to the nonsymmetric nature of the allylic radical **55** formed and whether the peroxy radical is prematurely quenched. The concentration of the thiol is a particularly important parameter for this transformation; only under very low concentrations of 4-methylbenzenethiol is the domino endoperoxide formed. This work also highlights the reversibility of the initial oxygen/allylic radical coupling step in determining product outcome.[47]

1.4.1 Peroxy Radical Additions **171**

Scheme 13 Thiol–Alkene Co-oxygenation Reaction of a Polyene[47]

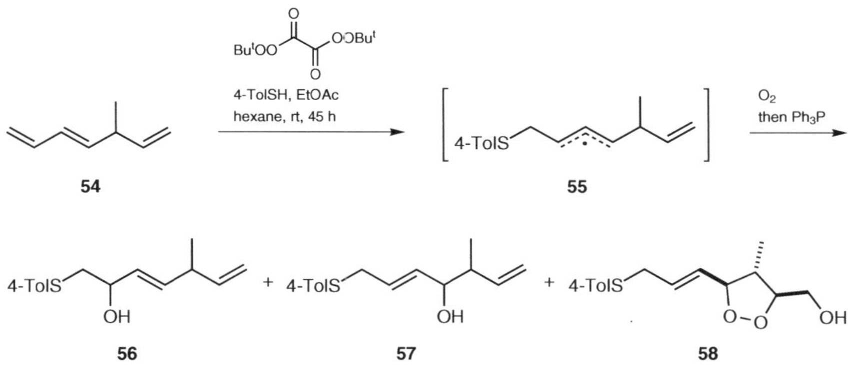

4-TolSH (M)	Yield (%) of **56**	Yield (%) of **57**	Yield (%) of **58**	Ref
0.04	47	29	<3	[47]
0.0045	44	21	26	[47]

4,8-Dimethyl-4-[(phenylsulfanyl)methyl]-2,3-dioxabicyclo[3.3.1]nonan-8-ol (50A/50B); Typical Procedure:[45]

> **CAUTION:** *Although no incidents occurred in the laboratory of the primary author,[45] particular attention should be given to potentially explosive mixtures of oxygen and vapors of organic solvents, particularly on large scales.*

O_2 gas was bubbled through a vigorously stirred soln of *R*-(+)-limonene [(*R*)-**28**; 6.13 g, 45 mmol, 3 equiv] and di-*tert*-butyl peroxyoxalate (105 mg, 0.45 mmol, 0.03 equiv) in heptane/benzene (5:2; 700 mL) (**CAUTION:** *carcinogen*) at rt. PhSH (1.65 g, 15 mmol, 1 equiv) in heptane (20 mL) was added over a period of 10 h to this mixture by syringe pump. After the addition was complete, the mixture was cooled to 0–5 °C. The mixture was then flushed with argon and diluted with CH_2Cl_2 (100 mL), and powdered Ph_3P (3.93 g, 15 mmol, 1 equiv) was added. The mixture was stirred for 2 h at 0–5 °C, warmed to rt, and stirred for an additional 1 h, and then the solvent was evaporated under reduced pressure. Flash chromatography (silica gel, EtOAc/hexane 1:49 to 3:7) afforded **50A/50B**; yield: 2.41 g (54%); and **51**; yield: 0.98 g (25%; 3:2 mixture of isomers). The excess amount of limonene was recovered by distillation; recovery: 72%.

1.4.1.3.1.2 Domino Transformations of Vinylcyclopropanes

Feldman and co-workers have devised peroxidation methodology based on a vinylcyclopropane ring-opening cascade.[48–51] These chalcogen-radical-induced processes are noteworthy in that the sulfur (or selenium) initiator is not incorporated into the final product. The overall process is shown in Scheme 14. The combination of diphenyl diselenide, 2,2′-azobisisobutyronitrile, and light produces a selenyl radical that adds to vinylcyclopropane **59**; the resulting radical intermediate undergoes cyclopropane scission to give **60**. This event is followed by recombination of **60** with oxygen to give **61**, and then 5-*exo* cyclization to give **62**. Rather than reacting further with oxygen, the newly formed carbon-centered radical (see compound **62**) ejects the selenium initiator, which leads to vinyldioxolane product **63** in 63% yield.

for references see p 184

Scheme 14 Vinylcyclopropane Oxygenation[48]

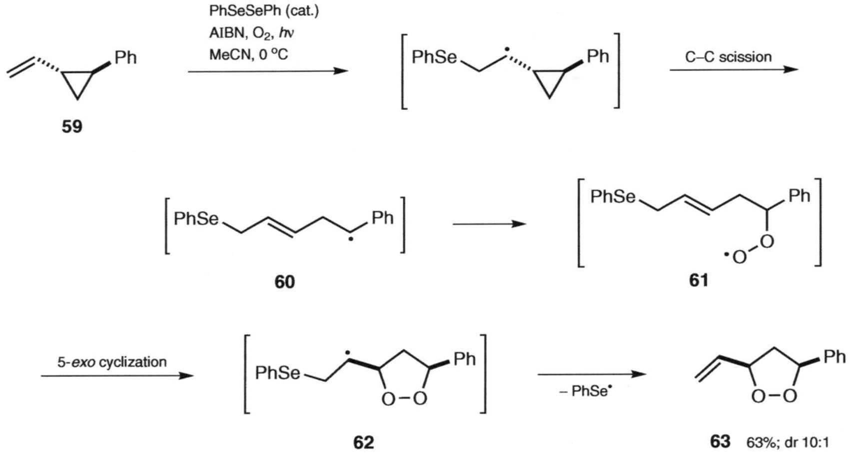

This process, which represents a formal insertion of oxygen into a cyclopropane, has been further developed into a general method for endoperoxide synthesis (Scheme 15).[49] A number of substituted vinylcyclopropanes **64** can be efficiently oxygenated to afford mixtures of isomeric dioxolane products **65A** und **65B** in good to excellent yields.

Scheme 15 1,2-Dioxolane Synthesis[49]

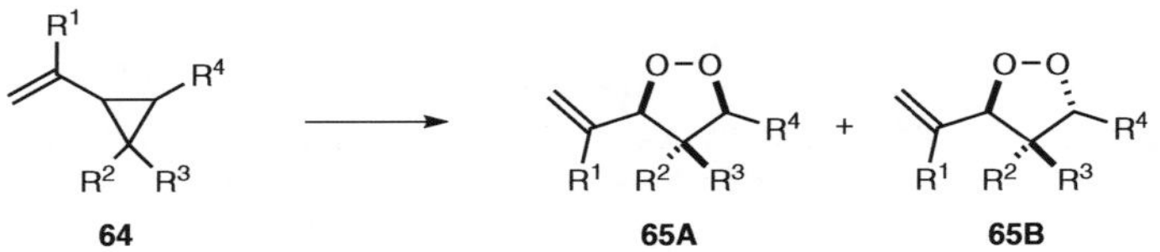

R^1	R^2	R^3	R^4	Conditions	Ratio (65A/65B)	Yield (%)	Ref
H	H	H	CO$_2$t-Bu	PhSeSePh (20 mol%), AIBN (10 mol%), MeCN, 0 °C	1:1.8	41	[49]
H	H	H	CH=CHCO$_2$Me	PhSeSePh (15 mol%), AIBN (8 mol%), MeCN, 0 °C	>19:1	88	[49]
H	H	Me	Ph	PhSeSePh (100 mol%), AIBN (10 mol%), MeOH, −50 °C	2.8:1	73	[49]
H	Me	Me	CH=CHCO$_2$Me	PhSeSePh (15 mol%), AIBN (8 mol%), MeCN, 0 °C	>19:1	73	[49,50]
H	Me	Me	Ph	PhSeSePh (20 mol%), AIBN (10 mol%), MeCN, 0 °C	>19:1	63	[49]
Me	H	H	CO$_2$t-Bu	PhSeSePh (20 mol%), AIBN (10 mol%), MeCN, 0 °C	1:1.7	53	[49]
t-Bu	H	H	CO$_2$t-Bu	PhSeSePh (20 mol%), AIBN (10 mol%), MeCN, 0 °C	1:1	52	[49]

Feldman has also studied domino processes in which multiple cyclopropane fragmentation events are strategically incorporated (Scheme 16).[49,50] These transformations lead to complex polydioxolanes in one operation; the yields are quite impressive given the structural complexity created in a single step.

Scheme 16 Polydioxolane Synthesis[49,50]

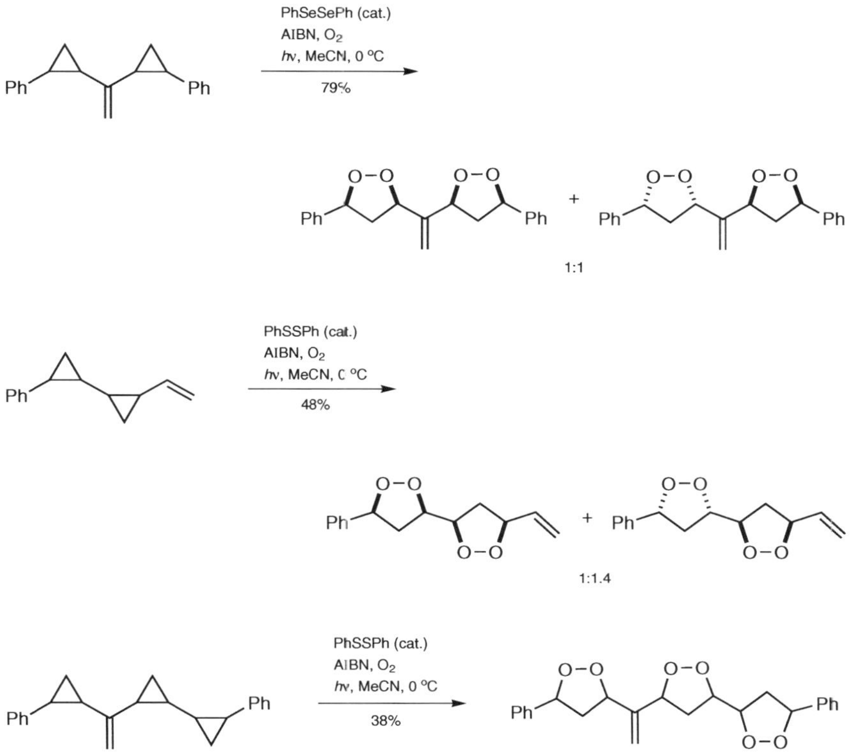

(3S,5R)-3-Phenyl-5-vinyl-1,2-dioxolane (63); Typical Procedure:[48,49]
A 12 mM soln of vinylcyclopropane **59** (1 equiv) in MeCN was cooled to 0 °C under an O_2 atmosphere. A 35 mM soln of PhSeSePh (0.1 equiv) in MeCN also containing AIBN (0.05 equiv) was added dropwise using a syringe pump while the mixture was irradiated with a sunlamp. Upon complete consumption of vinylcyclopropane **59** (TLC), the solvent was removed under reduced pressure and the *syn/anti* ratio and yield were determined by ¹H NMR spectroscopy. Purification by flash chromatography (silica gel) afforded the title compound; yield: 63%.

1.4.1.3.2 Carbon-Centered Radical Additions

The addition of carbon-centered radicals to alkenes can be utilized to initiate peroxy radical formation. However, the high reactivity of most radicals of this type toward oxygen greatly limits the scope of potential initiators. Typically, only highly electron deficient carbon-centered radicals can resist initial consumption by oxygen. In particular, α-keto radicals derived from 1,3-dicarbonyl-containing compounds efficiently add to conjugated alkenes; this process results in benzylic radicals that are rapidly trapped with oxygen. Multiple strategies exist to initiate formation of the first radical, including C—H abstraction as well as both chemical and electrochemical oxidation.

for references see p 184

1.4.1.3.2.1 By C–H Abstraction

Yoshida and Isoe have reported the 2,2'-azobisisobutyronitrile-mediated coupling of cyclic 1,3-diketones **66**, alkenes **67**, and oxygen to afford bicyclic peroxyketals **68** (Scheme 17).[52] 2,2'-Azobisisobutyronitrile serves as a radical initiator and promotes the formation of electron-deficient α-keto radical **69**. This intermediate adds smoothly to electron-rich alkene **67**, which results in peroxy radical **71** after reaction of **70** with oxygen. Radical **71** presumably abstracts a C–H bond from starting diketone **66**, and this closes the catalytic cycle and forms product **68** after ketalization. This transformation tolerates some structural diversity with respect to the alkene component, but styrenes afford the highest yields (Scheme 17). This coupling can also be elicited by electrochemical oxidation.[52]

Scheme 17 Radical Initiation by C–H Abstraction[52]

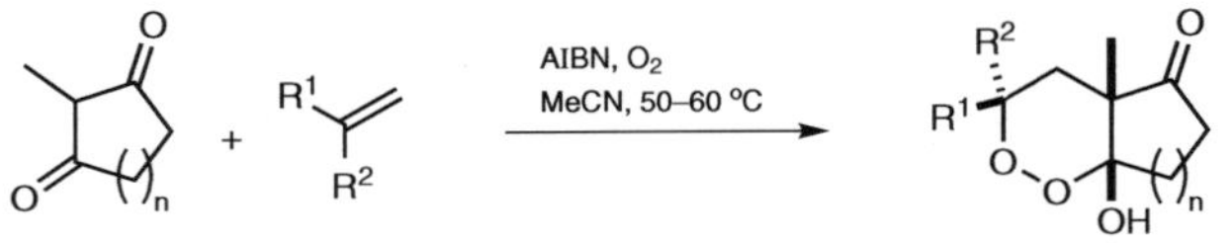

n	R[1]	R[2]	Alkene (Equiv)	AIBN (Equiv)	Yield[a] (%)	Ref
1	Ph	H	2	0.11	66	[52]
1	Ph	Me	2	0.11	90	[52]
1	CH₂TMS	H	6	0.14	73	[52]
1	OEt	H	19	0.13	65	[52]
1	CH=CH₂	Me	9	0.12	12	[52]
2	Ph	H	3	0.10	40	[52]

[a] Yields are based on 1,3-diketone **66**.

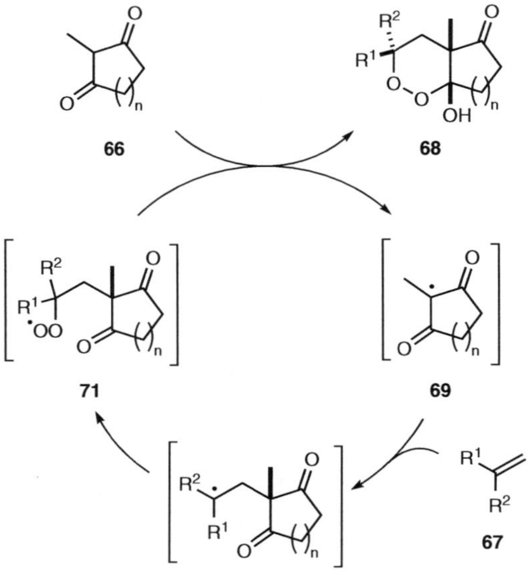

1.4.1 Peroxy Radical Additions **175**

(4aR,7aS)-7a-Hydroxy-4a-methylhexahydro-5H-cyclopenta[c][1,2]dioxin-5-ones 68; General Procedure:[52]

1,3-Diketone **66** (1 mmol) and alkene **67** were dissolved in MeCN (10 mL). O_2 gas was bubbled through the soln and AIBN (10 mol%) was added. The mixture was heated to 50–60 °C for 1 h. (Note: For cases in which the reaction was not complete after this time, an additional amount of AIBN was added and the mixture was heated. This cycle was repeated until the 1,3-diketone was consumed.) The mixture was cooled and partitioned between brine and Et_2O. The aqueous layer was extracted several times with Et_2O, and the combined organic layer was dried (Na_2SO_4). The solvent was removed under reduced pressure, and the residue was purified by flash chromatography (silica gel) to afford the title compound.

1.4.1.3.2.2 **Manganese(III)-Mediated Oxidation of 1,3-Dicarbonyls**

Manganese(III) salts have found widespread use in organic synthesis for the preparation of radicals from 1,3-dicarbonyls, and the addition of these electron-deficient species to alkenes has been exploited in numerous synthetic contexts.[53,54] Nishino and co-workers first demonstrated that this mode of reactivity could be utilized for the synthesis of endoperoxides (e.g., **74**) if the reaction was performed under aerobic conditions (Scheme 18).[55] Tris(acetylacetonato)manganese(III) serves as both the oxidant and the substrate; it provides radical **72**, which participates in chemistry analogous to that outlined in Section 1.4.1.3.2.1. As expected, this powerful transformation performs well with efficient radical acceptors, namely, 1,1-diarylalkenes **73** (Scheme 18). Exogenous 1,3-dicarbonyls and catalytic manganese(III) acetate can also initiate this process.[56,57] Although other metal oxidants can perform this transformation, none show the generality of manganese(III)-based systems.[58,59]

for references see p 184

Scheme 18 Initiation by Oxidation with Manganese(III)[55]

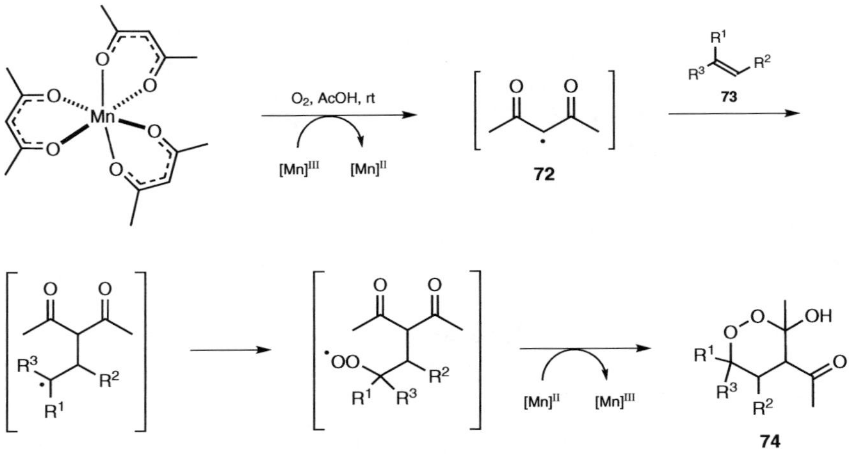

R[1]	R[2]	R[3]	Ratio [73/Mn(acac)$_3$]	Time (h)	Yield[a] (%)	Ref
Ph	H	Ph	1:1	11	92	[55]
4-ClC$_6$H$_4$	H	4-ClC$_6$H$_4$	1:2	12	90	[55]
4-MeOC$_6$H$_4$	H	4-MeOC$_6$H$_4$	1:2	12	87	[55]
(CH$_2$)$_5$Me	H	H	5:1	14	35	[55]
Ph	H	H	1:1	12	34[b]	[55]
(CH$_2$)$_4$		H	4:1	12	11[b]	[55]

[a] Yield based on alkene **73** except where stated.
[b] Yield based on Mn(acac)$_3$.

Majetich and co-workers have recently employed an intramolecular manganese(III)-mediated peroxide-forming cyclization en route to a highly elegant and efficient total synthesis of the terpene salvadione B (Scheme 19).[60] Treatment of polycyclic enone **75** with manganese(III) acetate induces formation of a carbon-centered radical that engages the neighboring alkene in a 6-*exo*-cyclization/oxygen-trapping cascade to lead to polycycle **76** in outstanding yield after peroxy radical reduction.

1.4.1 Peroxy Radical Additions

177

Scheme 19 Application of Magnesium(III)-Mediated Peroxide-Forming Cyclization in the Total Synthesis of Salvadione B[60]

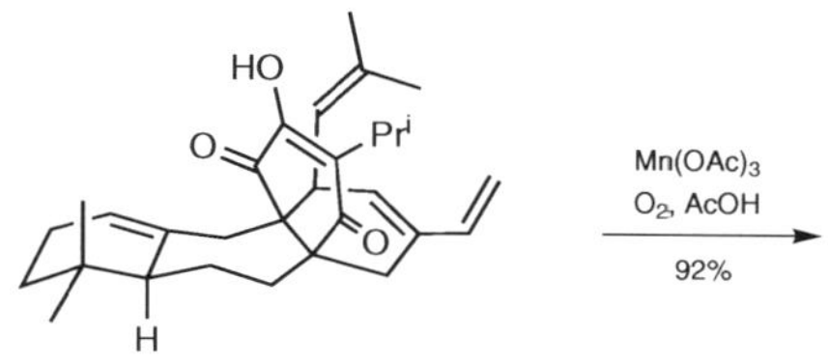

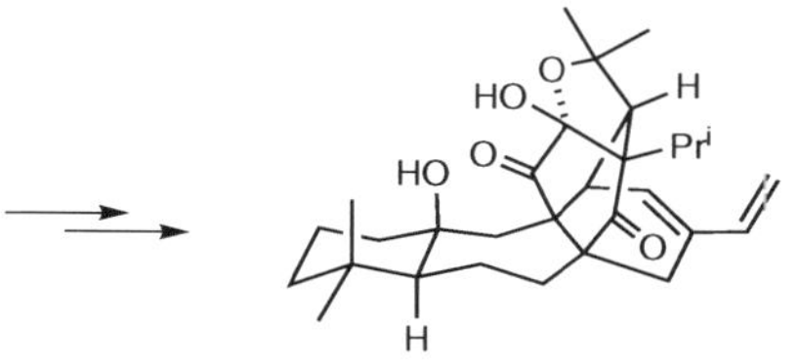

1-(3-Hydroxy-3-methyl-6,6-diphenyl-1,2-dioxan-4-yl)ethanone (74, R¹ = R³ = Ph; R² = H):[55] 1,1-Diphenylethene (**73**, R¹ = R³ = Ph; R² = H; 1 mmol) was dissolved in AcOH (25 mL) in a round-bottomed flask equipped with a CaCl$_2$ drying tube. Mn(acac)$_3$ (1 mmol) was added at rt, and the mixture was stirred under air until it changed from brown [indicative of Mn(III)] to pale yellow. The solvent was removed under reduced pressure, and the residue was triturated with 2 M HCl (30 mL), which was followed by extraction with CHCl$_3$. Purification by preparative TLC (Wakogel B-10, CHCl$_3$) followed by recrystallization (EtOH), afforded the title compound; yield: 287 mg (92%).

1.4.1.4 Heteroatom Oxidation/Cyclopropane Cleavage Pathways

Heteroatom oxidation provides an additional driving force for cleavage of pendent cyclopropane rings, and, if performed under aerobic conditions, peroxy radicals can be generated. Accordingly, the formation of 1,2-dioxolanes through the oxidation of cyclopropanols was documented nearly half a century ago (Scheme 20).[61] Recently, Wimalasena has disclosed that a variety of oxidative conditions can readily convert aminocyclopropane **77** into dioxolane **78** (as a mixture of isomers).[62] For the example shown, in which tris(1,10-phenanthroline)iron(III) hexafluorophosphate is used, amine **77** is presumably oxidized to radical cation **79**, which undergoes cyclopropyl ring opening, reaction with oxygen, and cyclization to radical cation **80**. Radical cation **80** is proposed to oxidize starting amine **77**, which thus regenerates the catalytic cycle. Analogous reactivity can also be achieved using electrochemical methods.[63] The heteroatom-centered radical does not need to be directly attached to the cyclopropane to facilitate ring opening, as shown in the facile conversion of cyclopropane **81** into dioxolane **83** by Creary and co-workers, which remarkably occurs upon standing in chloroform-*d*.[64] The intermediacy of phenoxy radical **82** is suspected in this process.

Scheme 20 Dioxolane Synthesis by Cyclopropyl Ring Opening[61,62,64]

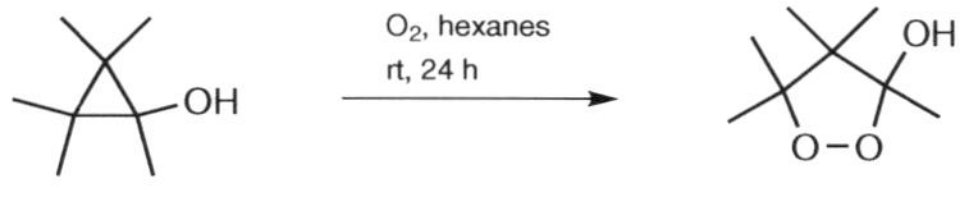

for references see p 184

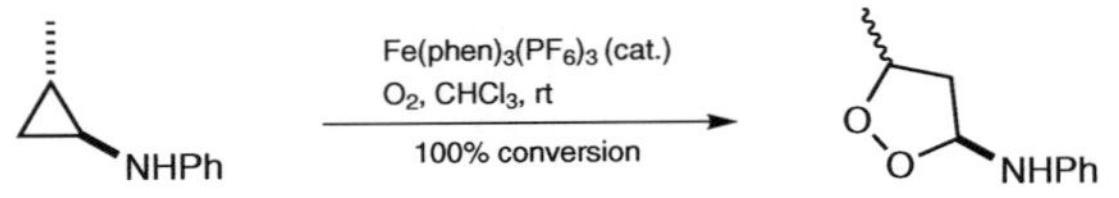

Domino Transformations 1.4 Radical Reactions

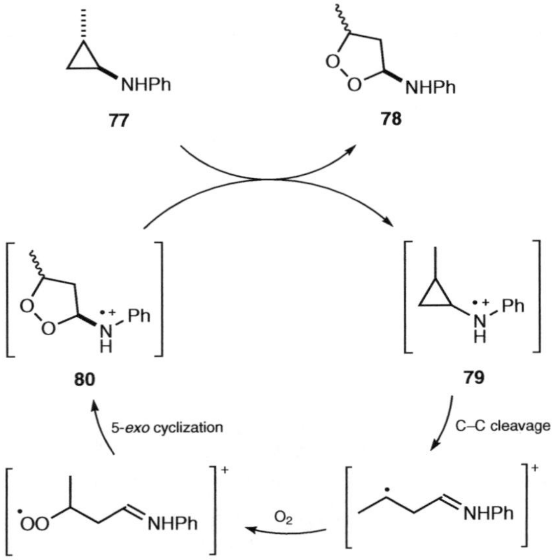

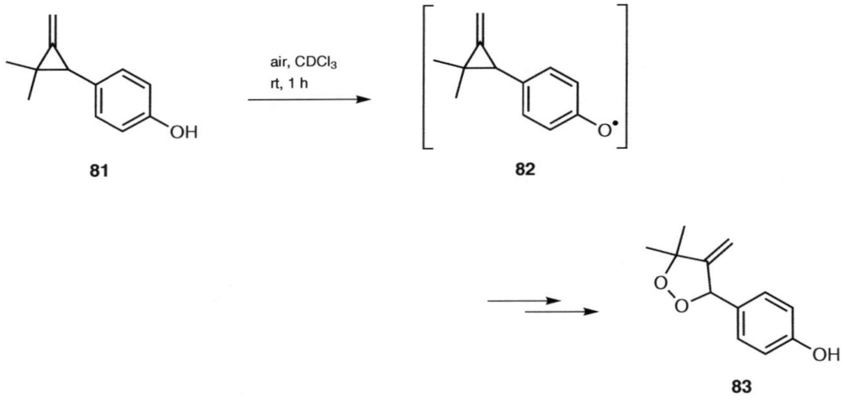

5-Methyl-N-phenyl-1,2-dioxolan-3-amine (78); Typical Procedure:[62]

A soln of *N*-cyclopropyl-*N*-phenylamine **77** (25 mg, 0.19 mmol) and a catalytic amount of [Fe(phen)$_3$(PF$_6$)$_3$] (ca. 1 mg, 0.6 mol%) in CHCl$_3$ (25 mL) was stirred for 1 h open to the atmosphere. The mixture was filtered through a plug of silica gel, concentrated under reduced pressure, and analyzed by ^{1}H NMR spectroscopy.

1.4.1.5 Radical Cation Intermediates

The oxidation of certain classes of electron-rich styrenes, dienes, and cyclopropanes produces radical cations. These radical cations can be intercepted with triplet oxygen thereby providing various 1,2-dioxolane and 1,2-dioxane products. These transformations can formally be viewed as cycloaddition reactions with oxygen. Photoinduced electron transfer has seen widespread use in initiating these oxygenation reactions.

1.4.1.5.1 1,2-Diarylcyclopropane Photooxygenation

The photooxygenation of 1,2-diarylcyclopropanes has been extensively studied by numerous groups from both mechanistic and synthetic perspectives.[65–68] Excited-state 9,10-dicyanoanthracene (DCA*), formed by irradiation with light, can efficiently oxidize certain classes of 1,2-diarylcyclopropanes to afford ring-opened, radical cation intermediates (e.g., reaction of **84** to give **86**) (Scheme 21). Reaction of **86** with triplet oxygen forms peroxy radical cation **87**, which after electron transfer with starting cyclopropane **84**, cyclizes to dioxolane product **85**. As shown in Scheme 21, this transformation (as well as other 9,10-dicyanoanthracene-sensitized photooxygenations) generally requires Ar¹ to be an electron-donating aromatic group.[69] Two modifications, however, can be employed to facilitate coupling: (1) the addition of biphenyl, which serves as an efficient one-electron shuttle between the excited state of 9,10-dicyanoanthracene and the substrate;[69] and (2), the incorporation of metal salts with weakly coordinating counterions [e.g., $Mg(ClO_4)_2$], which facilitate ionization of the initially formed complex between **86** and the radical anion of 9,10-dicyanoanthracene.

Scheme 21 9,10-Dicyanoanthracene-Sensitized Photooxygenation of Cyclopropanes[69]

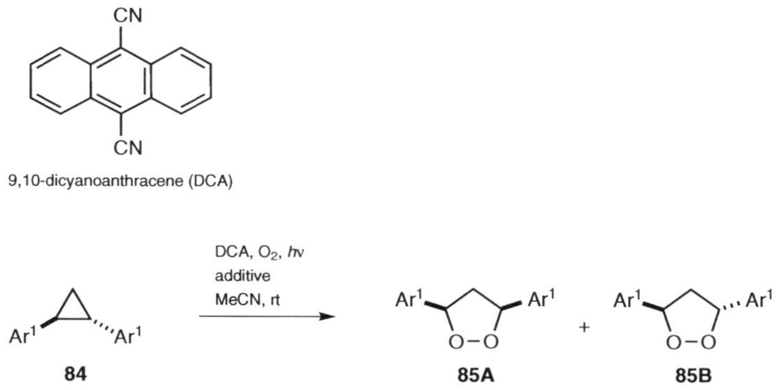

9,10-dicyanoanthracene (DCA)

Ar¹	Additive	Time (min)	Yield[a] (%) of **85A**	Yield[a] (%) of **85B**	Ref
Ph	none	410	0	0	[69]
Ph	biphenyl (0.5 equiv)	150	0	0	[69]
Ph	$Mg(ClO_4)_2$ (0.5 equiv)	305	25	15	[69]
4-MeOC$_6$H$_4$	none	100	70	30	[69]
4-MeOC$_6$H$_4$	biphenyl (0.5 equiv)	30	70	30	[69]
4-MeOC$_6$H$_4$	$Mg(ClO_4)_2$ (0.5 equiv)	30	70	30	[69]

[a] Yields were determined by ¹H NMR spectroscopy.

for references see p 184

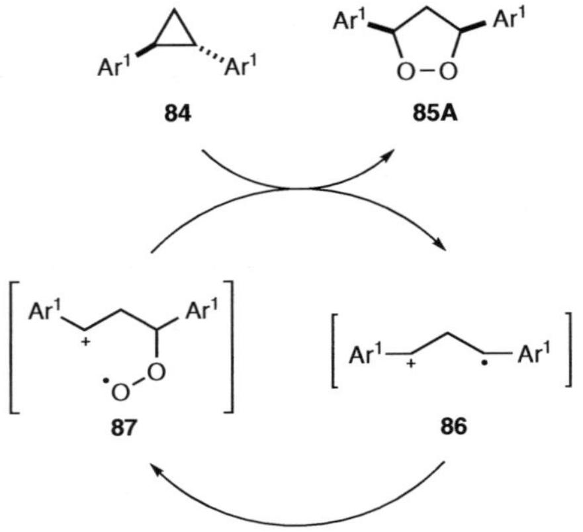

3,5-Disubstituted 1,2-Dioxolanes 85; General Procedure:[69]
A soln of cyclopropane **84** (0.2 mmol) and DCA (0.01 mmol) in dry MeCN (8 mL) was irradiated at rt with a 500-W high-pressure Hg arc through an aq NH_3/$CuSO_4$ filter soln in a stream of O_2 in the presence or absence of the indicated additives [biphenyl or $Mg(ClO_4)_2$]. The progress of the reaction was followed by GLC analysis. After consumption of cyclopropane **84**, the solvent was removed under reduced pressure, and the residue was extracted with hexane. The extract was washed with H_2O, dried (Na_2SO_4), and concentrated, and the residue was analyzed by ^{1}H NMR spectroscopy.

1.4.1.5.2 Alkene/Oxygen [2+2+2] Cycloaddition

Substituted styrenes can participate in a related photosensitized process to afford 1,2-dioxanes (e.g., reaction of **88** to give **89**) (Scheme 22).[70–73] Formally, this transformation represents a [2+2+2] cycloaddition between two alkenes and molecular oxygen. Mechanistically, this process is believed to proceed by initial excited 9,10-dicyanoanthracene induced styrene oxidation to produce radical cation **92**. Radical alkene addition and reaction with triplet oxygen produces radical cation **91**, which goes on to product **89** through an electron-transfer/cyclization pathway in analogy to the cyclization of **87** to give **85A** (Section 1.4.1.5.1). Gollnick and co-workers have conducted an extensive study of substituent effects on the performance of this transformation; to obtain synthetically useful yields of the 1,2-dioxane product, at least one of the aryl groups must typically possess an electron-donating group in the *ortho* or *para* position.[74] For cases in which this criterion is not met, alkene cleavage, likely through superoxide-mediated pathways, can predominate to afford diaryl ketones **90**.

1.4.1 Peroxy Radical Additions

Scheme 22 9,10-Dicyanoanthracene-Sensitized Alkene/Oxygen [2+2+2] Cycloaddition[74]

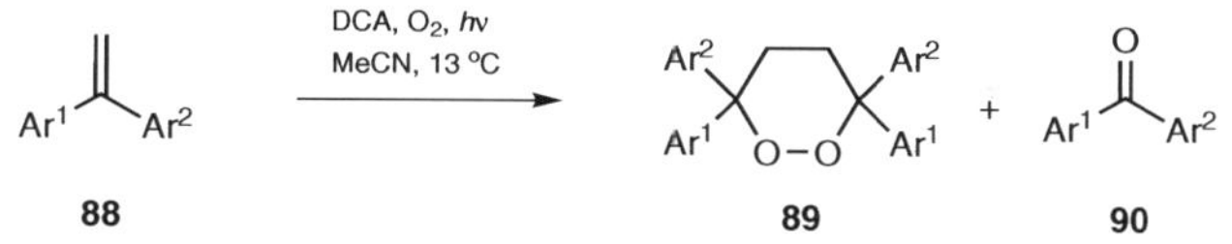

Ar^1	Ar^2	Yield[a] (%) of **89**	Yield[a] (%) of **90**	Ref
4-Me$_2$NC$_6$H$_4$	4-Me$_2$NC$_6$H$_4$	100	0	[74]
4-MeOC$_6$H$_4$	4-MeOC$_6$H$_4$	97	3	[74]
4-MeOC$_6$H$_4$	4-Tol	95	5	[74]
4-Tol	4-Tol	90	10	[74]
4-Tol	Ph	85	15	[74]
4-MeOC$_6$H$_4$	4-ClC$_6$H$_4$	90	10	[74]
3-MeOC$_6$H$_4$	3-MeOC$_6$H$_4$	<5	50	[74]
4-ClC$_6$H$_4$	4-ClC$_6$H$_4$	20	70	[74]

[a] Determined by ^{1}H NMR spectroscopy.

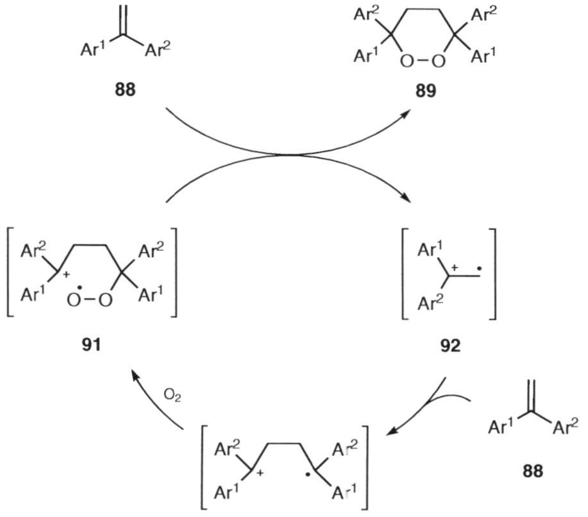

Intramolecular variants of this, and related, peroxidation reactions have been reported; these events can forge complex bicyclic endoperoxides rapidly (Scheme 23).[75–80] Miyashi and co-workers have reported the cyclization of diene **94** into bicyclic endoperoxide **95** by using 9,10-dicyanoanthracene as a photocatalyst.[75] This transformation is only efficient if the attached arene group (Ar^1) is electron rich, although magnesium perchlorate additives can improve this outcome. Recently, Nicewicz and Gesmundo, by employing pyrylium photocatalyst **93**, have explored a substantially larger class of dienes for participation in this and related peroxidation reactions.[30] Cyclic peroxides **95** or **97** are formed, depending on the substitution pattern of the dienes employed. As with Miyashi's system, electron-donating aryl rings are optimal for this chemistry. Substituents on the tether (R^1 in **96**) have also been explored, in addition to heteroatom linkages (not shown).

for references see p 184

Scheme 23 Intramolecular Cyclization/Endoperoxidation Employing Organic Photocatalysts[75,80]

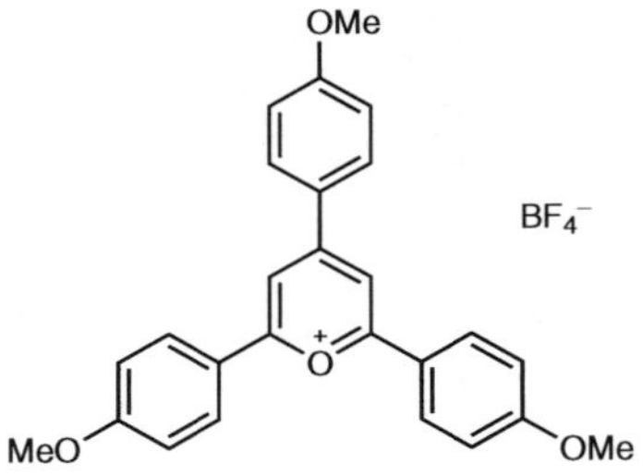

Ar^1	Photocatalyst	Yield (%)	Ref
Ph	DCA	6	[75]
4-MeOC$_6$H$_4$	DCA	72	[75]
4-MeOC$_6$H$_4$	**93**	50	[80]

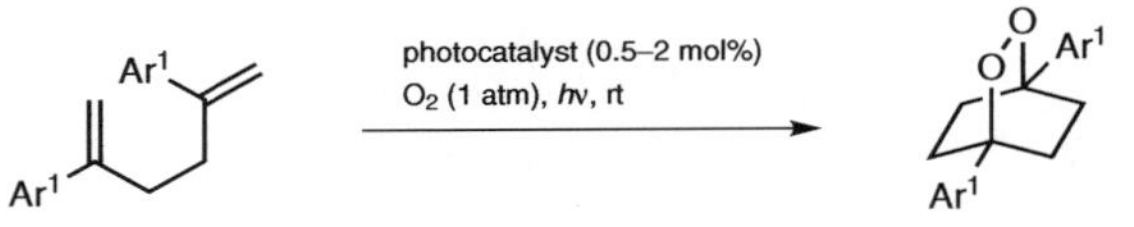

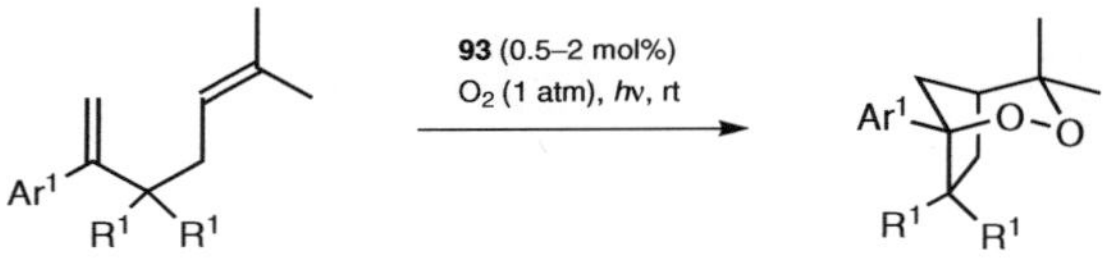

R^1	Ar^1	Photocatalyst	Yield (%)	Ref
H	4-MeOC$_6$H$_4$	**93**	32	[80]
Me	4-MeOC$_6$H$_4$	**93**	66	[80]

Very recently, Yoon and co-workers demonstrated that the tris(bipyrazyl)ruthenium(II) hexafluorophosphate complex {[Ru(bpz)$_3$(PF$_6$)$_2$]; bpz = bipyrazyl; **98**) functions as an efficient photocatalyst in the intramolecular [2+2+2] coupling of diene **99** and oxygen to afford bicyclic endoperoxides **100** (Scheme 24).[81] Analogously to prior transformations, an excited-state species of the catalyst formed through irradiation serves as the oxidizing agent, initiating radical cation formation. Visible light from a simple 200-W incandescent bulb is sufficient to initiate this transformation. At least one of the substituents must be an aryl group with an electron-donating substituent at the *ortho* or *para* position for efficient coupling to proceed (Scheme 24). A significant number of different substitution patterns are tolerated. Although diene **99** possesses an oxygen tether between the two alkenes, tethers based on carbon and nitrogen are also tolerated.

1.4.1 Peroxy Radical Additions

183

Scheme 24 Endoperoxide Synthesis by Ruthenium Photoredox Catalysis[81]

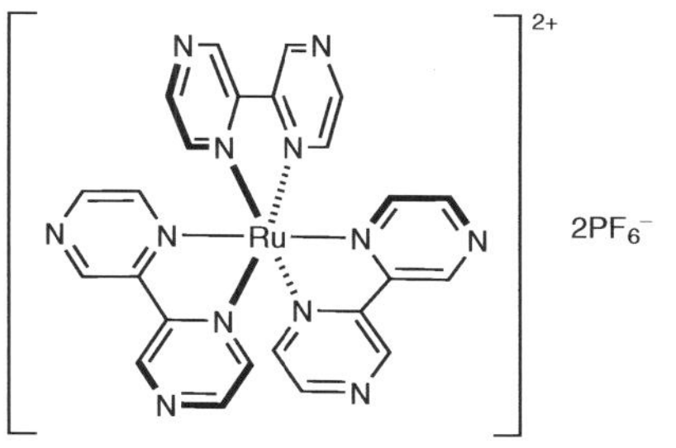

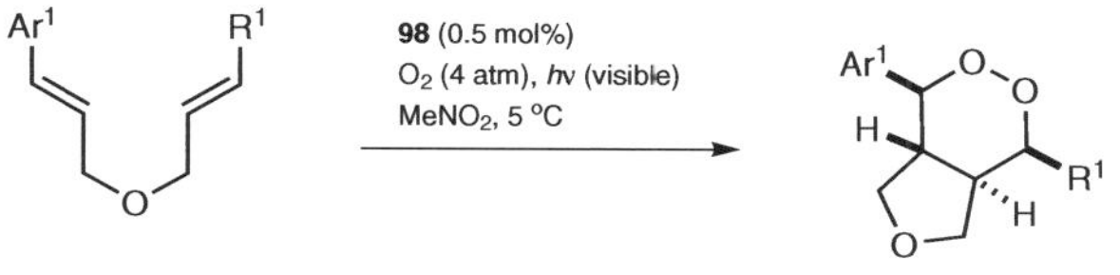

R^1	Ar^1	Time (h)	dr	Yield (%)	Ref
Ph	4-MeOC$_6$H$_4$	0.5	6:1	92	[81]
Ph	2-MeOC$_6$H$_4$	24	10:1	45^a	[81]
Ph	3-MeOC$_6$H$_4$	24	–	0	[81]
Ph	Ph	24	–	0	[81]
Me	4-MeOC$_6$H$_4$	24	–	<5	[81]
C≡CPh	4-MeOC$_6$H$_4$	2	2:1	82	[81]

^{a} 2 mol% of catalyst was used.

1,4-Bis(4-methoxyphenyl)-2,3-dioxabicyclo[2.2.2]octane (95, Ar^1 = 4-MeOC$_6$H$_4$); Typical Procedure:[80]

A flame-dried, 1-dram vial equipped with a magnetic stirrer bar was charged with catalyst **93** (0.25 mg) and then sealed with a septum cap; dry MeCN (0.5 mL) was added under an atmosphere of N$_2$. A soln of diene **94** (Ar^1 = 4-MeOC$_6$H$_4$; 30 mg) in MeCN (0.5 mL) was added to the stirring mixture. A balloon of O$_2$ was fitted to the vial and bubbled through the stirring soln for 2 min. After bubbling was suspended, an atmosphere of O$_2$ was maintained over the soln, and the mixture was irradiated at ambient temperature for 2 h by using 470-nm light-emitting diodes. Upon completion of the reaction, the crude material was purified by column chromatography (silica gel, Et$_2$O/hexanes 1:2) to afford the title compound; yield: 16.6 mg (50%).

(1S*,4R*,4aR*,7aR*)-1-(4-Methoxyphenyl)-4-phenyltetrahydro-1H,4H-furo[3,4-d][1,2]dioxin (100, R^1 = Ph; Ar^1 = 4-MeOC$_6$H$_4$); Typical Procedure:[81]

An oven-dried, 135-mL glass pressure vessel containing a magnetic stirrer bar was charged with the alkene substrate **99** (R^1 = Ph; Ar^1 = 4-MeOC$_6$H$_4$; 100 mg, 0.36 mmol), Ru(bpz)$_3$(PF$_6$)$_2$ (**98**; 0.5 mol%), and MeNO$_2$ (17.8 mL). The vessel was fitted with a regulator and pressurized with O$_2$ (4 atm). The vessel was cooled to 5 °C in a water bath and irradiated with a 200-W incandescent light bulb. After 30 min, the mixture was eluted through a short pad of silica gel (EtOAc/CH$_2$Cl$_2$). The solvent was removed under reduced pressure, and the crude material was purified by chromatography (silica gel, hexanes/EtOAc 8:1 to 2:1) to afford the title compound as a yellow oil; yield: 92% (6:1 mixture of diastereomers).

for references see p 184

References

[1] Spiteller, G., *Free Radical Biol. Med.*, (2006) **41**, 362.

[2] Liu, D.-Z.; Liu, J.-K., *Nat. Prod. Bioprospect.*, (2013) **3**, 161.

[3] Korshin, E. E.; Bachi, M. D., *Synthesis of Cyclic Peroxides, Patai's Chemistry of Functional Groups* (Online), Wiley: New York, (2009); pp 1–117.

[4] Dussault, P., *Synlett*, (1995), 997.

[5] Zabicky, J., In *The Chemistry of the Peroxide Group*, Rappoport, Z., Ed.; Wiley: Chichester, UK, (2006); Part 2, Vol. 2, pp 597–773.

[6] Shanley, E. S., In *Organic Peroxides*, Swern, D., Ed.; Wiley: New York, (1970); Vol 3, p 341.

[7] Funk, M. O.; Isaac, R.; Porter, N. A., *J. Am. Chem. Soc.*, (1975) **97**, 1281.

[8] Porter, N. A.; Funk, M. O.; Gilmore, D.; Isaac, R.; Nixon, J., *J. Am. Chem. Soc.*, (1976) **98**, 6000.

[9] Tokuyasu, T.; Kunikawa, S.; Abe, M.; Masuyama, A.; Nojima, M.; Kim, H.-S.; Begum, K.; Wataya, Y., *J. Org. Chem.*, (2003) **68**, 7361.

[10] Bartlett, P. D.; Benzing, E. P.; Pincock, R. E., *J. Am. Chem. Soc.*, (1960) **82**, 1762.

[11] Mendenhall, G. D., *Tetrahedron Lett.*, (1983) **24**, 451.

[12] Porter, N. A.; Zuraw, P. J., *J. Org. Chem.*, (1984) **49**, 1345.

[13] Spellmeyer, D. C.; Houk, K. N., *J. Org. Chem.*, (1987) **52**, 959.

[14] Beckwith, A. L. J.; Schiesser, C. H., *Tetrahedron*, (1985) **41**, 3925.

[15] Carless, H. A. J.; Batten, R. J., *Tetrahedron Lett.*, (1982) **23**, 4735.

[16] Bloodworth, A. J.; Curtis, R. J.; Mistry, N., *J. Chem. Soc., Chem. Commun.*, (1989), 954.

[17] Cointeaux, L.; Berrien, J.-F.; Mayrargue, J., *Tetrahedron Lett.*, (2002) **43**, 6275.

[18] Roe, A. N.; McPhail, A. T.; Porter, N. A., *J. Am. Chem. Soc.*, (1983) **105**, 1199.

[19] Porter, N. A.; Roe, A. N.; McPhail, A. T., *J. Am. Chem. Soc.*, (1980) **102**, 7574.

[20] Haynes, R. K.; Vonwiller, S. C., *J. Chem. Soc., Chem. Commun.*, (1990), 1102.

[21] Corey, E. J.; Wang, Z., *Tetrahedron Lett.*, (1994) **35**, 539.

[22] Fielder, S.; Rowan, D. D.; Sherburn, M. S., *Tetrahedron*, (1998) **54**, 12 907.

[23] Boukouvalas, J.; Pouliot, R.; Fréchette, Y., *Tetrahedron Lett.*, (1995) **36**, 4167.

[24] Tokuyasu, T.; Kunikawa, S.; Masuyama, A.; Nojima, M., *Org. Lett.*, (2002) **4**, 3595.

[25] Isayama, S.; Mukaiyama, T., *Chem. Lett.*, (1989), 1071.

[26] Mukaiyama, T.; Isayama, S.; Inoki, S.; Kato, K.; Yamada, T.; Takai, T., *Chem. Lett.*, (1989), 449.

[27] Inoki, S.; Kato, K.; Takai, T.; Isayama, S.; Yamada, T.; Mukaiyama, T., *Chem. Lett.*, (1989), 515.

[28] Magnus, P.; Payne, A. H.; Waring, M. J.; Scott, D. A.; Lynch, V., *Tetrahedron Lett.*, (2000) **41**, 9725.

[29] Magnus, P.; Waring, M. J.; Scott, D. A., *Tetrahedron Lett.*, (2000) **41**, 9731.

[30] Magnus, P.; Scott, D. A.; Fielding, M. R., *Tetrahedron Lett.*, (2001) **42**, 4127.

[31] Leggans, E. K.; Barker, T. J.; Duncan, K. K.; Boger, D. L., *Org. Lett.*, (2012) **14**, 1428.

[32] Okamoto, T.; Oka, S., *J. Org. Chem.*, (1984) **49**, 1589.

[33] Slack, R. D.; Jacobine, A. M.; Posner, G. H., *Med. Chem. Commun.*, (2012) **3**, 281.

[34] O'Neill, P. M.; Barton, V. E.; Ward, S. A., *Molecules*, (2010) **15**, 1705.

[35] Posner, G. H.; O'Neill, P. M., *Acc. Chem. Res.*, (2004) **37**, 397.

[36] Waser, J.; Carreira, E. M., *Angew. Chem.*, (2004) **116**, 4191; *Angew. Chem. Int. Ed.*, (2004) **43**, 4099.

[37] Iwasaki, K.; Wan, K. K.; Oppedisano, A.; Crossley, S. W. M.; Shenvi, R. A., *J. Am. Chem. Soc.*, (2014) **136**, 1300.

[38] King, S. M.; Ma, X.; Herzon, S. B., *J. Am. Chem. Soc.*, (2014) **136**, 6884.

[39] Hu, X.; Maimone, T. J., *J. Am. Chem. Soc.*, (2014) **136**, 5287.

[40] Posner, T., *Ber. Dtsch. Chem. Ges.*, (1905) **38**, 646.

[41] Walling, C.; Helmreich, W., *J. Am. Chem. Soc.*, (1959) **81**, 1144.

[42] Kharasch, M. S.; Nudenberg, W.; Mantell, G. J., *J. Org. Chem.*, (1951) **16**, 524.

[43] Beckwith, A. L. J.; Wagner, R. D., *J. Am. Chem. Soc.*, (1979) **101**, 7099.

[44] Szpilman, A. M.; Korshin, E. E.; Rozenberg, H.; Bachi, M. D., *J. Org. Chem.*, (2005) **70**, 3618.

[45] Korshin, E. E.; Hoos, R.; Szpilman, A. M.; Konstantinovski, L.; Posner, G. H.; Bachi, M. D., *Tetrahedron*, (2002) **58**, 2449.

[46] Beckwith, A. L. J.; Wagner, R. D., *J. Chem. Soc., Chem. Commun.*, (1980), 485.

[47] Barker, P. J.; Beckwith, A. L. J.; Fung, Y., *Tetrahedron Lett.*, (1983) **24**, 97.

[48] Feldman, K. S.; Simpson, R. E.; Parvez, M., *J. Am. Chem. Soc.*, (1986) **108**, 1328.

[49] Feldman, K. S.; Simpson, R. E., *J. Am. Chem. Soc.*, (1989) **111**, 4878.

[50] Feldman, K. S., *Synlett*, (1995), 217.

[51] Feldman, K. S.; Kraebel, C. M., *J. Org. Chem.*, (1992) **57**, 4574.

References

[52] Yoshida, J.-I.; Nakatani, S.; Sakaguchi, K.; Isoe, S., *J. Org. Chem.*, (1989) **54**, 3383.

[53] Snider, B. B., *Chem. Rev.*, (1996) **96**, 339.

[54] Mondal, M.; Bora, U., *RSC Adv.*, (2013) **3**, 18716.

[55] Tategami, S.-i.; Yamada, T.; Nishino, H.; Korp, J. D.; Kurosawa, K., *Tetrahedron Lett.*, (1990) **31**, 6371.

[56] Kumabe, R.; Nishino, H., *Tetrahedron Lett.*, (2004) **45**, 703.

[57] Kumabe, R.; Nishino, H.; Yasutake, M.; Nguyen, V.-H.; Kurosawa, K., *Tetrahedron Lett.*, (2001) **42**, 69.

[58] Yamada, T.; Iwahara, Y.; Nishino, H.; Kurosawa, K., *J. Chem. Soc., Perkin Trans. 1*, (1993), 609.

[59] Kajikawa, S.; Nishino, H.; Kurosawa, K., *Heterocycles*, (2001) **54**, 171.

[60] Majetich, G.; Zou, G.; Hu, S., *Org. Lett.*, (2013) **15**, 4924.

[61] Gibson, D. H.; DePuy, C. H., *Tetrahedron Lett.*, (1969) **10**, 2203.

[62] Wimalasena, K.; Wickman, H. B.; Mahindaratne, M. P. D., *Eur. J. Org. Chem.*, (2001), 3811.

[63] Madelaine, C.; Six, Y.; Buriez, O., *Angew. Chem.*, (2007) **119**, 8192; *Angew. Chem. Int. Ed.*, (2007) **46**, 8046.

[64] Creary, X.; Wolf, A.; Miller, K., *Org. Lett.*, (1999) **1**, 1615.

[65] Mizuno, K.; Kamiyama, N.; Otsuji, Y., *Chem. Lett.*, (1983), 477.

[66] Mizuno, K.; Kamiyama, N.; Ichinose, N.; Otsuji, Y., *Tetrahedron*, (1985) **41**, 2207.

[67] Shim, S. C.; Song, J. S., *J. Org. Chem.*, (1986) **51**, 2817.

[68] Gollnick, K.; Xiao, X.-L.; Paulmann, U., *J. Org. Chem.*, (1990) **55**, 5945.

[69] Tamai, T.; Mizuno, K.; Hashida, I.; Otsuji, Y., *J. Org. Chem.*, (1992) **57**, 5338.

[70] Haynes, R. K.; Probert, M. K. S.; Wilmot, I. D., *Aust. J. Chem.*, (1978) **31**, 1737.

[71] Gollnick, K.; Schnatterer, A., *Tetrahedron Lett.*, (1984) **25**, 185.

[72] Mattes, S. L.; Farid, S., *J. Am. Chem. Soc.*, (1986) **108**, 7356.

[73] Kojima, M.; Ishida, A.; Takamuku, S., *Chem. Lett.*, (1993), 979.

[74] Gollnick, K.; Schnatterer, A.; Utschick, G., *J. Org. Chem.*, (1993) **58**, 6049.

[75] Miyashi, T.; Konno, A.; Takahashi, Y., *J. Am. Chem. Soc.*, (1988) **110**, 3676.

[76] Mizuno, K.; Tamai, T.; Hashida, I.; Otsuji, Y.; Kuriyama, Y.; Tokumaru, K., *J. Org. Chem.*, (1994) **59**, 7329.

[77] Griesbeck, A. G.; Sadlek, O.; Polborn, K., *Liebigs Ann.*, (1996), 545.

[78] Kamata, M.; Ohta, M.; Komatsu, K.-I.; Kim, H.-S.; Wataya, Y., *Tetrahedron Lett.*, (2002) **43**, 2063.

[79] Miyashi, T.; Ikeda, H.; Takahashi, Y., *Acc. Chem. Res.*, (1999) **32**, 815.

[80] Gesmundo, N. J.; Nicewicz, D. A., *Beilstein J. Org. Chem.*, (2014) **10**, 1272.

[81] Parrish, J. D.; Ischay, M. A.; Lu, Z.; Guo, S.; Peters, N. R.; Yoon, T. P., *Org. Lett.*, (2012) **14**, 1640.

1.4.2   **Radical Cyclizations**

J. J. Devery, III, J. J. Douglas, and C. R. J. Stephenson

General Introduction

One of the main contributing classes to the rise of domino reactions in recent years has been that of radical cyclizations. The product from a radical cyclization is another radical, offering great potential for the use of such a new species in additional chemistry to lead to domino processes (Scheme 1). Historically, the use of radicals in multi-bond-forming reactions was limited by the supposition that these processes are unselective and uncontrollable. However, recent efforts, largely directed toward understanding their reactivity, have increased the power of the transformations for which they can be employed.[1–3] By harnessing the high reactivity that is associated with radicals, challenging bond constructions, such as the formation of highly substituted ring systems, can be achieved. The generation of new rings from linear precursors leads to an increase in molecular and stereochemical complexity, fueling the use of such domino cyclization approaches, particularly for natural product and complex molecule synthesis where such strategies can dramatically enhance efficiency.

Scheme 1   General Scheme of Domino Radical Cyclizations

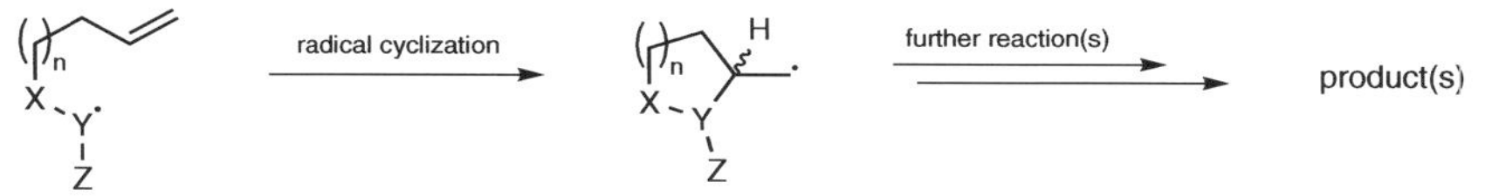

A further benefit of a domino radical process is that the high reactivity of a radical can be directed through substrates predisposed for cyclization, thus limiting unwanted intermolecular reactions. However, these reactions typically require a high level of substrate design, presenting one of the main drawbacks of typical radical domino cyclization approaches. Overall, these factors have contributed to the wealth of literature reporting the use of domino radical cyclizations in natural product synthesis, whereas general methods remain less well developed.

Recent comprehensive accounts have described the wide spectrum of radical domino reactions in considerable detail.[4,5] This chapter will focus on domino reactions that begin with intramolecular radical addition, after initial radical generation. It is arranged by method of radical generation: tin-, reductive-, oxidative-, and visible-light-mediated processes. References have been chosen following analysis of a range of criteria such as their general utility, efficiency, and substrate scope, with a particular emphasis on recent examples given the more historical perspective of the following chapter from Parker and co-authors (Section 1.4.3).

for references see p 214

1.4.2.1 Tin-Mediated Radical Cyclizations

Despite growing environmental concerns over the use of tin, its utility in organic chemistry means that it continues to provide the basis for novel research. The use of tin to initiate radical cyclizations in the synthesis of complex molecules has an especially rich history, owing principally to its functional group tolerance and ability to initiate reactivity through addition to unactivated alkenes and alkynes, amongst other species that often display low reactivity. Recent total syntheses from the group of Ishibashi,[6–9] such as the synthesis of (−)-cephalotaxine and (±)-stemonamide, have highlighted the continued interest in this approach.[10–12] Equally, the extension of radical cyclizations to N-centered radicals has been shown to be a versatile method for the construction of diverse alkaloid-type scaffolds. This method was elegantly employed by Zard and co-workers for the synthesis of (±)-aspidospermidine via a key 5-*exo-trig*/6-*endo-trig* radical cascade cyclization.[13,14]

1.4.2.1.1 Tin-Mediated Synthesis of Hexahydrofuropyrans

While the use of tin-mediated radical domino processes is well documented in total synthesis, their wider applicability to the synthesis of a diverse set of products has also been shown.[15] Fensterbank and Malacria[16] have disclosed a 5-*exo*/5-*exo*/3-*exo* cyclization cascade for the formation of fused polycyclic vinylcyclopropanes, which employs a (dichloromethyl)dimethylsilyl radical trigger. A recent report from the group of Wulff[17] outlines an elegant tin-mediated 6-*endo*/5-*exo* cyclization cascade from bis(vinyl ethers) that results in a range of complex hexahydrofuropyrans, a motif highly relevant for medicinal chemistry. The bis(vinyl ether) precursors **1** were efficiently prepared as part of a larger study into the iterative synthesis of higher order oligovinyl ethers. The reaction is proposed to proceed via addition of a triphenyltin radical to the pendant alkyne, with subsequent 6-*endo-trig*/5-*exo-trig* radical cyclizations furnishing the hexahydrofuropyrans **3** with moderate to high diastereoselectivity (2:1 to 8:1 dr) and in good yields (up to 81%), as shown in Scheme 2 (path B). An alternate mechanism involves initial 5-*exo-trig* cyclization followed by rearrangement to the more stable radical, a mechanism that has been proposed for other 6-*endo-trig* reactions.

In instances where the second vinyl ether is trisubstituted (i.e., $R^2 \neq H$), the cyclization sequence is prone to termination following the first cyclization (path A), yielding a mixture of pyran **2** and hexahydrofuropyran products **3** or exclusively the pyran product **2**. Further product utility was shown via efficient protodestannylation upon treatment with hydrochloric acid to provide the exocyclic methylene moiety.

1.4.2 Radical Cyclizations

Scheme 2 Synthesis of Hexahydrofuropyrans via 6-*endo-trig*/5-*exo-trig* Radical Cyclizations[17]

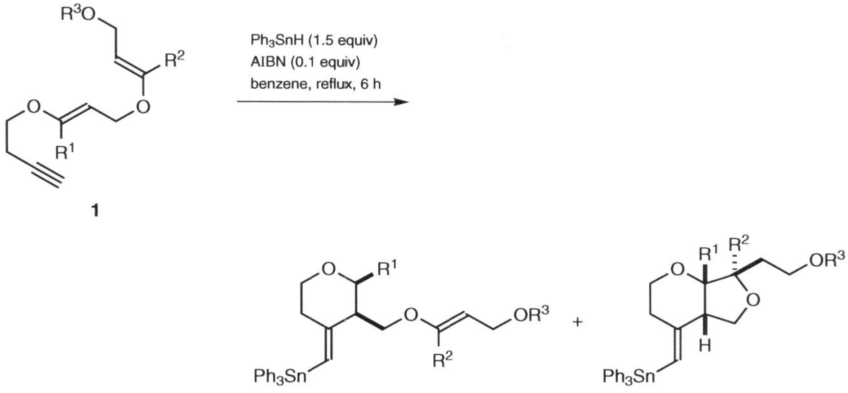

R[1]	R[2]	R[3]	dr of **2**	Yield (%) of **2**	dr of **3**[a,b]	Yield[b] (%) of **3**	Ref
Me	H	Me	–	–	4:1	77	[17]
Et	H	Me	–	–	8:1	76	[17]
Ph	H	Me	–	–	4:1	80	[17]
CF$_3$	H	Me	–	–	7:1	64	[17]
CH$_2$OTHP	H	Me	–	–	2:1	67	[17]
Me	Me	Me	4:1	54	5:1	33	[17]
Me	Me	Me	3:1	27	5:1	55[c]	[17]
Et	Et	Me	2:1	67	n.d.	n.d.	[17]
Ph	Me	Me	12:1	72	n.d.	n.d.	[17]
Me	H	H	–	–	1.4:1	81	[17]
Me	H	Bn	–	–	2.7:1	74	[17]

[a] Major diastereomer shown. Values reported refer to the ratio between the major observed product and the next most abundant species.

[b] n.d. = not detected.

[c] Experiment run at a concentration of 1.6 mM, other reactions run at 8 mM.

for references see p 214

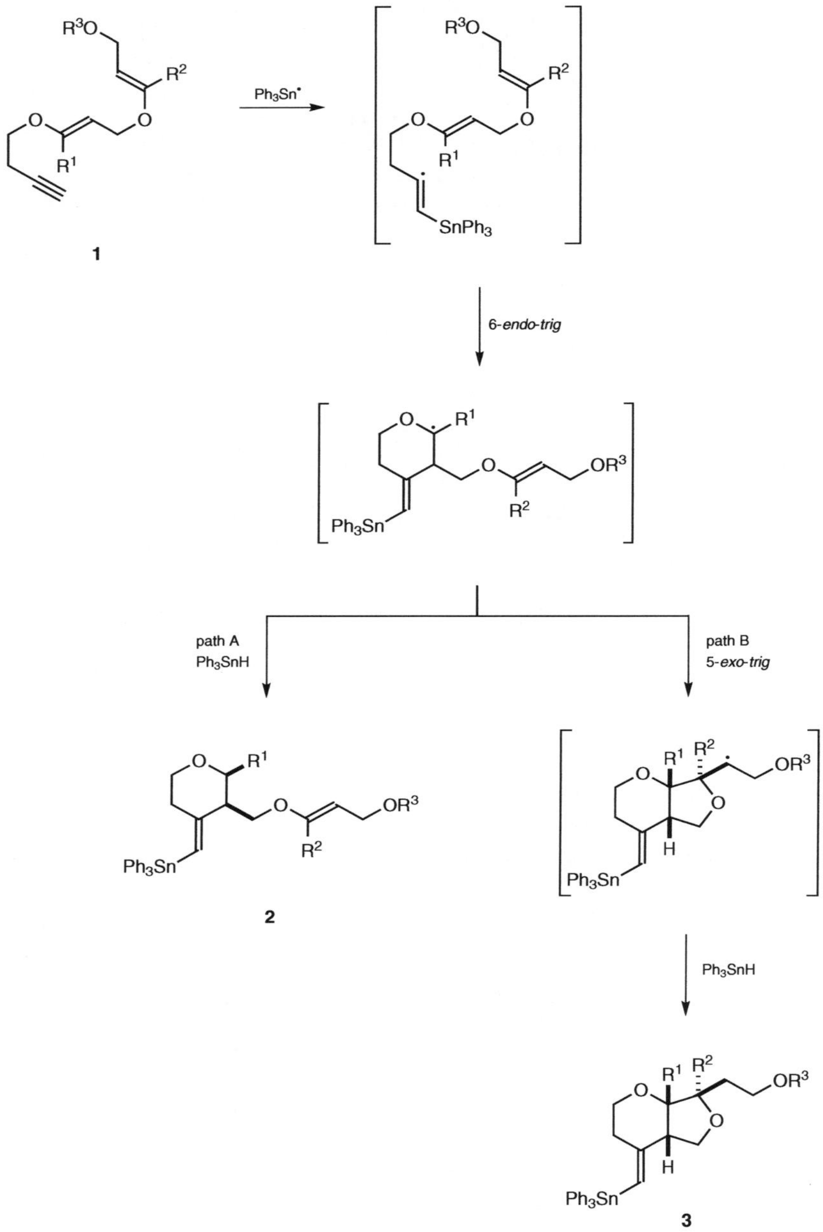

4-[(Triphenylstannyl)methylene]-3-[(vinyloxy)methyl]tetrahydropyrans 2 and 4-[(Triphen-ylstannyl)methylene]hexahydrofuro[3,4-b]pyrans 3; General Procedure:[17]

> **CAUTION:** *2,2′-Azobisisobutyronitrile (AIBN) is an explosion risk, avoid exposure of the reagent to light and heat; store protected from light in a refrigerator or freezer.*

A soln of Ph₃SnH (0.527 g, 1.50 mmol) and AIBN (0.0164 g, 0.10 mmol) in deoxygenated benzene (10 mL) (**CAUTION:** *carcinogen*) was added via syringe pump over 6 h to a refluxing soln of bis(vinyl ether) **1** (1.00 mmol) in deoxygenated benzene (115 mL). The resulting soln was stirred at reflux for an additional 1.5 h. The mixture was then concentrated

1.4.2 Radical Cyclizations

under reduced pressure and purified by flash column chromatography (silica gel) to afford the cyclized product.

1.4.2.1.2 Tin-Mediated Radical [3+2] Annulation

An elegant example of how a specific radical domino cyclization can be applied to the synthesis of a broader product selection was recently reported by Curran and co-workers.[18] These authors initially targeted (±)-epimeloscine (**8**) and (±)-meloscine,[19] alkaloid natural products that have attracted significant attention from the synthetic community. Curran and Zhang employed a radical annulation of divinylcyclopropane **4** to achieve rapid access to the ABCD ring system, which would then require minimal manipulation to access the completed natural product (Scheme 3). The proposed mechanism proceeds via addition of a tributyltin radical to one of the terminal alkenes of vinyl cyclopropane **4**, followed by cyclopropane fragmentation to yield radical **5**. Subsequent 6-*exo-trig* cyclization onto the tethered electron-rich alkene to give **6**, followed by a second cyclization and ultimately ejection of the tin radical, provides the final tetracyclic framework **7**. Although this process is achieved in a modest 55% yield, this step allows rapid access to the core of the natural products, with four additional steps providing (±)-epimeloscine (**8**) in 6% overall yield. (±)-Meloscine was obtained via a one-step epimerization of the hydrogen at C3.

Scheme 3 Domino Cyclizations En Route to (±)-Epimeloscine[18]

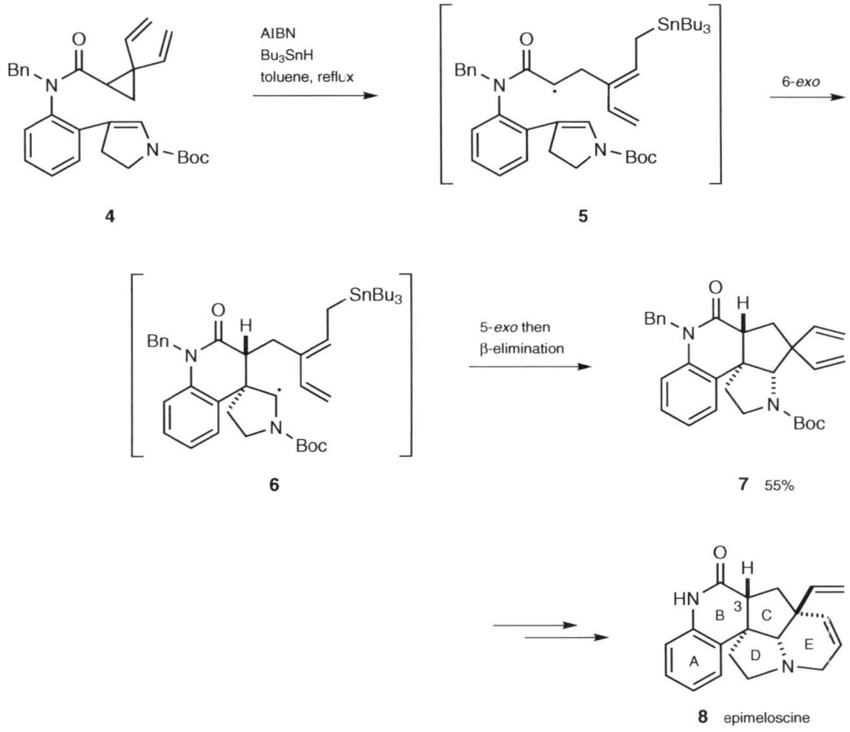

The modular synthesis of the radical cyclization precursor allows for the generalization of this [3+2]-annulation strategy to encompass a moderate range of divinylcyclopropanes of structure **9** (Scheme 4). Importantly, the dihydropyrrole motif at the core of these architectures can be altered to incorporate six- or seven-membered rings without any significant loss of efficiency. An isolated example also shows the method to be effective for an

for references see p 214

N-benzyl-*N*′-sulfonyldihydropyrrole derivative **9** (R^1 = Bn; R^2 = SO$_2$Ph; n = 1). The key reactions proceed in good yield (35–58%) given the number of C—C bond-forming/breaking events and the complexity of the final product. The versatility and practicality of this method have been demonstrated by elaboration of the products into 12 meloscine analogues with basic drug properties within typical Lipinski ranges.

Scheme 4 Domino Cyclizations En Route to (±)-Epimeloscine and Meloscine Analogues[18]

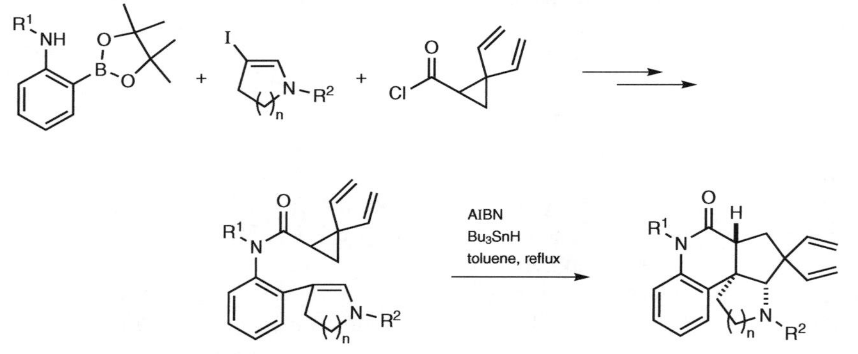

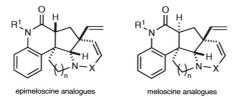

R^1	R^2	n	dr	Yield^a (%) of **10**	Ref
Boc	Boc	1	>20:1	53	[18]
Boc	Boc	2	>20:1	52	[18]
Boc	Boc	3	>20:1	35	[18]
Bn	SO$_2$Ph	1	>20:1	58	[18]

^{a} From **9**.

epimeloscine analogues   meloscine analogues

5-(*tert*-Butoxycarbonyl)-2,2-divinyl-1,2,3,3a,5,9b-hexahydro-4*H*-cyclopenta[*c*]quinolin-4-ones 10; General Procedure:[18]

CAUTION: *2,2′-Azobisisobutyronitrile (AIBN) is an explosion risk, avoid exposure of the reagent to light and heat; store protected from light in a refrigerator or freezer.*

A soln of AIBN (0.108 mmol, 0.5 equiv) and Bu$_3$SnH (0.448 mmol, 2.07 equiv) in toluene (6 mL) was added via syringe pump to a soln of divinylcyclopropane **9** (0.216 mmol, 1.0 equiv) in refluxing toluene (9 mL) over a period of 4 h. The mixture was cooled to rt, and the solvent was evaporated. The residue was purified by flash chromatography (silica gel) to provide the tetracyclic system, typically as a mixture of rotamers.

1.4.2.2 Reductive Radical Domino Cyclizations

There have been a number of recent methods utilizing reductants other than tin.[20–22] These developments have, in part, been driven by the desire to eliminate the toxicity concerns associated with tin but also to access novel reaction methodologies. In particular, titanium(III) and, more recently, samarium(II) iodide mediated[23] processes have been documented in radical–radical and radical–anionic cascade cyclizations. Both stoichiometric[24] and catalytic[25,26] titanium(III) epoxypolyene radical–radical domino cyclizations have found appreciable use in the construction of a wide array of cyclic terpenes, such as (+)-*seco*-C-oleanane and fomitellic acid. This methodology remains somewhat limited by the complexity requirements of the starting polyene, but due to its often-distinct reactivity relative to traditional cationic polyene cyclizations, should continue to find use and further development for the synthesis of complex molecules.

1.4.2.2.1 Samarium(II) Iodide Mediated Radical Cyclizations

The use of samarium(II) iodide as a reducing agent, first introduced by Kagan in 1977, has become important in organic synthesis, particularly in cyclization reactions en route to complex natural products.[27,28] An elegant demonstration of the distinct reactivity and selectivity of samarium(II) iodide induced radical–radical cascade cyclizations was recently reported by Procter and co-workers. Previous studies from the research group had documented the first addition reactions of radical anions, generated via samarium iodide directed lactone reduction.[29]

Extension of this methodology to cyclic lactones bearing two pendant chains with strategically positioned unsaturation led to biologically significant, complex molecular scaffolds, with the construction of up to four contiguous stereocenters.[30] Four-electron reductive cascade cyclizations of *cis*-disposed lactones with either two alkenes (i.e., **11**) or an alkene and an alkyne (i.e., **13**) were initially investigated. The method tolerates a moderate number of lactones with aryl-substituted *E*-alkenes, providing reasonable to excellent yields (50–90%) and good to excellent diastereoselectivities (4:1:1 to 8:1 dr; only the major diastereomer was stereochemically assigned) of readily separable bicyclic products **12** and **14** (Scheme 5).

Scheme 5 Alkene/Alkene and Alkene/Alkyne Samarium(II) Iodide/Water Mediated Radical Cyclization of Lactones[30]

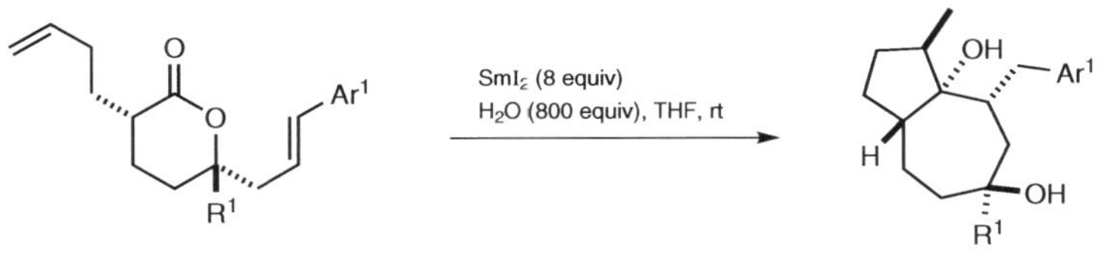

Ar¹	R¹	dr[a]	Yield (%)	Ref
Ph	Me	4:1:1	90	[30]
Ph	H	3:1	71	[30]
4-BrC₆H₄	Me	4:1:1	71	[30]
2-naphthyl	Me	3:1:1	79	[30]

[a] Major diastereomer shown, minor diastereomers not stereochemically assigned.

for references see p 214

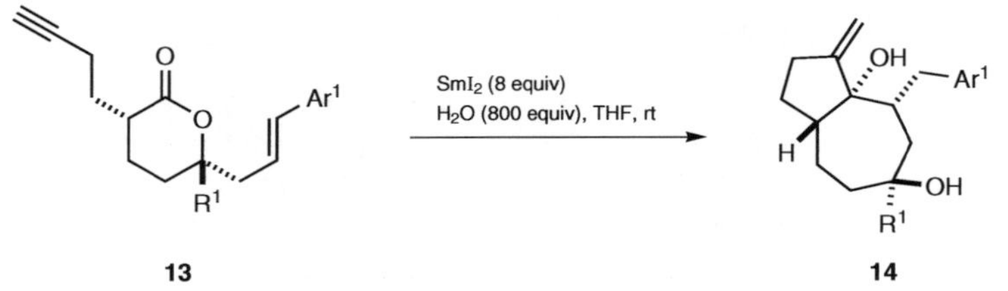

1.4 Radical Reactions

13 → **14**

Ar[1]	R[1]	dr[a]	Yield (%)	Ref
Ph	Me	10:2:1	54	[30]
4-ClC₆H₄	Me	8:1	50	[30]

[a] Major diastereomer shown, minor diastereomers not stereochemically assigned.

Calculations suggest an axial radical anion initializes the cascade cyclization in a "right then left" manner, leading to bicyclic products via the proposed mechanism and stereochemical rationale shown in Scheme 6. Electron transfer to the lactone carbonyl from samarium(II) iodide/water generates the axial radical anion **15**, which then undergoes cyclization with the styrenyl alkene. Following further reduction and protonation, opening of the hemiketal **16** provides ketone **17**. The second reductive cyclization leads to bicyclic diol **18**.

Scheme 6 Proposed Mechanism for the Radical Cyclization[30]

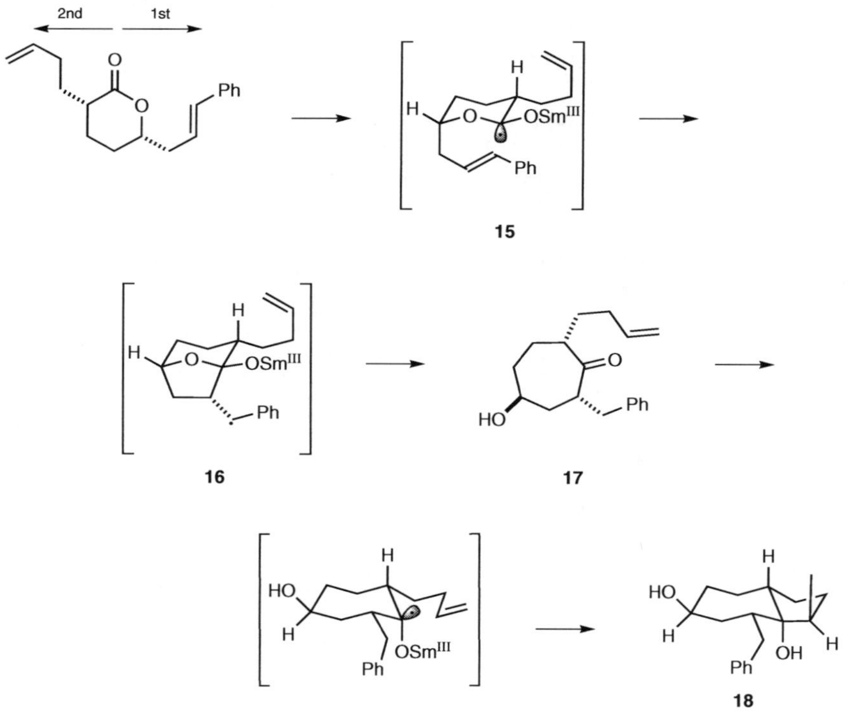

The method is also amenable to either *cis* or *trans* allenyl lactones **19**, giving structurally similar [5.3.0] bicyclic scaffolds **20** (Scheme 7). Good to excellent yields (63–98%) and diastereoselectivities (2:1 to >20:1 dr) are reported for a limited number of *cis* and *trans* allen-

yl lactones, along with two examples demonstrating the ability to generate a [4.2.1] bicyclic system.

Scheme 7 Allene/Alkene Samarium(II) Iodide/Water Mediated Radical Cyclization of Lactones[30]

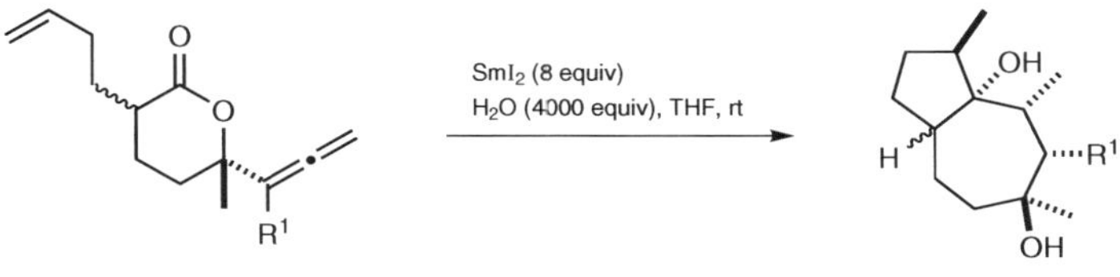

Config of **19**	R[1]	Config of **20**	dr[a]	Yield (%)	Ref
trans	Ph	cis	4:1	98	[30]
trans	4-MeOC$_6$H$_4$	cis	5:1	98	[30]
trans	3-ClC$_6$H$_4$	cis	2:1	63	[30]
cis	Ph	trans	>20:1	98	[30]
cis	4-MeOC$_6$H$_4$	trans	3:1	64	[30]
cis	3-ClC$_6$H$_4$	trans	>20:1	71	[30]

[a] Major diastereomer shown.

This method is defined by its one-pot assembly of highly elaborated carbocyclic skeletons in synthetically useful yields. Although only the *cis* stereoisomer is functional in the alkene/alkene or alkene/alkyne cyclizations, the use of either a *cis* or *trans* allenyl lactone allows substrate control over the [5.3.0] ring junction configuration. The use of differentially substituted Meldrum's acid derivatives has also been investigated; the radical cyclizations in these cases yield a range of complex carbocyclic frameworks in moderate to good yields (43–67%).[31]

4-Benzyl-3,6-dimethyloctahydroazulene-3a,6(1*H*)-diols 12 or 4-Benzyl-6-methyl-3-methyleneoctahydroazulene-3a,6(1*H*)-diols 14; General Procedure:[30]
To a stirred 0.1 M soln of SmI$_2$ in THF (3.1 mL, 0.31 mmol, 8 equiv) under N$_2$ at rt was added distilled H$_2$O (0.60 mL). This addition resulted in a color change from blue to red, at which point a soln of lactone **11** or **13** (0.04 mmol, 1.0 equiv) in THF (4 mL) was added quickly and the flask was sealed under N$_2$. The mixture was left to stir until decolorization had occurred. The reaction was quenched by opening the flask to air and adding sat. aq potassium sodium tartrate (15 mL). The aqueous layer was extracted with Et$_2$O (3 × 15 mL), and the organic extracts were combined, dried (Na$_2$SO$_4$), and concentrated under reduced pressure to yield the crude product. Purification by column chromatography (silica gel) gave the desired compound as a mixture of diastereomers.

3,4,6-Trimethyloctahydroazulene-3a,6(1*H*)-diols 20; General Procedure:[30]
An oven-dried flask was evacuated and refilled with argon three times before the addition of a soln of lactone **19** (0.10 mmol, 1.0 equiv) in THF (1 mL). To this flask was added distilled H$_2$O (7.30 mL) followed by the addition of 0.1 M SmI$_2$ in THF (8.10 mL, 0.81 mmol, 8 equiv) dropwise over 30 min. Once the addition was complete, the flask was sealed under a slight pressure of N$_2$ (or argon) before being left to stir until decolorization had occurred. The reaction was quenched by opening the flask to air and adding sat. aq tartaric acid (10 mL). The aqueous layer was extracted with Et$_2$O (3 × 15 mL), and the organic

for references see p 214

196 Domino Transformations 1.4 Radical Reactions

extracts were combined, dried (Na_2SO_4), and concentrated under reduced pressure to yield the crude product. Purification by column chromatography (silica gel) gave the desired compound as a mixture of diastereomers.

1.4.2.2.2 Samarium(II) Iodide Mediated Radical–Anionic Cyclizations

The samarium(II) iodide mediated radical–anionic dialdehyde cyclization cascade, which has been reported by Procter and co-workers, allows for rapid access to fused five- and six-membered carbocycles with an appended spirolactone unit (Scheme 8).[32,33] Although the dialdehyde starting materials 21 are not trivial to obtain, the subsequent one-pot, domino transformation leads to complex frameworks 22 containing four contiguous stereocenters, typically with high stereocontrol. The method is applicable to a range of cyclohexene starting materials bearing alkyl or alkenyl substituents in typically good yield (66–90%) and diastereoselectivity (typically >20:1 dr). Unfortunately, extension to the formation of a larger ring size in the second cyclization was unsuccessful in the reported protocol.

Scheme 8 Samarium(II) Iodide Mediated Radical–Anionic Cyclization of Aldehydes[32,33]

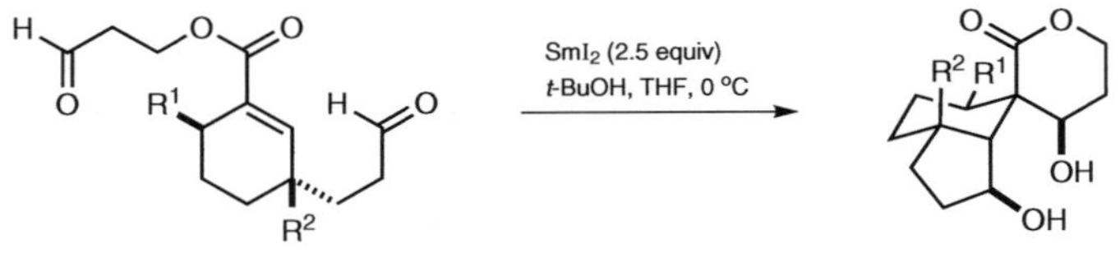

R^1	R^2	dr[a]	Yield (%)	Ref
H	CMe=CH$_2$	>20:1	90	[32,33]
Me	CMe=CH$_2$	3:1	70[b]	[32,33]
H	(CH$_2$)$_2$CH=CH	>20:1	69	[32,33]
Me	(CH$_2$)$_2$CH=CH	>20:1	66[b]	[32,33]
H	Me	>20:1	77	[32,33]

[a] Major diastereomer shown.
[b] Starting material 21 used as a 3:1 mixture of diastereomers.

The mechanism is proposed to proceed through an *anti*-selective ketyl–alkene cyclization via pre-transition state assembly 23 to give samarium enolate 24. A subsequent samarium(III) chelated diastereoselective aldol cyclization through assembly 25 furnishes the spirocyclic lactone (Scheme 9).

1.4.2 Radical Cyclizations **197**

Scheme 9 Samarium(II) Iodide Mediated Radical-Anionic Cyclization of Aldehydes[32,33]

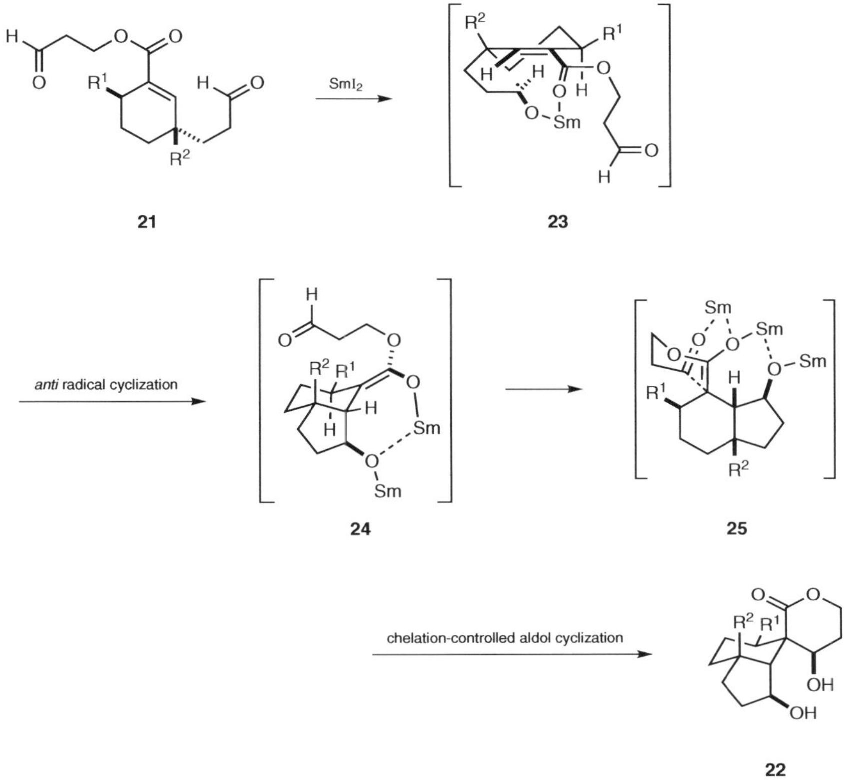

Although the high step count associated with the synthesis of the dialdehydes reduces the general applicability of this procedure, the efficient construction of molecular complexity with typically excellent diastereoselectivities has led to its employment in a range of complex molecule syntheses. Elegant syntheses of pleuromutilin by Procter and co-workers,[34,35] as well as (−)-maoecrystal Z by Reisman and co-workers,[36] have employed the methodology as key steps, demonstrating its ability to provide a useful synthetic disconnection.

3,4′-Dihydroxydecahydro-2′H,4′H-spiro[indene-4,3′-pyran]-2′-ones 22;
General Procedure:[32,33]
A 0.1 M soln of SmI$_2$ in THF (6.92 mL, 0.69 mmol, 2.5 equiv) was added to degassed t-BuOH (1.8 mL) and the resulting soln was stirred under a N$_2$ atmosphere for 20 min before being cooled to 0 °C. Next, the dialdehyde **21** (0.277 mmol, 1 equiv) was added dropwise as a soln in THF (2 mL) and the mixture was stirred for 30 min at 0 °C before the excess SmI$_2$ was quenched by allowing air to reach the mixture. Once the soln was yellow, sat. aq potassium sodium tartrate was added and the mixture was extracted with Et$_2$O (3 × 15 mL). The combined organic fractions were washed with H$_2$O and brine, dried (Na$_2$SO$_4$), and concentrated under reduced pressure. The crude products were purified by chromatography (silica gel) to give the desired product as a mixture of diastereomers.

for references see p 214

1.4.2.3 Oxidative Radical Cyclizations

Single-electron oxidative reactions, established by the seminal work of Faraday,[37] have played a role in organic synthesis for almost two centuries, predating Thomson's discovery of the electron by more than 60 years.[38] Radicals and radical cations generated via single-electron oxidation are routinely formed by metal-mediated, hypervalent iodine mediated, or electrochemically mediated oxidations. One of the benefits of these processes is that multiple oxidative methods are capable of performing the desired chemical transformation. However, it is always important to consider discrete interactions that can occur between oxidant, substrate, and solvent.

Despite the longstanding presence of single-electron oxidations in organic synthesis, utilization of radical cations as the key reactive intermediates in complex transformations has lagged behind anions, cations, and neutral radicals due to the perceived complexity of these high-energy species.[39] However, over the last 10–15 years, radical cations, and the radicals they decay to, have been applied in an expanding number of C—heteroatom and C—C bond-forming reactions. In particular, the oxidations of carbonyl-containing compounds have been applied in both the early- and late-stage syntheses of intricate natural products and biologically active targets.

1.4.2.3.1 Organo-SOMO-Activated Polyene Cyclization

Enamines result from the equilibrium condensation of an amine and a carbonyl-containing compound. These structures are electron rich and have traditionally been utilized as nucleophiles in two-electron chemistry. In 1992, Narasaka and co-workers achieved the oxidative coupling of enamines with silyl enol ethers[40] and in 2007, MacMillan and co-workers developed organo-SOMO activation of an aldehyde through the in situ addition of a chiral secondary amine, forming an enamine.[41] The cerium(IV)-mediated oxidative coupling of an aldehyde and an allylsilane in the presence of 20 mol% of an imidazolidinone catalyst resulted in the enantioselective formation of an α-substituted aldehyde in high yield through asymmetric addition to an intermediate radical cation. It was later demonstrated that this radical cation can readily couple with a variety of allylsilanes, pendant allylsilanes,[42] silyl enol ethers,[43] and other radicophiles,[44–46] providing access to a diverse set of chiral structures. The Nicolaou and MacMillan labs also independently reacted the organo-SOMO intermediate with pendant aromatic rings.[47–50] This approach provides access to bicyclic arenes and hetarenes. The SOMO intermediate can be generated from either ammonium cerium(IV) nitrate (CAN) mediated oxidation,[47] or oxidation by tris(1,10-phenanthroline)iron(III) hexafluorophosphate (Fephen).[48] Emphasizing the utility of this method, these groups applied it to the total synthesis of the natural products demethyl calamenene[47] and (–)-tashiromine, respectively.[48]

MacMillan and co-workers further extended the reactivity of the radical intermediate through unsaturations in polyenes to form multiple fused six-membered rings in one reaction with high enantioselectivity (Scheme 10).[51] Reaction of polyene **26** in the presence of catalyst **27** and copper(II) as an oxidant forms hexacyclic product **28** in 63% yield and 93% ee. This domino reaction undergoes four consecutive 6-*endo-trig* cyclizations. The mechanism is initiated by condensation of **26** with **27** to form an enamine, which is oxidized by copper(II) to form distonic radical cation **29**. This radical cation then undergoes rapid and successive 6-*endo-trig* cyclizations, terminating in the formation of conjugated radical **30**. Oxidation and elimination of a proton yield iminium **31**, which, upon hydrolysis, releases catalyst **27** and forms **28** with excellent enantioselectivity. This method provides synthetically useful yields of products bearing electron-deficient aromatic rings (Table 1, entries 1 and 4), electron-rich aromatic rings (entries 2 and 5), and heteroaromatics (entry 3), and can provide fused, hexacyclic systems in a single, enantioselective step.

1.4.2 Radical Cyclizations

199

Scheme 10 Organo-SOMO Polyene Cyclization[51]

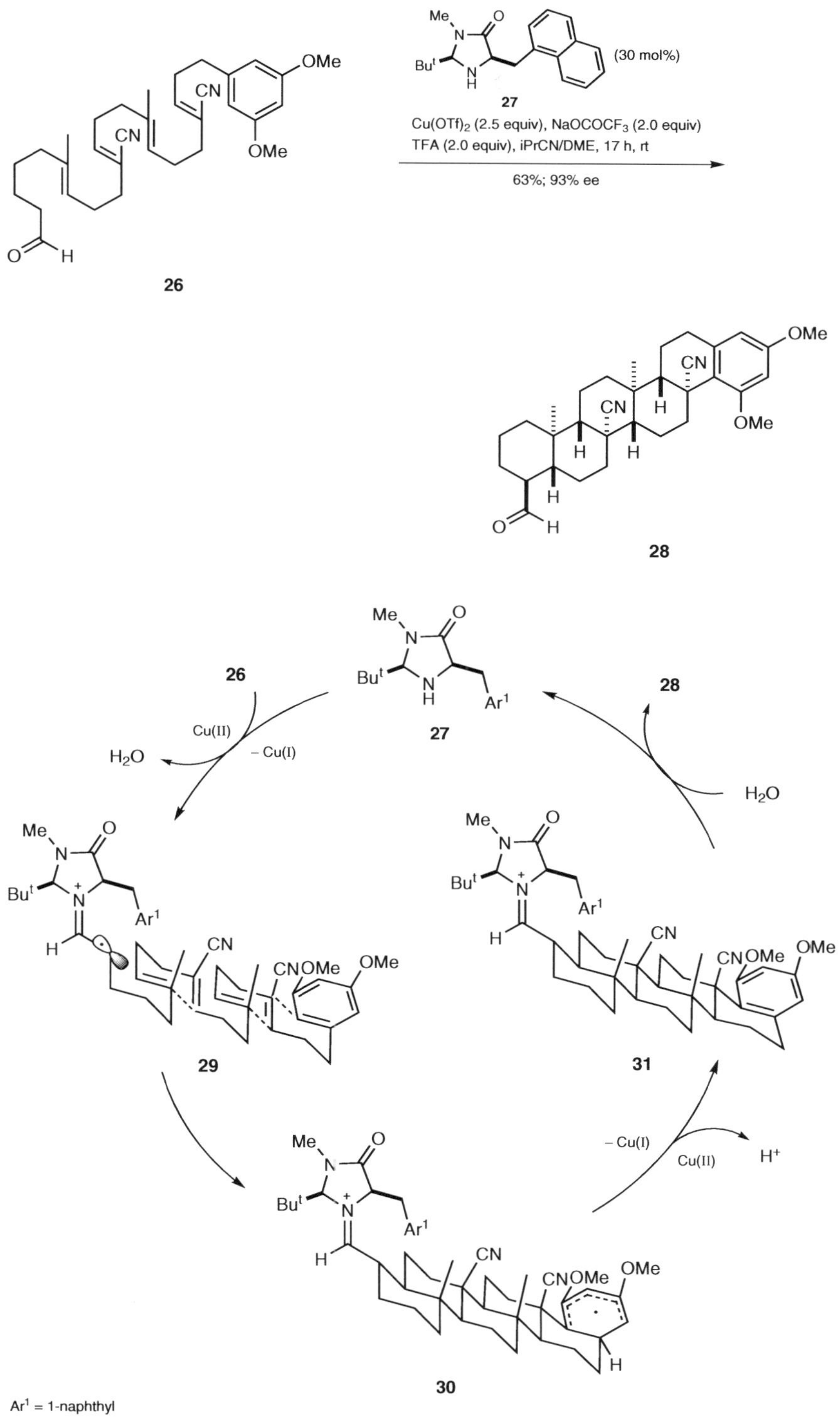

Ar¹ = 1-naphthyl

for references see p 214

200 Domino Transformations 1.4 Radical Reactions

Table 1 Organo-SOMO Polyene Cyclizations[51]

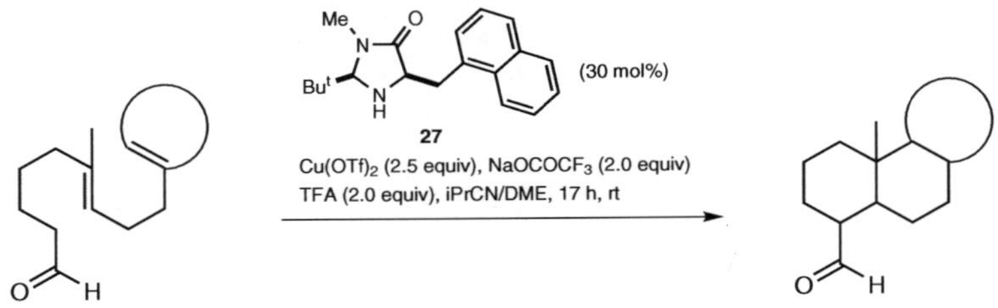

Entry	Substrate	Product	ee[a] (%)	Yield[b] (%)	Ref
1	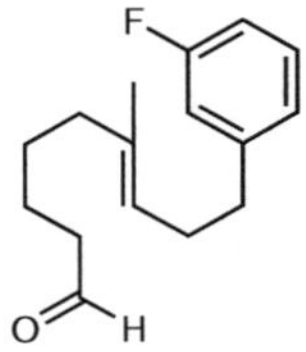	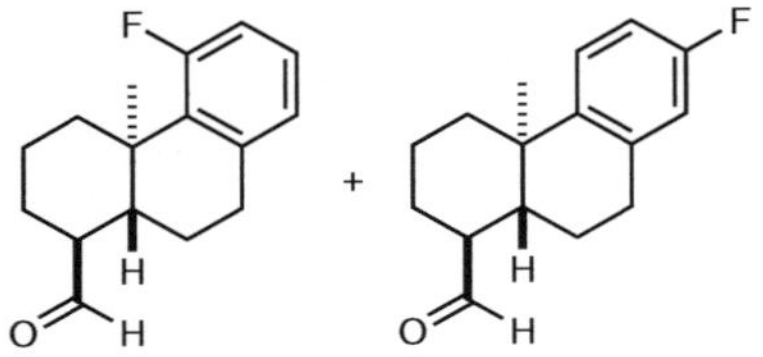	90	65	[51]
2	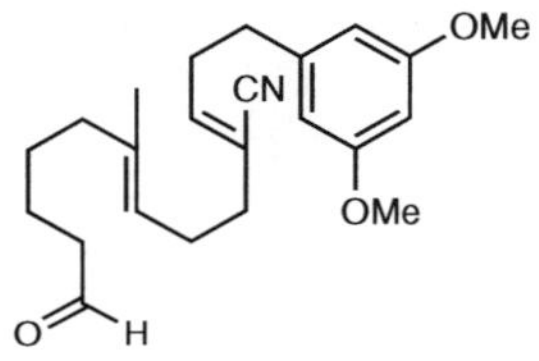	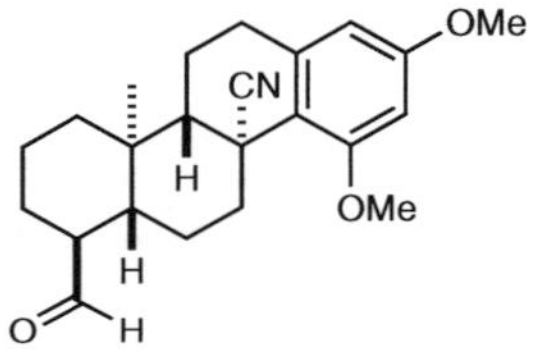	91	61	[51]
3	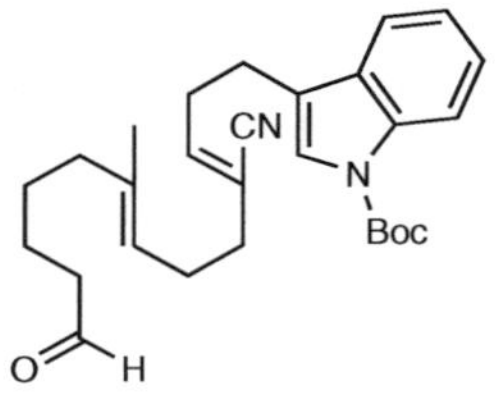	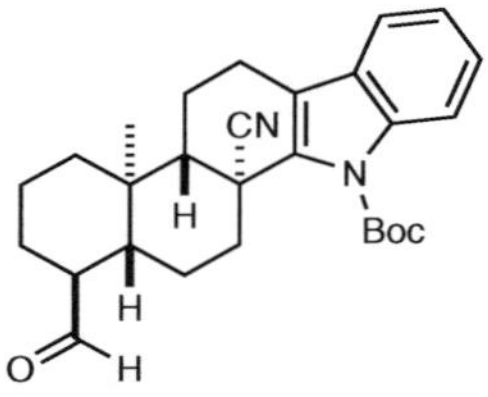	92	71	[51]
4	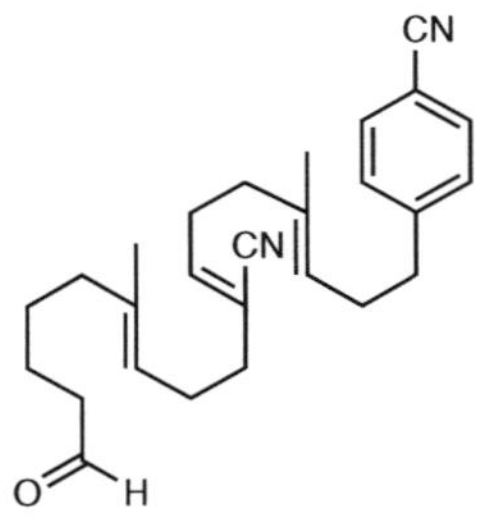	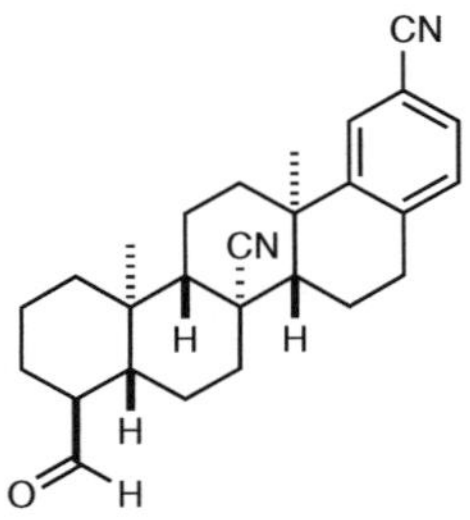	92[c]	56	[51]

[a] Determined by chiral HPLC analysis.
[b] Isolated yield.
[c] Determined by chiral supercritical fluid chromatography.

Polycyclic Aldehydes, e.g. 28 and Table 1; General Procedure:[51]

An oven-dried vial equipped with a magnetic stirrer bar and a rubber septum was charged with a soln of the TFA salt of catalyst **27** (0.060 mmol, 0.30 equiv) and the polyenal (0.200 mmol, 1.00 equiv) in iPrCN/DME (1:2; 1.5 mL) under an argon atmosphere. To this mixture was added slowly a soln of Cu(OTf)$_2$ (0.500 mmol, 2.50 equiv), NaOCOCF$_3$

(0.400 mmol, 2.00 equiv), and TFA (0.400 mmol, 2.00 equiv) in iPrCN (1.0 mL) over 7 h using a syringe pump at rt. After stirring had been continued for a further 17 h, the light-green soln was subjected to aqueous workup and purified by flash chromatography (silica gel) to afford the cyclization product.

1.4.2.3.2   **Oxidative Rearrangement of Silyl Bis(enol ethers)**

Silyl enol ethers are neutral structural analogues of metal enolates. They can be easily isolated and stored, and, even better, many are commercially available. The main structural difference between silyl enol ethers and lithium enolates is that silyl enol ethers are monomeric in nature. Ruzziconi and co-workers demonstrated an oxidative heterocoupling of silyl enol ethers in an ammonium cerium(IV) nitrate (CAN) mediated process.[52] They also observed that yields of the heterocoupled product decreased as the ratio of the silyl enol ethers approached 1:1. In 1998, Schmittel and co-workers reported the oxidative rearrangement of silyl bis(enol ethers).[53,54] Recently, Thomson and co-workers extended this method to the oxidative rearrangement of unsymmetrical silyl bis(enol ethers) **34** to generate 1,4-diketones **35** (Scheme 11).[55] These tethered substrates can be efficiently prepared via condensation of an enolate derived from ketone **32** with a chlorosilyl enol ether **33**. When ethers **34** are exposed to ammonium cerium(IV) nitrate in the presence of sodium hydrogen carbonate at reduced temperatures, the bis(enol ether) rearranges to form **35**. This transformation provides good yields for enol ethers bearing alkyl (R^1 = Me) and styryl (R^1 = CH=CHPh) groups as well as electron-neutral (R^1 = Ph), electron-deficient (R^1 = 4-ClC$_6$H$_4$), and electron-rich (R^1 = 4-Tol) aryl groups.

Scheme 11   Formation and Oxidative Rearrangement of Silyl Bis(enol ethers)[55]

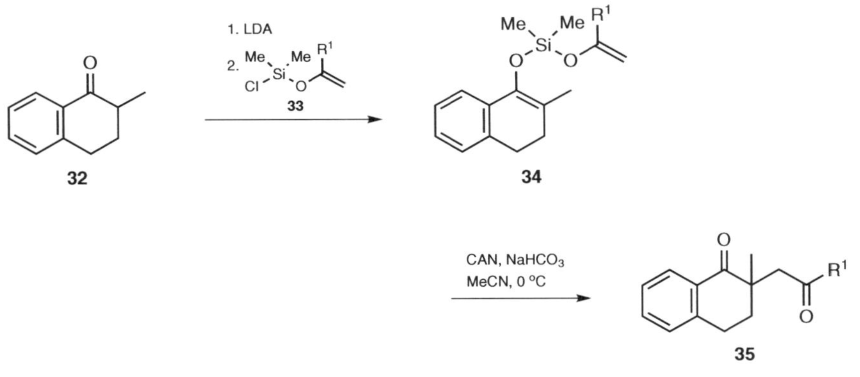

R^1	Yield[a] (%)	Ref
Me	71	[55]
Ph	81	[55]
4-ClC$_6$H$_4$	72	[55]
4-Tol	73	[55]
CH=CHPh	75	[55]

[a] Isolated yield over two steps.

In addition to the substitution pattern on the methyl ketone derived enol ether, the effect of ring size of the cyclic ketone derived enol ether has been investigated (Scheme 12). Suc-

for references see p 214

cessful formation of the 1,4-diketone is achieved for substrates bearing five-, six-, and seven-membered rings. Both the substitution pattern on the acyclic silyl enol ether and the ring size have a significant impact on the yield.

Scheme 12 Formation and Oxidative Rearrangement of Silyl Bis(enol ethers) with Various Ring Sizes[55]

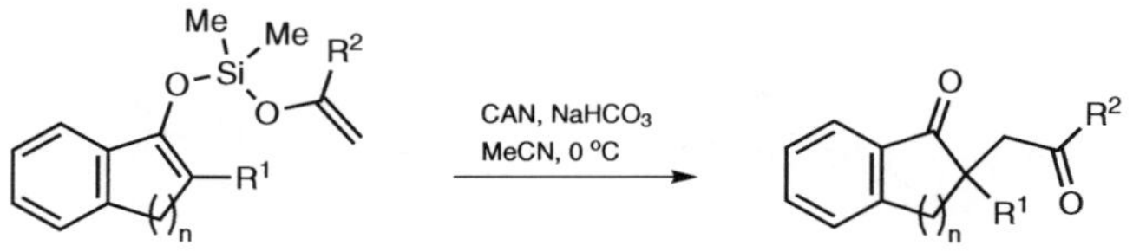

n	R^1	R^2	Yield^a (%)	Ref
1	Me	Me	52	[55]
1	Me	Ph	72	[55]
2	Et	Me	30	[55]
2	Et	Ph	76	[55]
3	Me	Me	0	[55]
3	Me	Ph	54	[55]

^{a} Isolated yield.

This process likely occurs via the mechanism displayed in Scheme 13. Cerium(IV)-mediated single-electron oxidation of one of the enol ethers of **36** results in the formation of a distonic radical cation **37**. This intermediate undergoes a *7-endo-trig* cyclization to form radical cation **38**. This radical cation then β-eliminates the silyl group through reaction with a nucleophile present in solution according to a known nucleophilic displacement mechanism[56] to form radical **39**. Cerium(IV)-mediated oxidation of the radical to give cation **40** and nucleophilic displacement of the silyl group then provides 1,4-diketone **41**. The Thomson lab demonstrated the utility of this process in their syntheses of metacycloprodigiosin and prodigiosin R1.[57]

1.4.2 Radical Cyclizations

Scheme 13 Mechanism of Oxidative Rearrangement of Silyl Bis(enol ethers)[55]

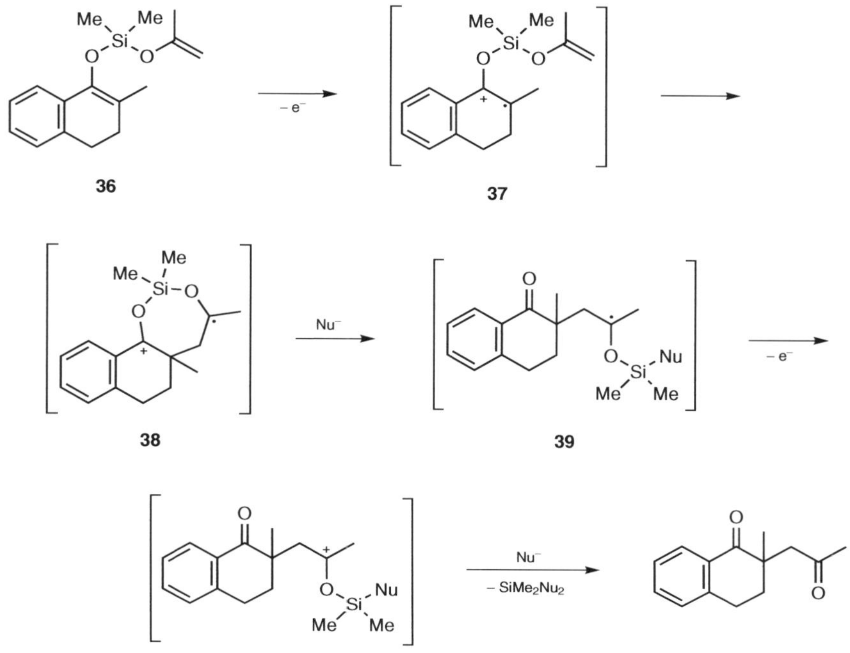

1-[Dimethyl(vinyloxy)siloxy]-2-methyl-3,4-dihydronaphthalenes 34; General Procedure:[55]
A soln of 2-methyltetrahydronaphthalen-1-one (**32**; 250 mg, 1.56 mmol) in THF (2 mL) was transferred via cannula (0.5 mL THF rinse to complete transfer) into a stirred soln of LDA [1.72 mmol, prepared by adding BuLi (1.72 mmol) to a stirred soln of iPr₂NH (1.75 mmol) in THF (12 mL) at 0 °C] in THF (12 mL) at 0 °C (as reported). The mixture was allowed to stir for 15 min at −78 °C and then warmed to 0 °C and stirred for 1 h at that temperature. In a separate round-bottomed flask, AcCl (122 µL, 1.72 mmol) was added dropwise to a stirred soln of an *N,N*-diethyl(dimethyl)(vinyloxy)silanamine (1.75 mmol) in THF (4 mL) at 0 °C. After stirring for 5 min at 0 °C and 10 min at rt, the soln of chlorosilane **33** was transferred by cannula to the soln of the lithium enolate. The reaction was monitored by TLC. Upon completion, the mixture was quenched with a soln of pH 7 buffer, warmed to rt, and extracted with pentanes (3 × 20 mL). The combined organic layers were washed with brine and dried (Na₂SO₄). Removal of the solvent under reduced pressure afforded the desired silyl bis(enol ether) in ≥95% purity as determined by ¹H NMR spectroscopy. The silyl bis(enol ethers) were used without further purification.

2-Methyl-2-(2-oxoalkyl)-3,4-dihydronaphthalen-1(2*H*)-ones 35; General Procedure:[55]
A soln of the silyl bis(enol ether) **34** (2.56 mmol) in MeCN (12 mL) was transferred by cannula (2 mL MeCN rinse to complete transfer) to a stirred suspension of CAN (1.71 g, 3.12 mmol) and NaHCO₃ (525 mg, 6.24 mmol) in MeCN (35 mL) at 0 °C. The reaction was monitored by TLC with reaction times of approximately 10 min. Upon completion, the soln was diluted with 1.0 M HCl (50 mL) and extracted with Et₂O (2 × 50 mL). The combined organic layers were washed with brine (100 mL) and dried (Na₂SO₄). Removal of the solvent under reduced pressure, followed by flash chromatography (silica gel), afforded the desired 1,4-diketone.

for references see p 214

1.4.2.3.3 **Diastereoselective Oxidative Rearrangement of Silyl Bis(enol ethers)**

Having demonstrated the propensity of structurally diverse silyl bis(enol ethers) to undergo oxidative rearrangement, Thomson and co-workers pursued the development of a diastereoselective process.[58] They found that when the enol ethers are linked by a sterically encumbered silyl group, as in **42**, a high degree of diastereoselectivity is imparted to the formation of the 1,4-diketones **43** (Table 2), consistent with Schmittel and co-workers' early results.[53,54]

This method provides good yields for the rearrangement of a variety of symmetric silyl bis(enol ethers). Importantly, the rearrangement is tolerant of substrates containing further degrees of unsaturation (entries 3 and 4). A mixture of acetonitrile and propanenitrile in the presence of 2 equivalents of dimethyl sulfoxide relative to the silyl bis(enol ether) greatly enhances the solubility of the lipophilic substrates and the polar oxidant and provides improved stereoselectivity over neat acetonitrile.

1.4.2 Radical Cyclizations **205**

Table 2 Cerium(IV)-Mediated Diastereoselective Intramolecular Oxidative Coupling of Silyl Bis(enol ethers)[58]

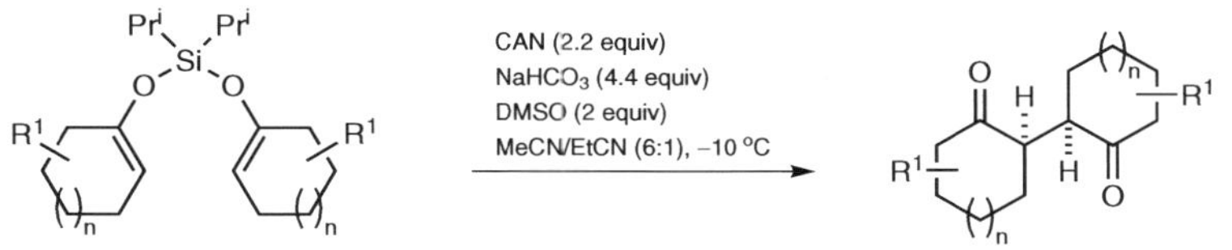

Entry	Substrate	Product	dr[a]	Yield[b] (%)	Ref
1	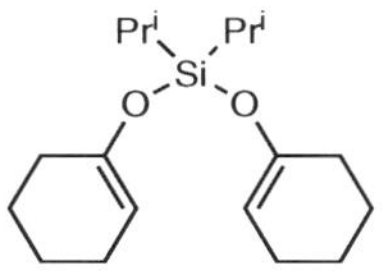	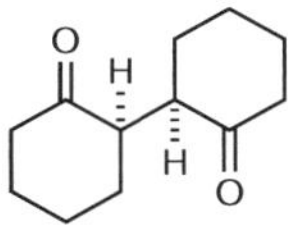	14:1	67	[58]
2	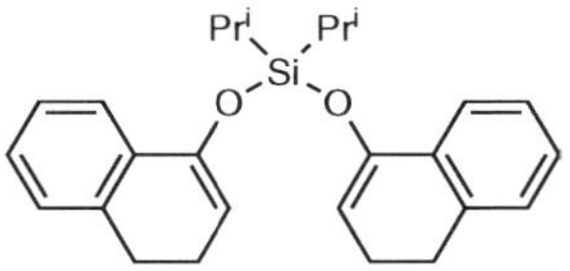	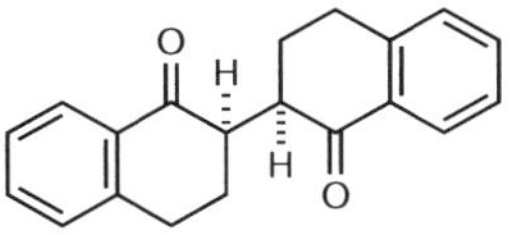	20:1	77	[58]
3	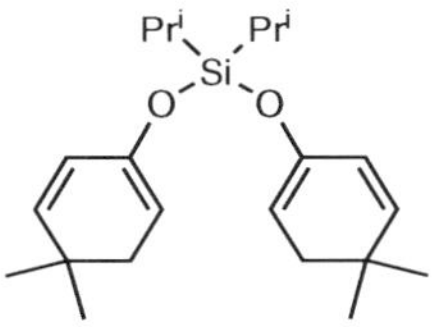	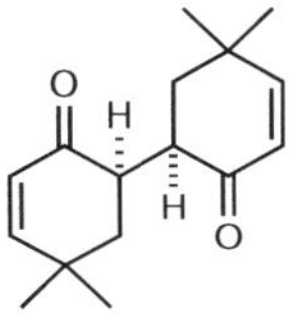	13:1	77	[58]
4	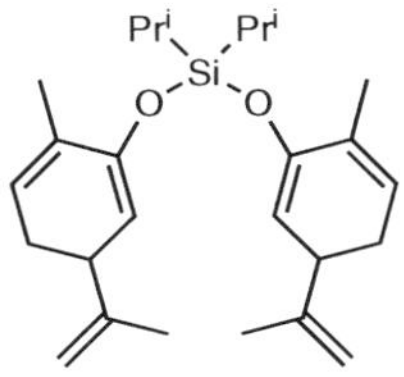	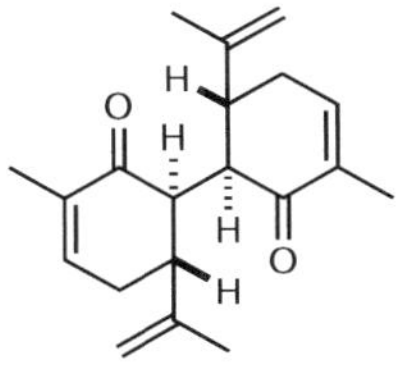	20:1	84	[58]
5	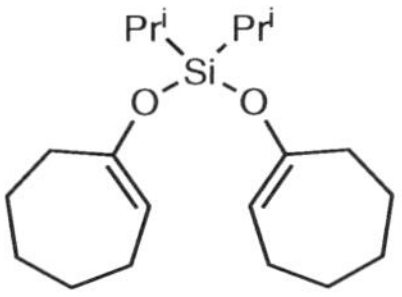	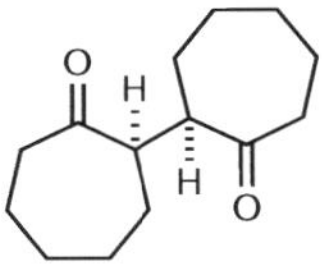	14:1	50	[58]

[a] Determined by ^{1}H NMR spectroscopy on unpurified reaction mixture.
[b] Isolated yield.

With conditions for a diastereoselective method in hand, Thomson and co-workers extended their method to oxidative cross coupling. Single-electron oxidation of unsymmetrical silyl bis(enol ethers) **44** affords the desired cross-coupled products **45** in varying yields, but with good diastereomeric ratios (Table 3).

for references see p 214

Table 3 Diastereoselectivity Cross-Coupling of Silyl Bis(enol ethers)[58]

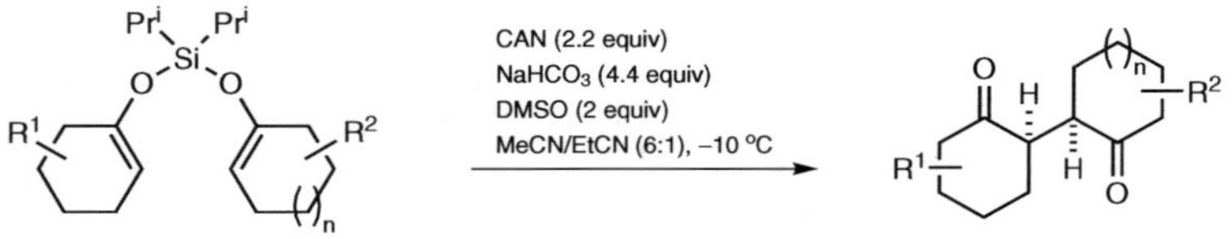

Entry	Substrate	Product	dr[a]	Yield[b] (%)	Ref
1	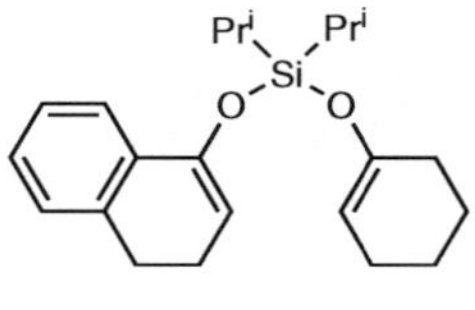	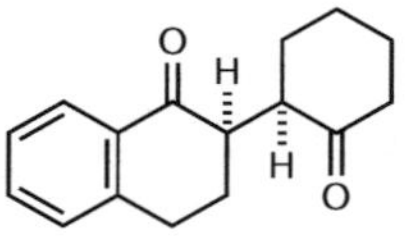	20:1	82	[58]
2	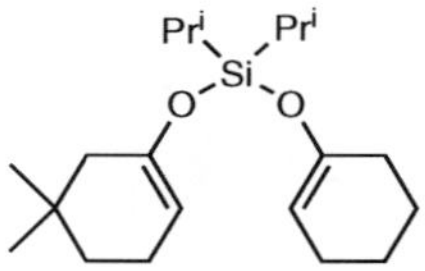	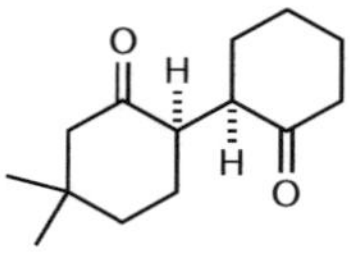	20:1[c]	50	[58]
3	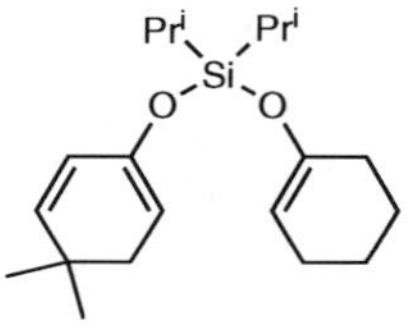	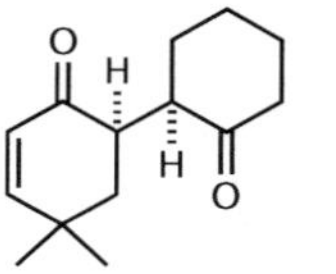	20:1	67	[58]
4	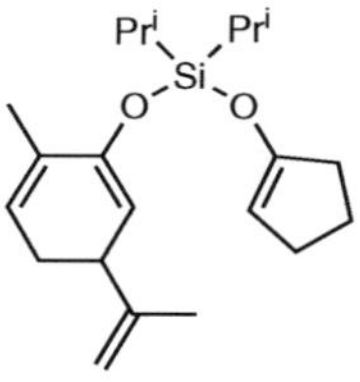	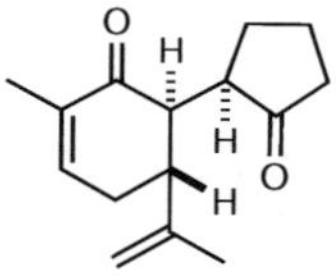	16:1	53	[58]
5	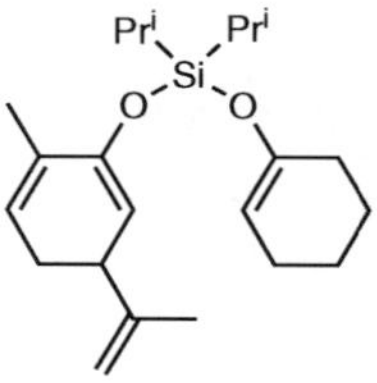	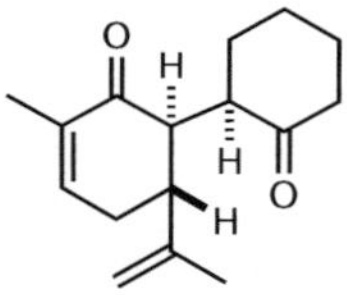	15:1	68	[58]

[a] Determined by ^{1}H NMR spectroscopy on unpurified reaction mixture.
[b] Isolated yield.
[c] Stereochemistry assigned by analogy.

The authors proposed a model for the stereoselectivity of this process consistent with that of Schmittel and co-workers (Scheme 14).[53,54] After single-electron oxidation of the bis-(enol ether), addition of the radical to the pendant enol ether through an *lk* approach (**46A**) would result in the formation of "chiral" product **47A**, which would be C_2-symmetric for symmetrical silyl bis(enol ethers); by contrast, the *ul* approach (**46B**) would form *meso* diketone **47B**.[58] When considering the transition state of C—C bond formation,

46A should be favored over **46B** due to steric interactions. This method has been used as one of the key steps in the synthesis of propolisbenzofuran B.[59]

Scheme 14 Rationale behind Diastereoselectivity[58]

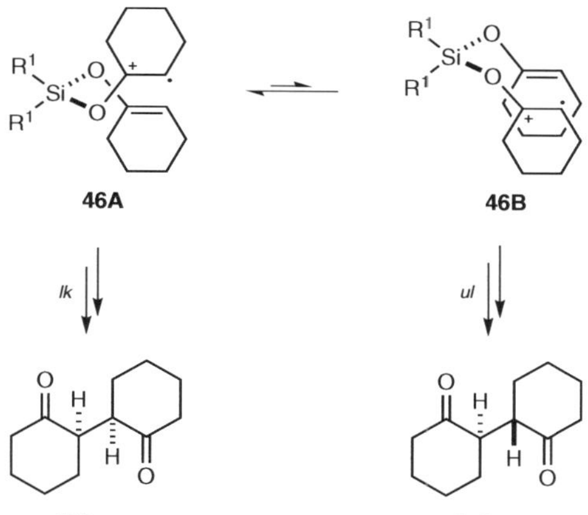

1,4-Diketones 43; **General Procedure:**[58]
A soln of the silyl bis(enol ether) **42** (1 mmol) in EtCN (4 mL) was transferred by cannula (1 mL EtCN rinse to complete transfer) to a stirred suspension of CAN (1.21 g, 2.2 mmol) and NaHCO$_3$ (396 mg, 4.4 mmol) in MeCN (30 mL) and DMSO (142 µL, 2.0 mmol) at −10 °C. The reaction was monitored by TLC with times ranging from 5 min to 24 h. Upon completion, the soln was diluted with sat. aq NaHCO$_3$ (35 mL) and extracted with CHCl$_3$ (3 × 35 mL). The combined organic layers were washed with brine (100 mL) and dried (Na$_2$SO$_4$). Removal of the solvent under reduced pressure, followed by flash chromatography (silica gel), afforded the desired 1,4-diketone.

1.4.2.4   Visible-Light-Mediated Reactions

Visible-light photoredox catalysis has established itself as a powerful technique for enacting free-radical transformations.[60–64] Typical photocatalysts are either transition-metal complexes or organic dyes that, upon photoexcitation by visible light to an excited triplet state, can be quenched through single-electron transfer or energy transfer to mediate free-radical transformations. This concept is emerging as a versatile and advantageous technique for mediating free-radical reactions for several reasons. First, photoredox catalysts are activated by visible light and reactions are conducted in typical borosilicate glassware with simple light fixtures. Second, the stoichiometric electron carrier is typically an inexpensive amine, as opposed to far more expensive stannanes or silanes as found in traditional radical processes. Finally, reactions are very robust, with typical transformations proceeding efficiently at low catalyst loadings.

1.4.2.4.1   Light-Mediated Radical Cyclization/Divinylcyclopropane Rearrangement

The Stephenson group reported a tin-free reductive dehalogenation domino reaction utilizing a visible-light-activated photocatalyst in 2009.[65] This photochemical process, which forms a radical intermediate, has been applied to the formation of tricyclic pyrrolidinones (Table 4).[66] When bromocyclopropanes **48** are irradiated with visible light in the presence of [Ir(ppy)$_2$(dtbbpy)]PF$_6$, acting as the photocatalyst, and triethylamine, tricyclic

for references see p 214

208   Domino Transformations **1.4** Radical Reactions

pyrrolidinones **49** are formed. The broad substrate scope of this method includes both electron-rich (entries 2 and 4) and electron-deficient (entry 3) arenes. Further, utilization of *trans*-1,2-diarylcyclopropanes provides the corresponding regioisomers (entries 5–7).

Table 4   Light-Mediated Radical Cyclization/Divinylcyclopropane Rearrangement[66]

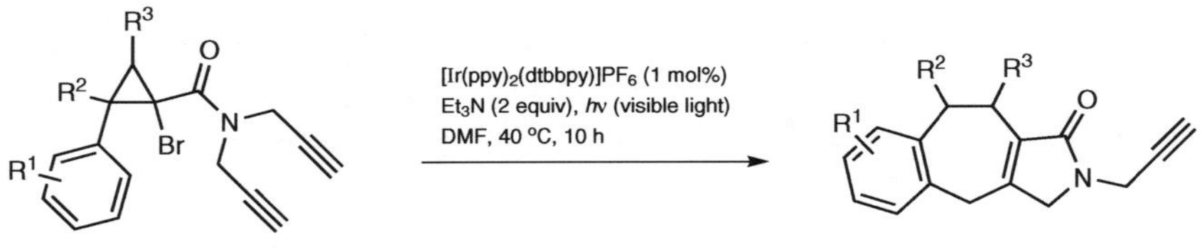

ppy = 2-(2-pyridyl)phenyl; dtbbpy = 4,4'-di-*tert*-butyl-2,2'-bipyridyl

Entry	Substrate	Product	Yield[a] (%)	Ref
1	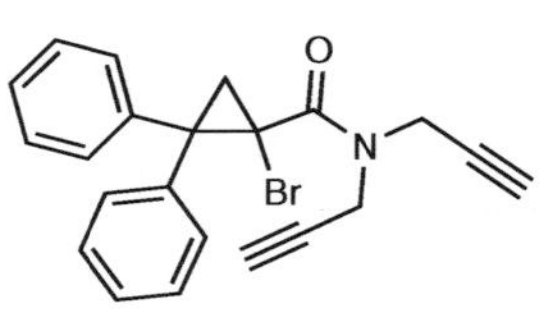	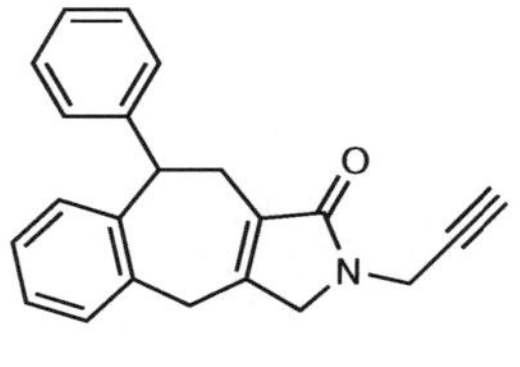	69	[66]
2	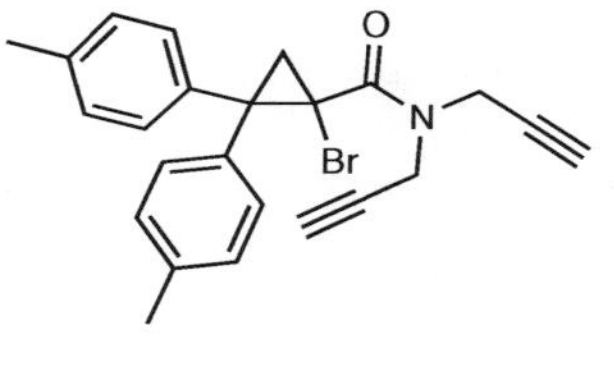	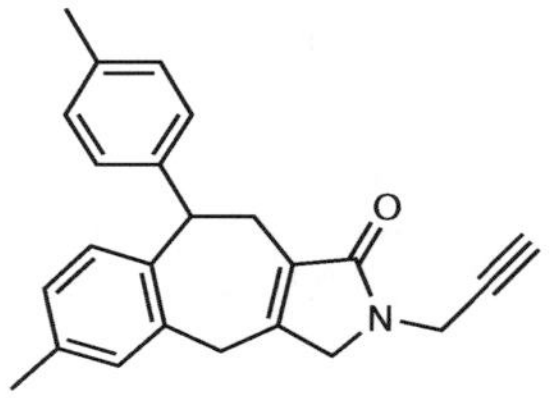	91	[66]
3	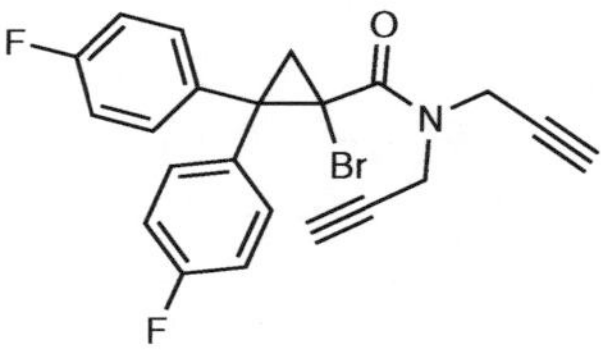	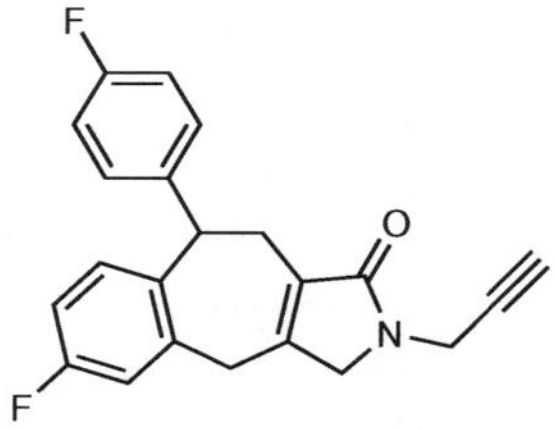	89	[66]
4	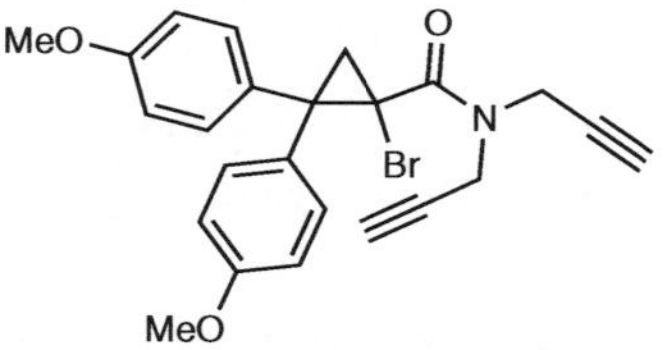	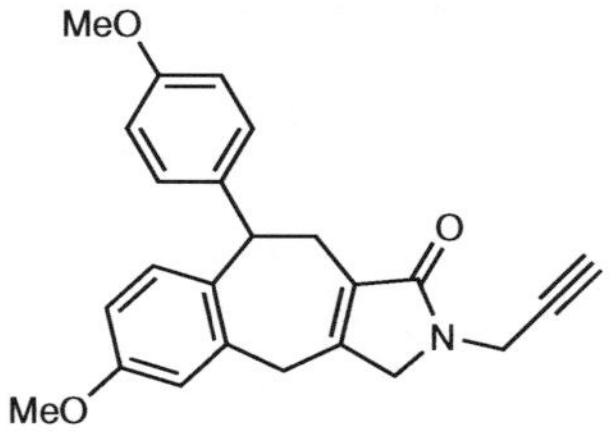	79	[66]

1.4.2 Radical Cyclizations **209**

Table 4 (cont.)

Entry	Substrate	Product	Yield[a] (%)	Ref
5	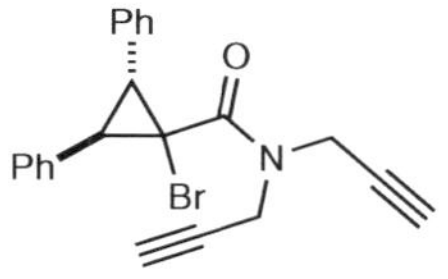	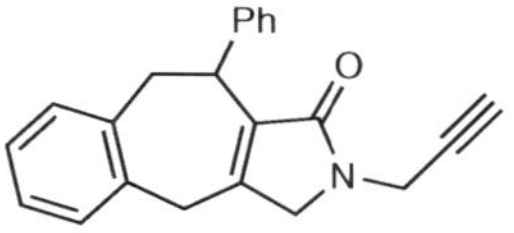	85	[66]
6	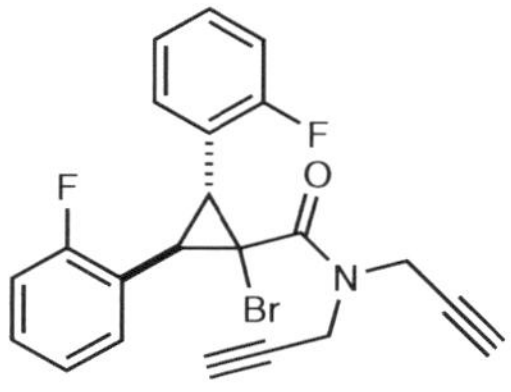	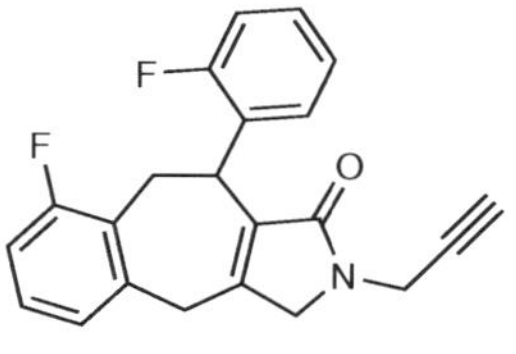	82	[66]
7	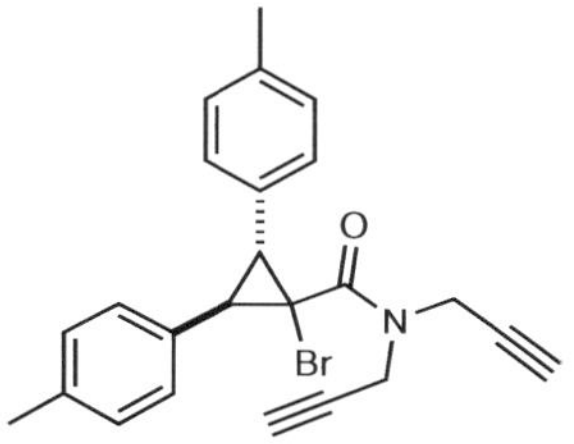	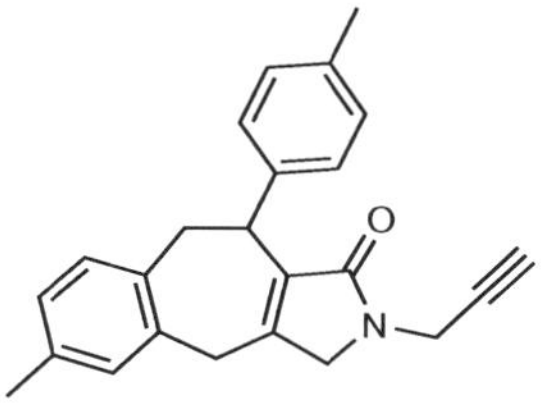	88	[66]

[a] Isolated yield.

This process begins with the reductive quenching of photoexcited iridium(III) by triethylamine to form an iridium(II) complex (Scheme 15). Single-electron transfer to **50** then provides α-radical **51**, which quickly undergoes a 5-*exo-dig* cyclization with the pendant alkyne to form lactam **52**. At this point, there are multiple possible paths. In path A, **52** can abstract a hydrogen atom to give **53**, followed by a thermal sigmatropic rearrangement that is driven by strain release of the cyclopropyl ring to yield **54**. Alternatively, in path B, radical **52** can undergo an additional cyclization onto the aromatic ring to give **55**, followed by β-scission to form radical **56**. Then, isomerization and hydrogen atom abstraction may lead to product **54**. Interestingly, trace quantities of **53** were isolated from the reaction mixture. To provide further mechanistic insight, intermediate **53** was heated at 40 °C in dimethylformamide in the absence of other redox conditions. Pyrrolidinone **54** was the only observed product, suggesting that path A is the likely route.

for references see p 214

Scheme 15 Visible-Light-Mediated Mechanism for Radical Cyclization/Divinylcyclopropane Rearrangement[66]

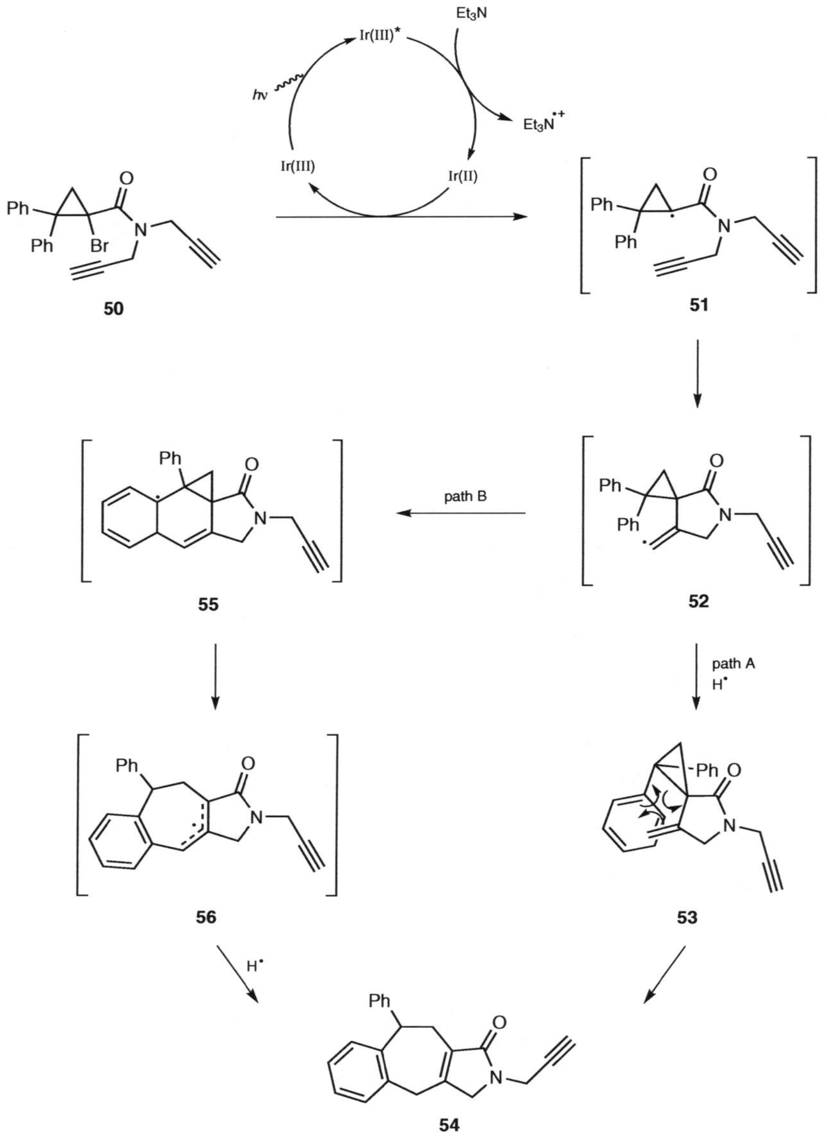

2-(Prop-2-ynyl)-3,4,9,10-tetrahydrobenzo[4,5]cyclohepta[1,2-c]pyrrol-1(2*H*)-ones 49; General Procedure:[66]

A 25-mL round-bottomed flask was equipped with a rubber septum and magnetic stirrer bar and was charged with bromocyclopropane **48** (1.0 mmol, 1.0 equiv), Et₃N (2.0 mmol, 2.0 equiv), DMF (15 mL), and [Ir(ppy)₂(dtbbpy)]PF₆ (0.010 mmol, 0.010 equiv). The mixture was then shielded from light and degassed by argon sparging for 10 min. The mixture was then irradiated by a 14-W compact fluorescent lamp under an atmosphere of argon for 6–24 h while the light and flask were surrounded by a layer of aluminum foil. After the reaction was complete (determined by TLC analysis), the mixture was poured into a separatory funnel containing EtOAc (25 mL) and H₂O (50 mL). The layers were separated and the aqueous layer was extracted with EtOAc (2 × 50 mL). The combined organic layers

were dried (Na_2SO_4) and concentrated. The residue was purified by flash column chromatography (silica gel) to afford the rearranged product.

1.4.2.4.2 Visible-Light-Mediated Radical Fragmentation and Bicyclization

In 2008, Yoon and co-workers reported a photocatalytic [2+2] reductive cyclization of a bis(enone) bearing a three-carbon tether.[67] A year later, they extended this methodology to an intermolecular process.[68] In particular, they noted that the addition of a soluble lithium ion source was necessary for successful conversion of the starting material, presumably through Lewis acid activation of the carbonyls of the enone-based substrates. They continued their efforts of using Lewis acid activation to facilitate photoredox processes with the development of a formal [3+2] cycloaddition of activated cyclopropanes with alkenes.[69] This process, facilitated by lanthanum(III) trifluoromethanesulfonate, results in the formation of fused [3.3.0] ring systems **58** from cyclopropane-containing tethered enones **57** (Scheme 16). Simple variation of the enone component demonstrates that a variety of esters, ketones, and thioesters participate in this cycloaddition. In particular, thioesters provide enhanced diastereoselectivity (e.g., **58**, R^1 = Me; R^2 = SEt). This method has been further extended beyond enones (Scheme 17), showing that additions to styrenes, alkenes, and alkynes proceed efficiently as well.

Scheme 16 Visible-Light-Mediated Radical Fragmentation and Bicyclization of Tethered Enones[69]

57 **58**

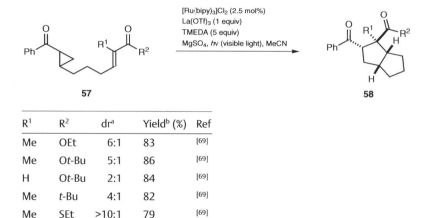

R^1	R^2	dr[a]	Yield[b] (%)	Ref
Me	OEt	6:1	83	[69]
Me	Ot-Bu	5:1	86	[69]
H	Ot-Bu	2:1	84	[69]
Me	t-Bu	4:1	82	[69]
Me	SEt	>10:1	79	[69]

[a] Ratio of major diastereomer to all other diastereomers as determined by GC.
[b] Averaged isolated yields from two reproducible experiments.

Scheme 17 Visible-Light-Mediated Radical Fragmentation and Bicyclization of Tethered Alkenes and Alkynes[69]

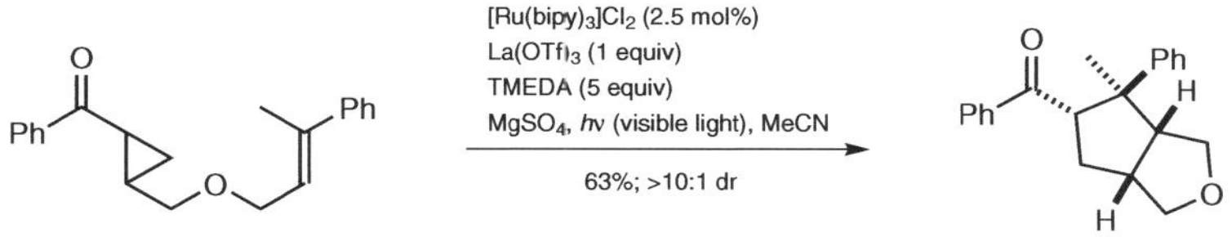

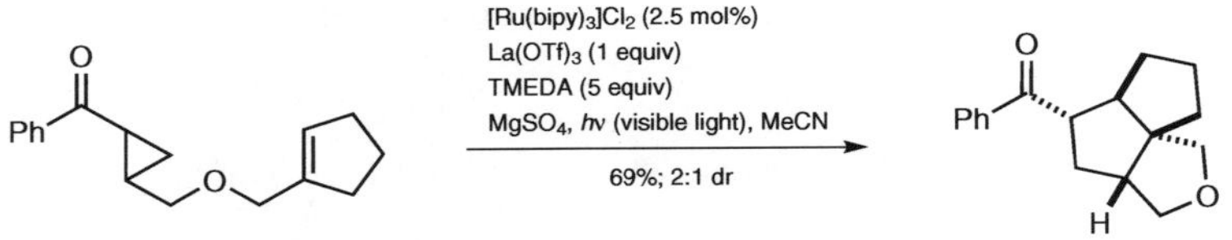

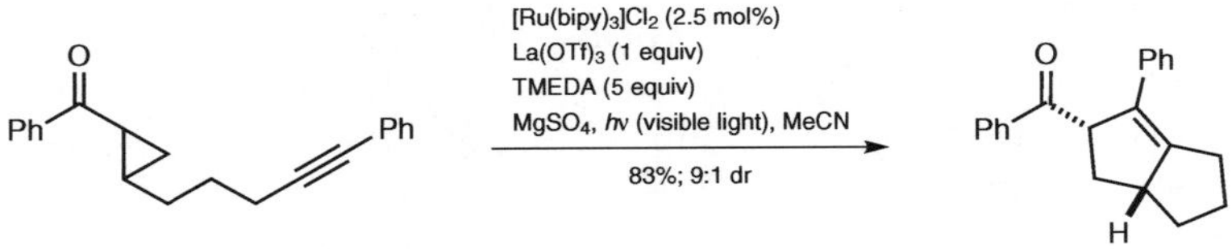

Domino Transformations   1.4 Radical Reactions

Initial single-electron reduction of Lewis acid activated species **59** by the reductively quenched photocatalyst results in formation of ketyl radical anion **60** (Scheme 18). Rapid retro 3-*exo-trig* cyclization leads to the formation of allyl radical **61**. Addition of the pendant alkene via a facile 5-*exo-trig* cyclization path yields tertiary radical intermediate **62**. Finally, ring closure via a second 5-*exo-trig* cyclization completes the formation of the [3.3.0] system to give ketyl **63**. Single-electron oxidation of this radical anion affords product **64**.

Scheme 18   Mechanism for the Visible-Light-Mediated Radical Fragmentation and Bicyclization[69]

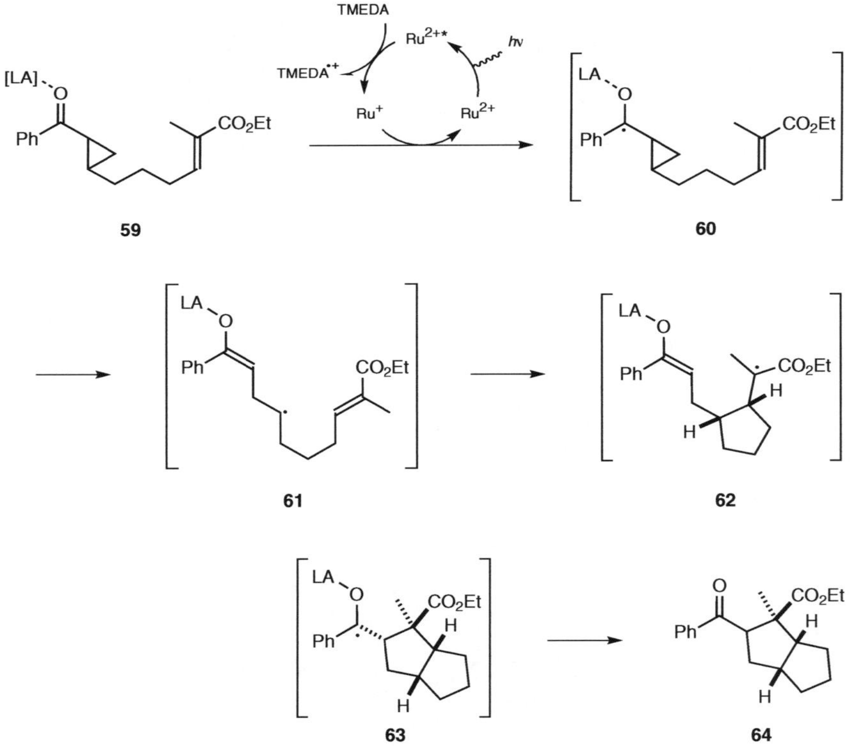

2-Benzoyloctahydropentalenes 58; **General Procedure:**[69]
A 25-mL Schlenk flask containing MgSO$_4$ (2% wt/wt with respect to substrate) was flame-dried under vacuum and cooled under dry N$_2$. The flask was then charged with substrate

57 (1.0 equiv), [Ru(bipy)$_3$]Cl$_2$•6H$_2$O (0.025 equiv), La(OTf)$_3$ (1 equiv), MeCN (0.05 M), and TMEDA (5 equiv). The soln was then degassed through three freeze–pump–thaw cycles under N$_2$. The reaction was placed in a water bath and stirred in front of a 23-W (1380 lumen) compact fluorescent lamp at a distance of 30 cm. Upon consumption of starting material, the mixture was diluted with Et$_2$O and passed through a short pad of silica gel using Et$_2$O. The filtrate was concentrated under reduced pressure and the residue was purified by column chromatography (silica gel) to give the desired product.

for references see p 214

References

[1] *Radicals in Organic Synthesis*, Renaud, P.; Sibi, M. P., Eds.; Wiley-VCH: Weinheim, Germany, (2001).

[2] Rowlands, G. J., *Tetrahedron*, (2009) **65**, 8603.

[3] Rowlands, G. J., *Tetrahedron*, (2010) **66**, 1593.

[4] Tietze, L. F.; Brasche, G.; Gericke, K. M., *Domino Reactions in Organic Synthesis*, Wiley-VCH: Weinheim, Germany, (2006); pp 219–279.

[5] An, G.; Li, G., In *Domino Reactions: Concepts for Efficient Organic Synthesis*, Tietze, L. F., Ed.; Wiley-VCH: Weinheim, Germany, (2014); pp 141–182.

[6] Takeuchi, K.; Ishita, A.; Matsuo, J.-i.; Ishibashi, H., *Tetrahedron*, (2007) **63**, 11101.

[7] Taniguchi, T.; Tanabe, G.; Muraoka, O.; Ishibashi, H., *Org. Lett.*, (2008) **10**, 197.

[8] Taniguchi, T.; Ishibashi, H., *Tetrahedron*, (2008) **64**, 8773.

[9] Taniguchi, T.; Ishibashi, H., *Org. Lett.*, (2008) **10**, 4129.

[10] Pattenden, G.; Stoker, D. A.; Thomson, N. M., *Org. Biomol. Chem.*, (2007) **5**, 1776.

[11] Pattenden, G.; Stoker, D. A.; Winne, J. M., *Tetrahedron*, (2009) **65**, 5767.

[12] Crick, P. J.; Simpkins, N. S.; Highton, A., *Org. Lett.*, (2011) **13**, 6472.

[13] Callier-Dublanchet, A.-C.; Cassayre, J.; Gagosz, F.; Quiclet-Sire, B.; Sharp, L. A.; Zard, S. Z., *Tetrahedron*, (2008) **64**, 4803.

[14] Sharp, L. A.; Zard, S. Z., *Org. Lett.*, (2006) **8**, 831.

[15] Khan, F. A.; Upadhyay, S. K., *Tetrahedron Lett.*, (2008) **49**, 6111.

[16] Maddess, M. L.; Mainetti, E.; Harrak, Y.; Brancour, C.; Devin, P.; Dhimane, A.-L.; Fensterbank, L.; Malacria, M., *Chem. Commun. (Cambridge)*, (2007), 936.

[17] Davies, K. A.; Wulff, J. E., *Org. Lett.*, (2011) **13**, 5552.

[18] Zhang, H.; Jeon, K. O.; Hay, E. B.; Geib, S. J.; Curran, D. P.; LaPorte, M. G., *Org. Lett.*, (2014) **16**, 94.

[19] Zhang, H.; Curran, D. P., *J. Am. Chem. Soc.*, (2011) **133**, 10376.

[20] Simpkins, N.; Pavlakos, I.; Male, L., *Chem. Commun. (Cambridge)*, (2012) **48**, 1958.

[21] Hierold, J.; Lupton, D. W., *Org. Lett.*, (2012) **14**, 3412.

[22] Zeng, R.; Fu, C.; Ma, S., *Angew. Chem. Int. Ed.*, (2012) **51**, 3888.

[23] Tu, Y.; Zhou, L.; Yin, R.; Lv, X.; Flowers, R. A., II; Choquette, K. A.; Liu, H.; Niu, Q.; Wang, X., *Chem. Commun. (Cambridge)*, (2012) **48**, 11026.

[24] Yamaoka, M.; Fukatsu, Y.; Nakazaki, A.; Kobayashi, S., *Tetrahedron Lett.*, (2009) **50**, 3849.

[25] Domingo, V.; Arteaga, J. F.; López Pérez, J. L.; Peláez, R.; Quílez del Moral, J. F.; Barrero, A. F., *J. Org. Chem.*, (2012) **77**, 341.

[26] Rosales, A.; Muñoz-Bascón, J.; Morales-Alcázar, V. M.; Castilla-Alcalá, J. A.; Oltra, J. E., *RSC Adv.*, (2012) **2**, 12922.

[27] Procter, D. J.; Flowers, R. A., II; Skrydstrup, T., *Organic Synthesis using Samarium Diiodide: A Practical Guide*, RSC: Cambridge, UK, (2010).

[28] Coote, S. C.; Flowers, R. A., II; Skrydstrup, T.; Procter, D. J., In *Encyclopedia of Radicals in Chemistry, Biology and Materials*, Chatgilialoglu, C.; Studer, A., Eds.; Wiley: Chichester, UK, (2012); pp 849–900.

[29] Parmar, D.; Price, K.; Spain, M.; Matsubara, H.; Bradley, P. A.; Procter, D. J., *J. Am. Chem. Soc.*, (2011) **133**, 2418.

[30] Parmar, D.; Matsubara, H.; Price, K.; Spain, M.; Procter, D. J., *J. Am. Chem. Soc.*, (2012) **134**, 12751.

[31] Sautier, B.; Lyons, S. E.; Webb, M. R.; Procter, D. J., *Org. Lett.*, (2012) **14**, 146.

[32] Helm, M. D.; Da Silva, M.; Sucunza, D.; Helliwell, M.; Procter, D. J., *Tetrahedron*, (2009) **65**, 10816.

[33] Helm, M. D.; Sucunza, D.; Da Silva, M.; Helliwell, M.; Procter, D. J., *Tetrahedron Lett.*, (2009) **50**, 3224.

[34] Helm, M. D.; Da Silva, M.; Sucunza, D.; Findley, T. J. K.; Procter, D. J., *Angew. Chem. Int. Ed.*, (2009) **48**, 9315.

[35] Fazakerley, N. J.; Helm, M. D.; Procter, D. J., *Chem.–Eur. J.*, (2013) **19**, 6718.

[36] Cha, J. Y.; Yeoman, J. T. S.; Reisman, S. E., *J. Am. Chem. Soc.*, (2011) **133**, 14964.

[37] Faraday, M., *Philos. Trans. R. Soc. London*, (1834) **124**, 77.

[38] Thomson, J. J., *Philos. Mag.*, (1897) **44**, 293.

[39] Todres, Z. V., In *Ion-Radical Organic Chemistry: Principles and Applications*, CRC: Boca Raton, FL, (2009).

[40] Narasaka, K.; Okaguchi, T.; Tanaka, K.; Murakami, M., *Chem. Lett.*, (1992), 2099.

[41] Beeson, T. D.; Mastracchio, A.; Hong, J.; Ashton, K.; MacMillan, D. W. C., *Science (Washington, D. C.)*, (2007) **316**, 582.

References

[42] Pham, P. V.; Ashton, K.; MacMillan, D. W. C., *Chem. Sci.*, (2011) **2**, 1470.

[43] Jang, H.; Hong, J.; MacMillan, D. W. C., *J. Am. Chem. Soc.*, (2007) **129**, 7004.

[44] Kim, H.; MacMillan, D. W. C., *J. Am. Chem. Soc.*, (2008) **130**, 398.

[45] Wilson, J. E.; Casarez, A. D.; MacMillan, D. W. C., *J. Am. Chem. Soc.*, (2009) **131**, 11332.

[46] Graham, T. H.; Jones, C. M.; Jui, N. T.; MacMillan, D. W. C., *J. Am. Chem. Soc.*, (2008) **130**, 16494.

[47] Nicolaou, K. C.; Reingruber, R.; Sarlah, D.; Bräse, S., *J. Am. Chem. Soc.*, (2009) **131**, 2086.

[48] Conrad, J. C.; Kong, J.; Laforteza, B. N.; MacMillan, D. W. C., *J. Am. Chem. Soc.*, (2009) **131**, 11640.

[49] Jui, N. T.; Lee, E. C. Y.; MacMillan, D. W. C., *J. Am. Chem. Soc.*, (2010) **132**, 10015.

[50] Jui, N. T.; Garber, J. A. O.; Finelli, F. G.; MacMillan, D. W. C., *J. Am. Chem. Soc.*, (2012) **134**, 11400.

[51] Rendler, S.; MacMillan, D. W. C., *J. Am. Chem. Soc.*, (2010) **132**, 5027.

[52] Paolobelli, A. B.; Latini, D.; Ruzziconi, R., *Tetrahedron Lett.*, (1993) **34**, 721.

[53] Schmittel, M.; Burghart, A.; Malisch, W.; Reising, J.; Söllner, R., *J. Org. Chem.*, (1998) **63**, 396.

[54] Schmittel, M.; Haeuseler, A., *J. Organomet. Chem.*, (2002) **661**, 169.

[55] Clift, M. D.; Taylor, C. N.; Thomson, R. J., *Org. Lett.*, (2007) **9**, 4667.

[56] Jiao, J.; Zhang, Y.; Devery, J. J., III; Xu, L.; Deng, L.; Flowers, R. A., II, *J. Org. Chem.*, (2007) **72**, 5486.

[57] Clift, M. D.; Thomson, R. J., *J. Am. Chem. Soc.*, (2009) **131**, 14579.

[58] Avetta, C. T., Jr.; Konkol, L. C.; Taylor, C. N.; Dugan, K. C.; Stern, C. L.; Thomson, R. J., *Org. Lett.*, (2008) **10**, 5621.

[59] Jones, B. T.; Avetta, C. T.; Thomson, R. J., *Chem. Sci.*, (2014) **5**, 1794.

[60] Ravelli, D.; Dondi, D.; Fagnoni, M.; Albini, A., *Chem. Soc. Rev.*, (2009) **38**, 1999.

[61] Narayanam, J. M. R.; Stephenson, C. R. J., *Chem. Soc. Rev.*, (2011) **40**, 102.

[62] Xuan, J.; Xiao, W.-J., *Angew. Chem. Int. Ed.*, (2012) **51**, 6828.

[63] Prier, C. K.; Rankic, D. A.; MacMillan, D. W. C., *Chem. Rev.*, (2013) **113**, 5322.

[64] Schultz, D. M.; Yoon, T. P., *Science (Washington, D. C.)*, (2014) **343**, 1239176.

[65] Narayanam, J. M. R.; Tucker, J. W.; Stephenson, C. R. J., *J. Am. Chem. Soc.*, (2009) **131**, 8756.

[66] Tucker, J. W.; Stephenson, C. R. J., *Org. Lett.*, (2011) **13**, 5468.

[67] Ischay, M. A.; Anzovino, M. E.; Du, J.; Yoon, T. P., *J. Am. Chem. Soc.*, (2008) **130**, 12886.

[68] Du, J.; Yoon, T. P., *J. Am. Chem. Soc.*, (2009) **131**, 14604.

[69] Lu, Z.; Shen, M.; Yoon, T. P., *J. Am. Chem. Soc.*, (2011) **133**, 1162.

1.4.3   **Tandem Radical Processes**

K. A. Parker

General Introduction

Tandem radical cyclizations (TRCs) continue to attract attention as chemists unravel their potential for the construction of a variety of ring systems. Indeed, the versatility of the general strategy is tantalizing.

The topic of tandem radical cyclizations (or radical domino or radical cascade cyclizations) has not gone unreviewed. Numerous accounts, devoted exclusively or in part to tandem radical cyclizations or to different aspects of radical cyclization, have appeared. The annotated list in Section 1.4.3.1 may be helpful for identifying those of particular interest.

The goal of this chapter is to present examples of the variety of structural types that have been generated by tandem radical cyclizations and by extensions of the concept. Its focus is on connectivity patterns and on reaction conditions. The goal is not to be comprehensive, but to highlight design principles that underlie some notable and prototypical examples of tandem radical cyclizations and related domino reactions. The reader should note that there is not necessarily a specific requirement for the ring system described to be prepared by the method presented; most methods have general utility.

For the purpose of the following discussion, a tandem radical cyclization is defined as a process in which a radical reacts in an intramolecular step to generate a new radical which then undergoes an additional intramolecular ring-closing step; this completes the formation of two new rings (or more than two new rings in cases in which three or more cyclizations occur) in a single reaction. This chapter presents methodologies for tandem radical cyclizations, for radical cyclizations that precede or follow a reaction with an external radical trap, and for tandem intermolecular radical reactions.

1.4.3.1   General and Specialized Reviews on Radical Cyclization Reactions

Hart's 1984 review[1] provided examples of free-radical C—C bond formation in organic synthesis. It shows the scope of the methodology at that time and suggests the expansion that was to follow. It features three classical tandem radical cyclizations.

Other general reviews have appeared regularly. As an introduction to the general topic, the chapter on free-radical cyclization reactions by Zimmerman, Halloway, and Sibi in the *Handbook of Cyclization Reactions* is particularly broad and informative.[2] A review of "Radical Reactions in Domino Processes", focused primarily on developments reported from 2008–2013, covers radical/cation, radical/anion, radical/radical, and radical/pericyclic domino processes.[3] An extensive review of radical cyclization reactions by Giese and co-workers in *Organic Reactions*[4] is organized according to method and ring size. The *Encyclopedia of Radicals in Chemistry, Biology and Materials* is a four-volume set that contains many relevant chapters,[5–29] including those entitled "Radical Cascade Reactions"[28] and "Main-Group Elements in Radical Chemistry".[29]

Numerous reviews focus on one or more aspects of radical cyclizations. Several are devoted to tandem radical cyclizations or other cascade processes, either entirely or partly.[30–32] Reviews devoted to nitrogen-centered[33–35] and oxygen-centered radicals[36] and reviews that contain sections on these topics have appeared.[37–39] Chatgilialoglu's reviews survey the use of silyl radicals,[40,41] while Rowland's reviews cover initiating reagents, in-

for references see p 239

218 Domino Transformations **1.4** Radical Reactions

termolecular addition reactions, cyclizations, rearrangements, and fragmentations.[42,43] Tandem radical cyclizations initiated at epoxides by reduction with titanocene(III) reagents have been reviewed,[44–46] and a mini-review from Stephenson discusses photochemical approaches.[47]

For specialized reviews that may be relevant in the context of domino processes, see "Mechanisms of Mn(OAc)$_3$-Based Oxidative Free-Radical Additions and Cyclizations",[48] "Selective Preparation of Complex Polycyclic Molecules from Acyclic Precursors via Radical-Mediated or Transition-Metal-Catalyzed Cascade Reactions",[49] "Applications of Free Radicals in Organic Synthesis",[50] "Bioinspired Terpene Synthesis: A Radical Approach",[51] "Polyene Cyclizations",[52] and "Radical Cascades Initiated by Intermolecular Radical Addition to Alkynes and Related Triple Bond Systems".[53]

Finally, in Section 1.4.2 Devery, Douglas, and Stephenson provide recent examples of radical cyclizations mediated by stannanes, samarium(II) iodide, ammonium cerium(IV) nitrate, SOMO (singly occupied molecular orbital) catalysts, and visible-light-initiated photoredox catalysis.

1.4.3.2 **A Brief History of Tandem Radical Cyclization Chemistry**

1.4.3.2.1 **The Tandem Radical Cyclization Concept: Fused Rings**

The earliest references to tandem cyclizations by saturated radicals are those of Julia, who isolated the products of "bicyclisation radicalaire" (translated as radical dicyclization in *Chemical Abstracts*) from α-cyano dienyl esters.[54,55] Julia's tandem radical cyclizations were initiated by heating the substrates with dibenzoyl peroxide (DBPO) in cyclohexane. A product from substrate **1** was isolated in 41% yield and tentatively assigned a *trans*-decalin structure. Decades later, in the context of studies with manganese(III) acetate initiated cyclizations, Snider had cause to reexamine this reaction; with modern NMR capabilities, he was able to identify four stereoisomers of the *cis*-fused hydrindane structure **2** in the product mixture (35% yield) (Scheme 1).[56]

Scheme 1 The Julia–Snider Tandem Radical Cyclization[56]

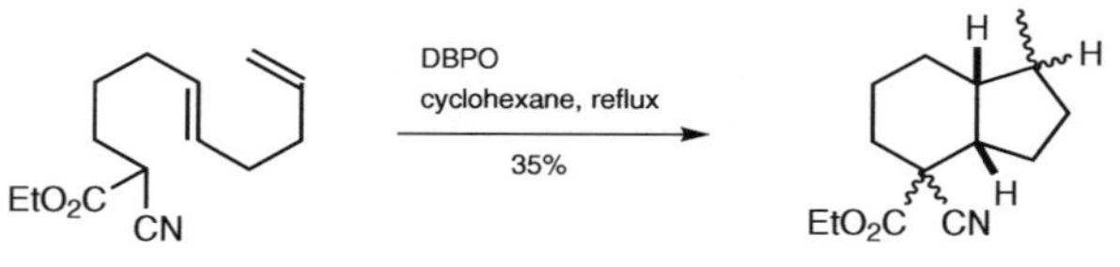

(3aR*,7aS*)-Octahydro-1H-indenes, e.g. 2; General Procedure for Cycloisomerization:[54–58] The following procedure is derived from the reports of Julia[54,55,57,58] and Snider.[56] To boiling cyclohexane (1.3 L) were added over 15 h, at the same time in six portions, a soln of cyano ester (0.11 mol) in cyclohexane (1.1 L) and dibenzoyl peroxide (6 g, 25 mmol). The resultant soln was stirred at reflux for an additional 20 h. Then, it was cooled and washed with aq FeSO$_4$, aq NaHCO$_3$, and H$_2$O. The organic phase was dried (CaCl$_2$) and concentrated. The residue was shaken with sat. aq KMnO$_4$ (to remove alkenic material) and the resulting mixture was filtered. The filter cake was washed with Et$_2$O and the washings were combined with the filtrate. The resulting mixture was partitioned between aq NaHCO$_3$ and Et$_2$O; the Et$_2$O soln was dried (CaCl$_2$) and concentrated. The residue was distilled.

1.4.3 Tandem Radical Processes **219**

1.4.3.2.2 **The Biomimetic Tandem Radical Cyclization Postulate**

In another early study, Breslow, Olin, and Groves[59] tested the reaction of farnesyl acetate (**3**) with dibenzoyl peroxide and copper(II) benzoate. They were interested in the premise that the oxidative biosynthetic cyclization of squalene might be radical-mediated. Although contemporaneous work by others showed this not to be the case, their results were of mechanistic and synthetic interest. For example, under the conditions shown in Scheme 2, they obtained 20–30% of an oxidized product to which they assigned the *trans*-decalin structure **4**.

Scheme 2 Breslow's Biomimetic Model[59]

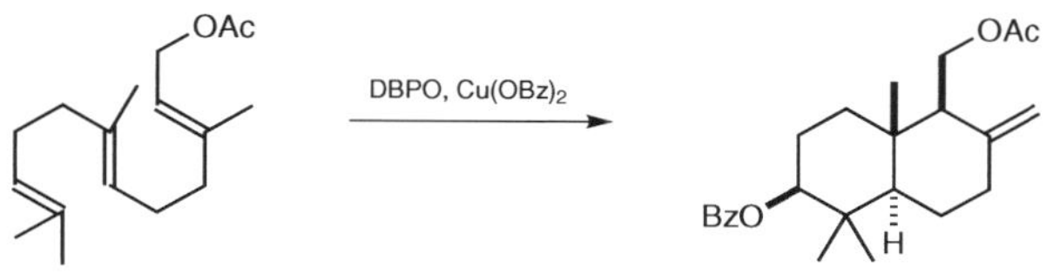

Conditions	Yield (%)	Ref
CuCl, heat	20–30	[59]
fluorescein, *hv*	~5	[59]

1.4.3.2.3 **A Vinyl Radical Tandem Radical Cyclization: A Product with Linked Rings**

Following studies that demonstrated the cyclization of vinyl radicals,[60,61] Stork reported the first dicyclization in which the intermediate radical was sp^2-hybridized.[62] This nice example of a reductive tandem radical cyclization, which generates product **5** containing an alkoxytetrahydrofuran ring linked to a *cis*-hydrindane system (Scheme 3), was considered a model for cardenolide synthesis. This application was later demonstrated by the total synthesis of (+)-digitoxigenin (**6**) with related radical cyclization chemistry.[63]

Scheme 3 Stork's Use of an Intermediate Vinyl Radical toward Digitoxigenin[62,63]

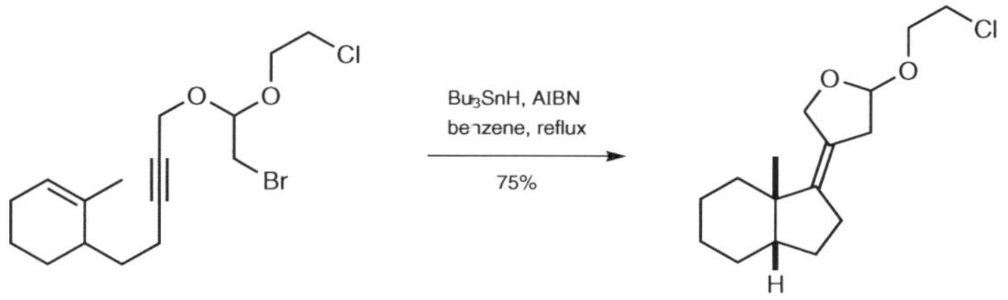

for references see p 239

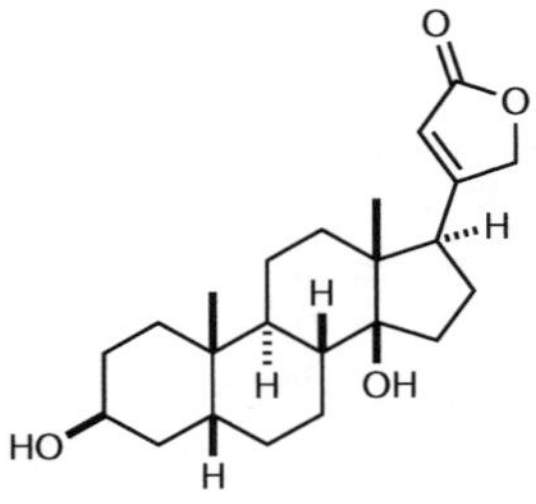

220   Domino Transformations **1.4** Radical Reactions

6 (+)-digitoxigenin

1.4.3.2.4   **Introduction to Selectivity: A Bridged Ring System**

Tandem radical cyclizations have proven noteworthy for increasing the complexity of existing ring systems, particularly ring systems derived from Diels–Alder reactions. Kraus and Kim described a tandem radical cyclization as the second step of a Diels–Alder/tandem radical cyclization sequence (Scheme 4).[64] The tandem radical cyclization is selectively initiated (Bu₃SnH, AIBN, refluxing benzene) at the tertiary sp^3 bromide of (unpurified) intermediate **7** in the presence of a vinyl bromide and a disubstituted alkyne. The bridged product **8** contains the tetracyclic ring system of cumbiasin (**9**).

Scheme 4   Kraus' "Over-the-Top" Tandem Radical Cyclization toward Cumbiasin[64]

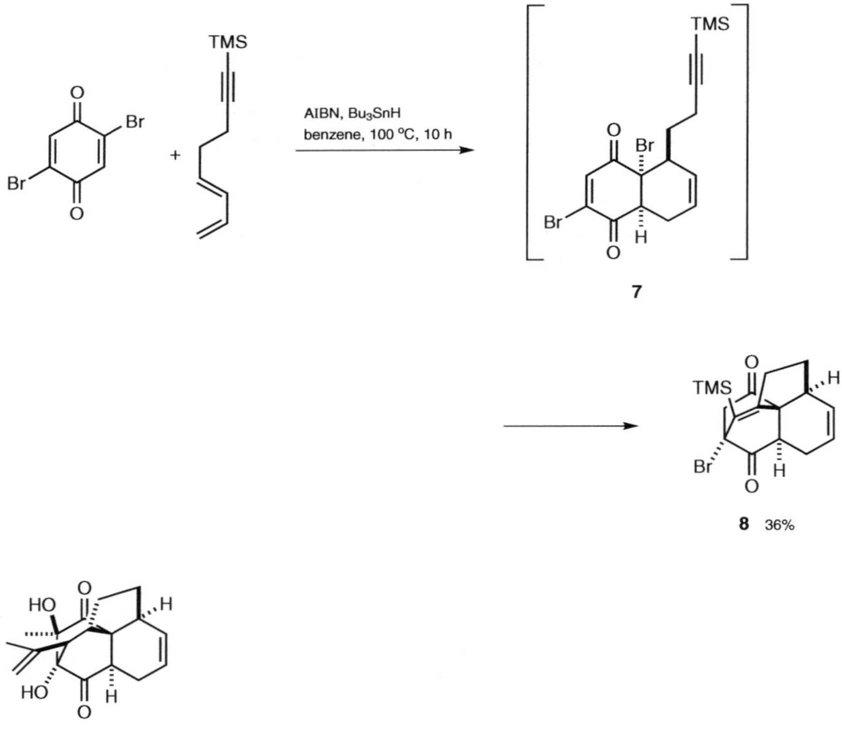

7

8 36%

9 cumbiasin

In summary, the diverse ring systems produced in the few examples shown here and in Sections 1.4.3.2.1–1.4.3.2.3 demonstrate the potential of the tandem radical cyclization concept for advancing the science of synthesis. Related strategies, such as coupling an in-

termolecular trapping reaction with a radical cyclization and coupling two intermolecular radical trapping reactions, have generated equally powerful transformations. Examples of both approaches for scaffold construction are featured in the sections devoted to reagents and methods that follow.

1.4.3.3 Alternative Reagents for Cascade Initiation: Getting Away from Tin, 2,2'-Azobisisobutyronitrile, and Peroxides

With the exception of the direct production of radicals by light or by thermolysis,[65] radical generation requires the use of one or more reagents/additives. Many prototypical radical cyclization reactions have been carried out with site-specific initiation by a peroxide or with the triorganotin hydride/2,2'-azobisisobutyronitrile system. However, alternative initiation and chain-carrying reagents are safer and more scalable, and they are currently considered more practical.

1.4.3.3.1 The Manganese(III) System

The first unambiguous examples of double radical carbocyclizations initiated by manganese(III) acetate were those described by Snider and co-workers.[66,67] The reagent oxidizes the α-position of a β-oxo ester or a malonic ester derivative, e.g. **10**, and the resulting stabilized radical begins the cascade. Some cyclizations give mixtures of products because the bicyclic radical partitions among pathways; in the reaction shown in Scheme 5, production of the exocyclic alkene **12** from oxidation of radical **11** is greatly improved by the presence of copper(II) acetate.

Scheme 5 Optimized Preparation of an Exocyclic Alkene[66,67]

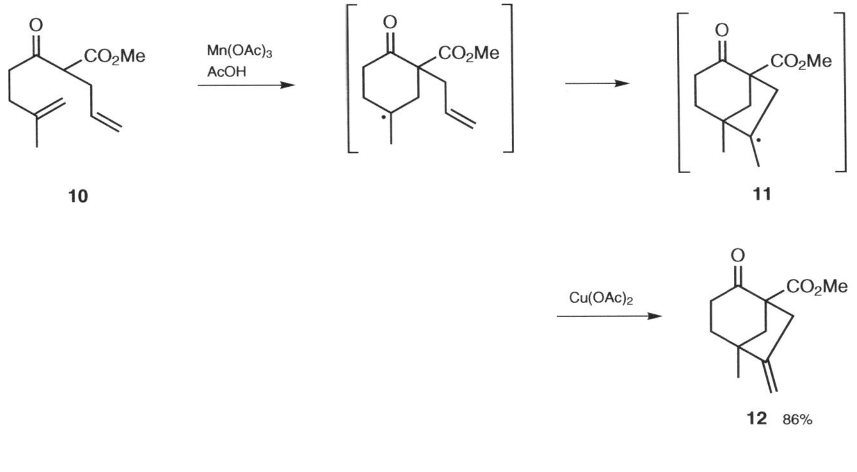

Cyclic Ketones, e.g. 12; General Procedure for Oxidative Cyclization with Manganese(III) Acetate and Copper(II) Acetate:[66]

To a stirred soln of $Mn(OAc)_3 \cdot 2H_2O$ (0.804 g, 3.0 mmol, 2 equiv) and $Cu(OAc)_2 \cdot H_2O$ (0.300 g, 1.5 mmol, 1 equiv) in glacial AcOH (13.5 mL) was added the β-oxo ester (1.5 mmol) in glacial AcOH (4 mL). The mixture was stirred at rt for 26 h and then H_2O (100 mL) was added. 10% aq $NaHSO_3$ was added dropwise to the mixture to decompose residual $Mn(OAc)_3$. The resulting soln was extracted with CH_2Cl_2 (3×30 mL). The combined organic soln was washed with sat. aq $NaHCO_3$ and dried (Na_2SO_4). Removal of the solvent under reduced pressure gave a crude product, which was purified by chromatography and/or recrystallization.

for references see p 239

222 Domino Transformations **1.4** Radical Reactions

1.4.3.3.2 **The Titanium(III) System**

A refinement of the radical cyclization approach to terpenoids is based on the introduction of a carbonyl along the chain of the acyclic substrate in order to interrupt the cyclization at a pre-determined position. Thus, Cuerva and co-workers[68] showed that epoxy ketone **13** affords tricyclic product **14** as shown in Scheme 6. The unexpected *cis* B,C ring junction stereochemistry is correlated with the radical addition to an enone rather than to an isolated alkene.

Scheme 6   Radical Cyclization of a Polyenyl Epoxy Ketone[68]

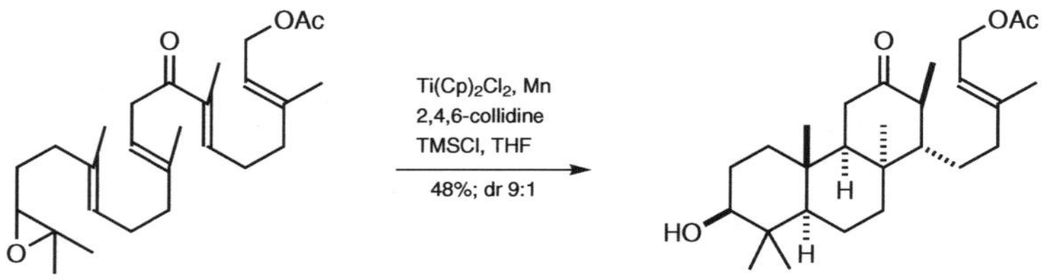

Polycyclic Terpenoids, e.g. 14; General Procedure for Chlorobis(η^5-cyclopentadienyl)titanium(III)-Catalyzed Cyclizations:[68]
Strictly deoxygenated THF (20 mL) was added to a mixture of Ti(Cp)$_2$Cl$_2$ (0.2 mmol) and Mn dust (8 mmol) under an argon atmosphere and the suspension was stirred at rt until it turned lime green (after about 15 min). Then, a soln of the epoxy polyprene substrate (1 mmol), 2,4,6-collidine (7 mmol) in THF (2 mL), and TMSCl (4 mmol) were added and the mixture was stirred for 16 h. The reaction was then quenched with 2 M HCl and the resulting mixture was extracted with EtOAc. The organic layer was washed with brine and dried (Na$_2$SO$_4$), and the solvent was removed. Products were isolated by flash chromatography of the residue.

1.4.3.3.3 **Using Silanes Rather than Stannanes**

Silanes can typically be used as the stoichiometric reducing agents in radical cyclization reactions. Of the silanes, tris(trimethylsilyl)silane [(TMS)$_3$SiH] is clearly the most popular.

1.4.3.3.3.1 **Carboxyarylation**

Sherburn's radical carboxyarylation strategy consists of a radical cyclization step followed by an aryl migration to the newly created radical. An example of this methodology served as the key step in a formal total synthesis of (−)-podophyllotoxin (**17**) (Scheme 7).[69] Here tris(trimethylsilyl)silane and 2,2′-azobisisobutyronitrile effect the cyclization/migration sequence that converts substrate **15** into the key synthetic intermediate **16**.

Scheme 7 Carboxyarylation toward Podophyllotoxin[69]

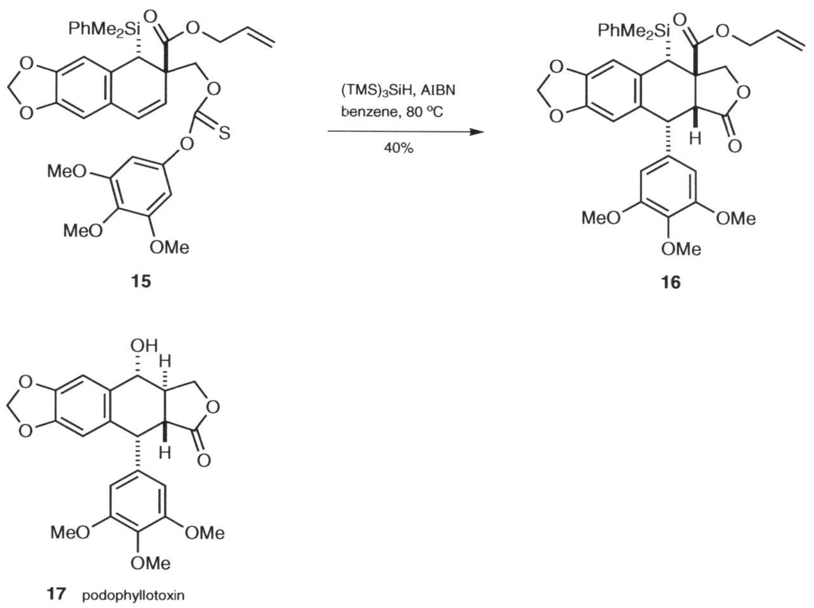

15

16

17 podophyllotoxin

Carboxyarylation Products, e.g. 16; General Procedure with Tris(trimethylsilyl)silane and 2,2′-Azobisisobutyronitrile:[69]

To a soln of thionocarbonate (0.1 mmol, 1.0 equiv) in benzene (4.7 mL) (**CAUTION:** *carcinogen*) was added (TMS)$_3$SiH (32 µL, 0.104 mmol, 1.1 equiv). The mixture was heated to reflux and treated with a soln of AIBN (9.3 mg, 0.57 mmol, 0.6 equiv) in benzene (0.30 mL) over 14 h using a syringe pump. Following the addition, the mixture was cooled to ambient temperature and concentrated under reduced pressure. The residue was subjected to chromatography (silica gel).

1.4.3.3.3.2 Reactions Terminated by Azide

Murphy and co-workers have developed a tandem radical cyclization method in which an aryl radical adds to an alkene, and the resulting alkyl radical, e.g. **18**, is quenched by intramolecular reaction with an azide (Scheme 8). Application of this methodology has provided a formal total synthesis of (±)-aspidospermidine[70] and a total synthesis of (±)-horsfiline (**19**).[71]

for references see p 239

224 Domino Transformations 1.4 Radical Reactions

Scheme 8 A Tandem Radical Cyclization Producing a Nitrogen-Containing Ring[71]

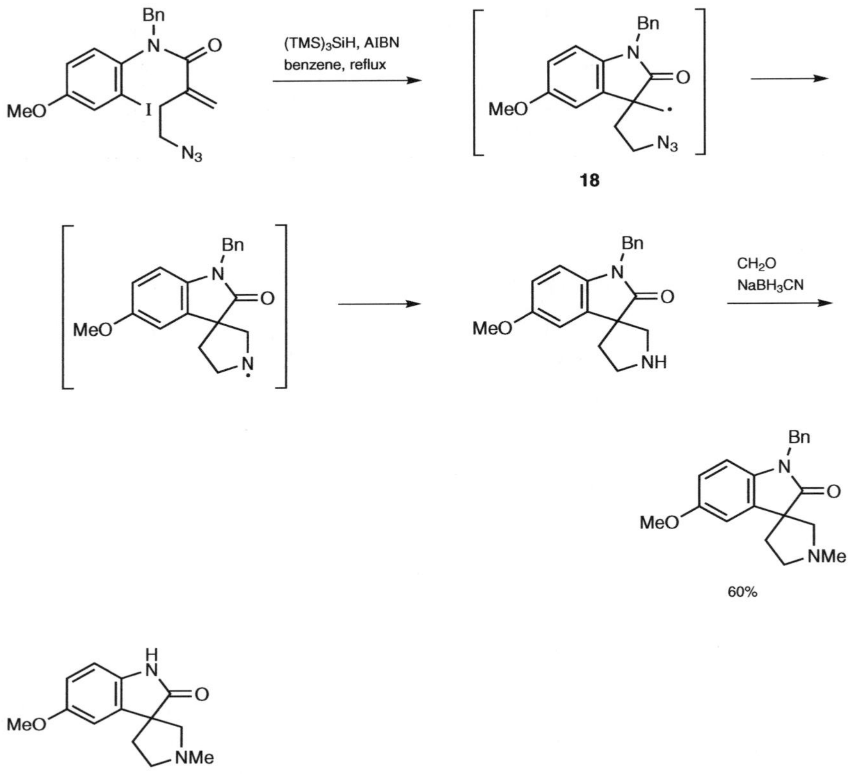

19 horsfiline

The cascade reaction can be intercepted by carbon monoxide so that a redesigned substrate **20** provides the diazaspiro[4.4]nonanedione **21** (Scheme 9).[72]

Scheme 9 The Cyclize/Carbonylate/Close-to-Azide Strategy[72]

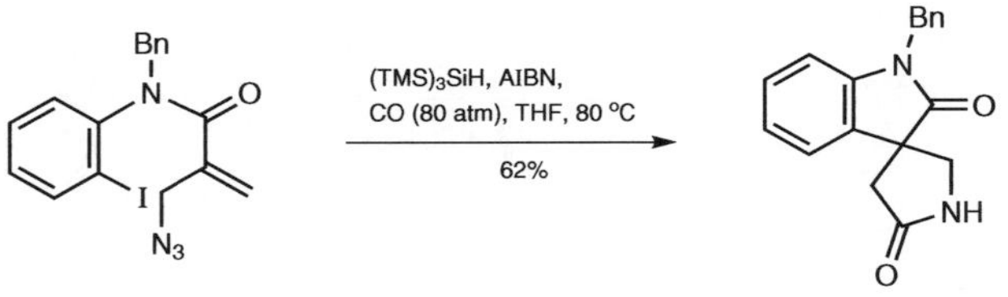

4,4-Spirocyclic γ-Lactams, e.g. 21; General Procedure for Tandem Radical Cyclization with Carbon Monoxide:[72]

> **CAUTION:** *Carbon monoxide is extremely flammable and toxic, and exposure to higher concentrations can quickly lead to a coma.*

A magnetic stirrer bar, a 2-(azidomethyl)-*N*-benzyl-*N*-(2-iodoaryl)acrylamide (0.36 mmol), AIBN (17.7 mg, 0.11 mmol), (TMS)$_3$SiH (178.3 mg, 0.72 mmol), and THF (17.9 mL) were placed in a 50-mL stainless steel autoclave. The autoclave was closed, purged with CO (3×), pressurized with CO (80 atm), and then heated at 80 °C (bath temperature) for 12 h.

Excess CO was discharged after the reaction. The mixture was concentrated under reduced pressure. The resulting residue was purified by column chromatography (silica gel).

1.4.3.3.4 **Reactions with Borane Initiators**

1.4.3.3.4.1 **Tin Hydrides with Triethylborane for Initiation and Fragmentation with Samarium(II) Iodide**

Among the variations on a Diels–Alder/radical cyclization/fragmentation strategy devised by Wood and co-workers for the total synthesis of phomoidrides (also known as the CP molecules) is one in which the bridged substrate **22** is converted into the tetracycle **27** in eight steps (Scheme 10).[73] The first key reaction involves addition of the tin radical to the alkyne and cyclization of the resulting vinyl radical onto the nearby ketone; the resulting alkoxyl radical is believed to be stabilized by complexation with the triethylborane.[74] The remainder of the conversion relies on the tandem radical cyclization of alkyl bromide **24** to polycyclic product **25**, and a fragmentation of dithiocarbonate **26** to give unsaturated product **27**, presumably triggered by a Barton–McCombie deoxygenation. Although later iterations may prove more practical for the total synthesis itself, this proof-of-principle sequence is interesting for its density of radical chemistry.

for references see p 239

Scheme 10 Key Radical Reactions that Rearrange Connections in a Complex Ring System in an Approach to Phomoidride B[73]

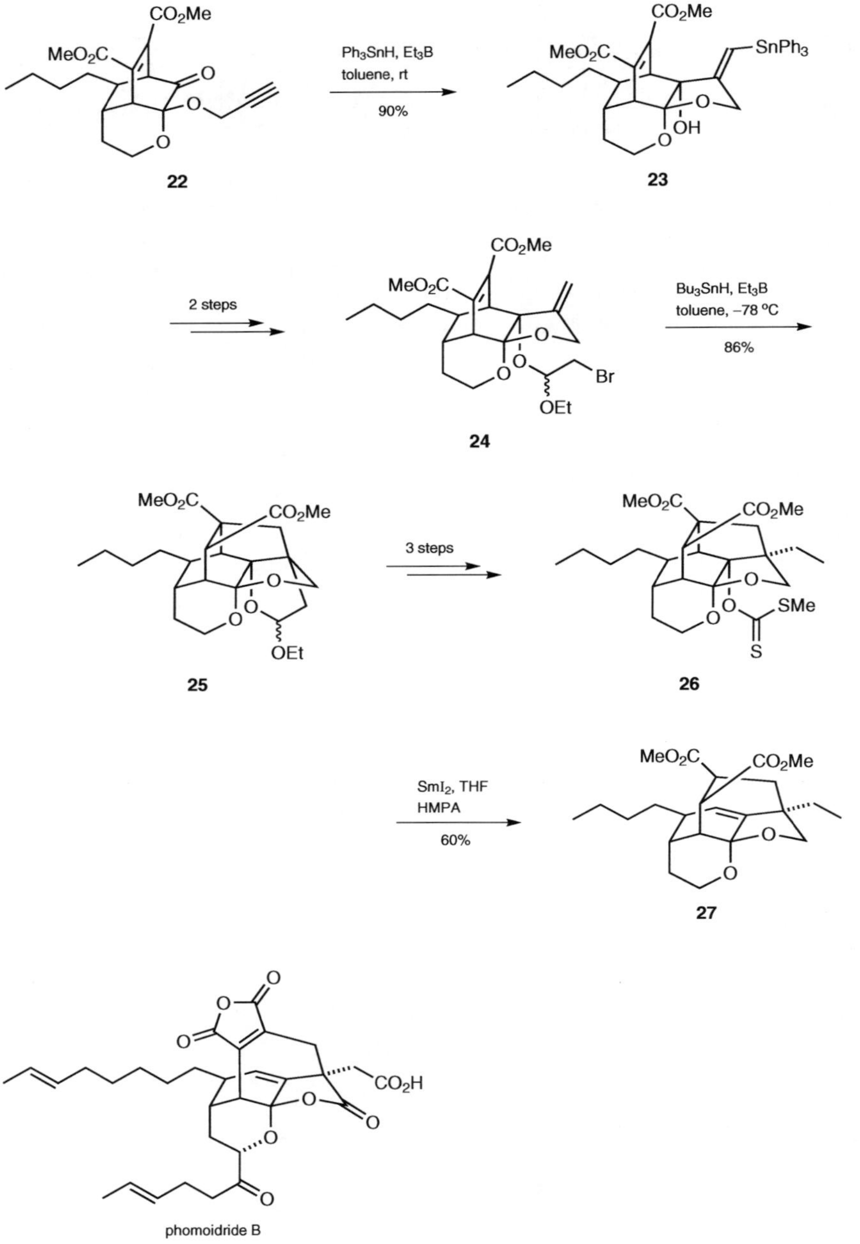

Like the Wood approach to the phomoidrides, that of Yoshimitsu[75,76] utilized three radical reactions as key steps in the modification of a Diels–Alder adduct to a phomoidride scaffold. These events are a radical-directed oxidation, a radical cyclization, and an alkoxyl radical β-cleavage.

Hydroxy(stannylmethylene)carbocycles, e.g. 23; General Procedure for Tin Radical Addition to an Alkyne Followed by Vinyl Radical Addition to a Ketone with Tin Hydride and Triethylborane:[73]

To a stirred, degassed soln of alkyne (0.36 mmol, 1 equiv) in toluene (22 mL) was added a mixture of Ph_3SnH (550 µL, 2.16 mmol, 6 equiv) and 1.0 M Et_3B in hexanes (1.08 mL, 1.08 mmol, 3 equiv) in 6 portions over 24 h at rt. The crude mixture was placed directly on a silica gel column and the product was isolated by elution.

Polycycles, e.g. 25; General Procedure for Tandem Radical Cyclization with Tin Hydride and Triethylborane:[73]

To a stirred soln of the bromo acetal (2.0 mmol, 1 equiv) and Bu_3SnH (1.37 mL, 5.1 mmol, 2.5 equiv) in toluene (55 mL) at −78 °C was added 1.0 M Et_3B in hexanes (6.1 mL, 6.1 mmol, 3 equiv). The flask was fitted with a balloon and the mixture was stirred under an atmosphere of N_2 at rt for 3 h. The crude mixture was poured onto a silica gel column and the product was isolated by elution and concentration.

Deoxygenated Polycyclic Systems, e.g. 27; General Procedure for the Barton–McCombie Radical Fragmentation with Samarium(II) Iodide:[73]

> **CAUTION:** *Hexamethylphosphoric triamide is a possible human carcinogen and an eye and skin irritant.*

To 0.1 M SmI_2 in THF (25 mL, 2.5 mmol, 17 equiv) was added HMPA (3 mL). After 15 min, a soln of the substrate (0.074 mmol, 1 equiv) in THF (1 mL) was added over 1 min. The mixture was stirred at rt for 30 min; it was then quenched with sat. aq NH_4Cl (20 mL) and diluted with EtOAc (100 mL). The organic layer was washed with H_2O (8 × 20 mL) and brine (20 mL), dried (Na_2SO_4), and concentrated. Chromatography furnished the product.

1.4.3.3.4.2 Triethylborane-Mediated Atom Transfer and Cobaloxime-Initiated Reductive Tandem Cyclization

Working in a system based on an acetal template, Hoffmann and co-workers[77] demonstrated the 5-*exo-trig*/6-*endo-dig* cyclizations of the diastereomeric mixture **29** (Scheme 11). Under triethylborane/oxygen conditions, a mixture of stereoisomeric tricyclic vinyl iodides **30A** and **30B** (X = I) is produced (50% yield). Although this atom-transfer protocol did not require the addition of exogenous oxygen,[78] many procedures do incorporate a step in which air is added to the reaction mixture as the reaction proceeds.[79] Under catalysis by cobaloxime **28**, the tricyclic reduction products **30A** and **30B** (X = H) are obtained (45–50% yield).

One should note, in the context of cobalt(I)-mediated radical cyclizations, that they, like those of other transition-metal-mediated reactions, can lead to product distributions that differ from those derived from more classical reductive cyclization procedures.

for references see p 239

Scheme 11 5-exo-trig/6-endo-dig Cyclizations[77]

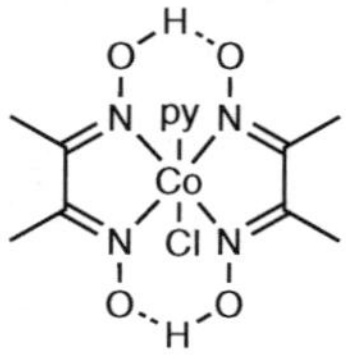

Substrate Concentration (M)	Conditions	X	Yield (%)	Ref
1	O_2 (cat.), Et_3B (2 equiv), toluene	I	50	[77]
0.2	**28**, $NaBH_4$, aq NaOH, EtOH, 55 °C, 2 h	H	45–50	[77]

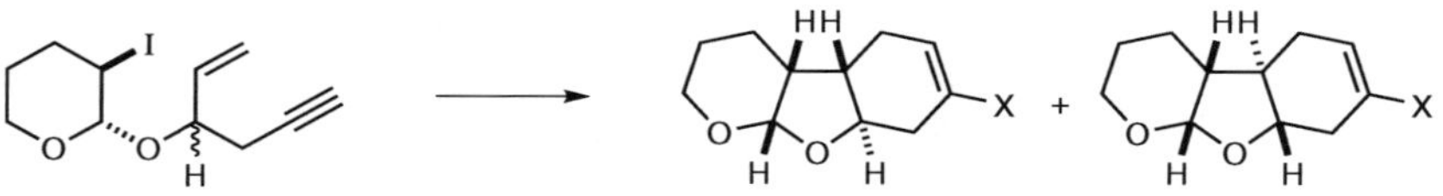

Iodocycloalkenes, e.g. 30 (X = I); General Procedure for Iodine Atom Transfer with Triethylborane/Oxygen:[77]
A well-dried, two-necked 10-mL flask, equipped with a reflux condenser, a $CaCl_2$ drying tube, and a septum, was charged with a soln of substrate (2.4 mmol) in toluene (2.4 mL) and heated to 100 °C. A 1 M soln of Et_3B in hexane (3.6 mL, 1.5 equiv) was slowly added by syringe over 15 min, and the reaction was monitored by TLC. After the mixture was stirred for 45 min, a further portion of Et_3B (0.5 equiv) was added. After the reaction was complete, as judged by TLC (~2.5 h), the solvent was evaporated and the oily residue was subjected to chromatography.

Cycloalkenes, e.g. 30 (X = H); General Procedure for Cobaloxime-Induced Cyclization:[77]
To a soln of substrate (0.65 mmol) in abs EtOH (5.6 mL) was added solid $NaBH_4$ (50 mg, 1.3 mmol) and 10 M aq NaOH (0.1 mL, 1 mmol). Argon was passed through the soln, and several portions of finely powdered chlorobis(dimethylglyoximato)cobalt(III)–pyridine complex (**28**; 14.6 mg, 0.033 mmol, 0.5 equiv) were added over the course of 1.5 h at 40 °C. The mixture was concentrated, and H_2O (10 mL) was added. The resulting mixture was extracted with Et_2O (3 × 10 mL) and the combined organic soln was washed with H_2O (5 mL) and brine (5 mL), and then dried ($MgSO_4$). After the solvent was removed, the residue was subjected to chromatography.

1.4.3.3.4.3 Tri-*sec*-butylborane/Oxygen/Tris(trimethylsilyl)silane Induced Reductive Cyclization

A tandem radical cyclization of aryl iodide **31** to give fused tetracycle **32** was employed as the key step in the total synthesis of (±)-bisabosqual A (**33**), as shown in Scheme 12.[80] Here, the second ring closure involves radical addition to the double bond of a vinylogous ester system. In the bisabosqual tandem radical cyclization, yields are noticeably better with the novel tri-*sec*-butylborane/oxygen initiating system than with the standard triethylborane/oxygen system.

1.4.3 Tandem Radical Processes

Scheme 12 Tandem Radical Cyclization to a Bisabosqual A Precursor[80]

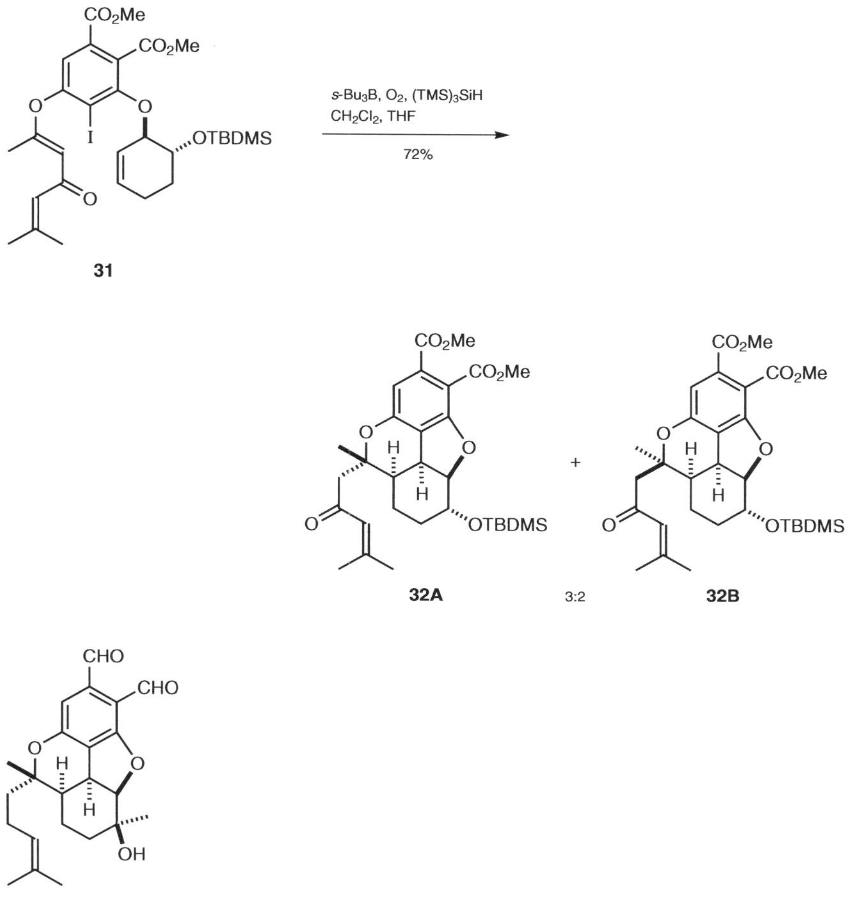

Polycycles, e.g. 32; General Procedure for Tandem Radical Closure with Tri-*sec*-butylborane, Tris(trimethylsilyl)silane, and Oxygen:[80]
To a stirred soln of substrate (3.36 mmol) and (TMS)₃SiH (1.25 g, 5.0 mmol, 1.5 equiv) in CH₂Cl₂ at rt were simultaneously added 1 M *s*-Bu₃B in THF (3.36 mL, 3.36 mmol, 1 equiv) and air (10 mL) via a syringe. The addition procedure took place over a period of 30 min. The mixture was stirred for an additional 15 min at rt and then the mixture was concentrated under reduced pressure. The crude residue was subjected to flash chromatography (silica gel) and, if necessary, preparative HPLC.

1.4.3.4 Nitrogen- and Oxygen-Centered Radicals

Synthetic chemists have exploited nitrogen- and oxygen-centered radicals much less than carbon-centered radicals; however, tandem radical cyclizations initiated at nitrogen and oxygen can also be powerful. The tandem radical cyclization shown in Scheme 13 was part of a study on alkaloid synthesis.[81] A related transformation was effected with irradiation in the presence of boron trifluoride.

for references see p 239

Scheme 13 Tandem Radical Cyclization of a Nitrogen-Centered Radical[81]

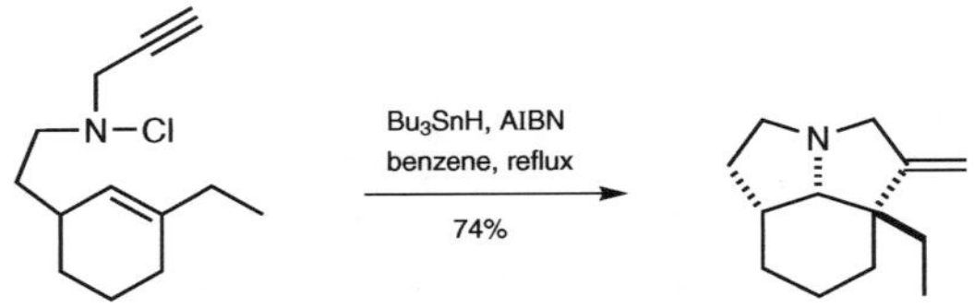

A tandem radical-cyclization-based formal synthesis of morphine (**37**) relies on a novel deprotection/hydroamination reaction for completion of the bridged ring system (Scheme 14).[82] The tandem radical cyclization of aryl bromide **34** to give tetracycle **35** is terminated by addition of the intermediate radical to the styrene alkene and subsequent elimination of the phenylthio radical. The reductive amination step to give nitrogen-containing polycyclic product **36**, which has not been shown to be general, involves cyclization of the conformationally restricted nitrogen radical at a rate that is competitive with quenching by the medium.

Scheme 14 Tandem Radical Cyclization and Reductive Amination To Give a Morphine Precursor[82]

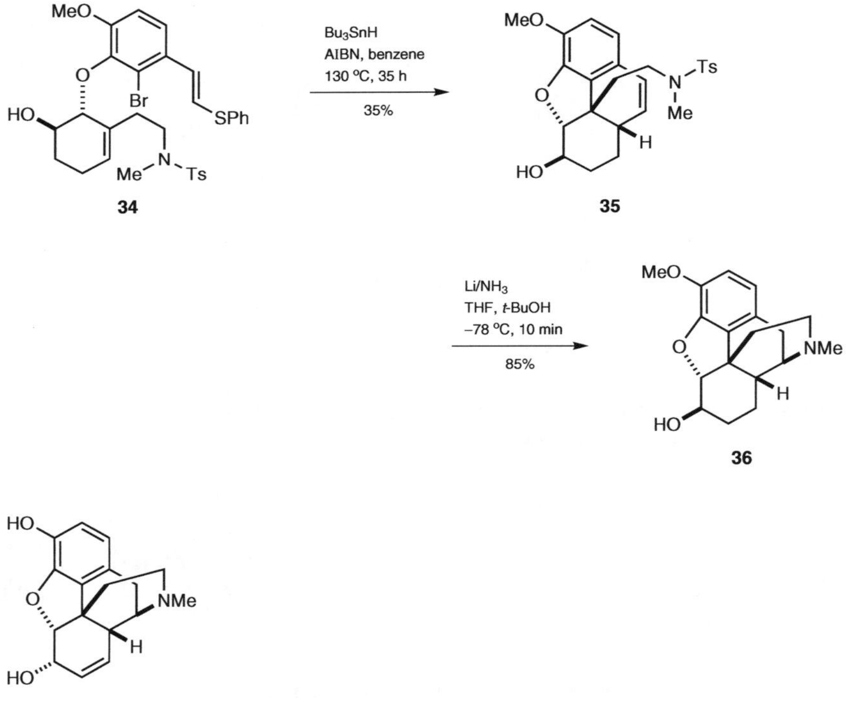

Oxygen-centered radicals have been used to initiate cascades that include a cyclization step.[36] For example, the chiral tetrahydrofuran **39** is obtained by exposure of the *N*-alkoxyphthalimide **38** to standard cyclization conditions (Scheme 15).[83]

Scheme 15 Cyclization Initiated by an Oxygen-Centered Radical[83]

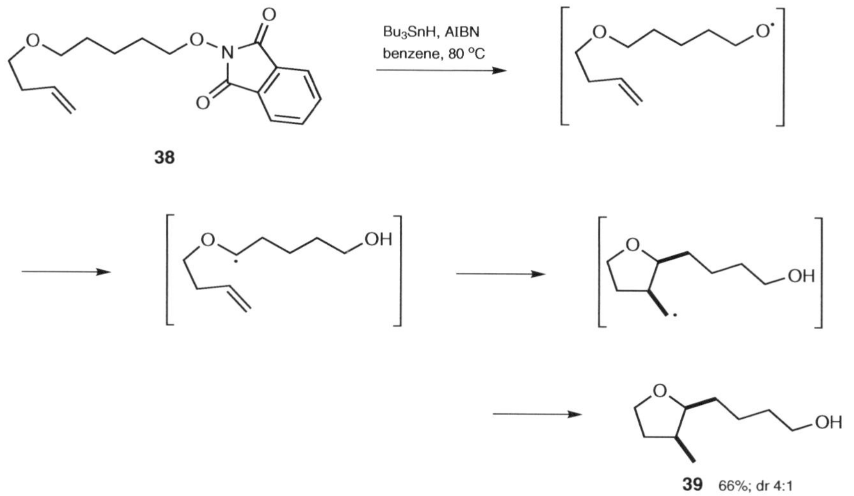

38

39 66%; dr 4:1

Oxygen-Substituted Cyclic Products, e.g. 39; General Procedure for Hydrogen Abstraction/Radical Cyclization Initiated by an O-Centered Radical:[83]

To a 0.02 M soln of cyclization substrate (1 equiv) in degassed benzene (**CAUTION:** *carcinogen*) at reflux was added, using a syringe pump (0.4 mL·h⁻¹), a 0.2 M soln of Bu₃SnH (1.2 equiv) and AIBN (0.15 equiv) in degassed benzene. The mixture was then stirred for an additional 1 h at reflux. The resulting soln was allowed to cool to rt, concentrated, and subjected to flash column chromatography to afford a mixture of cyclized products and linear alcohol. This product mixture was dissolved in CH_2Cl_2 (0.3 M) and the soln was cooled to 0 °C. MCPBA (3 equiv) was then added in a single portion. The resulting mixture was allowed to warm to rt and stirred overnight. The reaction was quenched with 2.0 M aq $Na_2S_2O_3$. The mixture was partitioned with sat. aq Na_2CO_3 (3 ×) and H_2O, dried (Na_2SO_4), and concentrated. The cyclized products were purified by flash column chromatography.

1.4.3.5 Intramolecular Plus Intermolecular Pathways

Intra- and intermolecular processes can be combined for novel and efficient syntheses of highly functionalized systems. Proper design of substrate and reaction conditions allows control of the order of reaction.

1.4.3.5.1 Cyclization/Trapping

The first trapping of a radical product of a radical cyclization in an intermolecular reaction was included in a 1975 paper by Pereyre and co-workers.[84] They reported that the major product from heating 1-iodohex-5-ene with allyltributylstannane (190 °C, 70 h) was but-3-enylcyclopentane. Keck studied the scope and limitations of the trapping of radicals with allyltin reagents;[85] this trapping is now known as the Keck reaction.

Stork and Sher used the radical cyclization/trapping concept as the basis of new methodology designed for the synthesis of prostaglandins. They showed that *tert*-butyl isocyanide was an effective trapping agent, delivering the cyanide group to the bicyclic radical derived from bromide **40** (X = Br) with hexaphenyldistannane (Ph₃SnSnPh₃) under photolysis. However, this photolysis procedure proved inefficient and inconvenient with stoichiometric tin reagent, prompting the development of a system for reductive cyclization that is catalytic in a tin reagent. When employed with iodo substrate **40**

for references see p 239

232 Domino Transformations **1.4** Radical Reactions

(X = I) in reactions driven by the catalytic tin system (Scheme 16), trapping is effective with a variety of electrophilic alkenes as well as with *tert*-butyl isocyanide, providing product **41**.[86]

Scheme 16 Radical Trapping with *tert*-Butyl Isocyanide[86]

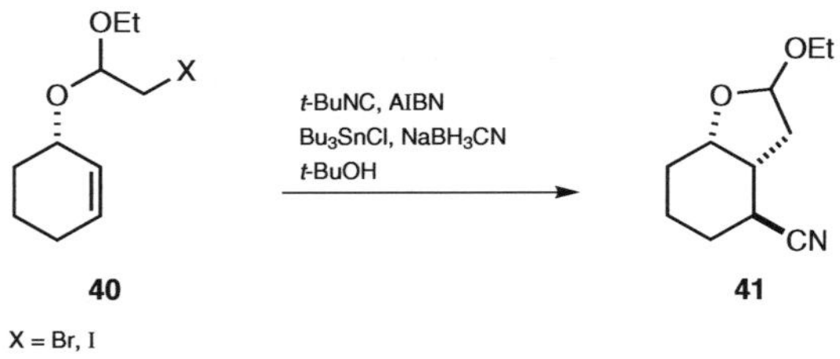

In a clever variation of this procedure, the same group introduced the completely functionalized prostaglandin side chain by trapping with 2-(trimethylsilyl)oct-1-en-3-one and oxidizing the intermediate enol silyl ether obtained following thermolysis of silyl ketone **42**, as shown in Scheme 17.[87] Elaboration of the resulting enone **43** completed a synthesis of $PGF_{2\alpha}$ (**44**). In this case, as in another in which the trapping agent was an enone, a photolysis-based procedure was developed; this protocol was presumably necessary because enones react directly with 2,2′-azobisisobutyronitrile.[88,89] With this methodology, the amount of tin byproduct was reduced.

Scheme 17 Radical Trapping with an α-Silyl Enone[87]

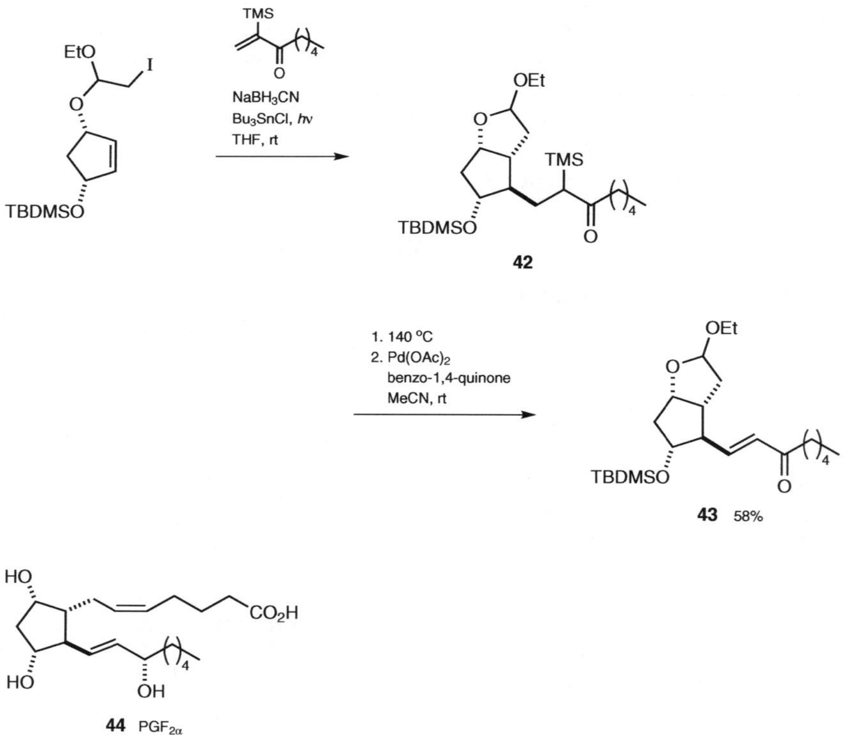

Amine *N*-oxide reagents are particularly attractive radical traps; reduction of the resulting alkoxyamine gives an alcohol at the position of trapping. Although these stable radicals had been used in rate (radical clock) studies since the 1980s,[90] it appears that their potential as trapping agents in synthesis was exploited only in the mid-1990s. At that time, Zard and co-workers[91] showed that they could be used to trap a radical derived from a cyclization, and Barrett showed that related products could be converted into the corresponding alcohols with activated zinc dust.[92] In Barrett's deiodination/trapping procedure with TEMPO (2,2,6,6-tetramethylpiperidin-1-oxyl) as the radical trapping reagent, the product distribution is consistent with steric approach control by the hindered *N*-oxide, e.g. favoring product epimer **45A** over **45B** in the reaction shown in Scheme 18. Subsequent reduction of **45A** gives the sugar derivative **46**.

Scheme 18 Trapping with TEMPO and Reduction of Alkoxyamine[92]

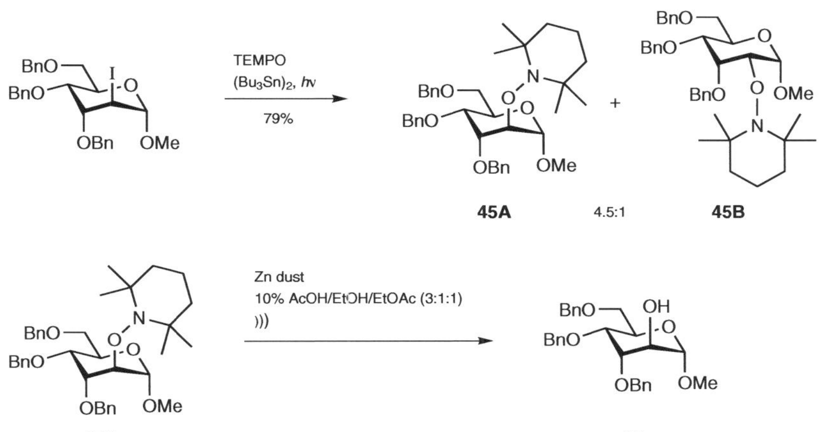

In a radical cyclization initiated by reaction of tris(trimethylsilyl)silane with aryl iodide **47**, Boger completed the desired transformation with a two-step TEMPO trapping/reduction procedure, resulting in alcohol **48** (Scheme 19).[93]

Scheme 19 Radical Cyclization/TEMPO Trapping/Reduction[93]

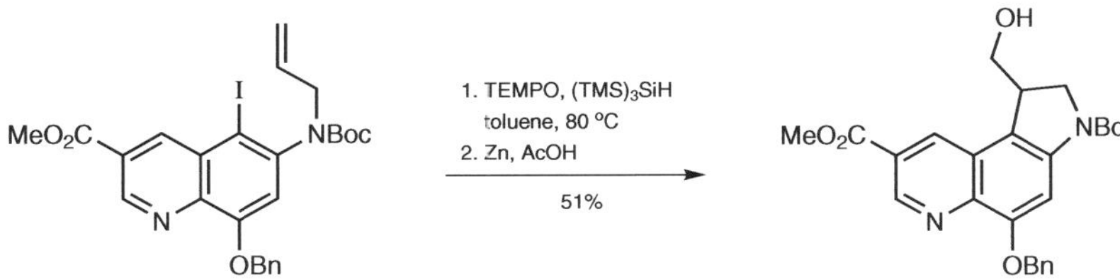

The power of the Diels–Alder/radical cyclization approach to cage molecules is illustrated in Snyder's total synthesis of (+)-scholarisine A (**54**).[94] Here the Diels–Alder-derived radical precursor **49** undergoes a triethylborane-initiated cyclization/trapping reaction (Scheme 20). Building on recent catalytic aerobic oxidation methodologies,[95,96] the authors invented a method for oxidation of the ketone **50** to the enone **51** in the presence of tetramethylguanidine and TEMPO (presumably a second radical-trapping reaction). Then, after functional-group manipulation, substrate **52** undergoes the third radical reac-

for references see p 239

234 Domino Transformations **1.4** Radical Reactions

tion of the sequence. A 1,5-hydrogen atom transfer followed by a homolytic aromatic sub-stitution (radical cyclization onto the aromatic ring and oxidation),[97–99] completes the construction of product **53** containing the scholarisine A ring system.

Scheme 20 Radical Reactions that Advance the Synthesis of Scholarisine A[94]

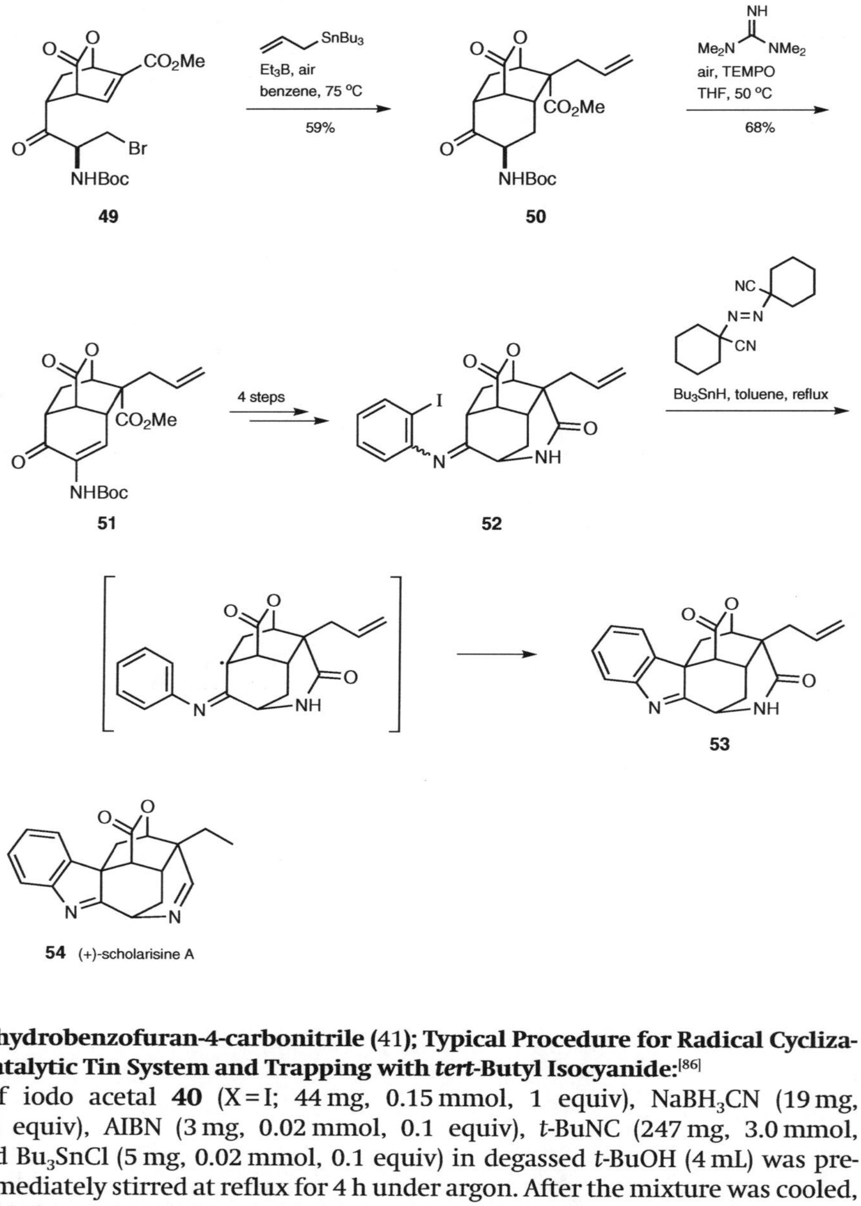

54 (+)-scholarisine A

2-Ethoxyoctahydrobenzofuran-4-carbonitrile (41); Typical Procedure for Radical Cyclization with a Catalytic Tin System and Trapping with *tert*-Butyl Isocyanide:[86]

A mixture of iodo acetal **40** (X = I; 44 mg, 0.15 mmol, 1 equiv), NaBH$_3$CN (19 mg, 0.30 mmol, 2 equiv), AIBN (3 mg, 0.02 mmol, 0.1 equiv), *t*-BuNC (247 mg, 3.0 mmol, 20 equiv), and Bu$_3$SnCl (5 mg, 0.02 mmol, 0.1 equiv) in degassed *t*-BuOH (4 mL) was pre-pared and immediately stirred at reflux for 4 h under argon. After the mixture was cooled, CH$_2$Cl$_2$ was added, and the resulting mixture was shaken vigorously with 3% aq NH$_3$. Then, brine was added and the organic phase was separated. The aqueous phase was subjected to two further extractions with CH$_2$Cl$_2$ and the combined organic soln was dried and con-centrated. The *t*-BuOH was removed by as an azeotrope with benzene (**CAUTION:** *carcin-ogen*) and the residue was subjected to chromatography.

1.4.3 Tandem Radical Processes

235

2-(Cycloalkyl)vinyl Ketones, e.g. 43; General Procedure for Radical Cyclization Using Catalytic Tin Reagent with Photolytic Initiation, Followed by Trapping with a Silyl Enone:[87]
The iodoalkene (1 equiv) was stirred under photolysis (254 nm) with Bu_3SnCl (0.1 equiv), $NaBH_3CN$ (2 equiv), and the silyl enone (7 equiv) in THF at rt. After 4 h, more Bu_3SnCl (0.1 equiv) was added. After stirring for a total of 10 h, the volatiles were removed (80 °C, high vacuum) and the silyl ketone residue was subjected to thermolysis (140 °C) to give crude enol silyl ether, which was subjected to the oxidation procedure of Saegusa.[100] Thus, to a stirring soln of $Pd(OAc)_2$ (0.5 mmol) and benzo-1,4-quinone (0.5 mmol) in MeCN (4 mL), the enol silyl ether (1.0 mmol) was added under N_2 at rt. The resultant mixture was stirred for 2–30 h. When gas chromatography of the mixture indicated that the desired α,β-unsaturated ketone had been produced, the product was isolated by column chromatography.

Cyclic Alcohols, e.g. 48; General Procedure for TEMPO Trapping and Reduction of the Adduct:[93]
A soln of aryl iodide (35 μmol, 1.0 equiv) in anhyd toluene (1.2 mL) was treated with a soln of TEMPO (16 mg, 0.11 mmol, 3.0 equiv) in toluene (0.11 mL) and $(TMS)_3SiH$ (11 μL, 37 μmol, 1.05 equiv). The soln was warmed to 80 °C, and TEMPO (2 × 14 mg, 5 equiv) in toluene (2 × 0.29 mL) and $(TMS)_3SiH$ (4 × 11 μL, 4 equiv) were added in portions over the next 4 h. After 16 h, the mixture was cooled to 25 °C and the volatiles were removed. Chromatography [silica gel (1.5 × 12 cm), EtOAc/hexane 1:4] provided the alkoxyamine. A soln of this intermediate (33 μmol, 1.0 equiv) in THF/H_2O (3:1; 0.90 mL) was treated with activated Zn powder (54 mg, 0.8 mmol, 25 equiv) and AcOH (0.20 mL), and the resulting suspension was warmed to 60 °C with vigorous stirring. After 7 h, additional Zn (54 mg) was added and the mixture was stirred for 4 h. The Zn powder was removed by filtration through Celite with a CH_2Cl_2 wash, and the mixture was concentrated. The resulting residue was dissolved in EtOAc and filtered through Celite, and the soln was concentrated under reduced pressure. Chromatography provided the product alcohol.

Allylcycloalkanes, e.g. 50; General Procedure for Radical Cyclization/Trapping with Allyltributylstannane:[94]
To a soln of the substrate bromide (4.91 mmol, 1.0 equiv) in benzene (55 mL) (**CAUTION:** *carcinogen*) under air (headspace of capped reaction vessel) was added allyltributylstannane (4.5 mL, 15 mmol, 3.0 equiv) followed by 1.0 M Et_3B in hexanes (1.0 mL, 1.0 mmol, 0.2 equiv) at 23 °C. The mixture was then heated to 75 °C and additional portions of 1.0 M Et_3B in hexanes (4 × 1.0 mL, 0.2 equiv each portion) were added every hour. After the final addition, the heating bath was switched off and the mixture was allowed to cool slowly to 23 °C overnight (13 h). The soln was then concentrated directly and the resultant crude product was purified by flash column chromatography.

Enones, e.g. 51; General Procedure for Air/TEMPO/1,1,3,3-Tetramethylguanidine Oxidation:[94]
To a soln of ketone (2.55 mmol, 1.0 equiv) in THF (34 mL) under an air atmosphere at 23 °C were sequentially added 1,1,3,3-tetramethylguanidine (0.35 mL, 2.80 mmol, 1.1 equiv) and TEMPO (0.438 g, 2.80 mmol, 1.1 equiv). The reaction flask was then capped and the soln was heated at 50 °C for 12.5 h. Upon completion of the reaction, the mixture was cooled to 23 °C and then quenched by the addition of sat. aq NH_4Cl (15 mL). The reaction contents were then poured into a separatory funnel, further diluting with H_2O (40 mL). The aqueous layer was extracted with EtOAc (3 × 60 mL). The combined organic soln was dried ($MgSO_4$), filtered, and concentrated. The crude product was purified by flash column chromatography.

for references see p 239

Cyclic Imines, e.g. 53; General Procedure for Hydrogen Atom Transfer/Cyclization:[94]

The (crude) imine (ca. 0.7 mmol, 1 equiv; from an oxidation/condensation sequence, typically a 2.4–2.8:1.0 mixture of geometric isomers) was dissolved in degassed toluene (10 mL) and heated at 110 °C under argon. A soln of Bu$_3$SnH (0.21 mL, 0.79 mmol, 1.2 equiv) and 1,1'-azobis(cyclohexanecarbonitrile) (194 mg, 0.79 mmol, 1.2 equiv) in degassed toluene (8 mL) was added to the heated soln using a syringe pump over 3.75 h. After the addition was complete, the mixture was stirred at 110 °C for an additional 45 min before being cooled to 23 °C and concentrated directly. The resultant residue was purified by flash column chromatography followed by further purification by preparative TLC.

1.4.3.5.2 Trapping/Cyclization

The remarkably concise synthesis of camptothecin (**56**) by Curran relies on an intermolecular trapping reaction by an isocyanide, which is followed by a tandem radical cyclization.[101,102] This three-radical sequence from substrate **55** effects a net 4 + 1 annulation, as shown in Scheme 21.

Scheme 21   The 4 + 1 Annulation Approach to Camptothecin[101]

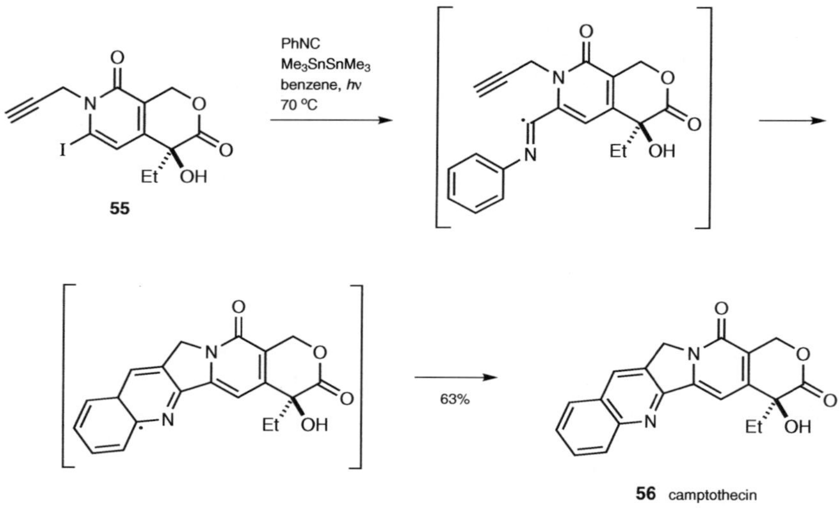

56 camptothecin

5,7-Dihydropyrido[2,3-a]indolizine Derivatives, e.g. 56; General Procedure for Photoinduced Annulation with Aryl Isocyanides:[101]

A soln of N-propargylated pyridinone derivative (0.02 mmol, 1 equiv) in benzene (**CAUTION:** *carcinogen*) was added to a cylindrical screw-capped glass vial (15 × 45 mm). The isocyanide (0.08 mmol, 4 equiv) and then hexamethyldistannane (0.03 mmol, 1.5 equiv) were added at rt. The vial was capped and the mixture was irradiated with a 275-W GE sunlamp for 8 h. If necessary, the vessel was tapped at regular intervals to allow the solid sticking on the walls to drop to the bottom of the flask. The solvent was then evaporated and the residue was purified by column chromatography. Hexamethyldisilane and tetrakis(trimethylsilyl)silane can be substituted for hexamethyldistannane with some (but not a major) loss in yield.

1.4.3.6 Intermolecular Trapping/Trapping Pathways

In progressing toward a total synthesis of resiniferatoxin (**62**), Inoue and co-workers[103] invoked the three-component sequential trapping strategy of Mizuno, Otsuji, and co-workers.[104] Thus, they generated the conformationally locked and sterically biased radical **60** from the costly but highly functionalized O/Se-acetal **57** in the presence of the "matched" enone **58** and allylic stannane **59** (Scheme 22). The resulting adduct **61** has notably increased stereochemical complexity.

Scheme 22 Three-Component Sequential Trapping[103]

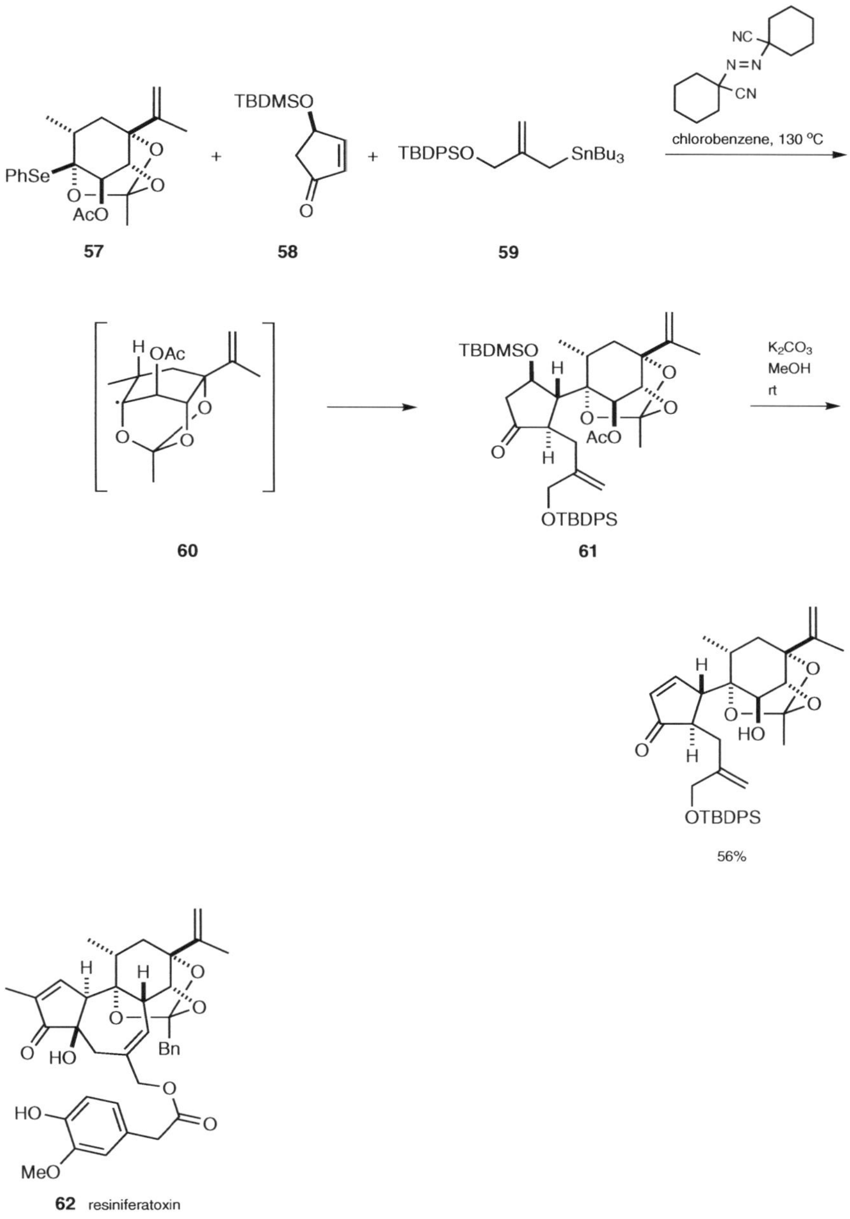

for references see p 239

238 Domino Transformations **1.4** Radical Reactions

Polycyclic Products, e.g. 61; General Procedure for Three-Component Trapping from an Se/O-Acetal:[103]

A two-necked round-bottomed flask equipped with a reflux condenser was charged with the Se/O-acetal (115 μmol, 1 equiv), enone (580 μmol, 5.0 equiv), 1,1'-azobis(cyclohexane-carbonitrile) (V-40; 7 mg, 29 μmol, 0.25 equiv), and toluene or chlorobenzene (0.8 mL). A separate flask was charged with the functionalized allyl stannane (580 μmol, 5.0 equiv), 1,1'-azobis(cyclohexanecarbonitrile) (7 mg, 29 μmol, 0.25 equiv), and toluene or chloro-benzene (0.4 mL). Both solns were degassed by the freeze–thaw procedure (3 ×). The latter soln was added to the refluxing former mixture by a syringe pump over 30 min, and then the mixture was concentrated to afford crude product, which was used for the next reaction without further purification.

1.4.3.7 Conclusions

The reactivities of radicals are orthogonal to those of most functional groups. Therefore, the development of their chemistry offers transformations that are complementary to those of more traditional organic reactions.

Proper design of radical precursors leads to products with predicted connectivities and ring systems. In fact, remarkable complexity can be generated. Exploitation of new strategies and methods is likely to expand further the application of domino radical processes in synthetic chemistry.

References

[1] Hart, D. J., *Science (Washington, D. C.)*, (1984) **223**, 883.

[2] Zimmerman, J.; Halloway, A.; Sibi, M. P., In *Handbook of Cyclization Reactions*, Ma, S., Ed.; Wiley-VCH, Weinheim: Germany, (2010); Vol. 2, pp 1099–1148.

[3] An, G.; Li, G., In *Domino Reactions: Concepts for Efficient Organic Synthesis*, Tietze, L. F., Ed.; Wiley-VCH: Weinheim, Germany, (2014); p 141.

[4] Giese, B.; Kopping, B.; Göbel, T.; Dickhaut, J.; Thoma, G.; Kulicke, K. J.; Trach, F., *Org. React. (N. Y.)*, (1995) **48**, 301.

[5] Sherburn, M. S., In *Encyclopedia of Radicals in Chemistry, Biology and Materials*, Chatgilialoglu, C.; Studer, A., Eds.; Wiley: Hoboken, NJ, (2012); p 57.

[6] Newcomb, M., In *Encyclopedia of Radicals in Chemistry, Biology and Materials*, Chatgilialoglu, C.; Studer, A., Eds.; Wiley: Hoboken, NJ, (2012); p 107.

[7] Sasaki, K.; Cumpstey, I.; Crich, D., In *Encyclopedia of Radicals in Chemistry, Biology and Materials*, Chatgilialoglu, C.; Studer, A., Eds.; Wiley: Hoboken, NJ, (2012); p 125.

[8] Pérez-Prieto, J.; Miranda, M. A., In *Encyclopedia of Radicals in Chemistry, Biology and Materials*, Chatgilialoglu, C.; Studer, A., Eds.; Wiley: Hoboken, NJ, (2012); p 275.

[9] Bardagí, J. I.; Vaillard, V. A.; Rossi, R. A., In *Encyclopedia of Radicals in Chemistry, Biology and Materials*, Chatgilialoglu, C.; Studer, A., Eds.; Wiley: Hoboken, NJ, (2012); p 333.

[10] Jasch, H.; Heinrich, M. R., In *Encyclopedia of Radicals in Chemistry, Biology and Materials*, Chatgilialoglu, C.; Studer, A., Eds.; Wiley: Hoboken, NJ, (2012); p 529.

[11] Chatgilialoglu, C.; Timokhin, V. I., In *Encyclopedia of Radicals in Chemistry, Biology and Materials*, Chatgilialoglu, C.; Studer, A., Eds.; Wiley: Hoboken, NJ, (2012); p 561.

[12] Renaud, P., In *Encyclopedia of Radicals in Chemistry, Biology and Materials*, Chatgilialoglu, C.; Studer, A., Eds.; Wiley: Hoboken, NJ, (2012); p 601.

[13] Kyne, S. H.; Schiesser, C. H., In *Encyclopedia of Radicals in Chemistry, Biology and Materials*, Chatgilialoglu, C.; Studer, A., Eds.; Wiley: Hoboken, NJ, (2012); p 629.

[14] Yang, Y.-H.; Sibi, M. P., In *Encyclopedia of Radicals in Chemistry, Biology and Materials*, Chatgilialoglu, C.; Studer, A., Eds.; Wiley: Hoboken, NJ, (2012); p 655.

[15] Gilmore, K.; Alabugin, I. V., In *Encyclopedia of Radicals in Chemistry, Biology and Materials*, Chatgilialoglu, C.; Studer, A., Eds.; Wiley: Hoboken, NJ, (2012); p 693.

[16] Murphy, J. A., In *Encyclopedia of Radicals in Chemistry, Biology and Materials*, Chatgilialoglu, C.; Studer, A., Eds.; Wiley: Hoboken, NJ, (2012); p 817.

[17] Riguet, E.; Hoffmann, N., In *Encyclopedia of Radicals in Chemistry, Biology and Materials*, Chatgilialoglu, C.; Studer, A., Eds.; Wiley: Hoboken, NJ, (2012); p 1217.

[18] Burton, J. W., In *Encyclopedia of Radicals in Chemistry, Biology and Materials*, Chatgilialoglu, C.; Studer, A., Eds.; Wiley: Hoboken, NJ, (2012); p 901.

[19] Coote, S. C.; Flowers, R. A., II; Skrydstrup, T.; Procter, D. J., In *Encyclopedia of Radicals in Chemistry, Biology and Materials*, Chatgilialoglu, C.; Studer, A., Eds.; Wiley: Hoboken, NJ, (2012); p 849.

[20] Spagnolo, P.; Nanni, D., In *Encyclopedia of Radicals in Chemistry, Biology and Materials*, Chatgilialoglu, C.; Studer, A., Eds.; Wiley: Hoboken, NJ, (2012); p 1019.

[21] Li, C., In *Encyclopedia of Radicals in Chemistry, Biology and Materials*, Chatgilialoglu, C.; Studer, A., Eds.; Wiley: Hoboken, NJ, (2012); p 943.

[22] Zard, S. Z., In *Encyclopedia of Radicals in Chemistry, Biology and Materials*, Chatgilialoglu, C.; Studer, A., Eds.; Wiley: Hoboken, NJ, (2012); p 965.

[23] Gansäuer, A.; Fleckhaus, A., In *Encyclopedia of Radicals in Chemistry, Biology and Materials*, Chatgilialoglu, C.; Studer, A., Eds.; Wiley: Hoboken, NJ, (2012); p 989.

[24] Vaillard, S. E.; Studer, A., In *Encyclopedia of Radicals in Chemistry, Biology and Materials*, Chatgilialoglu, C.; Studer, A., Eds.; Wiley: Hoboken, NJ, (2012); p 1059.

[25] Tansakul, C.; Braslau, R., In *Encyclopedia of Radicals in Chemistry, Biology and Materials*, Chatgilialoglu, C.; Studer, A., Eds.; Wiley: Hoboken, NJ, (2012); p 1095.

[26] Pérez-Martín, I.; Suárez, E., In *Encyclopedia of Radicals in Chemistry, Biology and Materials*, Chatgilialoglu, C.; Studer, A., Eds.; Wiley: Hoboken, NJ, (2012); p 1131.

[27] Bohn, M. A.; Paul, A.; Hilt, G., In *Encyclopedia of Radicals in Chemistry, Biology and Materials*, Chatgilialoglu, C.; Studer, A., Eds.; Wiley: Hoboken, NJ, (2012); p 1175.

[28] Baralle, A.; Baroudi, A.; Daniel, M.; Fensterbank, L.; Goddard, J.-P.; Lacôte, E.; Larraufie, M.-H.; Maestri, G.; Malacria, M.; Ollivier, C., In *Encyclopedia of Radicals in Chemistry, Biology and Materials*, Chatgilialoglu, C.; Studer, A., Eds.; Wiley: Hoboken, NJ, (2012); p 729.

[29] Baralle, A.; Baroudi, A.; Daniel, M.; Fensterbank, L.; Goddard, J.-P.; Lacôte, E.; Larraufie, M.-H.; Maestri, G.; Malacria, M.; Ollivier, C., In *Encyclopedia of Radicals in Chemistry, Biology and Materials*, Chatgilialoglu, C.; Studer, A., Eds.; Wiley: Hoboken, NJ, (2012); p 767.

[30] Nicolaou, K. C.; Edmonds, D. J.; Bulger, P. G., *Angew. Chem. Int. Ed.*, (2006) **45**, 7134.

[31] Chatgilialoglu, C.; Renaud, P., In *General Aspects of the Chemistry of Radicals*, Alfassi, Z. B., Ed.; Wiley: Chichester, UK, (1999); pp 501–538.

[32] Dhimane, A.-L.; Fensterbank, L.; Lacôte, E.; Malacria, M., *Actual. Chim.*, (2003) **265**, 46.

[33] Zard, S. Z., *Chem. Soc. Rev.*, (2008) **37**, 1603.

[34] Minozzi, M.; Nanni, D.; Spagnolo, P., *Chem.–Eur. J.*, (2009) **15**, 7830.

[35] Fallis, A. G.; Brinza, I. M., *Tetrahedron*, (1997) **53**, 17543.

[36] Hartung, J.; Gottwald, T.; Špehar, K., *Synthesis*, (2002), 1469.

[37] Curran, D. P., In *Comprehensive Organic Synthesis*, Trost, B. M.; Fleming, I., Eds.; Pergamon: Oxford, (1991); Vol. 4, pp 779–831.

[38] Loertscher, B. M.; Castle, S. L., In *Comprehensive Organic Synthesis*, 2nd ed., Knochel, P.; Molander, G. A., Eds.; Elsevier: Amsterdam, (2014); Vol. 4, pp 742–809.

[39] Quiclet-Sire, B.; Zard, S. Z., *Beilstein J. Org. Chem.*, (2013) **9**, 557.

[40] Chatgilialoglu, C.; Timokhin, V. I., *Adv. Organomet. Chem.*, (2008) **57**, 117.

[41] Chatgilialoglu, C., *Chem.–Eur. J.*, (2008) **14**, 2310.

[42] Rowlands, G. J., *Tetrahedron*, (2009) **65**, 8603.

[43] Rowlands, G. J., *Tetrahedron*, (2010) **66**, 1593.

[44] Cuerva, J. M.; Justicia, J.; Oller-López, J. L.; Bazdi, B.; Oltra, J. E., *Mini-Rev. Org. Chem.*, (2006) **3**, 23.

[45] Rosales, A.; Rodríguez-García, I.; Muñoz-Bascón, J.; Roldan-Molina, E.; Padial, N. M.; Morales, L. P.; García-Ocaña, M.; Oltra, J. E., *Eur. J. Org. Chem.*, (2015), 4567; corrigendum: *Eur. J. Org. Chem.*, (2015), 4592.

[46] Justicia, J.; Rosales, A.; Buñuel, E.; Oller-López, J. L.; Valdivia, M.; Haïdour, A.; Oltra, J. E.; Barrero, A. F.; Cárdenas, D. J.; Cuerva, J. M., *Chem.–Eur. J.*, (2004) **10**, 1778.

[47] Wallentin, C.-J.; Nguyen, J. D.; Stephenson, C. R. J., *Chimia*, (2012) **66**, 394.

[48] Snider, B. B., *Tetrahedron*, (2009) **65**, 10738.

[49] Malacria, M., *Chem. Rev.*, (1996) **96**, 289.

[50] Boger, D. L., *Isr. J. Chem.*, (1997) **37**, 119.

[51] Justicia, J.; Álvarez de Cienfuegos, L.; Campaña, A. G.; Miguel, D.; Jakoby, V.; Gansäuer, A.; Cuerva, J. M., *Chem. Soc. Rev.*, (2011) **40**, 3525.

[52] Snyder, S. A.; Levinson, A. M., In *Comprehensive Organic Synthesis*, 2nd ed., Knochel, P.; Molander, G. A., Eds.; Elsevier: Amsterdam, (2014); Vol. 3, pp 268–292.

[53] Wille, U., *Chem. Rev.*, (2013) **113**, 813.

[54] Julia, M.; Le Goffic, F.; Katz, L., *Bull. Soc. Chim. Fr.*, (1964), 1122.

[55] Julia, M.; Le Goffic, F., *Bull. Soc. Chim. Fr.*, (1964), 1129.

[56] Snider, B. B.; Armanetti, L.; Baggio, R., *Tetrahedron Lett.*, (1993) **34**, 1701.

[57] Julia, M.; Surzur, J. M.; Katz, L., *Bull. Soc. Chim. Fr.*, (1964), 1109.

[58] Julia, M.; Surzur, J. M.; Katz, L.; Le Goffic, F., *Bull. Soc. Chim. Fr.*, (1964), 1116.

[59] Breslow, R.; Olin, S. S.; Groves, J. T., *Tetrahedron Lett.*, (1968), 1837.

[60] Stork, G.; Baine, N. H., *J. Am. Chem. Soc.*, (1982) **104**, 2321.

[61] Stork, G.; Baine, N. H., *Tetrahedron Lett.*, (1985) **26**, 5927.

[62] Stork, G.; Mook, R., Jr., *J. Am. Chem. Soc.*, (1983) **105**, 3720.

[63] Stork, G.; West, F.; Lee, H. Y.; Isaacs, R. C. A.; Manabe, S., *J. Am. Chem. Soc.*, (1996) **118**, 10660.

[64] Kraus, G. A.; Kim, J., *Tetrahedron Lett.*, (2006) **47**, 7797.

[65] Mohamed, R. K.; Peterson, P. W.; Alabugin, I. V., *Chem. Rev.*, (2013) **113**, 7089.

[66] Dombroski, M. A.; Kates, S. A.; Snider, B. B., *J. Am. Chem. Soc.*, (1990) **112**, 2759.

[67] Kates, S. A.; Dombroski, M. A.; Snider, B. B., *J. Org. Chem.*, (1990) **55**, 2427.

[68] Morcillo, S. P.; Miguel, D.; Resa, S.; Martín-Lasanta, A.; Millán, A.; Choquesillo-Lazarte, D.; García-Ruiz, J. M.; Mota, A. J.; Justicia, J.; Cuerva, J. M., *J. Am. Chem. Soc.*, (2014) **136**, 6943.

[69] Reynolds, A. J.; Scott, A. J.; Turner, C. I.; Sherburn, M. S., *J. Am. Chem. Soc.*, (2003) **125**, 12108.

[70] Patro, B.; Murphy, J. A., *Org. Lett.*, (2000) **2**, 3599.

[71] Lizos, D. E.; Murphy, J. A., *Org. Biomol. Chem.*, (2003) **1**, 117.

[72] Ueda, M.; Uenoyama, Y.; Terasoma, N.; Doi, S.; Kobayashi, S.; Ryu, I.; Murphy, J. A., *Beilstein J. Org. Chem.*, (2013) **9**, 1340.

[73] Njardarson, J. T.; McDonald, I. M.; Spiegel, D. A.; Inoue, M.; Wood, J. L., *Org. Lett.*, (2001) **3**, 2435.

[74] Devin, P.; Fensterbank, L.; Malacria, M., *Tetrahedron Lett.*, (1998) **39**, 833.

References

[75] Yoshimitsu, T., *Chem. Rec.*, (2014) **14**, 268.

[76] Yoshimitsu, T.; Yanagisawa, S.; Nagaoka, H., *Org. Lett.*, (2000) **2**, 3751.

[77] Albrecht, U.; Wartchow, R.; Hoffmann, H. M. R., *Angew. Chem. Int. Ed. Engl.*, (1992) **31**, 910.

[78] Woltering, T. J.; Hoffmann, H. M. R., *Tetrahedron*, (1995) **51**, 7389.

[79] Gao, S.; Tzeng, T.; Lan, Y.-T.; Liu, J.-T.; Yao, C.-F., *ARKIVOC*, (2009) xii, 347.

[80] am Ende, C. W.; Zhou, Z.; Parker, K. A., *J. Am. Chem. Soc.*, (2013) **135**, 582.

[81] Banwell, M. G.; Lupton, D. W., *Heterocycles*, (2006) **68**, 71.

[82] Parker, K. A.; Fokas, D., *J. Am. Chem. Soc.*, (1992) **114**, 9688.

[83] Zhu, H.; Leung, J. C. T.; Sammis, G. M., *J. Org. Chem.*, (2015) **80**, 965.

[84] Grignon, J.; Servens, C.; Pereyre, M., *J. Organomet. Chem.*, (1975) **96**, 225.

[85] Keck, G. E.; Enholm, E. J.; Yates, J. B.; Wiley, M. R., *Tetrahedron*, (1985) **41**, 4079.

[86] Stork, G.; Sher, P. M., *J. Am. Chem. Soc.*, (1986) **108**, 303.

[87] Stork, G.; Sher, P. M., *J. Am. Chem. Soc.*, (1986) **108**, 6384.

[88] Kamimura, A.; Gunjigake, Y.; Mitsudera, H.; Yokoyama, S., *Tetrahedron Lett.*, (1998) **39**, 7323.

[89] Shi, M.; Zhao, G.-L., *Tetrahedron*, (2004) **60**, 2083.

[90] Bowry, V.; Lusztyk, J.; Ingold, K. U., *Pure Appl. Chem.*, (1990) **62**, 213.

[91] Boivin, J.; Yousfi, M.; Zard, S. Z., *Tetrahedron Lett.*, (1994) **35**, 5629.

[92] Barrett, A. G. M.; Rys, D. J., *J. Chem. Soc., Chem. Commun.*, (1994), 837.

[93] Boger, D. L.; Boyce, C. W., *J. Org. Chem.*, (2000) **65**, 4088.

[94] Smith, M. W.; Snyder, S. A., *J. Am. Chem. Soc.*, (2013) **135**, 12964.

[95] Cao, Q.; Dornan, L. M.; Rogan, L.; Hughes, N. L.; Muldoon, M. J., *Chem. Commun. (Cambridge)*, (2014) **50**, 4524.

[96] Ryland, B. L.; Stahl, S. S., *Angew. Chem. Int. Ed.*, (2014) **53**, 8824.

[97] Chan, T. L.; Wu, Y.; Choy, P. Y.; Kwong, F. Y., *Chem.–Eur. J.*, (2013) **19**, 15802.

[98] Tiecco, M., In *Free Radical Reactions*, Hey, D. H.; Waters, W. A., Eds.; Butterworths: London, (1975); pp 25–45.

[99] Vaillard, S. E.; Schulte, B.; Studer, A., In *Modern Arylation Methods*, Ackermann, L., Ed.; Wiley-VCH: Weinheim, Germany, (2009); pp 475–511.

[100] Ito, Y.; Hirao, T.; Saegusa, T., *J. Org. Chem.*, (1978) **43**, 1011.

[101] Josien, H.; Ko, S.-B.; Bom, D.; Curran, D. P., *Chem.–Eur. J.*, (1998) **4**, 67.

[102] Tangirala, R. S.; Antony, S.; Agama, K.; Pommier, Y.; Anderson, B. D.; Bevins, R.; Curran, D. P., *Bioorg. Med. Chem.*, (2006) **14**, 6202.

[103] Murai, K.; Katoh, S.-i.; Urabe, D.; Inoue, M., *Chem. Sci.*, (2013) **4**, 2364.

[104] Mizuno, K.; Ikeda, M.; Toda, S.; Otsuji, Y., *J. Am. Chem. Soc.*, (1988) **110**, 1288.

243

1.5   Non-Radical Skeletal Rearrangements

1.5.1   Protic Acid/Base Induced Reactions

D. Adu-Ampratwum and C. J. Forsyth

General Introduction

This chapter covers synthetic domino processes that are induced by protic acid or base. These events are broadly classified into capitalizing on (1) the release of oxirane ring strain under both acidic and basic conditions and (2) carbocyclic ring expansion and contraction under protic acid or basic conditions. The focus is on monocomponent, rather than multicomponent, domino processes. These reactions typically involve singular substrates that are poised to undergo thermodynamically driven sequential bond forming processes under the same reaction conditions. This sequence may simply involve intramolecular transetherification of hydroxy oxiranes, wherein a hydroxy O—H bond is exchanged for an oxirane C—O bond and another hydroxy O—H bond (three bond changes), or more extensive arrays of sequential bond reordering within a single substrate.

There are creative, base-mediated multicomponent domino processes that are useful to rapidly increase molecular complexity, and which involve nucleophilic initiation, such as anionic conjugate addition to generate nucleophilic enolate-type adducts.[1,2] However, these types of processes are omitted from the scope of this focused coverage of monocomponent unimolecular base-induced domino processes. There are notable recent monographs and a seminal review that cover domino reaction processes broadly but which only narrowly overlap with the coverage of this chapter.[3–5]

Acid-induced domino processes are particularly well suited for polyene- and oxirane-containing substrates, as well as carbocyclic ring expansions and contractions. For oxiranes, both *endo* and *exo* ring openings may occur, depending on the substrates. Thus, these openings are treated separately. Important acid-induced carbocyclic rearrangements include the Wagner–Meerwein, pinacol, and semipinacol rearrangements. Non-nucleophilic base-induced domino processes are more prominently represented in the carbocyclic ring expansions and contractions, including the benzilic acid and Favorskii rearrangements. Mechanistically related variants that are covered separately include the retro-benzilic acid and homo-Favorskii rearrangements. Final coverage is given to α-hydroxy ketone and related rearrangements.

The occurrence of architecturally complex natural products has inspired the design and development of acid- and base-induced domino processes in the synthetic chemistry laboratory. Alkaloids, steroids, ionophores, and ladder polyethers have all contributed to these developments. Important examples include Robinson's tropinone total synthesis through a double Mannich process and the cationic assembly of progesterone from an alkynyl polyene/tertiary allylic alcohol.[6,7] The Cane–Celmer–Wesley hypothesis for the biogenesis of monensin and related polyether antibiotics invokes a cationic polyepoxide cascade.[8] The biosynthesis of ladder polyether marine natural products such as the brevetoxins was originally proposed to occur through the domino opening of hypothetical polyepoxide intermediates derived from polyketides (see also Section 1.2).[9,10] However, no such

for references see p 265

biosynthetic polyepoxides have been encountered. This lends support to the idea that a single monooxygenase with broad substrate specificity may iteratively epoxidize linear polyene substrates and also catalyze intramolecular ring-expanding transetherification one alkene at a time.[11,12] Nonetheless, the notion that polyepoxide-opening cascades would result in fused polyether natural products has inspired the development and application of synthetic strategies and underlying methods. The stereospecific Shi epoxidation of unactivated alkenes has been particularly enabling in this regard.[13]

1.5.1.1 Intramolecular Epoxide-Opening Cyclizations

Epoxides constitute one of the functional groups most widely used in organic synthesis and, as such, epoxide-opening reactions represent a well-known class of organic transformation that has been well exploited in synthesis. This situation is due to the many efficient methods, including asymmetric methods, that have been developed to permit the facile introduction of the epoxide functionality and to predict the ease and directionality of epoxide opening with various nucleophiles with high levels of regio- and stereocontrol.[14-25] The intramolecular opening of an epoxide by an internal oxygen nucleophile is one of the most efficient ways to access cyclic ethers with ring sizes ranging from small to medium and, in the case of multiple epoxides, this approach provides direct entry to polycyclic polyethers through an intramolecular epoxide-opening domino process (Scheme 1).[26-32]

Scheme 1 Intramolecular Epoxide Opening by an Internal Oxygen Nucleophile[26-32]

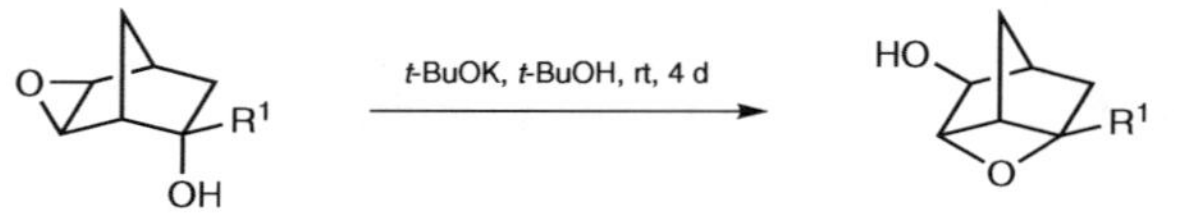

R[1]	Yield (%)	Ref
H	82	[26]
Bu	92	[27]

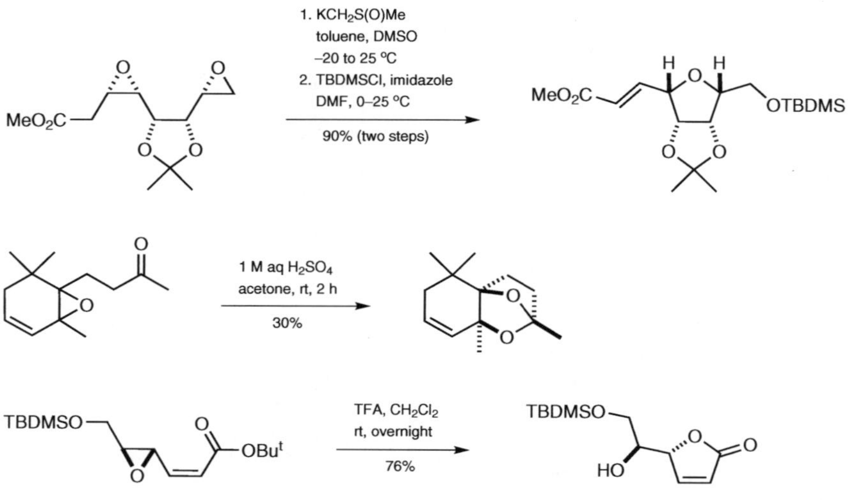

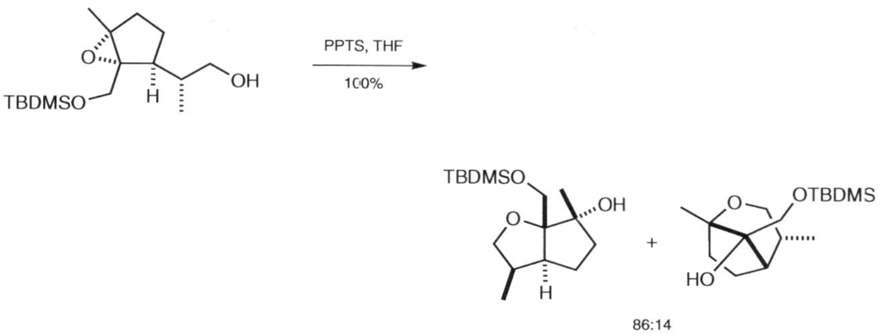

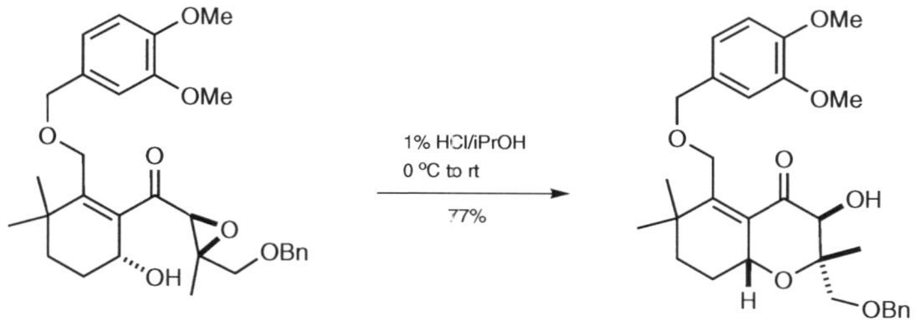

1.5.1 Protic Acid/Base Induced Reactions

245

1.5.1.1.1 Protic Acid Induced Intramolecular Epoxide Openings

In acid-catalyzed, intramolecular epoxide-opening reactions, two main issues must be addressed: the regioselectivity of the transformation and the stereoselectivity of the transformation. Since the vast majority of epoxide ring-opening reactions proceed with inversion of configuration, the stereoselectivity of the transformation is of less concern to synthetic chemists. However, the regioselectivity is far more variable and case dependent. For example, intramolecular opening of hydroxy epoxides (e.g., **1**) (Scheme 2) generally favors the *exo* mode of isomerization (path a) to yield smaller cyclic ethers (e.g., **3**),[33–35] owing to a less-strained, in-line transition state (e.g., **2**). By contrast, the *endo* mode (path b) leading to the larger ring (e.g., **5**) is relatively disfavored because of a more strained transition state (e.g., **4**).

Scheme 2 Regioselectivity of Intramolecular Hydroxy Epoxide Transetherification[33]

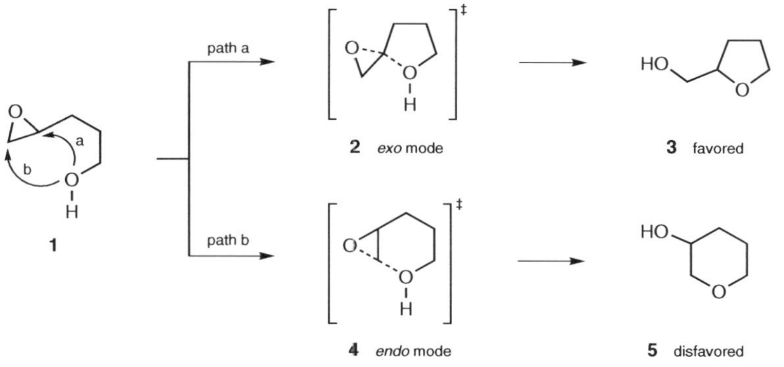

for references see p 265

1.5.1.1.1.1 *exo* Epoxide Ring Expansions

As acid-catalyzed intramolecular hydroxy epoxide isomerization by the *exo* mode can be facile, it is the most widely studied and applied process to access five- and six-membered cyclic ethers (Scheme 3). A broad range of epoxy alcohol and protected alcohol substrates can be successfully transformed into tetrahydrofuran and tetrahydropyran targets. For example, epoxy alcohol **6** can be readily converted into tetrahydrofuran **7** and tetrahydropyran **8** in near quantitative yield in a 91:9 ratio. Even tetraol **9** undergoes *exo*-selective epoxide opening to yield tetrahydrofuran **10** in high yield.[36]

Scheme 3 Protic Acid Catalyzed *exo*-Selective Intramolecular Epoxide Opening[36–42]

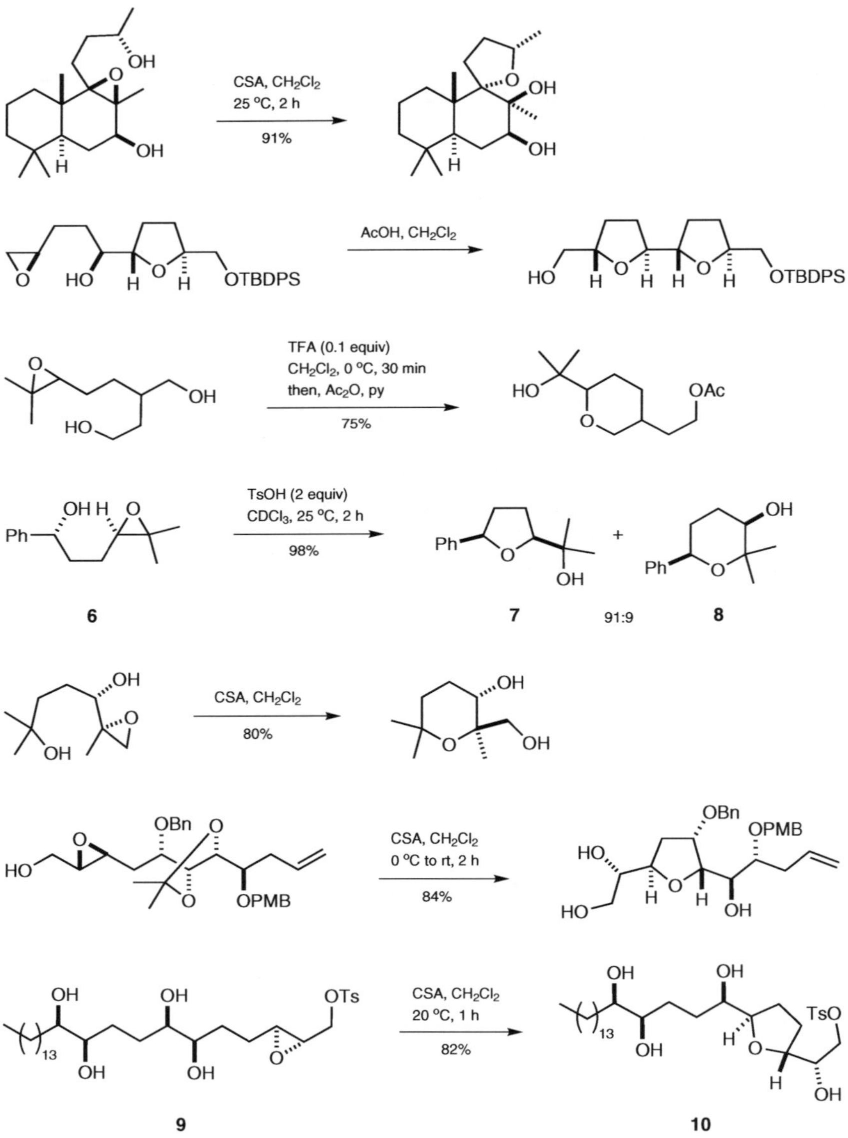

One major advantage of this regioselective intramolecular epoxide opening is that it provides an excellent opportunity to incorporate the *exo*-cyclization mode into a domino

transformation. This process allows direct access to polycyclic polyether frameworks of certain natural and unnatural products that feature 2,5-linked tetrahydrofuran and 2,6-linked tetrahydropyran ring systems. For example, under protic acid conditions, 1,5-diepoxide 11[43] and 1,6,11-triepoxide 13[44] undergo smooth all-*exo* epoxide-opening domino processes to afford polytetrahydrofuran 12 and polytetrahydropyran 14 motifs, respectively, in modest yields (Scheme 4). Molecules containing these types of motifs possess interesting cation-binding properties.

Scheme 4 Protic Acid Catalyzed Intramolecular All-*exo* Epoxide-Opening Domino Processes To Form Polytetrahydrofurans and Polytetrahydropyrans[43,44]

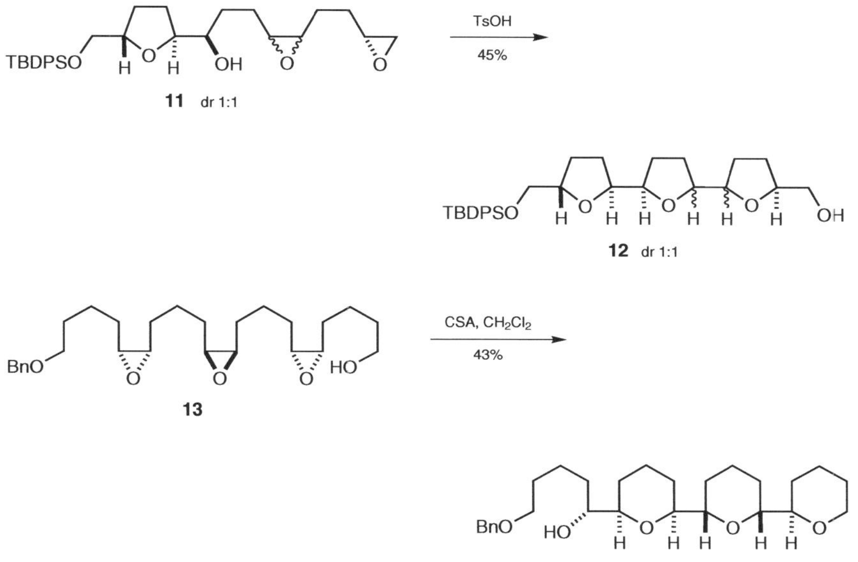

exo-Mode epoxide-opening domino transformations have been applied to the total synthesis of natural products. For example, Paterson observed that diastereomeric triepoxide 15 undergoes an epoxide-opening domino cascade catalyzed by 10-camphorsulfonic acid to provide the CDE ring system 16 of etheromycin and its diastereomer in moderate yield (Scheme 5).[45] Corey and Xiong, in their effort to confirm the proposed structure of glabrescol, employed an all-*exo* mode of domino epoxide openings as a key step in their synthesis. The authors recognized that, under protic acid conditions, tetraepoxide 17 undergoes a bidirectional all-*exo*-selective epoxide-opening domino cascade to yield 18 in modest yield. The choice of acidic conditions is crucial to ensure that the formation of the AB and A'B' rings proceeds preferentially and at a faster rate than cyclization to form the C ring.[46]

for references see p 265

Scheme 5 Application of Intramolecular All-*exo* Polyepoxide Opening in Natural Product Synthesis[45,46]

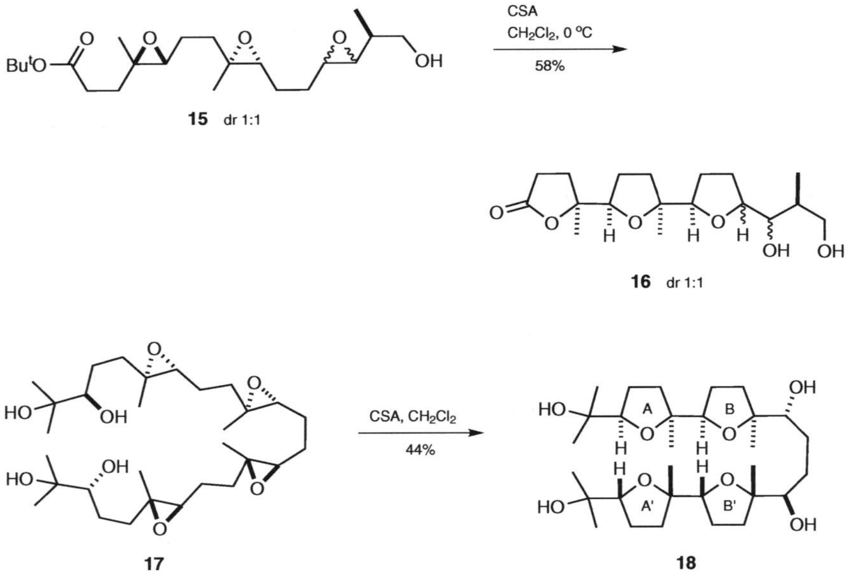

2-[(2S,5R)-5-Phenyltetrahydrofuran-2-yl]propan-2-ol (7) and (3R,6R)-2,2-Dimethyl-6-phenyltetrahydro-2H-pyran-3-ol (8); Typical Procedure for *exo* Opening of Epoxy Alcohols:[42]

TsOH (34.4 mg, 0.20 mmol) was added to a soln of epoxide **6** (19.4 mg, 94 µmol) in CDCl₃ (0.7 mL), and the resulting soln was stirred at 25 °C for 2 h. After removal of the solvent under reduced pressure (25 kPa/40 °C), the products were filtered through a short pad of alumina using Et₂O as the eluent. The organic phase was concentrated under reduced pressure to provide an oil, which was purified by column chromatography (silica gel, petroleum ether/acetone 4:1) to give the title compounds as a colorless liquid; yield: 19.0 mg (98%); ratio (**7/8**) 91:9 (as determined by ¹H NMR spectroscopy).

1.5.1.1.1.2 *endo* Epoxide Ring Expansions

Acid-catalyzed *endo*-selective intramolecular epoxide-opening cyclizations are generally disfavored, but it is possible to achieve regioselective epoxide opening through an *endo* mode by using directing groups. In the 10-camphorsulfonic acid catalyzed epoxide-opening cyclization, if the side chain in generic epoxy alcohol **19** is an aliphatic group [e.g., $R^1 = (CH_2)_2CO_2Me$], then attack of the nucleophile occurs preferentially through the *exo* mode (attack at carbon a) in accordance with Baldwin's rules, to yield tetrahydrofuran **20** in good yield (Scheme 6). However, placing an alkenyl group adjacent to the epoxide alters the selectivity and may lead preferentially to *endo*-product **21**. For example, *trans*-epoxide **19** with a vinyl substituent (e.g., $R^1 = CH=CH_2$) gives predominantly the *endo*-product, although the same is not true of the *cis*-epoxide, which gives an essentially 1:1 mixture of *exo/endo*-products. Adding an electron-withdrawing group to the alkene [e.g., $R^1 = (E)$-CH=CHCO₂Me] further complicates matters, with *trans*-epoxides giving variable product mixtures dependent upon the size of the ring formed and *cis*-epoxides returning the *exo*-product exclusively. The observed selectivity is attributed to stabilization of the positive charge that builds up on carbon b in the transition state by electron donation from the adjacent π bond. Without such incipient positive charge stabilization, *exo* cyclization is normally favored.[47,48]

1.5.1 Protic Acid/Base Induced Reactions **249**

Scheme 6 Protic Acid Catalyzed Intramolecular Epoxide Opening[47,48]

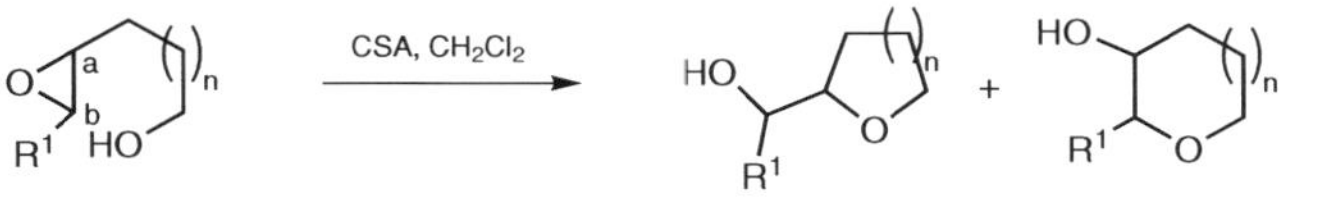

Stereochemistry of **19**	R[1]	n	Ratio [**20/21** (exo/endo)]	Yield (%)	Ref
trans	(CH$_2$)$_2$CO$_2$Me	1	100:0	94	[47]
trans	(CH$_2$)$_2$CO$_2$Me	2	100:0	70	[48]
trans	CH=CH$_2$	1	0:100	95	[47]
trans	CH=CH$_2$	2	18:82	75	[48]
cis	CH=CH$_2$	1	56:44	95	[47]
cis	CH=CH$_2$	2	50:50	73	[48]
trans	(E)-CH=CHCO$_2$Me	1	40:60	96	[47]
trans	(E)-CH=CHCO$_2$Me	2	78:22	75	[48]
cis	(E)-CH=CHCO$_2$Me	1	100:0	86	[47]
cis	(E)-CH=CHCO$_2$Me	2	100:0	76	[48]

Several other directing groups have been reported which favor *endo*-selective intramolecular epoxide transetherification. These include alkynyl,[49–52] alkyl,[53–55] silyl,[56–58] sulfonyl,[59–61] and methoxymethyl substituents,[62–64] and selected examples are shown in Scheme 7.[65–67] However, the need to use directing groups is one of the limitations of acid-catalyzed *endo*-selective intramolecular epoxide opening, and this demerit has hampered the use of this method (e.g., in domino transformations to access the polycyclic polyether structural motifs prevalent in ladder polyethers). While the use of directing groups may deliver the desired cyclic ether it does so with residual unwanted functionality. Thus, the utilization of this method in total synthesis requires removal or other synthetic elaboration of the residual substituents, and these limitations have stimulated an active area of research. Nevertheless, this directing-group approach has made it possible to implement the *endo*-selective mode in natural product synthesis, despite requiring superfluous subsequent transformations.

Scheme 7 Selected Examples of Intramolecular *endo* Epoxide Opening[65–67]

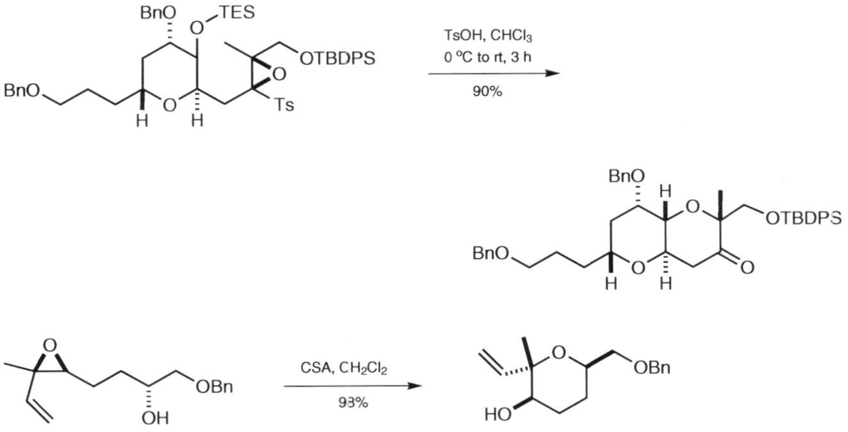

for references see p 265

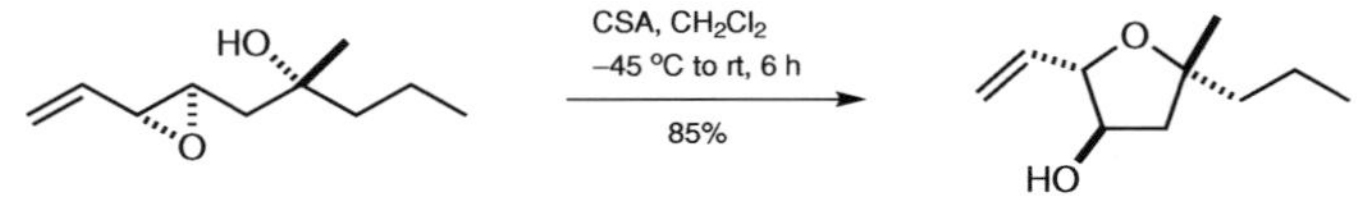

250 Domino Transformations **1.5** Non-Radical Skeletal Rearrangements

2-Vinyltetrahydro-2H-pyran-3-ol [21, n = 1; R¹ = CH=CH₂]; Typical Procedure for the *endo* Opening of Hydroxy Epoxides:[47]

CSA (9.0 mg, 0.04 mmol) was added in one portion to a stirred soln of hydroxy epoxide *trans*-**19** [n = 1; R¹ = CH=CH₂; 50 mg, 0.39 mmol] in dry CH₂Cl₂ (4 mL) at −40 °C. After stirring for 1 h (−40 to 25 °C), Et₃N (0.02 mL, 0.1 mmol) was added, the volatiles were evaporated, and the residue was subjected to flash chromatography (silica gel, Et₂O/petroleum ether 2:3) to afford the title compound; yield: 47 mg (95%).

1.5.1.1.2   **Base-Induced Intramolecular Epoxide Openings**

The most commonly encountered base-catalyzed intramolecular epoxide-opening reaction is the "epoxide-migration" reaction of 2,3-epoxy alcohols, popularly known as the Payne rearrangement.[68] This reversible process is known to proceed through an S_N2 pathway and benefits from the use of polar, protic solvents. The thermodynamic equilibrium generally lies in the direction of the more highly substituted epoxide, but rearrangement toward less-substituted epoxides may be influenced by steric factors. Base-catalyzed intramolecular epoxide openings are less common, but they can be used to synthesize medium-sized cyclic ethers and in some cases polycyclic ethers if polyepoxide substrates are used. Notable examples are shown in Scheme 8, including Dounay's conversion of diol epoxide **22** into substituted tetrahydrofuran **23**.[69–75]

Scheme 8   Synthesis of Cyclic Ethers by Base-Catalyzed Intramolecular Epoxide Opening[69–75]

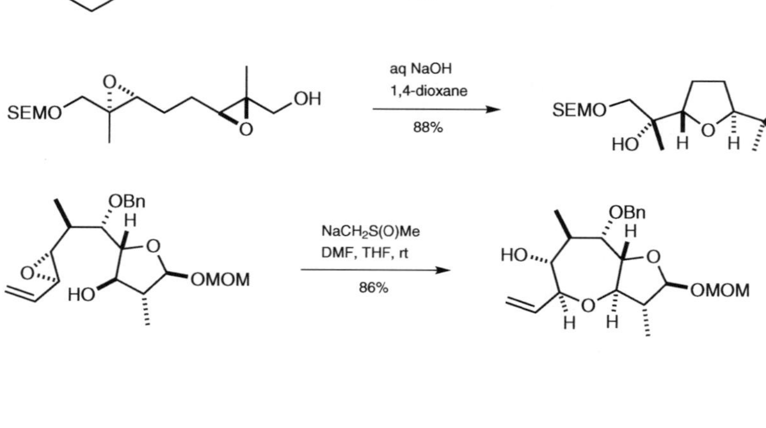

1.5.1 Protic Acid/Base Induced Reactions

Base-catalyzed intramolecular epoxide opening shows a general preference for *exo* selectivity, as can be seen in the examples already shown; however, *endo* selectivity can also be achieved through the use of directing groups (Scheme 9). As an instructive example, hydroxy oxiranes **24** possess a trimethylsilyl group directly on the epoxide, and this directs efficient *endo*-selective epoxide opening to yield fused bistetrahydropyran **25**. The use of cesium fluoride is important to effect removal of the trimethylsilyl group. Jamison has used this approach to synthesize, in 20% yield, fused tetrahydropyran tetramer **26**, a motif found in many ladder polyether natural products such as the brevetoxins.[76]

Scheme 9 Synthesis of Face-Fused Tetrahydropyrans by Base-Catalyzed Intramolecular Epoxide Opening[76]

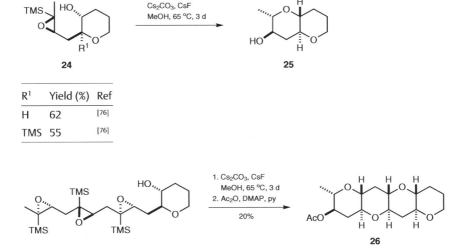

R^1	Yield (%)	Ref
H	62	[76]
TMS	55	[76]

for references see p 265

(2S,3S)-2-[3-(Benzyloxy)propyl]-5-(hydroxymethyl)tetrahydrofuran-3-ol (23):[74]
K_2CO_3 (228 mg, 1.65 mmol) was added to a stirred soln of epoxide **22** (44 mg, 0.17 mmol) in
MeOH (2.0 mL) at 0 °C. After 2.5 h at rt, sat. aq NH_4Cl (5 mL) was added, and the mixture
was extracted with EtOAc (3 × 15 mL). The combined organic phases were washed with
brine (10 mL), dried (Na_2SO_4), filtered, and concentrated under reduced pressure. Flash
chromatography (silica gel, EtOAc/MeOH/CH_2Cl_2 1:0:0 to 0:1:19) gave the title compound
as a mixture of isomers; yield: 43 mg (99%, as reported).

**(2S,3R,4aS,8aR)-2-Methyloctahydropyrano[3,2-*b*]pyran-3-ol (25); Typical Procedure for the
endo Cyclization of Epoxides:**[76]
Cs_2CO_3 (267 mg, 0.82 mmol) and CsF (125 mg, 0.82 mmol) were weighed into a flame-dried
Schlenk tube in a glovebox under an atmosphere of argon. A soln of epoxide **24** (R[1] = H;
8 mg, 32.7 µmol) in MeOH (427 µL) was added and the tube was sealed. The resulting slurry
was heated to 65 °C for 3 d, the MeOH was removed under reduced pressure, and the mixture was partitioned between sat. aq NH_4Cl (10 mL) and EtOAc (10 mL). The layers were
separated and the aqueous layer was extracted with EtOAc (4 × 10 mL). The combined organic layers were washed with brine, dried ($MgSO_4$), filtered, and concentrated under reduced pressure. The crude material was purified by column chromatography (silica gel,
EtOAc/hexane 1:1 to 3:1) to give the title compound; yield: 3.5 mg (62%); R_f = 0.34 (EtOAc/
hexane 3:2); $[\alpha]_D^{25}$ −11.2 (c 0.26, $CHCl_3$).

1.5.1.2 Carbocyclic Ring Expansions/Ring Contractions

Carbocyclic ring expansions/ring contractions are common features in synthetic organic
chemistry.[77] They provide powerful methods to construct complicated structures that
can be difficult to generate by alternative methods. These classes of organic transformation include acid-induced ring expansions/contractions, base-induced ring expansions/
contractions, oxidative rearrangements, and photochemical rearrangements. For the purpose of the topic under discussion, only protic acid induced and base-induced reactions
are presented, beginning with acidic modes.

1.5.1.2.1 Acid-Induced Carbocyclic Ring Expansions/Ring Contractions

Acid-induced ring expansions and ring contractions are well-known, facile reaction
classes in bicyclic systems that occur via the intermediacy of carbocations or incipient
carbocations. For example, treatment of cyclopropyl alcohol **27** with perchloric acid affords lactone **30** smoothly (Scheme 10). The rearrangement is thought to proceed via intermediate **28** with carbocationic character adjacent to the strained cyclopropane ring.
Ring expansion ensues, leading to allylic carbocation **29**, which finally undergoes intramolecular trapping by the pendant carboxy group to afford lactone **30** in 81% yield.[78] This
strategy has been used to synthesize (±)-confertin.[79]

1.5.1 Protic Acid/Base Induced Reactions 253

Scheme 10 Selected Examples of Carbocyclic Rearrangement and Application to the Synthesis of (±)-Confertin[78–82]

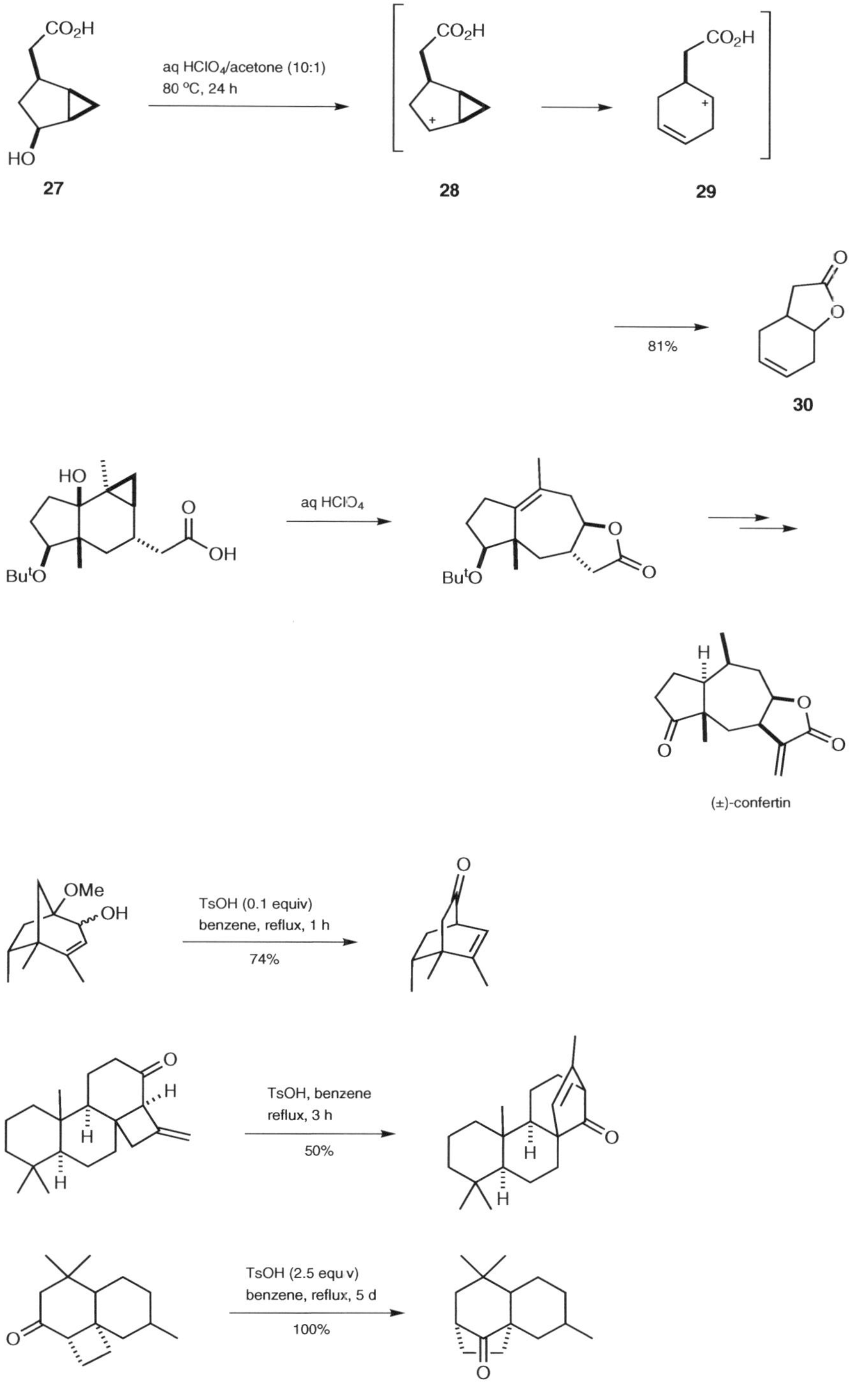

Many acid-induced carbocyclic ring expansions/contractions have been reported, catalyzed by either protic or Lewis acids. Two classical examples that are usually catalyzed by protic acids are the Wagner–Meerwein rearrangement and the pinacol rearrangement.

for references see p 265

254   Domino Transformations **1.5** Non-Radical Skeletal Rearrangements

3a,4,7,7a-Tetrahydrobenzo[*b*]furan-2(3*H*)-one (30):[78]

A soln of hydroxy acid **27** (0.19 g, 1.2 mmol) in 7% aq HClO$_4$ (10 mL) and acetone (1 mL) was stirred at 80 °C for 24 h. The product was isolated with Et$_2$O and distilled to afford the title compound; yield: 0.14 g (81%); bp 85 °C (bath temperature)/6.7 Pa.

1.5.1.2.1.1   **Wagner–Meerwein Rearrangements**

The Wagner–Meerwein rearrangement is a well-known classical organic transformation involving the acid-catalyzed generation of a carbocation followed by 1,2-migration of an adjacent group (hydride, aryl, or alkyl) to generate a new, usually more stable, carbocation.[83,84] One major limitation of the Wagner–Meerwein rearrangement is that the carbocationic nature of the rearrangement sometimes leads to complicated mixtures of products. Thus, the Wagner–Meerwein rearrangement is not often employed in target-oriented, complex molecule synthesis.

1.5.1.2.1.1.1   **Ring-Expansion Rearrangements**

The Wagner–Meerwein rearrangement is best known in polycyclic terpenes and terpenoids where it allows expansion of strained small rings to afford both common and medium-sized carbocycles. For example, cyclobutane **31** undergoes smooth ring expansion to afford cyclopentene-embedded (±)-isocomene (**34**) in excellent yield (Scheme 11).[85]

The observed ring expansion is readily rationalized by considering the release of cyclobutane ring strain in **32** on transition to the less-strained five-membered ring in **33**. These observations reinforce the concept that tertiary carbocations are favored over secondary carbocations. However, tertiary to secondary carbocation migration might be favored if the rearrangement allows a significant release of ring strain or steric energy.

Scheme 11   Selected Examples of Wagner–Meerwein Ring Expansion[85–89]

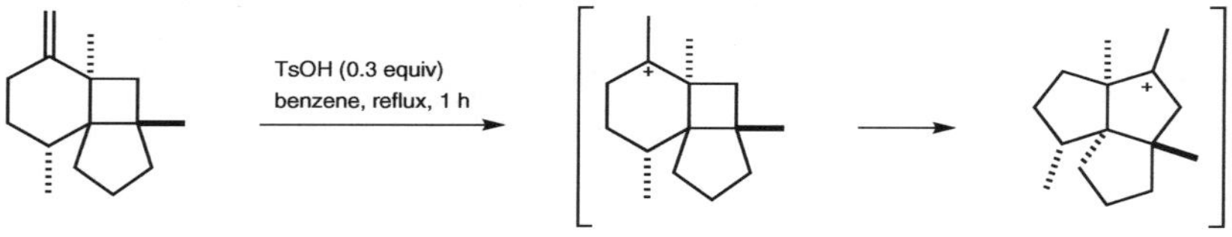

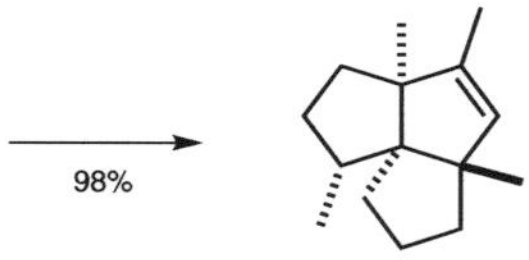

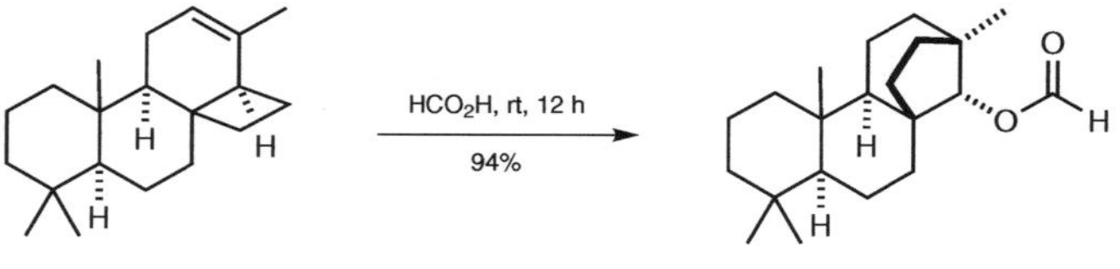

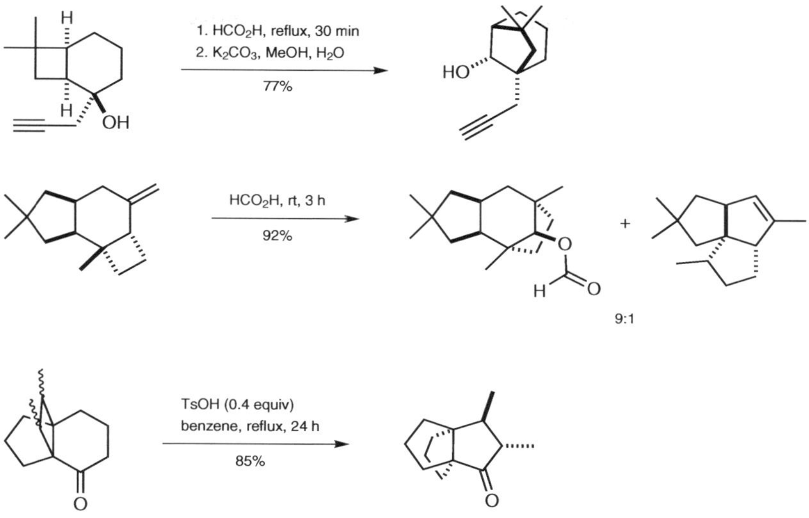

1.5.1 Protic Acid/Base Induced Reactions **255**

(1R,3aS,5aS,8aR)-1,3a,4,5a-Tetramethyl-1,2,3,3a,5a,6,7,8-octahydrocyclopenta[c]penta-lene (34):[85]

A 50-mL, round-bottomed flask was charged with **31** (1.52 g, 7.45 mmol), benzene (25 mL) (**CAUTION**: *carcinogen*), and TsOH (0.42 g, 2.2 mmol). The soln was heated at reflux for 1 h, cooled, and poured into H₂O. The layers were separated, and the aqueous phase was extracted with Et₂O. The combined organic phases were washed sequentially with sat. aq NaHCO₃ and NaCl, dried (MgSO₄), and filtered. Removal of the solvents under reduced pressure gave the title compound; yield: 1.488 g (98%).

1.5.1.2.1.1.2 **Ring-Contraction Rearrangements**

Carbocyclic ring contractions under Wagner–Meerwein conditions are less common, and this situation may be because the contraction is energetically costly. However, if rearrangement leads to a stable carbocationic intermediate, then this approach may offer a reliable protocol to construct smaller rings or derivatives. As an example, cyclobutane **35** presumably undergoes ring contraction to cyclopropane **36** (Scheme 12). Under the reaction conditions, the contracted cyclopropane undergoes a facile rearrangement process to yield ring-opened secondary alcohol **37** in good yield (88%). Even higher yields are observed at 50 °C with a pH 4 buffer.[90]

for references see p 265

Scheme 12 An Example of a Wagner–Meerwein Rearrangement Involving Ring Contraction[90]

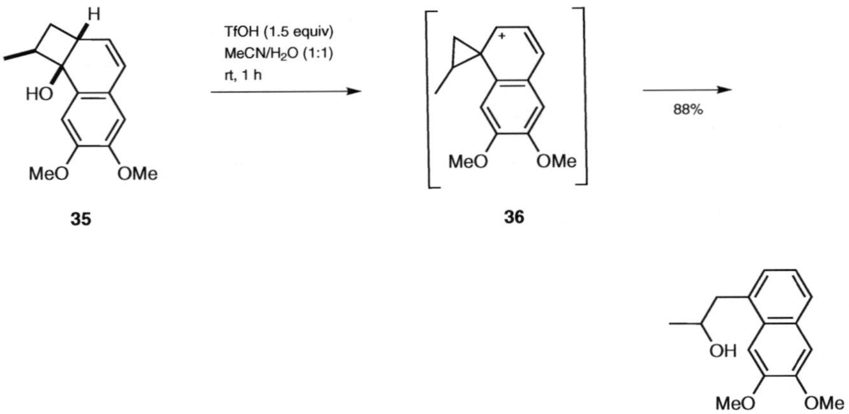

1-(6,7-Dimethoxynaphthalen-1-yl)propan-2-ol (37):[90]

TfOH (28 µL, 0.32 mmol) was added dropwise to a soln of **35** (52.6 mg, 214 µmol) in MeCN/H$_2$O (1:1; 1 mL). After stirring for 1 h at ambient temperature, sat. aq NaHCO$_3$ was added, the layers were separated, and the aqueous layer was extracted with EtOAc (2 ×). The combined organic layers were washed with brine, dried (Na$_2$SO$_4$), filtered, and concentrated under reduced pressure. The residue was purified by chromatography (silica gel, hexane/EtOAc 2:1 to 2:3) to afford the title compound; yield: 46.5 mg (88%).

1.5.1.2.1.2 Pinacol Rearrangements

Acid-induced rearrangement reactions involving vicinal diols are categorized as pinacol rearrangements, which also constitute a classic type of organic transformation. Although the process is related to the Wagner–Meerwein rearrangement, the driving force for the pinacol rearrangement is the formation of a new carbonyl functionality. Thus, vicinal tertiary cyclohexyl diol **38** readily undergoes smooth ring expansion to produce cycloheptanone **41** in good yield through carbocation rearrangement from tertiary carbocation **39** to a hydroxy-stabilized secondary carbocation, that is, oxacarbenium ion **40** (Scheme 13).[91]

Scheme 13 Selected Examples of the Pinacol Rearrangement[91–95]

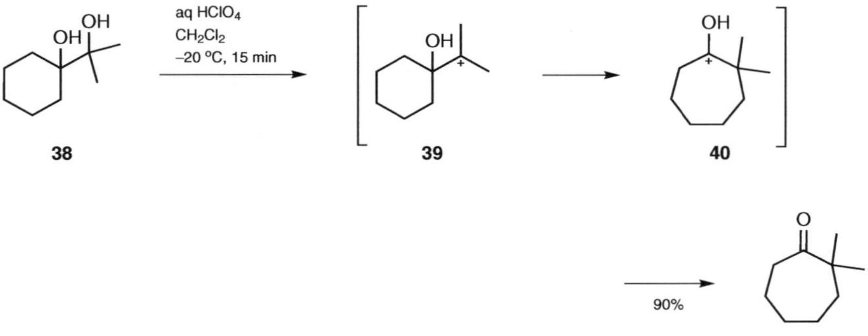

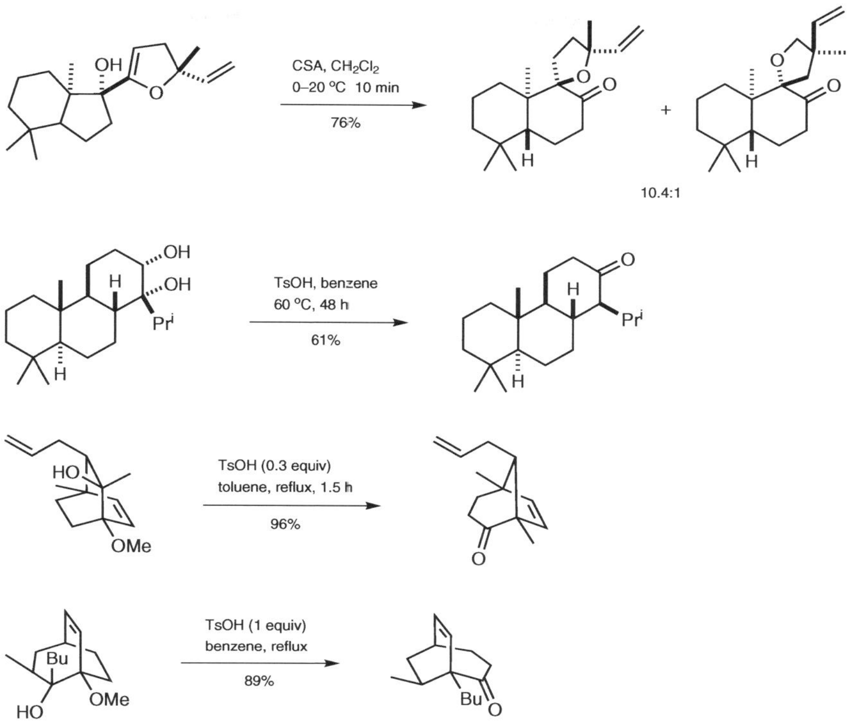

1.5.1 Protic Acid/Base Induced Reactions

As with the Wagner–Meerwein rearrangement, the pinacol rearrangement has seen limited synthetic application, although it can be a useful alternative to other standard methods for the synthesis of aldehydes and ketones.

2,2-Dimethylcycloheptanone (41):[91]
70% aq HClO$_4$ (15 mL) was added to a soln of diol **38** (158 mg, 1.0 mmol) in CH$_2$Cl$_2$ (5 mL) at –20 °C under an atmosphere of argon. The resulting soln was stirred at –20 °C for 15 min, H$_2$O was added, and this soln was extracted with Et$_2$O. The combined organic phases were washed with sat. aq NaHCO$_3$ and sat. NaCl, dried (MgSO$_4$), filtered, and concentrated to yield a pale yellow oil. Filtration through silica gel (Et$_2$O/petroleum ether 5:95) gave the title compound as a colorless oil; yield: 126 mg (90%).

for references see p 265

258   Domino Transformations   **1.5** Non-Radical Skeletal Rearrangements

1.5.1.2.1.3   **Semipinacol Rearrangements**

A very useful and commonly employed variant of the pinacol rearrangement is the semi-pinacol protocol, owing to the fact that the initiating carbocation can be easily triggered. This method generally gives high regio- and stereoselectivity and is high yielding. Notable examples of the semipinacol rearrangement in organic synthesis are shown in Scheme 14[96–100] (see Section 1.5.2 for additional examples with the use of Lewis acids as initiators).

Scheme 14   Selected Examples of the Semipinacol Rearrangement Reaction[96–100]

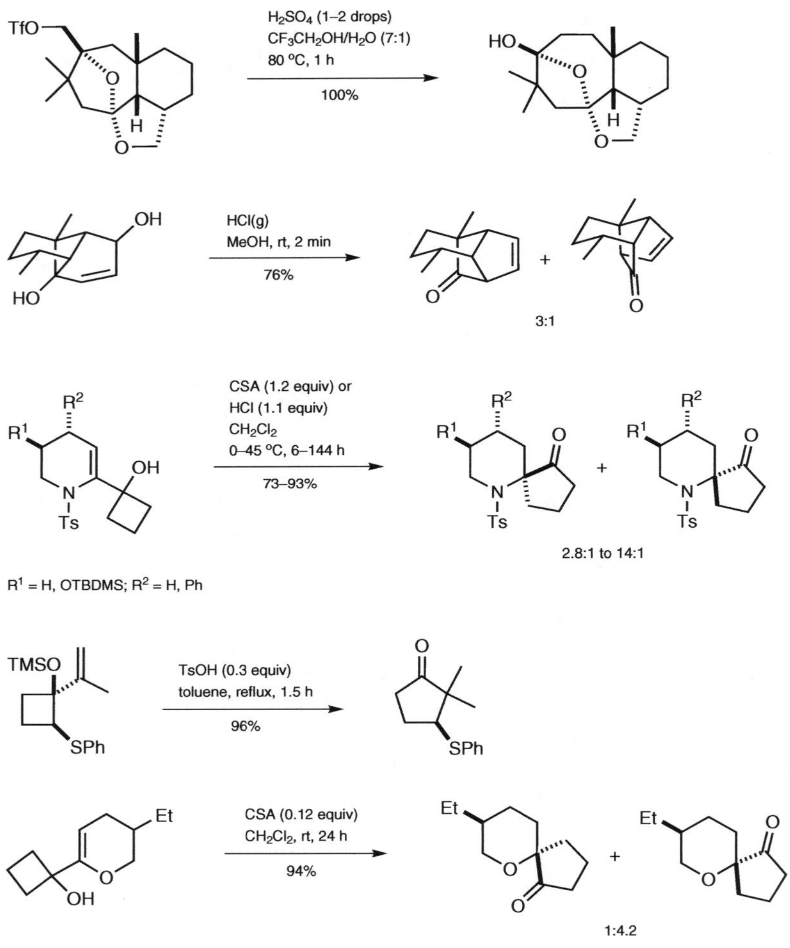

1.5.1 Protic Acid/Base Induced Reactions

259

Despite the limitations associated with the above-mentioned reactions, their versatility permits their use in combination with other organic transformations in domino processes such as the Prins/semipinacol rearrangement of protected cyclobutyl allylic alcohol **42** into spiro-fused cyclopentanone **43** (Scheme 15).[101]

Scheme 15   An Example of a Prins/Semipinacol Rearrangement[101]

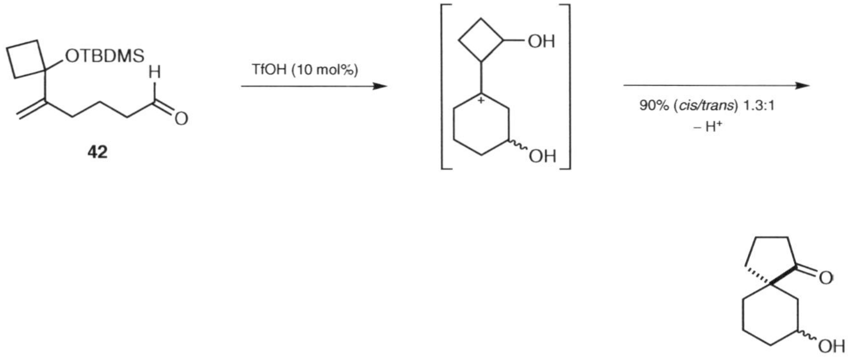

1.5.1.2.2   Base-Induced Carbocyclic Ring Expansions/Ring Contractions

Ring expansions and ring contractions also occur under basic reaction conditions. Commonly encountered examples are benzilic acid rearrangements, Favorskii rearrangements, and α-hydroxy ketone rearrangements. These events are usually performed under alkali metal hydroxide or alkoxide base conditions. The benzilic and Favorskii rearrangements both result in ring contractions in the case of cyclic systems, and are therefore synthetically useful complements to ring-expansion rearrangements, which usually occur under acidic conditions.

1.5.1.2.2.1   Benzilic Acid Rearrangements

Benzilic acid rearrangements occur readily with both aliphatic and aromatic α-diketones in moderate to excellent yield (depending on the substrate) to give α-hydroxy acids and their derivatives (depending on the type of base used). However, α-diketones having enolizable α-protons usually give poorer yields because of unavoidable aldol reaction pathways. As previously mentioned in Section 1.5.1.2.2, cyclic α-diketones undergo ring contraction to yield useful ring-contracted products. Notable examples are shown in Scheme 16.[102–106]

Scheme 16   Examples of the Benzilic Acid Type Rearrangement[102–106]

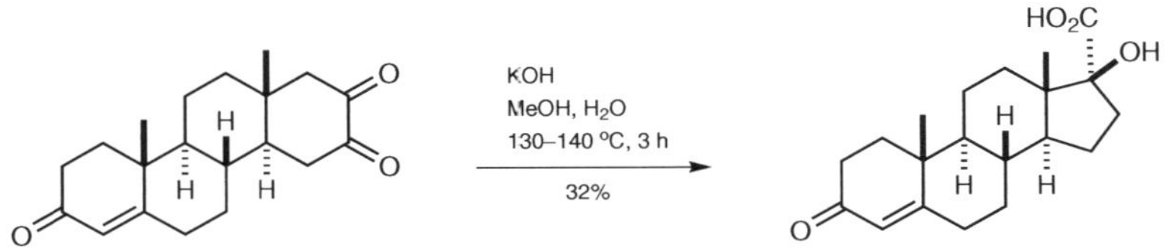

for references see p 265

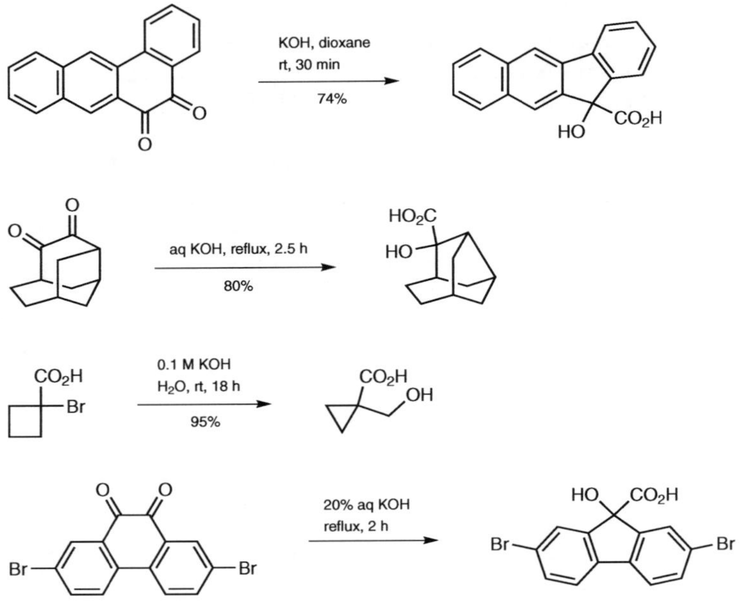

1.5.1.2.2.2 Retro-Benzilic Acid Rearrangements

For cases in which strained rings are encountered, the α-hydroxy carbonyl products of benzilic acid rearrangement may readily undergo a retro-benzilic acid rearrangement; this can be very synthetically useful. For example, Takeshita and co-workers have used this method to devise a short synthetic route to (±)-hinesol (**48A**) and (±)-agarospirol (**48B**), as shown in Scheme 17.[107] The diastereomeric mixture of α-hydroxy esters **46A/46B**, obtained readily by photoinduced [2+2] cycloaddition of **44** and **45**, undergoes a retro-benzilic acid rearrangement upon treatment with aqueous sodium carbonate to generate cyclopentenones **47A/47B**. These advanced intermediates can be further elaborated into **48A** and **48B**.

1.5.1 Protic Acid/Base Induced Reactions **261**

Scheme 17 Synthesis of the Spirocyclic Core of (±)-Hinesol and (±)-Agarospirol by a Retro-Benzilic Acid Rearrangement[107]

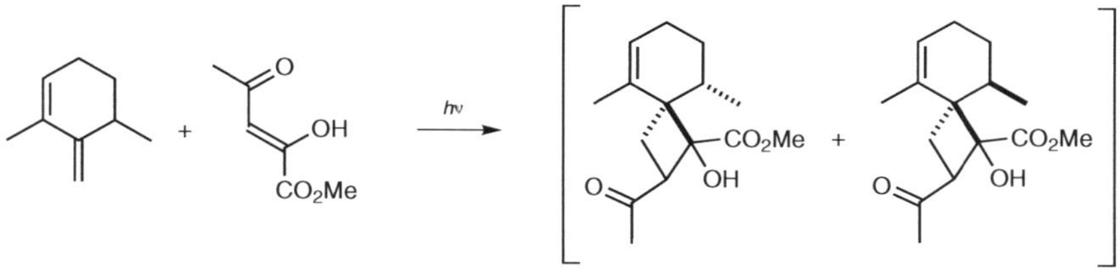

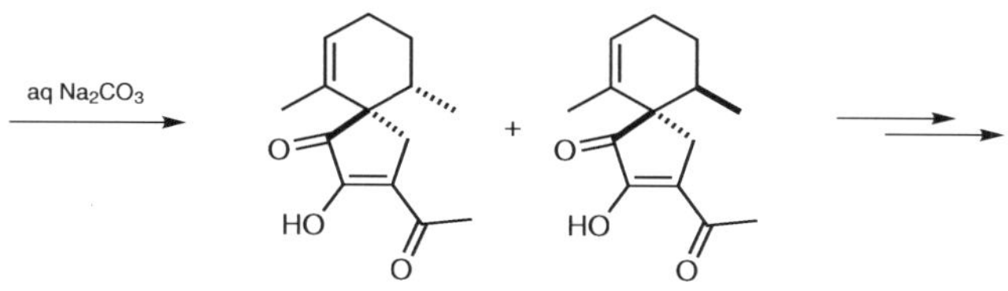

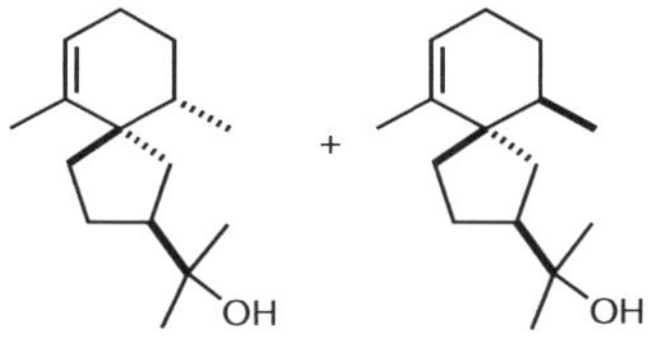

1.5.1.2.2.3 **Favorskii Rearrangements**

α-Halo ketones readily undergo Favorskii skeletal rearrangement. This reaction typically involves initial treatment of a substrate with hydroxide or an alkoxide base and proceeds via a cyclopropanone intermediate to yield α-branched esters and acids.[108,109] This rearrangement represents a very useful method for the synthesis of highly branched carboxylic acids and derivatives. Favorskii rearrangement of cyclic α-halo ketones usually results in ring contraction. In most cases, the yields are in the range of 40 to 80%, although higher yielding examples are known. The regio- and stereoselectivity are generally excellent and Scheme 18 shows notable examples of the process.[110–116] Ketones without enolizable protons readily undergo a quasi-Favorskii variant, as exemplified in the conversions of **49** and **51** into **50** and **52**, respectively.[110,115]

Scheme 18 Selected Examples of the Favorskii Rearrangement[110–116]

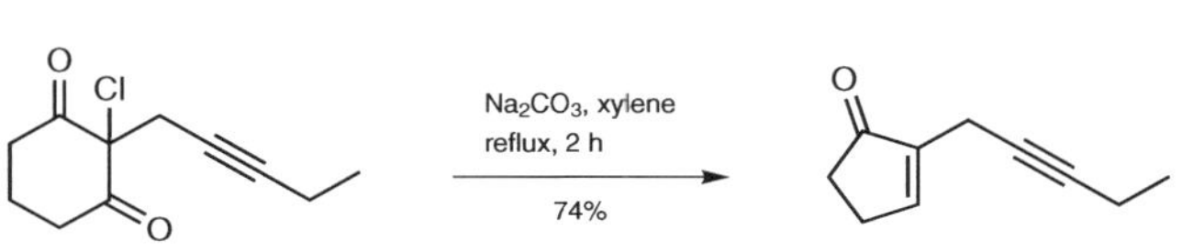

for references see p 265

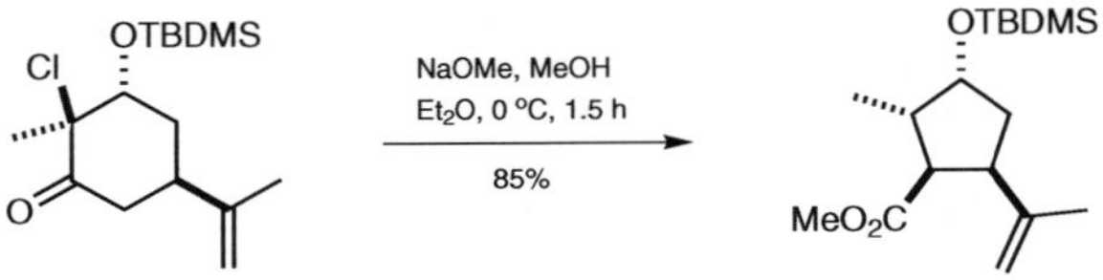

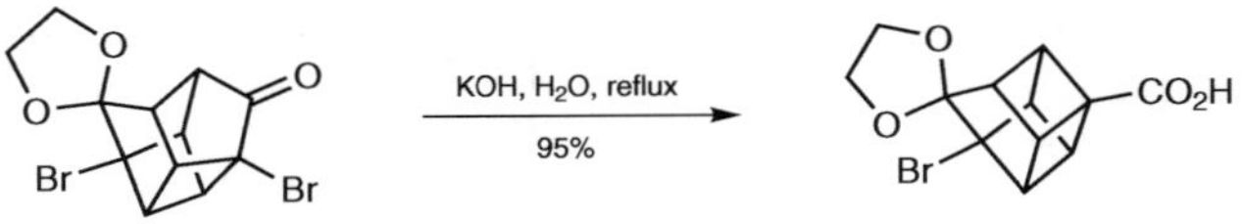

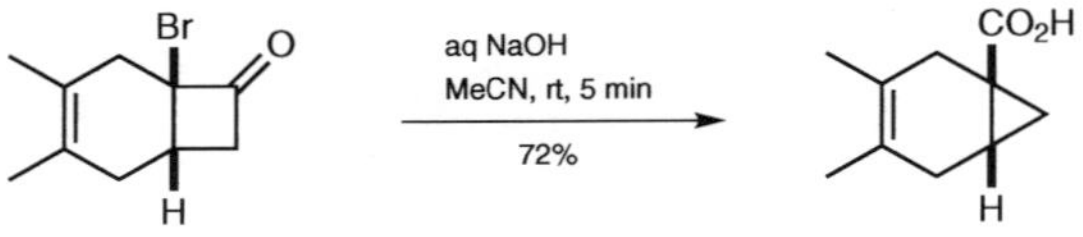

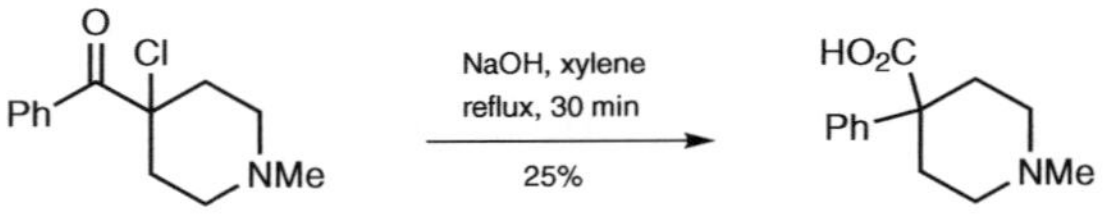

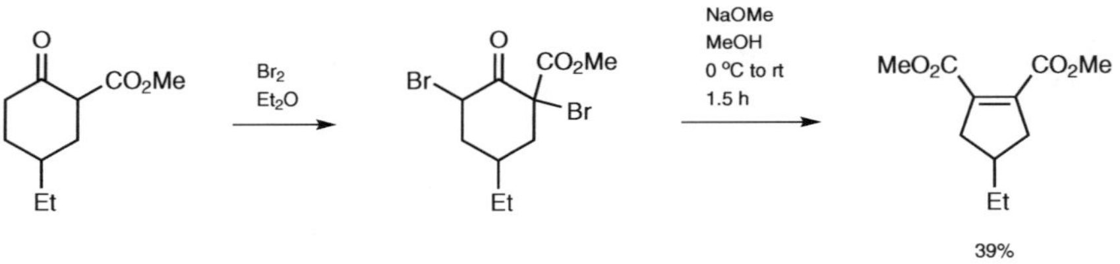

1-Methyl-4-phenylpiperidine-4-carboxylic Acid (50):[110]

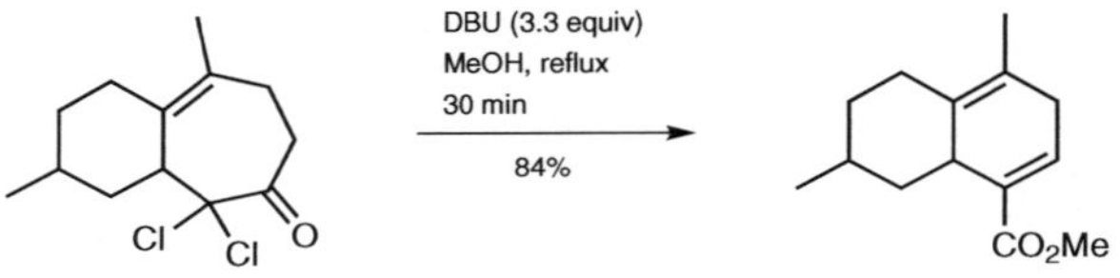

A soln of **49** (3.57 g, 15 mmol) in xylene (50 mL) was added over 30 min to a suspension of finely powdered, dried NaOH (18.0 g, 0.45 mol) in xylene (200 mL) heated at reflux. The mixture was then cooled and extracted with portions (25 mL) of H_2O until the pH of the extracts approached neutrality. The combined aqueous extracts were washed with Et_2O (3 × 25 mL) and then adjusted to pH 8.0 with HCl. The aqueous soln was concentrated to 50 mL, filtered, and acidified to pH 6.43 with HCl. This soln was cooled and the collected crystals were washed with H_2O, acetone, and Et_2O and then dried. Recrystallization (H_2O) gave the title compound as fine, white needles; yield: 0.81 g (25%); mp 309–310 °C.

Methyl 4,7-Dimethyl-3,5,6,7,8,8a-hexahydronaphthalene-1-carboxylate (52):[115]

A soln of **51** (16 mg, 0.061 mmol, 1.0 equiv) and DBU (33 mg, 0.21 mmol) in MeOH (1.0 mL) was heated to reflux under an atmosphere of N_2 for 30 min. After cooling to rt, the solvent was removed under reduced pressure, and the residue was purified by flash chromatography (silica gel, petroleum ether/EtOAc 30:1) to give the title compound; yield: 11 mg (84%).

1.5.1.2.2.3.1 Homo-Favorskii Rearrangements

Another commonly encountered variation is the homo-Favorskii rearrangement that typically involves β-halo ketones. Koreeda and co-workers have used a homo-Favorskii method as the key step in the synthesis of the tricyclic core of (±)-kelsoene (Scheme 19). In this example, treatment of β-sulfonyloxy ketone **53** with an excess of potassium *tert*-butoxide induces a homo-Favorskii rearrangement in less than two minutes at room temperature to yield **54** and **55** in a combined 95% yield. The relatively bulky, non-nucleophilic solvent cannot open the cyclobutanone intermediates, and both products are used in subsequent steps to synthesize (±)-kelsoene.[117]

Scheme 19 Synthesis of a (±)-Kelsoene Intermediate by Homo-Favorskii Rearrangement[117]

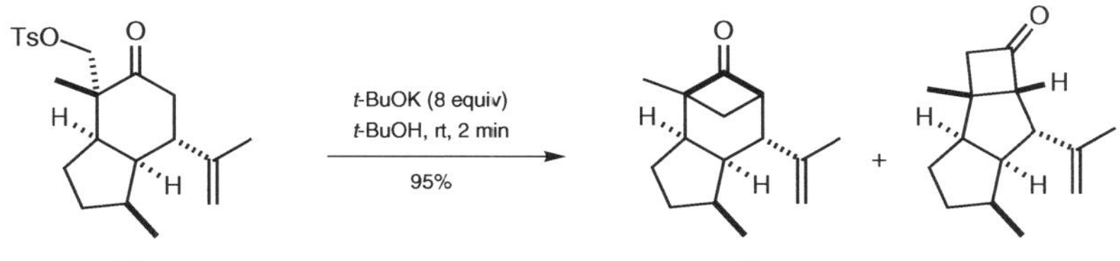

(1S,2R,5S,6S,7R,8R)-1,5-Dimethyl-7-(prop-1-en-2-yl)tricyclo[6.1.1.0^{2,6}]decan-9-one (54) and (2aS,2bR,5S,5aS,6R,6aR)-2a,5-Dimethyl-6-(prop-1-en-2-yl)octahydrocyclobuta[a]pentalen-1(2H)-one (55):[117]
A 1 M soln of *t*-BuOK in *t*-BuOH (0.70 mL, 0.70 mmol) was added in one portion to a soln of 4-toluenesulfonate **53** (33.8 mg, 87 µmol) in *t*-BuOH (2 mL) at rt under an atmosphere of argon. The soln immediately turned yellowish and, after 2 min, TLC analysis showed full consumption of **53**. Sat. aq NH$_4$Cl (2 mL) and H$_2$O (20 mL) were then added, and the resulting mixture was extracted with Et$_2$O (3 × 10 mL). The combined Et$_2$O layers were dried (MgSO$_4$), filtered, and concentrated under reduced pressure. The residue was purified by flash column chromatography (silica gel, hexanes/EtOAc 30:1) to give **54**; yield: 9.9 mg; and **55**; yield: 8.2 mg; combined yield: 95%.

1.5.1.2.2.4 α-Hydroxy Ketone Rearrangements

The α-hydroxy ketone rearrangement provides a complementary method to base-catalyzed carbocyclic rearrangement because it usually results in ring expansion of smaller and common rings to medium and large rings. An extensive volume of information on this classic organic transformation has been compiled by Paquette.[118] The rearrangement usually results in high yields with excellent regio- and stereoselectivity. However, a major limitation of the process is the use of nucleophilic hydroxides and alkoxides.

A recent application of the α-hydroxy ketone rearrangement in natural product synthesis is shown in Scheme 20. The cyclopenta[b]benzofuran core **58** of (−)-silvestrol (**59**) is obtained cleanly in high yield with excellent regioselectivity from regioisomeric α-hydroxy ketones **56** and **57** (obtained by photoinduced [3+2] cycloaddition, not shown). Ester **58** can then be further elaborated into the natural product **59**.[119]

for references see p 265

Scheme 20 Application of the α-Hydroxy Ketone Rearrangement in the Synthesis of the Cyclopenta[b]benzofuran Core of (−)-Silvestrol[119]

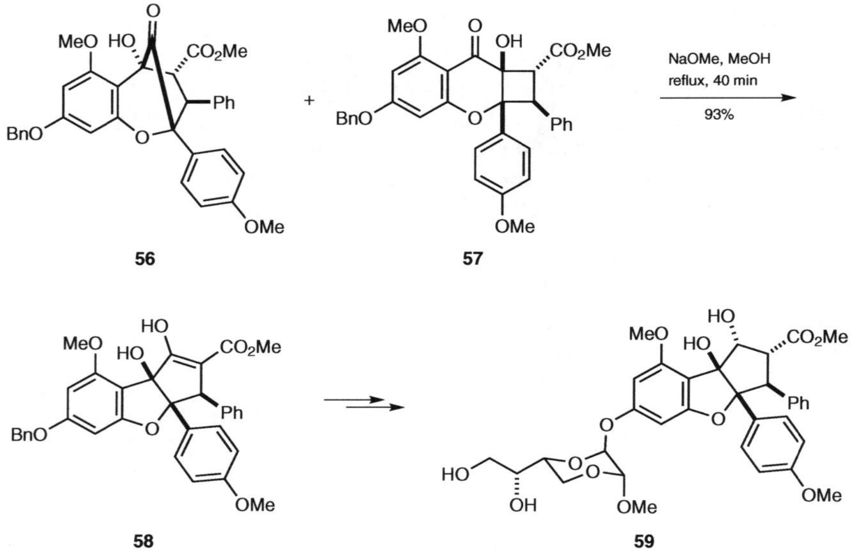

Methyl (3aR,8bS)-6-(Benzyloxy)-1,8-dihydroxy-8-methoxy-3a-(4-methoxyphenyl)-3-phenyl-3a,8b-dihydro-3H-cyclopenta[b]benzofuran-2-carboxylate (58):[119]
A 0.5 M soln of NaOMe in MeOH (7.1 mL, 3.55 mmol) was added to a soln of **56/57** (716 mg, 1.21 μmol) in MeOH (35 mL). The mixture was heated to reflux for 40 min before sat. aq NH₄Cl was added. The usual workup with EtOAc gave the title compound as a brown glassy oil containing a mixture of keto–enol tautomers; yield: 664 mg (93%).

References

[1] Coquerel, Y.; Filippini, M.-H.; Bensa, D.; Rodriguez, J., *Chem.–Eur. J.*, (2008) **14**, 3078.

[2] Brooks, J. L.; Frontier, A. J., *J. Am. Chem. Soc.*, (2013) **135**, 19362.

[3] Tietze, L. F., *Chem. Rev.*, (1996) **96**, 115.

[4] Tietze, L. F.; Brasche, G.; Gericke, K. M., *Domino Reactions in Organic Synthesis*, Wiley-VCH: Weinheim, Germany, (2006); pp 11, 48.

[5] *Domino Reactions: Concepts for Efficient Organic Synthesis*, Tietze, L. F., Ed.; Wiley-VCH: Weinheim, Germany, (2014).

[6] Robinson, R., *J. Chem. Soc., Trans.*, (1917) **111**, 762.

[7] Johnson, W. S.; Gravestock, M. B.; McCarry, B. E., *J. Am. Chem. Soc.*, (1971) **93**, 4332.

[8] Cane, D. E.; Celmer, W. D.; Westley, J. W., *J. Am. Chem. Soc.*, (1983) **105**, 3594.

[9] Nakanishi, K., *Toxicon*, (1985) **23**, 473.

[10] Lee, M. S.; Qin, G.; Nakanishi, K.; Zagorski, M. G., *J. Am. Chem. Soc.*, (1989) **111**, 6234.

[11] Gallimore, A. R.; Spencer, J. B., *Angew. Chem. Int. Ed.*, (2006) **45**, 4406.

[12] Wong, F. T.; Hotta, K.; Chen, X.; Fang, M.; Watanabe, K.; Kim, C.-Y., *J. Am. Chem. Soc.*, (2015) **137**, 86.

[13] Zhu, Y.; Wang, Q.; Cornwall, R. G.; Shi, Y., *Chem. Rev.*, (2014) **114**, 8199.

[14] Katsuki, T.; Sharpless, K. B., *J. Am. Chem. Soc.*, (1980) **102**, 5974.

[15] Johnson, R. A.; Sharpless, K. B., In *Catalytic Asymmetric Synthesis*, Ojima, I., Ed.; Wiley: Hoboken, NJ, (1993); p 103.

[16] Katsuki, T.; Martin, V. S., *Org. React. (N. Y.)*, (1995) **48**, 1.

[17] Zhang, W.; Loebach, J. L.; Wilson, S. R.; Jacobsen, E. N., *J. Am. Chem. Soc.*, (1990) **112**, 2801.

[18] Jacobsen, E. N.; Zhang, W.; Muci, A. R.; Ecker, J. R.; Deng, L., *J. Am. Chem. Soc.*, (1991) **113**, 7063.

[19] Brandes, B. D.; Jacobsen, E. N., *J. Org. Chem.*, (1994) **59**, 4378.

[20] Chang, S.; Galvin, J. M.; Jacobsen, E. N., *J. Am. Chem. Soc.*, (1994) **116**, 6937.

[21] Tu, Y.; Wang, Z.-X.; Shi, Y., *J. Am. Chem. Soc.*, (1996) **118**, 9806.

[22] Wang, Z.-X.; Tu, Y.; Frohn, M.; Zhang, J.-R.; Shi, Y., *J. Am. Chem. Soc.*, (1997) **119**, 11224.

[23] Shi, Y., *Acc. Chem. Res.*, (2004) **37**, 488.

[24] Wong, O. A.; Shi, Y., *Chem. Rev.*, (2008) **108**, 3958.

[25] Aggarwal, V. K.; Badine, D. M.; Moorthie, V. A., In *Aziridines and Epoxides in Organic Synthesis*, Yudin, A. K., Ed.; Wiley-VCH: Weinheim, Germany, (2006); p 1.

[26] Henbest, H. B.; Nicholls, B., *J. Chem. Soc.*, (1959), 221.

[27] Waddell, T. G., *Tetrahedron Lett.*, (1985) **26**, 6277.

[28] Dolle, R. E.; Nicolaou, K. C., *J. Am. Chem. Soc.*, (1985) **107**, 1691.

[29] Näf, R.; Velluz, A.; Decorzant, R.; Näf, F., *Tetrahedron Lett.*, (1991) **32**, 753.

[30] Urano, H.; Enomoto, M.; Kuwahara, S., *Biosci., Biotechnol., Biochem.*, (2010) **74**, 152.

[31] Kato, N.; Kamitamari, M.; Naganuma, S.; Arita, H.; Takeshita, H.; *Heterocycles*, (1990) **30**, 341.

[32] Mohr, P. J.; Halcomb, R. L., *Org. Lett.*, (2002) **4**, 2413.

[33] Baldwin, J. E., *J. Chem. Soc., Chem. Commun.*, (1976), 734.

[34] Coxon, J. M.; Hartshorn, M. P.; Swallow, W. H., *Aust. J. Chem.*, (1973) **26**, 2521.

[35] Vilotijevic, I.; Jamison, T. F., *Angew. Chem. Int. Ed.*, (2009) **48**, 5250.

[36] Wang, T.-L.; Hu, X. E.; Cassady, J. M., *Tetrahedron Lett.*, (1995) **36**, 9301.

[37] Basabe, P.; Estrella, A.; Marcos, I. S.; Díez, D.; Lithgow, A. M.; White, A. J. P.; Williams, D. J.; Urones, J. G., *Synlett*, (2001), 153.

[38] Koert, U.; Stein, M.; Wagner, H., *Liebigs Ann.*, (1995), 1415.

[39] Chen, R.; Rowand, D. A., *J. Am. Chem. Soc.*, (1980) **102**, 6609.

[40] Díez, D.; Moro, R. F.; Lumeras, W.; Rodríguez, L.; Marcos, I. S.; Basabe, P.; Escarcena, R.; Urones, J. G., *Synlett*, (2001), 1335.

[41] Sabitha, G.; Sandeep, A.; Rao, A. S.; Yadav, J. S., *Eur. J. Org. Chem.*, (2013), 6702.

[42] Hartung, J.; Drees, S.; Greb, M.; Schmidt, P.; Svoboda, I.; Fuess, H.; Murso, A.; Stalke, D., *Eur. J. Org. Chem.*, (2003), 2388.

[43] Koert, U.; Wagner, H.; Stein, M., *Tetrahedron Lett.*, (1994) **35**, 7629.

[44] Iimori, T.; Still, W. C.; Rheingold, A. L.; Staley, D. L., *J. Am. Chem. Soc.*, (1989) **111**, 3439.

[45] Paterson, I.; Boddy, I.; Mason, I., *Tetrahedron Lett.*, (1987) **28**, 5205.

[46] Xiong, Z.; Corey, E. J., *J. Am. Chem. Soc.*, (2000) **122**, 9328.

[47] Nicolaou, K. C.; Prasad, C. V. C.; Somers, P. K.; Hwang, C.-K., *J. Am. Chem. Soc.*, (1989) **111**, 5330.

[48] Nicolaou, K. C.; Prasad, C. V. C.; Somers, P. K.; Hwang, C.-K., *J. Am. Chem. Soc.*, (1989) **111**, 5335.

[49] Mukai, C.; Ikeda, Y.; Sugimoto, Y.-i.; Hanaoka, M., *Tetrahedron Lett.*, (1994) **35**, 2179.
[50] Mukai, C.; Sugimoto, Y.-i.; Ikeda, Y.; Hanaoka, M., *Tetrahedron Lett.*, (1994) **35**, 2183.
[51] Mukai, C.; Sugimoto, Y.-i.; Ikeda, Y.; Hanaoka, M., *Tetrahedron*, (1998) **54**, 823.
[52] Mukai, C.; Yamaguchi, S.; Kim, I. J.; Hanaoka, M., *Chem. Pharm. Bull.*, (2001) **49**, 613.
[53] McDonald, F. E.; Wang, X.; Do, B.; Hardcastle, K. I., *Org. Lett.*, (2000) **2**, 2917.
[54] Bravo, F.; McDonald, F. E.; Neiwert, W. A.; Do, B.; Hardcastle, K. I., *Org. Lett.*, (2003) **5**, 2123.
[55] Morimoto, Y.; Nishikawa, Y.; Ueba, C.; Tanaka, T., *Angew. Chem. Int. Ed.*, (2006) **45**, 810.
[56] Hudrlik, P. F.; Holmes, P. E.; Hudrlik, A. M., *Tetrahedron Lett.*, (1988) **29**, 6395.
[57] Adiwidjaja, G.; Flörke, H.; Kirschning, A.; Schaumann, E., *Tetrahedron Lett.*, (1995) **36**, 8771.
[58] Heffron, T. P.; Jamison, T. F., *Org. Lett.*, (2003) **5**, 2339.
[59] Mori, Y.; Yaegashi, K.; Furukawa, H., *J. Am. Chem. Soc.*, (1996) **118**, 8158.
[60] Mori, Y., *Chem.–Eur. J.*, (1997) **3**, 849.
[61] Mori, Y.; Furuta, H.; Takase, T.; Mitsuoka, S.; Furukawa, H., *Tetrahedron Lett.*, (1999) **40**, 8019.
[62] Fujiwara, K.; Tokiwano, T.; Murai, A., *Tetrahedron Lett.*, (1995) **36**, 8063.
[63] Fujiwara, K.; Saka, K.; Takaoka, D.; Murai, A., *Synlett*, (1999), 1037.
[64] Tokiwano, T.; Fujiwara, K.; Murai, A., *Chem. Lett.*, (2000) **29**, 272.
[65] Chen, Y.; Jin, J.; Wu, J.; Dai, W.-M., *Synlett*, (2006), 1177.
[66] Mori, Y.; Yaegashi, K.; Furukawa, H., *J. Am. Chem. Soc.*, (1997) **119**, 4557.
[67] González, I. C.; Forsyth, C. J., *J. Am. Chem. Soc.*, (2000) **122**, 9099.
[68] Payne, G. B., *J. Org. Chem.*, (1962) **27**, 3819.
[69] Salomon, R. G.; Basu, B.; Roy, S.; Sachinvala, N. D., *J. Am. Chem. Soc.*, (1991) **113**, 3096.
[70] Murai, A.; Ono, M.; Masamune, T., *J. Chem. Soc., Chem. Commun.*, (1976), 864.
[71] Morimoto, Y.; Nishikawa, Y.; Takaishi, M., *J. Am. Chem. Soc.*, (2005) **127**, 5806.
[72] Sasaki, M.; Inoue, M.; Takamatsu, K.; Tachibana, K., *J. Org. Chem.*, (1999) **64**, 9399.
[73] Banwell, M. G.; Chand, S.; Savage, G. P., *Tetrahedron: Asymmetry*, (2005) **16**, 1645.
[74] Dounay, A. B.; Florence, G. J.; Saito, A.; Forsyth, C. J., *Tetrahedron*, (2002) **58**, 1865.
[75] Hoye, T. R.; Suhadolnik, J. C., *Tetrahedron*, (1986) **42**, 2855.
[76] Simpson, G. L.; Heffron, T. P.; Merino, E.; Jamison, T. F., *J. Am. Chem. Soc.*, (2006) **128**, 1056.
[77] Kirchen, R. P.; Sorensen, T. S.; Wagstaff, K. E., *J. Am. Chem. Soc.*, (1978) **100**, 5134.
[78] Marshall, J. A.; Ellison, R. H., *J. Org. Chem.*, (1975) **40**, 2070.
[79] Marshall, J. A.; Ellison, R. H., *J. Am. Chem. Soc.*, (1976) **98**, 4312.
[80] Uyehara, T.; Sugimoto, M.; Suzuki, I.; Yamamoto, Y., *J. Chem. Soc., Chem. Commun.*, (1989), 1841.
[81] Do Khac Manh, D.; Fetizon, M.; Lazare, S., *J. Chem. Soc., Chem. Commun.*, (1975), 282.
[82] Do Khac Manh, D.; Fetizon, M.; Kone, M., *Tetrahedron*, (1978) **34**, 3513.
[83] Wagner, G., *J. Russ. Phys.-Chem. Soc.*, (1899) **31**, 690.
[84] Meerwein, H., *Justus Liebigs Ann. Chem.*, (1914) **405**, 129.
[85] Pirrung, M. C., *J. Am. Chem. Soc.*, (1981) **103**, 82.
[86] Do Khac Manh, D.; Fetizon, M.; Flament, J. P., *Tetrahedron*, (1975) **31**, 1897.
[87] Takeda, K.; Shimono, Y.; Yoshii, E., *J. Am. Chem. Soc.*, (1983) **105**, 563.
[88] Ohfune, Y.; Shirahama, H.; Matsumoto, T., *Tetrahedron Lett.*, (1976), 2869.
[89] Peet, N. P.; Cargill, R. L.; Bushey, D. F., *J. Org. Chem.*, (1973) **38**, 1218.
[90] Nagamoto, Y.; Hattori, A.; Kakeya, H.; Takemoto, Y.; Takasu, K., *Chem. Commun. (Cambridge)*, (2013) **49**, 2622.
[91] Corey, E. J.; Danheiser, R. L.; Chandrasekaran, S., *J. Org. Chem.*, (1976) **41**, 260.
[92] Paquette, L. A.; Wang, H.-L., *J. Org. Chem.*, (1996) **61**, 5352.
[93] Marcos, I. S.; Cubillo, M. A.; Moro, R. F.; Díez, D.; Basabe, P.; Sanz, F.; Urones, J. G., *Tetrahedron Lett.*, (2003) **44**, 8831.
[94] Uyehara, T.; Onda, K.; Nozaki, N.; Karikomi, M.; Ueno, M.; Sato, T., *Tetrahedron Lett.*, (2001) **42**, 699.
[95] Uyehara, T.; Furuta, T.; Kabawawa, Y.; Yamada, J.; Kato, T.; Yamamoto, Y., *J. Org. Chem.*, (1988) **53**, 3669.
[96] Magnus, P.; Diorazio, L.; Donohoe, T. J.; Giles, M.; Pye, P.; Tarrant, J.; Thom, S. J., *Tetrahedron*, (1996) **52**, 14147.
[97] Wenkert, E.; Bookser, B. C.; Arrhenius, T. S., *J. Am. Chem. Soc.*, (1992) **114**, 644.
[98] Fenster, M. D. B.; Patrick, B. O.; Dake, G. R., *Org. Lett.*, (2001) **3**, 2109.
[99] Cohen, T.; Yu, L.-C.; Daniewski, W. M., *J. Org. Chem.*, (1985) **50**, 4596.
[100] Paquette, L. A.; Lanter, J. C.; Johnston, J. N., *J. Org. Chem.*, (1997) **62**, 1702.
[101] Trost, B. M.; Chen, D. W. C., *J. Am. Chem. Soc.*, (1996) **118**, 12541.

References

[102] Georgian, V.; Kundu, N., *Tetrahedron*, (1963) **19**, 1037.

[103] Patra, A.; Ghorai, S. K.; De, S. R.; Mal, D., *Synthesis*, (2006), 2556.

[104] Trost, B. M.; Hiemstra, H., *Tetrahedron*, (1986) **42**, 3323.

[105] Estieu, K.; Ollivier, J.; Salaün, J., *Tetrahedron Lett.*, (1996) **37**, 623.

[106] Schmidt, J.; Bauer, K., *Ber. Dtsch. Chem. Ges.*, (1905) **38**, 3737.

[107] Hatsui, T.; Wang, J.-J.; Takeshita, H., *Bull. Chem. Soc. Jpn.*, (1995) **68**, 2393.

[108] Favorskii, A., *J. Russ. Phys.-Chem. Soc.*, (1894) **26**, 559.

[109] Favorskii, A., *J. Prakt. Chem./Chem.-Ztg.*, (1895) **51**, 533.

[110] Smissman, E. E.; Hite, G., *J. Am. Chem. Soc.*, (1959) **81**, 1201.

[111] Büchi, G.; Egger, B., *J. Org. Chem.*, (1971) **36**, 2021.

[112] Elford, T. G.; Hall, D. G., *J. Am. Chem. Soc.*, (2010) **132**, 1488.

[113] Eaton, P. E.; Cole, T. W., *J. Am. Chem. Soc.*, (1964) **86**, 3157.

[114] Ross, A. G.; Townsend, S. D.; Danishefsky, S. J., *J. Org. Chem.*, (2013) **78**, 204.

[115] Li, Z.; Alameda-Angulo, C.; Quiclet-Sire, B.; Zard, S. Z., *Tetrahedron*, (2011) **67**, 9844.

[116] White, J. D.; Kim, J.; Drapela, N. E., *J. Am. Chem. Soc.*, (2000) **122**, 8665.

[117] Zhang, L.; Koreeda, M., *Org. Lett.*, (2002) **4**, 3755.

[118] Paquette, L. A.; Hofferberth, J. E., *Org. React. (Hoboken, NJ, U. S.)*, (2003) **62**, 477.

[119] Adams, T. E.; El Sous, M.; Hawkins, B. C.; Hirner, S.; Holloway, G.; Khoo, M. L.; Owen, D. J.; Savage, G. P.; Scammells, P. J.; Rizzacasa, M. A., *J. Am. Chem. Soc.*, (2009) **131**, 1607.

1.5.2   **Lewis Acid/Base Induced Reactions**

S.-H. Wang, Y.-Q. Tu, and M. Tang

General Introduction

The efficient construction of complex molecular skeletons is always a hot topic in organic synthesis, especially in the synthesis of natural products, where many cyclic structural motifs can be found. Under the assiduous efforts of synthetic chemists, more and more methodologies are being developed to achieve the construction of cyclic skeletons. In particular, the beauty and high efficiency of organic synthesis are expressed vividly among those methods realized through a domino strategy. Based on these important methodologies, selected Lewis acid/base induced domino reactions leading to ring expansion, contraction, and closure will be presented.

1.5.2.1   **Ring Expansions**

1.5.2.1.1   **Semipinacol Rearrangement of 2,3-Epoxy Alcohols and Their Derivatives**

A variety of Lewis acid induced semipinacol rearrangements of 2-heterosubstituted alcohols, allylic alcohols, 2,3-epoxy alcohols, and their derivatives are well documented in the chemical literature.[1–3] When applied to cyclic compounds, the sequence may lead to ring-expansion products, and usually the starting ring is expanded by one carbon atom.

Although Lewis acid (e.g., $TiCl_4$ and $SnCl_4$) promoted semipinacol rearrangements of 2,3-epoxy alcohols and derivatives have been widely reported, they generally start from the hydroxy-protected epoxides (e.g., epoxy silyl ethers) and/or need an equivalent or excess amount of Lewis acid. Tu and co-workers have reported that a catalytic amount of anhydrous zinc(II) bromide (2–8 mol%) can effectively promote the diastereoselective rearrangement of unprotected 2,3-epoxy alcohols **1** into β-hydroxy ketones **2A** and **2B** (Scheme 1).[4]

Scheme 1   Zinc(II) Bromide Catalyzed Rearrangement of 2,3-Epoxy Alcohols[4]

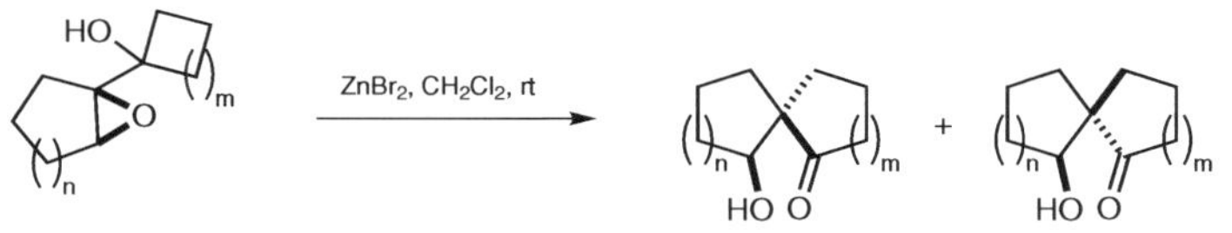

n	m	Ratio (**2A/2B**)	Yield (%)	Ref
1	1	57:43	94	[4]
2	2	>99:1	91	[4]
1	2	64:36	60	[4]
2	3	77:23	41	[4]
1	3	84:16	55	[4]
2	4	>99:1	83	[4]
1	4	72:28	68	[4]

for references see p 351

270 Domino Transformations **1.5** Non-Radical Skeletal Rearrangements

From Scheme 1, it can be seen that most of the substrates give satisfactory results with good yields. In particular, some substrates exhibit excellent stereoselectivity and give diastereomerically pure 2,3-*syn*-β-hydroxy ketones. It appears that the substrates **1** (n = 2) with a cyclohexene oxide moiety afford substantially better results than the corresponding substrates **1** (n = 1) containing a cyclopentene oxide moiety. Overall, though, a series of spirane skeletons with various ring sizes are successfully constructed.

Using a semipinacol rearrangement of enamide-derived 2,3-epoxy silyl ether **3** as a key step, Dake and co-workers have completed a formal total synthesis of the polycyclic alkaloid fasicularin (Scheme 2).[5–7] Treatment of epoxide **3** with titanium(IV) chloride leads to the desired ring-expanded product **4** in 96% yield with clean stereochemical control of the quaternary carbon center formed at the B,C-ring junction. Subsequent transformations, including methanesulfonylation, elimination (to give **5**), and enolization, yield vinyl trifluoromethanesulfonate **6**, which undergoes a Suzuki coupling reaction with borane **7** to give compound **8**. An eventual intramolecular S_N2 reaction completes the final ring of fasicularin.

Scheme 2 Formal Total Synthesis of Fasicularin[5–7]

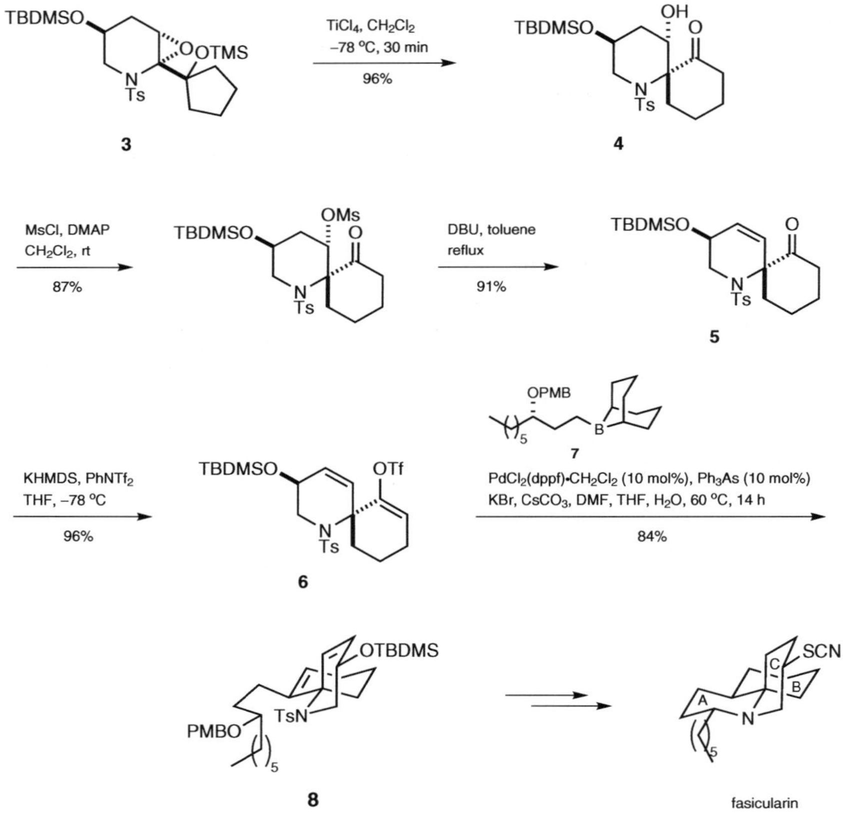

Although this domino sequence has usually been applied for the formation of four- to seven-membered rings from the corresponding three- to six-membered rings, expansion from an eight- to a nine-membered ring is also possible and has been used by the Tu group in the synthesis of (+)-alopecuridine, (+)-sieboldine A, and (−)-lycojapodine A (Scheme 3).[8,9] In this case, coupling of bromoalkene **9** and ketone **10** through the intermediacy of the vinylcerium species generated from the lithium salt of **9**, followed by direct epoxidation,

1.5.2 Lewis Acid/Base Induced Reactions

271

gives 2,3-epoxy alcohol **11** (dr 6:1) in 71% yield from **9**. Upon treatment of epoxide **11** with boron trifluoride–diethyl ether complex, the semipinacol rearrangement takes place to produce ketone **12** in 51% yield and in >99% ee. Conversion of enantiopure ketone **12** into (+)-alopecuridine (as its TFA salt) is then accomplished in six further steps. The biomimetic oxidation of the (+)-alopecuridine salt to (+)-sieboldine A can be achieved in a one-pot manner with 3-chloroperoxybenzoic acid and mercury(II) oxide in 60% yield. With manganese(IV) oxide as the oxidant, the biomimetic synthesis of (−)-lycojapodine A has also been completed in 82% yield.

Scheme 3   Total Syntheses of (+)-Alopecuridine, (+)-Sieboldine A, and (−)-Lycojapodine A[8,9]

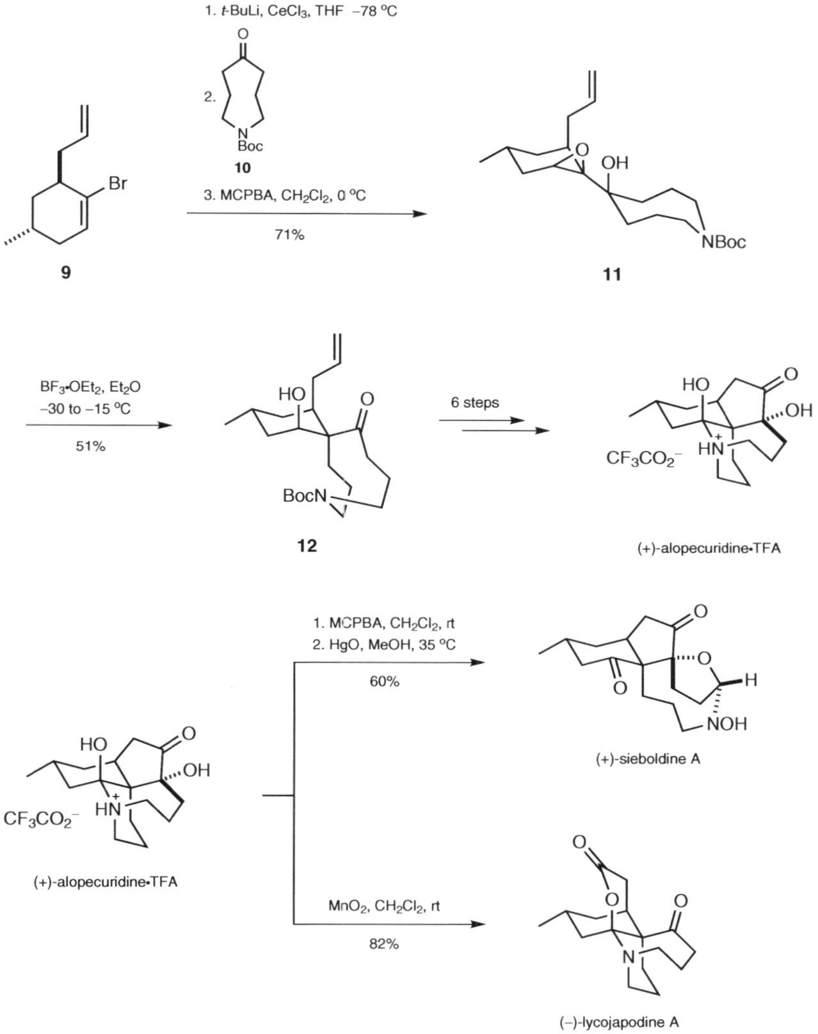

Spirocyclic β-Hydroxy Ketones 2A and 2B; General Procedure:[4]

ZnBr$_2$ (2–8 mol%) was added to a soln of 2,3-epoxy alcohol **1** (1 mmol) in dry CH$_2$Cl$_2$ (10 mL) under argon. The mixture was stirred at rt and monitored by TLC until the starting material disappeared. The mixture was partitioned with H$_2$O, and the organic phase was dried (MgSO$_4$) and concentrated under reduced pressure. The residue was purified by column chromatography (silica gel).

for references see p 351

1.5.2.1.2 Reductive Rearrangement of 2,3-Epoxy Alcohols with Aluminum Triisopropoxide

The preparation of spirocyclic diols has attracted increasing interest from organic chemists, possibly because some chiral spirocyclic diols have been used as ligands for asymmetric catalysis. Normally, such diols are prepared by reducing the corresponding spirocyclic dione, a process which typically is not very effective. Therefore, an efficient one-step approach to stereospecific spirocyclic diols from 2,3-epoxy alcohols has been developed. Upon treatment with aluminum triisopropoxide, racemic 2,3-epoxy alcohols **13** are transformed into the ring-expanded spirocyclic diol products **14A** and **14B** in one simple operation through a reductive rearrangement process (Scheme 4).[10,11]

Scheme 4 Synthesis of Spirocyclic Diols from 2,3-Epoxy Alcohols with Aluminum Triisopropoxide[10,11]

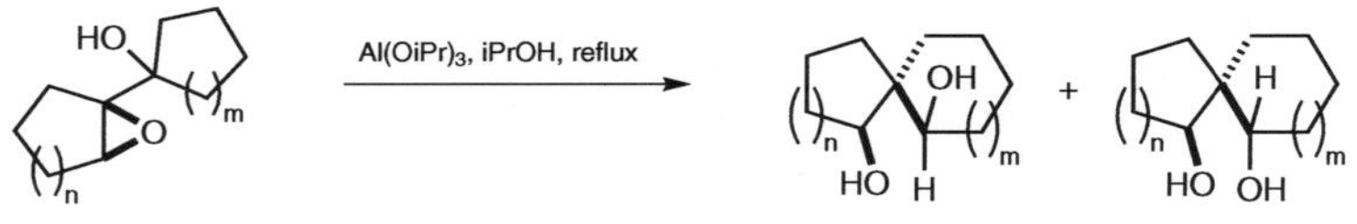

n	m	Ratio (**14A/14B**)	Yield (%)	Ref
2	1	>99:1	98	[10,11]
1	1	68:32	80	[10]
2	2	>99:1	60	[10]
1	2	76:24	50	[10]
2	3	>99:1	98	[10]
1	3	>99:1	77	[10]

A key strength of this methodology is the stereoselective construction of three contiguous carbon stereocenters, with one being quaternary. Based on the stereochemical outcome and deuteration experiments, a mechanism involving a semipinacol rearrangement and a subsequent Meerwein–Ponndorf–Verley reduction was proposed (Scheme 5).[11] In this transformation, aluminum triisopropoxide serves a dual role as both a Lewis acid to promote the semipinacol rearrangement and as a reducing agent for hydride transfer. Of the two possible pathways for hydride transfer, path a is more favorable in terms of steric hindrance than path b, and affords the C1–C2 *syn*-diastereomer **14A** as the major product.

1.5.2 Lewis Acid/Base Induced Reactions **273**

Scheme 5 Proposed Mechanism for the Reductive Rearrangement of 2,3-Epoxy Alcohols with Aluminum Triisopropoxide[11]

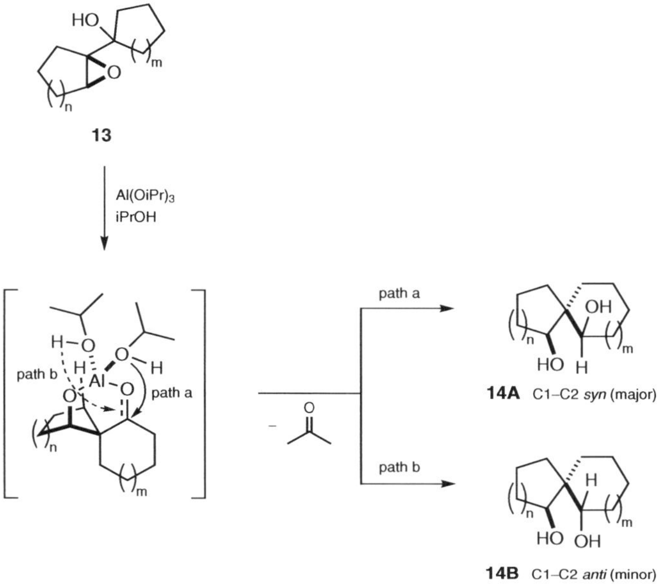

(1R*,6S*,7S*)-Spiro[5.5]undecane-1,7-diol (14A, n = 2; m = 1); **Typical Procedure:**[11]
A mixture of racemic 2,3-epoxy alcohol **13** (n = 2; m = 1; 0.5 g, 2.7 mmol), Al(OiPr)$_3$ (1.1 g, 5.4 mmol), and iPrOH (10 mL) was refluxed with stirring under argon for 4 h. The solvent was removed under reduced pressure and the gel residue obtained was partitioned between Et$_2$O and 10% aq NaOH. The aqueous layer was separated and re-extracted with Et$_2$O. The combined Et$_2$O layers were washed with H$_2$O and brine and then dried (Na$_2$SO$_4$). The solvent was evaporated under reduced pressure, and the residue was chromatographed (silica gel); yield: 0.49 g (98%).

1.5.2.1.3 **Tandem Semipinacol/Schmidt Reaction of α-Siloxy Epoxy Azides**

C-Aza quaternary center units are widespread in alkaloids, with the nitrogen atom often located at a highly sterically congested tertiary carbon. Tu and co-workers have designed and successfully developed a tandem semipinacol/Schmidt reaction of α-siloxy epoxy azides **15** for the efficient construction of many such aza quaternary carbon units.[12] In most cases, a new pyrrolidine ring is formed and the starting ring is usually expanded by two atoms (including a nitrogen atom). The reaction likely proceeds through the intermediate **16**, where chelation of titanium(IV) with two oxygen atoms induces epoxide ring opening and a synchronous antiperiplanar migration that leads to the diastereoselective formation of intermediate **17**. Next, the subsequent antiperiplanar migration of the quaternary carbon in intermediate **18** to the electron-deficient nitrogen proceeds with retention of configuration to provide the final polycyclic product **19** (Scheme 6).

for references see p 351

274 Domino Transformations 1.5 Non-Radical Skeletal Rearrangements

Scheme 6 Proposed Mechanism for a Tandem Semipinacol/Schmidt Domino Process[12]

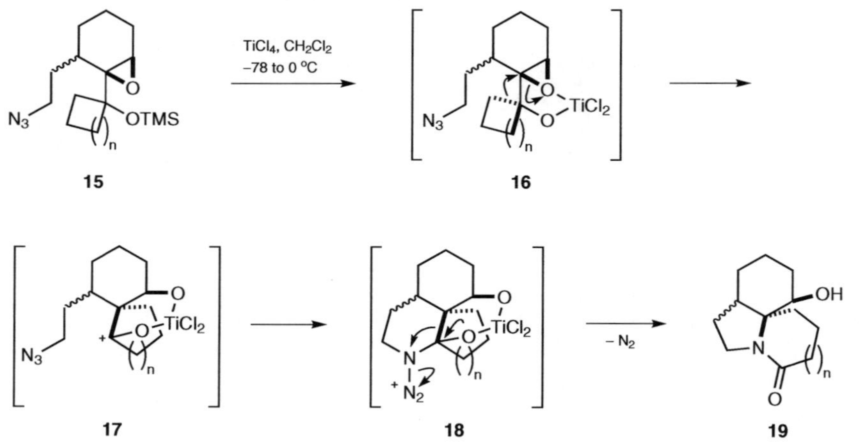

This domino process is effective for a wide range of substrates and is a powerful mode of construction for functionalized *C*-aza quaternary carbon centers. Moreover, the final reaction mixture is typically clean and readily purified; more importantly, the stereochemistry of this procedure is well controlled (Table 1).[12]

Table 1 Titanium(IV) Chloride Promoted Tandem Reaction of α-Siloxy Epoxy Azides[12]

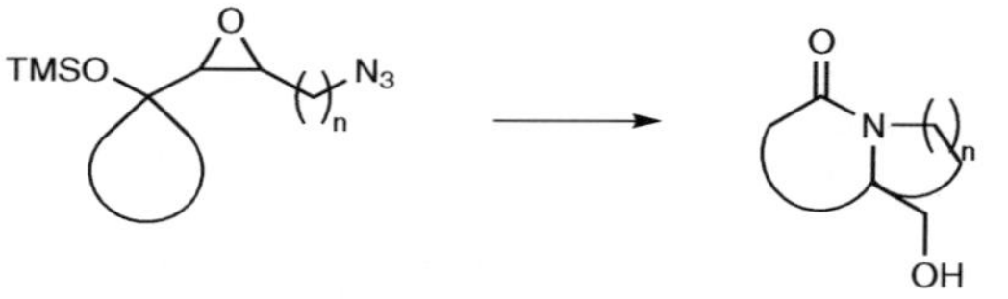

Substrate	Product	Yield (%)	Ref
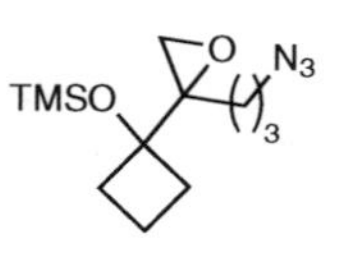	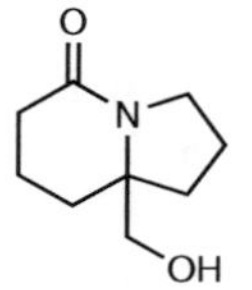	77	[12]
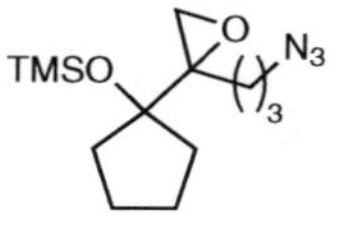	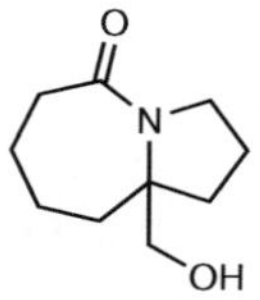	67	[12]
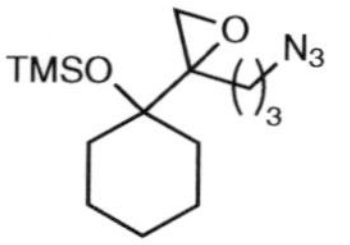	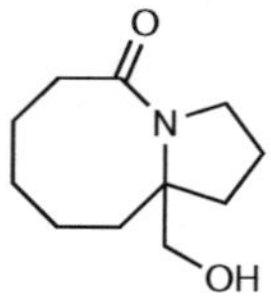	66	[12]

Table 1 (cont.)

Substrate	Product	Yield (%)	Ref
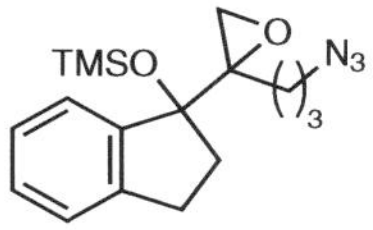	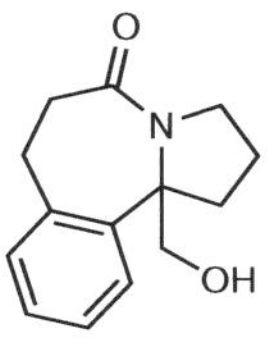	61	[12]
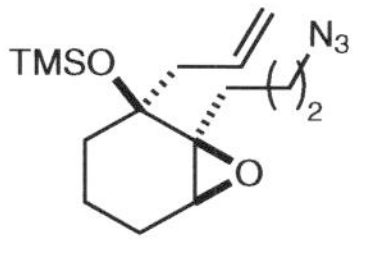	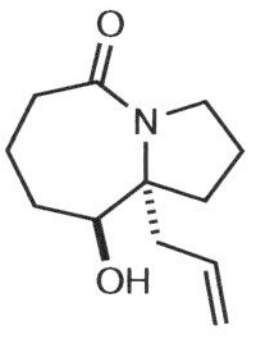	92	[12]
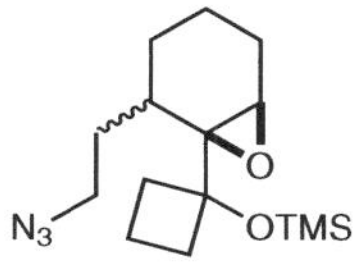	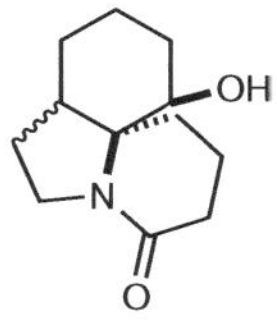	78	[12]
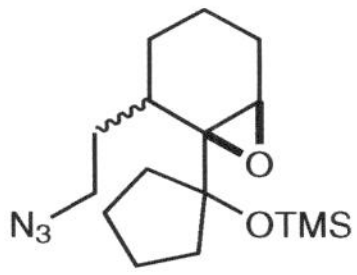	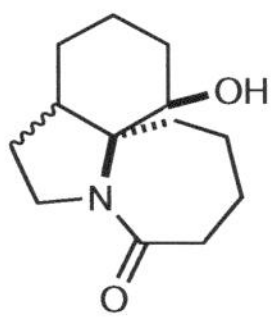	92	[12]

In one application of this method, the first total synthesis of (±)-stemonamine was achieved (Scheme 7).[13] Here, treating epoxy azide **21** with titanium(IV) chloride in dichloromethane at −78 to 0 °C transforms it into bicyclic lactam **24** as a single isomer in 68% yield via a tandem semipinacol/Schmidt process. Successive, stereospecific 1,2-migrations, as shown in the chelate transition states **22** and **23**, ensure the formation of the desired ring skeleton. Subsequent pyridinium chlorochromate oxidation, ozonolysis, and an aldol reaction generate tricyclic lactam **25**, which is then converted into (±)-stemonamine in just a few steps. In a similar way, (±)-cephalotaxine, (±)-maistemonine, (±)-stemonamide, and (±)-isomaistemonine have also been synthesized using this key domino process.[14–17]

for references see p 351

Scheme 7 Total Synthesis of (±)-Stemonamine[13]

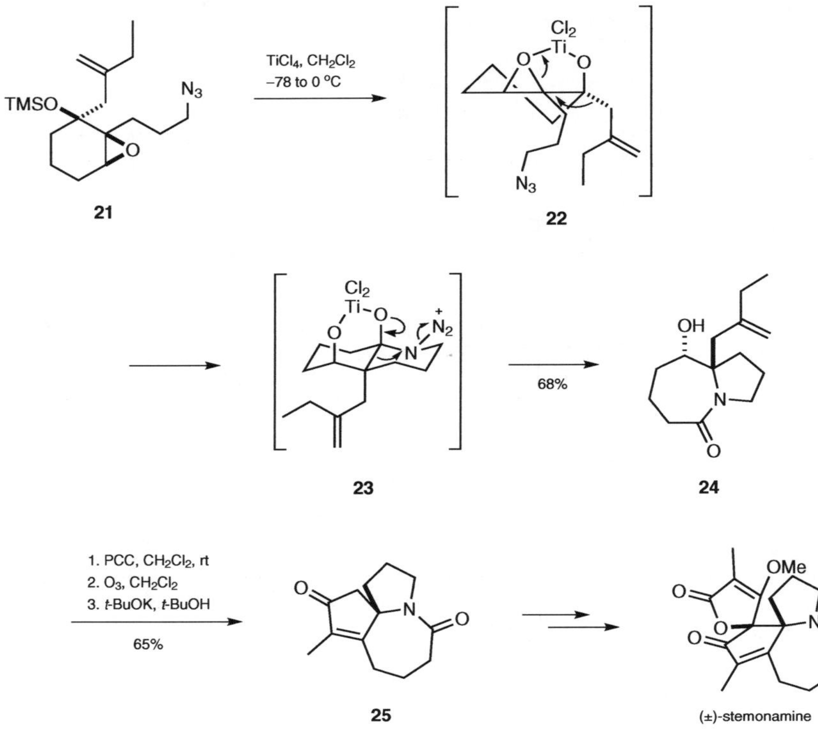

Fused (Hydroxymethyl)lactams 20; General Procedure:[12]

To a cooled (−78 °C) soln of the α-siloxy epoxy azide starting material (0.3 mmol) in CH₂Cl₂ (3 mL) under argon was added 1 M TiCl₄ in CH₂Cl₂ (0.66 mL, 0.66 mmol). The resultant mixture was stirred at −78 °C for 15 min and then warmed slowly to 25 °C for additional time before the reaction was quenched with H₂O (1 mL). The mixture was partitioned between H₂O and CH₂Cl₂ (50 mL). The organic layer was collected, dried (Na₂SO₄), filtered, and concentrated to give a crude product, which was purified by chromatography (silica gel).

1.5.2.1.4   **Prins–Pinacol Rearrangement**

The Overman group has developed an outstanding sequential series of acid-catalyzed cationic cyclization–rearrangements, which have been used to synthesize stereochemically complex substituted pyrrolidines,[18–20] substituted tetrahydrofurans,[21–24] and carbocyclic rings.[25,26] As a potentially useful annulation sequence, when cyclic substrates of type **26** are used, the rearrangements result in an unusual reaction in which elaboration of a new five-membered ring is coupled with a one-carbon ring enlargement of the starting carbocyclic ring (Scheme 8).

Scheme 8   Sequential Acid-Catalyzed Cationic Cyclization–Rearrangement

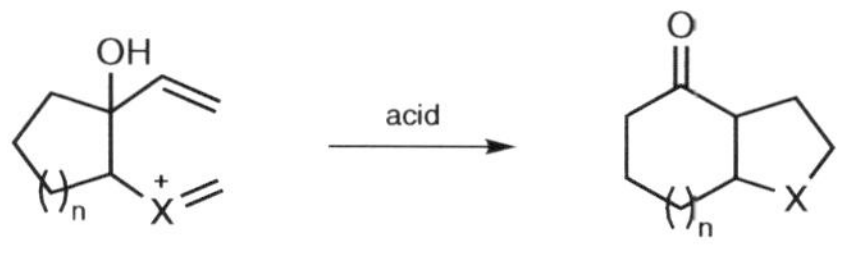

The coupling of a Prins cyclization with a pinacol rearrangement leads to the development of a useful family of reactions for forming five-membered rings. For this process, the synthesis of tetrahydrofurans is most extensively developed, having been shown to solve stereochemical problems in the assembly of complex cyclic ethers. When the starting allylic diol is cyclic, a ring-enlarging furan annulation leading to materials of type **27** can be realized, providing convenient access to a variety of *cis*-1-oxabicyclic ring systems (Table 2).[22] A variety of substituted *cis*-fused octahydrocyclohepta[*b*]furans and octahydrobenzofurans can be prepared in a stereocontrolled fashion from 1-alkenylcyclohexane-1,2-diol or 1-alkenylcyclopentane-1,2-diol precursors. In all cases, only a single bicyclic product is isolated (Table 2).[22]

for references see p 351

Domino Transformations 1.5 Non-Radical Skeletal Rearrangements

Table 2 Ring-Enlarging Tetrahydrofuran Annulations of Fused Acetals[22]

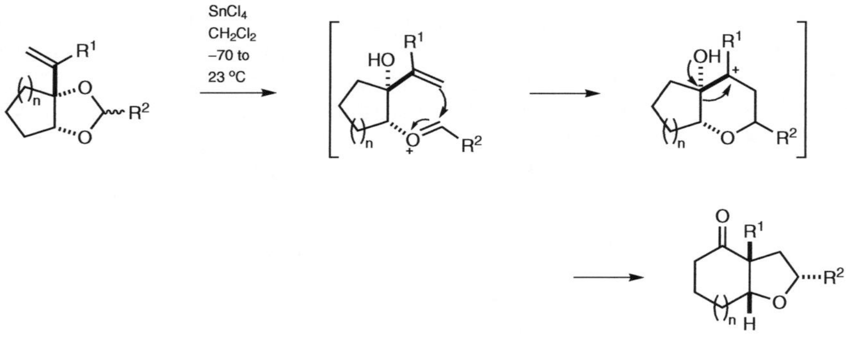

Entry	Substrate	Product	Yield (%)	Ref
1	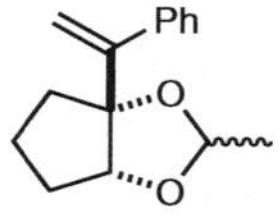	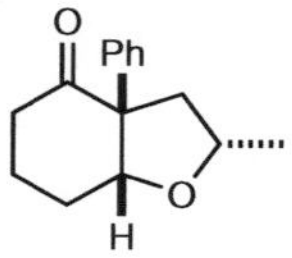	80	[22]
2	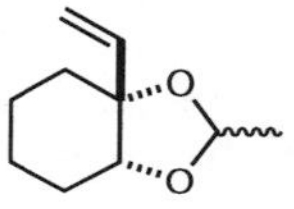	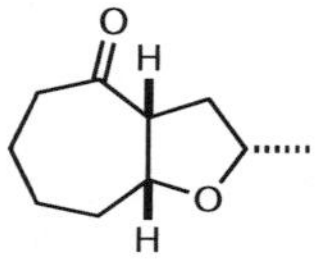	74	[22]
3	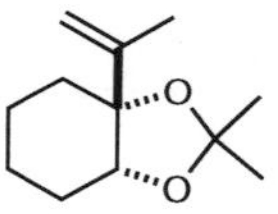	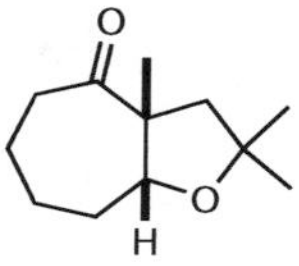	94	[22]
4	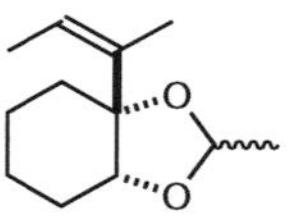	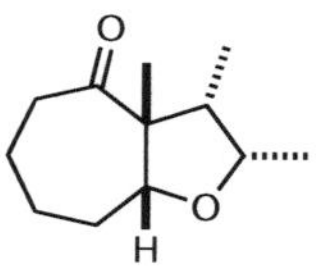	56	[22]
5	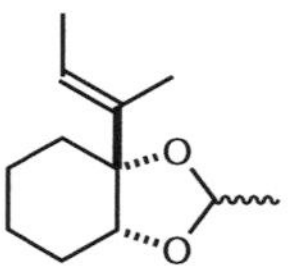	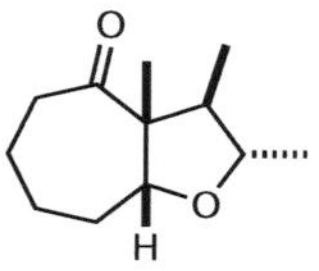	81	[22]

Based on the above mechanistic paradigm, the Overman group has developed what they term a "ring-enlarging cyclopentane annulation." Bicyclic and tricyclic ring systems containing five-, six-, seven-, and eight-membered carbocyclic rings can be assembled efficiently with high stereocontrol; moreover, *cis*-fused hydroindanes, hydroazulenes, and bicyclo[6.3.0]undecanes that contain functionality within both carbocyclic rings can be prepared readily, as shown in Table 3.[25]

1.5.2 Lewis Acid/Base Induced Reactions

279

Table 3 Stereocontrolled Construction of Carbocyclic Rings by Sequential Cationic Cyclization–Pinacol Rearrangements[25]

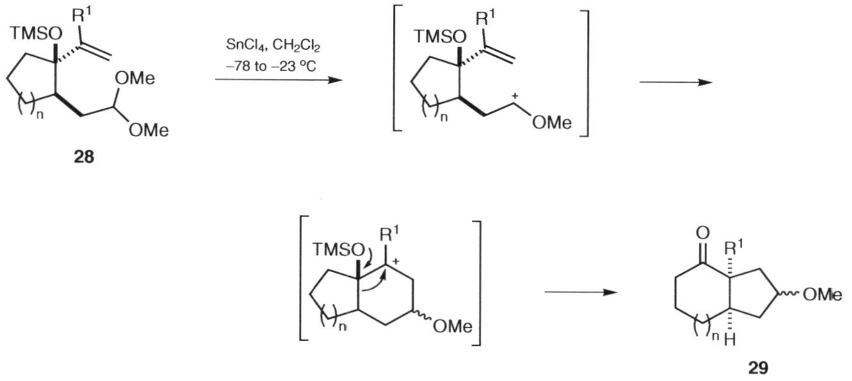

Entry	Substrate	Product	dr	Yield (%)	Ref
1	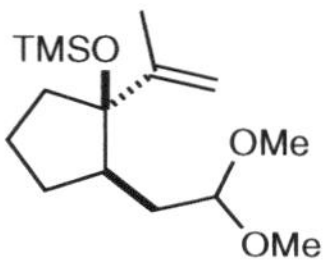	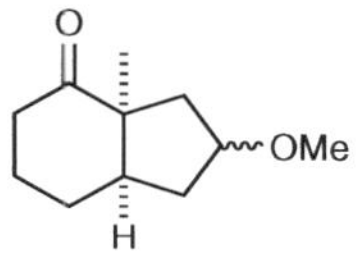	4:1	82	[25]
2	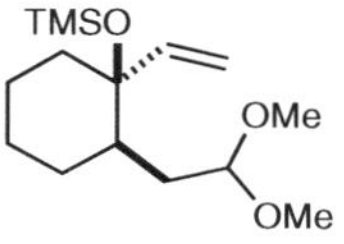	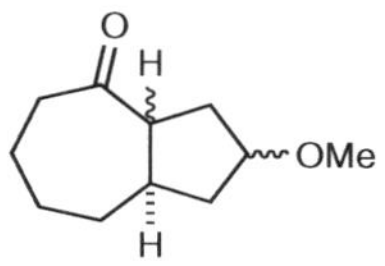	50:30:10:7	80	[25]
3	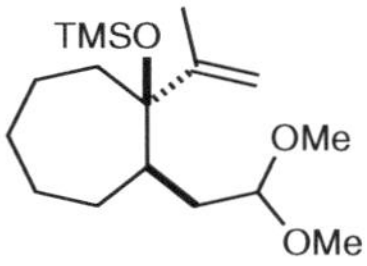	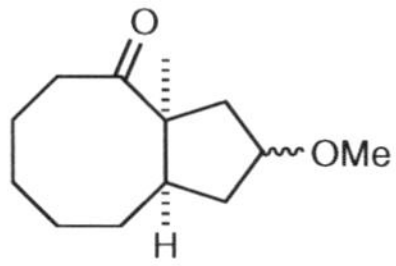	1:5	75	[25]
4	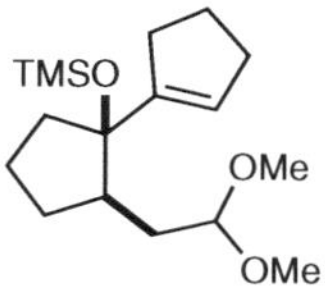	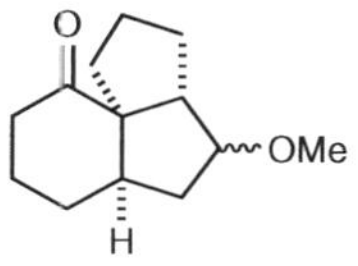	2:1	70	[25]

In the synthesis of (±)-kumausallene,[27] boron trifluoride–diethyl ether complex promotes condensation of diol **30** and 2-(benzoyloxy)acetaldehyde to deliver hydrobenzofuran **33** through transition states **31** and **32** in 71% yield. In this key step, a new five-membered ring is formed and the starting cyclopentane ring is expanded by one carbon. Next, a three-step sequence of standard reactions (Baeyer–Villiger oxidation, selenation, and selenoxide elimination) transforms ketone **33** into α,β-unsaturated lactone **34**. Treatment of the new product **34** with sodium methoxide then affects methanolysis of the lactone. Subsequent 1,4-addition of the resulting C7 alkoxide to the tethered α,β-unsaturated methyl ester and cleavage of the benzoyl protecting group provides *cis*-dioxabicyclo[3.3.0]octane **35** with high stereoselectivity. This compound is then converted into

for references see p 351

280 Domino Transformations **1.5** Non-Radical Skeletal Rearrangements

(±)-kumausallene in several steps (Scheme 9). By modifying the manner in which the hydrobenzofuranone intermediate **33** is elaborated, (±)-kumausyne can be synthesized in a similar way.[28]

Scheme 9 Total Synthesis of (±)-Kumausallene[27]

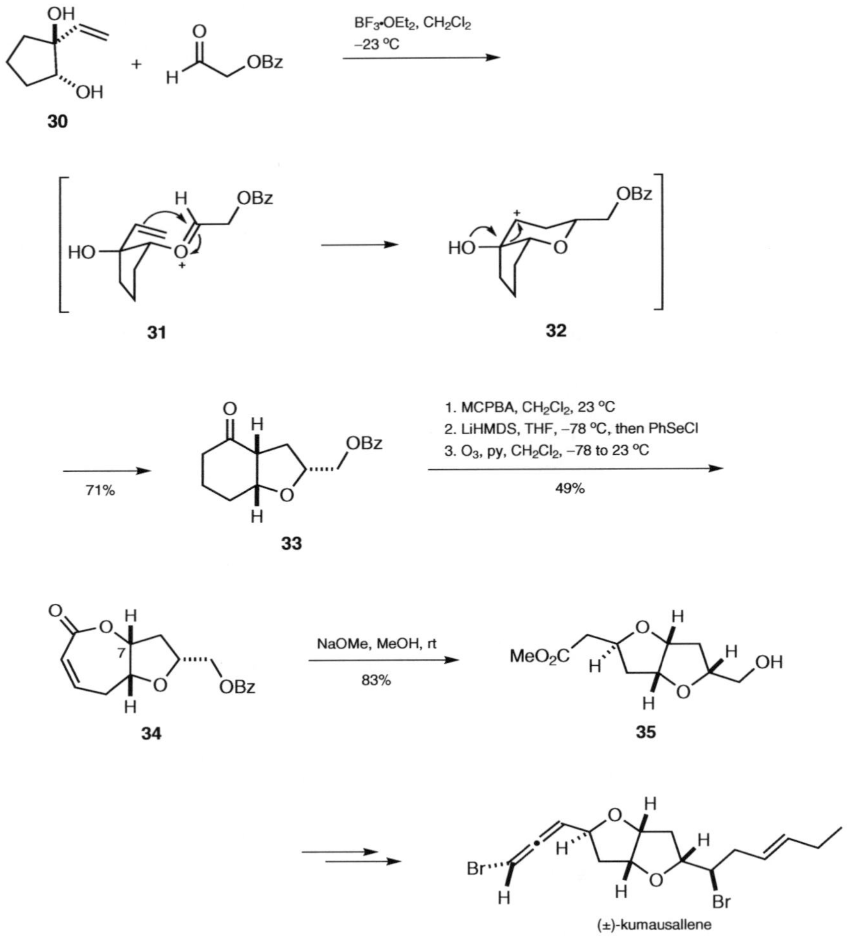

Overman and co-workers have applied the Prins–pinacol reaction of siloxy dienyl acetal **36** in the synthesis of (−)-magellanine and (+)-magellaninone as shown in Scheme 10.[29] Here, treatment of siloxy dienyl acetal **36** with 1.1 equivalents of tin(IV) chloride in dichloromethane provides carbotetracycle **37** in 57% yield as a 2:1 mixture of β- and α-methoxy epimers. Critically, this pivotal conversion establishes five of the six stereocenters of magellanine with complete stereocontrol. Oxidative cleavage of the cyclopentane ring followed by double reductive amination then furnishes the desired azatetracycle **38** in 60% overall yield from **37**.

1.5.2 Lewis Acid/Base Induced Reactions

281

Scheme 10 Enantioselective Total Syntheses of (−)-Magellanine and (+)-Magellaninone[29]

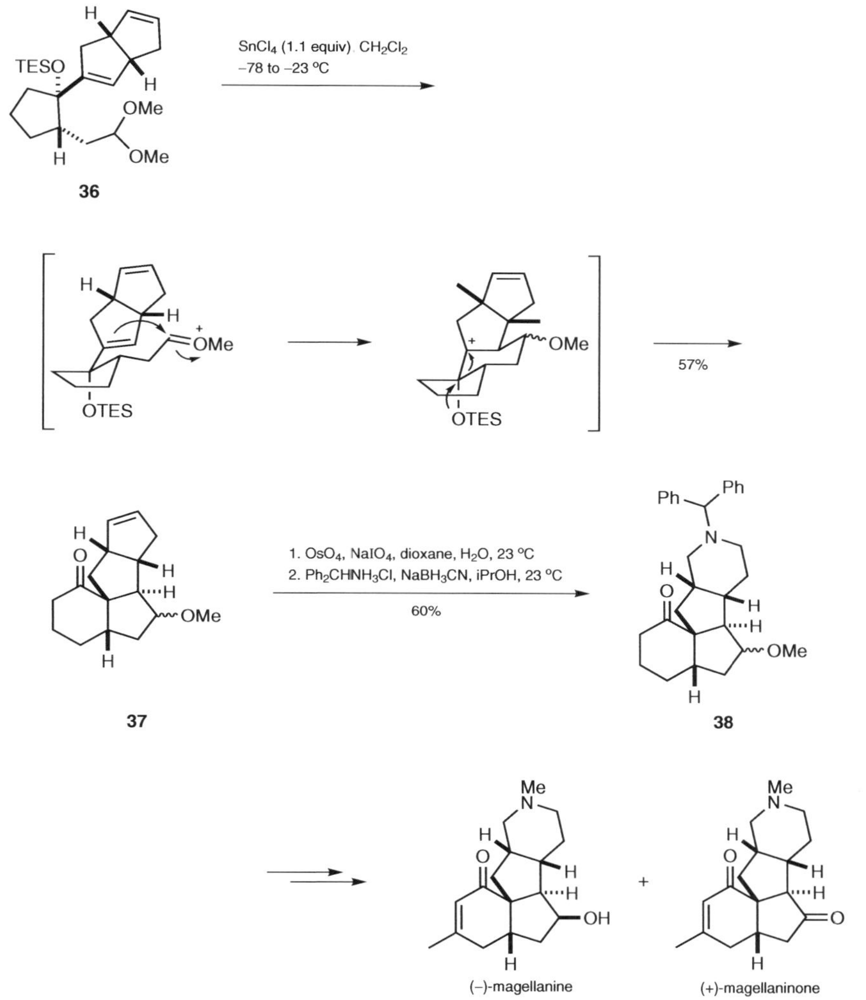

(−)-magellanine

(+)-magellaninone

(2S*,3aR*,8aR*)-2-Methyloctahydro-4H-cyclohepta[b]furan-4-one (Table 2, Entry 2); Typical Procedure:[22]

To a soln of 2-methyl-3a-vinylhexahydrobenzo[d][1,3]dioxole (43 mg, 0.26 mmol) in CH_2Cl_2 (1.5 mL) cooled to −70 °C in a dry ice/acetone bath was added $SnCl_4$ (66.4 mg, 0.255 mmol) and the mixture was stirred at −70 °C for 1 h. Et_3N (0.5 mL) was added followed by MeOH (1 mL) and, after warming to rt, the resulting mixture was poured into EtOAc (10 mL) and 1 M aq HCl (5 mL). The organic phase was separated, washed with sat. aq $NaHCO_3$ (5 mL), dried ($MgSO_4$), and concentrated to provide a yellow oil. The crude product was purified by chromatography [silica gel (230–400 mesh; 5 g), 15-mm o.d., Et_2O/hexane 1:4, 2-mL fractions] to give a colorless oil; yield: 32 mg (74%).

Fused Cyclopentanes 29 (n = 1–3); General Procedure:[25]

The l-alkenyl-2-(2,2-dimethoxyethyl)-l-(trimethylsiloxy)cycloalkane **28** (0.5 mmol) was dissolved in CH_2Cl_2 (1.5 mL) and cooled to −78 °C under an argon atmosphere. $SnCl_4$ (0.5 mmol) was added and the soln was stirred at −78 °C for 5 min. The soln was then warmed to −23 °C and was generally maintained at −23 °C for ca. 5 min before being

for references see p 351

quenched with Et$_3$N (0.5 mL) and then MeOH (1 mL). The mixture was then partitioned between 1 M HCl (10 mL) and EtOAc (30 mL), and the organic phase was washed with sat. aq NaHCO$_3$ (10 mL) and brine (10 mL). Concentration and purification of the residue by flash column chromatography (silica gel, EtOAc/hexane) gave the rearrangement product **29**.

1.5.2.2 **Ring Contractions**

1.5.2.2.1 **Rearrangement of Epoxides**

Lewis acid promoted rearrangement of epoxides, via 1,2-migration induced by an epoxide ring opening, has been shown to have wide utility in natural product synthesis. When the epoxide is within a cyclic substrate, the rearrangement may lead to a ring-contraction product in a highly efficient manner (Scheme 11).[30,31] Often, the presence of an electron-withdrawing group in the substrate directs the epoxide opening, allowing a good level of regioselectivity in the reaction.

Scheme 11 Synthesis of Carbonyls by Rearrangement of Cyclic Epoxides[30,31]

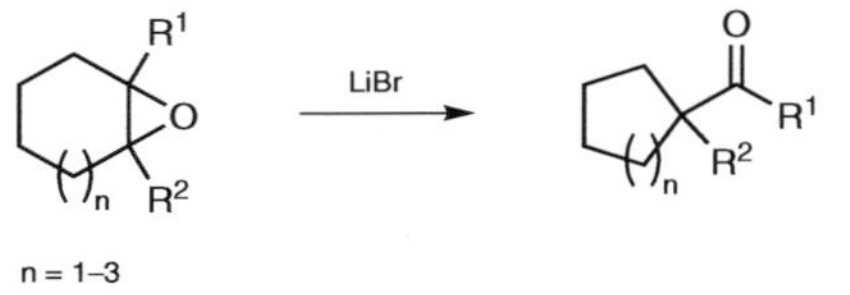

n = 1–3

The ring contraction of readily available cyclohexene oxides leads to cyclopentanecarbaldehydes or cyclopentyl ketones in good yields. These in situ generated aldehydes or ketones can further react with other reagents to give more complex cyclopentyl units. Tu and co-workers have reported the tandem rearrangement/reduction reactions of α-hydroxy epoxides promoted by samarium(II) iodide (Scheme 12)[32]or triethylaluminum (Table 4)[33–35] to construct 2-quaternary 1,3-diol units, which are widespread motifs within natural products.

Scheme 12 Samarium(II) Iodide Promoted Tandem Rearrangement/Reduction of α-Hydroxy Epoxides[32]

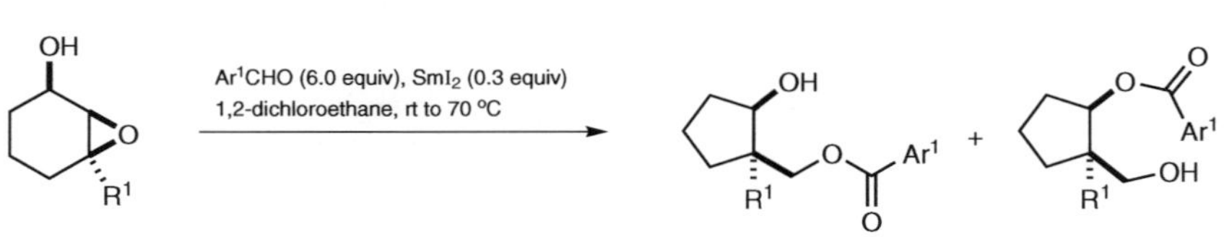

As shown in Table 4, triethylaluminum promoted tandem rearrangement/reduction of α-hydroxy (or α-benzyloxy) epoxides is applicable and versatile in the synthesis of natural products containing five- and/or six-membered-ring structures bearing a quaternary carbon stereocenter.

Table 4 Triethylaluminum Promoted Stereoselective Tandem Rearrangement/Reduction of α-Hydroxy or α-Benzyloxy Epoxides[33–35]

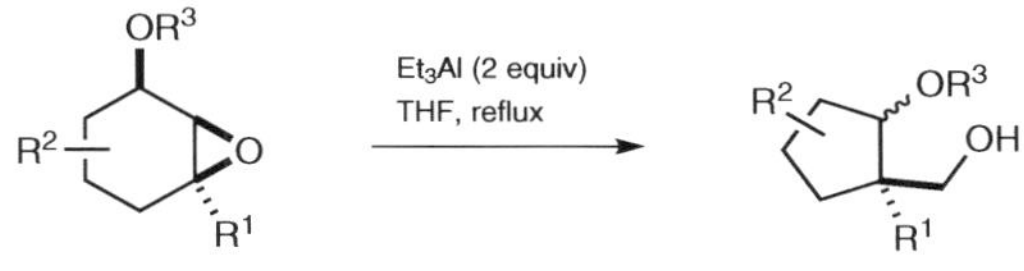

Entry	Substrate	Product	Yield (%)	Ref
1	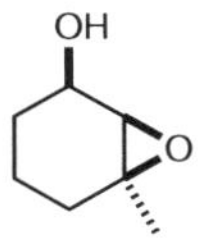	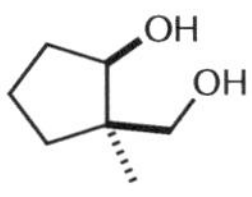	58	[34,35]
2	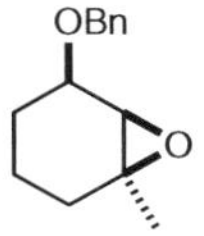	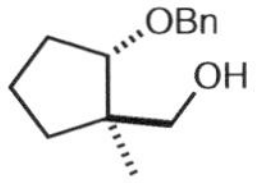	92	[33,35]
3	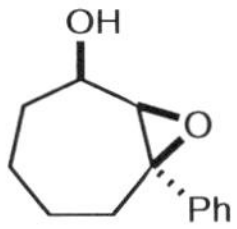	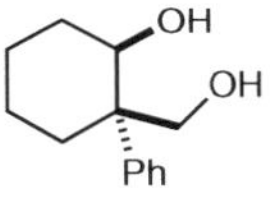	84	[34,35]
4	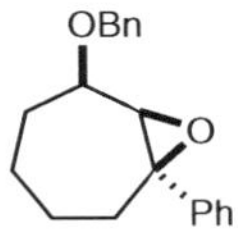	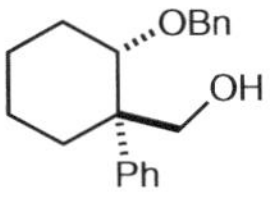	80	[33,35]

A particularly remarkable feature of this reaction is the high level of diastereoselectivity. Additionally, the protecting group on the hydroxy group affects the migration pattern. For instance, if unprotected substrates are used (entries 1 and 3), the *cis*-diol products are obtained. This outcome can be explained by means of the mechanism shown in Scheme 13. By contrast, *trans*-1,3-diols are obtained with benzyl-protected substrates under the same conditions (entries 2 and 4). Even though the mechanism of the transformation of benzyl-protected substrates remains undefined, the relative configuration of the product can be controlled by this strategy.

for references see p 351

Scheme 13 Proposed Mechanism for the Tandem Rearrangement/Reduction Reaction of α-Hydroxy Epoxides[34]

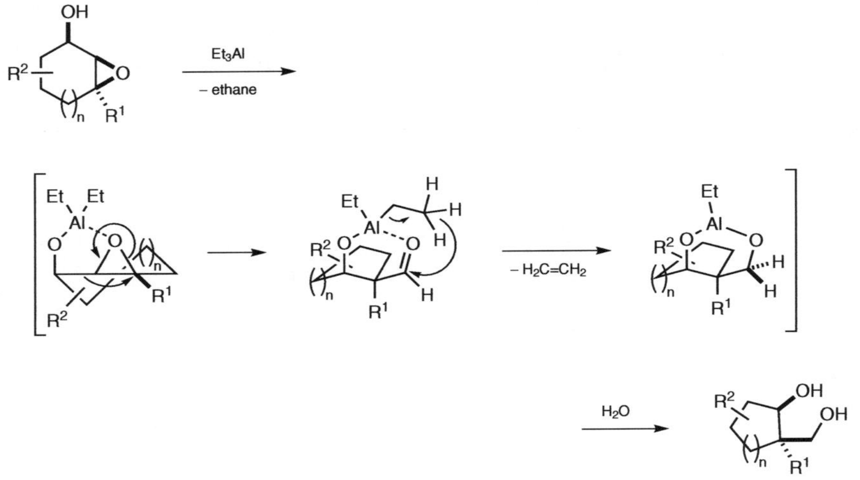

A samarium(II) iodide promoted rearrangement of compounds **39** and **40** allows for the total synthesis of the quaternary-carbon-containing aromatic sesquiterpenoids (±)-cuparene and (±)-herbertene to be achieved (Scheme 14).[36]

Scheme 14 Total Synthesis of (±)-Cuparene and (±)-Herbertene[36]

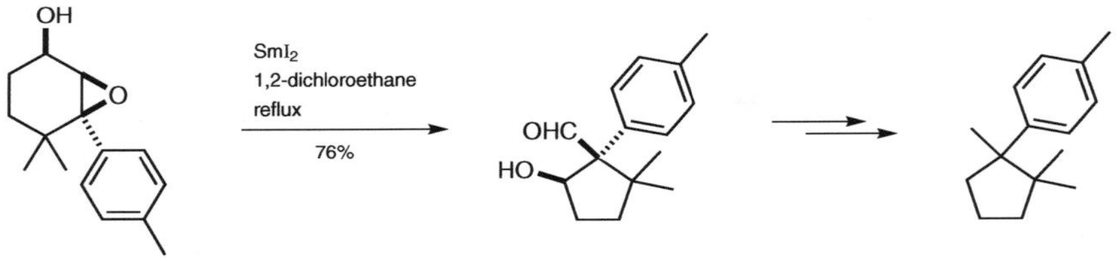

The rearrangement of tetrasubstituted epoxides that contain electron-withdrawing substituents has been broadly applied in organic synthesis to construct ring-contracted products in good yields. As shown through synthesis of microbiotol-type sesquiterpenes (Scheme 15), compound **41** can undergo a reliable, regioselective epoxide opening, affording cyclopentyl ketone **42** in 90% yield as a result of an α,β-unsaturated ester bonded to the epoxide. Further elaboration of ester **42** leads to the formation of diazo ketone **43**, which undergoes intramolecular cyclopropanation and Wittig alkenation to give (+)-β-microbiotene in 56% yield.[37] This method has also been used to synthesize other terpenes featuring a cyclopentane ring bearing multiple quaternary carbon centers.[38,39]

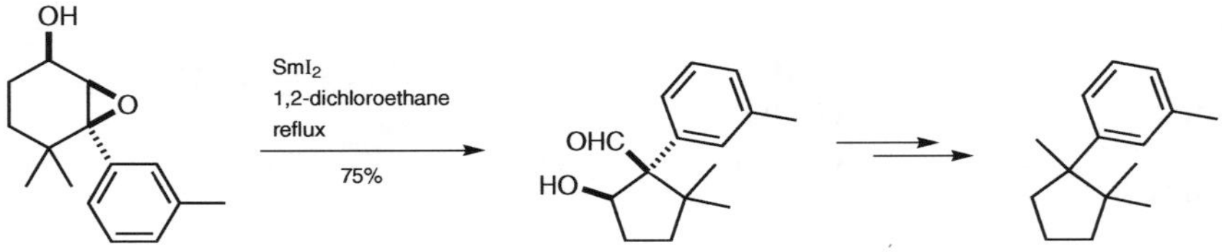

1.5.2 Lewis Acid/Base Induced Reactions

285

Scheme 15 Total Synthesis of (+)-β-Microbiotene and Other Sesquiterpenes[37–39]

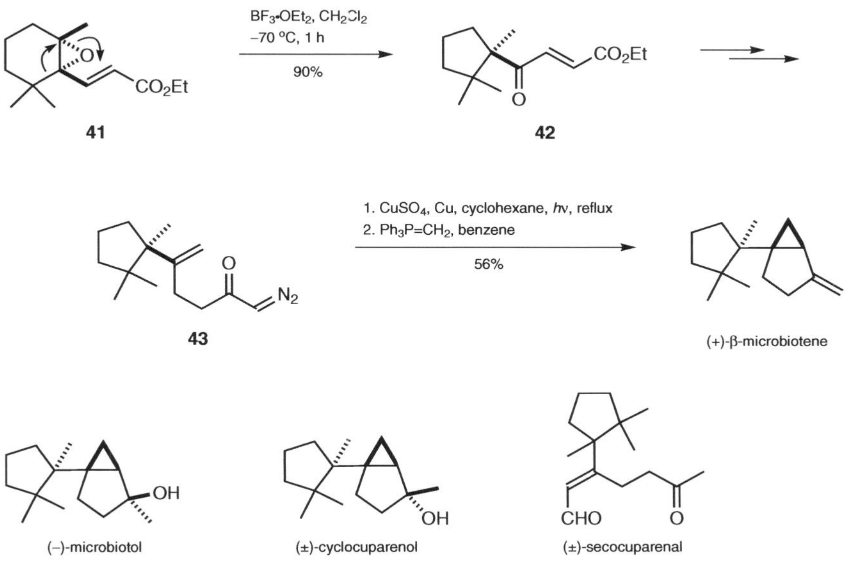

In addition to the construction of cyclopentanes in a highly selective manner, the rearrangement of tetrasubstituted epoxides has also been used to prepare spirocyclic compounds, including optically active molecules. Based on the results of some model studies, Kita and co-workers completed the total synthesis of fredericamycin A (Scheme 16).[40,41] In this case, the asymmetric reduction of ketone **44** using a chiral borane reagent and standard Sharpless epoxidation affords *cis*-epoxy alcohol **45** stereoselectively, which is then treated with (−)-camphanic acid under Mitsunobu conditions to give optically pure **46**. The rearrangement of the epoxy camphanoate **46** using boron trifluoride–diethyl ether complex gives the optically pure spiro CDEF-ring core **47**, which is subsequently transformed into fredericamycin A.

for references see p 351

Scheme 16 Total Synthesis of Fredericamycin A[40,41]

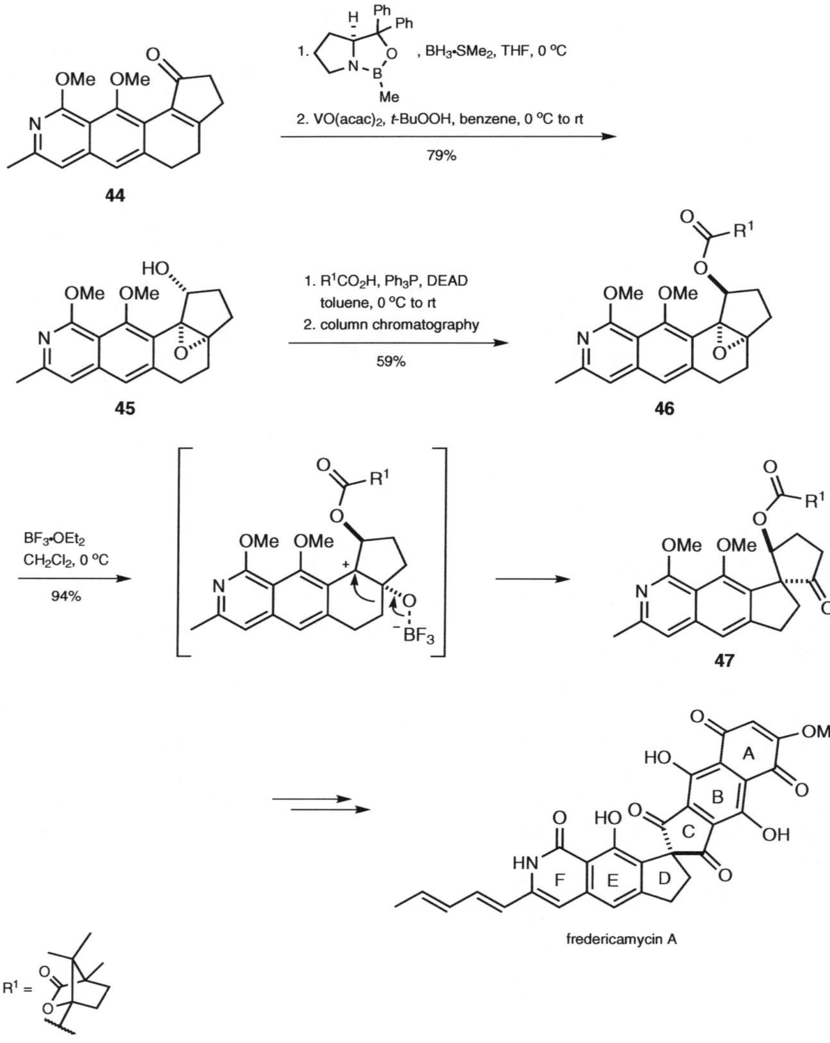

***cis*-2-(Hydroxymethyl)-2-methylcyclopentan-1-ol (Table 4, Entry 1); General Procedure:**[34]
Under an argon atmosphere, a 3 M soln of Et₃Al in THF (1.0 mL, 3.0 mmol) was added dropwise to a stirred soln of (1R*,2R*,6S*)-6-methyl-7-oxabicyclo[4.1.0]heptan-2-ol (190 mg,
1.5 mmol) in dry THF (10 mL). The mixture was refluxed under argon. After TLC analysis
showed the reaction was complete, the mixture was poured into 2 M aq HCl at 0 °C and
extracted several times with Et₂O. The combined extracts were dried (Na₂SO₄), concentrated under reduced pressure, and purified on a silica gel column to give a colorless oil; yield:
110 mg (58%).

1.5.2.2.2 Favorskii Rearrangement and Quasi-Favorskii Rearrangement

As a valuable synthetic procedure the Favorskii rearrangement, and its variant the quasi-
Favorskii rearrangement, has been known for more than a hundred years. The Favorskii
rearrangement occurs by treating α-halo ketones with bases and converting the formed
intermediates into carboxylic acids or their derivatives through formation of cyclopro-

panones, which normally give the thermodynamically more stable carbanions following cleavage of the three-membered ring (Scheme 17).

Scheme 17   Favorskii Rearrangement

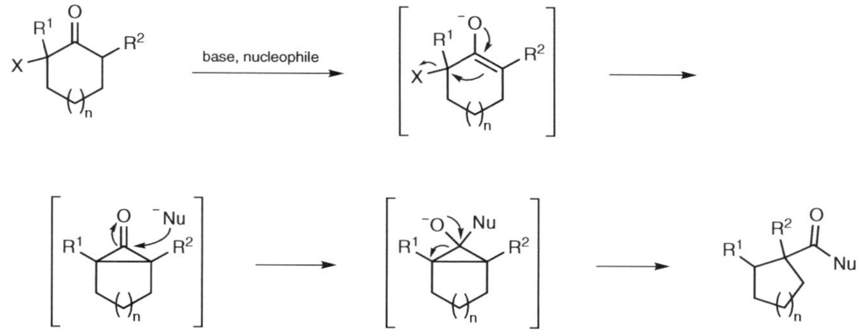

When the cyclopropanone intermediate cannot be formed because of the lack of an enolizable hydrogen or special configurational concerns of the substrate, the quasi-Favorskii rearrangement is operational. During this sequence, the carbon without the halogen atom migrates toward the one with the halogen (Scheme 18).

Scheme 18   Quasi-Favorskii Rearrangement

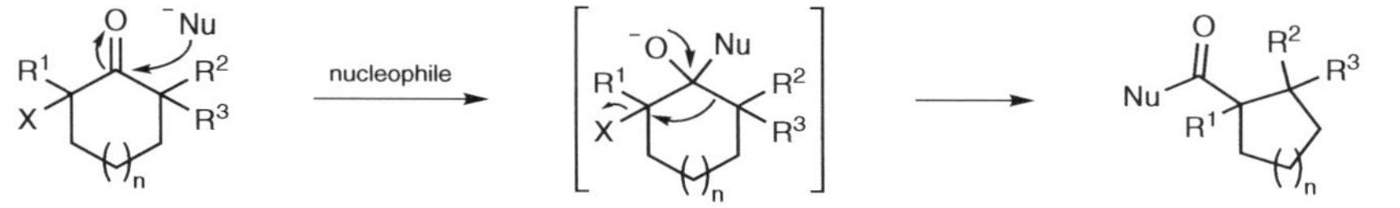

To promote the Favorskii rearrangement, the most commonly used bases are alkoxides (and hydroxides). In some cases, amines are suitable bases as well, and may lead to better results than alkoxides. The behavior of 2-chloro-6-phenylcyclohexanone (**48**) in the presence of two different bases exemplifies this statement particularly well (Scheme 19).[42]

Scheme 19   Favorskii Rearrangement of 2-Chloro-6-phenylcyclohexanone[42]

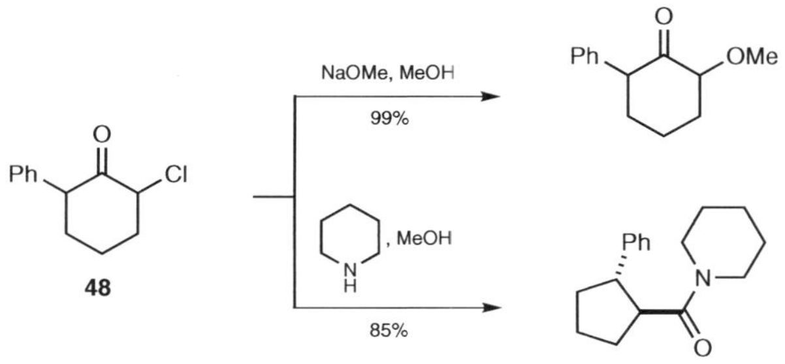

As shown in Scheme 20, during efforts to prepare the upper part of retigeranic acid A **56**, Frasier-Reid and Llera found that treatment of α-bromo ketone **49** with a methanolic solution of sodium methoxide affords a 2:1 mixture of α-methoxy ketones **50A** and **50B** in 80% yield.[43] The formation of these undesired products may be attributed to steric hindrance in removing the α-hydrogen, the first step in the desired Favorskii rearrangement, by the adjacent isopropyl group. Thus, treatment of substrate **49** with pyrrolidine affords

for references see p 351

the Favorskii product **55** in 75% yield. There is an explanation to account for the ability of secondary amines to produce high yields of Favorskii rearrangement amides from ketone **49** despite the failure of even high concentrations of methoxide ion to produce Favorskii esters. First, the enamine allylic bromide **51** is formed, with the formation of the three-membered ring and the departure of bromide leading to intermediate **52**. Next, opening of aminal intermediate **53** should occur with retention of configuration at the activating group for enamine formation to generate amidinium ion **54**. This species can be hydrolyzed to amide **55**. Clearly, the ability of piperidine to produce Favorskii products where methoxide ion is unable to do so is of synthetic and mechanistic significance.

Scheme 20 Preparation of the Upper Part of Retigeranic Acid A[43]

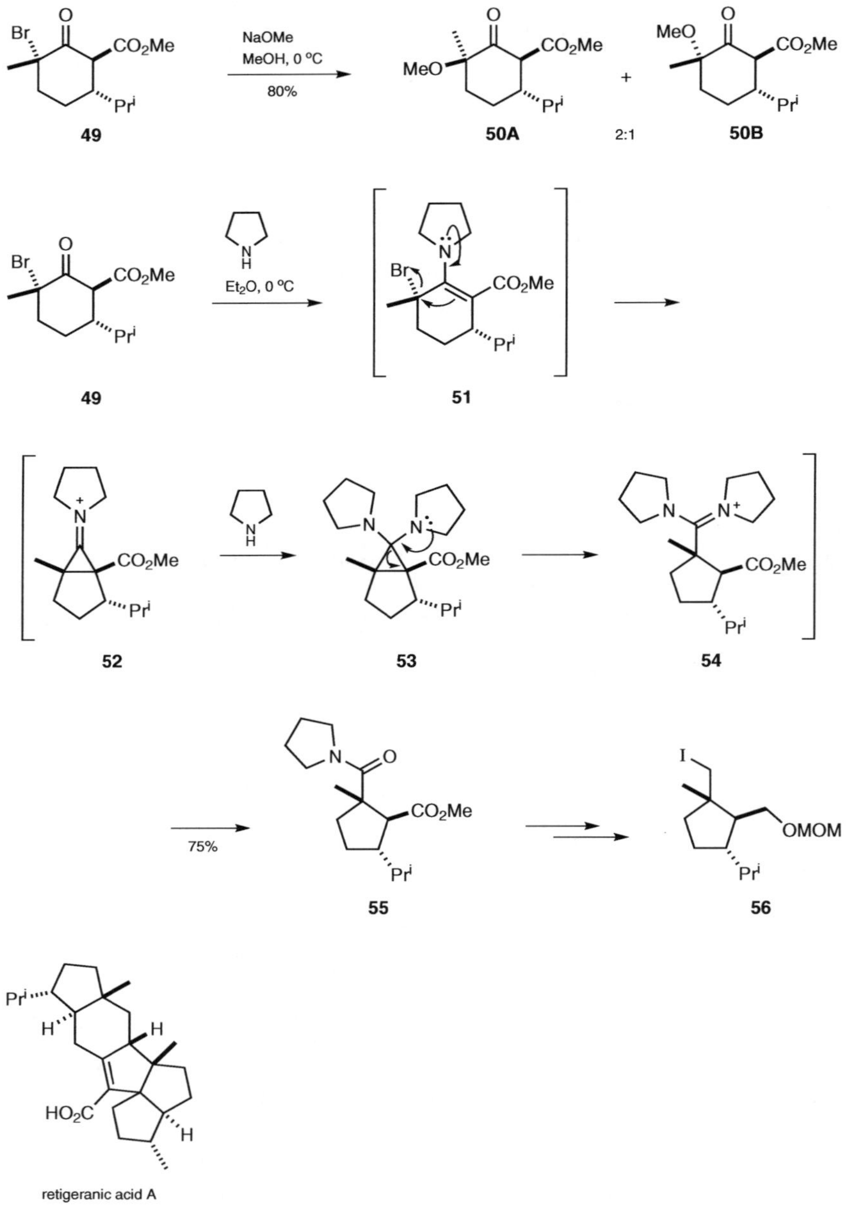

retigeranic acid A

1.5.2 Lewis Acid/Base Induced Reactions

In a total synthesis of the female sex pheromone (±)-sirenin, a quasi-Favorskii rearrangement has been used to construct the substituted cyclopropane moiety, which is fused to a cyclohexene ring and has a neopentyl-like quaternary carbon center. In this case, a silver(I) nitrate mediated rearrangement of α-chlorocyclobutanone **57** generates the ring-contracted product **58** diastereoselectively in 53% yield (Scheme 21).[44,45]

Scheme 21   Total Synthesis of (±)-Sirenin[44,45]

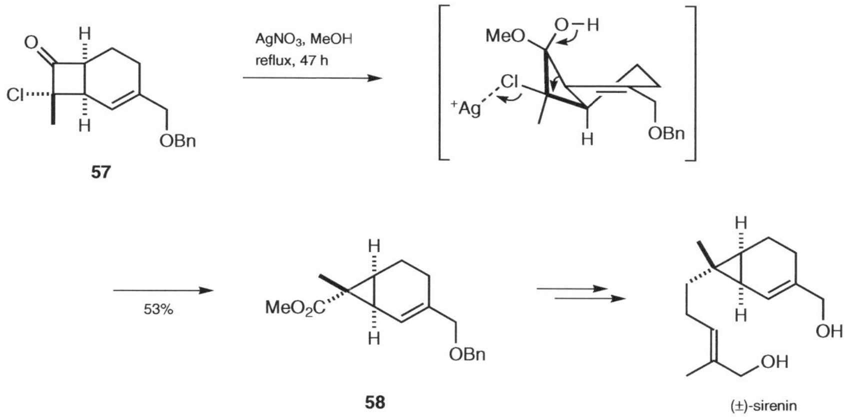

1.5.2.3 Ring Closures

Mono- or multicyclic skeletons feature in the majority of important natural products and bioactive compounds. Since the appearance of organic synthesis, the question of how to efficiently and rapidly construct such cyclic structures has been a hot topic and a key factor for the development of synthetic chemistry at large. As a result of tremendous efforts, a diverse array of methodologies have risen to the challenge of synthesizing cyclic compounds. Among these, domino-reaction-based strategies represent one of the most concise ways to construct target skeletons. This statement reflects the fact that domino reactions not only shorten the number of synthetic steps, but also help afford, in some cases, a better understanding of biosynthesis. In the long run, it is also of great importance to the sustainability of natural systems and the environment to generate less waste through more-efficient processes. Normally, in the presence of a Lewis acid or base, the domino procedure for the ring closure can be initiated by an electrophilic step, a pericyclic step, or a nucleophilic step. Some interesting procedures along these lines will be introduced in this section. In addition, some additional examples promoted by π-acids and other reagent choices are also presented.

1.5.2.3.1 Induction by an Electrophilic Step

Cation-induced domino cyclizations represent the major group of processes used for the construction of cyclic skeletons when initiated by an electrophilic step. The intrinsic properties of Lewis acids (LA) mean that they tend to accept an electron pair from a Lewis base (LB) to form a new adduct. The Lewis base part of the adduct will be electron deficient, and, if generated in the presence of certain functional groups such as epoxides or tertiary alcohols, carbocations will be generated. Consequently, the electron deficiency or the carbocation can further react with a nucleophile to trigger a domino transformation (Scheme 22).

for references see p 351

290   Domino Transformations   **1.5** Non-Radical Skeletal Rearrangements

Scheme 22   Cation-Induced Domino Reaction

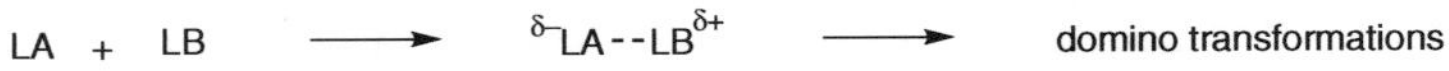

$$LA + LB \longrightarrow {}^{\delta^-}LA\text{-}\text{-}LB^{\delta^+} \longrightarrow \text{domino transformations}$$

1.5.2.3.1.1   **Initiation by Epoxide Ring Opening**

Due to their special structure, epoxides have been widely used in organic synthesis. Normally, under the activation of a Lewis acid, the epoxide moiety will be opened to give a carbocation intermediate which can trigger subsequent transformations such as skeletal rearrangements and/or cyclizations. As the selectivity of the epoxide opening can be controlled through the careful design of the substrate or choice of Lewis acid, this motif has been broadly embedded in domino transformations.

1.5.2.3.1.1.1   **Termination with a Carbon Nucleophile**

The first example that needs to be mentioned is the Lewis acid induced cationic polyalkene cyclization (see also Section 1.1.1) for the synthesis of pentacyclic triterpenes of the β-amyrin family (Scheme 23). In nature, these materials are derived from (S)-2,3-oxidosqualene through an enzymatic cyclization.[46]

Scheme 23   Typical Pentacyclic Triterpenes of the β-Amyrin Family

β-amyrin

erythrodiol

oleanolic acid

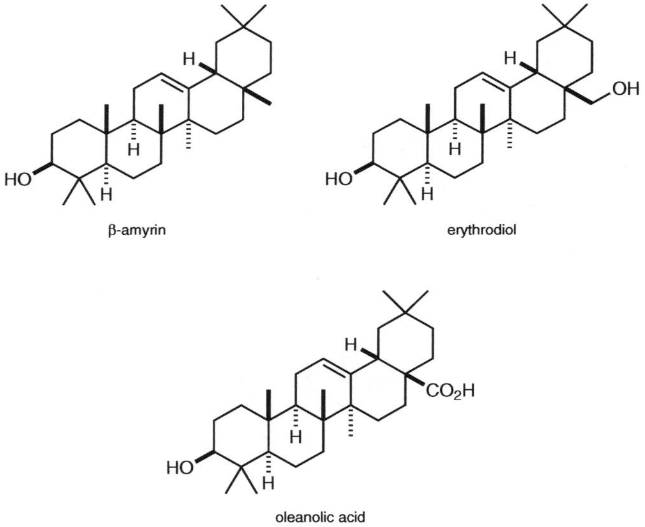

Inspired by the proposed biosynthesis mechanism, van Tamelen and co-workers first tried a biomimetic synthesis of this type of compound (Scheme 24).[47] By using tin(IV) chloride in nitromethane they successfully obtained the desired cyclization product **59** with a yield of 4% and subsequently accomplished the formal synthesis of β-amyrin. Similar types of reaction have also been carried out by Corey and co-workers.[48,49] Treatment of epoxide **60** with methylaluminum dichloride affords the pentacyclic products **61** and **62** (1.5:1 ratio) in a total yield of 39%. The yield of this reaction is further increased to 52% when using substrate **63** to give tricycle **64**. Based on these results, the enantioselective total synthesis of oleanolic acid, β-amyrin, erythrodiol, and other pentacyclic triterpenes has been realized.

1.5.2 Lewis Acid/Base Induced Reactions   **291**

Scheme 24   Lewis Acid Induced Tricyclizations[47–49]

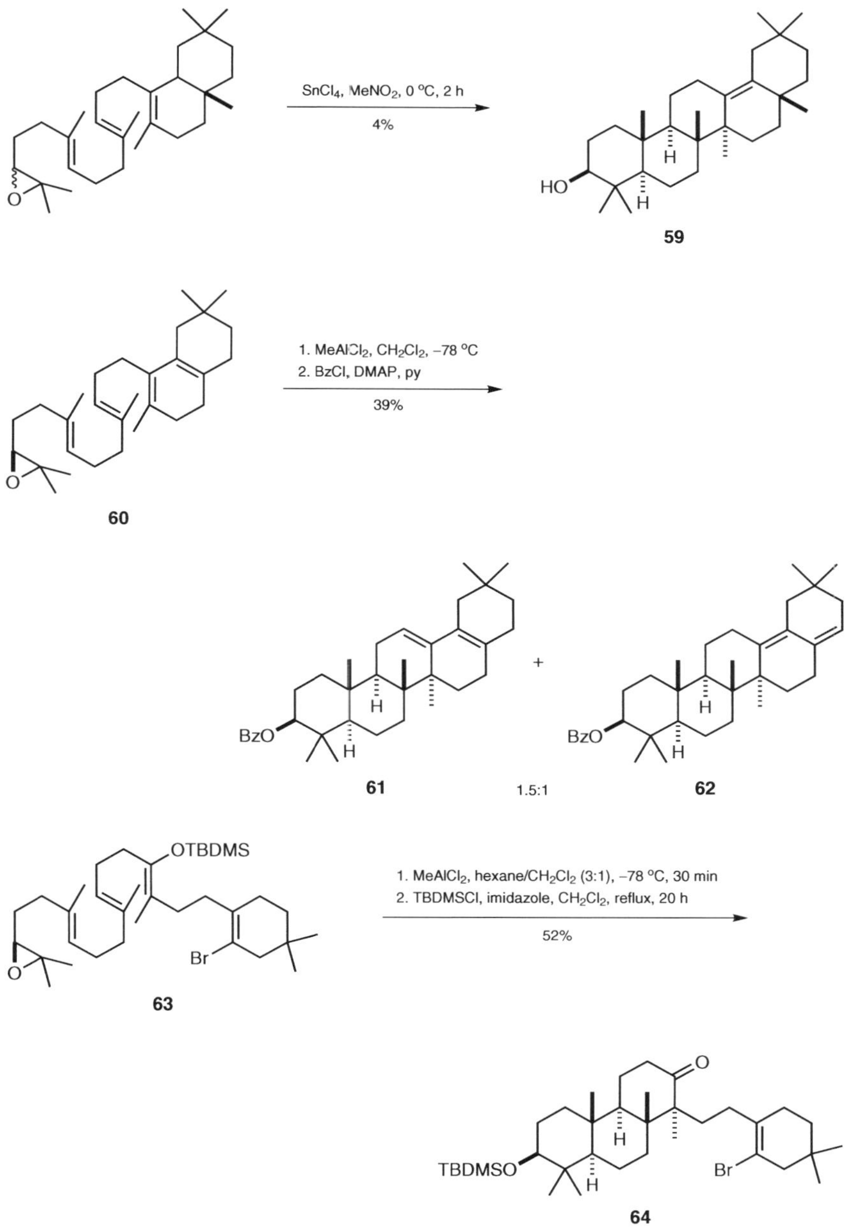

It is clear that this kind of reaction is affected by the Lewis acid, the substrate, and the reaction temperature. For example, the use of weaker Lewis acids improves the yield dramatically at lower temperature. Although there is one small issue with this transformation, which is selectivity for the formation of the double bond in ring C and the configuration of the resultant C14 methyl group, these are not significant problems because of the known transformations between these compounds.

Applying similar transformations, Corey and co-workers also completed the enantioselective total syntheses of dammarenediol II, scalarenedial, and (+)-α-onocerin (Scheme 25).[50–52]

for references see p 351

Scheme 25 Domino Cyclization for the Synthesis of Dammarenediol, Scalarenedial, and (+)-α-Onocerin[50–52]

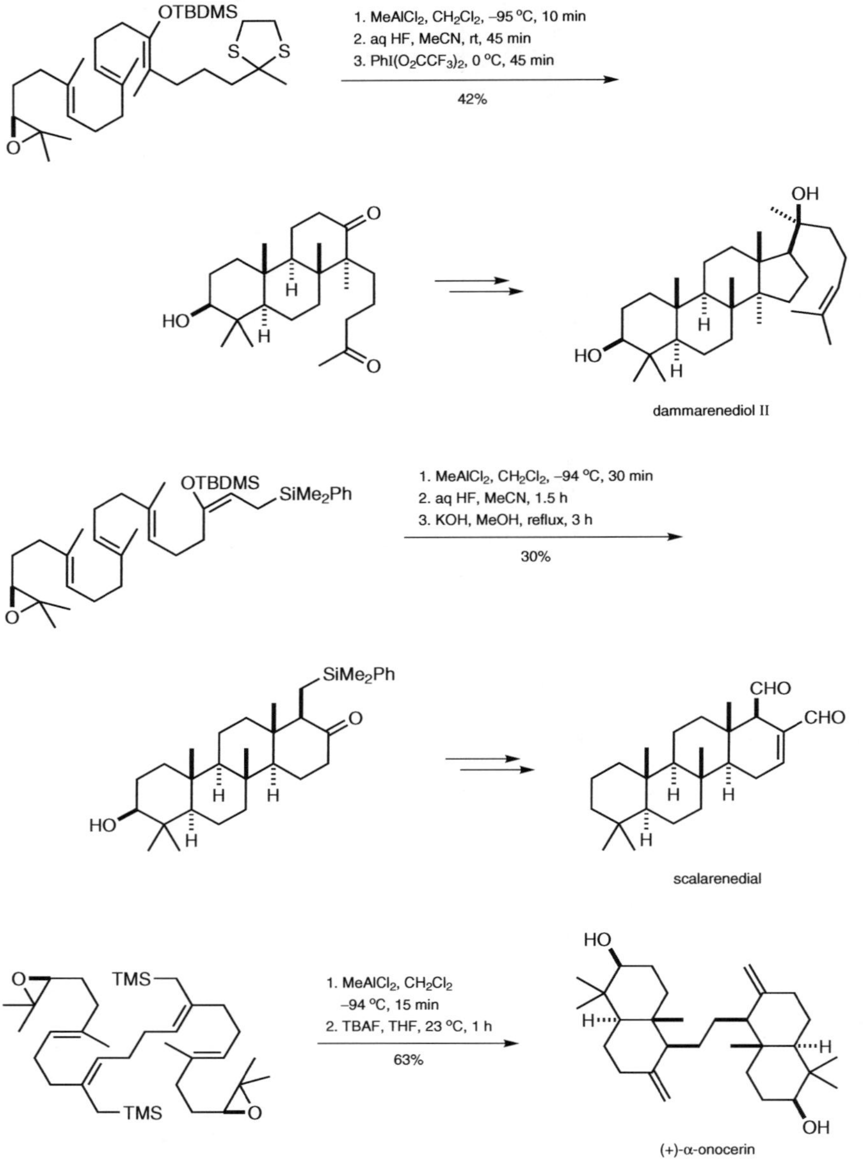

dammarenediol II

scalarenedial

(+)-α-onocerin

It should be noted that Johnson and co-workers have extensively investigated a pentacyclization strategy toward this type of natural product (Scheme 26).[53,54] Considering the potential problem of the epoxide-initiated cyclization, they prepared four substrates **65–68** with different initiators for this domino process. The four compounds share a common fluorine atom at the pro-C13 position (oleanane numbering). The presence of the fluorine is crucial for the cyclization, as it can effectively stabilize the carbocation at the C13 position during the reaction, affording the pentacyclic product in better yield with the desired ring system generated in situ. However, the reaction with Lewis acid is not successful, leading to the desired domino cyclization along with unwanted hydrogen fluoride elimination, an outcome caused by the steric congestion of the reaction intermediate. Al-

though the elimination gives the correct ring C alkene, it engenders difficulty for achieving any subsequent functionalization. Interestingly, the use of a protic acid such as trifluoroacetic acid can form the desired product **69**, albeit with lower yield, and leads to the synthesis of sophoradiol and some other pentacyclic triterpenoids of the ole-anane series (Scheme 27).[54]

Scheme 26 Pentacyclization Reactions in the Synthesis of Triterpenoid Natural Products[53,54]

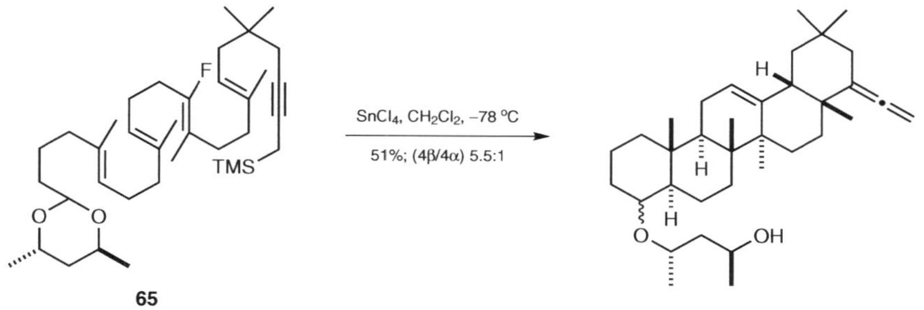

for references see p 351

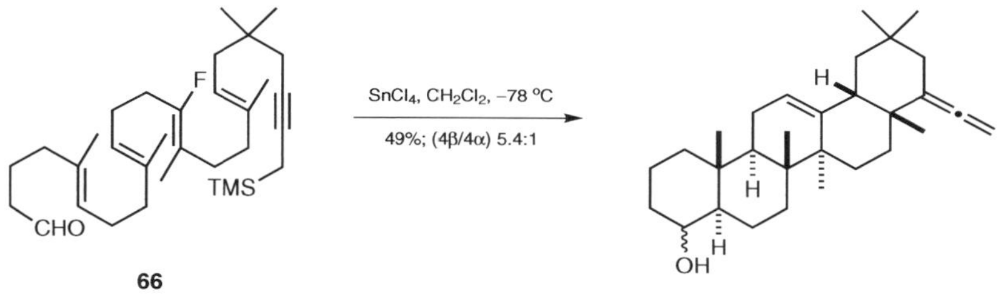

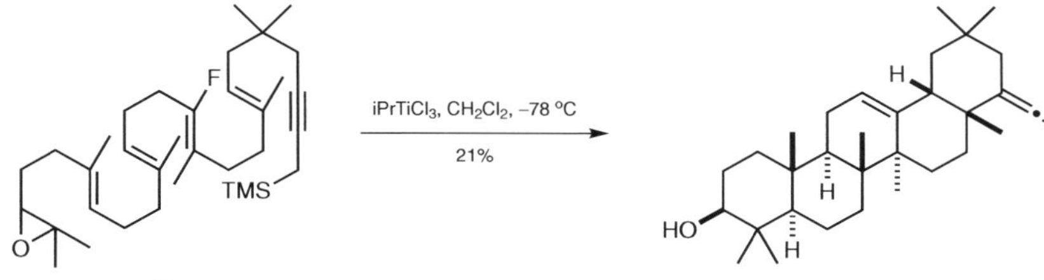

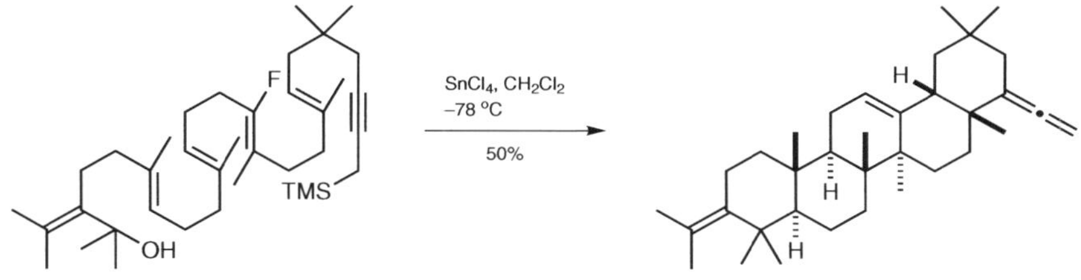

Scheme 27 Synthesis of Sophoradiol[54]

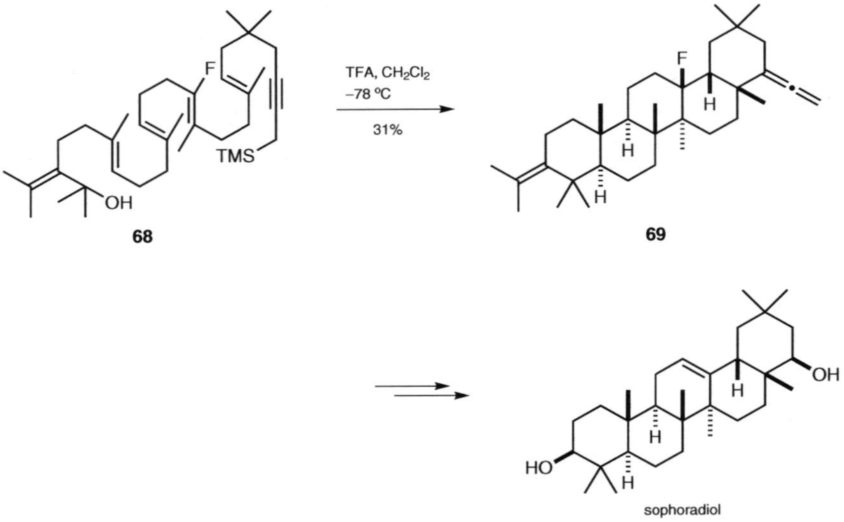

Besides polyenes, substrates with multiple epoxide moieties can also undergo such domino cyclizations in the presence of Lewis acids and protic acids (see also Section 1.1.1 and Section 1.2, respectively). McDonald and co-workers have applied such transformations to the total syntheses of abudinol B and related oxidative degradation products, which were isolated from Red Sea sponges of *Ptilocaulis spiculifer* (Scheme 28).[55] Inspired by the biosynthetic proposal for abudinols A and B, they used the Lewis acid catalyzed cyclization of bis-epoxides **70** and **73** to obtain key intermediates **72** and **75**, respectively, and used them to complete the total synthesis of *ent*-abudinol B. During the Lewis acid catalyzed domino procedures, the use of trimethylsilyl trifluoromethanesulfonate or *tert*-butyldimethylsilyl trifluoromethanesulfonate and 2,6-di-*tert*-butyl-4-methylpyridine (DTBMP) proves essential for the exclusive formation of intermediates **71** and **74**. Other Lewis acids, such as boron trifluoride–diethyl ether complex and dimethylaluminum chloride, cause partial desilylation of the products.

Scheme 28 Total Synthesis of *ent*-Abudinol B[55]

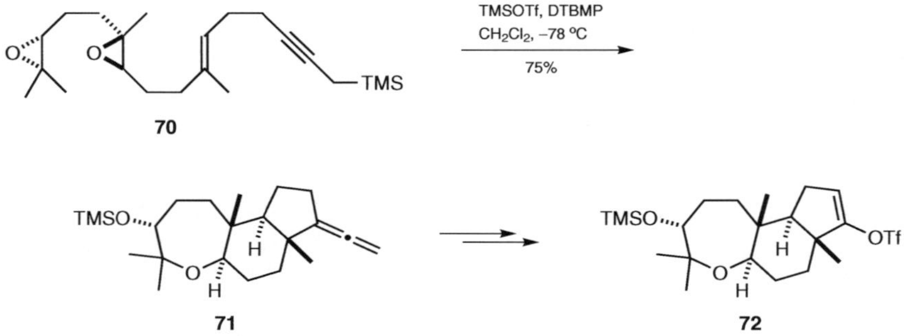

1.5.2 Lewis Acid/Base Induced Reactions

DTBMP = 2,6-di-*tert*-butyl-4-methylpyridine

Besides Lewis acids and protic acids, similar types of epoxide-opening cascade can also be initiated by a halonium ion. During studies toward the total synthesis of dioxepandehydrothyrsiferol, Jamison and co-workers developed a bromonium-promoted domino cyclization reaction.[56] As shown in Table 5, it is clear that the bromonium reagent does not significantly affect the yield, whereas the choice of nucleophilic trapping moiety has a large impact [i.e., a *tert*-butyl carbonate (entries 1 and 2) or a *tert*-butyl ester (entries 3 and 4) instead of a primary alcohol (entries 5 and 6)]. The absence of diastereoselectivity, however, requires further optimization. Finally, the total synthesis of *ent*-dioxepandehydrothyrsiferol was realized by the application of this method as the key step (Scheme 29).

Table 5 Bromonium-Induced Domino Cyclization[56]

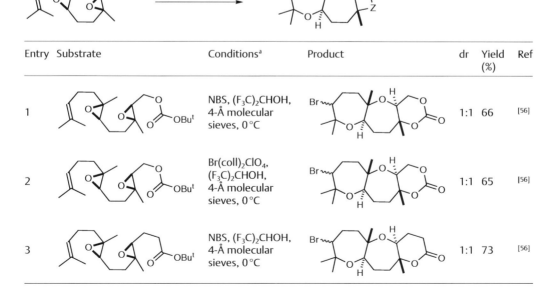

for references see p 351

Table 5 (cont.)

Entry	Substrate	Conditions[a]	Product	dr	Yield (%)	Ref
4	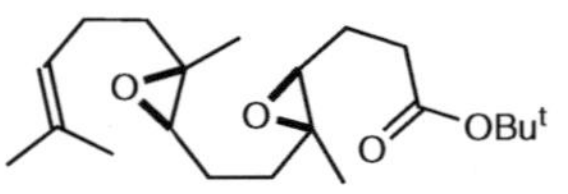	Br(coll)$_2$ClO$_4$, (F$_3$C)$_2$CHOH, 4-Å molecular sieves, 0 °C	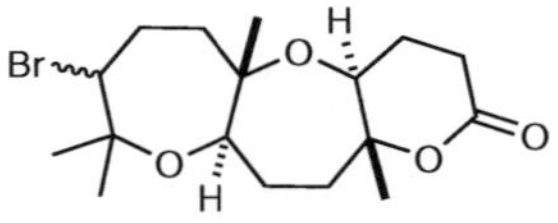	1:1	61	[56]
5	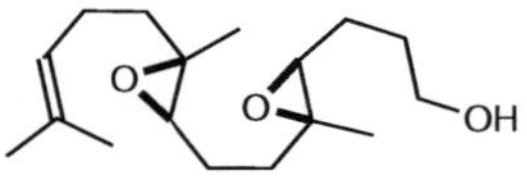	NBS, (F$_3$C)$_2$CHOH, 4-Å molecular sieves, 0 °C	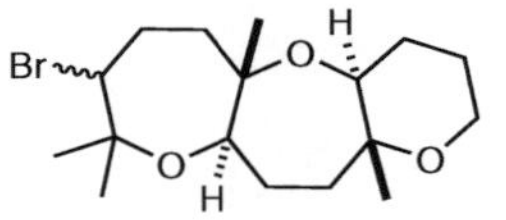	1:1	58	[56]
6	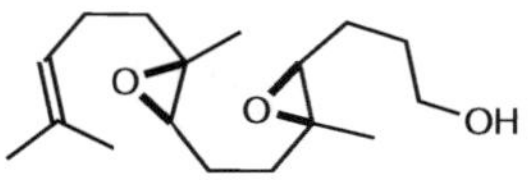	Br(coll)$_2$ClO$_4$, (F$_3$C)$_2$CHOH, 4-Å molecular sieves, 0 °C	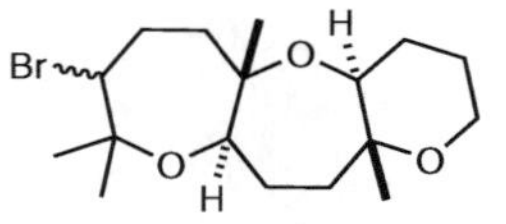	1:1	52	[56]

[a] Br(coll)$_2$ClO$_4$ = bis(collidine)bromonium perchlorate.

Scheme 29 Synthesis of *ent*-Dioxepandehydrothyrsiferol[56]

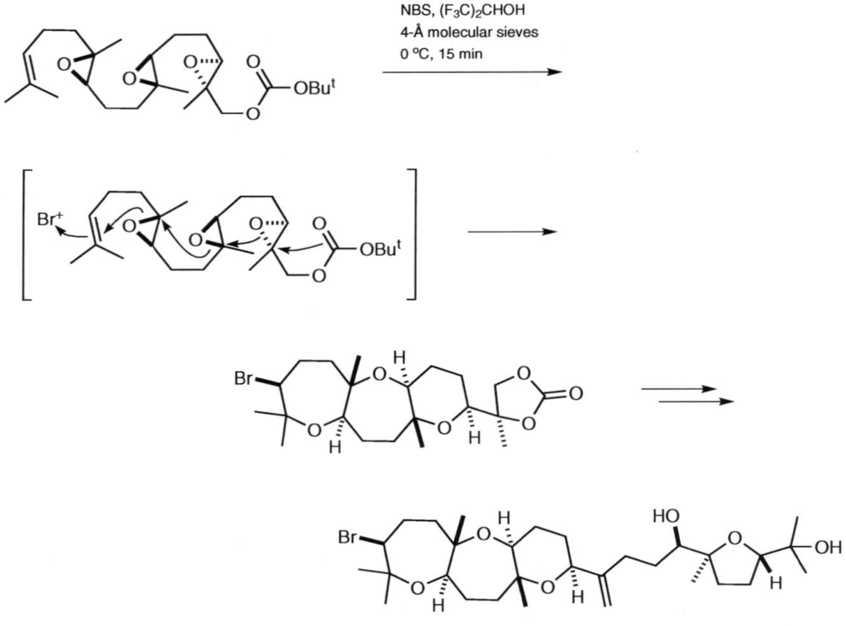

ent-dioxepandehydrothyrsiferol

(1S,4aR,4bR,7S,8aR,10aR)-1-[2-(2-Bromo-4,4-dimethylcyclohex-1-enyl)ethyl]-7-(*tert*-butyl-dimethylsiloxy)-1,4b,8,8,10a-pentamethyldodecahydrophenanthren-2(1H)-one (64); Typical Procedure:[49]

Compound **63** (30 mg, 48 μmol) was dissolved in a mixture of hexanes (3.6 mL) and CH$_2$Cl$_2$ (1.2 mL), and the soln was cooled to −78 °C. A 1.0 M soln of MeAlCl$_2$ in hexanes (56 μL, 56 μmol) was diluted with hexanes (2.1 mL) and CH$_2$Cl$_2$ (0.7 mL) and cooled to −78 °C. The diluted soln of MeAlCl$_2$ was added to the soln of **63** through a cannula. After the addition was complete, the mixture was stirred at −78 °C for another 30 min. Et$_3$N (57 mg,

0.56 mmol) was added, followed by MeOH/H$_2$O (4:1; 0.5 mL). The mixture was warmed to rt and poured into H$_2$O (5 mL). After separation of the layers, the aqueous phase was extracted with Et$_2$O (3×5 mL). The combined organic phases were washed with brine, dried (Na$_2$SO$_4$), and concentrated under reduced pressure to give the crude product as a white foam, which was azeotropically dried by evaporating with toluene. To a soln of the dried crude product in CH$_2$Cl$_2$ (1.0 mL) was added TBDMSCl (11 mg, 72 μmol) and imidazole (6.6 mg, 96 μmol). The resulting mixture was refluxed for 20 h. After the mixture had been cooled to 23 °C, the whole contents were passed through a short column of silica gel, which was further eluted with CH$_2$Cl$_2$ (20 mL). The eluent was concentrated under reduced pressure to give the crude product as a white foam, which was purified by preparative TLC [eluting with benzene (**CAUTION:** *carcinogen*)]; yield: 15.6 mg (52%).

1.5.2.3.1.1.2 **Termination with an Oxygen Nucleophile**

Another way to terminate the epoxide-initiated domino process described in Section 1.5.2.3.1.1.1 is through the use of a nucleophilic group that contains an oxygen atom. Accordingly, natural products with oxa ring skeletons can be constructed effectively. Among such molecules, the natural bioactive schweinfurthins and vedelianin are perfect targets (Scheme 30),[57–60] especially with the presence of the geminal methyl groups.

Scheme 30 Some Natural Schweinfurthins and Vedelianin[57–60]

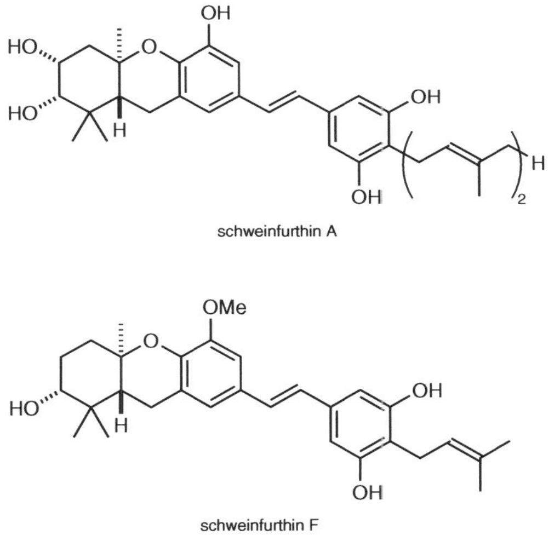

schweinfurthin A

schweinfurthin F

for references see p 351

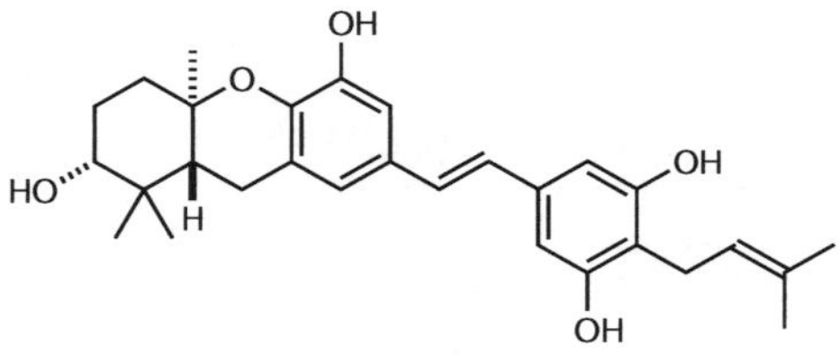

schweinfurthin G

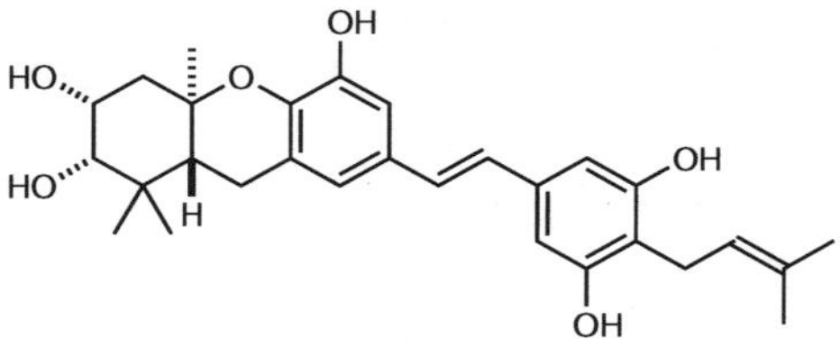

vedelianin

Aiming at the synthesis of these types of compound, Wiemer and co-workers have developed a Lewis acid mediated domino cyclization of epoxides terminated by a phenolic oxygen.[61–64] This reaction is highly regio- and stereoselective; i.e., the use of enantiopure substrates will give enantiopure products as a single diastereomer, which suggests that a pseudo-chair, chair-like transition state is probably involved. As the corresponding asymmetric epoxides are readily accessible, this method becomes even more practical for the synthesis of the related natural products. This cyclization can be induced by protic acids or Lewis acids, of which boron trifluoride–diethyl ether complex has proved to be the best choice. Generally, in the presence of a protic acid, the desired product is obtained in relatively low yield (30–40%). Among the Lewis acids screened, both very strong ones (e.g., TiCl$_4$) and weak ones are ineffective, leading either to decomposition or the recovery of the starting material. Furthermore, as shown in Table 6, the protecting group (R^1) is another key factor. First of all, the group should be labile to the Lewis acid, otherwise a bridged A-ring ether will be the major product (Table 6, entries 4 and 5). Accordingly, alkoxymethyl and benzyl-type protecting groups are commonly used. Secondly, because an electrophile is always released during the cyclization, the existence of a suitable aromatic ring will result in a domino process involving an additional electrophilic aromatic substitution (entry 2), a process which has been successfully applied using substrate **76** to give product **77** in the total synthesis of (+)-angelichalcone (Scheme 31).[63] Although the formation of products without aromatic substitution can be an issue for this reaction, it is still of special interest to organic chemists due to the assembly of an extra C—C bond.

1.5.2 Lewis Acid/Base Induced Reactions

299

Table 6 Domino Cyclization of Substituted Phenols[62–64]

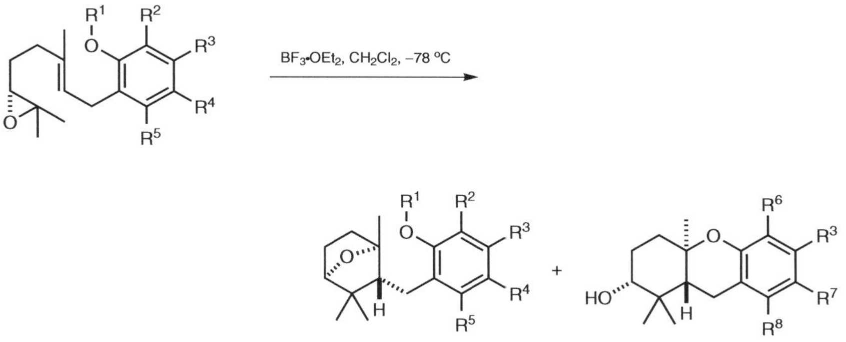

Entry	Substrate	Product	Yield (%)	Ref
1	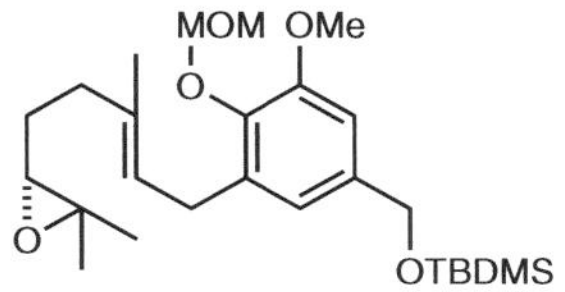	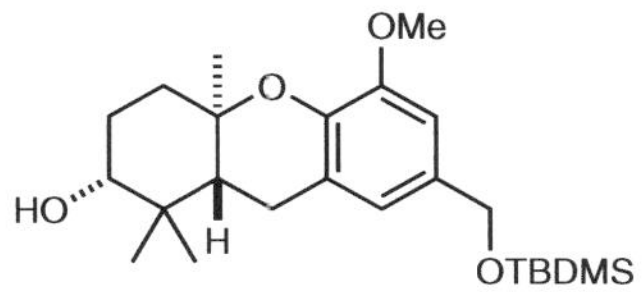	68	[62]
2	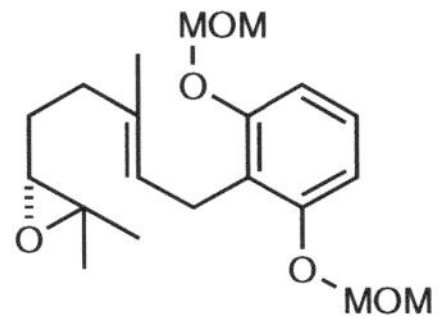	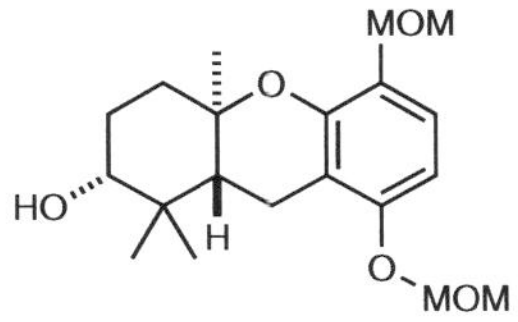	71	[53]
3	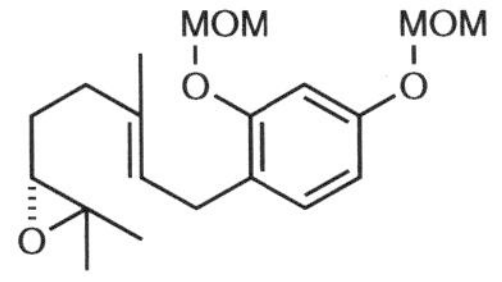	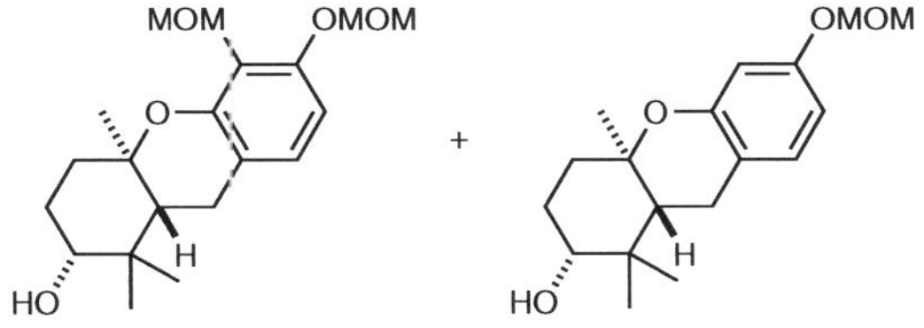	82	[64]
4	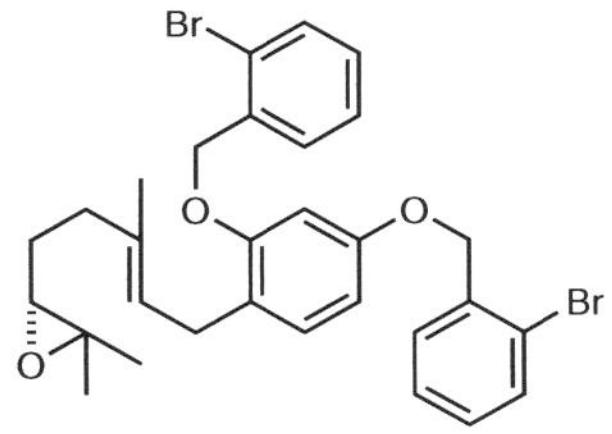	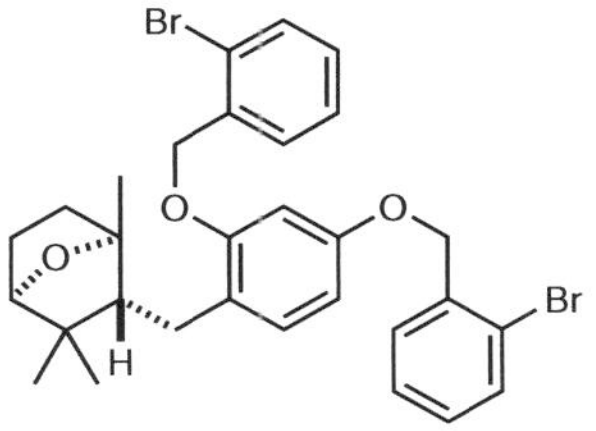	49	[64]
5	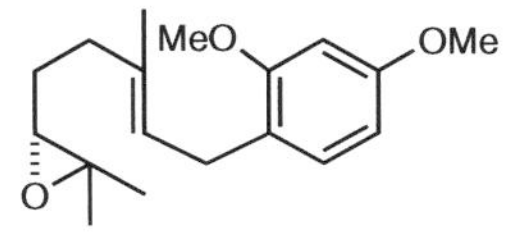	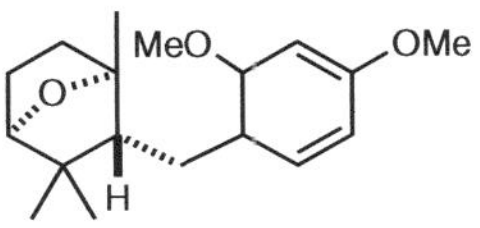	72	[64]

for references see p 351

Scheme 31 Total Synthesis of (+)-Angelichalcone[63]

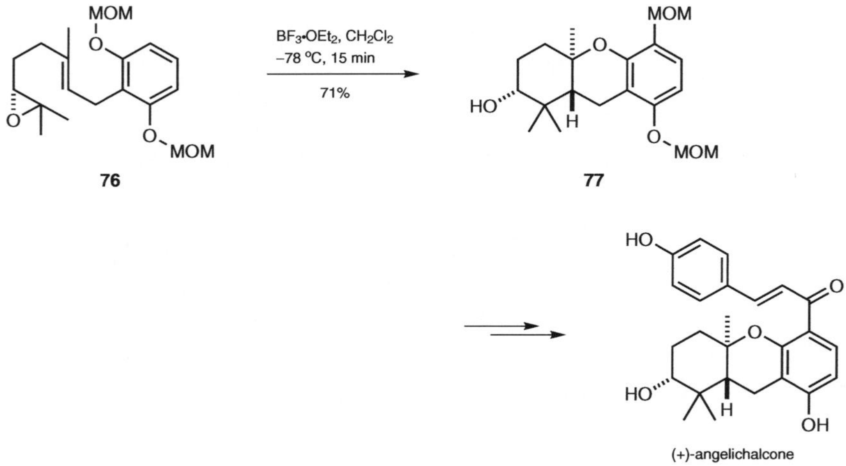

(2R,4aR,9aR)-5,8-Bis(methoxymethyl)-1,1,4a-trimethyl-2,3,4,4a,9,9a-hexahydro-1H-xanthen-2-ol (77); Typical Procedure:[63]

To a soln of epoxide **76** (130 mg, 0.37 mmol) in freshly distilled CH_2Cl_2 (50 mL) at −78 °C was added $BF_3 \cdot OEt_2$ (0.25 mL, 2.0 mmol). After 15 min, the reaction was quenched by addition of Et_3N (0.5 mL, 3.6 mmol). The resulting soln was concentrated under reduced pressure to near dryness and extracted with EtOAc. The combined organic phases were washed with 1 M HCl and brine, dried ($MgSO_4$), and concentrated under reduced pressure. Final purification by column chromatography (EtOAc/hexanes 3:7) afforded compound **77** as a colorless oil; yield: 92 mg (71%).

1.5.2.3.1.1.3 Termination with a Rearrangement

Although most epoxide-initiated (and even cation-induced) cyclizations are terminated by a simple nucleophilic addition to the related carbocation intermediate, terminations with a rearrangement transformation are also appealing because of the subsequent change to the molecular skeleton and the wide range of cation-induced rearrangements that are possible.

The first example introduced here is the intramolecular Schmidt reaction,[65] a transformation that uses an azide group as the terminator of the cation intermediate. The azide group can be considered as a nitrogen nucleophile. In particular, the intramolecular Schmidt reaction is a very useful transformation for the following reasons: (i) since the azide group is a remarkable cation terminator, functional groups other than epoxide also can be used, such as carbonyls and alkenes; (ii) as this reaction is always accompanied by a ring-expansion, ring-contraction, or ring-closure process to give nitrogen-containing cyclic compounds, it has been used broadly in the syntheses of alkaloids; and (iii) because a large number of Lewis acids and protic acids can be used to promote this transformation, it is likely that the desired rearrangement will be effected as long as the substrate is stable under acidic conditions.

The indolizidine alkaloids, possessing a typical azabicyclic ring skeleton, are an important class of bioactive natural products.[66–71] Their pharmaceutical applications and interesting biological activities have attracted extensive attention from organic chemists. In contrast to the traditional stepwise annulation strategies, a domino electrophilic cyclization/intramolecular Schmidt reaction provides a more efficient way to construct the indolizidine skeleton. Following the attempts with azido ketones and azidoalkenes by Pearson

1.5.2 Lewis Acid/Base Induced Reactions

and Aubé,[72,73] Baskaran and co-workers have developed an epoxide-initiated cationic cyclization of azides toward the synthesis of these molecules (Scheme 32).[74] Like most Schmidt reactions, a series of acids can be applied in this reaction, with the use of stoichiometric ethylaluminum dichloride giving the best yield. Also, substrates with different ring sizes are amendable to this reaction. It is noteworthy that between the *cis*- and *trans*-isomers of the starting material, only the *trans*-disposed substrate can go through this transformation to give the final product, an outcome which indicates that the Lewis acid promoted epoxide-opening step follows an S_N2 mechanism. Additionally, after in situ reduction of the reaction intermediate, this domino reaction gives the desired product as a single diastereomer. This outcome can be explained by the known stereoselective reduction of the similar iminium ion intermediate.

Scheme 32   Epoxide-Initiated Intramolecular Schmidt Reaction[74]

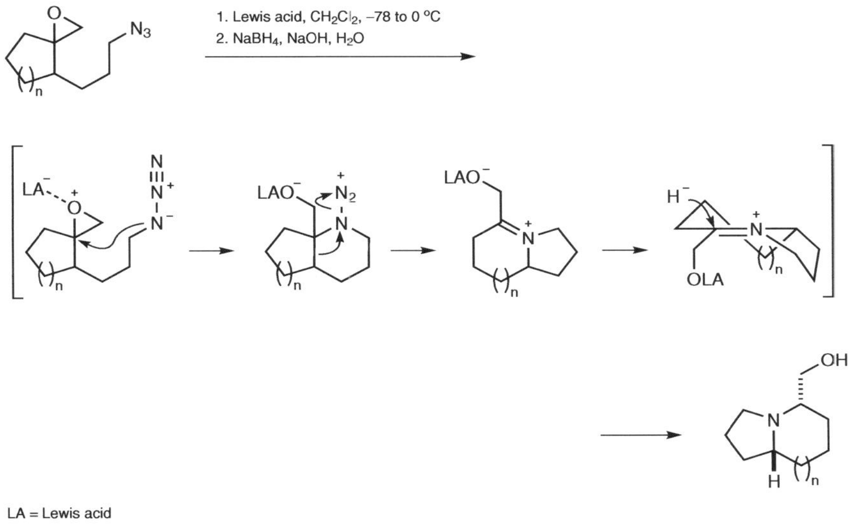

LA = Lewis acid

n	Lewis Acid	Temp (°C)	Yield (%)	Ref
1	TfOH	−40	25	[74]
1	TMSOTf	−78	21	[74]
1	$BF_3 \cdot OEt_2$	−78	50	[74]
1	$TiCl_4$	−25	43	[74]
1	$InCl_3$	0	25	[74]
1	$EtAlCl_2$	−78	63	[74]
2	$EtAlCl_2$	−78	42	[74]
3	$EtAlCl_2$	−78	47	[74]

Key is that the product **79A** has the same relative configuration as the C5 and C9 of the indolizidine alkaloids. Thus, using the Sharpless asymmetric dihydroxylation to introduce the expected chiral oxa-quaternary center, Baskaran and co-workers have applied this domino reaction in the enantioselective total synthesis of (+)- and (−)-indolizidines 167B ($R^1 = Pr$) and 209D [$R^1 = (CH_2)_5Me$] **80A** and **80B** from epoxides **78A** and **78B** via products **79A** and **79B**, respectively (Scheme 33).[74]

for references see p 351

Scheme 33 Total Synthesis of (+)- and (−)-Indolizidine 167B and 209D[74]

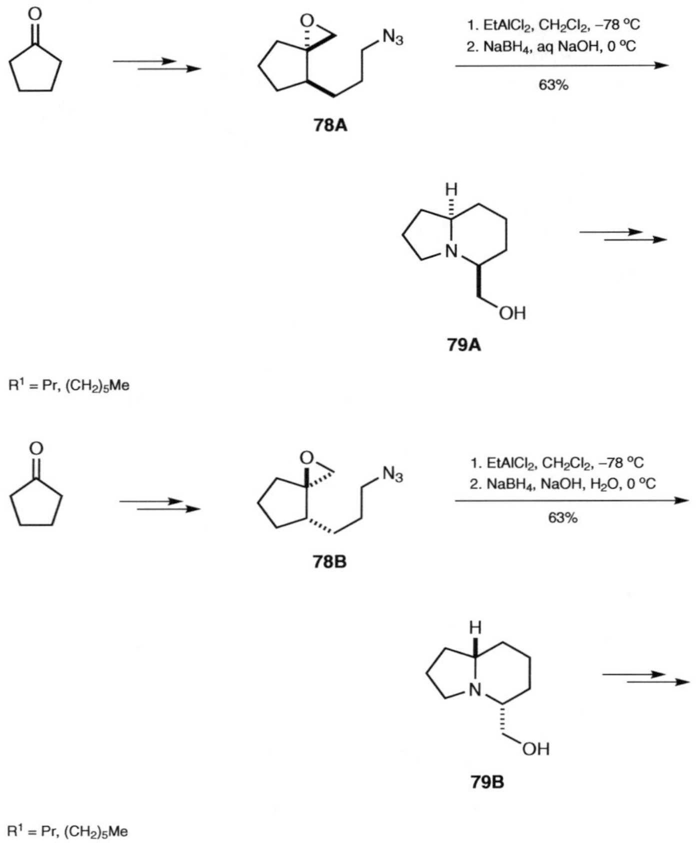

R¹ = Pr, (CH₂)₅Me

R¹ = Pr, (CH₂)₅Me

Another reaction that has been used to terminate the corresponding carbocation is the Wagner–Meerwein rearrangement,[75] a process that covers a large class of cation-initiated rearrangement reactions. Although such rearrangements usually occur in ring-expansion and contraction reactions, they can also give ring-closure products with proper termination of the resultant carbocation. One such domino reaction has been used in the synthesis of stachyflin, a potent anti-influenza A virus agent.[76,77] Stachyflin belongs to a small class of sesquiterpene natural products that possess similar bioactivity against influenza A virus (Scheme 34). These compounds all share the common tetracyclic benzo[d]xanthene skeleton **81**, a crucial element for the synthesis of these targets.

1.5.2 Lewis Acid/Base Induced Reactions

303

Scheme 34 Benzo[d]xanthene Sesquiterpene Natural Products

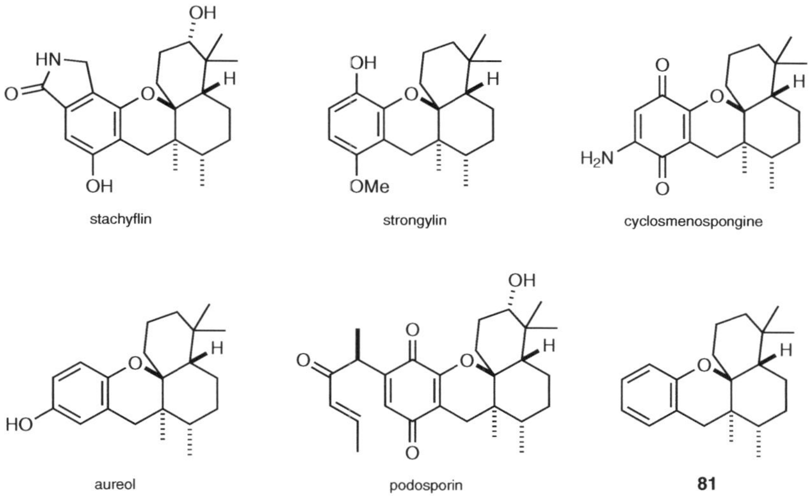

stachyflin

strongylin

cyclosmenospongine

aureol

podosporin

81

During their study of stachyflin, Katoh and co-workers designed an acid-induced epoxide-opening/rearrangement/cyclization reaction to construct the tetracyclic core.[78–80] As shown in Scheme 35, this reaction can be promoted by both Lewis acids and protic acids, specifically boron trifluoride–diethyl ether complex and trifluoroacetic acid, with the better yield obtained using the former. Other Lewis acids such as titanium(IV) chloride, aluminum trichloride, and dimethylaluminum chloride all lead to substrate decomposition, while stronger protic acids such as 4-toluenesulfonic acid and 10-camphorsulfonic acid do not promote this reaction at all. This transformation is highly stereospecific and only one enantiomer with the expected relative configuration is produced with the use of enantiopure substrate. As observed in most Wagner–Meerwein-type processes, rearrangement to a thermodynamically more stable structure is the main driving force for this procedure, an outcome which is also supported by computational analysis. Since the substituents on the phenyl ring seem to have no influence on this cyclization, the installation of the lactam moiety can be performed either before or after the key step. Furthermore, this element also provides the possibility for additional derivatization.

for references see p 351

Scheme 35 Lewis Acid Induced Epoxide Opening/Rearrangement/Cyclization[78–80]

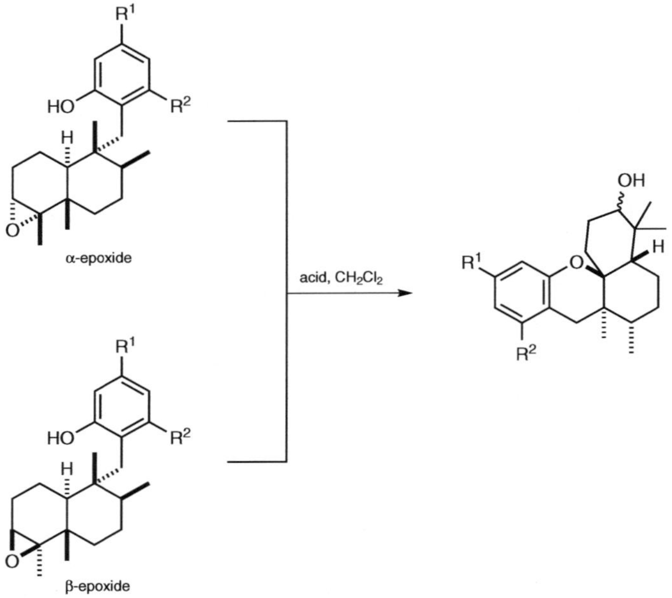

R^1	R^2	Acid	Yield (%)	Ref
H	H	$BF_3 \cdot OEt_2$	72^a	[78,80]
H	H	$BF_3 \cdot OEt_2$	76^b	[78,80]
H	H	$TiCl_4$	0^c	[80]
H	H	$AlCl_3$	0^c	[80]
H	H	Me_2AlCl	0^c	[80]
H	H	$MeAlCl_2$	0^c	[80]
H	H	TFA	43^b	[80]
H	H	TsOH	$-^d$	[80]
H	H	CSA	$-^d$	[80]
CO_2Me	OMe	$BF_3 \cdot OEt_2$	84^e	[80]

[a] Substrate was the α-epoxide; the hydroxy group in the product has β-configuration.
[b] Substrate was the β-epoxide; the hydroxy group in the product has α-configuration.
[c] Substrate decomposes.
[d] No reaction.
[e] Substrate was an 8:1 mixture of α- and β-epoxides; the product was a 6.8:1 mixture of β- and α-hydroxy epimers.

Based on this domino reaction, the Katoh group has completed the enantioselective total synthesis of stachyflin via advanced intermediate **82** (Scheme 36).[79,80] Later, the Cramer group attempted a ruthenium(III)-catalyzed domino cyclization for this tetracyclic skeleton.[81] Unfortunately, the selectivity of the corresponding reaction is not optimal.

Scheme 36 Total Synthesis toward Stachyflin[79,80]

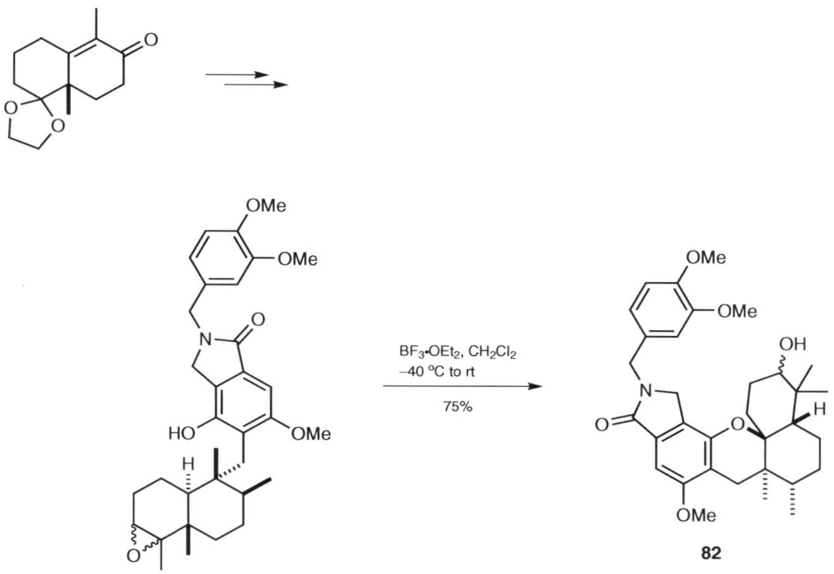

[(5S,8aR)-Octahydroindolizin-5-yl]methanol (79A); Typical Procedure:[74]

To a stirred soln of **78A** (181 mg, 1 mmol) in dry CH_2Cl_2 (5 mL) at −78 °C was added 1.8 M $EtAlCl_2$ in toluene (0.611 mL, 1.1 mmol) dropwise. The resultant mixture was stirred for 45 min at −78 °C and then allowed to warm to rt. After being stirred for an additional 5 min at rt, the mixture was cooled to 0 °C and treated with a soln of $NaBH_4$ (266 mg, 7 mmol) in 15% aq NaOH (3 mL). The mixture was allowed to warm to rt and stirred for 1.5 h. The layers were separated, and the aqueous layer was extracted with CH_2Cl_2. The combined organic layers were washed with H_2O, dried (Na_2SO_4), and concentrated under reduced pressure to give a crude compound, which was purified by column chromatography (basic alumina, hexane to EtOAc/hexane 1:9) to afford a colorless syrup; yield: 97 mg (63%).

1.5.2.3.1.1.4 Termination with a Pericyclic Reaction

In some cases, a domino process initiated by an epoxide ring-opening step can also be terminated by a pericyclic reaction, a process which is arguably the most popular way to form a ring-closure product. One significant advantage of pericyclic reactions over more standard cationic cyclizations is greater functional group tolerance, meaning that this type of domino reaction is highly practical for the synthesis of molecules with complex architectures.

One excellent example is the total synthesis of hirsutellone B (Scheme 37) as reported by Nicolaou and co-workers.[82] This molecule belongs to a class of fungal secondary metabolites, and its structure features a 6,5,6-fused tricyclic moiety and a γ-lactam embedded in a 12-membered ring as well as ten stereogenic centers. Obviously, the first major step toward the synthesis of hirsutellone B is the construction of the tricyclic core, for which a stereoselective Lewis acid induced domino epoxide opening/[4+2]-cycloaddition sequence has been designed. (R)-(+)-Citronellal is converted into the key epoxide precursor **83** for the domino process by means of classic transformations. In the critical event, treatment of epoxide **83** with diethylaluminum chloride smoothly affords tricyclic product **85** as a single diastereomer through the postulated mechanism shown in Scheme 37. This mechanism is supported by isolation of the protonated reaction intermediate **84** by

for references see p 351

quenching the reaction at −30 °C. Similarly, construction of the tricyclic core utilizing a Diels–Alder reaction has also been reported by Sorensen, Uchiro, Roush, and Kobayashi, albeit in a stepwise manner.[83]

Scheme 37 Total Synthesis of Hirsutellone B[82]

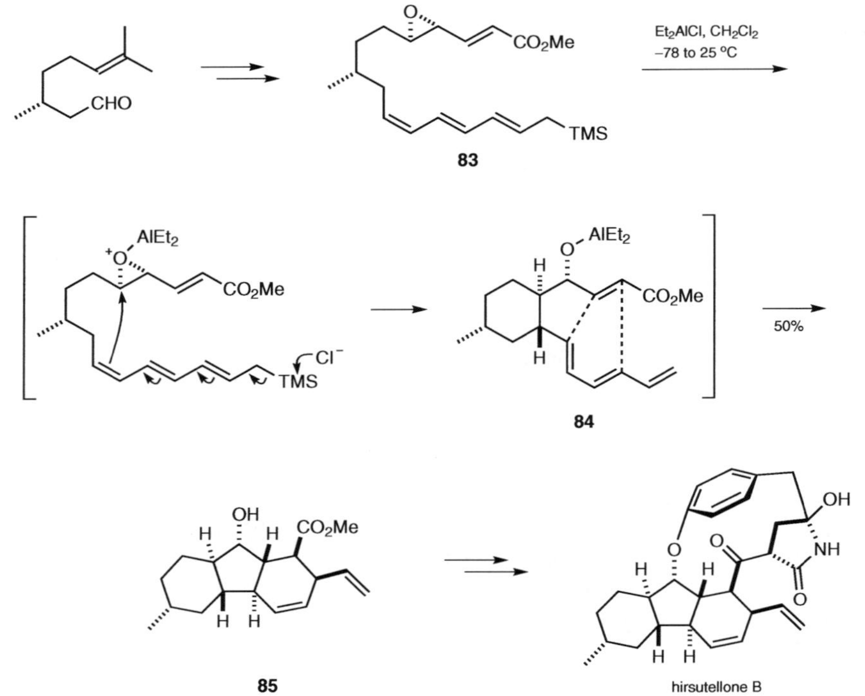

Methyl (1S,2S,4aS,4bS,6R,8aR,9S,9aS)-9-Hydroxy-6-methyl-2-vinyl-2,4a,4b,5,6,7,8,8a,9,9a-decahydro-1H-fluorene-1-carboxylate (85); Typical Procedure:[82]

To a stirred soln of epoxide **83** (100 mg, 0.276 mmol) in CH_2Cl_2 (10 mL) at −78 °C was added 1.0 M Et_2AlCl in hexanes (2.8 mL, 2.8 mmol, 10.0 equiv) dropwise. The resulting homogeneous soln was allowed to warm to −50 °C for 30 min, and then slowly to rt overnight (12 h). The reaction was then quenched with sat. aq $NaHCO_3$ (10 mL) and the mixture was extracted with EtOAc (3 × 10 mL). The combined organic phases were washed with brine (20 mL), dried ($MgSO_4$), and concentrated. Purification of the residue was carried out by flash column chromatography (silica gel, EtOAc/hexanes 1:3) to give a white solid; yield: 40 mg (50%).

1.5.2.3.1.2 Initiation with a Carbonyl and Its Derivatives

The second type of functional group that can be used as the initiator for ring-closure procedures are carbonyls and their derivatives, such as O,O- and O,N-acetals (Scheme 38). Due to the presence of the electronegative oxygen or nitrogen atom, such functional groups can interact with Lewis acids to give a cation either as an oxocarbenium or an aminocarbenium species: reactive intermediates that are capable of inducing subsequent transformations. In contrast to those initiated by epoxide opening, this type of reaction can take place either intermolecularly or intramolecularly. For the intramolecular cases, one advantage of this type of reaction is that carbocyclic or heterocyclic rings with an oxygen or nitrogen atom can be formed selectively, critically with proper placement of the

1.5.2 Lewis Acid/Base Induced Reactions

307

heteroatom. Generally, such cation-initiated domino reactions involving a ring-closure procedure can be terminated by a nucleophile or a rearrangement process. The latter variant is now broadly known as the Prins–pinacol rearrangement.[84,85] Despite the broad application of this type of transformation, the corresponding enantioselective versions are still rare.

Scheme 38 Carbonyl-Type Initiators

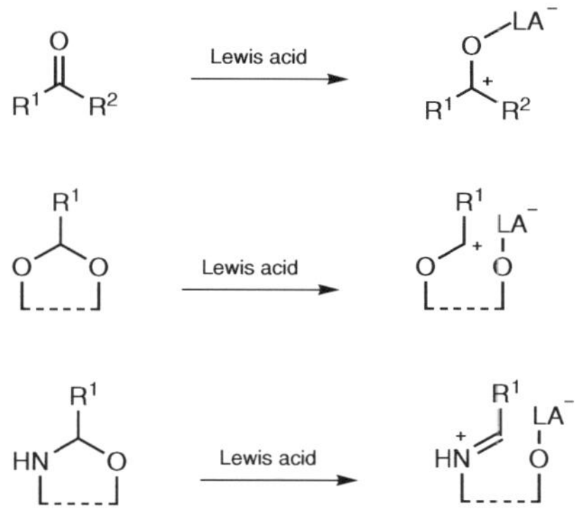

1.5.2.3.1.2.1 **Termination with a Nucleophile**

The most common way to terminate a domino cyclization initiated by a carbonyl or its derivatives is through the use of a nucleophile. Johnson and Kinnel first reported acetal-initiated cationic polyene cyclization.[86] Since their pioneering work, such reactions have been further developed for the efficient construction of polycycles.

In the case of intramolecular reactions, as mentioned in the previous section, both fully carbocyclic and oxacyclic rings can be forged selectively. One typical example introduced here is the first construction of 3-oxaterpenoids through the Prins–polyene cyclization as developed by Loh and co-workers.[87] As shown in Table 7, this transformation has very wide generality with respect to the reaction components in dichloromethane when promoted by indium(III) bromide. Firstly, a number of carbonyl compounds are compatible with this reaction; both aromatic and aliphatic aldehydes can be applied to afford the desired cyclic product in good to excellent yield regardless of the existence of a bulky substituent. Also, α,β-unsaturated aldehydes as well as ketones such as acetone and cyclohexanone are suitable substrates. Secondly, this domino reaction is also applicable to several homoallylic alcohol components (either primary or secondary). For those substrates with a phenyl ring, the electronic effect of its substituents does not show any clear influence on the reaction yield. Meanwhile, when other aromatic rings, such as furan and indole, are used, the results are also satisfactory. Additionally, an internal alkyne can also be used as the terminating group to give a bromoalkene product. Finally, just like those events that have been fulfilled in other cationic polycyclization reactions, stereoselective tetracyclic and pentacyclic skeletons are accessible by proper introduction of the diene and triene moieties, respectively.

for references see p 351

Table 7 Indium(III) Bromide Promoted Prins–Polyene Cyclizations[87]

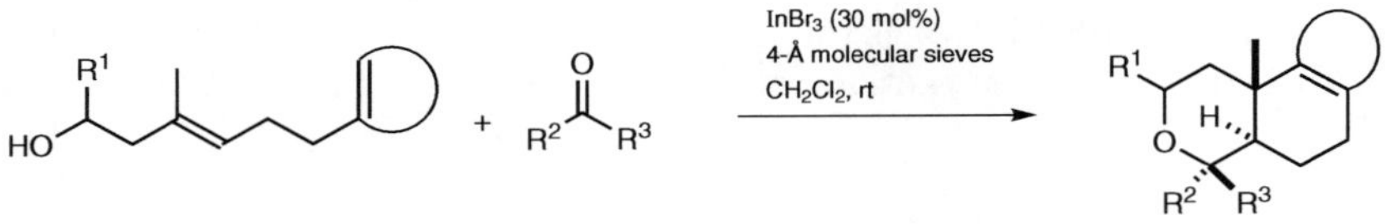

Substrates		Product	Yield (%)	Ref
Homoallylic Alcohol	Carbonyl			
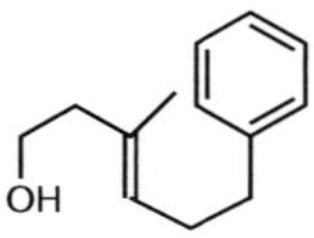	PhCHO	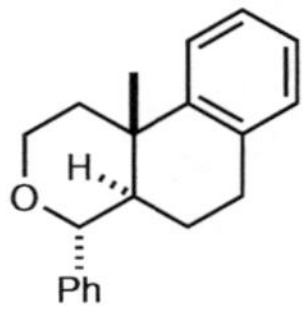	88	[87]
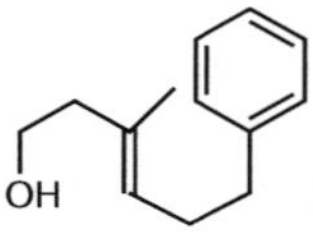	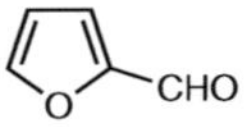 [furan]CHO	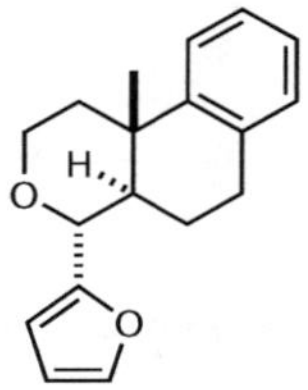	79	[87]
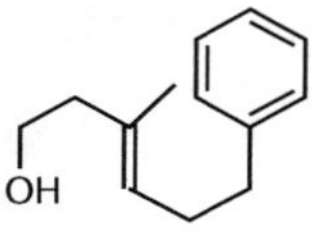	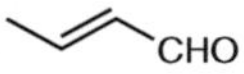 [CH=CH]CHO	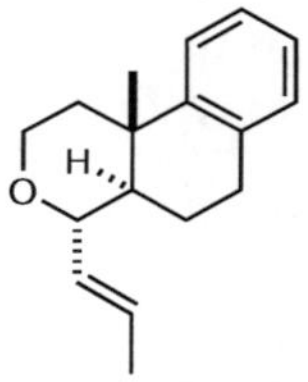	90	[87]
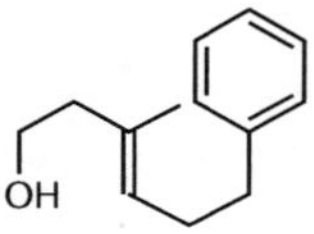	t-BuCHO	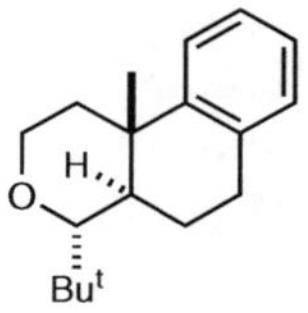	79	[87]
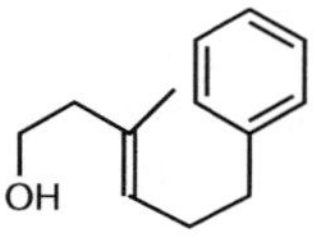	 acetone	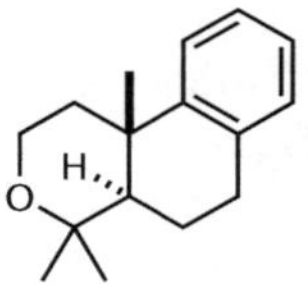	73	[87]
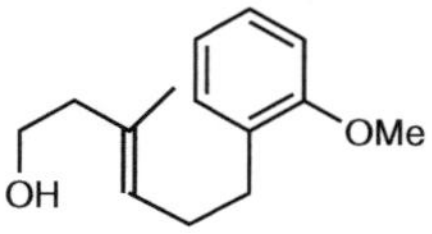	PhCHO	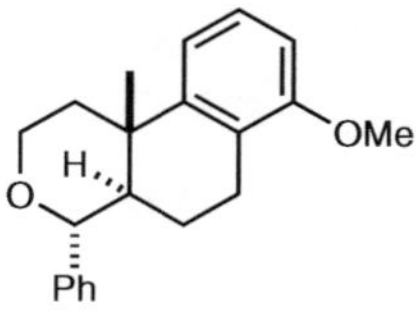	94	[87]

1.5.2 Lewis Acid/Base Induced Reactions **309**

Table 7 (cont.)

Substrates		Product	Yield (%)	Ref
Homoallylic Alcohol	Carbonyl			
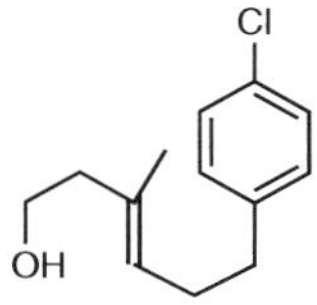	PhCHO	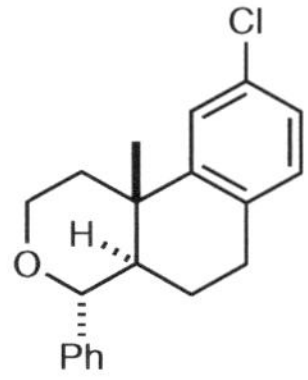	81	[87]
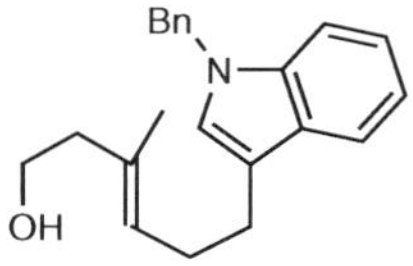	PhCHO	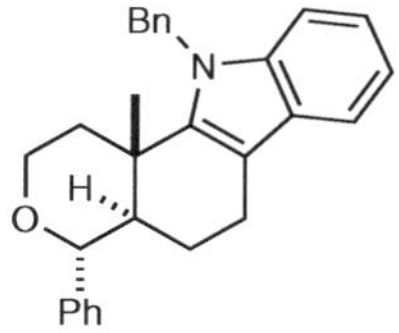	82	[87]
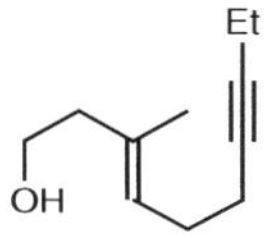	PhCHO	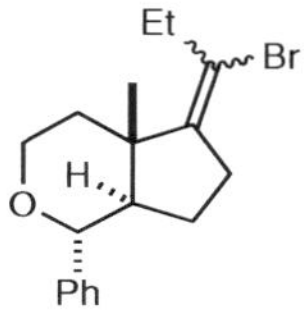	68	[87]
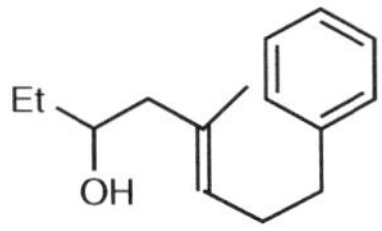	PhCHO	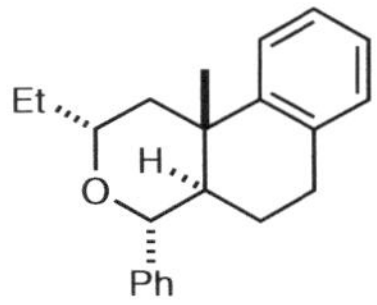	75	[87]
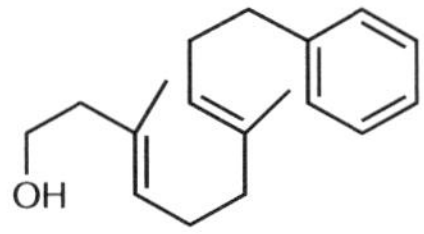	PhCHO	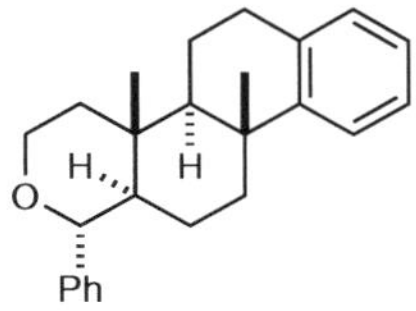	81	[87]
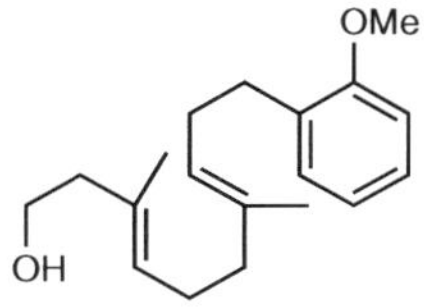	PhCHO	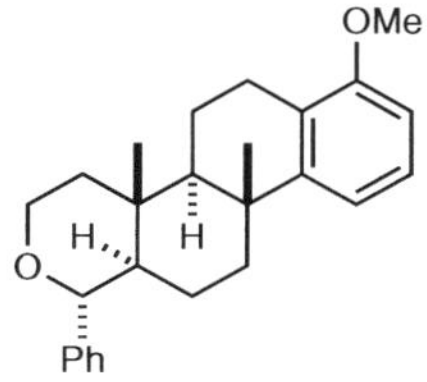	87	[87]

for references see p 351

Table 7 (cont.)

Substrates		Product	Yield (%)	Ref
Homoallylic Alcohol	Carbonyl			
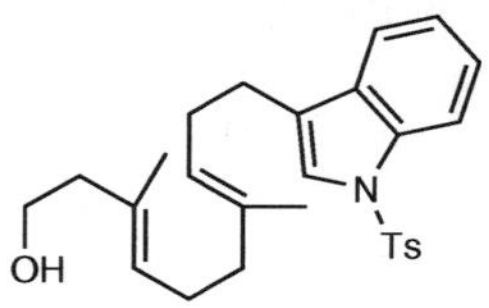	PhCHO	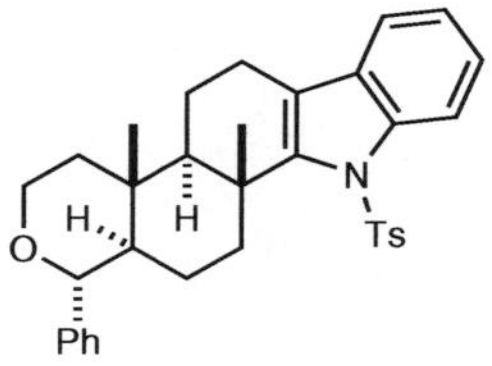	66	[87]
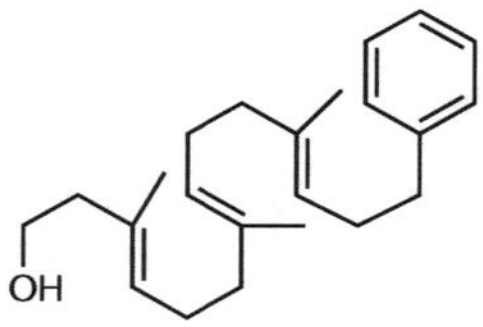	PhCHO	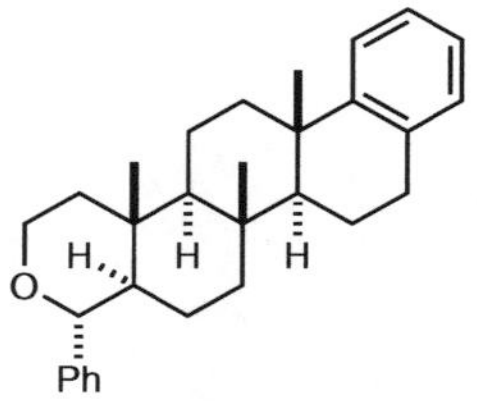	59	[87]

Using this Prins–polyene cyclization to give product **86** as the key step, Loh and co-workers completed the total synthesis of (±)-moluccanic acid methyl ester (Scheme 39).[87]

Scheme 39 Total Synthesis of (±)-Moluccanic Acid Methyl Ester[87]

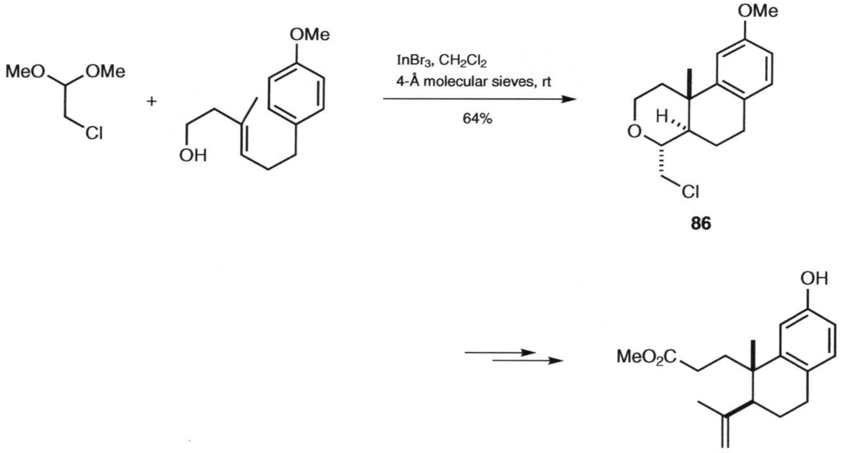

(±)-moluccanic acid methyl ester

Theoretically, intermolecular Prins–polyene cyclizations do have some potential advantages over intramolecular variants. Firstly, independent preparation of the acetal moiety can simplify the overall synthesis. Secondly, without having to connect the acetal part di-

1.5.2 Lewis Acid/Base Induced Reactions

rectly, the acyclic precursor becomes more flexible toward the desired cyclization; however, the corresponding flexibility can also be a problem for the diastereoselectivity. Loh's group has performed systematic studies on this kind of reaction (Table 8).[88–90] As indicated, the acetal used clearly affects the ratio between the desired polycyclic product and the monocyclized product. Among the acetals tested, cyclic variants give better results than the acyclic ones. Moreover, the asymmetric version of this reaction also has been attempted by using an enantiopure acetal component and applied in the enantioselective total synthesis of antiochic acid (Scheme 40).[91] Nonetheless, the moderate diastereomeric ratios of the products still affect the reaction's application to other contexts.

Table 8 Intermolecular Prins–Polyene Cyclizations[88]

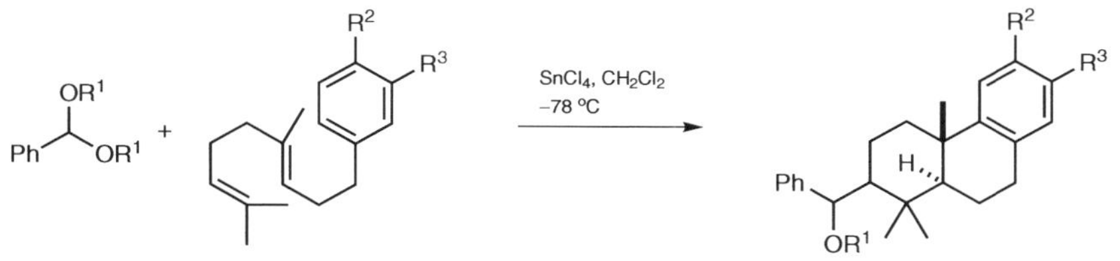

Substrates		Product(s)	Yield (%)	Ref
Acetal	Diene			
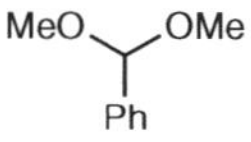 MeO OMe (Ph)	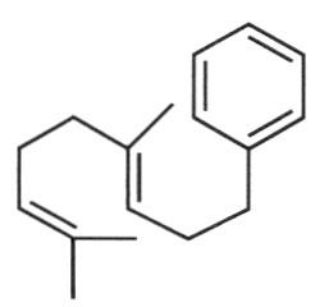	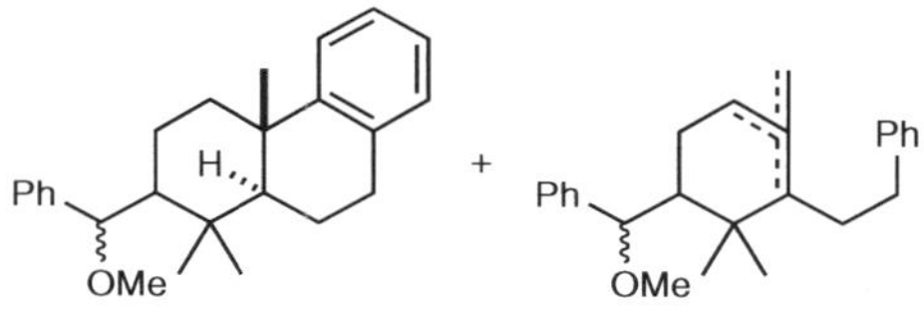 77:10	87	[88]
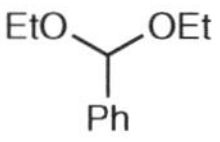 EtO OEt (Ph)	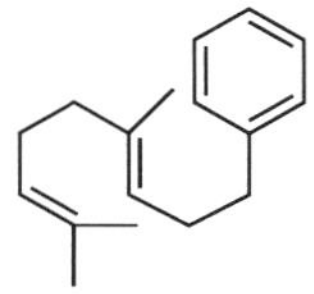	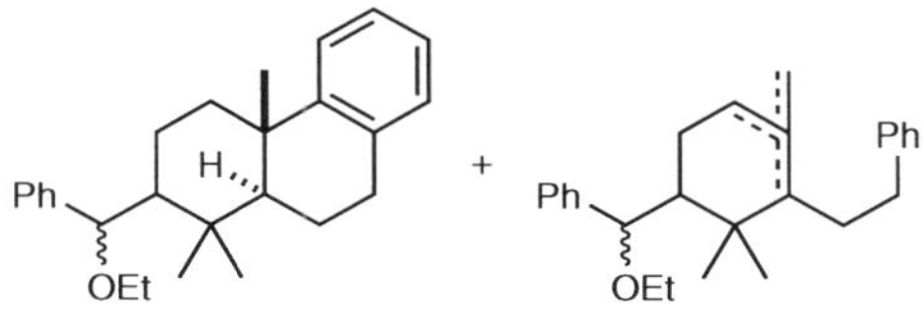 77:13	90	[38]
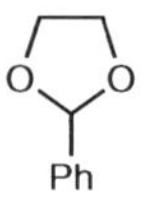 (dioxolane, Ph)	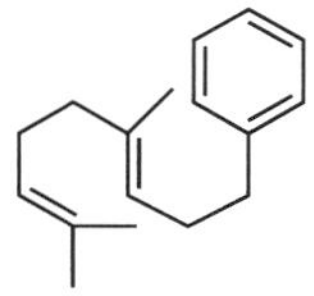	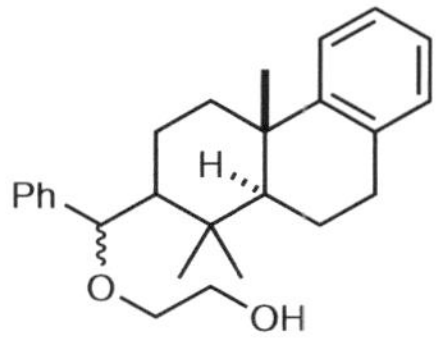	72	[88]
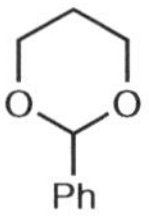 (dioxane, Ph)	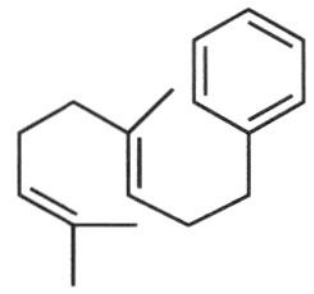	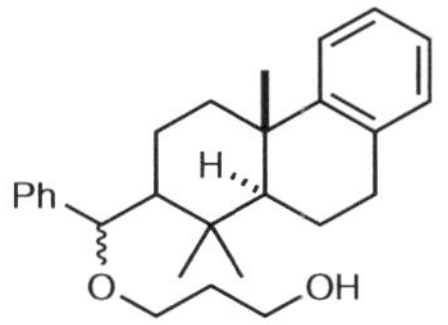	76	[88]

for references see p 351

Table 8 (cont.)

Substrates		Product(s)	Yield (%)	Ref
Acetal	Diene			
			89	[88]
		59:14:16		
			87	[88]
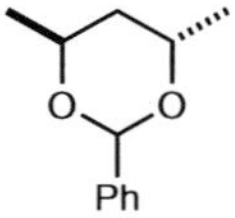	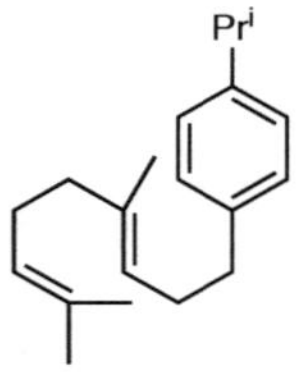	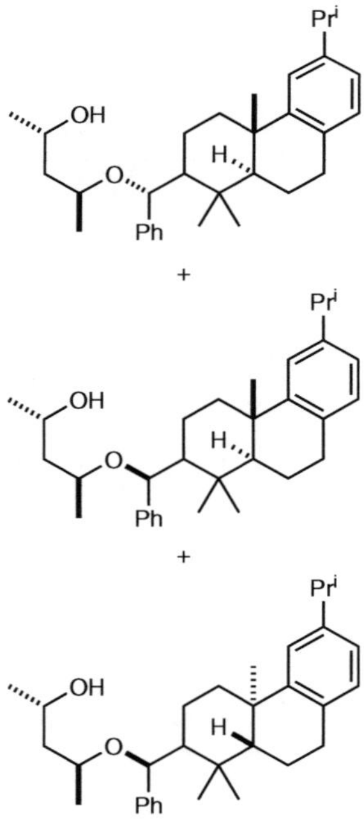 64:11:12		

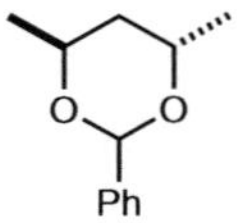

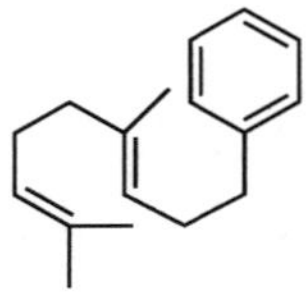

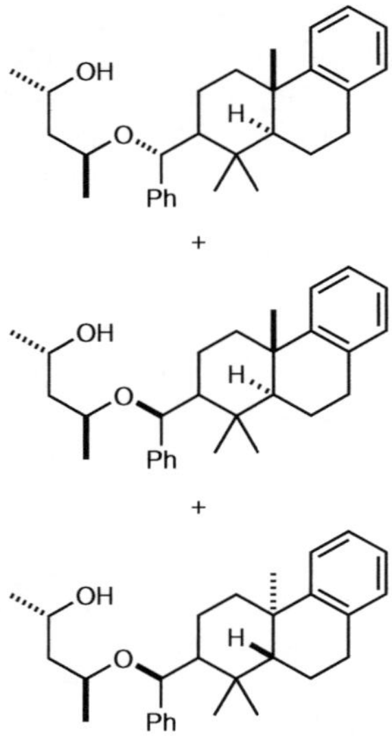

1.5.2 Lewis Acid/Base Induced Reactions

313

Scheme 40 Total Synthesis of Antiochic Acid[91]

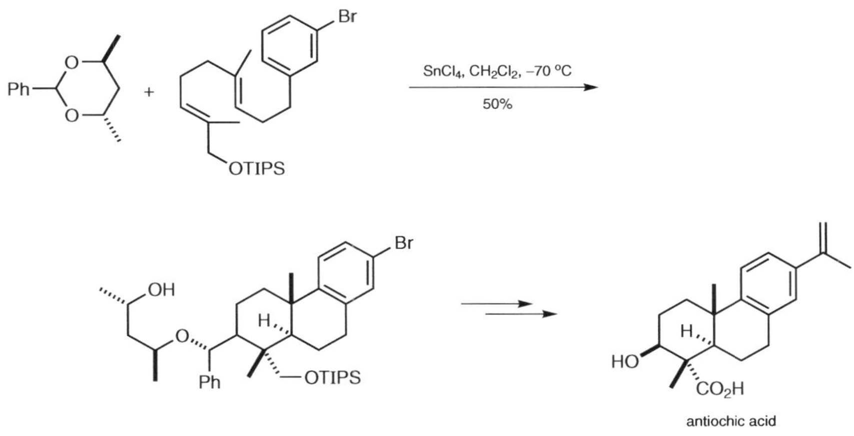

antiochic acid

Another type of carbonyl-initiated domino cyclization is the construction of oxa-polycy-clic skeletons. As acetal or ketal moieties are commonly made from an oxocarbenium spe-cies, such transformations can sometimes be used in a domino manner for the formation of structural skeletons containing ketal-type motifs. One typical example is the total syn-thesis of azaspiracid-1 (Scheme 41).[92,93] In this interesting and informative work, notwith-standing the importance of the total synthesis for structure elucidation,[94–101] Nicolaou and co-workers also developed many useful transformations, one of which is the Lewis acid induced domino cyclization leading to the trioxadispiroketal ABCD-ring domain of azaspiracid-1. During their attempts to identify the correct structure of natural azaspira-cid-1, several substrates were evaluated in this domino cyclization. Both protic and Lewis acids can trigger this reaction, with trimethylsilyl trifluoromethanesulfonate giving the best result (Table 9). Also, it should be noted that changing the reaction temperature leads to variation of the diastereomeric ratio of the products.

Scheme 41 Structure of Azaspiracid-1[92,93]

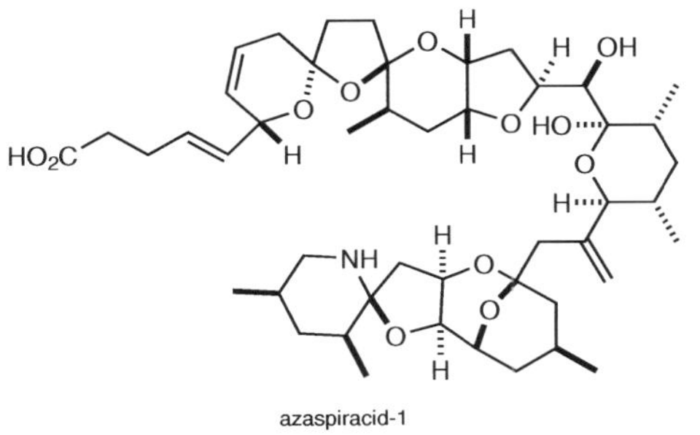

azaspiracid-1

for references see p 351

Table 9 Trimethylsilyl Trifluoromethanesulfonate Promoted Domino Cyclizations[94,101]

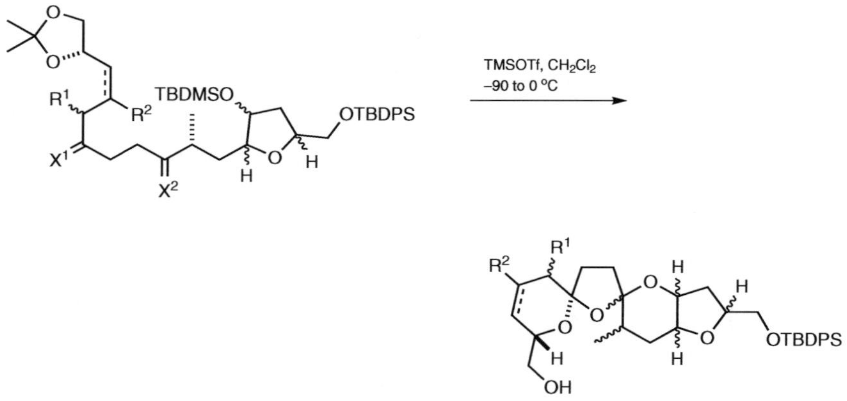

Substrate	Product	Yield (%)	Ref
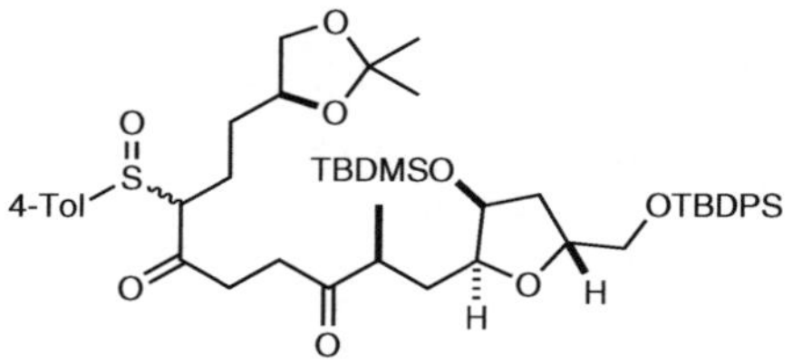	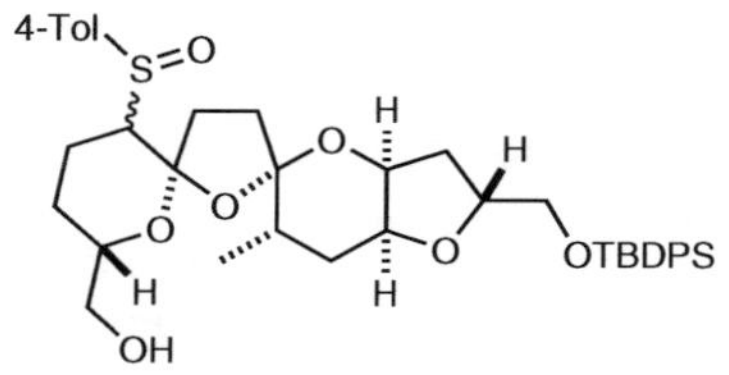	62	[94]
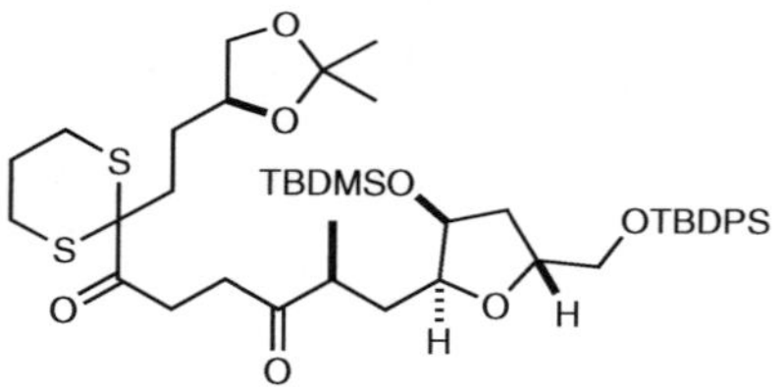	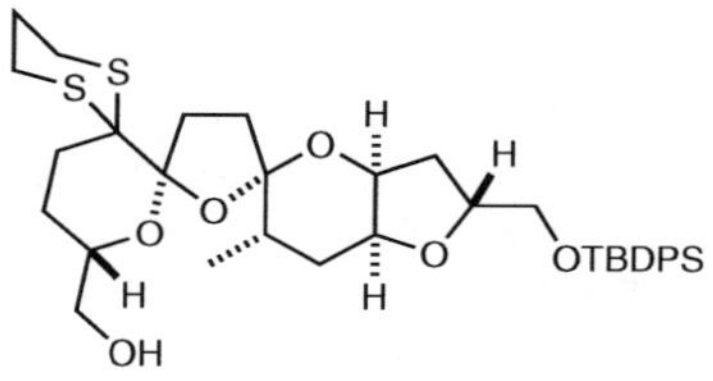	85	[94]
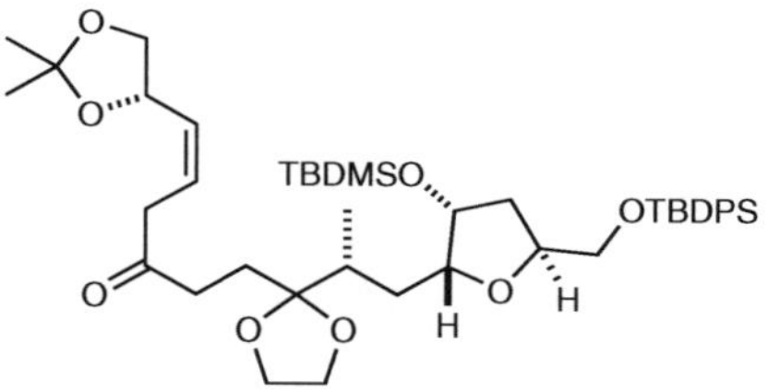	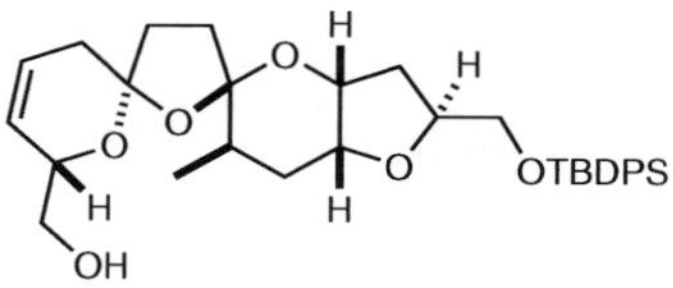	65	[101]
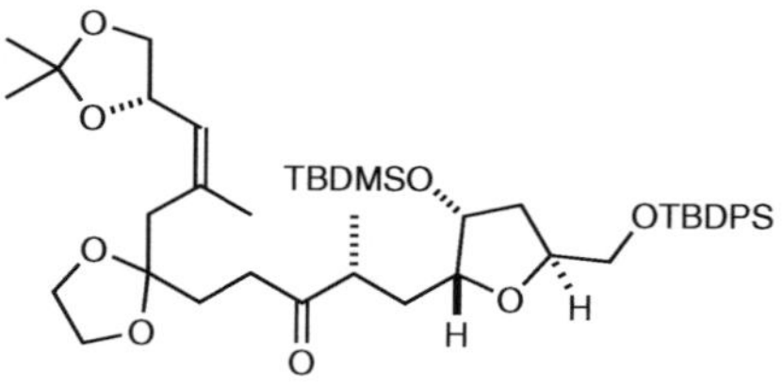	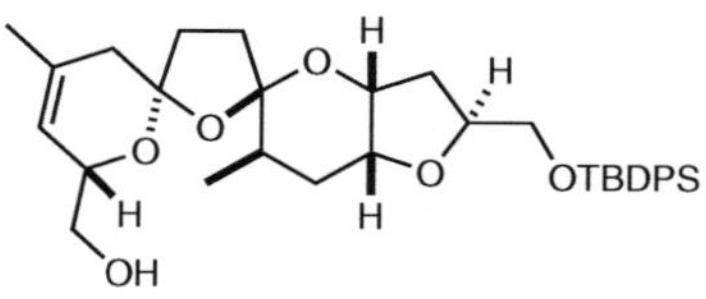	72	[101]

Another similar example is the assembly of the pseudoester cage ring system of (+)-CP-263,114 as reported by Shair and co-workers (Scheme 42).[102] In a very efficient total synthesis of the target molecule, treatment of advanced intermediate **87** with trimethyl

1.5.2 Lewis Acid/Base Induced Reactions

orthoformate and trimethylsilyl trifluoromethanesulfonate leads to the desired domino reaction, affording product **88** in nearly quantitative yield. During this reaction, along with the installation of the cage ring system and C14 quaternary carbon center, a free carboxy group is also released for the following homologation.

Scheme 42 Total Synthesis of (+)-CP-263,114[102]

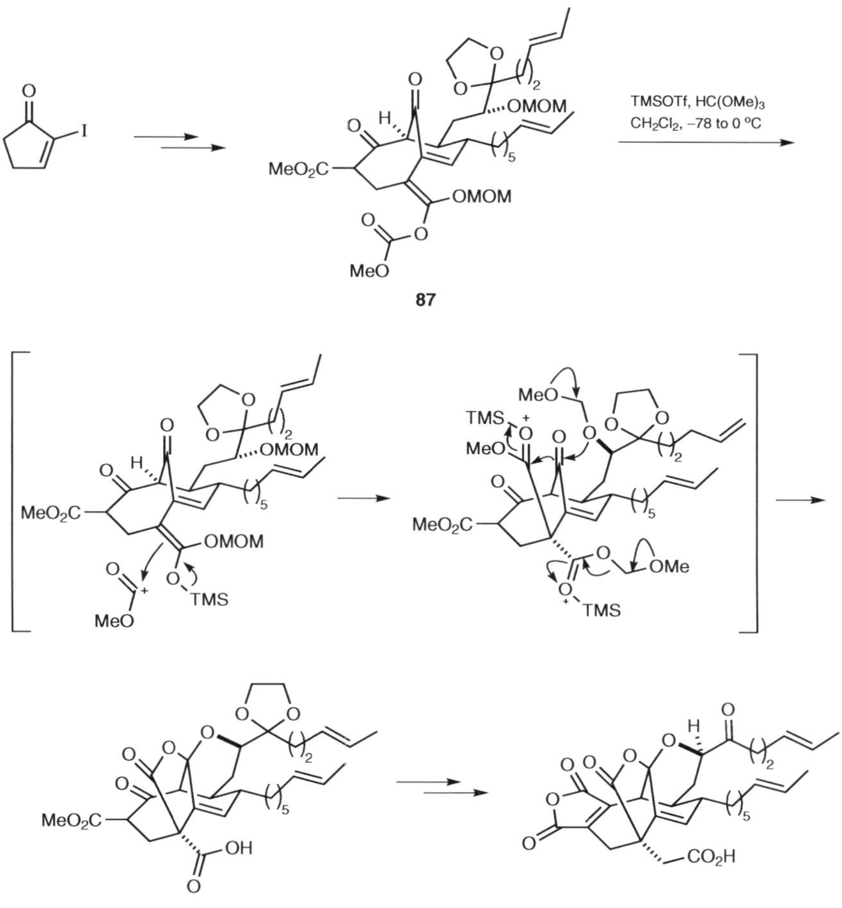

(4R*,4aR*,10bS*)-4-(Chloromethyl)-9-methoxy-10b-methyl-2,4,4a,5,6,10b-hexahydro-1H-benzo[f]-2-benzopyran (86); Typical Procedure:[87]
An oven-dried 10-mL round-bottomed flask equipped with a magnetic stirrer bar was charged with 4-Å molecular sieves (100 mg) and InBr$_3$ (0.06 mmol) and sealed with a rubber septum. Then, (E)-6-(4-methoxyphenyl)-3-methylhex-3-en-1-ol (0.20 mmol) and 2-chloro-1,1-dimethoxyethane (0.24 mmol, 1.2 equiv) were dissolved in dry CH$_2$Cl$_2$ (2 mL) and added via syringe. The soln was stirred at rt for 24–96 h. The reaction was quenched with sat. aq NaHCO$_3$ (5 mL) and the mixture was extracted with CH$_2$Cl$_2$ (3 × 20 mL). The combined organic layers were washed with brine (30 mL), dried (Na$_2$SO$_4$), and concentrated under reduced pressure. The crude product was purified by flash column chromatography (silica gel, hexane/EtOAc); yield: 64%.

for references see p 351

1.5.2.3.1.2.2 **Termination with a Rearrangement**

As mentioned in Section 1.5.2.1.4 on Prins–pinacol rearrangements, the Prins cyclization can also be followed in situ by a pinacol-type (or semipinacol-type) rearrangement to give a cyclic product. This type of domino transformation was first reported by Mousset and co-workers and further developed by the Overman group.[84,85] Now it is broadly named as the Prins–pinacol rearrangement. Similar to Prins–polyene cyclizations, it can be initiated intermolecularly or intramolecularly. For the intramolecular cases, a ring-closure process is generally involved, while the intermolecular processes remain underexplored and often suffer from mediocre stereoselectivity.

Since this type of transformation as initiated by an oxocarbenium ion has been mentioned earlier (Section 1.5.2.3.1.2), here another versatile cyclization reaction involving an iminium species is introduced that has been extensively studied and applied in the total synthesis of a number of natural products, namely the cationic aza-Cope–Mannich reaction (Scheme 43).[103] From simply looking at the reaction starting material and product, one might think it is very likely an extension of the Prins–pinacol rearrangement; however, this possibility has been ruled out mechanistically by related experiments.[104]

Scheme 43 Aza-Cope–Mannich Reaction

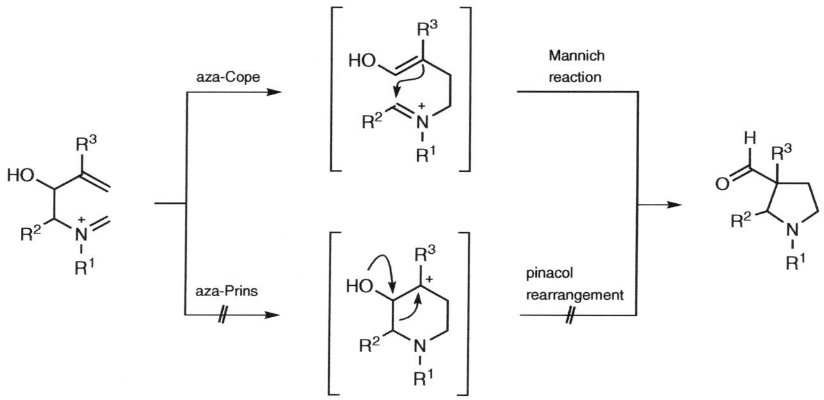

There are two key factors for the success of this domino reaction. First, the use of an iminium ion can significantly accelerate the 3,3-sigmatropic rearrangement process, a feature which has been proved by Geissman and Horowitz.[105] Secondly, incorporation of a hydroxy group at the 4-position can give a nucleophilic enol moiety after the rearrangement, and the appearance of an enol and iminium will lead to an irreversible Mannich reaction that finally terminates the domino process to give acyl-substituted pyrrolidines. The beauty of this transformation is that it can be promoted by different conditions, such as heating or exposure to Lewis and protic acids. Furthermore, because the aza-Cope rearrangement normally goes through a chair-like transition state, most of these reactions have good diastereoselectivity. As a result, it has found broad application in the total synthesis of many natural products, especially alkaloids (Scheme 44).[106–109]

1.5.2 Lewis Acid/Base Induced Reactions **317**

Scheme 44 Representative Alkaloids

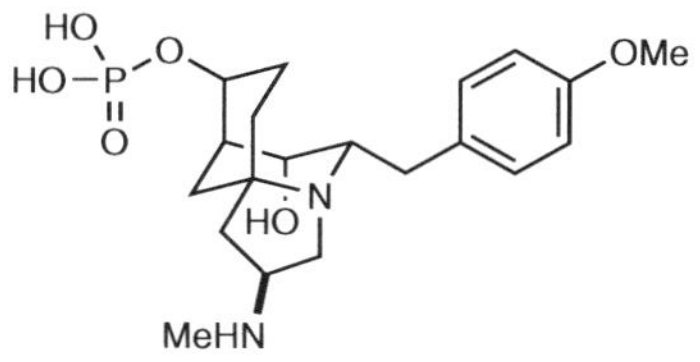

FR901483

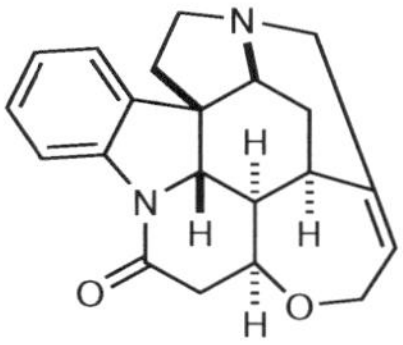

(−)-strychnine

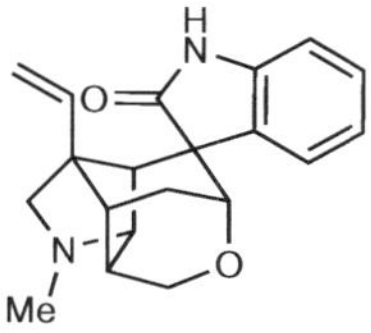

gelsemine

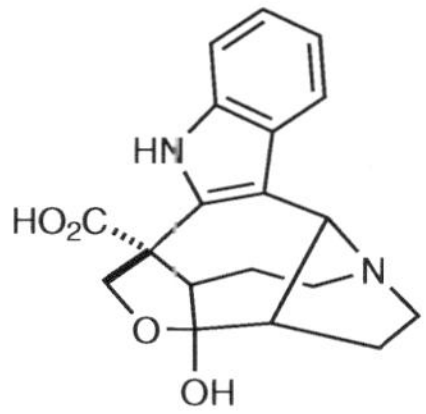

actinophyllic acid

Although the cationic aza-Cope–Mannich reaction is usually promoted by a protic acid or silver(I) salt, the use of Lewis acids can also provide satisfactory results. For example, during the synthesis of (±)-pancracine (Scheme 45),[110] such a rearrangement was designed for the stereoselective establishment of the hydroindolone skeleton. After it was determined that the use of protic acids does not afford the desired product in acceptable yield, Lewis acids were found to be suitable promoters. Among the Lewis acids screened, boron trifluoride–diethyl ether complex is the most effective, giving the desired hydroindolone **89** in nearly quantitative yield. By applying this domino methodology, Overman and co-workers have achieved the total synthesis of (±)-pancracine in 13 steps and 14% overall yield.

for references see p 351

Scheme 45 Total Synthesis of (±)-Pancracine[110]

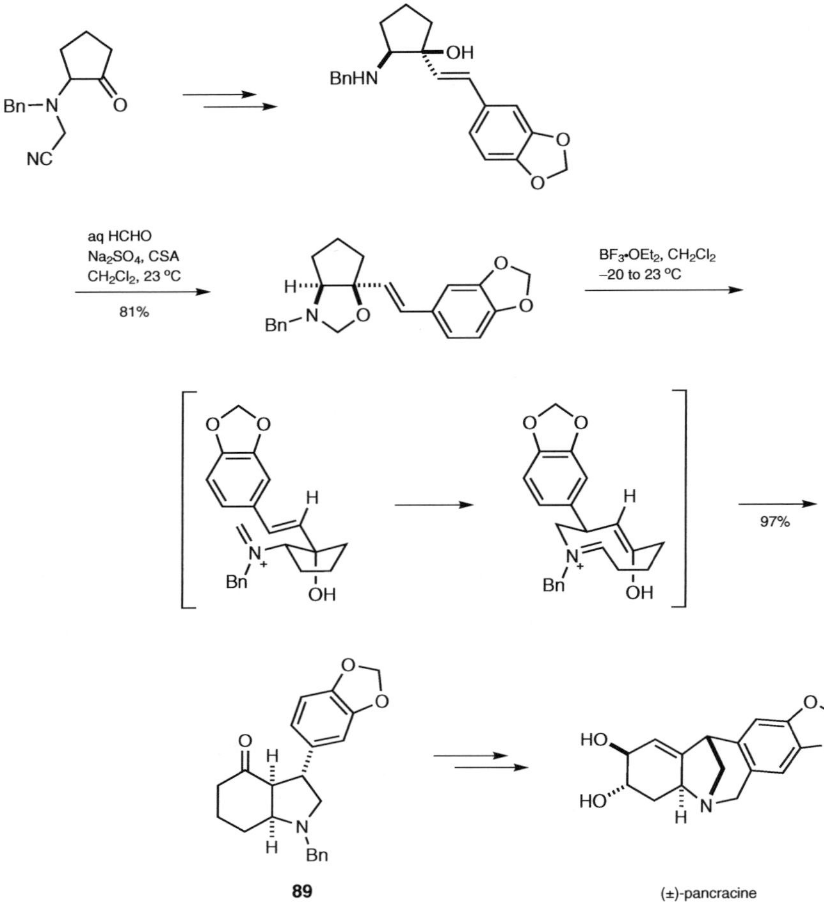

89

(±)-pancracine

1.5.2.3.1.3 Initiation by Activation of a π-Bond

Besides the epoxide and carbonyl functional group, a π-bond also can be used to generate a carbocation. Due to its special nucleophilicity, the π-bond is capable of reacting with various electrophiles such as a proton, Lewis acid, π-acid, or halonium to initiate a domino transformation. Similarly, such a process involving a ring-closure procedure can be terminated with a nucleophile or a rearrangement.

1.5.2.3.1.3.1 Initiation by a Lewis Acid

Biomimetic polyene cyclizations have attracted the broad attention of chemists, and are covered in more detail in Sections 1.1.1.1 and 1.2.3.1. As well as protons, the use of Lewis acids and π-acids as promoters has been investigated for a long time. As a result, in addition to mercury(II) salts, some other Lewis-acid-type reagents have proven to be effective in this type of domino reaction. Additionally, some such reactions can be carried out enantioselectively in the presence of a chiral ligand.

Yamamoto and co-workers first developed the enantioselective cyclization of polyenes induced by a Lewis acid assisted chiral Brønsted acid (LBA).[111–114] Due to the interaction between the two acids, not only does the acidity of the proton in the Brønsted acid increase, but also the way in which the LBA approaches the substrate is restricted, which

is sometimes ideal for enantioinduction. During a systematic study of this topic, a number of chiral LBAs and substrates have been employed. As shown in Scheme 46, this reaction is efficient for the enantioselective construction of polycyclic compounds. The yield and selectivity are clearly affected by the solvent used. Among them, dichloromethane and toluene are generally best. The former gives better yields and the latter affords better enantioselectivity. As for the substrates, both hydroxy and aryl groups can be used as terminators. Although the use of a less reactive nucleophile such as an aryl group usually gives the cyclization products in lower yield, the corresponding diastereoselectivity is normally higher than those for the sequences terminated by a hydroxy group. Critically, the lower yield can be overcome by the application of stepwise cyclizations promoted by chiral LBAs and stronger achiral LBAs. Additionally, for substrates with a phenyl ring as terminator, the substituents on the aromatic ring can affect the stereoselectivity of the second cyclization. For those with an electron-donating group *para* to the final ring-closure position, lower selectivity is usually observed. In contrast, substrates without the electron-donating substituent will lead to better selectivity. Also, it is worth mentioning that an abnormal cyclization involving a Claisen rearrangement has been observed with an aryl terminator, a process which provides a more efficient way to access the same product as that derived from a hydroxy terminator. Overall, this method has provided an alternative approach to polycyclic compounds, while the moderate enantioselectivity still gives plenty of room for further investigation and enhancements.

Scheme 46   Lewis Acid Assisted Chiral Brønsted Acid Catalyzed Cyclization of Polyenes[111,112]

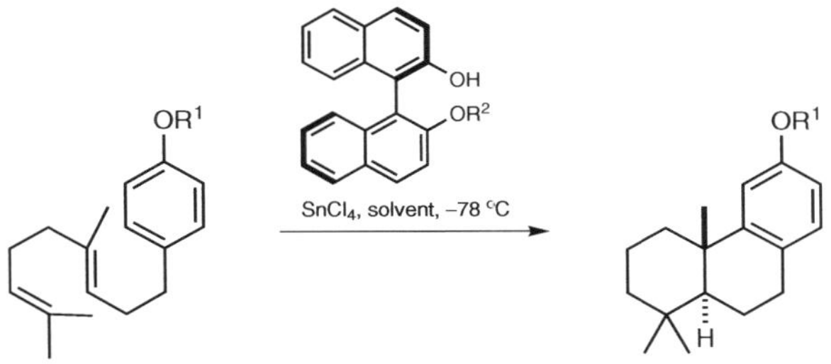

R¹	R²	Catalyst (Equiv)	Solvent	ee (%)	Yield (%)	Ref
H	Me	1.1	CH₂Cl₂	38	87	[112]
H	Me	1.1	toluene	49	28	[112]
Me	Bn	2.0	toluene	61	13	[112]

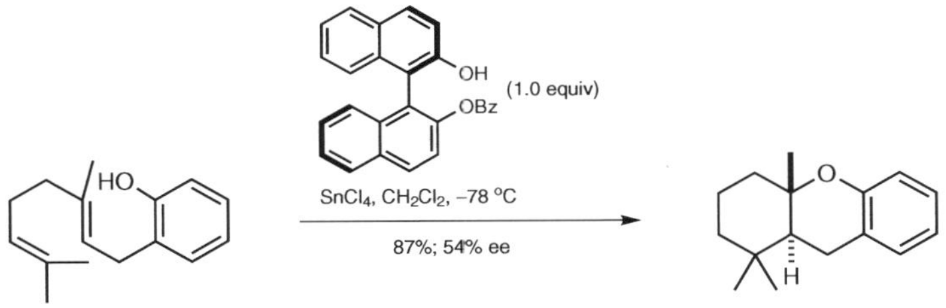

for references see p 351

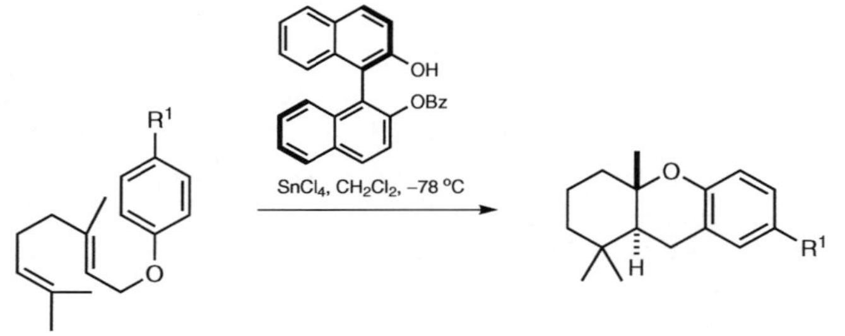

320 Domino Transformations **1.5** Non-Radical Skeletal Rearrangements

R¹	Catalyst (Equiv)	ee (%)	Yieldᵃ (%)	Ref
H	1.1	69	98	[111]
Br	0.2	87	85	[111]

ᵃ Determined by GC.

Applying this methodology, Yamamoto and co-workers have completed the total synthesis of various natural products, such as (−)-chromazonarol which is formed from cyclization product **91** generated from polyene **90** (Scheme 47).[114]

Scheme 47 Total Synthesis of (−)-Chromazonarol[114]

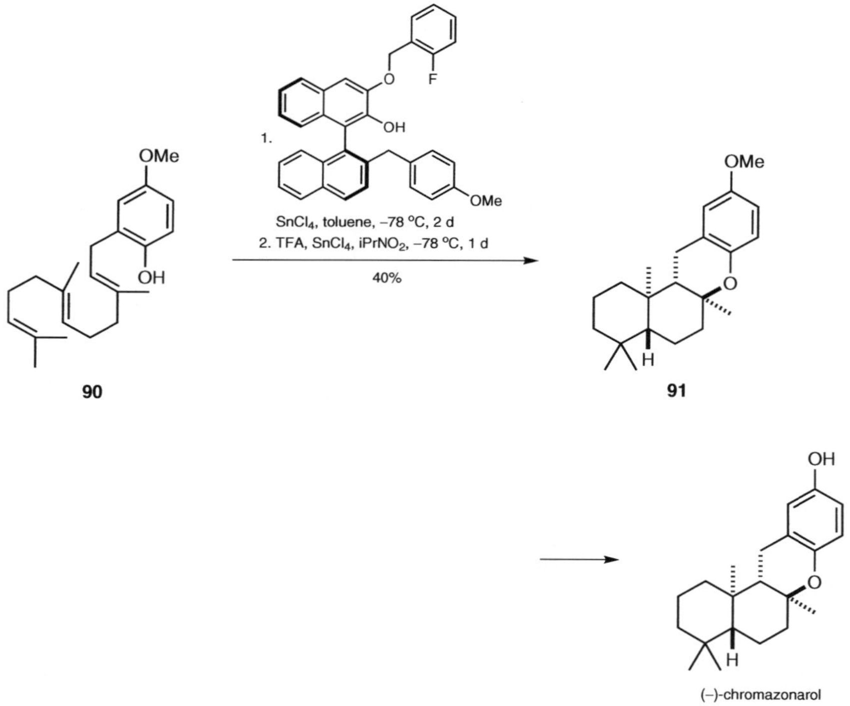

(4aR,6aS,12aS,12bR)-10-Methoxy-4,4,6a,12b-tetramethyl-2,3,4,4a,5,6,6a,12,12a,12b-deca-hydro-1H-benzo[a]xanthene (91); Typical Procedure:[114]
To a soln of (S)-3-(2-fluorobenzyloxy)-2′-(4-methoxybenzyl)-1,1′-binaphthyl-2-ol (100 mg, 0.20 mmol) in toluene (2 mL) was added 1.0 M SnCl₄ in hexane (0.10 mL, 0.10 mmol) at rt, and the mixture was stirred for 5 min. After the soln was cooled to −78 °C, polyene **90** (37 mg, 0.1 mmol) was added. The mixture was stirred for 2 d, the reaction was quenched

1.5.2 Lewis Acid/Base Induced Reactions **321**

with sat. aq $NaHCO_3$, and the resulting mixture was extracted with Et_2O. The combined organic phases were dried ($MgSO_4$) and concentrated. (*S*)-3-(2-Fluorobenzyloxy)-2′-(4-methoxybenzyl)-1,1′-binaphthyl-2-ol was removed from the crude products by column chromatography (silica gel, hexane/Et_2O). The mixture of cyclic products was dissolved in $iPrNO_2$ (2 mL), and 1.0 M $SnCl_4$ in hexane (0.20 mL, 0.20 mmol) and TFA (77 μL, 1.0 mmol) were added at −78 °C. After being stirred for 1 d, the reaction was quenched with sat. aq $NaHCO_3$ and the resulting mixture was extracted with Et_2O. The combined organic phases were dried ($MgSO_4$) and concentrated. The crude product was purified by column chromatography (silica gel, hexane/Et_2O); yield: 15 mg (40%).

1.5.2.3.1.3.2 Initiation by a π-Acid

In the past several decades, the application of π-acids in organic synthesis has drawn broad attention from chemists.[115–120] Due to their ability to activate different types of π-system and their high functional-group tolerance, they have been used in a number of domino reactions. As expected, the use of different π-system will lead to different reaction patterns.

1.5.2.3.1.3.2.1 Activation of Alkenes

Generally, reactions between π-acids and substrates with a C=C bond are the simplest amongst the three types of π-system most explored. Like those initiated by a Lewis acid, a carbocation or electropositive carbon center will be generated upon the coordination of the metal center to the double bond to induce the subsequent domino transformations.

Gagné and co-workers have performed a systematic study of the π-acid-catalyzed domino cyclization of polyprenoids (Table 10).[121,122] As the β-hydride elimination is usually the final step of the reaction cycle, the regioselective formation of the final double bond can be problematic. This is true in the presence of a palladium(II) catalyst but, fortunately, the use of [1,3-bis(diphenylphosphino)ethane]diiodoplatinum(II), silver(I) tetrafluoroborate, and resin-bound methyl trityl ether can give the desired product as a single diastereomer in a nitro-containing solvent. The authors attribute the excellent selectivity to the involvement of a β-agostic interaction in the reaction intermediate. It should be noted that the methyl trityl ether resin is a key factor for catalyst reoxidation, which requires the absorption of the H^+ and H^- generated from the reaction. With the optimal conditions in hand, the authors have successfully applied a series of dienols and trienols to this reaction. Like the domino cyclization of polyenes, internal *E*-alkenes always lead to *trans*-fused ring systems. In the case of the initiating alkene moieties, both mono- and disubstituted forms are applicable in this domino reaction. Additionally, for the 1,2-disubstituted initiating alkenes, the *E*- or *Z*-isomer can be stereospecifically converted into the corresponding epimer at the C4 position (Table 10, entries 7 and 9). However, substrates with trisubstituted initiating alkenes are not applicable in this reaction. Subsequently, Gagné's group has also attempted the asymmetric version of this transformation.[122] Chiral xylyl-PhanePhos **92** is found to be the best amongst the ligands screened. Interestingly, this catalyst system only works for substrates with monosubstituted or disubstituted initiating *Z*-alkene moieties, providing the desired product with moderate to good enantiomeric excess as a single diastereomer. By contrast, substrates with disubstituted initiating *E*-alkenes give poor enantioselectivities (entries 6 and 8). During the investigations, it was found that the initial domino cyclization is not the stereochemistry-determining step.

for references see p 351

Table 10 Platinum-Catalyzed Cyclization of Polyenes[121,122]

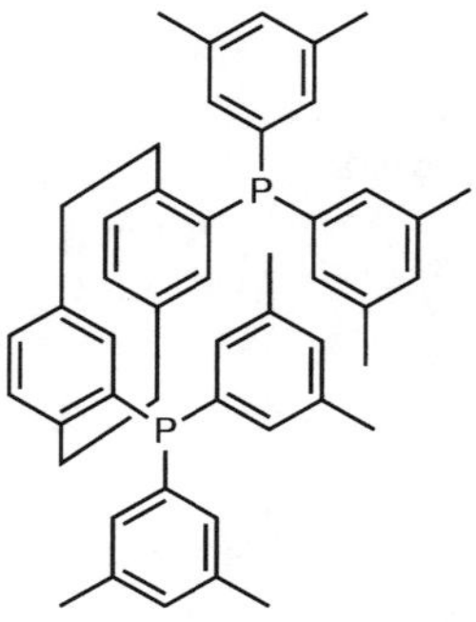

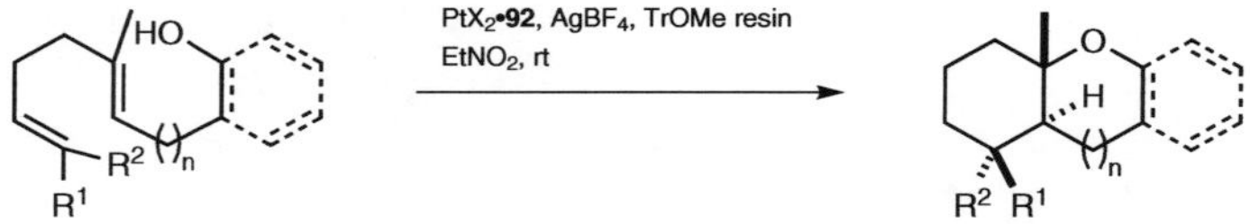

Entry	Substrate	Ligand	Product	ee[a] (%)	Yield[b] (%)	Ref
1	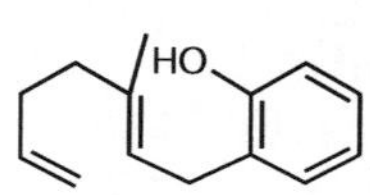	dppe	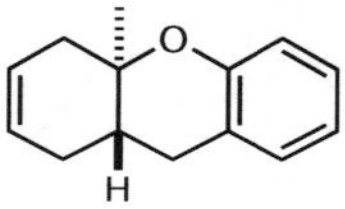	n.a.	73	[121]
2	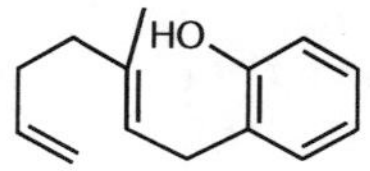	**92**	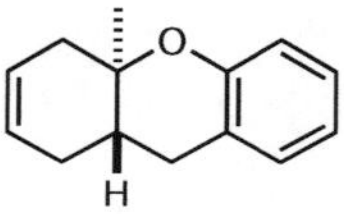	75	73	[122]
3	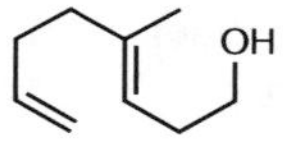	dppe	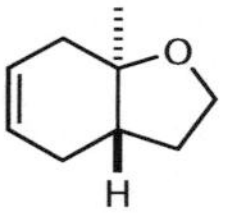	n.a.	84	[121]
4	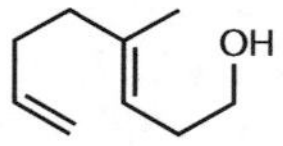	**92**	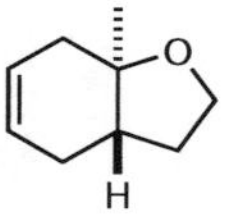	79	75	[122]
5	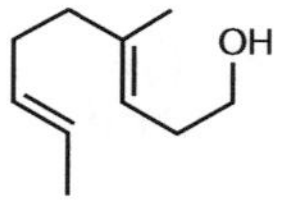	dppe	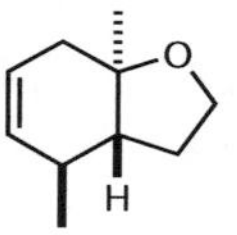	n.a.	67	[121]
6	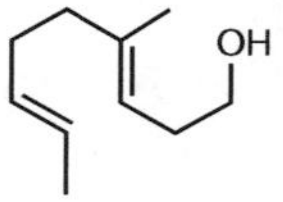	**92**	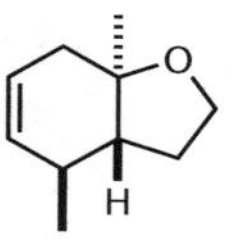	12	n.d.	[122]

1.5.2 Lewis Acid/Base Induced Reactions **323**

Table 10 (cont.)

Entry	Substrate	Ligand	Product	ee[a] (%)	Yield[b] (%)	Ref
7	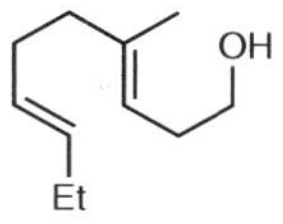	dppe	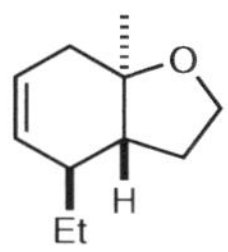	n.a.	65	[121]
8	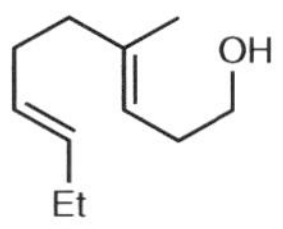	**92**	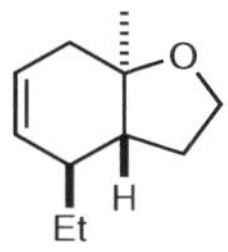	10	n.d.	[122]
9	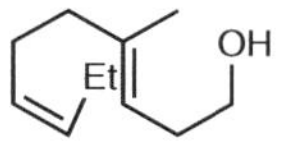	dppe	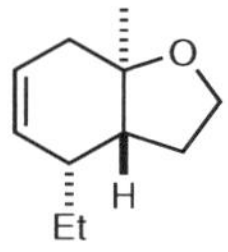	n.a.	72	[121]
10	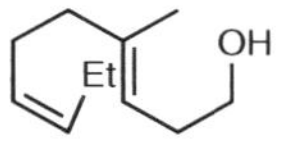	**92**	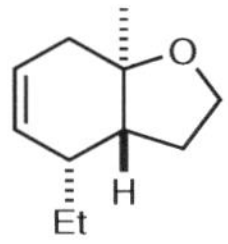	87	61	[122]
11	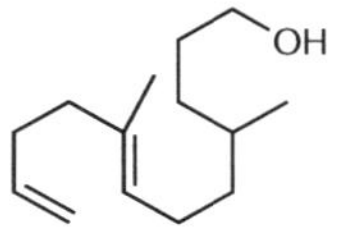	dppe	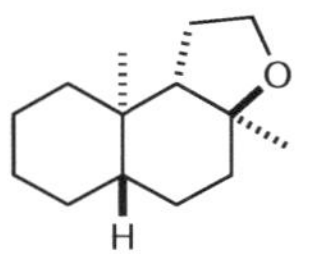	n.a.	90	[121]
12	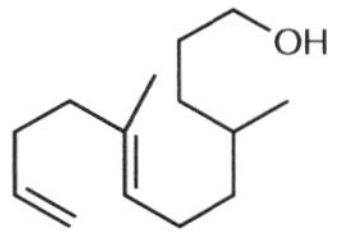	**92**	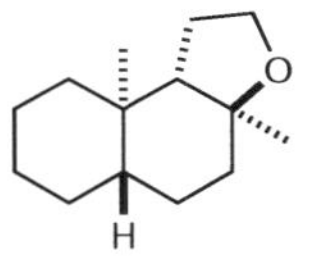	64	76	[122]

[a] n.a. = not applicable.
[b] n.d. = not determined.

Besides the domino cyclization, the carbocation generated from the activation of the C=C bond can also induce skeletal rearrangements. Orellana and co-workers have developed a palladium-catalyzed tandem semipinacol rearrangement/direct arylation reaction along these lines.[123] During this domino procedure, the coordination of the electrophilic palladium reagent to the alkene and hydroxy motif of substrate **93** readily promotes the expected 1,2-C–C migration to give a palladium homoenolate intermediate **94** selectively, which will afford the direct arylation product **95** due to the absence of a β-hydride. Subsequent investigation has found that a variety of substrates with different substituents are viable for this transformation, generating the desired products with moderate yield (Scheme 48). As well as the foreseeable low yield caused by the bromo substituent in substrate **93** (R^1 = Br), the regioselectivity of the arylation step is not always satisfactory based on the substitution pattern of the arene, for example using substrate **93** (R^1 = OEt; R^2 = OMe).

for references see p 351

Scheme 48 Palladium-Catalyzed Tandem Semipinacol Rearrangement/Direct Arylation Reaction[123]

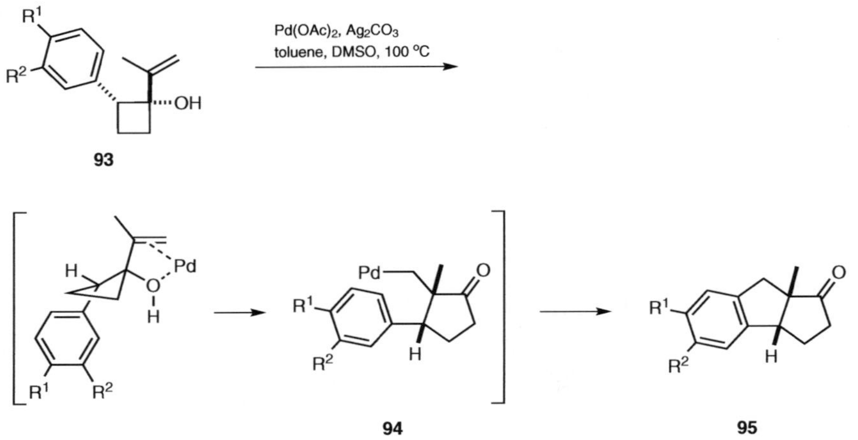

R¹	R²	Yield (%) of **95**	Ref
H	H	48	[123]
Me	H	66	[123]
F	H	45	[123]
Cl	H	48	[123]
Br	H	28	[123]
OEt	OMe	48ᵃ	[123]

ᵃ Obtained in a 3.4:1 ratio with the regioisomer (3aR*,8aS*)-6-ethoxy-7-methoxy-8a-methyl-3,3a,8,8a-tetrahydrocyclopenta[a]inden-1(2H)-one.

Using a very similar strategy, Nemoto and Ihara have accomplished the total synthesis of (+)-equilenin from diene **96** (Scheme 49).[124] In the key step, an alkene insertion is used to terminate the domino process. More importantly, the stereochemical control of the key reaction can be tuned by the solvent. As the coordination between the palladium catalyst and the hydroxy group is inhibited by the use of polar solvents such as hexamethylphosphoric triamide and tetrahydrofuran, the isomer **97B** with the desired *trans*-fused ring system becomes the major product together with the minor *cis*-fused form **97A**.

1.5.2 Lewis Acid/Base Induced Reactions **325**

Scheme 49 Total Synthesis of (+)-Equilenin[124]

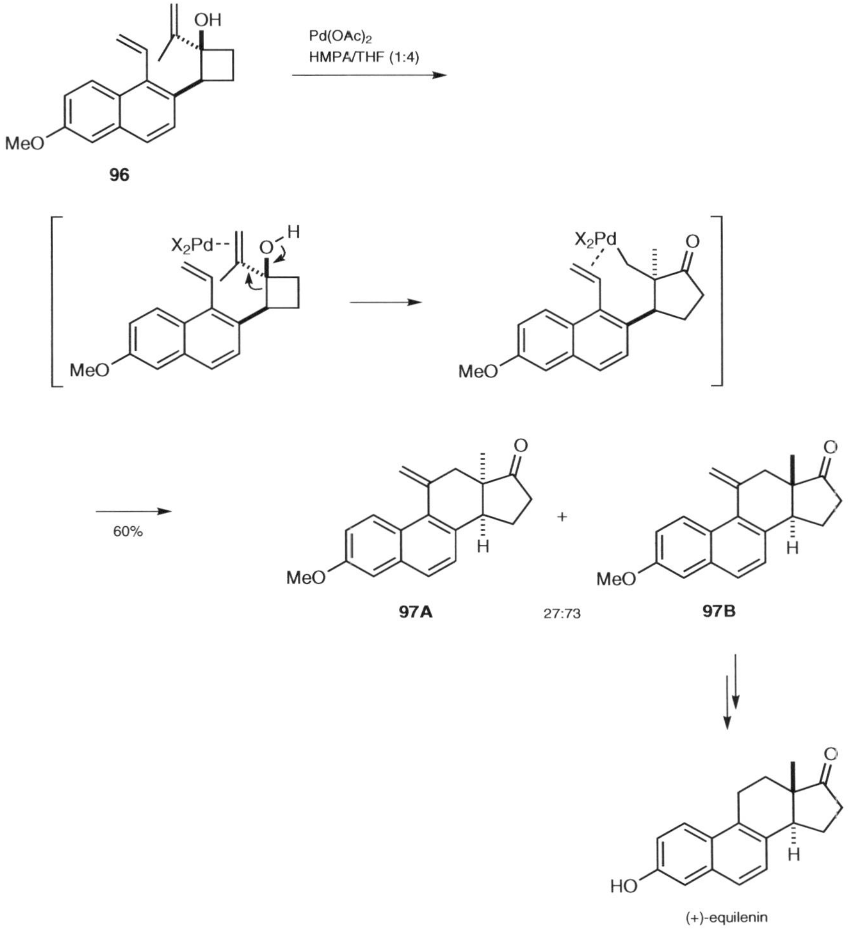

Yu and co-workers have developed a tandem rhodium(I)-catalyzed [(5+2)+1]-cycloaddition/aldol reaction.[125] During this tandem process, the first step is actually a rhodium(I)-catalyzed domino cyclization. In this one-pot reaction, a series of dienylcyclopropanes can be converted into synthetically useful intermediates with linear triquinane skeletons (Table 11). Based on the experimental results, it is clear that the reaction yield is affected by the structure of the substrates. The substitution pattern of the alk-1-enylcyclopropane moiety is of particular importance, with trisubstituted variants leading to much better yields than those that are disubstituted. Interestingly, the tethering group does not affect this reaction much; for example, the use of tosylamide and geminal diester tethered substrates all give very similar yields. For substrates with a trisubstituted alk-1-enylcyclopropane motif, the double bond with a Z configuration always gives the *cis*-fused cycloaddition intermediates selectively, which can be further transformed into the expected products with the *cis-anti-cis* configuration of the linear triquinanes (entries 3–5). This result can be explained by comparison of the possible reaction transition states. On the other hand, as the intermediate with a *trans*-fused configuration cannot go through the second aldol reaction, a substrate with an *E*-alk-1-enylcyclopropane will only give the bicyclic cyclooctanedione product (entry 6).

for references see p 351

Domino Transformations 1.5 Non-Radical Skeletal Rearrangements

Table 11 Rhodium(I)-Catalyzed [(5 + 2) + 1]-Cycloaddition/Aldol Reaction[125]

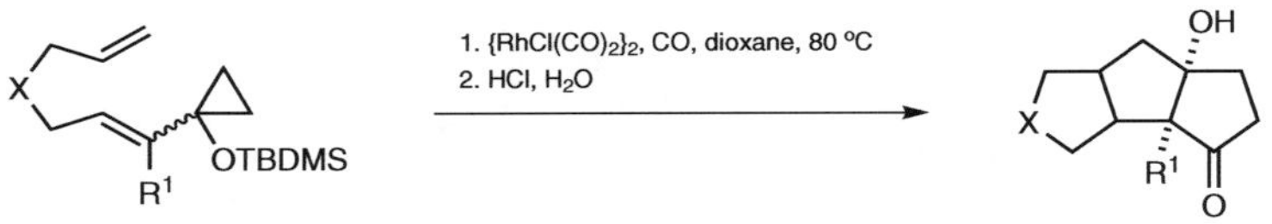

Entry	Substrate	Product	Yield (%)	Ref
1	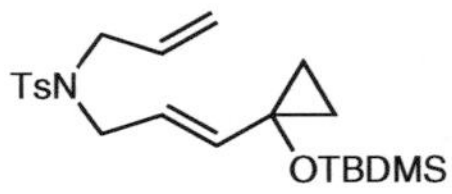	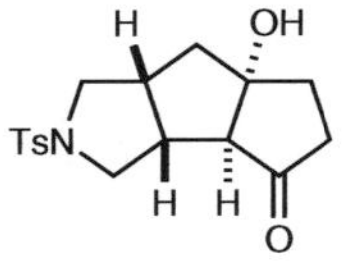	22	[125]
2	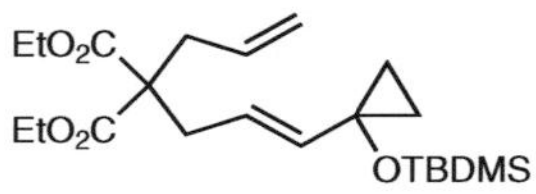	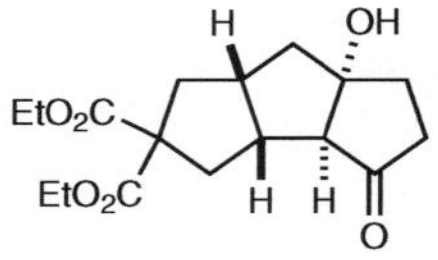	28	[125]
3	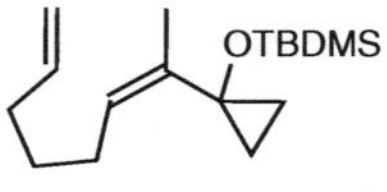	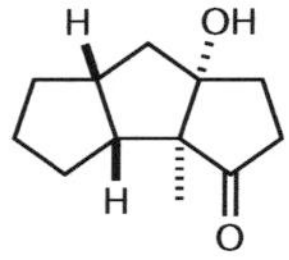	26	[125]
4	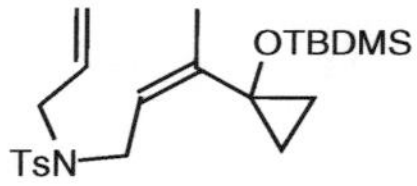	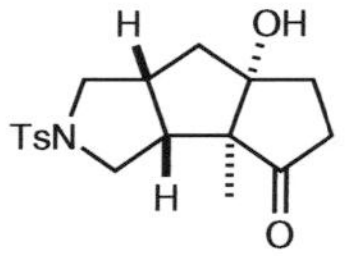	50	[125]
5	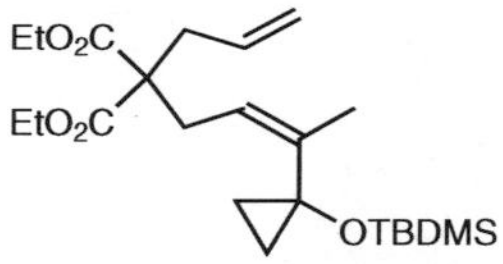	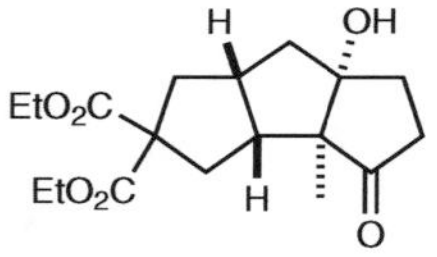	67	[125]
6	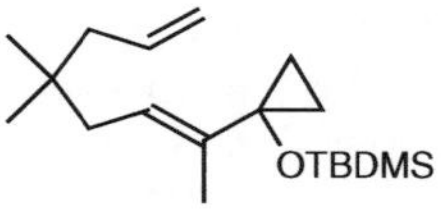	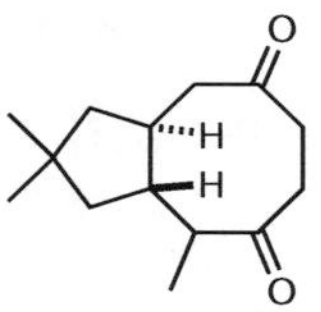	50	[125]

Some of the tricyclic products already contain the required configuration and quaternary center for the synthesis of linear triquinanes such as hirsutene, 1-desoxyhypnophilin, and coriolin. Accordingly, Yu and co-workers have been able to finish the total synthesis of (±)-hirsutene and (±)-1-desoxyhypnophilin from common intermediate **98** (Scheme 50).[125]

1.5.2 Lewis Acid/Base Induced Reactions

Scheme 50 Total Synthesis of (±)-Hirsutene and (±)-1-Desoxyhypnophilin[125]

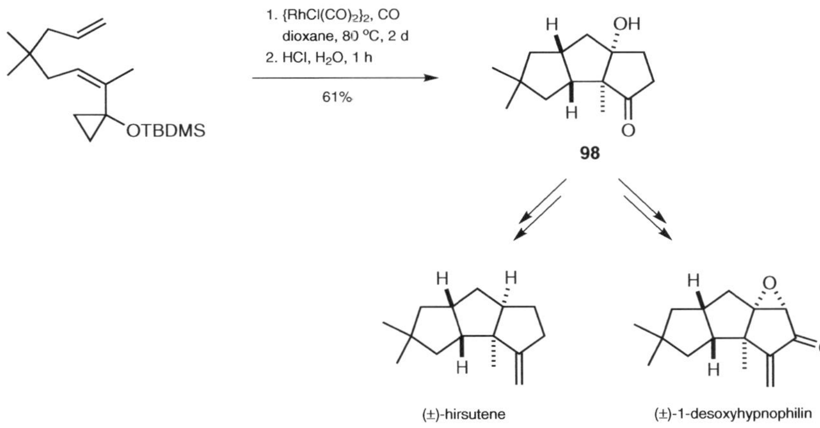

(±)-hirsutene   (±)-1-desoxyhypnophilin

(3aR,4R,7aR)-4-Ethyl-7a-methyl-2,3,3a,4,7,7a-hexahydrobenzofuran (Table 10, Entry 10); Typical Procedure:[122]

AgBF$_4$ (0.044 mmol) was added to a 13.3 mM soln of PtCl$_2$•**92** (typically 0.02 mmol) in EtNO$_2$. The mixture was stirred for 1 h in the dark and then TrOMe (0.42 mmol) on polystyrene resin and (3E,7Z)-4-methyldeca-3,7-dien-1-ol (0.2 mmol) were added. The resulting mixture was stirred at rt in the dark until the reaction was complete by GC analysis (typically 6–14 h). The reaction was then quenched by passage through a plug of silica gel, eluting with Et$_2$O. The solvent was then removed under reduced pressure and the residue was purified by column chromatography; yield: 61%; 87% ee.

cis-6,8a-Dimethyl-3,3a,8,8a-tetrahydrocyclopenta[a]inden-1(2H)-one (95, R^1 = Me; R^2 = H); Typical Procedure:[123]

An oven-dried 25-mL test tube equipped with a stirrer bar was charged with alcohol **93** (R^1 = Me; R^2 = H; 0.14 mmol), Ag$_2$CO$_3$ (0.28 mmol, 2.0 equiv), and Pd(OAc)$_2$ (3.1 mg, 0.014 mmol, 0.1 equiv). The reaction vessel was sealed with a rubber septum, an atmosphere of argon was introduced, and freshly distilled toluene (950 µL) and anhyd DMSO (50 µL) were added via syringe. The reaction vessel was placed in a hot oil bath (100 °C) and the progress of the reaction was monitored by TLC analysis. Upon completion (~1 h), the mixture was filtered through a pad of silica gel, eluting with EtOAc. The filtrate was washed with H$_2$O and brine, dried (MgSO$_4$), and concentrated under reduced pressure. Flash column chromatography (silica gel, EtOAc/hexanes 1:20) provided the product; yield: 66%.

for references see p 351

328 Domino Transformations **1.5** Non-Radical Skeletal Rearrangements

Diethyl (3aR*,3bS*,6aS*,7aR*)-6a-Hydroxy-3b-methyl-4-oxodecahydro-2H-cyclopenta-[a]pentalene-2,2-dicarboxylate (Table 11, Entry 5); Typical Procedure:[125]

> **CAUTION:** *Carbon monoxide is extremely flammable and toxic, and exposure to higher concentrations can quickly lead to a coma.*

A soln of diethyl (Z)-2-allyl-2-{3-[1-(*tert*-butyldimethylsiloxy)cyclopropyl]but-2-enyl}malonate (0.2 mmol) in anhyd dioxane (8 mL) was degassed with bubbling CO/N_2 (1:4 v/v) for 5 min. The catalyst $\{RhCl(CO)_2\}_2$ (8.8 mg, 22 μmol, 11 mol%) was added in one portion and a light yellow soln formed, through which was further bubbled the above gas mixture for 5 min. The resulting soln was heated to 80 °C in an oil bath with stirring under a positive pressure of the gas mixture. After 11 h, TLC indicated the absence of the starting material and the resulting brown soln was cooled to rt. The mixture was hydrolyzed by adding 1% HCl in EtOH (0.3 mL) and H_2O (0.1 mL) and stirred at rt for 20 min. The solvent was evaporated and the residue was purified by flash column chromatography [silica gel (5 g), petroleum ether/EtOAc 5:1 to 3:1]; yield: 67%.

1.5.2.3.1.3.2.2 Activation of Alkynes

Domino reactions induced by activation of an alkyne or allene moiety with π-acids are a little different from those cases mentioned in Section 1.5.2.3.1.3.2.1, largely because the carbenic intermediate formed in situ can be trapped in a number of ways.

Generally, in the presence of a π-acid, compounds with both C=C and C≡C bonds can sometimes go through domino cycloisomerization processes to give synthetically useful intermediates. Sarpong and co-workers have developed such a domino reaction of alkynyl indenes to construct tricyclic skeletons that commonly exist in the icetexane diterpenoids.[126–128] This reaction has the following features: firstly, it can be catalyzed by several transition-metal reagents, such as platinum(II) chloride, dicarbonyldichlororuthenium(II) dimer $[\{Ru(CO)_2Cl_2\}_2]$, and gallium trichloride. Secondly, it has relatively broad substrate scope. A variety of alkynyl indene compounds can be subjected to this reaction to afford the desired fused cycloheptadiene products (Table 12). However, small changes in the substrate structure lead to the use of different metal catalysts as the optimal promoter. For structurally simpler substrates, gallium trichloride is the best choice and leads to the expected compound as a single product (Table 12, entry 1 and Scheme 51). For structurally more complex materials, platinum(II) chloride can afford the desired products in good to excellent yield (Table 12, entries 2–5). One noteworthy observation is that carrying out the reaction at higher temperature can result in the formation of side products, likely by isomerization of the starting material.

1.5.2 Lewis Acid/Base Induced Reactions

Table 12 Transition-Metal-Catalyzed Cycloisomerization[126,128]

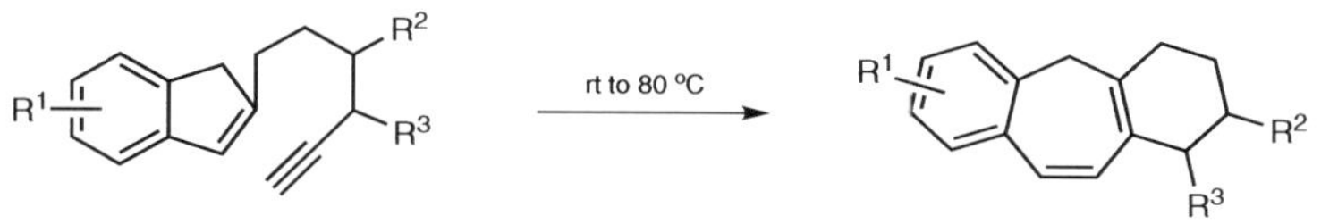

En-try	Substrate	Conditions	Product	Yield (%)	Ref
1	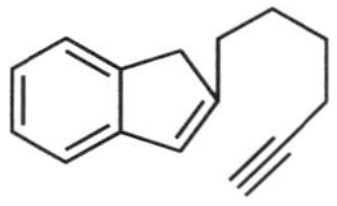	GaCl$_3$, toluene	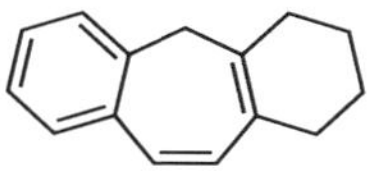	_a	[126]
2	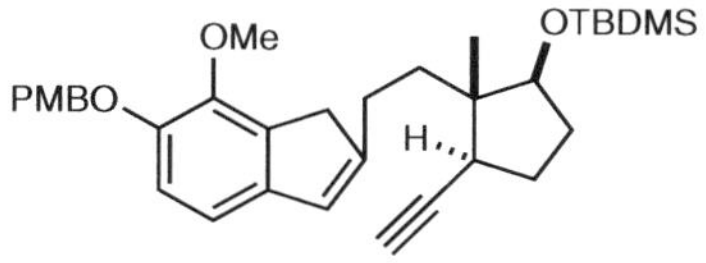	PtCl$_2$, benzene	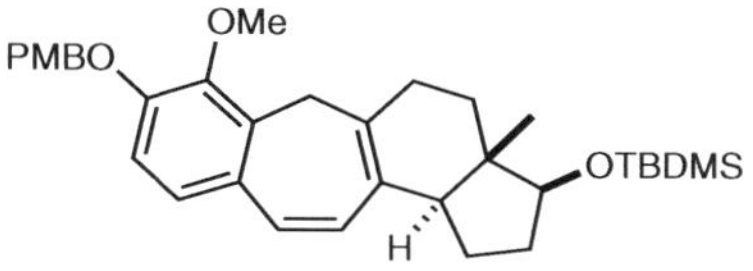	61	[128]
3	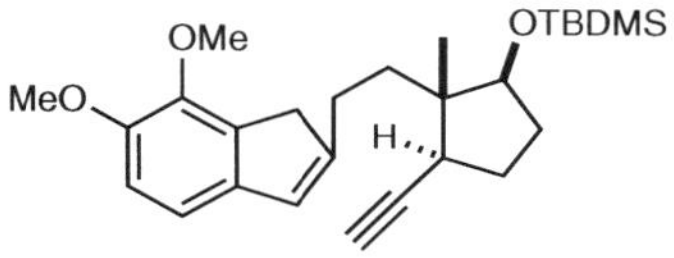	PtCl$_2$, benzene	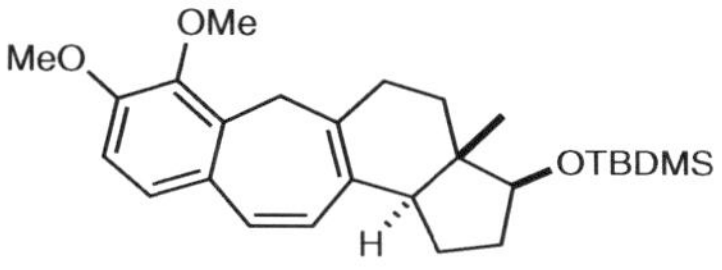	86	[128]
4	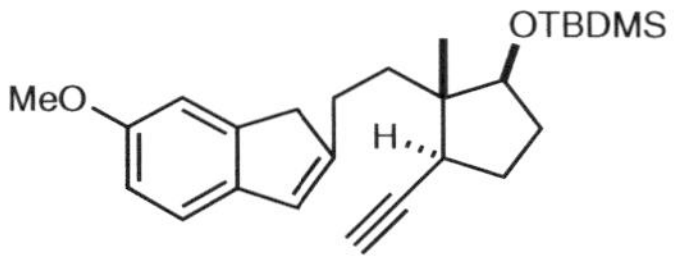	PtCl$_2$, benzene	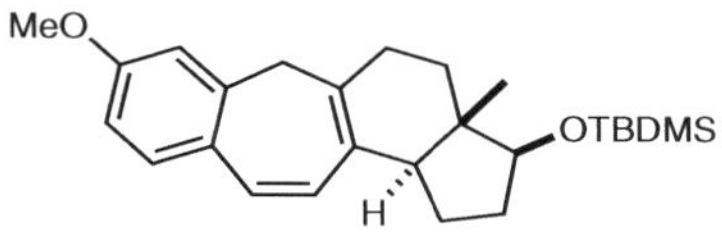	88	[128]
5	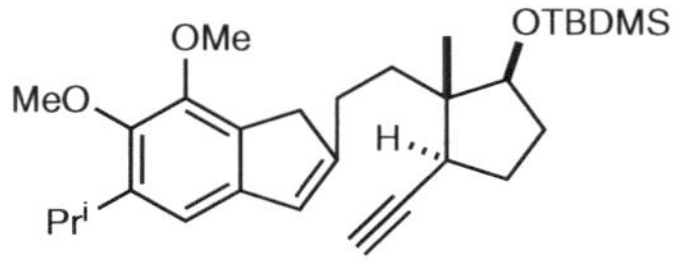	PtCl$_2$, benzene	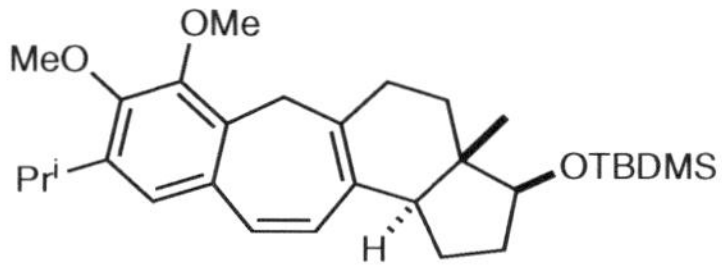	76	[128]

[a] Yield not reported.

Using this method, the Sarpong group has cyclized alkynylindene **99** to give tricyclic product **100** (Scheme 51) and completed total syntheses of (±)-salviasperanol, (±)-5,6-dihydro-6α-hydroxysalviasperanol, (±)-brussonol, and (±)-abrotanone as well as the construction of the pentacyclic core of cortistatin.[126–128]

for references see p 351

330   Domino Transformations   **1.5** Non-Radical Skeletal Rearrangements

Scheme 51   Total Synthesis of (±)-Salviasperanol, (±)-5,6-Dihydro-6α-hydroxy-salviasperanol, (±)-Brussonol, and (±)-Abrotanone[126–128]

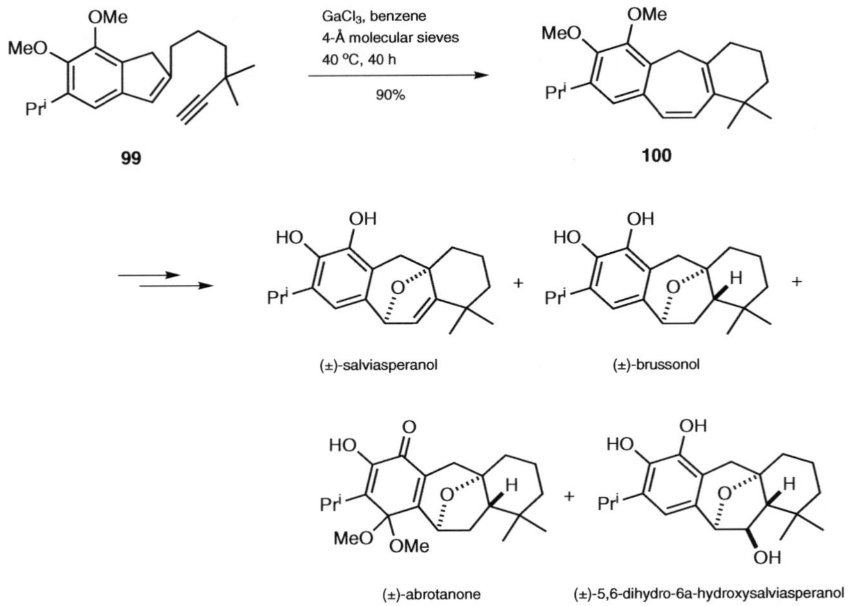

(±)-salviasperanol   (±)-brussonol

(±)-abrotanone   (±)-5,6-dihydro-6a-hydroxysalviasperanol

Domino nucleophilic addition reactions are another common way to proceed upon activation of the alkyne moiety, as an electrophilic center is always generated by the coordination of the π-acid with the C≡C bond. Following this kind of strategy, a number of important domino transformations have been realized by proper substrate design. One typical example is the gold-catalyzed domino reaction developed by Yang and co-workers (Table 13).[129] In the presence of [1,3-bis(2,6-diisopropylphenyl)imidazol-2-ylidene]chlorogold(I) [AuCl(IPr)] and silver(I) hexafluoroantimonate, substrates with two alkynes and a nucleophilic group strategically installed can undergo a domino cyclization terminated by an external nucleophile to give a tricyclic product. The nucleophilic group of the substrate is a key factor for this transformation. As substrates with a carboxylic acid will give an intermediate with a less nucleophilic alkene, the use of such substrates will lead to either a lower yield or no product. In contrast, higher yields and broader external nucleophile scope are observed with the application of a hydroxy group. Furthermore, for cyclic substrates, the ring size and the relative configuration between the two alkynes are also important for the success of the second cyclization. For example, in substrates with a six-membered-ring system, a *cis* relationship for the two alkynes will only give the monocyclization product (Table 13, entry 2), while the combination of the same configuration with a seven-membered ring can give the desired tricyclic product, albeit with a lower yield of 32% (entry 3). Furthermore, a series of external nucleophilic reagents are applicable for this tandem reaction.

1.5.2 Lewis Acid/Base Induced Reactions

Table 13 Gold-Catalyzed Domino Cyclization of Diynes[129]

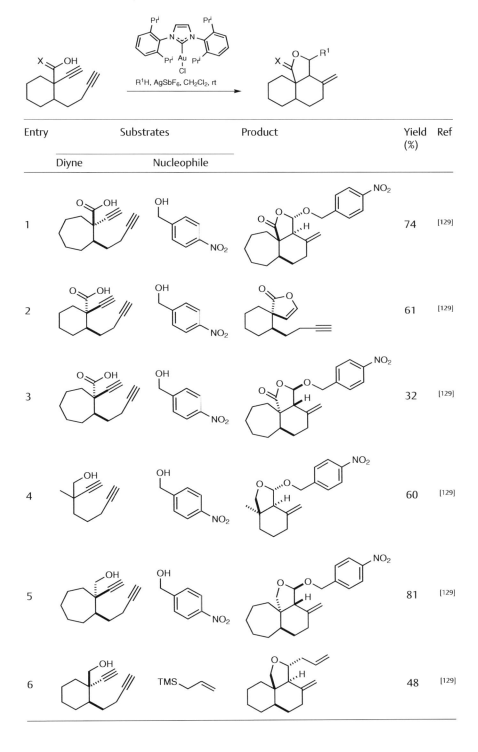

for references see p 351

Applying this tandem reaction as a key step, Yang and co-workers have completed the total synthesis of marasmene and three sesquiterpenoids (antrocin, anhydromarasmone, and kuehneromycin A) as shown in Scheme 52.[129]

Scheme 52 Total Synthesis of Marasmene, Kuehneromycin A, Anhydromarasmone, and Antrocin[129]

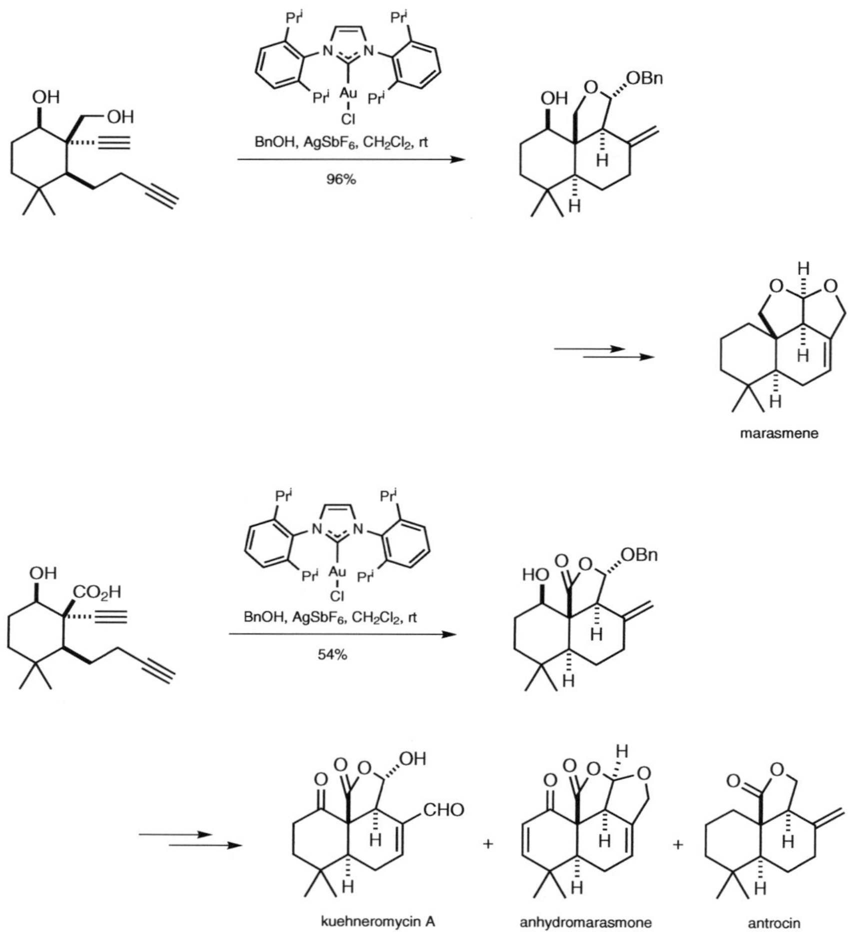

marasmene

kuehneromycin A   anhydromarasmone   antrocin

Another important type of tandem cyclization that has been extensively studied uses a 1,6-enyne as the substrate. Due to its special structural features, both metal carbenes and alkenylmetal intermediates could be involved in the reaction, which ultimately leads to the construction of various skeletons. One such representative example introduced here is the systematic study reported by the Echavarren group. Using a 1,6-enyne as the key moiety of the substrates, a series of tandem cyclization reactions was developed leading to synthetically useful intermediates containing a five-membered ring (Table 14).[130–134] In particular, these methodologies provide several practical strategies for the synthesis of fused 5,7-ring systems. During these transformations, the design of the substrates and the careful screening of the gold(I) catalysts (e.g., **101** and **102**) are the keys to the production of the desired product. Firstly, along with the prerequisite 1,6-enyne moiety, various functional groups can be used to trap the related metal carbene intermediate. When a carbonyl is used, the reaction affords the corresponding oxatricyclic products (Table 14, entries 5–8), while the use of a cyclopropyl ether can give tricyclic products

1.5.2 Lewis Acid/Base Induced Reactions **333**

with an octahydrocyclobuta[*a*]pentalene skeleton (entry 4). In the presence of a protected hydroxy group, which can potentially migrate, and an additional alkene moiety, compounds with a fused tricyclic 5,7,3-ring system can be produced (entries 1–3). Secondly, the relative configuration of the products can be tuned by the configuration of the C=C bond in the 1,6-enyne moiety. For example, in the case of substrates with a carbonyl group, products with different relative configuration at the oxabicyclo[3.2.1]octane motif are obtained selectively by varying the configuration of the C=C bond (entries 7 and 8). Furthermore, it should be mentioned that the overall stereoselectivity of these types of reaction is not excellent due to the presence of different active functional groups as well as the involvement of reversible transformations.

Table 14 Gold-Catalyzed Cyclization of 1,6-Enynes[130–132]

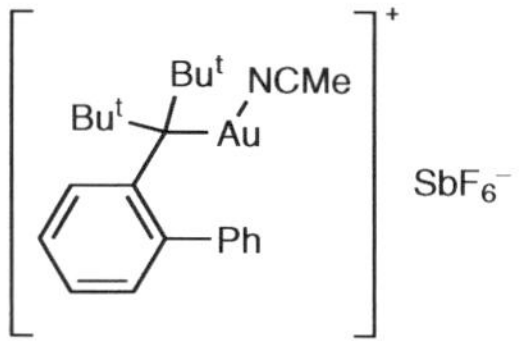

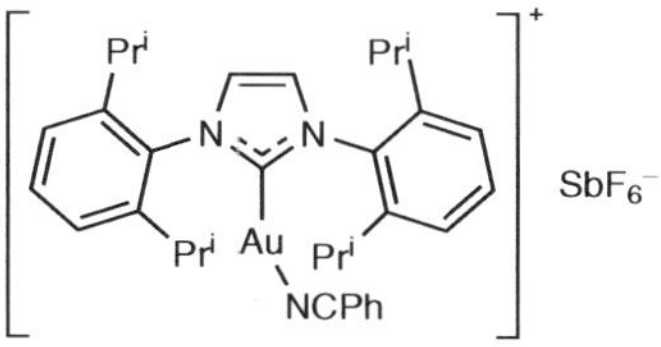

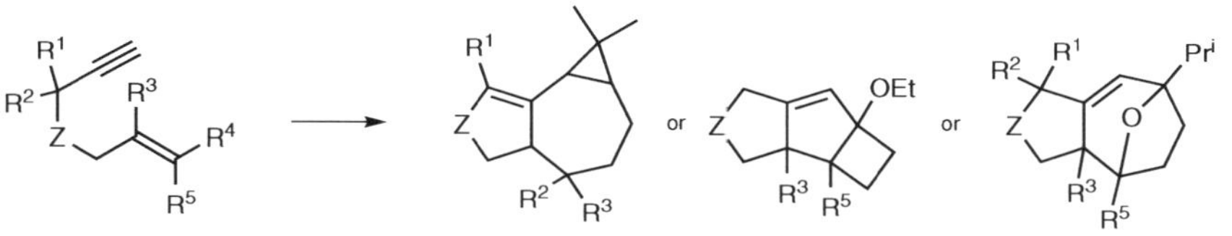

Entry	Substrate	Conditions	Product(s)	Yield (%)	Ref
1	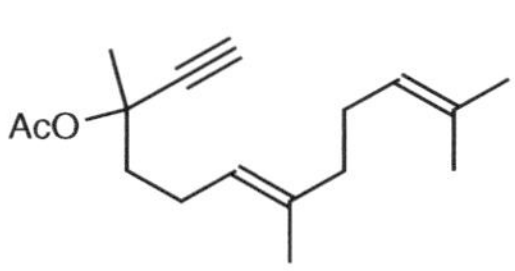	**101**, CH$_2$Cl$_2$, rt	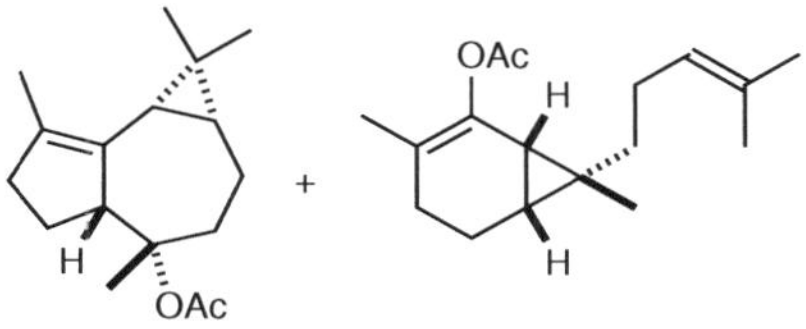 54:42	96	[132]
2	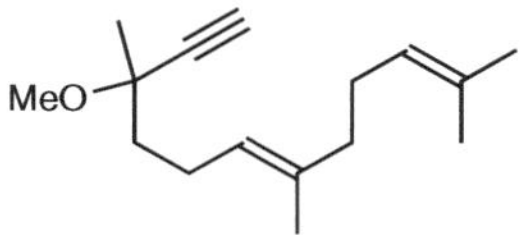	**101**, CH$_2$Cl$_2$, rt	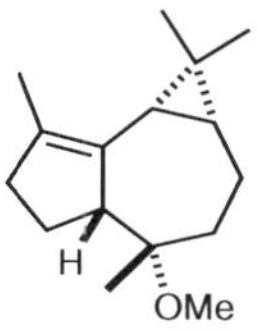	84	[132]
3	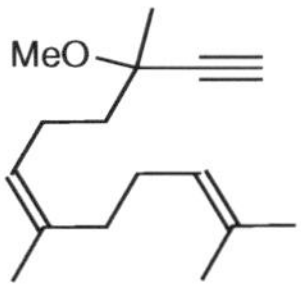	**101**, CH$_2$Cl$_2$, rt	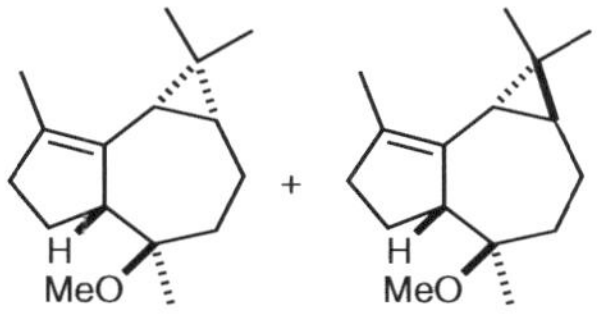 73:8	81	[132]

for references see p 351

Table 14 (cont.)

Entry	Substrate	Conditions	Product(s)	Yield (%)	Ref
4	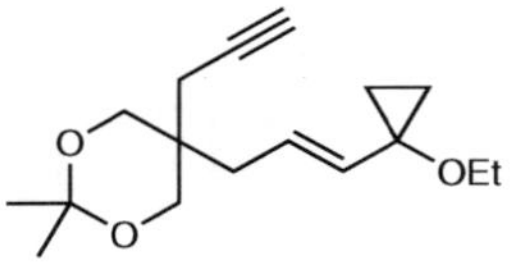	AuCl, CH_2Cl_2, rt	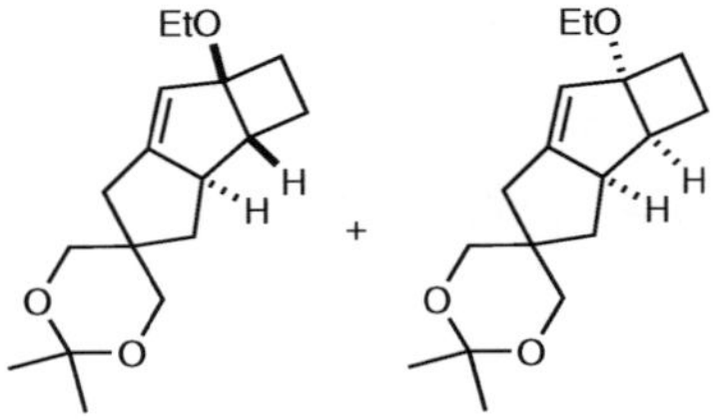 (84:7)	91	[130]
5	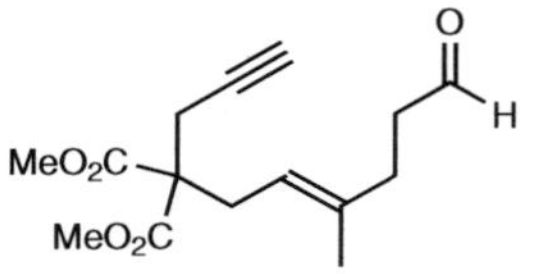	AuCl, CH_2Cl_2, rt	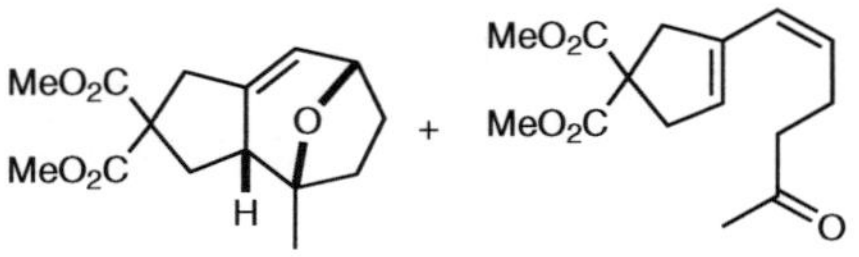 (58:18)	76	[130]
6	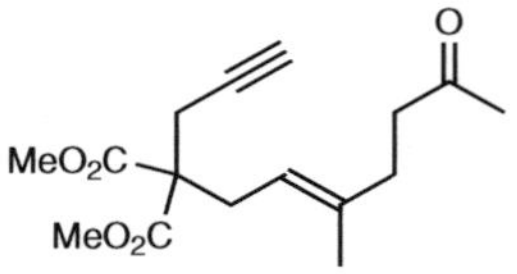	AuCl, CH_2Cl_2, rt	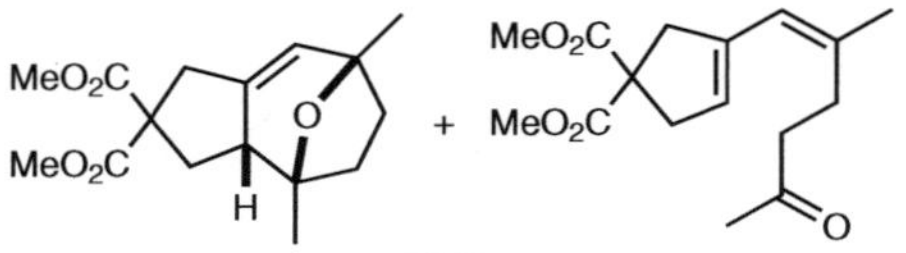 (79:10)	89	[130]
7	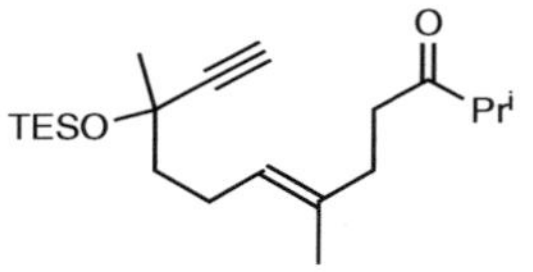	**102**, CH_2Cl_2, rt	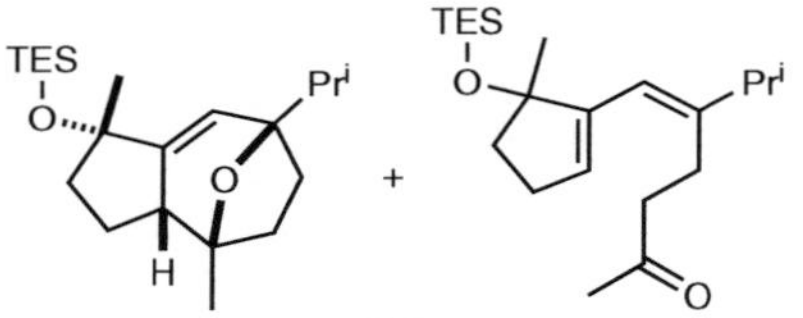 (65:4)	69	[131]
8	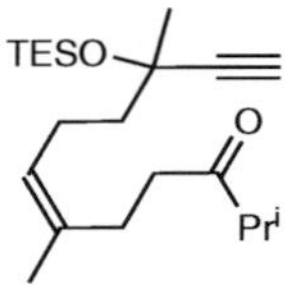	**101**, CH_2Cl_2, rt	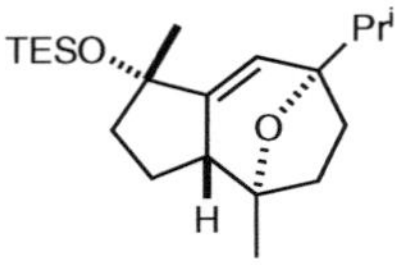	32	[131]

Using this method as the key step, Echavarren and co-workers have completed the total synthesis of (+)-orientalol F,[131] corrected the structure of pubinernoid B,[131] and achieved an asymmetric total synthesis of (−)-englerin A (Scheme 53).[135] This last target has also been synthesized by the Ma group through a similar strategy.[136]

1.5.2 Lewis Acid/Base Induced Reactions

335

Scheme 53 Total Syntheses of (+)-Orientalol F, Pubinernoid B, and (−)-Englerin A[131,135,136]

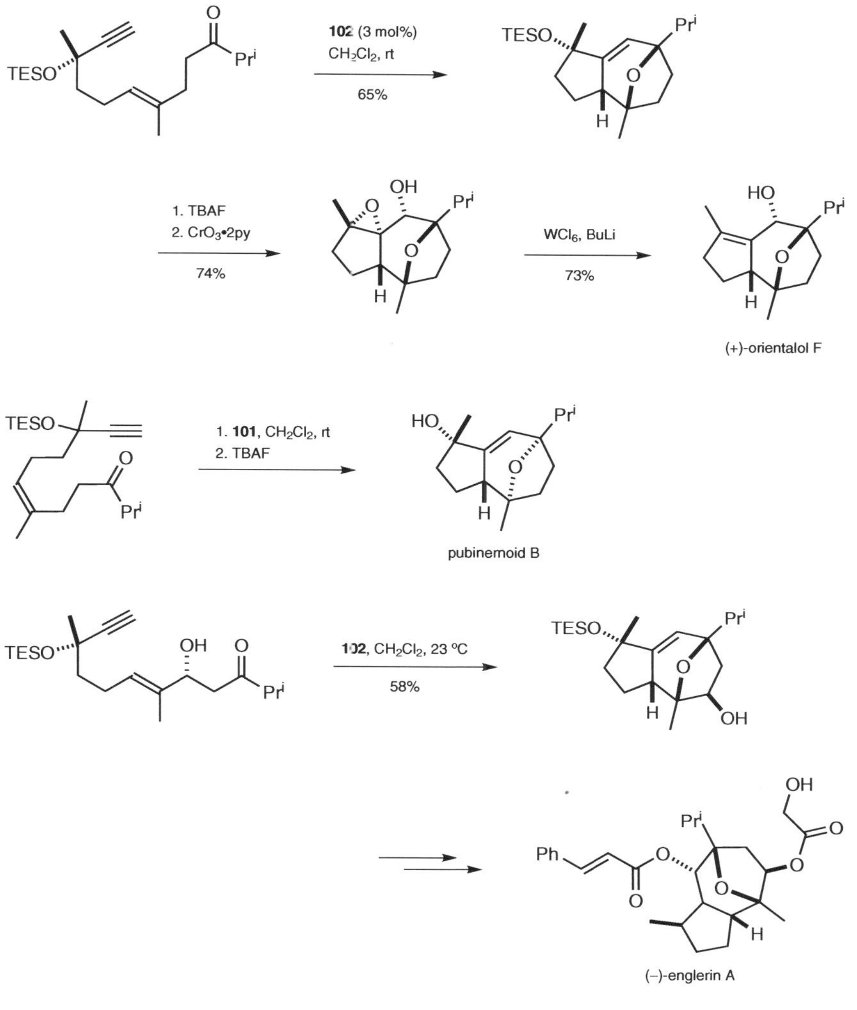

(+)-orientalol F

pubinernoid B

(−)-englerin A

Besides the reaction patterns mentioned above, the electrophilic center generated by the coordination of the C≡C bond to a metal catalyst can be used to induce a semipinacol rearrangement to give interesting skeletons.[137,138] Based on their systematic study of the semipinacol rearrangement, the Tu group has developed a silver(I)-catalyzed asymmetric hydroamination/semipinacol rearrangement of (4-aminobut-1-ynyl)cyclobutanols that affords synthetically useful azaspiro ketones **103** in excellent yield and with moderate to good enantioselectivity (Scheme 54). It should be noted that this method shows relatively narrow substrate scope, as only the use of the sulfonyl protecting group is successful.[139]

An asymmetric formal synthesis of (−)-cephalotaxine has been accomplished using this method via intermediate **104** (Scheme 55).[139]

for references see p 351

Scheme 54 Silver(I)-Catalyzed Asymmetric Hydroamination/Semipinacol Rearrangement of (4-Aminobut-1-ynyl)cyclobutanols[139]

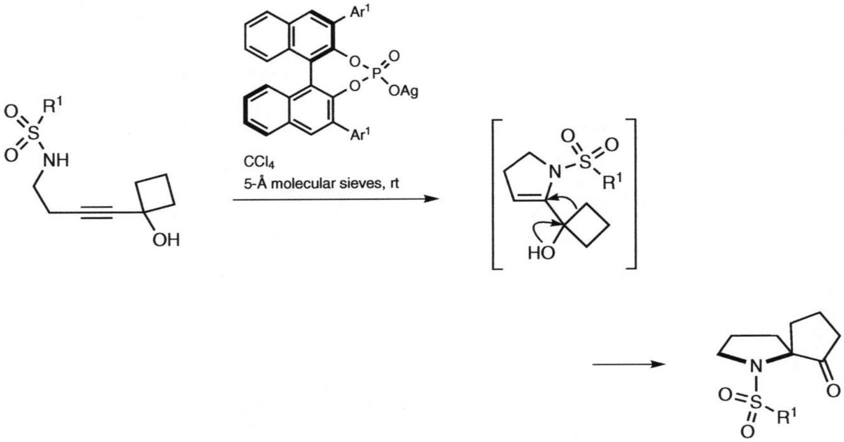

103

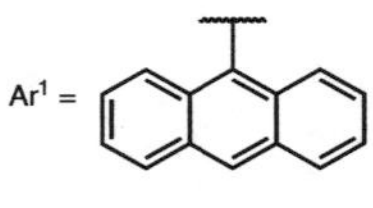

R^1	ee (%)	Yield (%)	Ref
4-Tol	82	99	[139]
4-BrC_6H_4	79	92	[139]
4-ClC_6H_4	82	98	[139]
4-*t*-BuC_6H_4	71	94	[139]
4-$F_3CC_6H_4$	72	98	[139]

Scheme 55 Formal Synthesis of (−)-Cephalotaxine[139]

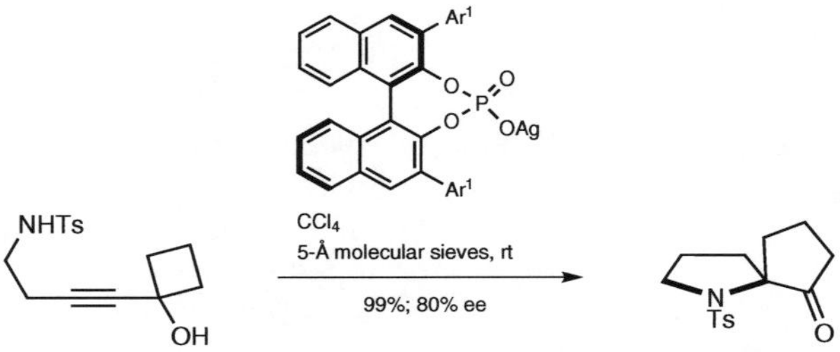

104

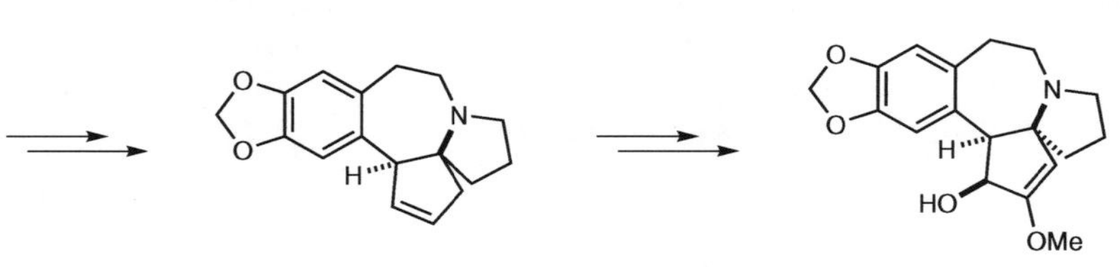

80% ee

(–)-cephalotaxine

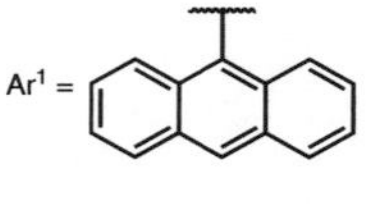

1.5.2 Lewis Acid/Base Induced Reactions

8-Isopropyl-6,7-dimethoxy-1,1-dimethyl-2,3,4,5-tetrahydro-1*H*-dibenzo[*a,d*][7]annulene (100); Typical Procedure:[127]

In a N_2-filled drybox, a Schlenk flask was charged with $GaCl_3$ (66.7 mg, 0.38 mmol) and powdered 4-Å molecular sieves (133 mg). The flask was removed from the drybox and a soln of enyne **99** (618 mg, 1.89 mmol) in freshly distilled benzene (47 mL) (**CAUTION:** *carcinogen*) was added via syringe under N_2 to give a dark orange-red soln. The flask was sealed and placed in an oil bath that was preheated to 40 °C. After being stirred at 40 °C for 40 h, the mixture was allowed to cool to rt. The soln was added to sat. aq $NaHCO_3$ (50 mL) and extracted with Et_2O (3 × 40 mL). The combined organic layers were washed with brine (50 mL), dried ($MgSO_4$), and concentrated to give a yellow oil (640 mg), which was purified by flash chromatography (hexanes/EtOAc 30:1) to give **100** as a viscous pale yellow oil; yield: 554 mg (90%). This crystallized to a white solid on standing overnight in a freezer. (In subsequent runs, it was found that the reaction was complete after 24 h. Alternatively, quantitative conversion could be achieved with only 10 mol% catalyst, though this required a reaction time of 72 h. Thus, in subsequent reactions 20 mol% catalyst was used and the mixture was heated for 24 h.)

(1*S*,3a*R*,7a*S*)-3a-Methyl-7-methylene-1-[(4-nitrobenzyl)oxy]octahydroisobenzofuran (Table 13, Entry 4); Typical Procedure:[129]

2-Ethynyl-2-methylhept-6-yn-1-ol (15 mg, 0.1 mmol) was dissolved in dry CH_2Cl_2 (0.05 M) and 4-nitrobenzyl alcohol (3 equiv), AuCl(IPr) (5 mol%), and $AgSbF_6$ (5 mol%) were added sequentially at rt. The mixture was stirred for 0.8–12 h and then the reaction was quenched by filtration through a pad of Celite, washing with CH_2Cl_2. The solvent was evaporated to dryness and the residue was purified by flash chromatography (silica gel, hexane/EtOAc 15:1); yield: 18 mg (60%).

(1a*R*,4*S*,4a*R*,7b*S*)-4-Methoxy-1,1,4,7-tetramethyl-1a,2,3,4,4a,5,6,7b-octahydro-1*H*-cyclopropa[*e*]azulene (Table 14, Entry 2); Typical Procedure:[132]

(*E*)-3-Methoxy-3,7,11-trimethyldodeca-6,10-dien-1-yne was dissolved in dry CH_2Cl_2 (0.1 M) containing activated 4-Å molecular sieves, the catalyst **101** was added, and the mixture was stirred at rt. When the starting material was consumed (TLC analysis) the reaction was stopped by addition of 0.1 M Et_3N in hexane. The soln was then filtered through a pad of silica gel. After removal of the solvent, the residue was purified by flash chromatography to give a colorless oil; yield: 84%.

for references see p 351

1.5.2.3.2 Induction by a Pericyclic Reaction

As pericyclic reactions are the most direct way to form a cyclic product, such processes have been commonly used in domino reactions, with the Diels–Alder reaction being perhaps the most common example (see Section 2.1.1). However, in the case of domino reactions induced by a Lewis acid/base catalyzed Diels–Alder process, only a limited number of results have been developed.

On the other hand, the emergence of organocatalysis,[140] the high efficiency in stereocontrol and environmental friendliness of which has drawn great attention, has led to the development of several interesting transformations in recent years.

One outstanding example introduced here is a catalytic organocascade reported by the MacMillan group. With the use of a catalyst they developed, a stereoselective tandem process involving a Diels–Alder/amine cyclization sequence was realized.[141–143] For this domino sequence, the screening of various organocatalysts **105–107** was necessary for the achievement of satisfactory yields and enantioselectivity, noting that the use of different counterions of the catalyst can lead to very different results (Scheme 56). Furthermore, the selectivity of the second cyclization can be controlled by the use of different alkene substituents. In the case of substrates with methylsulfanyl-substituted or terminal alkenes (e.g., **108** and **112**, respectively), protonation of the intermediate of type **109** would facilitate the 5-*exo* amine heterocyclization of **110** to give the products with a bridged ring system, e.g. **111** and **113**. By contrast, a β-elimination of intermediate **115** is preferred in the case of the methylselanyl-substituted substrate **114**. This case leads to an iminium ion intermediate, which affords product **116** with a fused tetracyclic ring system (Scheme 56).

Scheme 56   Secondary Amine Catalyzed Tandem Diels–Alder/Amine Cyclization[141–143]

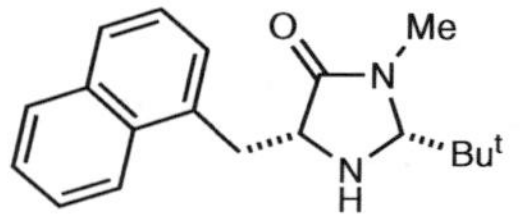

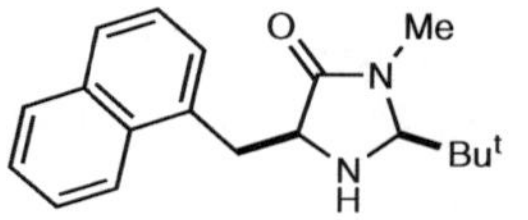

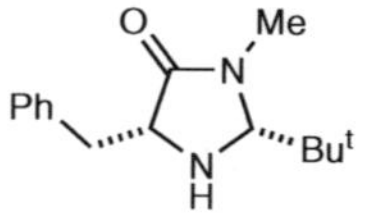

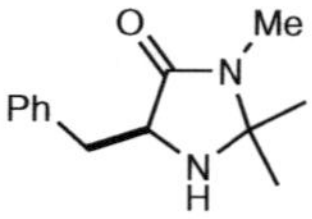

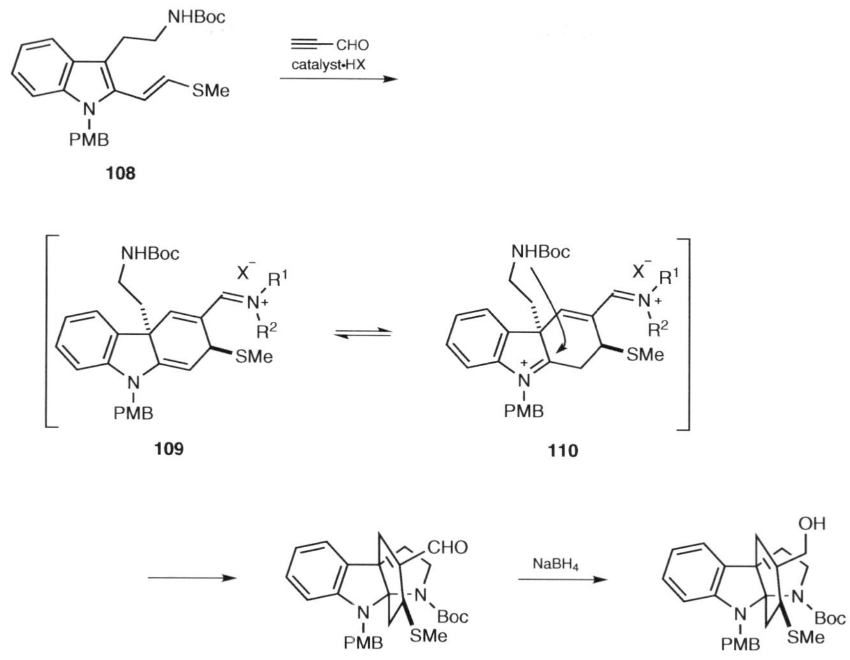

1.5.2 Lewis Acid/Base Induced Reactions

339

Catalyst	HX	ee (%)	Yield (%) of **111**	Ref
106	TFA	75	84	[141]
106	Br₃CCO₂H	88	81	[141]
105A	Br₃CCO₂H	96	87	[141]

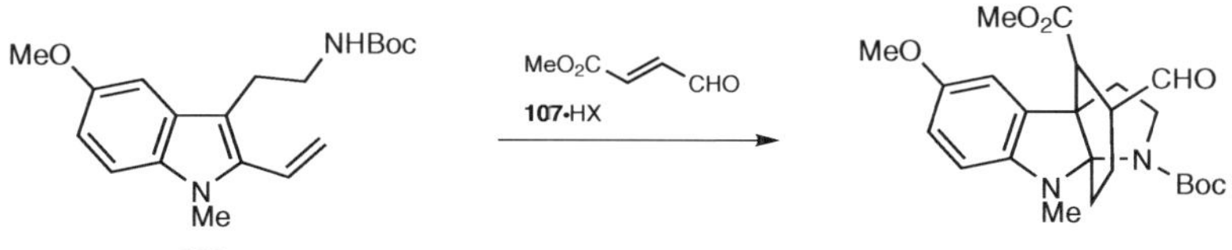

HX	ee (%)	Yield (%)	Ref
HCl	73	25	[143]
HBF₄	93	71	[143]

for references see p 351

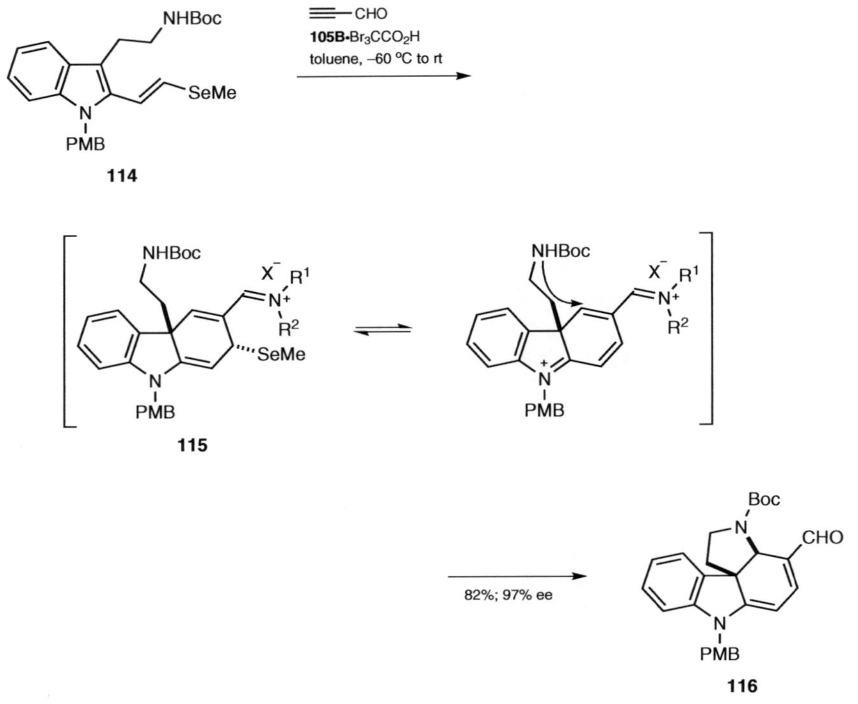

340 Domino Transformations **1.5** Non-Radical Skeletal Rearrangements

X = Br₃CCO₂⁻

Using this method, the asymmetric total synthesis of several indole alkaloids has been accomplished very efficiently from common intermediates **117** and **118** (Scheme 57).[142]

Scheme 57 Asymmetric Total Synthesis of Indole Alkaloids[142]

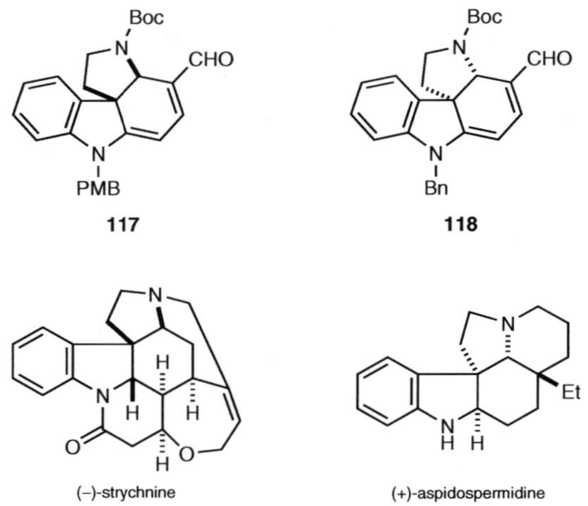

1.5.2 Lewis Acid/Base Induced Reactions **341**

(−)-akuammicine   (+)-vincadifformine

Another type of reaction presented here, reported in studies by Aubé and co-workers, features an intramolecular Schmidt reaction instead of a nucleophilic addition, in combination with a Diels–Alder cyclization.[144–147] As all the functional groups that are essential for the two transformations already exist in the substrates, there are two important issues to be resolved to successfully achieve this tandem process: (i) establishing reaction conditions that can efficiently promote the two transformations, and (ii) how to carry out the tandem reaction in the desired Diels–Alder/Schmidt reaction sequence. For the first issue, because both transformations can be promoted separately by Lewis and/or Brønsted acids, it is likely that such a tandem process can be realized by the use of the same promoter. As for the second, it is also feasible for several reasons. Firstly, due to the lower accessibility of intermolecular Schmidt reactions compared with the intramolecular version, the two reaction sites of the Schmidt reaction can be placed on two different substrates leading to an intermolecular Diels–Alder/intramolecular Schmidt reaction sequence. Secondly, even if the same two reaction sites are on the same substrate, the desired reaction sequence would still be viable through the deactivation of the corresponding carbonyl group. Accordingly, either an intermolecular or an intramolecular Diels–Alder reaction can be used in the corresponding tandem protocol. Based on the above analysis, three different reaction models have been investigated by the same group. Of the Lewis acids tested, methylaluminum dichloride, tin(IV) chloride, and boron trifluoride–diethyl ether complex give the best outcomes. As a result, most of the substrates can give the desired products with useful skeletons in good yields. In particular, for hexahydroindole products, the relative configuration of the fused positions can be tuned by using two different strategies (Table 15, entries 1 and 6). Also, a mixture of products can be obtained for the reactions in which two competitive intermediates are involved (entries 3 and 5). Stoichiometric amounts of the Lewis acid are normally required during this procedure because of the involvement of the Schmidt reaction.

for references see p 351

Table 15 Lewis Acid Promoted Diels–Alder/Schmidt Reactions[144,145,147]

Entry	Substrate(s)	Lewis Acid	Product(s)	Yield (%)	Ref
1	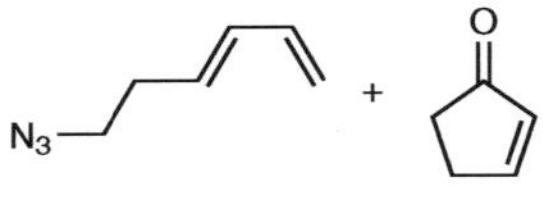	MeAlCl₂	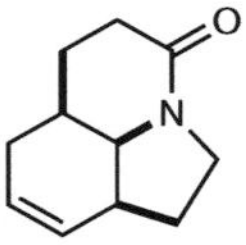	38	[145]
2	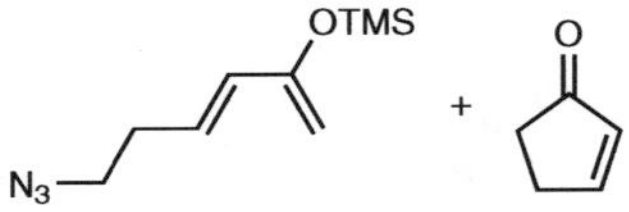	SnCl₄	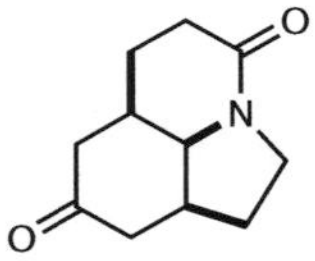	52	[145]
3	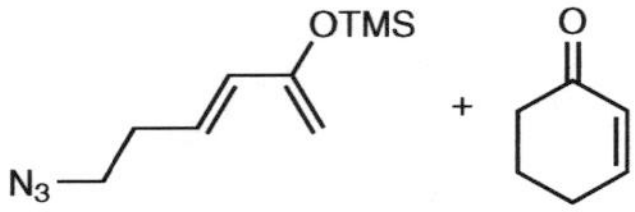	SnCl₄	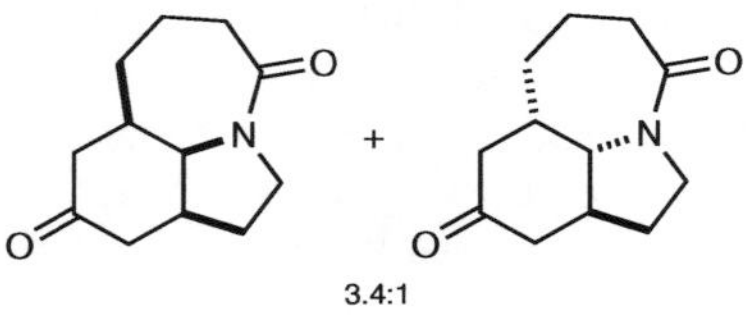 3.4:1	82	[145]
4	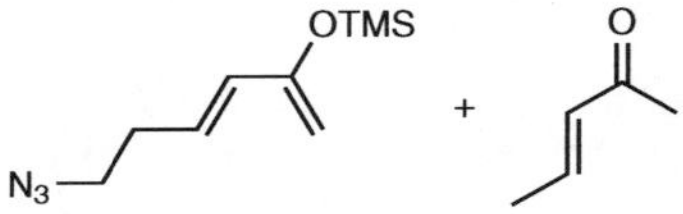	SnCl₄	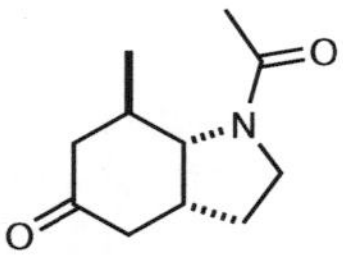	59	[145]
5	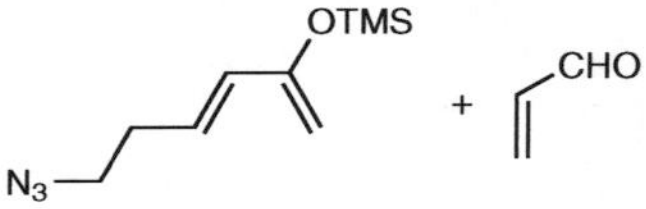	MeAlCl₂	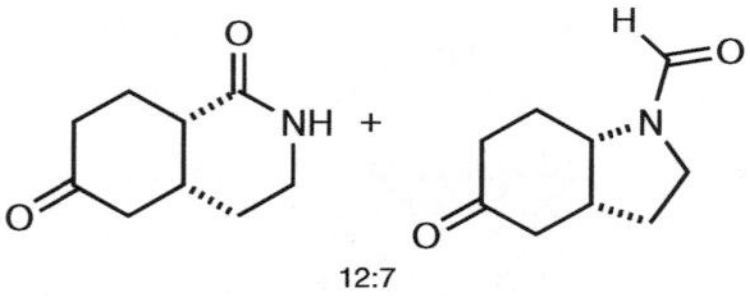 12:7	57	[145]
6	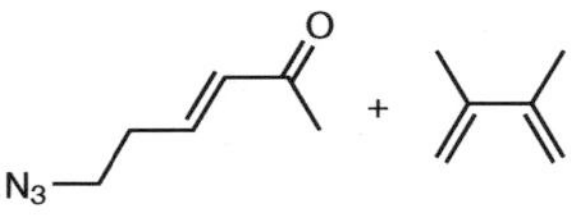	MeAlCl₂	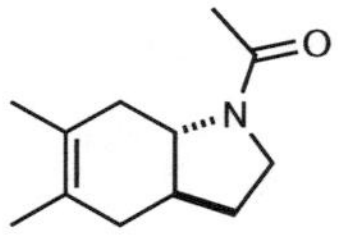	75	[145]
7	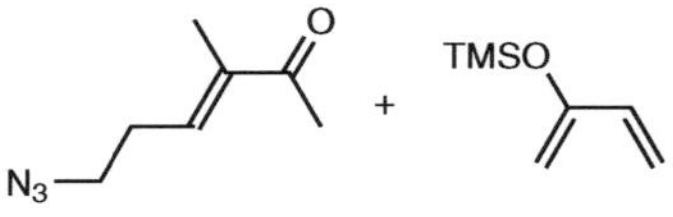	MeAlCl₂	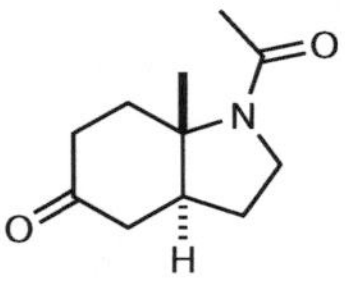	61	[145]
8	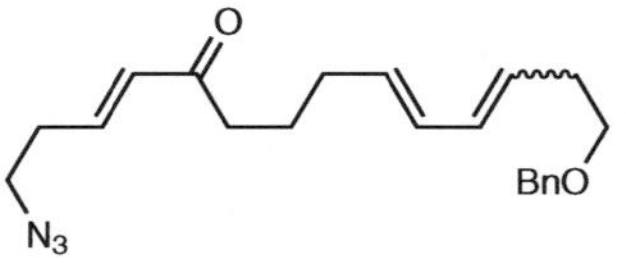	MeAlCl₂	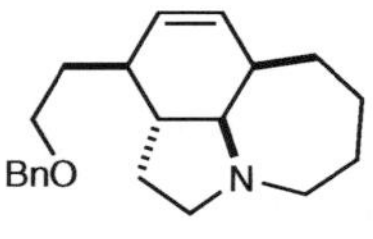	43	[144,147]

Subsequently, the total synthesis of (−)-stenine and neostenine have been accomplished by the Aubé group using this methodology (Scheme 58).[147]

1.5.2 Lewis Acid/Base Induced Reactions

Scheme 58 Total Synthesis of (−)-Stenine and Neostenine[147]

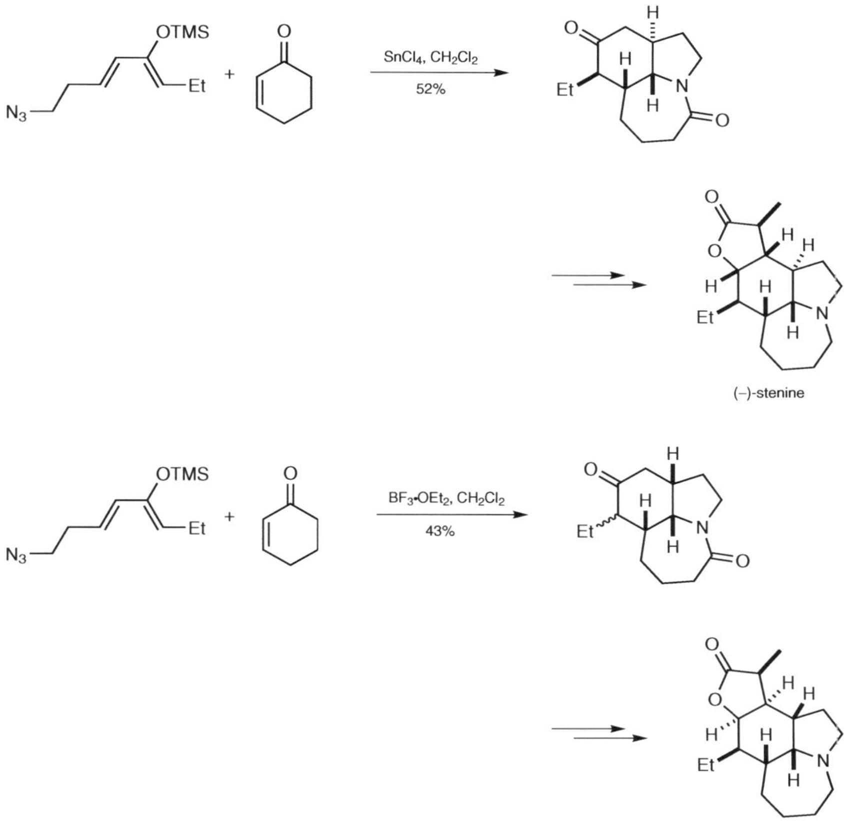

tert-Butyl (3aR,11bR)-4-Formyl-7-(4-methoxybenzyl)-1,2,3a,7-tetrahydro-3H-pyrrolo[2,3-d]carbazole-3-carboxylate (116); Typical Procedure:[142]
A 25-mL flask was charged with indole **114** (300 mg, 0.601 mmol, 1.00 equiv), organocatalyst **105B** (35.5 mg, 0.120 mmol, 0.20 equiv), and tribromoacetic acid (35.5 mg, 0.120 mmol, 0.20 equiv). Toluene (12 mL) was added and the resulting soln was cooled to −60 °C. Propynal (0.120 mL, 1.80 mmol) was added dropwise via syringe, whereupon the flask was placed in a −40 °C bath for 13 h. The resulting soln was allowed to warm to rt for 10 h, concentrated, and purified via flash column chromatography (hexanes/EtOAc 10:0 to 5:5) to give a dark orange foam; yield: 225.5 mg (82%); 97% ee.

1-[(3aR*,7aS*)-5,6-Dimethyl-2,3,3a,4,7,7a-hexahydro-1H-indol-1-yl]ethanone (Table 15, Entry 6); Typical Procedure:[145]
To a soln of (E)-6-azidohex-3-en-2-one (200 mg, 1.4 mmol) and 2,3-dimethylbuta-1,3-diene (3 equiv) in CH_2Cl_2 (15–20 mL) at 0 °C was added $MeAlCl_2$ (3 equiv). The resulting mixture was stirred at 0 °C for 2 h and then at rt for 12 h, and then the reaction was quenched with aq $NaHCO_3$. The mixture was partitioned between H_2O and $CHCl_3$. The organic layer was collected, dried (Na_2SO_4), filtered, and concentrated to give an oil. Chromatography (EtOAc/hexane 1:9 followed by $MeOH/CHCl_3$ 1:99 to 2:98) afforded the desired product as a yellow oil; yield: 208 mg (75%).

for references see p 351

344 Domino Transformations **1.5** Non-Radical Skeletal Rearrangements

1.5.2.3.3 **Induction by a Nucleophilic Step**

Other than the two major reaction types outlined in Sections 1.5.2.3.1 and 1.5.2.3.2, a domino process involving a ring closure can also be initiated by a nucleophilic step, which is normally induced by a Lewis basic reagent.

In 2004, Sorensen's group reported a concise synthesis of (+)-harziphilone,[148] an HIV-1 Rev/Rev-responsive element inhibitor with a challenging polyunsaturated structure. During their study, a 1,4-diazobicyclo[2.2.2]octane-catalyzed tandem cyclization was used as the key step for the construction of the crucial bicyclic system. Due to the existence of multiple reaction sites for the initial addition step, the feasibility of such a strategy was initially tested by a model study on the key reaction skeleton (Scheme 59). Accordingly, substrates **119** and **121** are treated with 1,4-diazobicyclo[2.2.2]octane, the most common promoter for the Morita–Baylis–Hillman reaction, and afford the desired products **120** and **122**, respectively. Interestingly, the presence of the unprotected 1,2-diol moiety in substrate **121** significantly facilitates the tandem reaction.

Scheme 59 1,4-Diazobicyclo[2.2.2]octane-Catalyzed Tandem Cyclization[148]

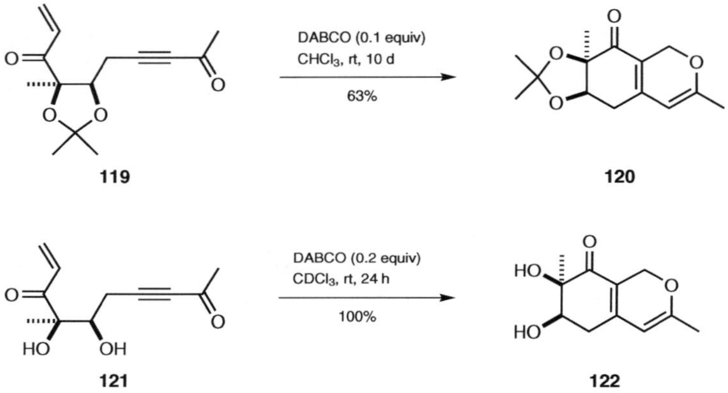

Applying this method, the total synthesis of (+)-harziphilone is accomplished through the cyclization of the more complex substrate **123**, which is prepared from commercially available 2-methylcyclopent-2-en-1-one in 17 steps (Scheme 60).[148]

1.5.2 Lewis Acid/Base Induced Reactions   **345**

Scheme 60   Total Synthesis of (+)-Harziphilone[148]

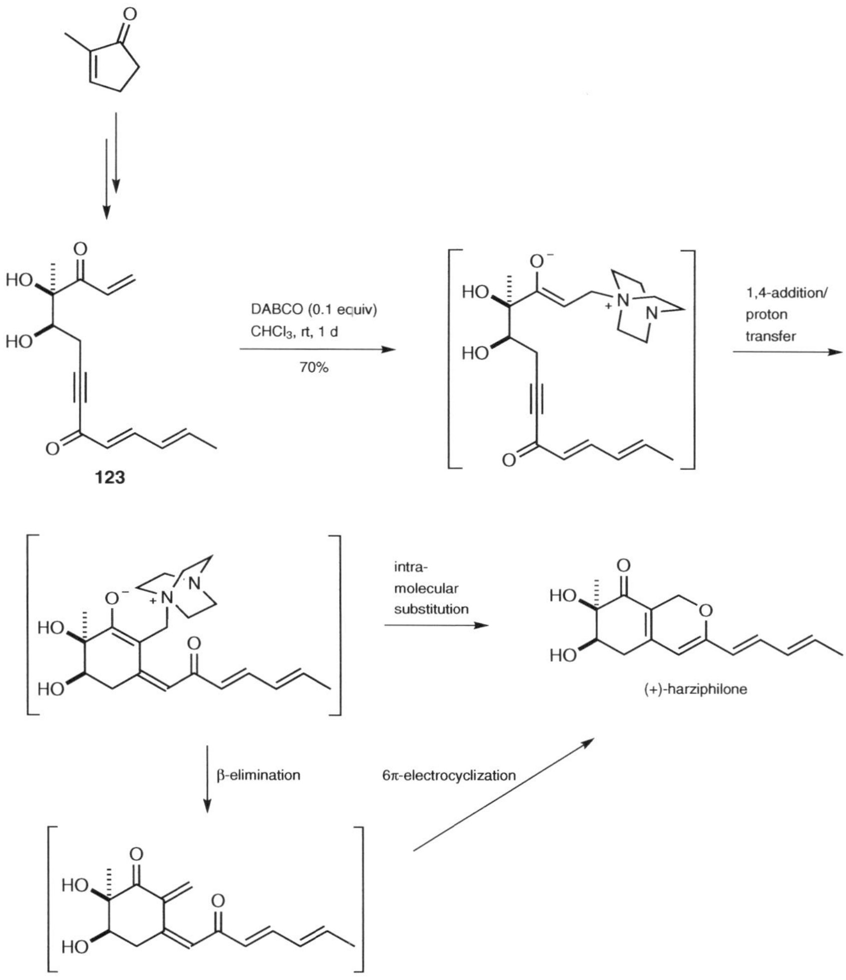

Cyclic secondary amines are another type of Lewis base catalyst that are used broadly in such domino reactions and also contribute to much of the current organocatalysis field. As such aminocatalysis usually proceeds through iminium and/or enamine activation, it has also been termed as iminium catalysis or enamine catalysis.[149–154]

For example, MacMillan and co-workers have developed an amine-catalyzed enantioselective tryptamine addition–cyclization reaction to give hexahydropyrrolo[2,3-b]indoles **126** (Scheme 61),[155] which are the key structural motif in many bioactive polyindoline alkaloids. As had been observed previously, different combinations of the amine and co-catalyst affect the outcome of the transformation. Here, the use of a tryptophan-derived catalyst **125** gives the best selectivity. Interestingly, the dielectric property of the solvent can also lead to changes in enantioselectivity. This reaction also shows broad substrate scope. Firstly, substrates **124** with different N1 and N10 substituents (R^3 and R^4) all give the desired product with excellent yield and enantiomeric excess. Secondly, the reaction is tolerant of variation in the electronic properties of the substituents on the indole ring (R^1 and R^2). Thirdly, for the aldehyde substrate with different substituents at the 4-position

for references see p 351

346 Domino Transformations **1.5** Non-Radical Skeletal Rearrangements

(R^5), both excellent absolute and relative stereocontrol can be achieved, albeit in dry dichloromethane. Furthermore, the corresponding tetrahydro-2*H*-furo[2,3-*b*]indole skeleton **127** can be constructed by the use of a tryptophol derivative (Scheme 61).

Scheme 61   Amine-Catalyzed Enantioselective Tryptamine Addition–Cyclization[155]

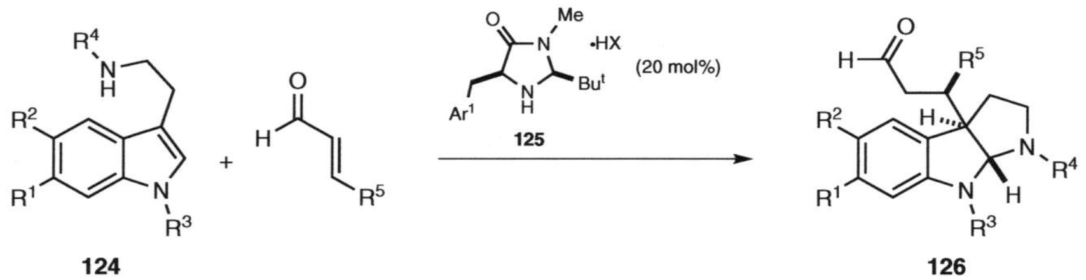

R^1	R^2	R^3	R^4	R^5	Ar^1	HX	Conditions	ee (%)	Yield (%)	Ref
H	H	CH$_2$CH=CH$_2$	Boc	H	Ph	TFA	CH$_2$Cl$_2$/H$_2$O, −85 °C	70	79	[155]
H	H	CH$_2$CH=CH$_2$	Boc	H	Ph	HBr	toluene/H$_2$O, 4 °C	84	50	[155]
H	H	CH$_2$CH=CH$_2$	Boc	H	indol-3-yl	TFA	CH$_2$Cl$_2$/H$_2$O, −85 °C	89	85	[155]
H	H	CH$_2$CH=CH$_2$	CO$_2$Et	H	indol-3-yl	TFA	CH$_2$Cl$_2$/H$_2$O, −80 °C	89	89	[155]
H	H	Bn	Boc	H	indol-3-yl	TFA	CH$_2$Cl$_2$/H$_2$O, −80 °C	90	82	[155]
H	H	CH$_2$CH=CH$_2$	Boc	CH$_2$OBz	Ph	TFA	CH$_2$Cl$_2$, −40 °C	91	66^a	[155]
H	Me	CH$_2$CH=CH$_2$	Boc	CO$_2$Me	Ph	TFA	CH$_2$Cl$_2$, −60 °C	92	94^b	[155]
Br	H	CH$_2$CH=CH$_2$	Boc	CO$_2$Me	Ph	TFA	CH$_2$Cl$_2$, −40 °C	97	86^c	[155]

^{a} dr 22:1.
^{b} dr 50:1.
^{c} dr 12:1.

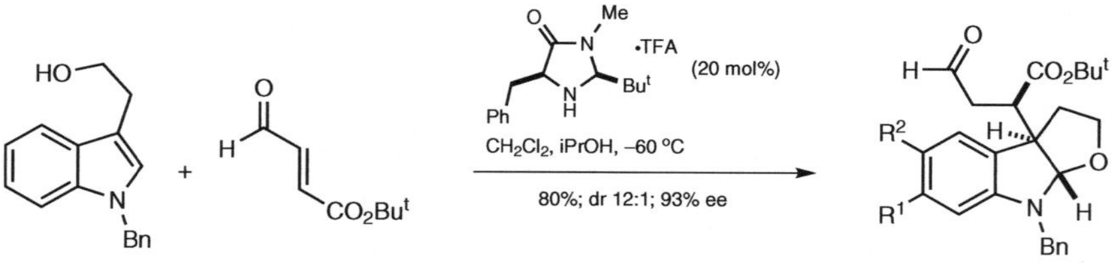

The synthetic utility of this method has also been demonstrated in the total synthesis of (−)-flustramine B (Scheme 62).[155]

Scheme 62 Total Synthesis of (−)-Flustramine B[155]

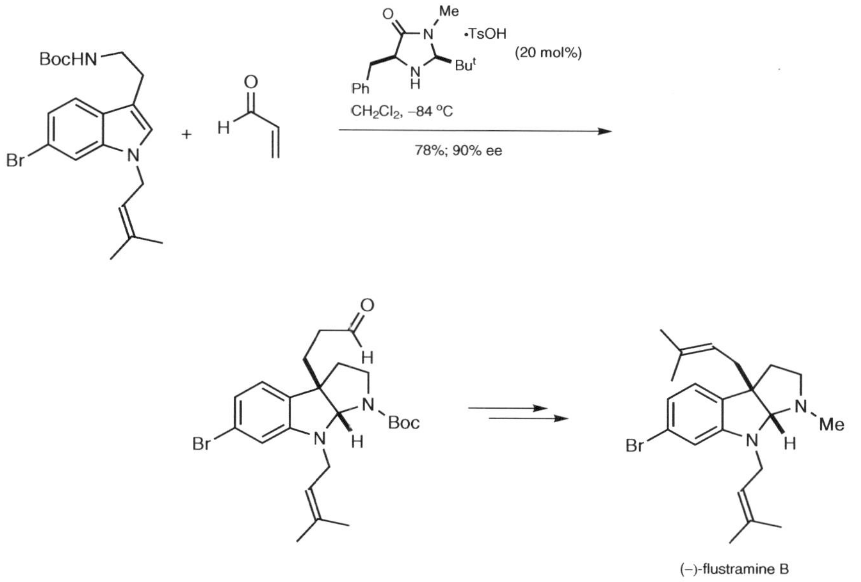

(−)-flustramine B

Besides the single use of two activation modes, numerous tandem transformations have been developed through the combination of these two activation modes with the design of substrates with different reactive sites. These transformations are in turn applied in the total synthesis of natural products and, among these reaction pathways, the iminium–enamine sequence is one of the most commonly applied.

At almost the same time, Wang and Córdova's groups reported an enantioselective domino aza-Michael/aldol reaction catalyzed by an (S)-diarylprolinol silyl ether **129** (Scheme 63).[156,157] This methodology shows very broad substrate scope. Both aliphatic and aromatic α,β-unsaturated aldehydes can be subjected to this transformation to give the desired products with excellent yield and enantioselectivity. For those aldehydes containing an aryl group, no electronic effects and no clear steric effects are observed. Furthermore, a variety of 2-aminobenzaldehydes **128** have been applied to this reaction irrespective of the electronic properties of the substituents on the phenyl ring and irrespective of the existence of an amine-protecting group. Interestingly, for substrates with a free amine, the use of an acidic additive such as benzoic acid gives satisfactory yields, whereas a basic additive is necessary for the same purpose for N-protected substrates.

for references see p 351

348 Domino Transformations **1.5** Non-Radical Skeletal Rearrangements

Scheme 63 Amine-Catalyzed Domino Aza-Michael/Aldol Reaction[156,157]

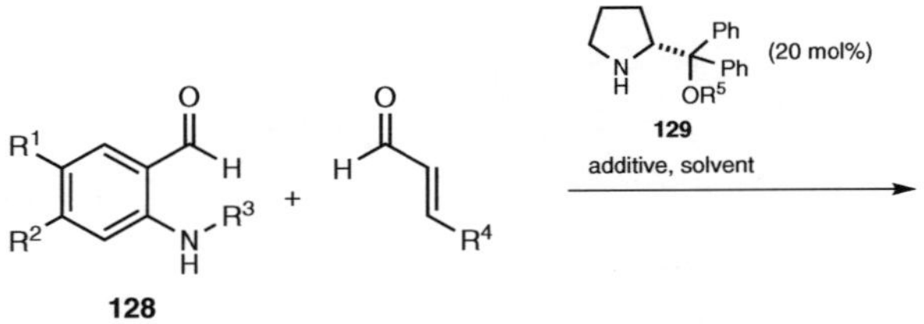

R¹	R²	R³	R⁴	R⁵	Additive	Solvent	Temp	ee (%)	Yield (%)	Ref
Cl	H	Cbz	4-O₂NC₆H₄	TES	NaOAc	1,2-dichloroethane	rt	94	95	[156]
OMe	OMe	Cbz	Ph	TES	NaOAc	1,2-dichloroethane	rt	95	84	[156]
H	H	Cbz	2-MeOC₆H₄	TES	NaOAc	1,2-dichloroethane	rt	92	96	[156]
H	F	H	CO₂Et	TMS	BzOH	DMF	−25 °C	90	83	[157]
H	H	H	Bu	TMS	BzOH	DMF	−25 °C	98	69	[157]
H	H	H	Pr	TMS	BzOH	DMF	−25 °C	96	78	[157]

Using this type of reaction as the key step, producing intermediates **130** and **131**, the Yao and Hamada groups completed the total syntheses of camptothecin and the chiral core **132** of martinelline, respectively (Scheme 64).[158,159]

Scheme 64 Synthesis of Camptothecin and the Chiral Core of Martinelline[158,159]

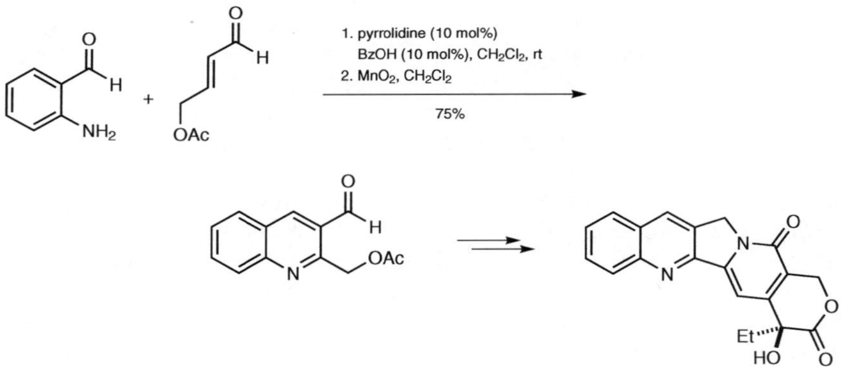

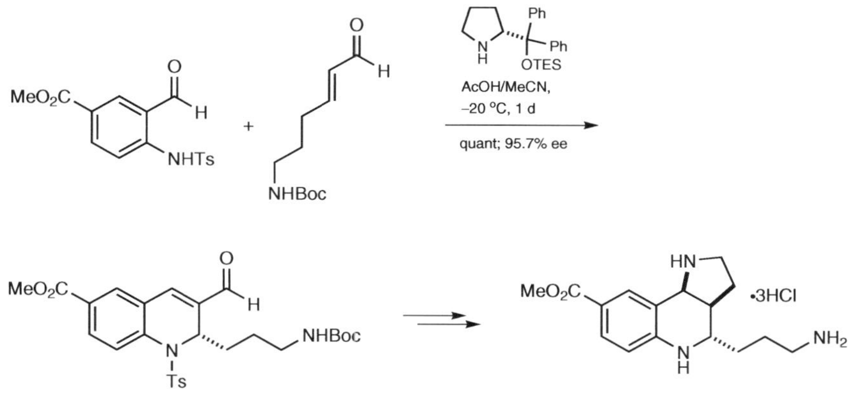

1.5.2 Lewis Acid/Base Induced Reactions

With an α,β-unsaturated aldehyde substrate two activation modes are possible upon reacting with a secondary amine catalyst, i.e. the formation of an iminium or a dienamine (Scheme 65). As there is an equilibrium between these two intermediates, dienamine activation is possible through proper screening of the reaction conditions.[160] Therefore, such an activation mode could lead to the enamine–iminium sequence, which is also effective for the construction of cyclic skeletons.

Scheme 65 Formation of an Iminium or Dienamine Intermediate

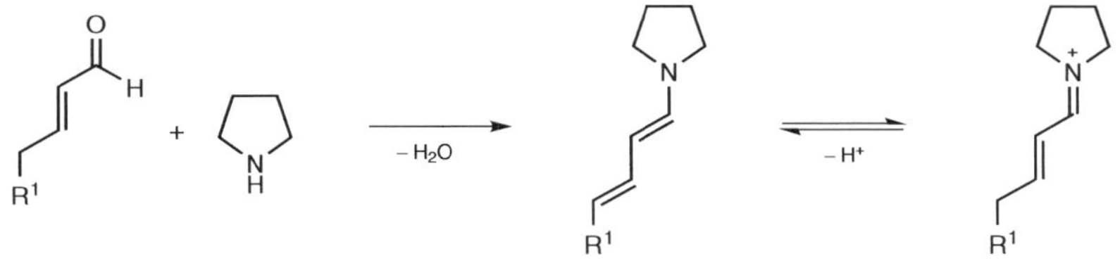

Using this sequence as the key step, Woggon and co-workers have reported a very efficient synthesis of α-tocopherol,[161] the most significant member of the vitamin E family. In their synthesis, the expected aldol/oxa-Michael sequence is realized in the presence of a proline-derived catalyst. In results similar to those observed by the Jørgensen group,[162] a catalyst with trifluoromethyl groups on the phenyl ring gives the best yield and stereoselectivity (Scheme 66).[161]

for references see p 351

350 Domino Transformations **1.5** Non-Radical Skeletal Rearrangements

Scheme 66 Total Synthesis of α-Tocopherol[161]

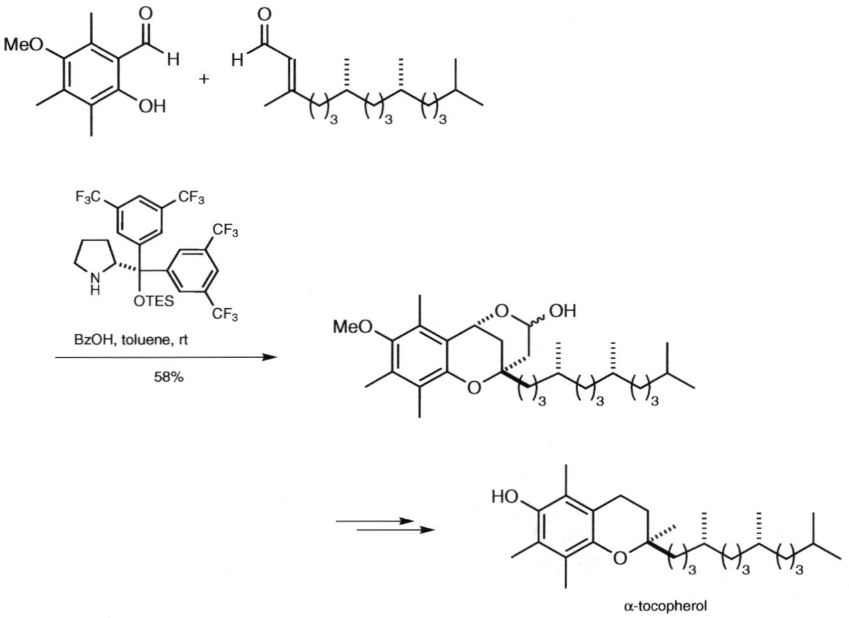

α-tocopherol

(6R,7R)-6,7-Dihydroxy-3,7-dimethyl-1,5,6,7-tetrahydro-2-benzopyran-8(8H)-one (122);
Typical Procedure:[148]
To a soln of enynone **121** (8 mg, 0.038 mmol) in $CDCl_3$ (0.75 mL) was added DABCO (1 mg, 0.009 mmol). The reaction was followed by ^1H NMR spectroscopy and reached completion after 24 h. The yellow soln was concentrated and purified on silica gel (hexanes/EtOAc 1:1) to give a bright orange solid; yield: 8 mg (100%).

tert-Butyl (3aS,8aS)-8-Benzyl-3a-(3-oxopropyl)-3,3a,8,8a-tetrahydropyrrolo[2,3-b]indole-
1(2H)-carboxylate (126, R¹ = R² = R⁵ = H; R³ = Bn; R⁴ = Boc); Typical Procedure:[155]
An amber 2-dram vial equipped with a magnetic stirrer bar, containing the catalyst **125** (Ar¹ = indol-3-yl; 114 mg, 0.40 mmol), TFA (31 μL, 0.40 mmol), and tryptamine derivative **124** (R¹ = R² = H; R³ = Bn; R⁴ = Boc; 700 mg, 2.0 mmol) was charged with CH_2Cl_2 (5.0 mL) and H_2O (1.0 mL) and then cooled to −80 °C. The soln was stirred for 5 min before addition of acrolein (0.54 mL, 8.0 mmol). The resulting suspension was stirred at −80 °C until complete consumption of the indole **124** was observed, as determined by TLC (24 h). To the mixture was then added pH 7.0 buffer. The contents were extracted with Et_2O and concentrated under reduced pressure. The resulting residue was purified by chromatography (silica gel, EtOAc/hexanes 1:9) to afford a colorless, viscous oil; yield: 670 mg (82%); 90% ee.

Methyl (S)-2-[3-(tert-Butoxycarbonylamino)propyl]-3-formyl-1-tosyl-1,2-dihydroquinoline-
6-carboxylate (131); Typical Procedure:[159]
To a stirred soln of (E)-6-(tert-butoxycarbonylamino)hex-2-enal (8.5 g, 0.04 mol) and (R)-diphenylprolinol triethylsilyl ether (370 mg, 1.0 mmol) in MeCN (40 mL) and AcOH (23 mL, 0.4 mmol) at −20 °C was added methyl 3-formyl-4-(4-toluenesulfonylamino)benzoate (7.6 g, 0.02 mol). After being stirred at −20 °C for 1 d under an argon atmosphere, the mixture was diluted with EtOAc, washed with H_2O and brine, dried (Na_2SO_4), filtered, and concentrated under reduced pressure. The residue was purified by column chromatography (silica gel, hexanes/EtOAc 2:1) to give a yellow, amorphous powder; yield: 11.7 g (quant); 95.7% ee.

References

[1] Song, Z.-L.; Fan, C.-A.; Tu, Y.-Q., *Chem. Rev.*, (2011) **111**, 7523.

[2] Wang, B.; Tu, Y. Q., *Acc. Chem. Res.*, (2011) **44**, 1207.

[3] Wang, S.-H.; Tu, Y.-Q.; Tang, M., In *Comprehensive Organic Synthesis II*, 2nd ed.; Knochel, P.; Molander, G. A., Eds.; Elsevier: Amsterdam, (2014); p 795.

[4] Tu, Y. Q.; Fan, C. A.; Ren, S. K.; Chan, A. S. C., *J. Chem. Soc., Perkin Trans. 1*, (2000), 3791.

[5] Fenster, M. D. B.; Dake, G. R., *Org. Lett.*, (2003) **5**, 4313.

[6] Fenster, M. D. B.; Dake, G. R., *Chem.–Eur. J.*, (2005) **11**, 639.

[7] Dake, G. R.; Fenster, M. D. B.; Fleury, M.; Patrick, B. O., *J. Org. Chem.*, (2004) **69**, 5676.

[8] Zhang, X.-M.; Tu, Y.-Q.; Zhang, F.-M.; Shao, H.; Meng, X., *Angew. Chem. Int. Ed.*, (2011) **50**, 3916.

[9] Zhang, X.-M.; Shao, H.; Tu, Y.-Q.; Zhang, F.-M.; Wang, S.-H., *J. Org. Chem.*, (2012) **77**, 8174.

[10] Tu, Y. Q.; Yang, L. M.; Chen, Y. Z., *Chem. Lett.*, (1998), 285.

[11] Tu, Y. Q.; Sun, L. D.; Wang, P. Z., *J. Org. Chem.*, (1999) **64**, 629.

[12] Gu, P.; Zhao, Y.-M.; Tu, Y. Q.; Ma, Y.; Zhang, F., *Org. Lett.*, (2006) **8**, 5271.

[13] Zhao, Y.-M.; Gu, P.; Tu, Y.-Q.; Fan, C.-A.; Zhang, Q., *Org. Lett.*, (2008) **10**, 1763.

[14] Zhao, Y.-M.; Gu, P.; Zhang, H.-J.; Zhang, Q.-W.; Fan, C.-A.; Tu, Y.-Q.; Zhang, F.-M., *J. Org. Chem.*, (2009) **74**, 3211.

[15] Chen, Z.-H.; Chen, Z.-M.; Zhang, Y.-Q.; Tu, Y.-Q.; Zhang, F.-M., *J. Org. Chem.*, (2011) **76**, 10173.

[16] Chen, Z.-H.; Zhang, Y.-Q.; Chen, Z.-M.; Tu, Y.-Q.; Zhang, F.-M., *Chem. Commun. (Cambridge)*, (2011) **47**, 1836.

[17] Chen, Z.-H.; Tu, Y.-Q.; Zhang, S.-Y.; Zhang, F.-M., *Org. Lett.*, (2011) **13**, 724.

[18] Overman, L. E.; Kakimoto, M.-a.; Okazaki, M.; Meier, G. P., *J. Am. Chem. Soc.*, (1983) **105**, 6622.

[19] Overman, L. E.; Mendelson, L. T.; Jacobsen, E. J., *J. Am. Chem. Soc.*, (1983) **105**, 6629.

[20] Overman, L. E.; Jacobsen, E. J.; Doedens, R. J., *J. Org. Chem.*, (1983) **48**, 3393.

[21] Hopkins, M. H.; Overman, L. E., *J. Am. Chem. Soc.*, (1987) **109**, 4748.

[22] Herrinton, P. M.; Hopkins, M. H.; Mishra, P.; Brown, M. J.; Overman, L. E., *J. Org. Chem.*, (1987) **52**, 3711.

[23] Hopkins, M. H.; Overman, L. E.; Rishton, G. M., *J. Am. Chem. Soc.*, (1991) **113**, 5354.

[24] Brown, M. J.; Harrison, T.; Herrinton, P. M.; Hopkins, M. H.; Hutchinson, K. D.; Mishra, P.; Overman, L. E., *J. Am. Chem. Soc.*, (1991) **113**, 5365.

[25] Hirst, G. C.; Howard, P. N.; Overman, L. E., *J. Am. Chem. Soc.*, (1989) **111**, 1514.

[26] Ando, S.; Minor, K. P.; Overman, L. E., *J. Org. Chem.*, (1997) **62**, 6379.

[27] Grese, T. A.; Hutchinson, K. D.; Overman, L. E., *J. Org. Chem.*, (1993) **58**, 2468.

[28] Brown, M. J.; Harrison, T.; Overman, L. E., *J. Am. Chem. Soc.*, (1991) **113**, 5378.

[29] Hirst, G. C.; Johnson, T. O., Jr.; Overman, L. E., *J. Am. Chem. Soc.*, (1993) **115**, 2992.

[30] Rickborn, B.; Gerkin, R. M., *J. Am. Chem. Soc.*, (1968) **90**, 4193.

[31] Suga, H.; Miyake, H., *Synthesis*, (1988), 394.

[32] Fan, C.-A.; Hu, X.-D.; Tu, Y.-Q.; Wang, B.-M.; Song, Z.-L., *Chem.–Eur. J.*, (2003) **9**, 4301.

[33] Li, D. R.; Xia, W. J.; Tu, Y. Q.; Zhang, F. M.; Shi, L., *Chem. Commun. (Cambridge)*, (2003), 798.

[34] Li, X.; Wu, B.; Zhao, X. Z.; Jia, Y. X.; Tu, Y. Q.; Li, D. R., *Synlett*, (2003), 623.

[35] Li, X.; Wang, B.-M.; Zhao, X.-Z.; Gao, S.-H.; Tu, Y.-Q.; Li, D.-R., *Chin. J. Chem.*, (2004) **22**, 1177.

[36] Zhao, X. Z.; Jia, Y. X.; Tu, Y. Q., *J. Chem. Res.*, (2003), 54.

[37] Srikrishna, A.; Nagamani, S. A.; Jagadeesh, S. G., *Tetrahedron: Asymmetry*, (2005) **16**, 1569.

[38] Srikrishna, A.; Nagamani, S. A., *J. Chem. Soc., Perkin Trans. 1*, (1999), 3393.

[39] Srikrishna, A.; Shaktikumar, L.; Satyanarayana, G., *ARKIVOC*, (2003), ix, 69.

[40] Kita, Y.; Higuchi, K.; Yoshida, Y.; Iio, K.; Kitagaki, S.; Akai, S.; Fujioka, H., *Angew. Chem. Int. Ed.*, (1999) **38**, 683.

[41] Kita, Y.; Higuchi, K.; Yoshida, Y.; Iio, K.; Kitagaki, S.; Ueda, K.; Akai, S.; Fujioka, H., *J. Am. Chem. Soc.*, (2001) **123**, 3214.

[42] Bordwell, F. G.; Almy, J., *J. Org. Chem.*, (1973) **38**, 571.

[43] Llera, J. M.; Fraser-Reid, B., *J. Org. Chem.*, (1989) **54**, 5544.

[44] Harding, K. E.; Trotter, J. W., *J. Org. Chem.*, (1977) **42**, 4157.

[45] Harding, K. E.; Strickland, J. B.; Pommerville, J., *J. Org. Chem.*, (1988) **53**, 4877.

[46] Corey, E. J.; Ortiz de Montellano, P. R., *J. Am. Chem. Soc.*, (1967) **89**, 3362.

[47] van Tamelen, E. E.; Seiler, M. P.; Wierenga, W., *J. Am. Chem. Soc.*, (1972) **94**, 8229.

[48] Corey, E. J.; Lee, J., *J. Am. Chem. Soc.*, (1993) **115**, 8873.

[49] Huang, A. X.; Xiong, Z.; Corey, E. J., *J. Am. Chem. Soc.*, (1999) **121**, 9999.

[50] Corey, E. J.; Lin, S., *J. Am. Chem. Soc.*, (1996) **118**, 8765.

[51] Corey, E. J.; Luo, G.; Lin, L. S., *J. Am. Chem. Soc.*, (1997) **119**, 9927.

[52] Mi, Y.; Schreiber, J. V.; Corey, E. J., *J. Am. Chem. Soc.*, (2002) **124**, 11290.

[53] Johnson, W. S.; Plummer, M. S.; Reddy, S. P.; Bartlett, W. R., *J. Am. Chem. Soc.*, (1993) **115**, 515.

[54] Fish, P. V.; Johnson, W. S., *J. Org. Chem.*, (1994) **59**, 2324.

[55] Tong, R.; Valentine, J. C.; McDonald, F. E.; Cao, R.; Fang, X.; Hardcastle, K. I., *J. Am. Chem. Soc.*, (2007) **129**, 1050.

[56] Tanuwidjaja, J.; Ng, S.-S.; Jamison, T. F., *J. Am. Chem. Soc.*, (2009) **131**, 12084.

[57] Beutler, J. A.; Shoemaker, R. H.; Johnson, T.; Boyd, M. R., *J. Nat. Prod.*, (1998) **61**, 1509.

[58] Beutler, J. A.; Jato, J.; Cragg, G. M.; Boyd, M. R., *Nat. Prod. Lett.*, (2000) **14**, 399.

[59] Thoison, O.; Hnawia, E.; Guiéritte-Voegelein, F.; Sévenet, T., *Phytochemistry*, (1992) **31**, 1439.

[60] Yoder, B. J.; Cao, S.; Norris, A.; Miller, J. S.; Ratovoson, F.; Razafitsalama, J.; Andriantsiferana, R.; Rasamison, V. E.; Kingston, D. G. I., *J. Nat. Prod.*, (2007) **70**, 342.

[61] Neighbors, J. D.; Beutler, J. A.; Wiemer, D. F., *J. Org. Chem.*, (2005) **70**, 925.

[62] Mente, N. R.; Neighbors, J. D.; Wiemer, D. F., *J. Org. Chem.*, (2008) **73**, 7963.

[63] Topczewski, J. J.; Callahan, M. P.; Neighbors, J. D.; Wiemer, D. F., *J. Am. Chem. Soc.*, (2009) **131**, 14630.

[64] Topczewski, J. J.; Kodet, J. G.; Wiemer, D. F., *J. Org. Chem.*, (2011) **76**, 909.

[65] Kürti, L.; Czakó, B., *Strategic Applications of Named Reactions in Organic Synthesis*, Elsevier: Burlington, (2005); p 396.

[66] Daly, J. W.; Sande, T. F., In *Alkaloids: Chemical and Biological Perspectives*, Pelletier, S. W., Ed.; Wiley: New York, (1986); Vol. 4, Chapter 1.

[67] Aronstam, R. S.; Daly, J. W.; Spande, T. F.; Narayanan, T. K.; Albuquerque, E. X., *Neurochem. Res.*, (1986) **11**, 1227.

[68] Michael, J. P., In *The Alkaloids: Chemistry and Biology*, Cordell, G. A., Ed.; Academic: New York, (2001); Vol. 55, p 91.

[69] Michael, J. P., *Nat. Prod. Rep.*, (2002) **19**, 719.

[70] Lourenço, A. M.; Máximo, P.; Ferreira, L. M.; Pereira, M. M. A., *Stud. Nat. Prod. Chem.*, (2002) **27**, 233.

[71] Michael, J. P., *Nat. Prod. Rep.*, (2003) **20**, 458.

[72] Pearson, W. H.; Walavalkar, R., *Tetrahedron*, (2001) **57**, 5081, and references cited therein.

[73] Wrobleski, A.; Sahasrabudhe, K.; Aubé, J., *J. Am. Chem. Soc.*, (2002) **124**, 9974, and references cited therein.

[74] Reddy, P. G.; Baskaran, S., *J. Org. Chem.*, (2004) **69**, 3093.

[75] Kürti, L.; Czakó, B., *Strategic Applications of Named Reactions in Organic Synthesis*, Elsevier: Burlington, (2005); p 476.

[76] Minagawa, K.; Kouzuki, S.; Yoshimoto, J. U. N.; Kawamura, Y.; Tani, H.; Iwata, T.; Terui, Y.; Nakai, H.; Yagi, S.; Hattori, N.; Fujiwara, T.; Kamigauchi, T., *J. Antibiot.*, (2002) **55**, 155.

[77] Minagawa, K.; Kouzuki, S.; Kamigauchi, T., *J. Antibiot.*, (2002) **55**, 165.

[78] Nakatani, M.; Nakamura, M.; Suzuki, A.; Inoue, M.; Katoh, T., *Org. Lett.*, (2002) **4**, 4483.

[79] Watanabe, K.; Sakurai, J.; Abe, H.; Katoh, T., *Chem. Commun. (Cambridge)*, (2010) **46**, 4055.

[80] Sakurai, J.; Kikuchi, T.; Takahashi, O.; Watanabe, K.; Katoh, T., *Eur. J. Org. Chem.*, (2011), 2948.

[81] Ngoc, D. T.; Albicker, M.; Schneider, L.; Cramer, N., *Org. Biomol. Chem.*, (2010) **8**, 1781.

[82] Nicolaou, K. C.; Sarlah, D.; Wu, T. R.; Zhan, W., *Angew. Chem. Int. Ed.*, (2009) **48**, 6870.

[83] Uchiro, H.; Kato, R.; Arai, Y.; Hasegawa, M.; Kobayakawa, Y., *Org. Lett.*, (2011) **13**, 6268, and references cited therein.

[84] Overman, L. E., *Acc. Chem. Res.*, (1992) **25**, 352.

[85] Kürti, L.; Czakó, B., In *Strategic Applications of Named Reactions in Organic Synthesis*, Elsevier: Burlington, (2005); p 366.

[86] Johnson, W. S.; Kinnel, R. B., *J. Am. Chem. Soc.*, (1966) **88**, 3861.

[87] Li, B.; Lai, Y.-C.; Zhao, Y.; Wong, Y.-H.; Shen, Z.-L.; Loh, T.-P., *Angew. Chem. Int. Ed.*, (2012) **51**, 10619.

[88] Zhao, Y.-J.; Chng, S.-S.; Loh, T.-P., *J. Am. Chem. Soc.*, (2007) **129**, 492.

[89] Zhao, Y.-J.; Loh, T.-P., *J. Am. Chem. Soc.*, (2008) **130**, 10024.

[90] Zhao, Y.-J.; Loh, T.-P., *Chem. Commun. (Cambridge)*, (2008), 1434.

[91] Zhao, Y.-J.; Loh, T.-P., *Org. Lett.*, (2008) **10**, 2143.

[92] Satake, M.; Ofuji, K.; Naoki, H.; James, K. J.; Furey, A.; McMahon, T.; Silke, J.; Yasumoto, T., *J. Am. Chem. Soc.*, (1998) **120**, 9967.

References

[93] Ofuji, K.; Satake, M.; McMahon, T.; Silke, J.; James, K. J.; Naoki, H.; Oshima, Y.; Yasumoto, T., *Nat. Toxins*, (1999) **7**, 99.

[94] Nicolaou, K. C.; Qian, W.; Bernal, F.; Uesaka, N.; Pihko, P. M.; Hinrichs, J., *Angew. Chem. Int. Ed.*, (2001) **40**, 4068.

[95] Nicolaou, K. C.; Li, Y.; Uesaka, N.; Koftis, T. V.; Vyskocil, S.; Ling, T.; Govindasamy, M.; Qian, W.; Bernal, F.; Chen, D. Y.-K., *Angew. Chem. Int. Ed.*, (2003) **42**, 3643.

[96] Nicolaou, K. C.; Chen, D. Y.-K.; Li, Y.; Qian, W.; Ling, T.; Vyskocil, S.; Koftis, T. V.; Govindasamy, M.; Uesaka, N., *Angew. Chem. Int. Ed.*, (2003) **42**, 3649.

[97] Nicolaou, K. C.; Vyskocil, S.; Koftis, T. V.; Yamada, Y. M. A.; Ling, T.; Chen, D. Y.-K.; Tang, W.; Petrovic, G.; Frederick, M. O.; Li, Y.; Satake, M., *Angew. Chem. Int. Ed.*, (2004) **43**, 4312.

[98] Nicolaou, K. C.; Koftis, T. V.; Vyskocil, S.; Petrovic, G.; Ling, T.; Yamada, Y. M. A.; Tang, W.; Frederick, M. O., *Angew. Chem. Int. Ed.*, (2004) **43**, 4318.

[99] Nicolaou, K. C.; Pihko, P. M.; Bernal, F.; Frederick, M. O.; Qian, W.; Uesaka, N.; Diedrichs, N.; Hinrichs, J.; Koftis, T. V.; Loizidou, E.; Petrovic, G.; Rodriquez, M.; Sarlah, D.; Zou, N., *J. Am. Chem. Soc.*, (2006) **128**, 2244.

[100] Nicolaou, K. C.; Chen, D. Y.-K.; Li, Y.; Uesaka, N.; Petrovic, G.; Koftis, T. V.; Bernal, F.; Frederick, M. O.; Govindasamy, M.; Ling, T.; Pihko, P. M.; Tang, W.; Vyskocil, S., *J. Am. Chem. Soc.*, (2006) **128**, 2258.

[101] Nicolaou, K. C.; Frederick, M. O.; Loizidou, E. Z.; Petrovic, G.; Cole, K. P.; Koftis, T. V.; Yamada, Y. M. A., *Chem.–Asian J.*, (2006) **1**, 245.

[102] Chen, C.; Layton, M. E.; Sheehan, S. M.; Shair, M. D., *J. Am. Chem. Soc.*, (2000) **122**, 7424.

[103] Overman, L. E.; Kakimoto, M.-a., *J. Am. Chem. Soc.*, (1979) **101**, 1310.

[104] Jacobsen, E. J.; Levin, J.; Overman, L. E., *J. Am. Chem. Soc.*, (1988) **110**, 4329.

[105] Horowitz, R. M.; Geissman, T. A., *J. Am. Chem. Soc.*, (1950) **72**, 1518.

[106] Knight, S. D.; Overman, L. E.; Pairaudeau, G., *J. Am. Chem. Soc.*, (1993) **115**, 9293.

[107] Martin, C. L.; Overman, L. E.; Rohde, J. M., *J. Am. Chem. Soc.*, (2008) **130**, 7568.

[108] Brummond, K. M.; Lu, J., *Org. Lett.*, (2001) **3**, 1347.

[109] Earley, W. G.; Jacobsen, J. E.; Madin, A.; Meier, G. P.; O'Donnell, C. J.; Oh, T.; Old, D. W.; Overman, L. E.; Sharp, M. J., *J. Am. Chem. Soc.*, (2005) **127**, 18046.

[110] Overman, L. E.; Shim, J., *J. Org. Chem.*, (1991) **56**, 5005.

[111] Ishihara, K.; Nakamura, S.; Yamamoto, H., *J. Am. Chem. Soc.*, (1999) **121**, 4906.

[112] Ishihara, K.; Ishibashi, H.; Yamamoto, H., *J. Am. Chem. Soc.*, (2002) **124**, 3647.

[113] Kumazawa, K.; Ishihara, K.; Yamamoto, H., *Org. Lett.*, (2004) **6**, 2551.

[114] Ishibashi, H.; Ishihara, K.; Yamamoto, H., *J. Am. Chem. Soc.*, (2004) **126**, 11122.

[115] Fürstner, A.; Davies, P. W., *Angew. Chem. Int. Ed.*, (2007) **46**, 3410.

[116] Hashmi, A. S. K., *Chem. Rev.*, (2007) **107**, 3180.

[117] Jiménez-Núñez, E.; Echavarren, A. M., *Chem. Commun. (Cambridge)*, (2007), 333.

[118] Arcadi, A.; Di Giuseppe, S., *Curr. Org. Chem.*, (2004) **8**, 795.

[119] Hashmi, A. S. K., *Angew. Chem. Int. Ed.*, (2005) **44**, 6990.

[120] Crone, B.; Kirsch, S. F., *Chem.–Eur. J.*, (2008) **14**, 3514.

[121] Mullen, C. A.; Gagné, M. R., *J. Am. Chem. Soc.*, (2007) **129**, 11880.

[122] Mullen, C. A.; Campbell, A. N.; Gagné, M. R., *Angew. Chem. Int. Ed.*, (2008) **47**, 6011.

[123] Schweinitz, A.; Chtchemelinine, A.; Orellana, A., *Org. Lett.*, (2011) **13**, 232.

[124] Yoshida, M.; Ismail, M. A. H.; Nemoto, H.; Ihara, M., *J. Chem. Soc., Perkin Trans. 1*, (2000), 2629.

[125] Jiao, L.; Yuan, C.; Yu, Z.-X., *J. Am. Chem. Soc.*, (2008) **130**, 4421.

[126] Simmons, E. M.; Sarpong, R., *Org. Lett.*, (2006) **8**, 2883.

[127] Simmons, E. M.; Yen, J. R.; Sarpong, R., *Org. Lett.*, (2007) **9**, 2705.

[128] Simmons, E. M.; Hardin, A. R.; Guo, X.; Sarpong, R., *Angew. Chem. Int. Ed.*, (2008) **47**, 6650.

[129] Shi, H.; Fang, L.; Tan, C.; Shi, L.; Zhang, W.; Li, C.-c.; Luo, T.; Yang, Z., *J. Am. Chem. Soc.*, (2011) **133**, 14944.

[130] Jiménez-Núñez, E.; Claverie, C. K.; Nieto-Oberhuber, C.; Echavarren, A. M., *Angew. Chem. Int. Ed.*, (2006) **45**, 5452.

[131] Jiménez-Núñez, E.; Molawi, K.; Echavarren, A. M., *Chem. Commun. (Cambridge)*, (2009), 7327.

[132] Jiménez-Núñez, E.; Raducan, M.; Lauterbach, T.; Molawi, K.; Solorio, C. R.; Echavarren, A. M., *Angew. Chem. Int. Ed.*, (2009) **48**, 6152.

[133] Gaydou, M.; Miller, R. E.; Delpont, N.; Ceccon, J.; Echavarren, A. M., *Angew. Chem. Int. Ed.*, (2013) **52**, 6396.

[134] López, S.; Herrero-Gómez, E.; Pérez-Galán, P.; Nieto-Oberhuber, C.; Echavarren, A. M., *Angew. Chem. Int. Ed.*, (2006) **45**, 6029.

[135] Molawi, K.; Delpont, N.; Echavarren, A. M., *Angew. Chem. Int. Ed.*, (2010) **49**, 3517.

[136] Zhou, Q.; Chen, X.; Ma, D., *Angew. Chem. Int. Ed.*, (2010) **49**, 3513.

[137] Klahn, P.; Duschek, A.; Liébert, C.; Kirsch, S. F., *Org. Lett.*, (2012) **14**, 1250.

[138] Canham, S. M.; France, D. J.; Overman, L. E., *J. Am. Chem. Soc.*, (2010) **132**, 7876.

[139] Zhang, Q.-W.; Xiang, K.; Tu, Y.-Q.; Zhang, S.-Y.; Zhang, X.-M.; Zhao, Y.-M.; Zhang, T. C., *Chem.– Asian J.*, (2012) **7**, 894.

[140] MacMillan, D. W. C., *Nature (London)*, (2008) **455**, 304.

[141] Jones, S. B.; Simmons, B.; MacMillan, D. W. C., *J. Am. Chem. Soc.*, (2009) **131**, 13606.

[142] Jones, S. B.; Simmons, B.; Mastracchio, A.; MacMillan, D. W. C., *Nature (London)*, (2011) **475**, 183.

[143] Horning, B. D.; MacMillan, D. W. C., *J. Am. Chem. Soc.*, (2013) **135**, 6442.

[144] Golden, J. E.; Aubé, J., *Angew. Chem. Int. Ed.*, (2002) **41**, 4316.

[145] Zeng, Y.; Reddy, D. S.; Hirt, E.; Aubé, J., *Org. Lett.*, (2004) **6**, 4993.

[146] Zeng, Y.; Aubé, J., *J. Am. Chem. Soc.*, (2005) **127**, 15712.

[147] Frankowski, K. J.; Golden, J. E.; Zeng, Y.; Lei, Y.; Aubé, J., *J. Am. Chem. Soc.*, (2008) **130**, 6018.

[148] Stark, L. M.; Pekari, K.; Sorensen, E. J., *Proc. Natl. Acad. Sci. U. S. A.*, (2004) **101**, 12064.

[149] List, B., *Tetrahedron*, (2002) **58**, 5573.

[150] Lelais, G.; MacMillan, D. W. C., *Aldrichimica Acta*, (2006) **39**, 79.

[151] Gaunt, M. J.; Johansson, C. C. C.; McNally, A.; Vo, N. T., *Drug Discovery Today*, (2007) **12**, 8.

[152] Erkkilä, A.; Majander, I.; Pihko, P. M., *Chem. Rev.*, (2007) **107**, 5416.

[153] Grondal, C.; Jeanty, M.; Enders, D., *Nat. Chem.*, (2010) **2**, 167.

[154] Ramachary, D. B.; Reddy, Y. V., *Eur. J. Org. Chem.*, (2012), 865.

[155] Austin, J. F.; Kim, S.-G.; Sinz, C. J.; Xiao, W.-J.; MacMillan, D. W. C., *Proc. Natl. Acad. Sci. U. S. A.*, (2004) **101**, 5482.

[156] Li, H.; Wang, J.; Xie, H.; Zu, L.; Jiang, W.; Duesler, E. N.; Wang, W., *Org. Lett.*, (2007) **9**, 965.

[157] Sundén, H.; Rios, R.; Ibrahem, I.; Zhao, G.-L.; Eriksson, L.; Córdova, A., *Adv. Synth. Catal.*, (2007) **349**, 827.

[158] Liu, G.-S.; Dong, Q.-L.; Yao, Y.-S.; Yao, Z.-J., *Org. Lett.*, (2008) **10**, 5393.

[159] Yoshitomi, Y.; Arai, H.; Makino, K.; Hamada, Y., *Tetrahedron*, (2008) **64**, 11568.

[160] Bertelsen, S.; Marigo, M.; Brandes, S.; Dinér, P.; Jørgensen, K. A., *J. Am. Chem. Soc.*, (2006) **128**, 12973.

[161] Liu, K.; Chougnet, A.; Woggon, W.-D., *Angew. Chem. Int. Ed.*, (2008) **47**, 5827.

[162] Marigo, M.; Fielenbach, D.; Braunton, A.; Kjærsgaard, A.; Jørgensen, K. A., *Angew. Chem. Int. Ed.*, (2005) **44**, 3703.

1.5.3   **Brook Rearrangement as the Key Step in Domino Reactions**

A. Kirschning, F. Gille, and M. Wolling

General Introduction

The Brook rearrangement,[1–4] named after A. G. Brook (1924–2013), refers to the intramolecular 1,2-anionic migration of a silyl group from carbon to oxygen in α-silylated carbinols **1** in the presence of a catalytic amount of a base or alkali metal (Scheme 1). By deprotonation of the hydroxy group, the migration is accelerated. Occasionally, the reverse process has been reported, which is named the retro-Brook rearrangement. If after silyl migration the newly formed carbanion **2** is trapped by an electrophile other than a proton, Brook rearrangements will become part of a cascade process (Scheme 1).[5]

Scheme 1   General Scheme of the Brook Rearrangement and Its Embedment into Cascade Processes[5]

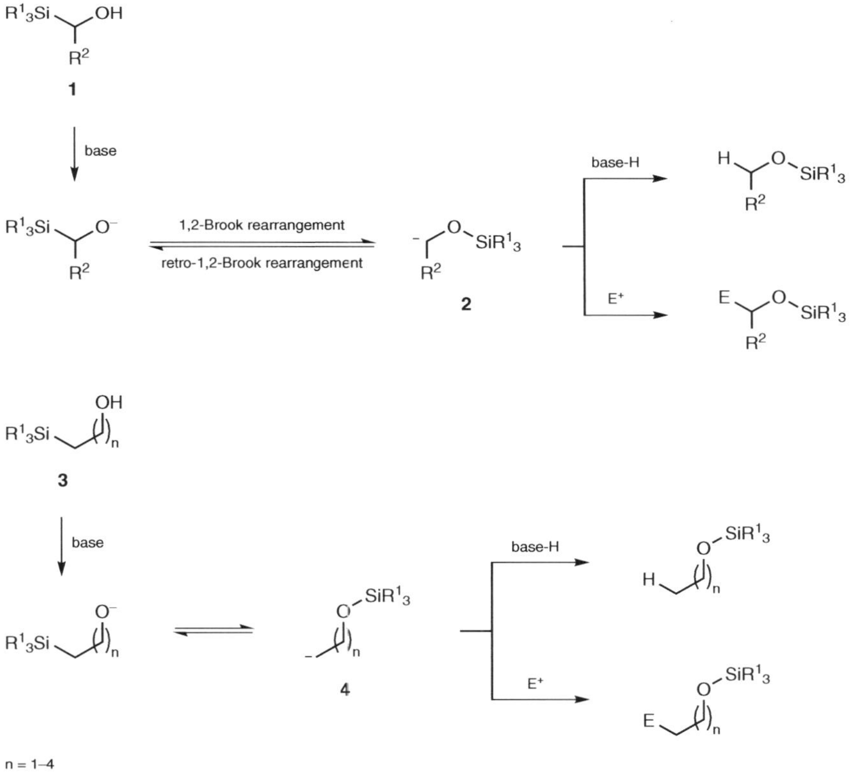

$n = 1–4$

Other variations of the Brook rearrangement are known (e.g., aza- and thia-Brook rearrangements) in which oxygen is exchanged by nitrogen or sulfur, respectively, and where silyl migrations occur across more bonds (1,3- to 1,5-Brook rearrangements[6,7]). The 1,3-Brook rearrangement of β-silylated carbinols **3** ($n = 1$) is most widely found as

for references see p 445

356 Domino Transformations **1.5** Non-Radical Skeletal Rearrangements

part of Peterson alkenations,[8] and therefore the number of cascade reactions that are based on trapping the intermediate carbanion **4** (n = 1) by an electrophile are rather restricted (Scheme 1). The competition between a homo-Peterson reaction leading to cyclopropanes and a cascade sequence is favored with respect to the latter option. Consequently, the number of examples and applications of cascade sequences initiated by 1,4-Brook rearrangement of γ-silylated carbinols **3** (n = 2) dwarfs those of 1,3-Brook rearrangements, whereas 1,5-Brook rearrangements are very rare.[6,7]

Synthetic routes toward alcoholates of silylated carbinols **1** or **3** are the key to develop new applications of domino Brook rearrangements. For example, these alcoholates are not only formed upon deprotonation of carbinols, but are also generated by attack of nucleophiles on carbonyl precursors or ring opening of epoxides.

This chapter will cover two major topics: (a) various methods to prepare precursors suited to undergoing Brook rearrangements as well as (b) their embedment in domino reactions including applications in natural product synthesis.

1.5.3.1 1,2-Brook Rearrangement

1,2-Brook rearrangements require access to α-silyl alkoxides. These compounds can be generated from aldehydes or ketones by nucleophilic addition of silyllithium species, from acylsilanes by addition of nucleophiles, from α-silyl carbinols by deprotonation, and from epoxysilanes if oxirane ring opening occurs at the non-silyl-substituted terminus. Retro-Brook rearrangements are also known, though less common, which unravels the reversibility of these rearrangements. Consequently, the retro-1,2-Brook rearrangement will also be covered in this chapter.

1.5.3.1.1 1,2-Brook Rearrangement with Aldehydes, Ketones, or Acyl Chlorides

Commonly, 1,2-Brook rearrangements with aldehydes, ketones, or acyl chlorides are initiated by the addition of silyl anions to the carbonyl moiety. A domino reductive McMurry-like alkenation is based on organochromium species and is initiated by nucleophilic addition of an in situ formed chromium–silane to aldehydes, followed by a Brook rearrangement (Scheme 2).[9] The intermediate organochromium species **5** reacts with a second aldehyde equivalent. Subsequent reductive elimination of chromium(IV) oxide provides the alkene. The chromium–silane species may be formed via one-electron transfer from the chromium(II) species to silicon, yielding a nucleophilic silyl anion and chromium(III).

Scheme 2 Reductive McMurry Alkenation of Aldehydes[9]

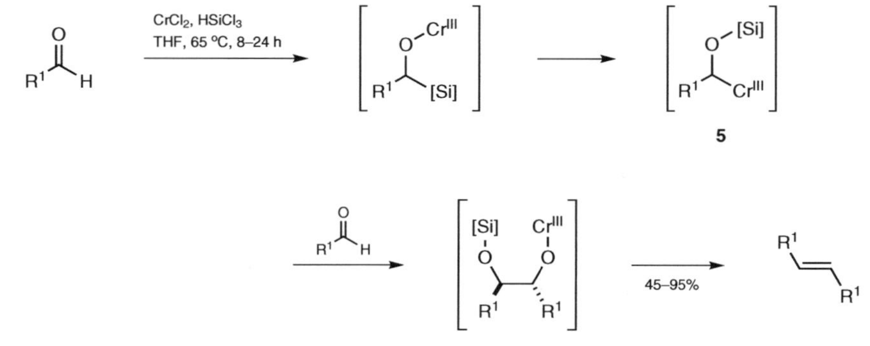

R^1 = Ph, 4-FC$_6$H$_4$, 4-BrC$_6$H$_4$, 4-Tol, 4-HOC$_6$H$_4$, 2-naphthyl, 4-BnO-3-MeOC$_6$H$_3$, (E)-CH=CHPh,

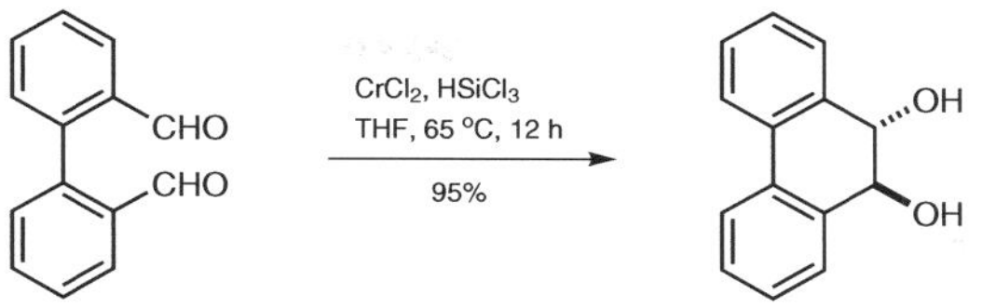

1.5.3 Brook Rearrangement as the Key Step in Domino Reactions **357**

Acyloins also find use in simple cascade reactions that include a 1,2-Brook rearrangement (Scheme 3).[10] After addition of an organosilyl group to the carbonyl group of acyloins, the following Brook rearrangement of intermediates **6** provides α-siloxy carbanions **7**, which undergo β-elimination to yield enol ethers, usually with high E selectivity.

Scheme 3   Preparation of Silyl Enol Ethers from Acyloins[10]

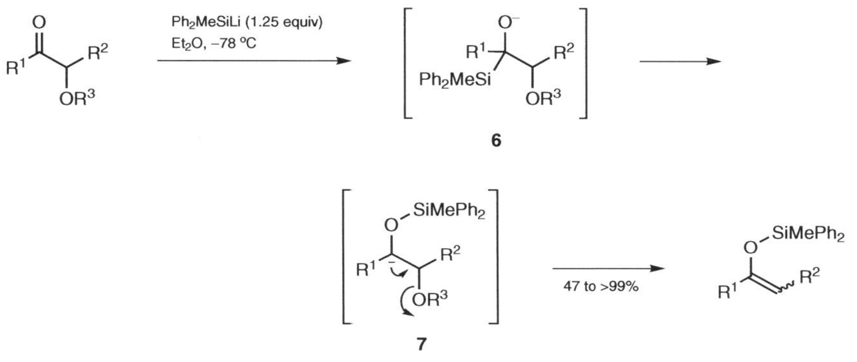

R^1 = Et, Pr, Ph; R^2 = H, Me, Et, Ph; R^3 = TMS, Ac, Me

2,3-Epoxy ketones also bear a substituent at the α-position that is prone to β-elimination. Consequently, the chemoselective, nucleophilic addition of (methyldiphenylsilyl)lithium to 2,3-epoxy ketones yields 3-hydroxy ketones **10**. After 1,2-Brook rearrangement of the alkoxide **8**, the resulting intermediate **9** undergoes ring opening of the oxirane (Scheme 4).[11] The resulting silyl enol ether is cleaved in situ by a second equivalent of (methyldiphenylsilyl)lithium. If the reaction is performed at −78 °C in toluene using only 1.7–2.0 equivalents of the silyl anion, the silyl enol ether can be isolated with low to high E/Z ratios (43:57 to 100:0).[12]

Scheme 4   Transformation of 2,3-Epoxy Ketones into β-Hydroxy Ketones via 1,2-Brook Rearrangement[11]

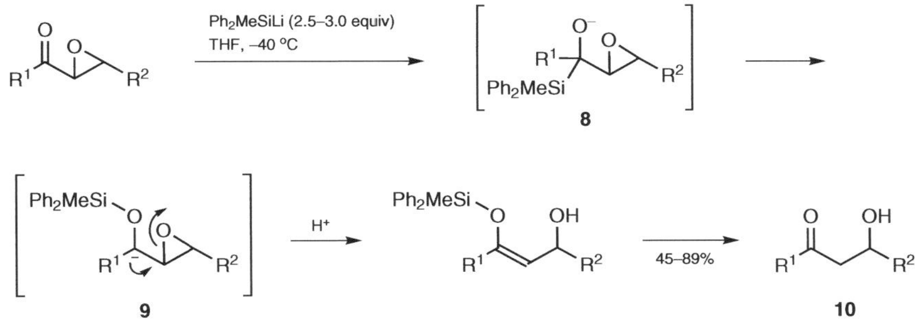

R^1 = Me, (CH$_2$)$_4$Me, 4-MeOC$_6$H$_4$, Ph; R^2 = H, Pr, Ph

for references see p 445

358   Domino Transformations   **1.5** Non-Radical Skeletal Rearrangements

The same cascade reaction can be conducted with the corresponding aziridines to yield β-amino ketones (Scheme 5).[13]

Scheme 5   Transformation of Acyl Aziridines to β-Amino Ketones via 1,2-Brook Rearrangement[13]

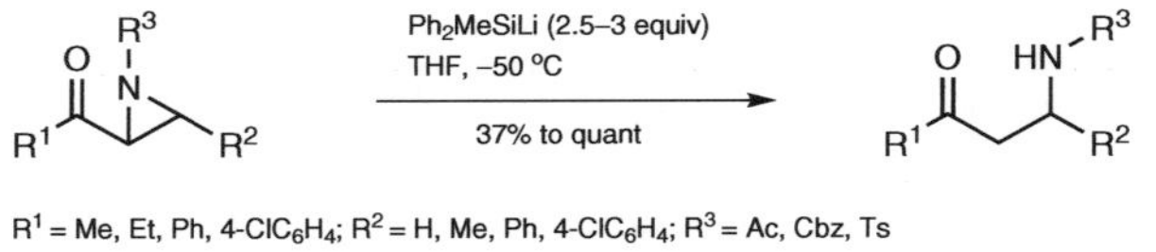

R^1 = Me, Et, Ph, 4-ClC$_6$H$_4$; R^2 = H, Me, Ph, 4-ClC$_6$H$_4$; R^3 = Ac, Cbz, Ts

Acyl chlorides have very rarely served as the carbonyl component.[14] Overall, they are versatile precursors to create α-siloxy carbanions after 1,2-Brook rearrangements, and the carbanions can be trapped with various electrophiles. For example, treatment of 2 equivalents of [dimethyl(phenyl)silyl]lithium with acyl chlorides provides 1,1-disilylalkoxide anions **11**. One silyl group undergoes silyl migration and the newly formed carbanions **12** react with various organic halides to yield α-silyl carbinols in the same pot (Scheme 6).[14]

Scheme 6   Acyl Chlorides as Precursors for 1,2-Brook Rearrangement after Double Silylation[14]

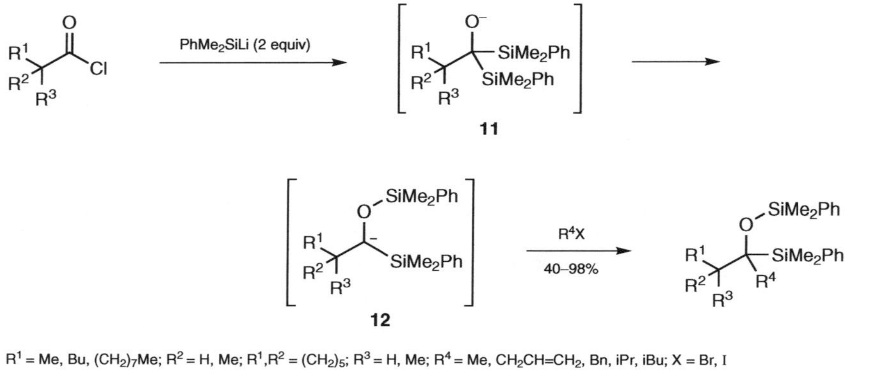

R^1 = Me, Bu, (CH$_2$)$_7$Me; R^2 = H, Me; R^1,R^2 = (CH$_2$)$_5$; R^3 = H, Me; R^4 = Me, CH$_2$CH=CH$_2$, Bn, iPr, iBu; X = Br, I

β-Hydroxy Ketones 10; General Procedure:[11]
A soln of an α,β-epoxy ketone (1 mmol) in anhyd THF (5 mL) was cooled to −40 °C under argon. A 1 M soln of Ph$_2$MeSiLi (2.5–3.0 equiv) in THF was added dropwise with stirring. The reaction was immediately quenched with sat. aq NH$_4$Cl. The mixture was extracted with CH$_2$Cl$_2$. The organic layer was dried (Na$_2$SO$_4$) and concentrated under reduced pressure to give the crude product.

1.5.3.1.2   **1,2-Brook Rearrangement with Acylsilanes**

1.5.3.1.2.1   **Domino Reactions of Acylsilanes by Addition of Nucleophiles**

A pioneering example of using acylsilanes for 1,2-Brook rearrangements in cascade sequences was reported well over 30 years ago.[15] Regio- and stereoselective synthesis of silyl enol ethers **15** is achieved by a nucleophilic attack of an organometallic reagent on α-phenylsulfanyl acylsilanes **13** under Felkin–Anh control. Subsequent base-induced 1,2-Brook rearrangement of alkoxide intermediate **14**, followed by elimination of the phenylsulfanyl group furnishes the product with good yield and selectivity (Scheme 7).[16,17]

1.5.3 Brook Rearrangement as the Key Step in Domino Reactions **359**

Scheme 7 Silyl Enol Ether Synthesis from an α-Phenylsulfanyl Acylsilane[16,17]

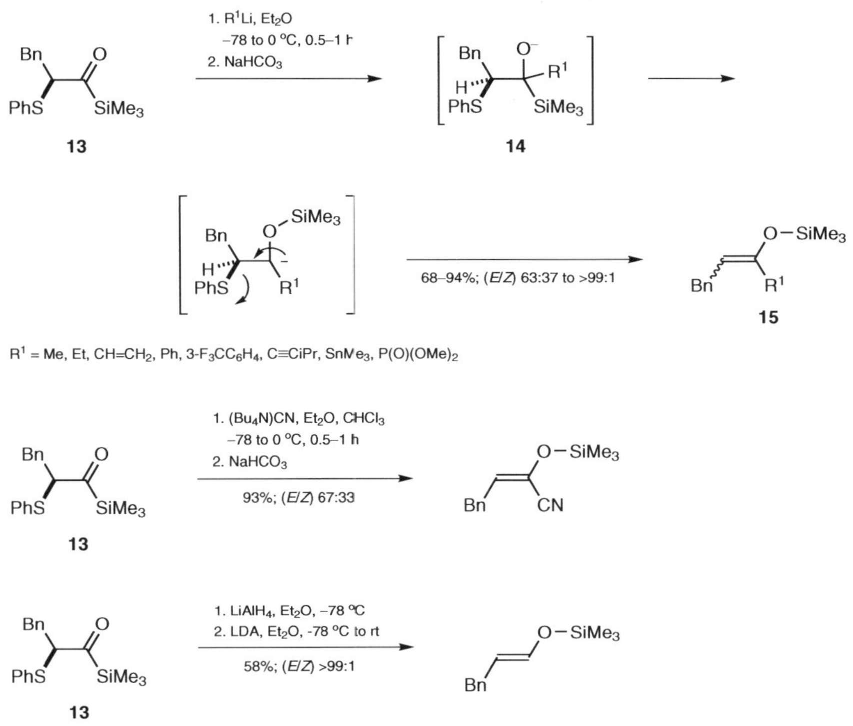

R^1 = Me, Et, CH=CH$_2$, Ph, 3-F$_3$CC$_6$H$_4$, C≡CiPr, SnMe$_3$, P(O)(OMe)$_2$

Closely related to this work are silyl enol ether syntheses that utilize acylsilanes and α-sulfinyl carbanions **16** as starting materials (Scheme 8).[18] Interestingly, the leaving group here is part of the nucleophile. The cascade is initiated by nucleophilic addition to the acylsilane, followed by Brook rearrangement of the alkoxide intermediate **17** and simultaneous formation of the product and elimination of an arylsulfinyl anion.

Scheme 8 Silyl Enol Ether Synthesis from Acylsilanes and α-Sulfinyl Carbanions[18]

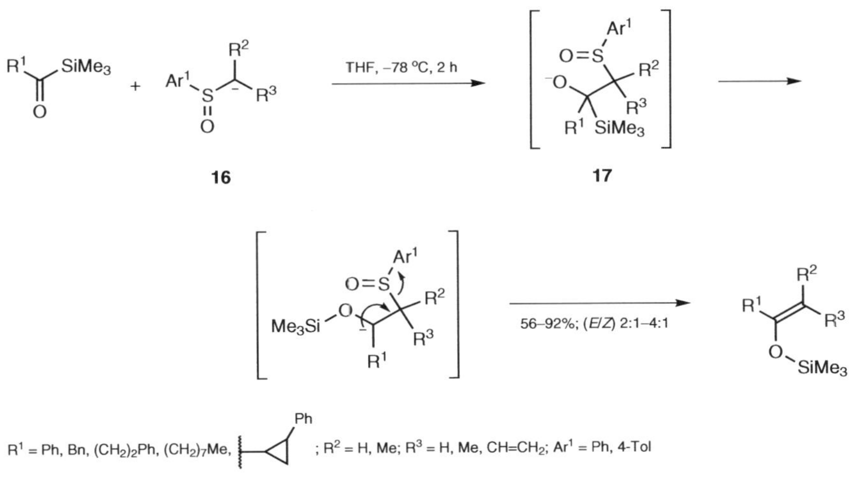

R^1 = Ph, Bn, (CH$_2$)$_2$Ph, (CH$_2$)$_7$Me, cyclopropyl-Ph; R^2 = H, Me; R^3 = H, Me, CH=CH$_2$; Ar^1 = Ph, 4-Tol

for references see p 445

360　Domino Transformations　**1.5** Non-Radical Skeletal Rearrangements

The addition of trimethylsilyl-substituted oxiranyl anions to acylsilanes leads to tetrasubstituted silyl enol ethers **19** (Scheme 9).[19] The addition results in the formation of an oxyanion **18** that undergoes 1,2-Brook rearrangement. This event leads to oxirane opening as a result of a β-elimination and formation of the allyl alcohol product **19**. The exclusive formation of the Z-isomer is supposedly controlled by dipole–dipole repulsion of the two oxygen atoms and steric interaction between the two silyl groups at the stage of nucleophilic addition. If the oxirane ring has a phenyl substituent instead of the trimethylsilyl group, this sequence proceeds in a similar manner.

Scheme 9　Silyl Enol Ether Synthesis from Acylsilanes and a Lithiated Epoxysilane[19]

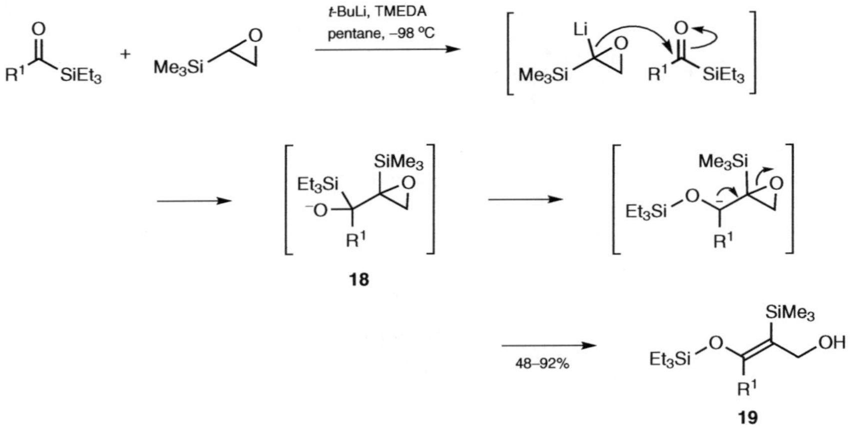

R^1 = 4-ClC$_6$H$_4$, 4-FC$_6$H$_4$, 4-Tol, 2-naphthyl, 3,4-(MeO)$_2$C$_6$H$_3$, 4-Me$_2$NC$_6$H$_4$, 4-MeOC$_6$H$_4$, 2-MeOC$_6$H$_4$, 3-furyl, 2-thienyl, CHMePh

The concept of silyl enol ether synthesis by a cascade based on 1,2-Brook rearrangement and β-elimination can be further extended by exploiting the reactive properties of silyl enol ethers, as exemplified in the preparation of geminal difluoro compounds **21** (Scheme 10).[20] The reaction starts with fluoride-catalyzed addition of trimethyl(trifluoromethyl)silane to an acylsilane. After 1,2-Brook rearrangement and β-elimination of fluoride, silyl enol ether **20** is generated. Without isolation, it is directly reacted with α,β-unsaturated carbonyl compounds as part of a Mukaiyama aldol protocol. After workup, the resulting 1,5-diketones **21** can be further cyclized under base promotion.[20] The protocol can also be utilized for the synthesis of 4,4-difluoroterpenes.[21] A similar method leads to geminal difluoro silyl enol ethers.[22]

1.5.3 Brook Rearrangement as the Key Step in Domino Reactions   **361**

Scheme 10   Preparation of Fluorinated 1,5-Diketones by a 1,2-Brook Rearrangement/β-Elimination/Mukaiyama Aldol Cascade[20]

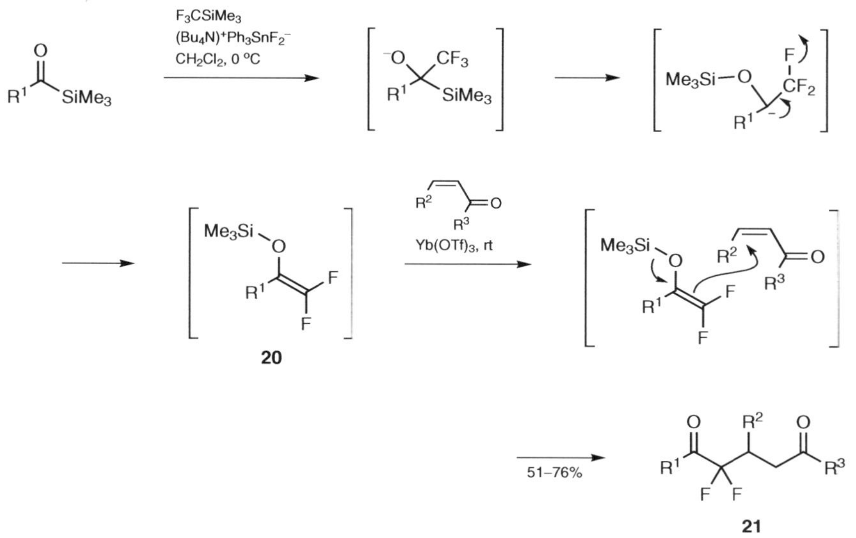

R^1 = Me, Ph, (CH$_2$)$_6$Me; R^2 = H; R^3 = Me; R^2,R^3 = (CH$_2$)$_3$

In addition, β-hydroxy α,α-difluoro ketone derivatives **24** can be prepared by Lewis acid promoted nucleophilic addition of an in situ formed Grignard reagent to an acyltriphenylsilane (e.g., **22**), followed by 1,2-Brook rearrangement and elimination of fluoride. The resulting silyl enol ether **23** is reacted in the same pot with the aldehydes (Scheme 11).[23]

Scheme 11   Preparation of Fluorinated β-Hydroxy Ketones by a 1,2-Brook Rearrangement/β-Elimination/Mukaiyama Aldol Cascade[23]

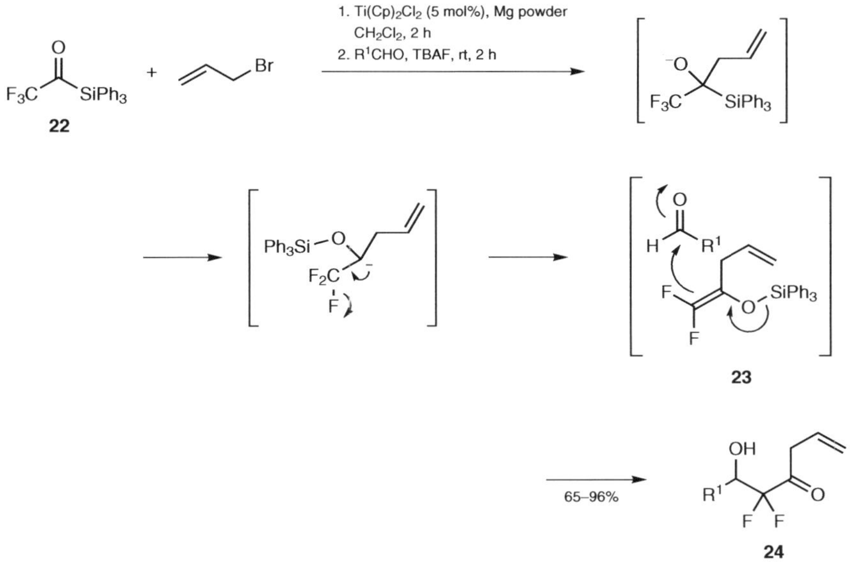

R^1 = 4-Tol, 4-MeOC$_6$H$_4$, 2-Tol, 4-Me$_2$NC$_6$H$_4$, 4-MeC$_2$CC$_6$H$_4$, 4-FC$_6$H$_4$, 4-O$_2$NC$_6$H$_4$, 2-O$_2$NC$_6$H$_4$, CH=CHPh, Cy

for references see p 445

The reaction of β-sulfanyl-substituted α,β-unsaturated acylsilanes, e.g. **25**, and ketone enolates **26** yields cyclopentanone silyl enol ethers such as **28**. Key features of this cascade reaction are nucleophilic addition to the acylsilane, 1,2-Brook rearrangement, and intramolecular attack of the allyl anion **27** to the carbonyl functionality to form a five-membered ring. Regiocontrol of the second C—C bond-forming process is achieved, in part, by the α-directing effect of the sulfanyl substituent (Scheme 12).[24–28] If this cascade reaction is conducted in the absence of a sulfanyl group, cyclopropanes **29** are formed via the alternative anionic position (Scheme 12).

Scheme 12 Reaction of a β-Sulfanyl-Substituted α,β-Unsaturated Acylsilane and Enolates Derived from Ketones To Yield Cyclopentanone Silyl Enol Ethers[25]

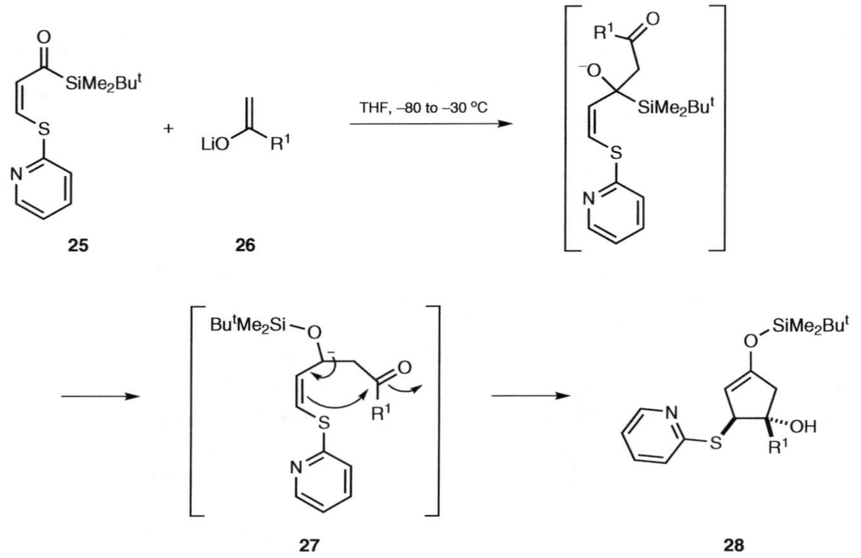

R[1]	dr	Yield (%)	Ref
Et	4:1	70	[25]
Pr	55:45	70	[25]
iPr	2:1	65	[25]
(CH$_2$)$_7$Me	2:1	54	[25]

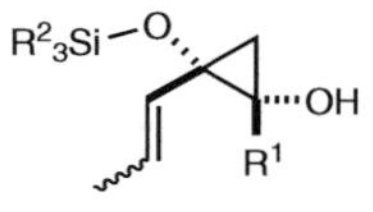

Likewise, β,β-dichloro-substituted α,β-unsaturated acylsilanes undergo a similar cascade reaction, but cleavage of the silyl ether results in loss of one chlorine substituent and reestablishment of the enone (Scheme 13).[25]

1.5.3 Brook Rearrangement as the Key Step in Domino Reactions **363**

Scheme 13 Reaction of β,β-Dichloro-Substituted α,β-Unsaturated Acylsilanes and Ketone Enolates To Yield Cyclopentenones[25]

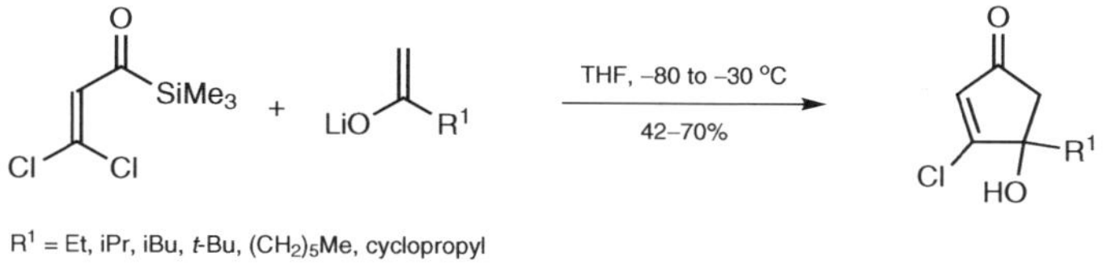

R^1 = Et, iPr, iBu, t-Bu, (CH$_2$)$_5$Me, cyclopropyl

A multicomponent cascade reaction toward pentasubstituted γ-butyrolactones **32** starts with a Reformatsky addition to silyl glyoxylates, which is followed by a 1,2-Brook rearrangement. The resulting carbanion **30** is trapped by a ketone followed by intramolecular transesterification of alcoholate **31** and lactone formation (Scheme 14).[29]

Scheme 14 1,2-Brook Rearrangement Initiated by Reformatsky Addition to a Silyl Glyoxylate and Synthesis of Pentasubstituted γ-Butyrolactones[29]

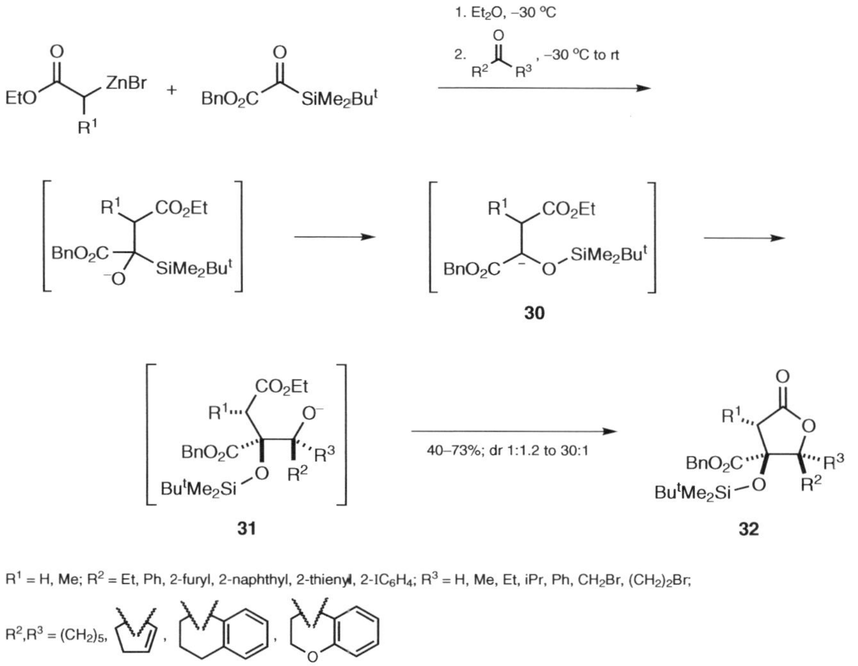

R^1 = H, Me; R^2 = Et, Ph, 2-furyl, 2-naphthyl, 2-thienyl, 2-IC$_6$H$_4$; R^3 = H, Me, Et, iPr, Ph, CH$_2$Br, (CH$_2$)$_2$Br;

R^2,R^3 = (CH$_2$)$_5$,

A [3+4]-cycloaddition approach provides cycloheptenones **35** and cycloheptenediones **36** (Scheme 15 and Scheme 16).[30–32] The sequence of events is initiated by nucleophilic attack of the enolate anion derived from a methyl vinyl ketone to the acylsilane, followed by a 1,2-Brook rearrangement and, in the case of the synthesis of cycloheptenones **35**, formation of a cyclopropane after intramolecular addition of the intermediate carbanion **33** to the carbonyl group. This step results in the formation of an alkoxy divinylcyclopropane intermediate **34** which sets the stage for an anionic oxy-Cope rearrangement. This elegant method has been successfully applied in the synthesis of the tricyclic skeleton of the cyathins.[33]

for references see p 445

Scheme 15 Synthesis of Cycloheptenones by a [3+4] Cycloaddition Triggered by a Brook Rearrangement[30]

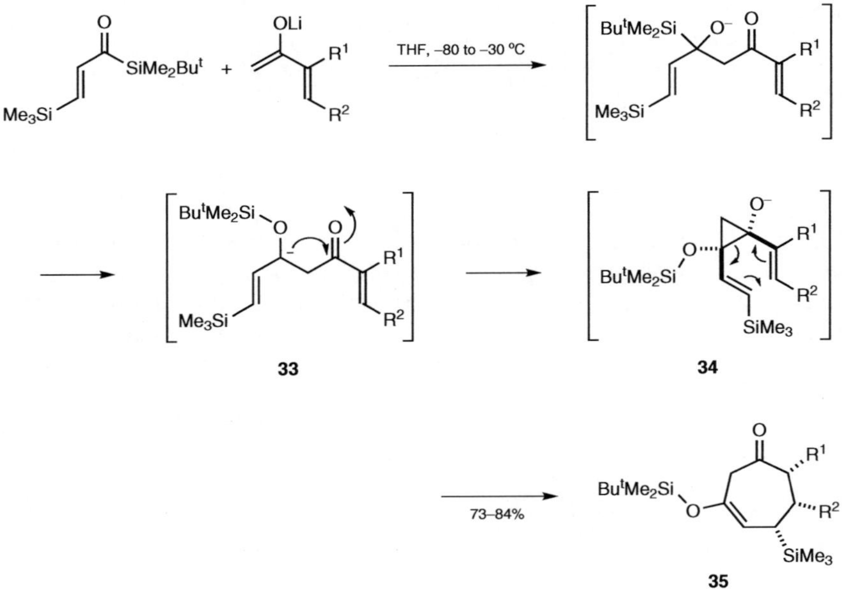

R[1] = H; R[2] = Pr, iPr, (CH$_2$)$_4$Me; R[1],R[2] = (CH$_2$)$_3$, (CH$_2$)$_4$

Scheme 16 Synthesis of Cycloheptenediones by a [3+4] Cycloaddition Triggered by a Brook Rearrangement[32]

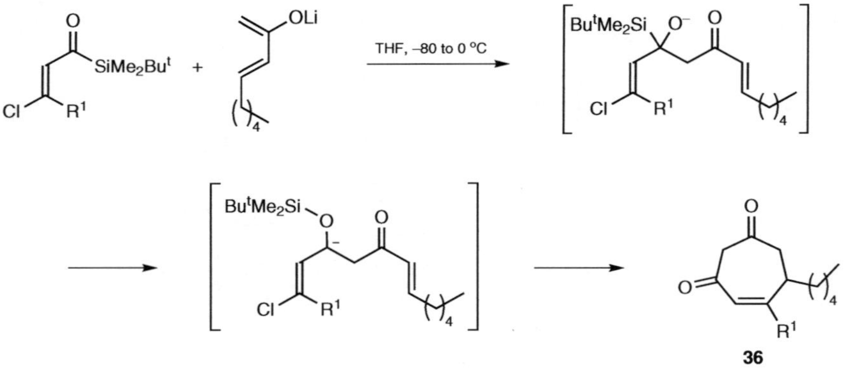

R[1]	Yield (%)	Ref
Me	75	[32]
Bu	41	[32]
(CH$_2$)$_5$Me	33	[32]
t-Bu	54[a]	[32]

[a] Reaction performed at rt.

1.5.3 Brook Rearrangement as the Key Step in Domino Reactions **365**

Bicyclo[3.3.2]decenone derivatives (e.g., **37**) are generated via a related [3+4]-cascade sequence starting from cycloheptenone enolates. α-Hydroxylation and oxidative cleavage using lead(IV) acetate provides cyclooctenes (Scheme 17).[34–37]

Scheme 17 Synthesis of a Bicyclo[3.3.2]decenone Derivative by a [3+4] Cycloaddition Triggered by a Brook Rearrangement[34–37]

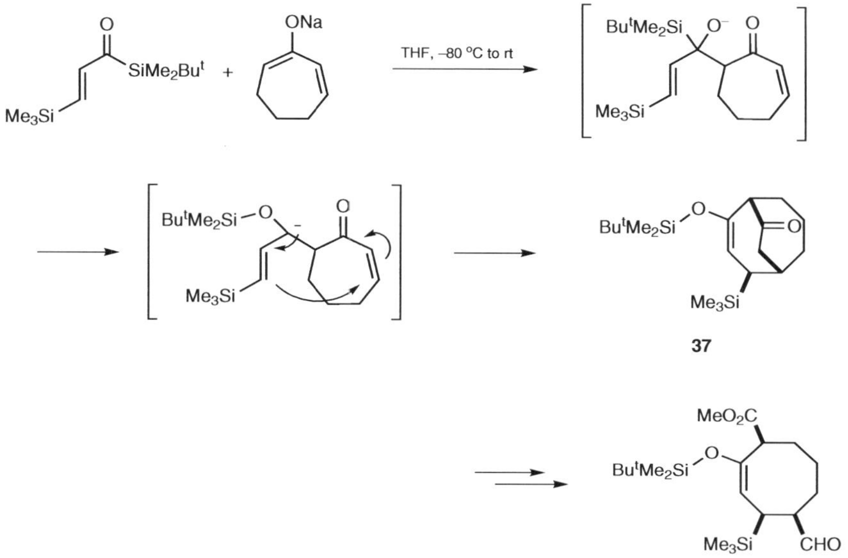

This concept can be extended to a cascade sequence that features two 1,2-Brook rearrangements (Scheme 18).[38] In this case, γ,δ-epoxy δ-silyl α,β-unsaturated acylsilane **38** is attacked by the enolate and a 1,2-Brook rearrangement of the alkoxide intermediate **39** provides allyl anion **40**, which is then followed by ring opening of the epoxysilane moiety. A second 1,2-Brook rearrangement provides new carbanionic intermediate **41**, which react in an intramolecular Michael addition to yield cycloheptenone **42**.

for references see p 445

Scheme 18 A Cascade Sequence to Cycloheptenones Involving Two 1,2-Brook Rearrangements[38]

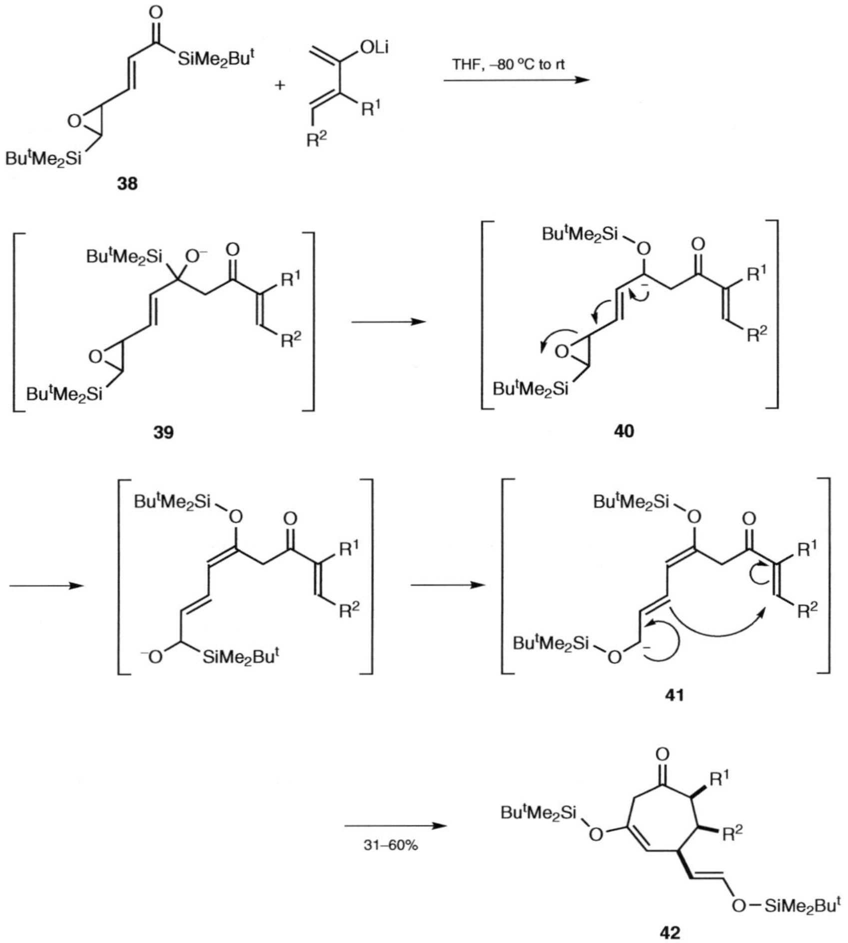

R^1 = H; R^2 = iPr, (CH$_2$)$_4$Me, OMe, Bn; R^1,R^2 = (CH$_2$)$_4$

The one-pot reaction of Reformatsky reagents with silyl glyoxylates followed by ring opening of the third component, a β-lactone or β-lactam, initiated by the carbanion **44** that is formed after 1,2-Brook rearrangement of alkoxide intermediate **43** yields ketones with 1,3- and 1,4-stereochemical control (Scheme 19).[39]

1.5.3 Brook Rearrangement as the Key Step in Domino Reactions **367**

Scheme 19 Three-Component Synthesis of Complex Ketones[39]

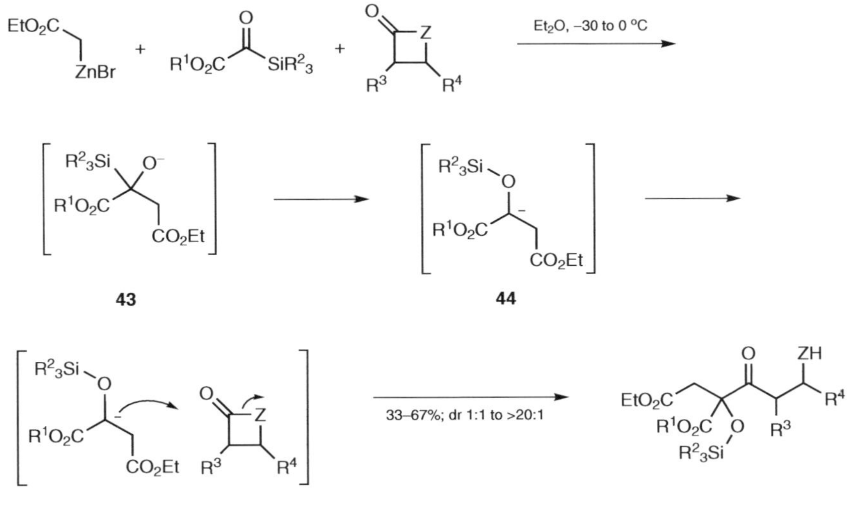

R^1 = Bn, t-Bu; SiR^2_3 = $SiEt_3$, $SiMe_2t$-Bu; Z = O, NBoc, NTs;
R^3 = H, Me; R^4 = Ph, C≡CTMS, $(CH_2)_2Ph$, CH_2OBn, $4-O_2NC_6H_4$, $CH_2OTBDPS$

A new synthesis of γ-amino β-hydroxy amides **46**, materials which are precursors for lactams, utilizes acylsilanes, imines, and amide-derived enolates (Scheme 20).[40] The cascade sequence starts with the addition of the enolate to acylsilanes, followed by 1,2-Brook rearrangement and addition of the newly formed carbanion **45** to the electrophilic imine.

Scheme 20 Three-Component Synthesis of Complex Amides[40]

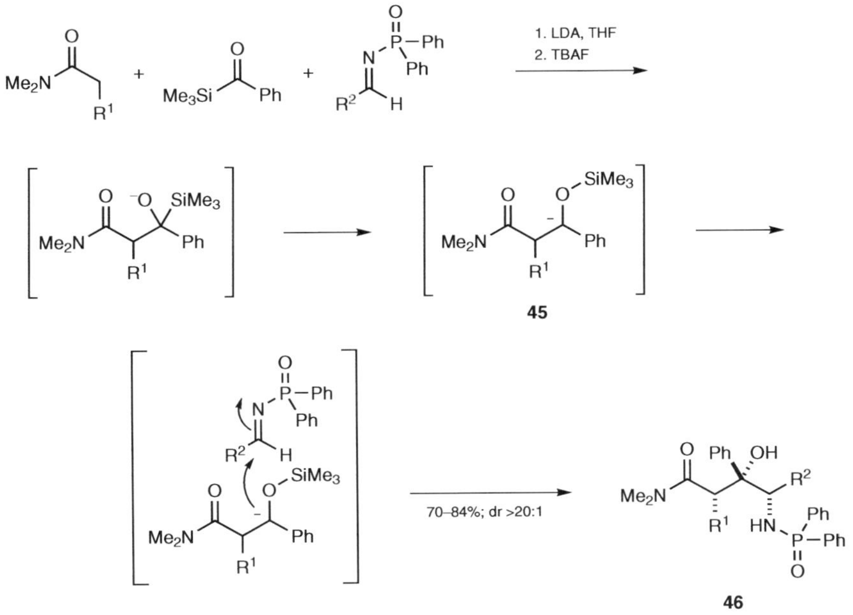

R^1 = H, Me, Bn; R^2 = Ph, $4-ClC_6H_4$, $4-BrC_6H_4$, $4-MeOC_6H_4$, 2-furyl

for references see p 445

The diastereoselective three-component coupling between glyoxylates, a nucleophile, and an electrophile in one pot can provide tetrasubstituted allenes **47** (Scheme 21). Key events of this domino sequence are metal chelation of the glyoxylate and a 1,2-Brook rearrangement. Various combinations of nucleophiles and electrophiles such as acetylides and aldehydes,[41] vinyl Grignard reagents and nitroalkenes,[42] and magnesium acetylides and nitroalkenes are possible.[43] This cascade sequence has been exploited in the total synthesis of leustroducsin B, using a Reformatsky reagent as nucleophile and a β-lactone as electrophile.[44]

Scheme 21 Three-Component Synthesis of Tetrasubstituted Allenes[41–43]

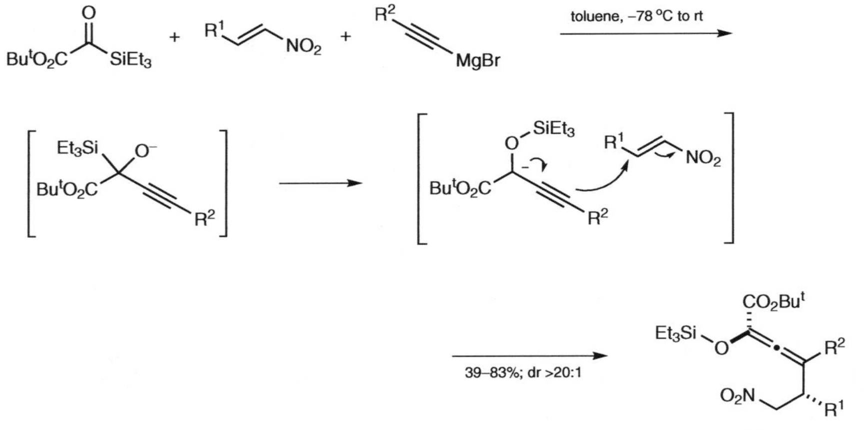

R^1 = Ph, 2-thienyl, 2-furyl, $(CH_2)_4Me$, iPr; R^2 = Me, Ph, $(CH_2)_4Me$, CMe=CH$_2$, TMS

Addition of nucleophiles to allyl esters of silyl glyoxylates provides an entry to a cascade sequence that proceeds via a 1,2-Brook rearrangement of alkoxide **48** followed by an Ireland–Claisen rearrangement leading to γ,δ-unsaturated α-siloxy acids **49** (Scheme 22).[45]

Scheme 22 Synthesis of γ,δ-Unsaturated α-Siloxy Acids by Combining 1,2-Brook Rearrangement with Ireland–Claisen Rearrangement[45]

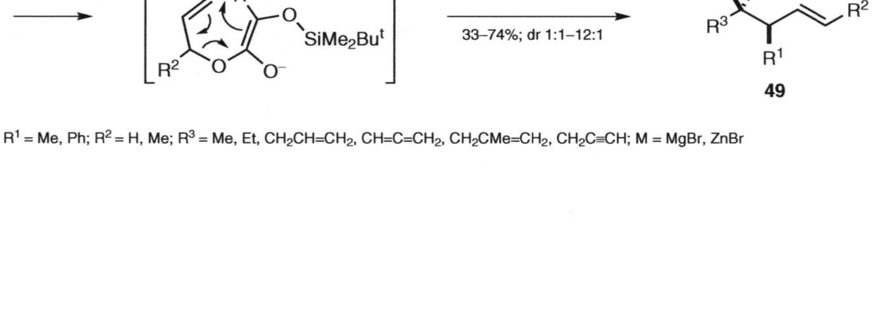

R^1 = Me, Ph; R^2 = H, Me; R^3 = Me, Et, CH$_2$CH=CH$_2$, CH=C=CH$_2$, CH$_2$CMe=CH$_2$, CH$_2$C≡CH; M = MgBr, ZnBr

1.5.3 Brook Rearrangement as the Key Step in Domino Reactions 369

An Oppenauer oxidation/aldol reaction cascade between alkoxides and silyl glyoxylates gives 2,3-dihydroxyglycolates **52** (Scheme 23).[46] Concerted oxidation of the alkoxide and reduction of the acylsilane is followed by 1,2-Brook rearrangement of intermediate **50** to generate the metal enolate **51**. This species reacts with the aldehyde or ketone in an aldol reaction via a Zimmerman–Traxler-like transition state to afford the aldol product; however, the transformation occurs with only low to moderate diastereoselectivity.

Scheme 23 Preparation of 2,3-Dihydroxyglycolates via Oppenauer Oxidation of Acylsilanes and Aldol Reaction[46]

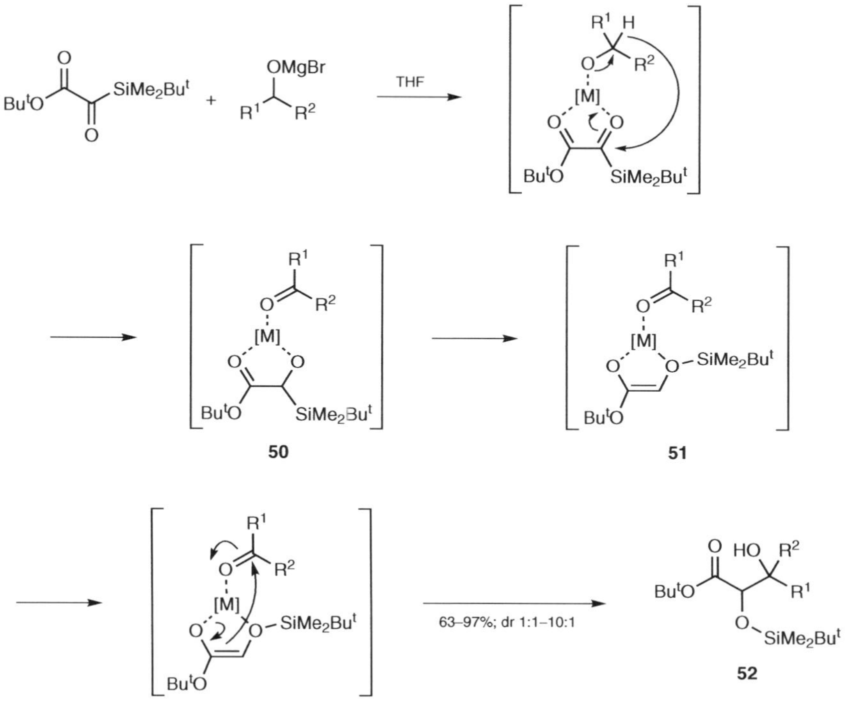

R^1 = Ph, iPr, $(CH_2)_4Me$, $(CH_2)_2TMS$, $(CH_2)_3CH=CH_2$, $4\text{-}ClC_6H_4$, $4\text{-}MeOC_6H_4$; R^2 = H, Me; R^1,R^2 = $(CH_2)_5$

Using lanthanide(III) isopropoxides as Lewis acid, silylated 2,3-dihydroxy esters **58** are obtained via an Oppenauer oxidation/aldol reaction cascade between aldehydes or ketones and silyl glyoxylates (Scheme 24).[47] The catalytic cycle starts by coordination of the Lewis acid to the silyl glyoxylate of form **54**, with one isopropoxy substituent serving as reductant in a Meerwein–Ponndorf–Verley reduction of the acylsilane. The resulting alkoxide **55** undergoes a 1,2-Brook rearrangement to yield a praseodymium enolate **56**, which undergoes an aldol reaction with the aldehyde or ketone. Subsequent liberation of the praseodymium Lewis acid from alkoxide **57** by exchange with the added vinyltrialkoxysilane **53** completes the catalytic cycle.

for references see p 445

370   Domino Transformations **1.5** Non-Radical Skeletal Rearrangements

Scheme 24   1,2-Brook Rearrangement as Part of an Oppenauer Oxidation/Aldol Reaction Cascade[47]

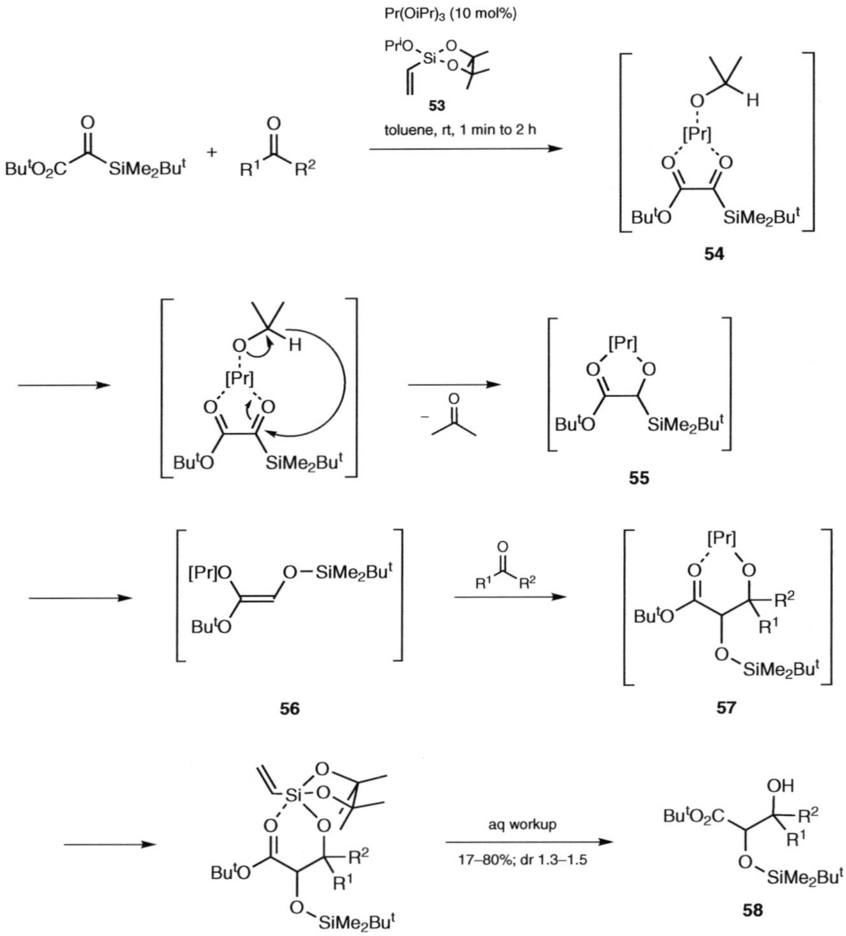

R^1 = Me, Ph, $(CH_2)_4$Me, 2-furyl, iPr; R^2 = H, Me

Related to cross-benzoin reactions, cyanoformate esters and acylsilanes can be coupled in a similar fashion (Scheme 25).[48,49] A key step of this domino reaction is the liberation of the cyanide anion, which adds to the acylsilane and initiates a 1,2-Brook rearrangement of alkoxide intermediate **60**. Using chiral aluminum Lewis acid **59**, the reaction proceeds enantioselectively.

1.5.3 Brook Rearrangement as the Key Step in Domino Reactions

371

Scheme 25 Benzoin-Type Coupling Reactions Using Cyanoformate Esters and Acylsilanes[49]

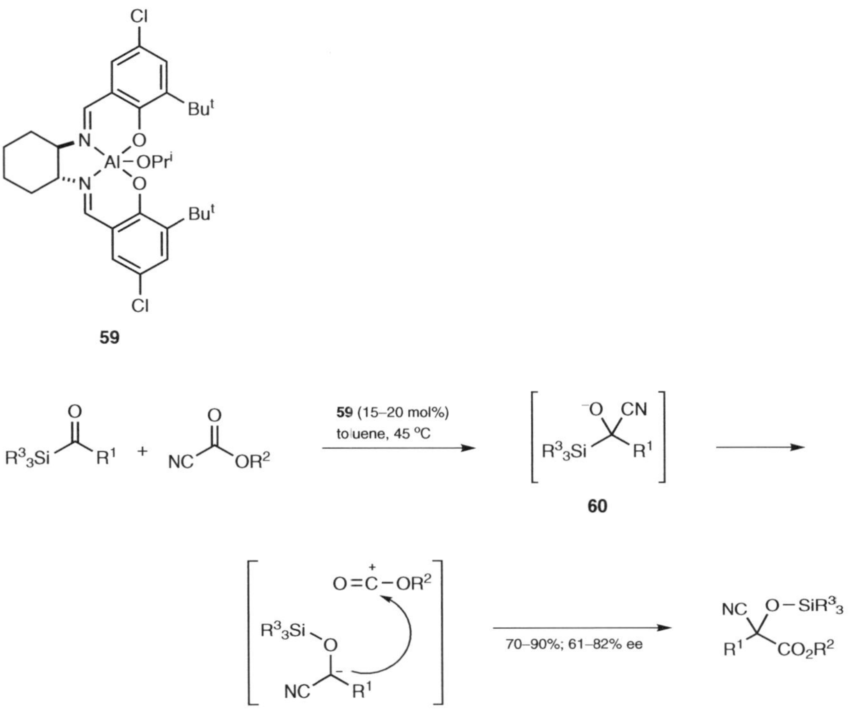

59

R¹ = Ph, 4-Tol, 4-MeOC₆H₄, 4-ClC₆H₄, 4-FC₆H₄, 4-NCC₆H₄, 2-naphthyl; R² = Et, Bn; SiR³₃ = SiEt₃, SiMe₂*t*-Bu

Addition of phenyllithium to acylsilanes carrying a Michael acceptor system at the β-, γ-, or δ-position allows the preparation of functionalized cyclobutanes, cyclopentanes, or cyclohexanes, respectively, with little relative stereocontrol after 1,2-Brook rearrangement and cyclization (Scheme 26).[50]

Scheme 26 Ring Formation by 1,2-Brook Rearrangement Followed by Intramolecular Michael Addition[50]

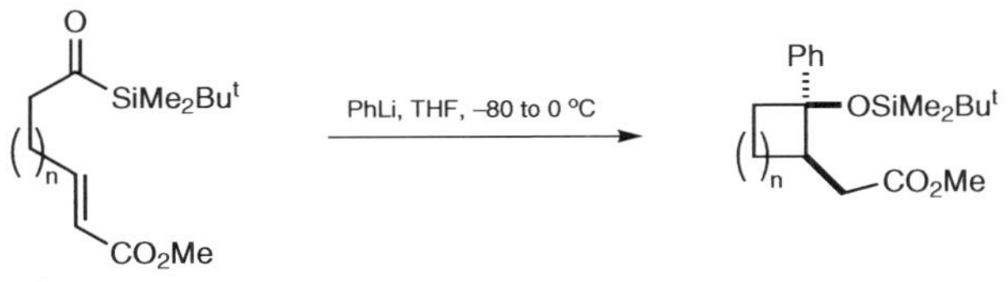

n	dr	Yield (%)	Ref
1	2:1	70	[50]
2	1.8:1	82	[50]
3	1.6:1	79	[50]

The cyanide anion enables a Brook rearrangement when reacted with a β-bromo acylsilane (Scheme 27).[51] The leaving group at the β-position is well located to form a cyclopropane after 1,2-Brook rearrangement of alkoxide **61** and intramolecular S$_N$2 displacement by the intermediate carbanion **62**.

for references see p 445

Scheme 27 Cyclopropane Formation by Cyanide-Initiated 1,2-Brook Rearrangement of a β-Bromo Acylsilane[51]

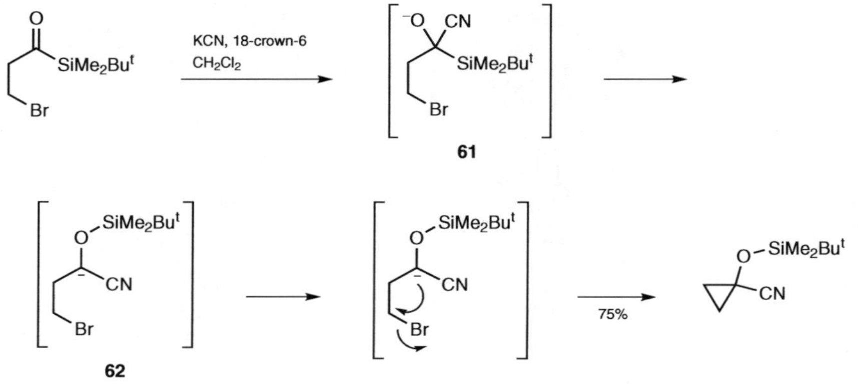

The 1,2-Brook rearrangement can also become part of a domino sequence initiated by formation of siloxyallenes followed by a [4+2] cycloaddition (Scheme 28).[52] This process starts with an enantioselective version of the Meerwein–Ponndorf–Verley-type reduction of alkynoylsilanes using chiral lithium amide **63** as base. The siloxyallene intermediates **64** are generated when the alkoxide undergoes a Brook rearrangement and S_E2' protonation.

Scheme 28 In Situ Preparation of Siloxy-Substituted Vinylallenes by 1,2-Brook Rearrangement Combined with Diels–Alder Cycloaddition[52]

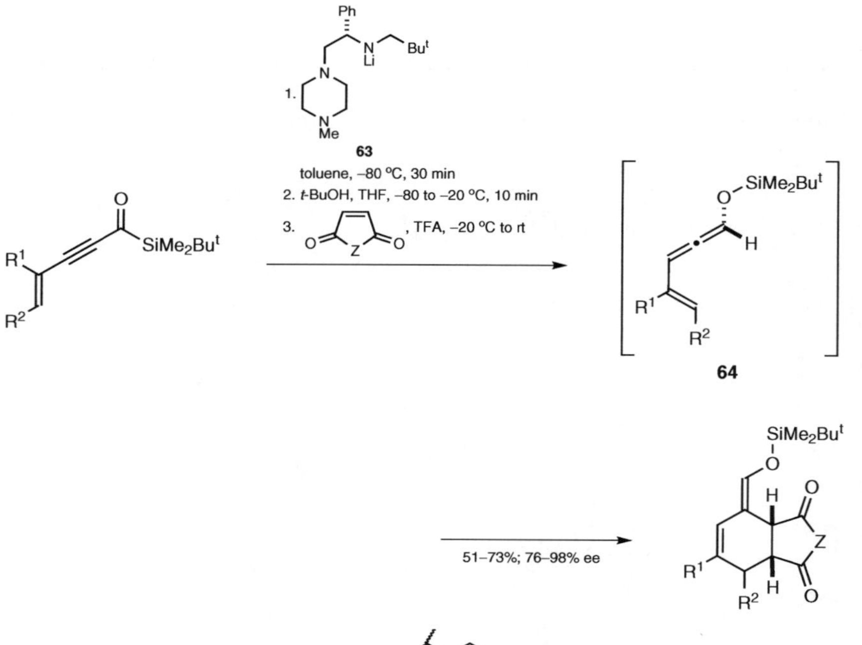

$R^1 = H$; $R^2 = (CH_2)_4OMe$; $R^1,R^2 = (CH_2)_3$, $(CH_2)_4$; $Z = O$, NMe,

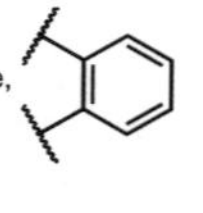

1.5.3 Brook Rearrangement as the Key Step in Domino Reactions **373**

A [6+2]-cyclization cascade gives cyclooctenone derivatives and starts with the addition of a vinyllithium species to an acylsilane, followed by a 1,2-Brook rearrangement of alkoxides **65**. Nucleophilic attack on the carbonyl group yields four-membered ring intermediates **66**. A subsequent anionic oxy-Cope rearrangement furnishes the products in moderate yield (Scheme 29).[53]

Scheme 29 A [6+2] Cyclization toward Cyclooctenones Based on a Brook Rearrangement/Anionic Oxy-Cope Rearrangement Cascade[53]

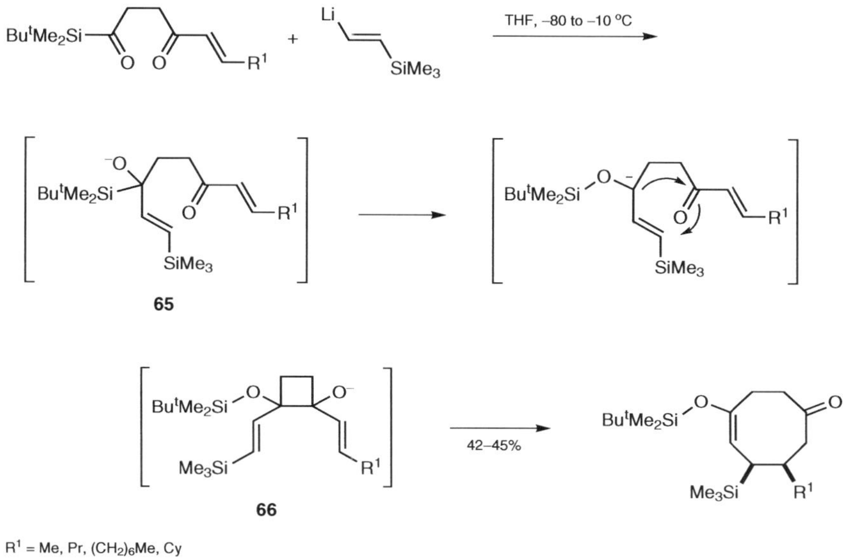

R^1 = Me, Pr, (CH$_2$)$_6$Me, Cy

A one-pot reaction sequence is initiated by the catalytic enantioselective alkynylation of acylsilanes in the presence of chiral ligand **67**, which is followed by a 1,2-Brook rearrangement of alkoxides **68** and an ene–allene carbocyclization of allenylzinc intermediates **69**. With propargyl-substituted acylsilanes, yne–allene carbocyclization takes place at this stage. The newly formed organozinc species are captured by an electrophile to give oxasilacyclopentane intermediates **70** before oxidative workup under Tamao–Fleming conditions terminates the sequence (Scheme 30 and Scheme 31).[54] The method is restricted to examples where no quaternary stereogenic center is formed.

Scheme 30 Enantioselective Alkynylation of Allyl-Substituted Acylsilanes To Induce a Cascade Reaction with 1,2-Brook Rearrangement[54]

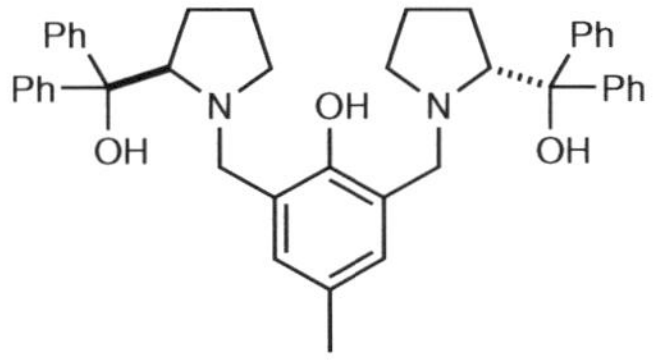

for references see p 445

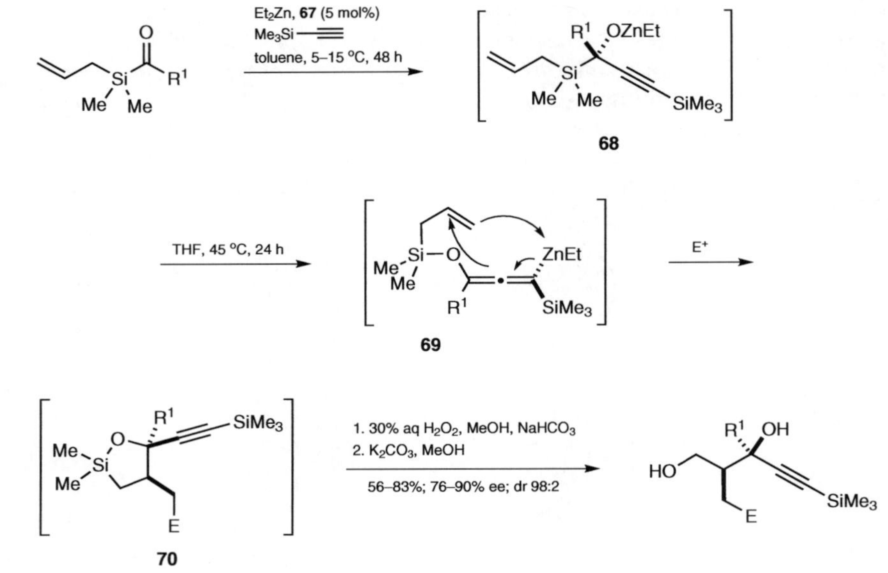

Scheme 31 reaction content.

Scheme 31 Enantioselective Alkynylation of Propargyl-Substituted Acylsilanes To Induce a Cascade Reaction with 1,2-Brook Rearrangement[54]

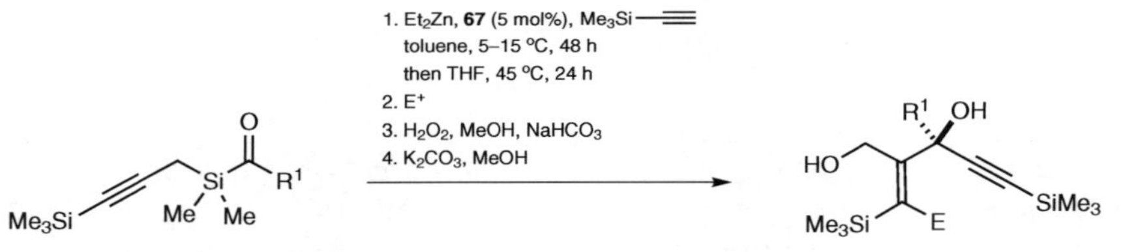

R¹ = Et, Bn, (CH₂)₅Me; E⁺ = I₂, Br₂, NH₄Cl; E = I, Br, H

R¹ = (CH₂)₅Me, Et, Bn; E⁺ = NH₄Cl, I₂; E = H, I

A diastereoselective method leads to *cis*-aminocyclopropanols **73**. The cascade sequence consists of three steps: (1) nucleophilic addition of an enolized *N*-sulfinyl ketimine to an acylsilane, (2) a 1,2-Brook rearrangement of intermediate **71**, and (3) ring closure of intermediate **72** through an intramolecular Mannich reaction (Scheme 32).[55]

1.5.3 Brook Rearrangement as the Key Step in Domino Reactions **375**

Scheme 32 Cyclopropane Formation by Ketimine-Initiated 1,2-Brook Rearrangement of Acylsilanes[55]

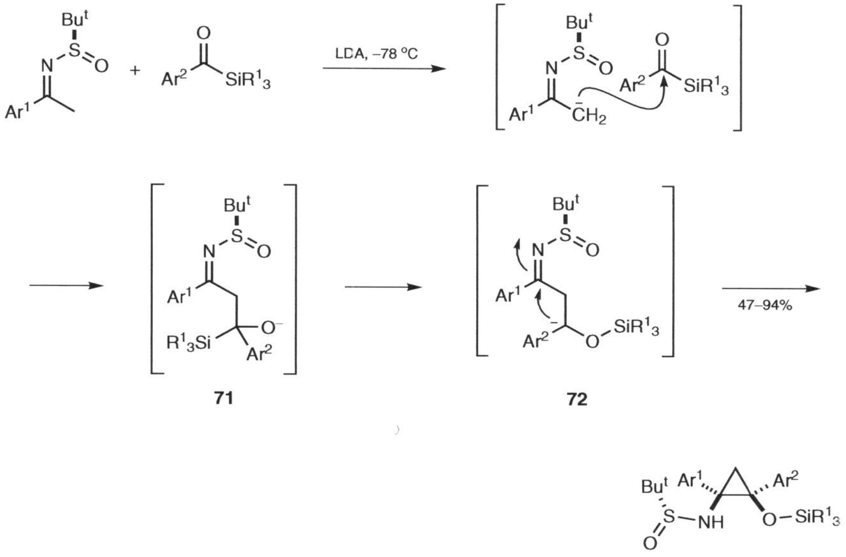

Ar[1] = Ph, 4-MeOC$_6$H$_4$, 2-MeOC$_6$H$_4$, 4-ClC$_6$H$_4$, 4-BrC$_6$H$_4$, 2-pyridyl, 3-pyridyl, 2-furyl; Ar[2] = Ph, 4-MeOC$_6$H$_4$, 4-Tol, 4-ClC$_6$H$_4$; SiR[1]$_3$ = SiEt$_3$, SiMe$_2$t-Bu

Based on the zinc-promoted Brook rearrangement/ene–allene carbocyclization cascade,[56,57] an enantioselective protocol using chiral imino alcohol **74** gives cyclopentanols **77** (Scheme 33).[58] The cascade is initiated by enantioselective, nucleophilic alkynylation of the acylsilane. Then, a zinc-mediated Brook rearrangement of intermediate **75** followed by suprafacial migration of the metal yields allenyl derivative **76**. This intermediate can further react in a diastereoselective zinc-mediated ene–allene cyclization. After coupling with the electrophile, a fluoride source is finally added to furnish cyclopentanols **77** in good yield and selectivity.

Scheme 33 Zinc-Promoted Brook Rearrangement/Ene–Allene Carbocyclization Cascade To Give Functionalized Cyclopentanols[58]

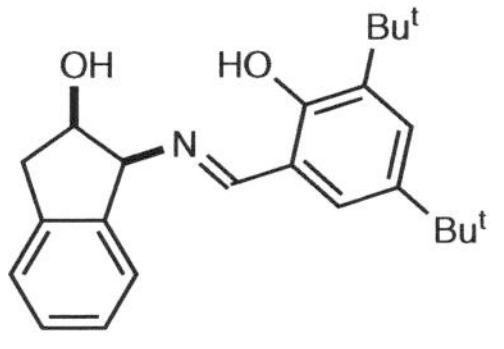

for references see p 445

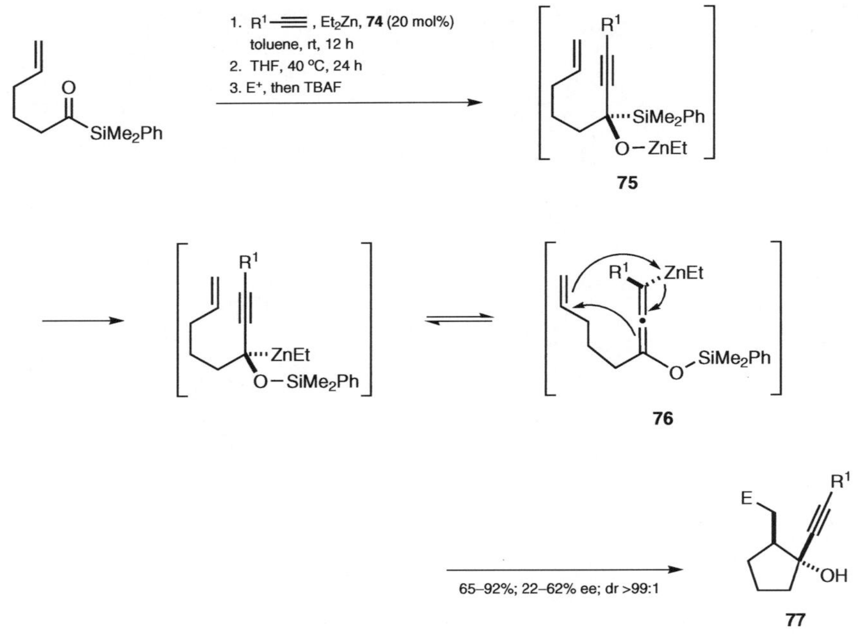
1. R1—≡, Et2Zn, 74 (20 mol%) toluene, rt, 12 h
2. THF, 40 °C, 24 h
3. E+, then TBAF
SiMe2Ph
O–ZnEt
75
ZnEt
O–SiMe2Ph
76
65–92%; 22–62% ee; dr >99:1
77

R^1 = Ph, $(CH_2)_5Me$, TMS; E^+ = I_2, NH_4Cl; E = I, H

Bis-silyl ketones are unusual and unexplored cases; they principally represent acyl dianion equivalents derived from formaldehyde. However, the chemistry of these bright-pink C1-building blocks remains bizarre. Sulfur nucleophiles are suited to add to the carbonyl carbon, which then initiates the desired 1,2-Brook rearrangement, but the newly formed silylated formyl anion has been trapped by only a few electrophiles so far (Scheme 34).[59] The use of other combinations of nucleophiles and electrophiles gives only low yield or leads to decomposition, a clear indication that the carbonyl moiety in bis-silyl ketones behaves differently to common carbonyl groups.

Scheme 34 1,2-Brook Rearrangement of Bis[dimethyl(phenyl)silyl] Ketone[59]

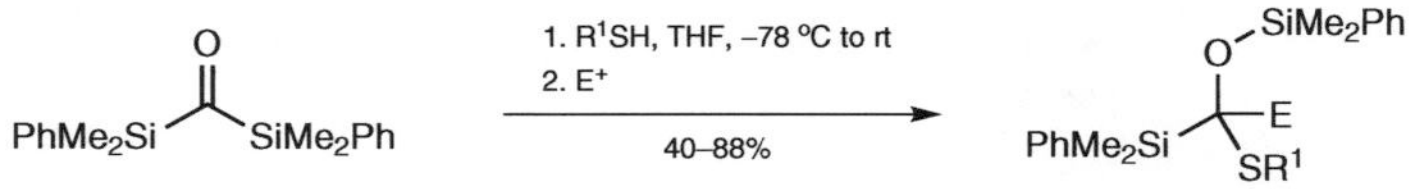
PhMe2Si
SiMe2Ph
1. R1SH, THF, –78 °C to rt
2. E+
40–88%
SR1

R^1 = Ph, Et; E^+ = H_2O, $H_2C{=}CHCH_2Br$; E = H, $CH_2CH{=}CH_2$

1-Alkynyl-2-methylcyclopentanols 77 (E = H); General Procedure:[58]

To a flame-dried, round-bottomed flask charged with an alkyne (1.0 mmol) in toluene (2 mL) was added a 1.5 M soln of Et_2Zn in toluene (0.66 mL, 1.0 mmol). The soln was stirred at rt for 1 h, and the ligand **74** (0.100 mmol) in toluene (1 mL) was added. After the soln had been stirred for an additional 1 h, an acylsilane (0.50 mmol) dissolved in toluene (1 mL) was added at rt. The yellow soln was stirred at rt until complete consumption of the acylsilane (8–12 h, TLC). THF (five times the volume of toluene) was added and the mixture was stirred at 40 °C for 24 h. The reaction was quenched with sat. aq NH_4Cl and the mixture was extracted with Et_2O (3 × 10 mL). The soln was concentrated to give the unpurified cyclic product, which was further desilylated. To a soln of the corresponding silyl ether in THF (10 mL) at rt was added a 1 M soln of TBAF (1.05 equiv). When the reaction was complete, the mixture was concentrated under reduced pressure. Brine was added and the

1.5.3 Brook Rearrangement as the Key Step in Domino Reactions   **377**

aqueous phase was extracted with Et_2O (5×). The combined organic phases were washed with brine and dried ($MgSO_4$).

1.5.3.1.2.2 **Domino Reactions of Acylsilanes Initiated by Nucleophiles Acting as Catalysts**

Several anions or nucleophiles can react catalytically in benzoin-type reactions with acylsilanes to promote silyl migration and carbon–carbon coupling reactions with electrophiles. These catalytically acting anions or nucleophiles, such as cyanide, phosphites, or N-heterocyclic carbenes, are commonly liberated at the end of the sequence.

Using the cyanide anion and an ytterbium complex as Lewis acid, silyl glyoxylates couple with aldehydes after addition of cyanide to the acylsilane (Scheme 35).[60] Subsequent Brook rearrangement of the alkoxide **78** and nucleophilic addition of the intermediate **79** to an aldehyde provides intermediates **80** that undergo 1,4-silyl migration and finally retrocyanation to yield the benzoin-type products.

Scheme 35 A Silyl Glyoxylate in Benzoin-Type Reactions Using Cyanide as Nucleophilic Catalyst[60]

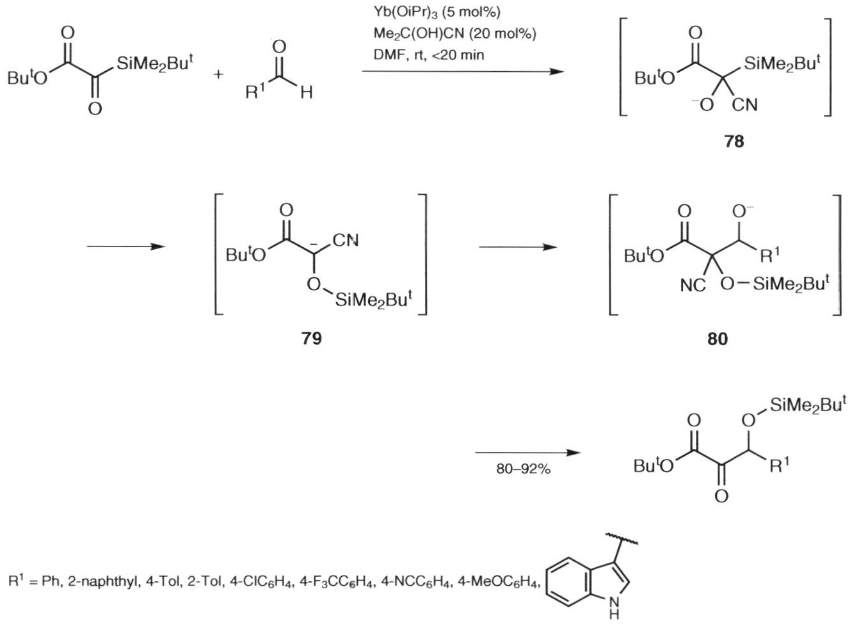

The scope of this cross-benzoin reaction between acylsilanes and aldehydes[61,62] has been extended to include ketones. Thus, the lanthanum(III) cyanide catalyzed Brook rearrangement has been successfully demonstrated with a variety of ketones, α,β-unsaturated ketones, and acylsilanes.[63] For example, 4-methoxyphenyl trimethylsilyl ketone reacts via intermediates **81** and **82** to afford the corresponding α-siloxy ketones (Scheme 36).[63] Chelation control via protected α-oxy ketones gives only moderate selectivities, whereas under Felkin–Anh control better results are obtained.

for references see p 445

Domino Transformations 1.5 Non-Radical Skeletal Rearrangements

Scheme 36 4-Methoxyphenyl Trimethylsilyl Ketone in Benzoin-Type Reactions Using Cyanide as Nucleophilic Catalyst[63]

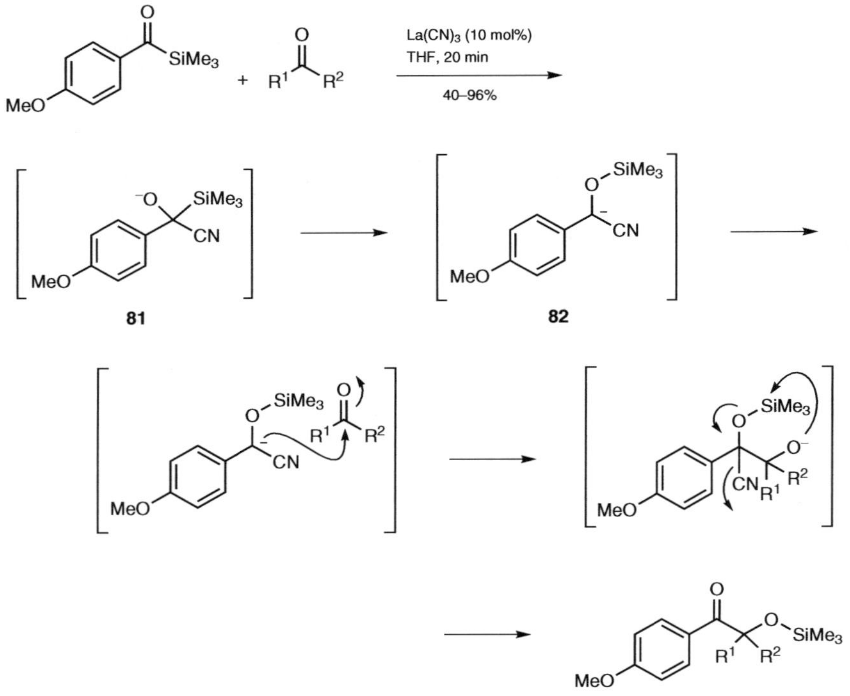

R^1 = Me, Ph; R^2 = Ph, 4-IC$_6$H$_4$, 2-naphthyl, 2-BrC$_6$H$_4$, Me; R^1,R^2 = (CH$_2$)$_5$, (CH$_2$)$_2$CH(t-Bu)(CH$_2$)$_2$, CH(Me)(CH$_2$)$_4$

A domino reaction converts γ,γ-difluoro-α,β-enoylsilanes into amide-type fluoroalkene isosteres (Scheme 37).[64] Attack of a cyanide anion on the acylsilane is followed by Brook rearrangement and elimination of fluoride, forming a vinylogous enolate anion. Protonation of the enolate using the camphor-derived sulfonamide **83** and subsequent nucleophilic attack of the formed amide anion on the simultaneously formed ketone gives the product after elimination of cyanide.

Scheme 37 Synthesis of an Amide-Type Fluoroalkene Isostere[64]

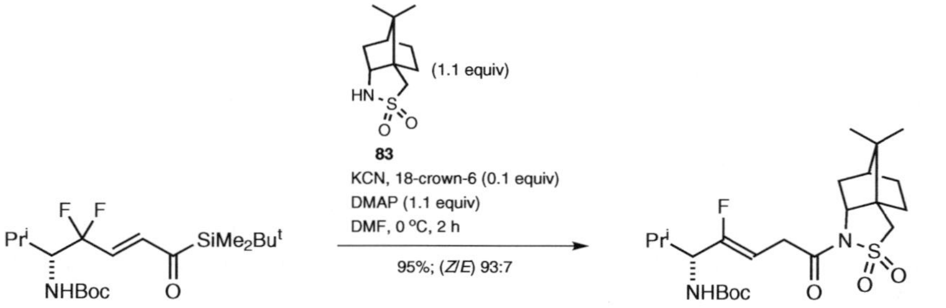

A sila-Stetter reaction employs a 1,2-Brook rearrangement. The cascade starts with attack of the N-heterocyclic carbene derived from thiazolium **84** on the acylsilane, forming a silyl enol ether after 1,2-Brook rearrangement of intermediate **85**. After Michael addition of the sila-Breslow intermediate **86** to the enone, cleavage of the carbene forms the product (Scheme 38).[65]

1.5.3 Brook Rearrangement as the Key Step in Domino Reactions **379**

Scheme 38 Acylsilanes in Stetter-Type Reactions Using an N-Heterocyclic Carbene as Nucleophilic Catalyst[65]

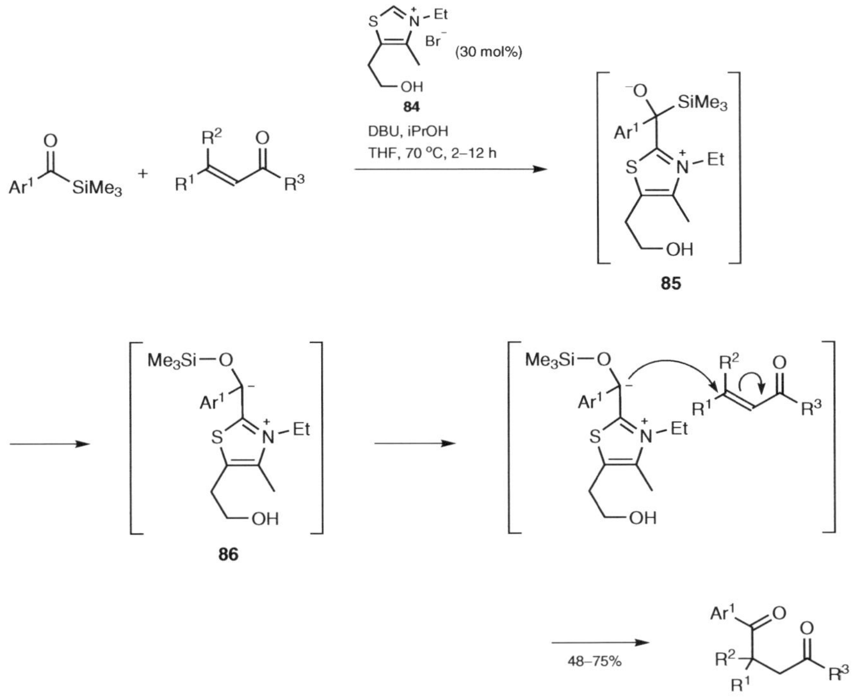

R^1 = H, Ph, CO_2Et; R^2 = H, CO_2Me; R^3 = Me, Et, *t*-Bu, OMe, OEt; Ar^1 = Ph, 4-ClC_6H_4

A cross-benzoin reaction is accomplished using chiral phosphite **87** as nucleophilic catalyst (Scheme 39).[66] Attack of the catalyst on the acylsilane is followed by Brook rearrangement of alkoxide **88** and addition of intermediate **89** to an aldehyde. The ketone product is formed after 1,4-silyl migration and liberation of the phosphite catalyst **87** from intermediate **90**.

for references see p 445

380 Domino Transformations **1.5** Non-Radical Skeletal Rearrangements

Scheme 39 Acylsilanes in Benzoin-Type Reactions Using a Chiral Phosphite as Nucleophilic Catalyst[66]

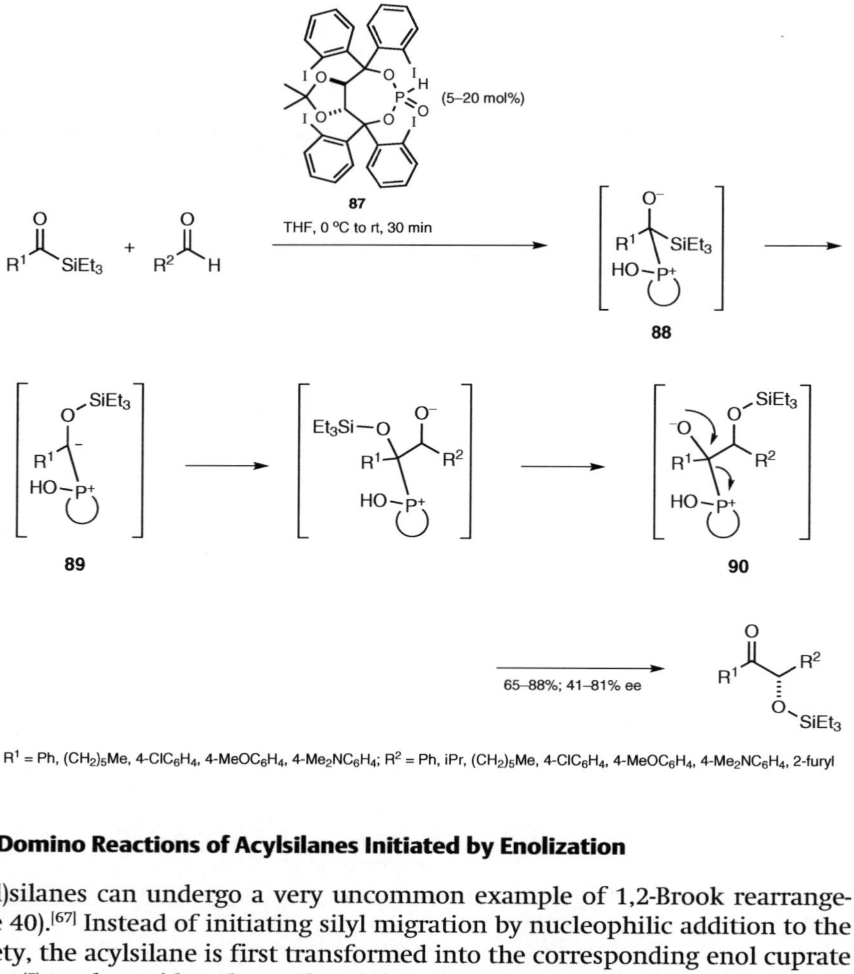

R^1 = Ph, (CH$_2$)$_5$Me, 4-ClC$_6$H$_4$, 4-MeOC$_6$H$_4$, 4-Me$_2$NC$_6$H$_4$; R^2 = Ph, iPr, (CH$_2$)$_5$Me, 4-ClC$_6$H$_4$, 4-MeOC$_6$H$_4$, 4-Me$_2$NC$_6$H$_4$, 2-furyl

1.5.3.1.2.3   **Domino Reactions of Acylsilanes Initiated by Enolization**

Acyl(triphenyl)silanes can undergo a very uncommon example of 1,2-Brook rearrangement (Scheme 40).[67] Instead of initiating silyl migration by nucleophilic addition to the carbonyl moiety, the acylsilane is first transformed into the corresponding enol cuprate **91** using copper(I) *tert*-butoxide as base. The triphenylsilyl group shows a strong tendency to rearrange to alkoxycopper centers. In the present case, copper-functionalized silyl enol ethers **92** are formed, which can be further processed in palladium-catalyzed cross-coupling reactions to yield trisubstituted silyl enol ethers.

1.5.3 Brook Rearrangement as the Key Step in Domino Reactions **381**

Scheme 40   Domino Reactions Initiated by Enolization of Acyl(triphenyl)silanes with Copper(I) *tert*-Butoxide[67]

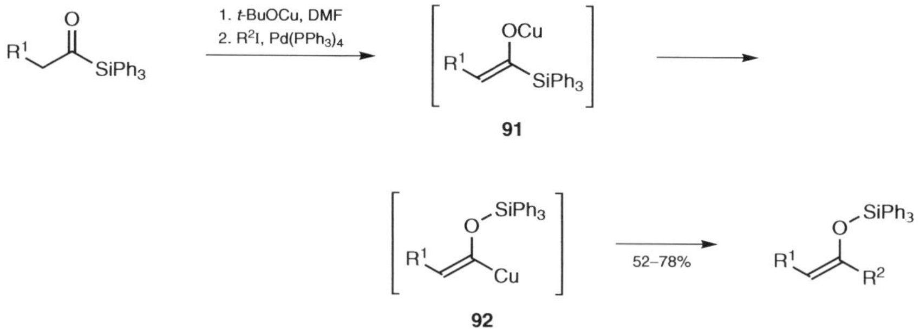

R^1 = Bn, Me, Et, CHBn$_2$; R^2 = Ph, 4-MeOC$_6$H$_4$, CH=CH(CH$_2$)$_5$Me

1.5.3.1.3   **1,2-Brook Rearrangement with α-Silyl Carbinols**

α-Silyl carbinols are also well suited for cascade reactions initiated by a 1,2-Brook rearrangement. For example, the 1,1-bis(silyl)alkoxide anion **93** formed after deprotonation of the corresponding alcohol undergoes rearrangement, and the newly formed carbanion **94** then reacts with various alkyl, allyl, and propargylic halides in good to moderate yields (Scheme 41).[14]

Scheme 41   Bis-silyl Carbinols as Precursors for 1,2-Brook Rearrangement[14]

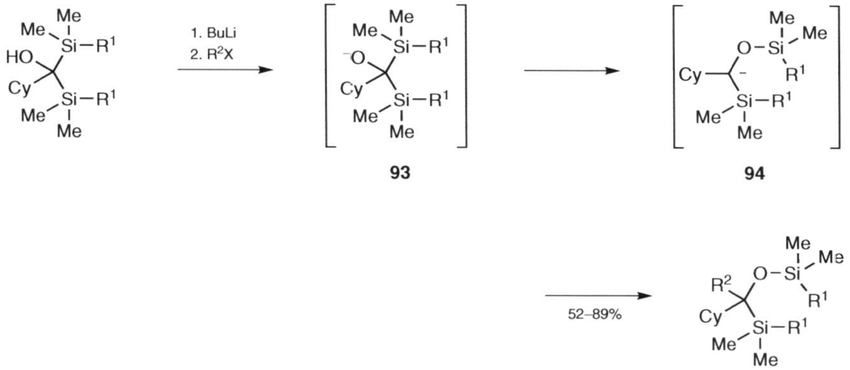

R^1 = Me, Ph; R^2 = Me, Et, iPr, iBu, Bn, CH$_2$CH=CH$_2$; X = I, Br

A modified Morita–Baylis–Hillman reaction starts by deprotonation of an α-silyl carbinol with butyllithium and is followed by a 1,2-Brook rearrangement of alkoxide **96**. The N-heterocyclic carbene **95** picks up the aldehyde to form Breslow adduct **98**, which catalyzes silyl migration from allene **97**. The formed enolate **100** undergoes a nucleophilic addition to the aldehyde to form aldol product **101**, which is O-silylated by silyl transfer from the silylated Breslow adduct **99**, a step that regenerates intermediate **98** (Scheme 42).[68]

for references see p 445

Scheme 42 Modified Morita–Baylis–Hillman Reaction Initiated by 1,2-Brook Rearrangement[68]

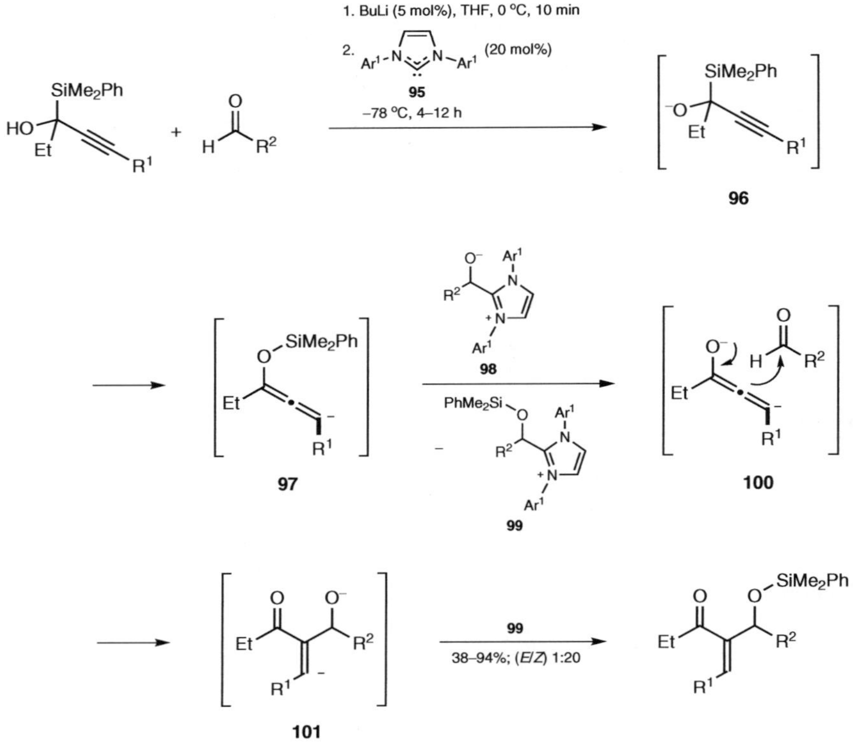

Ar[1] = 2,6-iPrC$_6$H$_3$; R[1] = Me, Bu, t-Bu, Ph, SiMe$_3$, (CH$_2$)$_3$OSiMe$_2$$t$-Bu;
R[2] = iPr, Ph, 4-ClC$_6$H$_4$, 4-MeOC$_6$H$_4$, 2-naphthyl, (CH$_2$)$_2$Ph

1.5.3.1.4 1,2-Brook Rearrangement with Epoxy Silanes

Several metalated O-silyl cyanohydrins derived from β-silyl-α,β-epoxy aldehydes undergo a cascade reaction initiated by deprotonation and oxirane opening of the resulting intermediate, e.g. **102**, followed by 1,2-Brook rearrangement of the alkoxide, e.g. **103**, and alkylation of the resulting allylic anion, e.g. **104** (Scheme 43).

1.5.3 Brook Rearrangement as the Key Step in Domino Reactions **383**

Scheme 43 Cascade Reaction Initiated by Intramolecular Ring Opening of an Epoxysilane Followed by 1,2-Brook Rearrangement[69,70]

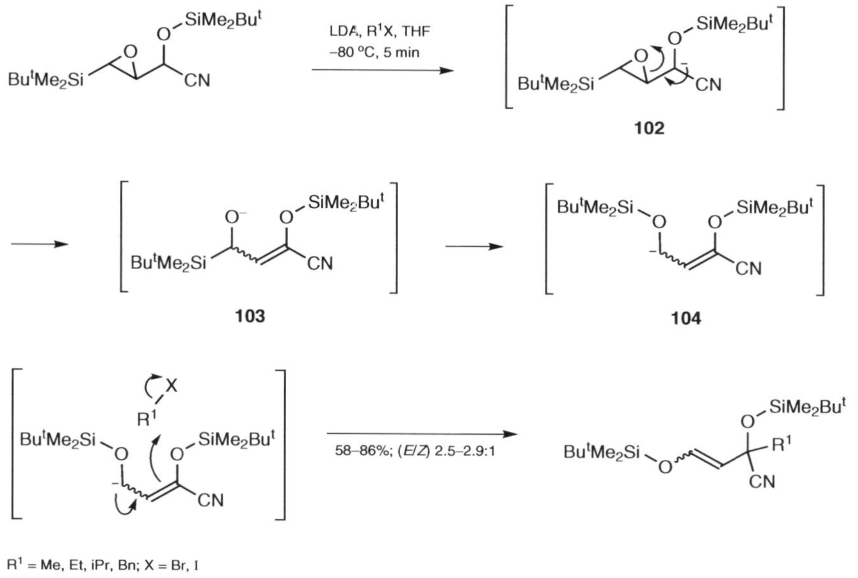

R^1 = Me, Et, iPr, Bn; X = Br, I

Detailed studies on the choice of base,[69,70] solvent,[69,70] and additives such as hexamethylphosphoric triamide[70] were conducted to improve yields and selectivity. In addition, the transformation was investigated with respect to chirality transfer.[71,72] The cascade is also suited for silyl-substituted epoxyalkanenitriles as starting material[73,74] or it can alternatively be initiated by addition of a cyanide anion to the acylsilane.[75] Finally, the protocol also allows utilization of bis-electrophiles (Table 1).[76] In the case of α,β-unsaturated esters (entries 1–5), the initially formed allylic anion **105** adds to the electrophile in a Michael-type addition and the newly formed enolate **106** reacts as a nucleophile with the second Michael acceptor system, as shown in Scheme 44 (or undergoes a nucleophilic displacement of the bromide). When the two electrophilic carbon atoms in the bis-electrophile are separated by two carbon atoms, cyclopentanes are formed (entries 1 and 2). If shorter or longer tethers are employed, the nucleophile is deprotonated after the first addition to give intermediate **107**, which then undergoes a second nucleophilic attack similar to the first one (Scheme 44 and Table 1, entries 3–8).[76]

for references see p 445

Table 1 Cascade Reactions Initiated by Intramolecular Ring Opening of α-Cyano-Functionalized Epoxysilanes[76]

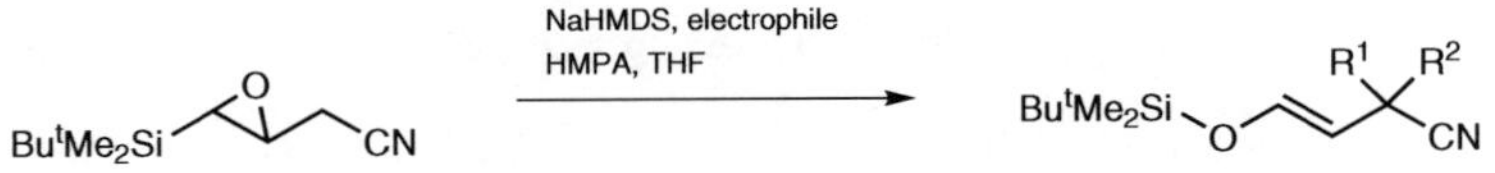

Entry	Electrophile	Conditions	Product	Yield (%)	Ref
1	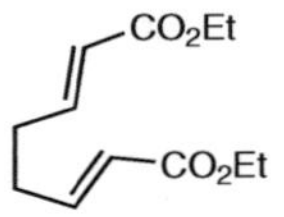 (diethyl ester diene)	−98 °C, 15 min	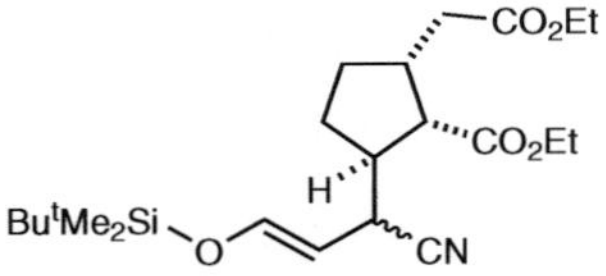 (cyclopentane product)	64	[76]
2	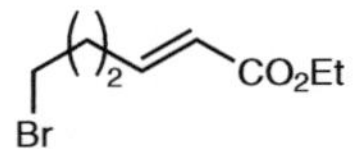 (bromo ester, n=2)	−98 °C	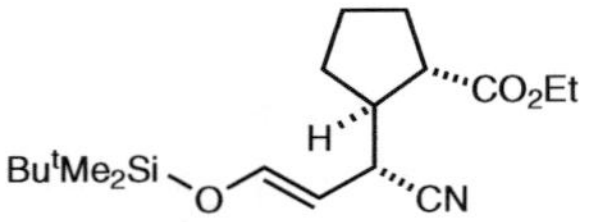 (cyclopentane product)	66	[76]
3	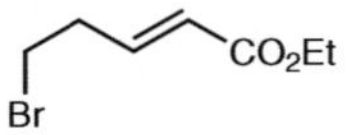 (bromo ester)	−80 °C	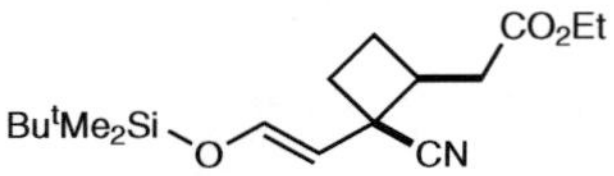 (cyclobutane product)	17	[76]
4	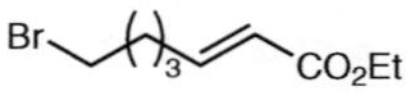 (bromo ester, n=3)	−80 °C	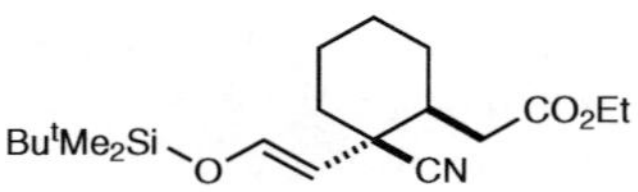 (cyclohexane product)	88	[76]
5	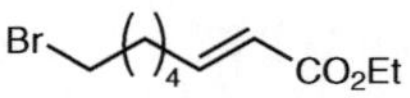 (bromo ester, n=4)	−80 °C	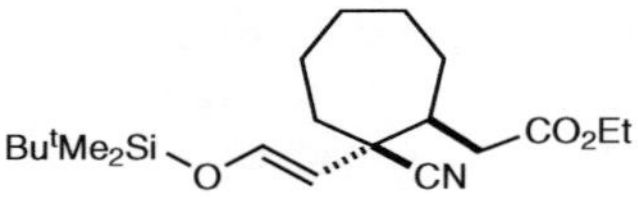 (cycloheptane product)	51	[76]
6	 I–(CH₂)–I	−80 °C, 120 min	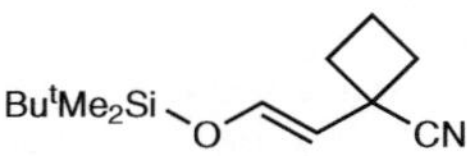 (cyclobutane product)	44	[76]
7	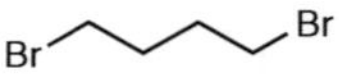 Br–(CH₂)–Br	−80 °C, 80 min	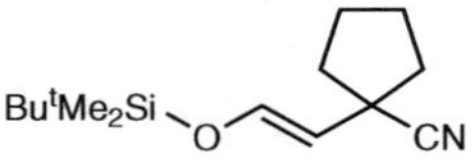 (cyclopentane product)	83	[76]
8	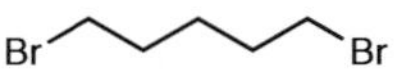 Br–(CH₂)–Br	−80 °C, 90 min	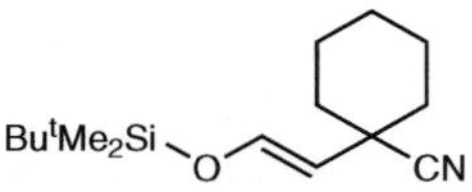 (cyclohexane product)	77	[76]

1.5.3 Brook Rearrangement as the Key Step in Domino Reactions **385**

Scheme 44 Intermediates in Cascade Reactions Initiated by Intramolecular Ring Opening of α-Cyano-Functionalized Epoxysilanes[76]

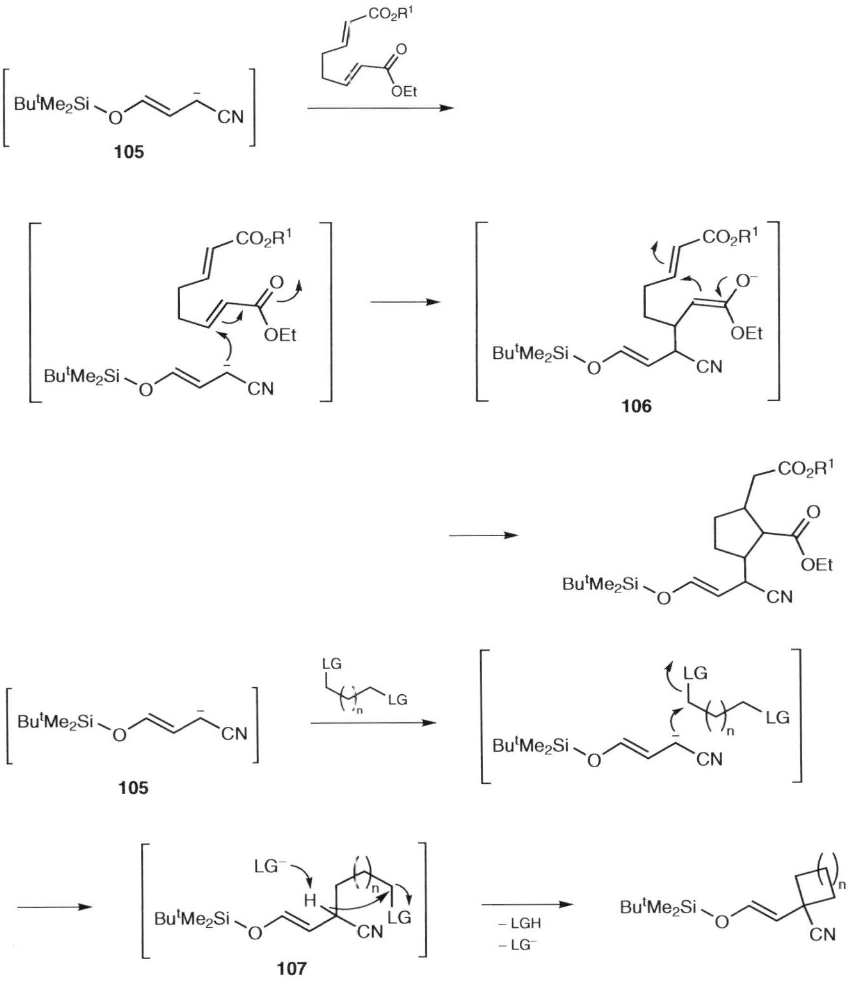

Highly functionalized cyclopropanes **110** can be prepared from epoxysilanes and the lithium enolate of a 2-chloroacetamide. The sequence is initiated by Michael addition of the enolate, which is followed by ring opening of the epoxide **108**, 1,2-Brook rearrangement of alkoxide **109**, and intramolecular alkylation (Scheme 45).[77]

for references see p 445

386 Domino Transformations **1.5** Non-Radical Skeletal Rearrangements

Scheme 45 Epoxysilanes as Tools for Cascade Reactions toward Functionalized Cyclopropanes[77]

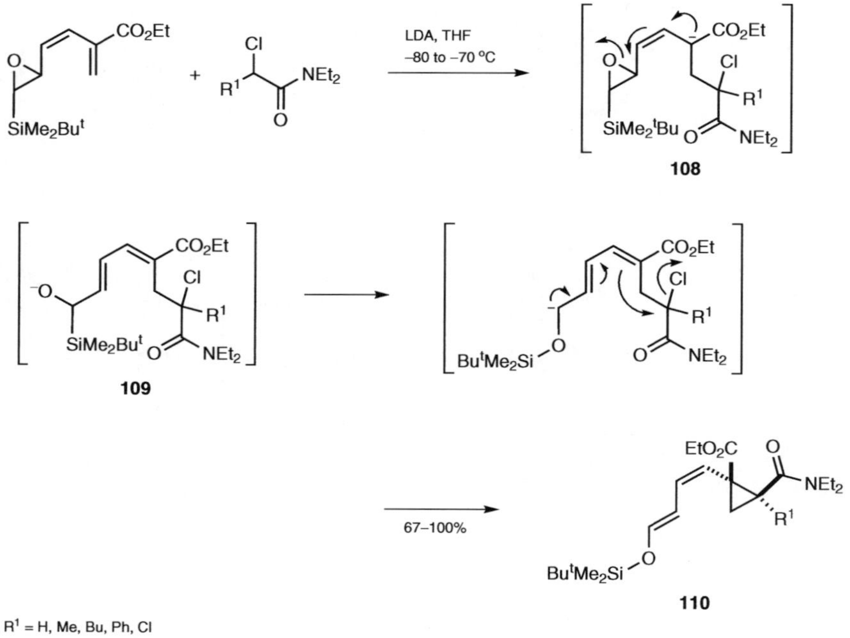

R[1] = H, Me, Bu, Ph, Cl

1.5.3.1.5 Miscellaneous Examples of 1,2-Brook Rearrangement

A few examples of 1,2-Brook rearrangements have been reported that do not fit the categories discussed above. For example, in contrast to established transformations of aryltrimethylsilanes with fluoride sources, acetyl(dimethyl)phenylsilane reacts unpredictably to give an unusual rearrangement product via pentacoordinated silyl species **111** (Scheme 46).[78] The proposed mechanism supposedly proceeds via a 1,2-phenyl shift and a 1,2-Brook rearrangement.

Scheme 46 1,2-Brook Rearrangement of Acetyl(dimethyl)phenylsilane after 1,2-Aryl Migration[78]

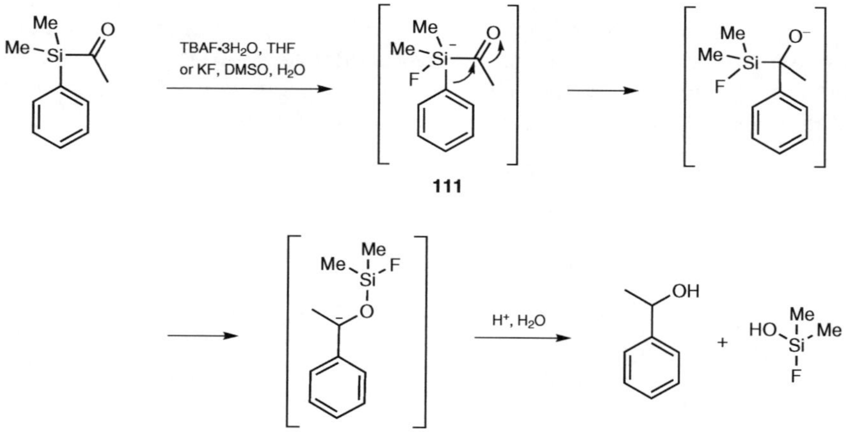

1.5.3 Brook Rearrangement as the Key Step in Domino Reactions

Silyl-substituted vinyloxiranes can be rearranged under palladium(0)-catalyzed conditions (Scheme 47).[79] Whereas aldehyde **113** results from a 1,2-shift after formation of allylpalladium complex **112**, siloxydiene **115** is formed via intermediate **114** derived after a 1,2-Brook rearrangement of allylpalladium complex **112**. The preference for either of the two products strongly depends on the chosen ligand system.

Scheme 47 1,2-Brook Rearrangement after Palladium(0)-Catalyzed Ring Opening of Vinylepoxysilanes[79]

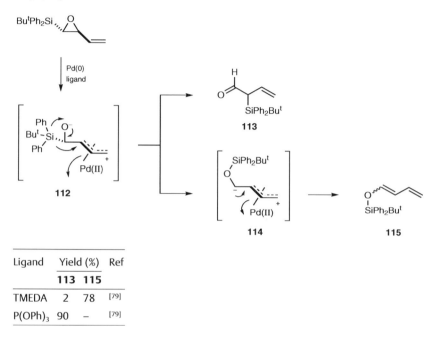

Ligand	Yield (%)		Ref
	113	115	
TMEDA	2	78	[79]
P(OPh)₃	90	–	[79]

1.5.3.1.6 Retro-1,2-Brook Rearrangement

Retro-1,2-Brook rearrangements are rare because of the larger binding affinity of silicon to oxygen compared to carbon. Still, aldehydes and ketones can serve as a starting point for retro-Brook rearrangements (Scheme 48).[80,81] When transformed into α-stannyl silyl ethers **116** by nucleophilic stannylation and O-silylation, these stannane derivatives (Bu₃Sn or Me₃Sn for sterically congested architectures) are transformed into lithium derivatives **117** by transmetalation. These organolithium species can undergo retro-1,2-Brook rearrangement to yield the α-silyl carbinols **118**.

Scheme 48 α-Stannyl Silyl Ethers as Precursors for Retro-1,2-Brook Rearrangement[80,81]

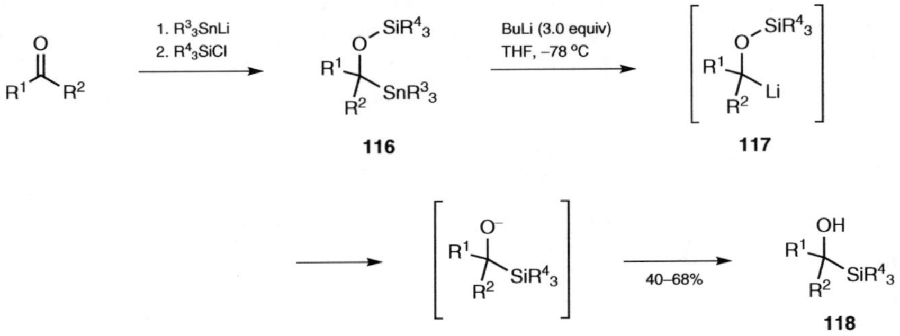

R^1 = Ph, (CH$_2$)$_4$Me, (CH$_2$)$_2$Ph, cyclopropyl, Cy, t-Bu; R^2 = H; R^1,R^2 = (CH$_2$)$_5$, (CH$_2$)$_2$CH(t-Bu)(CH$_2$)$_2$;
R^3 = Me, Bu; SiR$^4_3$ = SiMe$_3$, SiEt$_3$, SiMe$_2$$t$-Bu, Si(iPr)$_3$, SiMe$_2$Pr, SiMe$_2$Ph

After tin–lithium exchange, [1-(triorganosiloxy)vinyl]tin derivatives undergo retro-1,2-Brook rearrangement to form lithium enolates **119**, which can either be alkylated or silylated (Scheme 49).[82] This retro-Brook rearrangement also proceeds well with bulky silyl substituents.

Scheme 49 (Siloxyvinyl)tin Derivatives as Precursors for Retro-1,2-Brook Rearrangement[82]

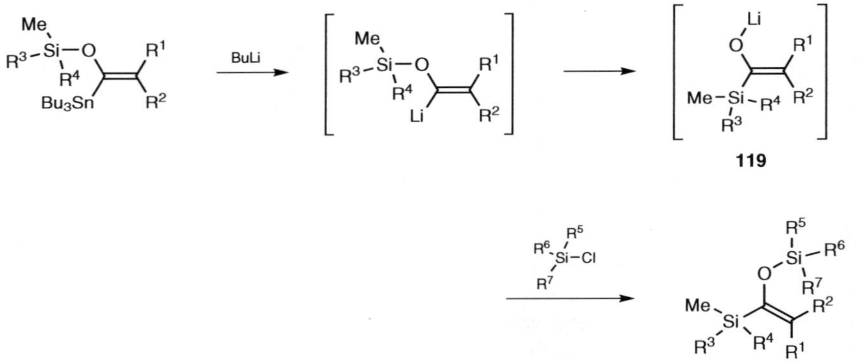

R^1	R^2	R^3	R^4	R^5	R^6	R^7	Yield (%)	Ref
H	H	Me	Me	Ph	Ph	Me	64	[82]
H	H	Me	t-Bu	Me	Me	Me	73	[82]
Me	Me	Me	Me	Et	Et	Et	66	[82]
H	H	Ph	Ph	Me	Me	Me	63	[82]

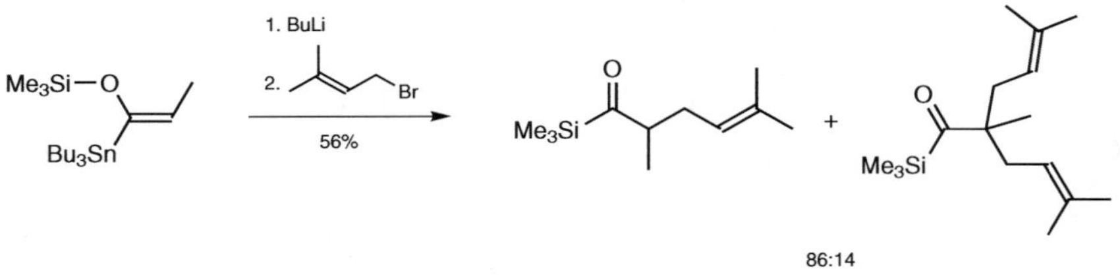

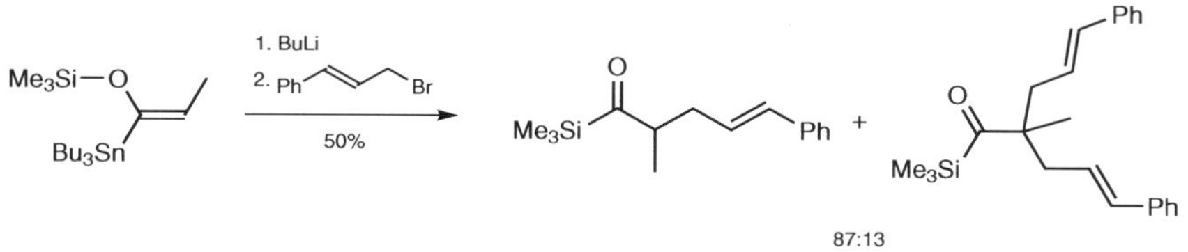

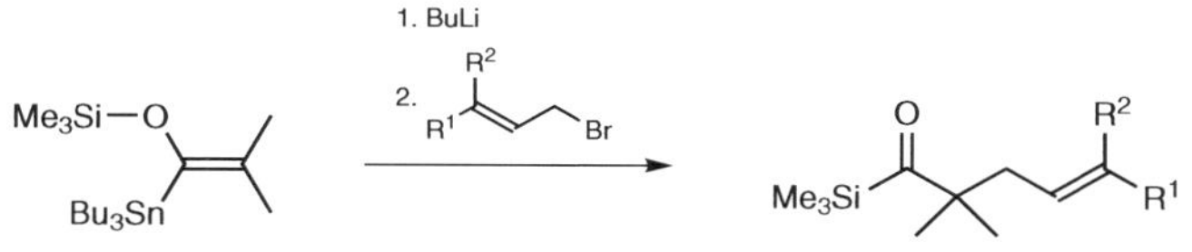

1.5.3 Brook Rearrangement as the Key Step in Domino Reactions

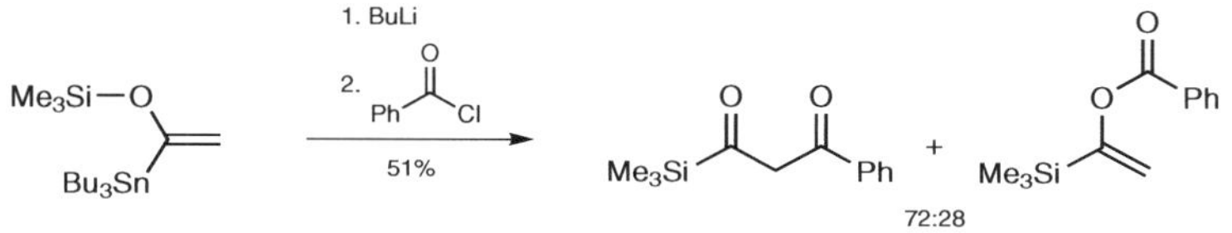

Monosilyl-protected vicinal diols can be prepared from silylated (α-hydroxyalkyl)stannanes **120** after tin–lithium exchange, a process which results in a reversible 1,2-Brook rearrangement (Scheme 50).[83] The carbanionic isomer **121** can be captured by electrophiles such as aldehydes or ketones to yield monoprotected diols **122**.

Scheme 50 α-Siloxy Stannanes as Precursors in Retro-1,2-Brook Rearrangement[83]

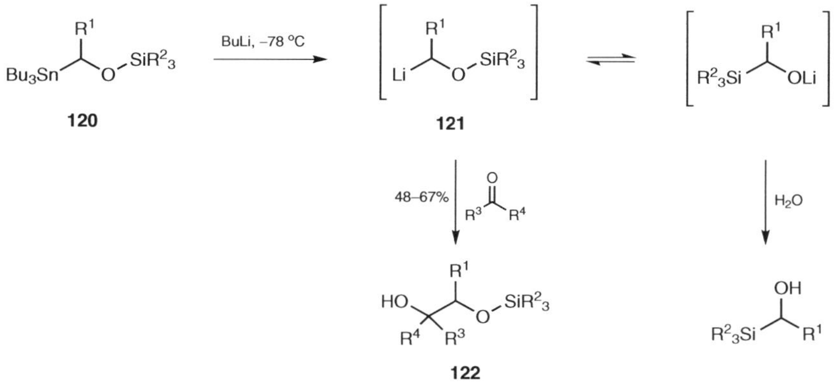

R^1 = H, Me, Ph; SiR^2_3 = SiMe$_2$t-Bu, SiPh$_2$t-Bu; R^3 = Ph; R^4 = H; R^3,R^4 = (CH$_2$)$_5$

Importantly, the retro-1,2-Brook rearrangement can also serve to prepare (hydroxyaryl)silanes starting from phenols, a process which provides access to silyl-substituted salen ligands (Scheme 51).[84] The required silyl-substituted salicylaldehyde precursors are obtained from the corresponding silylated 2,6-dibromophenols after double lithiation. This event initiates intramolecular silyl transfer from intermediate **123** and the second aryl anion **124** is trapped with dimethylformamide. Subsequent condensation of the arylalde-

for references see p 445

hyde product **125** with a chiral diamine terminates the synthetic sequence. The use of electrophiles other than dimethylformamide provides facile access to silyl-substituted phenolic esters, ketones, and boronic acids.

Scheme 51 Preparation of *ortho*-Silylphenols by Retro-1,2-Brook Rearrangement[84]

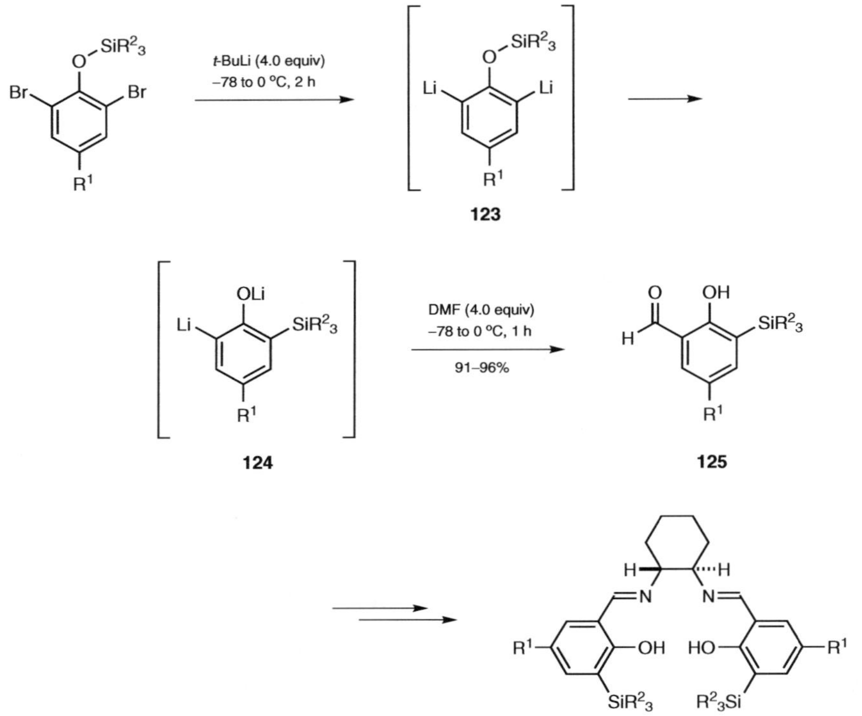

R^1 = Me, *t*-Bu, F; SiR^2_3 = SiMe₃, Si(iPr)₃, SiMe₂*t*-Bu, SiPh₂*t*-Bu

Propargylic silyl ethers are also good precursors for retro-1,2-Brook rearrangements (Scheme 52).[85] After α-lithiation, 1,2-silyl migration occurs to yield trialkyl(α-hydroxyalkynyl)silanes **126**. These propargylic silanes undergo a second silyl migration upon treatment with a Lewis acid to yield α,β-bis-silylated enals and enones.

1.5.3 Brook Rearrangement as the Key Step in Domino Reactions **391**

Scheme 52 Propargylic Silyl Ethers as Precursors for Retro-1,2-Brook Rearrangement[85]

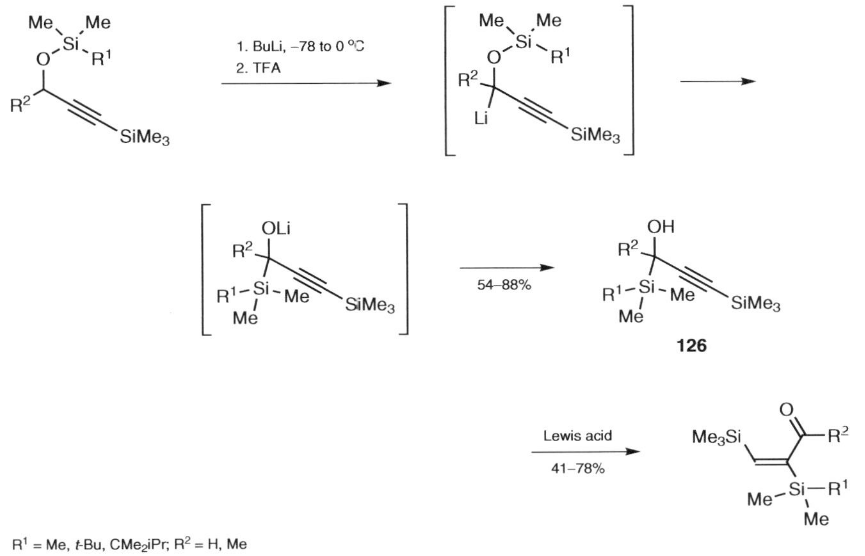

R[1] = Me, t-Bu, CMe₂iPr; R[2] = H, Me

When the retro-Brook rearrangement of propargylic silyl ethers does not proceed to completion, trialkyl(α-hydroxyalkynyl)silanes **127** are formed together with allenes **128** as byproducts (Scheme 53).[86] In situ formation of the starting silyl ethers from the corresponding propargyl alcohols leads to the same result (e.g., silanes **129** and allenes **130**) after deprotonation at the α-position.

Scheme 53 Propargyl Alcohols and Silyl Ethers as Precursors for Retro-1,2-Brook Rearrangement[86]

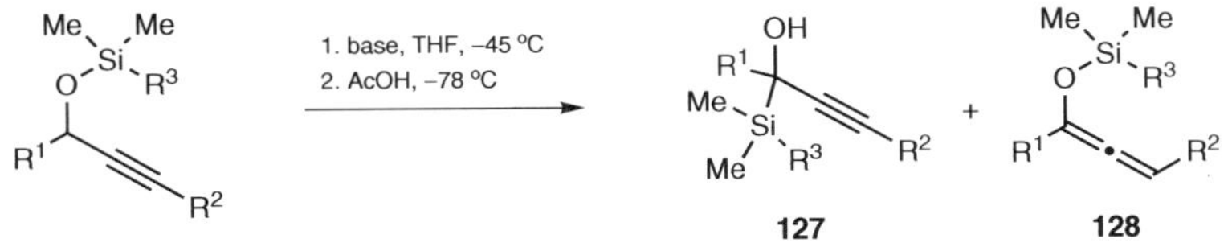

R[1]	R[2]	R[3]	Base (Equiv)	Time (h)	Yield (%)		Ref
					127	**128**	
H	H	t-Bu	s-BuLi (3)	22	16[a]	8	[86]
H	Me	t-Bu	BuLi (1.2)	1.5	45[b]	<2[c]	[86]
H	Me	t-Bu	BuLi (3.0)	1.5	78	<2[c]	[86]
H	Bu	Me	t-BuLi (3.0)	1.5	35	<2[c]	[86]
Me	Bu	Me	t-BuLi (3.0)	1.5	67[d]	31	[86]

[a] 22% of the starting material was recovered.
[b] 35% of the starting material was recovered.
[c] Determined by ¹H NMR spectroscopy.
[d] Reaction was carried out at −20 °C.

for references see p 445

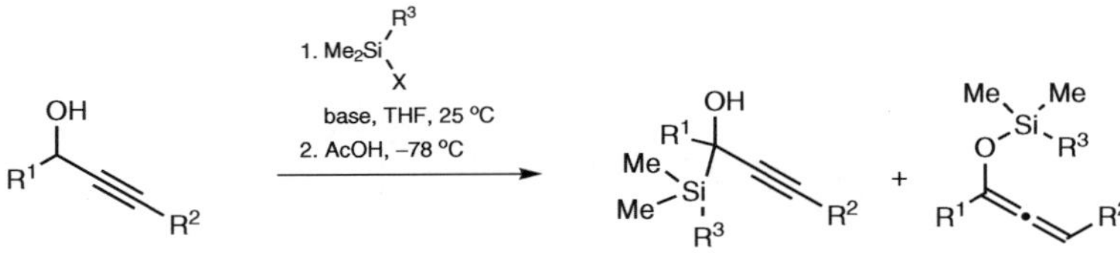

Domino Transformations 1.5 Non-Radical Skeletal Rearrangements

R¹	R²	R³	X	Base (Equiv)	Time (h)	Yield (%) 129	Yield (%) 130	Ref
H	H	t-Bu	Cl	s-BuLi (3)	22	70	4	[86]
H	Me	t-Bu	Cl	BuLi (1.2)	2	86	–	[86]
H	Me	t-Bu	OTf	BuLi (1.2)	2	32	–	[86]
H	Bu	Me	Cl	t-BuLi (1.2)	2	45	–	[86]
Me	Bu	Me	Cl	t-BuLi (3.0)	3	34[a]	–	[86]

[a] Reaction was carried out at –20 °C; 28% of the starting material was recovered.

Based on these results, an efficient method for the synthesis of tetrasubstituted allenyl ethers **132** has been developed. Retro-1,2-Brook rearrangement of propargyl silyl ethers provides propargylic alcohols **131**, which are captured by methoxymethyl chloride (Scheme 54).[87] The sequence is terminated by a second lithiation step followed by reaction of the allenyl anion with electrophiles.

Scheme 54 Propargyl Silyl Ethers as Precursors for Tetrasubstituted Allenes[87]

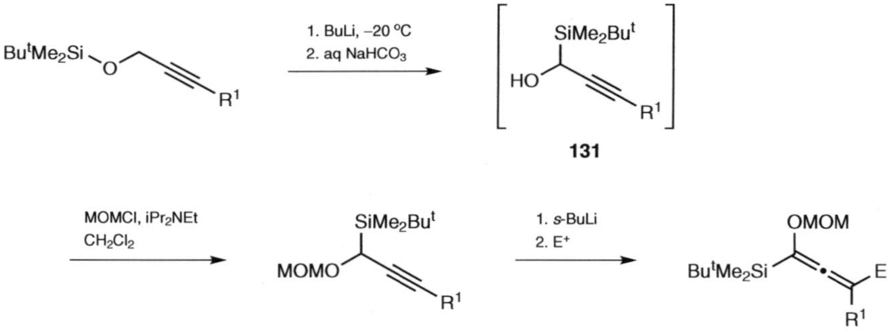

R¹ = Cy, (CH₂)₅Me; E⁺ = EtOH, MeI, H₂C=CHCH₂Br, BnBr, PhC(O)NMe(OMe), pentan-3-one, Ph₂CO, PhSSO₂Ph, iPr₃SiCl;
E = H, Me, CH₂CH=CH₂, Bn, Bz, C(OH)Et₂, C(OH)Ph₂, SPh, Si(iPr)₃

1.5.3.2 1,3-Brook Rearrangement

Synthetically, the 1,3-Brook rearrangement has had the largest impact on organic synthesis, because it is the last key step in the Peterson alkenation as first reported by A. G. Brook (Scheme 55).[1] However, Peterson alkenations are not commonly part of domino sequences. If the carbanion **134** formed after 1,3-silyl migration of alkoxide **133** does not expel the neighboring siloxy group, this silyl transfer step can be exploited to introduce an electrophile instead so that cascade reactions become feasible. Alternatively, a well-positioned second leaving group in the form of Z as shown in Scheme 55 can override the leaving group properties of the siloxy group so that β-elimination leads to (allyloxy)si-

lanes. In a few cases, silyl migration can proceed with vinylsilanes and the resulting carbanion is located at an sp²-hybridized center.

Scheme 55 The 1,3-Brook Rearrangement

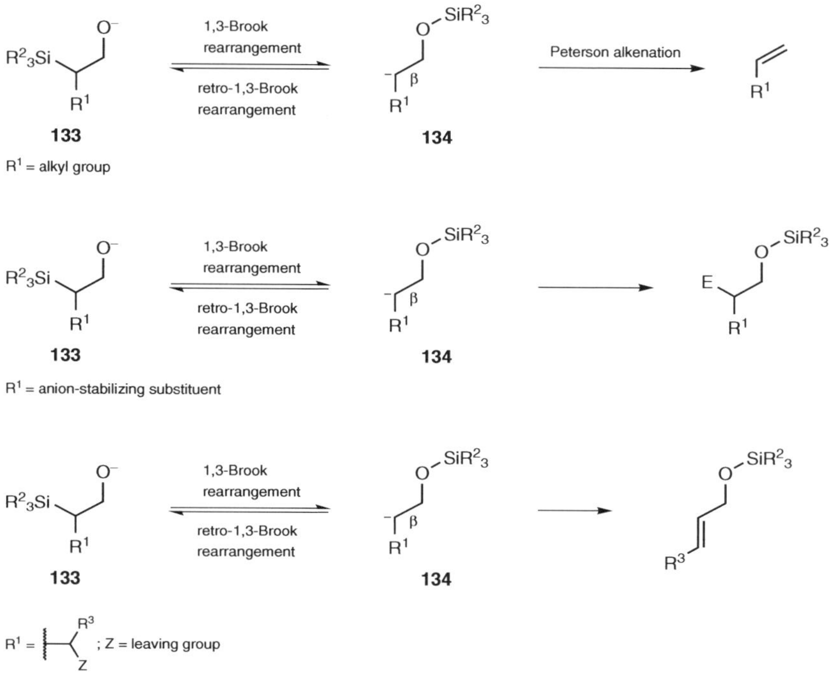

Several factors play a crucial role in terms of whether Peterson-type eliminations are hampered or how the elimination can be suppressed. The R^1 group needs to be able to stabilize the carbanion, which facilitates silyl migration, but should not stabilize it too much so that it retains enough nucleophilicity to react with an electrophile before Peterson elimination occurs. The process is further facilitated if the electrophile is part of the molecule and the final step proceeds in an intramolecular fashion.[2]

1.5.3.2.1 Addition of Silyl-Substituted Stabilized Organolithium Agents to Carbonyl Groups

After the addition of [(tert-butyldimethylsilyl)dichloromethyl]lithium, a versatile dianion synthon, to aldehydes, a 1,3-Brook rearrangement of alkoxides **135** occurs (Scheme 56). The resulting organolithium species **136** do not undergo Peterson alkenation due to the stabilizing effect of the two chloro substituents.[88] Therefore, they can be captured by a second electrophile to yield α,α-dichloroalkyl silyl ethers **137**.

for references see p 445

394 Domino Transformations **1.5** Non-Radical Skeletal Rearrangements

Scheme 56 1,3-Brook Rearrangement To Give Stable Carbanions after Nucleophilic Addition of [(*tert*-Butyldimethylsilyl)dichloromethyl]lithium to Aldehydes[88]

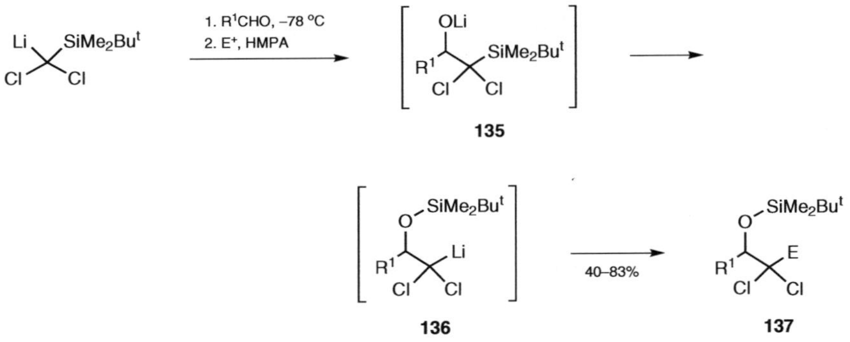

R[1] = Bu, Ph, CH=CHPh, Pr, 4-MeOC$_6$H$_4$; E[+] = MeI, H$_2$C=CHCH$_2$Br, PhCHO; E = Me, CH$_2$CH=CH$_2$, CH(OH)Ph

The sila-Morita–Baylis–Hillman reaction using aryl vinyl ketones also relies on a 1,3-Brook rearrangement (Scheme 57).[89] The presence of the trimethylsilyl group is essential for these types of enones because without the bulky substituent, dimerization of the enones prevails.

Scheme 57 1,3-Brook Rearrangement in Sila-Morita–Baylis–Hillman Reactions[89]

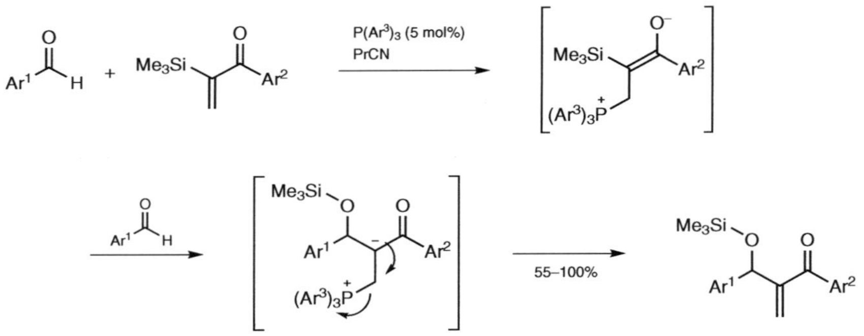

Ar[1] = Ph, 4-ClC$_6$H$_4$, 3-ClC$_6$H$_4$, 4-BrC$_6$H$_4$, 4-FC$_6$H$_4$, 4-MeOC$_6$H$_4$, 4-Tol, 2-Tol, 3-Tol, 4-TBDMSOC$_6$H$_4$;
Ar[2] = Ph, 4-Tol, 4-ClC$_6$H$_4$, 4-BrC$_6$H$_4$; Ar[3] = 2,4,6-(MeO)$_3$C$_6$H$_2$

Another anion-catalyzed Morita–Baylis–Hillman reaction uses dienolate intermediates **138** as nucleophiles (Scheme 58).[90] The products of the Brook rearrangement form allenes **139** via elimination of the nucleophilic catalyst.

1.5.3 Brook Rearrangement as the Key Step in Domino Reactions **395**

Scheme 58 1,3-Brook Rearrangement in Sila-Morita–Baylis–Hillman Reactions Using Allenoates as Nucleophiles[90]

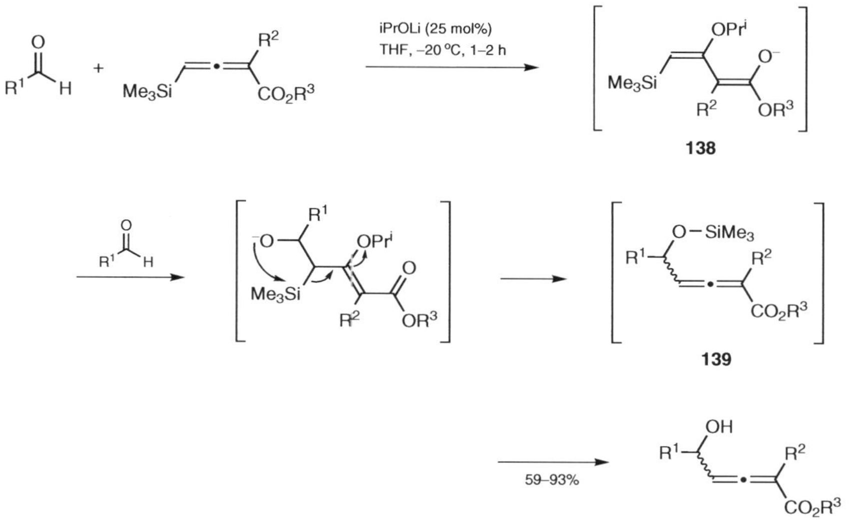

R[1] = Cy, *t*-Bu, C≡C(CH₂)₅Me, Ph, 4-MeOC₆H₄, 4-To , 4-*t*-BuC₆H₄, 4-ClC₆H₄, 4-O₂NC₆H₄, 4-FC₆H₄, 4-pyridyl; R[2] = Me, iPr, Bn; R[3] = Et, Ph

A number of Z-configured silyl enol ethers can be prepared by the Lewis acid catalyzed reaction of alkyl aryl ketones with diazo(trimethylsilyl)methane (Scheme 59).[91] After nucleophilic attack of diazo(trimethylsilyl)methane to the carbonyl group, a 1,2-aryl shift occurs and nitrogen gas is liberated. Then, the resulting α-silyl ketone undergoes a 1,3-Brook rearrangement to yield a Z-configured enol ether **140**. These conditions can also be applied to cyclic ketones to ultimately yield ring-expanded silyl enol ethers.

Scheme 59 1,3-Brook Rearrangement after Nucleophilic Addition of Diazo(trimethylsilyl)-methane to Aryl Ketones and 1,2-Aryl Migration[91]

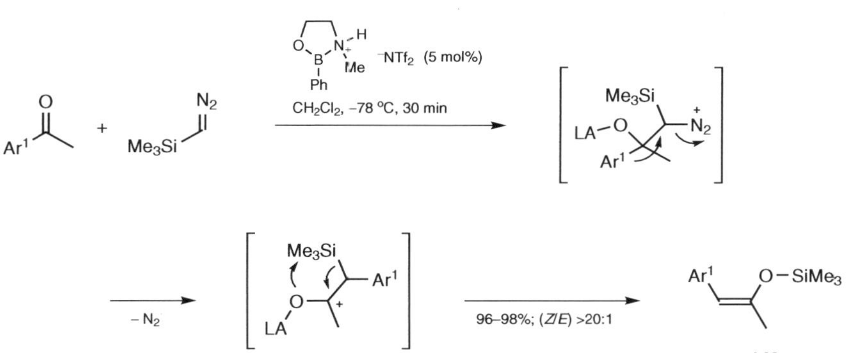

Ar[1] = Ph, 4-Tol, 2-Tol, 4-MeOC₆H₄, 4-BrC₆H₄, 4-F₃CC₆H₄, 4-O₂NC₆H₄, 1-naphthyl, 2-naphthyl, 2-furyl, 2-thienyl

When there is no aryl substituent present that can undergo 1,2-migration, the diazonium group can act as a 1,3-dipole in intramolecular dipolar cycloadditions (Table 2).[92] Thus, the base-catalyzed reaction of γ,δ-unsaturated aldehydes or ketones with diazo(trimethylsilyl)methane leads to bi- and tricyclic 4,5-dihydro-3*H*-pyrazoles. The reaction proceeds in

for references see p 445

a sequence composed of nucleophilic addition, 1,3-Brook rearrangement of alkoxide **141**, and a chemoselective intramolecular 1,3-dipolar cycloaddition of intermediate **142** overriding the competing Peterson-type alkenation.

Table 2 1,3-Brook Rearrangement after Nucleophilic Addition of Diazo(trimethylsilyl)-methane to γ,δ-Unsaturated Ketones Followed by Intramolecular 1,3-Dipolar Cycloaddition[92]

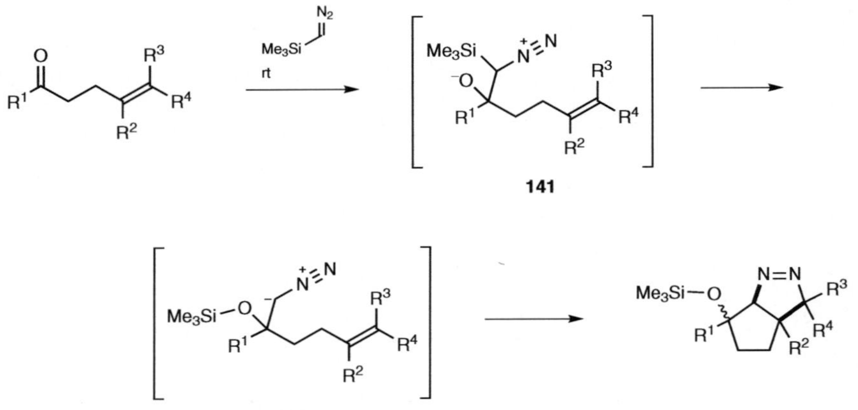

Substrate	Conditions[a]	Product	dr	Yield[b] (%)	Ref
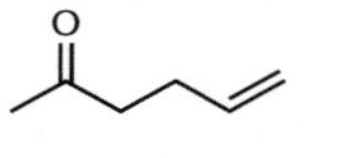	TBAT (2 mol%), THF	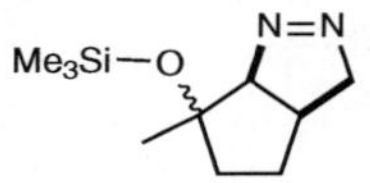	1.6:1	77	[92]
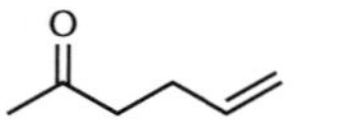	t-BuOK (10 mol%), THF	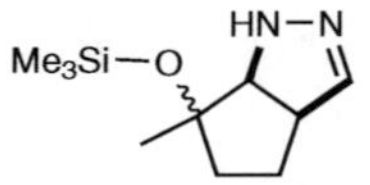	1:1	92	[92]
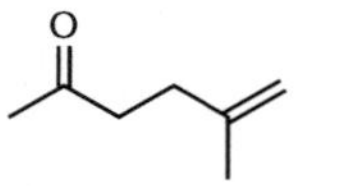	TBAT (2 mol%), THF	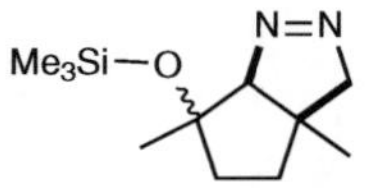	1:1	72	[92]
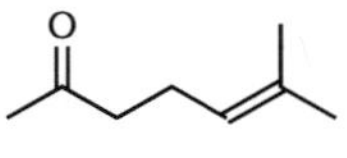	TBAT (2 mol%), THF	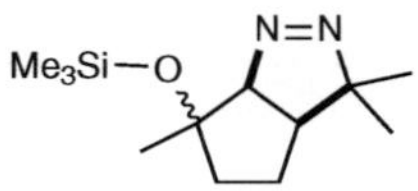	1:1	62	[92]
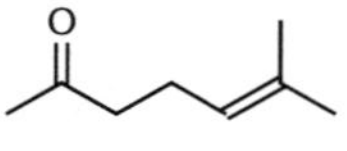	t-BuOK (10 mol%), THF	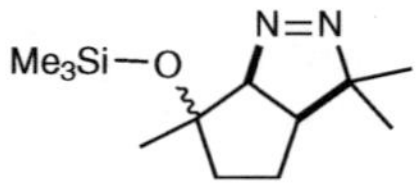	1:1	78	[92]
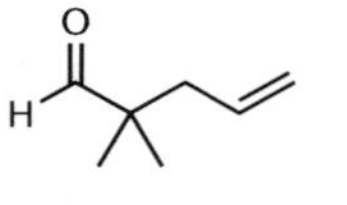	TBAT (2 mol%), THF	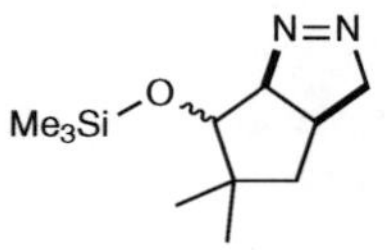	5:1	90	[92]

1.5.3 Brook Rearrangement as the Key Step in Domino Reactions **397**

Table 2 (cont.)

Substrate	Conditions[a]	Product	dr	Yield[b] (%)	Ref
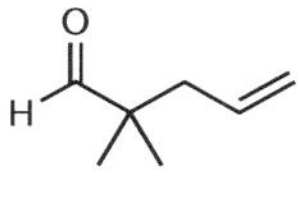	t-BuOK (10 mol%), THF	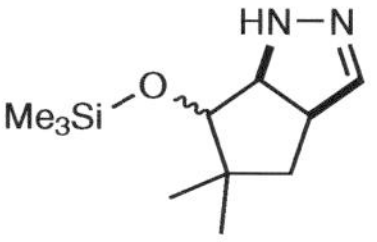	7:1	88	[92]
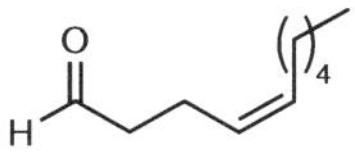	TBAT (2 mol%), THF	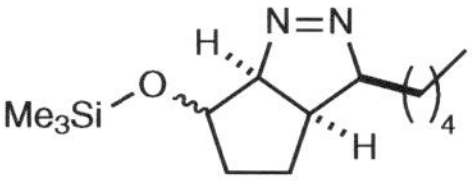	3:1	76	[92]
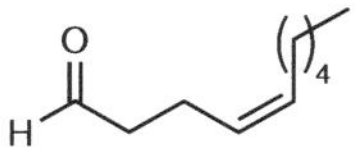	t-BuOK (10 mol%), THF	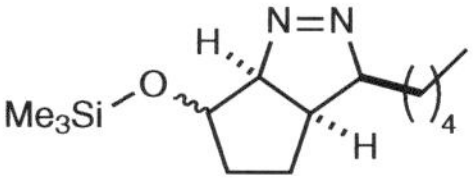	2:1	65	[92]
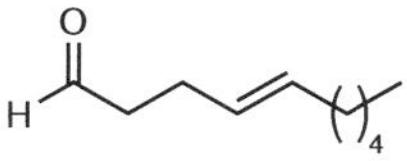	TBAT (2 mol%), THF	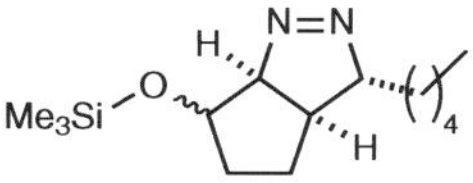	2.7:1	37	[92]
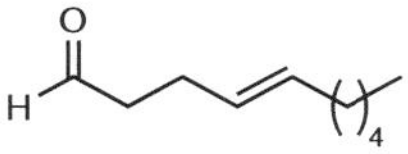	t-BuOK (10 mol%), THF	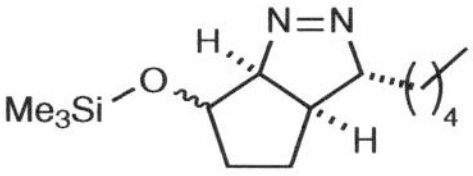	1:0	69	[92]
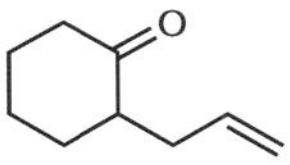	t-BuOK (10 mol%), THF	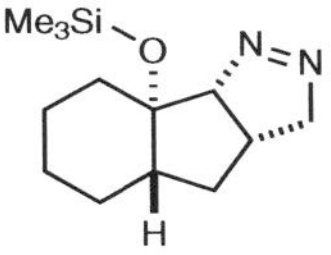	–	81	[92]
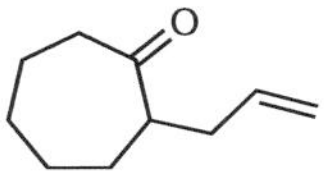	t-BuOK (10 mol%), THF	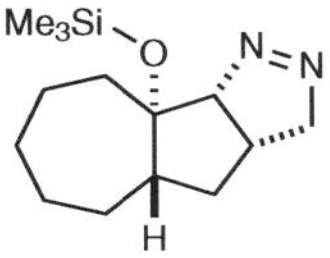	–	43[c]	[92]
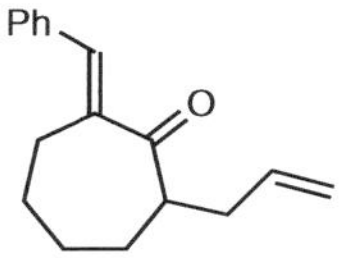	t-BuOK (10 mol%), THF	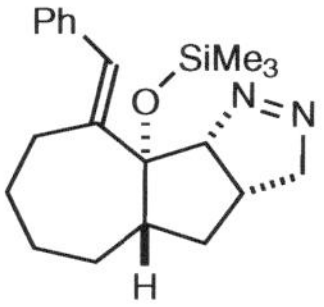	–	45	[92]
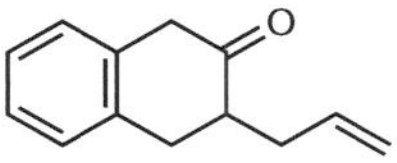	t-BuOK (10 mol%), THF	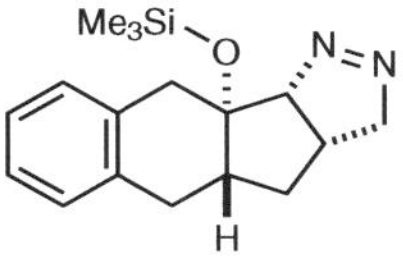	–	75	[92]
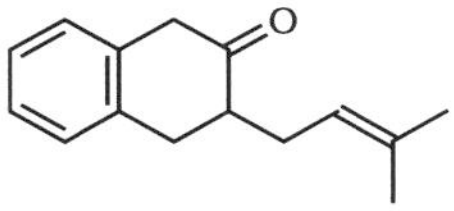	t-BuOK (10 mol%), THF	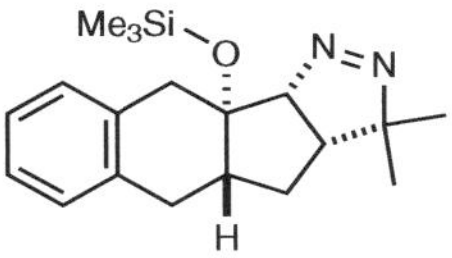	–	78	[92]

for references see p 445

Domino Transformations 1.5 Non-Radical Skeletal Rearrangements

Table 2 (cont.)

Substrate	Conditions[a]	Product	dr	Yield[b] (%)	Ref
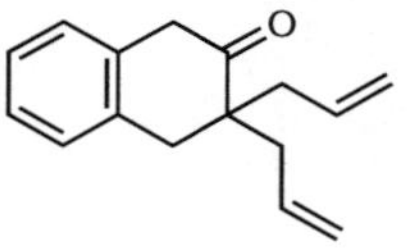	t-BuOK (10 mol%), THF	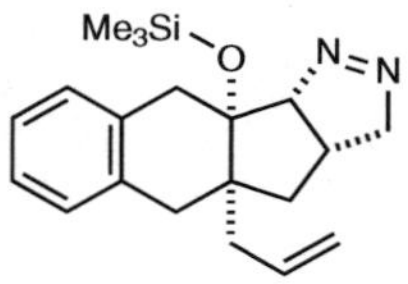	–	61	[92]
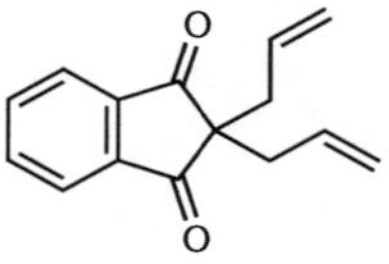	t-BuOK (10 mol%), THF	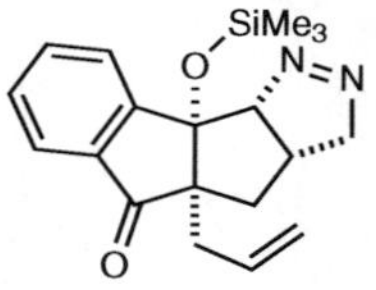	–	56	[92]
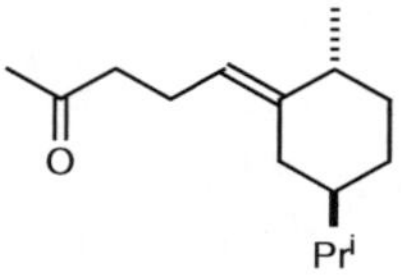	t-BuOK (10 mol%), THF	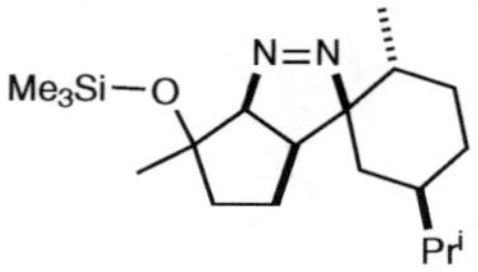	1:1	94	[92]
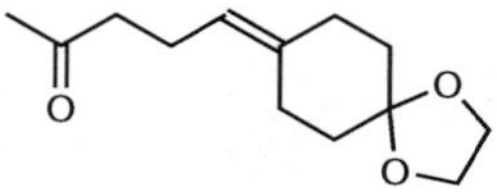	t-BuOK (10 mol%), THF	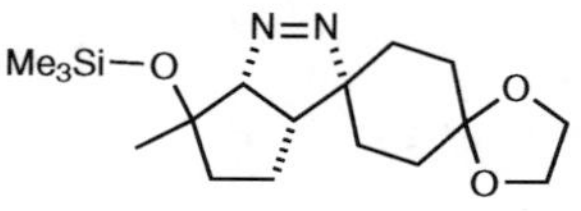	1:1	90	[92]
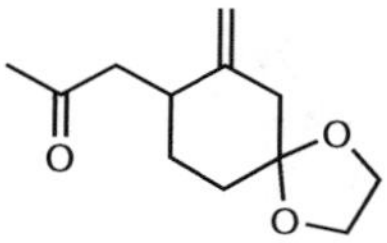	t-BuOK (10 mol%), THF	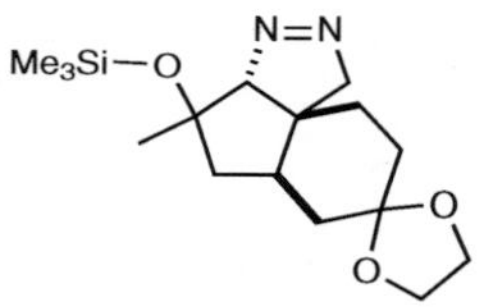	2.4:1	71	[92]
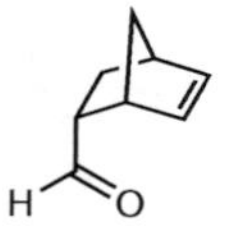	TBAT (2 mol%), THF	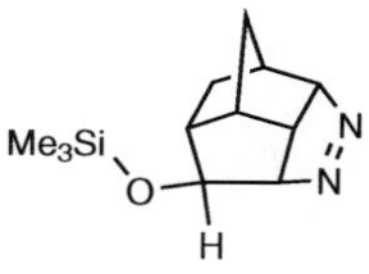	2:1	74	[92]
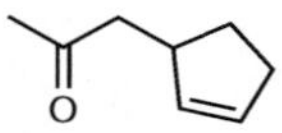	t-BuOK (10 mol%), THF	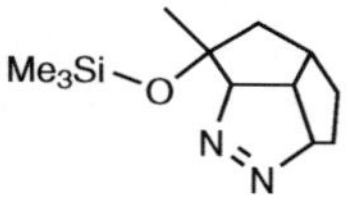	1:1	82	[92]
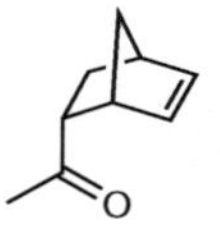	t-BuOK (10 mol%), THF	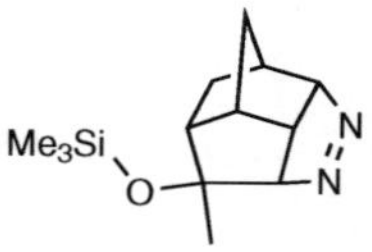	2:1	83	[92]

[a] TBAT = tetrabutylammonium triphenyldifluorosilicate.
[b] Isolated yield.
[c] 82% yield at −10 °C.

1.5.3 Brook Rearrangement as the Key Step in Domino Reactions

399

1.5.3.2.2   **1,3-Brook Rearrangement at sp²-Hybridized Carbon Atoms**

Based on principal investigations[93,94] into the structural requirements that hamper the Peterson-type elimination of carbanions positioned β to silyl ethers, a domino sequence starting from 2-(triphenylsilyl)allyl alcohols was developed (Scheme 60).[95] Copper(I) *tert*-butoxide serves as base in this well-designed system. The triphenylsilyl group in alkoxide intermediate **143** migrates to the oxygen center and the resulting soft organocopper species **144** does not undergo elimination but instead smoothly reacts with the added allyl halide.

Scheme 60   Formation of Vinylcuprates after 1,3-Brook Rearrangement and Subsequent Trapping with Electrophilic Allyl Systems[95]

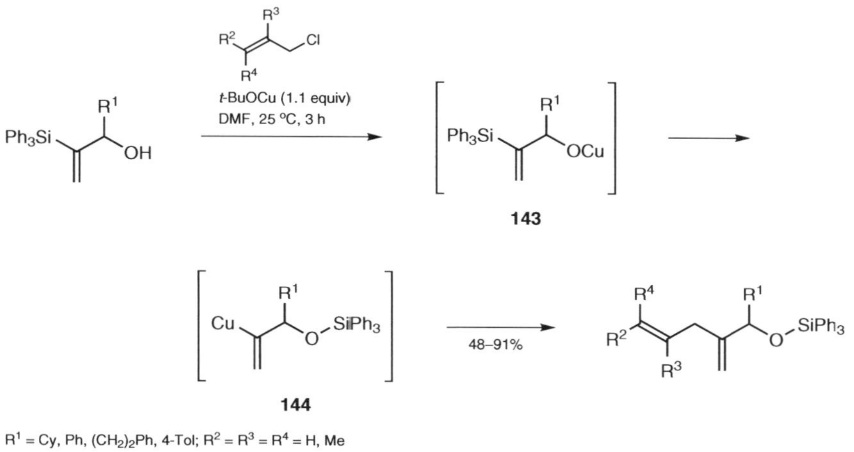

$R^1 = Cy, Ph, (CH_2)_2Ph, 4$-Tol; $R^2 = R^3 = R^4 = H, Me$

1.5.3.2.3   **1,3-Brook Rearrangement Accompanied by β-Elimination**

Another method to initiate a 1,3-Brook rearrangement is based on nucleophilic attack of [(phenylsulfanyl)(trimethylsilyl)methyl]lithium on epoxysilanes (e.g., **145**), which occurs on the silyl-terminus of the oxirane ring (Scheme 61).[96] After 1,3-Brook rearrangement of alkoxide **146**, a β-elimination of carbanion **147** yields the vinylsilane in 65% yield with an E/Z ratio of 3.3:1. It is noteworthy that the 1,3-silyl migration prevails over the alternative 1,4-Brook rearrangement.

Scheme 61   Cascade Reactions with 1,3-Brook Rearrangement Using an Epoxysilane[96]

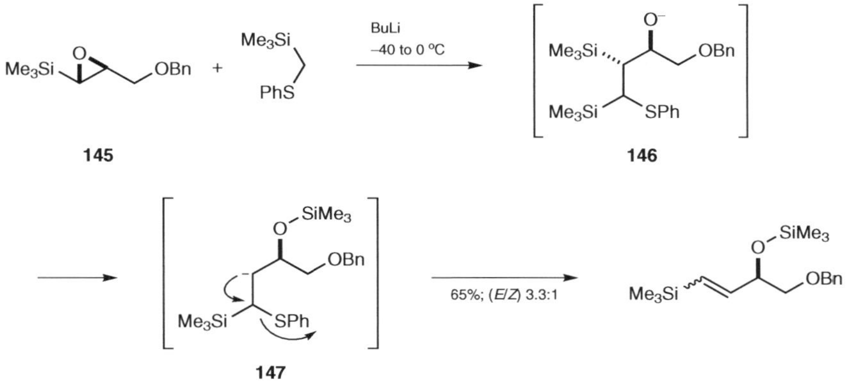

for references see p 445

A 1,3-Brook rearrangement has been successfully employed in the total synthesis of the alkaloid FR66 979 (**150**).[97,98] Following a 1,3-silyl migration induced by deprotonation of alcohol **148**, the neighboring aziridine is ring opened and alkene **149** is formed (Scheme 62).

Scheme 62   1,3-Brook Rearrangement as Key Step in the Synthesis of Alkaloid FR66 979[97,98]

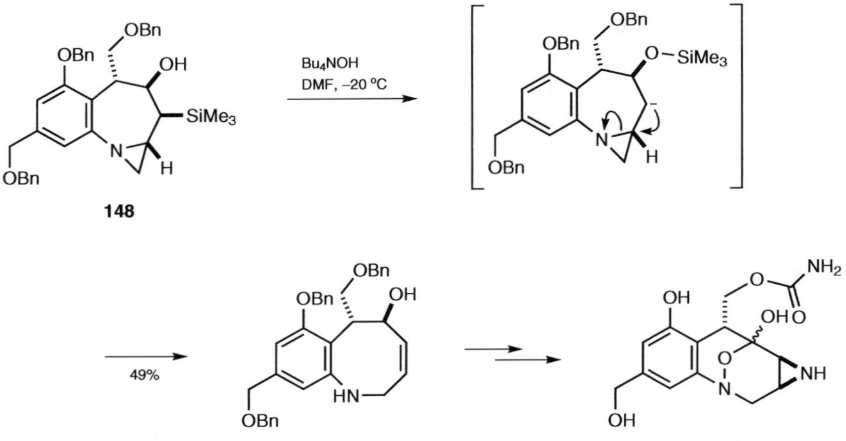

(5R*,6S*,Z)-7-(Benzyloxy)-6,9-bis[(benzyloxy)methyl]-1,2,5,6-tetrahydrobenzo[b]azocin-5-ol (149):[97]
40% aq Bu4NOH (5.50 mL, 8.4 mmol) was added dropwise to a soln of aziridine **148** (3.33 g, 5.61 mmol) in DMF (30 mL) at −20 °C. The mixture was stirred for 30 min at this temperature and then partitioned between H_2O and EtOAc. The combined organic layers were washed with brine, dried ($MgSO_4$), and concentrated. Purification by chromatography (silica gel, EtOAc/hexane 3:7 to 2:3) gave the product; yield: 1.46 g (49%).

1.5.3.2.4   Carbon to Nitrogen Rearrangement

Aza-Brook rearrangements are rare due to the lower affinity of silicon for nitrogen compared to oxygen (Scheme 63).[99] However, the synthesis of lactams is possible by addition of lithiated disulfanyl silylmethanes to isocyanates with an additional leaving group such as bromide. The domino sequence consists of a chemoselective, nucleophilic attack on the bromo isocyanate, 1,3-silyl carbon to nitrogen migration, and formation of a dithioacetal-stabilized carbanion **151** which finally undergoes ring closure by bromide displacement.

1.5.3 Brook Rearrangement as the Key Step in Domino Reactions **401**

Scheme 63 Carbon to Nitrogen 1,3-Brook Rearrangement En Route to Lactams[99]

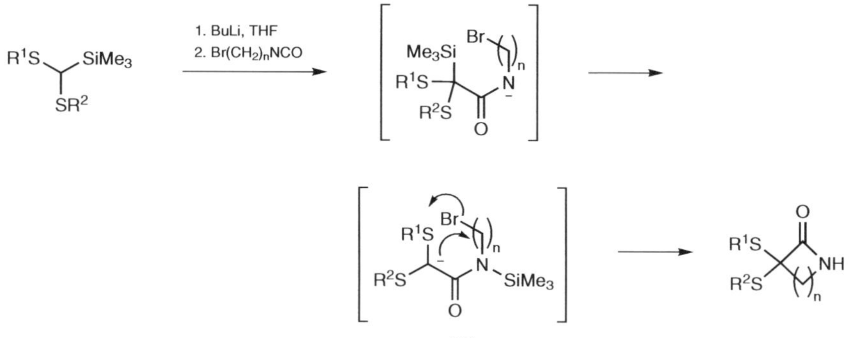

$R^1 = R^2 = Me; R^1,R^2 = (CH_2)_3; n = 2, 3$

1.5.3.2.5 **Carbon to Sulfur Rearrangement**

A thia-Brook rearrangement takes place when [tris(trimethylsilyl)methyl]lithium reacts with carbon disulfide to give the unusual sulfanylbis(trimethylsilyl)alkanethiones **155** (Scheme 64).[100] These thiones are stable at 0 °C for several months. A lithiated thiirane intermediate **152** (in equilibrium with bis-silyl dithioester **153**) is postulated and this unstable intermediate can successfully be captured with various alkyl iodides as well as activated organic bromides. The alkylation products **154** finally ring open to give the thiones **155**.

Scheme 64 Carbon to Sulfur Rearrangement after Nucleophilic Addition of [Tris(trimethylsilyl)methyl]lithium to Carbon Disulfide[100]

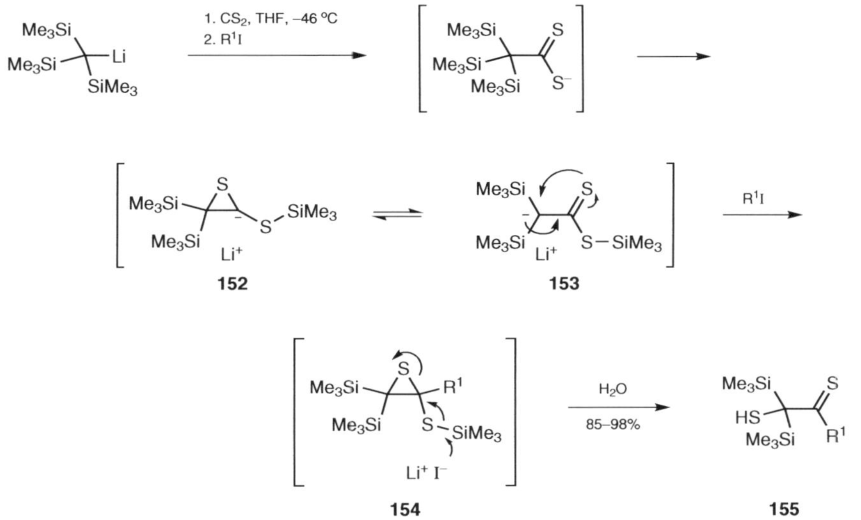

$R^1 = Me, Et, Pr, (CH_2)_2Ph, Bu, (CH_2)_4Me, (CH_2)_5Me$

for references see p 445

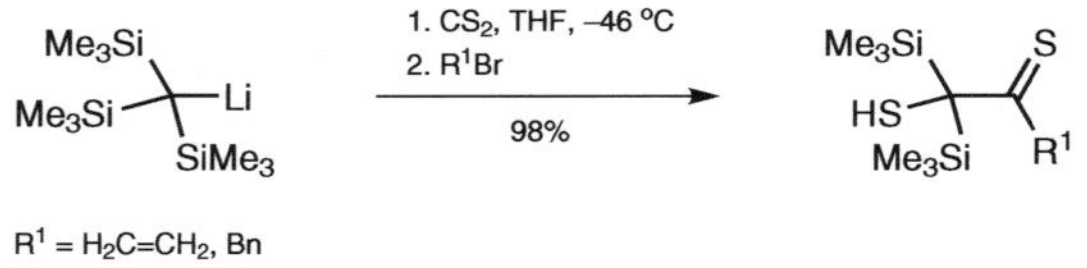

1.5.3.2.6 Retro-1,3-Brook Rearrangement

The key step in the synthesis of tea catechin lactone derivatives is the ring-opening reaction of epoxy benzyl ethers by an aryllithium species, which is generated by lithiation of the corresponding aryl bromide. The equilibrium between the lithiated species **156** and alkoxide **157**, arising from a rarely observed retro-1,3-Brook rearrangement, is highly solvent dependent (Scheme 65).[101] In tetrahydrofuran, as well as a mixture of tetrahydrofuran and diethyl ether (1:1), the rearrangement product **159** is formed in 85 and 92% yield, respectively. In contrast, the carbon coupling product **158** is formed exclusively in 88% yield when a 1:9 solvent mixture of tetrahydrofuran and diethyl ether is employed.

Scheme 65 Solvent Dependence of Retro-1,3-Brook Rearrangement in an *ortho*-Lithiated Siloxyarene[101]

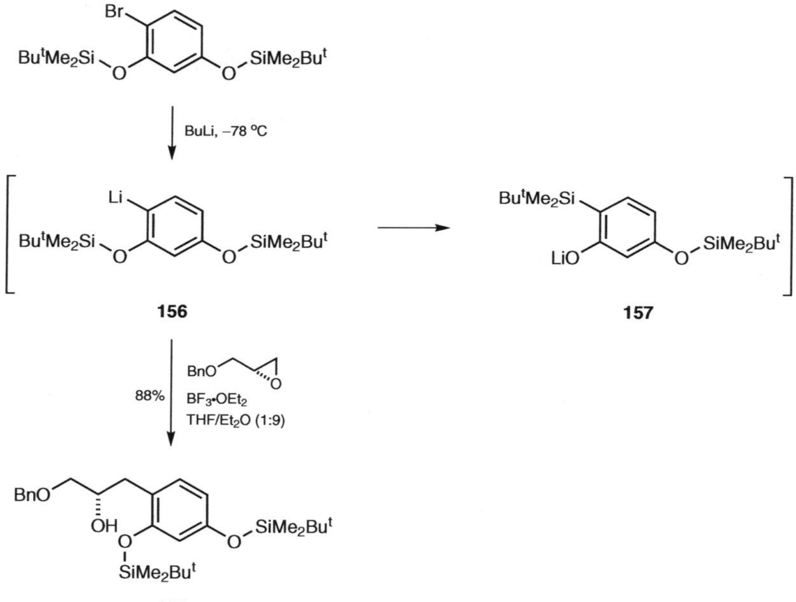

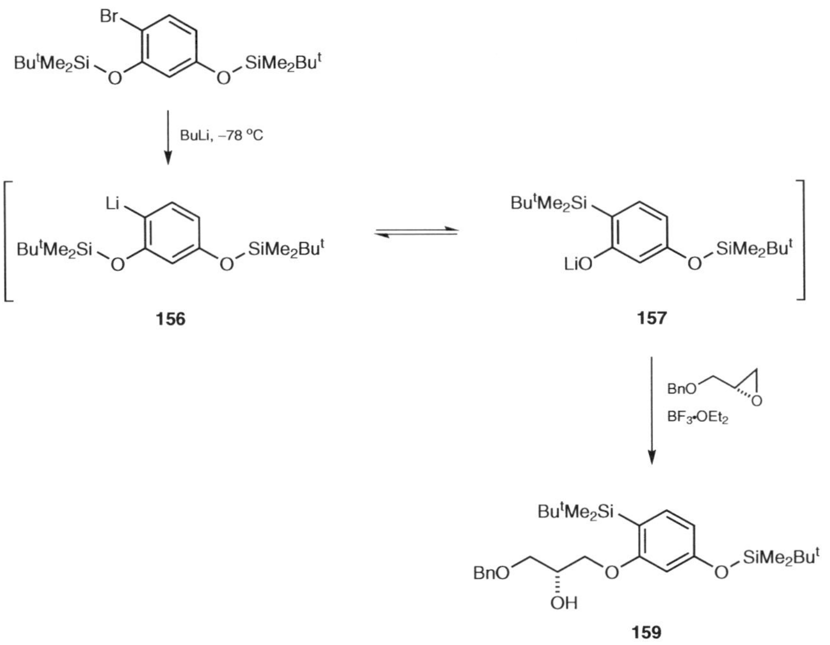

1.5.3 Brook Rearrangement as the Key Step in Domino Reactions

Solvent	Yield (%)	Ref
THF	85	[101]
THF/Et$_2$O (1:1)	92	[101]

1.5.3.3 1,4-Brook Rearrangement

1,4-Brook rearrangements have been extremely useful for the development of domino reactions, and recently several very impressive applications, mainly for the total synthesis of polyketide natural products, have been reported. These studies established that the carbanion **161** formed after 1,4-silyl migration of alkoxide **160** is less prone to homo-Peterson elimination leading to cyclopropanes compared to the scenario of the Peterson alkenation (see Section 1.5.3.2). Therefore, the formed carbanion **161** can be exploited for further transformations in a flexible manner (Scheme 66). Oxiranes can be regarded as homo-carbonyl equivalents and by formally substituting the carbonyl component employed in Peterson alkenations by oxiranes 1,4-Brook rearrangements become possible. However, these are only well induced if the resulting carbanion is stabilized by additional functional groups, labeled here as Z. Still, a delicate balance between stabilization of the carbanion and retaining nucleophilic reactivity for further transformations, such as with electrophiles, is crucial when selecting Z. Consequently, most 1,4-Brook rearrangements rely on nucleophilic ring opening of epoxides with silyl-substituted, Z-stabilized carbanions. Alternatively, γ-silyl-substituted carbinols can also serve as precursors for 1,4-Brook rearrangements after deprotonation. As in Section 1.5.3.2, where the Peterson alkenation was excluded, the homo-Peterson reaction is not discussed here in detail.

for references see p 445

404 Domino Transformations **1.5** Non-Radical Skeletal Rearrangements

Scheme 66 The 1,4-Brook Rearrangement

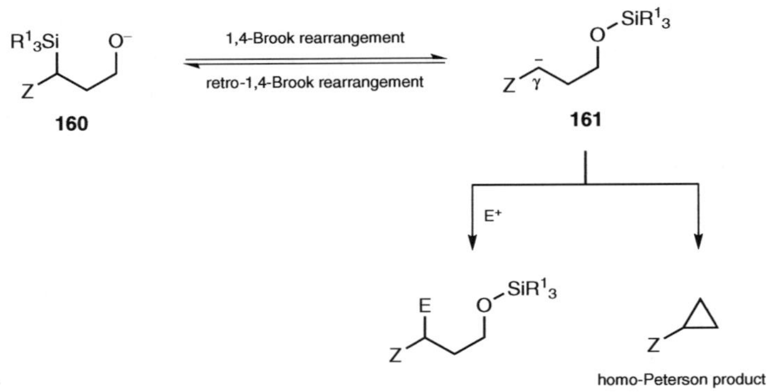

1.5.3.3.1 **1,4-Brook Rearrangement of Silyl-Substituted Carbanions with Epoxides**

A 1,4-silyl migration takes place in the reaction between lithiated 2-(trimethylsilyl)aceto-nitrile and epoxides. After ring opening, the 1,4-silyl migration yields a new nitrile-stabi-lized carbanion **162**. This intermediate can either be protonated and then converted into γ-lactones, or it can be involved in a second epoxide-ring-opening process to form com-plex nitriles **163** and then also converted into the corresponding γ-lactones (Scheme 67).[102]

Scheme 67 2-(Trimethylsilyl)acetonitrile as Dianion Equivalent in 1,4-Brook Rearrangement[102]

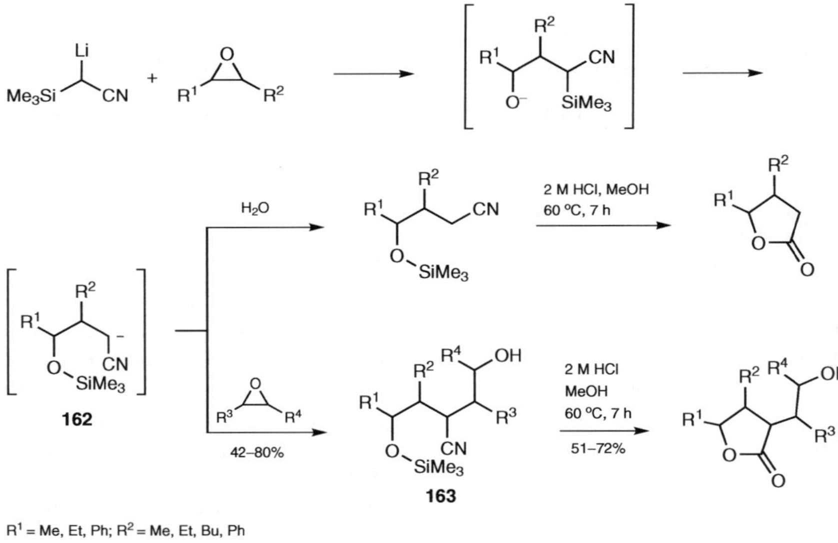

R^1 = Me, Et, Ph; R^2 = Me, Et, Bu, Ph

When [tris(trimethylsilyl)methyl]lithium is reacted with 2-phenyloxirane, the alkoxide **164** formed after ring opening leads to 1,4-silyl transfer, which is followed by a cycliza-tion forming the cyclopropane ring (Scheme 68).[103] A byproduct results from the reaction of the intermediate carbanion **165** with a second equivalent of 2-phenyloxirane to yield a 1,5-diol (after aqueous workup). These examples inspired several groups, especially Schaumann and co-workers, to develop a broadly applicable domino sequence.[104,105]

1.5.3 Brook Rearrangement as the Key Step in Domino Reactions **405**

Scheme 68 First Homo-Peterson Reaction[103]

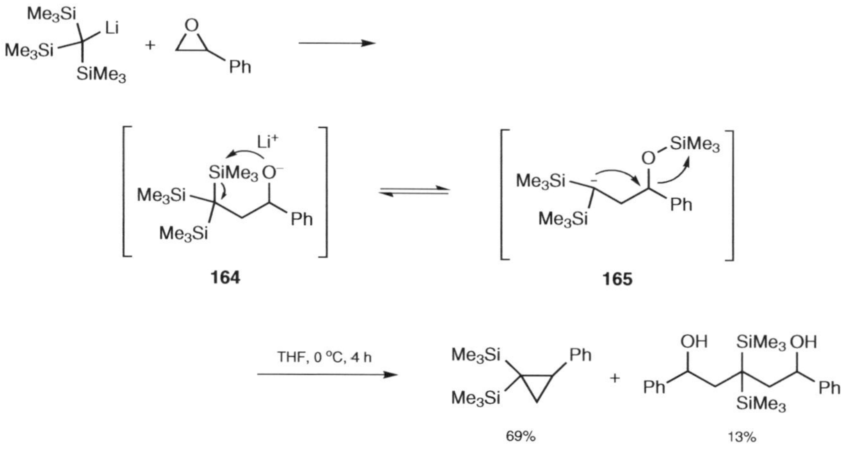

Based on the chemical properties of these C1 dianion equivalents, a series of applications for the preparation of different ring systems using bis-electrophiles have been developed.[104,105] The first example is a formal [4+1] cycloaddition leading to functionalized cyclopentanes. The lithium derivatives obtained from monosilylated dithioacetals react chemoselectively with epoxy sulfonates to yield silyl ethers **167** (Table 3). 1,4-Brook rearrangement provides a new carbanion **166**, which substitutes the 4-toluenesulfonate group. This example not only demonstrates the synthetic power of these dianion equivalents, but also reveals reactivity differences between 1,2-oxiranes and toluenesulfonates.

for references see p 445

Table 3 One-Pot [4 + 1] Cycloaddition Based on the 1,4-Brook Rearrangement To Give Cyclopentanes[104,105]

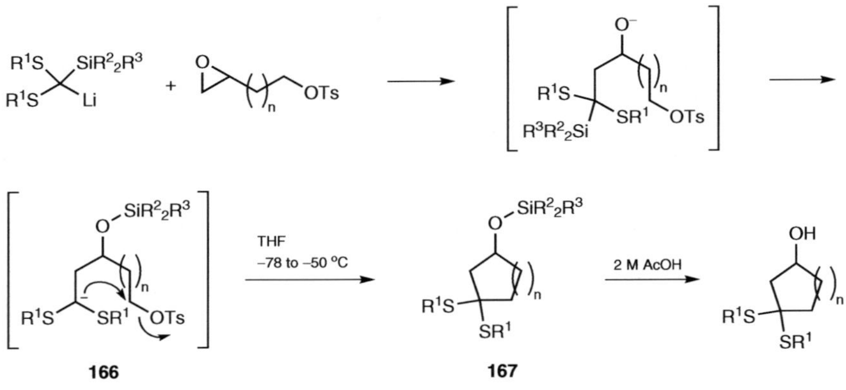

n = 1–3

Substrates		Product	Yield (%)	Ref
Organolithium	Oxirane			
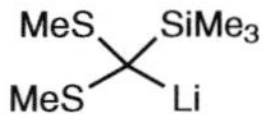 MeS, SiMe₃ / MeS, Li	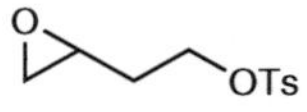 O epoxide OTs	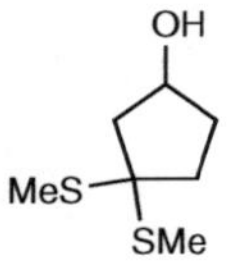 OH cyclopentane MeS SMe	80	[104]
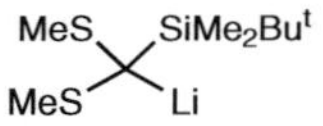 MeS, SiMe₂Buᵗ / MeS, Li	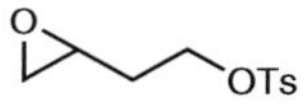 O epoxide OTs	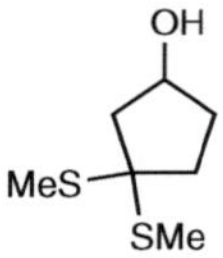 OH cyclopentane MeS SMe	46	[104]
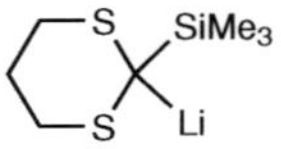 S,S SiMe₃ / Li dithiane	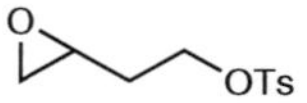 O epoxide OTs	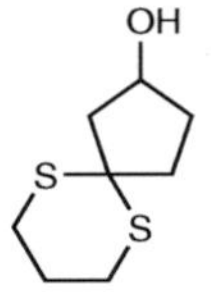 OH spirocycle S S	65	[104]
Me₃Si SiMe₃ / Li	O epoxide OTs	OH cyclopentane Me₃Si	55ᵃ	[104]
MeS, SiMe₃ / MeS, Li	H O epoxide cyclohexane OTs	H OH bicyclic MeS SMe	49ᵇ	[104]

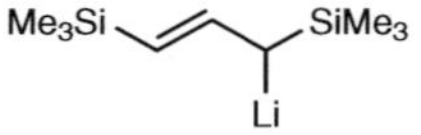

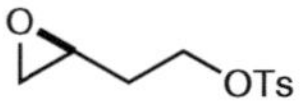

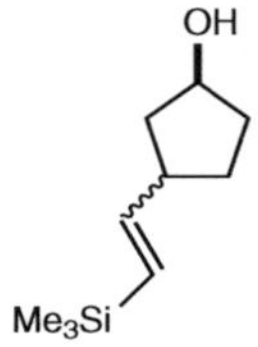

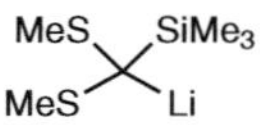

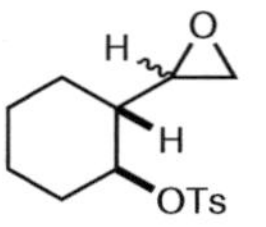

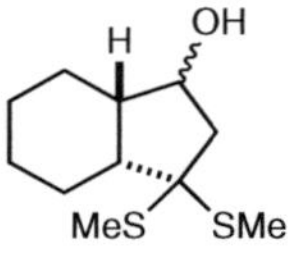

1.5.3 Brook Rearrangement as the Key Step in Domino Reactions **407**

Table 3 (cont.)

Substrates		Product	Yield (%)	Ref
Organolithium	Oxirane			

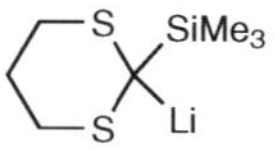 [structure: 2-(trimethylsilyl)-1,3-dithiane lithium]	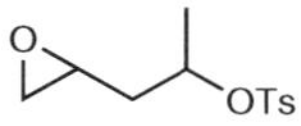 [oxirane with OTs chain]	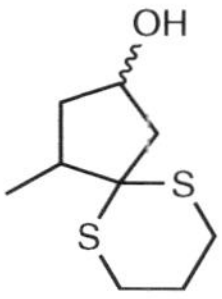 [spirocyclopentanol dithiane with OH]	65[a]	[105]
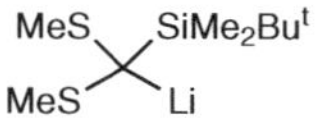 [structure: (MeS)₂C(SiMe₂Bu^t)Li]	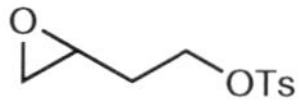 [oxirane with OTs chain]	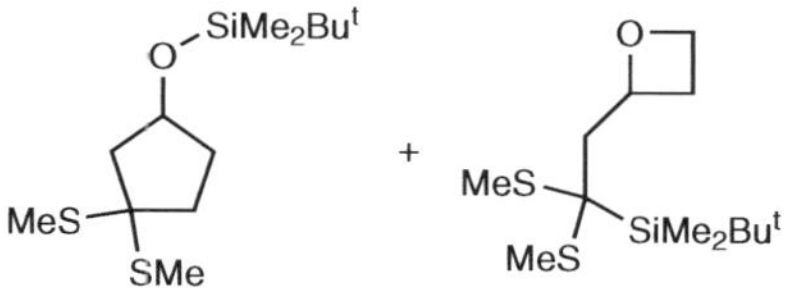 [cyclopentane OSiMe₂Bu^t with MeS, SMe] + [oxetane with MeS, MeS, SiMe₂Bu^t] 51:11	62[c]	[105]
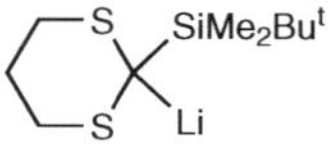 [structure: 2-(tert-butyldimethylsilyl)-1,3-dithiane lithium]	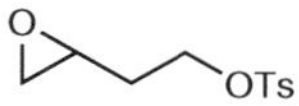 [oxirane with OTs chain]	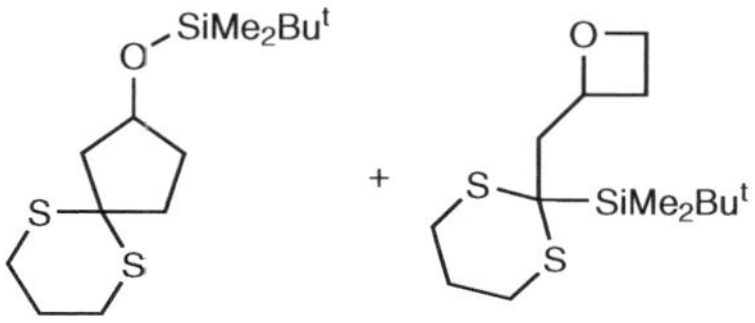 [cyclopentane OSiMe₂Bu^t with dithiane] + [oxetane with dithiane, SiMe₂Bu^t] 75:17	92[d]	[105]
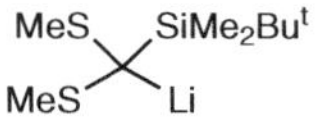 [structure: (MeS)₂C(SiMe₂Bu^t)Li]	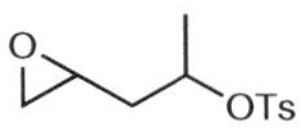 [oxirane with OTs chain]	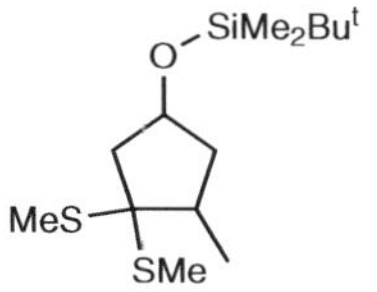 [cyclopentane OSiMe₂Bu^t with MeS, SMe, Me]	69[a,e]	[105]
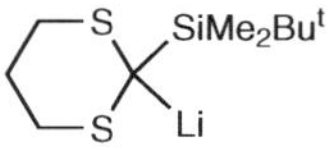 [structure: 2-(tert-butyldimethylsilyl)-1,3-dithiane lithium]	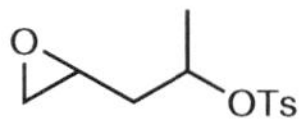 [oxirane with OTs chain]	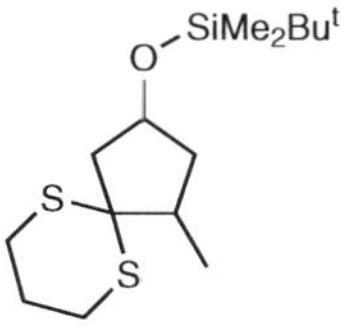 [cyclopentane OSiMe₂Bu^t with dithiane, Me]	41[a,f]	[105]

[a] Product was obtained as a mixture of diastereomers (ca. 1:1).
[b] Epoxide was employed as single diastereomer; product was obtained as a single diastereomer.
[c] The dithioacetal cyclopentanol silyl ether reacts to give the ketone in 63% under standard conditions.[105,106]
[d] The dithioacetal cyclopentanol silyl ether reacts to give the ketone in 64% under standard conditions.[105,106]
[e] The dithioacetal cyclopentanol silyl ether reacts to give the ketone in 34% under standard conditions.[105,106]
[f] The dithioacetal cyclopentanol silyl ether reacts to give the ketone in 27% under standard conditions.[105,106]

In principal, this cascade reaction can also serve to form cyclohexanes and heptanes; however, the formation of byproducts such as furan formation or substitution of the 4-toluenesulfonate group by a second equivalent of the organolithium species is difficult to suppress (Scheme 69).[105]

for references see p 445

Scheme 69 One-Pot [5+1] and [6+1] Cycloadditions Based on the 1,4-Brook Rearrangement To Access Cyclohexanes and Cycloheptanes[105]

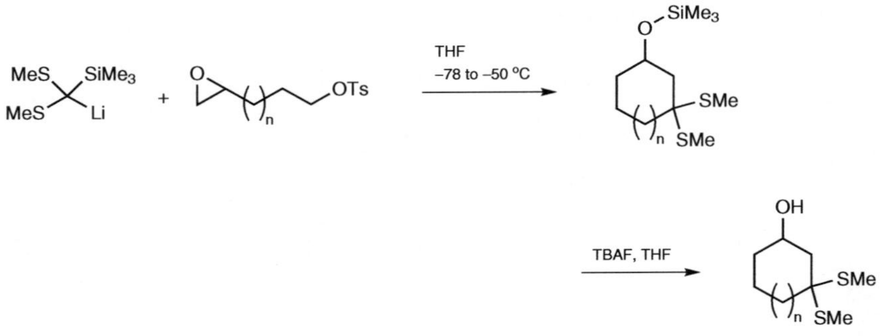

n	Yield (%)	Ref
1	49	[105]
2	47	[105]

Instead of two sulfur substituents, cases of one thioether and two silyl groups provide (silylmethyl)lithium species that can undergo 1,4-Brook rearrangements after nucleophilic ring opening of oxiranes followed by nucleophilic substitution of an internal 4-toluenesulfonate group.[107] Preliminary insights into migratory properties of different silyl groups in these Brook rearrangements obtained by competition experiments suggest the following trend: $SiMe_2Ph > SiMe_3 > SiMePh_2$. The phenyl ring has a stabilizing effect on the migrating group and this electronic effect compensates steric hindrance to some degree. However, this also implies that the bulkiness of the methyldiphenylsilyl group overrules the favorable electronic effect.

Vinyloxiranes (e.g., **168**) can also be used in these [4+1] cycloadditions. When the oxirane ring is first attacked by the organolithium species, a 1,4-Brook rearrangement occurs (Scheme 70).[108] As expected, the second intramolecular carbon–carbon coupling step only proceeds in a Michael-type fashion if the R^1 group is able to stabilize the carbanion. In the case of 2-[1-(phenylsulfanyl)vinyl]oxirane (**168**, $R^1 = SPh$), the main product is formed by vinylogous nucleophilic attack of the organolithium species.

1.5.3 Brook Rearrangement as the Key Step in Domino Reactions **409**

Scheme 70 1,4-Brook Rearrangement Initiated after Ring Opening of Vinyloxiranes with [Bis(methylsulfanyl)(trimethylsilyl)methyl]lithium[108]

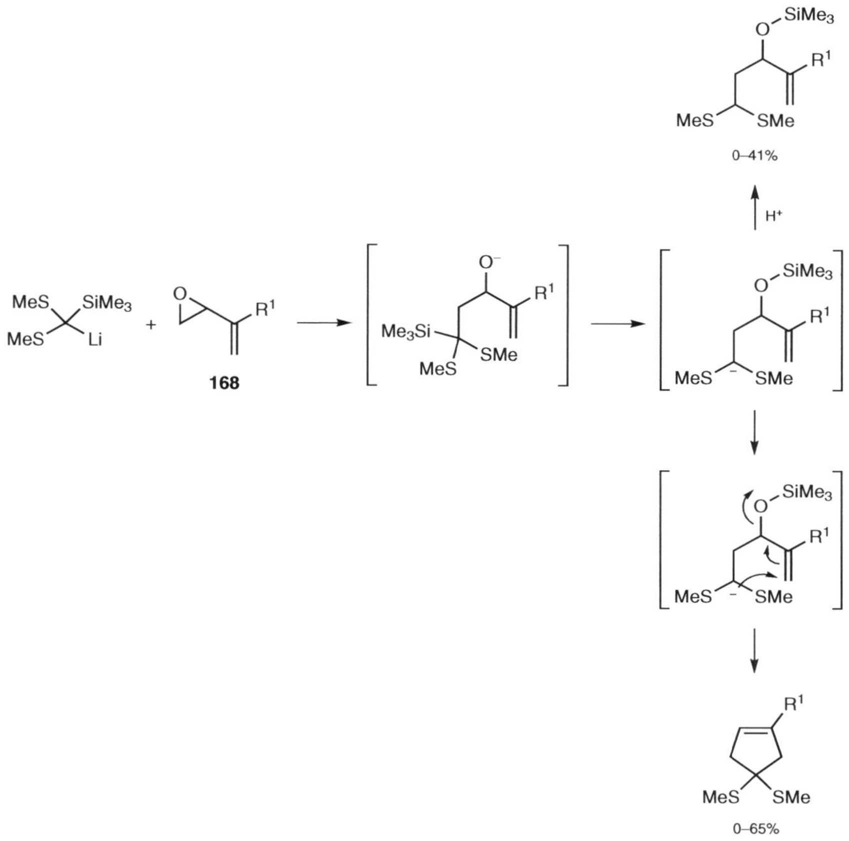

R^1 = H, Me, SiMe$_3$, SiMe$_2$Ph, SiPh$_3$

Bis-epoxides derived from carbohydrates can react as bis-electrophiles in these cascade reactions[109] to access highly functionalized carbocycles (Table 4). However, the chosen conditions provide mixtures of carbocycles **170** and **171** through 6-*exo* and 7-*exo* ring closures, respectively, and in some cases, an acyclic product **169** is also formed.

for references see p 445

410 Domino Transformations **1.5** Non-Radical Skeletal Rearrangements

Table 4 1,4-Brook Rearrangement Initiated after Ring Opening of Bisoxiranes with Lithiated Dithiosilyl-Sub-stituted Methanes[109]

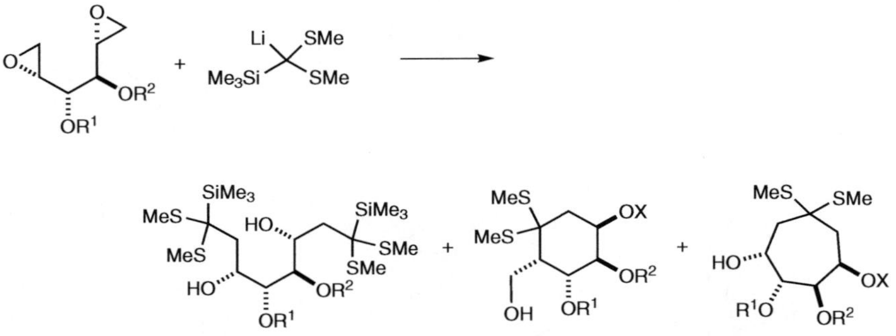

X = H, SiMe₃

Substrate	Products	Yield (%)	Ref
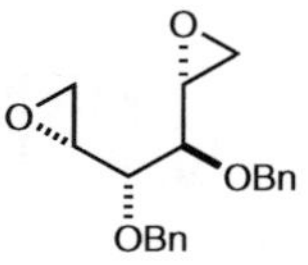	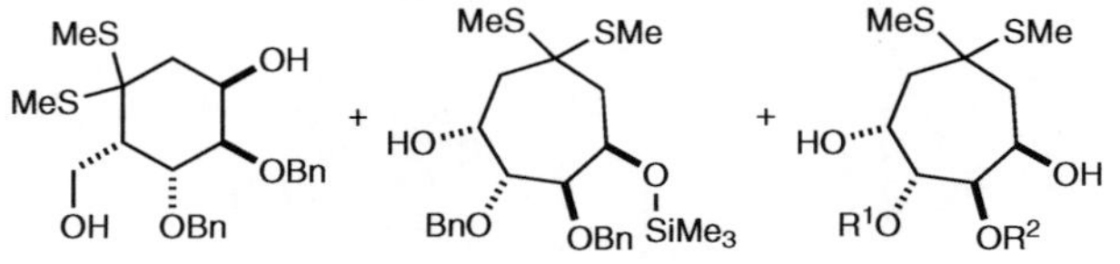 23:29:11	63[a]	[109]
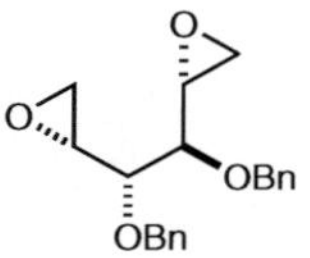	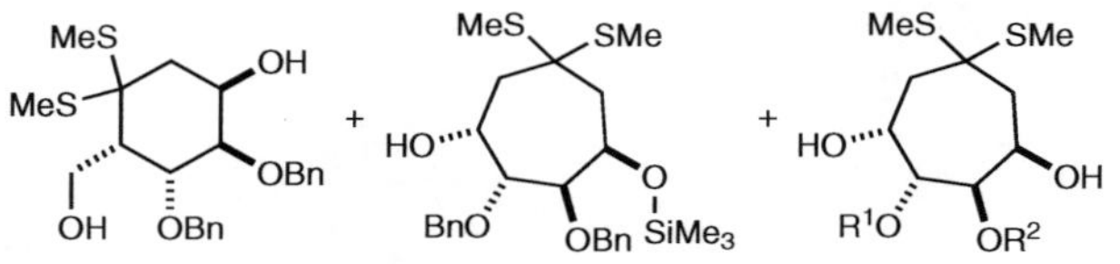 29:20:10	59[b]	[109]
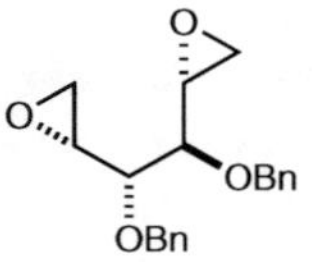	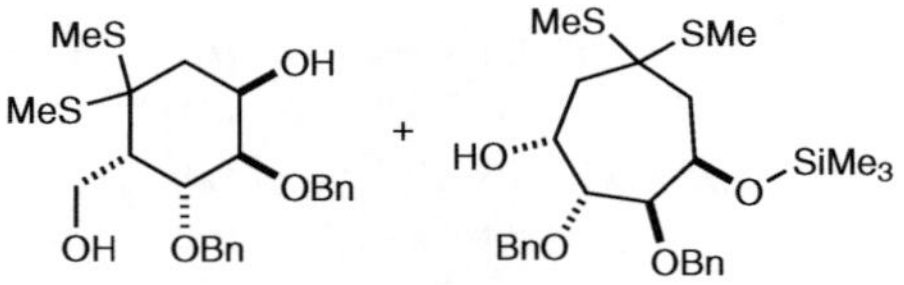 30:16	46[c]	[109]

1.5.3 Brook Rearrangement as the Key Step in Domino Reactions   **411**

Table 4   (cont.)

Substrate	Products			Yield (%)	Ref
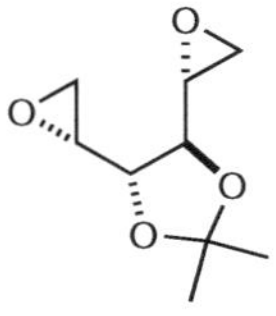	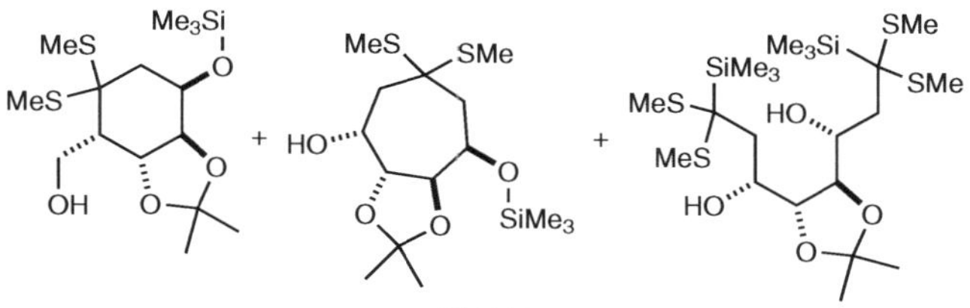 21:10:11			42[b]	[109]
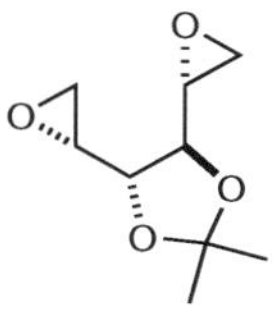	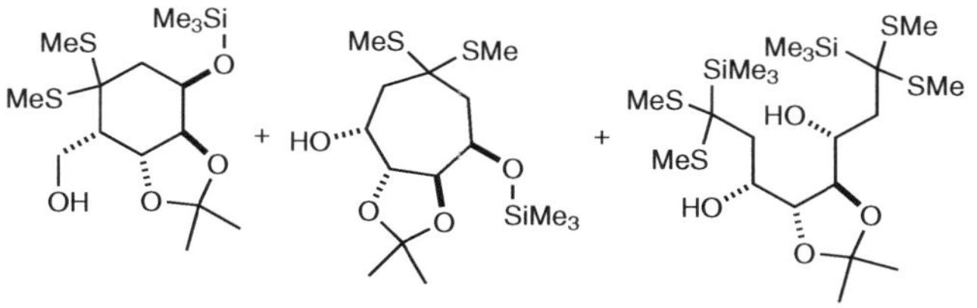 19:16:12			47[c]	[109]

[a] −40 to −20 °C, 24 h.
[b] −80 to −20 °C, 24 h.
[c] −80 °C, 24 h.

Similar cascade cyclizations based on these C1 dianion equivalents with various bis-electrophiles give enantiopure carbasugars (e.g., **172–175**) (Scheme 71).[110–112]

Scheme 71   1,4-Brook Rearrangement Initiated after Ring Opening of Bisoxiranes with Lithiated Disulfanyl-Substituted Methanes[110,112]

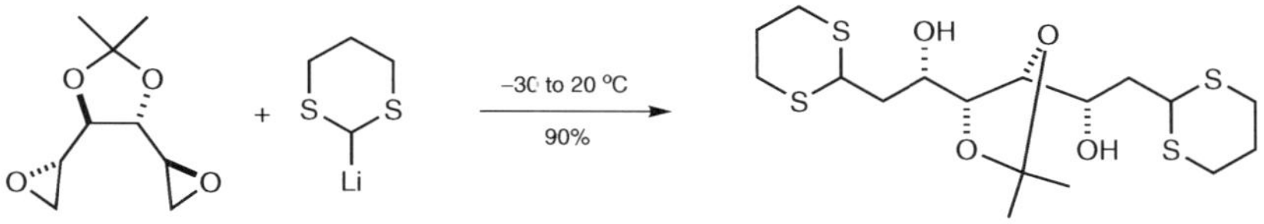

for references see p 445

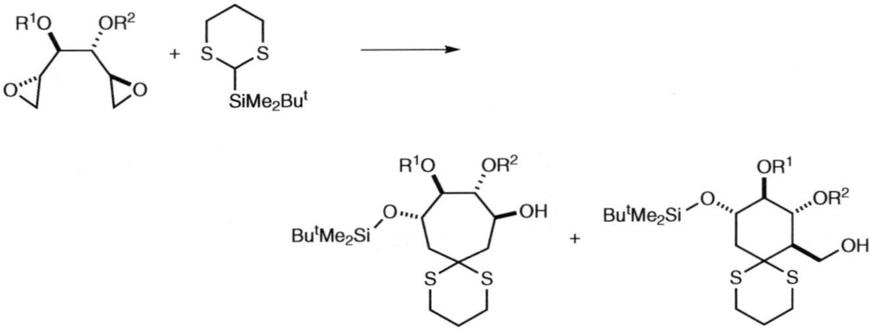

412 Domino Transformations 1.5 Non-Radical Skeletal Rearrangements

R¹	R²	Conditions	Yield (%) 172	Yield (%) 173	Ref
CMe₂		t-BuLi, THF/HMPA (9:1), −78 to −30 °C	72	15	[112]
CMe₂		BuLi, Bu₂Mg, THF, HMPA, rt	52	30	[112]
Bn	Bn	t-BuLi, THF/HMPA (9:1), −78 to −30 °C	21	43	[112]
Bn	Bn	BuLi, Bu₂Mg, THF, HMPA, rt	33	63	[112]

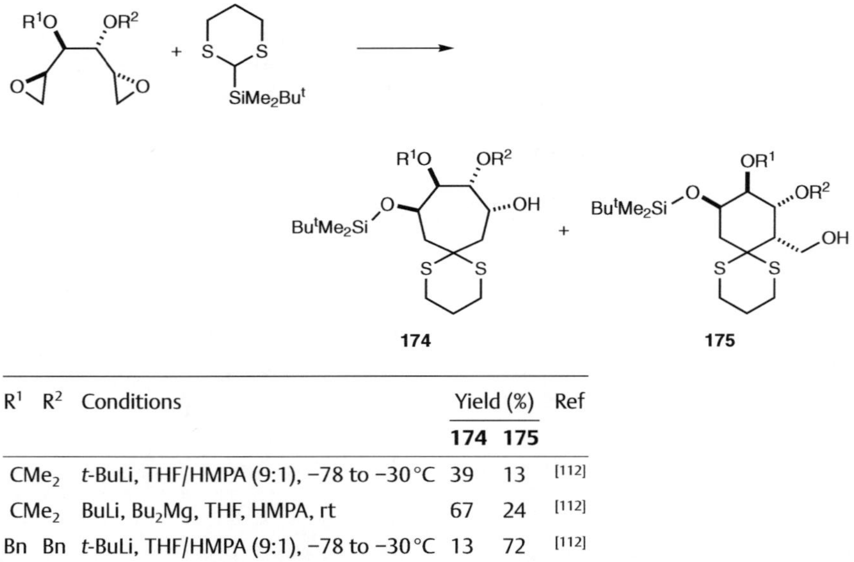

R¹	R²	Conditions	Yield (%) 174	Yield (%) 175	Ref
CMe₂		t-BuLi, THF/HMPA (9:1), −78 to −30 °C	39	13	[112]
CMe₂		BuLi, Bu₂Mg, THF, HMPA, rt	67	24	[112]
Bn	Bn	t-BuLi, THF/HMPA (9:1), −78 to −30 °C	13	72	[112]
Bn	Bn	BuLi, Bu₂Mg, THF, HMPA, rt	10	82	[112]

A domino sequence toward functionalized cyclopentane derivatives relies on a 1,4-silyl migration. Here, [2-(*tert*-butyldimethylsilyl)-1,3-dithian-2-yl]lithium and homochiral 1,4-bis-epoxides serve as starting materials to prepare carbanucleosides **176** and **177** (Scheme 72).[113,114] Selectivity with respect to mode of ring closure was improved compared to those examples that either led to cyclohexanes or cycloheptanes, respectively.

1.5.3 Brook Rearrangement as the Key Step in Domino Reactions **413**

Scheme 72 1,4-Brook Rearrangement Initiated after Ring Opening of Bisoxiranes with [2-(*tert*-Butyldimethylsilyl)-1,3-dithian-2-yl]lithium[113,114]

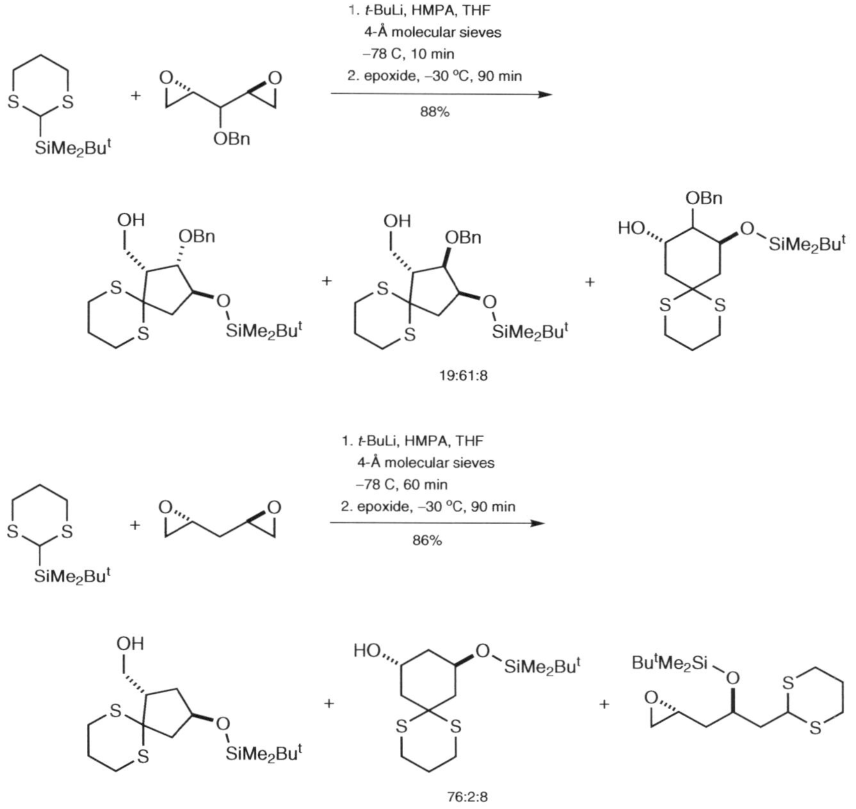

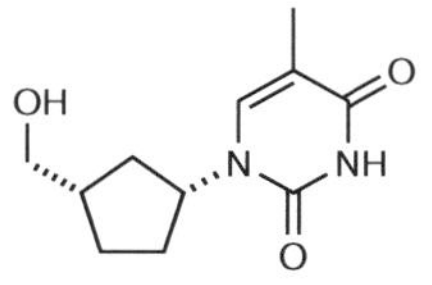

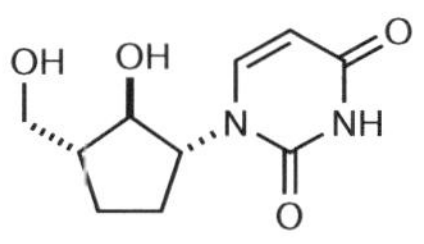

Finally, 1,2-epimino-3,4-epoxy-(*N*-toluenesulfonyl)butanes (e.g., **178**) can also be used as a homochiral bis-electrophile. These cascade reactions make use of the different relative electrophilicity of oxiranes versus *N*-tosylaziridines toward the C1 organolithium species (Scheme 73).[115] The expected cyclopentane formation occurs by initial attack of the oxirane ring and 1,4-Brook rearrangement. Addition of boron trifluoride is necessary to activate the aziridine ring for the second nucleophilic ring-opening step.

for references see p 445

Scheme 73 1,4-Brook Rearrangement Initiated after Ring Opening of an Oxirane Bearing an Additional Aziridine[115]

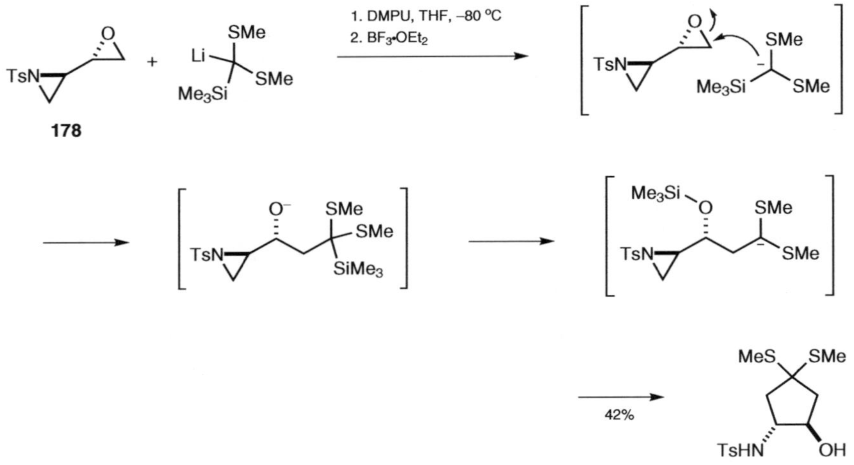

Enantiopure 1,5-diols **179** can be prepared by a tandem reaction involving a bisalkylation of lithiated 2-(trimethylsilyl)-1,3-dithiane with epoxides in the presence of a crown ether (Scheme 74).[116]

Scheme 74 Sequential Reaction of Lithiated 2-(Trimethylsilyl)-1,3-dithiane with Two Epoxides[116]

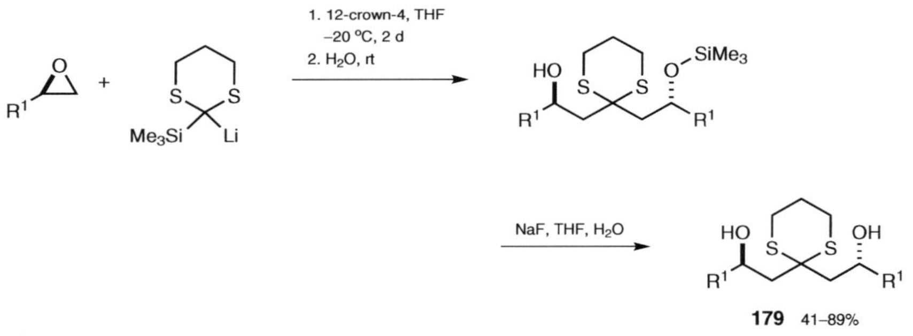

R^1 = Ph, 4-ClC$_6$H$_4$, 4-Tol, 4-MeOC$_6$H$_4$, Me

Bisalkylated products **180** are obtained in a one-pot cascade reaction of 2-(trialkylsilyl)-1,3-dithianes with two different oxiranes that are added sequentially (Table 5).[117] Lithiation of the C1 reagent in diethyl ether, addition of the first oxirane, addition of hexamethylphosphoric triamide, which facilitates 1,4-Brook rearrangement, and finally addition of the second oxirane (or benzyl bromide) affords the unsymmetrical bisalkylated products.

1.5.3 Brook Rearrangement as the Key Step in Domino Reactions **415**

Table 5 Sequential Reaction of Two Epoxides with Lithiated 2-(*tert*-Butyldimethylsilyl)-1,3-dithiane as a Dianion Equivalent[117]

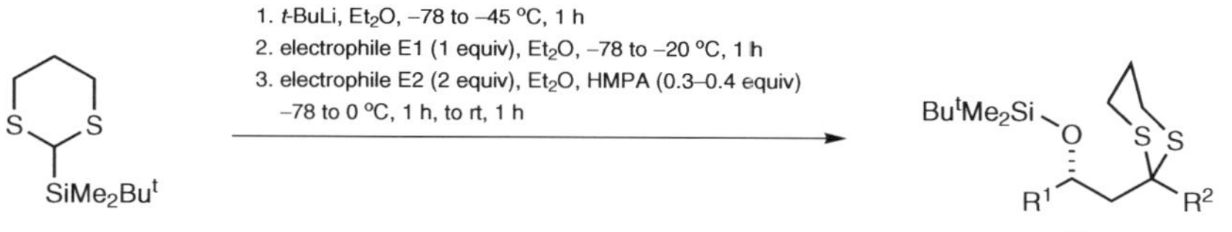

1. *t*-BuLi, Et$_2$O, −78 to −45 °C, 1 h
2. electrophile E1 (1 equiv), Et$_2$O, −78 to −20 °C, 1 h
3. electrophile E2 (2 equiv), Et$_2$O, HMPA (0.3–0.4 equiv)
   −78 to 0 °C, 1 h, to rt, 1 h

180

Substrates		Product	Yield (%)	Ref
Electrophile E1	Electrophile E2			
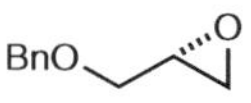 BnO~ epoxide	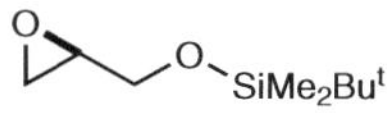 O epoxide–O–SiMe$_2$Bu^t	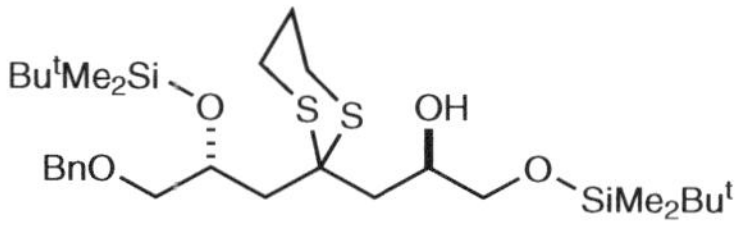 Bu^{t}Me$_2$Si–O, dithiane, OH, BnO...O–SiMe$_2$Bu^t	56	[117]
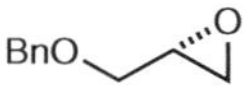 BnO~ epoxide	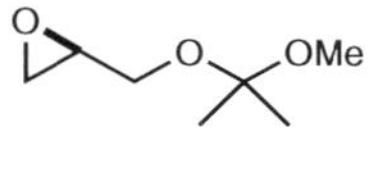 O epoxide–O–C(Me)$_2$–OMe	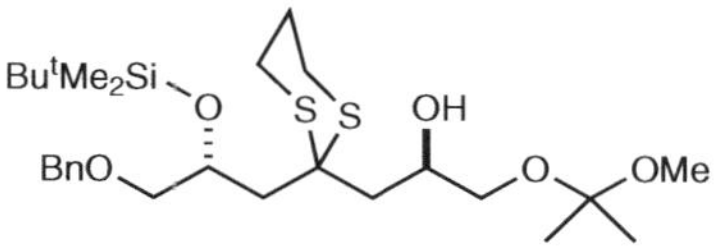 Bu^{t}Me$_2$Si–O, dithiane, OH, BnO...O–C(Me)$_2$–OMe	74	[117]
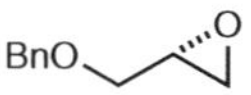 BnO~ epoxide	 Br~Ph	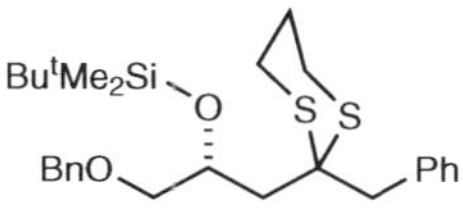 Bu^{t}Me$_2$Si–O, dithiane, BnO...Ph	62	[117]
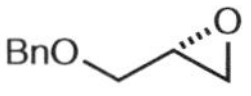 BnO~ epoxide	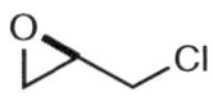 O epoxide~Cl	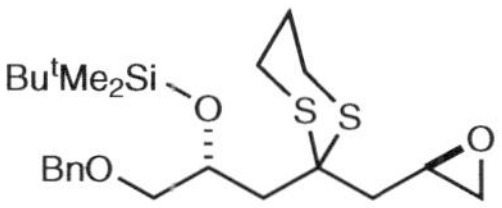 Bu^{t}Me$_2$Si–O, dithiane, BnO...epoxide	60	[117]
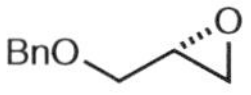 BnO~ epoxide	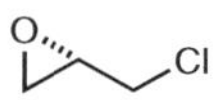 O epoxide~Cl	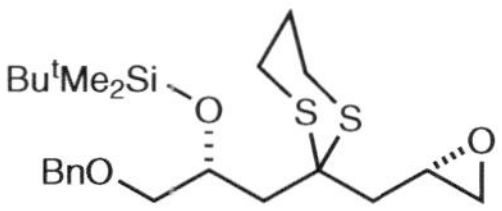 Bu^{t}Me$_2$Si–O, dithiane, BnO...epoxide	71[a]	[117]
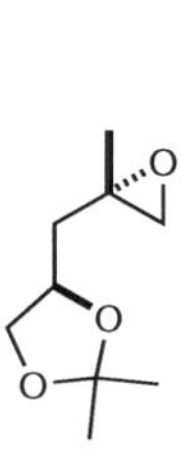 (epoxide with dioxolane)	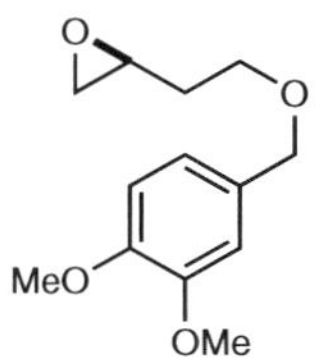 O epoxide~~O–CH$_2$–Ar(MeO)(OMe)	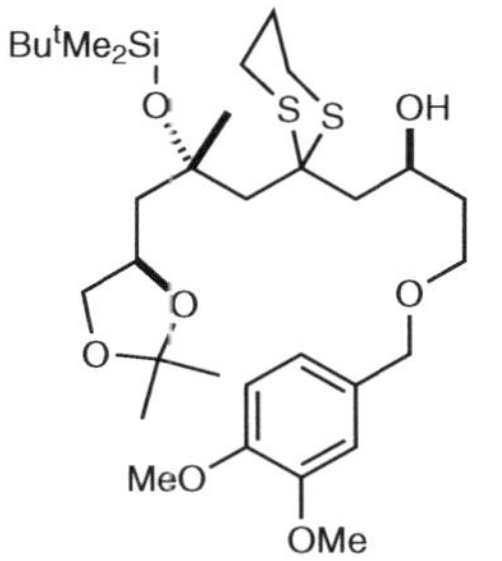 Bu^{t}Me$_2$Si–O, dithiane, OH, (macrocyclic product) MeO, OMe	59	[117]

[a] 1 equiv of electrophile E2 was used.

2-Silyl-functionalized vinyloxiranes bearing a spiro 1,3-dithian-2-yl group at the 2-position are also suited to undergoing 1,4-Brook rearrangements after S$_N$2′ oxirane ring opening with organolithium or organocuprate nucleophiles (Scheme 75).[118] The resulting alkox-

for references see p 445

416 Domino Transformations **1.5** Non-Radical Skeletal Rearrangements

ides undergo 1,4-Brook rearrangement in the presence of hexamethylphosphoric triamide and the newly formed dithioacetal-stabilized carbanions are captured by electrophiles. It is noteworthy that the *trans*-disubstituted vinyloxiranes yield a mixture of *E*- and *Z*-isomers, whereas *cis*-disubstituted analogues give excellent *E* selectivity. When lithiated dithianes are reacted with trisubstituted vinyloxiranes, excellent *E* selectivity is encountered; however, the reaction cannot be driven to completion and the Brook and non-Brook rearranged products are isolated in a 1:1 mixture.

Scheme 75 A 2-Silyl-Functionalized Vinyloxirane in Cascade Reactions with 1,4-Brook Rearrangement[118]

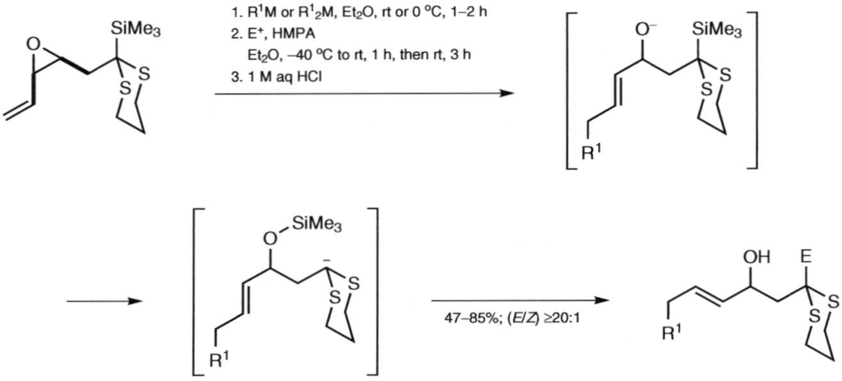

M = Li, CuLi; E+ = H₂C=CHCH₂Br, BnBr, Me₂C=CHCH₂Br; E = CH₂CH=CH₂, Bn, CH₂CH=CMe₂

1.5.3.3.2 **1,4-Brook Rearrangement with Dihalosilyl-Substituted Methyllithium**

Doubly halogenated trialkylsilyl-substituted methanes are also well suited to undergo 1,4-Brook rearrangement after lithiation and oxirane ring opening. For example, when a [(*tert*-butyldimethylsilyl)dihalomethyl]lithium is reacted with oxiranes, 1,4-silyl migration provides a new carbanion **181**, which smoothly reacts with a second electrophile such as iodomethane, benzaldehyde, or formic esters (Scheme 76).[119]

Scheme 76 [(*tert*-Butyldimethylsilyl)dihalomethyl]lithiums as Dianion Equivalents in Cascade Reactions[119]

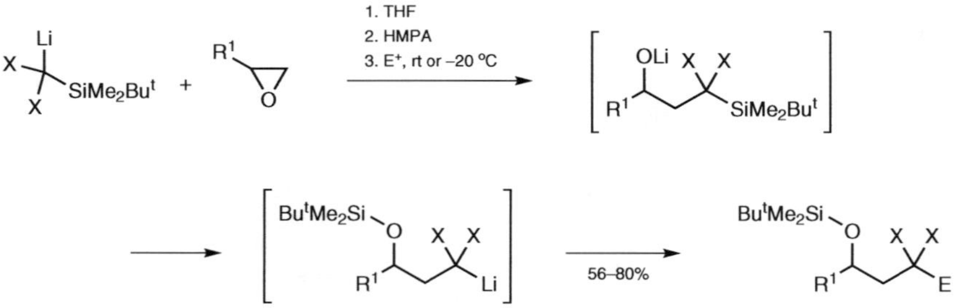

181

R¹ = H, Me; X = Cl, Br; E+ = MeI, PhCHO, HCO₂iPr; E = Me, CH(OH)Ph, CHO

1.5.3.3.3 1,4-Brook Rearrangement with Allylsilanes

Lithiated allylsilanes or in situ formed lithiated allylsilanes are also well suited to undergo 1,4-Brook rearrangements and consequently they can also serve as dianion equivalents in reactions with oxiranes. The allyl group serves as an anion-stabilizing "substituent" after 1,4-silyl migration to the lithiated alkoxide formed after oxirane ring opening. For example, lithiated bis-silylpropene **183** is generated after 1,4-Brook rearrangement of alkoxide **182** but it does not undergo homo-Peterson reaction to give the vinylcyclopropane **184** or the vinylogous cyclopentane product **185** (Scheme 77).[120] The anion **183** can, however, be trapped by an external electrophile; protonation or methylation yields the two regioisomeric silylated alkenols.

Scheme 77 A 1,3-Bis-silylpropene as a Precursor for Cascade Reactions with 1,4-Brook Rearrangement as the Key Step, and the (Not Formed) Homo-Peterson and Vinylogous Products[120]

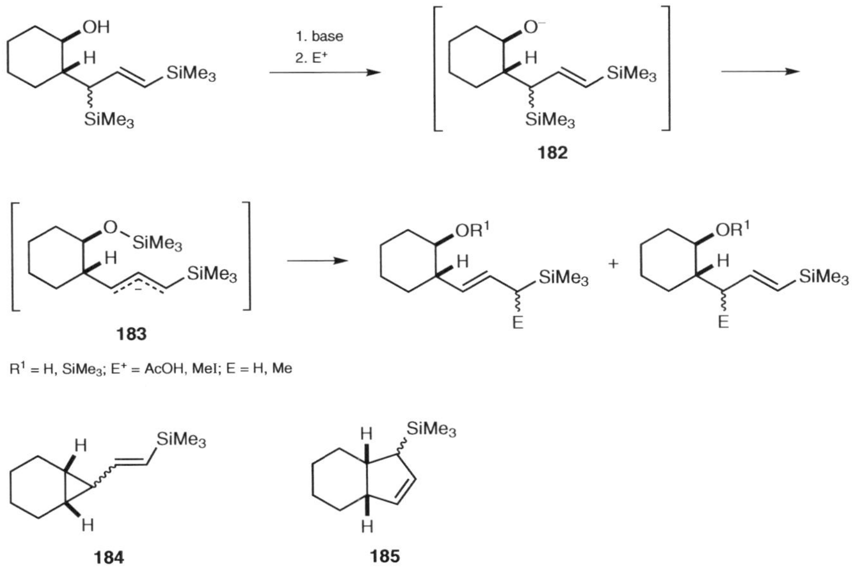

The triphenylsilyl group is particularly well suited to undergo 1,4-Brook rearrangements (Scheme 78 and Scheme 79).[121] For example, the allyllithium species **186** reacts in a one-pot procedure with oxiranes and electrophiles to yield the corresponding adducts as mixtures of regioisomers.

for references see p 445

Scheme 78 Cascade C–C Formation after Nucleophilic Addition of Lithiated Allyltriphenylsilane to Oxiranes[121]

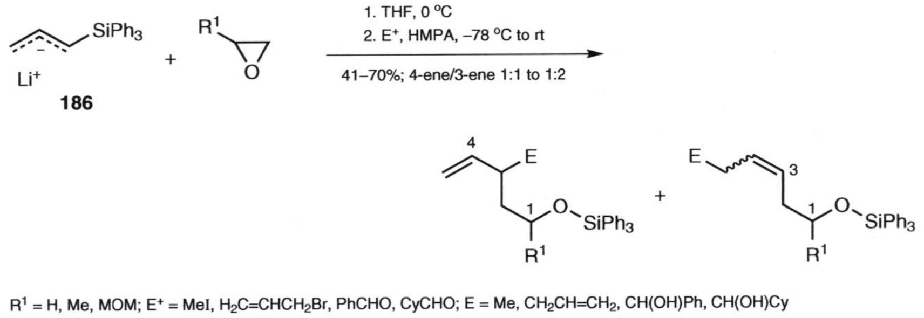

SiPh3
Li+
186
R1
O
1. THF, 0 °C
2. E+, HMPA, –78 °C to rt
41–70%; 4-ene/3-ene 1:1 to 1:2
4
E
1
O
SiPh3
R1
E
3
1
O
SiPh3
R1
R1 = H, Me, MOM; E+ = MeI, H2C=CHCH2Br, PhCHO, CyCHO; E = Me, CH2CH=CH2, CH(OH)Ph, CH(OH)Cy

The regiocontrol is improved in favor of the alk-4-enol product if lithiated (*E*)-prop-1-ene-1,3-diylbis(triphenylsilane) is employed and the intermediate anion is treated with electrophiles (Scheme 79).[121]

Scheme 79 Cascade C–C Formation with Lithiated (*E*)-Prop-1-ene-1,3-diylbis(triphenylsilane)[121]

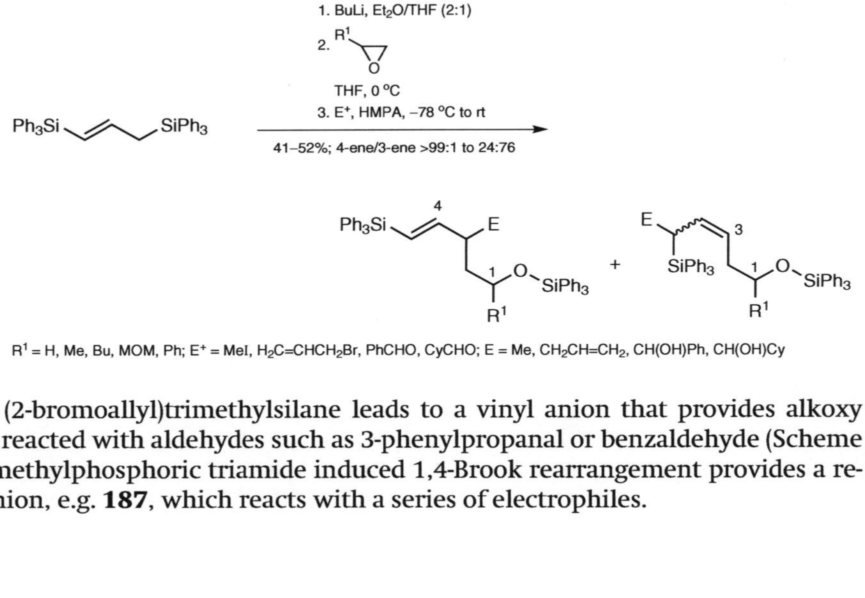

1. BuLi, Et2O/THF (2:1)
2. R1
O
THF, 0 °C
3. E+, HMPA, –78 °C to rt
Ph3Si
SiPh3
41–52%; 4-ene/3-ene >99:1 to 24:76
Ph3Si
4
E
1
O
SiPh3
R1
E
3
SiPh3
1
O
SiPh3
R1
R1 = H, Me, Bu, MOM, Ph; E+ = MeI, H2C=CHCH2Br, PhCHO, CyCHO; E = Me, CH2CH=CH2, CH(OH)Ph, CH(OH)Cy

Lithiation of (2-bromoallyl)trimethylsilane leads to a vinyl anion that provides alkoxy anions when reacted with aldehydes such as 3-phenylpropanal or benzaldehyde (Scheme 80).[122] Hexamethylphosphoric triamide induced 1,4-Brook rearrangement provides a reactive allyl anion, e.g. **187**, which reacts with a series of electrophiles.

1.5.3 Brook Rearrangement as the Key Step in Domino Reactions **419**

Scheme 80 [3-(Trimethylsilyl)prop-1-en-2-yl]lithium as Precursor in Cascade Reactions with 1,4-Brook Rearrangement[122]

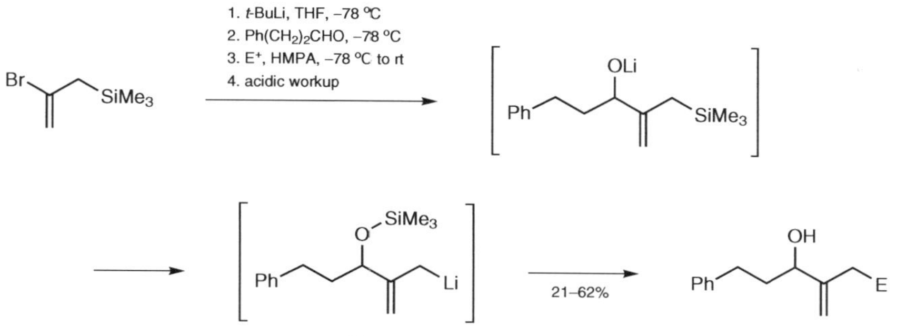

1. *t*-BuLi, THF, –78 °C
2. Ph(CH$_2$)$_2$CHO, –78 °C
3. E$^+$, HMPA, –78 °C to rt
4. acidic workup

187

E$^+$ = H$_2$C=CHCH$_2$Br, BnBr, MeI, PhC(O)Me; E = CH$_2$CH=CH$_2$, Bn, Me, C(OH)(Me)Ph

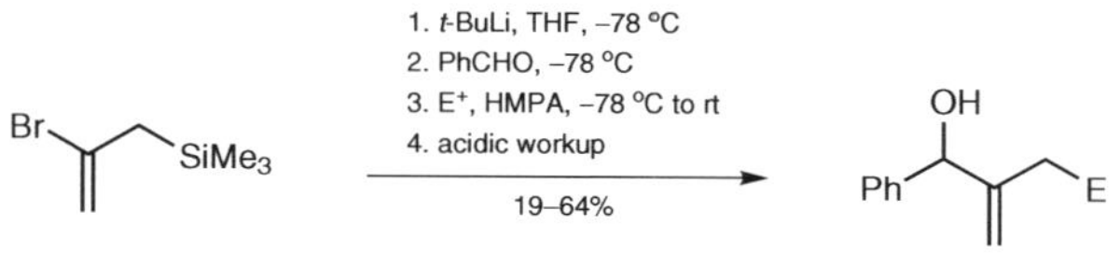

1. *t*-BuLi, THF, –78 °C
2. PhCHO, –78 °C
3. E$^+$, HMPA, –78 °C to rt
4. acidic workup

19–64%

E$^+$ = H$_2$C=CHCH$_2$Br, BnBr, BuBr, (PhS)$_2$, 4-ClC$_6$H$_4$CHO, 4-TolCHO, *t*-BuCHO, PhC(O)Me, EtC(O)Me, cyclohex-2-en-1-one;
E = CH$_2$CH=CH$_2$, Bn, Bu, SPh, 4-ClC$_6$H$_4$CH(OH), CH(OH)4-Tol, CH(OH)*t*-Bu, C(OH)MePh, C(OH)Et,

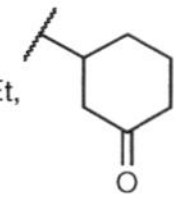

α-Disilylmethyl enals **188** are effective precursors in three-component domino reactions (Scheme 81).[123] The starting materials are accessible from the corresponding 3,3-disilyl aldehydes by Mannich reaction. After nucleophilic addition of the organolithium to the aldehyde, the lithium alkoxide formed undergoes a 1,4-Brook rearrangement to yield a siloxy-substituted allyl anion **189**, which can be captured by electrophiles to provide *E*-configured vinylsilanes **190** and α-addition products **191**, often with excellent regio- and stereocontrol.

for references see p 445

Scheme 81 α-Disilylmethyl Enal Precursors in Cascade Reactions with 1,4-Brook Rearrangement[123]

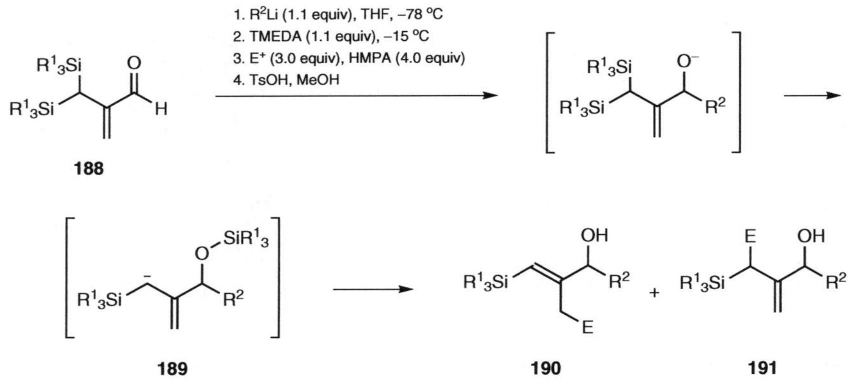

R¹	R²	E⁺	E	Ratio[a]		Ref
				(190/191)	**(E/Z)**	
Et	Bu	4-MeOC₆H₄CH₂Br	4-MeOC₆H₄CH₂	72:28	23:1	[123]
Et	Bu	(E)-PhCH=CHCH₂Br	(E)-CH₂CH=CHPh	78:22	100:0	[123]
Et	Bu	(PhS)₂	SPh	100:0	65:35	[123]
Me	Bu	PhCHO	CH(OH)Ph	100:0[b]	100:0	[123]
Me	Bu	4-MeOC₆H₄CHO	4-MeOC₆H₄CH(OH)	100:0[b]	100:0	[123]
Me	Bu	iPrCHO	CH(OH)iPr	100:0[b]	100:0	[123]
Me	Bu	t-BuCHO	CH(OH)t-Bu	100:0[c]	100:0	[123]
Me	Bu	PhC(O)Me	C(OH)MePh	100:0[d]	100:0	[123]
Me	Bu	Ph₂CO	C(OH)Ph₂	100:0	96:4	[123]
Me	cyclohex-1-enyl	Ph₂CO	C(OH)Ph₂	100:0	91:9	[123]
Et	C≡CPr	BnBr	Bn	70:30	13:1	[123]
Me	Ph	BnBr	Bn	72:28	17:1	[123]
Me	2-pyridyl	BnBr	Bn	100:0	100:0	[123]
Me	2-furyl	Ph₂CO	C(OH)Ph₂	100:0	100:0	[123]

[a] Ratios were determined by ¹H NMR spectroscopy.
[b] dr 2:1.
[c] dr 85:15.
[d] dr 58:42.

The allylsilane moiety can also be formed in situ before the 1,4-Brook rearrangement is initiated. For instance, the one-pot preparation of siloxy-substituted dienes relies on the combination of a 1,4-Brook rearrangement and Wittig alkenation (Scheme 82).[124] It uses β-silyl-substituted enones as the starting substrate. After Michael addition of tributylphosphine, intramolecular 1,4-silyl migration leads to a phosphorus ylide **192**, which initiates the alkenation in the presence of various aldehydes.

1.5.3 Brook Rearrangement as the Key Step in Domino Reactions **421**

Scheme 82 Cascade Reactions Based on 1,4-Brook Rearrangement and Wittig Alkenation[124]

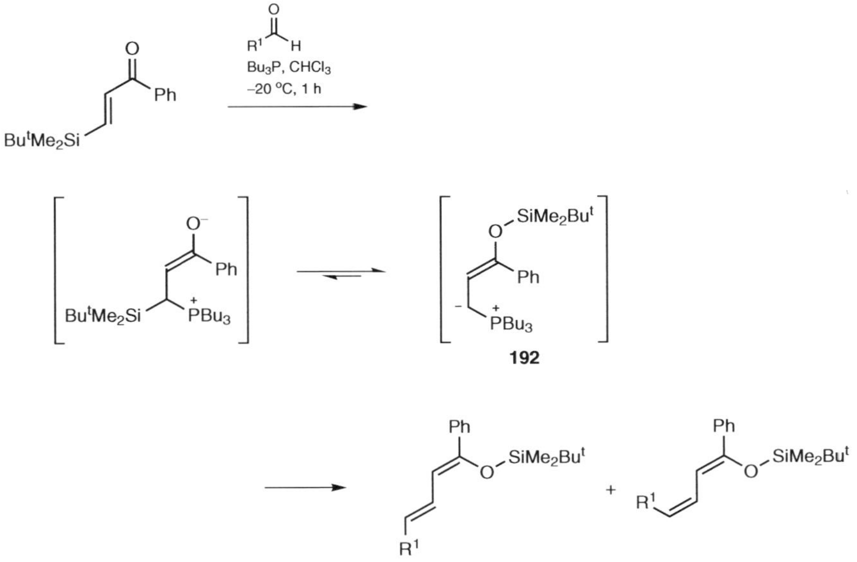

R^1	Ratio (E/Z)	Yield[a] (%)	Ref
Ph	69:31	quant	[124]
$4\text{-}ClC_6H_4$	71:29	92	[124]
$4\text{-}F_3CC_6H_4$	71:29	98	[124]
$4\text{-}O_2NC_6H_4$	–[b]	96	[124]
$4\text{-}MeOC_6H_4$	67:33	89	[124]
$2\text{-}MeOC_6H_4$	62:38	97	[124]
$2,6\text{-}Me_2C_6H_3$	82:18[c]	39	[124]
2-naphthyl	78:22	66	[124]
CH=CHPh	–[d]	82	[124]
Pr	74:26	87	[124]
$(CH_2)_2Ph$	72:21[e]	quant	[124]
Me	72:21	quant[f]	[124]

[a] All reactions were carried out using 1 equiv of aldehyde and 1 equiv of phosphine. All ratios were determined by ^1H NMR spectroscopy.
[b] The ratio was hard to determine due to rapid isomerization under the reaction conditions.
[c] A small amount of a second isomer was detected.
[d] Ratio was not determined.
[e] A second geometric isomer (7%) was observed.
[f] Excess aldehyde (3.0 equiv) was used.

for references see p 445

422 Domino Transformations **1.5** Non-Radical Skeletal Rearrangements

1,5-Amino alcohol derivatives can be obtained by ring opening of *tert*-butylsulfonyl-protected aziridines with the lithiated dithiane **193** (Scheme 83). The latter is generated from deprotonation of a carbinol located at the β-position of the dithioacetal group followed by 1,4-Brook rearrangement.[125]

Scheme 83 Cascade Reactions Initiated by Deprotonation of a Carbinol Located at the β-Position of a Dithioacetal Group[125]

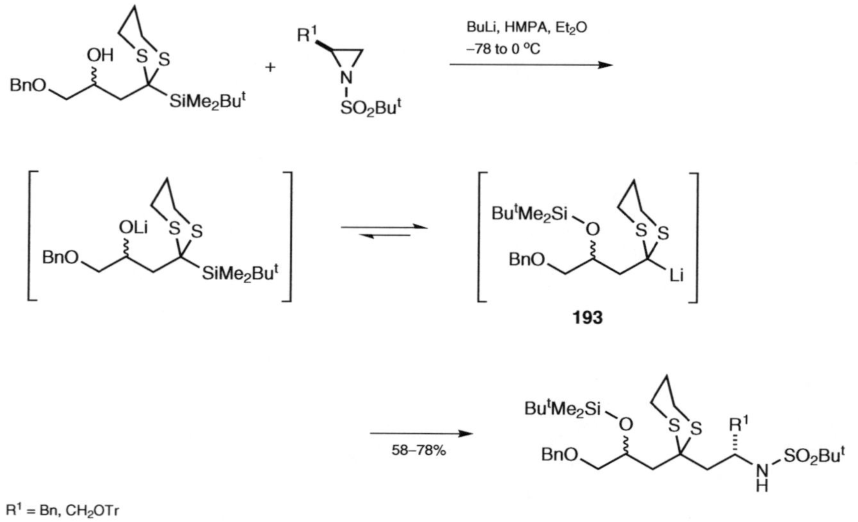

R¹ = Bn, CH₂OTr

1.5.3.3.4 1,4-Brook Rearrangement with Silylated Benzaldehydes

The tricarbonylchromium complex of 2-(trimethylsilyl)benzaldehyde undergoes efficient stereoselective bis-functionalization via carbonyl addition, 1,4-Brook rearrangement of an alkoxide, and alkylation of the resulting aryl anion to afford products **194** and **195** (Scheme 84).[126] This strategy also gives polycyclic and spirocyclic compounds **196**, and the cascade in this case involves aldol addition, 1,4-Brook rearrangement, and cyclization.[127]

1.5.3 Brook Rearrangement as the Key Step in Domino Reactions **423**

Scheme 84 Functionalization of Chromium Carbonyl Activated *ortho*-Silylated Benzaldehyde through 1,4-Brook Rearrangement[126,127]

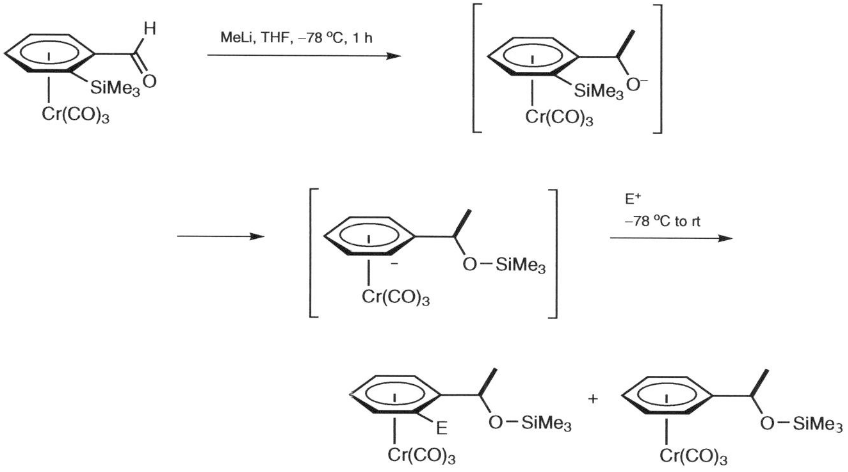

E+	E	Yield (%)		Ref
		194	**195**	
$H_2C=CHCH_2Br$	$CH_2CH=CH_2$	58	36	[126]
BnBr	Bn	52	33	[126]
MeI	Me	65	15	[126]
PhCHO	CH(OH)Ph	72	<5	[126]
Ph_2CO	$C(OH)Ph_2$	77	<5	[126]
$(PhS)_2$	SPh	83	<5	[126]
BrF_2CCF_2Br	Br	66	<5	[126]

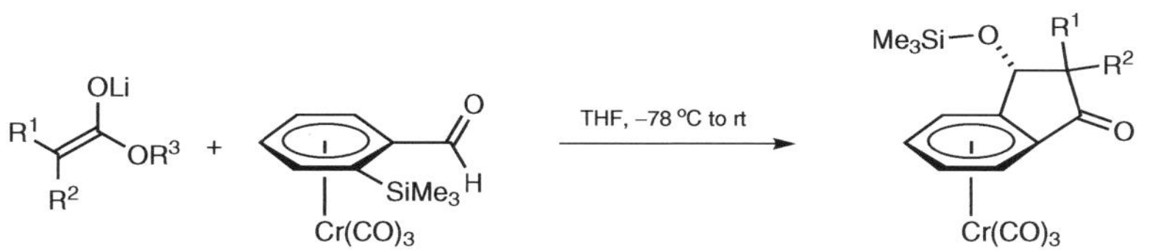

R¹	R²	R³	Yield (%)	Ref
H	H	Et	20	[127]
Me	Me	Me	70	[127]
$(CH_2)_4$		Me	72	[127]
$(CH_2)_5$		Me	84	[127]

ortho-Silylated benzaldehydes effectively act as linchpins for 1,4-silyl migrations after addition of various nucleophiles to the carbonyl group and formation of the alkoxide anion, a prerequisite for the 1,4-Brook rearrangement (Scheme 85).[128] This process paves the way for *ortho*-functionalization of benzaldehydes with various electrophiles.

for references see p 445

Scheme 85 ortho-Functionalization of Silylated Benzaldehydes through 1,4-Brook Rearrangement[128]

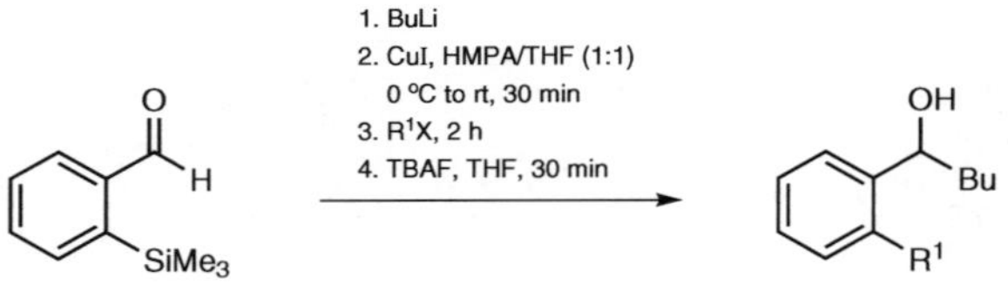

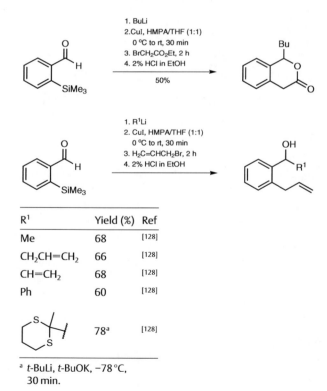

R[1]	X	Yield (%)	Ref
CH$_2$CH=CH$_2$	Br	69	[128]
Bn	Br	55	[128]
CH$_2$C≡CH	Br	58[a]	[128]
CH$_2$C(Me)=CH$_2$	Cl	64	[128]
CH=CH$_2$	Br	56[b]	[128]
SPh	SPh	71	[128]
SnBu$_3$	Cl	65	[128]

[a] Reaction time was 12 h.
[b] In the alkylation step, vinyl bromide, Pd(PPh$_3$)$_4$ (3 mol%), and THF were used.

R[1]	Yield (%)	Ref
Me	68	[128]
CH$_2$CH=CH$_2$	66	[128]
CH=CH$_2$	68	[128]
Ph	60	[128]
(dithiane)	78[a]	[128]

[a] t-BuLi, t-BuOK, −78 °C, 30 min.

2-(*tert*-Butyldimethylsilyl)furan-3-carbaldehyde (**197**) and 2-(*tert*-butyldimethylsilyl)thiophene-3-carbaldehyde (**199**) serve as bifunctional precursors for a solvent-controlled [HMPA/Et$_2$O (1:1) in the case of the furan derivative and DMPU/THF (1:1) in the case of

1.5.3 Brook Rearrangement as the Key Step in Domino Reactions **425**

the thiophene derivative] C(sp²)—O 1,4-Brook rearrangement of the alkoxide (e.g., **198**). Again, organolithium compounds (Scheme 86) and α-disubstituted ester enolates are appropriate nucleophiles (Scheme 87 and Scheme 88).[129] In the latter case, 2-(*tert*-butyldimethylsilyl)thiophene-3-carbaldehyde can give annulated products **200**, whereas with 2-(*tert*-butyldimethylsilyl)furan-3-carbaldehyde no cyclization takes place.

Scheme 86 *ortho* Functionalization of Silylated 3-Formylfurans and -thiophenes through 1,4-Brook Rearrangement[129]

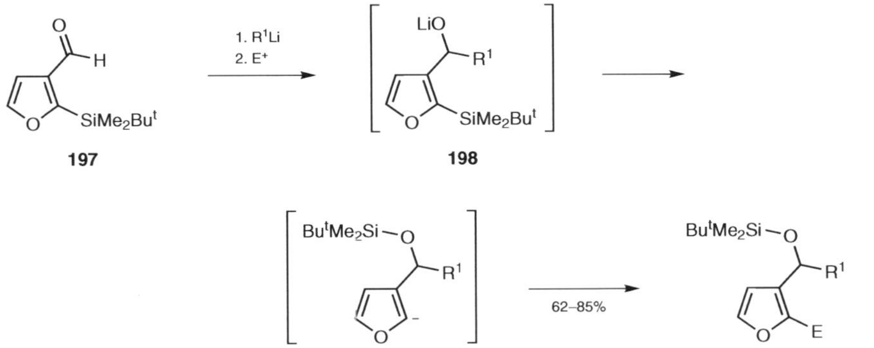

R¹ = Bu, Ph, 3-thienyl; E⁺ = MeI, iPrCHO, *t*-BuCHO, PhCHO, Ph₂CO;
E = Me, CH(OH)iPr, CH(OH)*t*-Bu, CH(OH)Ph, C(OH)Ph₂

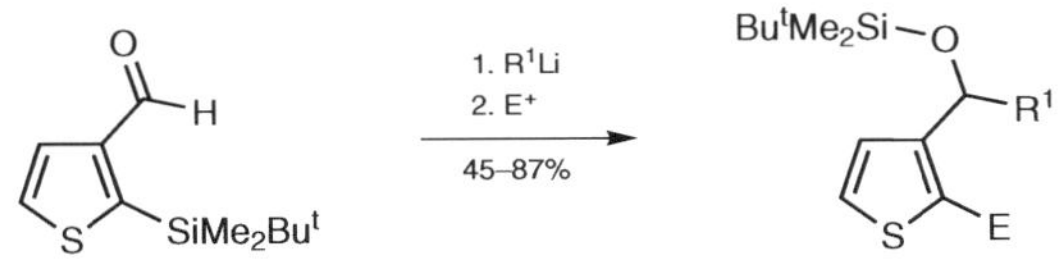

R¹ = Bu, Ph, 3-thienyl; E⁺ = MeI, iPrCHO, *t*-BuCHO, PhCHO, Ph₂CO;
E = Me, CH(OH)iPr, CH(OH)*t*-Bu, CH(OH)Ph, C(OH)Ph₂

Scheme 87 Annulation of 2-(*tert*-Butyldimethylsilyl)thiophene-3-carbaldehyde Promoted by 1,4-Brook Rearrangement[129]

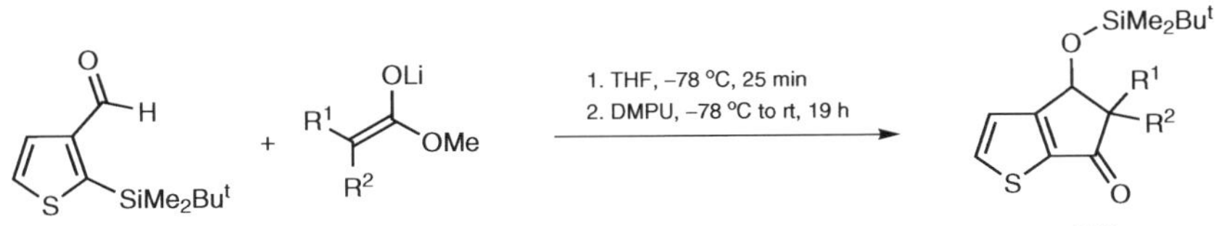

R¹	R²	Yield (%)	Ref
Me	Me	69	[129]
(CH₂)₄		88	[129]
(CH₂)₅		73	[129]

for references see p 445

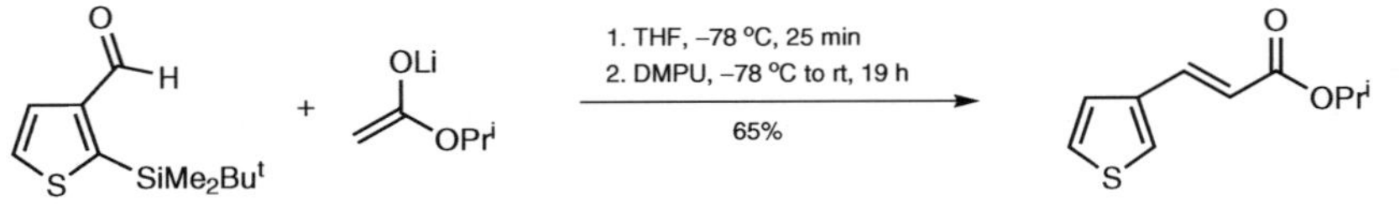

Domino Transformations 1.5 Non-Radical Skeletal Rearrangements

Scheme 88 Synthesis of 3-Substituted Furans[129]

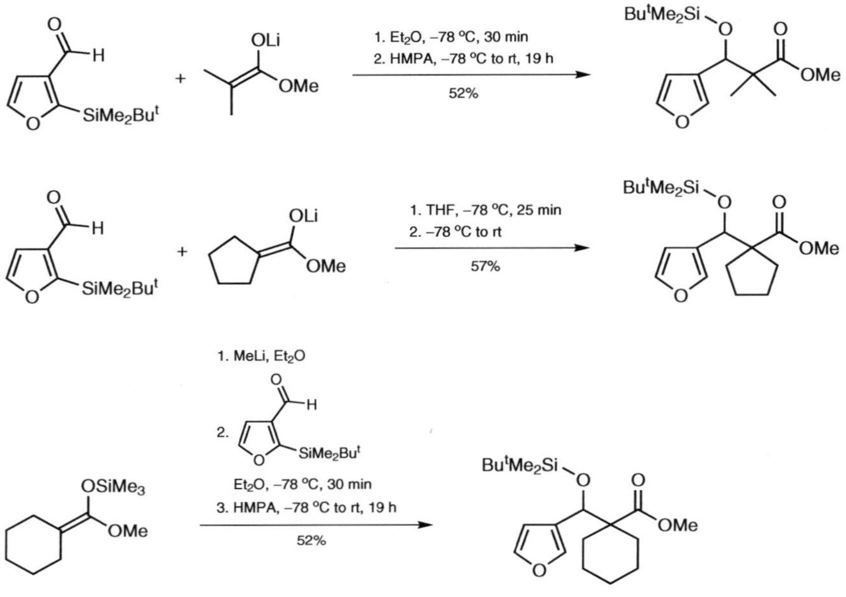

1.5.3.3.5 **1,4-Brook Rearrangement with Vinylsilanes**

Related to these 1,4-Brook rearrangements are reactions using vinylsilanes as precursors which lead to intermediate vinylmetal species that can be trapped by electrophiles. Geminal bis(silyl)enals are versatile precursors for this kind of cascade reaction (Scheme 89).[130] Overall, the process allows for the stereoselective preparation of trisubstituted vinylsilanes **201** and **202**. The cascade features a copper(I) cyanide promoted C(sp²)—O 1,4-silyl migration, and a large variety of nucleophiles and electrophiles can be employed.

Scheme 89 Bis(silyl)enals as Versatile Precursors for 1,4-Brook Rearrangement[130]

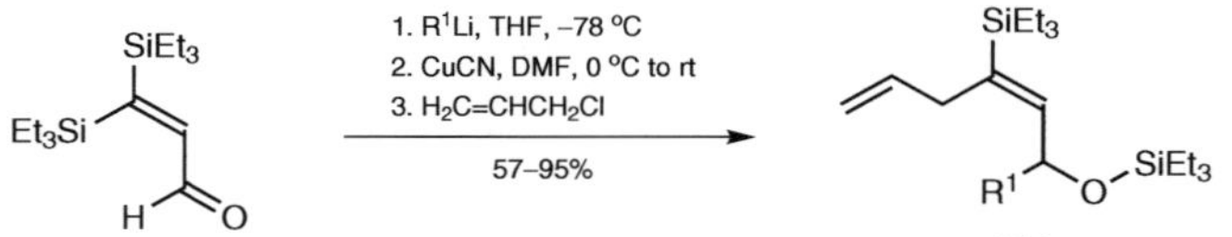

R^1 = Me, CH₂TMS, cyclohex-1-enyl, CH(Ph)CH=CH₂, 2-furyl, Ph, 2-pyridyl, 2-thienyl, C≡CPr, C≡CCH₂OTHP

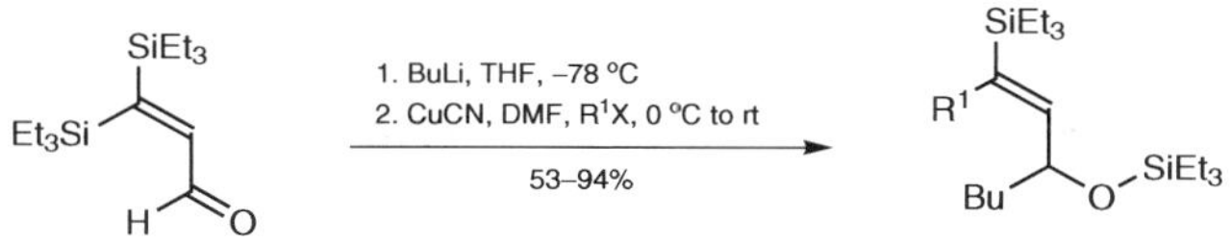

1.5.3 Brook Rearrangement as the Key Step in Domino Reactions

$R^1 = (E)\text{-}CH_2CH=CHSiEt_3, CH_2C(Me)=CH_2, CH_2C(iPr)=CH_2, CH_2C(Ph)=CH_2, CH_2C(CO_2Me)=CH_2,$
$CH_2C(CH_2TMS)=CH_2, CH_2C(Br)=CH_2, CH_2C\equiv CSiEt_3; X = OTs, OMs, Br$

Various Z-configured γ-trimethylsilyl allylic alcohols undergo a copper(I) *tert*-butoxide promoted 1,4-C(sp^2)—O silyl migration reaction (Scheme 90).[131] The intermediates **203** react with various allylic halides to afford the corresponding dienyl alcohols **204** with retention of configuration of the double bond. Aryl and vinyl halides can be attached to these intermediates through palladium-catalyzed cross-coupling chemistry in the final step (Scheme 91).[132]

Scheme 90 Ipso Substitution of 3-Silyl-Substituted Allyl Alcohols with 1,4-Brook Rearrangement as the Key Step[131]

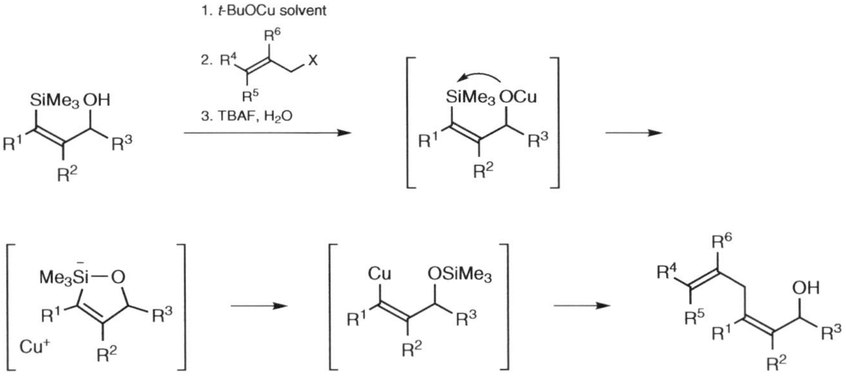

R^1	R^2	R^3	R^4	R^5	R^6	X	Solvent	Time (h)	Yield (%)	Ref
(CH$_2$)$_2$Ph	(CH$_2$)$_4$		H	H	Me	Cl	THF	20	80	[131]
(CH$_2$)$_2$Ph	(CH$_2$)$_4$		H	H	Me	Cl	DMF	2	84	[131]
(CH$_2$)$_2$Ph	(CH$_2$)$_4$		H	H	H	Cl	THF	16	88	[131]
(CH$_2$)$_2$Ph	(CH$_2$)$_4$		H	H	H	Br	THF	16	89	[131]
(CH$_2$)$_2$Ph	Me	Et	H	H	H	Cl	THF	14	80	[131]
(CH$_2$)$_2$Ph	Me	Et	H	H	Me	Cl	THF	21	85	[131]
(CH$_2$)$_2$Ph	Me	Et	Me	H	Me	Cl	THF	16	66	[131]
(CH$_2$)$_2$Ph	Me	Et	Me	H	Me	Cl	DMF	4	75	[131]
iBu	H	Ph	H	H	H	Cl	THF	30	51	[131]
iBu	H	Ph	H	H	H	Cl	DMF	10	76	[131]
Ph	H	H	H	H	Me	Cl	DMF	15	74	[131]
(CH$_2$)$_2$Ph	H	H	H	H	Me	Cl	THF	20	46	[131]
(CH$_2$)$_2$Ph	H	H	H	H	Me	Cl	DMF	12	73	[131]

for references see p 445

Scheme 91 Palladium-Catalyzed Ipso Substitution of 3-Silyl-Substituted Allyl Alcohols[132]

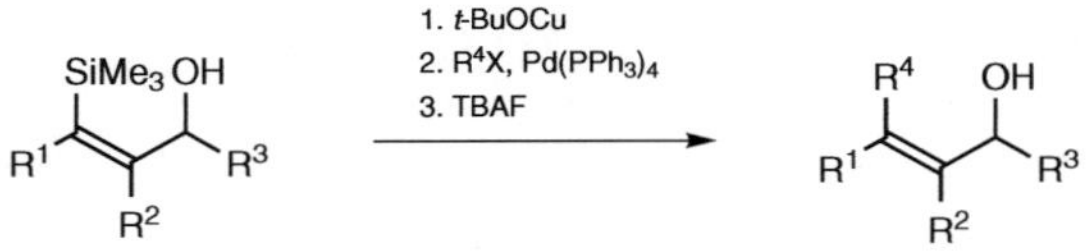

R^1	R^2	R^3	R^4	X	Yield[a] (%)	Ref
$(CH_2)_2Ph$	$(CH_2)_4$		Ph	I	78	[132]
$(CH_2)_2Ph$	$(CH_2)_4$		4-MeOC_6H_4	I	60	[132]
$(CH_2)_2Ph$	$(CH_2)_4$		$4\text{-O}_2NC_6H_4$	I	67	[132]
$(CH_2)_2Ph$	Me	Et	Ph	I	77	[132]
iBu	H	Ph	Ph	I	70	[132]
$(CH_2)_2Ph$	H	H	Ph	I	77	[132]
$(CH_2)_2Ph$	H	H	4-MeOC_6H_4	I	66	[132]
$(CH_2)_2Ph$	H	H	$4\text{-O}_2NC_6H_4$	I	63	[132]
$(CH_2)_2Ph$	$(CH_2)_4$		$CH{=}CH_2$	Br	71[b]	[132]
$(CH_2)_2Ph$	$(CH_2)_4$		$(E)\text{-CH}{=}CHPh$	Br	72	[132]
$(CH_2)_2Ph$	$(CH_2)_4$		$CH{=}CHMe$	Br	61[b,c]	[132]
iBu	H	Ph	$(E)\text{-CH}{=}CHPh$	Br	64[b]	[132]
$(CH_2)_2Ph$	H	H	$(E)\text{-CH}{=}CHPh$	Br	63	[132]

[a] The configurations of the double bonds were confirmed by NOE experiments.
[b] The product was isolated without final hydrolysis.
[c] Ratio (E/Z) 75:25 for starting allyl bromide; ratio (E/Z) 70:30 for product.

Bis-silylated allyl alcohols **205** undergo 1,4-silyl migration after deprotonation of the carbinol moiety using a copper alkoxide base (Scheme 92).[133] The resulting vinylcopper species **206** reacts with a wide range of electrophiles to furnish trisubstituted alkenes.

1.5.3 Brook Rearrangement as the Key Step in Domino Reactions

Scheme 92 1,5-Anion Relay/2,3-Wittig Rearrangement of Bis-silylated Allyl Alcohols[133]

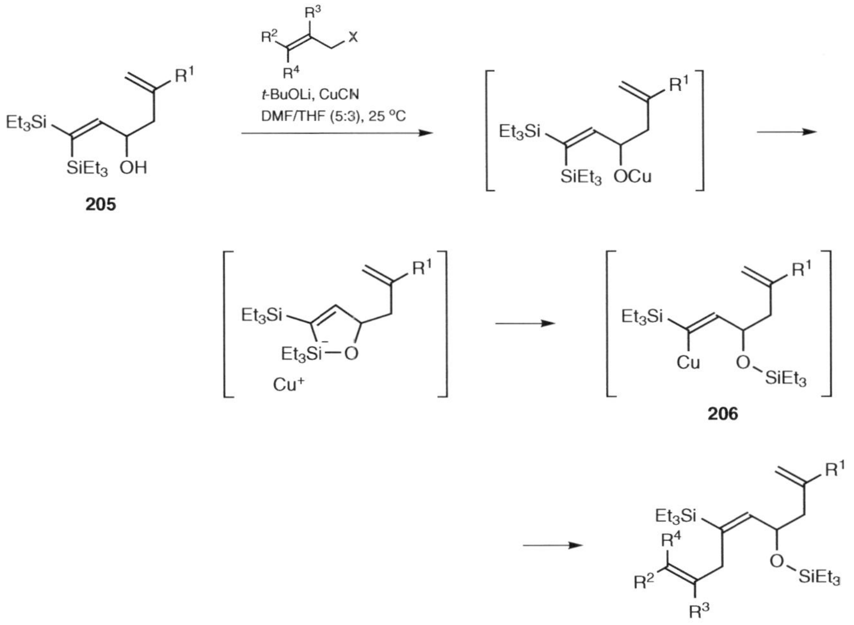

R¹	R²	R³	R⁴	X	Yield (%)	Ref
H	H	H	H	Cl	89	[133]
H	H	H	H	Br	76	[133]
H	H	H	H	OTs	80	[133]
H	H	Br	H	Br	69	[133]
H	H	Me	H	OTs	65	[133]
H	H	CH₂TMS	H	OTs	80	[133]
H	SiEt₃	H	H	OTs	68	[133]
H	Me	H	Me	Br	55	[133]
Me	H	H	H	Cl	75	[133]
iPr	H	H	H	Cl	83	[133]

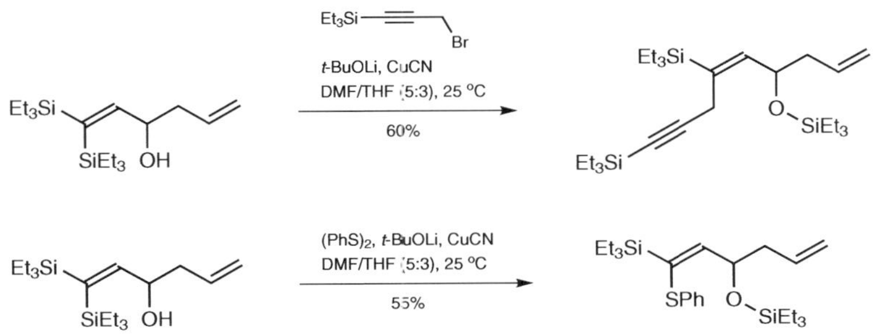

The observation that copper(I) salts promote silyl migrations from sp²-centers, especially when triphenylsilyl groups are involved, was further extended; 1-oxa-2-silacyclopentanes and -pentenes can be activated for this reaction by addition of organolithium species onto

for references see p 445

430   Domino Transformations   **1.5** Non-Radical Skeletal Rearrangements

silicon (Scheme 93).[134] In the presence of a copper source, the "ate" complex formed (e.g., **207**; supposedly in equilibrium with alkoxide **208**) opens up to an arylcopper species in the form of **209**, which is captured by various electrophiles.

Scheme 93   Substitution of 1-Oxa-2-silacyclopentane Derivatives[134]

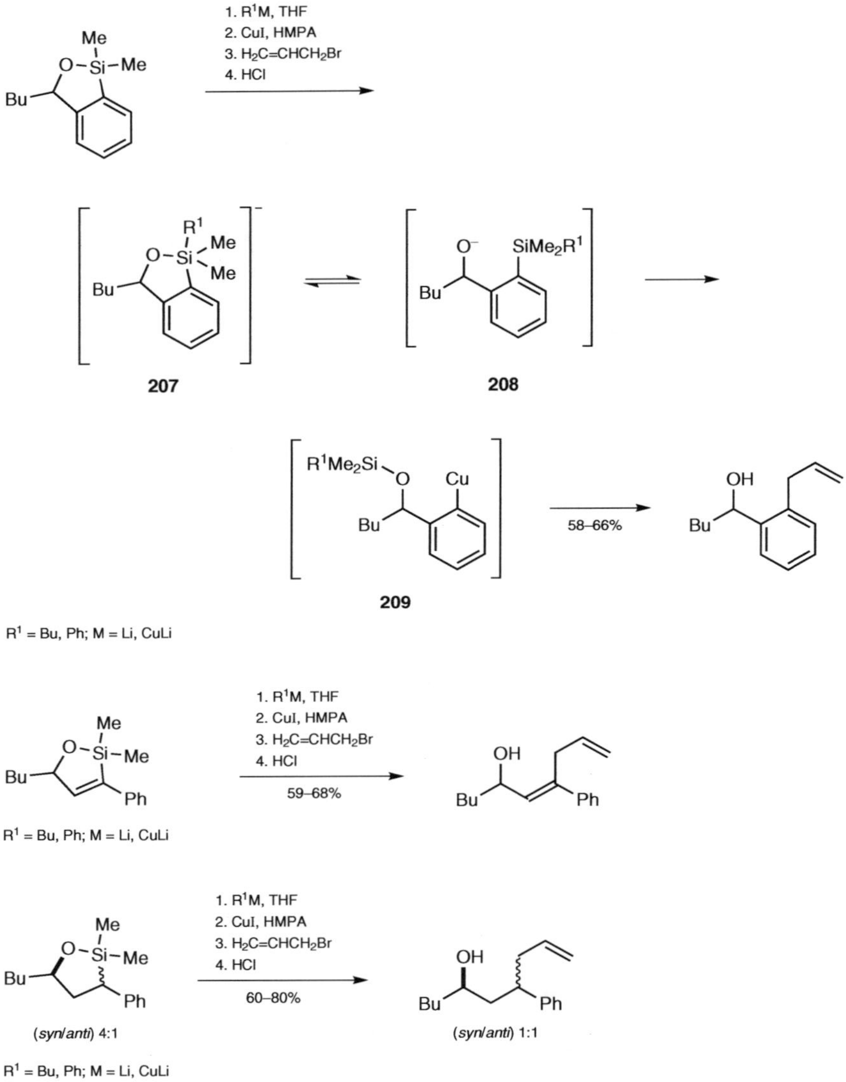

1.5.3.3.6   Sulfur to Oxygen Rearrangement

In a few instances, 1,4-Brook rearrangement can occur from heteroatoms such as sulfur to oxygen. A sulfur to oxygen 1,4-silyl migration of α-silylsulfanyl ketones (e.g., **210**) via enolate **211** yields Z-configured silyl enol thioethers with various electrophiles with excellent selectivity [ratio (Z/E) ≥95:5] (Scheme 94).[135] The Z-silyl enol thioethers formed are versatile substrates in Mukaiyama aldol reactions or Prins cyclizations.

1.5.3 Brook Rearrangement as the Key Step in Domino Reactions **431**

Scheme 94 Sulfur to Oxygen 1,4-Silyl Migration of α-Silylsulfanyl Ketones[135]

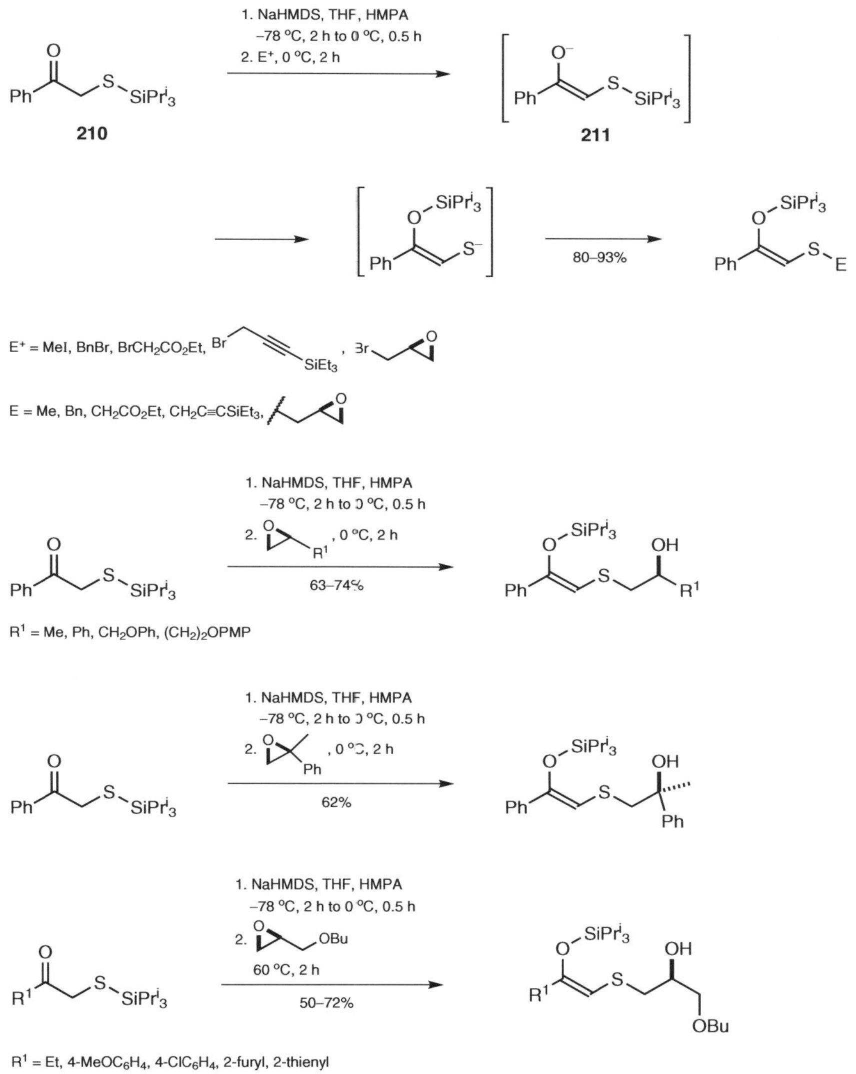

E^+ = MeI, BnBr, BrCH$_2$CO$_2$Et, Br⎯≡⎯SiEt$_3$, Br⎯△O

E = Me, Bn, CH$_2$CO$_2$Et, CH$_2$C≡CSiEt$_3$, △O

R^1 = Me, Ph, CH$_2$OPh, (CH$_2$)$_2$OPMP

R^1 = Et, 4-MeOC$_6$H$_4$, 4-ClC$_6$H$_4$, 2-furyl, 2-thienyl

1.5.3.3.7 **Retro-1,4-Brook Rearrangement**

Retro-1,4-Brook rearrangements have been encountered only in very few cases. For example, 1,3-dipolar cycloaddition of bis(trimethylsilyl) acetylenedicarboxylate with cyclooctyne yields the zwitterionic intermediate **212** which is stabilized in situ by a retro-1,4-silyl migration to yield lactone **213** (Scheme 95).[136] Cyclooct-1-en-5-yne reacts in a similar manner.

for references see p 445

432 Domino Transformations **1.5** Non-Radical Skeletal Rearrangements

Scheme 95 Retro-1,4-Brook Rearrangement Combined with 1,3-Dipolar Cycloaddition[136]

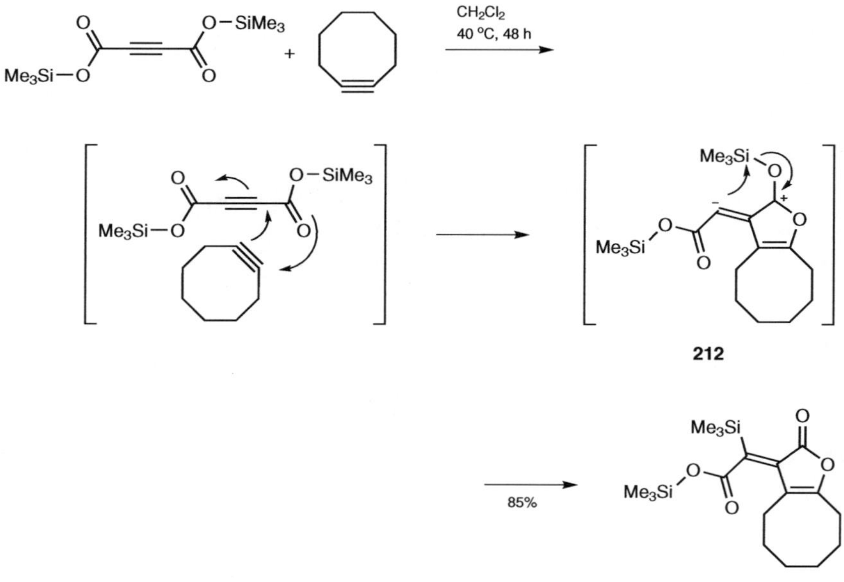

212

213

1.5.3.4 Applications in the Total Synthesis of Natural Products

Brook rearrangements embedded in cascade sequences as part of total syntheses of complex natural products reveal the power of Brook rearrangements in terms of general applicability and chemoselectivity. Cascade reactions of C1 dianion equivalents with two different oxiranes as electrophiles are particularly useful for applications in the total synthesis of polyketide backbones, because this cascade sequence provides products with three oxygen functionalities that are positioned in a 1,3-relationship. Importantly, the configuration of two carbinols can be controlled by the choice of the absolute configuration embedded in the two different terminal oxiranes. Brook rearrangements are also used in the synthesis of terpenes and alkaloids.

1.5.3.4.1 The 1,2-Brook Rearrangement in Natural Product Synthesis

In the enantioselective synthesis of scalarenedial (**214**), reaction of an acylsilane with an α-sulfonyl lithium derivative gives a silyl enol ether via a sequence of carbonyl addition, 1,2-Brook rearrangement, and β-sulfonyl elimination (Scheme 96).[137]

1.5.3 Brook Rearrangement as the Key Step in Domino Reactions **433**

Scheme 96 1,2-Brook Rearrangement in the Total Synthesis of Scalarenedial[137]

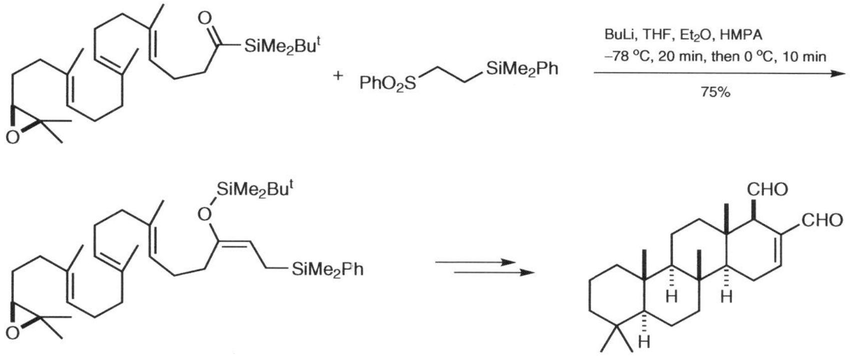

214 scalarenedial

1.5.3.4.2 The 1,4-Brook Rearrangement in Natural Product Synthesis

1.5.3.4.2.1 Synthesis of Polyketides

The 1,4-Brook rearrangement has seen the widest use in the total synthesis of polyketides using lithiated silyldisulfanyl-substituted methane as the dianion equivalent (see Section 1.5.3.3.1). The Smith group has had the largest impact on implementing the 1,4-Brook rearrangement in natural product synthesis. For instance, they utilized this cascade sequence to prepare a series of key intermediates in the total synthesis of several complex polyketide natural products such as roflamycoin (**219**); lithiated *tert*-butyl(1,3-dithian-2-yl)dimethylsilane (**215**) and the two epoxides **216** and **217** are utilized to generate the 1,3-polyol fragment **218** (Scheme 97).[117]

for references see p 445

434 Domino Transformations **1.5** Non-Radical Skeletal Rearrangements

Scheme 97 1,4-Brook Rearrangement in the Total Synthesis of Roflamycoin[117]

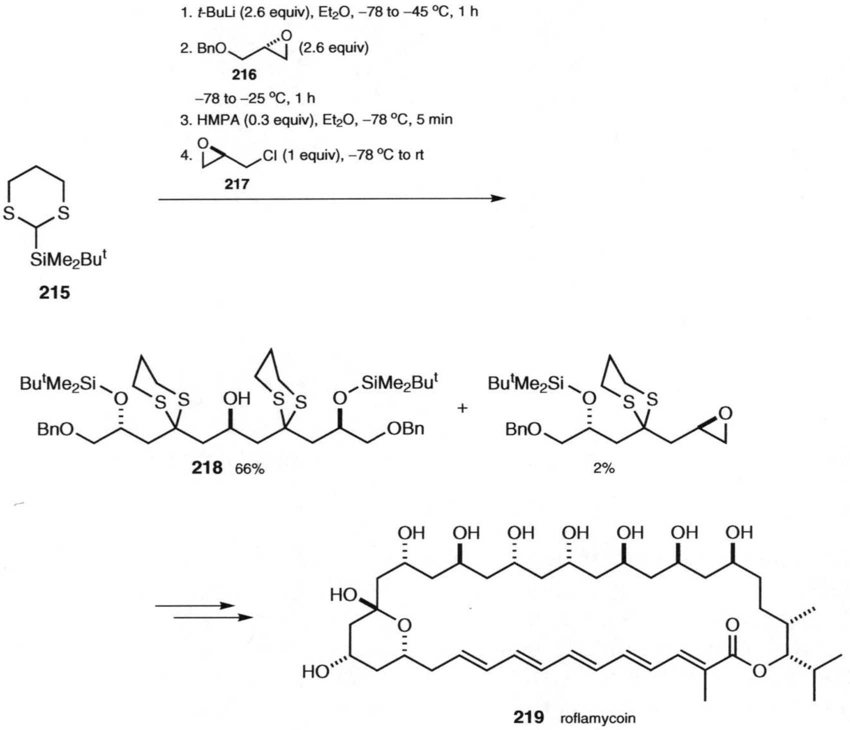

Another excellent illustration of this method is provided in the total synthesis of the spongistatins **220**. Several synthetic approaches utilize the 1,4-Brook rearrangement as part of a key domino sequence (Scheme 98).[138–142]

1.5.3 Brook Rearrangement as the Key Step in Domino Reactions **435**

Scheme 98 1,4-Brook Rearrangement in the Total Synthesis of the Spongistatins[138–142]

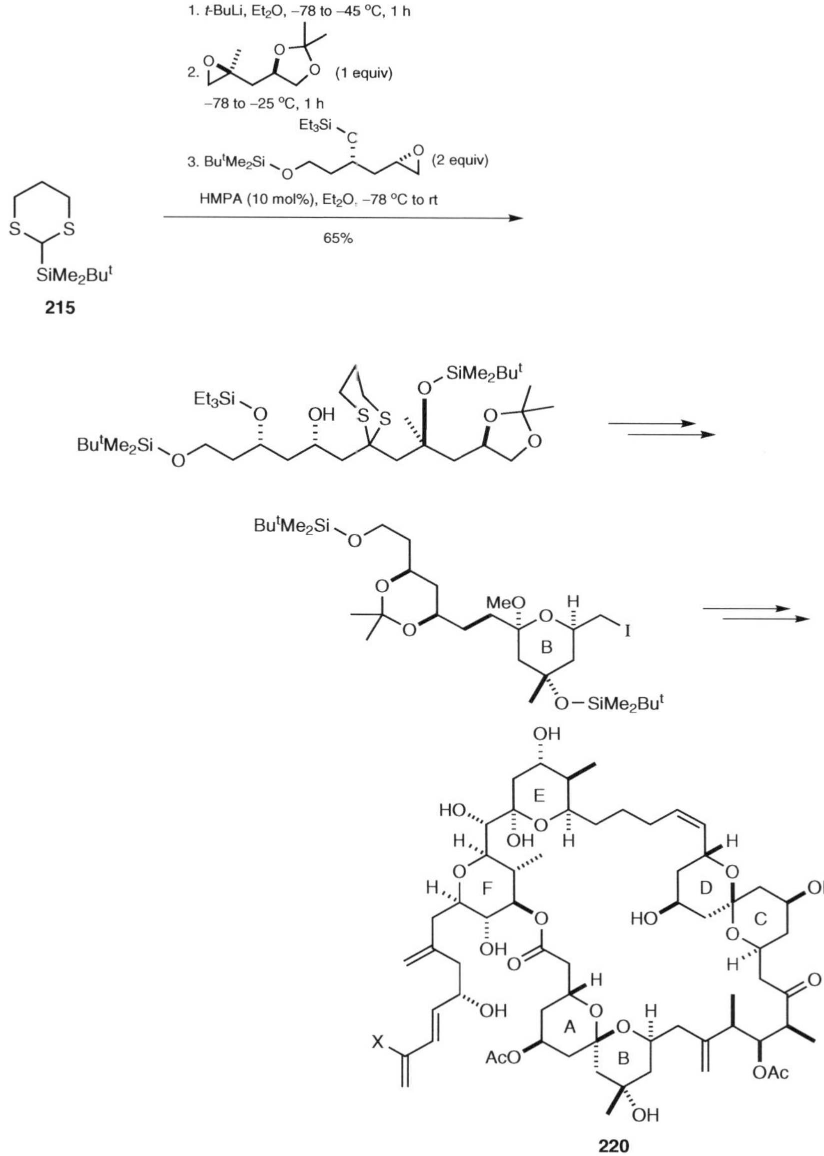

220

X = Cl, H

for references see p 445

436 Domino Transformations **1.5** Non-Radical Skeletal Rearrangements

1.5.3 Brook Rearrangement as the Key Step in Domino Reactions **437**

This cascade reaction has also been employed in the one-pot synthesis of a protected poly-ol, a key precursor for mycoticins A (**223**, R^1 = H) and B (**223**, R^1 = Me). The highlight of this synthesis is the use of bis-epoxide **221**, which allows the 1,4-Brook rearrangement to occur twice and generates advanced intermediate **222** in one pot (Scheme 99).[143]

Scheme 99 1,4-Brook Rearrangement in the Total Synthesis of Mycoticins A and B[143]

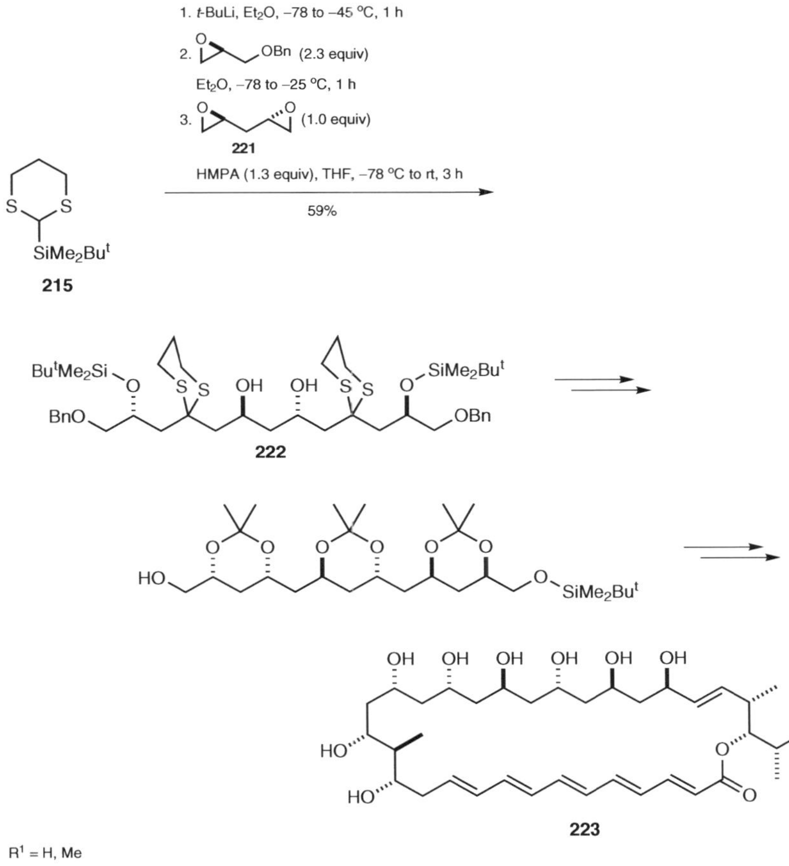

This cascade was also successfully employed in the total synthesis of the antibiotic ripostatin B (**227**), which is a potent inhibitor of the bacterial RNA polymerase (Scheme 100). The reduced basic character of the lithiated dithiane derivative is beneficial for these transformations, especially when multifunctionalized epoxides are employed. Here, epoxides **224** and **225** serve as electrophiles and the advanced synthetic precursor **226** is formed in good yield.[144]

for references see p 445

438 Domino Transformations **1.5** Non-Radical Skeletal Rearrangements

Scheme 100 1,4-Brook Rearrangement in the Total Synthesis of Ripostatin B[144]

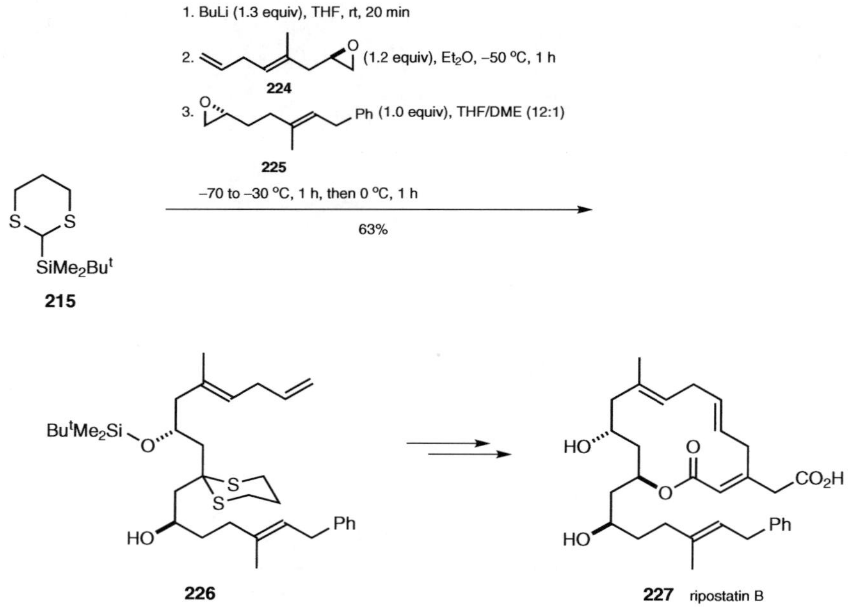

One key step in the stereocontrolled asymmetric synthesis of an advanced structural intermediate of the B-ring of bryostatin 1 (**230**) is the bisalkylation of the homochiral epoxide **228** with lithiated dithiane **215**, a process which results in the formation of homochiral product **229** (Scheme 101).[145]

1.5.3 Brook Rearrangement as the Key Step in Domino Reactions

Scheme 101 1,4-Brook Rearrangement in the Total Synthesis of a Bryostatin Precursor[145]

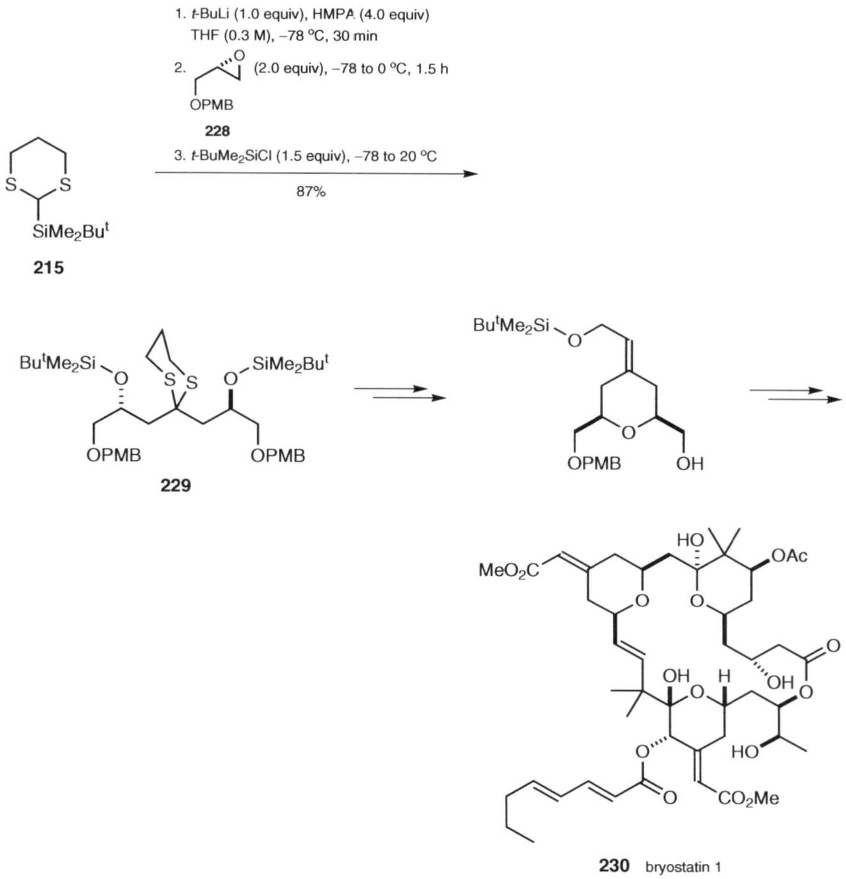

230 bryostatin 1

Likewise, a carbon to oxygen 1,4-silyl migration is used in the total synthesis of the cytotoxic V-ATPase inhibitor apicularen A (**232**) which provides the key intermediate dithiane **231** (Scheme 102).[146]

for references see p 445

Scheme 102 1,4-Brook Rearrangement in the Total Synthesis of Apicularen A[146]

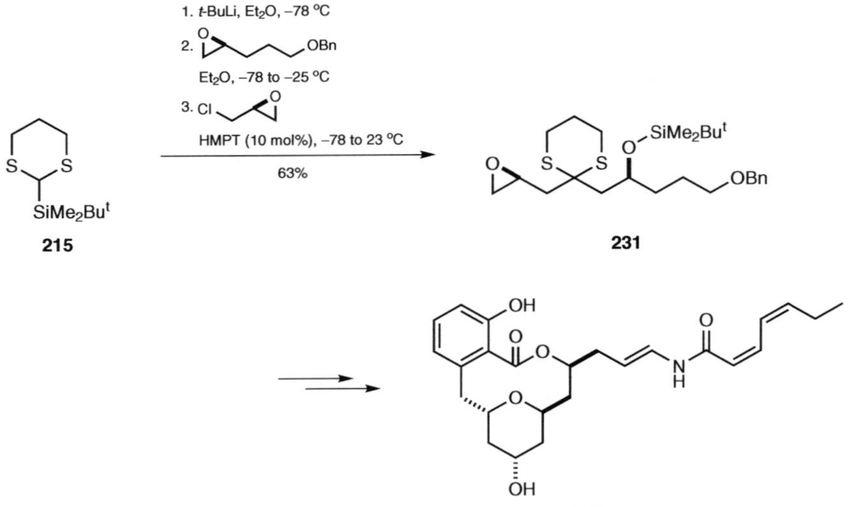

215

231

232 apicularen A

A 1,4-Brook rearrangement is used to build up the C1–C19 skeleton of (–)-cochleamycin A (**234**) (Scheme 103).[147] Here, deprotonation of the carbinol moiety in **233** followed by silyl migration provides a dithioacetal-stabilized carbanion that reacts with 2-bromo-1,1-diethoxyethane.

Scheme 103 1,4-Brook Rearrangement En Route to (–)-Cochleamycin A[147]

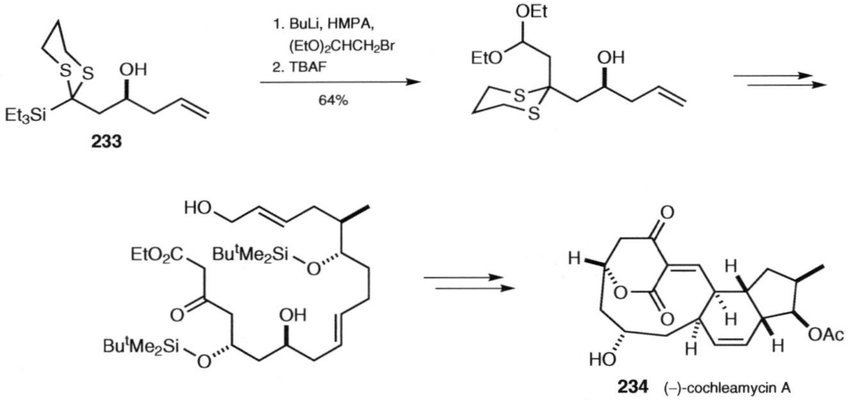

233

234 (–)-cochleamycin A

A biomimetically inspired total synthesis of (12S)-12-hydroxymonocerin (**236**) and the epimer (12R)-12-hydroxymonocerin, metabolites isolated from *Exserohilum rostratum*, a fungal strain endophytic in *Stemona* sp., uses a carbon to oxygen 1,4-silyl migration as one of its key steps (Scheme 104).[148] The reaction could be accelerated by the addition of hexamethylphosphoric triamide after *tert*-butyl(1,3-dithian-2-yl)dimethylsilane (**215**) had reacted with (S)-2-methyloxirane. Coupling with the second electrophile **235** terminates the domino sequence.

1.5.3 Brook Rearrangement as the Key Step in Domino Reactions **441**

Scheme 104 1,4-Brook Rearrangement in the Total Synthesis of (12S)-12-Hydroxymonocerin[148]

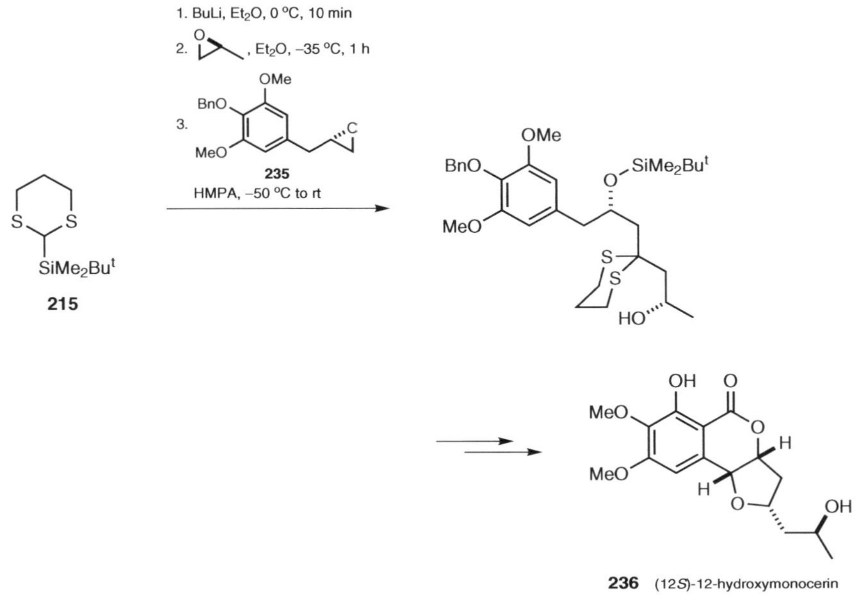

236 (12S)-12-hydroxymonocerin

A 1,4-Brook rearrangement was applied in the first total synthesis of the macrolactone dolabelide D (**239**) (Scheme 105).[149] Treatment of carbinol **237** with butyllithium and addition of copper(I) bromide–dimethyl sulfide complex in the presence of 1,3-dimethyl-3,4,5,6-tetrahydropyrimidin-2(1H)-one initiated the sp²-carbon to oxygen 1,4-silyl migration, and the resulting vinylcopper species was alkylated with iodomethane to provide the trisubstituted alkene **238** in excellent yield.

Scheme 105 1,4-Brook Rearrangement in the Total Synthesis of Dolabelide D[149]

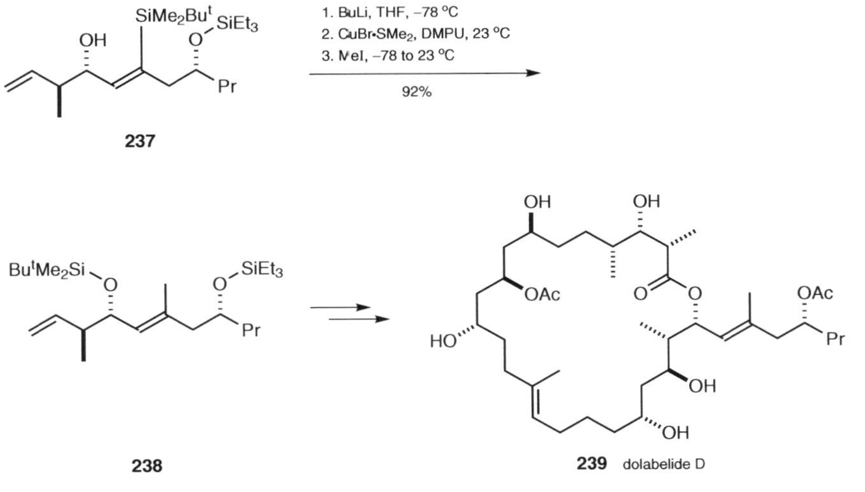

238

239 dolabelide D

for references see p 445

1.5.3.4.2.2 Synthesis of Terpenes

Until now, the 1,4-Brook rearrangement has rarely been used for the total synthesis of terpenes. One illustrative example was reported in the total synthesis of a gorgonian sesquiterpene **241**.[150] Deprotonation of 3-methylpenta-1,4-diene by Schlosser's base yields an organolithium species, which is added to aldehyde **240** (Scheme 106). Subsequent carbon to oxygen 1,4-silyl migration provides a new allyl anion, which is captured by 1-bromo-3-methylbut-2-ene.

Scheme 106 1,4-Brook Rearrangement in the Total Synthesis of a Gorgonian Sesquiterpene[150]

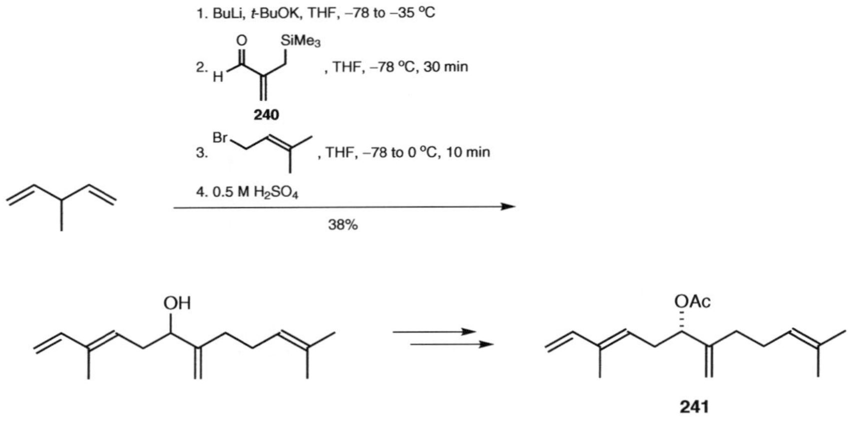

1.5.3.4.2.3 Synthesis of Alkaloids

Based on earlier findings that N-acylated aziridines can also be employed, N-tosylated aziridines **242** and **244** have been utilized in cascade reactions for the total synthesis of (−)-indolizidine 223AB (**243**) and alkaloid (−)-205B (**245**), respectively (Scheme 107 and Scheme 108).[151–153] Again, the key step is based on carbon to oxygen 1,4-silyl migration after nucleophilic ring opening of an epoxide by the established C1 dianion equivalent *tert*-butyl(1,3-dithian-2-yl)dimethylsilane, which is followed by nucleophilic attack of the aziridine-containing building block by the newly generated carbanion. This sequence elaborates protected 1,5-amino alcohols.

1.5.3 Brook Rearrangement as the Key Step in Domino Reactions **443**

Scheme 107 1,4-Brook Rearrangement in the Total Synthesis of (−)-Indolizidine 223AB[151,153]

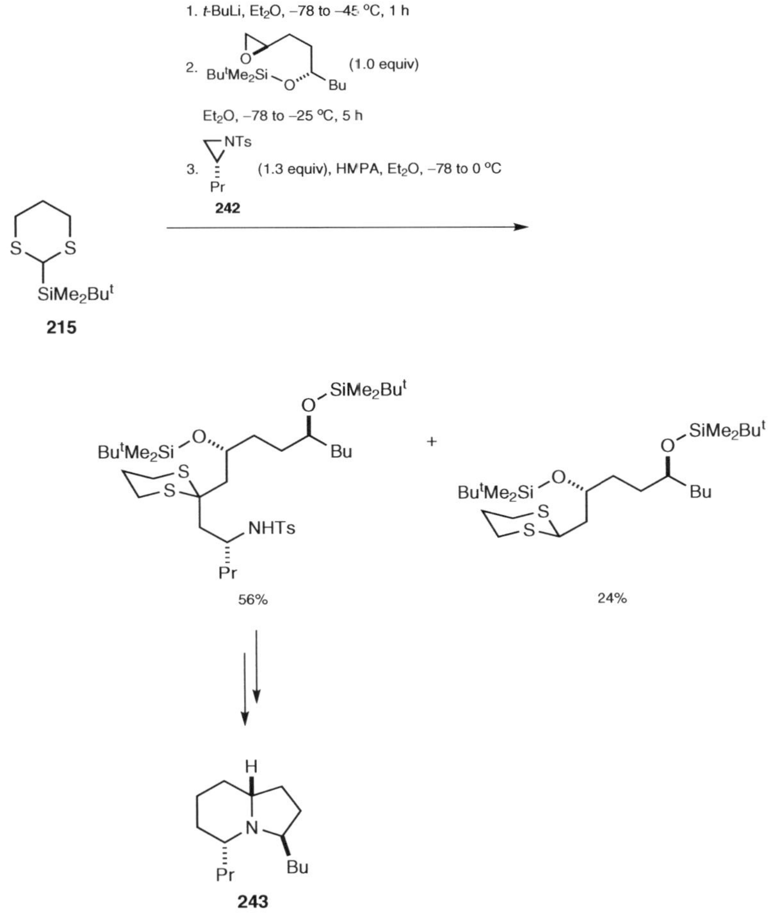

for references see p 445

Scheme 108 1,4-Brook Rearrangement in the Total Synthesis of Alkaloid (−)-205B[151,152]

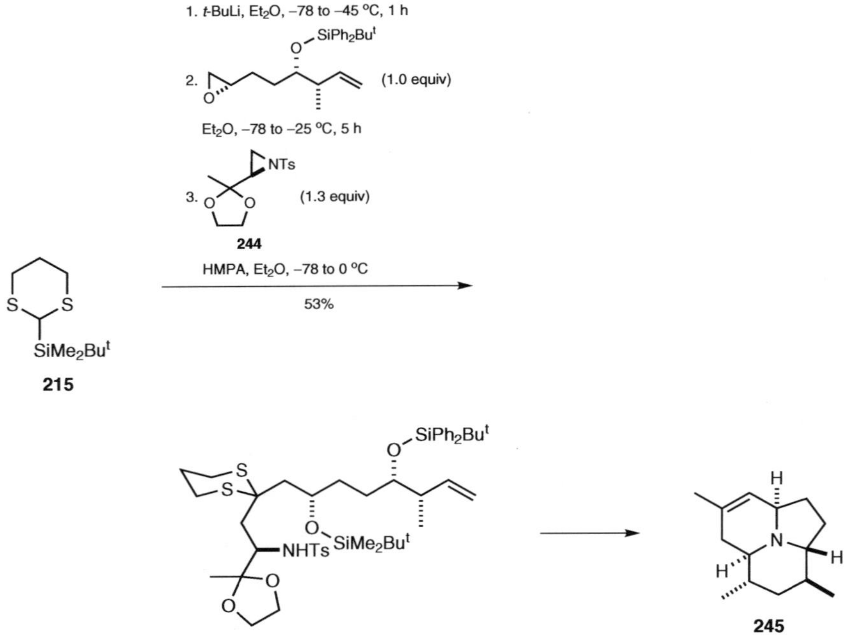

1.5.3.5 Conclusions

The Brook rearrangement can be categorized as classic and the transformation itself seems to be rather simple. Originally it was thought to have little impact on organic synthesis. Probably as a consequence, only few mechanistic studies have been reported over the decades.[154,155] These mainly cover the retro-1,2-Brook rearrangement. Since the early 1990s, the Cinderella status of the Brook rearrangement has changed because the Brook rearrangement has been embedded into domino reaction sequences. The value of this reaction is connected with the fact that an easily available oxyanion is transformed into a valuable carbanionic species which can further be exploited in carbon–carbon coupling reactions. Now, the large variety of different examples and applications available, including complex natural product syntheses, are due to a deeper understanding of how to efficiently initiate silyl migration (the well-balanced stabilization of the newly formed carbanionic species is crucial here) and the large number of methods to create an alkoxide anion. This has changed the relevance of the Brook rearrangement drastically. Indeed, one can expect many sophisticated applications in the future.

References

[1] Brook, A. G., *Acc. Chem. Res.*, (1974) **7**, 77.

[2] Parsons, P. J.; Penkett, C. S.; Shell, A. J., *Chem. Rev.*, (1996) **96**, 195.

[3] Tietze, L. F., *Chem. Rev.*, (1996) **96**, 115.

[4] Tietze, L. F.; Modi, A., *Med. Res. Rev.*, (2000) **20**, 304.

[5] Tietze, L. F.; Beifuss, U., *Angew. Chem.*, (1993) **105**, 137; *Angew. Chem. Int. Ed. Engl.*, (1993) **32**, 131.

[6] Smith, A. B., III; Xian, M.; Kim, W.-S.; Kim, D.-S., *J. Am. Chem. Soc.*, (2006) **128**, 12368.

[7] Gao, L.; Lu, J.; Song, Z.; Lin, X.; Xu, Y.; Yin, Z., *Chem. Commun. (Cambridge)*, (2013) **49**, 8961.

[8] Ager, D. J., *Synthesis*, (1984), 384.

[9] Baati, R.; Mioskowski, C.; Barma, D.; Kache, R.; Falck, J. R., *Org. Lett.*, (2006) **8**, 2949.

[10] Robertson, B. D.; Hartel, A. M., *Tetrahedron Lett.*, (2008) **49**, 2088.

[11] Reynolds, S. C.; Wengryniuk, S. E.; Hartel, A. M., *Tetrahedron Lett.*, (2007) **48**, 6751.

[12] Baker, H. K.; Hartel, A. M., *Tetrahedron Lett.*, (2009) **50**, 4012.

[13] Davis, A. L.; Korous, A. A.; Hartel, A. M., *Tetrahedron Lett.*, (2013) **54**, 3673.

[14] Fleming, I.; Lawrence, A. J.; Richardson, R. D.; Surry, D. S.; West, M. C., *Helv. Chim. Acta*, (2002) **85**, 3349.

[15] Reich, H. J.; Olson, R. E.; Clark, M. C., *J. Am. Chem. Soc.*, (1980) **102**, 1423.

[16] Reich, H. J.; Holtan, R. C.; Borkowsky, S. L., *J. Org. Chem.*, (1987) **52**, 312.

[17] Reich, H. J.; Holtan, R. C.; Bolm, C., *J. Am. Chem. Soc.*, (1990) **112**, 5609.

[18] Honda, M.; Nakajima, T.; Okada, M.; Yamaguchi, K.; Suda, M.; Kunimoto, K.-K.; Segi, M., *Tetrahedron Lett.*, (2011) **52**, 3740.

[19] Song, Z.; Kui, L.; Sun, X.; Li, L., *Org. Lett.*, (2011) **13**, 1440.

[20] Lefebvre, O.; Brigaud, T.; Portella, C., *Tetrahedron*, (1998) **54**, 5939.

[21] Lefebvre, O.; Brigaud, T.; Portella, C., *J. Org. Chem.*, (2001) **66**, 4348.

[22] Higashiya, S.; Chung, W. J.; Lim, D. S.; Ngo, S. C.; Kelly, W. H., IV; Toscano, P. J.; Welch, J. T., *J. Org. Chem.*, (2004) **69**, 6323.

[23] Wu, L., *J. Fluorine Chem.*, (2011) **132**, 367.

[24] Takeda, K.; Fujisawa, M.; Makino, T.; Yoshii, E.; Yamaguchi, K., *J. Am. Chem. Soc.*, (1993) **115**, 9351.

[25] Takeda, K.; Ohtani, Y.; Ando, E.; Fujimoto, K.-i.; Yoshii, E.; Koizumi, T., *Chem. Lett.*, (1998), 1157.

[26] Takeda, K.; Ubayama, H.; Sano, A.; Yoshii, E.; Koizumi, T., *Tetrahedron Lett.*, (1998) **39**, 5243.

[27] Takeda, K.; Yamawaki, K.; Hatakeyama, N., *J. Org. Chem.*, (2002) **67**, 1786.

[28] Takeda, K.; Nakayama, I.; Yoshii, E., *Synlett*, (1994), 178.

[29] Greszler, S. N.; Johnson, J. S., *Angew. Chem. Int. Ed.*, (2009) **48**, 3689.

[30] Takeda, K.; Takeda, M.; Nakajima, A.; Yoshii, E., *J. Am. Chem. Soc.*, (1995) **117**, 6400.

[31] Takeda, K.; Nakajima, A.; Takeda, M.; Okamoto, Y.; Sato, T.; Yoshii, E.; Koizumi, T.; Shiro, M., *J. Am. Chem. Soc.*, (1998) **120**, 4947.

[32] Takeda, K.; Ohtani, Y., *Org. Lett.*, (1999) **1**, 677.

[33] Takeda, K.; Nakane, D.; Takeda, M., *Org. Lett.*, (2000) **2**, 1903.

[34] Takeda, K.; Sawada, Y.; Sumi, K., *Org. Lett.*, (2002) **4**, 1031.

[35] Sawada, Y.; Sasaki, M.; Takeda, K., *Org. Lett.*, (2004) **6**, 2277.

[36] Sasaki, M.; Oyamada, K.; Takeda, K., *J. Org. Chem.*, (2010) **75**, 3941.

[37] Sasaki, M.; Hashimoto, A.; Tanaka, K.; Kawahata, M.; Yamaguchi, K.; Takeda, K., *Org. Lett.*, (2008) **10**, 1803.

[38] Nakai, Y.; Kawahata, M.; Yamaguchi, K.; Takeda, K., *J. Org. Chem.*, (2007) **72**, 1379.

[39] Greszler, S. N.; Malinowski, J. T.; Johnson, J. S., *J. Am. Chem. Soc.*, (2010) **132**, 17393.

[40] Lettan, R. B., II; Galliford, C. V.; Woodward, C. C.; Scheidt, K. A., *J. Am. Chem. Soc.*, (2009) **131**, 8805.

[41] Nicewicz, D. A.; Johnson, J. S., *J. Am. Chem. Soc.*, (2005) **127**, 6170.

[42] Boyce, G. R.; Johnson, J. S., *Angew. Chem. Int. Ed.*, (2010) **49**, 8930.

[43] Boyce, G. R.; Liu, S.; Johnson, J. S., *Org. Lett.*, (2012) **14**, 652.

[44] Greszler, S. N.; Malinowski, J. T.; Johnson, J. S., *Org. Lett.*, (2011) **13**, 3206.

[45] Schmitt, D. C.; Johnson, J. S., *Org. Lett.*, (2010) **12**, 944.

[46] Linghu, X.; Satterfield, A. D.; Johnson, J. S., *J. Am. Chem. Soc.*, (2006) **128**, 9302.

[47] Greszler, S. N.; Johnson, J. S., *Org. Lett.*, (2009) **11**, 827.

[48] Linghu, X.; Nicewicz, D. A.; Johnson, J. S., *Org. Lett.*, (2002) **4**, 2957.

[49] Nicewicz, D. A.; Yates, C. M.; Johnson, J. S., *J. Org. Chem.*, (2004) **69**, 6548.

[50] Takeda, K.; Tanaka, T., *Synlett*, (1999), 705.

[51] Takeda, K.; Ohnishi, Y., *Tetrahedron Lett.*, (2000) **41**, 4169.

[52] Sasaki, M.; Kondo, Y.; Kawahata, M.; Yamaguchi, K.; Takeda, K., *Angew. Chem. Int. Ed.*, (2011) **50**, 6375.

[53] Takeda, K.; Haraguchi, H.; Okamoto, Y., *Org. Lett.*, (2003) **5**, 3705.

[54] Smirnov, P.; Katan, E.; Mathew, J.; Kostenko, A.; Karni, M.; Nijs, A.; Bolm, C.; Apeloig, Y.; Marek, I., *J. Org. Chem.*, (2014) **79**, 12122.

[55] Liu, B.; Lu, C.-D., *J. Org. Chem.*, (2011) **76**, 4205.

[56] Unger, R.; Cohen, T.; Marek, I., *Org. Lett.*, (2005) **7**, 5313.

[57] Unger, R.; Cohen, T.; Marek, I., *Eur. J. Org. Chem.*, (2009), 1749.

[58] Unger, R.; Cohen, T.; Marek, I., *Tetrahedron*, (2010) **66**, 4874.

[59] Kirschning, A.; Luiken, S.; Migliorini, A.; Loreto, M. A.; Vogt, M., *Synlett*, (2009), 429.

[60] Steward, K. M.; Johnson, J. S., *Org. Lett.*, (2010) **12**, 2864.

[61] Bausch, C. C.; Johnson, J. S., *J. Org. Chem.*, (2004) **69**, 4283.

[62] Linghu, X.; Bausch, C. C.; Johnson, J. S., *J. Am. Chem. Soc.*, (2005) **127**, 1833.

[63] Tarr, J. C.; Johnson, J. S., *J. Org. Chem.*, (2010) **75**, 3317.

[64] Yamaki, Y.; Shigenaga, A.; Li, J.; Shimohigashi, Y.; Otaka, A., *J. Org. Chem.*, (2009) **74**, 3278.

[65] Mattson, A. E.; Bharadwaj, A. R.; Scheidt, K. A., *J. Am. Chem. Soc.*, (2004) **126**, 2314.

[66] Linghu, X.; Potnick, J. R.; Johnson, J. S., *J. Am. Chem. Soc.*, (2004) **126**, 3070.

[67] Tsubouchi, A.; Onishi, K.; Takeda, T., *J. Am. Chem. Soc.*, (2006) **128**, 14268.

[68] Reynolds, T. E.; Stern, C. A.; Scheidt, K. A., *Org. Lett.*, (2007) **9**, 2581.

[69] Takeda, K.; Kawanishi, E.; Sasaki, M.; Takahashi, Y.; Yamaguchi, K., *Org. Lett.*, (2002) **4**, 1511.

[70] Sasaki, M.; Kawanishi, E.; Nakai, Y.; Matsumoto, T.; Yamaguchi, K.; Takeda, K., *J. Org. Chem.*, (2003) **68**, 9330.

[71] Sasaki, M.; Shirakawa, Y.; Kawahata, M.; Yamaguchi, K.; Takeda, K., *Chem.–Eur. J.*, (2009) **15**, 3363.

[72] Sasaki, M.; Fujiwara, M.; Kotomori, Y.; Kawahata, M.; Yamaguchi, K.; Takeda, K., *Tetrahedron*, (2013) **69**, 5823.

[73] Sasaki, M.; Takeda, K., *Org. Lett.*, (2004) **6**, 4849.

[74] Okugawa, S.; Masu, H.; Yamaguchi, K.; Takeda, K., *J. Org. Chem.*, (2005) **70**, 10515.

[75] Tanaka, K.; Takeda, K., *Tetrahedron Lett.*, (2004) **45**, 7859.

[76] Matsumoto, T.; Masu, H.; Yamaguchi, K.; Takeda, K., *Org. Lett.*, (2004) **6**, 4367.

[77] Okamoto, N.; Sasaki, M.; Kawahata, M.; Yamaguchi, K.; Takeda, K., *Org. Lett.*, (2006) **8**, 1889.

[78] Zilch, H.; Tacke, R., *J. Organomet. Chem.*, (1986) **316**, 243.

[79] Le Bideau, F.; Malacria, M., *Phosphorus, Sulfur Silicon Relat. Elem.*, (1995) **107**, 275.

[80] Linderman, R. J.; Ghannam, A., *J. Org. Chem.*, (1988) **53**, 2878.

[81] Linderman, R. J.; Ghannam, A., *J. Am. Chem. Soc.*, (1990) **112**, 2392.

[82] Verlhac, J.-B.; Kwon, H.; Pereyre, M., *J. Organomet. Chem.*, (1992) **437**, C13.

[83] Antonsen, Ø.; Benneche, T.; Undheim, K., *Acta Chem. Scand.*, (1992) **46**, 757.

[84] Thadani, A. N.; Huang, Y.; Rawal, V. H., *Org. Lett.*, (2007) **9**, 3873.

[85] Mergardt, B.; Weber, K.; Adiwidjaja, G.; Schaumann, E., *Angew. Chem. Int. Ed. Engl.*, (1991) **30**, 1687.

[86] Sakaguchi, K.; Fujita, M.; Suzuki, H.; Higashino, M.; Ohfune, Y., *Tetrahedron Lett.*, (2000) **41**, 6589.

[87] Tokeshi, B. K.; Tius, M. A., *Synthesis*, (2004), 786.

[88] Shinokubo, H.; Miura, K.; Oshima, K.; Utimoto, K., *Tetrahedron Lett.*, (1993) **34**, 1951.

[89] Trofimov, A.; Gevorgyan, V., *Org. Lett.*, (2009) **11**, 253.

[90] Maity, P.; Lepore, S. D., *J. Am. Chem. Soc.*, (2009) **131**, 4196.

[91] Kang, B. C.; Shim, S. Y.; Ryu, D. H., *Org. Lett.*, (2014) **16**, 2077.

[92] Liu, H.; O'Connor, M. J.; Sun, C.; Wink, D. J.; Lee, D., *Org. Lett.*, (2013) **15**, 2974.

[93] Wilson, S. R.; Georgiadis, G. M., *J. Org. Chem.*, (1983) **48**, 4143.

[94] Yamamoto, K.; Kimura, T.; Tomo, Y., *Tetrahedron Lett.*, (1985) **26**, 4505.

[95] Tsubouchi, A.; Itoh, M.; Onishi, K.; Takeda, T., *Synthesis*, (2004), 1504.

[96] Raubo, P.; Wicha, J., *Tetrahedron: Asymmetry*, (1996) **7**, 763.

[97] Ducray, R.; Ciufolini, M. A., *Angew. Chem. Int. Ed.*, (2002) **41**, 4688.

[98] Ducray, R.; Cramer, N.; Ciufolini, M. A., *Tetrahedron Lett.*, (2001) **42**, 9175.

[99] Jung, A.; Koch, O.; Ries, M.; Schaumann, E., *Synlett*, (2000), 92.

[100] Safa, K. D.; Ghorbanpour, K., *J. Sulfur Chem.*, (2014) **35**, 170.

[101] Hamada, M.; Naruse, S.; Wada, M.; Kishimoto, T.; Nakajima, N., *Synthesis*, (2014) **46**, 1779.

[102] Matsuda, I.; Murata, S.; Ishii, Y., *J. Chem. Soc., Perkin Trans. 1*, (1979), 26.

[103] Fleming, I.; Floyd, C. D., *J. Chem. Soc., Perkin Trans. 1*, (1981), 969.

References

[104] Fischer, M.-R.; Kirschning, A.; Michel, T.; Schaumann, E., *Angew. Chem. Int. Ed. Engl.*, (1994) **33**, 217.

[105] Michel, T.; Kirschning, A.; Beier, C.; Bräuer, N.; Schaumann, E.; Adiwidjaja, G., *Liebigs Ann.*, (1996), 1811.

[106] Seebach, D., *Synthesis*, (1969), 17.

[107] Bräuer, N.; Michel, T.; Schaumann, E., *Tetrahedron*, (1998) **54**, 11481.

[108] Tries, F.; Schaumann, E., *Eur. J. Org. Chem.*, (2003), 1085.

[109] Bräuer, N.; Dreeßen, S.; Schaumann, E., *Tetrahedron Lett.*, (1999) **40**, 2921.

[110] Le Merrer, Y.; Gravier-Pelletier, C.; Maton, W.; Numa, M.; Depezay, J.-C., *Synlett*, (1999), 1322.

[111] Gravier-Pelletier, C.; Maton, W.; Le Merrer, Y., *Synlett*, (2003), 333.

[112] Gravier-Pelletier, C.; Maton, W.; Dintinger, T.; Tellier, C.; Le Merrer, Y., *Tetrahedron*, (2003) **59**, 8705.

[113] Leung, L. M. H.; Gibson, V.; Linclau, B., *J. Org. Chem.*, (2008) **73**, 9197.

[114] Leung, L. M. H.; Boydell, A. J.; Gibson, V.; Light. M. E.; Linclau, B., *Org. Lett.*, (2005) **7**, 5183.

[115] Harms, G.; Schaumann, E.; Adiwidjaja, G., *Synthesis*, (2001), 577.

[116] Tietze, L. F.; Geissler, H.; Gewert, J. A.; Jakobi, U., *Synlett*, (1994), 511.

[117] Smith, A. B., III; Boldi, A. M., *J. Am. Chem. Soc.*, (1997) **119**, 6925.

[118] Chen, M. Z.; Gutierrez, O.; Smith, A. B., III, *Angew. Chem. Int. Ed.*, (2014) **53**, 1279.

[119] Shinokubo, H.; Miura, K.; Oshima, K.; Utimoto, K., *Tetrahedron*, (1996) **52**, 503.

[120] Schaumann, E.; Kirschning, A.; Narjes, F., *J. Org. Chem.*, (1991) **56**, 717.

[121] Takaku, K.; Shinokubo, H.; Oshima, K., *Tetrahedron Lett.*, (1998) **39**, 2575.

[122] Smith, A. B., III; Duffey, M. O., *Synlett*, (2004), 1363.

[123] Gao, L.; Lin, X.; Lei, J.; Song, Z.; Lin, Z., *Org. Lett.*, (2012) **14**, 158.

[124] Matsuya, Y.; Koiwai, A.; Minato, D.; Sugimoto, K.; Toyooka, N., *Tetrahedron Lett.*, (2012) **53**, 5955.

[125] Sakakibara, K.; Nozaki, K., *Org. Biomol. Chem.*, (2009) **7**, 502.

[126] Moser, W. H.; Endsley, K. E.; Colyer, J. T., *Org. Lett.*, (2000) **2**, 717.

[127] Moser, W. H.; Zhang, J.; Lecher, C. S.; Frazier, T. L.; Pink, M., *Org. Lett.*, (2002) **4**, 1981.

[128] Smith, A. B., III; Kim, W.-S.; Wuest, W. M., *Angew. Chem. Int. Ed.*, (2008) **47**, 7082.

[129] Devarie-Baez, N. O.; Kim, W.-S.; Smith, A. B., III; Xian, M., *Org. Lett.*, (2009) **11**, 1861.

[130] Yan, L.; Sun, X.; Li, H.; Song, Z.; Liu, Z., *Org. Lett.*, (2013) **15**, 1104.

[131] Taguchi, H.; Ghoroku, K.; Tadaki, M.; Tsubouchi, A.; Takeda, T., *Org. Lett.*, (2001) **3**, 3811.

[132] Taguchi, H.; Ghoroku, K.; Tadaki, M.; Tsubouchi, A.; Takeda, T., *J. Org. Chem.*, (2002) **67**, 8450.

[133] Sun, X.; Lei, J.; Sun, C.; Song, Z.; Yan, L., *Org. Lett.*, (2012) **14**, 1094.

[134] Smith, A. B., III; Tong, R.; Kim, W.-S.; Maio, W. A., *Angew. Chem. Int. Ed.*, (2011) **50**, 8904.

[135] Sun, C.; Zhang, Y.; Xiao, P.; Li, H.; Sun, X.; Song, Z., *Org. Lett.*, (2014) **16**, 984.

[136] Banert, K.; Bochmann, S.; Ihle, A.; Plefka, O.; Taubert, F.; Walther, T.; Korb, M.; Rüffer, T.; Lang, H., *Molecules*, (2014) **19**, 14022.

[137] Corey, E. J.; Luo, G.; Lin, L. S., *J. Am. Chem. Soc.*, (1997) **119**, 9927.

[138] Smith, A. B., III; Zhuang, L.; Brook, C. S.; Lin, Q.; Moser, W. H.; Trout, R. E. L.; Boldi, A. M., *Tetrahedron Lett.*, (1997) **38**, 8671.

[139] Smith, A. B., III; Doughty, V. A.; Lin, Q.; Zhuang, L.; McBriar, M. D.; Boldi, A. M.; Moser, W. H.; Murase, N.; Nakayama, K.; Sobukawa, M., *Angew. Chem. Int. Ed.*, (2001) **40**, 191.

[140] Smith, A. B., III; Doughty, V. A.; Sfouggatakis, C.; Bennett, C. S.; Koyanagi, J.; Takeuchi, M., *Org. Lett.*, (2002) **4**, 783.

[141] Smith, A. B., III; Lin, Q.; Nakayama, K.; Boldi, A. M.; Brook, C. S.; McBriar, M. D.; Moser, W. H.; Sobukawa, M.; Zhuang, L., *Tetrahedron Lett.*, (1997) **38**, 8675.

[142] Smith, A. B., III; Zhu, W.; Shirakami, S.; Sfouggatakis, C.; Doughty, V. A.; Bennett, C. S.; Sakamoto, Y., *Org. Lett.*, (2003) **5**, 761.

[143] Smith, A. B., III; Pitram, S. M., *Org. Lett.*, (1999) **1**, 2001.

[144] Winter, P.; Hiller, W.; Christmann, M., *Angew. Chem. Int. Ed.*, (2012) **51**, 3396.

[145] Hale, K. J.; Hummersone, M. G.; Bhatia, G. S., *Org. Lett.*, (2000) **2**, 2189.

[146] Petri, A. F.; Bayer, A.; Maier, M. E., *Angew. Chem. Int. Ed.*, (2004) **43**, 5821.

[147] Mukherjee, S.; Lee, D., *Org. Lett.*, (2009) **11**, 2916.

[148] Fang, B.; Xie, X.; Jing, P.; Zhao, C.; Li, H.; Ma, H.; She, X., *Tetrahedron*, (2013) **69**, 11025.

[149] Park, P. K.; O'Malley, S. J.; Schmidt, D. R.; Leighton, J. L., *J. Am. Chem. Soc.*, (2006) **128**, 2796.

[150] Smith, A. B., III; Kim, D.-S.; Xian, M., *Org. Lett.*, (2007) **9**, 3307.

[151] Smith, A. B., III; Kim, D.-S., *J. Org. Chem.*, (2006) **71**, 2547.

[152] Smith, A. B., III; Kim, D.-S., *Org. Lett.*, (2005) **7**, 3247.

[153] Smith, A. B., III; Kim, D.-S., *Org. Lett.*, (2004) **6**, 1493.
[154] Boche, G.; Opel, A.; Marsch, M.; Harms, K.; Haller, F.; Lohrenz, J. C. W.; Thümmler, C.; Koch, W., *Chem. Ber.*, (1992) **125**, 2265.
[155] Kapeller, D. C.; Brecker, L.; Hammerschmidt, F., *Chem.–Eur. J.*, (2007) **13**, 9582.

449

1.6 # Metal-Mediated Reactions

1.6.1 ## Palladium-Mediated Domino Reactions

E. A. Anderson

General Introduction

Among many metals that have been employed in domino transformations, palladium arguably holds a privileged position. The breadth of palladium-catalyzed transformations available to the organic chemist (and the accompanying depth of mechanistic understanding), combined with the wide tolerance by such reactions of many useful functionalities (carbonyls, heteroatoms, etc.), facilitates the design of sequenced orthogonal reaction processes, and has led to the development of a wide range of domino transformations that incorporate palladium catalysis. A necessary consideration for the design of any domino transformation[1–4] is the relative rates of the processes involved in the reaction sequence, and the compatibility of each reaction step with a common set of conditions (reagents, solvents, temperature, etc.). As mentioned, orthogonality between each segment of the cascade is a useful (but not essential) factor, and palladium-catalyzed transformations are well suited for combination with other reaction manifolds.

In this section, the utility of palladium as a cascade-initiating component is discussed with a focus on applications in synthesis that exemplify reaction generality and wide substrate scope. The section is categorized according to the nature of the organopalladium intermediate initially generated in the cascade sequence, namely alkenylpalladium, arylpalladium, allylpalladium, allenylpalladium, and alkylpalladium complexes, with a degree of subcategorization according to the nature of subsequent transformations. As with any coverage of such a broadly useful topic, decisions were required as to the inclusion or omission of transformations. The chapter therefore aims to provide an informative and representative selection of reaction types as applied in a total or partial synthesis setting,[5] rather than a comprehensive coverage of the field, for which the reader is referred to more detailed reviews.[6–9]

1.6.1.1 ### Reactions Initiating with Alkenylpalladium Intermediates

The rich field of alkenylpalladium-initiated cascades owes much to the pioneering efforts of Grigg,[10–13] Negishi,[14–17] Trost,[18] and others[19–21] who established the feasibility of multiple C—C bond formations, followed by palladium-mediated termination steps (e.g., reductive elimination, β-hydride elimination, etc.), in reactions such as "zipper" cascades and related processes. Applications in synthesis are widespread, due to the ease of synthesis of reaction substrates together with the broad opportunities for the design of domino processes.

The Overman group's synthesis of scopadulcic acid A (**1**) (Scheme 1)[22,23] remains one of the prime examples of palladium-catalyzed cascade cyclization reactions that initiate with an alkenylpalladium complex; the transformation of iodoalkene **2** into tricyclic carbocycle **5** in a remarkable 90% yield stands as a landmark achievement in the field. The

for references see p 506

reaction proceeds by initial formation of an alkenylpalladium species, which engages in a highly facially selective (dr >20:1) 6-*exo-trig* carbopalladation onto the *exo*-methylene group of the seven-membered ring. The resulting alkylpalladium complex **3**, being unable to undergo a β-hydride elimination, instead participates in a second Heck reaction, a 5-*exo-trig* cyclization onto the cycloheptene. Finally, β-hydride elimination from intermediate **4** gives the observed product **5** (following desilylation). The inclusion of silver(I) carbonate in this reaction is essential to prevent palladium(II) hydride mediated isomerization of the newly formed alkene, by accelerating reductive elimination.

Scheme 1   Key Steps in the Synthesis of (−)-Scopadulcic Acid A[23]

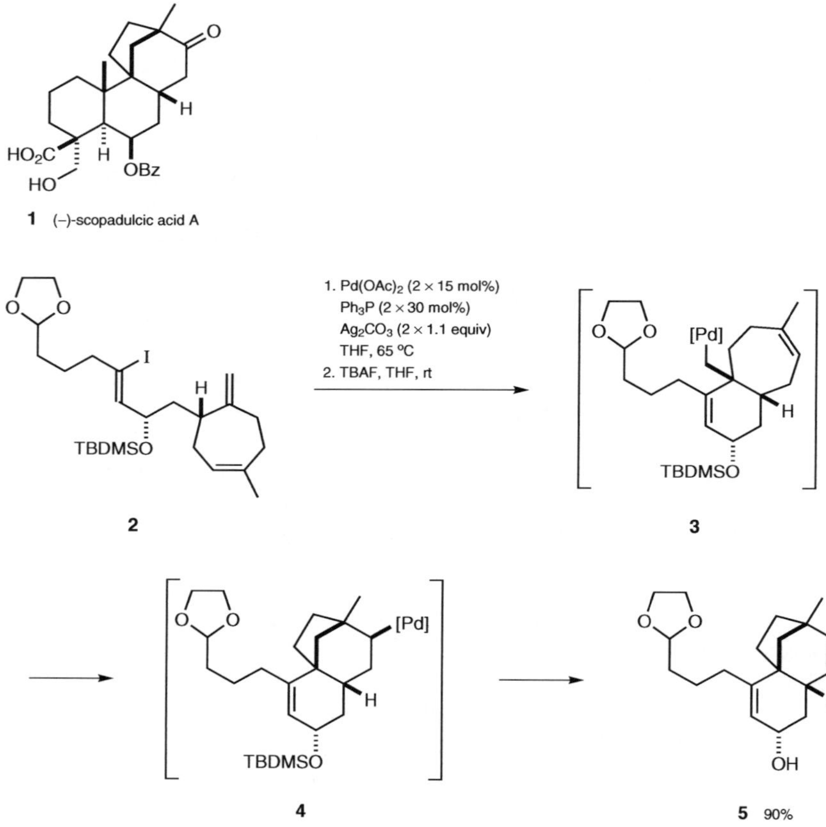

1   (−)-scopadulcic acid A

Many researchers have recognized the potential to construct substrates for pericyclic processes, in particular electrocyclizations, from alkenylpalladium intermediates. The rich history of this field owes much to biosynthetic electrocyclization pathways,[24] which have inspired numerous polyene syntheses for the purpose of exploring such biomimetic cascades. Although the Nicolaou group's landmark syntheses of the endiandric acids[25] indeed involves initiation by a palladium-catalyzed diyne semireduction, it is the Stille couplings used to assemble the tetraene precursors to the polyketides SNF4435C and SNF4435D, ocellapyrone A and ocellapyrone B, the elysiapyrones, and the tridachiapyrones that are most pertinent to this review. The groups of Trauner and Baldwin contemporaneously reported excellent seminal contributions in this arena;[26,27] here, a selection of pioneering work in the field is presented that emphasizes the flexibility of palladium catalysis to initiate these cascades.

1.6.1 Palladium-Mediated Domino Reactions

451

The Parker group was the first to achieve a synthesis of SNF4435C (**9**) and SNF4435D (**10**) (Scheme 2). Reaction of iodide **6** with stannane **7** using bis(acetonitrile)dichloropalladium(II) gives the presumed tetraene intermediate **8**, which undergoes the $8\pi/6\pi$-electrocyclization sequence to afford SNF4435C (**9**) and SNF4435D (**10**) in 53% yield in a 4:1 ratio.[28] Interestingly, the Trauner group showed that the efficiency of this transformation could be improved by reversing the polarity of the coupling partners: reaction of stannane **11** with iodide **12** affords SNF4435C (**9**) and SNF4435D (**10**) in 89% yield in a 3:1 ratio.[29] It is testament to the reliability of this sequence that many other such natural products have been approached in this manner.[24]

Scheme 2 Syntheses of SNF4435C and SNF4435D by Stille Coupling of Dienes and Electrocyclization[28,29]

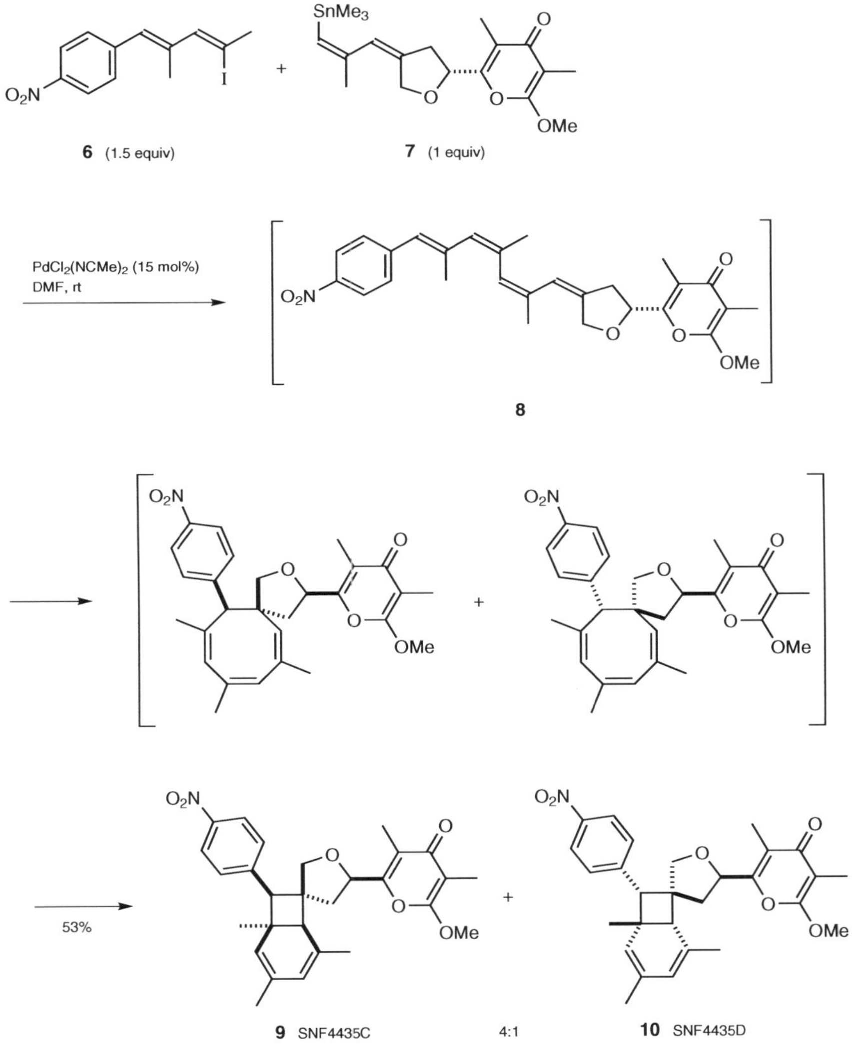

for references see p 506

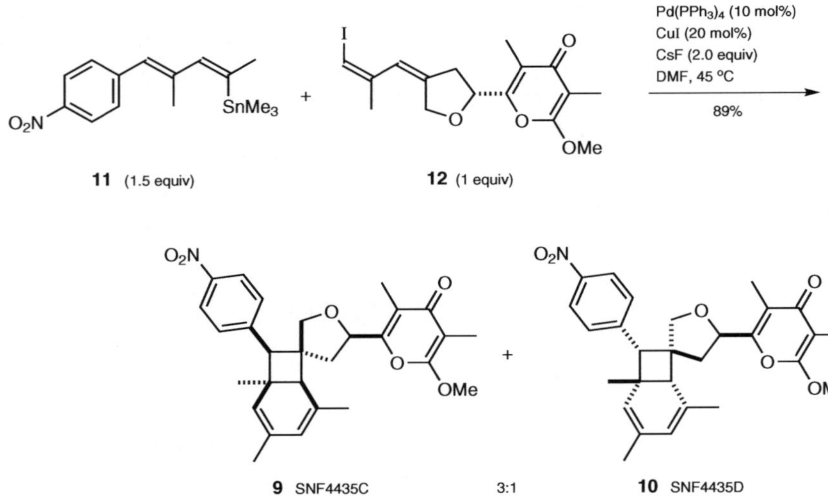

Intermediate **8** (Scheme 2) is, in fact, a double alkene stereoisomer of the natural product spectinabilin. The Baldwin approach to the SNF molecules recognized this relationship, and used a palladium(II) catalyst in a markedly different manner to initiate the cascade via alkene isomerization. Whilst application of this method in a model system met with some success,[27,30] the setting of the natural product itself revealed interesting, but complicated, reactivity.[31] Isomerization,[32] and then cyclization, of tetraene **13** on treatment with bis(acetonitrile)dichloropalladium(II) affords four products (Scheme 3): SNF4435C (**9**) and SNF4435D (**10**) (2.5:1 ratio, 22%), and byproducts **14** and **15** (2.1:1 ratio, 18%) arising from unexpected "over-isomerization" of spectinabilin prior to electrocyclization. Despite this mixture of products, this different, but effective, use of palladium catalysis demonstrates the flexibility of the metal in effecting cascade-initiating processes.

1.6.1 Palladium-Mediated Domino Reactions **453**

Scheme 3 Synthesis of SNF4435C and SNF4435D from a Tetraene[31]

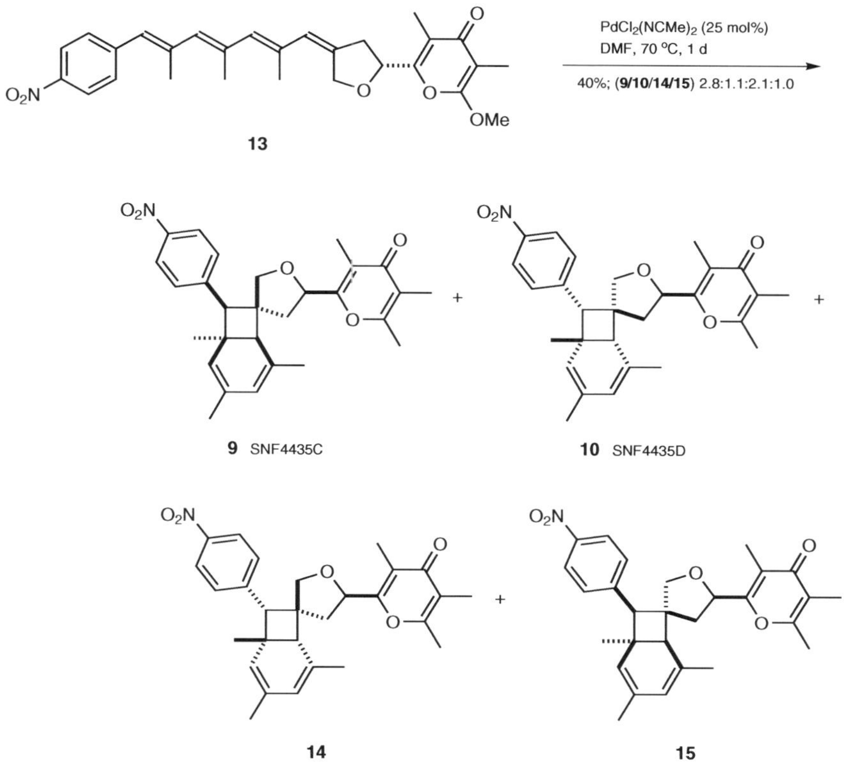

A twist to this tale of palladium-initiated $8\pi/6\pi$ cascades is found in recent efforts by the Trauner group to divert the 8π-electrocyclization product toward a Diels–Alder reaction, rather than a further electrocyclic ring closure, in work toward the insecticidal polyketide (−)-PF-1018 (**16**) (Scheme 4).[33] Although not yet rewarded with a total synthesis, success has been met with the 8π/Diels–Alder sequence. Stille coupling of stannane **17** with iodide **18**, followed by 8π-cyclization, affords **19** as a presumed intermediate, which indeed undergoes an intramolecular Diels–Alder cycloaddition in preference to the competing 6π-cyclization. This event yields product **20**, bearing the entire natural product core, in a highly efficient manner.

for references see p 506

454 Domino Transformations **1.6** Metal-Mediated Reactions

Scheme 4 Approach to (−)-PF-1018[33]

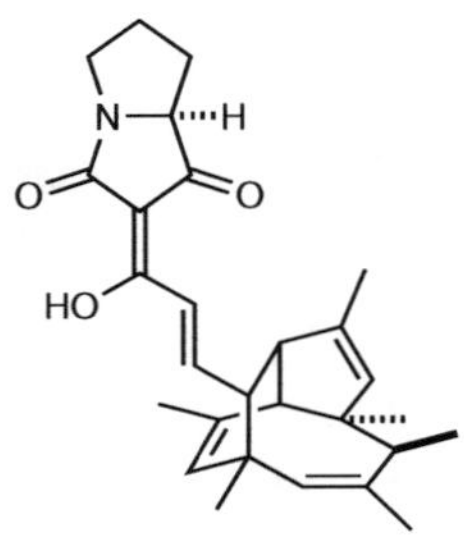

16 (−)-PF-1018

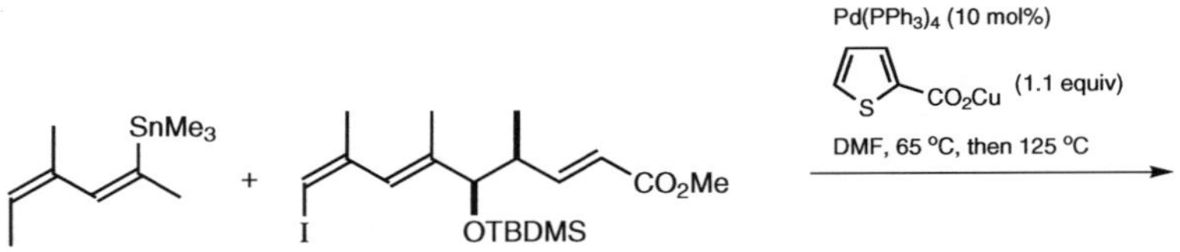

Pd(PPh3)4 (10 mol%)

CO2Cu (1.1 equiv)

DMF, 65 °C, then 125 °C

17   **18**

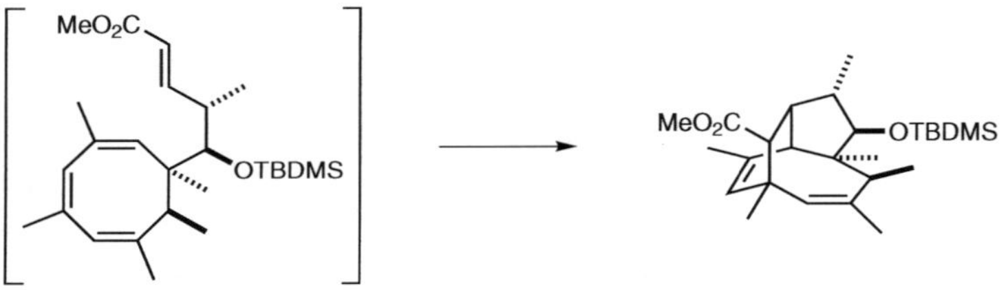

19   **20** 32%

A beautiful multicomponent cascade by the Katsumura group deploys a cross coupling/ electrocyclization in the setting of an impressive asymmetric synthesis of the alkaloid (−)-hippodamine (**21**) (Scheme 5).[34] This reaction initiates with a condensation of **22** and **23**, and Stille coupling with **24** to afford azatriene **25** (which is presumably in equilibrium with the cyclic aminal). This azatriene **25** then undergoes a highly efficient cyclization to the tetrahydropyridine product **26**, a transformation that constructs one of the hippodamine rings, with the stereoselectivity of the electrocyclization being controlled by the indane ring. Interestingly, the oxygen substituent of the indane plays a further stereocontrolling role in the synthesis by later mediating the facial selectivity of addition of a Grignard reagent to an iminium ion derived from ring opening of the aminal.

Scheme 5 Key Steps in the Synthesis of (−)-Hippodamine[34]

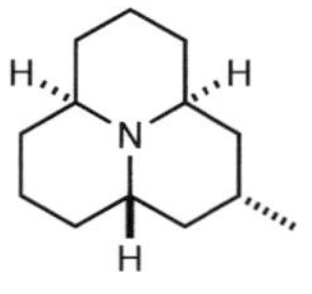

21 (−)-hippodamine

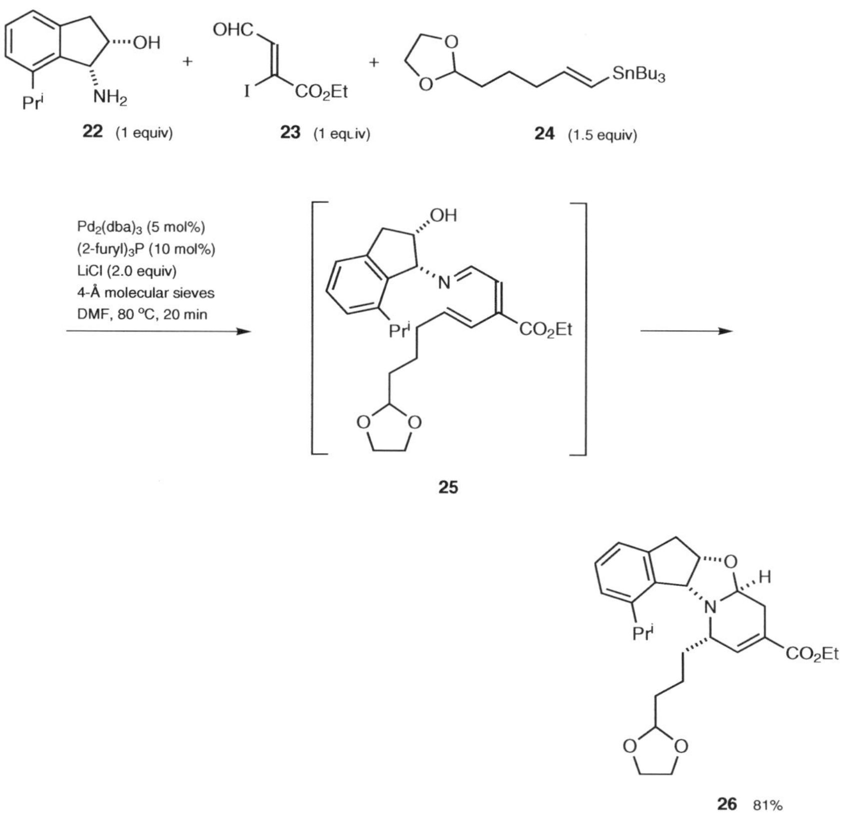

1.6.1 Palladium-Mediated Domino Reactions

An additional branch of pericyclic-terminating cascades involves reactions that feature an intramolecular carbopalladation step before an intermolecular cross coupling, and then a pericyclic process. This enables the formation of additional bonds, and in particular rings, before termination of the cascade. The Suffert group has been particularly active in this field, with a focus on complexity-inducing polycyclizations terminating in 6π- and 8π-electrocyclizations (see also Section 2.1.3).[35–38] Although several products can arise from these complex cascades, good control has been achieved in several "synthetic" settings. An example of particular relevance from the synthetic perspective of the natural product ophiobolin A (**27**)[39] is the carbopalladation/cross coupling of bromide **28** with stannane **29**. This is followed by 8π-electrocyclization to afford the 5,8,5-ring system **31**, corresponding to the carbon framework of ophiobolin A (Scheme 6). The reaction sequence involves oxidative addition into the C—Br bond of **28**, then *syn*-carbopalladation of the proximal alkynylsilane. The resulting dienylpalladium(II) complex undergoes Stille cross coupling with stannane **29**, which is followed by an 8π-electrocyclization to deliver the putative intermediate **30**. This material is unstable under the reaction conditions, with the highly reactive dihydroxycyclobutene motif then undergoing 4π-electrocyclic ring-opening to give a bis-enol, which, following tautomerization via a 1,5-hydrogen atom shift, delivers the observed product **31** and constructs the core of ophiobolin A in a single step. Although the yield for this transformation is modest, the formation of the 5,8,5-tricyclic core of the natural product in a single step is clearly impressive.

for references see p 506

Scheme 6 Approach toward Ophiobolin A[39]

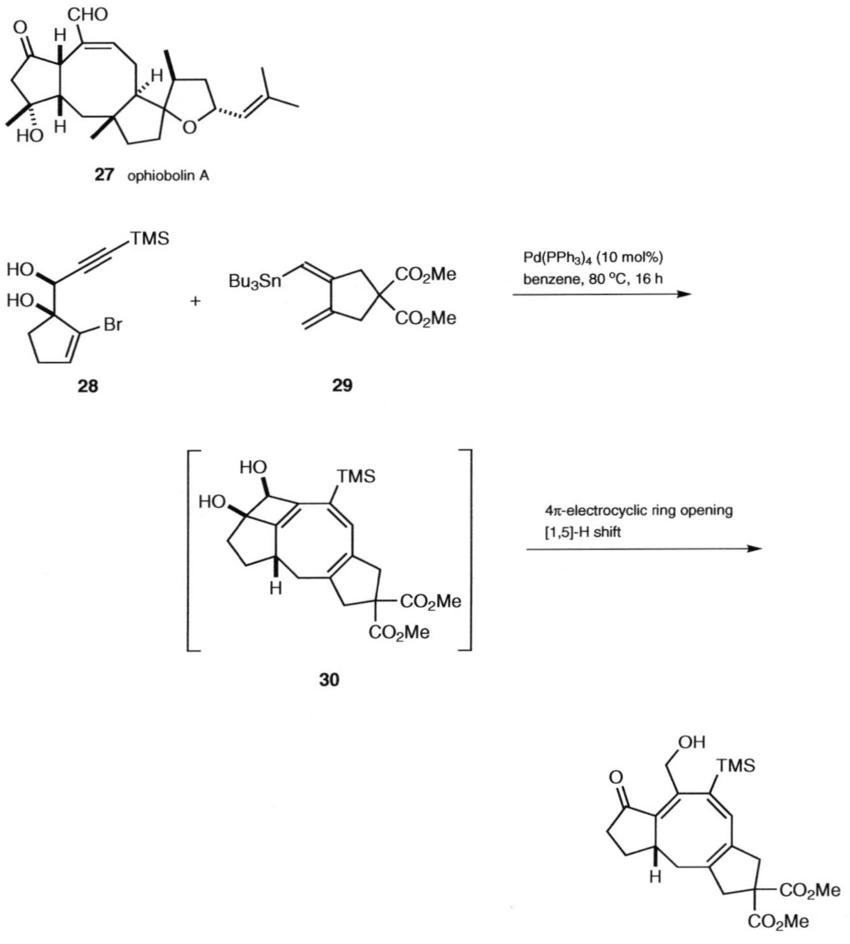

27 ophiobolin A

Pd(PPh₃)₄ (10 mol%)
benzene, 80 °C, 16 h

28   +   **29**

30

4π-electrocyclic ring opening
[1,5]-H shift

31 20%

The Anderson group has reported a related cyclization in the formation of the 7,8,5-CDE rings **35** of lancifodilactone G (**32**), starting from bromoenyne **33** (Scheme 7).[40] Treatment of **33** with stannane **34** led to an equivalent 8π/6π cascade to afford **35** in 62% yield. Importantly, a dihydrofuran coupling partner proved crucial in this reaction; use of an aromatic furylstannane partner led to further pericyclic processes, such as a 6π-electrocyclization of the 8π-electrocyclization product, presumably driven by accompanying furan rearomatization.

1.6.1 Palladium-Mediated Domino Reactions

Scheme 7 Approach toward Lancifodilactone G[40]

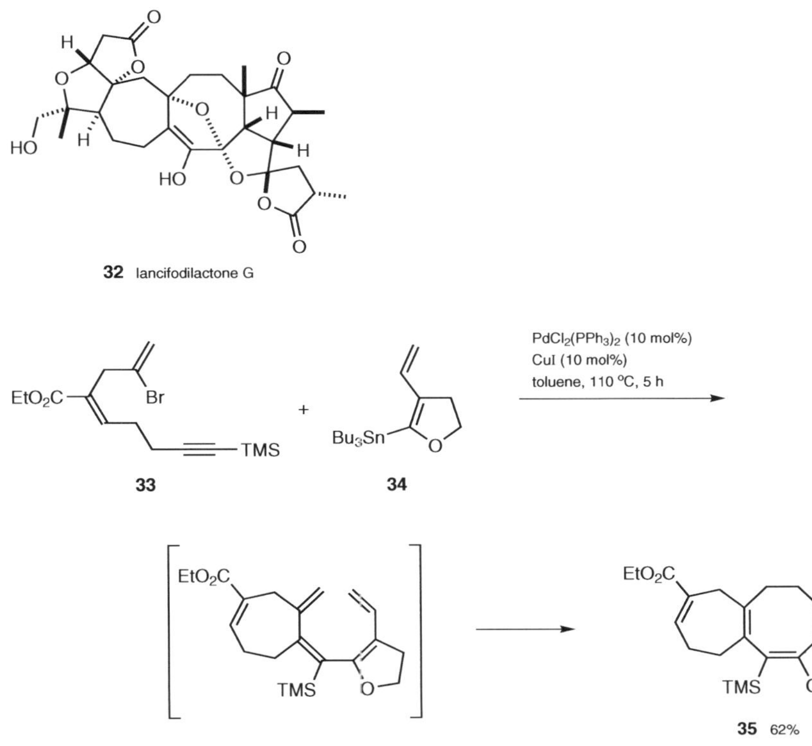

32 lancifodilactone G

PdCl2(PPh3)2 (10 mol%)
CuI (10 mol%)
toluene, 110 °C, 5 h

33 + **34**

35 62%

More unusual in the realm of carbopalladation/cross-coupling/electrocyclic cascades is the use of trialkylborate ions as coupling partners. The Ishikura group has worked extensively in this area,[41] with a particular focus on the use of triethyl(indolyl)borate salts as cascade-terminating agents. These efforts have culminated in a total synthesis of the antimalarial and anticancer natural products calothrixin A and calothrixin B (**41**) (Scheme 8).[42,43] Deprotonation of 1-methoxy-1*H*-indole (**36**) and trapping of the C2 carbanion with triethylborane affords triethyl(indolyl)borate anion **37**, which is not isolated but coupled directly with alkenyl bromide **38** using a tri-2-tolylphosphine-complexed palladium catalyst. The use of this particular ligand reduces the extent of direct (premature) coupling of the borate anion with the alkenyl bromide. The resultant product **39** does not undergo an in situ 6π-electrocyclization as is required in the synthesis, presumably due to the aromaticity of the indole component. Although a number of methods to achieve this cyclization have been examined, including photochemical promotion as used previously by the Ishikura group,[44] further investigation revealed that copper(II) trifluoromethanesulfonate, in either stoichiometric (83%) or catalytic quantities (72%), is best able to effect electrocyclization to the pentacyclic indole product **40**; four further steps furnished calothrixin B (**41**).

for references see p 506

Scheme 8 Synthesis of Calothrixin B[42,43]

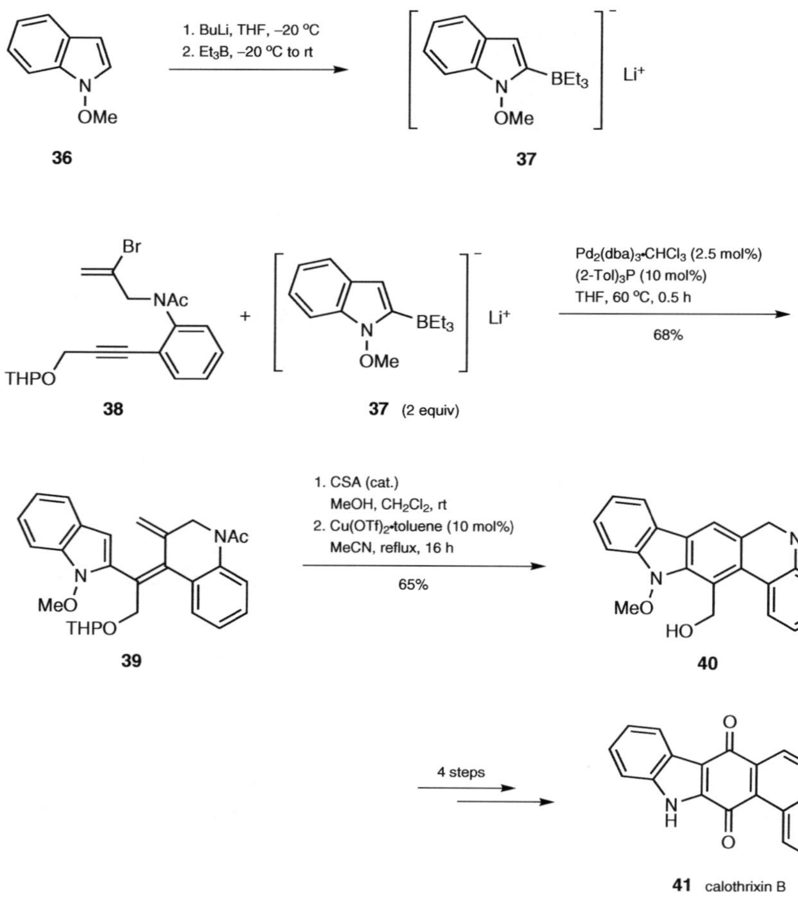

Palladium-catalyzed cycloisomerization provides an excellent, highly atom-economic means to forge complex molecular frameworks. This chemistry, pioneered by the Trost group,[45–47] is employed to great effect in a domino reaction in the setting of a formal asymmetric synthesis of echinopine A (**42**) and echinopine B by the Chen group (Scheme 9).[48] The key transformation involves the cyclization of dienyne **43** (synthesized as a 3:1 mixture of diastereomers) to give the intermediate 1,3-diene **45**, which undergoes a spontaneous Diels–Alder reaction with the pendent enoate to give cycloadduct **46** (dr 3:1 at the silyl ether bearing stereocenter). The mechanism of this impressive transformation, which forms three rings including an *ansa*-bridged system in a single step, is somewhat uncertain, but may well proceed via an initial hydropalladation of the terminal alkyne to form an alkenylpalladium intermediate **44**, which then engages with the proximal electron-rich alkene in a Heck-type reaction to form diene **45**.

1.6.1 Palladium-Mediated Domino Reactions

459

Scheme 9 Formal Synthesis of Echinopine A[48]

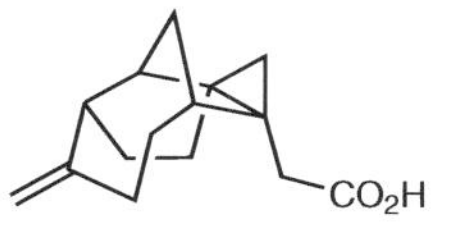

42 echinopine A

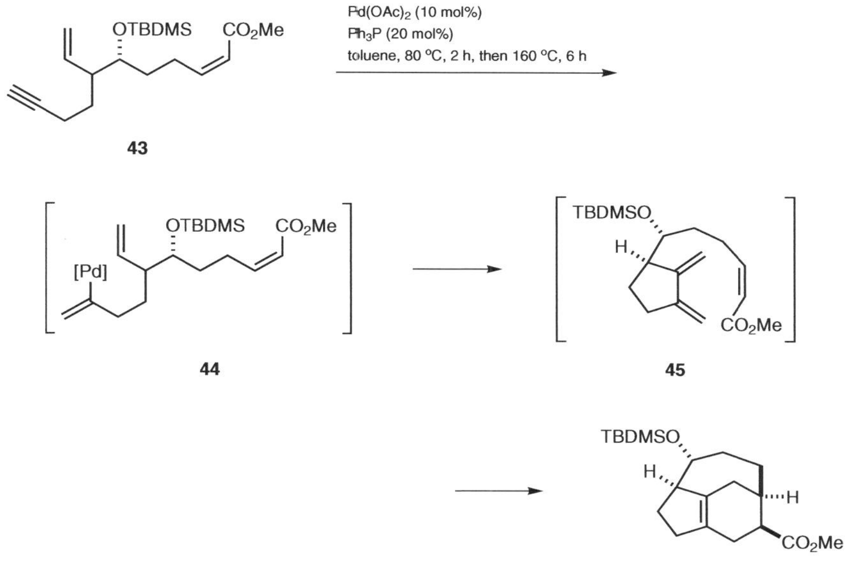

Pd(OAc)$_2$ (10 mol%)
Ph$_3$P (20 mol%)
toluene, 80 °C, 2 h, then 160 °C, 6 h

43

44

45

46 75%

In most of the examples discussed, alkenylative cross couplings usually set up a polyene system that is primed for electrocyclization. However, in the context of certain bioactive targets, this electrocyclization is undesired. Such a non-electrocyclizing cascade is employed in the synthesis of 1α,25-dihydroxyvitamin D$_3$ (**49**),[49] where reaction of the enol trifluoromethanesulfonate **47** with the alkenylpinacol boronic ester **48** affords 1α,25-dihydroxyvitamin D$_3$ (**49**) in an outstanding 81% yield (Scheme 10). This reaction is notable for the use of an alkenyl Suzuki cross-coupling reaction, which improves the environmental appeal of the transformation compared to earlier equivalent work using the Stille coupling,[50] and also for its ability to access other vitamin D$_3$ analogues under mild conditions that do not lead to isomerization of the somewhat delicate triene product (which is observed, for example, on standing in methanolic solution).

for references see p 506

460 Domino Transformations **1.6** Metal-Mediated Reactions

Scheme 10 Synthesis of 1α,25-Dihydroxyvitamin D₃ by Suzuki Cross Coupling[49]

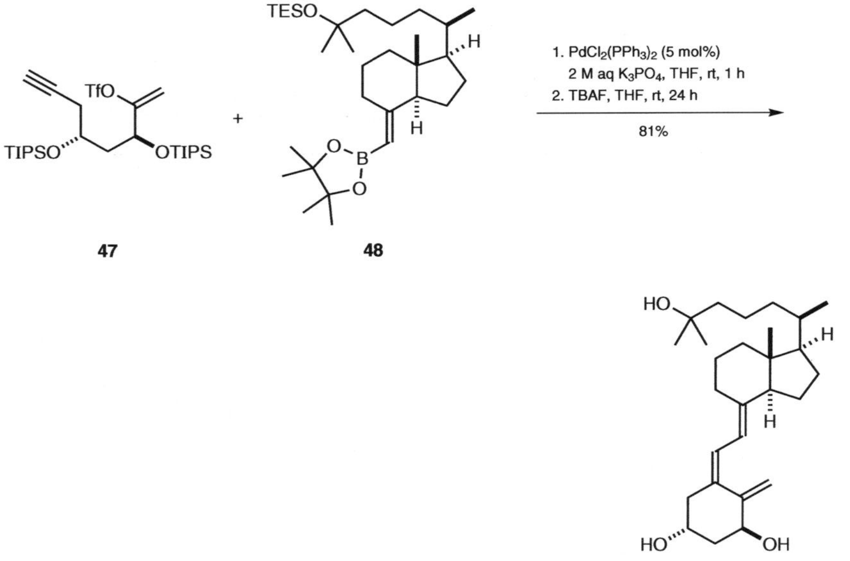

1. PdCl₂(PPh₃)₂ (5 mol%)
2 M aq K₃PO₄, THF, rt, 1 h
2. TBAF, THF, rt, 24 h

81%

47

48

49 1α,25-dihydroxyvitamin D3

Interestingly, this example is not the first approach to the vitamin D₃ framework using palladium catalysis. In earlier work, the Trost group had reversed the polarity of this reaction by employing alkenyl bromide **51** as a substrate to engage with enyne **50** (Scheme 11).[51,52] While the reaction conditions employed in this approach do lead to some isomerization of the product triene, it is still impressive that palladium catalysis is able to mediate bond formation in both "directions".

Scheme 11 Trost's Synthesis of 1α,25-Dihydroxyvitamin D₃[52]

1. Pd₂(dba)₃·CHCl₃ (3 mol%)
Ph₃P (30 mol%)
Et₃N, toluene, 110 °C
2. TBAF, THF, rt, 30 h

52%

50

51

49 1α,25-dihydroxyvitamin D3

1.6.1 Palladium-Mediated Domino Reactions

In addition to cross-coupling/electrocyclization cascades, fully intramolecular three-component cascade processes have been extensively developed following from seminal work by the de Meijere and Negishi groups.[53–56] These reactions usually feature a bromoalkene as the initiation point for the cascade, and then a combination of alkyne and/or alkene partners for the ensuing cyclization. With such a well-established synthetic methodology in place, synthetic applications toward natural products would seem inevitable. Surprisingly, such applications are relatively scarce.

Bromoene-diyne cyclizations have been applied in two synthetic contexts. The first involves construction of the tricyclic 7,6,5-core of the rubriflordilactone nortriterpenoid natural products, taking bromoene-diyne **52** as a starting point (Scheme 12).[57] This example is significant in that the formation of a seven-membered ring in such a cascade is a particularly challenging ring size, and in that the product **53** could be transformed into the core of either rubriflordilactone A (e.g., **54**) or rubriflordilactone B in two further steps.

Scheme 12 Approach toward Rubriflordilactone A[57]

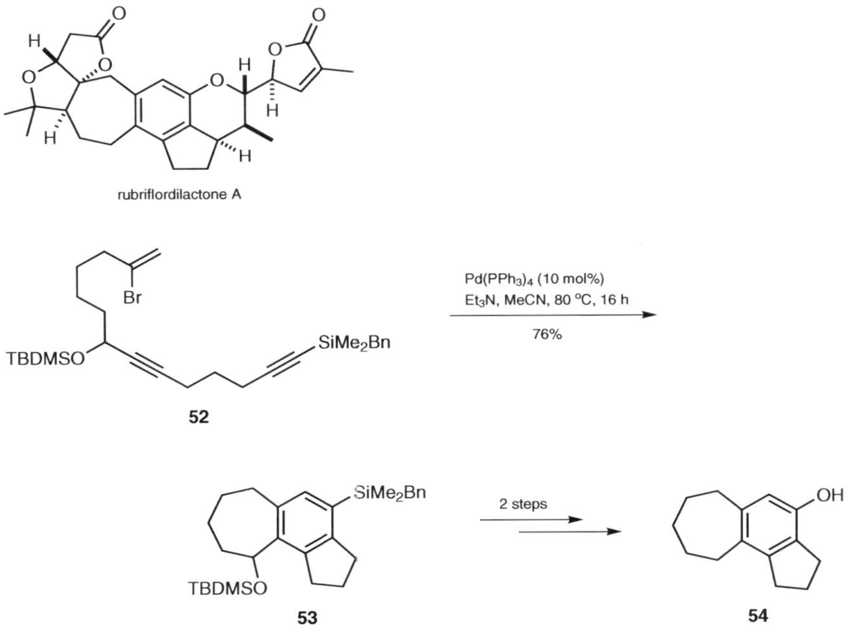

This fully intramolecular cascade cyclization has been extended to an ynamide setting, with a view toward the preparation of trikentrin- and herbindole-like frameworks (Scheme 13).[58] Taking bromoene-amide-diyne **55**, exposure to standard cyclization conditions affords the 5,6,5-azatricycle **56** in excellent yield; further synthetic manipulations, including a Friedel–Crafts *ipso*-desilylative acylation to introduce the trikentrin side chain, extended this product to didesmethyltrikentrin (**57**).

for references see p 506

Scheme 13 Synthesis of Didesmethyltrikentrin[58]

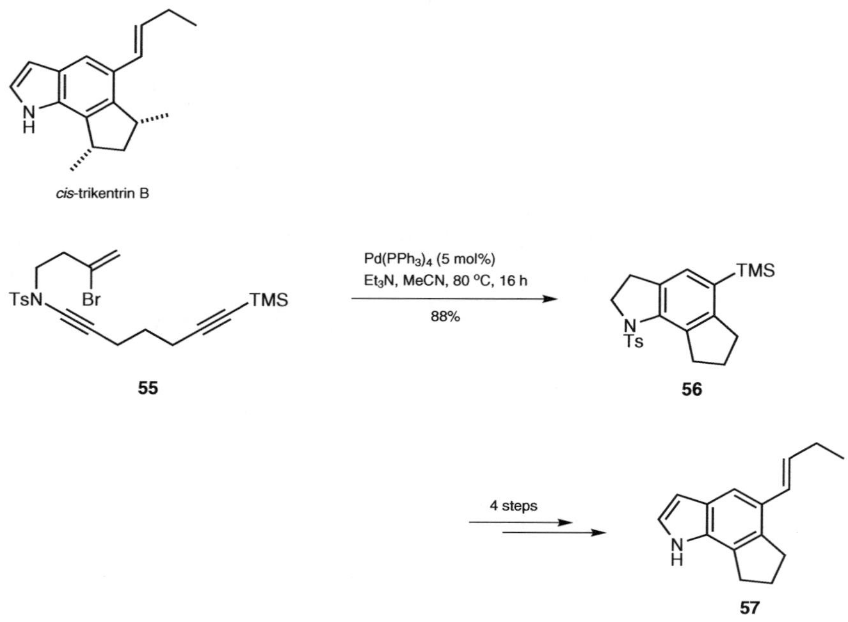

The Suffert group reported a more complex, elegant example of a bromoene-yne-ene cyclization that efficiently accesses the taxane framework (Scheme 14).[59] As with other work from this group, this involves an initial 4-*exo-dig* cyclization of the bromoalkene contained within the bromoene-yne-ene substrate **58**. Further 6-*exo-trig* cyclization onto the pendent alkene, and then β-hydride elimination, delivers a triene that undergoes 6π-electrocyclization to deliver product **59**. Notably, the stereochemistry at the newly formed stereocenters depends crucially on the nature and stereochemistry of the substituent at the alkene terminus, and generally leads to a choice of *cis*- (e.g., **59**) or *trans*-isomers in the product. With the diol unit in **59** constrained as an acetonide, subsequent 4π-electrocyclic ring opening is not possible, which enables a separate oxidative cleavage to reveal the taxane-like core **60**.

1.6.1 Palladium-Mediated Domino Reactions

463

Scheme 14 Approach to the Taxane Framework[59]

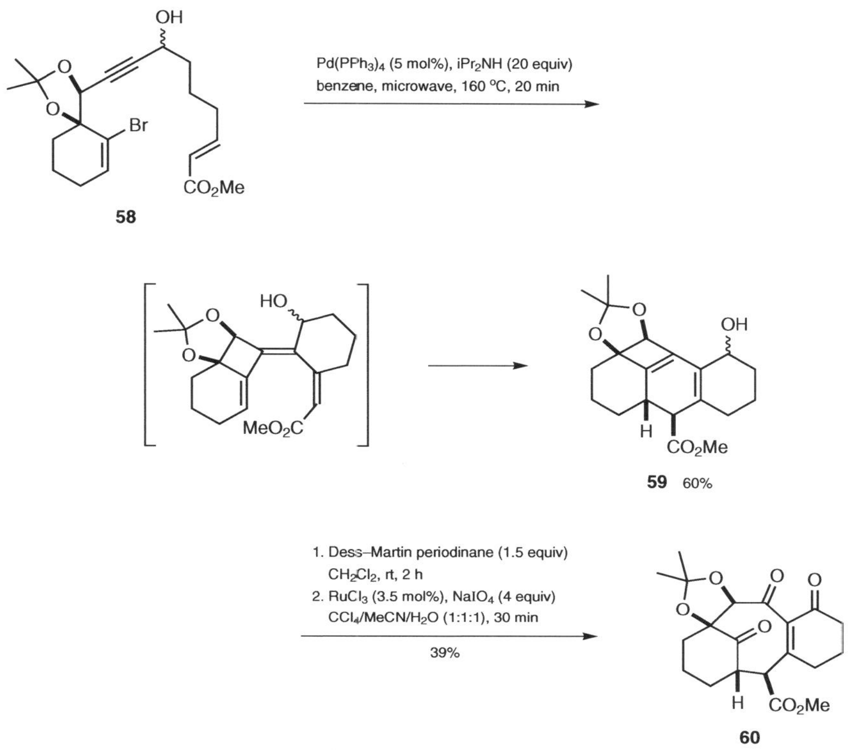

Oxopalladation and aminopalladation reactions provide a useful means to propagate palladium-catalyzed cascade pathways. This method has been extensively reviewed,[60–64] and has been applied in a number of synthetic contexts. A particularly elegant example is that of a synthesis of (+)-merobatzelladine B (**61**) reported by the Wolfe group (Scheme 15),[65] which features two aminopalladations as key ring-forming steps, each initiated by alkenylpalladium complexes. Both reactions follow an equivalent mechanism: following oxidative addition of the alkenyl halide, complexation of the alkenylpalladium complex to the alkene and nitrogen atom of substrate **62** or **65** enables a *syn*-aminopalladation reaction (e.g., via intermediate **63**).[66] Reductive elimination of the resultant alkylpalladium complex (not shown) gives the observed coupled products **64** and **66**, respectively. Notably, both cyclization reactions proceed with outstanding levels of diastereoselectivity.

Scheme 15 Key Steps in the Synthesis of (+)-Merobatzelladine B[65]

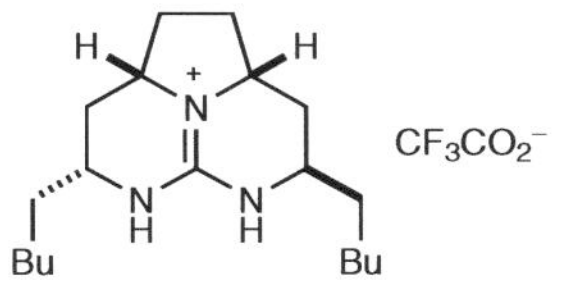

61 (+)-merobatzelladine B

for references see p 506

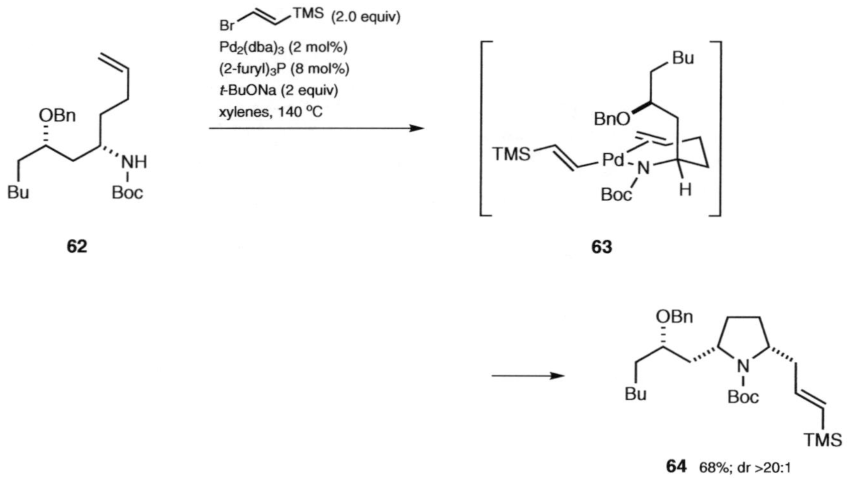

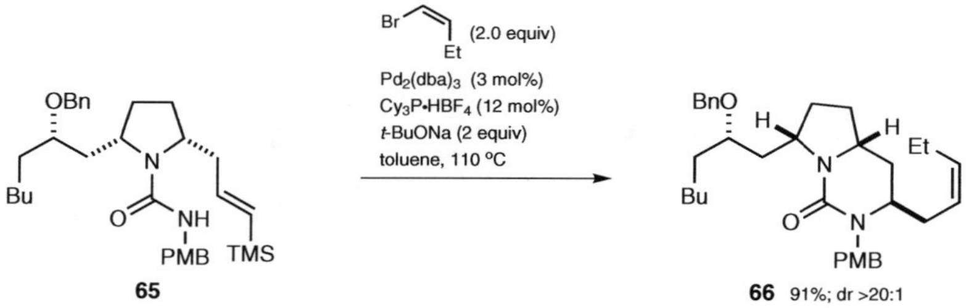

An alternative entry to alkenylpalladium species via nucleopalladation involves intramolecular nucleophilic attack onto a palladium(II)-activated alkyne. This chemistry, pioneered by Semmelhack[67] in an alkene setting and extended to alkynes by Marshall,[68] is put to excellent use in the MacMillan group's synthesis of callipeltoside C (**67**) (Scheme 16).[69] In an efficient and complexity-inducing transformation, treatment of alkynyl alcohol **68** with bis(acetonitrile)dichloropalladium(II) under a carbon monoxide atmosphere and oxidative conditions, with methanol as solvent, affords the pyranyl ester **70** in excellent yield and stereoselectivity. The reaction proceeds by initial oxopalladation of the alkyne, followed by carbon monoxide insertion and methyl ester formation. Under the mildly acidic reaction conditions, this enol ether **69** is then solvolyzed to give a kinetic mixture of pyran products;[68] however, in MacMillan's work, exposure to catalytic acid ensures that only the thermodynamic isomer **70** is formed.

1.6.1 Palladium-Mediated Domino Reactions

Scheme 16   Key Steps in the Synthesis of Callipeltoside C[69]

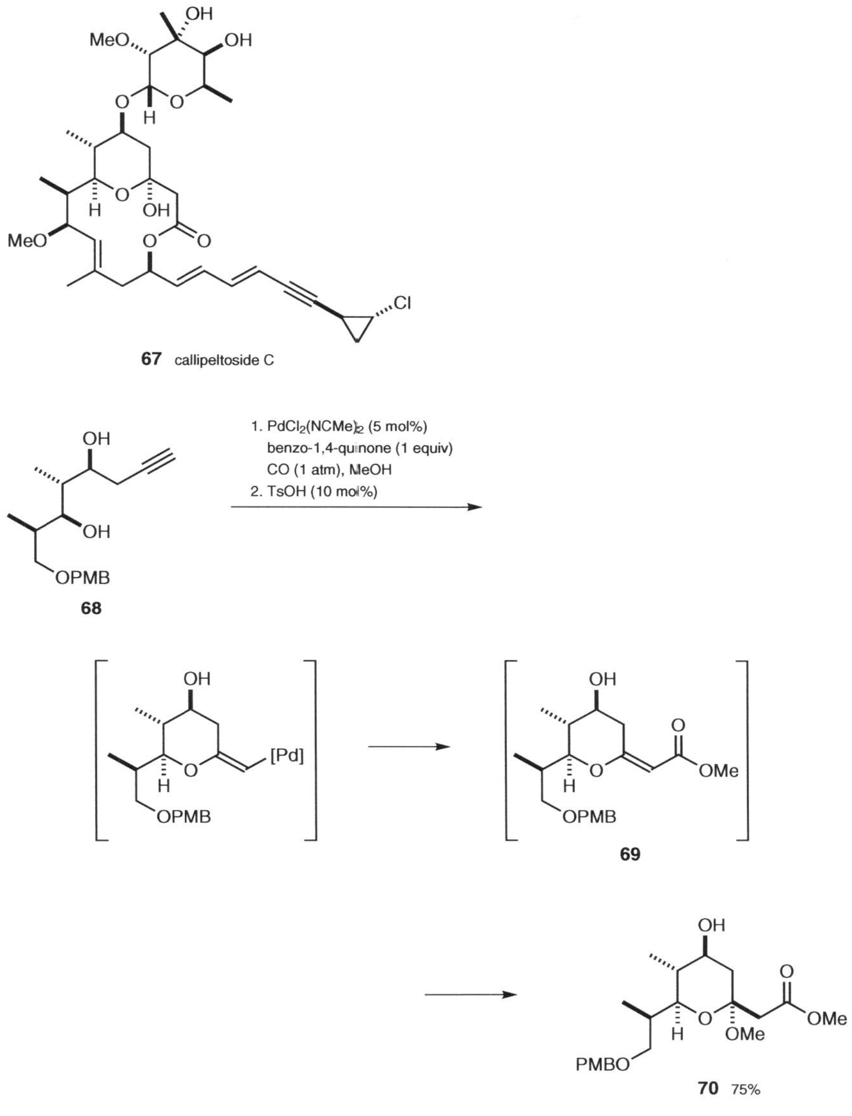

67 callipeltoside C

1. PdCl$_2$(NCMe)$_2$ (5 mol%)
   benzo-1,4-quinone (1 equiv)
   CO (1 atm), MeOH
2. TsOH (10 mol%)

68

69

70 75%

Aminopalladation can also be effected using the same principle of palladium-mediated alkyne activation to give alkenylpalladium intermediates. An excellent example of this idea is found in a formal synthesis of aspidospermidine (**71**) by the group of Han and Lu (Scheme 17),[70] in which aminopalladation of alkyne **72** leads to indolylpalladium complex **73**. This complex then undergoes an addition into the appended nitrile group, giving rise to tricycle **74**. As well as being optimized for cyclization to form a wide variety of tricycles, the specific reaction setting illustrated affords a product that could be advanced to aspidospermidine in five further steps.

for references see p 506

Domino Transformations 1.6 Metal-Mediated Reactions

Scheme 17 Formal Synthesis of Aspidospermidine[70]

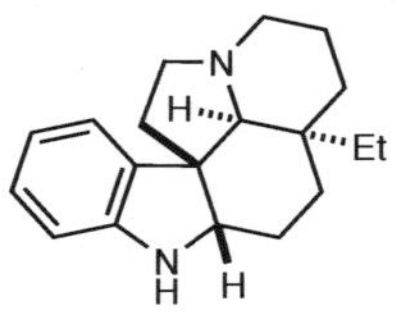

71 aspidospermidine

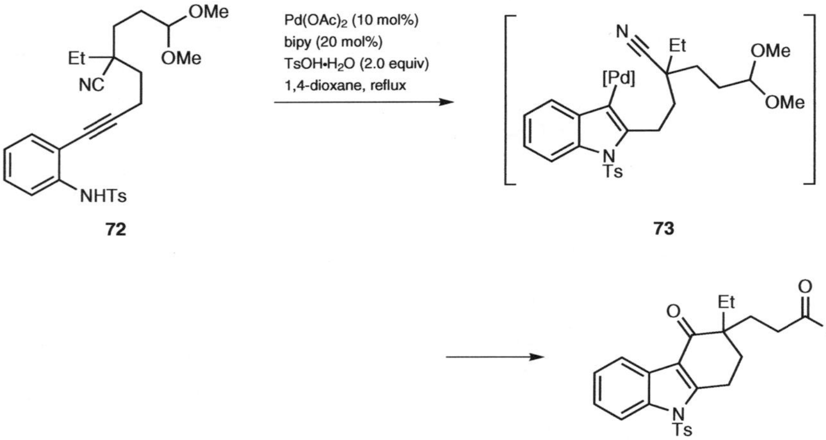

In the above chemistry, carbon monoxide insertion into an alkenylpalladium complex introduced an additional complexity-inducing element into the reaction cascade. This is an important and extremely well-established means to incorporate additional steps in such processes, and has been well used with virtually all types of palladium complex.[71,72] A particularly interesting setting is found in the Leighton group's approach to the phomoidrides, for example phomoidride B (**75**), also known as the CP-molecules (Scheme 18).[73] A wide variety of substrates were explored in this chemistry, the most advanced of which is illustrated here. The reaction involves a cyclization of alkenyl trifluoromethanesulfonate **76** in a carbonylative–spiroketalization process, forming two rings in one step. However, the resultant spiroketal lactone **77** is not the final product from the reaction, as it then undergoes a [3,3]-sigmatropic rearrangement to give **78**. Although this chemistry has not yet resulted in a completed total synthesis, its success in this challenging setting is noteworthy.

Scheme 18 An Approach toward the Phomoidrides[73]

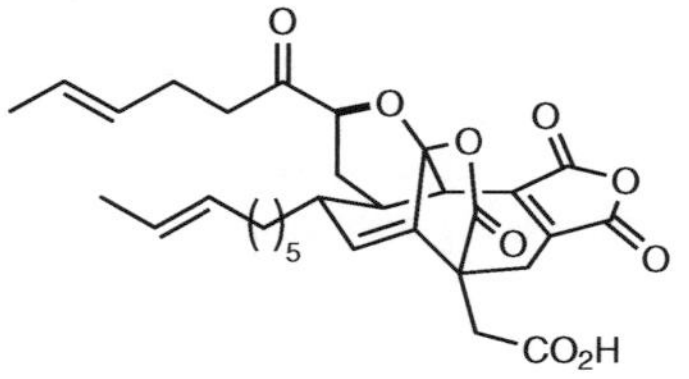

75 phomoidride B

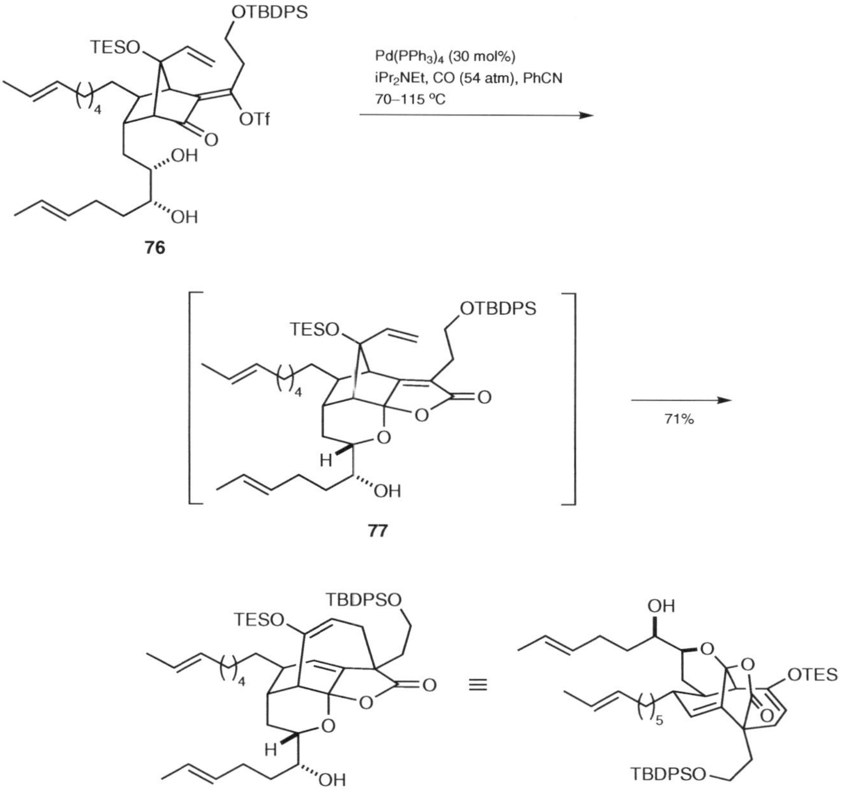

1.6.1 Palladium-Mediated Domino Reactions

A final mode of alkenylpalladium-initiating domino transformation that has been used to access pharmaceuticals or natural products is the formation of heterocycles from dihalogenated starting materials via double (or triple) C—X/C—C bond formation. This chemistry, developed extensively by the Lautens and Willis groups,[74,75] is typified by the syntheses shown in Scheme 19. In the first of these examples, the Willis group showed that reaction of dihalide **79** with hindered amine **80** smoothly affords indole **82**.[76] Although both alkenyl- and arylpalladium complexes are involved in this process, it is the former that initiates the cyclization, as evidenced by the fact that the stereochemistry of the alkenyl halide is not important: this is due to the facile isomerization of the enamine intermediate **81** following the first amination process. Indole **82** is an intermediate en route to demethylasterriquinone A1 (**83**). The second example, reported by the Lautens group, depicts the cyclization/coupling of 2-(dibromovinyl)aniline **84** and boronic acid **85** catalyzed by palladium(II) acetate and 2-(dicyclohexylphosphino)-2′,6′-dimethoxybiphenyl (SPhos).[77] This reaction likely proceeds via initial indole formation, followed by Suzuki coupling of the intermediate 2-bromoindole **86**; the product **87** was advanced to the Merck KDR kinase inhibitor **88**.

for references see p 506

468 Domino Transformations **1.6** Metal-Mediated Reactions

Scheme 19 Syntheses of Demethylasterriquinone A1 and Merck KDR Kinase Inhibitor[76,77]

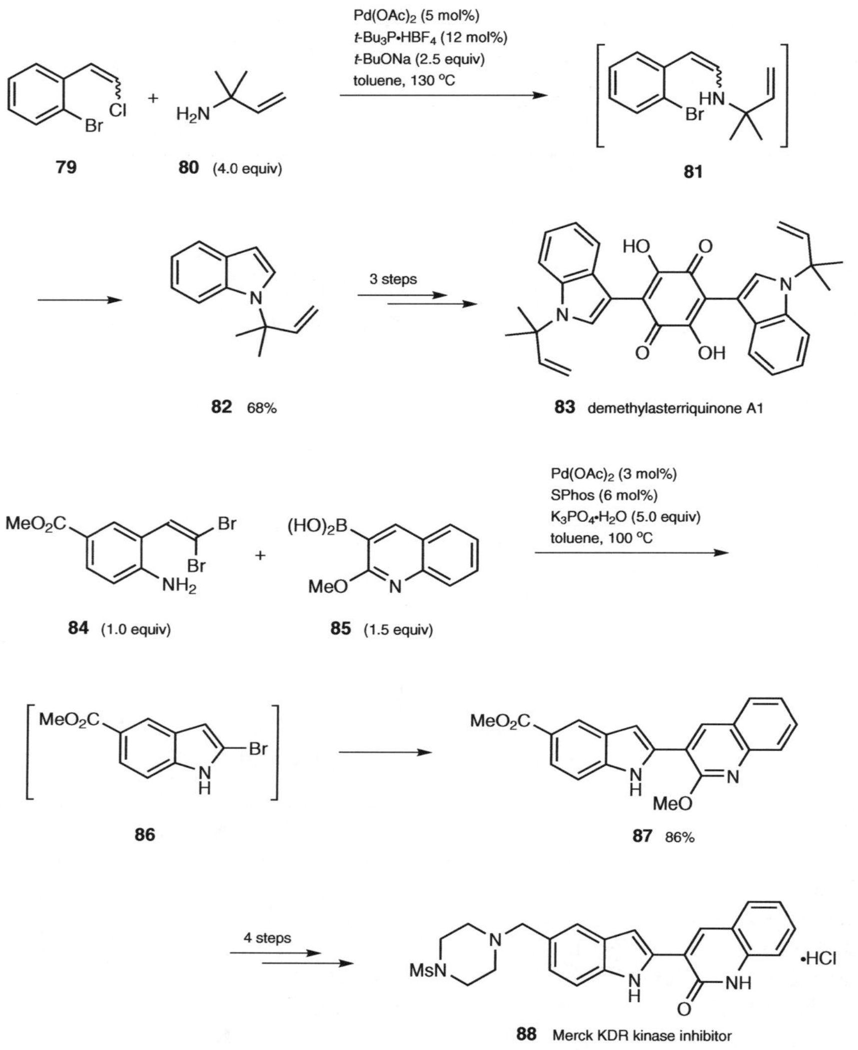

79   **80** (4.0 equiv)   **81**

82 68%   **83** demethylasterriquinone A1

84 (1.0 equiv)   **85** (1.5 equiv)

86   **87** 86%

88 Merck KDR kinase inhibitor

(2S,4aS,7S,9aS)-4-[3-(1,3-Dioxolan-2-yl)propyl]-7-methyl-1,2,5,6,7,9a-hexahydro-4a,7-methanobenzo[7]annulen-2-ol (5):[23]

To a degassed soln of vinyl iodide **2** (740 mg, 1.35 mmol, 1 equiv) in THF (75 mL) were added Ph$_3$P (107 mg, 0.41 mmol, 30 mol%), Ag$_2$CO$_3$ (410 mg, 1.49 mmol, 1.1 equiv), and Pd(OAc)$_2$ (46 mg, 0.20 mmol, 15 mol%). The resulting suspension was stirred at rt for 15 min, and then heated at 65 °C in a sealed tube for 12 h. A black suspension resulted after 10–20 min at 65 °C. After GC analysis of a filtered aliquot showed that the reaction had not proceeded to completion, additional Ph$_3$P (107 mg, 0.41 mmol, 30 mol%), Ag$_2$CO$_3$ (410 mg, 1.49 mmol, 1.1 equiv), and Pd(OAc)$_2$ (46 mg, 0.20 mmol, 15 mol%) were added, and the black suspension was stirred in the sealed tube at 65 °C for an additional 6 h. The suspension was then cooled to rt and filtered through a plug of silica gel (EtOAc), and the filtrate was concentrated to give the crude Heck product as a yellow oil. This sample was dissolved in THF (4 mL), and 1.0 M TBAF in THF (2.0 mL, 1.5 equiv) was added. The resulting soln was stirred at rt for 20 h, and then quenched with sat. aq NH$_4$Cl (20 mL). The resulting mixture was extracted with CH$_2$Cl$_2$ (3 × 20 mL), the combined organic layers

1.6.1 Palladium-Mediated Domino Reactions

were dried (Na_2SO_4) and concentrated, and the residue was purified by flash chromatography (hexanes/EtOAc 4:1) to provide the product as a pale yellow oil; yield: 370 mg (90%).

SNF4435C (9) and SNF4435D (10) from Iododiene 6 and Stannane 7:[28]

To a soln of iododiene **6** (62.9 mg, 0.191 mmol, 1.5 equiv) in dry DMF (1.3 mL) at rt were added vinylstannane **7** (56.0 mg, 0.127 mmol) and $PdCl_2(NCMe)_2$ (3.3 mg, 12.7 µmol, 10 mol%). The reaction flask was wrapped in Al foil and the mixture was stirred for 16 h at rt. Then, additional catalyst (1.65 mg, 6.35 µmol, 5 mol%) was added, and the mixture was stirred for a further 6 h at rt. To remove Me_3SnI, the mixture was stirred with KF on Celite (50 wt%) for 2 h, and then filtered through Celite. The resulting soln was washed with sat. aq $NaHCO_3$ and the aqueous layer was extracted with Et_2O. The combined organic phases were washed with H_2O and brine, dried ($MgSO_4$), and concentrated. The crude product was purified by flash chromatography (Et_2O/hexane 1:1) to give SNF4435C and SNF4435D as a yellow solid; yield: 32.3 mg (53%); ratio (**9/10**) 4:1 (^{1}H NMR).

SNF4435C (9) and SNF4435D (10) from Iododiene 12 and Stannane 11:[29]

To a soln of iododiene **12** (21 mg, 0.052 mmol, 1.0 equiv) and stannane **11** (29 mg, 0.078 mmol, 1.5 equiv) in DMF (1 mL) at rt were added CsF (16 mg, 0.10 mmol, 2.0 equiv), CuI (2 mg, 0.01 mmol, 20 mol%), and $Pd(PPh_3)_4$ (6 mg, 0.005 mmol, 10 mol%). The mixture was heated to 45 °C for 3 h, then it was cooled to rt and diluted with EtOAc (15 mL). The organic layer was washed with sat. aq NH_4Cl (3 × 10 mL), and the combined aqueous layers were extracted with EtOAc (3 × 15 mL). The combined organic layers were dried ($MgSO_4$) and concentrated. Purification by flash chromatography (hexanes/EtOAc 1:1) gave SNF4435C and SNF4435D, which could be separated by chiral HPLC; yield: 22 mg (89%); ratio (**9/10**) 3:1.

SNF4435C (9) and SNF4435D (10) from Tetraene 13:[31]

To tetraene **13** (311 mg, 0.65 mmol, 1.0 equiv) were added $PdCl_2(NCMe)_2$ (42 mg, 0.163 mmol, 25 mol%) and then dry DMF (10 mL). The resulting red-brown soln was heated in the dark at 70 °C for 23 h. At the end of the reaction, a precipitate of palladium black had formed and the soln had turned a pale yellowish color. The mixture was cooled to rt and then concentrated; the residue was purified by flash chromatography (pentanes/EtOAc 2:1) to afford a mixture of **9**, **10**, **14**, and **15** (160 mg). The mixture was subjected to preparative TLC (hexanes/Et_2O 1:1, multiple elutions) to afford a mixture of **14** and **15** [56 mg (18%); ratio (**14/15**) 2.1:1] as a yellowish stiff foam, followed by a mixture of **9** and **10** [68.5 mg (22%); ratio (**9/10**) 2.5:1] as a yellowish stiff foam.

Methyl (2S,3R,3aS,4S,Z)-3-(tert-Butyldimethylsiloxy)-2,3a,4,5,7,9-hexamethyl-2,3,3a,4,7,9a-hexahydro-1H-1,7-methanocyclopenta[8]annulene-10-carboxylate (20):[33]

Note: this reaction was performed in the absence of light. A soln of vinylstannane **17** (1.43 g, 5.55 mmol, 1.4 equiv) in benzene (12 mL) (**CAUTION:** *carcinogen*) and vinyl iodide **18** (1.84 g, 3.96 mmol, 1.0 equiv) were combined and concentrated (rotovap bath temperature kept below 35 °C) to azeotropically remove traces of H_2O, and then redissolved in dry DMF (40 mL). This soln was added to a mixture of $Pd(PPh_3)_4$ (458 mg, 0.40 mmol, 10 mol%) and copper(I) thiophene-2-carboxylate (831 mg, 4.36 mmol, 1.1 equiv) in a flame-dried Schlenk flask under N_2, wrapped in Al foil to shield it from light. The resulting mixture was immediately heated to 65 °C (external oil bath temperature) and stirred for 1 h. Following this period, the temperature was increased to 125 °C and the mixture was stirred for an additional 17 h. The mixture was cooled to rt and diluted with hexane/Et_2O (7:3; 50 mL); the reaction volume was filtered through a plug of silica gel [hexane/Et_2O 7:3 (2 × 100 mL)]. The solvent was evaporated, the residue was purified by flash chromatography (hexane/EtOAc 30:1), and the fractions containing the desired product were com-

for references see p 506

470 Domino Transformations **1.6** Metal-Mediated Reactions

bined and concentrated. An additional purification by flash chromatography was performed (hexane/EtOAc 30:1), after which a white solid was obtained. The solid was recrystallized (MeOH) to yield the product as colorless needles; yield: 550 mg (32%).

Ethyl (4b*R*,6*S*,9a*S*,10a*S*)-6-[3-(1,3-Dioxolan-2-yl)propyl]-4-isopropyl-4b,9,9a,10a-tetrahydro-6*H*,11*H*-indeno[1′,2′:4,5]oxazolo[3,2-*a*]pyridine-8-carboxylate (26):[34]

To a suspension of **23** (2.65 g, 10.45 mmol, 1.0 equiv) and 4-Å molecular sieves (10.45 g) in DMF (40 mL) was added amine **22** (2.00 g, 10.45 mmol, 1.0 equiv), and the mixture was stirred for 1 h at rt. To this suspension were then added LiCl (885 mg, 20.9 mmol, 2.0 equiv), tri(2-furyl)phosphine (243 mg, 1.05 mmol, 10 mol%), and Pd$_2$(dba)$_3$ (479 mg, 0.523 mmol, 5 mol%), and the mixture was stirred for 5 min at rt. A soln of stannane **24** (6.76 g, 15.68 mmol, 1.5 equiv) in DMF (10 mL) was then added to this suspension, and the mixture was heated to 80 °C for 20 min. The mixture was cooled to rt, H$_2$O was added, and the mixture was extracted with Et$_2$O. The organic layers were combined, washed with brine, dried (MgSO$_4$), and concentrated to give the crude product, which was purified by flash chromatography (EtOAc/hexane 5:95 to 13:87) to give the product as a pale yellow oil; yield: 3.72 g (81%).

Dimethyl (*S*,4*E*,6*E*)-6-(Hydroxymethyl)-7-oxo-5-(trimethylsilyl)-3,7,8,9,9a,10-hexahydrodicyclopenta[*a*,*d*][8]annulene-2,2(1*H*)-dicarboxylate (31):[39]

To a soln of bromide **28** (100 mg, 0.35 mmol, 1 equiv) in anhyd benzene (10 mL) (**CAUTION:** *carcinogen*) under argon were added sequentially Pd(PPh$_3$)$_4$ (40.4 mg, 0.035 mmol, 10 mol%) and stannane **29** (225 mg, 0.45 mmol, 1.3 equiv). The mixture was heated at reflux for 16 h, and then it was cooled to rt and concentrated. Purification by flash chromatography (Et$_2$O/hexane 4:1) afforded the product; yield: 29.0 mg (20%).

Ethyl (3a*Z*,6a*E*,12*Z*)-12-(Trimethylsilyl)-3,5,6,7,10,11-hexahydro-2*H*-cyclohepta[6,7]cycloocta[1,2-*b*]furan-8-carboxylate (35):[40]

To a degassed soln of bromoenyne **33** (30 mg, 0.0875 mmol, 1 equiv) in toluene (1 mL) were added PdCl$_2$(PPh$_3$)$_2$ (6 mg, 8.75 µmol, 10 mol%) and CuI (1.7 mg, 8.75 µmol, 10 mol%). The mixture was heated to 110 °C, and a degassed soln of stannane **34** (57 mg, 0.148 mmol, 1.7 equiv) in toluene (0.5 mL) was added via syringe pump over 1 h. The mixture was heated for a further 4 h, and then it was cooled and concentrated. The residue was purified by flash chromatography (petroleum ether/Et$_2$O 15:1 containing 0.5% Et$_3$N) to afford the product as a white solid; yield: 19.4 mg (62%).

(*Z*)-1-Acetyl-4-{1-(1-methoxy-1*H*-indol-2-yl)-2-[(tetrahydro-2*H*-pyran-2-yl)oxy]ethylidene}-3-methylene-1,2,3,4-tetrahydroquinoline (39):[43]

To a soln of 1-methoxy-1*H*-indole (**36**; 1.60 g, 11.0 mmol, 2.0 equiv) in THF (100 mL), at −20 °C under an argon atmosphere, was added 1.6 M BuLi in hexane (8.25 mL, 13.2 mmol, 2.4 equiv). The mixture was stirred for 30 min, and then 1.0 M Et$_3$B in hexane (13.2 mL, 13.2 mmol, 2.4 equiv) was added. The mixture was stirred for 30 min at −20 °C, and then it was gradually warmed to rt and stirred for 1 h. Enyne **38** (2.15 g, 5.50 mmol, 1.0 equiv), (2-Tol)$_3$P (167 mg, 0.55 mmol, 10 mol%), and Pd$_2$(dba)$_3$•CHCl$_3$ (142 mg, 0.138 mmol, 2.5 mol%) were then added to the mixture and it was stirred at rt for 10 min and then heated at 60 °C for 30 min. The mixture was then cooled in ice, and 10% aq NaOH (20 mL) and 30% aq H$_2$O$_2$ (10 mL) were added. The mixture was stirred for 20 min, then diluted with EtOAc (300 mL), and washed with brine, and the organic extract was dried (MgSO$_4$) and concentrated. The residue was separated by MPLC (hexane/EtOAc 1.5:1) to give the product; yield: 1.70 g (68%).

1.6.1 Palladium-Mediated Domino Reactions **471**

Methyl (5*R***,8***R***,8a***R***,9***S***)-8-(*tert*-Butyldimethylsiloxy)-1,2,4,5,6,7,8,8a-octahydro-5,3-ethano-azulene-9-carboxylate (46):**[48]
To a stirred soln of Pd(OAc)$_2$ (32 mg, 0.14 mmol, 10 mol%) in toluene (10 mL) at rt was added Ph$_3$P (77 mg, 0.28 mmol, 20 mol%). The resulting mixture was stirred for 30 min and then it was added to a soln of enyne **43** (500 mg, 1.40 mmol, 1.0 equiv) in toluene (47 mL). This mixture was heated to 80 °C and stirred for 2 h; it was then heated to 160 °C and stirred for a further 6 h. The mixture was then cooled to rt and concentrated. Flash chromatography (hexanes/EtOAc 98:2 to 95:5) afforded the product as a colorless oil; yield: 375 mg (75%); ca. 3:1 mixture of C10 epimers (^{1}H NMR).

1α,25-Dihydroxyvitamin D$_3$ (49) from Boronate 48 and Enol Trifluoromethanesulfonate 47:[49]
A 2.0 M aq soln of K$_3$PO$_4$ (3.0 mL) and PdCl$_2$(PPh$_3$)$_2$ (0.0081 g, 0.0115 mmol, 5 mol%) were successively added to a soln of boronate **48** (89 mg, 0.231 mmol, 1.0 equiv) and enol trifluoromethanesulfonate **47** (0.163 g, 0.271 mmol, 1.17 equiv) in THF (5 mL). The mixture was stirred vigorously at rt for 1 h. The mixture was then diluted with H$_2$O (5 mL) and extracted with Et$_2$O (3 × 10 mL). The combined organic phases were dried and concentrated; the residue was purified by flash chromatography (Et$_2$O/hexanes 1:99) to afford the protected hormone, which was dissolved in THF (5 mL). Then, 1 M TBAF in THF (1.38 mL, 1.38 mmol, 6 equiv) was added. The mixture was stirred for 24 h at rt and then diluted with sat. aq NH$_4$Cl (5 mL). The mixture was extracted with EtOAc (5 × 10 mL). The combined organic phases were dried and concentrated, and the residue was purified by flash chromatography (EtOAc/hexanes 1:1) to afford the product as a white solid (crystallized from CHCl$_3$/EtOAc); yield: 0.078 g (81%).

1α,25-Dihydroxyvitamin D$_3$ (49) from Enyne 50 and Bromide 51:[52]
A mixture of Pd$_2$(dba)$_3$•CHCl$_3$ (20 mg, 0.0192 mmol, 3 mol%) and Ph$_3$P (47 mg, 0.179 mmol, 30 mol%) in toluene (2 mL) and Et$_3$N (3 mL) was stirred at rt for 10 min. Then, a soln of bromide **51** (327 mg, 0.916 mmol, 1.5 equiv) and enyne **50** (224 mg, 0.610 mmol, 1.0 equiv) in toluene (1 mL) was added. The mixture was heated at reflux for 2 h, and then it was cooled to rt and diluted with pentane (5 mL). The mixture was filtered through a pad of silica gel (Et$_2$O). Rapid flash chromatography (Et$_2$O/pentane 2:1) gave a mixture of products (650 mg), which was treated with 0.5 M TBAF in THF (6 mL, 3.0 mmol). The mixture was stirred at rt for 30 h, and then purified by flash chromatography (EtOAc) to give the product; yield: 132 mg (52%).

4-(Benzyldimethylsilyl)-10-(*tert*-butyldimethylsiloxy)-1,2,3,6,7,8,9,10-octahydrocyclohepta[*e*]indene (53):[57]
Bromoenediyne **52** (150 mg, 0.275 mmol, 1.0 equiv) was dissolved in dry MeCN (0.7 mL), and the soln was degassed with argon bubbling for 30 min. Et$_3$N was separately degassed with argon bubbling for 30 min. A vial equipped with a stirrer bar was charged with Pd(PPh$_3$)$_4$ (32.0 mg, 0.0275 mmol, 10 mol%) in a glovebox, and placed under argon. The degassed soln of starting material was added to the catalyst by syringe, followed by degassed Et$_3$N (0.23 mL, 1.65 mmol, 6.0 equiv). The mixture was heated to 80 °C for 16 h and then it was cooled to rt, and the solvent carefully was removed under reduced pressure. Purification of the residue by flash column chromatography [petroleum ether (bp 40–60 °C)/Et$_2$O 99:1] afforded the product as a colorless oil; yield: 97 mg (76%).

1-Tosyl-5-(trimethylsilyl)-1,2,3,6,7,8-hexahydrocyclopenta[*g*]indole (56):[58]
To an oven-dried vial, equipped with a stirrer bar, was added Pd(PPh$_3$)$_4$ (37.2 mg, 0.032 mmol, 5 mol%) in a glovebox, and the vial was sealed with a rubber septum. To this were added a degassed (argon bubbling, 15 min) soln of the bromoenynamide **55** (300 mg, 0.643 mmol, 1.0 equiv) and Et$_3$N (0.54 mL, 3.86 mmol, 6.0 equiv) in anhyd MeCN (4.0 mL).

for references see p 506

472 Domino Transformations **1.6** Metal-Mediated Reactions

The rubber septum was replaced rapidly with a screw cap, and the mixture was heated to 80 °C until the reaction was complete (TLC monitoring; 16 h). On completion, the mixture was cooled to rt and concentrated under reduced pressure. Purification by column chromatography [petroleum ether (bp 40–60 °C)/Et$_2$O 3:1] afforded the product as a colorless crystalline solid; yield: 217 mg (88%).

Methyl (3aR,6aS,7S,11cS)-11-Hydroxy-2,2-dimethyl-5,6,6a,8,9,10,11,11c-octahydroanthra-[9′,1′:3,4,1]cyclobuta[1,2-d][1,3]dioxole-7-carboxylate (59):[59]

A soln of substrate **58** (100 mg, 0.234 mmol), Pd(PPh$_3$)$_4$ (17.0 mg, 0.014 mmol), and iPr$_2$NH (750 μL, 5.62 mmol) in benzene (3 mL) (**CAUTION:** *carcinogen*) was placed in a 3-mL microwaveable vial, which was sealed and degassed (argon purge) for 5 min. The mixture was then heated to 160 °C in a microwave (vertically focused IR temperature sensor monitoring) and maintained at this temperature for 20 min. The mixture was cooled, the solvent was evaporated, and the residue was purified directly by column chromatography (silica gel, pentane/Et$_2$O 3:2) to afford the product as a pale yellow oil; yield: 48 mg (60%).

(+)-tert-Butyl (E,2S,5R)-2-[(2S)-2-(Benzyloxy)heptyl]-5-[3-(trimethylsilyl)allyl]pyrrolidine-1-carboxylate (64):[65]

A flame-dried Schlenk flask was cooled under a stream of N$_2$ and charged with Pd$_2$(dba)$_3$ (18.3 mg, 0.02 mmol, 2 mol%), tri(2-furyl)phosphine (18.6 mg, 0.08 mmol, 8 mol%), and *t*-BuONa (200 mg, 2.08 mmol, 2.0 equiv). The flask was purged with N$_2$, then a soln of **62** (406 mg, 1.04 mmol, 1.0 equiv) in xylenes (5.2 mL) was added via syringe, and the resulting mixture was stirred at rt for 5 min. (E)-(2-Bromovinyl)trimethylsilane (319 μL, 2.08 mmol, 2.0 equiv) was added, and the flask was heated to 140 °C and stirred overnight. The mixture was cooled to rt, and sat. aq NH$_4$Cl (5 mL) and EtOAc (5 mL) were added. The mixture was filtered through a plug of silica gel (EtOAc, 20 mL). The filtrate was transferred to a separatory funnel, H$_2$O was added (10 mL), the layers were separated, and the aqueous layer was extracted with EtOAc (3 × 10 mL). The combined organic layers were dried (Na$_2$SO$_4$) and concentrated. The crude material was purified by flash chromatography to afford the product as a pale brown oil; yield: 347 mg (68%); mixture of rotamers (^{1}H and ^{13}C NMR).

Methyl 2-{(2S,4S,5R,6S)-4-Hydroxy-2-methoxy-6-[(R)-1-(4-methoxybenzyloxy)propan-2-yl]-5-methyltetrahydro-2H-pyran-2-yl}acetate (70):[69]

> **CAUTION:** *Carbon monoxide is extremely flammable and exposure to higher concentrations can quickly lead to a coma.*

A 25-mL round-bottomed flask containing a soln of alkynyl diol **68** (402 mg, 1.31 mmol, 1.0 equiv) in anhyd MeOH (8.75 mL) was purged with CO for 1 min, and then cooled to 0 °C under 1 atm CO pressure. Benzo-1,4-quinone (142 mg, 1.31 mmol, 1.0 equiv) and PdCl$_2$(NCMe)$_2$ (17.0 mg, 0.066 mmol, 5 mol%) were added, and the soln was re-purged with CO for 1 min and then stirred for an additional 3 h at 0 °C. TsOH•H$_2$O (25.0 mg, 0.131 mmol, 10 mol%) was added, and the soln was stirred for 30 min at 0 °C and then warmed to rt for 15 min. The mixture was then diluted with EtOAc (125 mL) and washed with 0.1 M NaOH (100 mL). The aqueous layer was back-extracted with EtOAc (2 × 50 mL), and the combined organic phases were dried (Na$_2$SO$_4$) and concentrated. Purification by flash chromatography (EtOAc/hexanes 2:3) afforded the product as a clear, colorless oil; yield: 389 mg (75%).

3-(3-Ethyl-4-oxo-9-tosyl-2,3,4,9-tetrahydro-1H-carbazol-3-yl)propanal (74):[70]

To a dry flask containing Pd(OAc)$_2$ (11 mg, 0.045 mmol, 10 mol%), 2,2′-bipyridine (15 mg, 0.090 mmol, 20 mol%), alkyne **72** (2.15 g, 4.59 mmol, 1.0 equiv), and TsOH•H$_2$O (1.75 g, 9.18 mmol, 2.0 equiv) was added 1,4-dioxane (40 mL), and the resulting mixture was heat-

1.6.1 Palladium-Mediated Domino Reactions

ed at reflux for 5 h. The mixture was cooled and diluted with H_2O (50 mL), and the aqueous soln was extracted with EtOAc (4×30 mL). The combined organic phases were dried (Na_2SO_4) and concentrated. The residue was purified by flash chromatography (petroleum ether/EtOAc 4:1 to 3:1) to give the product as a pale yellow oil; yield: 1.71 g (89%).

(2S,4R,4aR,8S,10aS)-8-[2-(*tert*-Butyldiphenylsilyloxy)ethyl]-2-[(*R*,*E*)-1-hydroxyhex-4-enyl]-13-[(*E*)-oct-6-enyl]-5-(triethylsilyloxy)-4,4a,7,8-tetrahydro-2*H*-8,10a,4-(prop[1]ene[1,1,3]triyl)pyrano[2,3-*b*]oxocin-9(3*H*)-one (78):[73]

> **CAUTION:** *Carbon monoxide is extremely flammable and exposure to higher concentrations can quickly lead to a coma.*

To a soln of trifluoromethanesulfonate **76** (38 mg, 0.039 mmol, 1.0 equiv) in benzonitrile (6 mL) was added iPr$_2$NEt (21 μL, 0.12 mmol, 3 equiv). The resulting soln was added by cannula to a Parr model 4700 pressure vessel, under an N_2 atmosphere, charged with Pd(PPh$_3$)$_4$ (14 mg, 0.012 mmol, 30 mol%). The pressure gauge assembly was then attached and the apparatus was charged with CO (54 atm), and vented twice. The apparatus was then pressurized to 54 atm CO and was heated at 70 °C (oil bath) for 5 h, and then at 115 °C (oil bath) for 2.5 h. The pressure vessel was cooled and vented, and the mixture was concentrated. The residue was purified by flash chromatography (Et$_2$O/hexanes 1:9) to give the product as an oil; yield: 23 mg (71%).

1-(2-Methylbut-3-en-2-yl)-1*H*-indole (82):[76]

To chloroalkene **79** (873 mg, 4.02 mmol, 1.0 equiv), Pd(OAc)$_2$ (45 mg, 0.20 mmol, 5 mol%), *t*-Bu$_3$P•HBF$_4$ (140 mg, 0.48 mmol, 12 mol%), and *t*-BuONa (964 mg, 10.04 mmol, 2.5 equiv) in a sealable reaction vessel under argon were added 2-methylbut-3-en-2-amine (**80**; 1.37 g, 16.06 mmol, 4 equiv) and toluene (8.5 mL). The reaction vessel was sealed and heated to 130 °C for 4 h, then the mixture was cooled to rt and filtered through a plug of Celite, and the filtrate was concentrated. The residue was purified by flash chromatography (Et$_2$O/petroleum ether 1:99) to give the product as a yellow oil; yield: 508 mg (68%).

Methyl 2-(2-Methoxyquinolin-3-yl)-1*H*-indole-5-carboxylate (87):[77]

To dibromoalkene **84** (168 mg, 0.50 mmol), boronic acid **85** (152 mg, 0.75 mmol, 1.5 equiv), Pd(OAc)$_2$ (3.4 mg, 0.015 mmol, 3 mol%), SPhos (12.3 mg, 0.030 mmol, 6 mol%), and K$_3$PO$_4$•H$_2$O (0.58 g, 2.50 mmol, 5.0 equiv) under argon was added toluene (2.5 mL). The resulting mixture was stirred at rt for 2 min, and then heated to 100 °C for 1.5 h. The mixture was cooled and diluted with EtOAc (10 mL) and H_2O, and the organic phase was separated, dried (Na_2SO_4), and concentrated. The residue was purified by flash chromatography (EtOAc/hexanes 1:4) to afford the product as a white solid; yield: 143 mg (86%).

1.6.1.2 Reactions Initiating with Arylpalladium Species

Arylpalladium complexes have found equally extensive applications as their alkenylpalladium counterparts, which is in great part likely due to their ready availability, and also the ability of transition metals to activate generally inert aryl halides through oxidative addition, thus facilitating a host of (carbon–carbon) bond formations under mild conditions.

As with alkenylpalladium intermediates, arylpalladium complexes serve as excellent starting points for carbopalladation cascades. An excellent example of an application of a carbopalladation-zipper that initiates with an arylpalladium complex is found in the Keay group's total synthesis of xestoquinone (**91**) (Scheme 20),[78] which is rendered all the more impressive for being an asymmetric variant of the zipper-Heck cyclization.[79] This challenging chemistry involves the formation of an unstable aryl trifluoromethanesulfonate from naphthol **89**; asymmetric zipper-Heck reaction of this trifluoromethanesulfo-

for references see p 506

nate, which includes a thermodynamically controlled 6-*endo-trig* cyclization, affords pentacycle **90** in 82% yield and 68% ee. This chemistry was later repeated by the Shibasaki group using a synthetically more convenient aryl bromide, to give comparable levels of enantioselectivity.[80]

Scheme 20 Synthesis of (+)-Xestoquinone[78]

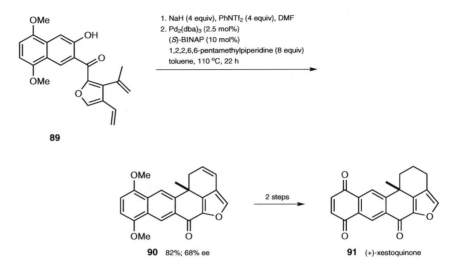

An intramolecular Heck/Tsuji–Trost cyclization was exploited by the Overman group in an elegant approach to spirotryprostatin B (**92**) (Scheme 21).[81,82] Key to this synthesis is the cyclization of aryl iodide **93** onto the proximal diene, a process that affords the π-allylpalladium intermediate **94** (and its epimer, with the palladium positioned on the opposite face of the allyl ligand). Cyclization of the diketopiperidine then takes place to give protected spirotryprostatin B (**95**) and its diastereomer **96** in good yield. Attempts to effect this transformation asymmetrically (using BINAP) were not successful.[82] Although this cyclization gives a mixture of diastereomers, it nevertheless provides a highly efficient entry to spirotryprostatin-like compounds.

Scheme 21 A Key Step in the Synthesis of (−)-Spirotryprostatin B[82]

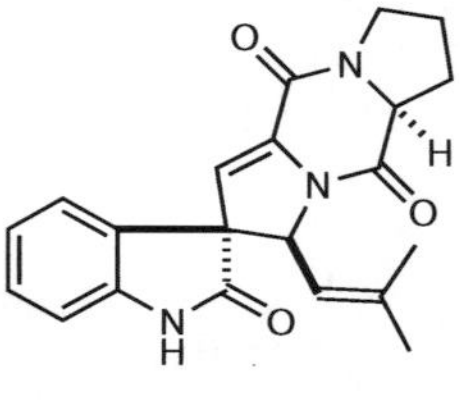

92 (−)-spirotryprostatin B

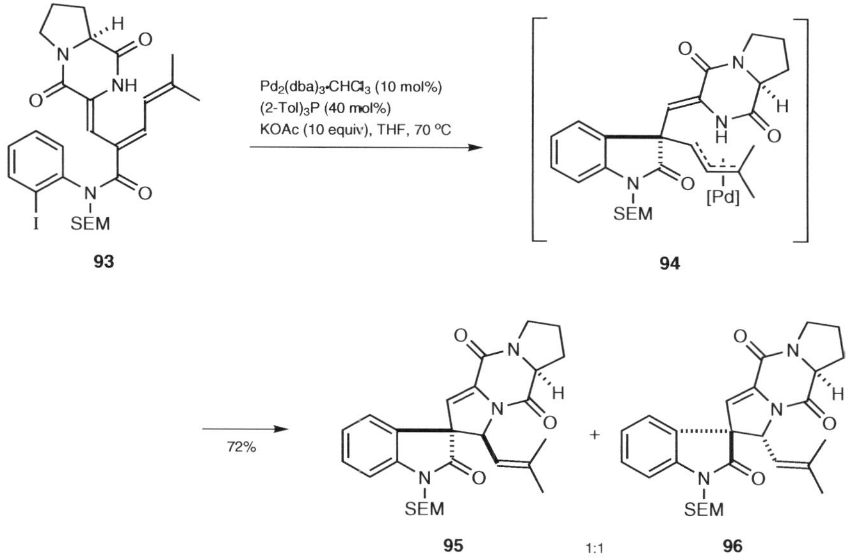

1.6.1 Palladium-Mediated Domino Reactions

475

Arylpalladium complexes are also readily able to participate in carbopalladation/cross-coupling sequences. In one of the first applications of ynamides in total synthesis, Meyer, Cossy, and co-workers demonstrated that an aryl bromide/terminal ynamide/aryl-boronic acid cascade could be used to forge key bonds in the framework of lennoxamine (**100**) (Scheme 22).[83] Thus, treatment of ynamide **97** and boronic acid **98** with palladium(II) acetate/triphenylphosphine afforded product **99** in excellent yield; this was advanced to lennoxamine in three further steps, including an acid-catalyzed Friedel–Crafts-type alkylation to construct the seven-membered benzazepine ring.

Scheme 22 Synthesis of Lennoxamine[83]

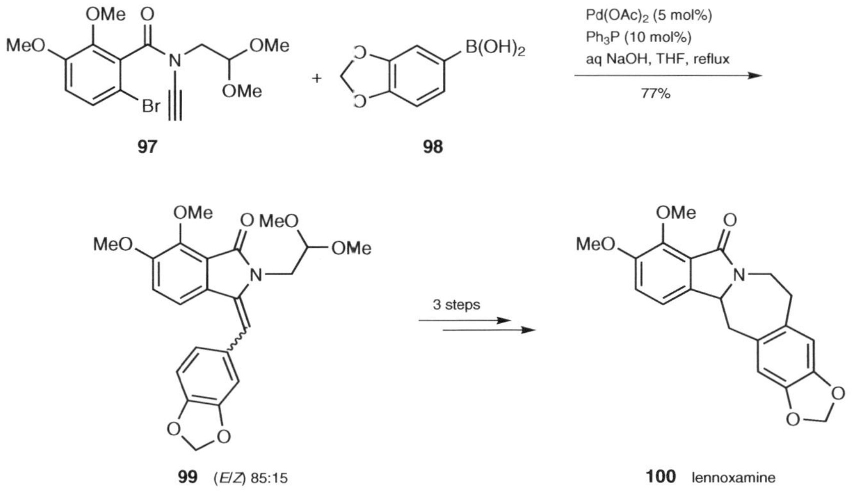

Equivalent to the intramolecular three-component cascades developed by de Meijere in an alkenylpalladium context (Section 1.6.1.1) are similar reactions that initiate with aryl halides.[84] A highly elegant example of this technology is found in the Tietze group's synthesis of the aryl dihydronaphthalene lignin linoxepin (**104**) (Scheme 23).[85] The pivotal

for references see p 506

476   Domino Transformations **1.6** Metal-Mediated Reactions

step of this concise synthesis involves an intramolecular domino cyclization of aryl bromide-alkyne-allylsilane **101**, using palladium(II) acetate and 2-(dicyclohexylphosphino)-2′,4′,6′-triisopropylbiphenyl (XPhos) as the catalyst, which affords pentacyclic product **102** in excellent yield. The allylsilane plays a key role in this sequence: it promotes high regioselectivity in the elimination of palladium(II) following the final carbopalladation of the alkene, and somewhat surprisingly gives desilylated alkene **102** as the major product of the reaction. The formation of **102** could be rationalized either through extensive protodesilylation of the initially formed alkenylsilane **103**,[86] or perhaps through a desilylative reductive elimination (rather than β-hydride elimination to give **103**) in this final step. In further work, Tietze showed that a vinylsilane would also serve as a suitable terminating partner in this cascade; this enabled a subsequent enantioselective hydroboration, and eventual enantioselective synthesis of linoxepin.[87]

Scheme 23   Synthesis of Linoxepin[85]

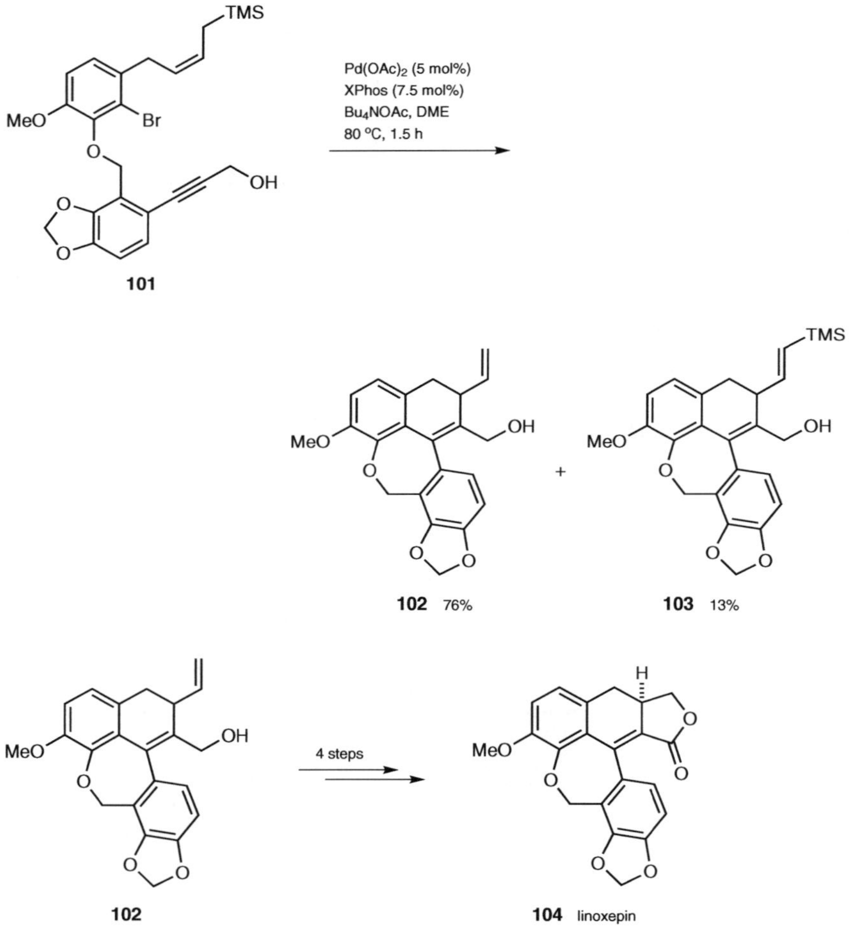

Interestingly, this approach to linoxepin is not the only one reported that employs a palladium-catalyzed cascade as a key strategy. The Lautens group utilized the venerable Catellani reaction[88,89] to rapidly assemble the elements of the natural product framework **111** from dihalide **105** (Scheme 24).[90] The reaction is thought to proceed via a Pd(0)/Pd(II)/Pd(IV) sequence whereby oxidative addition into the aryl iodide **105**, followed by norbornene carbopalladation, gives intermediate **107**. This positions the alkylpalladi-

1.6.1 Palladium-Mediated Domino Reactions **477**

um group in proximity to the arene and enables a C—H activation to give palladacycle **108**. One possible mechanism then involves oxidative addition with alkyl iodide **106** to generate Pd(IV) intermediate **109**, which by reductive elimination to **110**, and then retro-carbopalladation followed by Heck reaction with *tert*-butyl acrylate, delivers the product **111** in high yield. The utility and elegance of this transformation is underlined by the need for only three further steps (including a palladium-catalyzed cyclization) to convert **111** into (+)-linoxepin (**104**).

Scheme 24   Synthesis of (+)-Linoxepin[90]

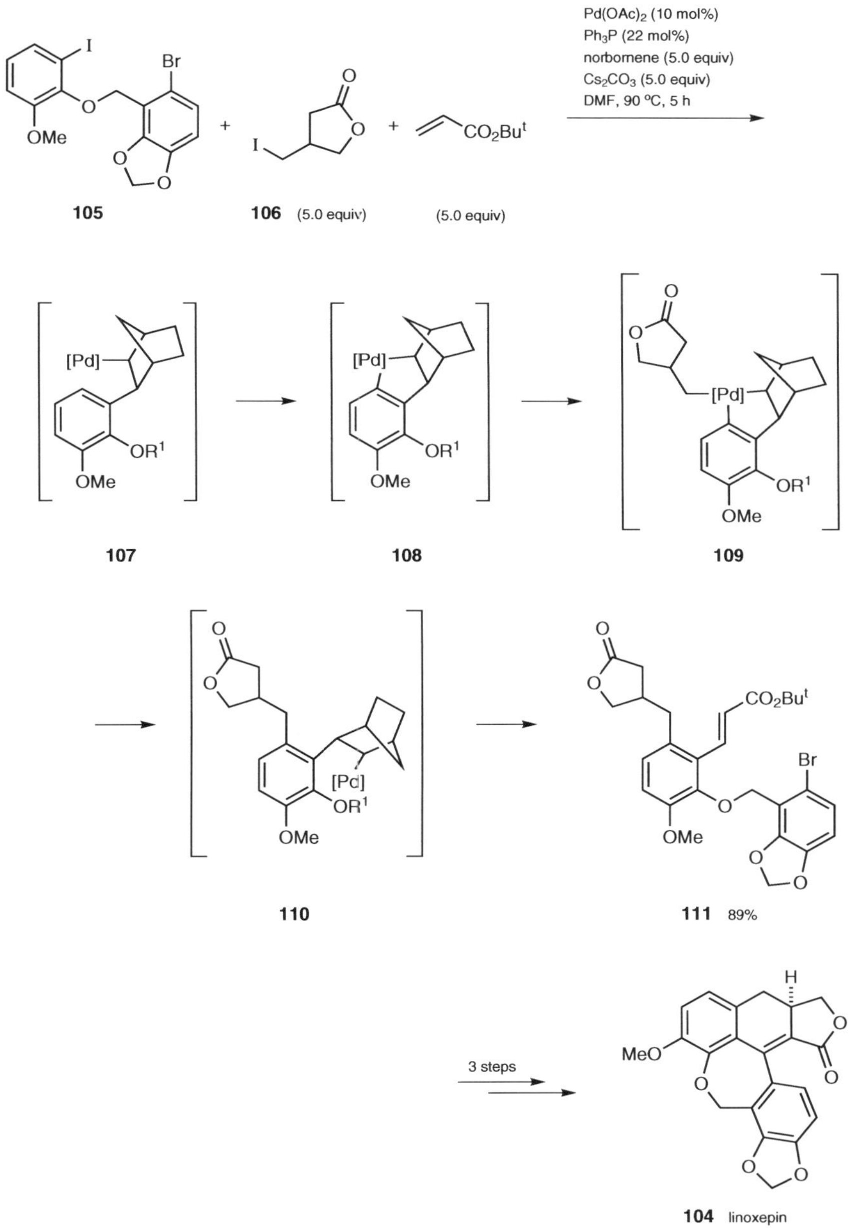

for references see p 506

An equally impressive application of the Catellani reaction is found in the Gu group's beautiful synthesis of rhazinal (**116**),[91] where treatment of iodopyrrole **112** with 1-bromo-2-nitrobenzene leads to the formation of product **115** in 85% yield (Scheme 25). The reaction presumably proceeds via norbornene carbopalladation/pyrrole C–H activation to give intermediate **113**, which, as shown in Scheme 24, undergoes coupling with 1-bromo-2-nitrobenzene, followed by retro-carbopalladation, to give pyrrolopalladium intermediate **114**. In this case, the ensuing Heck reaction takes place in an intramolecular sense, forming the six-membered pyrrolopiperidine ring required for the natural product. It is notable that carbopalladation of norbornene is so efficient that it proceeds to the exclusion of a premature intramolecular Heck cyclization, which is crucial for the successful installation of the nitrophenyl group at the indole C3-position. The power of this chemistry is further underlined by the fact that only two additional transformations are required to access rhazinal (**116**).

Scheme 25   Synthesis of Rhazinal[91]

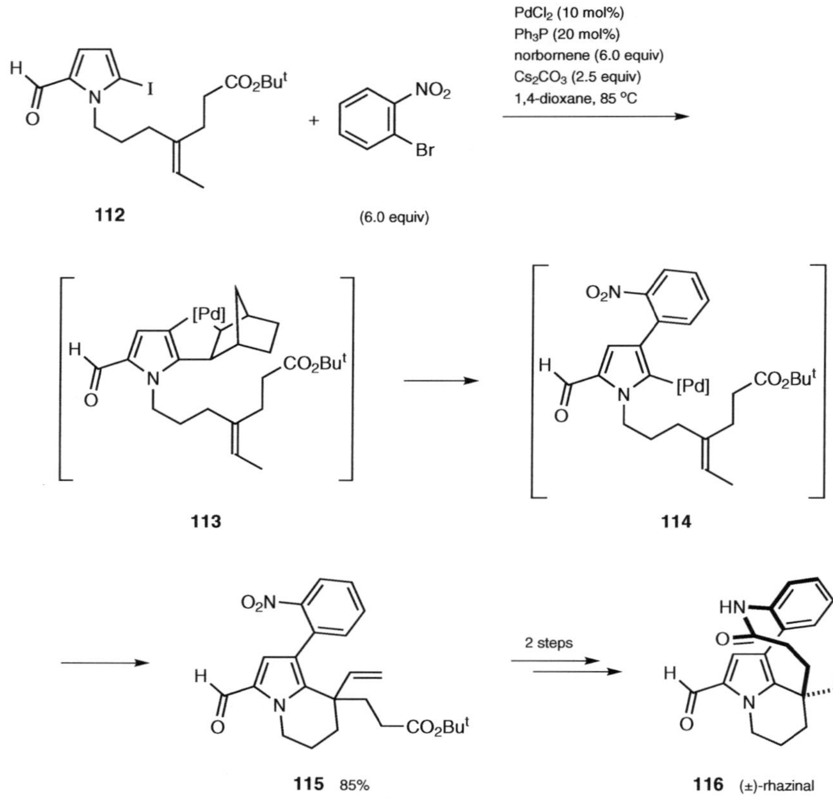

112

(6.0 equiv)

PdCl$_2$ (10 mol%)
Ph$_3$P (20 mol%)
norbornene (6.0 equiv)
Cs$_2$CO$_3$ (2.5 equiv)
1,4-dioxane, 85 °C

113

114

115   85%

2 steps

116   (±)-rhazinal

As mentioned in Section 1.6.1.1, aminopalladation provides a useful mechanism to propagate palladium-catalyzed cascade pathways. This method has been exploited by the Wolfe group to achieve a synthesis of the natural product preussin (**119**) (Scheme 26).[92] Key to this transformation is the *syn*-aminopalladation step, which sets the stereochemistry of the pyrrolidine ring in the course of ring formation. In the event, in the presence of palladium(II) acetate and bis[2-(diphenylphosphino)phenyl] ether (DPEphos), carbamate **117** is smoothly converted into pyrrolidine **118** in good yield, a product that is just two steps from preussin.

1.6.1 Palladium-Mediated Domino Reactions

479

Scheme 26 Synthesis of (+)-Preussin[92]

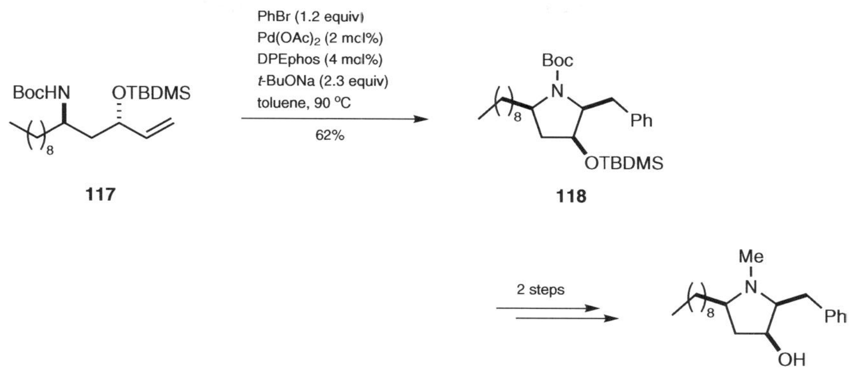

117

PhBr (1.2 equiv)
Pd(OAc)$_2$ (2 mcl%)
DPEphos (4 mcl%)
t-BuONa (2.3 equiv)
toluene, 90 °C
62%

118

2 steps

119 (+)-preussin

In a further advance in this field, the Wolfe group have developed an asymmetric aminopalladation that provides access to enantioenriched azacycles. This chemistry (Scheme 27) is elegantly applied to a synthesis of the antitumor/antiviral natural product (−)-tylophorine (**124**), where reaction of carbamate **121** and bromophenanthrene **122** delivers pyrrolidine **123**.[93,94]

Scheme 27 Synthesis of (−)-Tylophorine[93]

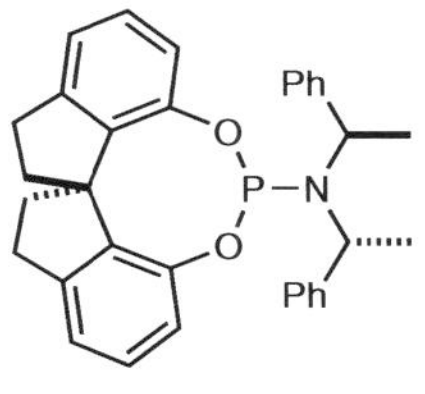

120 (*R*)-Siphos-PE

for references see p 506

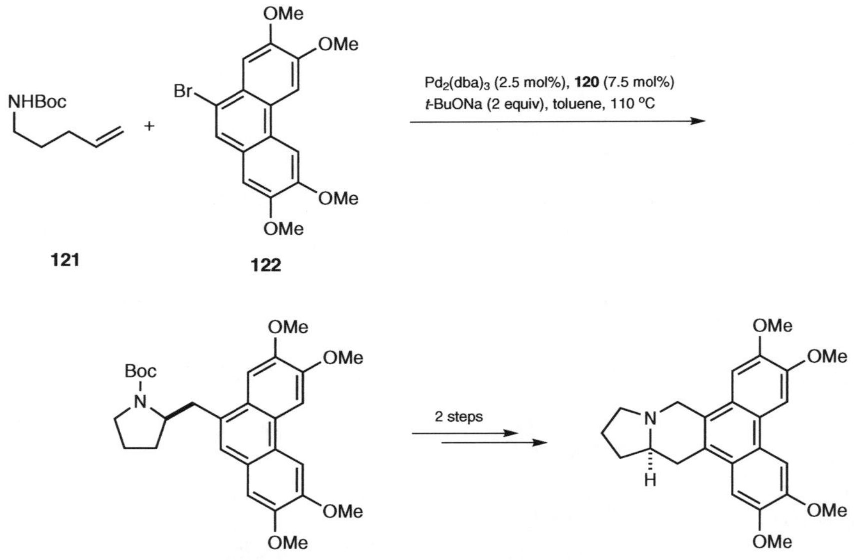

480 Domino Transformations **1.6** Metal-Mediated Reactions

Aminopalladation reactions can also proceed with an *anti* arrangement of the nucleophile and palladium activator. This is illustrated in a most elegant setting in the Ohno and Fujii group's synthesis of (+)-lysergic acid (**125**) and derivatives (Scheme 28).[95,96] In this case, arylpalladium complex **128** (formed from oxidative addition into aryl bromide **126**) complexes the allene motif, activating it toward nucleophilic attack by the sulfonamide and leading to product **127A** (along with a small amount of its diastereomer **127B**). This process forms two rings and a congested stereogenic center, where the latter is proposed to be controlled by the order of reaction events: aminopalladation (to give **129**), followed by reductive elimination, leads to the major (desired) product **127A**, whereas competing carbopalladation at the point of intermediate **128** (which forms a π-allylpalladium intermediate, not shown), followed by nucleophilic attack by the sulfonamide, leads to the minor stereoisomer **127B**.[95] In both cases, it is the geometrical definition of the allene that enables precise positioning of the two reacting components in the ring-forming sequence.

Scheme 28   Synthesis of (+)-Lysergic Acid[96]

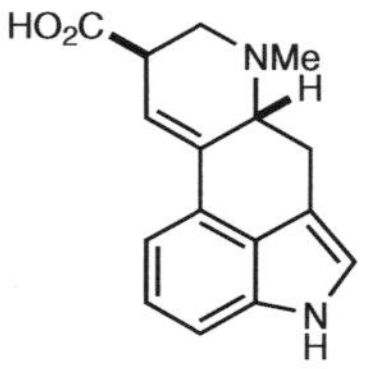

125   (+)-lysergic acid

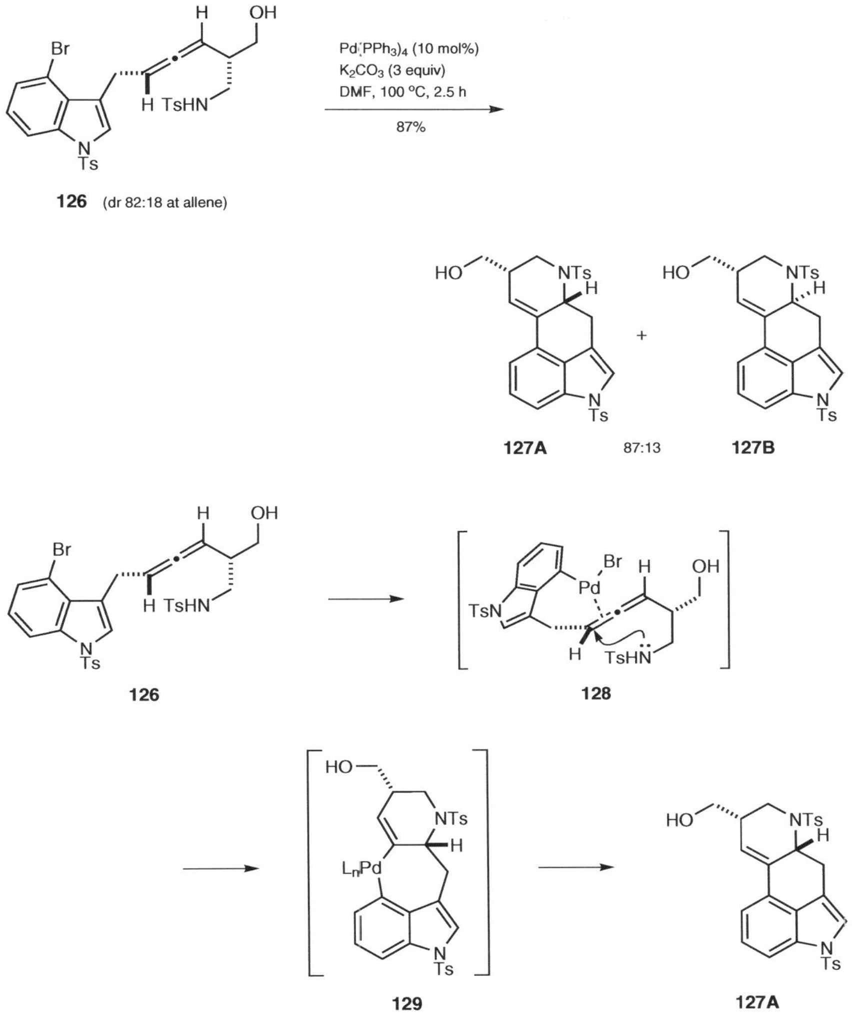

1.6.1 Palladium-Mediated Domino Reactions

481

126 (dr 82:18 at allene)

127A 87:13 **127B**

126 → **128** → **129** → **127A**

In Ohno and Fujii's work, nucleopalladation by a sulfonamide onto an arylpalladium-activated allene outcompetes allene carbopalladation, followed by nucleophilic attack. However, the balance between these processes can by tipped by variation of the nucleophile, as seen in the Dixon group's exploration of the cyclization of malonate-allene derivatives promoted by arylpalladium complexes (Scheme 29).[97] In the reaction of allene **131** with aryl iodide **130**, allene carbopalladation by the initially formed arylpalladium species gives π-allylpalladium complex **132**, which then undergoes cyclization by the pendent 1,3-dicarbonyl nucleophile to give product **133** as a single diastereomer. Although the yield for this specific transformation is low, it represents an application in a synthetic context (toward the manzamine A framework) of a generally high-yielding methodology, as developed in previous work by the same group.[98,99]

for references see p 506

Scheme 29 Synthesis of the Manzamine A Core[97]

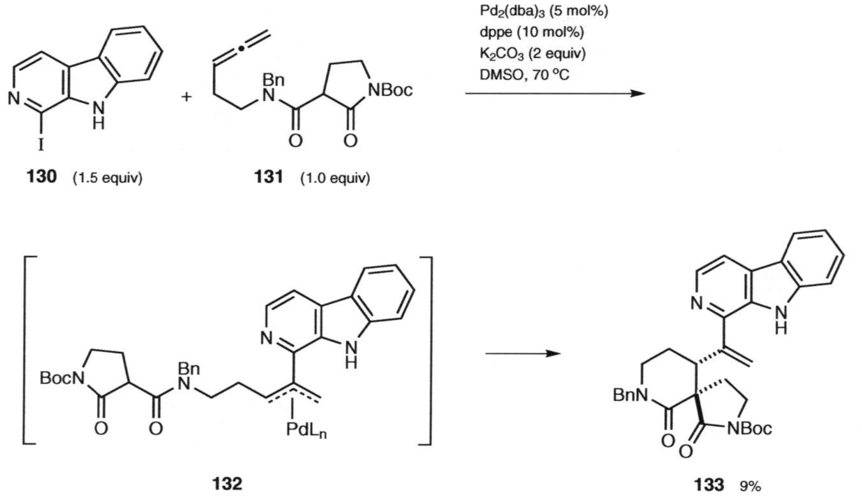
Pd2(dba)3 (5 mol%)
dppe (10 mol%)
K2CO3 (2 equiv)
DMSO, 70 °C
130 (1.5 equiv)
131 (1.0 equiv)
132
133 9%

Aminopalladation can be a reversible process, an attribute that proved synthetically useful in the construction of a number of benzophenanthridine alkaloids, such as sanguinarine (**139**) by the Xu group (Scheme 30).[100] Reaction of aryl iodide **134** with bridged azacycle **135** proceeds via an initial carbopalladation of the aza-norbornene ring. The C—Pd and adjacent C—N bonds in intermediate **136** are now well-aligned to undergo a strain-relieving retro-aminopalladation, delivering palladium amide **137**. This material, or the corresponding zinc amide, then undergoes lactamization to give the benzophenanthridinone **138**.

1.6.1 Palladium-Mediated Domino Reactions

483

Scheme 30 Synthesis of Sanguinarine[100]

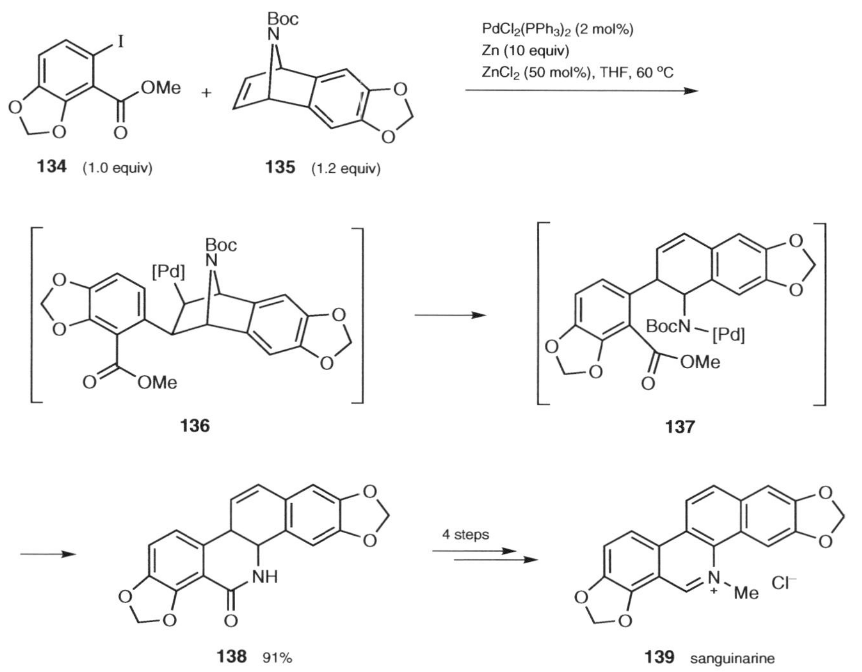

134 (1.0 equiv)   **135** (1.2 equiv)

PdCl$_2$(PPh$_3$)$_2$ (2 mol%)
Zn (10 equiv)
ZnCl$_2$ (50 mol%), THF, 60 °C

136

137

138 91%

4 steps

139 sanguinarine

A final example of a palladium-mediated C—C/C—N bond formation across an alkene is found in an unexpected outcome from the attempted Heck cyclization of aryl iodide **142** (Scheme 31), in work by the Trost group toward the synthesis of FR900482 (**140**).[101] Instead of isolating the eight-membered ring exocyclic alkene **141**, the group found that tetracycle **144** is the only observed product. This is proposed to arise from carbopalladation of the pendent alkene (following oxidative addition into the aryl iodide) giving azacycle **143**. However, instead of the expected β-hydride elimination, intramolecular nucleophilic attack is suggested, a process that corresponds to reductive elimination, and which gives product **144**. Thus, although unsuccessful in the context of the natural product, this cyclization reveals yet another mode of efficient organopalladium reactivity.

for references see p 506

Scheme 31 Unexpected Nucleophile-Mediated Reductive Elimination toward FR900482[101]

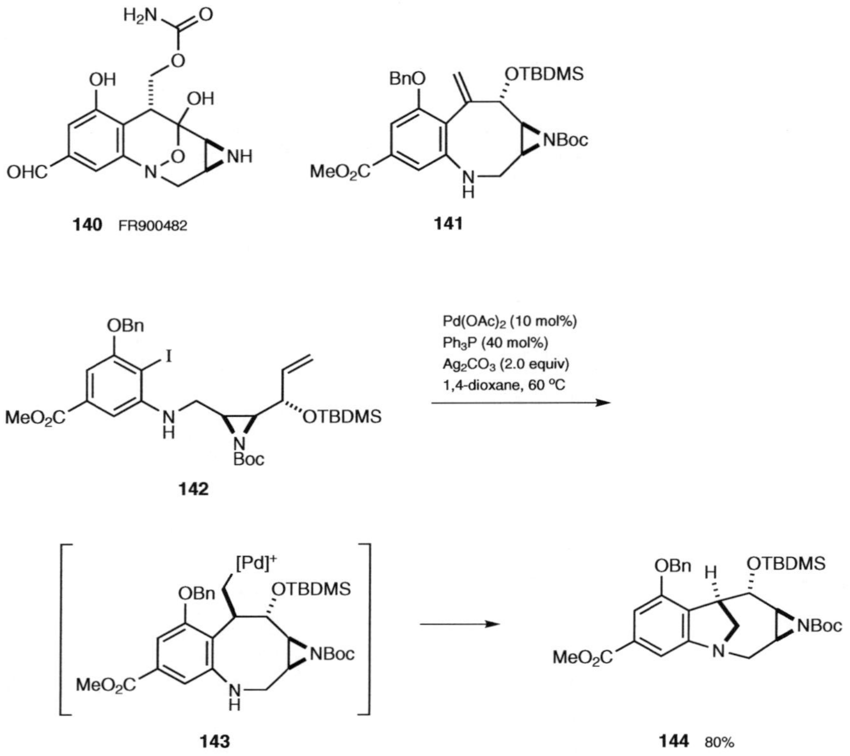

140 FR900482

141

142

Pd(OAc)$_2$ (10 mol%)
Ph$_3$P (40 mol%)
Ag$_2$CO$_3$ (2.0 equiv)
1,4-dioxane, 60 °C

143

144 80%

As discussed in Section 1.6.1.1, the insertion of small molecules such as carbon monoxide or equivalents within a cascade process can lead to impressive multicomponent assemblies. Analogous to work by the Willis group toward indole syntheses (Scheme 19) is the carbonylative cyclization employed in the same group's synthesis of thunberginol A (**149**) (Scheme 32).[102] In a sequence of two palladium-catalyzed steps, aryl iodide **146** is first coupled with the enolate of ketone **145** catalyzed by tris(dibenzylideneacetone)dipalladium and 2-(dicyclohexylphosphino)-2′-(dimethylamino)biphenyl (DavePhos). The product **147** is then subjected to carbonylative cyclization, catalyzed by tris(dibenzylideneacetone)dipalladium and bis[2-(diphenylphosphino)phenyl] ether (DPEphos), via the aryl bromide and the oxygen atom of the enolate of intermediate **147** to give 1*H*-2-benzopyran-1-one **148**. This highly efficient union leads to a product that requires simple methyl ether deprotection to afford thunberginol A.

1.6.1 Palladium-Mediated Domino Reactions

485

Scheme 32 Synthesis of Thunberginol A[102]

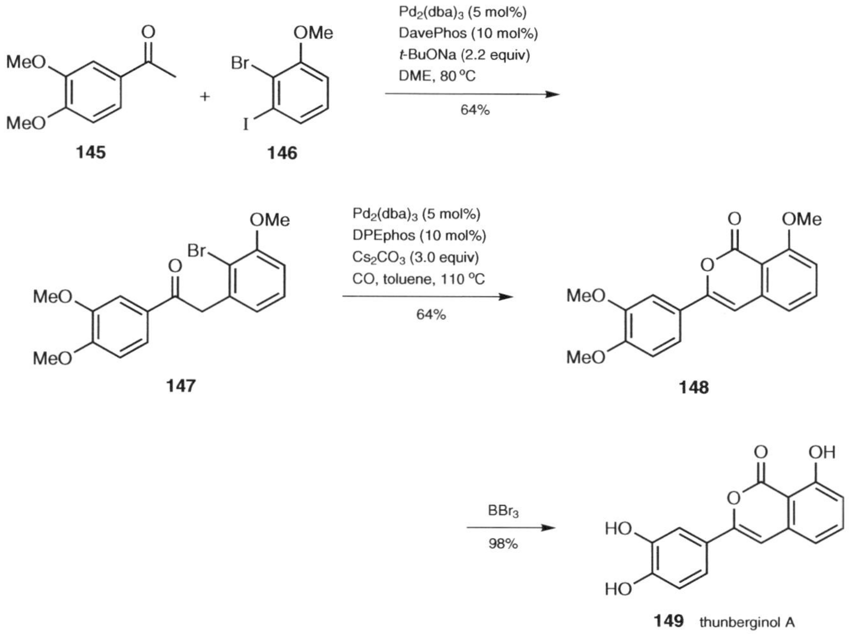

A particularly elegant example of an isocyanide insertion[103] is employed by the Curran group in a synthesis of the camptothecin analogue DB-67 (Scheme 33).[104] In this chemistry, reaction of pyridinone iodide **151** with isocyanide **152**, catalyzed by palladium(II) acetate and in the presence of 2.2 equivalents of silver(I) carbonate, leads to an excellent 70% yield of the cascade cyclization product **155**. This reaction likely proceeds by a migratory insertion of the isocyanide into the C—Pd bond generated through oxidative addition to give intermediate **153**; this process enables subsequent alkyne carbopalladation to give **154** followed by cyclization. Given that this reaction is conducted in the presence of silver(I) cations, it seems likely that a cationic palladium(II) species is generated that might well promote an electrophilic palladation/reductive elimination to form the quinoline ring. One further acidic deprotection step was required to form DB-67.

for references see p 506

Scheme 33 A Key Step in the Synthesis of DB-67[104]

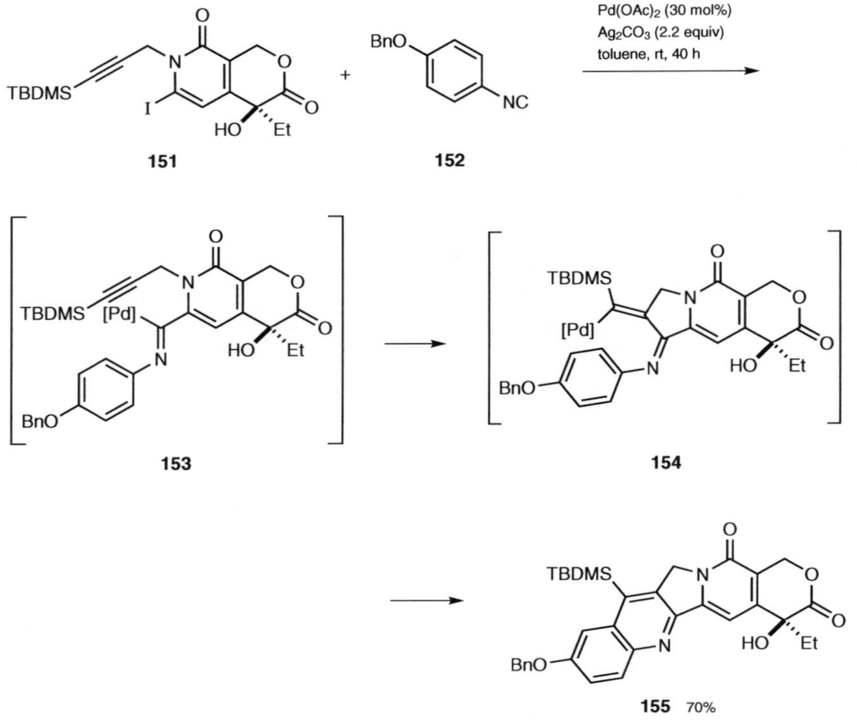

Halide-transfer reactions have emerged as an exciting area of palladium catalysis.[105,106] These processes, which in effect involve palladium-catalyzed carbohalogenation of alkenes, are most accessible when a β-hydride elimination is not possible from the first-generated alkylpalladium intermediate (although this has been shown to be a not insurmountable issue).[107] A very nice demonstration of the synthetic potential of this methodology is given in a formal synthesis of (+)-corynoline by the Lautens group (Scheme 34).[108] The key step of the synthesis is the palladium-catalyzed halide-transfer reaction of aryl iodide **155**, which presumably proceeds via alkylpalladium species **156**. The resultant iodide **157** arising from reductive elimination was transformed in a further six steps to **158**, which represents a formal asymmetric synthesis of (+)-corynoline.

Scheme 34 Formal Synthesis of (+)-Corynoline[108]

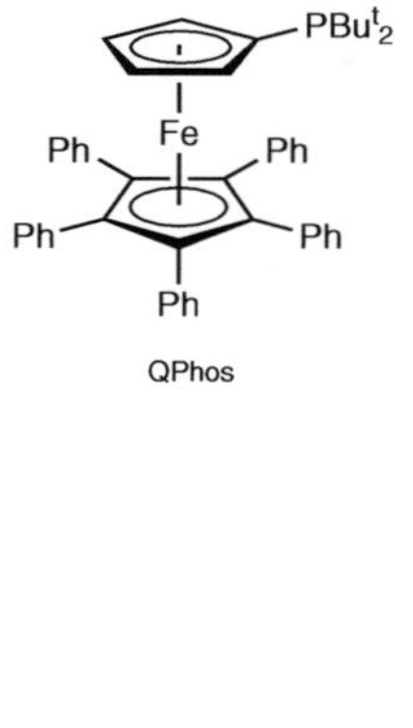

1.6.1 Palladium-Mediated Domino Reactions

155

156 → **157** 84%; dr 92:8; er >99:1

6 steps →

158

(S)-8,11-Dimethoxy-12b-methyl-1,12b-dihydro-6H-tetrapheno[5,4-bc]furan-6-one (90):[78]
A soln of naphthol **89** (14.4 mg, 0.039 mmol, 1.0 equiv) in dry DMF (2 mL) was cannulated into an ice-cooled flask containing NaH (3.8 mg, 0.156 mmol, 4 equiv). The resulting dark red soln was stirred at 0 °C for 15 min under N_2, followed by the addition of $PhNTf_2$ (56 mg, 0.156 mmol, 4 equiv), which rapidly decolorized the soln. The resultant colorless soln was stirred for 3.5 h, then diluted with CH_2Cl_2 (20 mL), and washed with brine (3 × 5 mL). The organic layer was dried (Na_2SO_4) and concentrated. Excess DMF and $PhNTf_2$ were removed by heating under vacuum (40 °C/0.1 Torr) to give the crude aryl trifluoromethanesulfonate, which was used in the polyene cyclization without further purification. The crude trifluoromethanesulfonate (~0.039 mmol) was dissolved in dry toluene (2 mL). 1,2,2,6,6-Pentamethylpiperidine (29 µL, 0.156 mmol, 4 equiv) and a toluene soln of Pd[(S)-BINAP]₂ [99 µL, 2.5 mol%; prepared by dissolving (+)-(S)-BINAP (13.9 mg) and $Pd_2(dba)_3$ (5.1 mg) in toluene (1 mL), 0.01 mmol/mL] were added. The soln was heated at 110 °C for 10 h under N_2. TLC analysis (Et_2O/CH_2Cl_2 9:1) indicated the presence of unreacted substrate. An additional portion of the catalyst soln (99 µL, 2.5 mol%) and 1,2,2,6,6-pentamethylpiperidine (29 µL, 0.156 mmol) were added, and the soln was heated at 110 °C for a further 12 h. The mixture was then cooled to rt, filtered through a pad of silica gel (hexanes/EtOAc 2:1), and purified by radial chromatography (hexanes/EtOAc 8:1) to afford the product as a yellow solid; yield: 10.7 mg (82% from naphthol **89**).

(2S,3S,5aS)-3-(2-Methylprop-1-enyl)-1′-{[2-(trimethylsilyl)ethoxy]methyl}-5a,6,7,8-tetrahydro-3H,5H,10H-spiro[dipyrrolo[1,2-a:1′,2′-d]pyrazine-2,3′-indoline]-2′,5,10-trione (95) and (2R,3R,5aS)-3-(2-Methylprop-1-enyl)-1′-{[2-(trimethylsilyl)ethoxy]methyl}-5a,6,7,8-tetrahydro-3H,5H,10H-spiro[dipyrrolo[1,2-a:1′,2′-d]pyrazine-2,3′-indoline]-2′,5,10-trione (96):[82]
A mixture of $Pd_2(dba)_3$•$CHCl_3$ (23 mg, 0.023 mmol, 10 mol%) and (2-Tol)₃P (27 mg, 0.090 mmol, 40 mol%) in THF (1.5 mL) was stirred at rt for 2 h, furnishing a bright red soln. This soln was added to a mixture of aryl iodide **93** (140 mg, 0.23 mmol, 1 equiv) and KOAc (220 mg, 2.3 mmol, 10 equiv) in a resealable tube. The mixture was sparged with

for references see p 506

argon for 10 min, then sealed, and heated at 70°C for 14 h. The brown mixture was allowed to cool to rt and then filtered through a pad of Celite (EtOAc). After concentration, the residue was purified by flash chromatography (EtOAc/hexanes 3:1) to afford a mixture of **95** and **96** as a light yellow solid; yield: 80 mg (72%); ratio (**95/96**) 1:1.

3-[(E/Z)-(1,3-Benzodioxol-5-yl)methylene]-2-(2,2-dimethoxyethyl)-6,7-dimethoxy-2,3-dihydro-1H-isoindol-1-one (99):[83]

Boronic acid **98** (300 mg, 1.81 mmol, 1.2 equiv), 1 M aq NaOH (2.3 mL, 2.3 mmol, 1.5 equiv), Pd(OAc)$_2$ (17 mg, 0.075 mmol, 5 mol%), and Ph$_3$P (40 mg, 0.15 mmol, 10 mol%) were successively added to a soln of ynamide **97** (561 mg, 1.51 mmol) in THF (60 mL). The mixture was heated to reflux for 2 h, and then it was cooled to rt, hydrolyzed with sat. aq NH$_4$Cl, and extracted with EtOAc. The combined extracts were washed with brine, dried (MgSO$_4$), filtered, and concentrated. The crude material was purified by flash chromatography (petroleum ether/EtOAc 3:2) to afford the product as a yellow oil; yield: 480 mg (77%); ratio (E/Z) 85:15.

(11-Methoxy-7-vinyl-7,13-dihydro-8H-[1,3]dioxolo[4′,5′:3,4]benzo[1,2-e]naphtho[1,8-bc]oxepin-6-yl)methanol (102):[85]

An oven-dried flask was charged with alkyne **101** (1.00 g, 1.93 mmol, 1.0 equiv), Pd(OAc)$_2$ (21.7 mg, 96.5 µmol, 5 mol%), XPhos (69.1 mg, 145 µmol, 7.5 mol%), and Bu$_4$NOAc (1.16 g, 3.86 mmol, 2.0 equiv). DME (20 mL) was added, and the mixture was heated to 80°C for 1.5 h (preheated oil bath). The mixture was then cooled to rt and diluted with CH$_2$Cl$_2$ (30 mL), and silica gel was added. The solvent was removed under reduced pressure and the residue was purified by flash chromatography (petroleum ether/EtOAc 4:1 to 3:1) to afford the product **102** as a colorless foam; yield: 535 mg (76%). In addition, vinylsilane **103** was isolated; yield: 112 mg (13%).

tert-Butyl (E)-3-{2-[(5-Bromo-1,3-benzodioxol-4-yl)methoxy]-3-methoxy-6-[(5-oxotetrahydrofuran-3-yl)methyl]phenyl}acrylate (111):[90]

Pd(OAc)$_2$ (38.5 mg, 0.17 mmol, 10 mol%), Ph$_3$P (99.1 mg, 0.38 mmol, 22 mol%), Cs$_2$CO$_3$ (2.80 g, 8.58 mmol, 5.0 equiv), aryl iodide **105** (797 mg, 1.72 mmol, 1.0 equiv), tert-butyl acrylate (1.26 mL, 8.58 mmol, 5.0 equiv), and lactone **106** (1.94 g, 8.58 mmol, 5.0 equiv) were added successively to a sealable pressure flask under argon. Degassed, anhyd DMF (2 mL) was added and the mixture was stirred at rt while argon was bubbled through the heterogeneous mixture. Norbornene (809 mg, 8.58 mmol, 5.0 equiv) was added, the flask was again purged with argon, and it was then sealed and heated at 90°C for 5 h. The mixture was cooled to rt and quenched by the addition of sat. aq NH$_4$Cl. H$_2$O and EtOAc/hexanes (1:1) were added, and the phases were separated. The organic layer was washed with brine and dried (MgSO$_4$), and the solvents were removed under reduced pressure. The crude material was purified by flash chromatography (hexanes/EtOAc 3:1 to 2:1) to give the product, which was recrystallized (EtOAc/hexanes) to afford a white solid; yield: 860 mg (89%).

tert-Butyl 3-[3-Formyl-1-(2-nitrophenyl)-8-vinyl-5,6,7,8-tetrahydroindolizin-8-yl]propanoate (115):[91]

A mixture of iodopyrrole **112** (0.297 g, 0.69 mmol, 1.0 equiv), PdCl$_2$ (12.2 mg, 0.069 mmol, 10 mol%), Ph$_3$P (36.2 mg, 0.138 mmol, 20 mol%), Cs$_2$CO$_3$ (0.562 g, 1.73 mmol, 2.5 equiv), norbornene (0.389 g, 4.14 mmol, 6.0 equiv), and 1-bromo-2-nitrobenzene (0.836 g, 4.14 mmol, 6.0 equiv) in 1,4-dioxane (7.0 mL) in a Schlenk tube was stirred at 85°C for 20 h. The mixture was cooled to rt and then filtered through a plug of Celite (EtOAc, 20 mL). The filtrate was washed with brine (15 mL), dried (Na$_2$SO$_4$), and concentrated. Purification by flash chromatography (EtOAc/hexanes 1:9 and CH$_2$Cl$_2$/hexanes 1:1) gave the product; yield: 0.249 g (85%).

tert-Butyl (2*S*,3*S*,5*R*)-2-Benzyl-3-(*tert*-butyldimethylsiloxy)-5-nonylpyrrolidine-1-carboxylate (118):[92]

A flame-dried Schlenk tube equipped with a magnetic stirrer bar was cooled under a stream of N_2 and charged with carbamate **117** (110 mg, 0.25 mmol, 1.0 equiv), Pd(OAc)$_2$ (1.2 mg, 0.0053 mmol, 2 mol%), DPEphos (5.4 mg, 0.01 mmol, 4 mol%), *t*-BuONa (56 mg, 0.575 mmol, 2.3 equiv), and bromobenzene (32 μL, 0.30 mmol, 1.2 equiv). The tube was purged with N_2 and toluene (1 mL) was added. The resulting mixture was heated to 90 °C with stirring for 5 h until the starting material had been consumed (GC analysis). The mixture was then cooled to rt and sat. aq NH$_4$Cl (1 mL) and EtOAc (1 mL) were added. The layers were separated, and the aqueous layer was extracted with EtOAc (3 × 5 mL). The combined organic layers were dried (Na$_2$SO$_4$) and concentrated. The crude product was purified by flash chromatography (EtOAc/hexanes 2.5:97.5) to afford the product as a colorless oil; yield: 80.2 mg (62%); mixture of rotamers 3:1 (¹H NMR).

tert-Butyl (*R*)-2-[(2,3,6,7-Tetramethoxyphenanthren-9-yl)methyl]pyrrolidine-1-carboxylate (123):[93]

A flame-dried Schlenk tube was cooled under a stream of N_2 and charged with Pd$_2$(dba)$_3$ (2.3 mg, 0.0025 mmol, 2.5 mol%), (*R*)-Siphos-PE (**120**; 3.8 mg, 0.0075 mmol, 7.5 mol%), *t*-BuONa (19 mg, 2.0 equiv [note: this quantity is not specified and has therefore been assumed based on the general procedure]), and bromide **122** (38 mg, 0.1 mmol, 1 equiv). The tube was purged with N_2, and then a soln of *N*-(*tert*-butoxycarbonyl)pent-4-enamine (24 mg, 0.13 mmol, 1.3 equiv) in toluene (0.2 M) was added via syringe. The mixture was heated to 110 °C with stirring until the starting material had been consumed, then it was cooled to rt, quenched with sat. aq NH$_4$Cl (2 mL), and diluted with EtOAc (5 mL). The layers were separated and the aqueous layer was extracted with EtOAc (2 × 5 mL). The combined organic layers were dried (Na$_2$SO$_4$) and concentrated. The crude product was purified by flash chromatography to give the product as an off-white solid; yield: 30 mg (62%)

{(6a*R*,9*S*)-4,7-Ditosyl-4,6,6a,7,8,9-hexahydroindolo[4,3-*fg*]quinolin-9-yl}methanol (127A) and {(6a*S*,9*S*)-4,7-Ditosyl-4,6,6a,7,8,9-hexahydroindolo[4,3-*fg*]quinolin-9-yl}methanol (127B):[96]

To a stirred mixture of aryl bromide **126** (dr 82:18; 20 mg, 0.032 mmol, 1.0 equiv) in DMF (0.87 mL) at rt under argon were added Pd(PPh$_3$)$_4$ (3.7 mg, 0.0032 mmol, 10 mol%) and K$_2$CO$_3$ (13.3 mg, 0.096 mmol, 3 equiv). The mixture was heated to 100 °C for 2.5 h and then cooled to rt and quenched by the addition of sat. aq NH$_4$Cl. The mixture was extracted with EtOAc; the extract was washed with H$_2$O and brine, and dried (MgSO$_4$). The filtrate was concentrated to give an oily residue, which was purified by flash chromatography (hexane/EtOAc 1:1) to give the product as a white amorphous solid; yield: 15.2 mg (87%); dr (**127A/127B**) 87:13.

tert-Butyl (5*R*,10*R*)-7-Benzyl-10-{1-(9*H*-pyrido[3,4-*b*]indol-1-yl)vinyl}-1,6-dioxo-2,7-diazaspiro[4.5]decane-2-carboxylate (133):[97]

Pd$_2$(dba)$_3$ (4.0 mg, 0.0040 mmol, 5 mol%) and dppe (3.0 mg, 0.0080 mmol, 10 mol%) in DMSO (0.2 mL) were stirred at rt for 30 min. To this mixture was added a suspension of amide **131** (32 mg, 0.080 mmol, 1.0 equiv), β-carboline **130** (35 mg, 0.12 mmol, 1.5 equiv), and K$_2$CO$_3$ (22 mg, 0.16 mmol, 2.0 equiv) in DMSO (0.2 mL). The resulting suspension was stirred at 70 °C in a sealed vial for 16 h and then cooled to rt and partitioned between Et$_2$O (1 mL) and H$_2$O (1 mL). The organic phase was separated, and the aqueous phase was extracted with Et$_2$O (3 × 2 mL). The combined organic phases were dried (MgSO$_4$) and concentrated. The residue was purified by flash chromatography (petroleum ether to EtOAc/petroleum ether 1:1 to EtOAc to MeOH/EtOAc 1:9) to yield the product as an orange oil; yield: 4.0 mg (9%).

for references see p 506

490   Domino Transformations   **1.6** Metal-Mediated Reactions

12b,13-Dihydro-[1,3]dioxolo[4′,5′:4,5]benzo[1,2-c][1,3]dioxolo[4,5-i]phenanthridin-14(5bH)-one (138):[100]

$PdCl_2(PPh_3)_2$ (14 mg, 0.020 mmol, 2 mol%), Zn powder (654 mg, 10 mmol, 10 equiv), and $ZnCl_2$ (68 mg, 0.50 mmol, 50 mol%) were added to a soln of azabicycle **135** (344 mg, 1.20 mmol, 1.2 equiv) and iodoarene **134** (306 mg, 1.00 mmol, 1.0 equiv) in THF (20 mL). The mixture was heated at 60 °C under a N_2 atmosphere until completion of the reaction (TLC monitoring). The mixture was cooled to rt and diluted with CH_2Cl_2 (15 mL), and then filtered through a short pad of Celite (CH_2Cl_2). After concentration, the crude product was purified by flash chromatography (EtOAc/petroleum ether 2:1) to give the product; yield: 305 mg (91%).

1-tert-Butyl 5-Methyl (1aS,3R,9S,9aS)-7-(Benzyloxy)-9-(tert-butyldimethylsiloxy)-1a,8,9,9a-tetrahydro-3,8-methanoazirino[2,3-f]benzo[b]azocine-1,5(2H)-dicarboxylate (144):[101]

A dry flask was charged with iodide **142** (185 mg, 0.28 mmol, 1.0 equiv), $Pd(OAc)_2$ (5.9 mg, 26 μmol, 10 mol%), Ph_3P (29.3 mg, 0.112 mmol, 40 mol%), and Ag_2CO_3 (144.3 mg, 0.522 mmol, 2.0 equiv). Freshly distilled 1,4-dioxane (7 mL) was added, and the mixture was warmed to 60 °C for 3 h until the starting material had been consumed (TLC analysis). The mixture was cooled to rt and filtered through a plug of Celite (EtOAc). The filtrate was concentrated and the residue was purified by flash chromatography (hexane/EtOAc/CH_2Cl_2 85:10:5) to give the product as a clear oil; yield: 133 mg (80%).

2-(2-Bromo-3-methoxyphenyl)-1-(3,4-dimethoxyphenyl)ethanone (147):[102]

Aryl iodide **146** (150 mg, 0.479 mmol, 1.5 equiv), acetophenone **145** (58 mg, 0.320 mmol, 1.0 equiv), and t-BuONa (67.6 mg, 0.703 mmol, 2.2 equiv) were added to an oven-dried flask charged with $Pd_2(dba)_3$ (14.6 mg, 0.016 mmol, 5 mol%) and DavePhos (12.6 mg, 0.032 mmol, 10 mol%) under argon. To this mixture was added anhyd DME (0.6 mL) and the reaction was heated at 80 °C for 5 h under argon. The mixture was cooled to rt, diluted with CH_2Cl_2 (~5 mL), and filtered through a pad of Celite, washing with CH_2Cl_2 (~40 mL). The filtrate was concentrated and the product was purified by flash column chromatography (toluene/Et_2O 93:7) to yield the product as a pale brown solid; yield: 74.6 mg (64%).

3-(3,4-Dimethoxyphenyl)-8-methoxy-1H-2-benzopyran-1-one (148):[102]

> **CAUTION:** *Carbon monoxide is extremely flammable and exposure to higher concentrations can quickly lead to a coma.*

Cs_2CO_3 (463 mg, 1.42 mmol, 3.0 equiv) was added to an oven-dried 30-mL Schlenk tube charged with $Pd_2(dba)_3$ (21.7 mg, 0.024 mmol, 5 mol%), DPEphos (25.5 mg, 0.047 mmol, 10 mol%), and ketone **147** (173 mg, 0.474 mmol) under argon. To this mixture was added anhyd toluene (1.0 mL). A balloon fitted with a glass tap attachment was filled with argon and evacuated (3 ×). The balloon was then filled with CO and attached to the top of the Schlenk tube. The inert atmosphere was then exchanged for CO by evacuation/refilling (3 ×). The vigorously stirred reaction was then heated to 110 °C for 15 h under CO. The mixture was cooled to rt and the CO balloon was removed. The mixture was then diluted with EtOAc (~10 mL), filtered through a pad of Celite [EtOAc (~40 mL)], and concentrated. The crude material was purified via flash chromatography (petroleum ether/EtOAc 1:1) to yield the product as a pale orange solid; yield: 93.0 mg (64%).

(S)-9-(Benzyloxy)-11-(tert-butyldimethylsilyl)-4-ethyl-4-hydroxy-1H-pyrano-[3′,4′:6,7]indolizino[1,2-b]quinoline-3,14(4H,12H)-dione (155):[104]

Iodopyridone **151** (13 mg, 0.027 mmol, 1.0 equiv), Ag_2CO_3 (11 mg, 0.040 mmol, 1.5 equiv), and $Pd(OAc)_2$ (1.2 mg, 0.0054 mmol, 20 mol%) were suspended in toluene (0.5 mL). A soln of isocyanide **152** (11 mg, 0.054 mmol, 2 equiv) in toluene (1 mL) was slowly added to this stirred suspension over 0.5 h. The mixture was stirred at rt for 20 h, and then filtered

through Celite. The mixture was concentrated, and the residue was mixed with Pd(OAc)$_2$ (0.6 mg, 0.0025 mmol, 10 mol%) and Ag$_2$CO$_3$ (5.5 mg, 0.020 mmol, 0.75 equiv) in toluene (0.5 mL). Then a soln of isocyanide **152** (5.5 mg, 0.025 mmol, 1.0 equiv) in toluene (0.5 mL) was added. The mixture was stirred at rt for a further 20 h after which time TLC analysis showed that **151** had been consumed. The mixture was then filtered through Celite, and the filtrate was concentrated. Purification by flash chromatography gave the product as a yellow solid; yield: 10.7 mg (70%).

(6R,7R)-7-(1,3-Benzodioxol-5-yl)-6-(iodomethyl)-6,8-dimethyl-7,8-dihydro-[1,3]dioxolo[4,5-h]isoquinolin-9(6H)-one (157):[108]

An oven-dried flask was charged with amide **155** (5.00 g, 10.4 mmol, 1.0 equiv), Pd(QPhos)$_2$ (1.65 g, 1.04 mmol, 10 mol%), and activated 4-Å molecular sieves (1.5 g) and purged with argon for 30 min. The solid components were taken up in dry, degassed toluene (208 mL, 5 freeze–pump–thaw cycles, transferred via cannula) and dry, degassed 1,2,2,6,6-pentamethylpiperidine (6.61 mL, 35.6 mmol, 3.5 equiv, 5 freeze–pump–thaw cycles). The reaction vessel was fitted with an oven-dried reflux condenser that had been cooled under a flow of argon for 10 min. The mixture was heated for 6 h at 100 °C, at which point TLC analysis indicated full consumption of the starting material. The mixture was cooled to rt and filtered through a plug of silica gel (EtOAc), and the filtrate was concentrated under reduced pressure. The crude mixture of dihydroquinolines was carefully purified using flash chromatography (hexanes/EtOAc 1:1) to yield the separate diastereomers; combined yield: 4.18 g (84%).

1.6.1.3 Reactions Initiating with Allylpalladium Intermediates

Allylpalladium(II) complexes are readily generated under mild conditions, and are also suitable for cascade processes incorporating a variety of secondary palladium-mediated reactions.[109,110] In a highly elegant piece of chemistry that demonstrates the preferential formation of a π-allylpalladium intermediate in the presence of an aryl iodide, the Tietze group has prepared tetracycline skeleton **161** (Scheme 35) from allyl acetate-aryl iodide **159** in a single synthetic operation that depends crucially on the order of palladium-catalyzed events. Heating **160** with sodium hydride and palladium(II) acetate/1,2-bis(diphenylphosphino)ethane in dimethylformamide effects an oxidative addition into the allyl acetate of **159**, then Tsuji–Trost cyclization of the deprotonated keto ester to form the first new six-membered ring in **160**. This process is followed by a slower Heck reaction in which the aryl iodide of **159** then cyclizes onto the pendent terminal alkene generated in the Tsuji–Trost process. This excellent transformation provides tetracycle **161** in 65% overall yield.[111]

for references see p 506

492 Domino Transformations **1.6** Metal-Mediated Reactions

Scheme 35 Approach to the Tetracycline Framework[111]

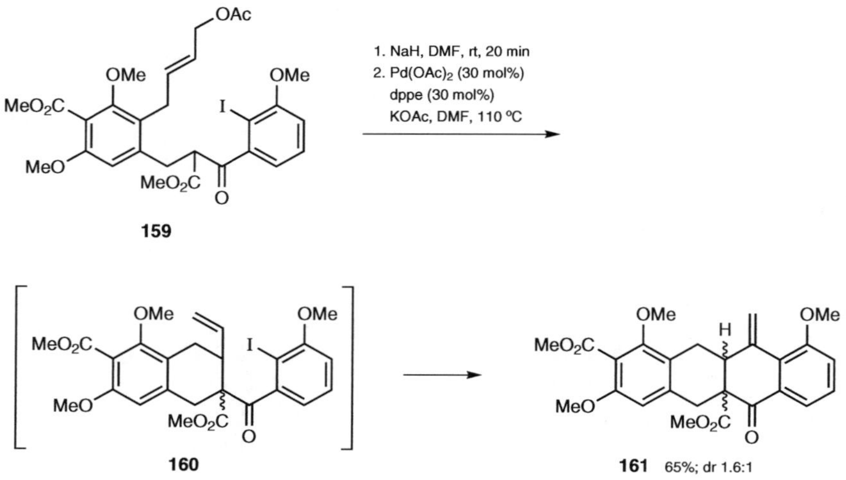

1. NaH, DMF, rt, 20 min
2. Pd(OAc)$_2$ (30 mol%)
   dppe (30 mol%)
   KOAc, DMF, 110 °C

159

160 → **161** 65%; dr 1.6:1

An allylpalladium formation/Tsuji–Trost/lactonization cascade has also been employed by the Fukuyama group toward a synthesis of the natural product jiadifenin (**162**) (Scheme 36).[112] The nature of the palladium catalyst is crucial to the success of this transformation. Namely, while reaction of allylic cyclic carbonate **163** with tetrakis(triphenylphosphine)palladium(0) as catalyst affords only bicycle **165**, switching to a bidentate ligand [(R)-BINAP] in the presence of lithium acetate completely reverses the product selectivity and affords the desired tricyclic lactone **167** in 57% yield. Under the former set of conditions, product **165** is proposed to arise from an invertive oxidative addition to give **164**, followed by β-hydride elimination and then aldol cyclization/dehydration. In the latter case, observation of the desired product may reflect a reduced propensity toward β-hydride elimination, allowing π–σ–π inversion of the π-allylpalladium intermediate, a theory that is supported by the stereochemistry at the newly formed quaternary stereocenter in the product, which must involve cyclization onto the diastereomer of **164** (i.e., methyl and hydroxymethyl groups interchanged, with palladium on the top face as drawn); alternatively, it may simply be that the conformation of the allylic carbonate is different at the point of oxidative addition. Either way, once cyclization of the β-oxo ester has been achieved in the desired pathway, lactonization of the residual primary alcohol in **166** (likely also requiring a degree of epimerization at the ester-bearing stereocenter) affords lactone **167**.

Scheme 36 Route toward Jiadifenin[112]

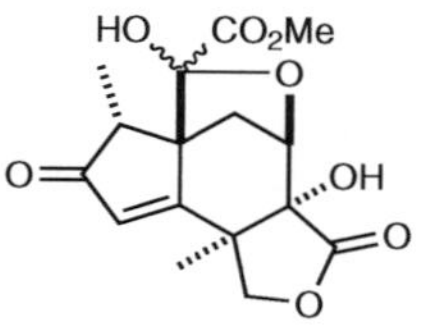

162 jiadifenin

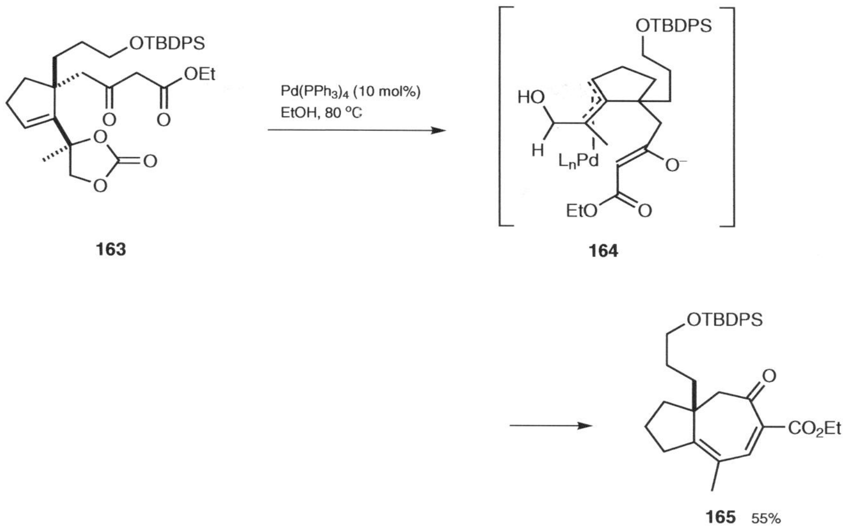

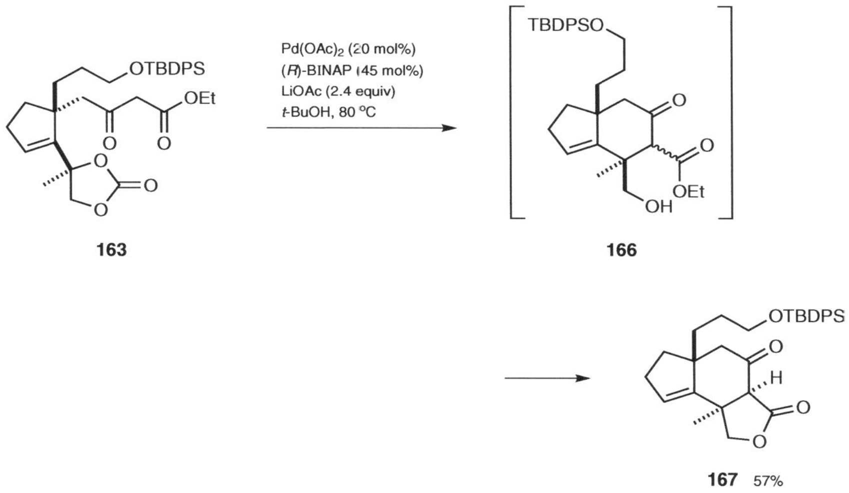

1.6.1 Palladium-Mediated Domino Reactions

An alternative entry to π-allylpalladium intermediates, i.e. a process other than oxidative addition into a heteroatom leaving group, is the reaction of a palladium(0) catalyst with a vinylcyclopropane. The efficiency of this process (originally developed by Tsuji)[113] is underlined in an approach by the Stoltz group toward members of the *Melodinus* alkaloids, such as scandine (**168**) (Scheme 37).[114] This chemistry relies on a formal [3+2] cycloaddition between vinylcyclopropane **170** and nitroalkene **169**, and is initiated by oxidative addition into the cyclopropane to generate π-allylpalladium species **171**. This step, which is presumably facilitated by the release of stabilized malonate anion, is followed by addition of this carbanion to the nitroalkene, generating a nitroalkane anion **172**. Being in proximity to the π-allylpalladium moiety, this anion undergoes cyclization to complete the formal [3+2] cascade and deliver cyclopentane **173**, albeit favoring the undesired diastereomer at the allylic stereocenter (dr 1:2). This product was advanced to tetracycle **174**, which bears a striking resemblance to the scandine framework.

for references see p 506

494 Domino Transformations **1.6** Metal-Mediated Reactions

Scheme 37 Route toward Scandine[114]

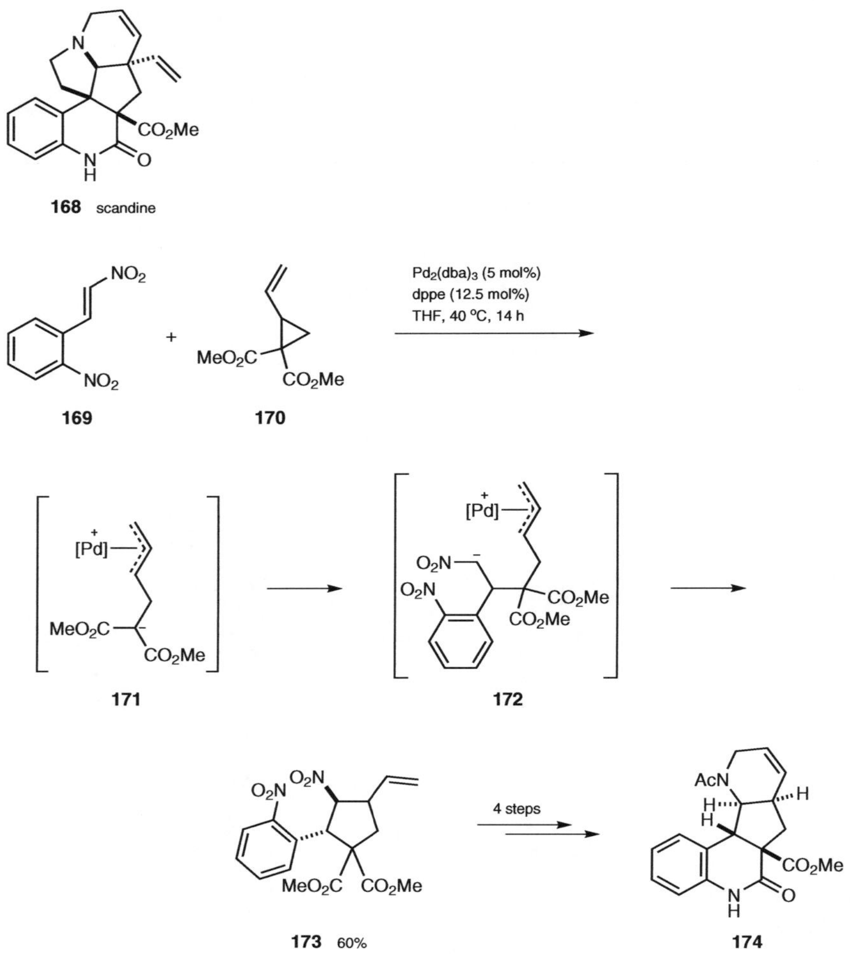

168 scandine

Pd$_2$(dba)$_3$ (5 mol%)
dppe (12.5 mol%)
THF, 40 °C, 14 h

169 **170**

171 **172**

4 steps

173 60% **174**

Finally, a highly elegant cyclization cascade initiated by an allylpalladium complex, reminiscent of the classic palladium-zipper reactions, is found in the Takahashi group's formal synthesis of dimethyl gloiosiphone A (**175**) (Scheme 38).[115] Treatment of allylic acetate **176** with an appropriate palladium catalyst leads to the formation of π-allylpalladium intermediate **177**. This intermediate, potentially due to the influence of the acetic acid solvent, is able to decomplex and undergo sequential Heck cyclization reactions to deliver product **178** (as an inconsequential mixture of diastereomers).

Scheme 38 Formal Synthesis of Dimethyl Gloiosiphone A[115]

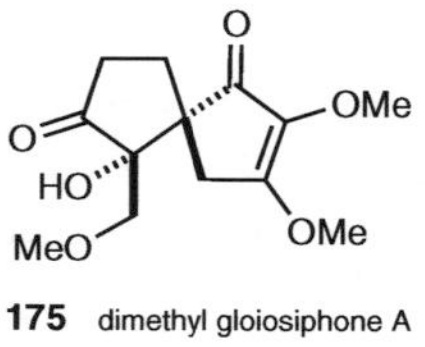

175 dimethyl gloiosiphone A

1.6.1 Palladium-Mediated Domino Reactions

176 → **177**

178 66%; dr 60:40

Dimethyl 1,3,10-Trimethoxy-11-methylene-6-oxo-5,5a,6,11,11a,12-hexahydrotetracene-2,5a-dicarboxylate (161):[111]

NaH (60% in mineral oil; 6.4 mg, 0.16 mmol, 1.1 equiv) was added in one portion to a soln of allyl acetate **159** (95 mg, 0.14 mmol, 1.0 equiv) in degassed DMF (0.9 mL) at 0 °C, and the mixture was warmed to rt and stirred for 20 min (soln A). In a separate flask, a soln of Pd(OAc)$_2$ (9.4 mg, 0.042 mmol, 30 mol%) and dppe (16.7 mg, 0.042 mmol, 30 mol%) in degassed DMF (0.6 mL) was stirred for 30 min at rt (soln B). Soln A was added dropwise via syringe to soln B, and the combined mixture was stirred for 10 min at rt. KOAc (28 mg, 0.29 mmol, 2.0 equiv) was then added, and the mixture was stirred for 2 h in a preheated oil bath at 110 °C. The mixture was cooled to rt, concentrated, and purified by flash chromatography (EtOAc/petroleum ether 1:4) to give the product as a yellow oil; yield: 42 mg (65%).

Ethyl (S)-3a-[3-(tert-Butyldiphenylsiloxy)propyl]-8-methyl-5-oxo-1,2,3,3a,4,5-hexahydro-azulene-6-carboxylate (165):[112]

Pd(PPh$_3$)$_4$ (2.3 mg, 2.0 µmol, 10 mol%) was added to a soln of carbonate **163** (13.1 mg, 22.1 µmol, 1.0 equiv) in EtOH (2.0 mL). The mixture was stirred for 5 h at 80 °C, and then concentrated and purified by flash chromatography (EtOAc/hexane 1:3) to give the product as a colorless oil; yield: 6.4 mg (55%).

(3aS,5aS,8bR)-5a-[3-(tert-Butyldiphenylsiloxy)propyl]-8b-methyl-3a,5,5a,6,7,8b-hexaahydro-1H-indeno[4,5-c]furan-3,4-dione (167):[112]

To a soln of carbonate **163** (14.6 mg, 24.4 µmol, 1.0 equiv) in t-BuOH (1 mL) were added LiOAc (3.86 µg, 58.6 µmol, 2.4 equiv), Pd(OAc)$_2$ (1.1 mg, 4.9 µmol, 20 mol%), and (R)-BINAP (6.9 mg, 11.1 µmol, 45 mol%). The mixture was stirred for 20 h at 80 °C. The solvent was removed under reduced pressure and the residue was purified by preparative TLC (Et$_2$O/hexane 2:1) to give the product as a colorless oil; yield: 7.0 mg (57%).

Dimethyl (2S,3S)-3-Nitro-2-(2-nitrophenyl)-4-vinylcyclopentane-1,1-dicarboxylate (173):[114]

Pd$_2$(dba)$_3$ (147 mg, 0.160 mmol, 5 mol%) and dppe (147 mg, 0.369 mmol, 12.5 mol%) were added to a flame-dried Schlenk tube. The Schlenk tube was evacuated and backfilled with argon (2×). In a separate, flame-dried, conical flask, THF was sparged with argon for 20 min. After this period, THF (25 mL) was added to the Schlenk tube, and the resulting purple soln was heated at 40 °C until the soln became bright orange, at which point it was allowed to cool to rt. To a separate, flame-dried conical flask were added 1-nitro-2-(2-nitrophenyl)ethene (**169**; 575 mg, 2.96 mmol, 1.0 equiv), THF (5 mL), and vinylcyclopropane **170** (500 µL, 2.96 mmol, 1.0 equiv). This soln was transferred via cannula to the Schlenk tube, and the mixture became red. The soln was heated to 40 °C and stirred for 14 h. Upon

for references see p 506

completion of the reaction, the mixture was concentrated and purified by flash chromatography (hexanes/EtOAc 10:1 to 5:1) yielding a yellow solid, which was further purified by trituration (Et$_2$O) to afford the product as a white solid; yield: 671 mg (60%); 1:2 mixture of diastereomers.

(1R,5S)-7-(4-Methoxybenzyloxy)-8-methylene-3,3-bis(phenylsulfonyl)-1-vinylspiro-[4.4]nonane (178):[115]

To a soln of allylic acetate **176** (2.50 g, 3.92 mmol) in AcOH (15 mL) were added Pd(OAc)$_2$ (88 mg, 0.392 mmol, 10 mol%) and Ph$_3$P (411 mg, 1.57 mmol, 40 mol%) at rt. The mixture was stirred at 90 °C for 2 h, and then it was concentrated. The residue was purified by flash chromatography (EtOAc/toluene 7:93) to give the product; yield: 1.50 g (66%); dr 60:40.

1.6.1.4 Reactions Initiating with Allenylpalladium Intermediates

Allenylpalladium complexes are among the more complicated organopalladium species in terms of structure and reactivity, as they can exist with controllable η^1- or η^3-coordination to the metal (**181** and **182**, respectively, Scheme 39),[116,117] with the latter representing a first step on the path of palladium migration between allenylpalladium and propargylpalladium species **181–184**. Allenylpalladium intermediates are easily formed, either by oxidative addition to an allenyl halide **179**, or, more commonly, by S$_N$2′-type oxidative addition to a propargylic electrophile **180**. Both of these approaches can reliably lead to enantioenriched allenylpalladium species when starting from chiral nonracemic electrophiles.

Scheme 39 Generation and Structure of Allenylpalladium(II) Complexes

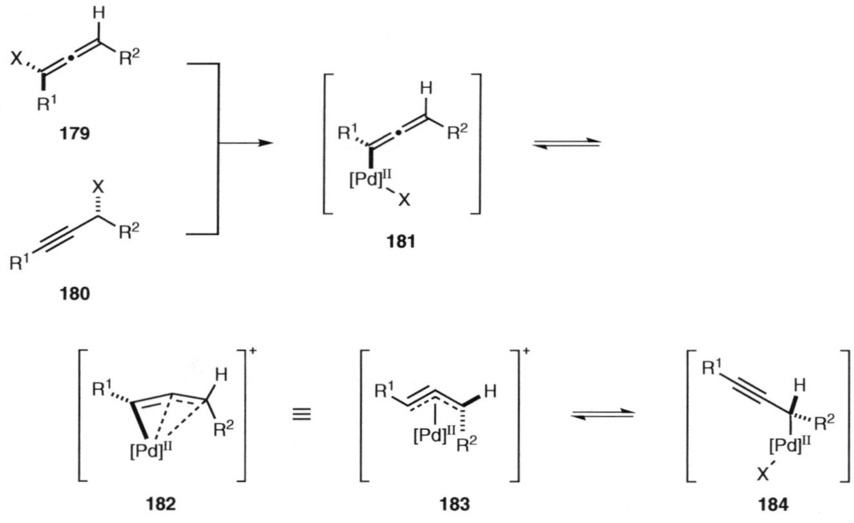

The most common form of capture of the allenylpalladium intermediate is by nucleophilic attack at the central carbon atom of the allene. This mode of reactivity, pioneered by Tsuji in the early 1980s using malonate nucleophiles,[118] has afforded a rich seam of reactivity in cascade processes as applied in synthesis, mainly due to the efforts of the Ohno and Fujii group,[8] who have elegantly demonstrated both approaches to allenylpalladium intermediates in the context of enantioselective syntheses of jaspine B (**188**) (Scheme 40). In their first-generation route,[119] bromoallene **184** serves as a precursor to bicycle **187**,

1.6.1 Palladium-Mediated Domino Reactions

497

where oxidative addition of palladium into the allenyl bromide and then nucleophilic attack on the resultant allenylpalladium species **185** affords an intermediate π-allyl complex **186**. This material undergoes a further cyclization by the benzamide carbonyl to give **187**. Interestingly, both diastereomers of bromoallene cyclize with equal efficiency, which suggests that the π-allyl complex **186** (and its diastereomer) can readily undergo π–σ–π epimerization to position the palladium atom on the opposite face to the benzamide nucleophile.

Scheme 40 First-Generation Approach to (+)-Jaspine B[119]

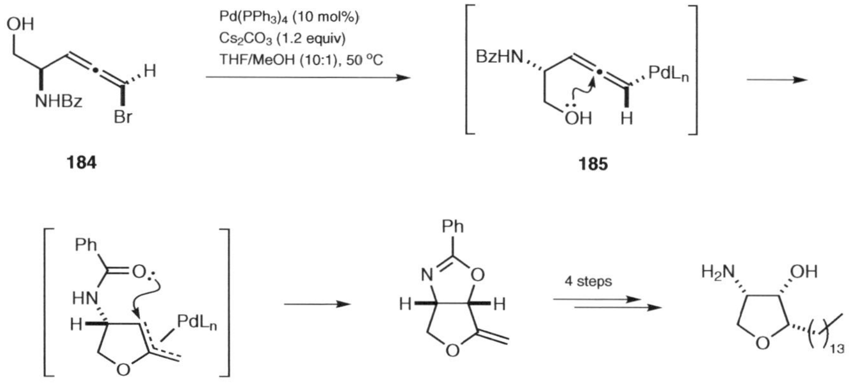

The group then later investigated the propargylic electrophile approach, which benefits from a pre-installation of the jaspine B side chain (Scheme 41).[120] Thus, treatment of either propargylic chloride **189** or propargylic carbonate **191** under similar conditions leads to efficient cyclization to bicyclic product **190**. In this case, however, the stereochemistry of the starting material (*syn* or *anti*) proved important, with the *syn* diastereomer (illustrated) undergoing a markedly more efficient cyclization. This outcome may be a consequence of the additional alkene substituent (i.e., the jaspine B side chain) leading to steric effects in the reaction pathway that prevent efficient π-allylpalladium epimerization or cyclization.

Scheme 41 Second-Generation Approach to (+)-Jaspine B[120]

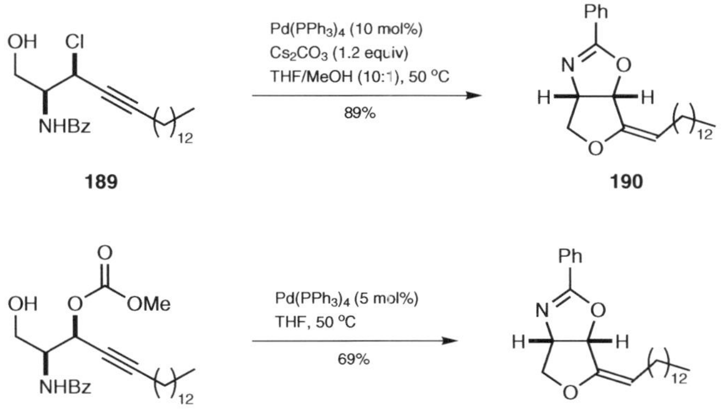

for references see p 506

Notably, the site of capture of allenylpalladium intermediates can be varied according to the positioning of the nucleophile, and also the nature of the other ligands around the metal. In related intramolecular cyclizations, the Anderson group has shown that the regiochemistry of intramolecular cyclization of nitrogen nucleophiles onto allenylpalladium intermediates (see **182**, Scheme 39) can be controlled by the bite angle of a bidentate phosphine ligand;[121,122] where a 1,2-bis(diphenylphosphino)ethane (dppe) ligated allenylpalladium species leads to central carbon attack as generally observed in such systems,[116] the use of the large bite angle ligand bis[2-(diphenylphosphino)phenyl] ether (DPEphos) results in nucleophilic attack at the distal carbon.

(3a*S*,6a*S*)-6-Methylene-2-phenyl-3a,4,6,6a-tetrahydrofuro[3,4-*d*]oxazole (187):[119]

To a stirred mixture of bromoallene **184** (40 mg, 0.142 mmol, 1.0 equiv) in THF/MeOH (10:1; 1.2 mL) at rt under argon were added $Pd(PPh_3)_4$ (8.2 mg, 0.0071 mmol, 5 mol%) and Cs_2CO_3 (55.5 mg, 0.170 mmol, 1.2 equiv). The mixture was stirred at 50 °C for 2.5 h and then cooled to rt and filtered through a short pad of silica gel (EtOAc). The filtrate was concentrated to give a yellow oil, which was purified by flash chromatography (hexane/EtOAc 3:1) to give the product as a white solid; yield: 25.5 mg (89%).

(3a*S*,6a*S*,*E*)-2-Phenyl-6-tetradecylidene-3a,4,6,6a-tetrahydrofuro[3,4-*d*]oxazole (190) from Propargylic Chloride 189:[120]

To a stirred mixture of propargylic chloride **189** (40 mg 0.095 mmol, 1.0 equiv) in THF/MeOH (10:1; 1.0 mL) at rt under argon were added $Pd(PPh_3)_4$ (11.0 mg, 0.0095 mmol, 10 mol%) and Cs_2CO_3 (37.1 mg, 0.114 mmol, 1.2 equiv). The mixture was stirred for 1 h at 50 °C and then cooled to rt and filtered through a short pad of silica gel (EtOAc). The filtrate was concentrated to give a yellow oil, which was purified by flash chromatography (hexane/EtOAc 4:1) to give the product as a white solid; yield: 32.4 mg (89%).

(3a*S*,6a*S*,*E*)-2-Phenyl-6-tetradecylidene-3a,4,6,6a-tetrahydrofuro[3,4-*d*]oxazole (190) from Propargylic Carbonate 191:[120]

To a stirred mixture of propargylic carbonate **191** (40 mg, 0.087 mmol, 1.0 equiv) in THF (0.9 mL) at rt under argon was added $Pd(PPh_3)_4$ (5.03 mg, 0.0044 mmol, 5 mol%). The mixture was stirred for 2 h at 50 °C and then cooled to rt and concentrated to give a yellow oil, which was purified by flash chromatography (hexane/EtOAc 4:1) to give the product as a white solid; yield: 23.1 mg (69%).

1.6.1.5 Reactions Initiating with Alkylpalladium Intermediates

Alkylpalladium complexes represent highly challenging intermediates to initiate, or convey through, domino reactions owing to a high propensity for β-hydride elimination, which is most usually avoided by a judicious choice of substrate that either lacks such hydrogen atoms, or the ability to adopt a conformation from which elimination can proceed. Nonetheless, alkylpalladium-initiated cascades have found considerable use in synthetic contexts.

Nucleopalladation reactions provide an excellent means to initiate cascade processes using alkylpalladium species. A highly elegant example of this principle is found in the Tietze group's synthesis of α-tocopherol (**192**), a constituent of vitamin E (Scheme 42). The key feature of the synthesis is the sequenced oxypalladation/intermolecular Heck reaction between phenol **194** and methyl acrylate, a reaction manifold pioneered by Semmelhack[123] that is effected in this setting with excellent enantioselectivity (96% ee) at room temperature under the influence of palladium(II) trifluoroacetate and chiral bis-oxazoline ligand **193**. This affords ester **195**, which could be converted into α-tocopherol using known chemistry. The reaction also proved successful using methyl vinyl ketone

as the Heck acceptor (not shown); the product from this process was produced in lower enantiomeric excess (65% ee) but could also be converted into α-tocopherol.[124]

Scheme 42 Synthesis of α-Tocopherol[124]

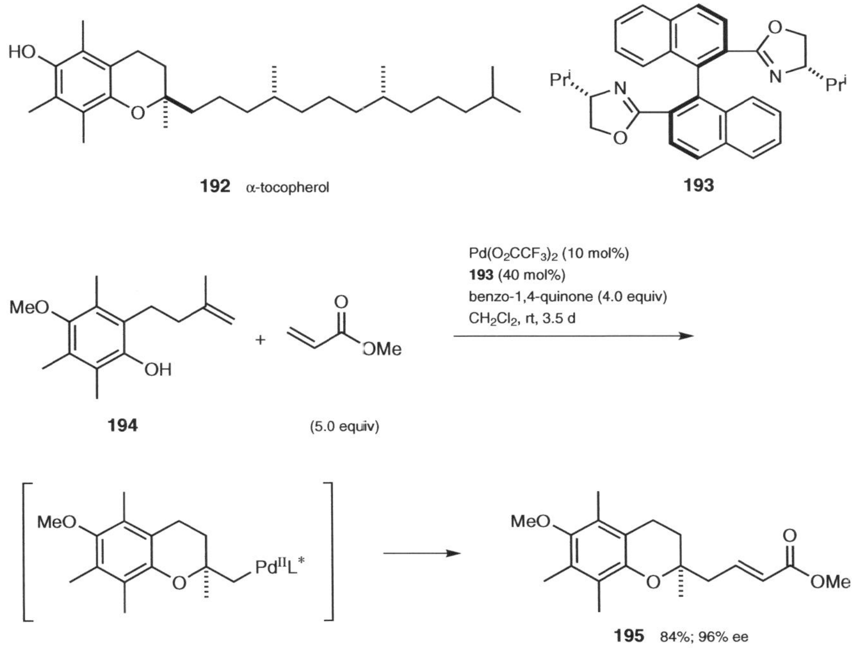
HO
O
Pr^i
N
O
N
Pr^i
O
192 α-tocopherol
193
MeO
OH
+
O
OMe
Pd(O_2CCF_3)$_2$ (10 mol%)
193 (40 mol%)
benzo-1,4-quinone (4.0 equiv)
CH_2Cl_2, rt, 3.5 d
194
(5.0 equiv)
MeO
O
Pd^{II}L*
MeO
O
O
OMe
195 84%; 96% ee

An extension of this reactivity has been reported by the France group, involving a termination of this oxypalladation cascade via an unusual β-halide elimination. Impressively, this chemistry has been applied to a synthesis of the antidepressant citalopram (**198**) (Scheme 43). The reaction sequence proceeds in the usual manner: oxypalladation of a pendent alkene in **196** leads to an intermediate alkylpalladium complex, which then undergoes carbopalladation with allyl chloride. Subsequent β-halide elimination delivers isobenzofuran **197** in 58% yield, a compound that is two steps away from citalopram itself.[125]

for references see p 506

Scheme 43 Synthesis of Citalopram[125]

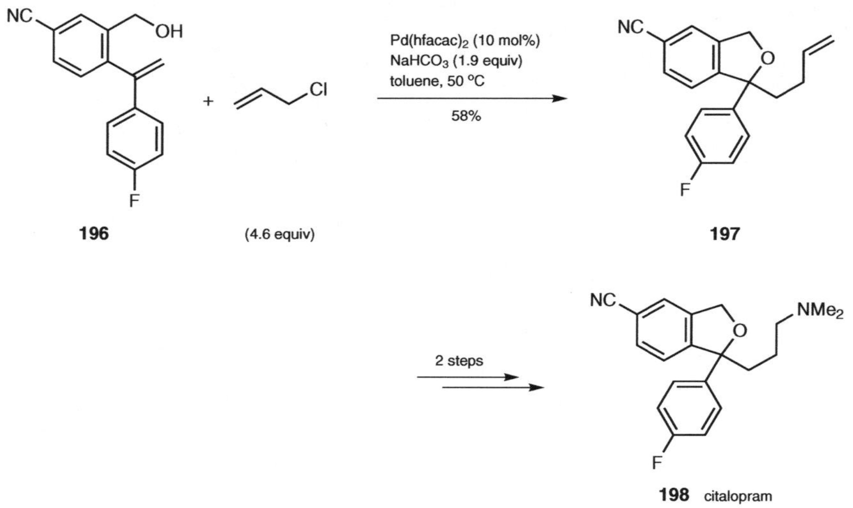

hfacac = 1,1,1,5,5,5-hexafluoropentane-2,4-dionate

As with many of the reactions discussed in Sections 1.6.1.1 and 1.6.1.2, the ability to insert additional small molecules (carbon monoxide, isocyanides) into alkylpalladium species provides further opportunities for bond formations. Carbonylative oxacyclization has proved particularly useful in the setting of complex molecule synthesis, with many examples of oxacycles, and oxabicycles, assembled using this technology. Scheme 44 illustrates some representative examples, dating back to early pioneering work by the Semmelhack group[67] toward the polyether ionophore tetronomycin.[126] In these studies, the group demonstrated that both complex furans (e.g., **200**) and pyrans could be accessed using simple reaction procedures, with excellent stereocontrol at the newly formed stereogenic center. The stereocontrol likely arises from a pseudoequatorial positioning of the activated alkene, and an *anti*-oxopalladation mechanism; a drawback is the requirement for a stoichiometric amount of the palladium species.

Soon after Semmelhack's seminal reports[127,128] was the disclosure by the Yoshida group of reaction conditions to effect the carbonylative oxacyclization cascade that are catalytic in palladium,[129] employing copper(II) chloride as a stoichiometric reoxidant. This variation has become widely adopted as the method of choice in synthetic applications. An example is illustrated by the cyclization of diol **201** to afford furan lactone **202** in high yield; this product was advanced to the natural product kumausallene.[130]

In certain cases, these conditions have been found to be less effective due to side reactions such as the incorporation of chloride ion, released from copper(II) chloride. The use of tetramethylthiourea as ligand improves this situation, with 2-methyloxirane acting as a chloride scavenger, and ammonium acetate preventing oxidative side reactions.[131–133] These conditions have been applied to the cyclization of diol **203** to give lactone **204**, a product that is an intermediate en route to pallambins C and D.[134] Further refinements in the nature of this thiourea ligand have been made, culminating in perhaps the most impressive example of a carbonylative cyclization, which installs the pyran-lactone rings **206** or **208** of schindilactone A.[135,136]

1.6.1 Palladium-Mediated Domino Reactions

501

Scheme 44 Representative Examples of Carbonylative Oxacyclization[126,130,134–136]

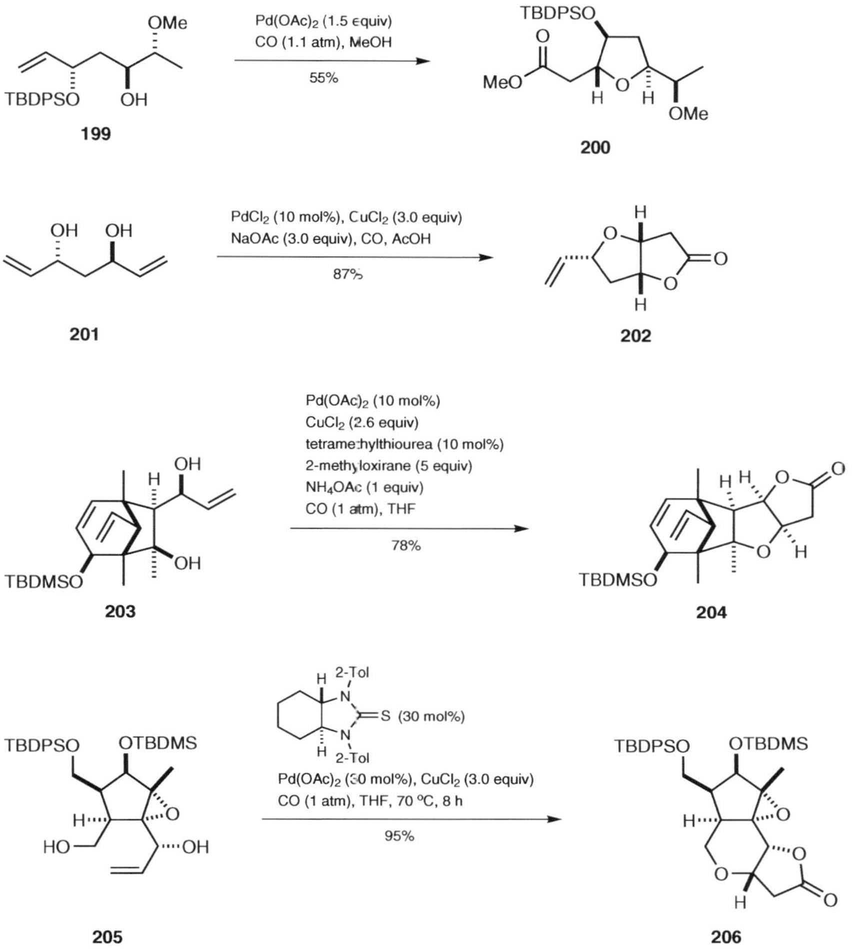

for references see p 506

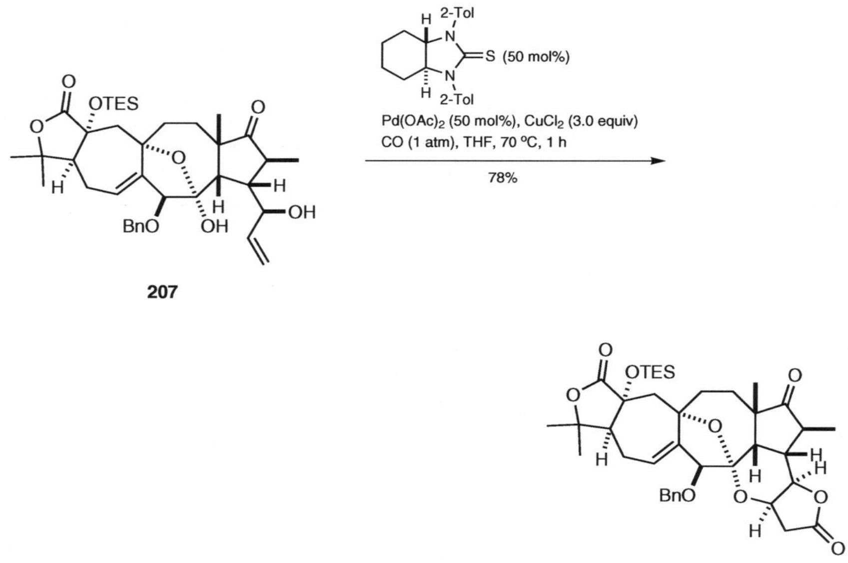

An example of such a process in the context of a phenol nucleophile is found in the synthesis of (−)-panacene (**209**) reported by Boukouvalas, Snieckus, and co-workers (Scheme 45).[137] The key step of this synthesis again involves a carbonylative-oxypalladation cascade, in which the enantioenriched phenol in **210** engages with the palladium(II)-activated alkene to give the bicyclic lactone **211** in a single synthetic operation. Unfortunately, this process requires a stoichiometric amount of palladium(II), as efforts to realize a catalytic process met with failure for reasons that are not clear. The stereoselectivity of this reaction is likely influenced by a preferred starting material conformation which minimizes 1,3-allylic strain; the product **211** was transformed to (−)-panacene in seven additional steps.

Scheme 45 Synthesis of Panacene[137]

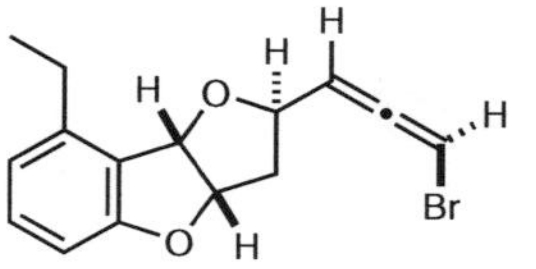

Finally, it is possible to achieve such cyclizations in an enantioselective sense using chiral palladium catalysts. In an elegant demonstration of this principle, the Tietze group showed that the phenol **214** undergoes carbonylative cyclization to give ester **215** in excellent yield and enantioselectivity (Scheme 46).[138] This transformation is all the more interesting given the failure of phenol **210** to cyclize using a substoichiometric amount of palladium catalyst (Scheme 45); in this case, the reaction proves effective with just

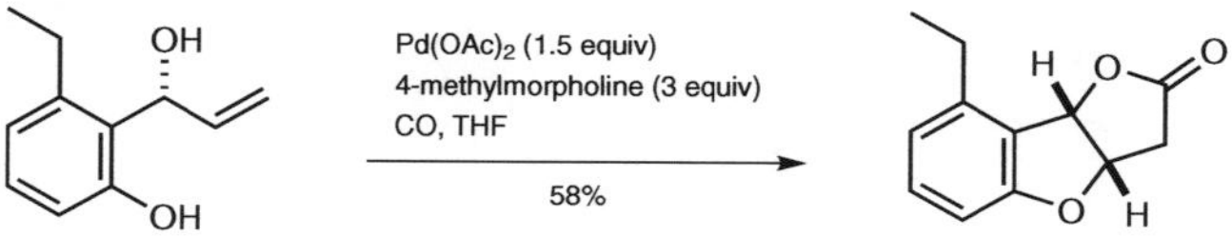

1.6.1 Palladium-Mediated Domino Reactions

503

5 mol% of palladium(II) trifluoroacetate, and enables a synthesis of the natural product (−)-blennolide C (**212**).

Scheme 46 A Key Step in the Synthesis of (−)-Blennolide C[138]

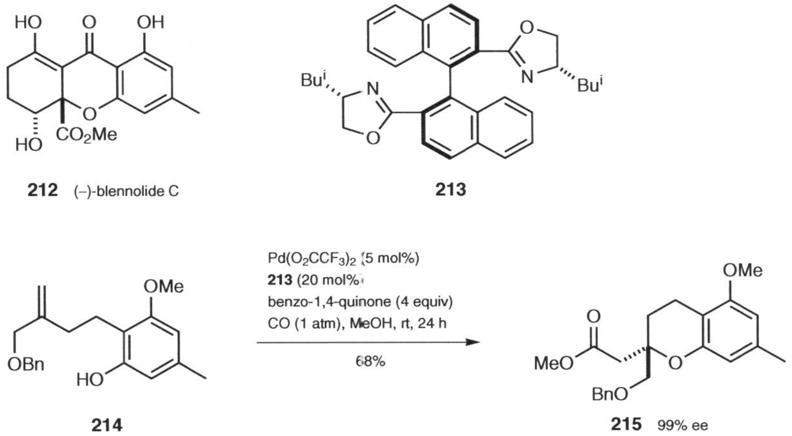

212 (−)-blennolide C

213

Pd(O₂CCF₃)₂ (5 mol%)
213 (20 mol%)
benzo-1,4-quinone (4 equiv)
CO (1 atm), MeOH, rt, 24 h

68%

214

215 99% ee

Methyl (*S,E*)-4-(6-Methoxy-2,5,7,8-tetramethyl-3,4-dihydro-2*H*-1-benzopyran-2-yl)but-2-enoate (195):[124]

A mixture of Pd(OCCF₃)₂ (7.1 mg, 0.0214 mmol, 10 mol%) and ligand **193** (40.7 mg, 0.0856 mmol, 40 mol%) in CH₂Cl₂ (0.1 mL, degassed) was stirred for 30 min at rt and then treated with benzo-1,4-quinone (93.1 mg, 0.855 mmol, 4.0 equiv), and stirred for a further 10 min. A soln of phenol **194** (50.0 mg, 0.214 mmol, 1.0 equiv) and methyl acrylate (91.9 mg, 1.07 mmol, 5.0 equiv) in CH₂Cl₂ (0.20 mL, degassed) was added to the suspension, and the mixture was stirred at rt for 3.5 d. At the end of the reaction (TLC), the mixture was treated with 1 M HCl (5 mL) and the aqueous phase was extracted with Et₂O (3 × 5 mL). The combined organic phases were washed with 1 M NaOH soln (3 × 5 mL) and dried (MgSO₄), and the solvent was removed under reduced pressure. The crude product was purified by column chromatography (pentane/Et₂O) to give the product as a clear oil; yield: 57 mg (84%).

1-(But-3-enyl)-1-(4-fluorophenyl)-1,3-dihydrobenzo[*b*]furan-5-carbonitrile (197):[125]

A 4-mL glass screw-top vial was charged with benzylic alcohol **196** (147 mg, 0.58 mmol, 1.0 equiv), allyl chloride (0.22 mL, 2.7 mmol, 4.6 equiv), toluene (2.2 mL), NaHCO₃ (92 mg, 1.1 mmol, 1.9 equiv), and Pd(hfacac)₂ (14 mg, 0.027 mmol, 5 mol%). The mixture was heated at 50 °C for 8.5 h, and then additional Pd(hfacac)₂ (14 mg, 0.027 mmol, 5 mol%) was added. Heating was continued for a further 18 h, and then the mixture was allowed to cool to rt and subjected directly to flash chromatography (petroleum ether/CH₂Cl₂ 1:1) to afford the product as a yellow oil; yield: 98 mg (58%).

Methyl 2-{(2*R*,3*S*,5*S*)-3-(*tert*-Butyldiphenylsiloxy)-5-[(*R*)-1-methoxyethyl]tetrahydrofuran-2-yl}acetate (200):[126]

CAUTION: *Carbon monoxide is extremely flammable and exposure to higher concentrations can quickly lead to a coma.*

A mixture of the alcohol **199** (48.4 mg, 0.122 mmol, 1.0 equiv) and Pd(OAc)₂ (41 mg, 0.183 mmol, 1.5 equiv) in MeOH (3 mL) under a CO atmosphere (1.1 atm) was stirred at

for references see p 506

504   Domino Transformations   **1.6** Metal-Mediated Reactions

23 °C for 3 h. The mixture was filtered and concentrated, and the residue was purified by preparative TLC (hexane/EtOAc 4:1) to afford the product as a colorless oil; yield: 30.8 mg (55%).

(3aR,5R,6aR)-5-Vinyltetrahydrofuro[3,2-b]furan-2(3H)-one (202):[130]

> **CAUTION:** *Carbon monoxide is extremely flammable and exposure to higher concentrations can quickly lead to a coma.*

$CuCl_2$ (3.34 g, 24.8 mmol, 3.0 equiv), $PdCl_2$ (147 mg, 0.827 mmol, 10 mol%), and NaOAc (2.04 g, 24.8 mmol, 3.0 equiv) were weighed into a flask. A soln of diene **201** (1.06 g, 8.27 mmol, 1 equiv) in AcOH (83 mL, 0.1 M) was added to the flask, which was evacuated and filled with CO (3×). The flask was fitted with a CO balloon and the contents were stirred for 24 h, at which point the color had changed from blue-green to tan. The mixture was filtered through Celite and concentrated. The residue was purified by flash chromatography (EtOAc/hexane 1:1) to give the product as a colorless oil; yield: 1.11 g (87%).

(3aR,4aR,5S,6S,9S,9aS,9bR,10S,Z)-6-(tert-Butyldimethylsiloxy)-4a,5,9-trimethyl-10-vinyl-3,3a,4a,5,6,9,9a,9b-octahydro-2H-5,9-methanocyclohepta[b]furo[2,3-d]furan-2-one (204):[134]

> **CAUTION:** *2-Methyloxirane is extremely flammable and forms explosive mixtures with air. It is an eye, skin, and respiratory tract irritant and a probable human carcinogen.*

> **CAUTION:** *Carbon monoxide is extremely flammable and exposure to higher concentrations can quickly lead to a coma.*

To a stirred mixture of $Pd(OAc)_2$ (1.0 mg, 0.005 mmol, 10 mol%) and $CuCl_2$ (17 mg, 0.13 mmol, 2.6 equiv) in THF (0.5 mL) was added 1,1,3,3-tetramethylthiourea (0.7 mg, 0.005 mmol, 10 mol%) at rt, and the resulting mixture was stirred for 30 min under an argon atmosphere. To this soln were added 2-methyloxirane (0.018 mL, 0.25 mmol, 5.0 equiv) and NH_4OAc (4.0 mg, 0.05 mmol, 1 equiv), and the mixture was purged with CO. A soln of enediol **203** (19.0 mg, 0.05 mmol, 1.0 equiv) in THF (0.1 mL) was added dropwise over 2 min, and then the mixture was stirred at 50 °C under CO for 12 h. The mixture was cooled to rt, then diluted with EtOAc (20 mL), and filtered through Celite. The filtrate was washed with sat. aq NH_4Cl (10 mL) and brine (10 mL), dried (Na_2SO_4), and concentrated. Purification of the residue by flash chromatography (hexane/EtOAc 10:1 to 5:1) afforded the product as a white semi-solid; yield: 16.0 mg (78%).

(1R,2S,2aR,4aS,7aR,7bS,8aS)-1-(tert-Butyldimethylsiloxy)-2-[(tert-butyldiphenylsiloxy)-methyl]-8a-methylhexahydro-1H-furo[3,2-b]oxireno[2′,3′:1,5]cyclopenta[1,2-d]pyran-6(2H)-one (206):[135]

To a soln of thiourea ligand (3aS,7aS)-1,3-di-2-tolyloctahydro-2H-benzimidazole-2-thione (34.6 mg, 0.103 mmol, 30 mol%) in THF (15 mL) were added $Pd(OAc)_2$ (23.1 mg, 0.103 mmol, 30 mol%) and $CuCl_2$ (138.3 mg, 1.03 mmol, 3.0 equiv), and the mixture was stirred under a N_2 atmosphere for 1 h at rt. To this soln was added dropwise a soln of diol **205** (200 mg, 0.34 mmol, 1.0 equiv) in THF (3 mL), and the mixture was then stirred at 70 °C for 8 h. The mixture was concentrated, and the residue was purified by flash chromatography (petroleum ether/EtOAc 6:1) to give the product as a white foam; yield: 196 mg (95%).

1.6.1 Palladium-Mediated Domino Reactions

505

Advanced Schindilactone A Intermediate 208:[136]

> **CAUTION:** *Carbon monoxide is extremely flammable and exposure to higher concentrations can quickly lead to a coma.*

To a soln of $Pd(OAc)_2$ (4.7 mg, 0.020 mmol, 50 mol%) and $CuCl_2$ (16 mg, 0.12 mmol, 3.0 equiv) in THF (2 mL) was added thiourea ligand (3aS,7aS)-1,3-di-2-tolyloctahydro-2H-benzimidazole-2-thione (7 mg, 0.02 mmol, 50 mol%), and the mixture was degassed with CO (balloon). To this soln was added a soln of allylic alcohol **207** (27 mg, 0.040 mmol, 1.0 equiv) in THF (2 mL), and the mixture was stirred at 70 °C for 1 h. The mixture was then concentrated, and the residue was purified by flash chromatography (petroleum ether/EtOAc 6:1) to give the product as a white solid; yield: 22 mg (78%).

(3aS,8bS)-8-Ethyl-3a,8b-dihydrofuro[3,2-b]benzofuran-2(3H)-one (211):[137]

> **CAUTION:** *Carbon monoxide is extremely flammable and exposure to higher concentrations can quickly lead to a coma.*

To a soln of diol **210** (50 mg, 0.28 mmol, 1.0 equiv) and 4-methylmorpholine (92 µL, 0.84 mmol, 3.0 equiv) in THF (3 mL) under a CO atmosphere (1 atm) was added a soln of $Pd(OAc)_2$ (94 mg, 0.42 mmol, 1.5 equiv) in THF (7 mL). The mixture was stirred at rt for 15 h and then filtered through a pad of Celite. The soln was concentrated and the residue was purified by flash column chromatography (hexanes/EtOAc 4:1) to afford the product as a white solid; yield: 33 mg (58%).

Methyl (S)-2-{2-[(Benzyloxy)methyl]-5-methoxy-7-methyl-3,4-dihydro-2H-1-benzopyran-2-yl}acetate (215):[138]

> **CAUTION:** *Carbon monoxide is extremely flammable and exposure to higher concentrations can quickly lead to a coma.*

A soln of $Pd(O_2CCF_3)_2$ (2.7 mg, 8.1 µmol, 5 mol%) and (S,S)-iBu-BOXAX (**213**; 16.2 mg, 32.1 µmol, 20 mol%) in MeOH (0.5 mL) was stirred at rt for 15 min and then added to alkene **214** (50 mg, 160 µmol, 1.0 equiv) by syringe (rinsing with 0.5 mL MeOH). Benzo-1,4-quinone (69 mg, 0.64 mmol, 4.0 equiv) was added and the mixture was stirred under a CO atmosphere (1 atm) at rt for 24 h. The mixture was then poured into 1 M HCl (10 mL) and the aqueous layer was extracted with t-BuOMe (3 × 5 mL). The combined organic phases were washed with 1 M NaOH (3 × 5 mL), dried (Na_2SO_4), and concentrated. Purification by flash chromatography (pentane/EtOAc 25:1 to 10:1) gave the product as a colorless oil that solidified under vacuum; yield: 40.5 mg (68%).

1.6.1.6 Conclusions

It is clear from the diversity of targets and methods described in this chapter that palladium catalysts continue to offer enormous opportunity for creativity and efficiency in the design of domino processes. Given the emergence of new methods and reaction paradigms in the field of palladium catalysis, it is surely beyond question that such innovation in reaction design will continue to hold a prominent position in the arena of cascade reactions.

for references see p 506

References

[1] Tietze, L. F.; Brasche, G.; Gericke, K. M., *Domino Reactions in Organic Synthesis*; Wiley-VCH: Weinheim, Germany, (2006).

[2] Nicolaou, K. C.; Edmonds, D. J.; Bulger, P. G., *Angew. Chem.*, (2006) **118**, 7292; *Angew. Chem. Int. Ed.*, (2006) **45**, 7134.

[3] Anderson, E. A., *Org. Biomol. Chem.*, (2011) **9**, 3997.

[4] Nicolaou, K. C.; Chen, J. S., *Chem. Soc. Rev.*, (2009) **38**, 2993.

[5] Majumdar, K. C.; Sinha, B., *Synthesis*, (2013) **45**, 1271.

[6] Vlaar, T.; Ruijter, E.; Orru, R. V. A., *Adv. Synth. Catal.*, (2011) **353**, 809.

[7] Nicolaou, K. C.; Bulger, P. G.; Sarlah, D., *Angew. Chem.*, (2005) **117**, 4516; *Angew. Chem. Int. Ed.*, (2005) **44**, 4442.

[8] Ohno, H., *Asian J. Org. Chem.*, (2013) **2**, 18.

[9] Muzart, J., *Tetrahedron*, (2013) **69**, 6735.

[10] Burns, B.; Grigg, R.; Ratananukul, P.; Sridharan, V.; Stevenson, P.; Sukirthalingam, S.; Worakun, T., *Tetrahedron Lett.*, (1988) **29**, 5565.

[11] Burns, B.; Grigg, R.; Sridharan, V.; Stevenson, P.; Sukirthalingam, S.; Worakun, T., *Tetrahedron Lett.*, (1989) **30**, 1135.

[12] Grigg, R.; Sridharan, V.; Sukirthalingam, S., *Tetrahedron Lett.*, (1991) **32**, 3855.

[13] Grigg, R.; Dorrity, M. J.; Malone, J. F.; Sridharan, V.; Sukirthalingam, S., *Tetrahedron Lett.*, (1990) **31**, 1343.

[14] Negishi, E.-i.; Noda, Y.; Lamaty, F.; Vawter, E. J., *Tetrahedron Lett.*, (1990) **31**, 4393.

[15] Owczarczyk, Z.; Lamaty, F.; Vawter, E. J.; Negishi, E.-i., *J. Am. Chem. Soc.*, (1992) **114**, 10091.

[16] Zhang, Y.; Negishi, E.-i., *J. Am. Chem. Soc.*, (1989) **111**, 3454.

[17] Zhang, Y.; Wu, G.-z.; Agnel, G.; Negishi, E.-i., *J. Am. Chem. Soc.*, (1990) **112**, 8590.

[18] Trost, B. M.; Lee, D. C., *J. Am. Chem. Soc.*, (1988) **110**, 7255.

[19] Wang, R. T.; Chou, F. L.; Luo, F. T., *J. Org. Chem.*, (1990) **55**, 4846.

[20] Nuss, J. M.; Rennels, R. A.; Levine, B. H., *J. Am. Chem. Soc.*, (1993) **115**, 6991.

[21] Negishi, E.-i.; Copéret, C.; Ma, S.; Liou, S.-Y.; Liu, F., *Chem. Rev.*, (1996) **96**, 365.

[22] Kucera, D. J.; O'Connor, S. J.; Overman, L. E., *J. Org. Chem.*, (1993) **58**, 5304.

[23] Fox, M. E.; Li, C.; Marino, J. P., Jr.; Overman, L. E., *J. Am. Chem. Soc.*, (1999) **121**, 5467.

[24] Beaudry, C. M.; Malerich, J. P.; Trauner, D., *Chem. Rev.*, (2005) **105**, 4757.

[25] Nicolaou, K. C.; Petasis, N. A.; Zipkin, R. E., *J. Am. Chem. Soc.*, (1982) **104**, 5560.

[26] Beaudry, C. M.; Trauner, D., *Org. Lett.*, (2002) **4**, 2221.

[27] Moses, J. E.; Baldwin, J. E.; Marquez, R.; Adlington, R. M.; Cowley, A. R., *Org. Lett.*, (2002) **4**, 3731.

[28] Parker, K. A.; Lim, Y.-H., *J. Am. Chem. Soc.*, (2004) **126**, 15968.

[29] Beaudry, C. M.; Trauner, D., *Org. Lett.*, (2005) **7**, 4475.

[30] Moses, J. E.; Baldwin, J. E.; Brückner, S.; Eade, S. J.; Adlington, R. M., *Org. Biomol. Chem.*, (2003) **1**, 3670.

[31] Jacobsen, M. F.; Moses, J. E.; Adlington, R. M.; Baldwin, J. E., *Org. Lett.*, (2005) **7**, 2473.

[32] Tan, E. H. P.; Lloyd-Jones, G. C.; Harvey, J. N.; Lennox, A. J. J.; Mills, B. M., *Angew. Chem.*, (2011) **123**, 9776; *Angew. Chem. Int. Ed.*, (2011) **50**, 9602.

[33] Webster, R.; Gaspar, B.; Mayer, P.; Trauner, D., *Org. Lett.*, (2013) **15**, 1866.

[34] Fujita, S.; Sakaguchi, T.; Kobayashi, T.; Tsuchikawa, H.; Katsumura, S., *Org. Lett.*, (2013) **15**, 2758.

[35] Suffert, J.; Salem, B.; Klotz, P., *J. Am. Chem. Soc.*, (2001) **123**, 12107.

[36] Salem, B.; Klotz, P.; Suffert, J., *Org. Lett.*, (2003) **5**, 845.

[37] Charpenay, M.; Boudhar, A.; Blond, G.; Suffert, J., *Angew. Chem.*, (2012) **124**, 4455; *Angew. Chem. Int. Ed.*, (2012) **51**, 4379.

[38] Bour, C.; Blond, G.; Salem, B.; Suffert, J., *Tetrahedron*, (2006) **62**, 10567.

[39] Salem, B.; Suffert, J., *Angew. Chem.*, (2004) **116**, 2886; *Angew. Chem. Int. Ed.*, (2004) **43**, 2826.

[40] Cordonnier, M.-C. A.; Kan, S. B. J.; Anderson, E. A., *Chem. Commun. (Cambridge)*, (2008), 5818.

[41] Ishikura, M., *Heterocycles*, (2011) **83**, 247.

[42] Abe, T.; Ikeda, T.; Yanada, R.; Ishikura, M., *Org. Lett.*, (2011) **13**, 3356.

[43] Abe, T.; Ikeda, T.; Choshi, T.; Hibino, S.; Hatae, N.; Toyata, E.; Yanada, R.; Ishikura, M., *Eur. J. Org. Chem.*, (2012), 5018.

[44] Ishikura, M.; Takahashi, N.; Yamada, K.; Abe, T.; Yanada, R., *Helv. Chim. Acta*, (2008) **91**, 1828.

[45] Trost, B. M.; Lautens, M., *J. Am. Chem. Soc.*, (1985) **107**, 1781.

References

[46] Trost, B. M.; Tanoury, G. J.; Lautens, M.; Chan, C.; MacPherson, D. T., *J. Am. Chem. Soc.*, (1994) **116**, 4255.

[47] Trost, B. M.; Romero, D. L.; Rise, F., *J. Am. Chem. Soc.*, (1994) **116**, 4268.

[48] Peixoto, P. A.; Severin, R.; Tseng, C.-C.; Chen, D. Y.-K., *Angew. Chem.*, (2011) **123**, 3039; *Angew. Chem. Int. Ed.*, (2011) **50**, 3013.

[49] Gogoi, P.; Sigüeiro, R.; Eduardo, S.; Mouriño, A., *Chem.–Eur. J.*, (2010) **16**, 1432.

[50] Nuss, J. M.; Murphy, M. M.; Rennels, R. A.; Heravi, M. H.; Mohr, B. J., *Tetrahedron Lett.*, (1993) **34**, 3079.

[51] Trost, B. M.; Dumas, J., *J. Am. Chem. Soc.*, (1992) **114**, 1924.

[52] Trost, B. M.; Dumas, J.; Villa, M., *J. Am. Chem. Soc.*, (1992) **114**, 9836.

[53] Negishi, E.-i.; Harring, L. S.; Owczarczyk, Z.; Mohamud, M. M.; Ay, M., *Tetrahedron Lett.*, (1992) **33**, 3253.

[54] Meyer, F. E.; de Meijere, A., *Synlett*, (1991), 777.

[55] Schweizer, S.; Tokan, W. M.; Parsons, P. J.; de Meijere, A., *Eur. J. Org. Chem.*, (2010), 4687.

[56] de Meijere, A.; von Zezschwitz, P.; Bräse, S., *Acc. Chem. Res.*, (2005) **38**, 413.

[57] Goh, S. S.; Baars, H.; Gockel, B.; Anderson, E. A., *Org. Lett.*, (2012) **14**, 6278.

[58] Campbell, C. D.; Greenaway, R. L.; Holton, O. T.; Chapman, H. A.; Anderson, E. A., *Chem. Commun. (Cambridge)*, (2014) **50**, 5187.

[59] Petrignet, J.; Boudhar, A.; Blond, G.; Suffert, J., *Angew. Chem.*, (2011) **124**, 3343; *Angew. Chem. Int. Ed.*, (2011) **50**, 3285.

[60] McDonald, R. I.; Liu, G.; Stahl, S. S., *Chem. Rev.*, (2011) **111**, 2981.

[61] Kočovský, P.; Bäckvall, J.-E., *Chem.–Eur. J.*, (2015) **21**, 36.

[62] Wolfe, J. P., *Synlett*, (2008), 2913.

[63] Wolfe, J. P., *Eur. J. Org. Chem.*, (2007), 571.

[64] Schultz, D. M.; Wolfe, J. P., *Synthesis*, (2012) **44**, 351.

[65] Babij, N. R.; Wolfe, J. P., *Angew. Chem.*, (2012) **124**, 4204; *Angew. Chem. Int. Ed.*, (2012) **51**, 4128.

[66] Bertrand, M. B.; Neukom, J. D.; Wolfe, J. P., *J. Org. Chem.*, (2008) **73**, 8851.

[67] Semmelhack, M. F.; Kim, C.; Zhang, N.; Bodurow, C.; Sanner, M.; Dobler, W.; Meier, M., *Pure Appl. Chem.*, (1990) **62**, 2035.

[68] Marshall, J. A.; Yanik, M. M., *Tetrahedron Lett.*, (2000) **41**, 4717.

[69] Carpenter, J.; Northrup, A. B.; Chung, d.; Wiener, J. J. M.; Kim, S.-G.; MacMillan, D. W. C., *Angew. Chem.*, (2008) **120**, 3624; *Angew. Chem. Int. Ed.*, (2008) **47**, 3568.

[70] Xia, G.; Han, X.; Lu, X., *Org. Lett.*, (2014) **16**, 2058.

[71] Wu, X.-F.; Neumann, H.; Beller, M., *Chem. Rev.*, (2013) **113**, 1.

[72] Wu, X.-F.; Neumann, H.; Beller, M., *Chem. Soc. Rev.*, (2011) **40**, 4986.

[73] Bio, M. M.; Leighton, J. L., *J. Org. Chem.*, (2003) **68**, 1693.

[74] Sadig, J. E. R.; Willis, M. C., *Synthesis*, (2011), 1.

[75] Ball, C. J.; Willis, M. C., *Eur. J. Org. Chem.*, (2013), 425.

[76] Fletcher, A. J.; Bax, M. N.; Willis, M. C., *Chem. Commun. (Cambridge)*, (2007), 4764.

[77] Fang, Y.-Q.; Karisch, R.; Lautens, M., *J. Org. Chem.*, (2007) **72**, 1341.

[78] Maddaford, S. P.; Andersen, N. G.; Cristofoli, W. A.; Keay, B. A., *J. Am. Chem. Soc.*, (1996) **118**, 10766.

[79] McCartney, D.; Guiry, P. J., *Chem. Soc. Rev.*, (2011) **40**, 5122.

[80] Miyazaki, F.; Uotsu, K.; Shibasaki, M., *Tetrahedron*, (1998) **54**, 13073.

[81] Overman, L. E.; Rosen, M. D., *Angew. Chem.*, (2000) **112**, 4768; *Angew. Chem. Int. Ed.*, (2000) **39**, 4596.

[82] Overman, L. E.; Rosen, M. D., *Tetrahedron*, (2010) **66**, 6514.

[83] Couty, S.; Liegault, B.; Meyer, C.; Cossy, J., *Tetrahedron*, (2006) **62**, 3882.

[84] Grigg, R.; Loganathan, V.; Sridharan, V., *Tetrahedron Lett.*, (1996) **37**, 3399.

[85] Tietze, L. F.; Duefert, S.-C.; Clerc, J.; Bischoff, M.; Maaß, C.; Stalke, D., *Angew. Chem.*, (2013) **125**, 3273; *Angew. Chem. Int. Ed.*, (2013) **52**, 3191.

[86] Tietze, L. F.; Kahle, K.; Raschke, T., *Chem.–Eur. J.*, (2002) **8**, 401.

[87] Tietze, L. F.; Clerc, J.; Biller, S.; Duefert, S.-C.; Bischoff, M., *Chem.–Eur. J.*, (2014) **20**, 17119.

[88] Catellani, M.; Frignani, F.; Rangoni, A., *Angew. Chem.*, (1997) **109**, 142; *Angew. Chem. Int. Ed.*, (1997) **36**, 119.

[89] Ferraccioli, R., *Synthesis*, (2013) **45**, 581.

[90] Weinstabl, H.; Suhartono, M.; Qureshi, Z.; Lautens, M., *Angew. Chem.*, (2013) **125**, 5413; *Angew. Chem. Int. Ed.*, (2013) **52**, 5305.

[91] Sui, X.; Zhu, R.; Li, G.; Ma, X.; Gu, Z., *J. Am. Chem. Soc.*, (2013) **135**, 9318.

[92] Bertrand, M. B.; Wolfe, J. P., *Org. Lett.*, (2006) **8**, 2353.

[93] Mai, D. N.; Wolfe, J. P., *J. Am. Chem. Soc.*, (2010) **132**, 12 157.

[94] Rossiter, L. M.; Slater, M. L.; Giessert, R. E.; Sakwa, S. A.; Herr, R. J., *J. Org. Chem.*, (2009) **74**, 9554.

[95] Inuki, S.; Oishi, S.; Fujii, N.; Ohno, H., *Org. Lett.*, (2008) **10**, 5239.

[96] Inuki, S.; Iwata, A.; Oishi, S.; Fujii, N.; Ohno, H., *J. Org. Chem.*, (2011) **76**, 2072.

[97] Hawkins, A.; Jakubec, P.; Ironmonger, A.; Dixon, D. J., *Tetrahedron Lett.*, (2013) **54**, 365.

[98] Li, M.; Dixon, D. J., *Org. Lett.*, (2010) **12**, 3784.

[99] Li, M.; Hawkins, A.; Barber, D. M.; Bultinck, P.; Herrebout, W.; Dixon, D. J., *Chem. Commun. (Cambridge)*, (2013) **49**, 5265.

[100] Lv, P.; Huang, K.; Xie, L.; Xu, X., *Org. Biomol. Chem.*, (2011) **9**, 3133.

[101] Trost, B. M.; O'Boyle, B. M.; Hund, D., *Chem.–Eur. J.*, (2010) **16**, 9772.

[102] Tadd, A. C.; Fielding, M. R.; Willis, M. C., *Chem. Commun. (Cambridge)*, (2009), 6744.

[103] Vlaar, T.; Ruijter, E.; Maes, B. U. W.; Orru, R. V. A., *Angew. Chem.*, (2013) **125**, 7222; *Angew. Chem. Int. Ed.*, (2013) **52**, 7084.

[104] Curran, D. P.; Du, W., *Org. Lett.*, (2002) **4**, 3215.

[105] Liu, H.; Li, C.; Qiu, D.; Tong, X., *J. Am. Chem. Soc.*, (2011) **133**, 6187.

[106] Newman, S. G.; Lautens, M., *J. Am. Chem. Soc.*, (2011) **133**, 1778.

[107] Monks, B. M.; Cook, S. P., *Angew. Chem.*, (2013) **125**, 14464; *Angew. Chem. Int. Ed.*, (2013) **52**, 14214.

[108] Petrone, D. A.; Yoon, H.; Weinstabl, H.; Lautens, M., *Angew. Chem.*, (2014) **126**, 8042; *Angew. Chem. Int. Ed.*, (2014) **53**, 7908.

[109] Trost, B. M.; Crawley, M. L., *Chem. Rev.*, (2003) **103**, 2921.

[110] Lu, Z.; Ma, S., *Angew. Chem.*, (2008) **120**, 264; *Angew. Chem. Int. Ed.*, (2008) **47**, 258.

[111] Tietze, L. F.; Redert, T.; Bell, H. P.; Hellkamp, S.; Levy, L. M., *Chem.–Eur. J.*, (2008) **14**, 2527.

[112] Harada, K.; Imai, A.; Uto, K.; Carter, R. G.; Kubo, M.; Hioki, H.; Fukuyama, Y., *Org. Lett.*, (2011) **13**, 988.

[113] Shimizu, I.; Ohashi, Y.; Tsuji, J., *Tetrahedron Lett.*, (1985) **26**, 3825.

[114] Goldberg, A. F. G.; Stoltz, B. M., *Org. Lett.*, (2011) **13**, 4474.

[115] Doi, T.; Iijima, Y.; Takasaki, M.; Takahashi, T., *J. Org. Chem.*, (2007) **72**, 3667.

[116] Guo, L.-N.; Duan, X.-H.; Liang, Y.-M., *Acc. Chem. Res.*, (2011) **44**, 111.

[117] Tsuji, J.; Mandai, T., *Angew. Chem.*, (1995) **107**, 2830; *Angew. Chem. Int. Ed.*, (1995) **34**, 2589.

[118] Tsuji, J.; Watanabe, H.; Minami, I.; Shimizu, I., *J. Am. Chem. Soc.*, (1985) **107**, 2196.

[119] Inuki, S.; Yoshimitsu, Y.; Oishi, S.; Fujii, N.; Ohno, H., *Org. Lett.*, (2009) **11**, 4478.

[120] Inuki, S.; Yoshimitsu, Y.; Oishi, S.; Fujii, N.; Ohno, H., *J. Org. Chem.*, (2010) **75**, 3831.

[121] Daniels, D. S. B.; Thompson, A. L.; Anderson, E. A., *Angew. Chem.*, (2011) **123**, 11 708; *Angew. Chem. Int. Ed.*, (2011) **50**, 11 506.

[122] Daniels, D. S. B.; Jones, A. S.; Thompson, A. L.; Paton, R. S.; Anderson, E. A., *Angew. Chem.*, (2014) **126**, 1946; *Angew. Chem. Int. Ed.*, (2014) **53**, 1915.

[123] Semmelhack, M. F.; Epa, W. R., *Tetrahedron Lett.*, (1993) **34**, 7205.

[124] Tietze, L. F.; Sommer, K. M.; Zinngrebe, J.; Stecker, F., *Angew. Chem.*, (2005) **117**, 262; *Angew. Chem. Int. Ed.*, (2005) **44**, 257.

[125] Hewitt, J. F. M.; Williams, L.; Aggarwal, P.; Smith, C. D.; France, D. J., *Chem. Sci.*, (2013) **4**, 3538.

[126] Semmelhack, M. F.; Epa, W. R.; Cheung, A. W.-H.; Gu, Y.; Kim, C.; Zhang, N.; Lew, W., *J. Am. Chem. Soc.*, (1994) **116**, 7455.

[127] Semmelhack, M. F.; Bozell, J. J.; Sato, T.; Wulff, W.; Spiess, E.; Zask, A., *J. Am. Chem. Soc.*, (1982) **104**, 5850.

[128] Semmelhack, M. F.; Bodurow, C., *J. Am. Chem. Soc.*, (1984) **106**, 1496.

[129] Tamaru, Y.; Kobayashi, T.; Kawamura, S.-i.; Ochiai, H.; Hojo, M.; Yoshida, Z.-i., *Tetrahedron Lett.*, (1985) **26**, 3207.

[130] Werness, J. B.; Tang, W., *Org. Lett.*, (2011) **13**, 3664.

[131] Li, Z.; Gao, Y.; Tang, Y.; Dai, M.; Wang, G.; Wang, Z.; Yang, Z., *Org. Lett.*, (2008) **10**, 3017.

[132] Li, Z.; Gao, Y.; Jiao, Z.; Wu, N.; Wang, D. Z.; Yang, Z., *Org. Lett.*, (2008) **10**, 5163.

[133] Tamaru, Y.; Hojo, M.; Yoshida, Z.-i., *J. Org. Chem.*, (1991) **56**, 1099.

[134] Xu, X.-S.; Li, Z.-W.; Zhang, Y.-J.; Peng, X.-S.; Wong, H. N. C., *Chem. Commun. (Cambridge)*, (2012) **48**, 8517.

[135] Tang, Y.; Zhang, Y.; Dai, M.; Luo, T.; Deng, L.; Chen, J.; Yang, Z., *Org. Lett.*, (2005) **7**, 885.

References

[136] Xiao, Q.; Ren, W.-W.; Chen, Z.-X.; Sun, T.-W.; Li, Y.; Ye, Q.-D.; Gong, J.-X.; Meng, F.-K.; You, L.; Liu, Y.-F.; Zhao, M.-Z.; Xu, L.-M.; Shan, Z.-H.; Shi, Y.; Tang, Y.-F.; Chen, J.-H.; Yang, Z., *Angew. Chem.*, (2011) **123**, 7511; *Angew. Chem. Int. Ed.*, (2011) **50**, 7373.

[137] Boukouvalas, J.; Pouliot, M.; Robichaud, J.; MacNeil, S.; Snieckus, V., *Org. Lett.*, (2006) **8**, 3597.

[138] Tietze, L. F.; Jackenkroll, S.; Hierold, J.; Ma, L.; Waldecker, B., *Chem.–Eur. J.*, (2014) **20**, 8628.

1.6.2   **Dirhodium-Catalyzed Domino Reactions**

X. Xu, P. Truong, and M. P. Doyle

General Introduction

Domino reactions (also known as tandem reactions or cascade reactions), in which two or more sequential bond-forming events occur inter- or intramolecularly under the same reaction conditions, are efficient and economical processes for the synthesis of organic compounds.[1,2] Over the past two decades, the development of cascade processes has advanced dramatically due to increasing demand for highly atom-economic and operationally simple methodologies that rapidly build up molecular complexity through catalysis.[3–5] In spite of the great advancements that have been achieved over the past few decades, major challenges remain. A comprehensive review by Padwa from 1996 highlights catalytic metal–carbene reactions as providing effective pathways for domino processes,[6] and dirhodium(II) catalysts have been identified as more suitable for these transformations than other transition-metal catalysts due to their robustness under the reaction conditions and the often low catalyst loadings that can be employed. Dirhodium(II) carboxylates (see examples in Scheme 1) form highly electrophilic rhodium–carbene intermediates with diazo compounds that undergo intramolecular addition to a C≡C bond to form a second metal–carbene which completes the cascade through traditional carbene transformations such as cyclopropanation, cyclopropenation, C—H functionalization, or ylide formation (Scheme 2). Given past contributions, this chapter only focuses on the most recent progress of rhodium-catalyzed domino reactions, with a strong emphasis on reports in which the initially formed reactive metal–carbene intermediates cascade intramolecularly into other rhodium-bound intermediates through bond-forming events that eventually produce a stable product (e.g., Scheme 2). This limitation excludes a number of well-known chemical transformations such as intermolecular ylide formation and subsequent intramolecular rearrangement, which can be found elsewhere,[7,8] but will allow for focus on some other impressive transformations.

Scheme 1   Dirhodium(II) Carboxylate Catalysts

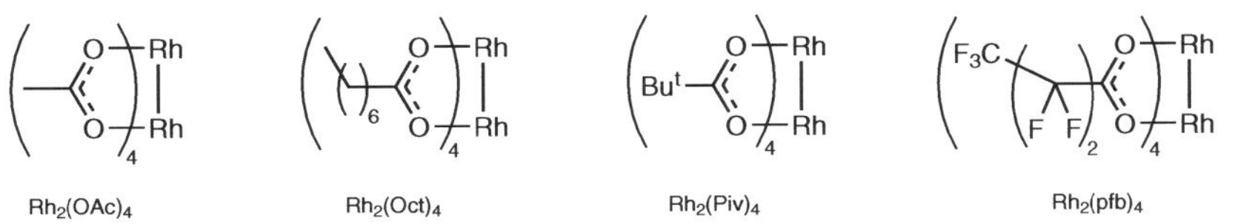

Rh₂(OAc)₄   Rh₂(Oct)₄   Rh₂(Piv)₄   Rh₂(pfb)₄

for references see p 533

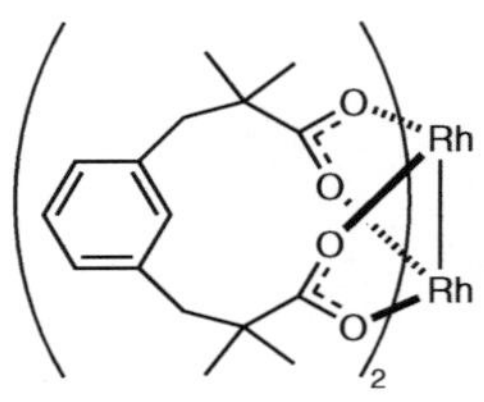

512 Domino Transformations 1.6 Metal-Mediated Reactions

Rh₂(esp)₂   Rh₂(S-TCPTTL)₄

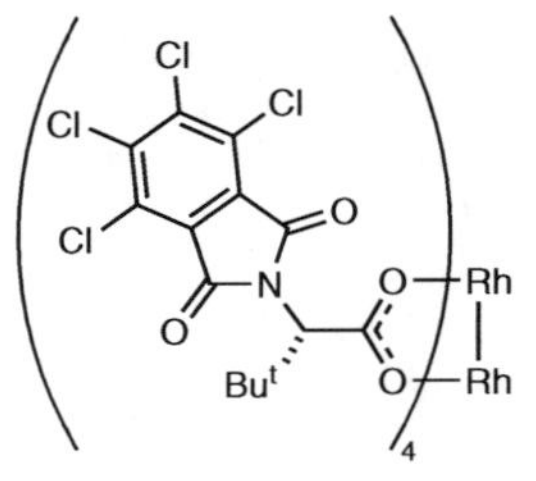

Scheme 2 Cascade Processes Involving Metal–Carbene to Carbene Transformations

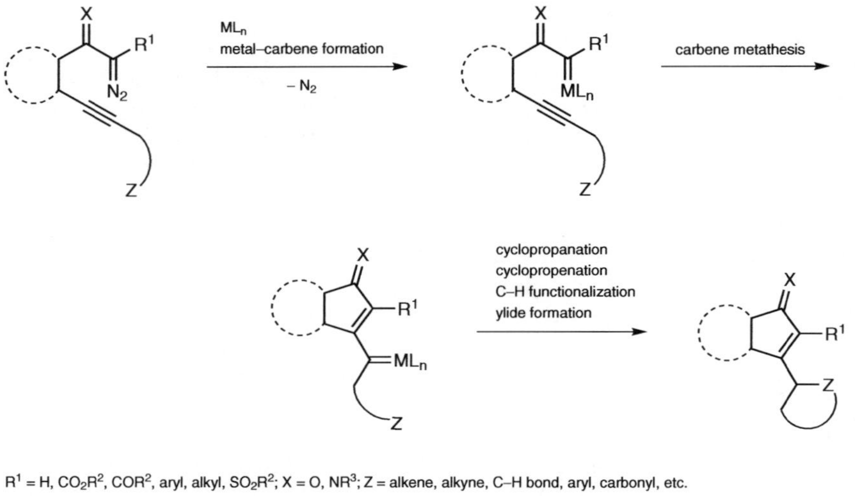

R^1 = H, CO_2R^2, COR^2, aryl, alkyl, SO_2R^2; X = O, NR^3; Z = alkene, alkyne, C–H bond, aryl, carbonyl, etc.

1.6.2.1   1-Sulfonyl-1,2,3-triazoles as (Azavinyl)carbene Precursors in Domino Reactions

α-Diazocarbonyl compounds **1** have been widely utilized in rhodium(II)-catalyzed dinitrogen extrusion reactions, forming versatile dirhodium(II)–carbene intermediates **2** (Scheme 3) that undergo a variety of novel transformations including insertion, cyclopropanation, cycloaddition, ylide formation, and other reactions.[9–12] These diazo compounds, ranging from diazo ketones to diazoacetates and -malonates, as well as assorted derivatives, are proven reactive templates for cascade transformations.[6,9,10,13–16]

Scheme 3 Access to Rhodium(II)–Carbenes from α-Diazocarbonyl Compounds[9–12]

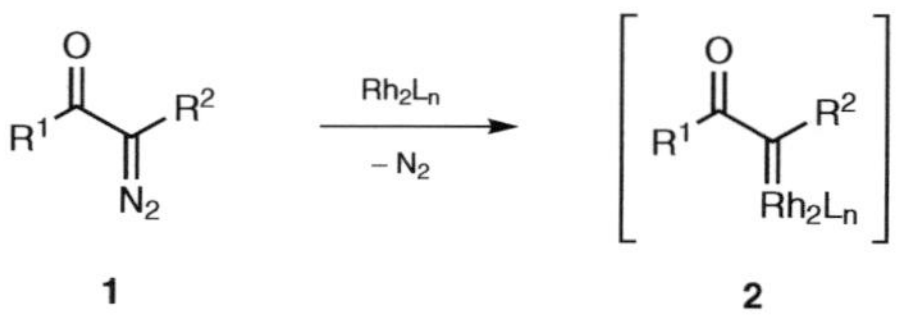

Reactions of the structurally related α-diazoimines **4** have not been as extensively explored due to their previous inaccessibility.[17,18] Only recently have Fokin, Gevorgyan, and others discovered that readily available and stable 1-sulfonyl-1,2,3-triazoles **3** are ef-

1.6.2 Dirhodium-Catalyzed Domino Reactions **513**

fective precursors to rhodium(II)–(azavinyl)carbenes **5** through rhodium(II)-catalyzed dinitrogen extrusion reactions (Scheme 4).[19,20] This discovery has generated great interest since the carbenes **5** serve as a versatile template for cascade reactions in the synthesis of complex nitrogen-containing heterocyclic compounds. This section will highlight the most recent work on rhodium(II)-catalyzed generation of metal–carbenes from 1-sulfonyl-1,2,3-triazoles that are involved in cascade processes.

Scheme 4 Access to Rhodium(II)–(Azavinyl)carbenes from 1-Sulfonyl-1,2,3-triazoles and Their Subsequent Use in the Synthesis of Nitrogen-Containing Heterocycles[19,20]

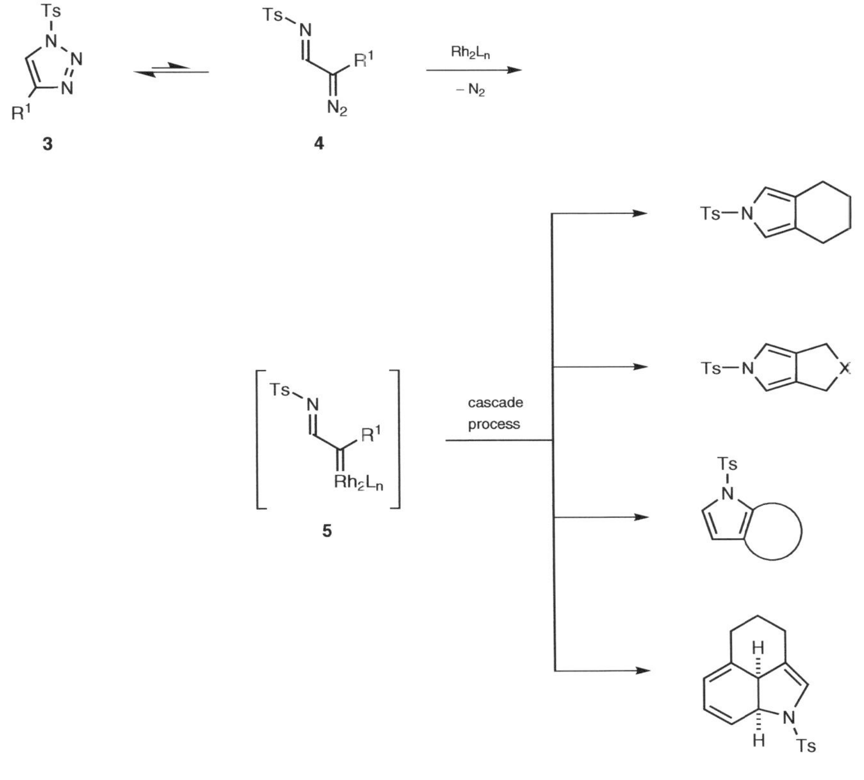

Taking advantage of the unique reactivity of rhodium–(azavinyl)carbenes **5** that has been established by Fokin and co-workers,[20–23] several groups have explored their intramolecular reactions with tethered unsaturated functional groups at C4 of the triazole to attain 3,4-fused pyrroles (Scheme 5). Sarpong reported that Rh₂(Oct)₄ (Scheme 1), assisted by microwave irradiation, catalyzes the dedinitrogenative rearrangement of allene-tethered 1-sulfonyl-1,2,3-triazoles **6** for the highly effective production of 3,4-fused pyrroles **7**.[24] Similarly, Gevorgyan reported that 3,4-fused pyrroles **9** are obtained by Rh₂(esp)₂-catalyzed decomposition of alkynyl-tethered 1-sulfonyl 1,2,3-triazoles **8**.[25] Although the proposed mechanisms for these two transformations are not identical, both involve initial dirhodium-catalyzed dinitrogen extrusion and rhodium–carbene formation. Rhodium–carbene formation is followed by an attack of the allene or alkyne at the electrophilic rhodium–carbene center with subsequent cyclization to provide the basic structural skeleton, and product formation is completed by aromatization that eventually furnishes 3,4-fused pyrroles.

for references see p 533

Scheme 5 Synthesis of 3,4-Fused Pyrroles from Allene- and Alkyne-Tethered 1-Sulfonyl-1,2,3-triazoles[24,25]

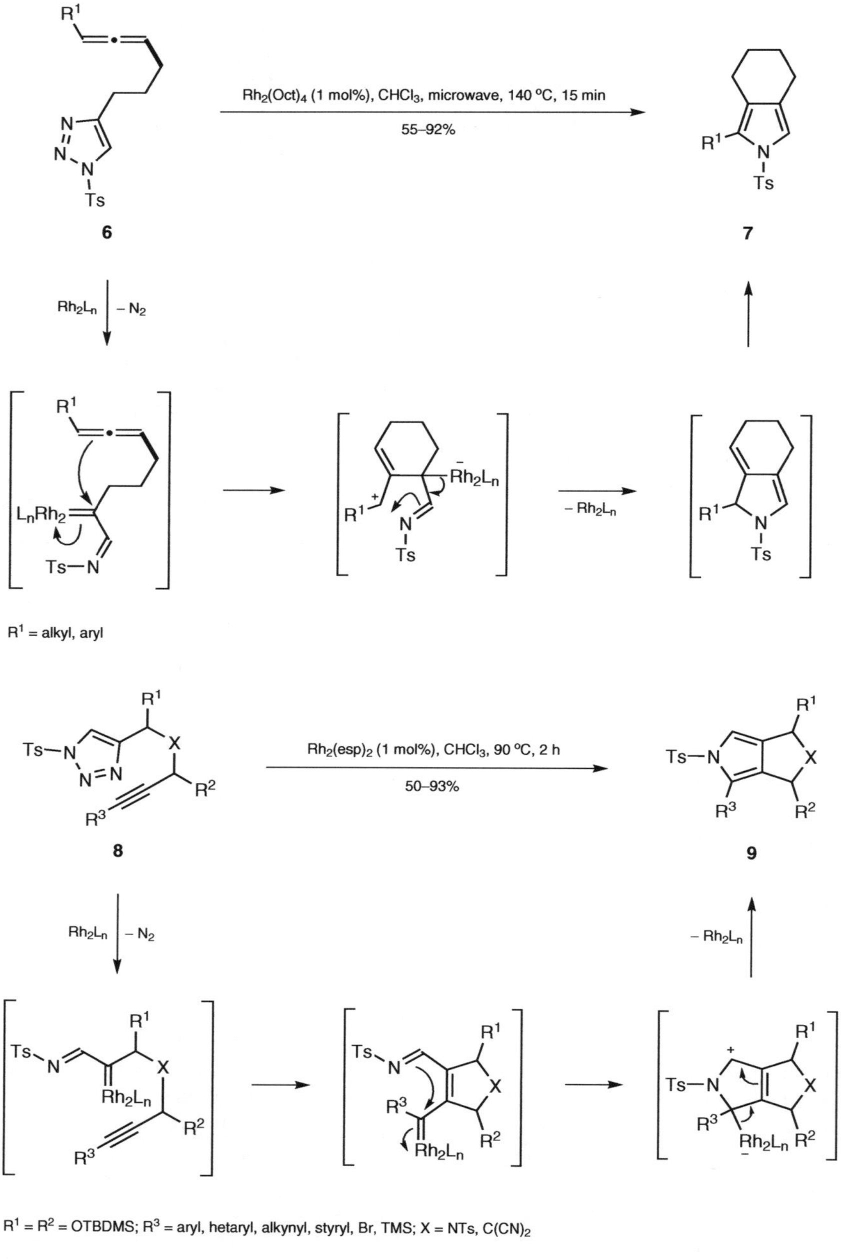

R^1 = alkyl, aryl

$R^1 = R^2$ = OTBDMS; R^3 = aryl, hetaryl, alkynyl, styryl, Br, TMS; X = NTs, C(CN)$_2$

Davies and co-workers have discovered that Rh$_2$(esp)$_2$ (Scheme 1) effectively promotes the rearrangement of 4-alkenyl-1-sulfonyl-1,2,3-triazoles **10** with the loss of dinitrogen to form 2,3-fused pyrroles **11** (Scheme 6).[26] They propose that the rhodium–carbene intermediate adopts the *s-trans* conformation **12**, which undergoes a 4π-electrocyclization (Nazarov-type cyclization) to form intermediate **13**. Subsequent proton elimination and aromatization leads to vinylrhodium **14** which, upon reprotonation, produces pyrroles **11**.

1.6.2 Dirhodium-Catalyzed Domino Reactions

Scheme 6 Synthesis of 2,3-Fused Pyrroles from Alkenyl-Tethered 1-Sulfonyl-1,2,3-triazoles[26]

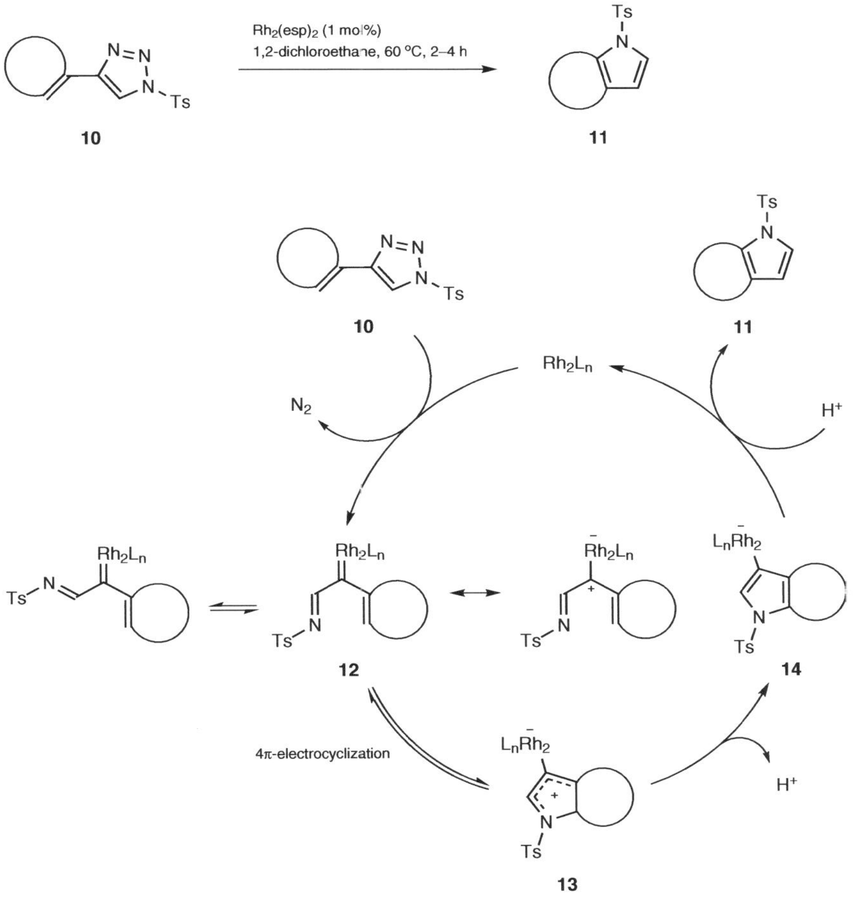

Murakami and co-workers have developed a $Rh_2(Piv)_4$-catalyzed intramolecular dearomatizing annulation of 4-(3-arylpropyl)-1-sulfonyl-1,2,3-triazoles **15** for the synthesis of 3,4-fused dihydroindoles **17** (Scheme 7).[27] Remarkably, this reaction tolerates both electron-donating and electron-withdrawing substituents on the phenyl ring. The authors showcase one example for asymmetric induction, achieving 81% ee through the use of $Rh_2(S\text{-TCPTTL})_4$ (Scheme 1) as the chiral catalyst (Scheme 7). As illustrated from the proposed mechanism, the key step to achieve the formal [3+2]-annulation with the formation of the dihydroindole is the cyclization of rhodium-bound zwitterionic intermediate **16**.

for references see p 533

Scheme 7 Synthesis of 3,4-Fused Indoles by Intramolecular Dearomatizing [3+2]-Annulation and Catalytic Asymmetric Intramolecular Dearomatizing [3+2]-Annulation of an α-Imino Carbene with a Phenyl Group[27]

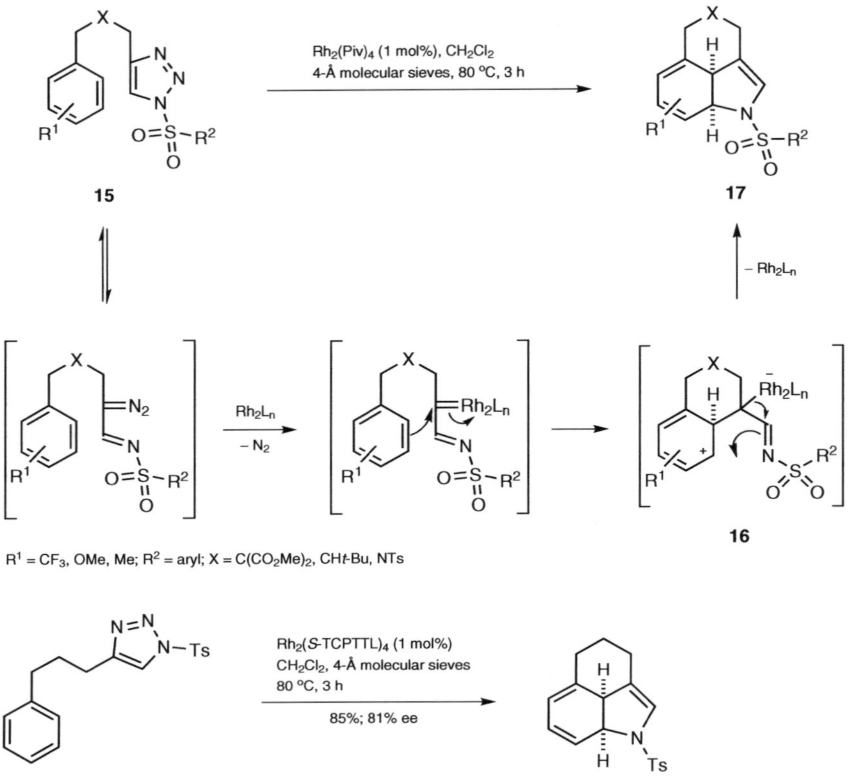

R^1 = CF_3, OMe, Me; R^2 = aryl; X = $C(CO_2Me)_2$, CHt-Bu, NTs

3,4-Fused 1-Tosylpyrroles 9; General Procedure:[25]

An oven-dried 3.0-mL vial equipped with a stirrer bar was charged with $Rh_2(esp)_2$ (1.5 mg, 1 mol%), triazole **8** (0.2 mmol), and $CHCl_3$ (3 mL) under a N_2 atmosphere. The reaction vessel was capped with a Mininert syringe valve and the mixture was stirred at 90 °C for 2 h. Upon completion of the reaction, the mixture was cooled to rt and concentrated under reduced pressure, and the crude product was purified by column chromatography to afford the corresponding 3,4-fused pyrrole **9**.

1.6.2.2 Dirhodium(II)-Catalyzed Generation of Rhodium–Carbenes from Cyclopropenes and Their Subsequent Reactions

Cyclopropenation reactions of alkynes with a catalytically generated rhodium–carbene intermediate formed from a diazo compound have been intensively studied, and several chiral catalytic systems have been devised to control the enantioselectivity of such a transformation.[28,29] Compared to the well-known cyclopropenation reaction, the use of substituted cyclopropenes as carbene precursors under dirhodium catalysis has not been fully explored, and their applications in organic synthesis are still primitive.

Müller and co-workers have systematically examined the rearrangement reaction of multisubstituted cyclopropenes in the presence of Lewis acidic dirhodium(II) heptafluorobutanoate [$Rh_2(pfb)_4$; Scheme 1] and contrasted the differential reactivity of $Rh_2(pfb)_4$ with other transition-metal complexes (Scheme 8).[30–32] Although some degree of similarity has been discovered for $Rh_2(pfb)_4$ and copper complexes, drastic differences in most cases

1.6.2 Dirhodium-Catalyzed Domino Reactions

517

have underscored the uniqueness of rhodium catalysts in these transformations. More importantly, a large number of reaction outcomes are only consistent with the formation of a rhodium–vinylcarbene intermediate of type **18** from a cyclopropene (Scheme 8). Such a carbene can then go on to perform traditional carbene transformations. For instance, treatment of the triphenylcyclopropenes **19** and **21** with a catalytic amount of Rh$_2$(pfb)$_4$ results in the formation of 1,2-diphenylindene **20** or α,β-unsaturated ketone **22**, depending on the substitution pattern on the cyclopropene ring. Heating the carboxylate-substituted cyclopropene **23** under Rh$_2$(pfb)$_4$ catalysis presumably generates a vinylcarbene; this species then likely forms an oxonium ylide via intramolecular addition of the ester carbonyl and the ylide rearranges to the final furan product **24**. The reaction sequence of rhodium–vinylcarbene formation and 1,2-hydride migration is observed when an alkyl-substituted cyclopropene **25** is employed, e.g. to give (E,Z)- and (E,E)-**26**. Interestingly, a small amount of cycloheptatriene product **29** is isolated along with the anticipated C—H insertion product **28** when cyclopropene **27** is used, an outcome which could be rationalized by an intermolecular Buchner reaction of the rhodium–vinylcarbene intermediate with the benzene solvent.

Scheme 8 Rhodium–Vinylcarbene Formation from Cyclopropenes under Dirhodium(II) Heptafluorobutanoate Catalysis[30–32]

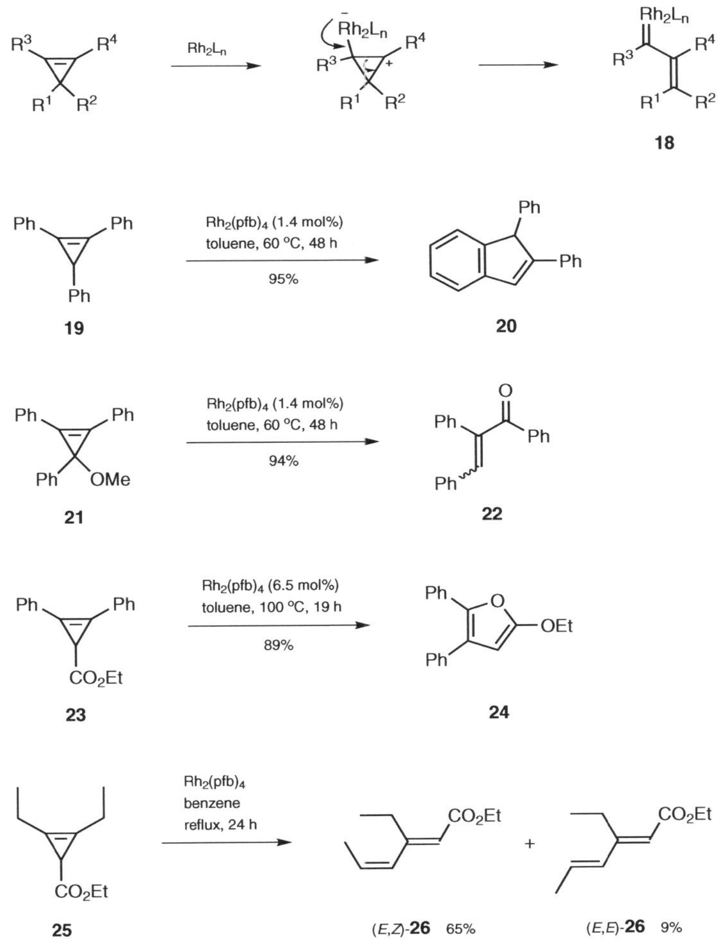

for references see p 533

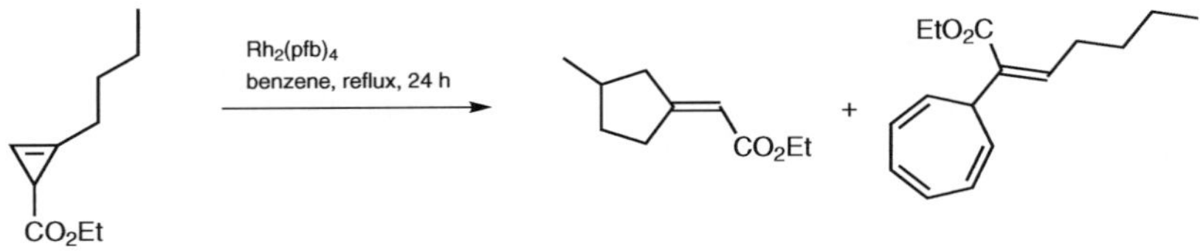

518 Domino Transformations **1.6** Metal-Mediated Reactions

Cossy and co-workers have recognized the unique advantage of using cyclopropenes as rhodium–vinylcarbene precursors to access donor-type carbenes. They have recently reported a very useful approach for the diastereoselective construction of multisubstituted carbocycles and oxygen heterocycles (Scheme 9).[33] For example, dimethylcyclopropenes **32** can be easily prepared by treating tribromocyclopropane **30** with 2 equivalents of butyllithium to afford the corresponding lithiated dimethylcyclopropene **31**, which subsequently reacts with an aliphatic aldehyde. Dimethylcyclopropenes **32** show very high reactivity toward dirhodium-catalyzed vinylcarbene formation; in fact, such a transformation occurs rapidly at room temperature in dichloromethane with only 0.5 mol% $Rh_2(OAc)_4$. While the reaction outcomes could potentially be complicated by the formation of the two regioisomeric vinylcarbene intermediates, rhodium–vinylcarbene **33**, which can be schematically understood by the initiation of dirhodium catalyst coordination to the less-substituted cyclopropene carbon, appears to be formed exclusively. Intramolecular 1,5-C–H insertion occurs with moderate diastereoselectivity when $X = CH_2$ (e.g., to give 2-alkylidenecyclopentanol **34**); intramolecular 1,6-C–H insertion occurs with excellent diastereocontrol when $X = O$ (e.g., to give 4-alkylidenetetrahydropyran-3-ol **36**). A deuterium labeling experiment suggests that the C–H insertion process proceeds in a concerted stereospecific manner, and the observed high diastereoselectivity is rationalized by a seven-membered cyclic boat transition state with minimal 1,3-allylic strain.

1.6.2 Dirhodium-Catalyzed Domino Reactions

519

Scheme 9 Diastereoselective Construction of Carbocycles and Oxygen Heterocycles through Dirhodium-Catalyzed Isomerization of Substituted Cyclopropenes[33]

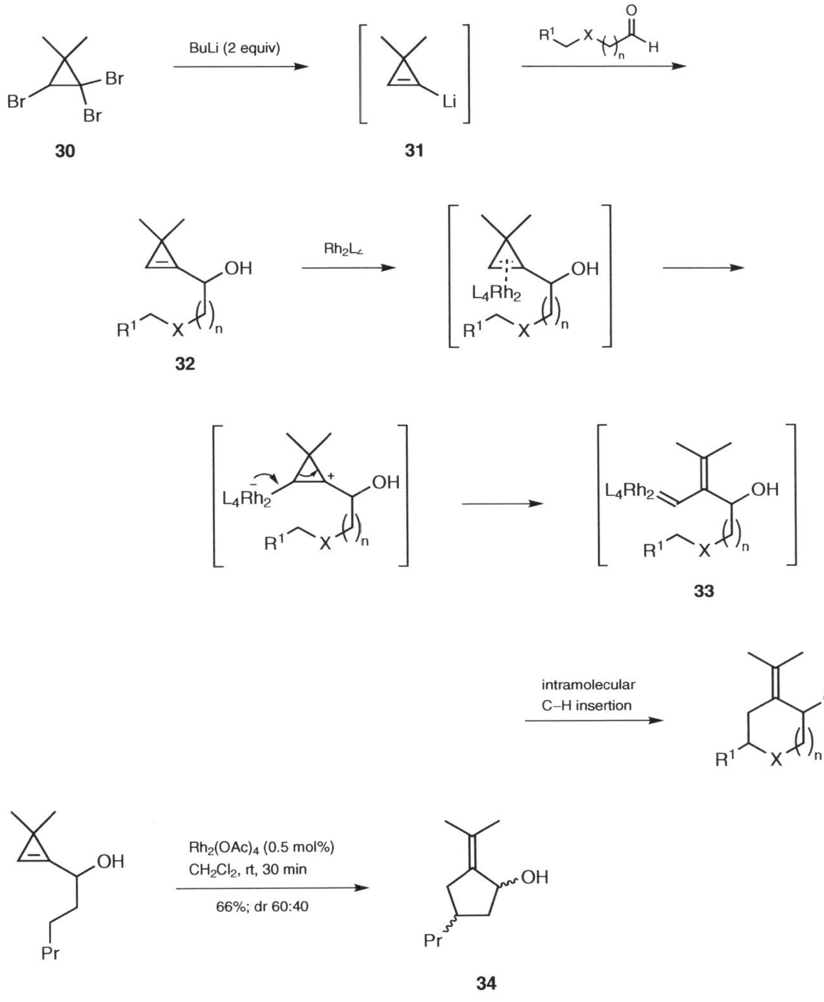

for references see p 533

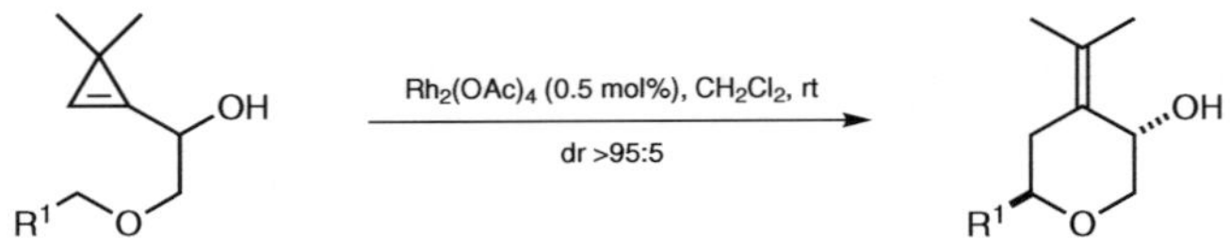

R[1]	Time (h)	Yield (%)	Ref
(CH$_2$)$_6$Me	8	46[a]	[33]
CO$_2$t-Bu	–[b]	0	[33]
Ph	3	98	[33]
3-MeOC$_6$H$_4$	2	99	[33]
4-MeOC$_6$H$_4$	2	86	[33]
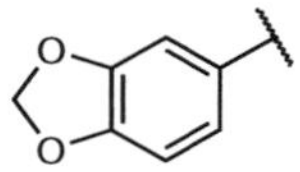	1.5	82	[33]
4-F$_3$CC$_6$H$_4$	1.5	83	[33]
4-FC$_6$H$_4$	1.5	90	[33]
2-BrC$_6$H$_4$	1.5	81	[33]
3-O$_2$NC$_6$H$_4$	1.75	77	[33]

[a] Using Rh$_2$(OAc)$_4$ (1.5 mol%).
[b] Not reported.

Sequential cyclopropene ring opening/rhodium–vinylcarbene formation and intramolecular electrophilic aromatic substitution have been reported by Vicente and co-workers.[34] Coordination of the π-Lewis acidic dirhodium catalyst[35] to the alkyne moiety provides the activation for the nucleophilic attack of the carbonyl oxygen, and subsequent rearrangement produces the rhodium–carbene intermediate that reacts with a terminal alkyne to generate a stable and isolable furan-substituted cyclopropene **37**. The overall result is ring enlargement of the cyclopropene ring. However, the final electrophilic aromatic substitution reaction is not regioselective and results in the formation of two regioisomers (**38** and **39**) with low selectivity (Scheme 10).

1.6.2 Dirhodium-Catalyzed Domino Reactions **521**

Scheme 10 Dirhodium(II) Carboxylate Catalyzed Cyclization, Cyclopropenation, and Ring-Expansion of Enynones[34]

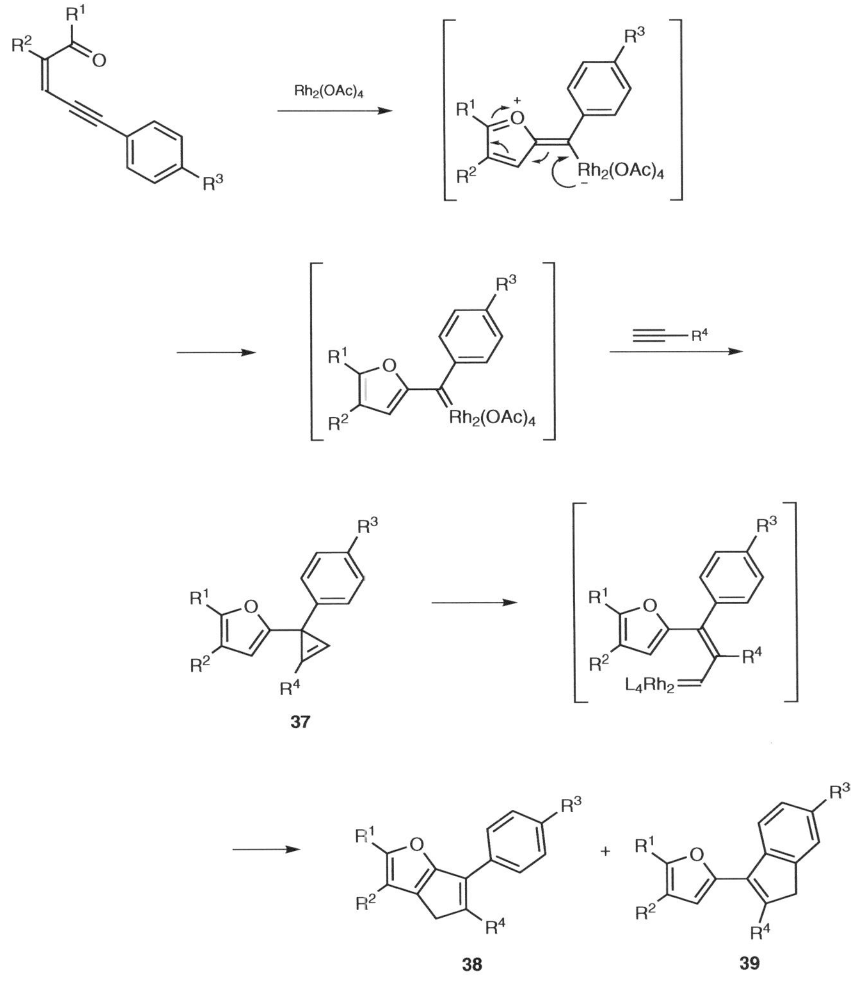

**6-Heptyl-4-(propan-2-ylidene)tetrahydro-2H-pyran-3-ol [36, R^1 = (CH$_2$)$_6$Me];
Typical Procedure:**[33]

To a soln of cyclopropenylcarbinol **35** (0.517 mmol) in CH$_2$Cl$_2$ (5 mL) at rt, was added Rh$_2$(OAc)$_4$ (1.1 mg, 2.6 μmol, 0.5 mol%). After 7 h, the reaction was not complete and more Rh$_2$(OAc)$_4$ was added (2.2 mg, 5.2 μmol, 1 mol%). After a further 1 h of stirring at rt, the mixture was concentrated under reduced pressure, and analysis of the crude material by ^{1}H NMR spectroscopy determined the diastereomeric ratio. Purification by flash chromatography (silica gel, petroleum ether/EtOAc 9:1 to 4:1) gave tetrahydropyran **36** [R^1 = (CH$_2$)$_6$Me] as a colorless oil; yield 57.3 mg (46%).

1.6.2.3   **Dirhodium(II)-Catalyzed Carbene/Alkyne Metathesis**

The metathesis reaction of a metallocarbene with an alkyne has been studied extensively and reviewed comprehensively over the last two decades.[6] Nevertheless, two recent examples are still very informative. May and co-workers have demonstrated that introduction of an alkyne functionality into the structural framework of a diazoacetate **40** derived

for references see p 533

522 Domino Transformations **1.6** Metal-Mediated Reactions

from a tertiary alcohol enables the construction of bridged polycyclic compound **41** with great ease.[36] Upon catalysis with Rh$_2$(esp)$_2$ (Scheme 1), the carbene/alkyne metathesis and C–H insertion proceed fairly well for most substrates; however, diastereoselectivity appears to be moderate to poor (Scheme 11).

Scheme 11   Carbene/Alkyne Metathesis and C–H-Insertion Reaction Cascade[36]

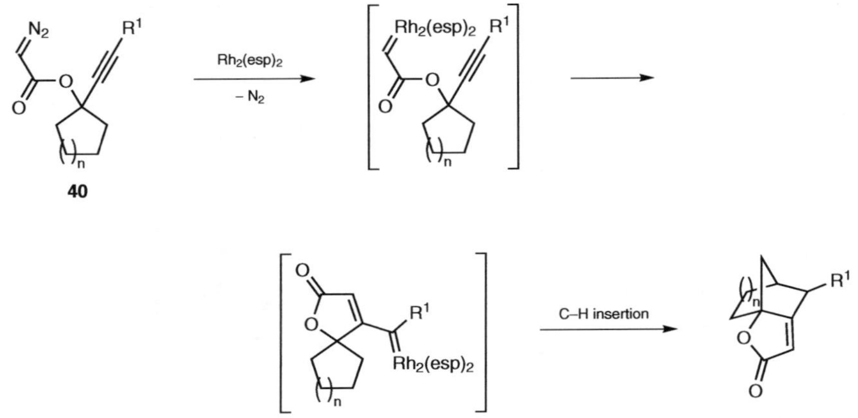

n	R^1	Yield (%)	Ref
1	4-F$_3$CC$_6$H$_4$	64	[36]
1	4-MeOC$_6$H$_4$	63	[36]
1	4-MeO$_2$CC$_6$H$_4$	65	[36]
1	TMS	78	[36]
2	TMS	69^a	[36]
3	TMS	72^a	[36]
4	TMS	65^a	[36]

^{a} Mixture of products including the bicyclo[2.2.n] isomer.

Encouraged by the recent successes of using enol diazoacetates in Lewis acid catalyzed reactions including Mukaiyama aldol,[37,38] Mannich,[37,39] and Michael addition,[40,41] Doyle and co-workers have reported a convenient approach for the coupling of a propargylic ester **43** with a silyl-protected diazo enol **42** (Scheme 12).[42] The resulting diazo compounds with tethered alkynyl functionality participate in cascade processes, and the reaction outcomes depend upon the Z substituent.[42] Reactions of β-keto-α-diazocarbonyl compounds **44** (R^1 = Ph, OMe) with a catalytic amount of Rh$_2$(Oct)$_4$ (Scheme 1) initiate domino events that form the secondary carbene **45** and then continue through formation of a carbonyl ylide with the proximal carbonyl group that, by elimination of Rh$_2$(Oct)$_4$, terminate at cyclopenta[c]furan-4-ones **46** (Scheme 13).

1.6.2 Dirhodium-Catalyzed Domino Reactions

Scheme 12 Lewis Acid Catalyzed Coupling of Silyl-Protected Diazo Enols and Propargylic Acetates[42]

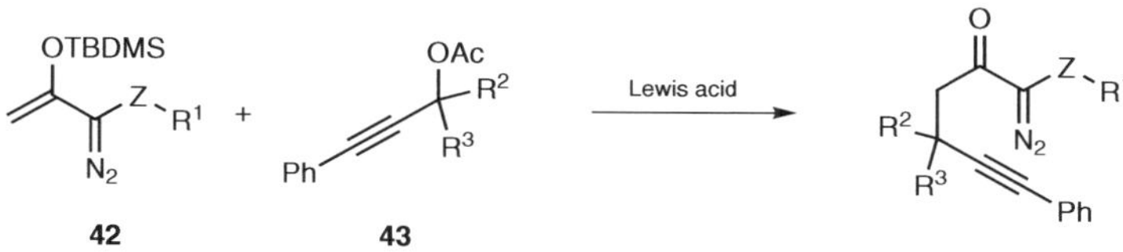

Z	R^1	R^2	R^3	Lewis Acid	Yield (%)	Ref
CO	OCH$_2$CH=CH$_2$	Me	Me	Sc(OTf)$_3$	90	[42]
CO	OMe	Me	Me	Sc(OTf)$_3$	82	[42]
CO	Ph	Me	Me	TMSOTf	82	[42]
CO	Ph	Et	Et	TMSOTf	78	[42]
SO$_2$	4-Tol	Me	Me	TMSOTf	83	[42]
SO$_2$	Mes	Me	Me	TMSOTf	72	[42]

Scheme 13 Carbene/Alkyne Metathesis and Carbonyl Ylide Formation[42]

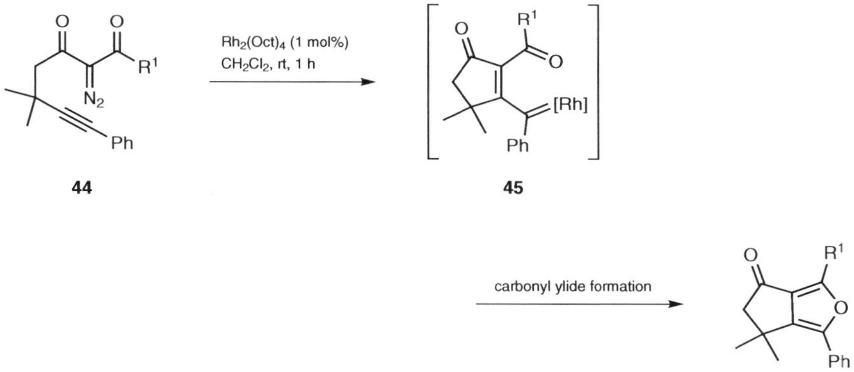

R^1	Yield (%)	Ref
Ph	86	[42]
OMe	68	[42]
OCH$_2$CH=CH$_2$	–^{a}	[42]

^{a} Not isolated.

While investigating the initial rhodium-catalyzed furan synthesis, it was found that the reaction of the allyl diazoacetoacetate **44** (R^1 = OCH$_2$CH=CH$_2$), unlike the methyl ester analogue **44** (R^1 = OMe), does not terminate at the alkoxyfuran **46** (R^1 = OCH$_2$CH=CH$_2$). Instead, the reaction provides mixtures of products which favor Claisen rearrangement product **47** {shown from the reaction catalyzed by copper(II) hexafluoroacetylacetonate [Cu(hfacac)$_2$], Scheme 14}. However, the Claisen product **47** is also found to undergo further rearrangement via a thermally induced Cope rearrangement to give butenolide **48**. Since competitive intramolecular cyclopropanation to provide **49** is not observed in this reaction, the carbene/alkyne metathesis occurs at a faster rate than cyclopropanation of

for references see p 533

524 Domino Transformations 1.6 Metal-Mediated Reactions

the starting diazoacetoacetate. This cascade process is synthetically notable in that the highly substituted bicyclic butenolide **48**, bearing fully substituted centers at the remote C3- and C4-positions, is formed in one step from a readily available alkynyl diazoacetoacetate.

Scheme 14 Carbene Metathesis/Claisen/Cope Cascade Reaction[42]

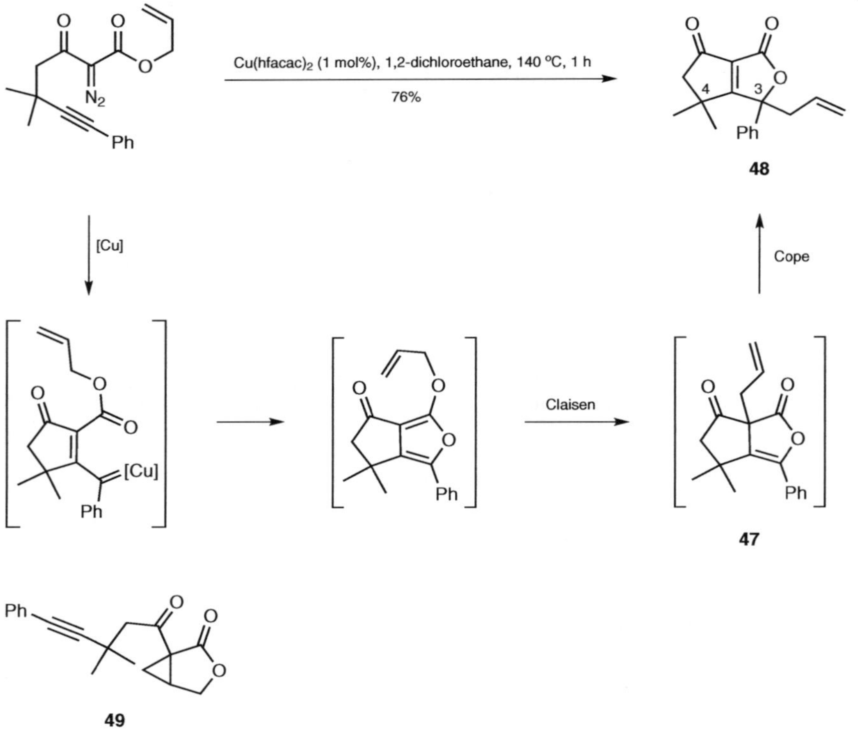

The outcome of the cascade reaction with the sulfonyl substituents at position Z (see Scheme 12) is notably different from those outlined in Scheme 14. With substrate **50**, the secondary carbene **51** reacts with the sulfone to provide an intermediate sulfoxonium ylide which undergoes O-atom transfer to deliver the racemic sulfoxide **52** in 90% yield (Scheme 15). Surprisingly, when the 4-tolyl sulfone is replaced with the mesityl sulfone, as in **53**, the O-atom transfer process no longer occurs and, instead, the secondary carbene **54** undergoes benzylic 1,7-C—H insertion to provide the benzothiepin ring system **55** in 80% yield. Although ylide formation is believed to be kinetically favored over other carbene reactions (especially to form five-membered-ring cyclic ylides), the preference for the cascade reaction of the mesityl sulfone to selectively target the benzylic C—H bond is surprising and indicates a conformational bias in the transition state brought about by the sterically encumbered mesityl group.

1.6.2 Dirhodium-Catalyzed Domino Reactions

Scheme 15 Effect of Aryl Sulfone Substitution on the Cascade Outcome[42]

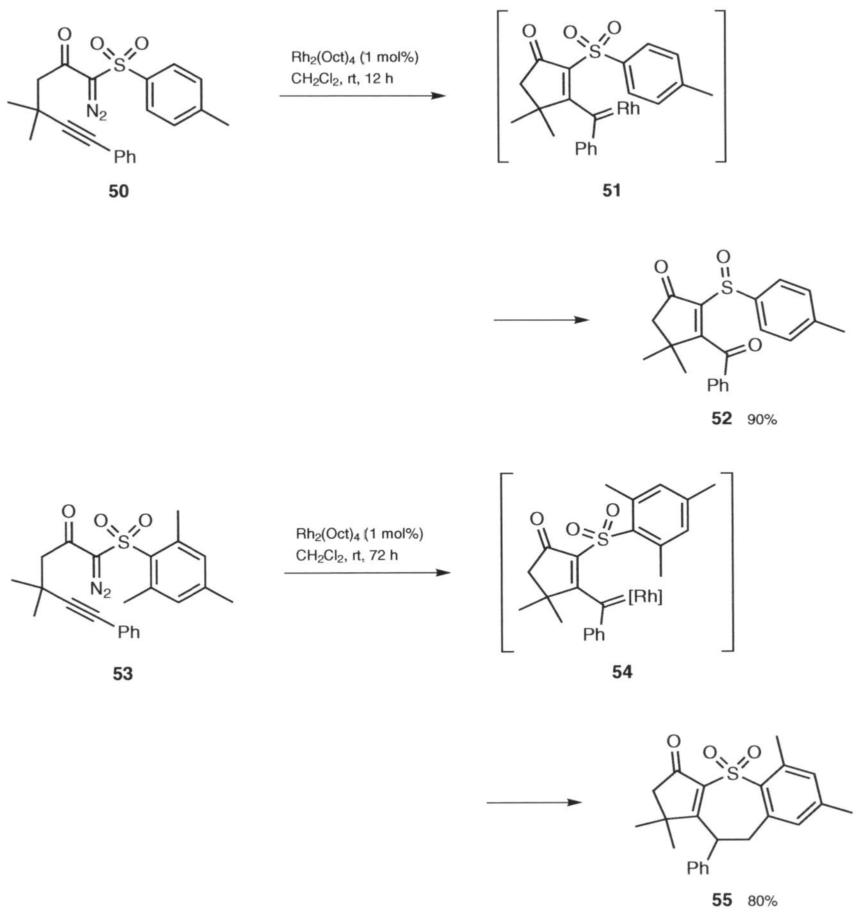

52 90%

55 80%

In the absence of a reactive functional group attached to the diazo-containing carbon, *gem*-diethyl diazo ketone **56** provides a unique opportunity to direct the cascade reaction with termination via intramolecular C—H insertion by either the first-formed or a secondary metallocarbene intermediate (Scheme 16). In this case, the reaction of ketone **56** with catalytic Rh$_2$(esp)$_2$ provides the cyclopentenone product **58** in 72% yield as a single diastereomer through insertion into a primary C—H bond. The reaction of the first-formed carbene of **56** is selective toward carbene/alkyne metathesis, forming the secondary phenyl(vinyl)carbene **57**. Compound **59**, which would result from insertion of the first-formed carbene into a secondary C—H bond, is not observed. The overall reaction presents a significant advantage over May's reported approach[36] in which mixtures of diastereomers are obtained with low selectivity.

for references see p 533

526 Domino Transformations **1.6** Metal-Mediated Reactions

Scheme 16 Carbene Metathesis/C–H Insertion Cascade Reaction[42]

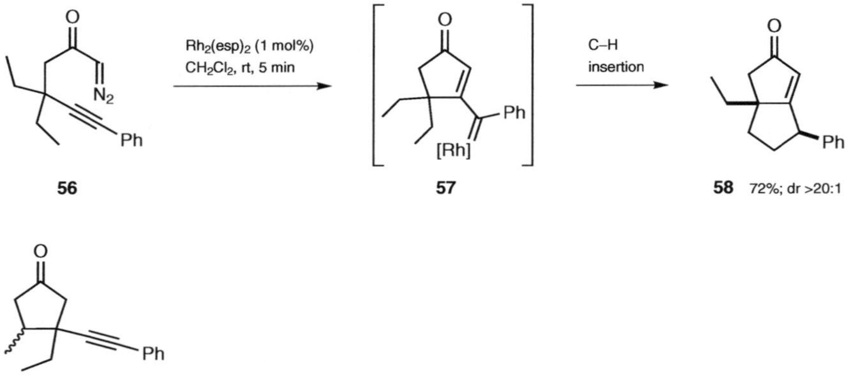

56   **57**   **58** 72%; dr >20:1

59

In an effort to establish a clearer picture of the rate of carbene/alkyne metathesis relative to intermolecular carbonyl ylide formation, the reaction of diazo ketone **60** with 4-methoxybenzaldehyde under Rh$_2$(esp)$_2$ catalysis was investigated; this reaction affords exclusively dihydrofuran **62** as a single diastereomer in 81% yield, presumably via an intramolecular 1,3-dipolar cycloaddition of ylide **61** (Scheme 17). This result demonstrates that intermolecular carbonyl ylide formation proceeds at a faster rate than carbene/alkyne metathesis, since the competing cycloaddition products **63** and **64** are not observed.

Scheme 17 Intermolecular Ylide Formation/Intramolecular Dipolar Cycloaddition[42]

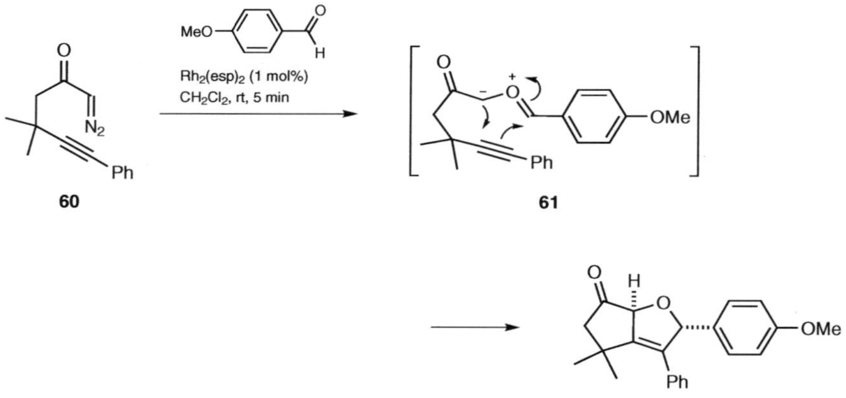

60   **61**

62 81%; dr >20:1

1.6.2 Dirhodium-Catalyzed Domino Reactions

527

3-Allyl-4,4-dimethyl-3-phenyl-4,5-dihydro-1H-cyclopenta[c]furan-1,6(3H)-dione (48); Typical Procedure:[42]

A mixture of allyl 2-diazo-5,5-dimethyl-3-oxo-7-phenylhept-6-ynoate (25 mg, 0.081 mmol) and $Cu(hfacac)_2$ (0.40 mg, 0.00081 mmol) in 1,2-dichloroethane (1.6 mL) was heated at 140 °C in a screw-capped vial for 1 h. The mixture was concentrated and the residue was purified by automated flash chromatography (hexane/EtOAc) to provide **48** as a colorless oil; yield: 76%.

1.6.2.4 Nitrene Cascade Reactions Catalyzed by a Dirhodium Complex

The incorporation of a nitrogen atom into an organic molecule with the formation of a C—N bond is one of the most performed transformations in organic synthesis.[43,44] Research dedicated to the development of highly selective methods for the construction of the C—N bond has attracted enormous attention.[45,46]

The neutral single-valent nitrene with six valence electrons has historically been regarded as a highly potent intermediate for the direct formation of a C—N bond.[47,48] However, due to the high reactivity and undiscriminating nature of free nitrenes, the efficiency and selectivity of these conversions are generally moderate to low.[45] The discovery that transition-metal complexes are capable of catalyzing and directing nitrene transfer,[49–52] in combination with the use of an iodine(III) oxidant[53–55] for the in situ generation of metallonitrene species, has stimulated the growth of nitrene chemistry.

Closely resembling their carbene siblings, nitrenes preferentially engage in electrophilic addition reactions, and both free nitrenes and metallonitrenes react with hydrocarbons and alkenes to afford the corresponding C—H amination and aziridination products.[46] Among transition-metal complexes that are capable of directing nitrene transfer, paddlewheel dirhodium complexes, and $Rh_2(esp)_2$ in particular, display exceptional reactivity and, more importantly, a high level of selectivity.[56] While metal carbenes readily undergo carbene/alkyne metathesis reactions both intra- and intermolecularly, the parallel nitrene/alkyne metathesis reactions have been largely underreported. Similarly to its carbene/alkyne analogue, nitrene/alkyne metathesis would result in the formation of a carbene intermediate or an equivalent (Scheme 18), which, in principle, could be trapped by nucleophiles internally or externally.

for references see p 533

Scheme 18 Alkyne/Nitrene Metathesis To Afford α-Imino Carbenes

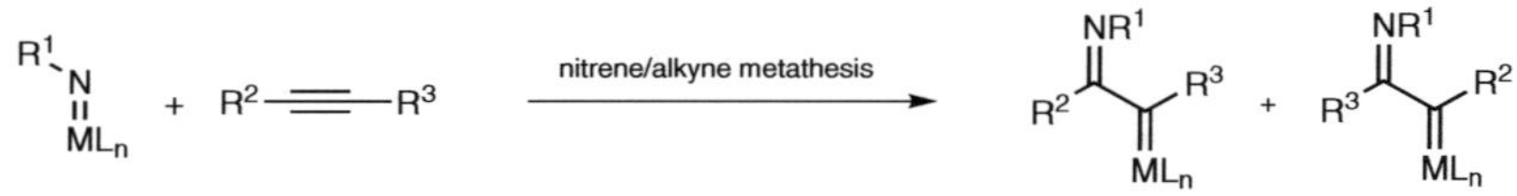

As shown by Blakey and co-workers, the intramolecular nitrene/alkyne metathesis process can be initiated by judicious placement of an alkyne functionality in proximity to a catalytically generated sulfamate-derived metallonitrene intermediate (e.g., **66**; Scheme 19).[57] The putative rhodium–carbene intermediate **67** is trapped intramolecularly by the electron lone pair of an ether oxygen to generate oxonium ylide intermediate **68**, a species which undergoes a facile [1,2]-Stevens or [2,3]-sigmatropic rearrangement, depending on the ether being alkyl or allyl, respectively. The resulting *N*-(sulfonyl)imine **69** can be reduced to sulfonamide **70** by sodium borohydride or undergo addition of a Grignard reagent. In the latter case, the overall reaction sequence constructs two vicinal quaternary chiral centers with complete diastereoselectivity. Mechanistically, the nitrene/alkyne metathesis of compounds of type **71** represents a key step in the overall rearrangement. Several electrophilic intermediates **72–74** have been proposed, but the exact identity of the intermediate remains elusive (Scheme 20).[57]

Scheme 19 Rh$_2$(esp)$_2$-Catalyzed Alkyne/Nitrene Metathesis, Oxonium Ylide Formation, and 2,3-Sigmatropic Rearrangement Reaction[57]

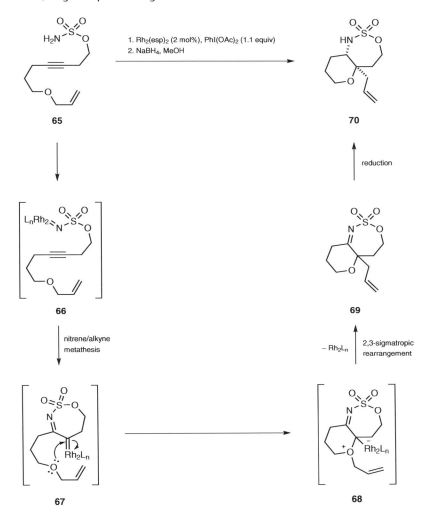

Scheme 20 Potential Reactive Intermediates Produced in the Alkyne/Nitrene Metathesis Reaction[57]

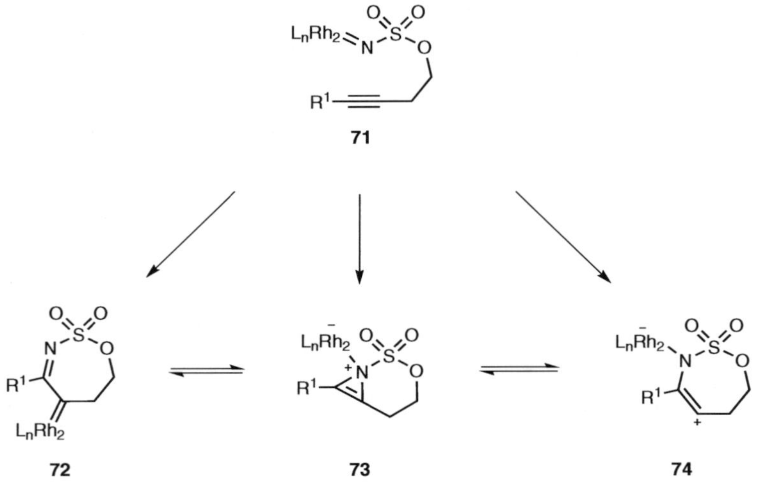

In an effort to elucidate the mechanistic detail for the nitrene/alkyne metathesis process, the alkyne-tethered sulfamate **75** was synthesized.[58] Treatment of the sulfamate with Rh$_2$(esp)$_2$ in the presence of (diacetoxyiodo)benzene results in the sole formation of the electrophilic aromatic substitution product **76** (Scheme 21). This evidence, together with the following substrate scope study, appears to corroborate a rhodium-bound zwitterionic vinyl cation intermediate of type **74** and discounts the rhodium–(α-imino)carbene intermediate **72**. However, when the terminal aromatic or heteroaromatic functionality is replaced by an alkene, stereospecific cyclopropanation occurs exclusively with Z- and E-alkenes to afford *cis*- and *trans*-cyclopropanes, respectively (Scheme 22). The concerted cyclopropanation with minimal isomerization is reminiscent of a rhodium–(α-imino)carbene intermediate **72**. Although the 7-*endo-dig* nature of the cyclization of the sulfamate-derived metallonitrene **71** and an alkyne is highly favorable, reaction products derived from 6-*exo-dig*-like cyclization are produced in significant quantities with selected substrates. While Blakey advocates a highly electrophilic rhodium-attached 1*H*-azirine **73** or a rhodium-bound zwitterionic vinyl cation **74**, this matter remains unresolved.

1.6.2 Dirhodium-Catalyzed Domino Reactions

531

Scheme 21 Electrophilic Aromatic Substitution is the Only Pathway for the Intermediate Generated from Alkyne/Nitrene Metathesis[58]

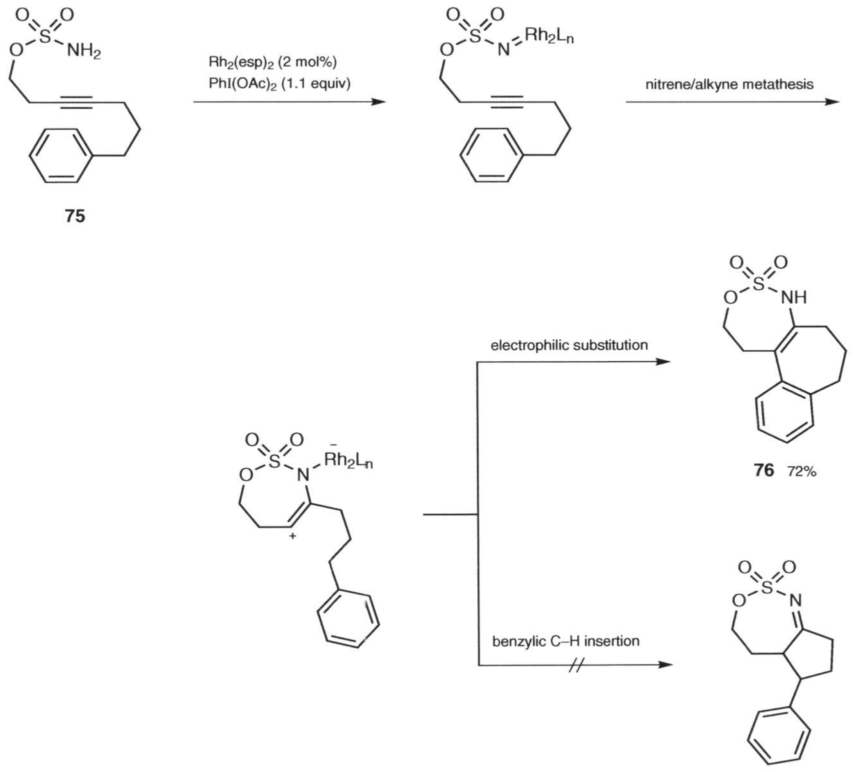

Scheme 22 Stereospecific Cyclopropanation Occurs for the Alkyne/Nitrene Metathesis Intermediate[58]

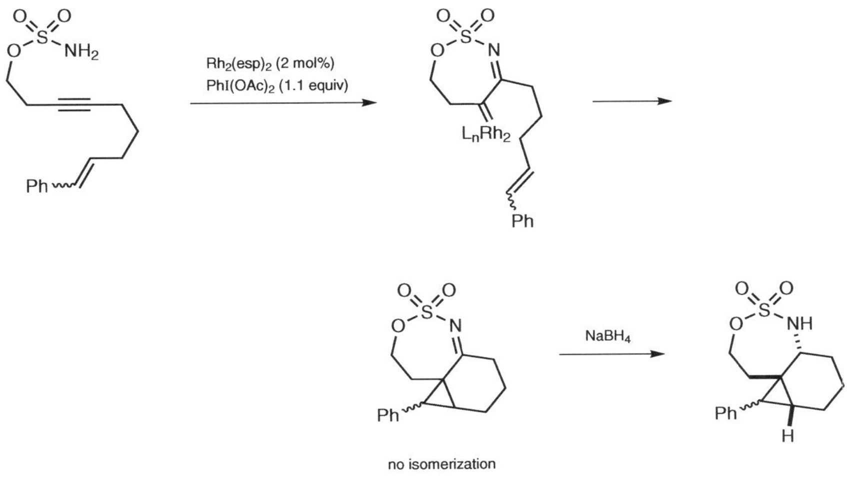

for references see p 533

(5a*R**,9a*S**)-5a-Allylhexahydro-1*H*,4*H*-pyrano[3,2-*d*][1,2,3]oxathiazepine 2,2-Dioxide (70); Typical Procedure:[57]

Sulfamate ester **65** (1.0 equiv), PhI(OAc)$_2$ (1.1 equiv), and Rh$_2$(esp)$_2$ (2 mol%) were combined in a 2-dram reaction vial, which was capped with a Teflon-lined septum. CH$_2$Cl$_2$ was added and the mixture was stirred under argon at 25 °C until complete consumption of starting material (15–30 min, TLC). MeOH (1.5 mL) and NaBH$_4$ (3 equiv) were added and the mixture was stirred for 1 h at 25 °C. Silica gel (ca. 250 mg) was added and the resulting mixture was concentrated. The silica gel was then eluted with CH$_2$Cl$_2$/Et$_2$O (1:1; 10 mL). The eluate was concentrated under reduced pressure and the residue was purified by flash chromatography (silica gel) to afford oxathiazepane **70**.

1.6.2.5 Conclusions

Overall, dirhodium catalysts have exhibited outstanding performance in generating reactive intermediates and directing cascade processes in metal–carbene and metal–nitrene chemistry. The synthetic utilities for these intriguing processes enable the construction of complex molecular architectures from simple and readily available substrates. It is because of these advantages that their uses in chemical catalysis will expand into even more diverse areas.

References

[1] Tietze, L. F., *Chem. Rev.*, (1996) **96**, 115.

[2] Fogg, D. E.; Dos Santos, E. N., *Coord. Chem. Rev.*, (2004) **248**, 2365.

[3] Grondal, C.; Jeanty, M.; Enders, D., *Nat. Chem.*, (2010) **2**, 167.

[4] Sebren, L. J.; Devery, J. J., III; Stephenson, C. R. J., *ACS Catal.*, (2014) **4**, 703.

[5] Volla, C. M. R.; Atodiresei, I.; Rueping, M., *Chem. Rev.*, (2014) **114**, 2390.

[6] Padwa, A.; Weingarten, M. D., *Chem. Rev.*, (1996) **96**, 223.

[7] Murphy, G. K.; Stewart, C.; West, F. G., *Tetrahedron*, (2013) **69**, 2667.

[8] Guo, X.; Hu, W., *Acc. Chem. Res.*, (2013) **46**, 2427.

[9] Doyle, M. P.; McKervey, M. A.; Ye, T., *Modern Catalytic Methods for Organic Synthesis with Diazo Compounds: From Cyclopropanes to Ylides*; Wiley-Interscience: New York, (1998).

[10] Doyle, M. P.; Duffy, R.; Ratnikov, M.; Zhou, L., *Chem. Rev.*, (2010) **110**, 704.

[11] Doyle, M. P.; McKervey, M. A., *Chem. Commun. (Cambridge)*, (1997), 983.

[12] Doyle, M. P.; Ratnikov, M.; Liu, Y., *Org. Biomol. Chem.*, (2011) **9**, 4007.

[13] Davies, H. M. L.; Morton, D., *Chem. Soc. Rev.*, (2011) **40**, 1857.

[14] Davies, H. M. L.; Beckwith, R. E. J., *Chem. Rev.*, (2003) **103**, 2861.

[15] Hansen, J.; Davies, H. M. L., *Coord. Chem. Rev.*, (2008) **252**, 545.

[16] Davies, H. M. L.; Manning, J. R., *Nature (London)*, (2008) **451**, 417.

[17] Gilchrist, T. L.; Gymer, G. E., *Adv. Heterocycl. Chem.*, (1974) **16**, 33.

[18] Harmon, R. E.; Stanley, F.; Gupta, S. K.; Johnson, J., *J. Org. Chem.*, (1970) **35**, 3444.

[19] Chuprakov, S.; Hwang, F. W.; Gevorgyan, V., *Angew. Chem. Int. Ed.*, (2007) **46**, 4757.

[20] Horneff, T.; Chuprakov, S.; Chernyak, N.; Gevorgyan, V.; Fokin, V. V., *J. Am. Chem. Soc.*, (2008) **130**, 14972.

[21] Chuprakov, S.; Kwok, S. W.; Fokin, V. V., *J. Am. Chem. Soc.*, (2013) **135**, 4652.

[22] Zibinsky, M.; Fokin, V. V., *Angew. Chem. Int. Ed.*, (2013) **52**, 1507.

[23] Chuprakov, S.; Worrell, B. T.; Selander, N.; Sit, R. K.; Fokin, V. V., *J. Am. Chem. Soc.*, (2014) **136**, 195.

[24] Schultz, E. E.; Sarpong, R., *J. Am. Chem. Soc.*, (2013) **135**, 4696.

[25] Shi, Y.; Gevorgyan, V., *Org. Lett.*, (2013) **15**, 5394.

[26] Alford, J. S.; Spangler, J. E.; Davies, H. M. L., *J. Am. Chem. Soc.*, (2013) **135**, 11712.

[27] Miura, T.; Funakoshi, Y.; Murakami, M., *J. Am. Chem. Soc.*, (2014) **136**, 2272.

[28] Goto, T.; Takeda, K.; Shimada, N.; Nambu, H.; Anada, M.; Shiro, M.; Ando, K.; Hashimoto, S., *Angew. Chem. Int. Ed.*, (2011) **50**, 6803.

[29] Briones, J. F.; Davies, H. M. L., *J. Am. Chem. Soc.*, (2012) **134**, 11916.

[30] Müller, P.; Pautex, N.; Doyle, M. P.; Bagheri, V., *Helv. Chim. Acta*, (1990) **73**, 1233.

[31] Müller, P.; Gränicher, C., *Helv. Chim. Acta*, (1993) **76**, 521.

[32] Müller, P.; Gränicher, C., *Helv. Chim. Acta*, (1995) **78**, 129.

[33] Archambeau, A.; Miege, F.; Meyer, C.; Cossy, J., *Angew. Chem. Int. Ed.*, (2012) **51**, 11540.

[34] González, M. J.; López, E.; Vicente, R., *Chem. Commun. (Cambridge)*, (2014) **50**, 5379.

[35] Doyle, M. P.; Colsman, M. R.; Chinn, M. S., *Inorg. Chem.*, (1984) **23**, 3684.

[36] Jansone-Popova, S.; May, J. A., *J. Am. Chem. Soc.*, (2012) **134**, 17877.

[37] Doyle, M. P.; Kundu, K.; Russell, A. E., *Org. Lett.*, (2005) **7**, 5171.

[38] Kundu, K.; Doyle, M. P., *Tetrahedron: Asymmetry*, (2006) **17**, 574.

[39] Xu, X.; Ratnikov, M. O.; Zavalij, P. Y.; Doyle, M. P., *Org. Lett.*, (2011) **13**, 6122.

[40] Liu, Y.; Bakshi, K.; Zavalij, P.; Doyle, M. P., *Org. Lett.*, (2010) **12**, 4304.

[41] Liu, Y.; Zhang, Y.; Jee, N.; Doyle, M. P., *Org. Lett.*, (2008) **10**, 1605.

[42] Qian, Y.; Shanahan, C. S.; Doyle, M. P., *Eur. J. Org. Chem.*, (2013), 6032.

[43] Babu, Y. S.; Chand, P.; Bantia, S.; Kotian, P.; Dehghani, A.; El-Kattan, Y.; Lin, T.-H.; Hutchison, T. L.; Elliott, A. J.; Parker, C. D.; Ananth, S. L.; Horn, L. L.; Laver, G. W.; Montgomery, J. A., *J. Med. Chem.*, (2000) **43**, 3482.

[44] Hili, R.; Yudin, A. K., *Nat. Chem. Biol.*, (2006) **2**, 284.

[45] Dequirez, G.; Pons, V.; Dauban, P., *Angew. Chem. Int. Ed.*, (2012) **51**, 7384.

[46] Müller, P.; Fruit, C., *Chem. Rev.*, (2003) **103**, 2905.

[47] Barton, D. H. R.; Morgan, L. R., *J. Chem. Soc.*, (1962), 622.

[48] Barton, D. H. R.; Starratt, A. N., *J. Chem. Soc.*, (1965), 2444.

[49] Breslow, R.; Gellman, S. H., *J. Am. Chem. Soc.*, (1983) **105**, 6728.

[50] Mansuy, D.; Mahy, J.-P.; Dureault, A.; Bedi, G.; Battioni, P., *J. Chem. Soc., Chem. Commun.*, (1984), 1161.

[51] Evans, D. A.; Faul, M. M.; Bilodeau, M. T., *J. Am. Chem. Soc.*, (1994) **116**, 2742.

[52] Müller, P.; Baud, C.; Jacquier, Y., *Tetrahedron*, (1996) **52**, 1543.

[53] Yu, X.-Q.; Huang, J.-S.; Zhou, X.-G.; Che, C.-M., *Org. Lett.*, (2000) **2**, 2233.

[54] Espino, C. G.; Du Bois, J., *Angew. Chem. Int. Ed.*, (2001) **40**, 598.

[55] Dauban, P.; Sanière, L.; Tarrade, A.; Dodd, R., *J. Am. Chem. Soc.*, (2001) **123**, 7707.

[56] Roizen, J. L.; Harvey, M. E.; Du Bois, J., *Acc. Chem. Res.*, (2012) **45**, 911.

[57] Thornton, A. R.; Blakey, S. B., *J. Am. Chem. Soc.*, (2008) **130**, 5020.

[58] Thornton, A. R.; Martin, V. I.; Blakey, S. B., *J. Am. Chem. Soc.*, (2009) **131**, 2434.

535

1.6.3   **Gold-Mediated Reactions**

E. Merino, A. Salvador, and C. Nevado

General Introduction

In this section, a broad definition of domino transformations has been applied; reactions are discussed in which several new bonds are generated in a single synthetic operation. The aim of this section is not to provide an exhaustive compilation of examples reported in this already mature field, but instead to target a careful selection of relevant examples of multicomponent reactions, cycloisomerizations, and cycloadditions catalyzed by electrophilic gold(I) and gold(III) complexes. Most of these processes are triggered by activation of the π-bonds in the presence of catalytic amounts of gold complexes, in line with their powerful carbophilic Lewis acid character.[1–6]

1.6.3.1   **Gold-Catalyzed Annulations**

1.6.3.1.1   **Using *ortho*-Alkynylbenzaldehydes**

Metal-catalyzed benzannulation reactions provide a very efficient route for the synthesis of carbo- and heterocycles fused to a benzene ring with different degrees of functionalization.[7] Different cascade processes can take place upon coordination of gold to the triple bond of an alkyne. For instance, substituted 3-aryl-2,3-dihydro-4H-1-benzopyran-4-ones (isoflavanones) are efficiently prepared by annulation of 2-hydroxybenzaldehydes and external alkynes in the presence of a gold catalyst.[8] The use of a very electron-rich ancillary ligand, such as tributylphosphine, is necessary to obtain the desired products in good yield whereas other phosphines (e.g., triphenylphosphine) inhibit the reaction. In an extension of this chemistry, 2-aminobenzaldehyde derivatives and alkynes produce 3-aryl-2,3-dihydroquinolin-4(1H)-ones (azaisoflavanones)[9] as well as quinolines[10] in moderate yields.

2-Alkynylbenzaldehydes react with external nucleophiles in the presence of palladium,[11] copper,[12–15] and gold complexes[12,14–19] to produce naphthalene related compounds. Benzannulation reactions involving 2-alkynylbenzaldehydes and vinyl ethers in the presence of cationic gold(I) complex **1** produce dihydronaphthalenes **2** or 2-benzopyrans **3** in moderate to good yields under mild conditions (Scheme 1).[20]

for references see p 573

Scheme 1 Synthesis of Dihydronaphthalene and 2-Benzopyran Derivatives from 2-Alkynyl-benzaldehydes and a Vinyl Ether in the Presence of a Bulky Cationic Gold(I) Catalyst[20]

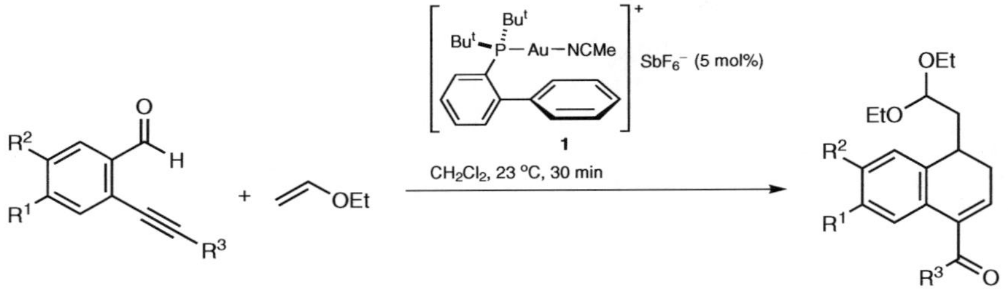

R[1]	R[2]	R[3]	Yield (%)	Ref
H	H	4-FC$_6$H$_4$	67	[20]
H	H	3-Tol	56	[20]
OMe	OMe	Ph	53	[20]

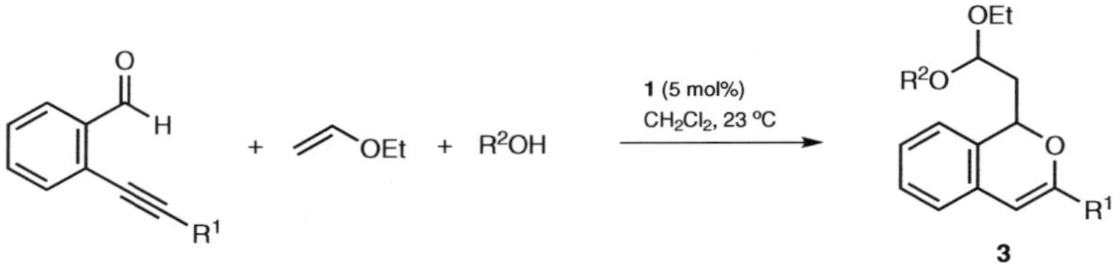

R[1]	R[2]	Yield (%)	Ref
4-FC$_6$H$_4$	Et	82	[20]
3-Tol	Et	71	[20]
Ph	Me	89	[20]
Ph	iPr	91	[20]

An enantioselective tandem addition/cycloisomerization of 2-alkynylbenzaldehydes with primary, secondary, and tertiary alcohols employing chiral acyclic diaminocarbene–gold(I) complexes **4–6** gives 1*H*-2-benzopyrans **7** in good yields and with 61 to >99% enantiomeric excess.[21] Compared to other chelating or bridging diaminocarbene-based complexes, the performance of complexes **4–6** depends dramatically on the 2′-aryl substituent on the 1,1′-binaphthalen-2-yl group and, as a result of a secondary reversible gold–arene interaction, the electron-deficient aryl substituent 3,5-bis(trifluoromethyl)phenyl appears to produce a dynamic chiral pocket. The presence of lithium bis(trifluoromethane-sulfonyl)imide is necessary as an additive to achieve acceptable levels of catalytic activity in these transformations (Scheme 2).

1.6.3 Gold-Mediated Reactions

Scheme 2 Enantioselective Synthesis of 2-Benzopyrans[21]

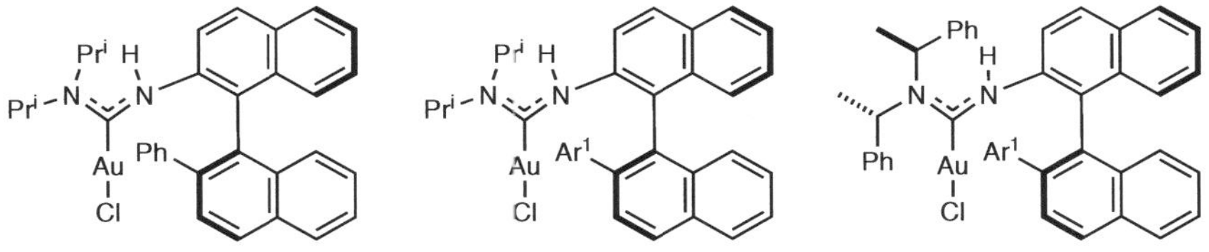

4   **5**   Ar^1 = 3,5-(F$_3$C)$_2$C$_6$H$_3$ **6**   Ar^1 = 3,5-(F$_3$C)$_2$C$_6$H$_3$

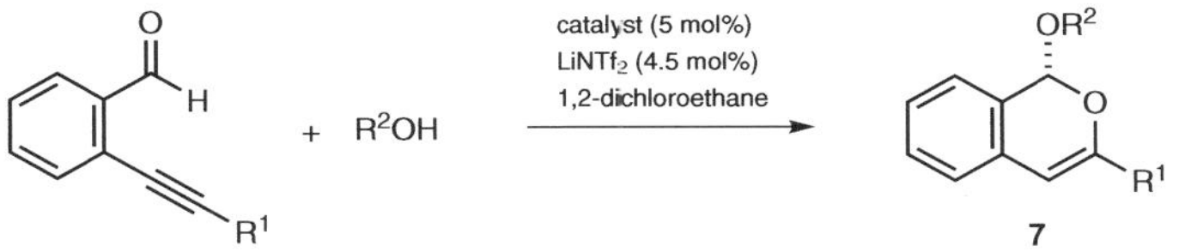

R^1	R^2	Catalyst	Temp (°C)	ee (%)	Yield (%)	Ref
Ph	iPr	**4**	25	61	28	[21]
Ph	t-Bu	**5**	60	96	66	[21]
Ph	(CH$_2$)$_7$Me	**5**	60	99	87	[21]
Ph	Bn	**5**	60	95	68	[21]
Ph	Cy	**5**	60	98	65	[21]
4-Tol	Cy	**5**	60	98	70	[21]
Pr	(CH$_2$)$_7$Me	**6**	25	99	86	[21]
Ph	iPr	**6**	25	>99	70	[21]

4-Acyl-1-(2,2-diethoxyethyl)-1,2-dihydronaphthalenes 2; General Procedure:[20]

Gold catalyst **1** (7.2 μmol, 0.05 equiv) was added to a soln of the 2-alkynylbenzaldehyde (0.145 mmol, 1.0 equiv) and ethyl vinyl ether (0.72 mmol, 5.0 equiv) in CH$_2$Cl$_2$ (1.0 mL). The mixture was stirred for 30 min at 23 °C and then the reaction was quenched with H$_2$O and this mixture was extracted with CH$_2$Cl$_2$. The solvent in the organic layer was removed under reduced pressure and the crude material was purified by column chromatography (silica gel) to afford the product.

3-Substituted 1-Alkoxy-1H-2-benzopyrans 7; General Procedure:[21]

Under dry and O$_2$-free conditions, chiral gold catalyst **4–6** (25 μmol, 0.05 equiv) and LiNTf$_2$ (22.5 μmol, 0.045 equiv) were dissolved in dry 1,2-dichloroethane (1.0 mL). The vial was heated at 45 °C, and the soln was stirred for 15 min. The mixture was then allowed to cool to rt (or heated at 60 °C if a preliminary trial indicated a slow reaction at 25 °C). A soln of the aldehyde (0.5 mmol, 1 equiv) and alcohol (0.5 mmol, 1.0 equiv) in dry 1,2-dichloroethane (1.0 mL) was then added via syringe. After 8 h, further alcohol (0.5 mmol, 1.0 equiv) was added. The crude material was purified by column chromatography (silica gel) to afford the product.

for references see p 573

1.6.3.1.2 Using Arylimines and Alkynes

The intermolecular three-component reaction of aldimines, terminal alkynes, and tosyl isocyanate (**8**) preferentially produces five-membered oxazolidin-2-imines **9** upon treatment with catalytic amounts of chloro(triphenylphosphine)gold(I) and silver(I) bis(trifluoromethanesulfonyl)imide (Scheme 3).[22] Products **10**, stemming from a 6-*endo-dig* cyclization, are also detected as minor regioisomers in the reaction. A two-component version of this transformation uses a preformed alkynylimine in combination with isocyanate **8**.[23]

Scheme 3 Three-Component Reaction for the Synthesis of Oxazolidin-2-imines[22]

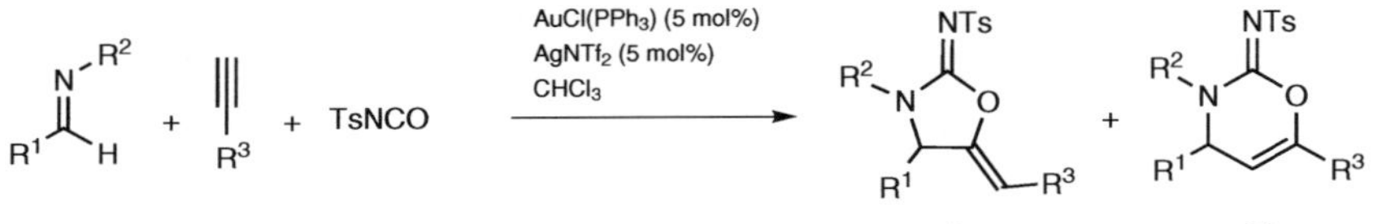

R¹	R²	R³	Temp (°C)	Time (h)	Ratio (9/10)	Yield (%) of 9	Ref
Ph	Ph	Ph	35	20	17:1	84	[22]
3-Tol	Ph	Ph	35	20	>20:1	83	[22]
2-naphthyl	Ph	Ph	35	20	19:1	78	[22]
Ph	4-MeOC₆H₄	Ph	35	20	>20:1	73	[22]
Ph	4-IC₆H₄	Ph	35	48	16:1	76	[22]
Ph	Ph	4-F₃CC₆H₄	50	48	>20:1	90	[22]

3,4-Diaryl-5-(arylmethylene)-*N*-tosyloxazolidin-2-imines 9; General Procedure:[22]
An amber vial was charged with the imine (0.28 mmol, 1 equiv), alkyne (0.34 mmol, 1.2 equiv), and tosyl isocyanate (**8**; 0.34 mmol, 1.2 equiv) in CHCl₃ (0.90 mL). A soln of Au(NTf₂)(PPh₃) [prepared from AuCl(PPh₃) and AgNTf₂; 14 µmol, 0.05 equiv] in CHCl₃ (0.40 mL) was added quickly and the vial was sealed (imine molarity ~0.2 M). The reaction was left to stand at the indicated temperature (35 or 50 °C) for the indicated time period. The mixture was then either loaded directly onto silica gel and purified by column chromatography or concentrated and the residue recrystallized.

1.6.3.1.3 Using Alcohols and Dienes

A highly efficient and completely atom-economical synthesis of dihydrobenzo[*b*]furans from phenols **11** and dienes **12** utilizes catalytic gold(III) chloride and silver(I) trifluoromethanesulfonate.[24] While gold(I) species are not effective in this transformation, cationic gold(III) catalysts deliver tricyclic compounds **13** in good yield. The reaction occurs in a stereoselective fashion, producing products with good-to-excellent *cis/trans* diastereomeric ratios (Table 1).

1.6.3 Gold-Mediated Reactions **539**

Table 1 Dihydrobenzo[*b*]furan Synthesis by Direct Phenol Addition to Cyclic Dienes[24]

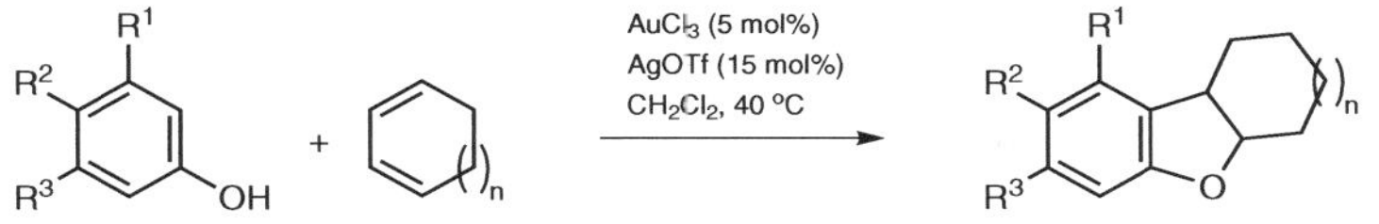

Entry	Substrate		Product	Ratio (*cis/trans*)	Yield (%)	Ref
	Phenol	Diene				
1	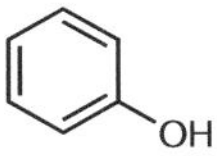		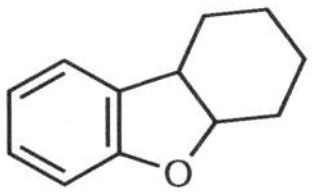	3:1	71	[24]
2	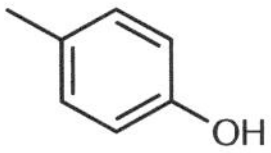		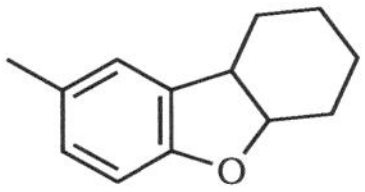	4:1	74	[24]
3	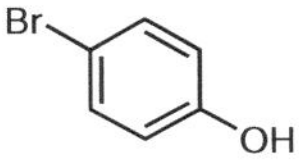		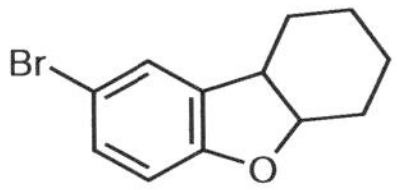	3:1	53	[24]
4	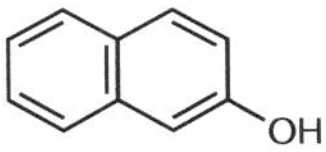		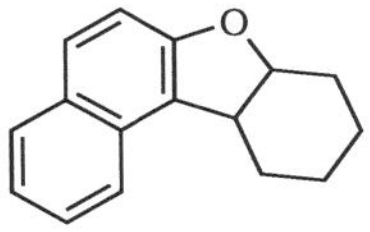	11:1	80	[24]
5	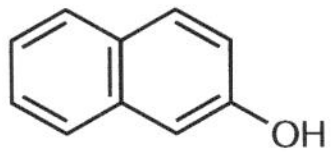		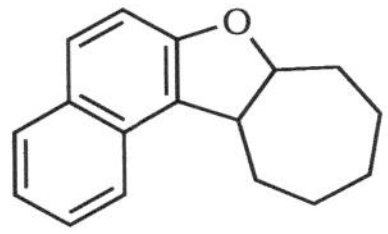	8:1	77	[24]
6	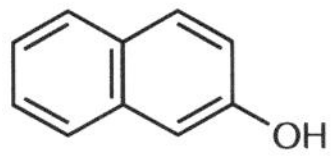		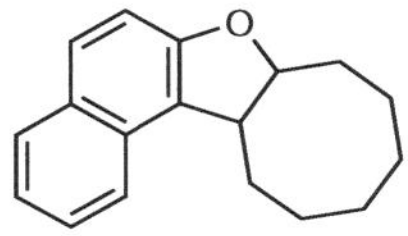	5:1	49	[24]

1,2,3,4,4a,9b-Hexahydrodibenzo[*b,d*]furan (Table 1, Entry 1); Typical Procedure:[24]
A soln of AuCl$_3$ (7.6 mg, 0.025 mmol) and AgOTf (19.3 mg, 0.075 mmol) was stirred in dry CH$_2$Cl$_2$ (1.5 mL) at 23 °C for 2 h. Phenol (47 mg, 0.5 mmol) was then added, which was followed by the dropwise (approximately 1 drop per second) addition of cyclohexa-1,3-diene (124 µL, 1.0 mmol) in dry CH$_2$Cl$_2$ (0.5 mL). The resulting soln was stirred for 16 h at 40–45 °C and then filtered through a cotton plug. The solvent was removed under reduced pressure to produce a dark oil and this crude material was purified by column chromatography (silica gel, hexane/CH$_2$Cl$_2$ 40:1 to 2:1 gradient) to afford the product as a colorless oil; yield: 62 mg (71%).

for references see p 573

1.6.3.1.4 Using Carbonyl Compounds, Alkynes, and Nitrogen-Containing Compounds

The combination of ketones or aldehydes with alkynes **14** or **16** and hydrazines in the presence of cationic gold(I) complexes provides access to complex nitrogen-containing scaffolds by a three-component reaction (Scheme 4). The catalytic system formed by [1,3-bis(2,6-diisopropylphenyl)imidazol-2-ylidene]chlorogold(I) [AuCl(IPr)]/silver(I) trifluoromethanesulfonate is utilized for the synthesis of dihydropyrazole derivatives **15**, arenefused dihydropyrazoles **17**,[25] and even pyrazolo[4,3-b]indoles[26] in good yields by careful combination of the starting materials.

Scheme 4 Three-Component Reaction for the Synthesis of Polyfunctionalized Dihydropyrazoles[25]

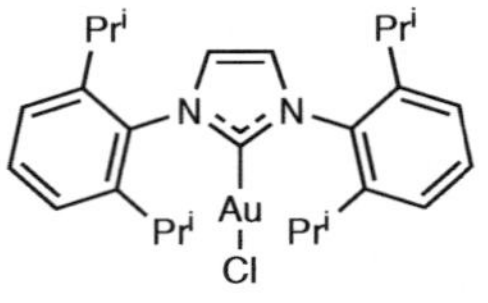

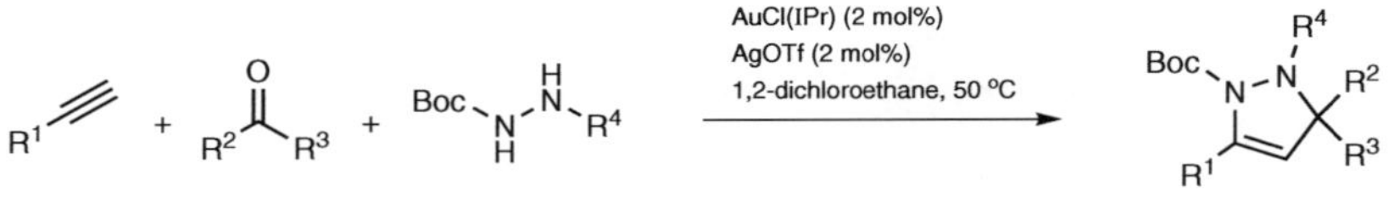

R^1	R^2	R^3	R^4	Yield (%)	Ref
t-Bu	iPr	H	Bn	79	[25]
Ph	iPr	H	Bn	96	[25]
4-O$_2$NC$_6$H$_4$	iPr	H	Bn	81	[25]
4-MeOC$_6$H$_4$	iPr	H	Bn	90	[25]
Ph	iPr	H	Ph	84	[25]
Ph	Et	Et	Bn	47	[25]

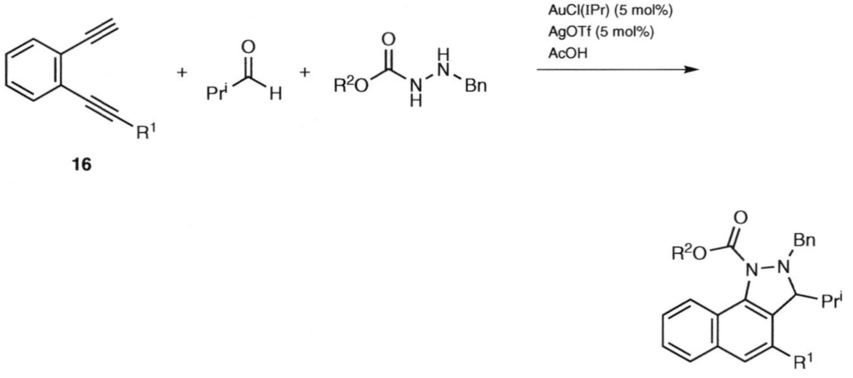

1.6.3 Gold-Mediated Reactions

R^1	R^2	Temp (°C)	Yield (%)	Ref
Pr	Me	80	65	[25]
Ph	t-Bu	50	84	[25]

In an analogous process, gold(III) bromide is capable of catalyzing the three-component reaction between arylglyoxals, secondary amines, and terminal alkynes to furnish 1,2,4-trisubstituted furans **18** bearing an amino group at the 3-position (Scheme 5).[27]

Scheme 5 Furan Synthesis from a Three-Component, Gold(III) Bromide Catalyzed Reaction[27]

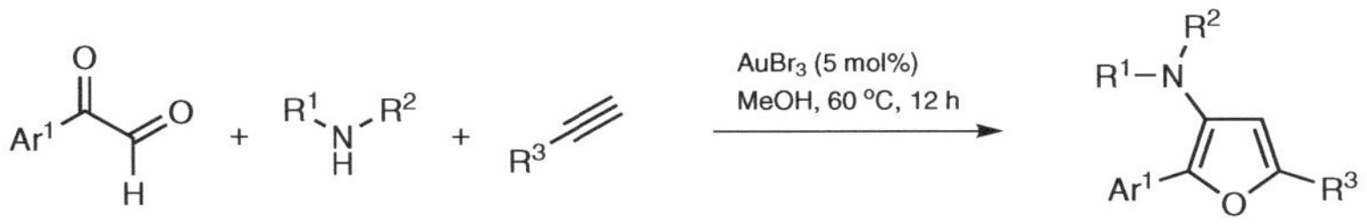

Ar^1	R^1	R^2	R^3	Yield (%)	Ref
Ph	Et	Et	Ph	90	[27]
Ph	Pr	Pr	Ph	93	[27]
Ph	(CH$_2$)$_2$O(CH$_2$)$_2$		Ph	91	[27]
Ph	(CH$_2$)$_2$O(CH$_2$)$_2$		4-FC$_6$H$_4$	93	[27]
4-ClC$_6$H$_4$	(CH$_2$)$_2$O(CH$_2$)$_2$		Ph	83	[27]
4-O$_2$NC$_6$H$_4$	(CH$_2$)$_2$O(CH$_2$)$_2$		Ph	85	[27]

Pyrazoles 15 (R^1 = Ph); General Procedure with Aldehydes:[25]
The hydrazine (0.18 mmol, 1 equiv), AuCl(IPr) (3.6 μmol, 0.02 equiv), and AgOTf (3.6 μmol, 0.02 equiv) were suspended in 1,2-dichloroethane (0.9 mL) under an argon atmosphere. The corresponding aldehyde (0.22 mmol, 1.2 equiv) and phenylacetylene (**14**, R^1 = Ph; 0.22 mmol, 1.2 equiv) were added to the stirred mixture and the soln was stirred at 50 °C for 1 h. The resulting mixture was filtered and concentrated under reduced pressure. The crude material was purified by column chromatography (silica gel) to afford the product.

2,5-Diarylfuran-3-amines 18; General Procedure:[27]
A soln of the arylglyoxal derivative (1 mmol, 1 equiv), amine (1.5 mmol, 1.5 equiv), AuBr$_3$ (0.05 mmol, 0.05 equiv), and alkyne (2 mmol, 2 equiv) in anhyd MeOH (1 mL) was stirred at 60 °C for 12 h under N$_2$. After completion of the reaction (TLC monitoring), the solvent was evaporated under reduced pressure to obtain a crude residue. This crude residue was purified by column chromatography (silica gel) to afford the product.

for references see p 573

542 Domino Transformations **1.6** Metal-Mediated Reactions

1.6.3.2 **Gold-Catalyzed Domino Reactions via Addition of Carbon Nucleophiles to π-Electrophiles**

1.6.3.2.1 **1,n-Enynes**

1,n-Enyne cycloisomerizations are one of the most relevant areas that utilize modern gold catalysis.[28–30] A representative example is the cyclization of 1,6-enynes, a reaction that rapidly builds up molecular complexity.[31] Tricyclic compounds **20A–20D** are obtained by a gold(I)-catalyzed intramolecular 1,5-migration reaction of the ether (OR^1) group of propargyl alcohols **19** (R^1 = H), ethers **19** (R^1 = alkyl), and silyl ethers **19** (R^1 = TMS) (Scheme 6).

Scheme 6 1,5-Migration of Hydroxy, Alkoxy, or Siloxy Groups in 1,6-Dienynes[31]

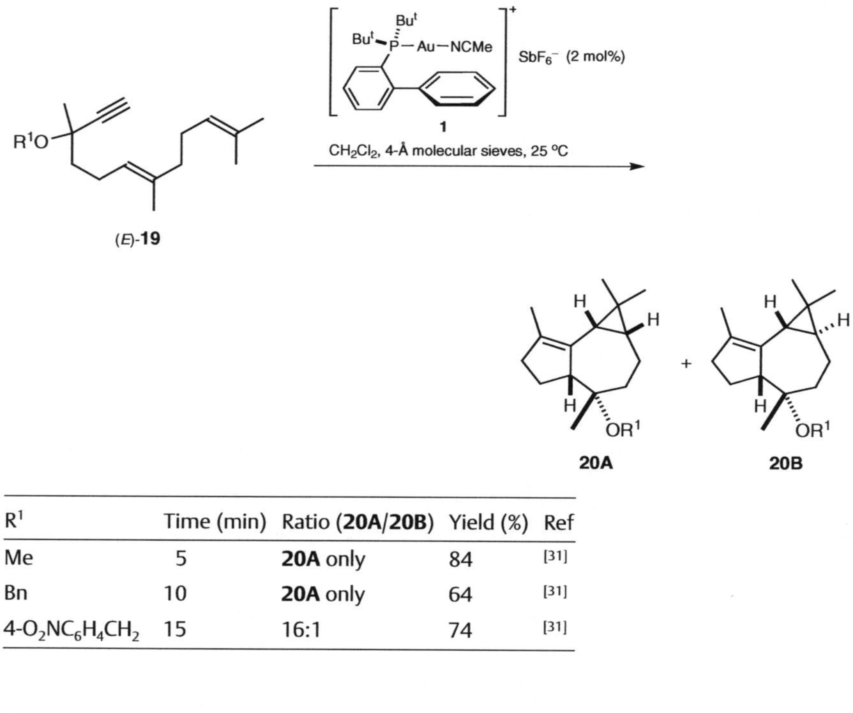

R^1	Time (min)	Ratio (**20A/20B**)	Yield (%)	Ref
Me	5	**20A** only	84	[31]
Bn	10	**20A** only	64	[31]
4-O$_2$NC$_6$H$_4$CH$_2$	15	16:1	74	[31]

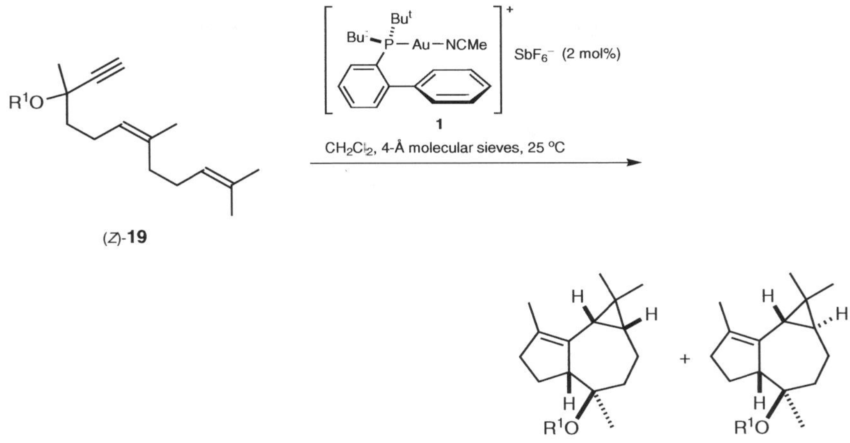

1.6.3 Gold-Mediated Reactions

543

R^1	Time (min)	Ratio (**20C**/**20D**)	Yield (%)	Ref
H	5	11:1	52	[31]
TMS	10	6:1	46	[31]
Me	5	9:1	81	[31]
MOM	10	7:1	72	[31]
4-$O_2NC_6H_4CH_2$	40	7:1	87	[31]

Deuterium-labeling experiments have confirmed that the 1,5-migration is an intramolecular process. The mechanistic proposal involves π-activation of the alkyne in (E)- or (Z)-**19** to generate intermediate **21** (Scheme 7). Formation of compounds **20A** and **20B** is justified via an allylic carbocation in the form of **22**, generated by intermolecular addition of a molecule of alcohol (*anti* addition, path a). Alternatively, the intramolecular migration of the OR¹ group (path b) furnishes the allylgold cation **23** which corresponds to a *syn* addition. Products **20C** and **20D** are ultimately obtained after intramolecular cyclopropanation with the alkene (Scheme 7). This methodology has been applied in the synthesis of (+)-schisanwilsonene A via tandem gold-catalyzed cyclization/1,5-migration/cyclopropanation on a related 1,6-dienyne.[32]

for references see p 573

544 Domino Transformations **1.6** Metal-Mediated Reactions

Scheme 7 Mechanism of 1,5-Migration of Hydroxy, Alkoxy, or Siloxy Groups in 1,6-Dien-ynes via an Allylgold Cation[31]

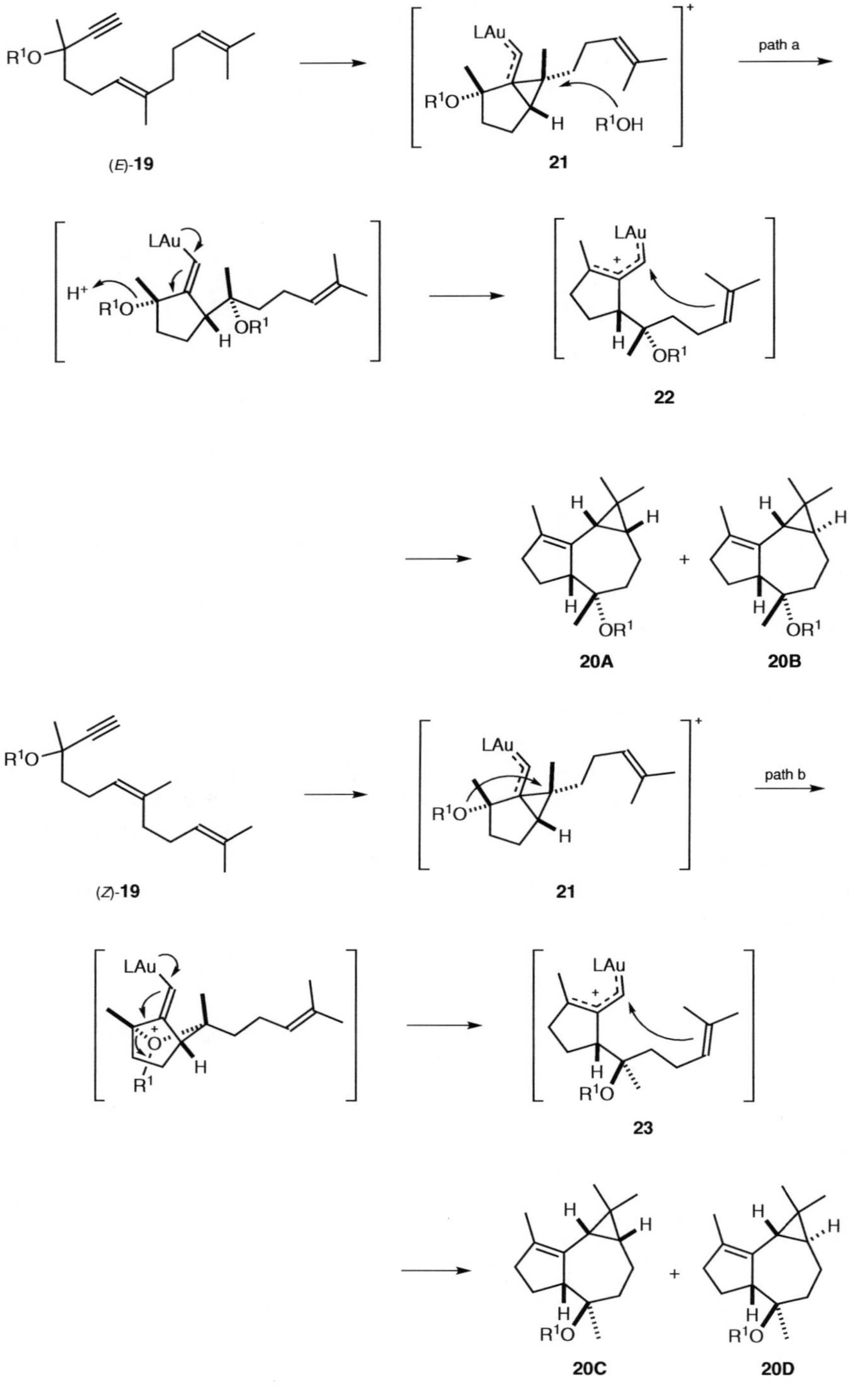

1,5-Enynes **24** can also undergo a gold-catalyzed cyclization followed by an oxidative [3+2]-cycloaddition process in the presence of monomeric nitrosobenzene derivatives such as **25**.[33] This efficient transformation yields functionalized isoxazolidines **26** in high yield (Scheme 8). The presence of an internal alkene in the 1,5-enyne is necessary, while the scope in the nitrosobenzene moiety includes both electron-rich and electron-deficient aromatic rings.

Scheme 8 1,5-Enyne Cyclization/Oxidative [3+2] Cycloaddition with Nitrosobenzene[33]

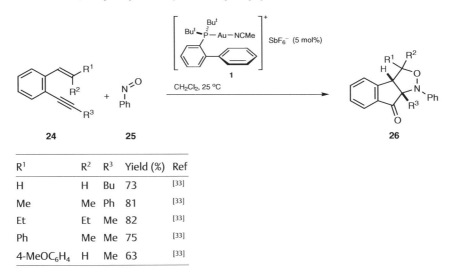

R[1]	R[2]	R[3]	Yield (%)	Ref
H	H	Bu	73	[33]
Me	Me	Ph	81	[33]
Et	Et	Me	82	[33]
Ph	Me	Me	75	[33]
4-MeOC$_6$H$_4$	H	Me	63	[33]

1.6.3.2.2 1,n-Diynes

1,5-Diynes **27** that possess a terminal aliphatic group can deliver benzofulvenes upon dual activation with [1,3-bis(2,6-diisopropylphenyl)imidazol-2-ylidene][bis(trifluoromethanesulfonyl)amido]gold(I) [Au(IPr)(NTf$_2$)] at high temperatures.[34,35] Under these conditions, tricyclic compounds **28** are obtained in high yields, with the process proving tolerant of both electron-donating and electron-withdrawing substituents on the aromatic ring. The *tert*-butyl group on the internal alkyne can be replaced by isopropyl, cyclopentyl, or cyclohexyl groups, as well as various aryl and hetaryl groups (Scheme 9).

for references see p 573

546 Domino Transformations **1.6** Metal-Mediated Reactions

Scheme 9 Synthesis of Benzofulvenes by Dual Gold Catalysis[34]

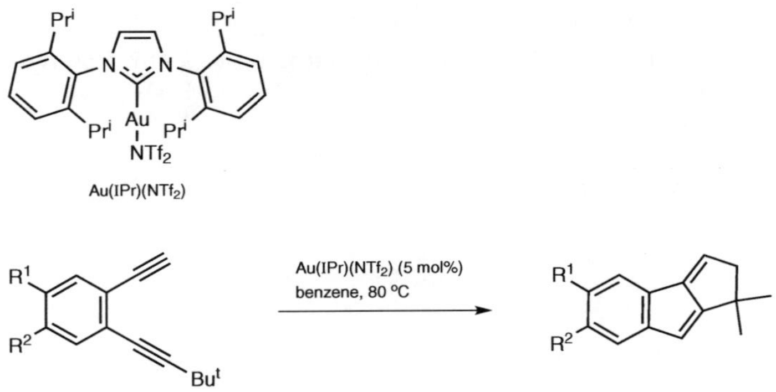

Au(IPr)(NTf$_2$)

27 → 28

Au(IPr)(NTf$_2$) (5 mol%)
benzene, 80 °C

R^1	R^2	Yield (%)	Ref
H	H	92	[34]
H	Me	43	[34]
F	H	73	[34]
OMe	H	41	[34]
OMe	OMe	73	[34]

Mechanistic studies invoke the key role of dinuclear gold species **29** [a gold(I) acetylide and a gold(I) π-coordinated alkyne] in the dual activation of the substrate. The π-activated triple bond undergoes nucleophilic attack by the gold(I) acetylide to give gold(I)–vinylidene-like species **30**. This highly electrophilic carbenoid then formally inserts into the C—H bond of the *tert*-butyl group delivering vinylgold(I) species **31**. Upon intramolecular trapping of the resultant carbocation **31**, final protodeauration by the alkyne delivers compound **28** (R^1 = R^2 = H) and regenerates the active catalytic species (Scheme 10).

1.6.3 Gold-Mediated Reactions   **547**

Scheme 10   Proposed Mechanistic Pathway for the Cyclization of 1,5-Diynes[34]

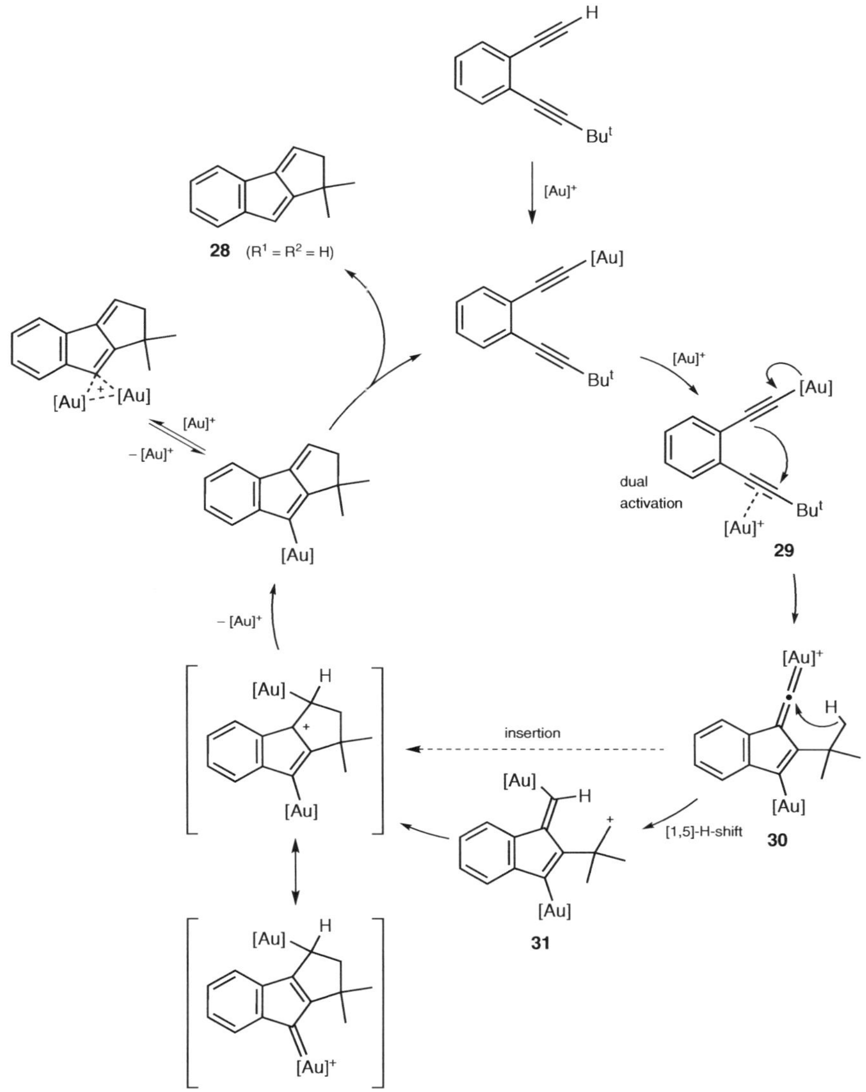

Substituted polycyclic structures, including indanes, heterocycle-fused benzenes, or phenols, are obtained through the dual gold-catalyzed cyclization of *cis*-enediynes **33** (Table 2).[36] In this domino process, it is proposed that gold first activates the internal alkyne, forming a "gold aryne" complex in situ. This intermediate then facilitates the 6-*endo-dig* cyclization with the terminal alkyne affording the corresponding products **34**.[36]

for references see p 573

548 Domino Transformations **1.6** Metal-Mediated Reactions

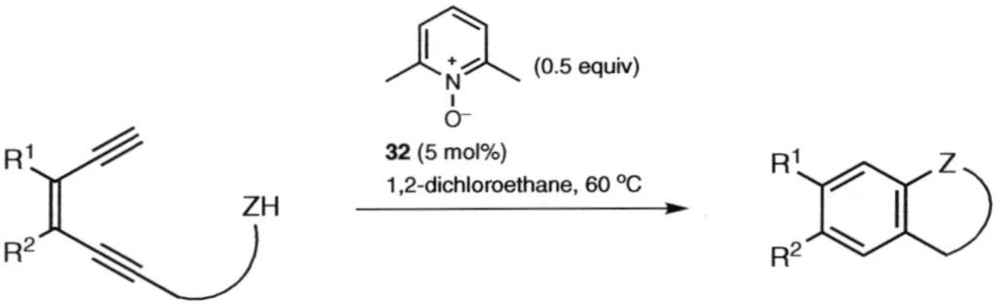

Table 2 Gold-Catalyzed Cyclization of *cis*-Enediynes[36]

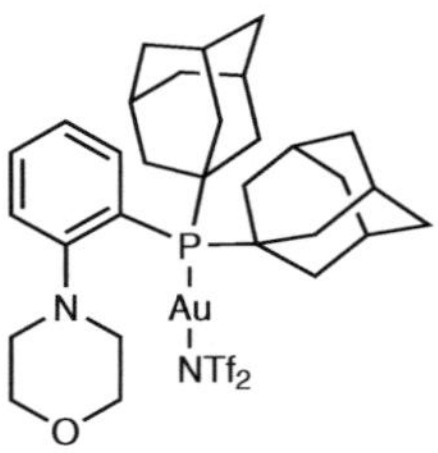

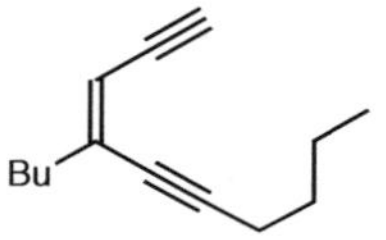

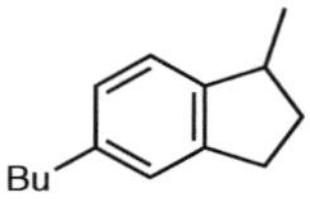

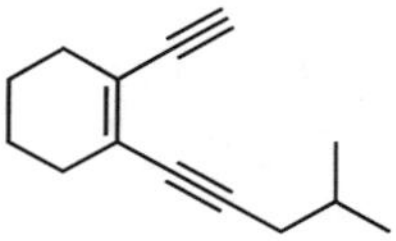

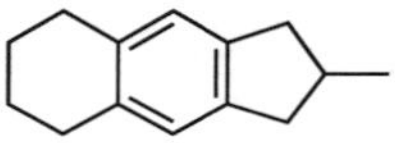

1,1-Dimethyl-1,2-dihydrocyclopenta[a]indene (28, $R^1 = R^2 = H$); Typical Procedure:[34]
1-(3,3-Dimethylbut-1-ynyl)-2-ethynylbenzene (**27**, $R^1 = R^2 = H$; 150 mg, 823 μmol) and
Au(IPr)(NTf₂) (35.6 mg, 41.1 μmol) were stirred in benzene (5.0 mL) (**CAUTION:** *carcinogen*)
at 80 °C for 6 h. After evaporation of the solvent, the crude material was purified by column chromatography (silica gel, petroleum ether) to afford the product as a yellow solid;
yield: 757 μmol (92%).

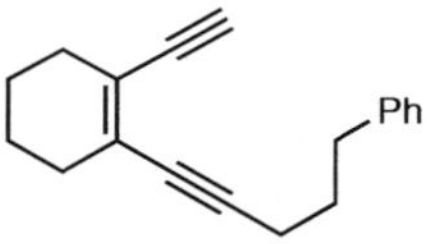

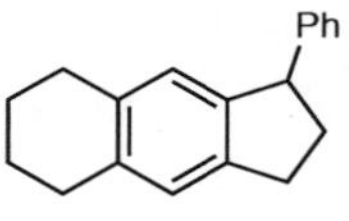

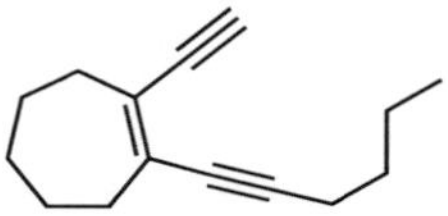

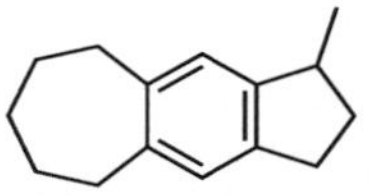

1.6.3 Gold-Mediated Reactions **549**

1-Methyl-1,2,3,5,6,7,8,9-octahydrocyclohepta[*f*]indene (Table 2, Entry 2);
Typical Procedure:[36]
2,6-Dimethylpyridine *N*-oxide (0.15 mmol), Au(NTf$_2$)(Mor-DalPhos) **32** (0.015 mmol), and
4-Å molecular sieves (75 mg) were added in this order to a soln of 1-ethynyl-2-(4-methyl-
pent-1-ynyl)cyclohexene (0.30 mmol) in 1,2-dichloroethane (6.0 mL) at 23 °C. The mixture
was stirred at 60 °C (TLC monitoring). Upon completion (30 h), the mixture was concentrated and the crude material was purified by column chromatography (silica gel, hex-
ane/EtOAc) to afford the product; yield: 60%.

1.6.3.2.3 1,n-Allenenes

Transition-metal-catalyzed cycloisomerization of 1,n-allenynes and 1,n-allenenes[37] and
the electrophilic activation of allenenes and allenynes[38] have been reviewed. 1,6-All-
enenes afford complex tetracyclic products **35** in a stereoselective manner in the presence of the catalytic system chloro(triphenylphosphine)gold(I)/silver(I) hexafluoroantimonate (Scheme 11).[39] Gold carbenes have been proposed as intermediates in these transformations, and these species can undergo a cyclopropanation reaction to deliver variants
of **35** as single diastereomers.

Scheme 11 Cycloisomerization of Allenenes[39]

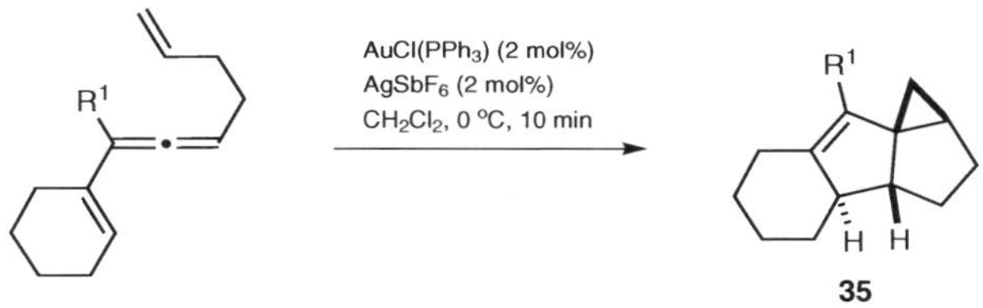

R^1	Yield (%)	Ref
H	87	[39]
Me	77	[39]
t-Bu	65	[39]

1,7-Allenenes undergo gold-catalyzed asymmetric [2+2] cycloadditions with a TADDOL-
based phosphoramidite–gold complex **36**.[40] The corresponding products **37** are obtained
in very good yields and with excellent enantioselectivity (Scheme 12).

Scheme 12 Asymmetric [2+2] Cycloadditions of 1,7-Allenenes[40]

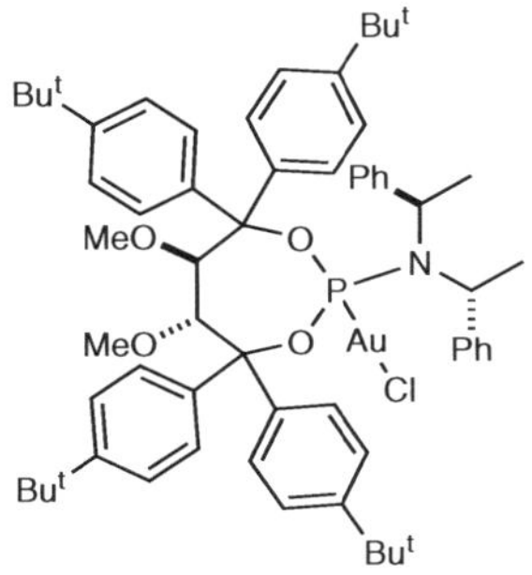

for references see p 573

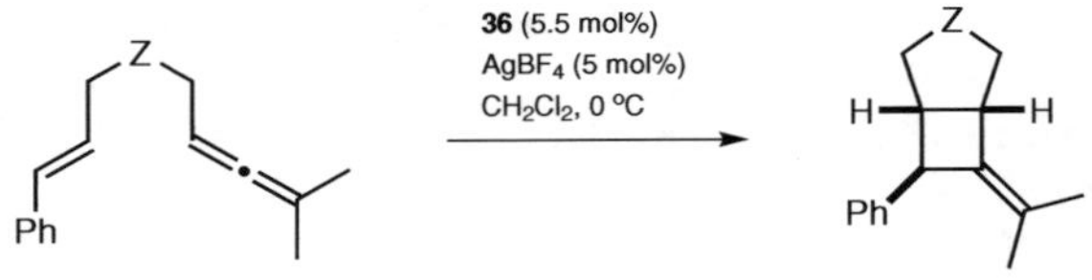

Domino Transformations **1.6** Metal-Mediated Reactions

Z	ee (%)	Yield (%)	Ref
C(CO$_2$Me)$_2$	99	91	[40]
C(CO$_2$Bn)$_2$	99	98	[40]
C(CO$_2$t-Bu)$_2$	99	94	[40]
C(SO$_2$Ph)$_2$	81	60	[40]
NTs	95	52	[40]

Decahydrocyclopropa[2,3]cyclopenta[1,2-a]indenes 35; General Procedure:[39]
AuCl(PPh$_3$) (6–20 µmol, 0.02 equiv) was added to a soln of AgSbF$_6$ (6–20 µmol, 0.02 equiv) in anhyd CH$_2$Cl$_2$ (0.025 M) at 0 °C. The mixture was stirred for 5 min and the substrate (0.3–1.0 mmol, 1 equiv) was added. Reaction progress was monitored by TLC analysis and, upon completion, the mixture was filtered through a short pad of silica gel. The solvent was removed under reduced pressure, and the crude material was purified by column chromatography (silica gel).

1.6.3.2.4 **1,n-Allenynes**

The gold-catalyzed hydrative carbocyclization of 1,7- and 1,5-allenynes in the presence of chloro(triphenylphosphine)gold(I)/silver(I) trifluoromethanesulfonate produces five-membered-ring exocyclic ketones **38** and **39**, respectively (Scheme 13).[41] Water is incorporated upon π-activation of the alkyne with gold in a highly regioselective Markovnikov fashion. Depending on the substitution pattern on the allenyne starting material, compounds with all-carbon quaternary stereocenters can be obtained. Investigation of the stereochemical aspects of this reaction with chiral allenynes shows that this process provides the corresponding ketones with only slight erosion of enantioselectivity.

Scheme 13 Hydrative Cyclization of 1,7- and 1,5-Allenynes[41]

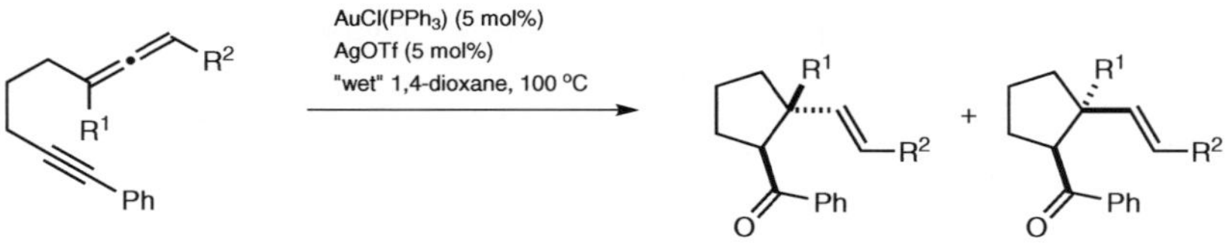

R^1	R^2	Time (h)	Ratio (trans/cis)	Yield (%)	Ref
H	Ph	36	5.1	73	[41]
H	Bu	48	11.4	35	[41]
Me	Me	6	1.4	70	[41]
Et	Ph	10	1.2	78	[41]

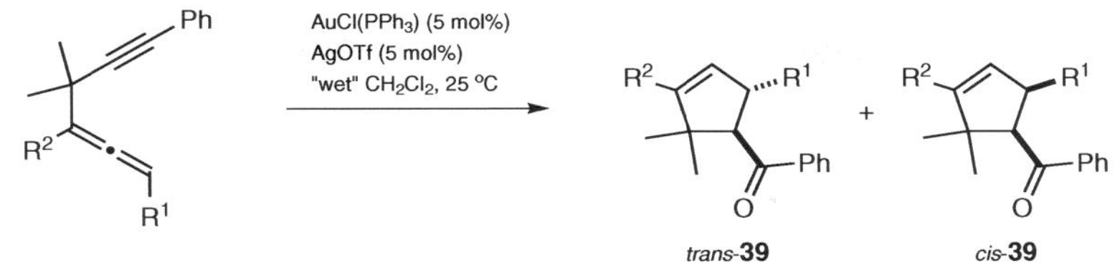

1.6.3 Gold-Mediated Reactions **551**

R^1	R^2	Time (h)	Ratio (*trans/cis*)	Yield (%)	Ref
Me	H	12	5.1	65	[41]
Bu	H	8	1.5	78	[41]
Me	Me	6	5.1	80	[41]
Bu	Me	6	6.7	75	[41]

Phenyl 2-Vinylcyclopentyl Ketones 38; **General Procedure:**[41]
A soln of Au(OTf)(PPh$_3$) (0.018 mmol, 0.05 equiv) was prepared by mixing AuCl(PPh$_3$) (0.018 mmol) and AgOTf (0.018 mmol) in 1,4-dioxane (1.9 mL). To this soln was added the acyclic precursor compound (0.37 mmol, 1 equiv) at 100 °C, and the mixture was stirred for the indicated time. The resulting soln was filtered through Celite, and the crude material was purified by column chromatography (silica gel) to afford the product.

1.6.3.3 Gold-Catalyzed Domino Reactions via Addition of Heteroatom Nucleophiles to π-Electrophiles

1.6.3.3.1 Addition of Nitrogen Nucleophiles to Alkynes

Among contemporary methods for the synthesis of nitrogen-containing heterocycles, gold-catalyzed domino transformations play a prominent role due to their efficiency and versatility. For example, an enantioselective indole formation/allylic substitution cascade reaction for the preparation of variously substituted oxazino-fused indoles **42** takes advantage of dinuclear gold–phosphine complex **40** (Scheme 14).[42] The reaction is highly atom-economical as two bonds (C—C and C—O) are formed simultaneously from *N*-(4-hydroxybut-2-enyl)-2-(3-hydroxyprop-1-ynyl)anilines **41**, with water as the only byproduct. The reaction scope is very broad and the enantioselectivity of the process is high, with enantiomeric excesses typically in the 85–90% range, particularly in the case of substrates bearing a 1-tosylpiperidin-4-yl group at the propargylic position.

Scheme 14 Enantioselective Formation of Oxazino-Fused Indoles[42]

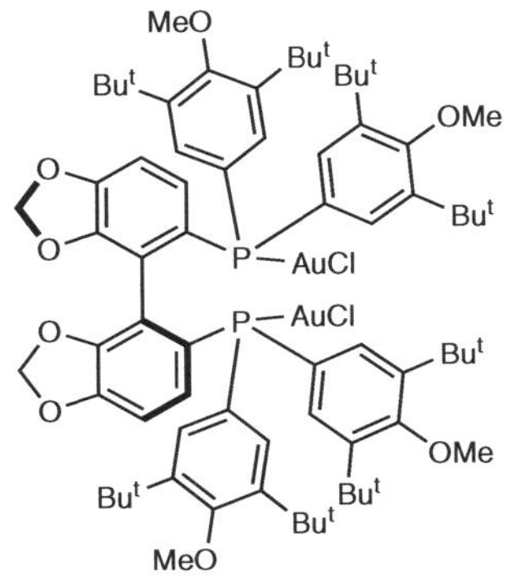

for references see p 573

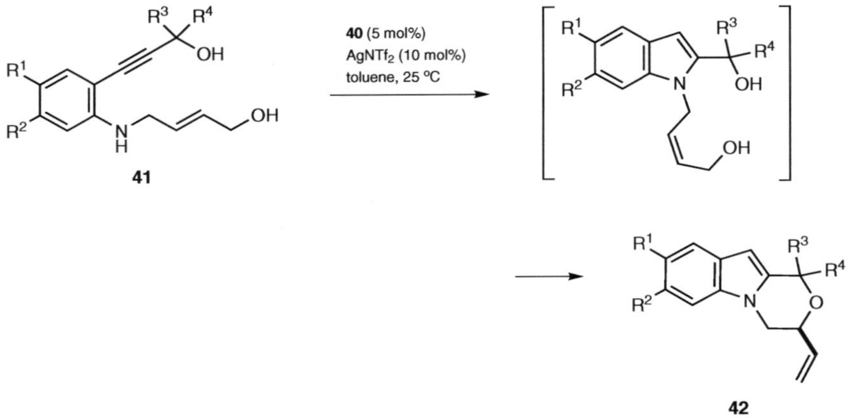

Reaction scheme with catalyst **40** (5 mol%), AgNTf$_2$ (10 mol%), toluene, 25 °C; substrate **41** converting to intermediate and product **42**.

R^1	R^2	R^3	R^4	ee (%)	Yield (%)	Ref
H	H	H	H	85	88	[42]
Cl	H	Me	Me	82	61	[42]
H	CF$_3$	Me	Me	86	84	[42]
H	H	(CH$_2$)$_5$		84	72	[42]
Cl	H	(CH$_2$)$_2$NTs(CH$_2$)$_2$		90	84	[42]
H	CF$_3$	(CH$_2$)$_2$NTs(CH$_2$)$_2$		95	62	[42]
H	Me	(CH$_2$)$_2$NTs(CH$_2$)$_2$		98	62	[42]

3,4-Dihydro-1H-[1,4]oxazino[4,3-a]indoles 42; General Procedure:[42]

(S)-DTBM-SEGPHOS (2.5 µmol, 0.05 equiv) was dissolved in anhyd CH$_2$Cl$_2$ (0.5 mL) in an oven dried, two-necked round-bottomed flask and AuCl(SMe$_2$) (5 µmol, 0.1 equiv) was added. The soln was stirred for 10 min and then the volatiles were removed, leaving the resultant complex **40** under vacuum for an extra 20 min for further drying. The complex was then dissolved in anhyd toluene (0.5 mL) and AgNTf$_2$ (5 µmol, 0.1 equiv) was added in the dark followed by a soln of the desired substrate **41** in toluene (0.5 mL). The reaction was stirred at 25 °C until complete consumption of the acyclic precursor. The crude material was purified by column chromatography (silica gel) to afford the product.

1.6.3.3.2 Addition of Oxygen Nucleophiles to Alkynes and Allenes

1.6.3.3.2.1 Alcohols as Nucleophiles

The nucleophilic attack of alcohols on gold π-activated unsaturated moieties has been extensively developed, both in an inter- and intramolecular fashion.[43,44] In a representative example, highly functionalized phenols **45** are obtained from enediynes **43** in the presence of an acyclic nitrogen carbene–gold complex **44** (Scheme 15).[45] In the first stage, attack of the alcohol in **43** onto the activated alkyne forms a furan intermediate that subsequently rearranges to deliver the observed phenol derivatives **45**.

1.6.3 Gold-Mediated Reactions

Scheme 15 Phenol Synthesis through a Furan Intermediate[45]

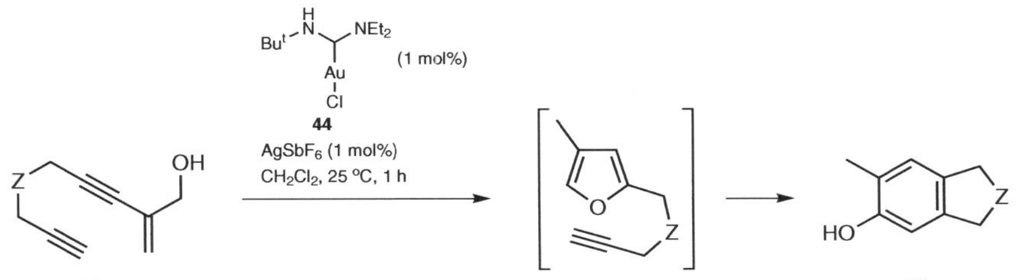

Z	Yield (%)	Ref
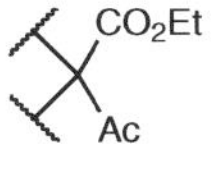 CO₂Et / Ac	64	[45]
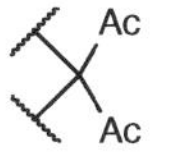 Ac / Ac	42	[45]
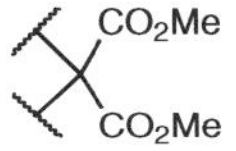 CO₂Me / CO₂Me	59	[45]
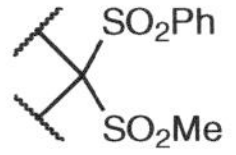 SO₂Ph / SO₂Me	49	[45]
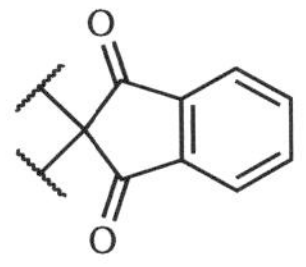 (indene-1,3-dione)	41	[45]
NTs	36	[45]
(CH₂)₂	>99	[45]

5′-Hydroxy-6′-methyl-1′,3′-dihydro-2,2′-spirobi[indene]-1,3-dione [45, Z = 1,3-Dioxo-2,3-di-hydro-1H-indene-2,2-diyl]; Typical Procedure:[45]

The enediyne **43** (350 mg, 1.26 mmol) was dissolved in CH_2Cl_2 (5 mL). To this mixture was added a soln of gold(I) catalyst **44** (4.89 mg, 12.6 µmol) and $AgSbF_6$ (4.32 mg, 12.6 µmol). The mixture was stirred at 25 °C for 1 h, the solvent was removed under reduced pressure, and the crude material was purified by column chromatography (silica gel, petroleum ether/EtOAc 3:1) to afford the product as a colorless solid; yield: 145 mg (41%).

1.6.3.3.2.2 Epoxides as Nucleophiles

Epoxyalkyne substrates **46** react in the presence of halogen-containing additives such as N-bromo-, N-chloro-, or N-iodosuccinimide and picolinate–gold(III) dichloride catalyst **47** to give halogenated eight-membered rings **48** in high yields through a double ring-expansion reaction.[46] The presence of an aryl- or hetaryl-substituted internal triple bond is required for a successful reaction outcome (Scheme 16).

for references see p 573

Scheme 16 Gold(III)-Catalyzed Synthesis of Macrocyclic Ethers[46]

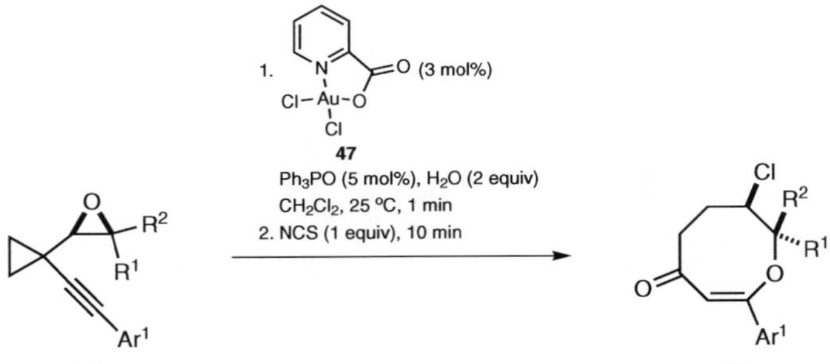

Ar¹	R¹	R²	Yield (%)	Ref
Ph	Me	H	83	[46]
Ph	Me	Me	57	[46]
Ph	(CH₂)₄Me	H	81	[46]
4-NCC₆H₄	(CH₂)₄Me	H	69	[46]
2-thienyl	(CH₂)₄Me	H	79	[46]
benzothiophen-2-yl	(CH₂)₄Me	H	82	[46]

Alternatively, epoxyallenes produce cyclic ethers with high diastereoselectivity (ranging from 4:1 to the formation of only a single diastereomer) in the presence of cationic phosphite-based gold catalyst chloro(triphenyl phosphite)gold(I)/silver(I) trifluoromethanesulfonate (5 mol%).[47] In the case of malonate-containing bridged starting material **49**, the product of a 9-*endo*-*trig* cyclization is observed in the form of **50**, while substrates bearing a nitrogen-containing tether, e.g. **51**, deliver 7-*exo* compounds, e.g. **52**, in moderate yield (Scheme 17).

Scheme 17 Medium- and Large-Sized Cyclic Ether Synthesis[47]

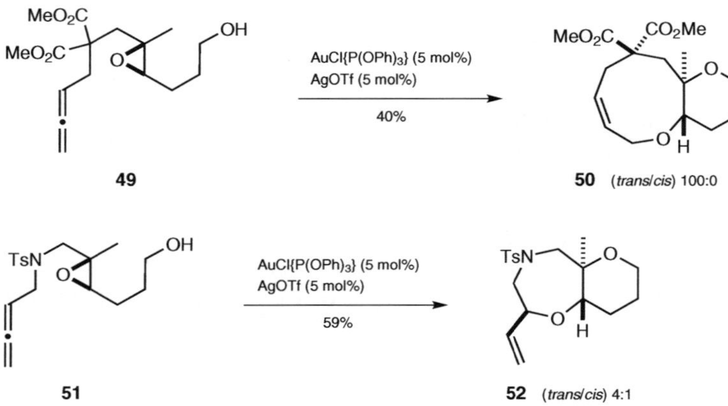

(2S*,5aR*,9aS*)-5a-Methyl-4-tosyl-2-vinyloctahydro-2H-pyrano[2,3-f][1,4]oxazepine (52); Typical Procedure:[47]

To a 20-mL scintillation vial preloaded with AgOTf (0.05 equiv) and AuCl{P(OPh)₃} (0.05 equiv) as white solids was added CH₂Cl₂ (1.0 mL). A white grey suspension formed within 5 min and the epoxyallene **51** was then added by pipet, the tip of which was

1.6.3 Gold-Mediated Reactions

washed with fresh CH_2Cl_2 (0.3 mL) into the reaction. The mixture was stirred until complete consumption of the acyclic precursor and then loaded directly onto a silica gel column for purification by chromatography to afford the product.

1.6.3.3.3 Addition of Heteroatom Nucleophiles to Alkenes

In addition to the gold-catalyzed hydrofunctionalization of alkenes with heteroatom-centered nucleophiles,[48] additional oxidative cross-couplings have been developed.[49–53] For example, a gold-catalyzed aminoarylation of alkenes uses a dinuclear gold catalyst [bis(diphenylphosphino)methane]dibromodigold(I) [dppm(AuBr)$_2$, 53] (Scheme 18).[54,55] In this process, arylboronic acids provide an efficient route for the synthesis of functionalized pyrrolidines 54 in high yield in the presence of Selectfluor [1-(chloromethyl)-4-fluoro-1,4-diazoniabicyclo[2.2.2]octane bis(tetrafluoroborate)] (Scheme 18). The reaction occurs under a gold(I)–gold(III) redox catalytic cycle that is promoted by Selectfluor as a two-electron oxidant, whereas intermolecular C—C bond formation is proposed to occur through bimolecular reductive elimination (Scheme 19).[54] The use of neutral N-heterocyclic carbene–gold complexes has also been reported for this transformation under similar reaction conditions.[56]

Scheme 18 Intramolecular Aminoarylation of Alkenes[54]

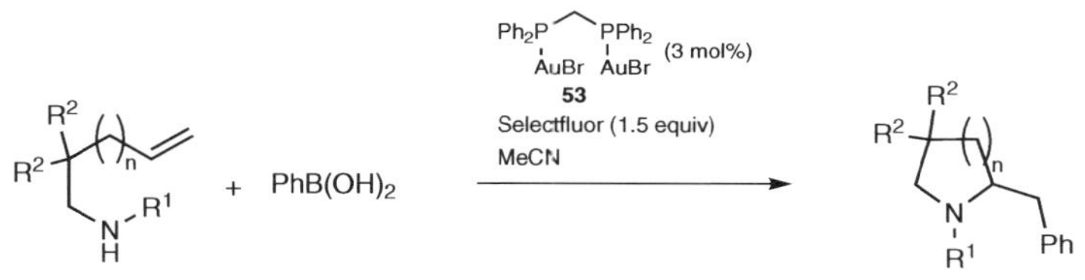

R^1	R^2	n	Temp (°C)	Yield (%)	Ref
COCF$_3$	Ph	1	60	51	[54]
Ts	H	1	25	70	[54]
Ts	H	2	40	82	[54]

Scheme 19 Mechanistic Proposal for the Aminoarylation of Alkenes[54]

for references see p 573

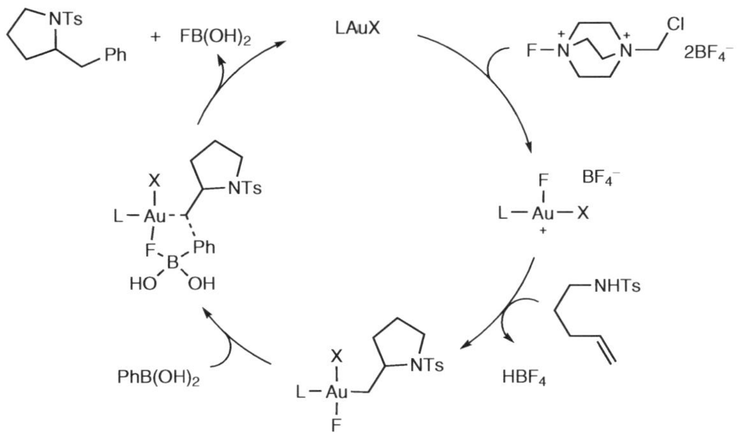

Alkenes can undergo intermolecular oxidative oxyarylation in a single synthetic step catalyzed by dinuclear gold(I) complex **53** to give O-substituted 1-arylalkan-2-ols **55** (Scheme 20).[57] In this three-component reaction, alcohols, carboxylic acids, and even water can be used as nucleophiles. Alkyl-, halo-, carboxymethyl-, and formyl-substituted arylboronic acids are also amenable to the process. This transformation can be explained by a gold(I)/gold(III) cycle, where Selectfluor is the oxidant.

Scheme 20 Oxidative Oxyarylation of Alkenes[57]

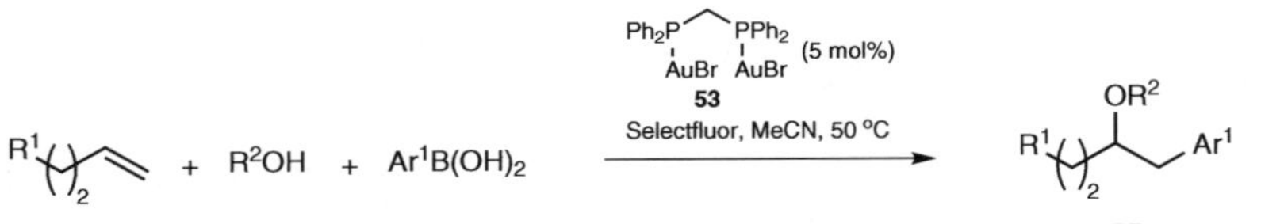

R¹	R²	Ar¹	Yield (%)	Ref
Bu	H	Ph	73	[57]
Bu	CH₂t-Bu	4-BrC₆H₄	73	[57]
2-BrC₆H₄	Me	2-MeO₂CC₆H₄	78	[57]
CH₂NPhth	Me	Ph	79	[57]
CH₂NPhth	iPr	Ph	90	[57]
CH₂NPhth	C(O)Et	Ph	69	[57]
CH₂N(Ts)CH₂CH=CH₂	H	3-FC₆H₄	72	[57]

The oxyarylation of alkenes may also use arylsilanes instead of arylboronic acids.[58] In the intramolecular version of this reaction, various substituted oxygen-containing heterocycles are obtained efficiently.[59]

2-Benzyl-1-tosylpyrrolidine (54, n = 1; R¹ = Ts; R² = H); Typical Procedure:[54]
The sulfonamide (24 mg, 0.1 mmol) was dissolved in anhyd MeCN (1.0 mL) with phenylboronic acid (24.4 mg, 0.2 mmol) and Au complex **53** (2.8 mg, 3 μmol). Selectfluor (0.15 mmol) was then added in one portion, and the soln was stirred in a sealed vial at 25 °C for 18 h. Sat. aq Na₂S₂O₃ (3.0 mL) was added to the mixture and the aqueous layer was extracted with Et₂O (3 × 5 mL). The combined organic layers were washed with brine (5 mL), dried (MgSO₄), filtered, and concentrated. The crude material was purified by column chromatography (silica gel, Et₂O/hexanes 1:9) to afford the product as a viscous oil that solidified upon standing; yield: 22 mg (70%).

O-Substituted 1-Arylalkan-2-ols 55; General Procedure:[57]
The alkene (0.1 mmol, 1 equiv), arylboronic acid (0.1 mmol, 1 equiv), and Au complex **53** (2.5 μmol, 0.025 equiv) were dissolved in a mixture of MeCN/nucleophile (9:1; 1.0 mL) at 25 °C in a vial equipped with a magnetic stirrer bar. Selectfluor (0.2 mmol) was added in one portion and the mixture was heated to 50 °C. After 2 h, additional arylboronic acid (0.1 mmol) and dppm(AuBr)₂ **53** (2.5 μmol, 0.025 equiv) were added to the mixture which was then stirred for an additional 12 h at 50 °C. The mixture was cooled to 25 °C and the reaction was quenched by the addition of sat. aq Na₂S₂O₃ (5 mL). The aqueous layer was extracted with Et₂O (3 × 5 mL) and the combined organic extracts were washed with brine, dried (MgSO₄), and concentrated. The crude material was purified by column chromatography (silica gel) to afford the product.

1.6.3.4 Gold-Catalyzed Domino Reactions Involving the Rearrangement of Propargyl Esters

The rearrangement of propargyl esters in the presence of a catalytic amount of a late-transition-metal complex is a versatile and valuable synthetic tool with diverse applications.[60,61] The starting materials are easily obtained in a few steps, with their subsequent reaction in the presence of a catalytic amount of metal, such as gold, copper, or platinum, producing complex molecular architectures through multiple domino reactions. Thus, in this vein, gold complexes can efficiently activate propargyl carboxylates toward 1,2- or 1,3-acyloxy migrations.[62–64] As seen in Scheme 21, it is widely accepted that terminal or electronically demanding alkynes react via path a[65–73] producing gold carbene intermediates **56**, whereas internal alkynes seem to prefer path b via allenyl acetate **57**[74–78] with only a few exceptions to this general reactivity pattern.[79,80]

Scheme 21 Mechanistic Proposals for the Gold-Catalyzed Rearrangement of Propargyl Acetates[65,74]

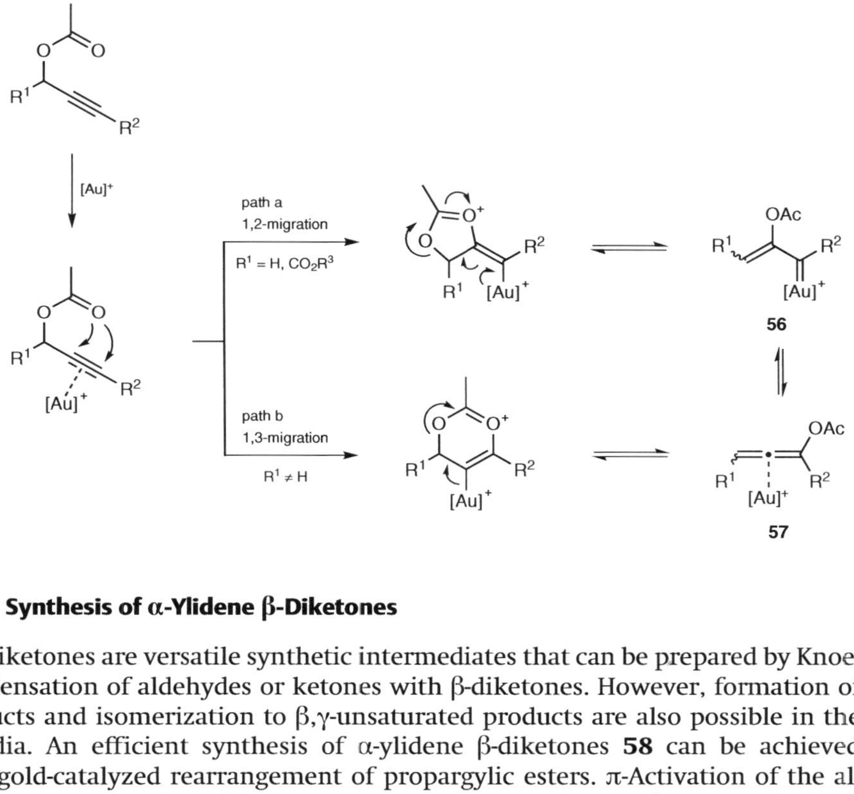

1.6.3.4.1 Synthesis of α-Ylidene β-Diketones

α-Ylidene β-diketones are versatile synthetic intermediates that can be prepared by Knoevenagel condensation of aldehydes or ketones with β-diketones. However, formation of Michael adducts and isomerization to β,γ-unsaturated products are also possible in the reaction media. An efficient synthesis of α-ylidene β-diketones **58** can be achieved through the gold-catalyzed rearrangement of propargylic esters. π-Activation of the alkyne by gold and 1,3-migration of the acetate group triggers an intramolecular acyl migration onto the nucleophilic vinyl–gold(III) intermediate delivering the corresponding α-ylidene β-diketones **58** in excellent yields and with moderate stereoselectivity (Scheme 22).[76] By contrast, the reaction of alkynyl benzoates in the presence of [bis(trifluoromethanesulfonyl)amido](triphenylphosphine)gold(I) [Au(NTf₂)(PPh₃)] and Selectfluor affords 1-(benzoyloxy)vinyl ketones by a gold-catalyzed, oxidative C—O bond-forming reaction.[81]

for references see p 573

558 Domino Transformations **1.6** Metal-Mediated Reactions

Scheme 22 Gold-Catalyzed Synthesis of 3-Alkylidenealkane-2,4-diones[76]

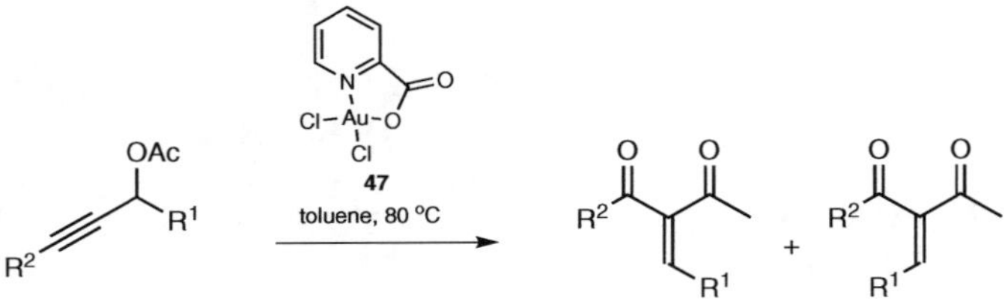

R^1	R^2	mol% of **47**	Time (h)	Yield (%)	Ref
Me	Bu	5	1	94	[76]
Me	Ph	5	1	91	[76]
Me	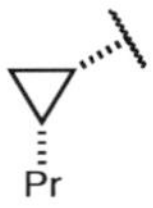 (cyclopropyl-Pr)	5	1	93	[76]
iPr	Bu	10	3	95	[76]
Ph	Cy	5	1	97	[76]
Ph	Ph	5	1	90	[76]

3-Alkylidenealkane-2,4-diones 58; **General Procedure:**[76]

To a soln of the propargylic ester (0.1 mmol) in dry toluene (2.0 mL) was added gold complex **47** (5–10 µmol, 0.05–0.10 equiv). The mixture was stirred at 80 °C and the reaction was monitored by TLC analysis until the ester was completely consumed. The mixture was then concentrated and the crude material was purified by column chromatography (silica gel) to afford the product.

1.6.3.4.2 **Synthesis of Dienes**

(1Z,3E)-2-(Pivaloyloxy)-1,3-dienes **60** are synthesized with excellent stereoselectivity when propargylic pivalic esters **59** containing electronically unbiased internal alkynes are reacted with a catalytic amount of [1,3-bis(2,6-diisopropylphenyl)imidazol-2-ylidene][bis(trifluoromethanesulfonyl)amido]gold(I) [Au(IPr)(NTf$_2$)] (Scheme 23).[80] In contrast to the classical pattern shown in Scheme 21, these substrates react via 1,2-acyloxy migration, which generates a carbene intermediate that evolves via hydrogen abstraction at the neighboring carbon atom to give the products **60**.

Scheme 23 Gold-Catalyzed Formation of (1Z,3E)-2-(Pivaloyloxy)-1,3-dienes from Propargylic Pivalates[80]

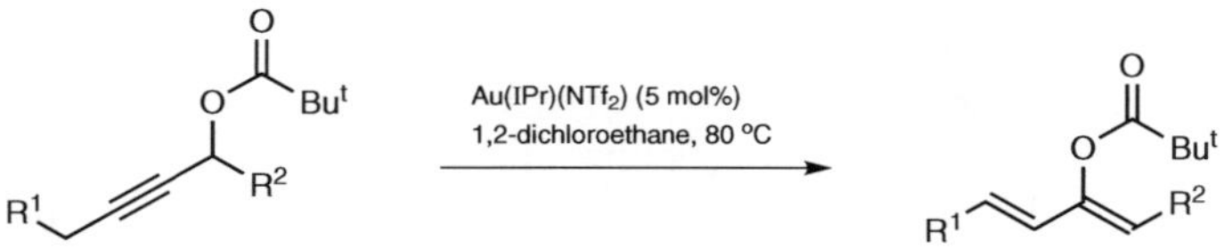

1.6.3 Gold-Mediated Reactions

R¹	R²	Time (h)	Yield (%)	Ref
Pr	(CH₂)₄Me	8	85	[80]
Pr	CH₂OBn	10	36	[80]
Pr	(CH₂)₂OTBDMS	11	67	[80]
(CH₂)₂OBn	Me	10	67	[80]
(CH₂)₂OTBDPS	Me	22	71	[80]

1,4-Diacetoxybut-2-ynes also undergo a highly selective tandem 1,2-/1,2-bis(acetoxy) migration to give 2,3-diacetoxy-1,3-dienes.[82] Although various stereoisomers can be formed, the reaction proceeds in a highly regio- and stereocontrolled manner dependent entirely on the catalyst used. A 1,2-/1,2-bis(acetoxy) migration of symmetrically substituted diacetates **61** in the presence of [1,3-bis(2,6-diisopropylphenyl)imidazol-2-ylidene][bis(trifluoromethanesulfonyl)amido]gold(I) [Au(IPr)(NTf₂)] selectively delivers (1Z,3Z)-**62**. In contrast, the use of the more cationic complex [bis(trifluoromethanesulfonyl)amido](triphenylphosphine)gold(I) [Au(NTf₂)(PPh₃)] enables the isolation of (1Z,3E)-**62**. The reaction can be extended to unsymmetrically substituted diacetates with good levels of both efficiency and selectivity (Scheme 24).

Scheme 24 Tandem 1,2-/1,2-Bis(acetoxy) Migration[82]

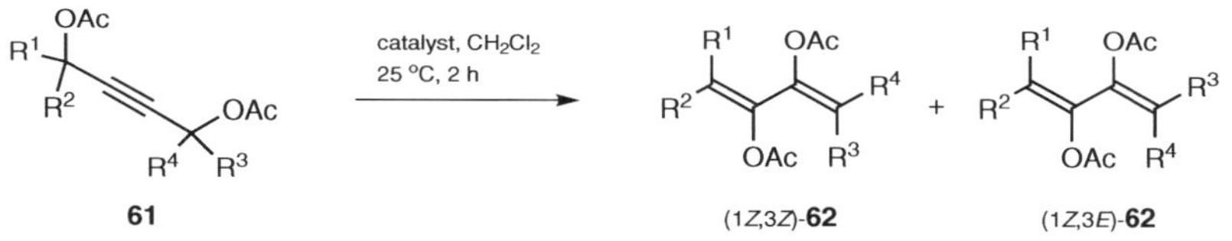

R¹	R²	R³	R⁴	Catalyst	Ratioᵃ (Z,Z/Z,E)	Yieldᵇ (%)	Ref
H	3,5-(MeO)₂C₆H₃	H	3,5-(MeO)₂C₆H₃	Au(IPr)(NTf₂)	13:1	92	[82]
H	3,5-(MeO)₂C₆H₃	H	3,5-(MeO)₂C₆H₃	Au(NTf₂)(PPh₃)	1:10	92	[82]
4-MeOC₆H₄	H	Me	Ph	Au(IPr)(NTf₂)	13:1	83	[82]
4-MeOC₆H₄	H	Me	Ph	Au(NTf₂)(PPh₃)	1:10	87	[82]
Me	Me	Ph	Me	Au(IPr)(NTf₂)	>20:1ᶜ	80	[82]
Me	Me	Ph	Me	Au(NTf₂)(PPh₃)	>1:20ᶜ	82	[82]

ᵃ Determined by ¹H NMR analysis.
ᵇ Isolated yield of the major isomer after column chromatography; the E,E-isomer could also be detected.
ᶜ Ratio (Z/E).

According to the mechanistic proposal depicted in Scheme 25, in this process the acetoxy moiety next to the group best able to stabilize the developing positive electron density migrates first. The resultant intermediate **63** determines the Z stereochemistry of the alkene linked to the gold carbenoid by avoiding the 1,3-allylic strain between gold and R¹. When the second acetoxy is also attached to a stabilizing group, the stereochemical outcome of the second acetoxy migration depends on the nature of the catalyst. For example, Au(IPr)(NTf₂) affords the Z-alkene whereas Au(NTf₂)(PPh₃) favors E-alkene formation.

for references see p 573

560 Domino Transformations **1.6** Metal-Mediated Reactions

Scheme 25 Mechanism of Tandem 1,2-/1,2-Bis(acetoxy) Migration[82]

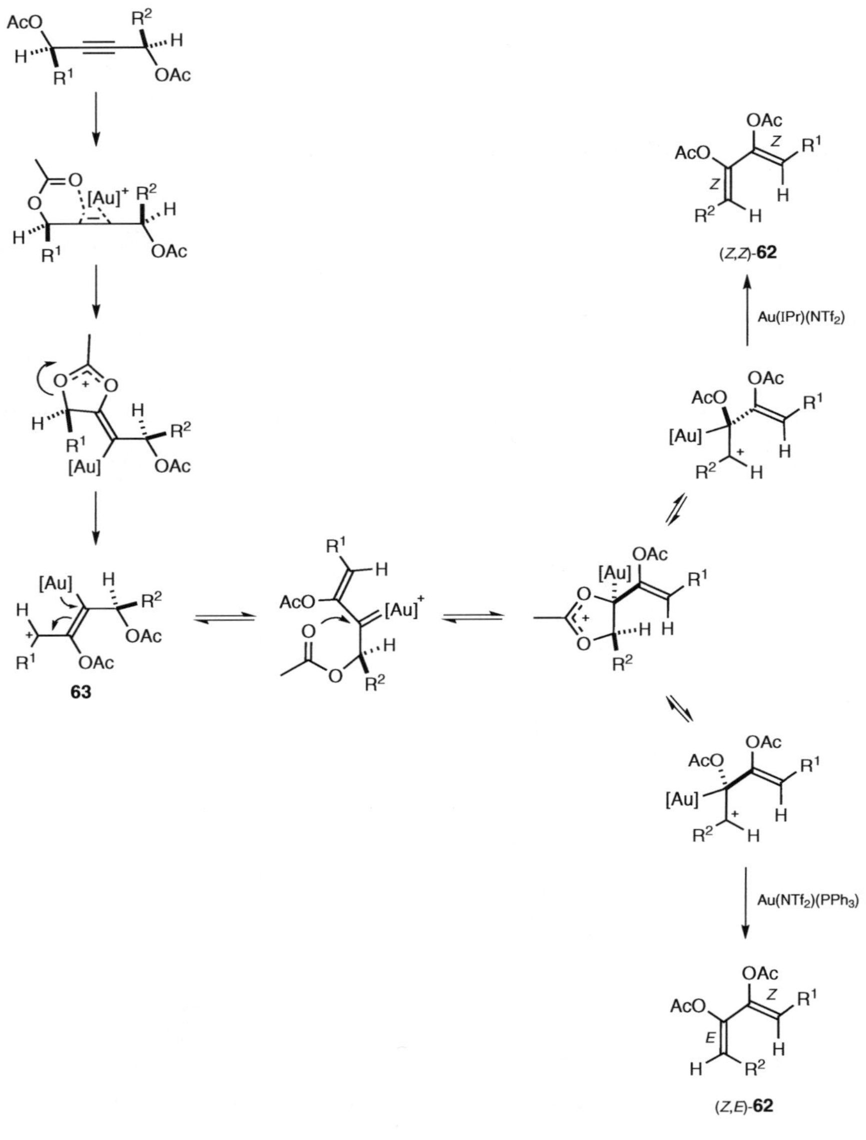

2-(Pivaloyloxy)alka-1,3-dienes 60; General Procedure:[80]

To a soln of a propargyl pivalate **59** (1 mol) in anhyd 1,2-dichloroethane (2 mL) in a flame-dried 2-dram vial was added Au(IPr)(NTf$_2$) (0.05 mol, 0.05 equiv). The vial was sealed tightly with Teflon tape inside and outside the cap and heated to 80 °C. Upon completion of the reaction, the solvent was removed under vacuum, and the residue was purified by flash column chromatography (Et$_3$N-deactivated silica gel) to provide the product.

2,3-Diacetoxyalka-1,3-dienes 62; General Procedure:[82]

Au(IPr)(NTf$_2$) (2–5 μmol, 0.02–0.05 equiv) or Au(NTf$_2$)(PPh$_3$) [2 μmol, 0.02 equiv; 0.05 M soln prepared from AuCl(PPh$_3$) and AgNTf$_2$ in freshly distilled CH$_2$Cl$_2$] was added to a soln of diacetate **61** (0.05 mmol, 1 equiv) in anhyd CH$_2$Cl$_2$ (2 mL). The mixture was stirred for 2 h

1.6.3 Gold-Mediated Reactions

at 25 °C and then the solvent was evaporated under reduced pressure. The crude material was purified by column chromatography (silica gel) to afford the product.

1.6.3.4.3 Synthesis of α-Substituted Enones

1.6.3.4.3.1 Synthesis of α-Halo-Substituted Enones

Propargylic acetates are suitable starting materials for the synthesis of α-haloenones. For example, (E)-α-iodoenones **66** (X = I) are obtained by a sequential propargyl acetate rearrangement followed by allene halogenation in the presence of catalytic amounts of triazole–gold complex **64** (Scheme 26).[83] The kinetic E-products are obtained in excellent yields and good stereoselectivity. Critically, the use of N-bromosuccinimide delivers the corresponding bromoenones **66** (X = Br), but the reaction with N-chlorosuccinimide is unsuccessful under these conditions due to its decreased reactivity. The E-isomers are transformed into the thermodynamically more stable Z-stereoisomers by treatment with protic acids. Alternatively, when tertiary propargyl acetates are used or [bis(trifluoromethanesulfonyl)amido](triphenylphosphine)gold(I) [Au(NTf₂)(PPh₃)] is employed as the reaction catalyst, Z-configured iodoenones are initially obtained from the reaction mixture.[84,85] Secondary propargyl acetates **65** in the presence of [1,3-bis(2,6-diisopropylphenyl)-4,5-dihydroimidazol-2-ylidene]chlorogold(I) [AuCl(SIPr)][86] or [1,3-bis(2,6-diisopropylphenyl)imidazol-2-ylidene][bis(trifluoromethanesulfonyl)amido]gold(I) [Au(IPr)(NTf₂)][87] and Selectfluor give α-fluoroenones **66** (X = F). In this case, when the reaction is performed with secondary propargyl acetates, E selectivity is observed preferentially.

Scheme 26 Synthesis of (E)-α-Halo-Substituted Enones[83,86,87]

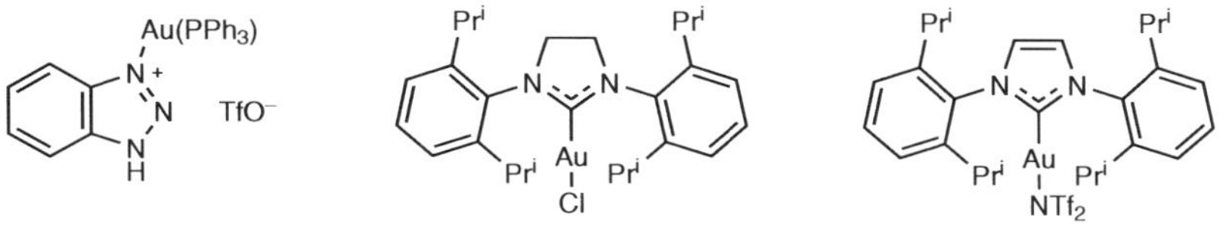

R¹	R²	R³	X	Conditions	Ratio (E/Z)	Yield (%)	Ref
2-O₂NC₆H₄	H	Bu	I	NIS (1.2 equiv), **64** (1 mol%), MeNO₂, 0 °C	3:1	92	[83]
4-FC₆H₄	H	Bu	I	NIS (1.2 equiv), **64** (1 mol%), MeNO₂, 0 °C	12:1	92	[83]
Me	Me	Bu	Br	NBS (1.2 equiv), **64** (1 mol%), MeNO₂, 25 °C	–	84	[83]
Ph	H	Bu	Br	NBS (1.2 equiv), **64** (1 mol%), MeNO₂, 25 °C	7:1	88	[83]
Ph	Ph	Ph	F	Selectfluor (2 equiv), AuCl(SIPr) (5 mol%), AgOTf (12.5 mol%), MeCN, 40 °C	–	58	[86]

for references see p 573

R^1	R^2	R^3	X	Conditions	Ratio (E/Z)	Yield (%)	Ref
Ph	H	Ph	F	Selectfluor (2 equiv), AuCl(SIPr) (5 mol%), AgOTf (12.5 mol%), MeCN, 40 °C	12.2:1	62	[86]
Me	Ph	Bu	F	Selectfluor (2 equiv), NaHCO₃ (1 equiv), Au(IPr)(NTf₂) (5 mol%), MeCN/H₂O (20:1), 80 °C	1.5:1	84	[87]
Me	Me	Bu	F	Selectfluor (2 equiv), NaHCO₃ (1 equiv), Au(IPr)(NTf₂) (5 mol%), MeCN/H₂O (20:1), 80 °C	–	77	[87]

2-Fluoroalk-2-en-1-ones 66 (X = F); General Procedure:[87]

The propargyl acetate **65** (1 equiv) and Au(IPr)(NTf₂) (0.05 equiv) were added sequentially to a soln of Selectfluor (2 equiv) and NaHCO₃ (1 equiv) in MeCN/H₂O (20:1; 0.02 M). The resulting mixture was stirred for 20 min at 80 °C and, upon completion, the mixture was diluted with CH₂Cl₂ (5 mL) and 5% aq Na₂S₂O₃ soln (2 mL) was added. The mixture was extracted with CH₂Cl₂ (3 × 5 mL) and the combined organic phases were dried (MgSO₄). The solvent was evaporated under reduced pressure and the crude residue was purified by column chromatography (silica gel) to afford the product.

1.6.3.4.3.2 Synthesis of α-Aryl-Substituted Enones

The gold-catalyzed oxidative cross coupling of propargylic acetates with arylboronic acids generates α-aryl enones **67** in one synthetic operation (Scheme 27).[88] The proposed mechanism for this transformation suggests a gold(I)/gold(III) catalytic cycle where, upon acyloxy migration, Selectfluor is able to oxidize the initial vinyl–gold(I) species into a gold(III) intermediate which then reacts with the boronic acid to give a new C(sp²)–C(sp²) bond via reductive elimination. The variety of substituents present on the phenylboronic acid reagent show that the process is somewhat sensitive to the steric bulk on the aromatic ring. *ortho*-Substituents give the corresponding cross-coupling products in lower yields; however, all the reactions proceed with excellent *E* selectivity.

Scheme 27 Gold-Catalyzed Oxidative Cross Coupling between Propargyl Acetates and Boronic Acids[88]

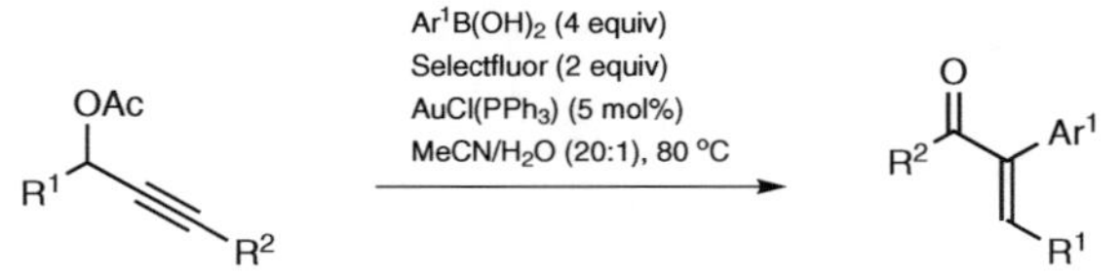

R^1	R^2	Ar^1	Yield (%)	Ref
Ph	Bu	Ph	62	[88]
Me	Ph	Ph	59	[88]
Cy	Bu	4-Tol	72	[88]
Cy	Bu	4-ClC₆H₄	58	[88]

2-Arylalk-2-en-1-ones 67; General Procedure:[88]

AuCl(PPh₃) (7.5 μmol, 0.05 equiv) was added to a soln of the propargylic acetate substrate (0.15 mmol, 1 equiv), Selectfluor (0.30 mmol, 2 equiv), and arylboronic acid (0.6 mmol, 4 equiv) in MeCN/H₂O (20:1) under N₂. The mixture was heated at 80 °C for 15 min and then cooled to 25 °C. The reaction was quenched with 5% aq Na₂S₂O₃ soln (10 mL) and

1.6.3 Gold-Mediated Reactions **563**

this mixture was extracted with Et$_2$O (3 × 10 mL). The combined organic layers were washed with brine (10 mL), dried (MgSO$_4$), and filtered. The filtrate was concentrated under reduced pressure, and the crude material was purified by column chromatography (silica gel).

1.6.3.4.4 **Synthesis of Cyclopentenones**

An interesting example of a one-pot domino synthesis of cyclopentenones **69** from enynyl acetates **68** uses a gold(I)-catalyzed tandem [3,3]-acyloxy rearrangement and Nazarov reaction (Table 3).[78] The combination of chloro(triphenylphosphine)gold(I) and silver(I) hexafluoroantimonate plays a dual role in the reaction by activating both the alkyne in the starting material **68** and the allenyl acetate intermediate. The reaction is carried out using only 1 mol% of gold catalyst and "wet" dichloromethane to favor the hydrolysis of cyclopentadienylic acetate to the corresponding ketone.

Table 3 Synthesis of Cyclopentenones via Tandem Gold(I)-Catalyzed 3,3-Rearrangement and Nazarov Reaction[78]

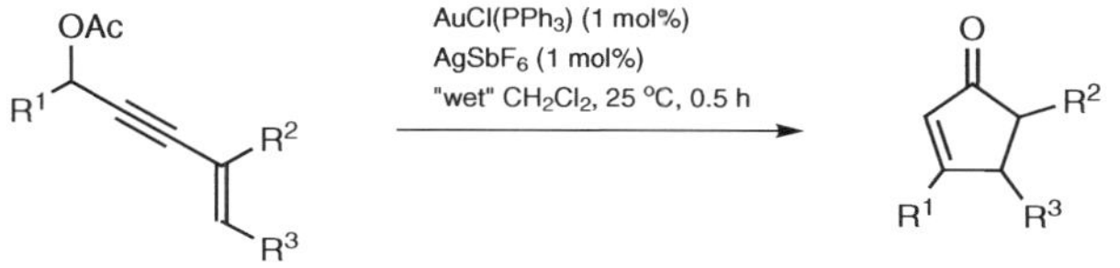

Entry	Substrate	Product	Yield (%)	Ref
1	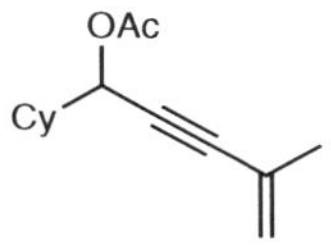	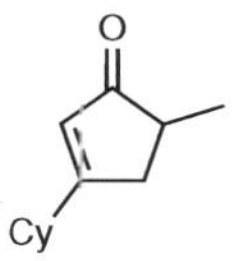	95	[78]
2	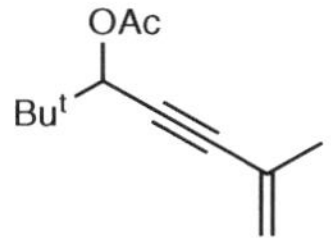	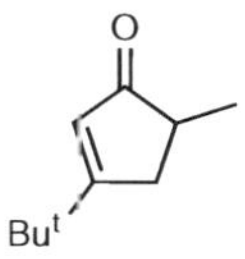	88	[78]
3	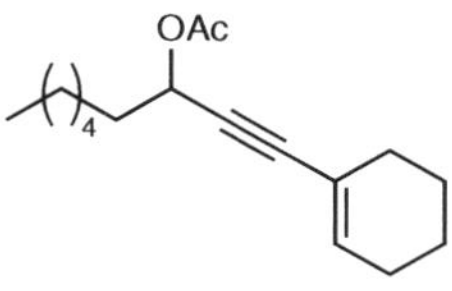	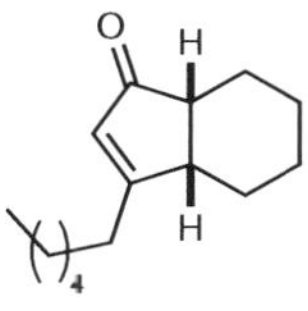	84	[78]

Cyclopentenones 69; **General Procedure:**[78]

To a soln of the enynyl acetate **68** (0.2 mmol, 1 equiv) in "wet" CH$_2$Cl$_2$ (4.0 mL) at 25 °C was added AuCl(PPh$_3$)/AgSbF$_6$ [0.2 mL; pre-generated as a 0.01 M soln in CH$_2$Cl$_2$ by mixing AuCl(PPh$_3$) with AgSbF$_6$] and the mixture was stirred for 30 min, treated with one drop of Et$_3$N, and concentrated. The crude material was purified by column chromatography (silica gel) to afford the product.

for references see p 573

564 Domino Transformations **1.6** Metal-Mediated Reactions

1.6.3.4.5   **Acetate Migration and Reaction with π-Electrophiles**

The gold intermediates generated upon migration of a carboxylate group in propargyl carboxylic substrates can be engaged in additional transformations with π-electrophiles (i.e., alkenes, alkynes) present in the reaction media.[6,89–91]

1.6.3.4.5.1   **Acetate Migration and Reaction with Alkynes**

The synthesis of 2,4a-dihydro-1H-fluorenes of general structure **72** is achieved by a domino gold(I)-catalyzed 1,2-acyloxy migration, cyclopropenation, and Nazarov cyclization of 1,6-diyne esters **70** using gold complex **71** (Table 4).[92]

Table 4   Gold-Catalyzed Cycloisomerization of 1,6-Diyne Esters to 2,4a-Dihydro-1H-fluorenes[92]

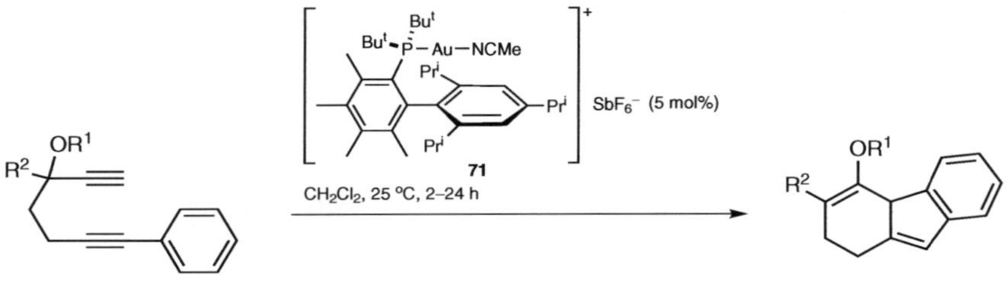

Entry	Substrate	Product	Yield (%)	Ref
1	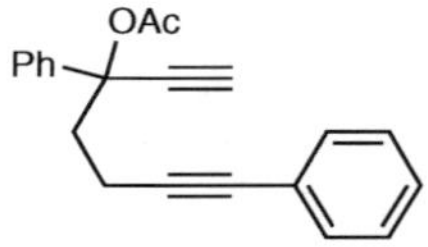	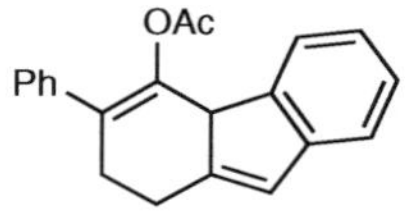	71	[92]
2	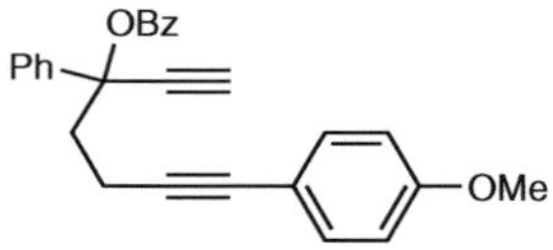	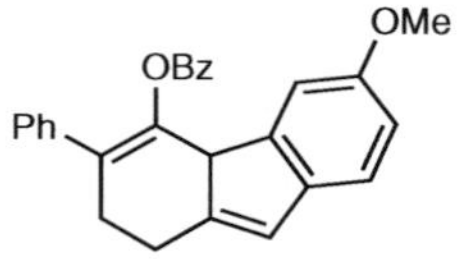	71	[92]
3	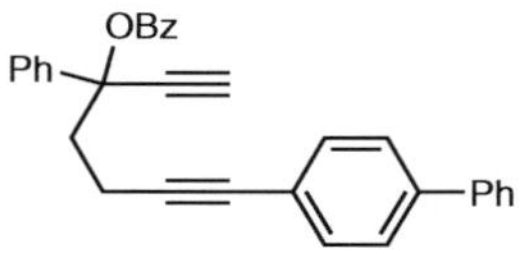	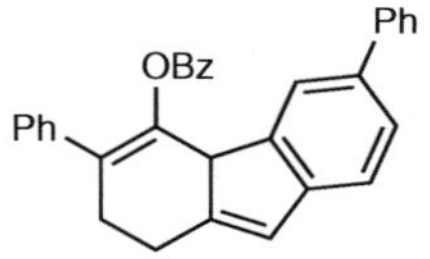	77	[92]

1.6.3 Gold-Mediated Reactions 565

Table 4 (cont.)

Entry	Substrate	Product	Yield (%)	Ref
4	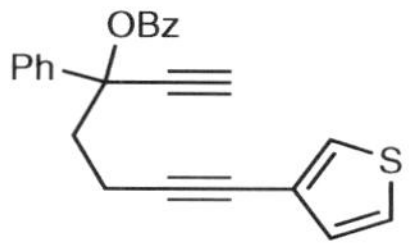	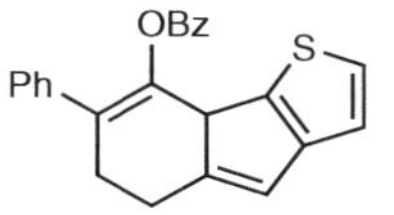	72	[92]
5	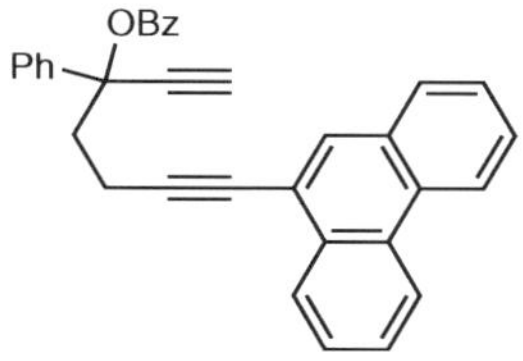	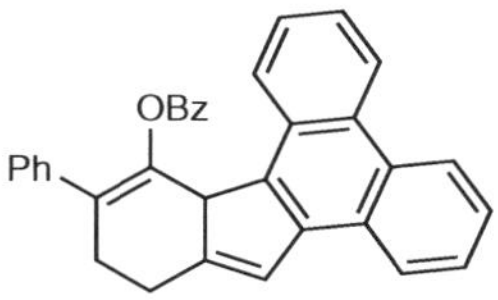	71	[92]
6	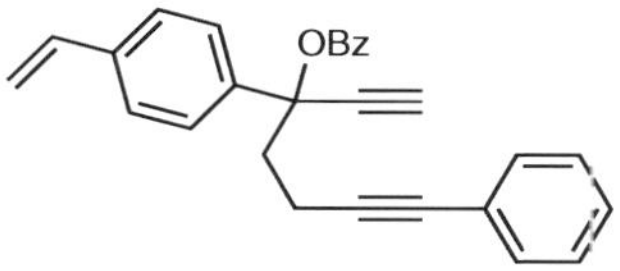	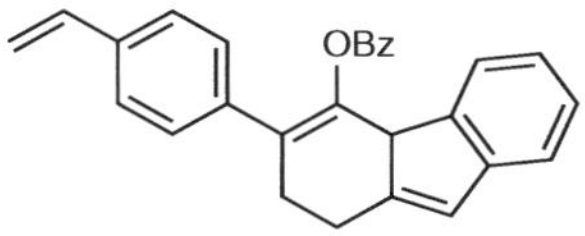	60	[92]

Alternatively, similar aromatic diynes can be used to construct naphthyl skeletons in excellent yields via 1,2-migration of the acetoxy group, intramolecular cyclopropenation, and gold-mediated ring opening of the cyclopropene unit.[93]

2,4a-Dihydro-1*H*-fluorenes 72; General Procedure:[92]
To a soln of the diynyl acetate **70** (0.2 mmol) in anhyd CH_2Cl_2 (4.0 mL) was added Au complex **71** (10 µmol) at 25 °C and the reaction was monitored by TLC analysis. Upon completion of the reaction, the soln was concentrated under reduced pressure, and the crude material was purified by column chromatography (silica gel) to afford the product.

1.6.3.4.5.2 **Acetate Migration and Reaction with Alkenes**

The synthesis of highly functionalized 3-azabicyclo[4.2.0]oct-5-enes **74** takes place by a tandem reaction sequence starting from 1,7-enyne benzoates **73** via a gold(I)-catalyzed benzoate 1,3-migration followed by [2+2] cycloaddition (Scheme 28).[94] The coordination of the gold to the alkene (rather than to the allene) in the reaction intermediate is proposed to precede the [2+2] cycloaddition after the initial 1,3-ester migration.

for references see p 573

566   Domino Transformations   **1.6** Metal-Mediated Reactions

Scheme 28   Gold-Catalyzed Tandem 1,3-Migration/[2 + 2] Cycloaddition of 1,7-Enyne Benzoates[94]

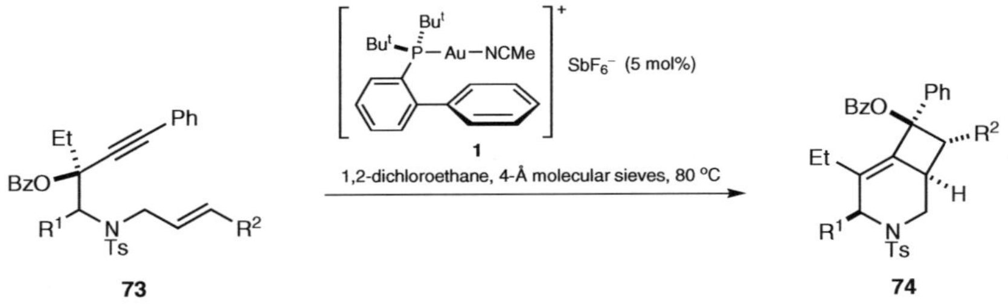

73

74

R^1	R^2	Yield (%)	Ref
Bn	H	75	[94]
CH$_2$SO$_2$Me	H	68	[94]
iBu	Ph	90	[94]
iPr	Ph	41	[94]

7-(Benzoyloxy)-5-ethyl-7-phenyl-3-tosyl-3-azabicyclo[4.2.0]oct-5-enes 74;
General Procedure:[94]

To a soln of 1,7-enyne benzoate **73** (0.15 mmol, 1 equiv) and 4-Å molecular sieves (150 mg) in anhyd 1,2-dichloroethane (1.5 mL) was added gold(I) complex **1** (7.5 μmol, 0.05 equiv) under an argon atmosphere. The mixture was stirred at 80 °C for 15–24 h, cooled to 25 °C, filtered through Celite, and washed with CH$_2$Cl$_2$. The solvent was removed under reduced pressure and the crude residue was purified by column chromatography (silica gel) to afford the product.

1.6.3.4.6   **Acetate Migration and Ring-Opening Reactions**

Gold-catalyzed ring-opening reactions are a well-defined strategy to achieve molecular complexity.[95] Selected examples are given in Sections 1.6.3.4.6.1 and 1.6.3.4.6.2.

1.6.3.4.6.1   **Cyclopentannulations**

Cyclopent-1-enyl ketones **76** are obtained efficiently under mild reaction conditions from 1-cyclopropylpropargyl esters **75** in the presence of catalytic amounts of (triphenylphosphine)gold(I) hexafluoroantimonate, prepared from chloro(triphenylphosphine)gold(I) and silver(I) hexafluoroantimonate (Table 5). The reaction involves a Rautenstrauch rearrangement (i.e., 1,2-migration of the carboxy group) followed by cyclopropyl ring opening and cyclization to produce cyclopentenyl ketones **76** in very good yields. Enantiomerically enriched ketones are synthesized by the gold-catalyzed cyclization of optically active propargyl acetates. The stereochemical information transfer observed in these reactions suggests that gold-stabilized nonclassical carbocations with a certain configurational stability might intervene along the reaction pathway.[96]

1.6.3 Gold-Mediated Reactions **567**

Table 5 Gold-Catalyzed Cyclopent-1-enyl Ketone Synthesis[96]

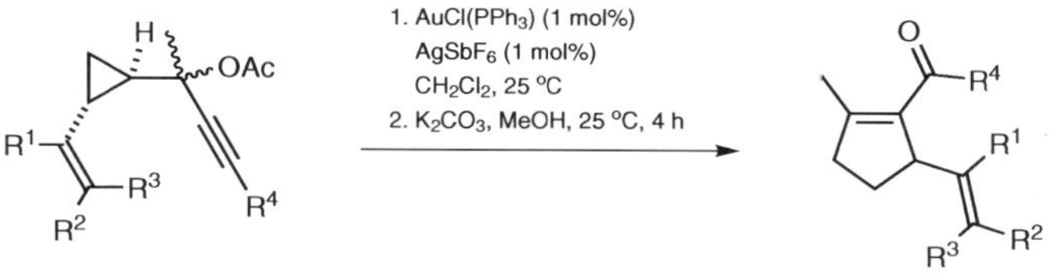

Entry	Substrate	Time (min) for Step 1	Product	Yield (%)	Ref
1	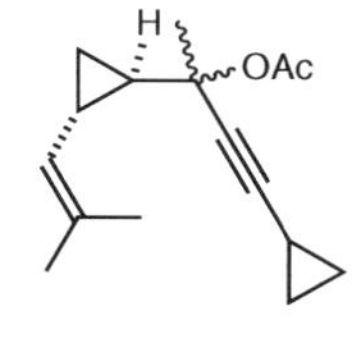	5	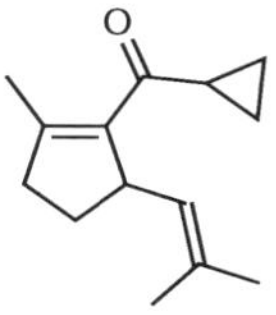	78	[96]
2	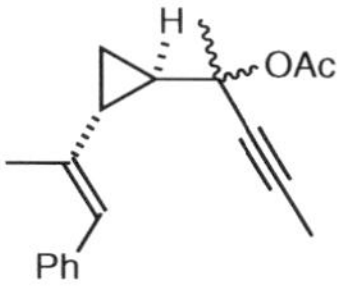	5	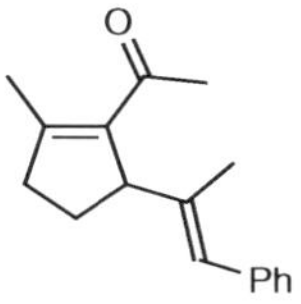	88	[96]
3	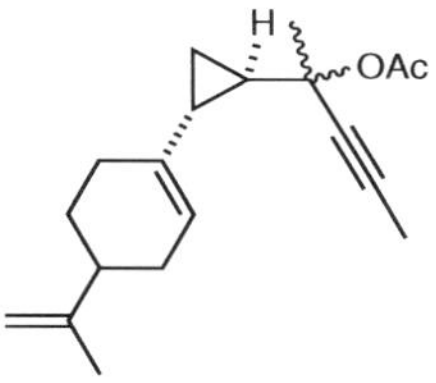	2	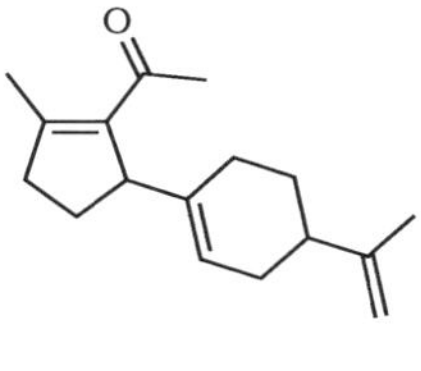	87	[96]
4	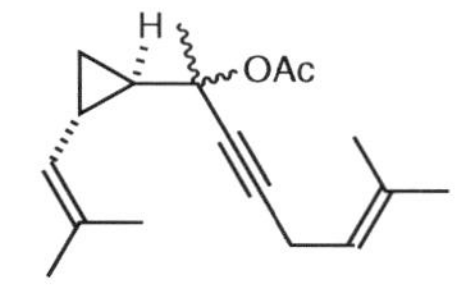	10	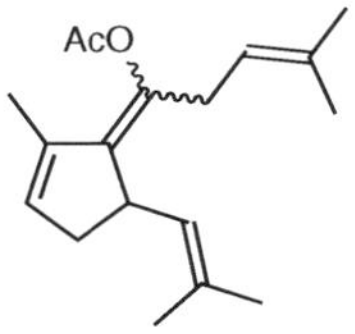	90	[96]

Interestingly, 3-cyclopropylpropargylic esters also undergo rearrangement under gold catalysis. Mechanistic aspects of this gold(I)-catalyzed [3,3]-rearrangement have been studied in detail, ultimately revealing the stereospecific nature of these transformations.[97,98] When the reaction is carried out with propargylic vinyl ethers, the [3,3]-rearrangement is irreversible and the reaction proceeds through a concerted pathway.[99]

An intermolecular version of this reaction utilizes gold-promoted 1,2-migration of the acetate group to trigger the intermolecular cyclopropanation of an alkene counterpart. Cycloisomerization of the cyclopropyl–vinyl acetate intermediate then produces the corresponding cyclopentanones or cyclopentenyl acetates in very good yields (59–90%). The reaction is highly diastereoselective, furnishing, in most cases, exclusively *trans*-2,3-disubstituted cyclopentenyl derivatives.[98]

for references see p 573

Cyclopropyl 2-Methyl-5-(2-methylprop-1-enyl)cyclopent-1-enyl Ketone (Table 5, Entry 1); Typical Procedure:[96]

CH_2Cl_2 (10 mL) was added to a mixture of $AuCl(PPh_3)$ (49.4 mg, 0.10 mmol) and $AgSbF_6$ (34.3 mg, 0.10 mmol), and the mixture was stirred for 10 min under an argon atmosphere. The resulting soln was filtered through Celite and a soln of the 1-cyclopropylpropargyl ester (2.46 g, 10 mol) in CH_2Cl_2 (10 mL) was added. This mixture was stirred for 5 min, MeOH (10 mL) and K_2CO_3 (2.76 g, 20 mmol) were added, and the resulting mixture was stirred for 4 h and concentrated. The residue was neutralized with 1 M HCl and extracted with t-BuOMe (3 × 30 mL). The combined organic layers were washed with brine (20 mL), dried ($MgSO_4$), and concentrated under reduced pressure. The crude material was purified by column chromatography (silica gel, hexane/t-BuOMe 10:1) to afford the product as a yellow oil; yield: 1.59 g (78%).

1.6.3.4.6.2 **Cyclohexannulations**

The gold (I)-catalyzed intramolecular cycloisomerization of 1-cyclopropylpropargyl esters **77** followed by hydrolysis in basic media gives cyclohex-2-enones **78** under very mild conditions. Different substitution patterns are tolerated (i.e., alkyl and aryl substituents on the double bond), ultimately leading to an array of products in good to excellent yields (Table 6).[96]

Table 6 Gold(I)-Catalyzed Cyclohex-2-enone Synthesis[96]

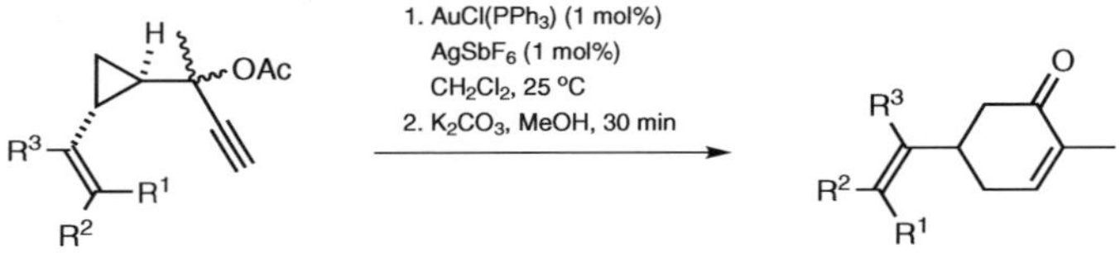

Entry	Substrate	Time (min) for Step 1	Product	Yield (%)	Ref
1	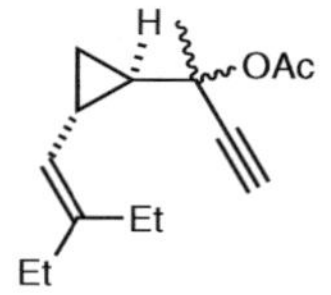	5	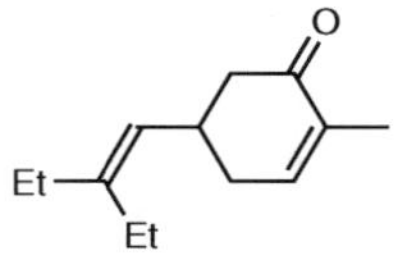	78	[96]
2	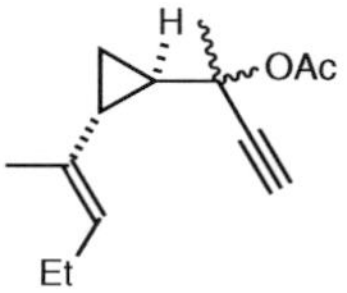	10	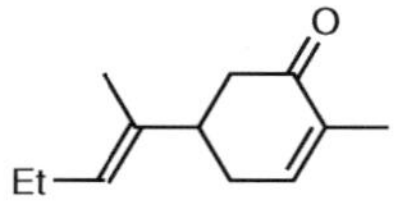	96	[96]
3	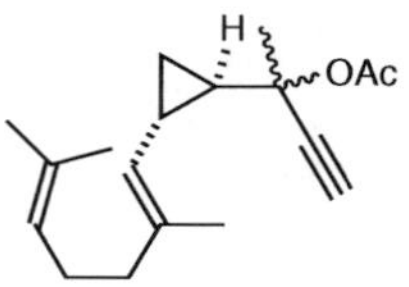	10	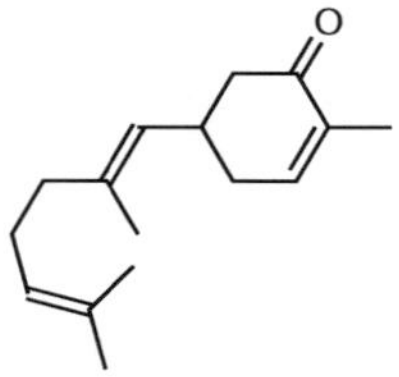	71	[96]

1.6.3 Gold-Mediated Reactions

Table 6 (cont.)

Entry	Substrate	Time (min) for Step 1	Product	Yield (%)	Ref
4	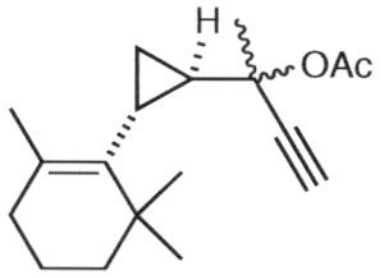	360	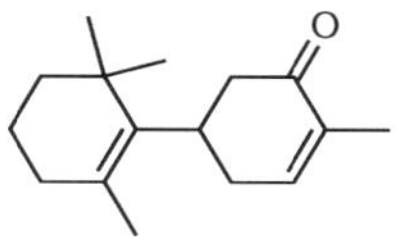	90	[96]
5	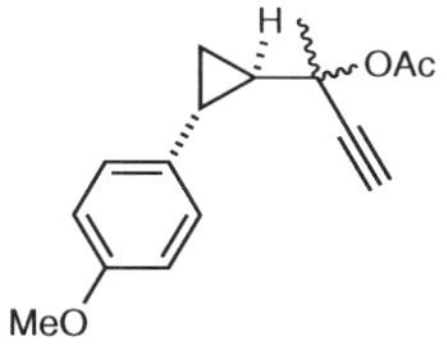	60	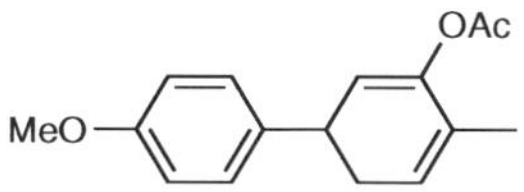	70	[96]

2-Methyl-5-vinylcyclohex-2-enones 78; General Procedure:[96]
CH$_2$Cl$_2$ (10 mL) was added to a mixture of AuCl(PPh$_3$) (0.10 mmol, 0.01 equiv) and AgSbF$_6$ (0.10 mmol, 0.01 equiv), and the mixture was stirred for 10 min under an argon atmosphere. The resulting soln was filtered through Celite and a soln of **77** (10 mmol, 1 equiv) in CH$_2$Cl$_2$ (10 mL) was added. This mixture was stirred for the indicated time, MeOH (10 mL) and K$_2$CO$_3$ (20 mmol) were added, and the resulting mixture was stirred for 30 min and then concentrated. The residue was neutralized with 1 M HCl and extracted with *t*-BuOMe (3 × 30 mL). The combined organic layers were washed with brine (20 mL), dried (MgSO$_4$), and concentrated under reduced pressure. The crude material was purified by column chromatography (silica gel) to afford the product.

1.6.3.4.6.3 Cycloheptannulations

Seven-membered-ring carbocycles (such as cycloheptenes) are present in a wide variety of natural products including members of the guanacastepenes and the frondosins (Scheme 29). They are also included in the ring systems of historically and medicinally important compounds such as ingenol.[100] Among the more common routes to obtain such cycloheptenes are inter- and intramolecular metal-mediated [5+2] and [4+3] reactions involving rhodium and copper carbenoids and dienes.

Scheme 29 Examples of a Guanacastepene and a Frondosin

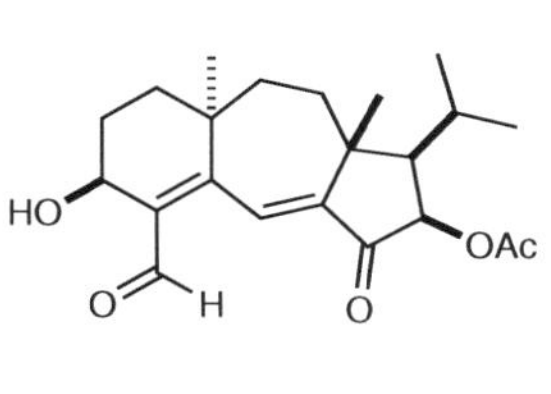

guanacastepene A

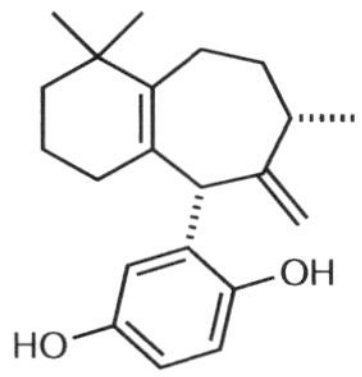

frondosin A

A formal [4+3] cycloaddition catalyzed by gold starts from propargyl acetates **79** and dienes **80** in a three-step cascade where the gold effects 1,2-acetoxy migration on the propargyl acetate substrate as well as the subsequent cyclopropanation of the diene counterpart. Finally, the gold also reactivates the vinyl acetate intermediate in situ, triggering a formal homo-Cope rearrangement to give the observed seven-membered-ring products,

for references see p 573

e.g. **81** (Table 7).[98] This cycloheptannulation process is highly stereoselective, affording only *cis*-cycloheptenyl acetates as the major product. This methodology has been used in a formal enantioselective synthesis of frondosin A.

Table 7 Gold-Catalyzed Cycloheptannulations[98]

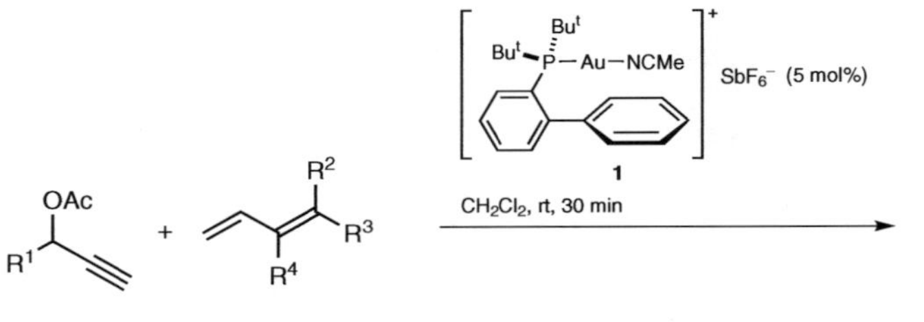

Entry	Substrates		Product	Yield (%)	Ref
	Alkyne	Diene			
1	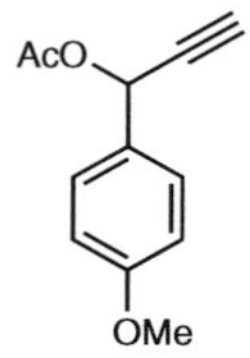	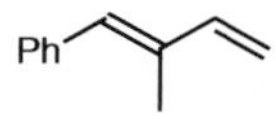	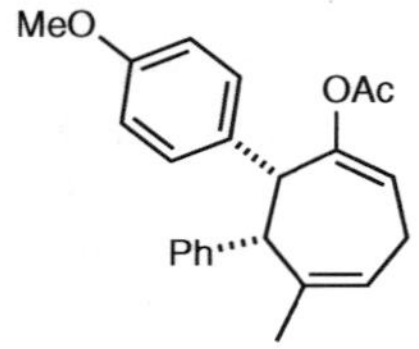	85	[98]
2	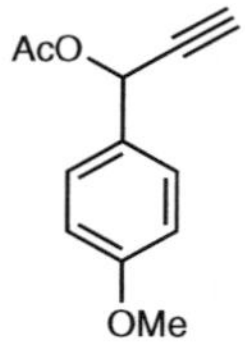	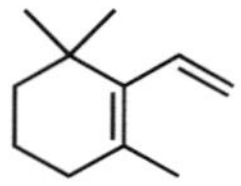	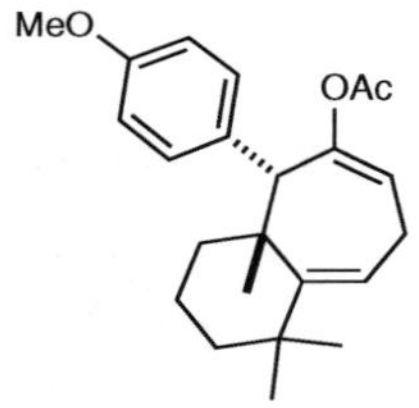	59	[98]
3	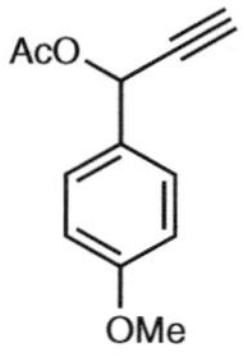	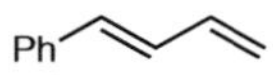	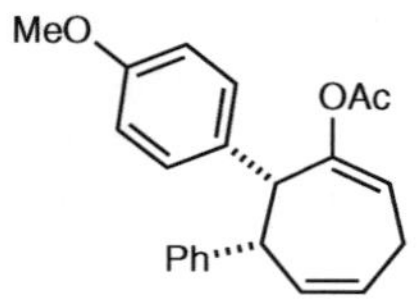	92	[98]
4	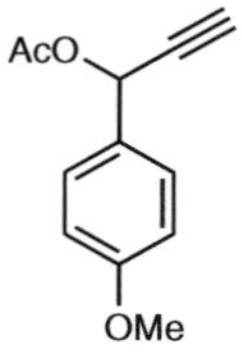	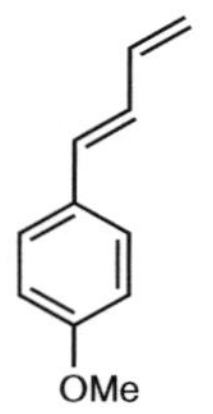	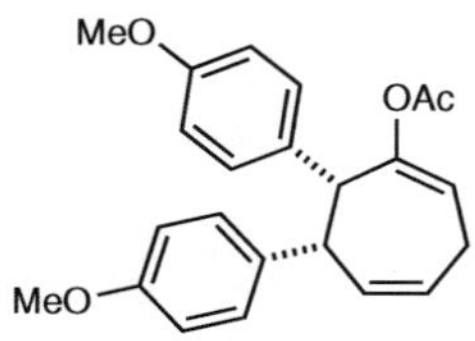	77	[98]

1.6.3 Gold-Mediated Reactions

Table 7 (cont.)

Entry	Substrates		Product	Yield (%)	Ref
	Alkyne	Diene			
5	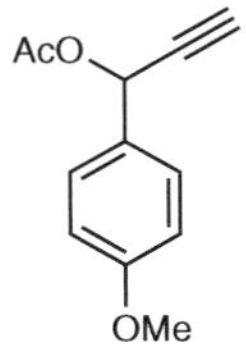	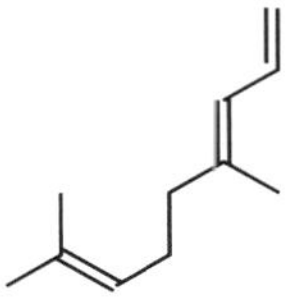	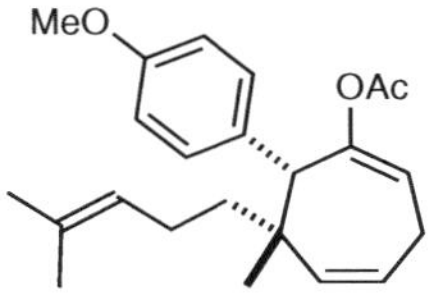	37	[98]
6	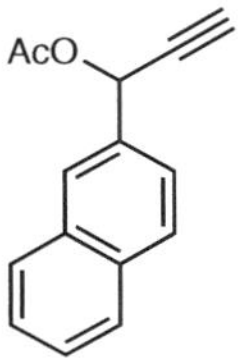	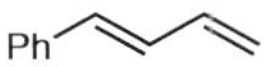	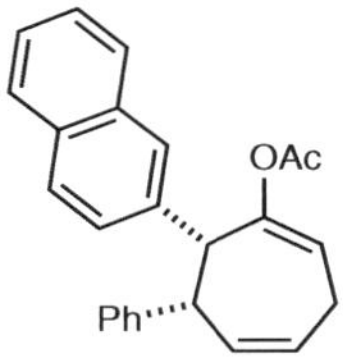	72	[98]
7	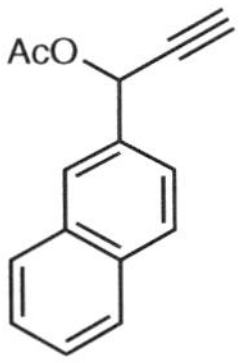	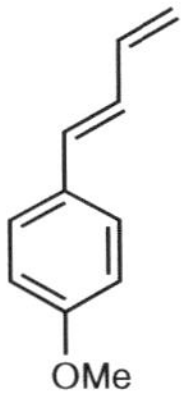	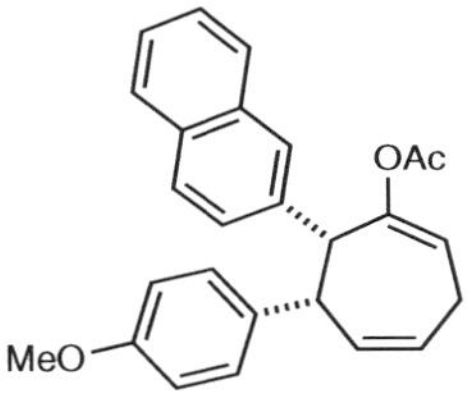	55	[98]
8	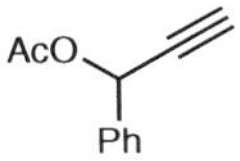	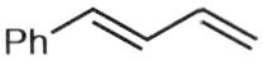	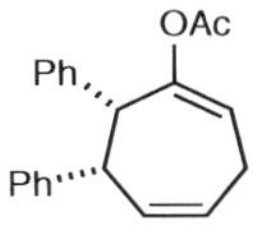	80	[98]
9	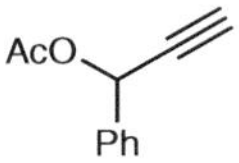	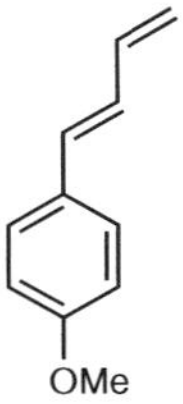	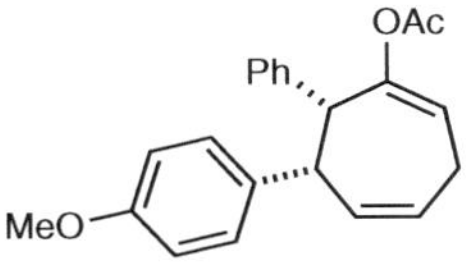	50	[98]
10	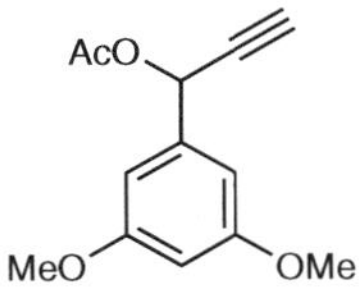	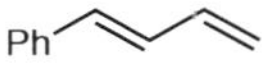	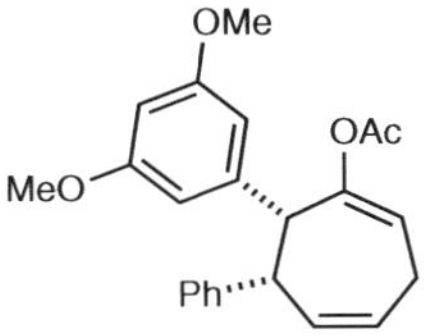	74	[98]
11	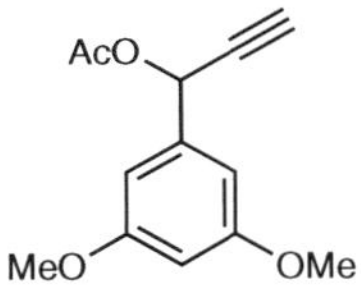	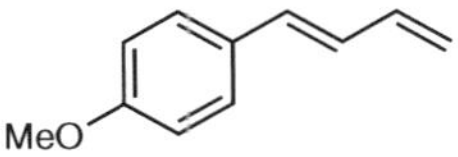	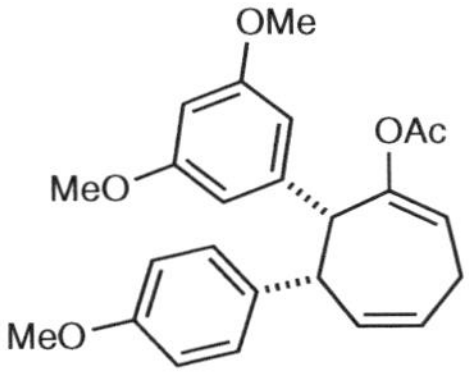	44	[98]

for references see p 573

(6R*,7R*)-1-Acetoxy-7-(4-methoxyphenyl)-5-methyl-6-phenylcyclohepta-1,4-diene (Table 7, Entry 1); Typical Procedure:[98]

To a soln of 1-(4-methoxyphenyl)prop-2-ynyl acetate (1.0 equiv) and (2-methylbuta-1,3-dienyl)benzene (1.1 equiv) in CH_2Cl_2 (0.1 M) was added catalyst **1** (0.05 equiv). The mixture was stirred at 25 °C for 30 min and then the reaction was quenched with Et_3N (0.05 equiv). The solvent was removed under reduced pressure, and the crude material was purified by column chromatography (silica gel, hexane/EtOAc) to afford the product; yield: 85%.

References

[1] Fürstner, A.; Davies, P. W., *Angew. Chem.*, (2007) **119**, 3478; *Angew. Chem. Int. Ed.*, (2007) **46**, 3410.

[2] Gorin, D. J.; Toste, F. D., *Nature (London)*, (2007) **446**, 395.

[3] Hashmi, A. S. K., *Chem. Rev.*, (2007) **107**, 3180.

[4] Hashmi, A. S. K., *Angew. Chem.*, (2008) **120**, 6856; *Angew. Chem. Int. Ed.*, (2008) **47**, 6754.

[5] Jiménez-Núñez, E.; Echavarren, A. M., *Chem. Rev.*, (2008) **108**, 3326.

[6] Michelet, V.; Toullec, P. Y.; Genêt, J.-P., *Angew. Chem.*, (2008) **120**, 4338; *Angew. Chem. Int. Ed.*, (2008) **47**, 4268.

[7] Kotha, S.; Misra, S.; Halder, S., *Tetrahedron*, (2008) **64**, 10775.

[8] Skouta, R.; Li, C.-J., *Angew. Chem.*, (2007) **119**, 1135; *Angew. Chem. Int. Ed.*, (2007) **46**, 1117.

[9] Skouta, R.; Li, C.-J., *Synlett*, (2007), 1759.

[10] Cai, S.; Zeng, J.; Bai, Y.; Liu, X.-W., *J. Org. Chem.*, (2012) **77**, 801.

[11] Asao, N.; Nogami, T.; Takahashi, K.; Yamamoto, Y., *J. Am. Chem. Soc.*, (2002) **124**, 764.

[12] Asao, N.; Nogami, T.; Lee, S.; Yamamoto, Y., *J. Am. Chem. Soc.*, (2003) **125**, 10921.

[13] Isogai, Y.; Menggenbateer; Nawaz Khan, F.; Asao, N., *Tetrahedron*, (2009) **65**, 9575.

[14] Asao, N.; Sato, K.; Menggenbateer; Yamamoto, Y., *J. Org. Chem.*, (2005) **70**, 3682.

[15] Asao, N., *Synlett*, (2006), 1645.

[16] Asao, N.; Aikawa, H.; Yamamoto, Y., *J. Am. Chem. Soc.*, (2004) **126**, 7458.

[17] Asao, N.; Menggenbateer; Seya, Y.; Yamamoto, Y.; Chen, M.; Zhang, W.; Inoue, A., *Synlett*, (2012) **23**, 66.

[18] Asao, N.; Takahashi, K.; Lee, S.; Kasahara, T.; Yamamoto, Y., *J. Am. Chem. Soc.*, (2002) **124**, 12650.

[19] Asao, N.; Yamamoto, Y., In *Modern Gold Catalyzed Synthesis*, Hashmi, A. S. K.; Toste, F. D., Eds.; Wiley-VCH: Weinheim, Germany, (2012); p 55.

[20] Malhotra, D.; Liu, L.-P.; Mashuta, M. S.; Hammond, G. B., *Chem.–Eur. J.*, (2013) **19**, 4043.

[21] Handa, S.; Slaughter, L. M., *Angew. Chem.*, (2012) **124**, 2966; *Angew. Chem. Int. Ed.*, (2012) **51**, 2912.

[22] Campbell, M. J.; Toste, F. D., *Chem. Sci.*, (2011) **2**, 1369.

[23] Pereshivko, O. P.; Peshkov, V. A.; Jacobs, J.; Meervelt, L. V.; Van der Eycken, E. V., *Adv. Synth. Catal.*, (2013) **355**, 781.

[24] Nguyen, R.-V.; Yao, X.; Li, C.-J., *Org. Lett.*, (2006) **8**, 2397.

[25] Suzuki, Y.; Naoe, S.; Oishi, S.; Fujii, N.; Ohno, H., *Org. Lett.*, (2012) **14**, 326.

[26] Hou, Z.; Oishi, S.; Suzuki, Y.; Kure, T.; Nakanishi, I.; Hirasawa, A.; Tsujimoto, G.; Ohno, H.; Fujii, N., *Org. Biomol. Chem.*, (2013) **11**, 3288.

[27] Li, J.; Liu, L.; Ding, D.; Sun, J.; Ji, Y.; Dong, J., *Org. Lett.*, (2013) **15**, 2884.

[28] Abu Sohel, S. M.; Liu, R.-S., *Chem. Soc. Rev.*, (2009) **38**, 2269.

[29] Echavarren, A. M.; Jiménez-Núñez, E., *Top. Catal.*, (2010) **53**, 924.

[30] Gorin, D. J.; Sherry, B. D.; Toste, F. D., *Chem. Rev.*, (2008) **108**, 3351.

[31] Jiménez-Núñez, E.; Raducan, M.; Lauterbach, T.; Molawi, K.; Solorio, C. R.; Echavarren, A. M., *Angew. Chem.*, (2009) **121**, 6268; *Angew. Chem. Int. Ed.*, (2009) **48**, 6152.

[32] Gaydou, M.; Miller, R. E.; Delpont, N.; Ceccon, J.; Echavarren, A. M., *Angew. Chem.*, (2013) **125**, 6524; *Angew. Chem. Int. Ed.*, (2013) **52**, 6396.

[33] Chen, C.-H.; Tsai, Y.-C.; Liu, R.-S., *Angew. Chem.*, (2013) **125**, 4697; *Angew. Chem. Int. Ed.*, (2013) **52**, 4599.

[34] Hashmi, A. S. K.; Braun, I.; Nösel, P.; Schädlich, J.; Wieteck, M.; Rudolph, M.; Rominger, F., *Angew. Chem.*, (2012) **124**, 4532; *Angew. Chem. Int. Ed.*, (2012) **51**, 4456.

[35] Hashmi, A. S. K.; Wieteck, M.; Braun, I.; Nösel, P.; Jongbloed, L.; Rudolph, M.; Rominger, F., *Adv. Synth. Catal.*, (2012) **354**, 555.

[36] Wang, Y.; Yepremyan, A.; Ghorai, S.; Todd, R.; Aue, D. H.; Zhang, L., *Angew. Chem.*, (2013) **125**, 7949; *Angew. Chem. Int. Ed.*, (2013) **52**, 7795.

[37] Aubert, C.; Fensterbank, L.; Garcia, P.; Malacria, M.; Simonneau, A., *Chem. Rev.*, (2011) **111**, 1954.

[38] Cañeque, T.; Truscott, F. M.; Rodriguez, R.; Maestri, G.; Malacria, M., *Chem. Soc. Rev.*, (2014) **43**, 2916.

[39] Lemière, G.; Gandon, V.; Cariou, K.; Hours, A.; Fukuyama, T.; Dhimane, A.-L.; Fensterbank, L.; Malacria, M., *J. Am. Chem. Soc.*, (2009) **131**, 2993.

[40] Teller, H.; Corbet, M.; Mantilli, L.; Gopakumar, G.; Goddard, R.; Thiel, W.; Fürstner, A., *J. Am. Chem. Soc.*, (2012) **134**, 15331.

[41] Yang, C.-Y.; Lin, G.-Y.; Liao, H.-Y.; Datta, S.; Liu, R.-S., *J. Org. Chem.*, (2008) **73**, 4907.

[42] Chiarucci, M.; Mocci, R.; Syntrivanis, L.-D.; Cera, G.; Mazzanti, A.; Bandini, M., *Angew. Chem.*, (2013) **125**, 11050; *Angew. Chem. Int. Ed.*, (2013) **52**, 10850.

[43] Muzart, J., *Tetrahedron*, (2008) **64**, 5815.

[44] Alcaide, B.; Almendros, P.; Alonso, J. M., *Org. Biomol. Chem.*, (2011) **9**, 4405.

[45] Hashmi, A. S. K.; Häffner, T.; Rudolph, M.; Rominger, F., *Chem.–Eur. J.*, (2011) **17**, 8195.

[46] Liao, H.-H.; Liu, R.-S., *Chem. Commun. (Cambridge)*, (2011) **47**, 1339.

[47] Tarselli, M. A.; Zuccarello, J. L.; Lee, S. J.; Gagné, M. R., *Org. Lett.*, (2009) **11**, 3490.

[48] de Haro, T.; Garayalde, D.; Nevado, C., In *Science of Synthesis Knowledge Updates*, (2012); Vol. 2012/3, Section 3.6.15.1, p 171.

[49] Ball, L. T.; Lloyd-Jones, G. C.; Russell, C. A., *Science (Washington, D. C.)*, (2012) **337**, 1644.

[50] Ball, L. T.; Lloyd-Jones, G. C.; Russell, C. A., *J. Am. Chem. Soc.*, (2014) **136**, 254.

[51] Brenzovich, W. E., Jr.; Brazeau, J.-F.; Toste, F. D., *Org. Lett.*, (2010) **12**, 4728.

[52] Levin, M. D.; Toste, F. D., *Angew. Chem.*, (2014) **126**, 6325; *Angew. Chem. Int. Ed.*, (2014) **53**, 6211.

[53] Wang, W.; Jasinski, J.; Hammond, G. B.; Xu, B., *Angew. Chem.*, (2010) **122**, 7405; *Angew. Chem. Int. Ed.*, (2010) **49**, 7247.

[54] Brenzovich, W. E., Jr.; Benitez, D.; Lackner, A. D.; Shunatona, H. P.; Tkatchouk, E.; Goddard, W. A., III; Toste, F. D., *Angew. Chem.*, (2010) **122**, 5615; *Angew. Chem. Int. Ed.*, (2010) **49**, 5519.

[55] Tkatchouk, E.; Mankad, N. P.; Benitez, D.; Goddard, W. A., III; Toste, F. D., *J. Am. Chem. Soc.*, (2011) **133**, 14293.

[56] Zhu, S.; Ye, L.; Wu, W.; Jiang, H., *Tetrahedron*, (2013) **69**, 10375.

[57] Melhado, A. D.; Brenzovich, W. E., Jr.; Lackner, A. D.; Toste, F. D., *J. Am. Chem. Soc.*, (2010) **132**, 8885.

[58] Ball, L. T.; Lloyd-Jones, G. C.; Russell, C. A., *Chem.–Eur. J.*, (2012) **18**, 2931.

[59] Zhang, G.; Cui, L.; Wang, Y.; Zhang, L., *J. Am. Chem. Soc.*, (2010) **132**, 1474.

[60] Marco-Contelles, J.; Soriano, E., *Chem.–Eur. J.*, (2007) **13**, 1350.

[61] Nevado, C., *Chimia*, (2010) **64**, 247.

[62] Fehr, C.; Galindo, J., *Angew. Chem.*, (2006) **118**, 2967; *Angew. Chem. Int. Ed.*, (2006) **45**, 2901.

[63] Fürstner, A.; Hannen, P., *Chem.–Eur. J.*, (2006) **12**, 3006.

[64] Marion, N.; Nolan, S. P., *Angew. Chem.*, (2007) **119**, 2806; *Angew. Chem. Int. Ed.*, (2007) **46**, 2750.

[65] Amijs, C. H. M.; López-Carrillo, V.; Echavarren, A. M., *Org. Lett.*, (2007) **9**, 4021.

[66] Crone, B.; Kirsch, S. F., *Chem.–Eur. J.*, (2008) **14**, 3514.

[67] Dudnik, A. S.; Schwier, T.; Gevorgyan, V., *Tetrahedron*, (2009) **65**, 1859.

[68] Gorin, D. J.; Dubé, P.; Toste, F. D., *J. Am. Chem. Soc.*, (2006) **128**, 14480.

[69] Johansson, M. J.; Gorin, D. J.; Staben, S. T.; Toste, F. D., *J. Am. Chem. Soc.*, (2005) **127**, 18002.

[70] Mamane, V.; Gress, T.; Krause, H.; Fürstner, A., *J. Am. Chem. Soc.*, (2004) **126**, 8654.

[71] Shi, X.; Gorin, D. J.; Toste, F. D., *J. Am. Chem. Soc.*, (2005) **127**, 5802.

[72] Witham, C. A.; Mauléon, P.; Shapiro, N. D.; Sherry, B. D.; Toste, F. D., *J. Am. Chem. Soc.*, (2007) **129**, 5838.

[73] Marion, N.; de Frémont, P.; Lemière, G.; Stevens, E. D.; Fensterbank, L.; Malacria, M.; Nolan, S. P., *Chem. Commun. (Cambridge)*, (2006), 2048.

[74] Buzas, A.; Gagosz, F., *J. Am. Chem. Soc.*, (2006) **128**, 12614.

[75] Marion, N.; Díez-González, S.; de Frémont, P.; Noble, A. R.; Nolan, S. P., *Angew. Chem.*, (2006) **118**, 3729; *Angew. Chem. Int. Ed.*, (2006) **45**, 3647.

[76] Wang, S.; Zhang, L., *J. Am. Chem. Soc.*, (2006) **128**, 8414.

[77] Zhang, L., *J. Am. Chem. Soc.*, (2005) **127**, 16804.

[78] Zhang, L.; Wang, S., *J. Am. Chem. Soc.*, (2006) **128**, 1442.

[79] Buzas, A.; Gagosz, F., *Org. Lett.*, (2006) **8**, 515.

[80] Li, G.; Zhang, G.; Zhang, L., *J. Am. Chem. Soc.*, (2008) **130**, 3740.

[81] Peng, Y.; Cui, L.; Zhang, G.; Zhang, L., *J. Am. Chem. Soc.*, (2009) **131**, 5062.

[82] Huang, X.; de Haro, T.; Nevado, C., *Chem.–Eur. J.*, (2009) **15**, 5904.

[83] Wang, D.; Ye, X.; Shi, X., *Org. Lett.*, (2010) **12**, 2088.

[84] Yu, M.; Zhang, G.; Zhang, L., *Org. Lett.*, (2007) **9**, 2147.

[85] Yu, M.; Zhang, G.; Zhang, L., *Tetrahedron*, (2009) **65**, 1846.

[86] Hopkinson, M. N.; Giuffredi, G. T.; Gee, A. D.; Gouverneur, V., *Synlett*, (2010), 2737.

[87] de Haro, T.; Nevado, C., *Chem. Commun. (Cambridge)*, (2011) **47**, 248.

[88] Zhang, G.; Peng, Y.; Cui, L.; Zhang, L., *Angew. Chem.*, (2009) **121**, 3158; *Angew. Chem. Int. Ed.*, (2009) **48**, 3112.

References

[89] Toullec, P. Y.; Genin, E.; Leseurre, L.; Genêt, J.-P.; Michelet, V., *Angew. Chem.*, (2006) **118**, 7587; *Angew. Chem. Int. Ed.*, (2006) **45**, 7427.

[90] Nieto-Oberhuber, C.; Pérez-Galán, P.; Herrero-Gómez, E.; Lauterbach, T.; Rodríguez, C.; López, S.; Bour, C.; Rosellón, A.; Cárdenas, D. J.; Echavarren, A. M., *J. Am. Chem. Soc.*, (2008) **130**, 269.

[91] Bae, H. J.; Baskar, B.; An, S. E.; Cheong, J. Y.; Thangadurai, D. T.; Hwang, I.-C.; Rhee, Y. H., *Angew. Chem.*, (2008) **120**, 2295; *Angew. Chem. Int. Ed.*, (2008) **47**, 2263.

[92] Rao, W.; Koh, M. J.; Li, D.; Hirao, H.; Chan, P. W. H., *J. Am. Chem. Soc.*, (2013) **135**, 7926.

[93] Lauterbach, T.; Gatzweiler, S.; Nösel, P.; Rudolph, M.; Rominger, F.; Hashmi, A. S. K., *Adv. Synth. Catal.*, (2013) **355**, 2481.

[94] Rao, W.; Susanti, D.; Chan, P. W. H., *J. Am. Chem. Soc.*, (2011) **133**, 15248.

[95] Garayalde, D.; Nevado, C., *ACS Catal.*, (2012) **2**, 1462.

[96] Zou, Y.; Garayalde, D.; Wang, Q.; Nevado, C.; Goeke, A., *Angew. Chem.*, (2008) **120**, 10264; *Angew. Chem. Int. Ed.*, (2008) **47**, 10110.

[97] Garayalde, D.; Gómez-Bengoa, E.; Huang, X.; Goeke, A.; Nevado, C., *J. Am. Chem. Soc.*, (2010) **132**, 4720.

[98] Garayalde, D.; Krüger, K.; Nevado, C., *Angew. Chem.*, (2011) **123**, 941; *Angew. Chem. Int. Ed.*, (2011) **50**, 911.

[99] Mauleón, P.; Krinsky, J. L.; Toste, F. D., *J. Am. Chem. Soc.*, (2009) **131**, 4513.

[100] Maimone, T. J.; Baran, P. S., *Nat. Chem. Biol.*, (2007) **3**, 396.

1.6.4   **Rare Earth Metal Mediated Domino Reactions**

T. Ohshima

General Introduction

The group of rare earth metals comprises 17 chemical elements in the periodic table, namely scandium, yttrium, and 15 lanthanides (La, Ce, Pr, Nd, Pm, Sm, Eu, Gd, Tb, Dy, Ho, Er, Tm, Yb, and Lu). Despite their name, rare earth elements are relatively abundant in the Earth's crust. The availability of lanthanides at quite inexpensive prices facilitates their use in diverse applications. For example, much progress has recently been achieved in the use of rare earth based catalysts in synthetic chemistry.[1–3] Rare earth elements exist almost exclusively in their trivalent state [i.e., M(III)] in coordination complexes. In the series of lanthanides, a systematic contraction of the ionic radii is observed when going from lanthanum to lutetium (often referred to as the lanthanide contraction), but this variation is so smooth and limited, with only a 15% contraction between lanthanum and lutetium and ca. 1% between two successive lanthanides, that it is possible to fine-tune ionic radii, Lewis acidity, and Brønsted basicity of rare earth complexes. As a result of the large size of the lanthanide ions compared to other metal ions, lanthanide ions have high coordination numbers varying from 6 to 12. In water, for example, large lanthanide(III) ions at the beginning of the series (La to Nd) adopt nine-coordinate geometries, which are gradually transformed into eight-coordinate geometries for small lanthanide(III) ions (Tb to Lu). In addition, the nature of the coordination sphere is controlled by subtle interplay between electrostatic interactions and inter-ligand steric constraints, and thus variable coordination numbers (6 to 12) and geometries are observed in lanthanide complexes. Due to the strong oxophilicity of rare earth elements, their metal ions have a hard Lewis acidic nature. Most particularly, rare earth metal trifluoromethanesulfonates have been regarded as new types of Lewis acids and have several useful characteristic features compared to common Lewis acids such as BX_3, AlX_3, TiX_4, and SnX_4. For example, rare earth(III) trifluoromethanesulfonates are stable and active in the presence of many Lewis bases containing nitrogen, oxygen, phosphorus, and sulfur atoms, and thus work as Lewis acids even in water. In many cases, catalytic amounts of the trifluoromethanesulfonates are enough to complete the Lewis acid mediated reactions. Moreover, once their role as catalysts is complete, most rare earth(III) trifluoromethanesulfonates can be recovered easily and reused without loss of activity.

Another important type of rare earth metal species, the rare earth metal alkoxides [M(III)(OR¹)₃], exhibit both Lewis acidity and Brønsted basicity. Such multifunctionality of rare earth metal alkoxides enables the synergistic function of the active sites, making substrates more reactive in the transition state and controlling their position. These collated characteristic features of rare earth based complexes, such as high coordination numbers, a hard Lewis acidic nature, high compatibility with various functional groups, ease of fine-tuning, and multifunctionality, have led to the development of a variety of domino reactions catalyzed largely by rare earth(III) trifluoromethanesulfonates and alkoxides. As mentioned above, catalytic amounts of rare earth metal species afford acceptable results in many cases; thus, this review focuses on rare earth metal catalyzed domino reactions up to early 2014. The reactions are classified into six sections, depending on the key and/or initial reaction of each domino process. In each section, the reactions are listed

for references see p 599

578   Domino Transformations   **1.6** Metal-Mediated Reactions

in chronological order, and in most cases one transformation is selected in terms of providing a representative experimental procedure that has high practicality and substrate generality.

1.6.4.1   **Addition to C=O or C=C—C=O as a Primary Step**

1.6.4.1.1   **Aldol-Type Reactions**

The aldol reaction is generally regarded as one of the most versatile and efficient C—C bond-forming reactions. Many efforts have been directed toward the development of catalytic asymmetric aldol reactions, but most of these reactions require a preconversion of the nucleophile (i.e., aldehyde, ketone, or ester) into a more reactive species, such as a silyl enol ether or ketene silyl acetal (Mukaiyama-type aldol reaction). The use of such activated substrates, however, causes the formation of more than stoichiometric amounts of unwanted chemical waste in both the preactivation and the aldol steps. Thus, a direct catalytic aldol reaction, one in which the reaction proceeds by proton exchange of substrates without generating any waste, has recently attracted much attention from an environmental and economic point of view. When preactivated substrates are used, usually a Lewis acid is the catalyst of choice. On the other hand, for direct aldol reaction, catalysts having a Brønsted basic nature, such as a rare earth metal alkoxide [M(OR^1)$_3$], are used because of the necessity of generating an enolate anion in situ from the substrate. After the primary aldol reaction step, the resulting alkoxide can promote deprotonation of the substrate or another type of reaction, allowing it to achieve domino reactions and access even greater complexity.

The first report of an aldol-initiated domino reaction was achieved by a chiral praseodymium trilithium tris[(R)-binaphthoxide] (PrLB; **1**) catalyst, which promotes asymmetric intermolecular nitroaldol reaction of nitromethane to diketo aldehyde **2** followed by intramolecular nitroaldol reaction and basic epimerization of the nitro group, generating bicyclic ketone **3** in 41% yield and 79% ee after crystallization (65% ee before crystallization) (Scheme 1).[4] In this reaction, the praseodymium ion works as a Lewis acid to activate the carbonyl group, while the lithium naphthoxide portion functions as a Brønsted base (multifunctional catalyst) to generate a lithium nitronate from the nitro compounds (nitromethane or intermolecular nitroaldol product).

Scheme 1   Domino Inter- and Intramolecular Nitroaldol Reaction[4]

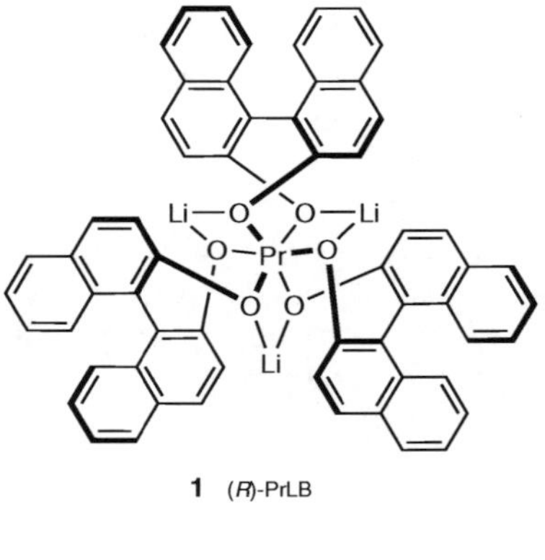

1   (R)-PrLB

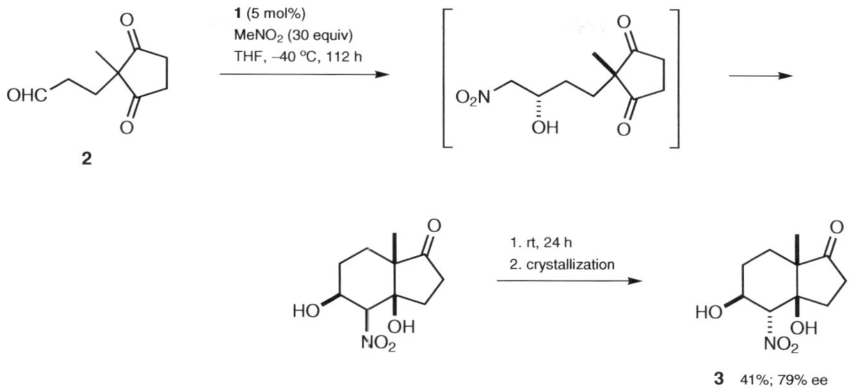

1.6.4 Rare Earth Metal Mediated Domino Reactions

579

3 41%; 79% ee

Another example of a domino reaction initiated by the aldol reaction is the aldol/Tishchenko reaction. In most cases of direct aldol reactions, substrates are limited to rather simple carbonyl donor compounds, such as methyl ketones, α-hydroxy ketones, and easily enolizable aliphatic aldehydes; moreover, it is well known that in the reactions of methylene ketones, such as ethyl ketone, it is quite difficult to obtain the product in high yield and high selectivity due to a strong tendency for retro-aldol reactions to predominate. One of the best solutions for this kind of problem is to couple an irreversible reaction, such as a Tishchenko reaction, to a reversible reaction, such as the aldol reaction of methylene ketones. The metalated aldolate delivered from the initial direct aldol reaction (reversible) is active toward the addition of another aldehyde molecule (reversible), and the resulting intermediate exhibits the appropriate orientation to rapidly undergo [3,3] bond reorganization (Tischenko reaction, irreversible) to provide monoacylated 1,3-diols. Indeed, the lanthanum–binaphthoxide catalyst prepared from lanthanum(III) trifluoromethanesulfonate, (R)-BINOL, and butyllithium, catalyzes a domino aldol/Tishchenko reaction of a variety of arylaldehydes **5** with aryl methylene ketones **4** to afford Tishchenko products **6**; after methanolysis using sodium methoxide in methanol, the corresponding *anti*-diols **7** are obtained in high yield, high enantiomeric excess, and nearly perfect diastereoselectivity. In these operations, the enantioselectivity is controlled in the aldol reaction step and the diastereoselectivity is controlled in the Tishchenko reaction step (Scheme 2).[5,6] In addition, the reaction of arylaldehydes with aliphatic ketones, such as diethyl ketone, can also be achieved using a chiral amino alcohol based ytterbium catalyst.[7,8]

for references see p 599

Scheme 2 Domino Aldol/Tishchenko Reaction[5]

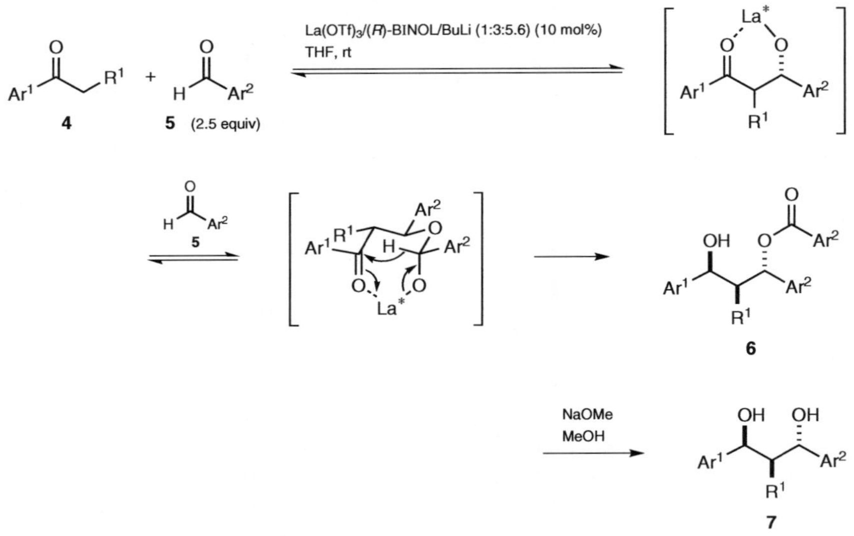

Ar^1	R^1	Ar^2	Time (h)	Yield (%) of **7**	ee (%)	Ref
4-F$_3$CC$_6$H$_4$	Me	4-ClC$_6$H$_4$	60	95	93	[5]
4-F$_3$CC$_6$H$_4$	Me	4-BrC$_6$H$_4$	48	96	95	[5]
4-F$_3$CC$_6$H$_4$	Me	4-FC$_6$H$_4$	72	85	92	[5]
4-F$_3$CC$_6$H$_4$	Me	4-Tol	94	67	92	[5]
4-F$_3$CC$_6$H$_4$	Me	Ph	84	95	91	[5]
4-F$_3$CC$_6$H$_4$	Me	3-BrC$_6$H$_4$	48	92	86	[5]
4-F$_3$CC$_6$H$_4$	Me	3-MeOC$_6$H$_4$	72	65	85	[5]
4-F$_3$CC$_6$H$_4$	Me	2-naphthyl	80	67	88	[5]
4-F$_3$CC$_6$H$_4$	Me	2-furyl	84	77	93	[5]
4-F$_3$CC$_6$H$_4$	Me	2-thienyl	84	82	94	[5]
4-BrC$_6$H$_4$	Me	4-BrC$_6$H$_4$	48	70	85	[5]
3-ClC$_6$H$_4$	Me	4-ClC$_6$H$_4$	48	60	84	[5]
3,4-Cl$_2$C$_6$H$_3$	Me	4-ClC$_6$H$_4$	48	81	88	[5]
3,5-Cl$_2$C$_6$H$_3$	Me	4-ClC$_6$H$_4$	48	73	85	[5]
3,5-F$_2$C$_6$H$_3$	Me	4-ClC$_6$H$_4$	48	77	87	[5]
4-F$_3$CC$_6$H$_4$	Et	4-BrC$_6$H$_4$	90	90	88	[5]
4-F$_3$CC$_6$H$_4$	Pr	4-BrC$_6$H$_4$	90	88	87	[5]

One-pot, sequential reactions, where a single catalyst promotes mechanistically different reactions by tuning of the catalyst component with appropriate additives, can be achived using a multifunctional yttrium catalyst.[9] For instance, the chiral yttrium trilithium tris-[(S)-binaphthoxide] catalyst (YLB; **8**) can promote both an asymmetric cyanation reaction and a nitroaldol reaction with high enantioselectivity; however, the best conditions for the former reaction (**8**/Ar^1$_3$P=O) are not suitable for the latter reaction. A highly enantio-selective, one-pot sequential cyanation/nitroaldol process was eventually achieved by tuning the chiral environment in the yttrium trilithium tris[(S)-binaphthoxide] catalyst for the individual steps by the addition of different achiral additives. For example, the more

1.6.4 Rare Earth Metal Mediated Domino Reactions

reactive aliphatic aldehyde group in dialdehyde **9** was selectively reacted with ethyl cyanoformate by tuning yttrium trilithium tris[(S)-binaphthoxide] (**8**) with water, butyllithium, and tris(2,6-dimethoxyphenyl)phosphine oxide, to give cyano carbonate **10** (Scheme 3). Cyano carbonate **10** was subjected to a nitroaldol reaction with nitromethane by tuning the catalyst with lithium tetrafluoroborate, leading to the desired nitroaldol product **11** in 78% yield with 2.6:1 diastereoselectivity and 93% ee (major diastereomer).

Scheme 3 One-Pot Sequential Cyanation/Nitroaldol Process[9]

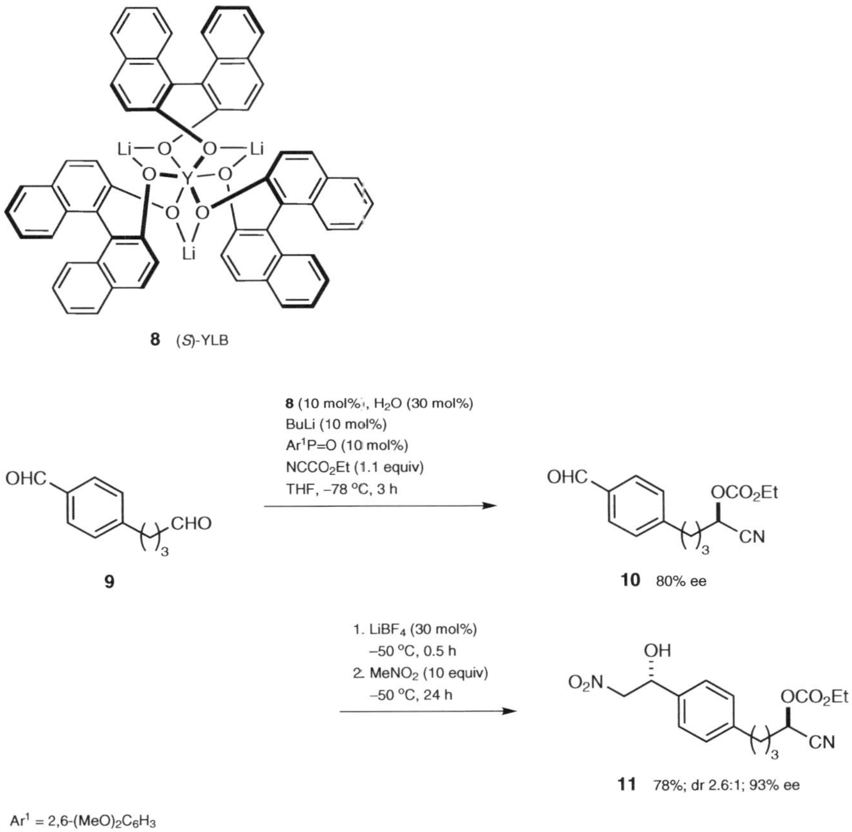

8 (S)-YLB

$Ar^1 = 2,6\text{-}(MeO)_2C_6H_3$

(1S,2S,3S)-1-(4-Chlorophenyl)-2-methyl-3-[4-(trifluoromethyl)phenyl]propane-1,3-diol (7, $Ar^1 = 4\text{-}F_3CC_6H_4$; $R^1 = Me$; $Ar^2 = 4\text{-}ClC_6H_4$); Typical Procedure:[5]

A soln of $La(OTf)_3/(R)$-BINOL/BuLi (1:3:5.6) complex in THF (0.2 M) was prepared as follows: To a suspension of (R)-BINOL (429 mg, 1.5 mmol) and $La(OTf)_3$ (293 mg, 0.5 mmol) in anhyd THF (2.5 mL) at 4°C was slowly added a 1.6 M soln of BuLi in hexane (1.75 mL, 2.8 mmol) over 2 min. The ice/water bath was then removed and the mixture was allowed to stir for 30 min at rt. The mixture was cooled to −78°C and then concentrated slowly via a needle attached to a vacuum pump. After 1 min, the mixture was slowly allowed (over the course of 30 min) to warm to rt under vacuum. When the mixture became a dry solid, the vessel was connected directly to vacuum. After 30 min, the vacuum was released with argon gas and anhyd THF (2.5 mL) was added to make a 0.2 M (based on La) stock soln of the catalyst.

A portion of the 0.2 M stock soln (0.25 mL, 0.05 mmol) was slowly added to a mixture of 1-[4-(trifluoromethyl)phenyl]propan-1-one (**4**, $Ar^1 = 4\text{-}F_3CC_6H_4$; $R^1 = Me$; 101 mg, 0.50 mmol) and 4-chlorobenzaldehyde **5** ($Ar^2 = 4\text{-}ClC_6H_4$; 170 mg, 1.25 mmol) in THF (0.25 mL) at 4°C. After stirring for 60 h at rt, the reaction was quenched by the addition

for references see p 599

of 1 M aq HCl (2 mL). The aqueous layer was extracted with EtOAc (2 × 10 mL). The combined organic layers were washed with brine, dried (Na_2SO_4), filtered, and concentrated under reduced pressure. The residue was purified by column chromatography (silica gel, hexane/Et_2O 5:1 to 4:1) to afford aldol/Tishchenko ester **6** as a colorless oil. The product was dissolved in MeOH (1 mL), NaOMe (27 mg, 0.5 mmol) was added, and the resulting mixture was stirred for 1 h at rt. The mixture was poured into brine and extracted with EtOAc. The combined organic layers were dried (Na_2SO_4), filtered, and concentrated under reduced pressure. The residue was purified by column chromatography (silica gel, hexane/EtOAc 10:1 to 8:1) to afford diol **7** as a single diastereomer; yield: 185 mg (95%, 2 steps); 93% ee.

1.6.4.1.2 1,4-Addition Reactions

A nucleophilic reagent can be exchanged for a labile ligand within a lanthanide complex to generate another nucleophilic complex, such as a lanthanide cyanide or azide complex, which can work as a highly nucleophilic catalyst. Based on the nature of the lanthanide complex, considering properties such as moderate Lewis acidity, multifunctionality, large coordination numbers, and fast ligand-exchange ability, subsequent addition of nucleophilic reagents would alter the structure and function of the lanthanide complex in situ through dynamic ligand exchange to promote different reactions successively. Indeed, lanthanide–BINOL complexes appear to be suitable catalysts.[10] Addition of an alkyl peroxide (R^1OOH) to lanthanide–BINOL complexes induces an exchange of the alkoxide ligand on the lanthanide metal to generate an active lanthanide–peroxide complex [e.g., (BINOL)Sm—OOR^1], a species which efficiently catalyzes an initial highly enantioselective epoxidation of α,β-unsaturated amides **12** (Scheme 4). The subsequent addition of another reagent, azidotrimethylsilane, generates a highly reactive nucleophilic lanthanide–azide complex in situ through dynamic ligand exchange (e.g., $X_2Sm—N_3$; X_2 = BINOL or other ligands); this complex catalyzes a regioselective epoxide opening in one reaction vessel to afford a variety of *anti*-β-azido-α-hydroxyamides **13** (20 examples) in good to excellent yield (70–99%) and nearly perfect enantioselectivity (96 to >99% ee). Reactions using trimethyl(phenylsulfanyl)silane or benzenethiol instead of azidotrimethylsilane also proceed to give *anti*-α-hydroxy-β-(phenylsulfanyl)amides in good to excellent yield (70–93%), high regioselectivity [ratio (β-SPh/α-SPh) 4:1 to >49:1], and near perfect to perfect enantioselectivity (96 to >99% ee).

Scheme 4 One-Pot, Sequential Epoxidation/Regioselective Epoxide Opening Process[10]

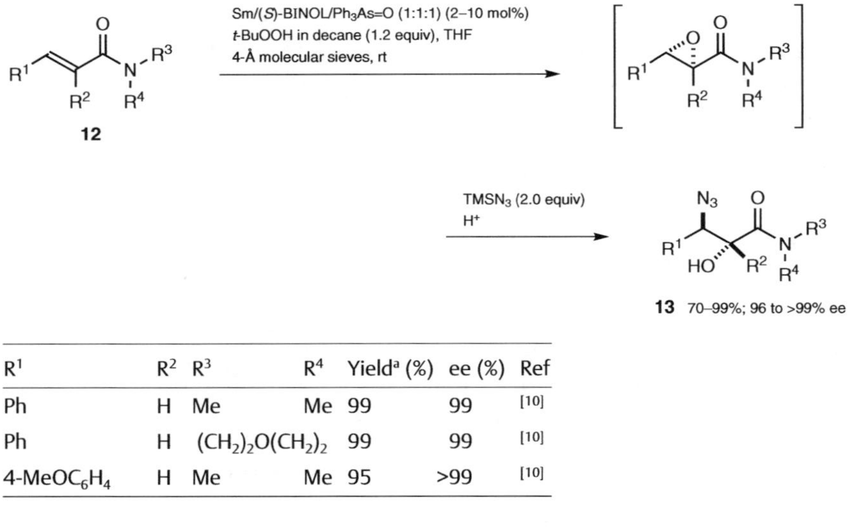

R^1	R^2	R^3	R^4	Yield[a] (%)	ee (%)	Ref
Ph	H	Me	Me	99	99	[10]
Ph	H	$(CH_2)_2O(CH_2)_2$	99	99	[10]
4-MeOC$_6$H$_4$	H	Me	Me	95	>99	[10]

1.6.4 Rare Earth Metal Mediated Domino Reactions

R^1	R^2	R^3	R^4	Yield[a] (%)	ee (%)	Ref
4-FC$_6$H$_4$	H	Me	Me	98	>99	[10]
2-naphthyl	H	Me	Me	99	>99	[10]
(E)-CH=CHPh	H	Me	Me	90	>99	[10]
(CH$_2$)$_3$		H	Bn	97	96	[10]
(CH$_2$)$_2$Ph	H	Me	Me	84	98	[10]
(CH$_2$)$_2$Ph	H	(CH$_2$)$_2$O(CH$_2$)$_2$		92	98	[10]
Cy	H	Me	Me	75	99	[10]

[a] Isolated yield; the regioselectivity was generally greater than the detection limit by 500 MHz ^{1}H NMR spectroscopy (>98:2).

1.6.4.2 Addition to C=N or C=C–C=N as a Primary Step

1.6.4.2.1 Strecker-Type Reactions

The three-component condensation reaction of aldehydes, amines (or ammonia), and cyanides through imine formation, the so-called Strecker reaction, is one of the most important synthetic methods for the preparation of α-amino cyanides, materials which can be readily converted into the corresponding α-amino acids, including chiral ones. A variety of catalysts, such as Brønsted acid catalysts, organocatalysts, and Lewis acid catalysts, have been developed for promoting this type of reaction. Lewis acid catalysts, in general, are utilized with preformed imines as the substrate for cyanation, because water derived from the condensation of aldehydes and amines often decomposes or deactivates such Lewis acid catalysts. However, due to their high stability in the presence of water, rare earth(III) trifluoromethanesulfonates can be applied as Lewis acid catalysts for the three-component Strecker reaction. Indeed, the highly Lewis acidic ytterbium(III) trifluoromethanesulfonate is a very efficient catalyst for the one-pot, three-component condensation reaction of cyclohexanone, α-amino esters **14** (derived from six natural α-amino acids), and trimethylsilyl cyanide to afford α,α-disubstituted α-amino cyanides **15** (8 examples reported) (Scheme 5).[11]

Later, this domino reaction was further coupled with another Brønsted acid catalyzed cyclization to afford tetracyclic indole derivatives **17** after treatment of intermediates **16** with phosphoric acid.[12]

Scheme 5 Three-Component Strecker Reactions[11,12]

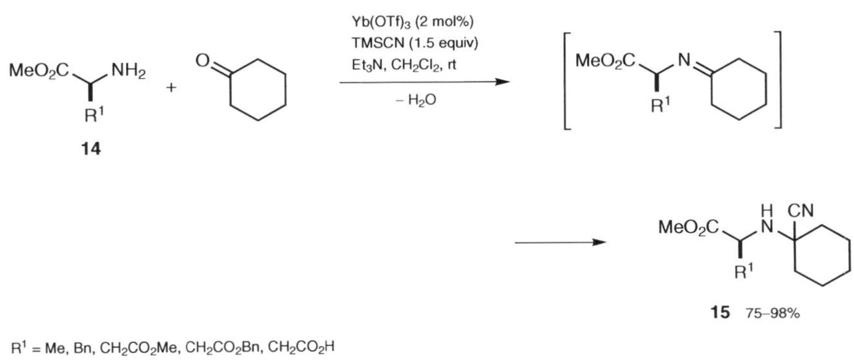

R^1 = Me, Bn, CH$_2$CO$_2$Me, CH$_2$CO$_2$Bn, CH$_2$CO$_2$H

for references see p 599

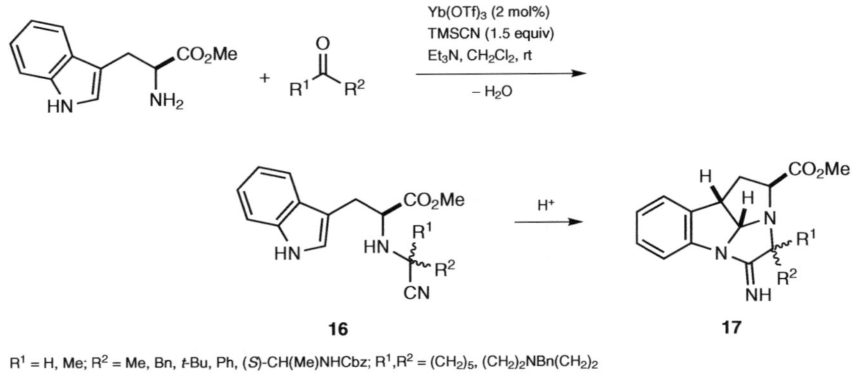

Domino Transformations 1.6 Metal-Mediated Reactions

$R^1 = H, Me; R^2 = Me, Bn, \textit{t}-Bu, Ph, (\textit{S})-CH(Me)NHCbz; R^1,R^2 = (CH_2)_5, (CH_2)_2NBn(CH_2)_2$

(S)-N-(1-Cyanocyclohexyl)amino Acid Methyl Esters 15; General Procedure:[11]

> **CAUTION:** *Trimethylsilyl cyanide and its hydrolysis products are extremely toxic.*

Et_3N (1 mmol) was added to a soln of amino acid derivative **14** (1 mmol) in anhyd CH_2Cl_2 (20 mL). After 30 min of stirring at rt under argon, cyclohexanone (103.6 µL, 1 mmol) was added, and stirring was maintained for 3 h. Then, the soln was cooled to $-23\,°C$ and $Yb(OTf)_3$ (13.6 mg, 0.02 mmol) and TMSCN (1.5 mmol) were added. After being stirred for 1 h at $-23\,°C$ and 20 h at rt, the mixture was concentrated to dryness. The residue was dissolved in CH_2Cl_2 (20 mL), and the soln was washed with H_2O (5 mL) and brine (5 mL), dried (Na_2SO_4), and concentrated. The residue was purified by flash chromatography (silica gel, EtOAc/hexane 1:4 to 1:1).

1.6.4.2.2   Other Reactions Initiated by Imine Formation

Because of the high reactivity of imines and iminium cations generated from aldehydes and amines, their formation can trigger various types of reactions. For example, lanthanide trifluoromethanesulfonates catalyze the reaction of furan-2-carbaldehyde (**18**) and amines **19** to result in the formation of *trans*-4,5-diaminocyclopent-2-enones **20** (Scheme 6).[13] The use of 10 mol% dysprosium(III) trifluoromethanesulfonate (secondary amines) or scandium(III) trifluoromethanesulfonate (primary anilines) successfully affords the desired diamine compounds **20** in high yield. The reaction likely proceeds through a pathway involving iminium cation formation, nucleophilic attack of the amine at the 5-position, furan ring opening to generate a linear zwitterion intermediate, and Nazarov-type thermal conrotatory 4π-electrocyclization.

1.6.4 Rare Earth Metal Mediated Domino Reactions **585**

Scheme 6 Domino Condensation/Ring-Opening/Electrocyclization Process[13]

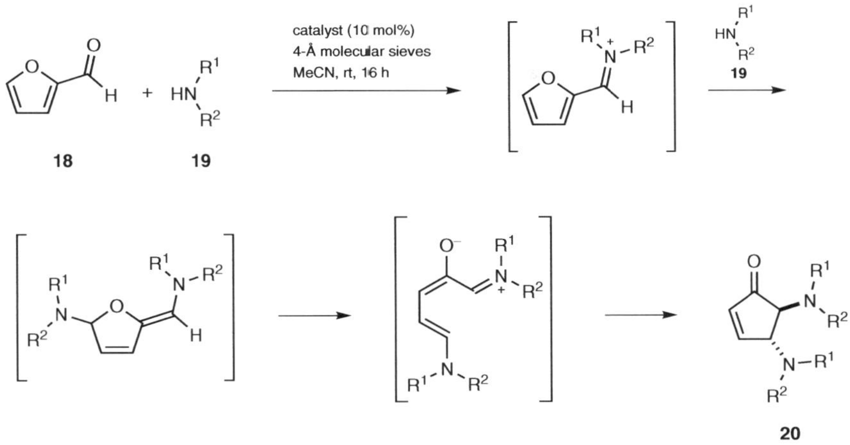

R¹	R²	Catalyst	Yield (%)	Ref
(CH₂)₂O(CH₂)₂		Dy(OTf)₃	quant	[13]
CH₂CH=CH₂	CH₂CH=CH₂	Dy(OTf)₃	82	[13]
Bn	Bn	Dy(OTf)₃	98	[13]
PMB	PMB	Dy(OTf)₃	quant	[13]
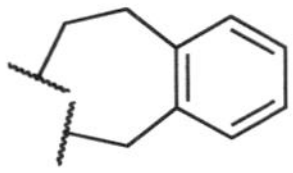 (ring structure)		Dy(OTf)₃	92	[13]
Me	Ph	Dy(OTf)₃	78	[13]
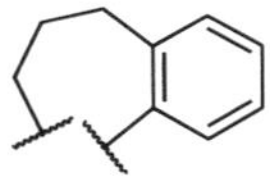 (ring structure)		Dy(OTf)₃	81	[13]
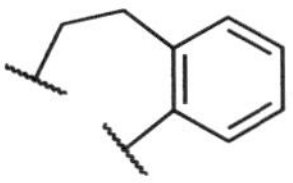 (ring structure)		Dy(OTf)₃	99	[13]
Ph	H	Sc(OTf)₃	78	[13]
2-MeOC₆H₄	H	Sc(OTf)₃	52	[13]

In the presence of lanthanum(III) trifluoromethanesulfonate, amine-substituted nickel porphyrins **21** react with cyclic enol ethers **22** [2,3-dihydrofuran ($n = 1$) or 3,4-dihydro-2H-pyran ($n = 2$)] to give pyrido[2,3-b]porphyrins **23** bearing two vicinal hydroxyalkyl groups (Scheme 7).[14] In this reaction, the cyclic enol ether **22** first reacts with water to generate a cyclic hemiacetal, and the following condensation reaction with amine **21** then generates the imine intermediate. A subsequent hetero-Diels–Alder reaction between the imine and cyclic enol ether **22** proceeds with complete regioselectivity; treatment with 4-toluenesulfonic acid in methanol results in the formation of pyrido[2,3-b]porphyrins **23** in good to moderate yields for the whole domino process.

for references see p 599

Scheme 7 Domino Condensation/Hetero-Diels–Alder Process[14]

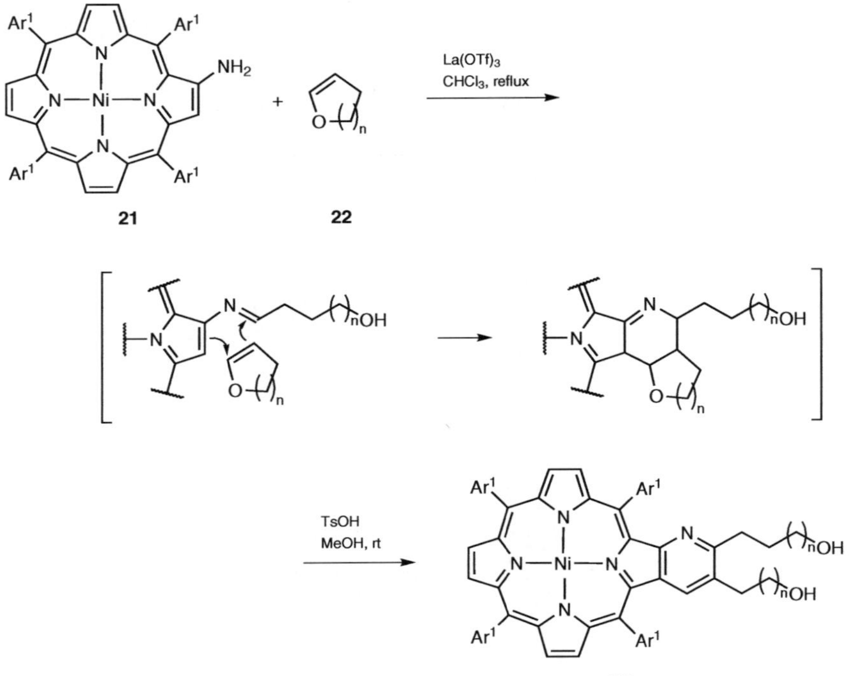

23 31–69%

Ar^1	n	Yield (%)	Ref
Ph	1	68	[14]
Ph	2	69	[14]
3-MeOC$_6$H$_4$	1	48	[14]
3-MeOC$_6$H$_4$	2	46	[14]
4-BrC$_6$H$_4$	1	62	[14]
4-BrC$_6$H$_4$	2	31	[14]
4-MeO$_2$CC$_6$H$_4$	1	50	[14]
4-MeO$_2$CC$_6$H$_4$	2	33	[14]

Quinoline is one of the ubiquitous heterocyclic scaffolds widely found in many natural products and pharmaceuticals. It is also frequently used as a building block in material science. Thus, a variety of synthetic methods for quinoline derivatives have been developed. A highly efficient synthesis of 2,4-disubstituted quinolines from N-arylprop-2-ynyl-amines using copper(I) chloride and silver(I) trifluoromethanesulfonate as catalysts has been reported.[15] Later, a three-component condensation reaction of aldehydes, aryl-amines, and terminal alkynes to afford quinolines **24** through a similar propynylamine intermediate was found to be catalyzed by ytterbium(III) (Scheme 8).[16,17]

1.6.4 Rare Earth Metal Mediated Domino Reactions

Scheme 8 Three-Component Synthesis of Substituted Quinolines[16,17]

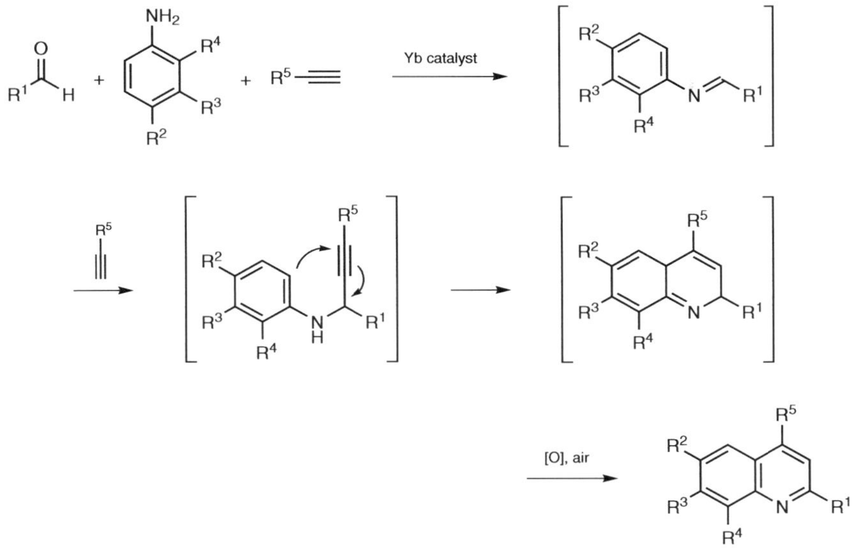

R¹	R²	R³	R⁴	R⁵	Conditions[a]	Yield (%)	Ref
3,4,5-(MeO)$_3$C$_6$H$_2$	H	H	H	Ph	Yb(OTf)$_3$ (10 mol%), [bmim]BF$_4$, microwave (80 W), 80 °C, 5.5 bar, 3 min	88	[16]
2-FC$_6$H$_4$	H	H	H	Ph	Yb(OTf)$_3$ (10 mol%), [bmim]BF$_4$, microwave (80 W), 80 °C, 5.5 bar, 3 min	76	[16]
4-O$_2$NC$_6$H$_4$	H	H	H	Ph	Yb(OTf)$_3$ (10 mol%), [bmim]BF$_4$, microwave (80 W), 80 °C, 5.5 bar, 3 min	89	[16]
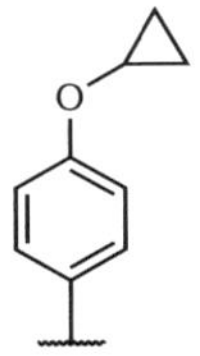 (cyclopropyloxy-phenyl)	Br	H	H	Ph	Yb(OTf)$_3$ (10 mol%), [bmim]BF$_4$, microwave (80 W), 80 °C, 5.5 bar, 3 min	91	[16]
3-ClC$_6$H$_4$	NO$_2$	H	H	Ph	Yb(OTf)$_3$ (10 mol%), [bmim]BF$_4$, microwave (80 W), 80 °C, 5.5 bar, 3 min	89	[16]
4-NCC$_6$H$_4$	H	H	H	4-Tol	Yb(OTf)$_3$ (10 mol%), [bmim]BF$_4$, microwave (80 W), 80 °C, 5.5 bar, 3 min	88	[16]
3,4,5-(MeO)$_3$C$_6$H$_2$	Br	H	H	4-O$_2$NC$_6$H$_4$	Yb(OTf)$_3$ (10 mol%), [bmim]BF$_4$, microwave (80 W), 80 °C, 5.5 bar, 3 min	88	[16]
Ph	H	H	H	Ph	Yb(OCOC$_6$F$_5$)$_3$ (2 mol%), neat, 80 °C, 12 h	88	[17]
Ph	OMe	H	H	Ph	Yb(OCOC$_6$F$_5$)$_3$ (2 mol%), neat, 80 °C, 12 h	90	[17]
Ph	Me	H	H	Ph	Yb(OCOC$_6$F$_5$)$_3$ (2 mol%), neat, 80 °C, 12 h	92	[17]
Ph	F	H	H	Ph	Yb(OCOC$_6$F$_5$)$_3$ (2 mol%), neat, 80 °C, 12 h	89	[17]
Ph	Cl	H	H	Ph	Yb(OCOC$_6$F$_5$)$_3$ (2 mol%), neat, 80 °C, 12 h	92	[17]
Ph	H	Cl	H	Ph	Yb(OCOC$_6$F$_5$)$_3$ (2 mol%), neat, 80 °C, 12 h	65	[17]
Ph	H	H	Cl	Ph	Yb(OCOC$_6$F$_5$)$_3$ (2 mol%), neat, 80 °C, 12 h	75	[17]
4-Tol	H	H	H	Ph	Yb(OCOC$_6$F$_5$)$_3$ (2 mol%), neat, 80 °C, 12 h	89	[17]
4-MeOC$_6$H$_4$	H	H	H	Ph	Yb(OCOC$_6$F$_5$)$_3$ (2 mol%), neat, 80 °C, 12 h	91	[17]

for references see p 599

588 Domino Transformations 1.6 Metal-Mediated Reactions

R¹	R²	R³	R⁴	R⁵	Conditions[a]	Yield (%)	Ref
4-ClC₆H₄	H	H	H	Ph	Yb(OCOC₆F₅)₃ (2 mol%), neat, 80 °C, 12 h	62	[17]
4-O₂NC₆H₄	H	H	H	Ph	Yb(OCOC₆F₅)₃ (2 mol%), neat, 120 °C, 12 h	70	[17]
2-O₂NC₆H₄	H	H	H	Ph	Yb(OCOC₆F₅)₃ (2 mol%), neat, 120 °C, 12 h	63	[17]
3-O₂NC₆H₄	H	H	H	Ph	Yb(OCOC₆F₅)₃ (2 mol%), neat, 120 °C, 12 h	trace	[17]

[a] [bmim] = 1-butyl-3-methylimidazolium.

Related reactions of ferrocenylacetylenes are also efficiently catalyzed by cerium(III) trifluoromethanesulfonate to give the corresponding ferrocene-containing (Fc) quinolines **25** (18 examples) in moderate to good yields (Scheme 9).[18]

Scheme 9 Three-Component Synthesis of Ferrocene-Containing Quinolines[18]

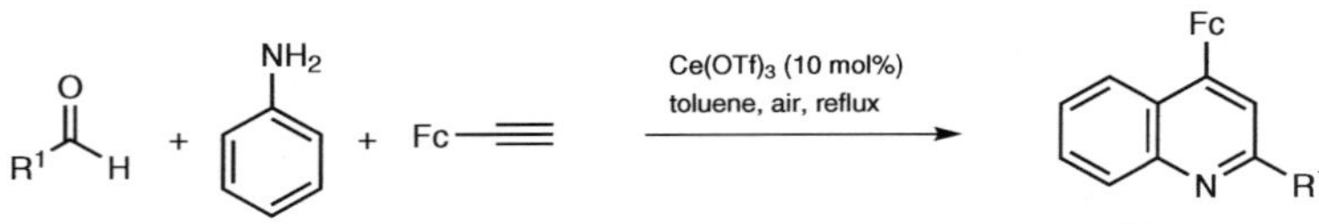

25 22–75%

R¹	Yield (%)	Ref
Ph	75	[18]
4-Tol	63	[18]
4-MeOC₆H₄	64	[18]
4-FC₆H₄	58	[18]
4-ClC₆H₄	69	[18]
2-ClC₆H₄	57	[18]
4-BrC₆H₄	75	[18]
3-O₂NC₆H₄	57	[18]
2-furyl	48	[18]
2-thienyl	53	[18]

Pyrazoles are also an important motif for pharmaceuticals and are found in several well-known medicines such as celecoxib and Viagra. Fully substituted pyrazoles **29** (14 examples) are efficiently synthesized by a scandium(III) trifluoromethanesulfonate catalyzed one-pot, three-component reaction of phenylhydrazine (**26**), aldehydes **27**, and ethyl acetoacetate (**28**) under solvent-free/microwave-irradiation conditions (Scheme 10).[19] Other Brønsted and Lewis acids also promote the reaction, but scandium(III) trifluoromethanesulfonate (5 mol%) was found to be the best. Hydrazine **26** can react with β-keto ester **28** to generate an enamine intermediate (see Section 1.6.4.3.1), but based on the position of the phenyl group in the pyrazole product, the reaction may proceed through initial hydrazone formation.

Scheme 10 Three-Component Domino Reaction of a Hydrazine, Aldehydes, and a β-Keto Ester To Give Pyrazoles[19]

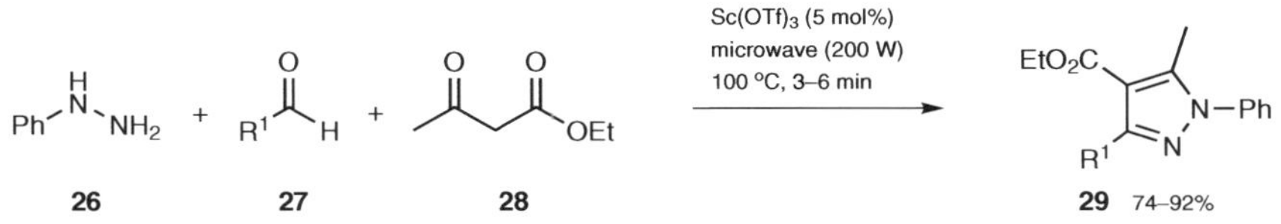

R^1	Yield (%)	Ref
Ph	83	[19]
4-BrC$_6$H$_4$	83	[19]
4-O$_2$NC$_6$H$_4$	88	[19]
3-O$_2$NC$_6$H$_4$	92	[19]
3-ClC$_6$H$_4$	86	[19]
3-BrC$_6$H$_4$	85	[19]
4-ClC$_6$H$_4$	84	[19]
4-FC$_6$H$_4$	90	[19]
2-furyl	82	[19]
2-thienyl	84	[19]

1.6.4.3 Enamine Formation as a Primary Step

1.6.4.3.1 Enamines from β-Keto Esters

A condensation reaction of aldehydes or ketones having an α-proton with secondary amines generates enamines via an iminium cation species. An enamine, of course, is the tautomeric form of an imine, and enamine–imine tautomerism is considered to be analogous to enol–keto tautomerism. Indeed, enamines behave in a similar way to enols, but are more reactive (nucleophilic), making them very useful chemical intermediates. When β-dicarbonyl compounds, such as diketones and keto esters, are used as a coupling partner, the reaction with even primary amines forms an enamine, primarily because the C=O bond can be in conjugation with the C=C bond in the enamine motif. For example, a three-component domino reaction between aromatic amines **30**, β-keto esters or thioesters **31**, and α,β-unsaturated aldehydes **32** is efficiently catalyzed by ammonium cerium(IV) nitrate to provide 1,4-dihydropyridines **33** (19 examples) in good yield (Scheme 11).[20] Ammonium cerium(IV) nitrate is normally employed as a one-electron oxidant for many types of carbon–carbon and carbon–heteroatom bond-forming reactions, but, based on several mechanistic studies, in this domino reaction it may behave as a Lewis acid.

The use of naphthoquinones **34** instead of α,β-unsaturated aldehydes **32** gives benzo[g]indoles **35** (17 examples).[21] Furthermore, when tryptamine (**36**) is used as the amine component, highly substituted indolo[2,3-a]quinolizines **37** (16 examples) are obtained in high yield in a process in which two rings are generated in a single synthetic operation through the creation of two C—C and two C—N bonds.[22] The reaction is highly diastereoselective and affords the product exclusively with a *trans* relationship between the protons at C2 and C12b. Although one cannot rule out the possibility that the reaction proceeds through initial imine formation between the amine and the α,β-unsaturated carbonyl compound (see Section 1.6.4.2.2), it is more likely that the mechanism follows an initial Michael addition of the in situ generated enaminone to the α,β-unsaturated carbonyl compound, followed by a nucleophilic cyclization of the amine in the enaminone to form a C—N bond.

for references see p 599

Scheme 11 Three-Component Domino Reactions of Amines, β-Dicarbonyl Compounds, and α,β-Unsaturated Carbonyl Compounds[20–22]

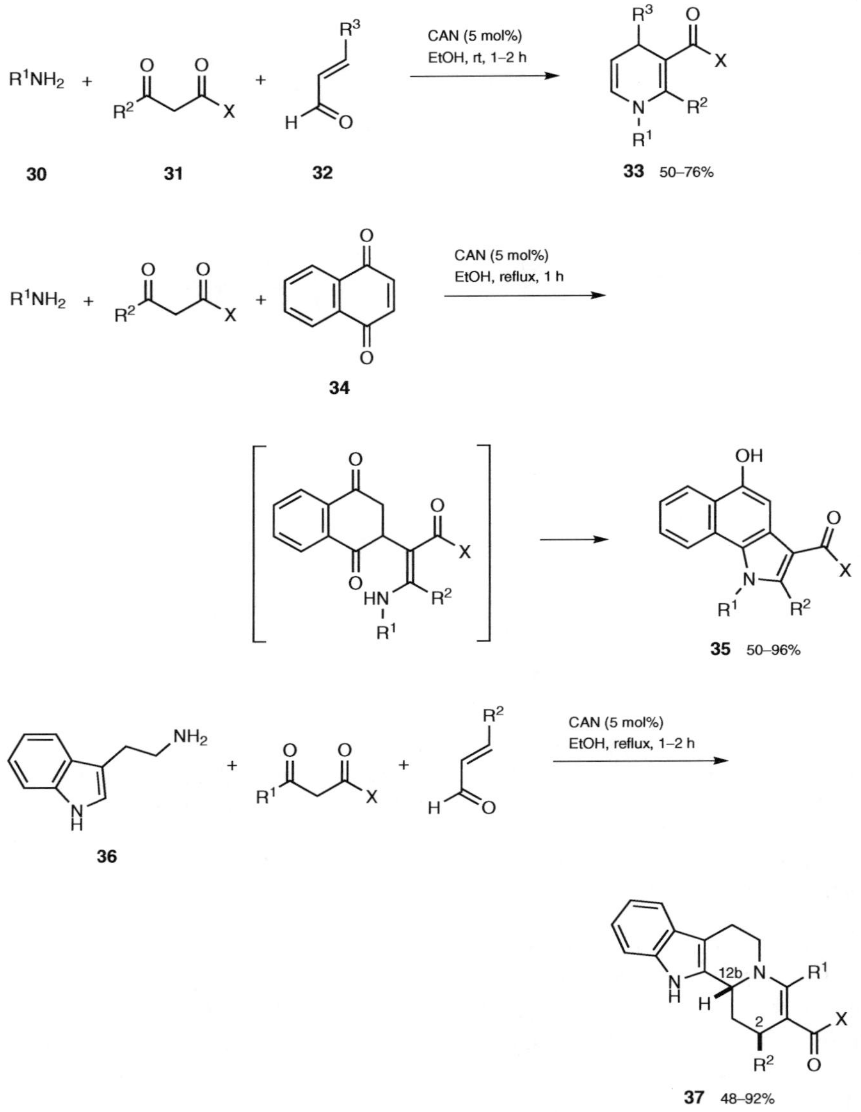

1.6.4.3.2 Enamines from Alkynes

An alternative approach to access enamines is through the hydroamination of alkynes. Hydroamination of highly electron-deficient alkynes, acetylenedicarboxylates **38**, with amines (mainly aniline derivatives) is efficiently catalyzed by ytterbium(III) trifluoromethanesulfonate; subsequent reaction with two equivalents of formaldehyde affords 3,4,5-trisubstituted 3,6-dihydro-2H-1,3-oxazines **39** (16 examples) in good yield (Scheme 12).[23] In this reaction, the hard properties of Yb³⁺ as a Lewis acid may result in a significant enhancement of the electrophilicity of the intermediates.

1.6.4 Rare Earth Metal Mediated Domino Reactions

591

Scheme 12 Three-Component Synthesis of 3,4,5-Trisubstituted 3,6-Dihydro-2H-1,3-oxazines[23]

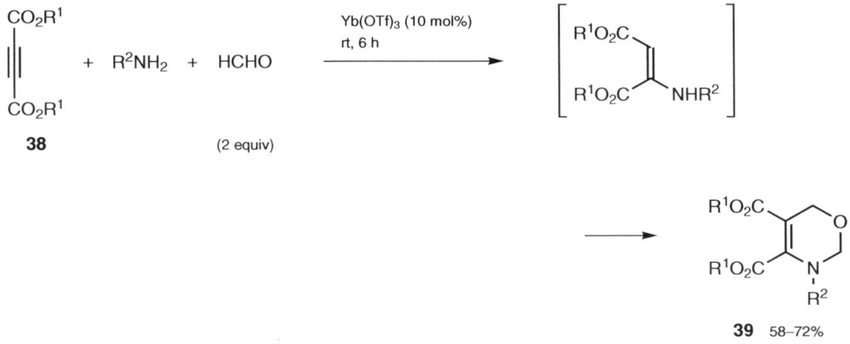

R^1 = Me, Et; R^2 = Ph, 4-O_2NC_6H$_4$, 4-MeOC_6H$_4$, 4-BrC_6H$_4$, 2-MeOC_6H$_4$, 2-O_2NC_6H$_4$, Bu, Cy

Pyrroles are important heterocycles because they are key skeletal components of many natural products including hemoglobin, the chlorophylls, and vitamin B_{12}. Once again, enamines are efficient intermediates for the synthesis of substituted pyrroles by transition-metal-catalyzed C—H activation of enamines with alkenes or alkynes. Alternatively, 1,2-diketones can be used as an alkyne equivalent; a four-component domino reaction of β-keto esters **40** (2 equivalents), primary alkylamines **41**, and α-diketones **42** is efficiently catalyzed by cerium(III) chloride heptahydrate, through activation of allylic $C(sp^3)$—H, vinylic $C(sp^2)$—H, and N—H bonds of an enamine ester intermediate, to afford pentasubstituted pyrroles **43** (8 examples) in high yield (Scheme 13).[24] Interestingly, most Lewis acids tested for this reaction, including ammonium cerium(IV) nitrate (see Section 1.6.4.3.1), do not give the desired products, while the combination of cerium(III) chloride heptahydrate with magnesium sulfate gives better results than cerium(III) chloride alone. The reaction may proceed through initial enamine formation from the β-keto ester and the primary amine, followed by a condensation with the α-diketone to generate an exocyclic alkene and finally 1,4-addition of a second β-keto ester. In these events, cerium(III) chloride may promote enamine formation, magnesium sulfate may act as a desiccant, and potassium iodide may have an important role in the deprotonation/protonation processes. By changing substrates and modifying the reaction procedure, three- and seven-component reactions also proceed smoothly. For example, when arylamines and diaryl-substituted 1,2-diketones are used as the substrates, N,N-bis[(1H-pyrrol-2-yl)methyl]amines **44** (7 examples), bearing seven aromatic moieties, are obtained in good yield.

for references see p 599

592   Domino Transformations   **1.6** Metal-Mediated Reactions

Scheme 13   Four- and Seven-Component Synthesis of Pentasubstituted Pyrroles[24]

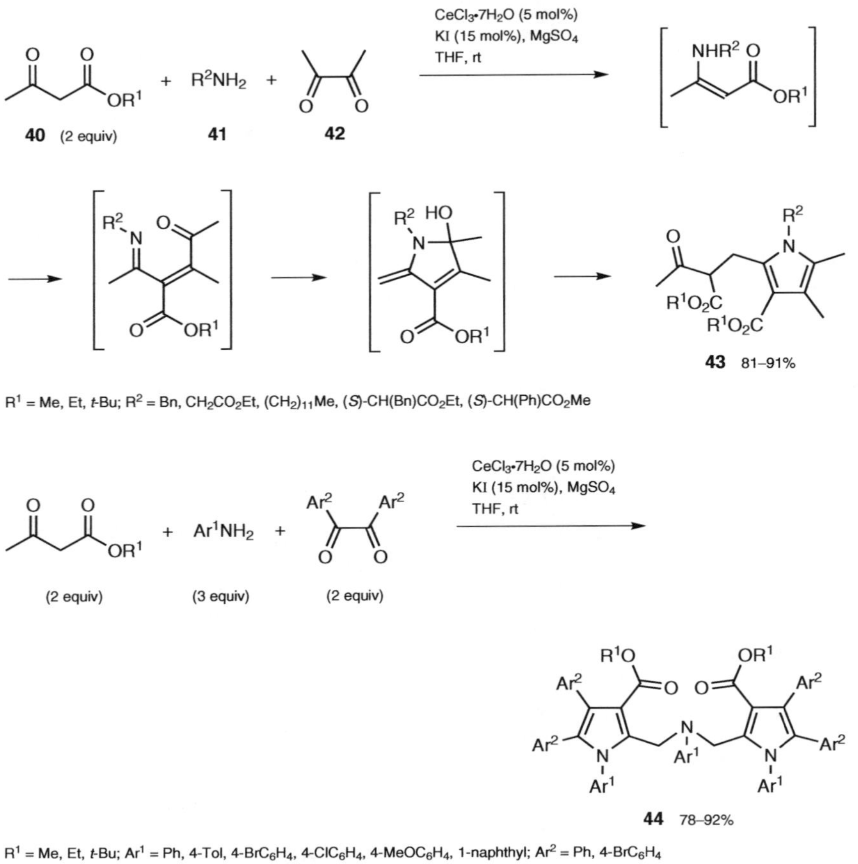

R^1 = Me, Et, t-Bu; R^2 = Bn, CH_2CO_2Et, $(CH_2)_{11}Me$, (S)-CH(Bn)CO$_2$Et, (S)-CH(Ph)CO$_2$Me

R^1 = Me, Et, t-Bu; Ar^1 = Ph, 4-Tol, 4-BrC$_6$H$_4$, 4-ClC$_6$H$_4$, 4-MeOC$_6$H$_4$, 1-naphthyl; Ar^2 = Ph, 4-BrC$_6$H$_4$

Alkyl 2-[2-(Alkoxycarbonyl)-3-oxobutyl]-4,5-dimethyl-1H-pyrrole-3-carboxylates 43;
General Procedure:[24]

β-Keto ester **40** (1.25 mmol), aliphatic amine **41** (1.25 mmol), CeCl$_3$•7H$_2$O (5 mol%), KI (100 mg, 15 mol%), and anhyd MgSO$_4$ in dry THF (30 mL) were stirred for 24 h at rt. 1,2-Diketone **42** (1 mmol) and β-keto ester **40** (1.25 mmol) were added and the resultant mixture was stirred at rt for 43–48 h. After removal of the solvent, the mixture was diluted with EtOAc (25 mL). The organic layer was washed with H$_2$O (3 × 10 mL), dried (Na$_2$SO$_4$), and concentrated under reduced pressure in a rotary evaporator at rt. The residue was then subjected to column chromatography [silica gel (60–120 mesh), petroleum ether/EtOAc 97:3] to afford fully substituted pyrrole **43**.

1.6.4.4   **Ring-Opening or Ring-Closing Reactions as a Primary Step**

1.6.4.4.1   **Ring-Opening Reactions**

Highly strained three-membered-ring compounds, such as cyclopropanes, epoxides, and aziridines, undergo a variety of ring-opening reactions because the relief of ring strain provides a potent thermodynamic driving force. Therefore, such small rings have been widely utilized as substrates in various domino reactions based on this property.

In the presence of scandium(III) trifluoromethanesulfonate (20 mol%), the reaction of 2-formylcyclopropane-1,1-dicarboxylates **45** and 1,3-diketones **46**, including cyclohex-

ane-1,3-dione, affords highly functionalized furans **47** (17 examples) through aldol/homo-Michael reactions (Scheme 14).[25] Several Lewis acids were tested, but scandium(III) trifluoromethanesulfonate gives the best results. The key to this reaction process is an intramolecular S_N2-type ring-opening reaction of the cyclopropane ring by the carbonyl oxygen of the aldol products derived from **45** and **46**.

Scheme 14 Domino Aldol and Ring-Opening Reaction[25]

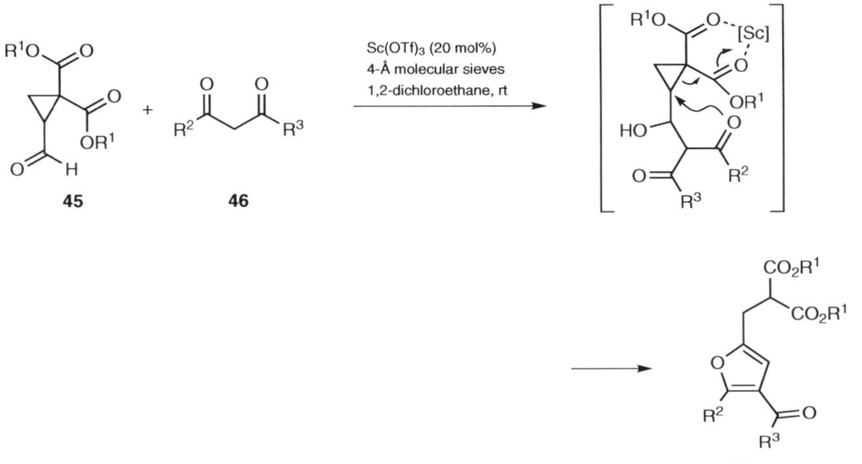

47 28–100%

R¹	R²	R³	Time (h)	Yield (%)	Ref
Et		(CH₂)₃	40	84	[25]
Me		(CH₂)₃	40	67	[25]
iPr		(CH₂)₃	40	61	[25]
Et	Me	Me	18	90	[25]
Et	Ph	Ph	40	56	[25]
Et	Ph	OEt	12	99	[25]
Me	Me	OMe	18	86	[25]
Et	CH₂CO₂Et	OEt	14	28	[25]

Aziridines have also been utilized as a building block for the synthesis of various nitrogen-containing heterocyclic compounds. For example, the chiral, substituted 4,5-dihydro-1H-pyrroles **49** (16 examples) are efficiently synthesized through an S_N2-type ring-opening reaction of the aziridine ring of **48** by the carbanion derived from malononitrile, followed by intramolecular cyclization (Scheme 15).[26] The best result was obtained using scandium(III) trifluoromethanesulfonate as the Lewis acid and potassium *tert*-butoxide as the base. When optically pure aziridines are used, the desired 4,5-dihydropyrrole products are obtained in high yield without any loss of enantiopurity. Not only monosubstituted aziridines, but also 2,3-disubstituted aziridines are amenable to this domino reaction. N-(4-Nitrophenylsulfonyl)aziridines are more reactive than N-tosylaziridines, and the 4-nitrophenylsulfonyl group of the product can be easily removed by a combination of potassium carbonate and benzenethiol to afford the corresponding N-unsubstituted product. Later, similar domino reactions using donor–acceptor cyclopropanes with malononitrile were developed, affording functionalized cyclopentene derivatives in good yield.[27]

for references see p 599

594　　Domino Transformations　**1.6** Metal-Mediated Reactions

Scheme 15　Domino Ring-Opening Reaction and Intramolecular Cyclization[26]

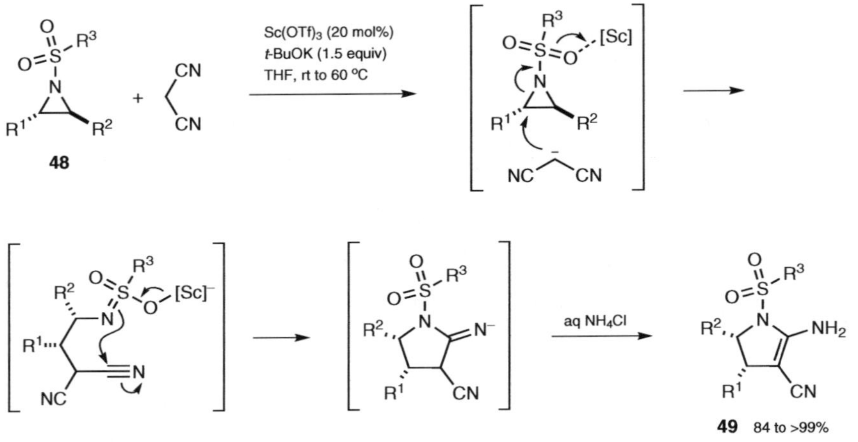

R[1] = H, Ph, 4-ClC$_6$H$_4$, 3-ClC$_6$H$_4$, 4-BrC$_6$H$_4$, 4-FC$_6$H$_4$, 4-Tol, 4-t-BuC$_6$H$_4$, 4-AcOC$_6$H$_4$;
R[2] = H, Me, iPr, Bn, Pr, CH$_2$CH=CH$_2$, CH$_2$OTBDMS; R[3] = 4-O$_2$NC$_6$H$_4$, 4-Tol, 4-t-BuC$_6$H$_4$

2-Amino-1-(arylsulfonyl)-4,5-dihydro-1*H*-pyrrole-3-carbonitriles 49; General Procedure:[26]
To a suspension of *t*-BuOK (1.5 equiv) in THF (2.0 mL) was added malononitrile (1.5 equiv) at rt under a N$_2$ atmosphere. Solutions of *N*-sulfonylaziridine **48** (1.0 equiv) and Sc(OTf)$_3$ (20 mol%) in THF were then added sequentially to the mixture. The mixture was stirred at 60 °C until completion of the reaction (monitored by TLC). Once complete, the reaction was quenched with sat. aq NH$_4$Cl (1.0 mL). After separating the organic phase, the aqueous phase was extracted with EtOAc (3 × 1.0 mL), and the combined organic extracts were washed with brine and dried (Na$_2$SO$_4$). After removal of the solvent under reduced pressure, the crude mixture was purified by flash column chromatography [silica gel (230–400 mesh), 20–25% EtOAc in petroleum ether] to give pure 4,5-dihydro-1*H*-pyrroles **49**.

1.6.4.4.2　Ring-Closing Reactions

The intramolecular Friedel–Crafts process is one of the most effective methods for the construction of polycyclic aromatic molecular architectures from relatively simple starting materials. For example, a variety of tricyclic compounds **51** can be efficiently synthesized from ω-aryl propargyl tosylates **50** through a double C—H functionalization of the arene and a selective 1,2-alkyl migration (Scheme 16).[28] When *N*-tosyl-tethered substrates (X = NTs) are used, both [1,3-bis(2,6-diisopropylphenyl)imidazol-2-ylidene](chloro)gold(I)/silver(I) tetrafluoroborate and scandium(III) trifluoromethanesulfonate give satisfactory results. By contrast, the gold catalyst system is not effective for oxygen- or carbon-tethered compounds **50** (X = O, CH$_2$), affording a complex mixture instead, with only scandium(III) trifluoromethanesulfonate catalyzing these reactions smoothly. The reaction may proceed via: (1) activation of the alkyne upon coordination to the electrophilic scandium(III) ion; (2) tosylate migration with a concomitant 1,2-shift of the R^2 group; (3) the first Friedel–Crafts-type reaction; (4) elimination of the 4-toluenesulfonyloxy group to generate a cationic intermediate; and (5) the second Friedel–Crafts-type reaction.

1.6.4 Rare Earth Metal Mediated Domino Reactions **595**

Scheme 16 Domino C–H Funtionalization of Arenes[28]

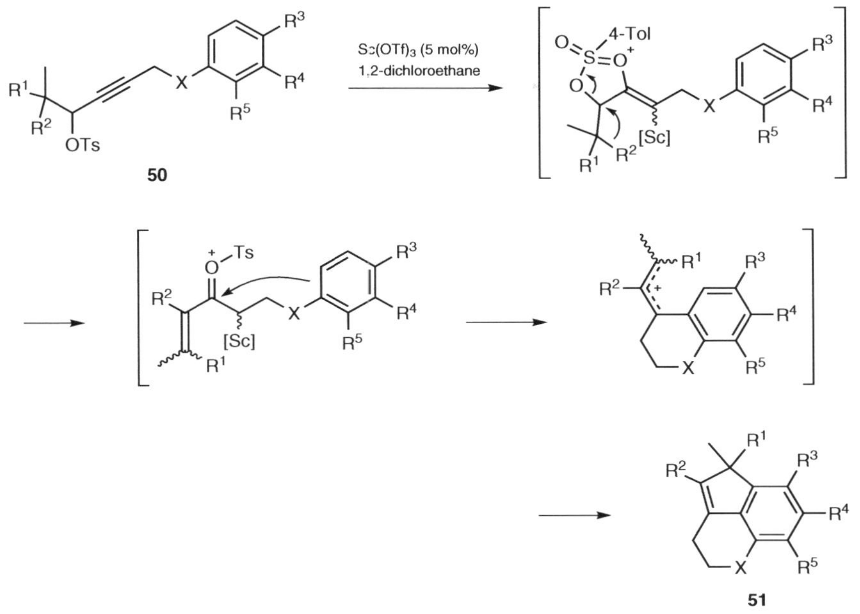

R¹	R²	X	R³	R⁴	R⁵	Conditions	Yield (%)	Ref
Me	Me	O	H	H	H	60°C, 40 min	60	[28]
Me	Me	CH₂	H	H	H	60°C, 30 min	66	[28]
(CH₂)₄		NTs	H	H	H	60°C, 40 min	61	[28]
(CH₂)₄		CH₂	H	H	H	60°C, 5 min	80	[28]
Me	Ph	NTs	Me	H	H	25°C, 3 h	92	[28]
Me	Me	NTs	F	H	H	60°C, 40 h	62	[28]
Me	Me	NTs	I	H	H	60°C, 32 h	54	[28]
Me	Me	NTs	Me	H	H	60°C, 22 h	68	[28]
Me	Me	NTs	CN	H	H	60°C, 24 h	60	[28]
Me	Me	O	Ph	H	H	60°C, 5 h	77	[28]
Me	Me	O	NO₂	H	H	60°C, 35 h	56	[28]
Me	Me	O	H	H	Ph	60°C, 24 h	70	[28]
Me	Me	O	H	H	Br	60°C, 5 h	91	[28]
Me	Me	CH₂	Br	H	H	60°C, 6 min	87	[28]
Me	Me	O	Me	Me	H	60°C, 21 h	36	[28]

Xanthenes are also important heteroaromatic compounds because of their diverse biological properties and use as dyes and fluorescent materials. For the synthesis of functionalized 9-substituted xanthene derivatives **54** (13 examples), a scandium(III) trifluoromethanesulfonate catalyzed domino intermolecular and intramolecular Friedel–Crafts-type arylation strategy was employed (Scheme 17).[29] Although other Lewis and Brønsted acids, such as iron(III) chloride, aluminum trichloride, and trifluoromethanesulfonic acid, also promote the reaction, they require strictly inert atmosphere conditions because they immediately react with moisture rather than the substrates. By contrast, utilization of scandium(III) trifluoromethanesulfonate makes this process convenient to handle,

for references see p 599

596   Domino Transformations **1.6** Metal-Mediated Reactions

and, under normal atmospheric conditions, 10 mol% of the catalyst is sufficient to obtain the desired compound in good yield. This domino process may include a double arylation reaction of aldehydes **52** with electron-rich arenes **53** (intermolecular Friedel–Crafts-type arylation) to generate triarylmethanes, with subsequent elimination of one aryl group and intramolecular Friedel–Crafts-type arylation completing the synthesis of **54**. In a similar way, 9-(arylsulfanyl)xanthenes can also be synthesized, using arenethiols instead of electron-rich arenes.

Scheme 17   Domino Inter- and Intramolecular C–H Functionalization of Arenes[29]

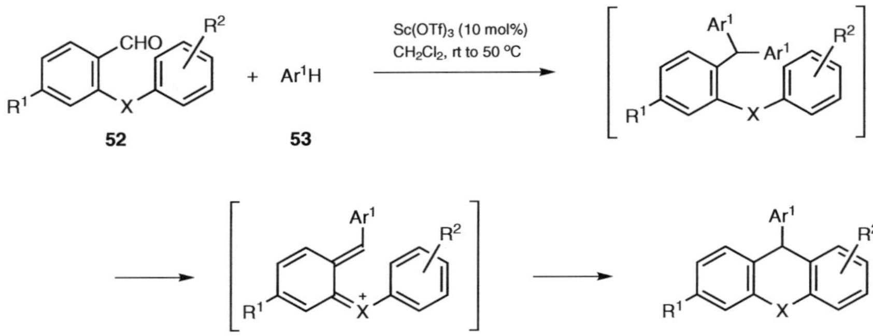

X = O, S

1.6.4.5   **Rearrangement Reactions**

Rearrangement reactions enable the alteration of a carbon skeleton to give a structural isomer of the original molecule, and thus are widely utilized for the syntheses of complex molecules. In fact, various sequential sigmatropic rearrangements have been developed by combinations of Claisen, Cope, oxy-Cope, or Ireland–Claisen rearrangements, etc. Bisphenolic products, such as **56** (8 examples), that incorporate two contiguous aryl—$C(sp^3)$ bonds have been synthesized in a highly stereocontrolled manner by the double aryl-Claisen/[3,3]-sigmatropic rearrangements of bis(aryl ethers) **55**. Although classical thermal conditions do not lead to the desired product, tris(6,6,7,7,8,8,8-heptafluoro-2,2-dimethyloctane-3,5-dionato)europium(III) [Eu(fod)$_3$] effectively promotes the reaction to give **56** in high yield with excellent chirality transfer (Scheme 18).[30] The corresponding mono-rearranged product was not observed, suggesting that the second rearrangement is faster than the first one and that both rearrangements are efficiently catalyzed by tris(6,6,7,7,8,8,8-heptafluoro-2,2-dimethyloctane-3,5-dionato)europium(III).

Scheme 18   Domino Aryl-Claisen/[3,3]-Sigmatropic Rearrangements[30]

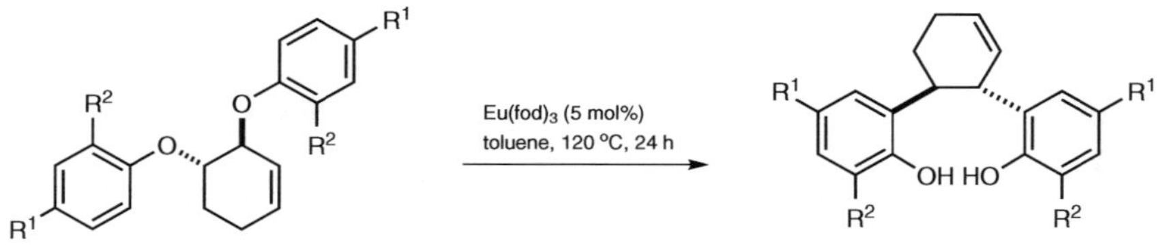

fod = 6,6,7,7,8,8,8-heptafluoro-2,2-dimethyloctane-3,5-dionato

1.6.4 Rare Earth Metal Mediated Domino Reactions

R^1	R^2	Yield (%)	Ref
H	H	92	[30]
SMe	H	94	[30]
OMe	H	96	[30]
F	H	96	[30]
NO$_2$	H	10	[30]
H	NO$_2$	27	[30]

Unique lanthanum(III) trifluoromethanesulfonate catalyzed domino syntheses of 1,4-benzoxazines **59** (X=O; 18 examples) from 2-furylcarbinols **57** and *o*-aminophenols **58** (X=O) have also been reported (Scheme 19).[31] A number of Lewis and Brønsted acids were examined, and only rare earth metal salts, trifluoromethanesulfonic acid, and iron(III) chloride showed good catalytic activity, with lanthanum (III) trifluoromethane-sulfonate being the best. The use of N-monoprotected benzene-1,2-diamines **58** (X=NTs, NMs, NAc; 4 examples) instead of *o*-aminophenols gives the 1,2,3,4-tetrahydroquinoxaline derivatives. The reaction mechanism is proposed to involve a Piancatelli rearrangement (transformation of the 2-furylcarbinol into a 4-hydroxycyclopentenone derivative through an acid-catalyzed rearrangement), C—N coupling, and Michael addition.

Scheme 19 Domino Piancatelli Rearrangement/C—N Coupling/Michael Addition[31]

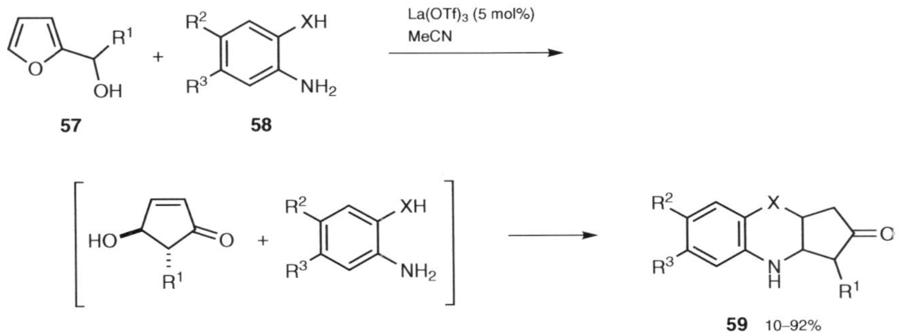

R^1	R^2	R^3	X	Conditions	Yield (%)	Ref
Ph	H	H	O	80 °C, 4 h	81	[31]
4-MeOC$_6$H$_4$	H	H	O	80 °C, 4 h	46	[31]
4-FC$_6$H$_4$	H	Cl	O	80 °C, 4 h	78	[31]
4-MeOC$_6$H$_4$	H	Me	O	80 °C, 4 h	60	[31]
Bu	NO$_2$	H	O	80 °C, 4 h	53	[31]
4-F$_3$CC$_6$H$_4$	NO$_2$	H	O	80 °C, 4 h	92	[31]
Ph	H	H	NTs	90 °C, 2 h	88	[31]
Ph	H	H	NMs	90 °C, 2 h	83	[31]
4-FC$_6$H$_4$	H	H	NTs	90 °C, 2 h	68	[31]

for references see p 599

598 Domino Transformations **1.6** Metal-Mediated Reactions

1.6.4.6 **Miscellaneous Reactions**

1.6.4.6.1 **Domino Reactions with Transition-Metal Catalysts**

Rare earth(III) trifluoromethanesulfonates are compatible with many transition-metal catalysts, and thus are employed as cocatalysts for transition-metal-catalyzed domino reactions (Scheme 20).[32–34] The key reactions of these domino processes are mainly catalyzed by the transition metals, not the rare earth(III) trifluoromethanesulfonate component. Therefore, although such reactions are included in this section for the sake of completeness, only reaction schemes are shown, without any detailed explanation.

Scheme 20 Domino Reactions Catalyzed by Transition-Metal Catalysts and Rare Earth(III) Trifluoromethanesulfonate Catalysts[32–34]

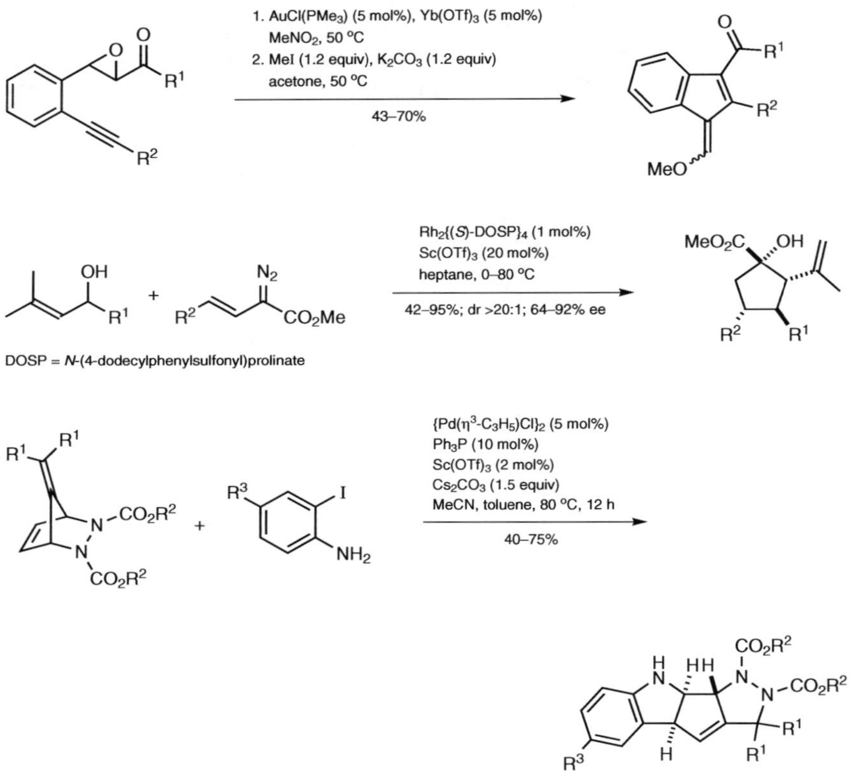

References

[1] Molander, G. A.; Romero, J. A. C., *Chem. Rev.*, (2002) **102**, 2161.
[2] Shibasaki, M.; Yoshikawa, N., *Chem. Rev.*, (2002) **102**, 2187.
[3] Kobayashi, S.; Sugiura, M.; Kitagawa, H.; Lam, W. W.-L., *Chem. Rev.*, (2002) **102**, 2227.
[4] Sasai, H.; Hiroi, M.; Yamada, Y. M. A.; Shibasaki, M., *Tetrahedron Lett.*, (1997) **38**, 6031.
[5] Gnanadesikan, V.; Horiuchi, Y.; Ohshima, T.; Shibasaki, M., *J. Am. Chem. Soc.*, (2004) **126**, 7782.
[6] Horiuchi, Y.; Gnanadesikan, V.; Ohshima, T.; Masu, H.; Katagiri, K.; Sei, Y.; Yamaguchi, K.; Shibasaki, M., *Chem.–Eur. J.*, (2005) **11**, 5195.
[7] Mlynarski, J.; Mitura, M., *Tetrahedron Lett.*, (2004) **45**, 7549.
[8] Mlynarski, J.; Rakiel, B.; Stodulski, M.; Suszczyńska, A.; Frelek, J., *Chem.–Eur. J.*, (2006) **12**, 8158.
[9] Tian, J.; Yamagiwa, N.; Matsunaga, S.; Shibasaki, M., *Angew. Chem. Int. Ed.*, (2002) **41**, 3636.
[10] Tosaki, S.-y.; Tsuji, R.; Ohshima, T.; Shibasaki, M., *J. Am. Chem. Soc.*, (2005) **127**, 2147.
[11] González-Vera, J. A.; García-López, M. T.; Herranz, R., *J. Org. Chem.*, (2005) **70**, 3660.
[12] González-Vera, J. A.; García-López, M. T.; Herranz, R., *J. Org. Chem.*, (2005) **70**, 8971.
[13] Li, S.-W.; Batey, R. A., *Chem. Commun. (Cambridge)*, (2007), 3759.
[14] Alonso, C. M. A.; Serra, V. I. V.; Neves, M. G. P. M. S.; Tomé, A. C.; Silva, A. M. S.; Paz, F. A. A.; Cavaleiro, J. A. S., *Org. Lett.*, (2007) **9**, 2305.
[15] Kuninobu, Y.; Inoue, Y.; Takai, K., *Chem. Lett.*, (2007) **36**, 1422.
[16] Kumar, A.; Rao, V. K., *Synlett*, (2011), 2157.
[17] Tang, J.; Wang, L.; Mao, D.; Wang, W.; Zhang, L.; Wu, S.; Xie, Y., *Tetrahedron*, (2011) **67**, 8465.
[18] Chen, S.; Li, L.; Zhao, H.; Li, B., *Tetrahedron*, (2013) **69**, 6223.
[19] Kumari, K.; Raghuvanshi, D. S.; Jouikov, V.; Singh, K. N., *Tetrahedron Lett.*, (2012) **53**, 1130.
[20] Sridharan, V.; Perumal, P. T.; Avendaño, C.; Menéndez, J. C., *Tetrahedron*, (2007) **63**, 4407.
[21] Suryavanshi, P. A.; Sridharan, V.; Menéndez, J. C., *Org. Biomol. Chem.*, (2010) **8**, 3426.
[22] Suryavanshi, P. A.; Sridharan, V.; Menéndez, J. C., *Chem.–Eur. J.*, (2013) **19**, 13 207.
[23] Epifano, F.; Pelucchini, C.; Rosati, O.; Genovese, S.; Curini, M., *Catal. Lett.*, (2011) **141**, 844.
[24] Dhara, D.; Gayen, K. S.; Khamarui, S.; Pandit, P.; Ghosh, S.; Maiti, D. K., *J. Org. Chem.*, (2012) **77**, 10441.
[25] Bao, J.; Ren, J.; Wang, Z., *Eur. J. Org. Chem.*, (2012), 2511.
[26] Ghorai, M. K.; Tiwari, D. P., *J. Org. Chem.*, (2013) **78**, 2617.
[27] Ghorai, M. K.; Talukdar, R.; Tiwari, D. P., *Chem. Commun. (Cambridge)*, (2013) **49**, 8205.
[28] Suárez-Pantiga, S.; Palomas, D.; Rubio, E.; González, J. M., *Angew. Chem. Int. Ed.*, (2009) **48**, 7857.
[29] Singh, R.; Panda, G., *Org. Biomol. Chem.*, (2010) **8**, 1097.
[30] Ramadhar, T. R.; Kawakami, J.-i.; Lough, A. L.; Batey, R. A., *Org. Lett.*, (2010) **12**, 4446.
[31] Liu, J.; Shen, Q.; Yu, J.; Zhu, M.; Han, J.; Wang, L., *Eur. J. Org. Chem.*, (2012), 6933.
[32] Dai, L.-Z.; Shi, M., *Chem.–Eur. J.*, (2010) **16**, 2496.
[33] Parr, B. T.; Li, Z.; Davies, H. M. L., *Chem. Sci.*, (2011) **2**, 2378.
[34] Chand, S. S.; Jijy, E.; Prakash, P.; Szymoniak, J.; Preethanuj, P.; Dhanya, B. P.; Radhakrishnan, K. V., *Org. Lett.*, (2013) **15**, 3338.

1.6.5   **Cobalt and Other Metal Mediated Domino Reactions:**
The Pauson–Khand Reaction and Its Use in Natural Product Total Synthesis

L. Shi and Z. Yang

General Introduction

Complex natural product total synthesis provides the ultimate test of reaction design and applicability. The selection of a powerful reaction for the total synthesis of specific target molecules with considerable structural and stereochemical complexity is both an intellectual challenge and a critical step in natural product synthesis. The Pauson–Khand reaction is a method used for the construction of a cyclopentenone skeleton from an alkene, an alkyne, and carbon monoxide based on a domino sequence of bond constructions. This reaction contributes immeasurably to both the science and art of total synthesis, resulting in improved practical efficiencies and an enhanced aesthetic appeal to synthetic planning.[1–3]

Developed in 1973 by Pauson and Khand at the University of Strathclyde in Glasgow, the original procedure involved using a stoichiometric amount of octacarbonyldicobalt(0) in a reaction with an alkyne, and then heating the hexacarbonyldicobalt–alkyne complex with an alkene to effect cyclization (Scheme 1). However, this reaction was usually carried out under harsh conditions with a low reaction yield, and this disadvantage initially retarded the development of further synthetic applications. More recently, with the discovery of the cobalt-mediated [2+2+1]-cyclization reaction, the Pauson–Khand reaction has undergone many improvements in terms of reactivity, selectivity, and catalysis.[4–14]

Scheme 1   General Pauson–Khand Reaction[1–3]

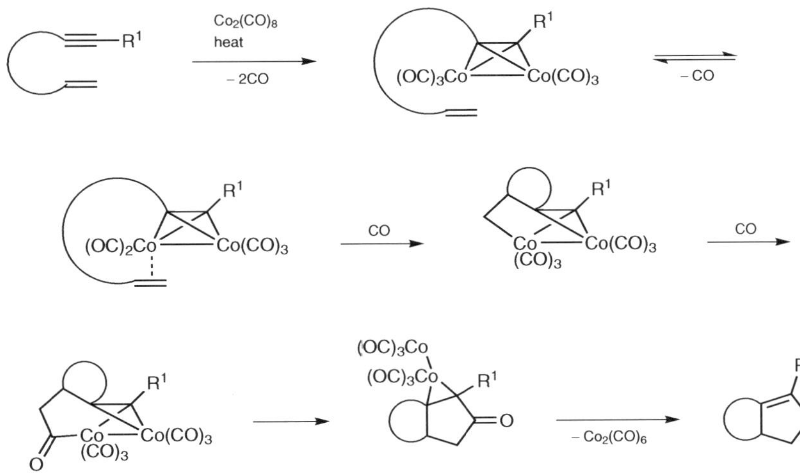

As one of the most powerful reactions in complex natural product total synthesis, the Pauson–Khand reaction is exceptionally tolerant of common organic functionality and of the local environment around the alkene and alkyne reaction partners. It can be used to dramatically increase molecular complexity when proceeding from starting material to product. Considerable efforts have been devoted to optimizing the reaction conditions, and

for references see p 628

several important improvements have been made when using transition-metal complexes based on titanium,[15] ruthenium,[16,17] rhodium,[18] and iridium[19] and additives such as tertiary amine N-oxides,[20–22] primary amines[23] [for a related paper that also describes a method for the in situ complexation of alkynes with cobalt(II) bromide, zinc, and carbon monoxide see ref[24]], and thiourea.[25,26] The reward has been the discovery of impressive catalytic activity. Asymmetric variants of the catalytic cycloaddition are now available,[27,28] while safer and more environmentally friendly procedures that avoid using toxic carbon monoxide are starting to appear. These improvements pave the way for the Pauson–Khand reaction to become one of the most important reactions in modern organic synthesis, and perhaps the most important cobalt-mediated domino process.

Given the impressive application of the Pauson–Khand reaction as an important step in complex natural product total synthesis, as well as the lack of a critical review of this topic over the last decade, we present here a personal selection of the examples we believe are the most interesting and relevant. These topics highlight the instrumental role of the Pauson–Khand reaction in the total synthesis of selected target molecules, and focus largely on cobalt-mediated processes, though a few examples achieved through the use of other promoters will be provided for didactic value and to give a sense of relative strengths and weaknesses.

The high value of the Pauson–Khand reaction in the synthesis of natural products and other complex compounds has been frequently demonstrated. For example, the synthesis of (+)-epoxydictymene (4),[29] which is a commercially important anti-inflammatory and analgesic agent,[30,31] shows the potential of the Pauson–Khand reaction in total synthesis.

The asymmetric total synthesis of (+)-epoxydictymene (4) started with the preparation of enyne 1 and then employed an intramolecular Nicholas reaction,[32] a Lewis acid mediated conversion of enyne 1, for the preparation of Pauson–Khand substrate 2. This substrate was then converted into the epoxydictymene precursor 3. The total synthesis of (+)-epoxydictymene was achieved from compound 2 in 13 steps (Scheme 2).[33,34]

1.6.5 The Pauson–Khand Reaction in Natural Product Total Synthesis **603**

Scheme 2 Total Synthesis of (+)-Epoxydictymene[33,34]

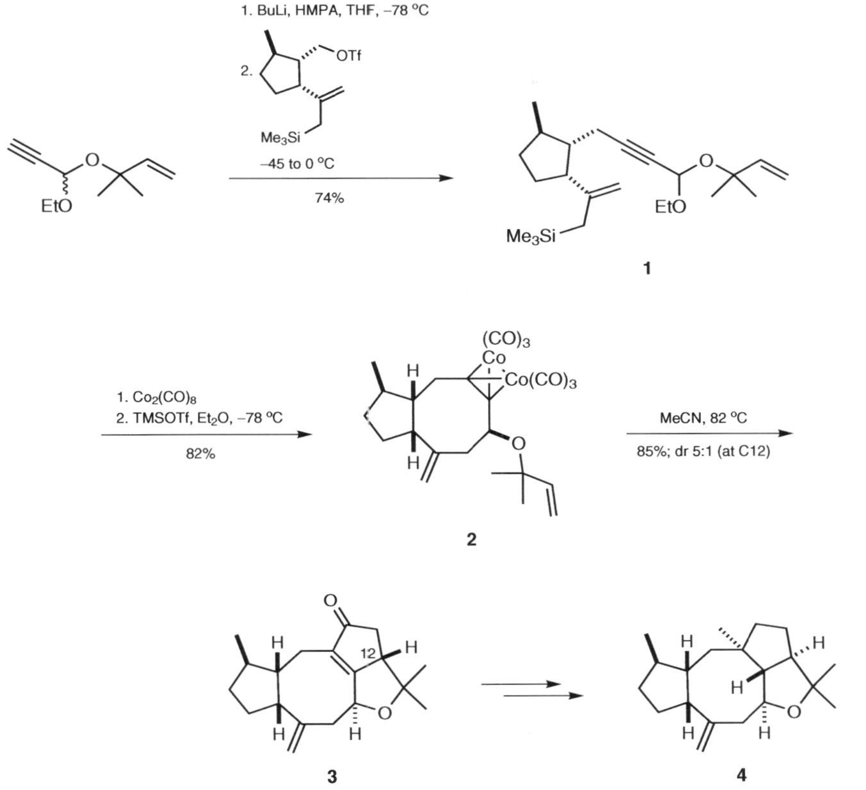

A second classic example is the application of an intramolecular Pauson–Khand reaction as a key step in the asymmetric preparation of an important intermediate in the enantioselective formal synthesis of hirsutene (**8**; Scheme 3).[35] In this report, the authors disclosed the first efficient asymmetric synthesis of cyclopentenone **6** based on chiral auxiliary directed face discrimination in acetylenic 0-alkyl enol ether–hexacarbonyldicobalt complexes. In the key event, enyne **5** was treated with octacarbonyldicobalt(0) (1.1 equiv) in hexane at 42 °C for 12 hours, producing enone **6** diastereoselectively in 55% yield. The sequential treatment of enone **6** with lithium in ammonia and samarium(II) iodide reductively afforded intermediate **7**, which was then employed in the enantioselective total synthesis of (+)-hirsutene (**8**).[36]

for references see p 628

604 Domino Transformations **1.6** Metal-Mediated Reactions

Scheme 3 Asymmetric Approach to the Formal Synthesis of (+)-Hirsutene[35]

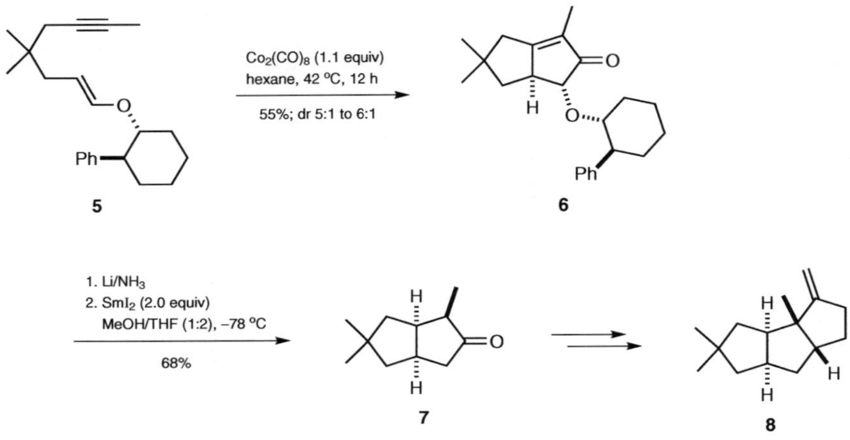

In 1999, Brummond and co-workers developed a novel synthetic strategy featuring the application of a hexacarbonylmolybdenum(0)-mediated allenic Pauson–Khand-type cyclo-addition providing facile entry into the cyclopentenone ring system **10** from allenyne **9** (Scheme 4).[37] The chemistry developed allowed the preparation of racemic hydroxymeth-ylacylfulvene (**11**),[38,39] a natural-product-derived antitumor agent, as well as a series of new analogues of that compound.

Scheme 4 Allenic Pauson–Khand Approach to Racemic Hydroxymethylacylfulvene[37]

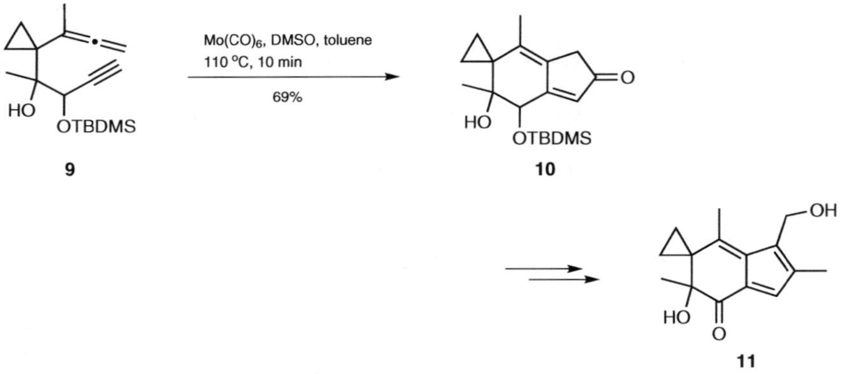

The fourth classic example, reported by Cassayre and Zard, is the short synthesis of (−)-dendrobine (**14**), which is the major alkaloid constituent of the Chinese ornamental orchid *Dendrobium nobile*[40,41] and exhibits interesting antipyretic and hypotensive activities.[42] A concise, enantioselective synthesis of (−)-dendrobine (**14**) has been achieved in 13 steps, with the important step in this synthesis being an octacarbonyldicobalt(0)-mediated Pauson–Khand reaction of enyne **12**, which constructs the crucial quaternary center at C11 of the tricyclic (−)-dendrobine core structure **13** (Scheme 5).[43]

1.6.5 The Pauson–Khand Reaction in Natural Product Total Synthesis

Scheme 5 Concise, Enantioselective Synthesis of (−)-Dendrobine[43]

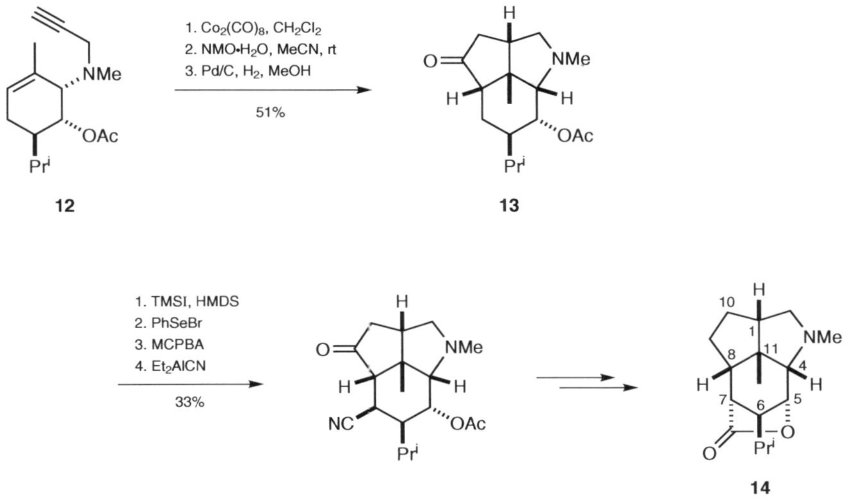

The syntheses illustrated in Schemes 2–5 should be familiar to most synthetic chemists. Therefore, rather than dwell further on the triumphs of the past, this chapter will highlight and discuss a selection of more recent examples to showcase the application of domino-based Pauson–Khand reactions in the total syntheses of natural products. Fresh evidence of the utility of this powerful reaction in the realm of chemical synthesis is offered. We do not aim to present a comprehensive survey of the literature, but rather emphasize a broad variety of strategies that have been developed for the construction of complex molecules using the Pauson–Khand reaction. Where possible, we have tried to avoid duplicating related scaffold constructions prepared by similar strategies/tactics. Only representative examples were selected from the extensive literature, and because the candidate pool was further restricted to application in the total synthesis of natural products, we will inevitably omit discussion about many impressive Pauson–Khand reactions that have been used for the construction of pharmaceuticals.

The very nature of Pauson–Khand reactions, which often involve many distinct steps, makes them difficult to classify. For convenience, we categorize selected examples into three sections: typical enyne-, heteroenyne-, and allenyne-based Pauson–Khand reactions. This categorization could be viewed as arbitrary, particularly in the case of examples that feature more than one class of reaction, but we have tried to place each Pauson–Khand reaction according to what can be argued as being the major theme of the sequence. Within each section, we ordered the examples to provide a continuous discussion, rather than chronologically.

Before discussing the chosen examples, however, some discussion regarding the terminology and organization of this chapter is warranted. The examples feature novel strategies, either by the interplay of several known reactions that would normally be executed independently into a one-pot domino sequence, or by the development of new reaction pathways. Nevertheless, the common feature of all these reactions is their inspired use of the Pauson–Khand reaction to generate molecular complexity in a concise fashion.

for references see p 628

606 Domino Transformations **1.6** Metal-Mediated Reactions

1.6.5.1 **Enyne-Based Pauson–Khand Reactions**

1.6.5.1.1 **A Short Synthesis of Racemic 13-Deoxyserratine**

Lycopodium alkaloids[44,45] exhibit fascinating structural complexity and have emerged as challenging synthetic targets in recent years. Little attention has been given to alkaloids in the serratinane subgroup with the exception of racemic serratinine[46,47] and its corresponding 8-deoxy derivative,[48,49] both of which have been synthesized previously by long and low-yielding routes.

Zard and co-workers developed a concise and efficient ten-step synthesis of racemic 13-deoxyserratine (**20**; 12% overall yield) using the Pauson–Khand reaction for the construction of cyclopentenone **16** from enyne **15**, and an amidyl-radical-initiated tandem annulation reaction of N-benzoyloxy amide **17** to create two adjacent quaternary centers at C4 and C12 of indolizidine **19** via intermediate **18** in one step, with the relative stereochemistry being correct. The presence of these centers in the serratine skeleton has considerably hampered previous approaches. The developed radical cascade can thus be directed at will to produce either an indolizidine or a pyrrolizidine framework, both of which are found in a large number of alkaloids (Scheme 6).[50]

Scheme 6 Highly Efficient Synthesis of Racemic 13-Deoxyserratine[50]

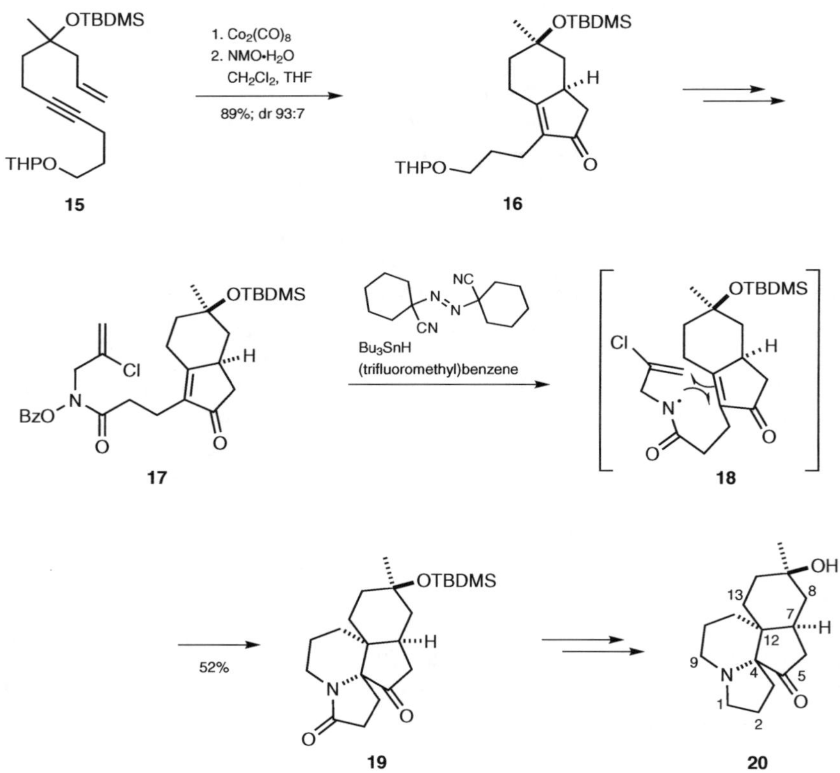

1.6.5.1.2 **Total Synthesis of Paecilomycine A**

Paecilomycine A (**24**) was isolated from *Isoria japonica*, and is capable of fostering neurite outgrowth in PC_{12} cells at 10 nM by enhancing the expression levels of the neurotrophic factors in previously conditioned human astrocytoma cells.[51] The first total synthesis of

1.6.5 The Pauson–Khand Reaction in Natural Product Total Synthesis   **607**

paecilomycine A (**24**) was achieved by Danishefsky and co-workers using a cobalt-mediated intramolecular Pauson–Khand reaction as a key step to form the critical cyclopentenone **22** from enyne **21**.[52] To attach a methyl group to C5, enone **22** is reduced to its corresponding allylic alcohol under Luche conditions,[53] enabling cyclopropanation with the in situ generated Furukawa reagent[54,55] followed by an oxidation of the hydroxy group and a dissolving metal reduction of the resultant cyclopropane to afford cyclopentanone **23** in 28% yield after four steps. The total synthesis of paecilomycine A (**24**) was eventually achieved by a stereoselective insertion of the hydroxy group at C12 followed by intramolecular ketalization (Scheme 7).[52]

Scheme 7   Total Synthesis of Paecilomycine A[52]

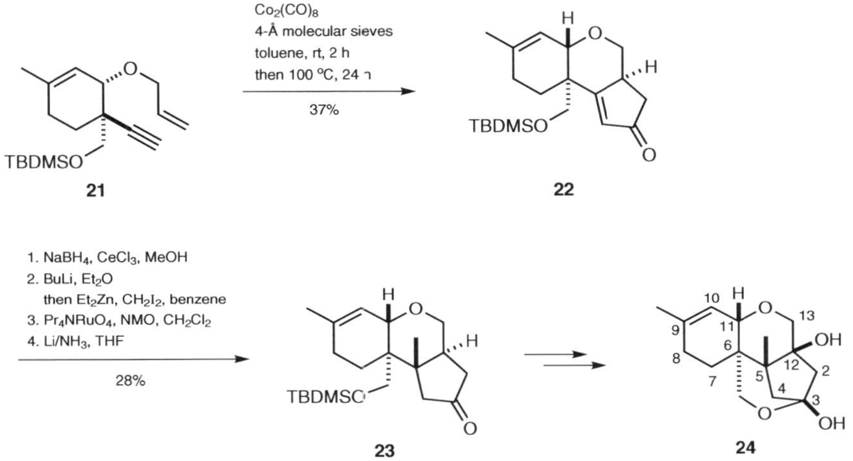

(3a*R**,5a*R**,9a*S**)-9a-[(*tert*-Butyldimethylsiloxy)methyl]-7-methyl-3a,4,5a,8,9,9a-hexahydrocyclopenta[*c*][1]benzopyran-2(3*H*)-one (22):[52]
To a soln of enyne **21** (52 mg, 0.16 mmol) in toluene (6 mL), which was degassed by bubbling with argon for 15 min, were added 4-Å molecular sieves (400 mg) and $Co_2(CO)_8$ (64 mg, 0.19 mmol). The mixture was stirred for 2 h at 23 °C and then heated to 100 °C. After 24 h, the mixture was filtered through a pad of Celite and the filtrate was concentrated under reduced pressure. The residue was purified by column chromatography (silica gel, hexane/EtOAc 3:1) to afford the product as a colorless oil; yield: 22 mg (37%).

1.6.5.1.3   Stereoselective Total Syntheses of (−)-Magellanine, (+)-Magellaninone, and (+)-Paniculatine

Magellanine (**30**),[56] magellaninone (**31**),[57] and paniculatine (**32**)[58,59] all have a 6/5/5/6-tetracyclic framework with five to seven stereogenic centers as a common structural feature. They were isolated from *Lycopodium paniculatum* and *Lycopodium magellanicum* in the mid-to-late 1970s. Because of their unique and challenging structures, several studies on the construction of the core tetracyclic skeleton of these alkaloids have been reported.[60–62] In 1994, Overman completed the first total synthesis of (−)-magellanine (**30**) and (+)-magellaninone (**31**) via a Prins-pinacol rearrangement.[63] A total synthesis of both alkaloids in racemic form based on a tandem Michael/Michael addition was accomplished in the same year by Paquette.[64,65] In addition, the masked benzo-1,2-quinone Diels–Alder protocol enabled Liao to complete the total synthesis of racemic magellanine (**30**).[66] Recently, a formal total synthesis of racemic magellanine (**30**) based on an intramolecular Pauson–

for references see p 628

608 Domino Transformations **1.6** Metal-Mediated Reactions

Khand reaction for the preparation of the two five-membered (B and C) rings was reported by Ishizaki.[67,68] The first total synthesis of (+)-paniculatine (**32**) was achieved by a tandem radical cyclization of α-carbonyl radicals by the Sha group.[69]

Mukai and co-workers completed the total syntheses of the three *Lycopodium* alkaloids (−)-magellanine (**30**), (+)-magellaninone (**31**), and (+)-paniculatine (**32**) from diethyl ʟ-tartrate in a stereoselective manner. A noteworthy tactical feature of the reported synthesis involves two Pauson–Khand reactions of the enynes **25** and **27** to afford the corresponding cyclopentenones **26** and **28**, respectively, and this distinguishes the reported synthesis from conventional strategies and enabled them to accomplish a new total synthesis of these three alkaloids via the common intermediate **29** (Scheme 8).[70]

Scheme 8   Synthetic Route to (−)-Magellanine, (+)-Magellaninone, and (+)-Paniculatine[70]

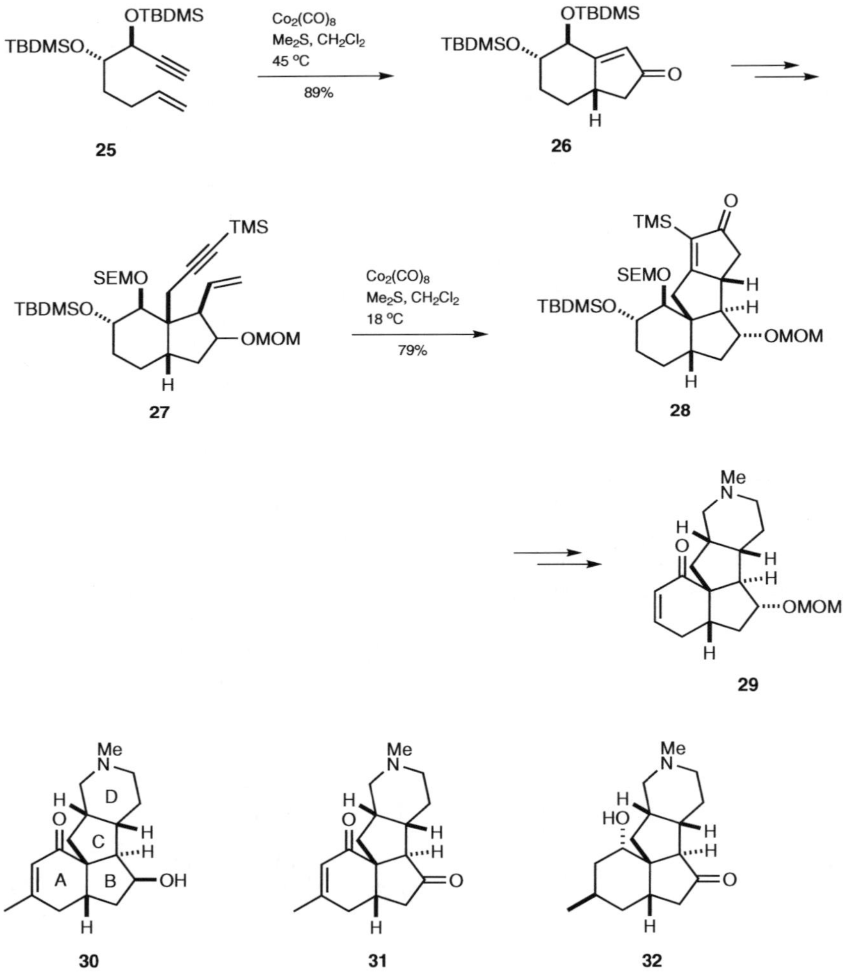

1.6.5.1.4   Concise, Enantioselective Total Synthesis of (−)-Alstonerine

The macroline/sarpagine class of indole alkaloids is a rich source of biologically active natural products, some of which exhibit hypotensive, antiamoebic, and antimalarial activities.[71–73] For example, alstonerine (**35**) has recently been reported to exhibit cytotoxic activity against two human lung cancer cell lines.[74]

1.6.5 The Pauson–Khand Reaction in Natural Product Total Synthesis **609**

Scheme 9 Enantioselective Total Synthesis of (−)-Alstonerine[75]

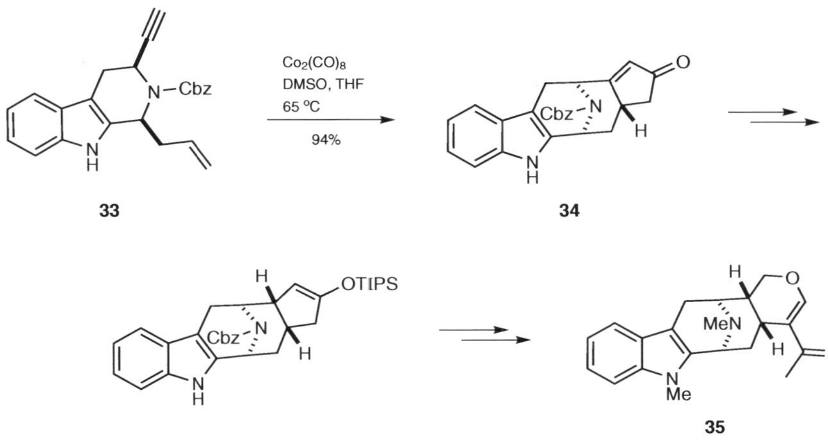

Martin and co-workers developed a concise, enantioselective total synthesis of (−)-alstonerine (**35**) starting from L-tryptophan employing the Pauson–Khand reaction of azaenyne **33** as the important step to give cyclopentenone **34** in 94% yield. The striking feature of this synthesis is that it is the first example of a Pauson–Khand reaction wherein an aza-bridged bicyclic structure is formed. Therefore, after several synthetic transformations, including a demanding 1,4-enone hydrosilylation followed by oxidative cleavage and a new, mild two-step protocol to acetylate a cyclic enol ether, the target molecule (−)-alstonerine (**35**) was obtained in 15 steps with an overall yield of 4.4% (Scheme 9).[75]

1.6.5.1.5 **Pauson–Khand Approach to the Hamigerans**

The hamigerans (e.g., **36–41**) are a small class of brominated terpenes and were isolated by Cambie and co-workers from the marine sponge *Hamigerans tarangaensis*.[76] These terpenes possess several unique structural features including the benzannulation of an indenye or an azulene fragment. In addition to these novel structural features, these natural products exhibit antiviral activity against several viruses, including polio.

Scheme 10 Pauson–Khand Approach to Hamigerans[77]

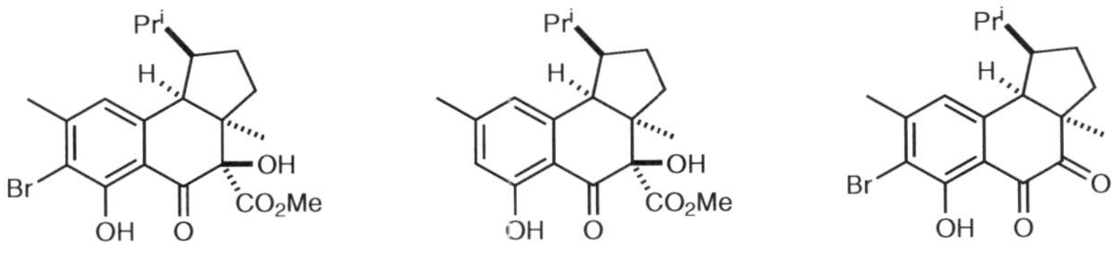

for references see p 628

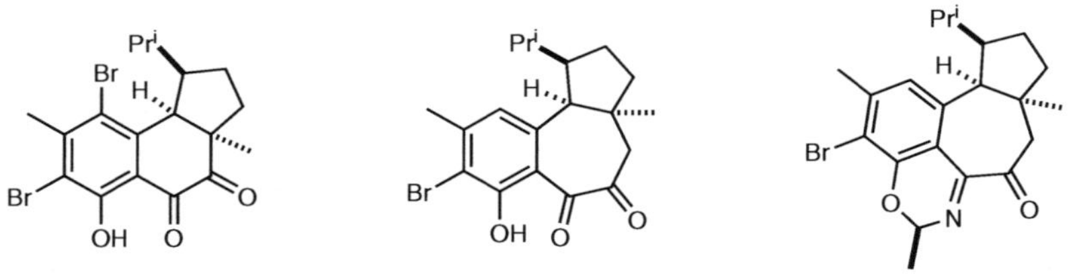

An efficient ten-step sequence starting from 4-methylsalicylic acid was developed by Lovely and co-workers to give the complete tricyclic ring product **43**, which is the core structure of several members of the hamigeran family of natural products. The important step involves an intramolecular Pauson–Khand reaction of arylenyne **42**. Critical to the success of this cyclization is the presence of a bulky silylene protecting group, which tethers the two reacting fragments together, thereby facilitating the reaction by reducing its entropic penalty and providing steric acceleration through buttressing effects (Scheme 10).[77]

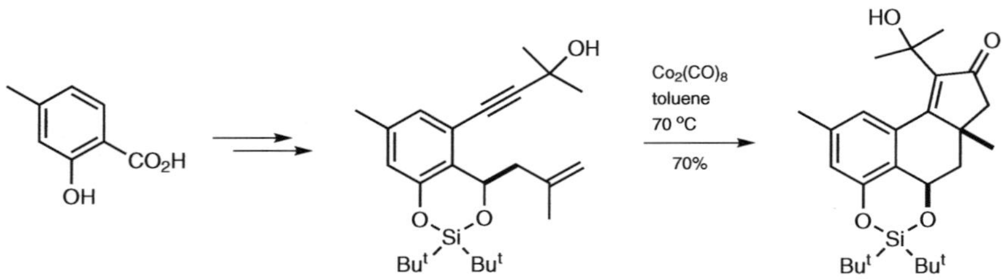

1.6.5.1.6 Enantioselective Synthesis of (−)-Pentalenene

Triquinane natural products are considered to be classic targets for total synthesis because of their biological activity and their compact, structurally complex architectures.[78–82] As the biosynthetic precursor to the pentalenolactone family of antibiotics,[83] pentalenene (**47**) has attracted intense and consistent interest from the synthetic community, culminating in numerous total and formal syntheses. However, most of these syntheses have produced racemic pentalenene.[84–88] Therefore, an enantioselective synthesis is highly desirable.

Fox and co-workers reported the first enantioselective synthesis of (−)-pentalenene (**47**) using a catalytic enantioselective cyclopropenation and a subsequent intramolecular Pauson–Khand reaction of cylopropenyne **44** to produce a quaternary center in the tricyclic cyclopentenone diastereomers **45**. Particular emphasis was laid on the tetramethylthiourea because it plays a critical role in ensuring that the desired Pauson–Khand annulation is carried out in acceptable yield. Only 0.6 equivalents of octacarbonyldicobalt(0) proved to be effective, avoiding a stoichiometric amount of this toxic reagent. Therefore, the asymmetric total synthesis of (−)-pentalenene (**47**) is complete after installation of the third cyclopentane ring onto intermediate **46**, which is derived from diastereomer **45B** (Scheme 11).[89]

1.6.5 The Pauson–Khand Reaction in Natural Product Total Synthesis **611**

Scheme 11 Enantioselective Synthesis of (−)-Pentalenene[89]

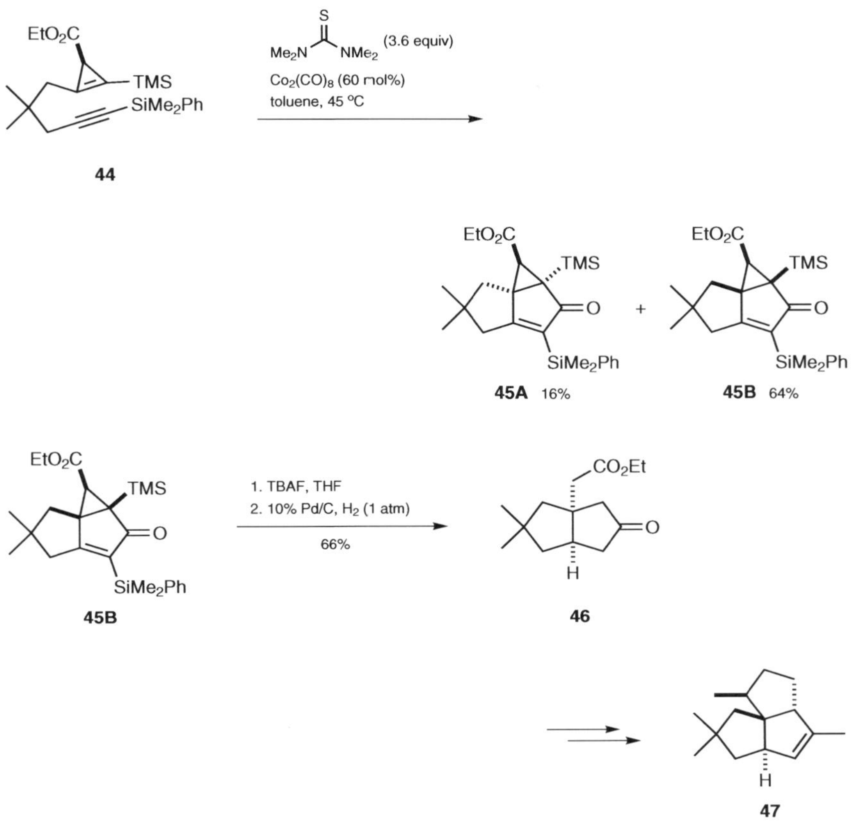

1.6.5.1.7 **Formal Synthesis of (+)-Nakadomarin A**

Nakadomarin A (**52**) is a representative manzamine-related alkaloid that was isolated from the Okinawan marine sponge *Amphimedon* sp. (SS-264) by Kobayashi.[90,91] Nakadomarin A (**52**) exhibits cytotoxicity against murine lymphoma L1210 cells (IC_{50} 1.3 µg/mL), inhibitory activity against cyclin-dependent kinase 4 (IC_{50} 9.9 µg/mL), and antimicrobial activity against a fungus (*Trichophyton mentagrophytes*, MIC 23 µg/mL) and a Gram-positive bacterium (*Corynebacterium xerosis*, MIC 11 µg/mL). This marine natural product has a unique fused hexacyclic framework consisting of a series of 6/5/5/5/8/15 rings. Intensive efforts have been directed toward its synthesis, leading thus far to four total syntheses,[92–95] and several synthetic studies.[96–103]

Mukai and co-workers achieved a formal synthesis of (+)-nakadomarin A (**52**) using an intramolecular Pauson–Khand reaction for the stereoselective formation of the diazatricyclo[6.4.0.0]dodecane skeleton **49** from enyne **48** bearing suitable substituents for further synthetic transformation in a total synthesis (Scheme 12).[104] After protecting-group manipulation, the vinyl residue in compound **49** is subjected to regioselective osmium(VIII) oxide/4-methylmorpholine N-oxide catalyzed dihydroxylation, followed by an intramolecular ketalization and dehydration to afford a furan ring in compound **50**. This intermediate is then subjected to several synthetic transformations to generate diene **51**, which is an intermediate also used in the total synthesis of nakadomarin A (**52**) by the Nishida and Kerr groups.[92–94]

for references see p 628

612　Domino Transformations　**1.6** Metal-Mediated Reactions

Scheme 12　Formal Synthesis of (+)-Nakadomarin A[104]

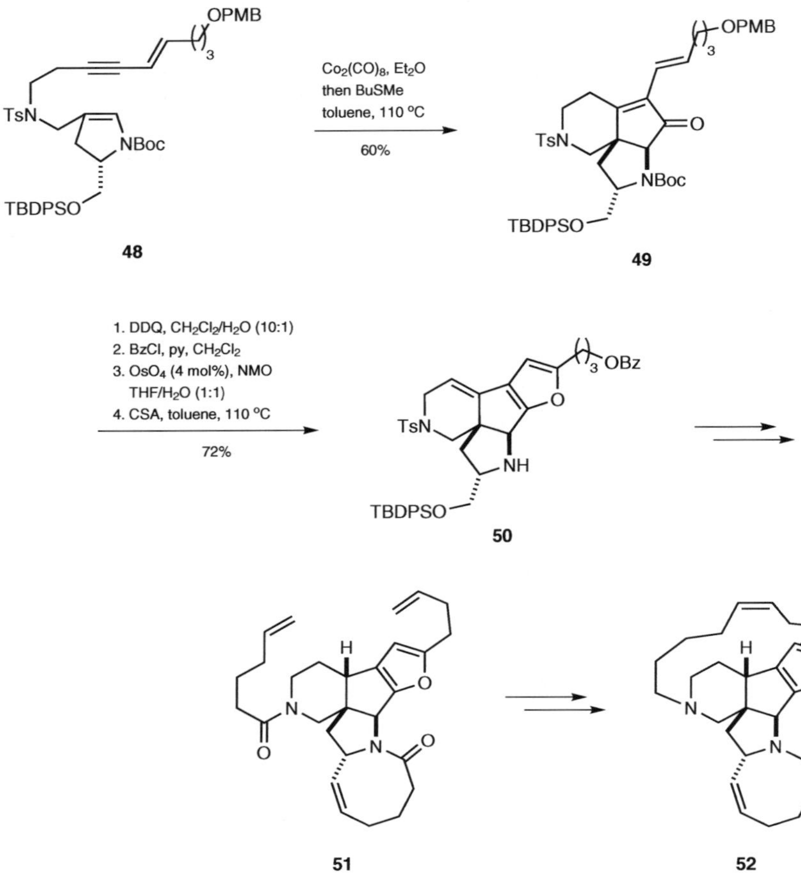

It is worth mentioning that Mukai and co-workers used a conjugated enyne **48** as a subunit in the Pauson–Khand reaction to broaden the substrate scope. Additionally, Wender has reported a dienyl-based Pauson–Khand reaction to give excellent yields of alkenylcyclopentenones under mild conditions with low catalyst loadings (1–2.5 mol%).[105,106]

1.6.5.1.8　Diastereoselective Total Synthesis of Racemic Schindilactone A

Schindilactone A (**58**) is a representative member of a novel group of nortriterpenoids isolated from *Schisandraceae* plants.[107] They have been used in China for the treatment of rheumatic lumbago and related diseases.[108] Preliminary biological assays have indicated that some of these compounds possess biological activity and inhibit hepatitis, tumors, and HIV-1.[109] The synthetic challenge posed by schindilactone A (**58**) stems from the complexity of its molecular structure, which is a highly oxygenated framework bearing 12 stereogenic centers, eight of which are contiguous chiral centers located in the FGH tricyclic ring system. Additionally, it possesses an oxa-bridged ketal within an unprecedented seven-/eight-membered ring fused carbocyclic core. Its structural complexity together with its attractive biological activities have rendered schindilactone A (**58**) an attractive target for synthetic studies.

Yang and co-workers accomplished the total synthesis of racemic schindilactone A (**58**) in 29 steps, and the relevant features of the synthetic route include: (a) an intermolec-

1.6.5 The Pauson–Khand Reaction in Natural Product Total Synthesis **613**

ular Diels–Alder reaction to produce the B ring system; (b) a silver-mediated cyclopropane rearrangement to generate the C ring; (c) a ring-closing metathesis reaction for the diastereoselective formation of the fully functionalized eight-membered CDE ring system; (d) a thiourea/cobalt-catalyzed Pauson–Khand reaction for the stereoselective construction of the F ring; (e) a thiourea/palladium-catalyzed carbonylative annulation for the stereoselective synthesis of the GH ring system; and (f) a Dieckmann-type condensation to generate the A ring. In the syntheses of intermediates **54** and **57** from their corresponding starting materials **53** and **55** via the cobalt-mediated Pauson–Khand reaction and the palladium-catalyzed carbonylative annulation reaction, respectively, both tetramethylthiourea[110,111] and thiourea **56**[112–114] play a critical role in the annulation reactions. Without these reagents, the desired reactions would not be possible (Scheme 13).[115]

Scheme 13 Total Synthesis of Racemic Schindilactone A[115]

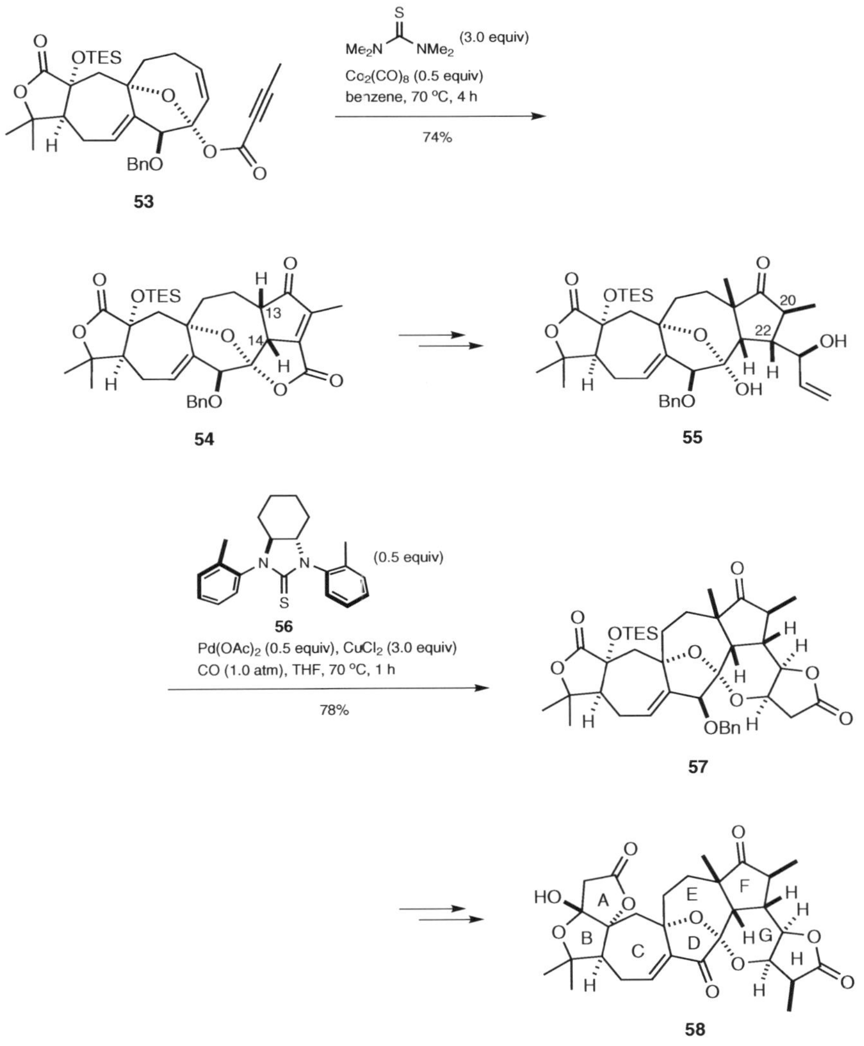

for references see p 628

614   Domino Transformations **1.6** Metal-Mediated Reactions

1.6.5.1.9   **Asymmetric Total Synthesis of (−)-Huperzine Q**

Huperzine Q (**63**) was isolated from *Huperzia serrata* in 2002,[116] and this type of natural product is reported to have a broad range of biological activity, such as the inhibition of acetylcholinesterase[117,118] and the ability to promote neurite outgrowth.[119] Huperzine Q (**63**) consists of a unique pentacyclic skeleton possessing a spiroaminal moiety and six stereogenic centers including a quaternary carbon center.

Scheme 14   Asymmetric Total Synthesis of (−)-Huperzine Q[120]

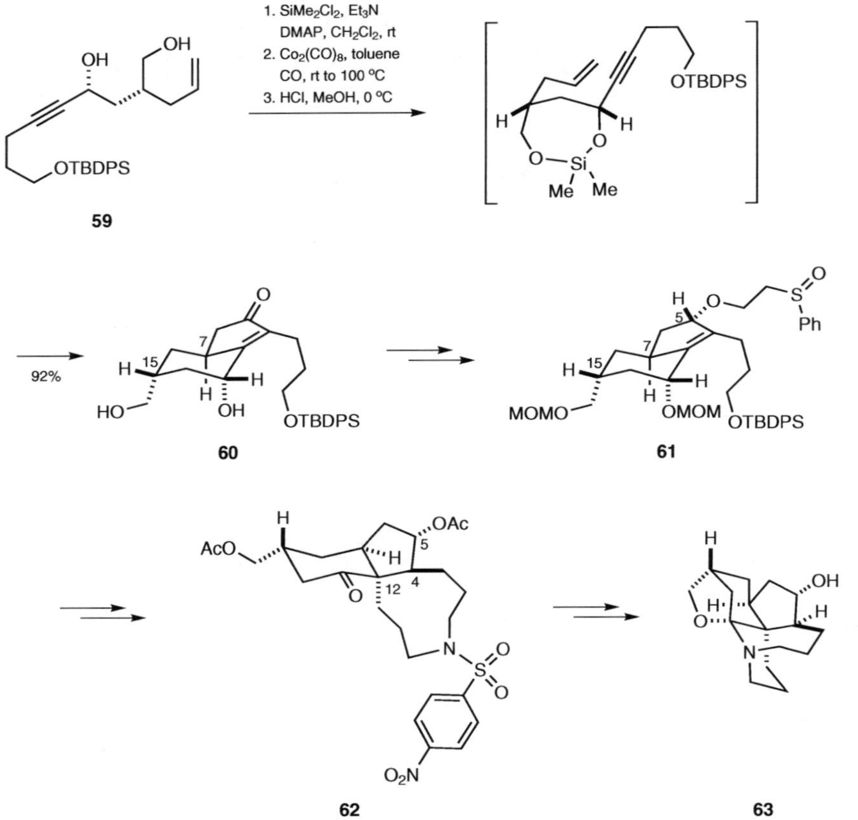

Takayama and co-workers completed the total synthesis of huperzine Q (**63**) using an octacarbonyldicobalt(0)-mediated stereoselective Pauson–Khand reaction as the critical step. Bicyclic pentenone **60** is obtained in 92% yield from chiral diol **59** through sequential protection, Pauson–Khand reaction, and desilylation. The C5 chiral center in compound **61** is constructed by Corey–Bakshi–Shibata reduction with (*R*)-(+)-2-methyloxazaborolidine reagent in good stereoselectivity. The C5, C4, and C12 chiral centers in **62** are successfully installed from intermediate **61** by a vinyl Claisen rearrangement and a subsequent hydroboration/oxidation process. The completion of the total synthesis of huperzine Q (**63**) is achieved by azonane ring formation from **62** via intramolecular Mitsunobu reaction as the important step (Scheme 14).[120]

1.6.5.1.10 Total Synthesis of (−)-Jiadifenin

Jiadifenin (**70**), a novel nonpeptidyl neurotrophic modulator, was isolated from the pericarps of *Illicium jiadifengpi* in 2002. This molecule significantly promotes neurite outgrowth in the primary cultures of fetal rat cortical neuronal cells at concentrations of 0.1~10 µM.[121] This cage-shaped molecule has six stereocenters, including two all-carbon quaternary centers (C5 and C9) and two oxo-functionalized fully substituted centers (C6 and C10). It features a unique secoprezizaane-type skeleton (A and B rings) decorated with a fused γ-lactone (C ring) and a bridged cyclic hemiacetal (D ring). Several groups have undertaken the total synthesis of this attractive molecule. For example, Danishefsky employed a Claisen-type condensation as the key step for the formation of the core structure[122,123] while Theodorakis reported a palladium(0)-mediated methylation and a palladium(0)-catalyzed carbomethoxylation to prepare rings A and C, respectively.[124,125] Another synthetic strategy by Fukuyama and co-workers used a cascade Tsuji–Trost cyclization/lactonization sequence as the important step.[126]

for references see p 628

Scheme 15 Total Synthesis of (−)-Jiadifenin[127]

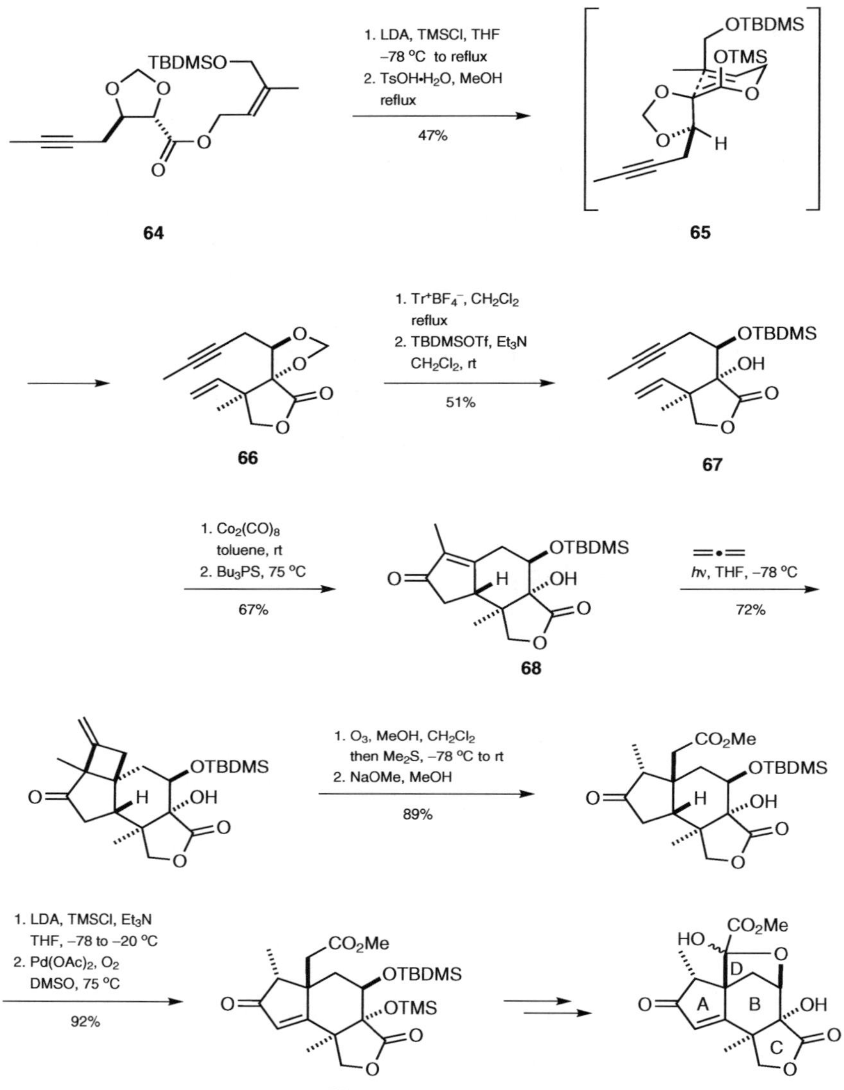

A successful asymmetric total synthesis of (−)-jiadifenin (**70**) by Zhai and co-workers uses a cobalt-mediated Pauson–Khand reaction as the key step. In this case, ester **64** was chosen as the asymmetric starting material to synthesize enyne alcohol **67**. Thus, treatment of ester **64** with lithium diisopropylamide/chlorotrimethylsilane in tetrahydrofuran afforded lactone **66** as the major diastereomer in 47% yield by [3,3]-sigmatropic rearrangement via the preferred chair-like transition state **65** followed by desilylation. After removal of the dioxolane ring, the generated diol was monoprotected as a *tert*-butyldimethylsilyl ether to give enyne alcohol **67**. In the key step, this enyne was treated with 1.2 equivalents of octacarbonyldicobalt(0) in anhydrous toluene under argon. After the mixture had been stirred at room temperature for 2 hours, 5.5 equivalents of tributylphosphine sulfide in toluene was added. The mixture was then heated at 75 °C for 16 hours to afford the corresponding cyclopentenone **68**. The total synthesis of the target molecule (−)-jiadifenin (**70**) was achieved from intermediate **69** in eight steps (Scheme 15).[127]

1.6.5 The Pauson–Khand Reaction in Natural Product Total Synthesis **617**

1.6.5.1.11 **Total Synthesis of Penostatin B**

Penostatin B (**74**) is a representative of the nine molecules in this family that have been isolated from a strain of *Penicillium* sp., originally separated from the marine alga *Enteromorpha intestinalis* by Numata and co-workers in 1996. The polyketide-derived penostatins all exhibit significant cytotoxicity against cultured P388 cells.[128–130]

The first total synthesis of penostatin B (**74**) was completed by Shishido and co-workers in a longest linear sequence of 16 steps with an overall yield of 12%. The unique features of this work included the use of a highly diastereoselective intramolecular Pauson–Khand cyclization of dienyne **71** to give cyclopentenone **72**, the successful application of an efficient relay ring-closing metathesis reaction of intermediate **73** to assemble the dihydropyranone moiety, and the diastereoselective introduction of an alkenyl side chain at C12. The synthetic route developed here is general and efficient, and can also be applied to the synthesis of other penostatins (Scheme 16).[131]

Scheme 16 Total Synthesis of Penostatin B[131]

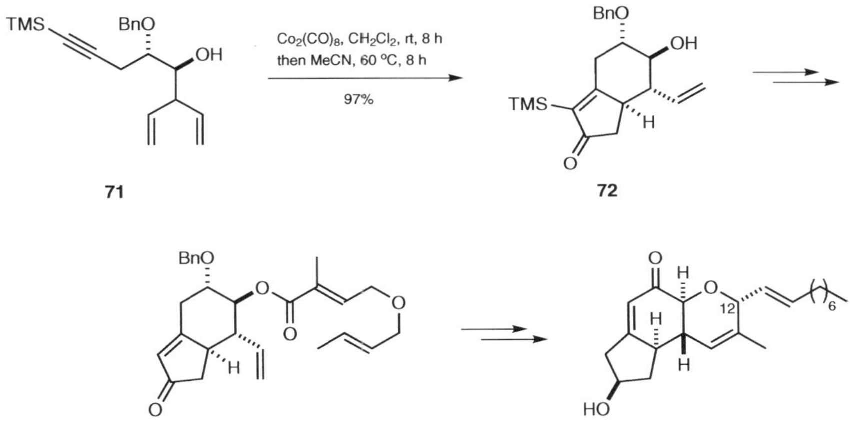

1.6.5.1.12 **Total Synthesis of Racemic Pentalenolactone A Methyl Ester**

Pentalenolactone A is a prominent member of a group of naturally occurring antibiotics produced by a prokaryotic organism. It exhibits a broad spectrum of activities against bacteria, fungi, viruses, and tumors.[132–135] The biological evaluation of pentalenolactones has shown that they can specifically inhibit glyceraldehyde-3-phosphate dehydrogenase in *Trypanosoma brucei*,[136–138] indicating that these compounds are potential candidates for the development of new antibiotics.

Yang and co-workers successfully developed a novel strategy for the stereoselective synthesis of pentalenolactones. One critical step in this strategy involves a cobalt-mediated intramolecular Pauson–Khand reaction of the trimethylsilyl-substituted enyne **75** that stereoselectively cyclizes to bicyclic cyclopentenone **76** upon treatment with 1 equivalent of octacarbonyldicobalt(0) under carbon monoxide atmosphere in benzene at 70 °C. Another critical step is a trimethylsilyl-promoted telescoped intramolecular Michael/alkenation reaction of ketophosphonate **77** for the stereoselective synthesis of α-methylene δ-pentyrolactone **78**. A striking increase in molecular complexity is achieved in a single step. The successful implementation of this novel strategy is reported in the first stereoselective total synthesis of pentalenolactone A methyl ester (**79**; Scheme 17).[139]

for references see p 628

618   Domino Transformations   **1.6** Metal-Mediated Reactions

Scheme 17   Total Synthesis of Racemic Pentalenolactone A Methyl Ester[139]

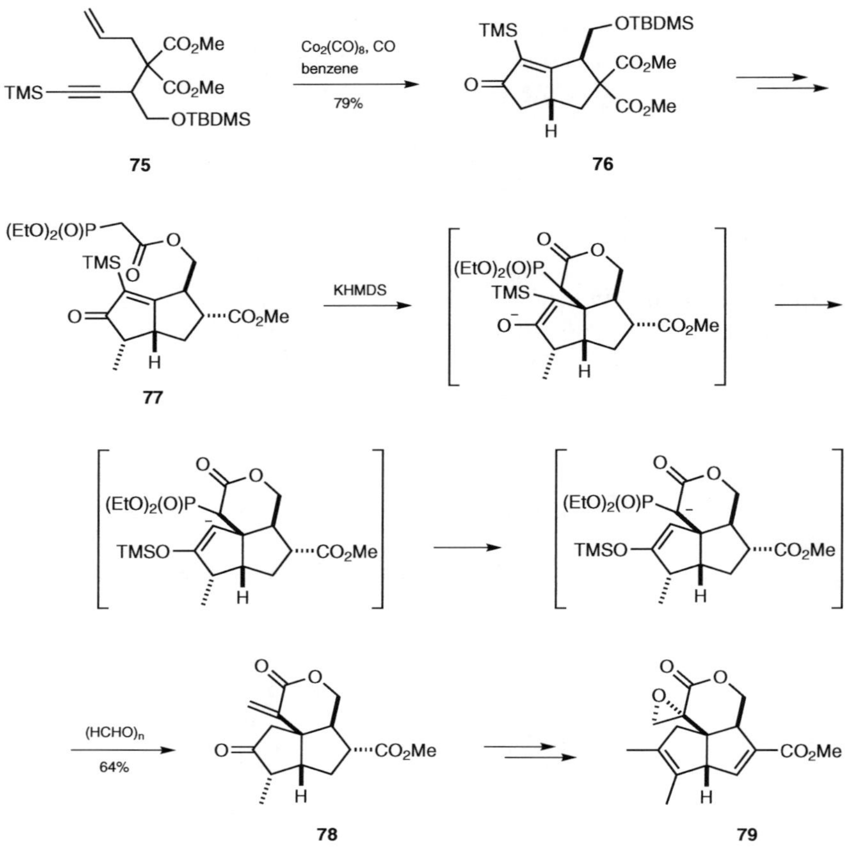

1.6.5.1.13   Asymmetric Total Synthesis of (+)-Fusarisetin A

(+)-Fusarisetin A (**83**)[140] represents a potential class of anticancer agents structurally related to the cytochalasin family of fungal metabolites.[141] This molecule has elicited considerable levels of biological interest because it exerts a potent inhibitory effect on metastasis in MDA-MB-231 breast cancer cells. Additionally, it inhibits acinar morphogenesis (77 μM), cell migration (7.7 μM), and cell invasion (26 μM) in the same cell line without any significant cytotoxicity. The structure of (+)-fusarisetin A (**83**) was established by NMR and X-ray diffraction analyses as well as by its total synthesis.[142]

Structurally, (+)-fusarisetin A (**83**) is composed of 10 stereocenters and it is characterized by an unprecedented carbon skeleton consisting of a pentacyclic ring system comprising decalin (6/6) and tricyclic (5/5/5) moieties that are both decorated with a variety of different functionalities. The novel structural features of these fungal metabolites have led to considerable synthetic efforts that have culminated in several successful total syntheses of (+)-fusarisetin A (**83**).[142–145]

1.6.5 The Pauson–Khand Reaction in Natural Product Total Synthesis **619**

Scheme 18 Asymmetric Total Synthesis of (+)-Fusarisetin A[146]

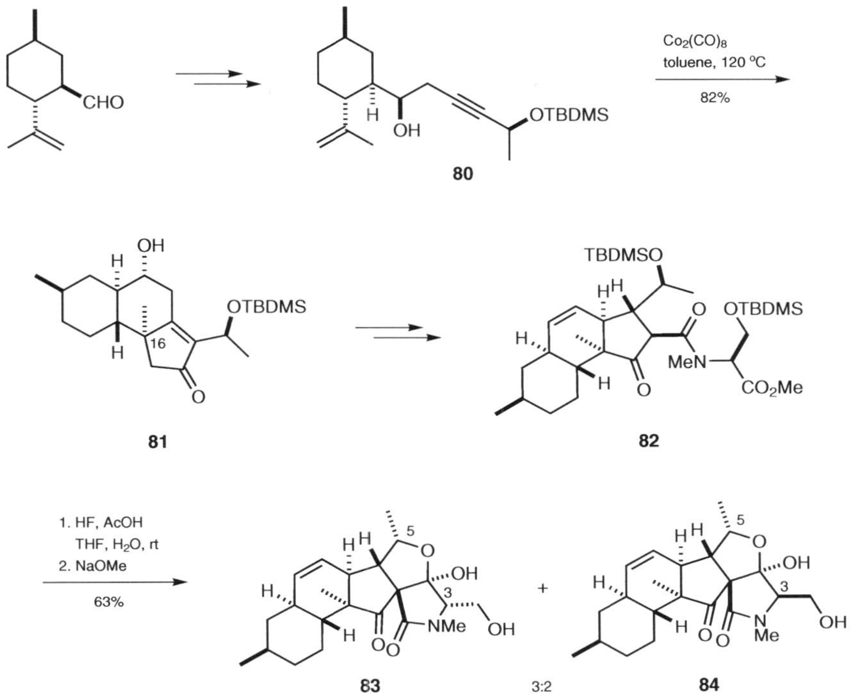

Yang and co-workers completed the total synthesis of (+)-fusarisetin A (**83**) using an intramolecular Pauson–Khand reaction as the key step. The enyne substrate **80** for the Pauson–Khand reaction is prepared from the commercially available starting material (1R,2R,5R)-5-methyl-2-(prop-1-en-2-yl)cyclohexane-1-carbaldehyde. Under the action of a stoichiometric amount of octacarbonyldicobalt(0) in toluene at reflux, the expected tricyclic product **81** is obtained in 82% yield with the desired stereochemistry at the C16 quaternary carbon. Using an 11-step sequence, enone **81** is then converted into intermediate **82**. The total synthesis of (+)-fusarisetin A (**83**) is finally achieved by desilylation followed by base-mediated condensation to afford the natural product together with its diastereomer **84** in 63% yield and a diastereomeric ratio of 3:2. The developed chemistry offers an alternative strategy to the intramolecular Diels–Alder reaction that was used by others for the synthesis of fusarisetin A (**83**), and is applicable to the synthesis of its analogues for biological evaluation (Scheme 18).[146]

1.6.5.2 Heteroatom-Based Pauson–Khand Reaction

The first hetero-Pauson–Khand-type reactions were independently carried out by the Buchwald[147,148] and Crowe[149] groups in 1996 via the intramolecular titanium-mediated [2+2+1] cycloaddition of α,β-unsaturated ketones and aldehydes with carbon monoxide. These processes result in the formation of bicyclic lactone species (oxa-Pauson–Khand-type reaction). Chatani and Murai[150,151] later discovered the dodecacarbonyltriruthenium(0)-catalyzed intramolecular oxa-Pauson–Khand reaction and the aza-Pauson–Khand reaction to provide unsaturated butenolides and α,β-unsaturated lactams, respectively.[150,151] In 2002, Kang and co-workers reported an efficient route to methylene lactones involving the dodecacarbonyltriruthenium(0)-catalyzed [2+2+1] cycloaddition between allenyl aldehydes and ketones with carbon monoxide.[152] In 2003, Saito developed a hexa-

for references see p 628

620 Domino Transformations **1.6** Metal-Mediated Reactions

carbonylmolybdenum(0)-mediated aza-Pauson–Khand reaction involving the cyclocarbonylation of an alkyne carbodiimide to generate diazabicyclic compounds.[153] In the following year, Yu and co-workers reported a convenient synthesis of α-methylene-γ-butyrolactones from allenyl carbonyl units mediated by hexacarbonylmolybdenum(0) through intramolecular cyclocarbonylation.[154] In 2007, Saito published a rhodium-catalyzed intramolecular alkyne carbodiimide Pauson–Khand-type reaction leading to the syntheses of structurally diverse 4,5-dihydro-1H-pyrrolo[2,3-b]pyrrolin-2-ones and 1H-pyrrolo[2,3-b]indol-2-ones.[155]

In addition, Adrio and Carretero developed a method for the synthesis of butenolides by the molybdenum-mediated hetero-Pauson–Khand reaction of alkynyl aldehydes. This novel hetero-Pauson–Khand reaction occurs under very mild conditions in the absence of a carbon monoxide gas atmosphere. Starting from readily available chiral aldehydes, highly valuable and enantiomerically pure fused butenolides are obtained.[156] In the presence of a stoichiometric amount of hexacarbonylmolybdenum(0) or octacarbonyldicobalt(0) or a catalytic amount of a rhodium catalyst under carbon monoxide at atmospheric pressure, a thiocarbonyl-induced heterocumulenic Pauson–Khand-type reaction can occur, leading to an effective process for the preparation 3-substituted 2H-thieno[2,3-b]indol-2-ones.[157] Finally, in 2009, Zhai and co-workers identified that tricarbonyltris(dimethylformamide)molybdenum(0) [Mo(CO)$_3$(DMF)$_3$] was also an effective agent for the formation of lactones from allenyl aldehydes, and this method has been successfully used in the total synthesis of (+)-mintlactone.[158]

1.6.5.2.1 Total Synthesis of Physostigmine

Physostigmine (**87**) was originally isolated from calabar beans in 1864,[159,160] and later identified from *Streptomyces* as an insecticidal compound. This type of alkaloid ring system has also been found in marine alkaloids such as the flustramines from the bryozoan *Flustra foliacea*. (–)-Physostigmine (**87**) is an inhibitor of acetylcholinesterase,[161] and is used clinically in the treatment of glaucoma and myasthenia gravis[162] and for protection against organophosphate poisoning.[163] The oral or intravenous administration of (–)-physostigmine (**87**) can affect memory in patients with Alzheimer's disease.[164–169]

In the total synthesis of racemic physostigmine (**87**), Mukai and co-workers developed a novel cobalt-catalyzed aza-Pauson–Khand-type reaction of alkyne carbodiimide **85** to give pyrrolo[2,3-b]indol-2-one **86** using the octacarbonyldicobalt(0)/tetramethylthiourea complex;[170] this represents the first application of octacarbonyldicobalt(0) in a hetero-Pauson–Khand reaction (Scheme 19).[26]

In 2007, using the same synthetic strategy, Mukai and co-workers achieved the total syntheses of (±)-flustramide B, (±)-flustramines B and E, (±)-debromoflustramides B and E, and (±)-debromoflustramines B and E.[171]

1.6.5 The Pauson–Khand Reaction in Natural Product Total Synthesis **621**

Scheme 19 Total Synthesis of Racemic Physostigmine[26]

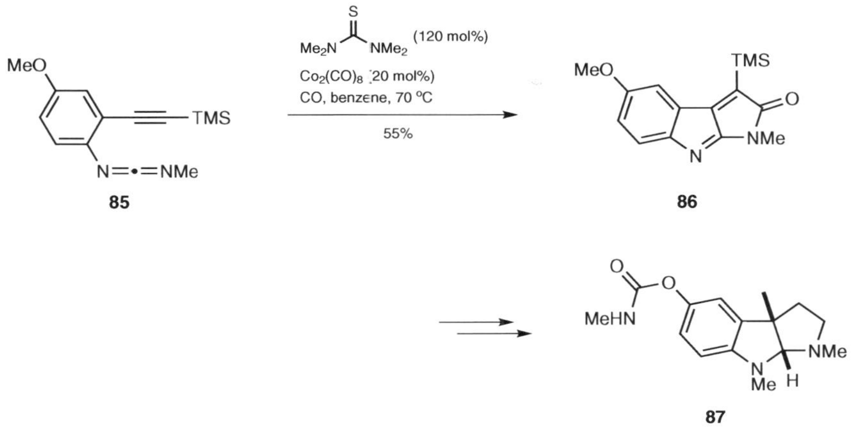

5-Methoxy-1-methyl-3-(trimethylsilyl)pyrrolo[2,3-*b*]indol-2(1*H*)-one (86):[26]

> **CAUTION:** *Carbon monoxide is extremely flammable and toxic, and exposure to higher concentrations can quickly lead to a coma.*

To a soln of N-{4-methoxy-2-[(trimethylsilyl)ethynyl]phenyl}-N-methylmethanediimine (**85**; 0.15 mmol) in benzene (1.5 mL) (**CAUTION:** *carcinogen*) were added tetramethylthiourea (24 mg, 0.18 mmol, 1.2 equiv) and $Co_2(CO)_8$ (11.8 mg, 30 µmol, 20 mol%). The mixture was heated at 70 °C (oil bath temperature) under CO until complete disappearance of the starting material (TLC) and then passed through a short pad of Celite with EtOAc, and the filtrate was concentrated to dryness. The residue was purified by chromatography (silica gel, hexane/EtOAc); yield: 24 mg (55%).

1.6.5.2.2 **Asymmetric Total Synthesis of Racemic Merrilactone A**

Merrilactone A (**93**) was isolated from the pericarps of *Illicium merrillianum* in 2000 by Fukuyama and co-workers, and it is a nonpeptidal neurotrophic factor that promotes neurite outgrowth in the culture of fetal rat cortical neurons.[172,173]

Merrilactone A (**93**) contains seven contiguous chiral centers, including five quaternary centers. It is a complex cage-shaped pentacyclic sesquiterpene that possesses an oxetane moiety, two lactone functionalities, and a highly substituted cyclopentane ring. The structure of merrilactone A (**93**) was established by NMR and X-ray crystallographic analyses, and the absolute configuration was determined using the Mosher protocol.

The total synthesis of racemic merrilactone A (**93**) by Zhai and co-workers incorporates a hetero-Pauson–Khand reaction for the early stage formation of the ABC tricyclic core of target **89**. This event is followed by a vinylogous Mukaiyama/Michael reaction and a samarium(II) iodide mediated reductive carbonyl alkene coupling to afford the ABCD tetracyclic system in **92**. Synthetically, ynal **88** is treated with tricarbonyltris(dimethylformamide)molybdenum(0) [$Mo(CO)_3(DMF)_3$, 1.1 equiv] in tetrahydrofuran and the resulting mixture is stirred at room temperature under carbon monoxide, leading to the formation of α,β-unsaturated lactone **89** with the ABC tricyclic ring system in 69% yield via an intramolecular ynal hetero-Pauson–Khand reaction. Treatment of α,β-unsaturated lactone **89** with *tert*-butyldimethylsilyl trifluoromethanesulfonate in the presence of triethylamine in dichloromethane affords siloxyfuran **90**, which is then reacted with methyl vinyl ketone in the presence of 1,1,3,3-tetrakis(trifluoromethanesulfonyl)propane[174–176] to furnish ketone **91** (61% yield) along with its C4 epimer (8% yield) through a

for references see p 628

622 Domino Transformations **1.6** Metal-Mediated Reactions

vinylogous Mukaiyama/Michael reaction. The good facial selectivity (dr 7.2:1) found in this reaction might be a result of the presence of the bulky *tert*-butyldimethylsilyl group. Further treatment of ketone **91** with samarium(II) iodide in tetrahydrofuran at room temperature leads to the formation of the coupled product **92** with the correct stereochemistry at the C1 and C9 positions as required for the target natural product. The total synthesis of racemic merrilactone A (**93**) is then completed by stereoselective installation of its E ring and incorporation of the C2 hydroxy group (Scheme 20).[177]

Scheme 20 Asymmetric Total Synthesis of Racemic Merrilactone A[177]

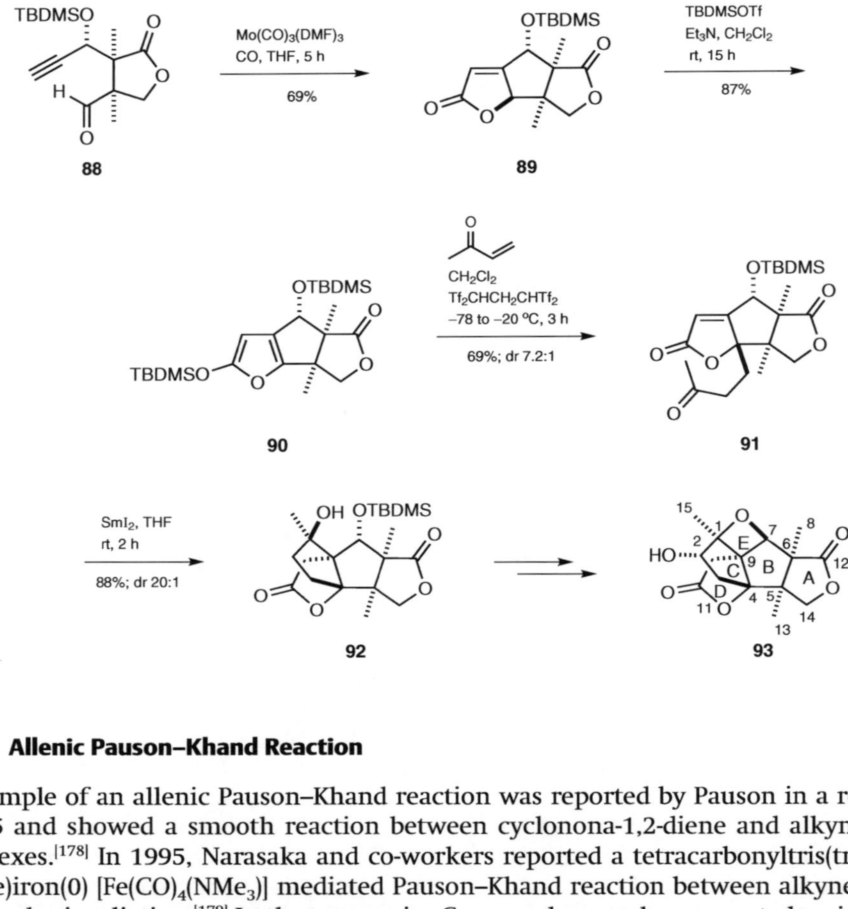

1.6.5.3 Allenic Pauson–Khand Reaction

The first example of an allenic Pauson–Khand reaction was reported by Pauson in a review in 1985 and showed a smooth reaction between cyclonona-1,2-diene and alkyne metal complexes.[178] In 1995, Narasaka and co-workers reported a tetracarbonyltris(trimethylamine)iron(0) [Fe(CO)$_4$(NMe$_3$)] mediated Pauson–Khand reaction between alkynes and allenes under irradiation.[179] In the same vein, Cazes and co-workers reported an intermolecular cobalt-mediated Pauson–Khand reaction between alkynes and allenes in the presence of 4-methylmorpholine *N*-oxide at room temperature to afford 4-alkylidene cyclopentenones.[180–183]

In 2001 and 2002, Narasaka and then the Mukai and Brummond groups reported a rhodium-catalyzed allenic Pauson–Khand-type reaction for the assembly of the carbon skeleton of bicyclo[5.3.0]dec-1,7-dien-9-ones.[184–186] Later, a molybdenum-mediated Pauson–Khand reaction by Brummond and co-workers demonstrated chirality transfer from a chiral nonracemic allene to an α-alkylidene and an α-silylidene cyclopentenone.[187] Since then, the allenic-type Pauson–Khand reaction has been used as an important step in the construction of the core structure of complex natural products such as guanacastepene A.[188] In 2012, Shi and co-workers reported a novel rhodium(I)-catalyzed Pauson–Khand-type [3+2+1]-cycloaddition reaction of enevinylidene cyclopropanes and carbon

1.6.5 The Pauson–Khand Reaction in Natural Product Total Synthesis **623**

monoxide to give a series of aza- and oxabicyclic compounds in moderate to good yields in a highly regio- and diastereoselective manner. This approach provides an alternative and efficient synthetic approach to access bicyclic cyclohexanone frameworks.[189] In the same year, Shi reported an intramolecular Pauson–Khand-type cycloaddition reaction between ene-vinylidenecyclopropanes and carbon monoxide using chloro(cycloocta-1,5-diene)rhodium(I) dimer as the catalyst. The reaction is highly efficient using 1,2-dichloroethane and 1,1,2,2-tetrachloroethane as solvent, affords excellent yields (90–99%), and provides easy access to a series of fused 6/5-ring structures containing spirocyclopropane units that are useful for drug design and development.[190]

1.6.5.3.1 Total Synthesis of (+)-Achalensolide

(+)-Achalensolide (**97**) was isolated in 1983 from the aerial parts of *Decachaeta thieleana* gathered in Turrucares, Costa Rica, and from *Stevia achalensis*, collected in Copina, Córdoba, Argentina,[191] and, similar to other guaianolides, is a potent inhibitor of aromatase enzyme in human placental microsomes.

Mukai and co-workers completed the first total synthesis of (+)-achalensolide (**97**) in 2008 starting from commercially available D-(−)-isoascorbic acid (**94**) using a rhodium-catalyzed Pauson–Khand-type reaction of allenyne **95**, which enabled direct construction of the bicyclo[5.3.0]decane core **96** in excellent yield (Scheme 21).[192] The stereoselective installation of the *cis*-fused γ-lactone moiety in (+)-achalensolide (**97**) was realized using the Ueno–Stork reaction.[193–195]

Scheme 21 Total Synthesis of (+)-Achalensolide[192]

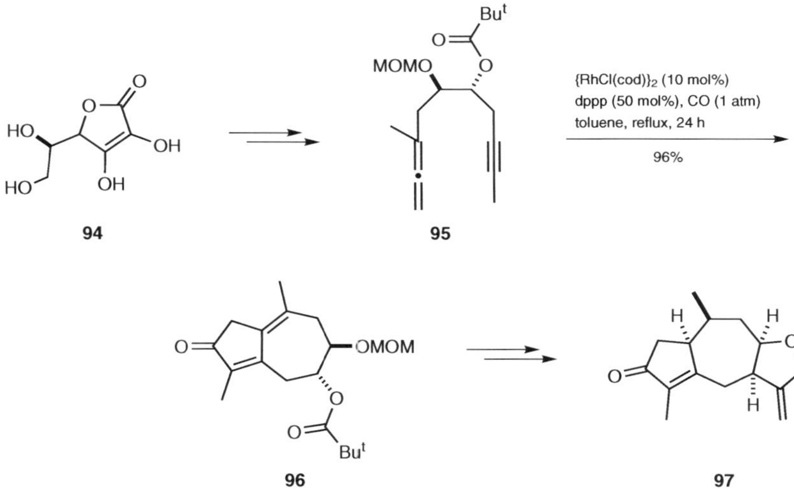

(−)-(5*R*,6*R*)-6-(Methoxymethoxy)-3,8-dimethyl-2-oxo-1,2,4,5,6,7-hexahydroazulen-5-yl Pivalate (96):[192]

> **CAUTION:** *Carbon monoxide is extremely flammable and toxic, and exposure to higher concentrations can quickly lead to a coma.*

To a soln of (5*R*,6*R*)-6-(methoxymethoxy)-8-methyldeca-8,9-dien-2-yn-5-yl pivalate (**95**; 100 mg, 0.324 mmol) in toluene (3.2 mL) were added dppp (66.8 mg, 0.162 mmol) and {RhCl(cod)}$_2$ (16.0 mg, 32.4 µmol), and the mixture was heated at reflux for 24 h under

for references see p 628

624 Domino Transformations **1.6** Metal-Mediated Reactions

CO. The solvent was evaporated, and the residue was purified by chromatography (silica gel, hexane/EtOAc 2:1) to afford the product as colorless needles; yield: 105 mg (96%); mp 73–74 °C (hexane/EtOAc); $[\alpha]_D^{25}$ −123.4 (c 0.97, CHCl$_3$).

1.6.5.3.2 Synthesis of 6,12-Guaianolide

The guaianolides constitute the largest class of sesquiterpene lactones known, many of which possess an α-methylene butyrolactone, a privileged moiety represented in 10% of all natural products.[196,197] The skeleton of this class of compounds is typified by a 5/7/5-fused ring system and can be further classified into four subgroups: 6,12-, 8,12-, pseudo-, and dimeric guaianolides; each subgroup is represented in Scheme 22 by arglabin (**98**), (+)-achalensolide (**97**), helenalin (**99**), and arteminolide (**100**), respectively. Those four guaianolides are densely oxygenated, as are many guaianolides; fulvenoguianolide (**101**) is an exception.[198]

The guaianolides have a wide range of biological activities.[199–206] The α-methylene butyrolactone, cyclopentenone, and acrylic ester moieties endow these molecules with bioactivity and selectivity.[207] Recently, the guaianolides have received attention from medicinal chemists as potential therapeutics because of their controlled and targeted specific covalent modification.[208,209]

Scheme 22 Representative Naturally Occurring Guaianolides[198]

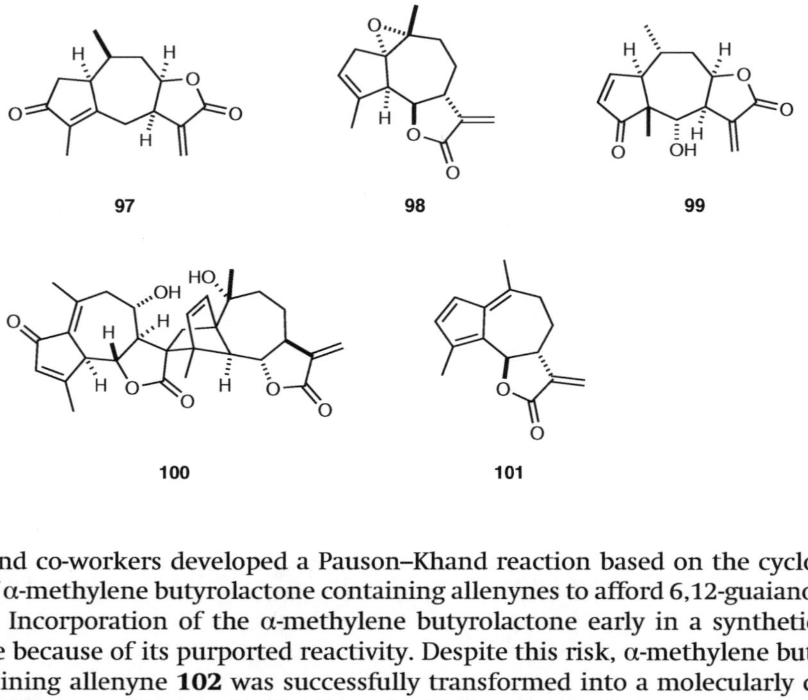

97

98

99

100

101

Brummond and co-workers developed a Pauson–Khand reaction based on the cyclocarbonylation of α-methylene butyrolactone containing allenynes to afford 6,12-guaianolide ring systems. Incorporation of the α-methylene butyrolactone early in a synthetic sequence is rare because of its purported reactivity. Despite this risk, α-methylene butyrolactone containing allenyne **102** was successfully transformed into a molecularly complex skeleton **103** in excellent yield. The three double bonds and the ketone in the resultant 5/7/5-ring system have significantly different reactivity and are ideally positioned for synthetic application to 6,12-guaianolides as well as to their analogues. This strategy is appealing because of the rapid entry into a highly complex skeleton as well as the diverse functionalities (Scheme 23).[198] Ring systems with 6/7/5 fusion such as **104** and **105** can also be obtained (Scheme 23). The enantioselective synthesis of 5/7-bicyclic ring systems from axially chiral allenes using a rhodium(I)-catalyzed cyclocarbonylation reaction was also reported more recently by the same group.[210]

1.6.5 The Pauson–Khand Reaction in Natural Product Total Synthesis **625**

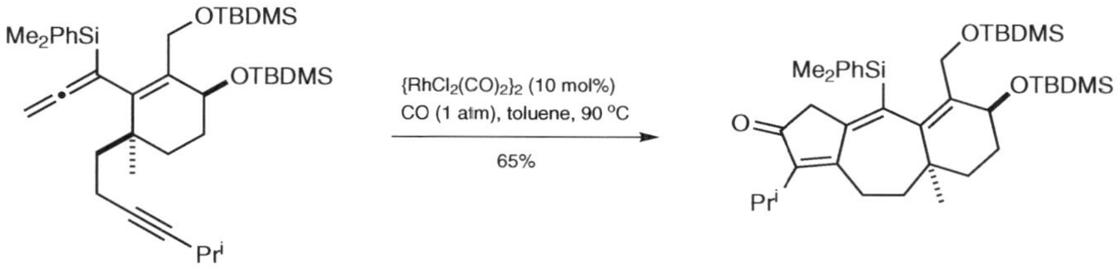

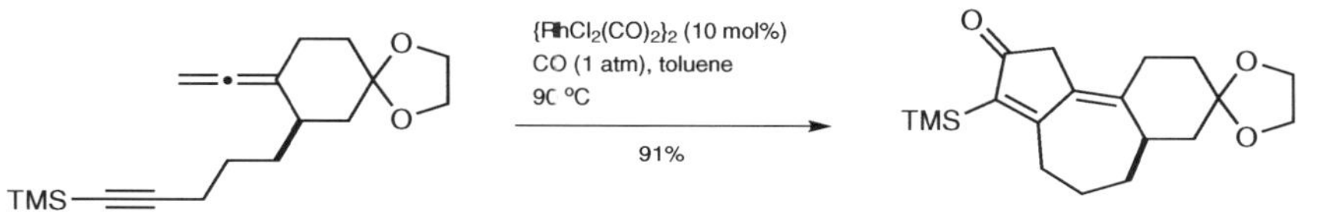

Scheme 23 Synthesis of 6,12-Guaianolides[198]

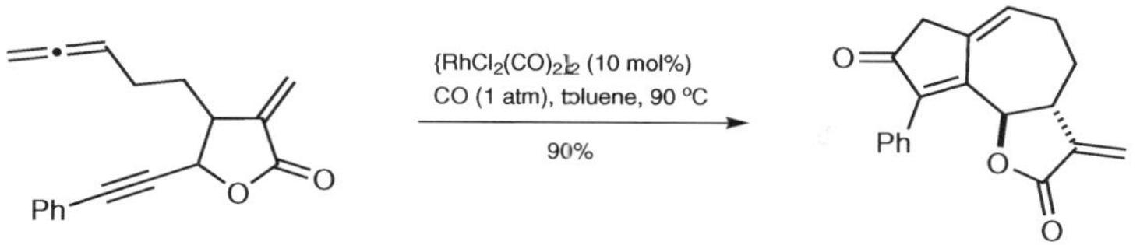

1.6.5.3.3 **Stereoselective Total Syntheses of Uncommon Sesquiterpenoids**

Two uncommon tricyclic sesquiterpenes **108** (R[1] = H, Ac), possessing the so-called cycloax-4(15)-ene skeleton, have been isolated from *Jatropha neopauciflora*.[211] Mukai and co-workers completed the total synthesis of these novel sesquiterpenoids from dimethyl D-tartrate in a stereoselective manner, The reaction features a rhodium(I)-catalyzed Pauson–Khand-type reaction of allenene derivative **106** leading to the exclusive formation of the bicyclo[4.3.0]nonenone framework **107**, which possesses an angular methyl group; it also relies on the highly stereoselective construction of the isopropylcyclopropane ring (Scheme 24).[212]

Scheme 24 Stereoselective Total Synthesis of Uncommon Sesquiterpenoids[212]

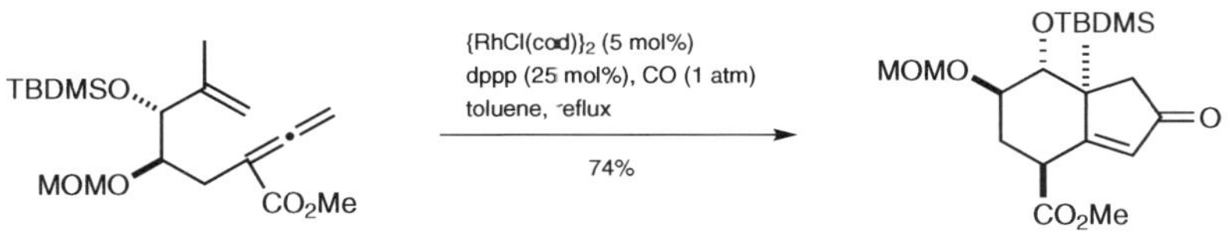

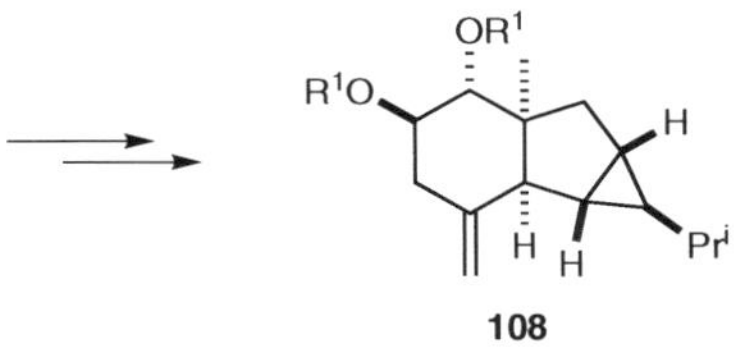

R[1] = H, Ac

for references see p 628

626   Domino Transformations **1.6** Metal-Mediated Reactions

1.6.5.3.4   **14-Step Synthesis of (+)-Ingenol from (+)-3-Carene**

Ingenol (**114**) is a diterpenoid with unique architecture, and its derivatives possess important anticancer activity. An example is Picato, a first-in-class drug that has recently been approved by the Food and Drug Administration for the treatment of the precancerous skin condition actinic keratosis.[213–217]

The structure of ingenol (**114**) features a unique *in,out*-[4.4.1]bicyclododecane core, which is considerably more strained than the typical *out,out*-configuration.[218] The fascinating structure and intriguing biological activity of ingenol has attracted tremendous interest from synthetic organic chemists over the last 30 years.[219] During that time, three total syntheses[170,220,221] and one formal synthesis have been completed,[110] along with numerous approaches toward the bicyclic core.[219] Although these landmark efforts have led to elegant syntheses of ingenol, the employed routes required between 37 and 45 steps. This length has precluded their practical use in accessing synthetic analogues in meaningful quantities.

Very recently, Baran and co-workers reported a 14-step synthesis of (+)-ingenol (**114**) from (+)-3-carene, which is more practical for the synthesis of this molecule and will certainly allow for the creation of fully synthetic analogues. It also provides a concise approach for its chemical synthesis.

Scheme 25   14-Step Synthesis of (+)-Ingenol from (+)-3-Carene[222]

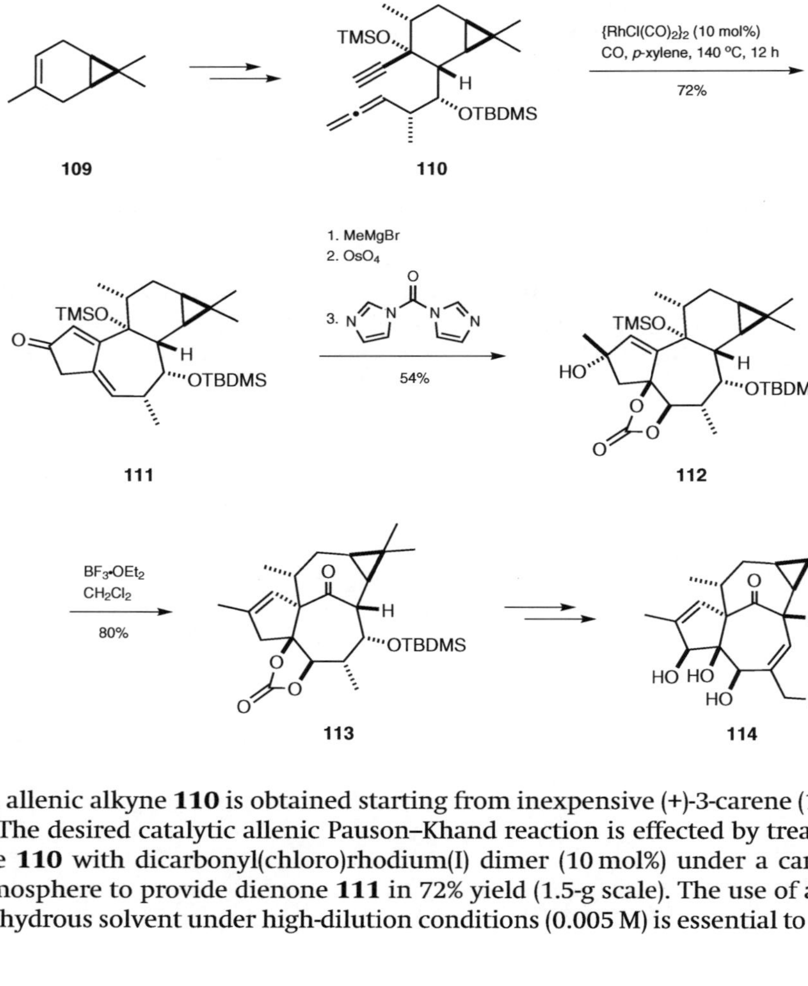

Synthetically, allenic alkyne **110** is obtained starting from inexpensive (+)-3-carene (**109**) in five steps. The desired catalytic allenic Pauson–Khand reaction is effected by treating allenic alkyne **110** with dicarbonyl(chloro)rhodium(I) dimer (10 mol%) under a carbon monoxide atmosphere to provide dienone **111** in 72% yield (1.5-g scale). The use of a degassed and anhydrous solvent under high-dilution conditions (0.005 M) is essential to pro-

vide a high yield in this reaction. Methylation of the ketone with methylmagnesium bromide (80% yield) is followed by osmium(VIII) oxide mediated dihydroxylation to afford the required diol. This diol is protected with 1,1′-carbonyldiimidazole leading to carbonate **112** in 54% yield in three steps (100-mg scale). The desired ingenol scaffold is thus generated by the treatment of intermediate **112** with boron trifluoride–diethyl ether complex in dichloromethane at −78 to −40 °C, followed by quenching with triethylamine and methanol to produce ketone **113**. This generates the strained *in,out*-stereochemistry of ingenol in 80% yield (191-mg scale), which eventually leads to the total synthesis of ingenol (**114**) in 14 steps (Scheme 25).[222]

1.6.5.4 Conclusions

The Pauson–Khand reaction is important for the synthesis of various kinds of molecules. The reactions presented in this chapter demonstrate the power of this reaction in the total synthesis of complex natural products, highlighting the array of complex frameworks that can result from its domino series of bond constructions. Therefore, the Pauson–Khand reaction has a prominent place in the repertoire of synthetic organic chemists. Its use enables the construction of complex molecules in a convergent and atom-economic way, starting from structurally simple precursors.

for references see p 628

References

[1] Khand, I. U.; Knox, G. R.; Pauson, P. L.; Watts, W. E., *J. Chem. Soc. D*, (1971), 36.

[2] Khand, I. U.; Knox, G. R.; Pauson, P. L.; Watts, W. E.; Foreman, M. I., *J. Chem. Soc., Perkin Trans. 1*, (1973), 977.

[3] Pauson, P. L.; Khand, I. U., *Ann. N. Y. Acad. Sci.*, (1977) **295**, 2.

[4] Blanco-Urgoiti, J.; Añorbe, L.; Pérez-Serrano, L.; Domínguez, G.; Pérez-Castells, L., *Chem. Soc. Rev.*, (2004) **33**, 32.

[5] Gibson, S. E.; Stevenazzi, A., *Angew. Chem. Int. Ed.*, (2003) **42**, 1800.

[6] Gibson, S. E.; Mainolfi, N., *Angew. Chem. Int. Ed.*, (2005) **44**, 3022.

[7] Brummond, K. M.; Kent, J. L., *Tetrahedron*, (2000) **56**, 3263.

[8] Boñaga, L. V. R.; Krafft, M. E., *Tetrahedron*, (2004) **60**, 9795.

[9] Fletcher, A. J.; Christie, S. D. R., *J. Chem. Soc., Perkin Trans. 1*, (2000), 1657.

[10] Chung, Y. K., *Coord. Chem. Rev.*, (1999) **188**, 297.

[11] Jeong, N., In *Transition Metals for Organic Synthesis*, Beller, M.; Bolm, C., Eds.; Wiley-VCH: Weinheim, Germany, (1998); p 560.

[12] Geis, O.; Schmalz, H.-G., *Angew. Chem. Int. Ed.*, (1998) **37**, 911.

[13] Ingate, S. T.; Marco-Contelles, J., *Org. Prep. Proced. Int.*, (1998) **30**, 121.

[14] Buchwald, S. L.; Hicks, F. A., In *Comprehensive Asymmetric Catalysis*, Jacobsen, E. N.; Pfaltz, A.; Yamamoto, H., Eds.; Springer: Berlin, (1999); Vol. 2, p 491.

[15] Grossman, R. B.; Buchwald, S. L., *J. Org. Chem.*, (1992) **57**, 5803.

[16] Morimoto, T.; Chatani, N.; Fukumoto, Y.; Murai, S., *J. Org. Chem.*, (1997) **62**, 3762.

[17] Kondo, T.; Suzuki, N.; Okada, T.; Mitsudo, T., *J. Am. Chem. Soc.*, (1997) **119**, 6187.

[18] Koga, Y.; Kobayashi, T.; Narasaka, K., *Chem. Lett.*, (1998) **27**, 249.

[19] Shibata, T.; Takagi, K., *J. Am. Chem. Soc.*, (2000) **122**, 9852.

[20] Shambayati, S.; Crowe, W. E.; Schreiber, S. L., *Tetrahedron Lett.*, (1990) **31**, 5289.

[21] Jeong, N.; Chung, Y. K.; Lee, B. Y.; Lee, S. H.; Yoo, S.-E., *Synlett*, (1991), 204.

[22] Gordon, A. R.; Johnstone, C.; Kerr, W. J., *Synlett*, (1996), 1083.

[23] Sugihara, T.; Yamada, M.; Ban, H.; Yamaguchi, M.; Kaneko, C., *Angew. Chem. Int. Ed.*, (1997) **36**, 2801.

[24] Rajesh, T.; Periasamy, M., *Tetrahedron Lett.*, (1998) **39**, 117.

[25] Tang, Y.; Deng, L.-J.; Zhang, Y.-D.; Dong, G.-B.; Chen, J.-H.; Yang, Z., *Org. Lett.*, (2005) **7**, 593.

[26] Mukai, C.; Yoshida, T.; Sorimachi, M.; Odani, A., *Org. Lett.*, (2006) **8**, 83.

[27] Murakami, M.; Itami, K.; Ito, Y., *J. Am. Chem. Soc.*, (1997) **119**, 2950.

[28] Jeong, N.; Sung, B. K.; Choi, Y. K., *J. Am. Chem. Soc.*, (2000) **122**, 6771.

[29] Enoki, N.; Furusaki, A.; Suehiro, K.; Ishida, R.; Matsumoto, T., *Tetrahedron Lett.*, (1983) **24**, 4341.

[30] Winter, C. A.; Risley, E. A.; Nuss, G. W., *Proc. Soc. Exp. Biol. Med.*, (1962) **111**, 544.

[31] Adeyemi, O. O.; Okpo, O. S.; Okpaka, O., *J. Ethnopharmacol.*, (2004) **90**, 45.

[32] Nicholas, K. M., *Acc. Chem. Res.*, (1987) **20**, 207.

[33] Jamison, T. F.; Shambayati, S.; Crowe, W. E.; Schreiber, S. L., *J. Am. Chem. Soc.*, (1994) **116**, 5505.

[34] Jamison, T. F.; Shambayati, S.; Crowe, W. E.; Schreiber, S. L., *J. Am. Chem. Soc.*, (1997) **119**, 4353.

[35] Castro, J.; Sörensen, H.; Riera, A.; Morin, C.; Moyano, A.; Pericàs, M. A.; Greene, A. E., *J. Am. Chem. Soc.*, (1990) **112**, 9388.

[36] Nozoe, S.; Furukawa, J.; Sankawa, U.; Shibata, S., *Tetrahedron Lett.*, (1976), 195.

[37] Brummond, K. M.; Lu, J.-L., *J. Am. Chem. Soc.*, (1999) **121**, 5087.

[38] MacDonald, J. R.; Muscoplat, C. C.; Dexter, D. L.; Mangold, G. L.; Chen, S.-F.; Kelner, M. J.; McMorris, T. C.; Von Hoff, D. D., *Cancer Res.*, (1997) **57**, 279.

[39] Kelner, M. J.; McMorris, T. C.; Estes, L. A.; Wang, W.; Samson, K. M.; Taetle, R., *Invest. New Drugs*, (1996) **14**, 161.

[40] Suzuki, H.; Keimatsu, I.; Ito, K., *J. Pharm. Soc. Jpn.*, (1932) **52**, 1049.

[41] Suzuki, H.; Keimatsu, I.; Ito, K., *J. Pharm. Soc. Jpn.*, (1934) **54**, 802.

[42] Porter, L. A., *Chem. Rev.*, (1967) **67**, 441.

[43] Cassayre, J.; Zard, S. Z., *J. Am. Chem. Soc.*, (1999) **121**, 6072.

[44] Ayer, W. A., *Nat. Prod. Rep.*, (1991) **8**, 455.

[45] Aver, W. A.; Trifonov, L. S., In *The Alkaloids: Chemistry and Pharmacology*, Cordell, G. A.; Brossi, A., Eds.; Academic: San Diego, (1994); Vol. 45, pp 233–266.

[46] Harayama, T.; Ohtani, M.; Oki, M.; Inubushi, Y., *Chem. Pharm. Bull.*, (1975) **23**, 1511.

[47] Harayama, T.; Ohtani, M.; Oki, M.; Inubushi, Y., *J. Chem. Soc., Chem. Commun.*, (1974), 827.

References

[48] Harayama, T.; Takatani, M.; Inubushi, Y., *Chem. Pharm. Bull.*, (1980) **28**, 1276.

[49] Harayama, T.; Takatani, M.; Inubushi, Y., *Tetrahedron Lett.*, (1979), 4307.

[50] Cassayre, J.; Gagosz, F.; Zard, S. Z., *Angew. Chem. Int. Ed.*, (2002) **41**, 1783.

[51] Kikuchi, H.; Miyagawa, Y.; Sahashi, Y.; Inatomi, S.; Haganuma, A.; Nakahata, N.; Oshima, Y., *Tetrahedron Lett.*, (2004) **45**, 6225.

[52] Min, S.-J.; Danishefsky, S. J., *Angew. Chem. Int. Ed.*, (2007) **46**, 2199.

[53] Luche, J. L., *J. Am. Chem. Soc.*, (1978) **100**, 2226.

[54] Furukawa, J.; Kawabata, N.; Nishimura, J., *Tetrahedron Lett.*, (1966), 3353.

[55] Furukawa, J.; Kawabata, N.; Nishimura, J., *Tetrahedron*, (1968) **24**, 53.

[56] Castillo, M.; Loyola, L. A.; Morales, G.; Singh, I.; Calvo, C.; Holland, H. L.; MacLean, D. B., *Can. J. Chem.*, (1976) **54**, 2893.

[57] Loyola, L. A.; Morales, G.; Castillo, M., *Phytochemistry*, (1979) **18**, 1721.

[58] Castillo, M.; Morales, G.; Loyola, L. A.; Singh, I.; Calvo, C.; Holland, H. L.; MacLean, D. B., *Can. J. Chem.*, (1975) **53**, 2513.

[59] Castillo, M.; Morales, G.; Loyola, L. A.; Singh, I.; Calvo, C.; Holland, H. L.; MacLean, D. B., *Can. J. Chem.*, (1976) **54**, 2900.

[60] Mehta, G.; Reddy, M. S., *Tetrahedron Lett.*, (1990) **31**, 2039.

[61] Sandham, D. A.; Meyers, A. I., *J. Chem. Soc., Chem. Commun.*, (1995), 2511.

[62] Mehta, G.; Reddy, M. S.; Thomas, A., *Tetrahedron*, (1998) **54**, 7865.

[63] Hirst, G. C.; Johnson, T. O., Jr.; Overman, L. E., *J. Am. Chem. Soc.*, (1993) **115**, 2992.

[64] Paquette, L. A.; Friedrich, D.; Pinard, E.; Williams, J. P.; St. Laurent, D.; Roden, B. A., *J. Am. Chem. Soc.*, (1993) **115**, 4377.

[65] Williams, J. P.; St. Laurent, D. R.; Friedrich, D.; Pinard, E.; Roden, B. A.; Paquette, L. A., *J. Am. Chem. Soc.*, (1994) **116**, 4689.

[66] Yen, C.-F.; Liao, C.-C., *Angew. Chem. Int. Ed.*, (2002) **41**, 4090.

[67] Ishizaki, M.; Niimi, Y.; Hoshino, O., *Tetrahedron Lett.*, (2003) **44**, 6029.

[68] Ishizaki, M.; Niimi, Y.; Hoshino, O.; Hara, H.; Takahashi, Y., *Tetrahedron*, (2005) **61**, 4053.

[69] Sha, C.-K.; Lee, F.-K.; Chang, C.-J., *J. Am. Chem. Soc.*, (1999) **121**, 9875.

[70] Kozaka, T.; Miyakoshi, N.; Mukai, C., *J. Org. Chem.*, (2007) **72**, 10147.

[71] Lounasmaa, M.; Hanhinen, P.; Westersund, M.; Halonen, N., In *The Alkaloids: Chemistry and Biology*, Cordell, G. A., Ed.; Academic: San Diego, (1999); Vol. 52, pp 103–195.

[72] Hamaker, L. K.; Cook, J. M., In *Alkaloids: Chemical and Biological Perspectives*, Pelletier, S. W., Ed.; Elsevier Science: New York, (1995); Vol. 9, pp 23–84.

[73] Keawpradub, N.; Kirby, G. C.; Steele, J. C. P.; Houghton, P. J., *Planta Med.*, (1999) **65**, 690.

[74] Keawpradub, N.; Eno-Amooquaye, E.; Burke, P. J.; Houghton, P. J., *Planta Med.*, (1999) **65**, 311.

[75] Miller, K. A.; Martin, S. F., *Org. Lett.*, (2007) **9**, 1113.

[76] Wellington, K. D.; Cambie, R. C.; Rutledge, P. S.; Bergquist, P. R., *J. Nat. Prod.*, (2000) **63**, 79.

[77] Madu, C. E.; Lovely, C. J., *Org. Lett.*, (2007) **9**, 4697.

[78] Mehta, G.; Srikrishna, A., *Chem. Rev.*, (1997) **97**, 671.

[79] Paquette, L. A.; Doherty, A. M., *Polyquinane Chemistry*, Springer: Berlin, (1987).

[80] Paquette, L. A., *Top. Curr. Chem.*, (1984) **119**, 1.

[81] Trost, B. M., *Chem. Soc. Rev.*, (1982) **11**, 141.

[82] Paquette, L. A., *Top. Curr. Chem.*, (1979) **79**, 41.

[83] Seto, H.; Yonehara, H., *J. Antibiot.*, (1980) **33**, 92.

[84] Trost, B. M.; Jiang, C., *Synthesis*, (2006), 369.

[85] Corey, E. J.; Guzman-Perez, A., *Angew. Chem. Int. Ed.*, (1998) **37**, 388.

[86] Fuji, K., *Chem. Rev.*, (1993) **93**, 2037.

[87] Christoffers, J.; Baro, A., *Adv. Synth. Catal.*, (2005) **347**, 1473.

[88] Zhu, Y.-Y.; Burnell, D. J., *Tetrahedron: Asymmetry*, (1996) **7**, 3295.

[89] Pallerla, M. K.; Fox, J. M., *Org. Lett.*, (2007) **9**, 5625.

[90] Kobayashi, J.; Watanabe, D.; Kawasaki, N.; Tsuda, M., *J. Org. Chem.*, (1997) **62**, 9236.

[91] Kobayashi, J.; Tsuda, M.; Ishibashi, M., *Pure Appl. Chem.*, (1991) **71**, 1123.

[92] Nagata, T.; Nakagawa, M.; Nishida, A., *J. Am. Chem. Soc.*, (2003) **125**, 7484.

[93] Ono, K.; Nakagawa, M.; Nishida, A., *Angew. Chem. Int. Ed.*, (2004) **43**, 2020.

[94] Young, I. S.; Kerr, M. A., *J. Am. Chem. Soc.*, (2007) **129**, 1465.

[95] Jakubec, P.; Cockfield, D. M.; Dixon, D. J., *J. Am. Chem. Soc.*, (2009) **131**, 16632.

[96] Fürstner, A.; Guth, O.; Rumbo, A.; Seidel, G., *J. Am. Chem. Soc.*, (1999) **121**, 11108.

[97] Fürstner, A.; Guth, O.; Duffels, A.; Seidel, G.; Liebl, M.; Gabor, B.; Mynott, R., *Chem.–Eur. J.*, (2001) **7**, 4811.

[98] Nagata, T.; Nishida, A.; Nakagawa, M., *Tetrahedron Lett.*, (2001) **42**, 8345.

[99] Magnus, P.; Fielding, M. R.; Wells, C.; Lynch, V., *Tetrahedron Lett.*, (2002) **43**, 947.

[100] Ahrendt, K. A.; Williams, R. M., *Org. Lett.*, (2004) **6**, 4539.

[101] Young, I. S.; Williams, J. L.; Kerr, M. A., *Org. Lett.*, (2005) **7**, 953.

[102] Nilson, M. G.; Funk, R. L., *Org. Lett.*, (2006) **8**, 3833.

[103] Deng, H.; Yang, X.; Tong, Z.; Li, Z.; Zhai, H., *Org. Lett.*, (2008) **10**, 1791.

[104] Inagaki, F.; Kinebuchi, M.; Miyakoshi, N.; Mukai, C., *Org. Lett.*, (2010) **12**, 1800.

[105] Wender, P. A.; Deschamps, N. M.; Gamber, G. G., *Angew. Chem. Int. Ed.*, (2003) **42**, 1853.

[106] Wender, P. A.; Deschamps, N. M.; Williams, T. J., *Angew. Chem. Int. Ed.*, (2004) **43**, 3076.

[107] Huang, S.-X.; Li, R.-T.; Liu, J.-P.; Lu, Y.; Chang, Y.; Lei, C.; Xiao, W.-L.; Yang, L.-B.; Zheng, Q.-T.; Sun, H.-D., *Org. Lett.*, (2007) **9**, 2079.

[108] Xiao, W.-L.; Li, R.-T.; Huang, S.-X.; Pu, J.-X.; Sun, H.-D., *Nat. Prod. Rep.*, (2008) **25**, 871.

[109] Sun, H.-D.; Qiu, S.-X.; Lin, L.-Z.; Wang, Z.-Y.; Lin, Z.-W.; Pengsuparp, T.; Pezzuto, J. M.; Fong, H. H. S., *J. Nat. Prod.*, (1996) **59**, 525.

[110] Watanabe, K.; Suzuki, Y.; Aoki, K.; Sakakura, A.; Suenaga, K.; Kigoshi, H., *J. Org. Chem.*, (2004) **69**, 7802.

[111] Wang, Y.-F.; Xu, L.-M.; Yu, R.-C.; Chen, J.-H.; Yang, Z., *Chem. Commun. (Cambridge)*, (2012) **48**, 8183.

[112] Tang, Y.-F.; Zhang, Y.-D.; Dai, M. J.; Luo, T. P.; Deng, L.-J.; Chen, J.-H.; Yang, Z., *Org. Lett.*, (2005) **7**, 885.

[113] Dai, M. J.; Liang, B.; Wang, C.-H.; Chen, J.-H.; Yang, Z., *Org. Lett.*, (2004) **6**, 221.

[114] Dai, M. J.; Liang, B.; Wang, C.-H.; You, Z.-J.; Xiang, J.; Dong, G.-B.; Chen, J.-H.; Yang, Z., *Adv. Synth. Catal.*, (2004) **346**, 1669.

[115] Xiao, Q.; Ren, W.-W.; Chen, Z.-X.; Sun, T.-W.; Li, Y.; Ye, Q.-D.; Gong, J.-X.; Meng, F.-K.; You, L.; Liu, Y.-F.; Zhao, M.-Z.; Xu, L.-M.; Shan, Z.-H.; Shi, Y.; Tang, Y.-F.; Chen, J.-H.; Yang, Z., *Angew. Chem. Int. Ed.*, (2011) **50**, 7373.

[116] Tan, C.-H.; Ma, X.-Q.; Chen, G.-F.; Zhu, D.-Y., *Helv. Chim. Acta*, (2002) **85**, 1058.

[117] Tang, X. C.; Han, Y. F.; Chen, X. P.; Zhu, X. D., *Acta Pharmacol. Sin.*, (1986) **7**, 507.

[118] Tang, X. C.; De Sarno, P.; Sugaya, K.; Giacobini, E., *J. Neurosci. Res.*, (1989) **24**, 276.

[119] Ishiuchi, K.; Kubota, T.; Hoshino, T.; Obara, Y.; Nakahata, N.; Kobayashi, J., *Bioorg. Med. Chem.*, (2006) **14**, 5995.

[120] Nakayama, A.; Kogure, N.; Kitajima, M.; Takayama, H., *Angew. Chem. Int. Ed.*, (2011) **50**, 8025.

[121] Yokoyama, R.; Huang, J.-M.; Yang, C.-S.; Fukuyama, Y., *J. Nat. Prod.*, (2002) **65**, 527.

[122] Cho, Y. S.; Carcache, D. A.; Tian, Y.; Li, Y. M.; Danishefsky, S. J., *J. Am. Chem. Soc.*, (2004) **126**, 14358.

[123] Carcache, D. A.; Cho, Y. S.; Hua, Z. H.; Tian, Y.; Li, Y. M.; Danishefsky, S. J., *J. Am. Chem. Soc.*, (2006) **128**, 1016.

[124] Xu, J.; Trzoss, L.; Chang, W. K.; Theodorakis, E. A., *Angew. Chem. Int. Ed.*, (2011) **50**, 3672.

[125] Trzoss, L.; Xu, J.; Lacoske, M. H.; Mobley, W. C.; Theodorakis, E. A., *Org. Lett.*, (2011) **13**, 4554.

[126] Harada, K.; Imai, A.; Uto, K.; Carter, R. G.; Kubo, M.; Hioki, H.; Fukuyama, Y., *Org. Lett.*, (2011) **13**, 988.

[127] Yang, Y.; Fu, X.; Chen, J.; Zhai, H., *Angew. Chem. Int. Ed.*, (2012) **51**, 9825.

[128] Takahashi, C.; Numata, A.; Yamada, T.; Minoura, K.; Enomoto, S.; Konishi, K.; Nakai, M.; Matsuda, C.; Nomoto, K., *Tetrahedron Lett.*, (1996) **37**, 655.

[129] Iwamoto, C.; Minoura, K.; Hagishita, S.; Nomoto, K.; Numata, A., *J. Chem. Soc., Perkin Trans. 1*, (1998), 449.

[130] Iwamoto, C.; Minoura, K.; Oka, T.; Hagishita, S.; Numata, A., *Tetrahedron*, (1999) **55**, 14353.

[131] Fujioka, K.; Yokoe, H.; Yoshida, M.; Shishido, K., *Org. Lett.*, (2012) **14**, 244.

[132] Spanevello, R. A.; Pellegrinet, S. C., *Curr. Top. Phytochem.*, (2000) **3**, 225.

[133] Takeuchi, S.; Ogawa, Y.; Yonehara, H., *Tetrahedron Lett.*, (1969), 2737.

[134] Martin, D. G.; Slomp, G.; Mizsak, S.; Duchamp, D. J.; Chidester, C. G., *Tetrahedron Lett.*, (1970), 4901.

[135] Cane, D. E.; Sohng, J.-K.; Williard, P. G., *J. Org. Chem.*, (1992) **57**, 844.

[136] Duszenko, M.; Mecke, D., *Mol. Biochem. Parasitol.*, (1986) **19**, 223.

[137] Willson, M.; Lauth, N.; Perie, J.; Callens, M.; Opperdoes, F. R., *Biochemistry*, (1994) **33**, 214.

[138] Cane, D. E.; Sohng, J.-K., *Biochemistry*, (1994) **33**, 6524.

References

631

[139] Liu, Q.; Yue, G.-Z.; Wu, N.; Lin, G.; Li, Y.-Z.; Quan, J. M.; Li, C.-C.; Wang, G.-X.; Yang, Z., *Angew. Chem. Int. Ed.*, (2012) **51**, 12072.

[140] Jang, J.-H.; Asami, Y.; Jang, J.-P.; Kim, S.-O.; Moon, D. O.; Shin, K.-S.; Hashizume, D.; Muroi, M.; Saito, T.; Oh, H.; Kim, B. Y.; Osada, H.; Ahn, J. S., *J. Am. Chem. Soc.*, (2011) **133**, 6865.

[141] Li, J. Y.; Strobel, G. A.; Harper, J. K.; Lobkovsky, E.; Clardy, J., *Org. Lett.*, (2000) **2**, 767.

[142] Deng, J.; Zhu, B.; Lu, Z.-Y.; Yu, H.-X.; Li, A., *J. Am. Chem. Soc.*, (2012) **134**, 920.

[143] Xu, J.; Caro-Diaz, J. E.; Trzoss, L.; Theodorakis, E. A., *J. Am. Chem. Soc.*, (2012) **134**, 5072.

[144] Xu, J.; Caro-Diaz, J. E.; Lacoske, M. H.; Hung, C.-I.; Jamora, C.; Theodorakis, E. A., *Chem. Sci.*, (2012) **3**, 3378.

[145] Yin, J.; Wang, C.; Kong, L.-L.; Cai, S.-J.; Gao, S.-H., *Angew. Chem. Int. Ed.*, (2012) **51**, 7786.

[146] Huang, J.; Fang, L.-C.; Long, R.; Shi, L.-L.; Shen, H.-J.; Li, C.-C.; Yang, Z., *Org. Lett.*, (2013) **15**, 4018.

[147] Kablaoui, N. M.; Hicks, F. A.; Buchwald, S. L., *J. Am. Chem. Soc.*, (1996) **118**, 5818.

[148] Kablaoui, N. M.; Hicks, F. A.; Buchwald, S. L., *J. Am. Chem. Soc.*, (1997) **119**, 4424.

[149] Crowe, W. E.; Vu, A. T., *J. Am. Chem. Soc.*, (1996) **118**, 1557.

[150] Chatani, N.; Morimoto, T.; Fukumoto, Y.; Murai, S., *J. Am. Chem. Soc.*, (1998) **120**, 5335.

[151] Chatani, N.; Morimoto, T.; Kamitani, A.; Fukumoto, Y.; Murai, S., *J. Organomet. Chem.*, (1999) **579**, 177.

[152] Kang, S.-K.; Kim, K.-J.; Hong, Y.-T., *Angew. Chem. Int. Ed.*, (2002) **41**, 1584.

[153] Saito, T.; Shiotani, T.; Otani, T.; Hasaba, S., *Heterocycles*, (2003) **60**, 1045.

[154] Yu, C.-M.; Hong, Y.-T.; Lee, J.-H., *J. Org. Chem.*, (2004) **69**, 8506.

[155] Saito, T.; Sugizaki, K.; Otani, T.; Suyama, T., *Org. Lett.*, (2007) **9**, 1239.

[156] Adrio, J.; Carretero, J. C., *J. Am. Chem. Soc.*, (2007) **129**, 778.

[157] Saito, T.; Nihei, H.; Otani, T.; Suyama, T.; Furukawa, N.; Saito, M., *Chem. Commun. (Cambridge)*, (2008), 172.

[158] Gao, P.; Xu, P.-F.; Zhai, H., *J. Org. Chem.*, (2009) **74**, 2592.

[159] Salway, A. H., *J. Chem. Soc., Trans.*, (1911) **99**, 2148.

[160] Jobst, J.; Hesse, O., *Justus Liebigs Ann. Chem.*, (1864) **129**, 115.

[161] Koelle, G. B., In *The Pharmacological Basis of Therapeutics*, 5th ed., Goodman, R. S.; Gilman, A., Eds.; Macmillan: New York, (1975); pp 445, 466.

[162] Walker, M. B., *Lancet*, (1934) **223**, 1200.

[163] Desphpande, S. S.; Viana, G. B.; Kauffman, F. C.; Rickett, D. L.; Albuquerque, E. X., *Fundam. Appl. Toxicol.*, (1986) **6**, 566.

[164] Mohs, R. C.; Davis, B. M.; Johns, C. A.; Mathé, A. A.; Greenmald, B. S.; Horvath, T. B.; Davis, K. L., *Am. J. Psychiatry*, (1985) **142**, 28.

[165] Mohs, R. C.; Davis, B. M.; Greenmald, B. S.; Mathé, A. A.; Johns, C. A.; Horvath, T. B.; Davis, K. L., *J. Am. Geriatr. Soc.*, (1985) **33**, 749.

[166] Beller, S. A.; Overall, J. E.; Swann, A. C., *Psychopharmacology (Berl.)*, (1985) **87**, 147.

[167] Caltagirone, C.; Gainotti, G.; Masullo, C., *Int. J. Neurosci.*, (1982) **16**, 247.

[168] Jatkowitz, S., *Ann. Neurol.*, (1983) **14**, 690.

[169] Agnoli, A.; Martucci, N.; Manna, V.; Conli, L.; Fioravanti, M., *Clin. Neuropharmacol.*, (1983) **6**, 311.

[170] Nickel, A.; Maruyama, T.; Tang, H.; Murphy, P. D.; Greene, B.; Yusuff, N.; Wood, J. L., *J. Am. Chem. Soc.*, (2004) **126**, 16300.

[171] Aburano, D.; Yoshida, T.; Miyakoshi, N.; Mukai, C., *J. Org. Chem.*, (2007) **72**, 6878.

[172] Huang, J.-M.; Yokoyama, R.; Yang, C.-S.; Fukuyama, Y., *Tetrahedron Lett.*, (2000) **41**, 6111.

[173] Huang, J.-M.; Yang, C.-S.; Tanaka, M.; Fukuyama, Y., *Tetrahedron*, (2001) **57**, 4691.

[174] Nozari, M. S., DE 2609148, (1976); *Chem. Abstr.*, (1976) **85**, 162115.

[175] Siefken, M. W., DE 2609150, (1976); *Chem. Abstr.*, (1976) **85**, 193902.

[176] Koshar, R. J.; Barber, L. L., Jr., US 4053519, (1977); *Chem. Abstr.*, (1978) **88**, 38575.

[177] Chen, J.; Gao, P.; Yu, F.; Yang, Y.; Zhu, S.; Zhai, H., *Angew. Chem. Int. Ed.*, (2012) **51**, 5897.

[178] Pauson, P. L., *Tetrahedron*, (1985) **41**, 5855.

[179] Shibata, T.; Koga, Y.; Narasaka, K., *Bull. Chem. Soc. Jpn.*, (1995) **68**, 911.

[180] Ahmar, M.; Antras, F.; Cazes, B., *Tetrahedron Lett.*, (1995) **36**, 4417.

[181] Ahmar, M.; Chabanis, O.; Gauthier, J.; Cazes, B., *Tetrahedron Lett.*, (1997) **38**, 5277.

[182] Antras, F.; Ahmar, M.; Cazes, B., *Tetrahedron Lett.*, (2001) **42**, 8153.

[183] Antras, F.; Ahmar, M.; Cazes, B., *Tetrahedron Lett.*, (2001) **42**, 8157.

[184] Brummond, K. M.; Chen, H.; Fisher, K. D.; Kerekes, A. D.; Rickards, B.; Sill, P. C.; Geib, S. J., *Org. Lett.*, (2002) **4**, 1931.

[185] Kobayashi, T.; Koga, Y.; Narasaka, K., *J. Organomet. Chem.*, (2001) **624**, 73.

[186] Mukai, C.; Nomura, I.; Yamanishi, K.; Hanaoka, M., *Org. Lett.*, (2002) **4**, 1755.

[187] Brummond, K. M.; Kerekes, A. D.; Wan, H., *J. Org. Chem.*, (2002) **67**, 5156.

[188] Brummond, K. M.; Gao, D., *Org. Lett.*, (2003) **5**, 3491.

[189] Lu, B.-L.; Wei, Y.; Shi, M., *Organometallics*, (2012) **31**, 4601.

[190] Yuan, W.; Dong, X.; Shi, M.; McDowell, P.; Li, G.-G., *Org. Lett.*, (2012) **14**, 5582.

[191] Castro, V.; Cicio, F.; Alvarado, S.; Bohlmann, F.; Schmeda-Hirschmann, G.; Jakupovic, J., *Liebigs Ann. Chem.*, (1983), 974.

[192] Hirose, T.; Miyakoshi, N.; Mukai, C., *J. Org. Chem.*, (2008) **73**, 1061.

[193] Ueno, Y.; Chino, K.; Watanabe, M.; Moriya, O.; Okawara, M., *J. Am. Chem. Soc.*, (1982) **104**, 5564.

[194] Wakamatsu, T.; Hara, H.; Ban, Y., *J. Org. Chem.*, (1985) **50**, 108.

[195] Ueno, Y.; Moriya, O.; Chino, K.; Watanabe, M.; Okawara, M., *J. Chem. Soc., Perkin Trans. 1*, (1986), 1351.

[196] Kitson, R. R. A.; Millemaggi, A.; Taylor, R. J. K., *Angew. Chem. Int. Ed.*, (2009) **48**, 9426.

[197] Hoffmann, H. M. R.; Rabe, J., *Angew. Chem. Int. Ed. Engl.*, (1985) **24**, 94.

[198] Grillet, F.; Huang, C.-F.; Brummond, K. M., *Org. Lett.*, (2011) **13**, 6304.

[199] Christensen, S.; Skytte, D.; Denmeade, S. R.; Dionne, C.; Møller, J.; Nissen, P.; Isaacs, J. T., *Adv. Anticancer Agents Med. Chem.*, (2009) **9**, 276.

[200] Wagner, S.; Hofmann, A.; Siedle, B.; Terfloth, L.; Merfort, I.; Gasteiger, J., *J. Med. Chem.*, (2006) **49**, 2241.

[201] Bruno, M.; Rosselli, S.; Maggio, A.; Raccuglia, R. A.; Bastow, K. F.; Lee, K.-H., *J. Nat. Prod.*, (2005) **68**, 1042.

[202] Lee, S.-H.; Kim, H.-K.; Seo, J.-M.; Kang, H.-M.; Kim, J.-H.; Son, K.-H.; Lee, H.; Kwon, B.-M.; Shin, J.; Seo, Y., *J. Org. Chem.*, (2002) **67**, 7670.

[203] Blanco, J. G.; Gil, R. R.; Bocco, J. L.; Meragelman, T. L.; Genti-Raimondi, S.; Flury, A., *J. Pharmacol. Exp. Ther.*, (2001) **297**, 1099.

[204] Dirsch, V. M.; Stuppner, H.; Vollmar, A. M., *Cancer Res.*, (2001) **61**, 5817.

[205] Yuuya, S.; Hagiwara, H.; Suzuki, T.; Ando, M.; Yamada, A.; Suda, K.; Kataoka, T.; Nagai, K., *J. Nat. Prod.*, (1999) **62**, 22.

[206] Giordano, O. S.; Pestchanker, M. J.; Guerreiro, E.; Saad, J. R.; Enriz, R. D.; Rodríguez, A. M.; Jáuregui, E. A.; Guzmán, J.; María, A. O. M.; Wendel, G. H., *J. Med. Chem.*, (1992) **35**, 2452.

[207] Konaklieva, M. I.; Plotkin, B. J., *Mini-Rev. Med. Chem.*, (2005) **5**, 73.

[208] Singh, J.; Petter, R. C.; Baillie, T. A.; Whitty, A., *Nat. Rev. Drug Discovery*, (2011) **10**, 307.

[209] Potashman, M. H.; Duggan, M. E., *J. Med. Chem.*, (2009) **52**, 1231.

[210] Grillet, F.; Brummond, K. M., *J. Org. Chem.*, (2013) **78**, 3737.

[211] García, A.; Delgado, G., *Helv. Chim. Acta*, (2006) **89**, 16.

[212] Hayashi, Y.; Miyakoshi, N.; Kitagaki, S.; Mukai, C., *Org. Lett.*, (2008) **10**, 2385.

[213] Ogbourne, S. M.; Suhrbier, A.; Jones, B.; Cozzi, S.-J.; Boyle, G. M.; Morris, M.; McAlpine, D.; Johns, J.; Scott, T. M.; Sutherland, K. P.; Gardner, J. M.; Le, T. T. T.; Lenarczyk, A.; Aylward, J. H.; Parsons, P. G., *Cancer Res.*, (2004) **64**, 2833.

[214] Fujiwara, M.; Ijichi, K.; Tokuhisa, K.; Katsuura, K.; Shigeta, S.; Konno, K.; Wang, G. Y.; Uemura, D.; Yokota, T.; Baba, M., *Antimicrob. Agents Chemother.*, (1996) **40**, 271.

[215] Vasas, A.; Rédei, D.; Csupor, D.; Molnár, J.; Hohmann, J., *Eur. J. Org. Chem.*, (2012), 5115.

[216] Hasler, C. M.; Acs, G.; Blumberg, P. M., *Cancer Res.*, (1992) **52**, 202.

[217] U. S. Food and Drug Administration, 2012 Novel New Drugs Summary; FDA Publication, (2013); available online at www.fda.gov/downloads/Drugs/DevelopmentApprovalProcess/DrugInnovation/UCM337830.pdf (accessed August 2015).

[218] Alder, R. W.; East, S. P., *Chem. Rev.*, (1996) **96**, 2097.

[219] Kuwajima, I.; Tanino, K., *Chem. Rev.*, (2005) **105**, 4661.

[220] Tanino, K.; Onuki, K.; Asano, K.; Miyashita, M.; Nakamura, T.; Takahashi, Y.; Kuwajima, I., *J. Am. Chem. Soc.*, (2003) **125**, 1498.

[221] Winkler, J. D.; Rouse, M. B.; Greaney, M. F.; Harrison, S. J.; Jeon, Y. T., *J. Am. Chem. Soc.*, (2002) **124**, 9726.

[222] Jørgensen, L.; McKerrall, S. J.; Kuttruff, C. A.; Ungeheuer, F.; Felding, J.; Baran, P. S., *Science (Washington, D. C.)*, (2013) **341**, 878.

Keyword Index

In this keyword index, which should be used in conjunction with the Table of Contents, starting material entries are indicated in an *italic font*, product entries are identified by an arrow (→), and all other entries are given in a roman font.

A

→ Abietic acid, dehydro-, precursor, chiral, from enantioselective polyene cyclization of a 1,3,7-triene, chiral binaphthol–Lewis acid complex catalyzed 20

→ (±)-Abrotanone, tricyclic intermediate, from al-kynylindene, via domino cycloisomerization, transition-metal catalyzed 329, 330

→ *ent*-Abudinol B, key intermediate, from diep-oxyalkenes, via *endo*-selective epoxide-open-ing cascade, silyl trifluoromethanesulfonate initiated, alkene terminated 58–60

→ *ent*-Abudinol B, key intermediate, from domino cyclization of enyne diepoxides, Lewis acid induced 294

Acetals, cyclic fused, ring-enlarging tetrahydrofuran annulation, tin(IV) chloride promoted, via Prins cyclization/pinacol rearrangement, 1-oxabicyclic ring system synthesis 277, 278

Acetates, diazo-, alkyne substituted, rhodium-cata-lyzed rearrangement/cyclization, bridged polycy-clic compound synthesis, via cascade sequence 521, 522

Acetates, diazo-, alkyne tethered, reaction with 4-methoxybenzaldehyde, rhodium catalyzed, cy-clopenta[b]furan-6-one synthesis, via intermolec-ular ylide formation/intramolecular dipolar cy-cloaddition cascade sequence 526

Acetates, enynyl, tandem [3,3]-acyloxy rearrange-ment/Nazarov cyclization sequence, gold/silver catalyzed, cyclopent-2-enone synthesis 563

Acetates, propargyl, cross coupling with arylboronic acids, gold complex catalyzed, α-arylenone syn-thesis, via domino acyloxy migration/oxidation/reductive elimination sequence 562

Acetates, propargyl, [4+3] cycloaddition with 1,3-dienes, cationic gold complex catalyzed, cyclo-hepta-1,4-diene synthesis, via domino 1,2-acetoxy migration/cyclopropanation/homo-Cope rear-rangement sequence 569–571

Acetates, propargyl, reaction with N-halosuccin-imides, gold complex catalyzed, α-haloenone syn-thesis, via domino rearrangement/allene halo-genation sequence 561

Acetoacetates, diazo-, allyl ester substituted/alkyne tethered, carbene metathesis/cyclization/Claisen/Cope cascade sequence, copper catalyzed, bicyclic butenolide synthesis 523, 524

Acetylenedicarboxylates, three-component domino condensation process with amines and formalde-hyde, ytterbium catalyzed, 3,4,5-trisubstituted 3,6-dihydro-1,3-oxazine synthesis 590, 591

→ (+)-Achalensolide, bicyclic cyclopentenone intermediate, from a functionalized allenyne and carbon monoxide, via intramolecular Pauson–Khand-type cyclization, rhodium/phosphine catalyzed 623

Acid chlorides, reaction with organosilyllithium reagents, via addition/1,2-Brook rearrangement cascade sequence, then organic halide addition, α-silyl carbinol synthesis 358

Acids – see also Carboxylic acids

→ Acids, cyclic, ring contracted, from cyclic α-halo ketones, via Favorskii rearrangement, base promoted 260

Acids, hydroxy – see α-Hydroxy acids

Acrylamides, azido-N-(2-iodoaryl)-, tandem radical cyclization, using tris(trimethylsilyl)silane/2,2'-azobisisobutyronitrile, (±)-horsfiline precursor synthesis 223, 224

→ (–)-Acylfulvene, key intermediate, from a dien-yne, via domino double enyne ring-closing metathesis, ruthenium alkylidene complex catalyzed, then tetrabutylammonium fluo-ride treatment 102, 103

Acyloins, reaction with organosilyllithium reagents, via addition/1,2-Brook rearrangement cascade sequence, silyl enol ether synthesis 357

→ (±)-Agarospirol, key intermediate, from a spiro-cyclobutyl α-hydroxy ester, via retro-benzilic acid rearrangement, base promoted 260, 261

→ (+)-Akuammicine, tetracyclic intermediate, from reaction of N-protected 3-(2-aminoeth-yl)-2-[(methylsulfanyl)vinyl]indoles with pro-pynal, via asymmetric tandem Diels–Alder/amine cyclization sequence, chiral secondary amine catalyzed 341

Alcohols, allyl – see Allyl alcohols

Alcohols, amino – see Amino alcohols

Alcohols, benzyl – see Benzyl alcohols

Alcohols, cyclopropyl, carbocyclic ring expansion, acid induced, lactone synthesis 252

Alcohols, dienyl – see Dienols; also Dienynols

Alcohols, epoxy, domino intramolecular open-ing/cyclization, acid catalyzed, endo versus exo selectivity 248, 249

Alcohols, epoxy, intramolecular epoxide opening/cy-clization, base catalyzed, cyclic ether synthesis 250

Alcohols, epoxy, endo-selective intramolecular ep-oxide opening/cyclization, base catalyzed, using directing groups, face-fused tetrahydropyran syn-thesis 251

Alcohols, 2,3-epoxy, reductive rearrangement/ring expansion, aluminum triisopropoxide catalyzed, spirocyclic diol synthesis 272, 273

634 Keyword Index

Alcohols, 2,3-epoxy, semipinacol rearrangement/ ring expansion, Lewis acid catalyzed, spirocyclic β-hydroxy ketone synthesis 269–271

→ Alcohols, secondary, from tricyclic cyclobutanols, via Wagner–Meerwein ring contraction/ ring opening, acid induced 255, 256

Aldehydes, aryl-, domino asymmetric aldol–Tishchenko reaction with aryl methylene ketones, chiral lanthanum complex catalyzed, then methanolysis, chiral 1,3-diol synthesis 579, 580

Aldehydes, aryl-, reaction with arenes, scandium catalyzed, via domino intermolecular/intramolecular C—H functionalization sequence, 9-substituted xanthene synthesis 595, 596

Aldehydes, aryl-, sila-Morita–Baylis–Hillman reaction with aryl 1-silylvinyl ketones, triarylphosphine catalyzed, via 1,3-Brook rearrangement, 2-[aryl(siloxy)methyl]-1-arylprop-2-enone synthesis 394

Aldehydes, diketo, domino asymmetric intermolecular nitroaldol reaction with nitromethane/intramolecular nitroaldol reaction/epimerization sequence, praseodymium complex catalyzed, chiral bicyclic nitro ketone synthesis 578

Aldehydes, nucleophilic O-silylation and stannylation, then domino lithium transmetalation/retro-1,2-Brook rearrangement sequence, α-silyl carbinol synthesis 387, 388

→ Aldehydes, polycyclic, chiral, from polyenes, via asymmetric domino oxidative radical cyclization, organo-SOMO activated, chiral imidazolidin-4-one catalyzed 198–201

Aldehydes, reaction with [(silyl)dichloromethyl]lithium, via domino nucleophilic addition/1,3-Brook rearrangement/electrophilic capture sequence, α,α-dichlorosilyl ether synthesis 393, 394

Aldehydes, reductive McMurry alkenation cascade, via 1,2-Brook rearrangement, using trichlorosilane, chromium catalyzed, alkene synthesis 356

Aldehydes, sila-Morita–Baylis–Hillman reaction with silylallenoates, lithium isopropoxide catalyzed, via 1,3-Brook rearrangement, trisubstituted allene synthesis 394, 395

Aldehydes, three-component domino condensation process with ferrocenylacetylenes and anilines, cerium catalyzed, 4-ferrocenylquinoline synthesis 588

Aldehydes, three-component reaction with terminal alkynes and hydrazines, gold/silver catalyzed, polyfunctionalized 2,3-dihydropyrazole synthesis, via cyclization domino reaction 540

Aldimines, three-component reaction with terminal alkynes and tosyl isocyanate, gold/silver catalyzed, oxazolidin-2-imine synthesis, via cyclization domino reaction 538

Alkadienes – see 1,3-Dienes, 1,4-Dienes, 1,5-Dienes, 1,6-Dienes

→ Alkaloid (−)-205B, key intermediate, from reaction of lithiated 2-silyl-1,3-dithianes with epoxide and N-tosylaziridine, via 1,4-Brook rearrangement cascade sequence 442, 444

→ Alkane-2,4-diones, 3-alkylidene-, from propargyl esters, via domino intramolecular rearrangement reaction, gold complex catalyzed 557, 558

→ Alkanethiones, sulfanylbis(silyl)-, from reaction of carbon disulfide with [tris(silyl)methyl]lithium/alkyl iodides, via nucleophilic addition/ 1,3-thia-Brook rearrangement/electrophilic capture/ring opening cascade sequence 401

→ Alkanols, β,β-dihalo-, from reaction of oxiranes with dihalo(silyl)methyllithium, via lithiation/ring opening/1,4-Brook rearrangement/ electrophilic capture cascade sequence 416

→ Alkan-2-ols, 1-aryl-, O-substituted, from three-component reaction of alkenes with alcohols and arylboronic acids, gold complex catalyzed, via intermolecular oxidative oxyarylation domino reaction 556

Alkenes, amino-, reaction with arylboronic acids, gold complex catalyzed, functionalized 2-benzylpyrrolidine synthesis, via intramolecular arylation domino reaction 555

Alkenes, diepoxy-, endo-selective epoxide-opening cascade, silyl trifluoromethanesulfonate initiated, alkene terminated, 3-siloxy-8-vinylidenedodecahydroindeno[5,4-b]oxepin synthesis 58, 59

Alkenes, epoxy-, aryl substituted, endo-selective epoxide-opening cascade, indium bromide initiated, arene terminated, octahydrophenanthren-2-ol synthesis 60–62

Alkenes, epoxy-, O-protected phenol substituted, endo-selective epoxide-opening cascade, Lewis acid initiated, protected phenol terminated, substituted hexahydroxanthen-2-ol synthesis 63

Alkenes, epoxy-, O-protected phenol substituted, tandem endo-selective epoxide-opening/electrophilic substitution cascade, Lewis acid initiated, protected phenol terminated, substituted hexahydroxanthen-2-ol synthesis 63, 64

→ Alkenes, from reductive McMurry aldehyde alkenation cascade, via 1,2-Brook rearrangement, using trichlorosilane, chromium catalyzed 356

Alkenes, Mukaiyama hydration/hydroperoxidation, using oxygen, metal complex catalyzed, then metathesis with triethylsilane, silylated peroxide synthesis 163, 164

→ Alkenes, siloxy-, from reaction of oxiranes with lithiated allyltriphenylsilane, via nucleophilic addition/1,4-Brook rearrangement/electrophilic capture cascade sequence 417, 418

Alkenes, three-component reaction with alcohols and arylboronic acids, gold complex catalyzed, O-substituted 1-arylalkan-2-ol synthesis, via intermolecular oxidative oxyarylation domino reaction 556

→ Alkenes, trisubstituted, from reaction of bis-silylated allyl alcohols with alkyl halides, copper(I) tert-butoxide promoted, via cupration/ 1,4-Brook rearrangement/electrophilic trapping cascade sequence 428, 429

Keyword Index

635

Alkenes, trisubstituted, reaction with arenecarbaldehydes, via carbonyl–alkene cross metathesis, trityl cation catalyzed, β-alkylstyrene synthesis 151, 152

→Alkenols, functionalized, from reaction of 2-silyl-1,3-dithianyl functionalized vinyloxiranes with organometallic reagents, via ring opening/1,4-Brook rearrangement/electrophilic capture cascade sequence 415, 416

→Alkenols, silylated, from 1,3-bis(silyl)propenes, via lithiation/reaction with epoxides/1,4-Brook rearrangement/electrophilic capture cascade sequence 417

→Alk-4-enols, from reaction of oxiranes with lithiated prop-1-ene-1,3-diylbis(triphenylsilane), via nucleophilic addition/1,4-Brook rearrangement/electrophilic capture cascade sequence 417, 418

→Alkynediols, chiral, from enantioselective reaction of allyl-substituted acylsilanes with alkynylsilanes, chiral diarylprolinol catalyzed, zinc promoted, via nucleophilic alkynylation/1,2-Brook rearrangement/ene–allene carbocyclization/electrophilic trapping/oxidation cascade sequence 373, 374

Alkynes, borylated, enyne cross metathesis with terminal alkene, ruthenium alkylidene complex catalyzed, (–)-amphidinolide K intermediate synthesis 91, 92

Alkynes, cross-metathesis reactions with alkenes, ruthenium alkylidene complex catalyzed, 1,3-diene synthesis, selectivity in natural product synthesis 86, 87

Alkynes, epoxy-, reaction with N-halosuccinimides, gold complex catalyzed, 7-chlorotetrahydrooxocin-4-one synthesis, via double ring-expansion domino reaction 553, 554

Alkynes, terminal, cross metathesis with ethene, then ring-closing metathesis with 2-methylpenta-1,4-diene, both ruthenium alkylidene complex catalyzed, (–)-amphidinolide E key intermediate synthesis 117–119

1,6-Allenenes, intramolecular cycloisomerization domino reaction, gold/silver catalyzed, tetracyclic carbocycle synthesis 549

1,7-Allenenes, intramolecular asymmetric [2+2] cycloaddition domino reaction, gold–chiral phosphoramidite/silver catalyzed, chiral bicyclic carbocycle synthesis 549

Allenes, epoxy-, intramolecular cyclization domino reaction, gold/silver catalyzed, cyclic ether synthesis 554

→Allenes, tetrasubstituted, from propargyl silyl ethers, via domino α-lithiation/retro-1,2-Brook rearrangement sequence, then methoxymethyl chloride capture, lithiation, and reaction with electrophiles 392

→Allenes, tetrasubstituted, from three-component reaction of silyl glyoxylates with magnesium acetylides and nitroalkenes, via nucleophilic addition/1,2-Brook rearrangement cascade sequence 368

→Allenes, trisubstituted, from sila-Morita–Baylis–Hillman reaction of aldehydes with silylallenoates, lithium isopropoxide catalyzed, via 1,3-Brook rearrangement 394, 395

Allenynes, functionalized, conversion into (hydroxymethyl)acylfulvene, via intramolecular Pauson–Khand cyclization, using hexacarbonylmolybdenum 604

1,5-Allenynes, intramolecular hydrative carbocyclization domino reaction, gold/silver catalyzed, 4-alkanoylcyclopentene synthesis 550

1,7-Allenynes, intramolecular hydrative carbocyclization domino reaction, gold/silver catalyzed, 1-alkanoyl-2-vinylcyclopentane synthesis 550

→(–)-Allonorsecurinine, from a dienyne, via domino double enyne ring-closing metathesis, ruthenium alkylidene complex catalyzed, then allylic bromination and deprotection-induced N-alkylation 110, 111

Allyl alcohols, bis-silylated, reaction with alkyl halides, copper(I) tert-butoxide promoted, via cupration/1,4-Brook rearrangement/electrophilic trapping cascade sequence, trisubstituted alkene synthesis 428, 429

Allyl alcohols, 3-silylated, ipso substitution, by reaction with alkyl halides, copper(I) tert-butoxide promoted, palladium catalyzed, via cupration/1,4-Brook rearrangement/alkylation cascade sequence, trisubstituted allyl alcohol synthesis 427, 428

Allyl alcohols, 3-silylated, ipso substitution, by reaction with allylic halides, copper(I) tert-butoxide promoted, via cupration/1,4-Brook rearrangement/alkylation cascade sequence, dienyl alcohol synthesis 427

→Allyl alcohols, substituted, from reaction of 3-phenylpropanals with [3-(trimethylsilyl)-prop-1-en-2-yl]lithium, via nucleophilic addition/1,4-Brook rearrangement/electrophilic capture cascade sequence 418, 419

Allyl alcohols, 2-(triphenylsilyl)-, reaction with allyl chlorides/copper tert-butoxide, via nucleophilic addition/1,3-Brook rearrangement/electrophilic trapping cascade sequence, 2-(siloxymethyl)-1,4-diene synthesis 399

→Allyl alcohols, trisubstituted, from ipso substitution of 3-silylated allyl alcohols, by reaction with alkyl halides, copper(I) tert-butoxide promoted, palladium catalyzed, via cupration/1,4-Brook rearrangement/alkylation cascade sequence 427, 428

Allylic alcohols, cyclobutyl substituted, domino Prins/semipinacol rearrangement, acid catalyzed, spiro-fused cyclopentanone synthesis 257, 258

636   Keyword Index

Allylic alcohols, enantioselective polyene cyclization, iridium/chiral dioxaphosphepin catalyzed, chiral hexahydrophenanthrene synthesis 24

→ Allylic alcohols, from reaction of acylsilanes with lithiated epoxysilanes, via nucleophilic addition/1,2-Brook rearrangement/ring opening cascade sequence 360

→ (+)-Alopecuridine, key intermediate, from a 2,3-epoxy alcohol, via semipinacol rearrangement/ring expansion, Lewis acid catalyzed 270, 271

→ (−)-Alstonerine, pentacyclic cyclopentenone intermediate, from a functionalized azaenyne and octacarbonyldicobalt, via intramolecular Pauson–Khand cyclization 608, 609

→ Amides, γ-amino-β-hydroxy, from three-component reaction of acylsilanes with imines and amide-derived enolates, via nucleophilic addition/1,2-Brook rearrangement/imine addition cascade sequence 367

→ Amides, β-azido-α-hydroxy-, chiral, from α,β-unsaturated amides, via one-pot sequential enantioselective epoxidation/ring opening process, using alkyl peroxides then azidosilanes, chiral samarium complex catalyzed 582

Amides, α,β-unsaturated, one-pot sequential enantioselective epoxidation/ring opening process, using alkyl peroxides then azidosilanes, chiral samarium complex catalyzed, chiral β-azido-α-hydroxyamide synthesis 582

Amines, aryl-, three-component domino condensation process with aldehydes and terminal alkynes, ytterbium catalyzed, 2,4-disubstituted quinoline synthesis 586, 587

Amines, aryl-, three-component domino condensation process with β-keto esters and α,β-unsaturated aldehydes, cerium catalyzed, 1,4-dihydropyridine synthesis 589, 590

→ 1,5-Amino alcohols, from reaction of 2-(hydroxyethyl)-2-silyl-1,3-dithianes with N-sulfonylaziridines, via lithiation/1,4-Brook rearrangement/ring opening cascade sequence 422

→ (−)-Amphidinolide E, key intermediate, from a terminal alkyne, via cross metathesis with ethene, then ring-closing metathesis with 2-methylpenta-1,4-diene, both ruthenium alkylidene complex catalyzed 117–119

→ (−)-Amphidinolide K, key intermediate, from enyne cross metathesis of a borylated alkyne with a terminal alkene, ruthenium alkylidene complex catalyzed 91, 92

→ Amphidinolide V, key intermediate, from an ester-tethered 1,15-diyne, via ring-closing metathesis, molybdenum complex catalyzed, then enyne cross metathesis with ethene, ruthenium alkylidene complex catalyzed 88, 89

→ Amphidinolide V, key intermediate, from proline-mediated aldol condensation of two aldehydes, one from enyne cross metathesis of an alkynol with ethene, the other from enyne ring-closing metathesis of an alkynylsilyl ether with ethene, both ruthenium alkylidene complex catalyzed 90, 91

→ (+)-Anatoxin-a, from a 1,8-enyne, via ring-closing metathesis, ruthenium alkylidene complex catalyzed, then oxidative cleavage and deprotection 97, 98

→ (+)-Angelichalcone, key intermediate, from distal O-protected phenol-substituted epoxides, via domino cyclization, Lewis acid induced 298, 300

→ Angucyclinone-type natural products, from 1,7-enynes, via ring-closing metathesis, ruthenium alkylidene complex catalyzed, then Diels–Alder cycloaddition with naphtho-1,4-quinones, elimination/aromatization, oxidation, and deprotection 125, 126

Anilines, N-(4-hydroxybut-2-enyl)-2-(3-hydroxyprop-1-ynyl)-, enantioselective intramolecular cyclization domino reaction, chiral gold complex catalyzed, chiral [1,4]oxazino[4,3-a]indole synthesis 551, 552

→ Antiochic acid, key intermediate, from reaction of an aryl-substituted diene with an acetal, via intermolecular domino Prins reaction/polyene cyclization sequence, tin(IV) chloride promoted 311, 313

→ Antrocin, tricyclic intermediate, from a diynol acid, via domino cycloisomerization, gold complex catalyzed 332

→ Apicularen A, key intermediate, from reaction of lithiated 2-silyl-1,3-dithianes with two epoxides, via 1,4-Brook rearrangement cascade sequence 439, 440

Aryl bromides, polysubstituted, tandem radical cyclization, using tributyltin hydride/2,2′-azobisisobutyronitrile, then reductive amination, using lithium/ammonia, morphine precursor synthesis 230

Aryl iodides, cyclohex-2-enyloxy substituted, tandem reductive radical cyclization, using tri-sec-butylborane/oxygen/tris(trimethylsilyl)silane, (±)-bisabosqual A precursor synthesis 228, 229

→ Asperolide C, from enantioselective polyene cyclization of allylic alcohols, iridium/chiral dioxaphosphepine catalyzed 24

→ (+)-Asperpentyn, from enediyne, via ring-closing metathesis/metallotropic [1,3]-shift domino sequence, ruthenium alkylidene complex catalyzed 120, 121

→ Aspidospermidine, precursor, from tandem radical cyclization of azido-N-(2-iodoaryl)-acrylamides, using tris(trimethylsilyl)silane/2,2′-azobisisobutyronitrile 223, 224

Keyword Index

→Aspidospermidine, tetracyclic intermediate, from reaction of N-protected 3-(2-aminoethyl)-2-[(methylsulfanyl)vinyl]indoles with propynal, via asymmetric tandem Diels–Alder/amine cyclization sequence, chiral secondary amine catalyzed 340

→Aspidospermidine, tricyclic intermediate, from intramolecular cyclization of an aminoalkyne, palladium/bipyridine catalyzed, via aminopalladation/oxidative addition cascade sequence 465, 466

→Asterriquinone A1, demethyl-, key intermediate, from cyclization of an alkenyl dihalide with an alkenylamine, palladium/phosphine catalyzed, via amination/isomerization/amination cascade sequence 467, 468

→3-Azabicyclo[4.2.0]oct-5-enes, functionalized, from cycloisomerization of 1,7-enyne benzoates, cationic gold complex catalyzed, via intramolecular domino 1,3-ester migration/[2+2] cycloaddition sequence 565, 566

→Azaspiracid-1, trioxadispiroketal intermediate, from intramolecular domino cyclization of polyfunctionalized acetals, Lewis acid promoted 313, 314

Azides, dienyl, reaction with enones, via tandem intermolecular Diels–Alder/intramolecular Schmidt reaction sequence, Lewis acid promoted, fused nitrogen heterocycle synthesis 341, 342

Azides, α-siloxy epoxy, tandem semipinacol/Schmidt domino process, titanium(IV) chloride promoted, fused hydroxy lactam synthesis 273, 274

Aziridines, 2-acyl-, reaction with organosilyllithium reagents, via addition/1,2-Brook rearrangement/ring-opening cascade sequence, β-amino ketone synthesis 358

Aziridines, 2-(oxiran-2-yl)-N-tosyl-, reaction with lithiated monosilyl dithioacetals, via ring opening/1,4-Brook rearrangement/ring closure cascade sequence, functionalized cyclopentanol synthesis 413, 414

Aziridines, N-protected, reaction with malononitrile, scandium catalyzed, via domino ring-opening/intramolecular cyclization sequence, 2-aminopyrrole-3-carbonitrile synthesis 593, 594

→Azulene-3a,6-diols, octahydro-, from allenyl/alkenyl-substituted lactones, via reductive radical cyclization cascade, samarium(II) iodide/water mediated 194, 195

→Azulenes, polysubstituted, from dialkenyl-substituted lactones, via alkene/alkene reductive radical cyclization cascade, samarium(II) iodide/water mediated 193, 194

→Azulen-6-ols, hexahydro-, O-protected, from polyene cyclization of O-protected dienynols, ruthenium carbene complex catalyzed, via cycloisomerization 36

B

Benzaldehydes, 2-alkynyl-, enantioselective reaction with alcohols, chiral diaminocarbene–gold complex catalyzed, via addition/cycloisomerization domino reaction, chiral 2-benzopyran synthesis 536, 537

Benzaldehydes, 2-alkynyl-, reaction with vinyl ethers, bulky cationic gold complex catalyzed, via benzannulation domino reaction, 1,2-dihydronaphthalene synthesis 535, 536

Benzaldehydes, 2-amino-, cyclization with enals, via enantioselective domino aza-Michael/aldol reaction, chiral diarylprolinol silyl ether catalyzed, chiral quinoline-3-carbaldehyde synthesis 347, 348

→Benzaldehydes, 2-hydroxy-3-silyl-, from O-silylated 2,6-dibromophenols, via domino double lithiation/silyl transfer/retro-1,2-Brook rearrangement/dimethylformamide trapping sequence 389, 390

Benzaldehydes, 2-silyl-, ortho-functionalization, via lithiation/1,4-Brook rearrangement/electrophilic capture cascade sequence, 1-(2-alkylphenyl)pentanol synthesis 423, 424

Benzaldehydes, 2-silyl-, tricarbonylchromium complexes, reaction with methyllithium and electrophiles, via carbonyl addition/1,4-Brook rearrangement/alkylation cascade sequence, 1-alkyl-2-(1-siloxyethyl)benzene–tricarbonylchromium complex synthesis 422, 423

→Benzenes, 1-alkyl-2-(1-siloxyethyl)-, tricarbonylchromium complexes, from reaction of 2-silylbenzaldehyde–tricarbonylchromium complexes with methyllithium and electrophiles, via carbonyl addition/1,4-Brook rearrangement/alkylation cascade sequence 422, 423

Benzenes, 1,2-dialkynyl-, three-component reaction with aldehydes and hydrazines, gold/silver catalyzed, polyfunctionalized tricyclic pyrazole synthesis, via cyclization domino reaction 540

Benzilic acid rearrangement, cyclic α-diketones to ring-contracted α-hydroxy acids, base promoted 258

→Benzofulvenes, from intramolecular cyclization domino reaction of 1,5-diynes, gold complex catalyzed 545–547

→Benzo[b]furan-4-carbonitriles, octahydro-, from radical cyclization/trapping of 3-(2-iodoethoxy)cyclohexenes, using tributyltin chloride/2,2′-azobisisobutyronitrile/tert-butyl isocyanide 231, 232

→Benzo[b]furan-2-ones, tetrahydro-, from bicyclic cyclopropyl alcohols, via carbocyclic ring expansion, acid induced 252, 253

→Benzo[b]furans, hexahydro-, chiral, from dienols, via enantioselective domino cyclization sequence, platinum/chiral ligand catalyzed 321, 322

638 Keyword Index

→ Benzo[*b*]furans, polycyclic, from reaction of phenols with cyclic 1,3-dienes, gold/silver catalyzed, via cyclization domino reaction 538, 539

→ Benzo[*c*]furan-1,3-diones, tetrahydro-, chiral, from enantioselective reaction of alkenynoyl-silanes with chiral lithium amides and furan-2,5-diones, via reduction/1,2-Brook rearrangement/Diels–Alder cycloaddition cascade sequence 372

Benzoic acids, 2,4-dihydroxy-, O-protected, conversion into acid chloride, coupling with substituted phenol, and Takai–Utimoto alkenation, then alkene metathesis, Schrock molybdenum complex catalyzed, and hydrogenation, phytoalexin synthesis 137

→ Benzo[g]indoles, from three-component domino condensation process of naphtho-1,4-quinones with β-keto esters and arylamines, cerium catalyzed 589, 590

→ Benzoin-type compounds, from reaction of silyl glyoxylates with aldehydes, cyanide ion/Lewis acid catalyzed, via cyanide addition/1,2-Brook rearrangement/nucleophilic addition/1,4-silyl migration/cyanide elimination cascade sequence 377

→ Benzo[*f*]isochromenes, hexahydro-, from reaction of distal aryl substituted homoallylic alcohols with aldehydes, via domino Prins reaction/polyene cyclization sequence, indium(III) bromide promoted 307–309

→ 2-Benzopyrans, chiral, from enantioselective reaction of 2-alkynylbenzaldehydes with alcohols, chiral diaminocarbene–gold complex catalyzed, via addition/cycloisomerization domino reaction 536, 537

→ 2-Benzopyrans, from reaction of 2-alkynylbenzaldehydes with vinyl ethers and alcohols, bulky cationic gold complex catalyzed, via benzannulation domino reaction 535, 536

→ Benzothiepins, tricyclic, from alkyne-tethered α-diazo-β-(mesitylsulfonyl) carbonyl compounds, via carbene metathesis/cyclization cascade sequence, rhodium catalyzed 524, 525

→ Benzo[*d*]xanthenes, from bicyclic epoxides, via domino epoxide opening/Wagner–Meerwein rearrangement/cyclization process, Lewis acid induced 302–304

→ 1,4-Benzoxazines, tricyclic, from reaction of 2-furylcarbinols with 2-aminophenols, lanthanum catalyzed, via domino Piancatelli rearrangement/C—N coupling/Michael addition sequence 597

→ Benzyl alcohols, 2-allyl-, from reaction of 1-oxa-2-silacyclopentanes with allyl halides, copper(I) *tert*-butoxide promoted, via organolithium addition/1,4-Brook rearrangement/electrophilic trapping cascade sequence 429, 430

→ Bicyclic 5,7-ring systems, fused, from domino cyclization of 1,6-enynes, gold complex catalyzed 332, 333

→ Bicyclo[3.2.1]octane-1-carboxylates, 6-methylene-2-oxo-, from radical polyene oxidative cyclization of 2-allyl-3-oxohept-6-enoates, manganese mediated 26, 27

→ (±)-Bisabosqual A, precursor, from tandem reductive radical cyclization of cyclohex-2-enyloxy-substituted aryl iodides, using tri-*sec*-butylborane/oxygen/tris(trimethylsilyl)silane 228, 229

Bis(enol ethers), silyl, symmetrical, diastereoselective intramolecular oxidative coupling, cerium(IV) mediated, chiral 1,4-diketone synthesis 204, 205

Bis(enol ethers), silyl, unsymmetrical, diastereoselective intramolecular oxidative cross-coupling, cerium(IV) mediated, chiral 1,4-diketone synthesis 204–207

Bis(enol ethers), silyl, unsymmetrical, domino oxidative radical rearrangement, cerium(IV) mediated, 1,4-diketone synthesis 201–203

Bis-epoxides – see Bisoxiranes

Bisoxiranes, reaction with lithiated monosilyl dithioacetals, via ring opening/1,4-Brook rearrangement/ring closure cascade sequence, functionalized cycloalkane synthesis 409, 410

Bisoxiranes, reaction with lithiated 2-silyl-1,3-dithianes, via ring opening/1,4-Brook rearrangement/ring closure cascade sequence, spirocyclic carbasugar synthesis 411

→ Bis(phenols), from bis(aryl ethers), via domino aryl-Claisen/[3,3]-sigmatropic rearrangement sequence, europium catalyzed 596

Bis(vinyl ethers), intramolecular radical cyclization cascade, triphenyltin radical mediated, hexahydrofuro[3,4-b]pyran synthesis 188

→ (−)-Blennolide C, key intermediate, from asymmetric cyclization of a 2-alkenylphenol, palladium/chiral bis-oxazoline catalyzed, via carbonylative oxypalladation cascade sequence 502, 503

→ Brevetoxin, structural fragment, from a hydroxy triepoxide, via *endo*-selective intramolecular epoxide opening/cyclization, base catalyzed, using directing groups 251

Brook rearrangement, overview 355

1,3-Brook rearrangement, overview 392, 393

1,3-Brook rearrangement, retro-, solvent dependence 402

1,4-Brook rearrangement, overview 403, 404

→ (±)-Brussonol, tricyclic intermediate, from alkynylindene, via domino cycloisomerization, transition-metal catalyzed 329, 330

→ Bryostatin 1, key intermediate, from reaction of lithiated 2-silyl-1,3-dithianes with an epoxide, via 1,4-Brook rearrangement cascade sequence 438, 439

Keyword Index

→ Bryostatin 1 fragment, intermediate, from an alkenyl ester, via carbonyl–alkene metathesis, using in situ generated titanium–ethylidene complex 147

→ Butane-1,4-diols, 1,4-bis(octahydro-2,2′-bifuran-5-yl)-, from tetraepoxides, via *exo*-selective epoxide-opening cascade, 10-camphorsulfonic acid promoted 43, 44

→ But-3-enals, 4-aryl-, from reaction of strained cyclopropenes with substituted benzaldehydes, via carbonyl–alkene ring-opening cross metathesis, bicyclic hydrazine catalyzed 150

But-2-enamines, N-(2-methylallyl)-, radical retrocycloisomerization, cobalt–salen complex catalyzed, 4-(prop-1-en-2-yl)pyrrolidine synthesis 30

→ But-3-enenitriles, 2-alkyl-2,4-disiloxy-, from *O*-silyl cyanohydrin-substituted epoxysilanes, by reaction with alkyl halides, via deprotonation/ring opening/1,2-Brook rearrangement/alkylation cascade sequence 382, 383

→ Butenolides, bicyclic, from allyl ester substituted/alkyne-tethered diazoacetoacetates, via carbene metathesis/cyclization/Claisen/Cope cascade sequence, copper catalyzed 523, 524

But-2-ynes, 1,4-diacetoxy-, intramolecular tandem 1,2-/1,2-bis(acetoxy) migration reaction, gold complex catalyzed, 2,3-diacetoxy-1,3-diene synthesis 559

→ γ-Butyrolactones, pentasubstituted, from silyl glyoxylates, via Reformatsky reagent addition/1,2-Brook rearrangement/ketone trapping/transesterification/lactonization cascade sequence 363

C

→ Callipeltoside C, key intermediate, from reaction of an alkynyl alcohol with carbon monoxide, palladium catalyzed, via oxopalladation/carbon monoxide insertion/ester formation/acidification cascade sequence 464, 465

→ Calothrixin B, pentacyclic intermediate, from reaction of an alkenyl bromide with a triethyl(indolyl)borate anion, palladium/phosphine catalyzed, then 6π-electrocyclization, copper catalyzed 457, 458

→ Camptothecin, from [4+1] annulation of 6-iodo-7-(prop-2-ynyl)pyrano[3,4-*c*]pyridines, via tandem radical trapping/cyclization, using phenyl isocyanide/hexamethyldistannane 234, 235

→ Camptothecin, key intermediate, from cyclization of 2-aminobenzaldehyde with an enal, via domino aza-Michael/aldol reaction, pyrrolidine catalyzed, then oxidation 348

→ Capnellene, key intermediate, from a tricyclic norbornene, via ring-opening alkene metathesis/carbonyl cyclization, using Tebbe reagent 139, 140

→ Carbanucleosides, precursors, from reaction of bisoxiranes with lithiated 2-silyl-1,3-dithianes, via ring opening/1,4-Brook rearrangement/ring closure cascade sequence 412, 413

→ Carbasugars, spirocyclic, from reaction of bisoxiranes with lithiated 2-silyl-1,3-dithianes, via ring opening/1,4-Brook rearrangement/ring closure cascade sequence 411

Carbenes, homotropic shift versus metallotropic [1,3]-shift 75, 76

Carbinols, α,α′-bis-silyl, domino deprotonation/1,2-Brook rearrangement sequence, then coupling with alkyl halides, [alkyl(siloxy)methyl]silane synthesis 381

Carbinols, 2-furyl-, reaction with 2-aminophenols, lanthanum catalyzed, via domino Piancatelli rearrangement/C—N coupling/Michael addition sequence, tricyclic 1,4-benzoxazine synthesis 597

→ Carbinols, α-silyl, from aldehydes, via nucleophilic stannylation and O-silylation, then domino lithium transmetalation/retro-1,2-Brook rearrangement sequence 387, 388

→ Carbinols, α-silyl, from reaction of acid chlorides with organosilyllithium reagents, via addition/1,2-Brook rearrangement cascade sequence, then organic halide addition 358

Carbinols, α-silyl, modified Morita–Baylis–Hillman reaction, using aldehydes and N-heterocyclic carbenes, via deprotonation/1,2-Brook rearrangement/silyl migration/nucleophilic addition/O-silylation cascade sequence, α-(1-siloxyalkyl) enone synthesis 381, 382

→ Carbocycles, bicyclic, chiral, from 1,7-allenenes, via intramolecular asymmetric [2+2] cycloaddition domino reaction, gold–chiral phosphoramidite/silver catalyzed 549, 550

→ Carbocycles, polycyclic, from intramolecular cyclization domino reaction of enediynes, gold complex catalyzed 547, 548

→ Carbocycles, tetracyclic, from intramolecular cycloisomerization domino reaction of 1,6-allenenes, gold/silver catalyzed 549

→ Carbocycles, tricyclic, from cycloisomerization domino reaction of 1,6-enynes, gold complex catalyzed, via 1,5-migration 542, 543

Carbon disulfide, reaction with [tris(silyl)methyl]-lithium/alkyl iodides, via nucleophilic addition/1,3-thia-Brook rearrangement/electrophilic capture/ring opening cascade sequence, sulfanyl-bis(silyl)alkanethione synthesis 401

Carbonyl–alkene metathesis, catalysts 136

Carbonyl–alkene metathesis, overview 135, 136

Carbonyl compounds – *see also* Aldehydes, Esters, Ketones

Carbonyl compounds, α-diazo-β-keto-, alkyne tethered, carbene metathesis/carbonyl ylide/cyclization cascade sequence, rhodium catalyzed, cyclopenta[c]furan-4-one synthesis 522, 523

640 Keyword Index

→ Carbonyl compounds, α-diazo-β-keto-, alkyne tethered, from coupling of silyl-protected diazo enols with propargylic esters, Lewis acid catalyzed 522, 523

Carbonyl compounds, α-diazo-β-(mesitylsulfonyl), alkyne tethered, carbene metathesis/cyclization cascade sequence, rhodium catalyzed, tricyclic benzothiepin synthesis 524, 525

→ Carbonyl compounds, from cyclic epoxides, via rearrangement/ring contraction, Lewis acid promoted 282

Carboxylic acids – *see also* Acids

Carboxylic acids, hydroxy – *see* α-Hydroxy acids

→ Carboxylic acids, γ,δ-unsaturated α-siloxy, from reaction of silyl glyoxylate allyl esters with organometallic reagents, via nucleophilic addition/1,2-Brook rearrangement/Ireland–Claisen rearrangement cascade sequence 368

→ Cardamom peroxide, from a dienedione, via domino hydroperoxidation/cyclization, using oxygen, manganese complex catalyzed 166, 167

→ (−)-Cephalotaxine, key intermediate, from (4-aminobut-1-ynyl)cyclobutanol, via asymmetric domino hydroamination/semipinacol rearrangement sequence, chiral silver complex catalyzed 335, 336

→ (−)-Chromazonarol, from an aryl-substituted triene, via enantioselective domino cyclization sequence, chiral Brønsted acid catalyzed, Lewis acid assisted 320

→ Chrysenes, octahydro-, from enantioselective polyene cyclization of 1,5-dienes, chiral binaphthol–Lewis acid complex catalyzed 19, 20

→ Citalopram, precursor, from reaction of a 1,1-diarylethene with allyl chloride, palladium catalyzed, via oxypalladation/carbopalladation/β-halide elimination/cyclization cascade sequence 499, 500

→ (−)-Cochleamycin A, key intermediate, from a dienyne, via domino double enyne ring-closing metathesis, ruthenium alkylidene complex catalyzed, then deprotection 105

→ (−)-Cochleamycin A, key intermediate, from reaction of lithiated γ-silyl carbinol with 2-bromo-1,1-diethoxyethane, via 1,4-Brook rearrangement cascade sequence 440

→ (±)-Confertin, key intermediate, from tricyclic cyclopropyl alcohol, via carbocyclic ring expansion, acid induced 252

→ (+)-Corynoline, key intermediate, from intramolecular cyclization of an aryl iodide/amide/alkene, palladium catalyzed, via halide transfer/cyclization/reductive elimination cascade sequence 486, 487

→ (+)-CP-263,114, pseudoester cage ring system, from intramolecular domino cyclization sequence of polyfunctionalized intermediate, Lewis acid promoted 314, 315

→ Cumbiasin, tetracyclic precursor, from Diels–Alder/tandem radical cyclization sequence of a 1,3,7-dienyne with a cyclic vinyl bromide, using tributyltin hydride/2,2′-azobisisobutyronitrile 220

→ (±)-Cuparene, key intermediate, from α-hydroxy epoxides, via tandem rearrangement/reduction/ring contraction, samarium(II) iodide promoted 284

→ Cyanides, α-amino, α,α-disubstituted, from α-amino esters, via three-component Strecker reaction with cyclohexanone and silyl cyanides, ytterbium catalyzed 583

→ Cyanthiwigin core, from a bis(enone)-substituted bridging cyclohexene, via metathesis polycyclization cascade, ruthenium carbene complex catalyzed 33, 34

Cycloalkanes, 1-alkenyl-2-(2,2-dialkoxyethyl)-1-siloxy-, ring-enlarging cyclopentane annulation, tin(IV) chloride promoted, via sequential cationic cyclization/pinacol rearrangement, fused cyclopentane synthesis 278, 279

→ Cycloalkanes, 1-cyano-1-(2-siloxyvinyl)-, from reaction of α-cyano-functionalized epoxysilanes with bis-electrophiles, via deprotonation/ring opening/1,2-Brook rearrangement/nucleophilic addition cascade sequence 383–385

→ Cycloalkanes, functionalized, from reaction of bisoxiranes with lithiated monosilyl dithioacetals, via ring opening/1,4-Brook rearrangement/ring closure cascade sequence 409–411

Cycloalkanols, semipinacol rearrangement, acid catalyzed, ring-expanded cycloalkanone synthesis 258

→ Cycloalkanones, ring expanded, from cycloalkanols, via semipinacol rearrangement, acid catalyzed 258

→ Cycloalkenes, iodo-, tricyclic, from 3-(2-iodopyranyloxy)-substituted 1,5-enynes, via tandem radical cyclization, oxygen catalyzed, triethylborane mediated 227, 228

→ Cycloalkenes, tricyclic, from 3-(2-iodopyranyloxy)-substituted 1,5-enynes, via reductive tandem radical cyclization, using sodium borohydride, cobaloxime catalyzed 227, 228

→ Cyclobutanes, functionalized, from reaction of β,γ-unsaturated acylsilanes with phenyllithium, via nucleophilic addition/1,2-Brook rearrangement/intramolecular Michael addition cascade sequence 371

Cyclobutanes, polycyclic, Wagner–Meerwein ring expansion, acid induced, polycyclic cyclopentane synthesis 254

Cyclobutanols, 1-alkenyl-2-aryl-, tandem semipinacol rearrangement/direct arylation reaction sequence, palladium catalyzed, tetrahydrocyclopenta[a]inden-1-one synthesis 323, 324

Keyword Index

Cyclobutanols, (4-aminobut-1-ynyl)-, asymmetric domino hydroamination/semipinacol rearrangement sequence, chiral silver complex catalyzed, chiral azaspiro ketone synthesis 335, 336

Cyclobutanols, tricyclic, Wagner–Meerwein ring contraction/ring opening, acid induced, bicyclic secondary alcohol synthesis 255, 256

→ Cyclobutenes, from 1,5-enynes, via ring-closing metathesis, ruthenium alkylidene complex catalyzed 93

→ Cycloheptadienes, fused, from alkynylindenes, via domino cycloisomerization, transition-metal catalyzed 328, 329

→ Cyclohepta-1,4-dienes, from [4+3] cycloaddition of propargyl acetates with 1,3-dienes, cationic gold complex catalyzed, via domino 1,2-acetoxy migration/cyclopropanation/homo-Cope rearrangement sequence 569–571

→ Cyclohepta[*b*]furans, octahydro-, from reaction of 1-alkenylcyclohexane-1,2-diol acetals, tin(IV) chloride promoted, via ring-enlarging tetrahydrofuran annulation, by Prins cyclization/pinacol rearrangement 277, 278

→ Cycloheptanols, functionalized, from [6+1] cycloaddition of epoxy sulfonates with lithiated monosilyl dithioacetals, via ring opening/1,4-Brook rearrangement/ring closure cascade sequence 407, 408

→ Cycloheptanones, from vicinal tertiary cyclohexyl diols, via pinacol rearrangement, acid catalyzed 256

→ Cycloheptenediones, from [3+4] cycloaddition of β-chloro-substituted acylsilanes with vinyl ketone enolates, via nucleophilic addition/1,2-Brook rearrangement/cyclization cascade sequence 363, 364

→ Cycloheptenones, from [3+4] cycloaddition of β-silyl-substituted acylsilanes with vinyl ketone enolates, via nucleophilic addition/1,2-Brook rearrangement/cyclopropanation/anionic oxy-Cope rearrangement cascade sequence 363, 364

→ Cycloheptenones, tetrasubstituted, from [3+4] cycloaddition of γ,δ-epoxy-δ-silyl α,β-unsaturated acylsilanes with vinyl ketone enolates, via nucleophilic addition/1,2-Brook rearrangement/ring opening/1,2-Brook rearrangement/intramolecular Michael addition cascade sequence 365, 366

Cyclohexane-1,2-diols, acetals, 1-alkenyl-, reaction with aldehydes, tin(IV) chloride promoted, ring-enlarging tetrahydrofuran annulation, via Prins cyclization/pinacol rearrangement, octahydrocyclohepta[b]furan synthesis 277, 278

→ Cyclohexanes, 1-alkylidene-3-methylene-, from polyene cyclization of 1,6-dienes, palladium/phosphine catalyzed, via π-allylpalladium complexes 34, 35

Cyclohexanes, (phenylselanyl)-, tricyclic, three-component sequential radical trapping with cyclopent-2-en-1-ones and allylstannanes, 1,1'-azobis(cyclohexanecarbonitrile) promoted, resiniferatoxin precursor synthesis 237

→ Cyclohexanols, functionalized, from [5+1] cycloaddition of epoxy sulfonates with lithiated monosilyl dithioacetals, via ring opening/1,4-Brook rearrangement/ring closure cascade sequence 407, 408

Cyclohexanones, 2-chloro-, Favorskii rearrangement, using amines, cyclopentanecarboxamide synthesis 287

Cyclohex-1-ene-1-carboxylates, 6-oxo-, intermolecular anionic polycyclization with 3-hydroxyalka-2,4-dienoate-substituted diketals, cesium carbonate promoted, 2,5-dioxooctahydronaphthalene-1,4a-dicarboxylate synthesis, via intramolecular Diels–Alder cycloaddition/aldol condensation 37

Cyclohexenes, bridging, bis(enone) substituted, metathesis polycyclization cascade, ruthenium carbene complex catalyzed, cyanthiwigin core synthesis 33, 34

Cyclohexenes, bridging, enol substituted, metathesis polycyclization cascade, ruthenium carbene complex catalyzed, hexahydroinden-1-ol synthesis 33, 34

Cyclohexenes, dialdehyde substituted, radical-anionic cyclization cascade, samarium(II) iodide mediated, carbocycle-fused spirolactone synthesis 196, 197

Cyclohexenes, 3-(2-iodoethoxy)-, radical cyclization/trapping, using tributyltin chloride/2,2'-azobisisobutyronitrile/tert-butyl isocyanide, octahydro-1-benzofuran-4-carbonitrile synthesis 231, 232

→ Cyclohex-2-enones, from 1-cyclopropylpropargyl esters, via intramolecular domino cycloisomerization sequence, gold/silver catalyzed, then hydrolysis 568, 569

Cyclohex-2-enones, 4-(pent-3-enyl)-, chiral, polyene cyclization, ethylaluminum dichloride catalyzed, chiral hexahydroinden-5-one synthesis, via Wagner–Meerwein shift 21

Cyclohex-2-enones, substituted, conversion into spirocyclic macrocycle, then carbonyl–alkene metathesis, Schrock molybdenum complex mediated, tricyclic huperzine Q intermediate synthesis 148, 149

→ Cyclooctenes, tetrafunctionalized, from [3+4] cycloaddition of β-silyl-substituted acylsilanes with cycloheptenone enolates, via nucleophilic addition/1,2-Brook rearrangement/cyclization cascade sequence, then α-hydroxylation and oxidative cleavage 365

→ Cyclooctenones, substituted, from [6+2] cyclization of silylated 1,4-diones with vinyllithium reagents, via nucleophilic addition/1,2-Brook rearrangement/ring closure/anionic oxy-Cope rearrangement cascade sequence 372, 373

642 Keyword Index

Cyclooctyne, reaction with bis-silyl acetylenedicar-boxylate, via 1,3-cycloaddition/retro-1,4-Brook rearrangement cascade sequence, bicyclic lactone synthesis 431, 432

→ Cyclopenta[*b*]benzofurans, from rearrangement of tricyclic α-hydroxy ketones, base catalyzed 263, 264

→ Cyclopenta[*c*][1,2]dioxin-5-ones, 7a-hydroxy-hexahydro-, from cyclic 1,3-diketones, via coupling with alkenes and oxygen by C—H radical abstraction, 2,2′-azobisisobutyronitrile initiated 174

→ Cyclopenta[*b*]furan-6-ones, from reaction of alkyne-tethered diazoacetates with 4-methoxybenzaldehyde, rhodium catalyzed, via intermolecular ylide formation/intramolecular dipolar cycloaddition cascade sequence 526

→ Cyclopenta[*b*]furans, 6-aryl-, from reaction of enynones with terminal alkynes, rhodium catalyzed, via cyclization/cyclopropenation/ring expansion cascade sequence 520, 521

→ Cyclopenta[*c*]furan-4-ones, from alkyne-tethered β-keto-α-diazocarbonyl compounds, via carbene metathesis/carbonyl ylide/cyclization cascade sequence, rhodium catalyzed 522, 523

→ Cyclopenta[*c*]furans, hexahydro-, from distal cyclopropane containing alkenes, via radical fragmentation/bicyclization, visible light mediated, lanthanum(III) facilitated, ruthenium complex catalyzed 211, 212

→ Cyclopenta[*a*]inden-1-ones, tetrahydro-, from 1-alkenyl-2-arylcyclobutanols, via tandem semipinacol rearrangement/direct arylation reaction sequence, palladium catalyzed 323, 324

→ Cyclopentanecarboxamides, from Favorskii rearrangement of 2-chlorocyclohexanones, using amines 287

→ Cyclopentane-1,1-dicarboxylates, from reductive radical polyene cyclization of 2,2-bis(2-methylallyl)malonates, manganese catalyzed 29

→ Cyclopentane-1,1-dicarboxylates, 3-(hydrazino-methyl)-, from radical polyene cyclization/azidation of 2,2-diallylmalonates, using azodicarboxylates, manganese catalyzed 28, 29

→ Cyclopentanes, 1-alkanoyl-2-vinyl-, from intramolecular hydrative carbocyclization domino reaction of 1,7-allenynes, gold/silver catalyzed 550

→ Cyclopentanes, fused, from 1-alkenyl-2-(2,2-dialkoxyethyl)-1-siloxycycloalkanes, via ring-enlarging cyclopentane annulation, tin(IV) chloride promoted, by sequential cationic cyclization/pinacol rearrangement 278, 279

→ Cyclopentanes, polycyclic, from polycyclic cyclobutanes, via Wagner–Meerwein ring expansion, acid induced 254

→ Cyclopentanols, 2-alkylidene-, from 1-(cyclo-prop-1-enyl)hexanols, via rhodium-catalyzed rearrangement/cyclization cascade sequence 518, 519

→ Cyclopentanols, 1-alkynyl-, chiral, from enantioselective reaction of alkenyl-substituted acylsilanes with alkynes, chiral imino alcohol catalyzed, zinc promoted, via nucleophilic alkynylation/1,2-Brook rearrangement/ene–allene carbocyclization/electrophilic trapping cascade sequence 375, 376

→ Cyclopentanols, functionalized, from [4+1] cycloaddition of epoxy sulfonates with lithiated monosilyl dithioacetals, via ring opening/1,4-Brook rearrangement/ring closure cascade sequence 405–407

→ Cyclopentanols, functionalized, from reaction of 2-(oxiran-2-yl)-*N*-tosylaziridines with lithiated monosilyl dithioacetals, via ring opening/1,4-Brook rearrangement/ring closure cascade sequence 413, 414

→ Cyclopentanones, silyl enol ethers, from reaction of β-sulfanyl-substituted α,β-unsaturated acylsilanes with ketone enolates, via nucleophilic addition/1,2-Brook rearrangement/cyclization cascade sequence 362

→ Cyclopentanones, spiro fused, from cyclobutyl-substituted allylic alcohols, via domino Prins/semipinacol rearrangement, acid catalyzed 258

→ Cyclopenta[*c*]oxepines, hexahydro-, from polyene cyclization of dienynols, gold complex catalyzed, via cycloisomerization 36

→ Cyclopenta[*b*]thiophen-6-ones, 4-siloxy-, from annulation of 2-silylthiophene-3-carbaldehydes, using α,α-disubstituted ester enolates, via lithiation/1,4-Brook rearrangement/cyclization cascade sequence 424, 425

→ Cyclopentenes, 4-alkanoyl-, from intramolecular hydrative carbocyclization domino reaction of 1,5-allenynes, gold/silver catalyzed 549, 551

→ Cyclopentenes, functionalized, from [4+1] cycloaddition of vinyloxiranes with lithiated monosilyl dithioacetals, via ring opening/1,4-Brook rearrangement/ring closure cascade sequence 408, 409

Cyclopentenes, 3-(2-iodoethoxy)-, photolytic radical cyclization/trapping, using tributyltin hydride/2-silyloct-1-en-3-one, prostaglandin PGF$_{2α}$ precursor synthesis 232

→ Cyclopentenones, bicyclic, from alkyne-tethered *gem*-diethyl diazoketones, via carbene metathesis/cyclization/C—H insertion cascade sequence, rhodium catalyzed 525, 526

→ Cyclopentenones, from reaction of β,β-dichloro-substituted α,β-unsaturated acylsilanes with ketone enolates, via nucleophilic addition/1,2-Brook rearrangement/cyclization cascade sequence 362, 363

→Cyclopentenones, spirocyclic, from spirocyclobutyl α-hydroxy esters, via retro-benzilic acid
rearrangement, base promoted 260, 261
→Cyclopent-2-enones, 4,5-diamino-, from reaction of furan-2-carbaldehyde with secondary
amines, via domino condensation/ring open-
ing/electrocyclization process, dysprosium
catalyzed 584, 585
→Cyclopent-2-enones, from enynyl acetates, via
tandem [3,3]-acyloxy rearrangement/Nazarov
cyclization sequence, gold/silver catalyzed
563
→Cyclopropa[e]azulenes, octahydro-, from domino cyclization of dienynes, gold complex catalyzed 332, 333
*Cyclopropane-1,1-dicarboxylates, 2-formyl-, reac-
tion with 1,3-diketones, scandium catalyzed, via
domino aldol/ring opening sequence, functionalized furan synthesis 592, 593*
*Cyclopropanes, amino-, heteroatom oxidation, using
oxygen, iron complex catalyzed, by oxidation to
radical cation/ring opening/peroxidation/cycliza-
tion, 1,2-dioxolan-3-amine synthesis 177, 178*
*Cyclopropanes, bromo(N,N-dipropargylamido),
domino radical cyclization/rearrangement, light
mediated, iridium complex photocatalyzed, tricyclic pyrrolidinone synthesis 207–210*
→Cyclopropanes, 1-cyano-1-siloxy-, from reaction
of β-bromo acylsilanes with cyanide ion, via
nucleophilic addition/1,2-Brook rearrange-
ment/intramolecular S_N2 displacement cascade sequence 371, 372
*Cyclopropanes, 1,2-diaryl-, photooxygenation,
9,10-dicyanoanthracene sensitized, 3,5-diaryl-
1,2-dioxolane synthesis 179*
*Cyclopropanes, dienyl-, domino [(5+2)+1] cycloaddi-
tion/aldol reaction sequence, rhodium catalyzed,
triquinane synthesis 325, 326*
*Cyclopropanes, divinyl-, intramolecular domino
radical cyclization, tributyltin radical mediated,
(±)-epimeloscine key intermediate synthesis 191*
→Cyclopropanes, phenyl-, from homo-Peterson
reaction of 2-phenyloxiranes with [tris(silyl)-
methyl]lithium, via 1,4-Brook rearrangement/
cyclization cascade sequence 404, 405
→Cyclopropanes, polyfunctionalized, from reaction of diene-functionalized epoxysilanes
with 2-chloroacetamides, via enolization/
Michael addition/ring opening/1,2-Brook
rearrangement/intramolecular alkylation cascade sequence 385, 386
*Cyclopropanes, vinyl-, domino hydroperoxidation,
using oxygen, cobalt complex catalyzed, then metathesis with triethylsilane, endoperoxide synthesis 165, 166*
*Cyclopropanes, vinyl-, domino ring opening/peroxi-
dation/cyclization cascade, using oxygen/light,
2,2′-azobisisobutyronitrile/diphenyl diselenide
catalyzed, 3-vinyl-1,2-dioxolane synthesis 171,
172*

→Cyclopropanols, amino-, from reaction of acylsilanes with enolized N-sulfinyl ketimines, via
nucleophilic addition/1,2-Brook rearrange-
ment/ring closure cascade sequence 374, 375
*Cyclopropanols, pentamethyl-, heteroatom oxida-
tion, using oxygen, pentamethyl-1,2-dioxolan-3-ol
synthesis 177*
*Cycloprop-2-ene-1-carboxylate, 2,3-diphenyl-, rhodi-
um-catalyzed rearrangement, 2-ethoxy-4,5-di-
phenylfuran synthesis, via cascade sequence 516,
517*
*Cyclopropenes, strained, reaction with substituted
benzaldehydes, via carbonyl–alkene ring-opening
cross metathesis, bicyclic hydrazine catalyzed,
substituted 4-arylbut-3-enal synthesis 150*
*Cyclopropenes, 1,2,3-triphenyl-, rhodium-catalyzed
rearrangement, 1,2-diphenylindene synthesis, via
cascade sequence 516, 517*

D

→Dammarenediol II, key intermediate, from
domino cyclization of a polyene epoxide,
Lewis acid induced 291, 292
→DB-67, pentacyclic intermediate, from reaction
of a pyridinone iodide with an aryl isocyanide, palladium catalyzed, via oxidative addi-
tion/migratory insertion/carbopalladation/cy-
clization cascade sequence 485, 486
→Decalins, from diene epoxides, using ethylaluminum dichloride, via cationic polyene cyclization 15, 16
→Decalins, polysubstituted, from farnesyl acetate, via tandem radical cyclization, using dibenzoyl peroxide/copper benzoate 219
→(−)-Dendrobine, tricyclic core, from a functionalized enyne and octacarbonyldicobalt, via
intramolecular Pauson–Khand cyclization
604, 605
→1-Desoxyhypnophilin, key triquinane intermediate, from dienylcyclopropane, via domino [(5+2)+1] cycloaddition/aldol reaction sequence, rhodium catalyzed 326, 327
*Dialdehydes, one-pot enantioselective sequential
cyanation/nitroaldol process with cyanoformate
and then nitromethane, chiral yttrium complex
catalyzed, chiral nitroaldol cyano carbonate synthesis 580, 581*
*Dialdehydes, radical-anionic cyclization cascade,
samarium(II) iodide mediated, carbocycle-fused
spirolactone synthesis 196, 197*
→Diazaspiro[4.4]nonanediones, from tandem radical cyclization of azido-N-(2-iodoaryl)acryl-
amides with carbon monoxide, using tris(trimethylsilyl)silane/2,2′-azobisisobutyroni-
trile 224
→Dibenzo[a,d][7]annulenes, tetrahydro-, from alkynylindenes, via domino cycloisomerization, transition-metal catalyzed 328, 329
*Dienediones, domino hydroperoxidation/cyclization,
using oxygen, manganese complex catalyzed, cardamom peroxide synthesis 166, 167*

Dienes, aryl substituted, enantioselective domino cyclization sequence, chiral Brønsted acid catalyzed, Lewis acid assisted, chiral octahydrophenanthrene synthesis 318, 319

Dienes, aryl substituted, reaction with acetals, via intermolecular domino Prins reaction/polyene cyclization sequence, tin(IV) chloride promoted, substituted octahydrophenanthrene synthesis 311, 312

→ Dienes, siloxy-, from vinyl-substituted epoxysilanes, via palladium complexation/1,2-Brook rearrangement cascade sequence 387

→ 1,3-Dienes, 2,3-diacetoxy-, from 1,4-diacetoxybut-2-ynes, via intramolecular tandem 1,2-/1,2-bis(acetoxy) migration reaction, gold complex catalyzed 559

→ 1,3-Dienes, from enynes, via intramolecular metathesis reaction, ruthenium alkylidene complex catalyzed 68

→ 1,3-Dienes, 2-(pivaloyloxy)-, from propargylic pivalates, via domino intramolecular rearrangement reaction, gold complex catalyzed 558

→ 1,3-Dienes, siloxy substituted, from reaction of β-silyl-substituted enones with aldehydes/tributylphosphine, via Michael addition/1,4-Brook rearrangement/Wittig alkenation cascade sequence 420, 421

→ 1,4-Dienes, 2-(siloxymethyl)-, from reaction of 2-(triphenylsilyl)allyl alcohols with allyl chlorides/copper tert-butoxide, via nucleophilic addition/1,3-Brook rearrangement/electrophilic trapping cascade sequence 399

1,4-Dienes, thiol–alkene co-oxygenation reaction, using di-tert-butyl peroxyoxalate/benzenethiol/oxygen, endoperoxide synthesis 168, 169

1,5-Dienes, cationic polyene cyclization, bromonium salt catalyzed, 4-isocymobarbatol precursor synthesis 17, 18

1,5-Dienes, enantioselective polyene cyclization, chiral binaphthol–Lewis acid complex catalyzed, chiral octahydrochrysene synthesis 19, 20

1,5-Dienes, enantioselective polyene cyclization, Lewis acid–Brønsted acid complex catalyzed, chiral hexahydroxanthene synthesis 19

1,5-Dienes, enantioselective polyene halocyclization, using N-iodosuccinimide, chiral phosphoramidite catalyzed, chiral 2-iodooctahydrophenanthrene synthesis 20, 21

1,5-Dienes, intramolecular cyclization/endoperoxidation, using oxygen, 9,10-dicyanoanthracene photosensitized, bicyclic endoperoxide synthesis 181, 182

1,5-Dienes, intramolecular cyclization/endoperoxidation, using oxygen, triarylpyrylium salt photosensitized, bicyclic endoperoxide synthesis 181, 182

1,5-Dienes, iodophenyl substituted, Heck polyene cyclization, palladium/phosphine catalyzed, spiro[cyclopentane-indene] synthesis 32, 33

1,5-Dienes, polyene cyclization, cationic palladium complex mediated, hexahydroxanthene synthesis 23, 24

1,5-Dienes, polyene cyclization, episulfonium ion initiated, 2-(phenylsulfanyl)hexahydrophenanthrene synthesis 23

1,5-Dienes, vinyl trifluoromethanesulfonate substituted, Heck polyene cyclization, palladium catalyzed, spiro[cyclopentane-naphthalene] synthesis 32, 33

1,6-Dienes, intramolecular cyclization/endoperoxidation, using oxygen, ruthenium complex photocatalyzed, bicyclic endoperoxide synthesis 182, 183

1,6-Dienes, polyene cyclization, palladium/phosphine catalyzed, via π-allylpalladium complexes, 1-alkylidene-3-methylenecyclohexane synthesis 34, 35

Dienols, enantioselective domino cyclization sequence, platinum/chiral ligand catalyzed, chiral hexahydrobenzo[b]furan synthesis 321, 322

→ Dienols, from ipso substitution of 3-silylated allyl alcohols, by reaction with allylic halides, copper(I) tert-butoxide promoted, via cupration/1,4-Brook rearrangement/alkylation cascade sequence 426, 427

Dienynes, aza, hydrochloride salt, domino double enyne ring-closing metathesis, ruthenium alkylidene complex catalyzed, (±)-erythrocarine synthesis 109, 110

Dienynes, domino double enyne ring-closing metathesis, ruthenium alkylidene complex catalyzed, (–)-englerin A key intermediate synthesis 100–102

Dienynes, domino double enyne ring-closing metathesis, ruthenium alkylidene complex catalyzed, guanacastepene A skeleton synthesis 99, 100

Dienynes, domino double enyne ring-closing metathesis, ruthenium alkylidene complex catalyzed, ent-lepadin F key intermediate synthesis 106–108

Dienynes, domino double enyne ring-closing metathesis, ruthenium alkylidene complex catalyzed, then allylic bromination and deprotection-induced N-alkylation, (–)-norsecurinine synthesis 110, 111

Dienynes, domino double enyne ring-closing metathesis, ruthenium alkylidene complex catalyzed, then allylic oxidation, allylic bromination, ring closure, and deprotection, (–)-securinine synthesis 108, 109

Dienynes, domino double enyne ring-closing metathesis, ruthenium alkylidene complex catalyzed, then deprotection, cochleamycin A key intermediate synthesis 105

Dienynes, domino double enyne ring-closing metathesis, ruthenium alkylidene complex catalyzed, then deprotection and acetylation, (+)-kempene-2 synthesis 103, 104

Dienynes, domino double enyne ring-closing metathesis, ruthenium alkylidene complex catalyzed, then hydrogenation, hydroboration, oxidation, and Mannich reaction with formaldehyde, lycoflexine synthesis 106

Dienynes, domino double enyne ring-closing metathesis, ruthenium alkylidene complex catalyzed, then tetrabutylammonium fluoride treatment, (−)-acylfulvene key intermediate synthesis 102, 103

Dienynes, reaction with but-3-en-2-ol, via domino ring-closing metathesis/cross metathesis sequence, ruthenium alkylidene complex catalyzed, (+)-panepophenanthrin key intermediate synthesis 114, 115

1,3-Dien-7-ynes, Diels–Alder/tandem radical cyclization sequence with a cyclic vinyl bromide, using tributyltin hydride/2,2′-azobisisobutyronitrile, tetracyclic cumbiasin precursor synthesis 220

→ *1,5-Dien-3-ynes, from dienediynes, via ring-closing metathesis/metallotropic [1,3]-shift domino sequence, ruthenium alkylidene complex catalyzed* 120

1,5-Dien-10-ynes, domino cyclization, gold complex catalyzed, fused tricyclic 5,7,3-ring system synthesis 332, 333

1,6-Dien-8-ynes, domino relay cross metathesis with a 2-allyl-3,9-dioxabicyclo[4.2.1]nonane, ruthenium alkylidene complex catalyzed, laureatin key intermediate synthesis 122, 123

Dienynols, polyene cyclization, gold complex catalyzed, hexahydrocyclopenta[c]oxepine synthesis, via cycloisomerization 36

Dienynols, O-protected, polyene cyclization, ruthenium carbene complex catalyzed, O-protected hexahydroazulen-6-ol synthesis, via cycloisomerization 36

Diepoxides, alkenyl, bromonium ion induced domino cyclization, dioxepandehydrothyrsiferol intermediate synthesis 295, 296

Diepoxides, bis(trimethylsilyl) containing, endo-selective epoxide-opening cascade, Brønsted base/fluoride ion promoted, then acetylation, decahydrodipyrano[3,2-b:2′,3′-e]pyran synthesis 51, 52

Diepoxides, bromo substituted, endo-selective epoxide-opening cascade, carbocation initiated via halide abstraction, using silver trifluoromethanesulfonate, octahydropyrano[3,2-b]pyran synthesis 55, 56

Diepoxides, N,N-dimethylcarbamate containing, endo-selective epoxide-opening cascade, Lewis acid promoted, 7-hydroxyhexahydropyrano[3,2-d][1,3]dioxin-2-one synthesis 47, 48

Diepoxides, diol, exo-selective epoxide-opening cascade, Brønsted base promoted, tetraol synthesis 45

Diepoxides, distal alkene containing, endo-selective epoxide-opening cascade, bromonium ion initiated, external nucleophile terminated, 3-bromodecahydrooxepino[3,2-b]oxepin synthesis 53, 54

Diepoxides, distal alkene containing, endo-selective epoxide-opening cascade, bromonium ion initiated, internal nucleophile terminated, 8-bromodecahydro[1,3]dioxino[5,4-b]oxepino[2,3-f]-oxepin-2-one synthesis 52, 53

Diepoxides, 5-hydroxy-1,3-dioxane containing, endo-selective epoxide-opening cascade, water promoted, octahydropyrano[2′,3′:5,6]pyrano[3,2-d][1,3]dioxin synthesis 57

Diepoxides, 3-hydroxytetrahydropyran containing, endo-selective epoxide-opening cascade, water promoted, decahydrodipyrano[3,2-b:2′,3′-e]pyran synthesis 56

Diepoxides, exo-selective epoxide-opening cascade, Brønsted base promoted, (+)-omaezakianol precursor synthesis 45

Diepoxides, exo-selective epoxide-opening cascade, 10-camphorsulfonic acid promoted, 2-(dodecahydro-2,2′:5′,2′′-terfuran-5-yl)propan-2-ol synthesis 43, 44

Diepoxides, endo-selective epoxide-opening cascade, oxocarbenium ion initiated, via photooxidative cleavage, decahydro[1,3]dioxino[5,4-b]-oxepino[2,3-f]oxepin-2-one synthesis 54, 55

Diepoxides, exo-selective epoxide-opening cascade, oxocarbenium ion initiated, via photooxidative cleavage, 2-(octahydro-2,2′-bifuran-5-yl)ethanol synthesis 46, 47

1,5-Diepoxides, hydroxy, exo-selective domino intramolecular opening/cyclization, acid catalyzed, poly(tetrahydrofuran) synthesis 247

→ (±)-*Differolide, from ester-linked 1,6-enyne, via domino enyne ring-closing metathesis/Diels–Alder dimerization, ruthenium alkylidene complex catalyzed* 124

→ (+)-*Digitoxigenin, key synthetic step, vinyl radical tandem radical cyclization of alkoxy-substituted 1,6-enyne, using tributyltin hydride/2,2′-azobisisobutyronitrile* 219

1,2-Diketones, cyclic, benzilic acid rearrangement, base promoted, ring-contracted α-hydroxy acid synthesis 259, 260

1,3-Diketones, cyclic, coupling with alkenes and oxygen by C—H radical abstraction, 2,2′-azobisisobutyronitrile catalyzed, peroxyketal synthesis 174

→ *1,4-Diketones, chiral, from symmetrical silyl bis(enol ethers), via diastereoselective intramolecular oxidative coupling, cerium(IV) mediated* 204, 205

→ *1,4-Diketones, chiral, from unsymmetrical silyl bis(enol ethers), via diastereoselective intramolecular oxidative cross-coupling, cerium(IV) mediated* 205–207

→ *1,4-Diketones, from sila-Stetter reaction of acylsilanes with enones, N-heterocyclic carbene catalyzed, via nucleophilic addition/1,2-Brook rearrangement/Michael addition/carbene cleavage cascade sequence* 378, 379

646 Keyword Index

→ 1,4-Diketones, from unsymmetrical silyl bis-(enol ethers), via domino oxidative radical rearrangement, cerium(IV) mediated 201–203

→ 1,5-Diketones, fluorinated, from reaction of acylsilanes with trimethyl(trifluoromethyl)silane and α,β-unsaturated carbonyl compounds, via nucleophilic addition/1,2-Brook rearrangement/β-elimination/Mukaiyama aldol addition cascade sequence 360, 361

Diol diepoxides, exo-selective epoxide-opening cascade, Brønsted base promoted, tetraol synthesis 45

Diol epoxides, intramolecular epoxide opening/cyclization, base catalyzed, 5-(hydroxymethyl)tetrahydrofuran-3-ol synthesis 250, 251

Diol pentaepoxides, exo-selective epoxide-opening cascade, Brønsted base promoted, (–)-glabrescol synthesis 45

→ Diols, monosilyl protected, from (α-siloxyalkyl)-stannanes and aldehydes, via domino tin–lithium exchange/retro-1,2-Brook rearrangement/electrophilic trapping sequence 389

→ Diols, spirocyclic, from 2,3-epoxy alcohols, via reductive rearrangement/ring expansion, aluminum triisopropoxide catalyzed 272, 273

Diols, vicinal, tertiary cyclohexyl, pinacol rearrangement, acid catalyzed, cycloheptanone synthesis 256

→ 1,3-Diols, chiral, from arylaldehydes, via domino asymmetric aldol–Tishchenko reaction with aryl methylene ketones, chiral lanthanum complex catalyzed, then methanolysis 579, 580

→ 1,3-Diols, chiral, from α-hydroxy epoxides, via stereoselective tandem rearrangement/reduction/ring contraction, triethylaluminum promoted 282–284

→ 1,3-Diols, from α-hydroxy epoxides, via tandem rearrangement/reduction/ring contraction, samarium(II) iodide promoted 282

→ 1,5-Diols, from sequential reaction of lithiated 2-silyl-1,3-dithiane with two epoxides, via ring opening/1,4-Brook rearrangement sequence 414

→ 2,3-Dioxabicyclo[2.2.2]octanes, from 1,5-dienes, via intramolecular cyclization/endoperoxidation, using oxygen, triarylpyrylium salt photosensitized 181, 182

→ 1,2-Dioxanes, substituted, from peroxy radical domino cyclization of polyene hydroperoxides, di-tert-butyl peroxyoxalate mediated 159, 160

→ 1,2-Dioxanes, substituted, from peroxy radical domino cyclization of polyene hydroperoxides, samarium(II) iodide/oxygen induced 162

→ 1,2-Dioxanes, 3,3,6,6-tetraaryl-, from α-aryl substituted styrenes, via [2+2+2] cycloaddition with oxygen, 9,10-dicyanoanthracene photosensitized 180, 181

→ ent-Dioxepandehydrothyrsiferol, key intermediate, from alkenyl triepoxide carbonate, via bromonium ion induced domino cyclization 295, 296

→ ent-Dioxepandehydrothyrsiferol, key intermediate, from tert-butyl ester containing triepoxides, via endo-selective epoxide-opening cascade, Lewis acid promoted, then silylation 49

→ ent-Dioxepandehydrothyrsiferol, key intermediate, from distal alkene containing triepoxides, via endo-selective epoxide-opening cascade, bromonium ion initiated, internal nucleophile terminated 53

→ [1,3]Dioxino[5,4-b]oxepino[2,3-f]oxepin-2-ones, 8-bromodecahydro-, from distal alkene containing diepoxides, via endo-selective epoxide-opening cascade, bromonium ion initiated, internal nucleophile terminated 52, 53

→ [1,3]Dioxino[5,4-b]oxepino[2,3-f]oxepin-2-ones, decahydro-, from endo-selective epoxide-opening cascade, oxocarbenium ion initiated, via photooxidative cleavage 54, 55

→ [1,3]Dioxino[5,4-b]oxepino[2,3-f]oxepin-2-ones, 8-hydroxydecahydro-, from tert-butyl carbonate containing triepoxides, via endo-selective epoxide-opening cascade, Lewis acid promoted 48–50

→ 1,2-Dioxolan-3-amines, from aminocyclopropanes, by heteroatom oxidation, using oxygen, iron complex catalyzed, via oxidation to radical cation/ring opening/peroxidation/cyclization 177, 178

→ 1,2-Dioxolanes, 3,5-diaryl-, from 1,2-diarylcyclopropanes, via photooxygenation, 9,10-dicyanoanthracene sensitized 179

→ 1,2-Dioxolanes, substituted, from peroxy radical domino cyclization of polyene hydroperoxides, copper trifluoromethanesulfonate/oxygen induced 161

→ 1,2-Dioxolanes, 3-vinyl-, from vinylcyclopropanes, via domino ring opening/peroxidation/cyclization cascade, using oxygen/light, 2,2′-azobisisobutyronitrile/diphenyl diselenide catalyzed 171, 172

→ 1,2-Dioxolan-3-ols, pentamethyl-, from heteroatom oxidation of pentamethylcyclopropanols, using oxygen 177

→ Dipyrano[3,2-b:2′,3′-e]pyrans, decahydro-, from bis(trimethylsilyl)-containing diepoxides, via endo-selective epoxide-opening cascade, Brønsted base/fluoride ion promoted, then acetylation 51, 52

→ Dipyrano[3,2-b:2′,3′-e]pyrans, decahydro-, from 3-hydroxytetrahydropyran-containing diepoxides, via endo-selective epoxide-opening cascade, water promoted 56

1,3-Dithianes, 2-(hydroxyethyl)-2-silyl-, reaction with N-sulfonylaziridines, via lithiation/1,4-Brook rearrangement/ring opening cascade sequence, 1,5-amino alcohol synthesis 422

Keyword Index

647

1,3-Dithianes, 2-silyl-, lithiated, sequential reaction with two epoxides, via ring opening/1,4-Brook rearrangement cascade sequence, 1,5-diol synthesis 414

1,6-Diyne esters, cycloisomerization, cationic gold complex catalyzed, dihydrofluorene synthesis, via intramolecular domino 1,2-acyloxy migration/cyclopropenation/Nazarov cyclization sequence 564

Diynes, domino cycloisomerization, gold complex catalyzed, tricyclic compound synthesis 330, 331

1,3-Diynes, intermolecular reactions with alkenes, ruthenium alkylidene complex catalyzed, initiation of metallotropic [1,3]-shift via cross metathesis 80–82

1,3-Diynes, intramolecular reactions with tethered alkenes, ruthenium alkylidene complex catalyzed, initiation of metallotropic [1,3]-shift via ring-closing metathesis 82, 83

1,3-Diynes, intramolecular reactions with tethered alkenes, via ring-closing metathesis, ruthenium alkylidene complex catalyzed, without metallotropic [1,3]-shift 82, 83

1,5-Diynes, intramolecular cyclization domino reaction, gold complex catalyzed, benzofulvene synthesis 545–547

1,15-Diynes, ester tethered, ring-closing metathesis, molybdenum complex catalyzed, then enyne cross metathesis with ethene, ruthenium alkylidene complex catalyzed, amphidinolide V key intermediate synthesis 88, 89

→ Dolabelide D, key intermediate, from reaction of a lithiated γ-silyl carbinol with copper(I) bromide–dimethyl sulfide complex and iodomethane, via 1,4-Brook rearrangement cascade sequence 441

E

→ Echinopine A, tricyclic intermediate, from a dienyne, via intramolecular cyclization/Diels–Alder cycloaddition cascade sequence, palladium/phosphine catalyzed 458, 459

Enals, bis(silyl), reaction with alkyl halides, copper(I) cyanide promoted, via nucleophilic addition of organolithiums/1,4-Brook rearrangement/alkylation cascade sequence, trisubstituted vinylsilane synthesis 426, 427

→ Enals, α,β-bis(silyl), from propargylic silyl ethers, via domino α-lithiation/retro-1,2-Brook rearrangement/silyl migration sequence, then Lewis acid treatment 390, 391

Enals, α-(disilylmethyl)-, reaction with electrophiles, via nucleophilic addition/1,4-Brook rearrangement/electrophilic capture cascade sequence, functionalized vinylsilane synthesis 419, 420

→ Enamides, bicyclic, from alkenyl lactams, via carbonyl–alkene metathesis, using an in situ generated titanium–alkylidene complex 146

→ Endoperoxides, bicyclic, from 1,5-dienes, via intramolecular cyclization/endoperoxidation, using oxygen, 9,10-dicyanoanthracene photosensitized 181, 182

→ Endoperoxides, bicyclic, from 1,5-dienes, via intramolecular cyclization/endoperoxidation, using oxygen, triarylpyrylium salt photosensitized 181, 182

→ Endoperoxides, bicyclic, from 1,6-dienes, via intramolecular cyclization/endoperoxidation, using oxygen, ruthenium complex photocatalyzed 182, 183

→ Endoperoxides, bicyclic, from domino hydroperoxidation of (S)-limonene, using oxygen, cobalt catalyzed, then metathesis with triethylsilane 164, 165

→ Endoperoxides, from conversion of tris(acetylacetonato)manganese(III) into 1,3-dicarbonyl radicals, then oxidative addition to alkenes/cyclization, manganese(III) mediated 175, 176

→ Endoperoxides, from 1,4-dienes, via thiol–alkene co-oxygenation reaction, using di-tert-butyl peroxyoxalate/benzenethiol/oxygen 168, 169

→ Endoperoxides, from domino hydroperoxidation/cyclization of (R)-limonene, via thiol–alkene co-oxygenation reaction, using di-tert-butyl peroxyoxalate/benzenethiol/oxygen 169, 170

→ Endoperoxides, from domino hydroperoxidation of vinylcyclopropanes, using oxygen, cobalt complex catalyzed, then metathesis with triethylsilane 165, 166

→ Endoperoxides, from peroxy radical domino cyclization of polyene hydroperoxides, di-tert-butyl peroxyoxalate mediated 159, 160

→ Endoperoxides, from vinylcyclopropanes, via domino ring opening/peroxidation/cyclization cascade, using oxygen/light, 2,2′-azobisisobutyronitrile/diphenyl diselenide catalyzed 171, 172

Enediynes, intramolecular cyclization domino reaction, gold complex catalyzed, polycyclic carbocycle synthesis 547, 548

Enediynes, reaction with 1,4-diacetoxybut-2-ene, via domino relay metathesis, double metallotropic [1,3]-shift, and cross metathesis sequence, ruthenium alkylidene complex catalyzed, panaxytriol key intermediate synthesis 121, 122

Enediynes, ring-closing metathesis/metallotropic [1,3]-shift domino sequence, ruthenium alkylidene complex catalyzed, 1,5-dien-3-yne synthesis 120

Enediynols, intramolecular cyclization domino reaction, carbene gold complex catalyzed, functionalized bicyclic phenol synthesis 552, 553

→ (−)-Englerin A, key intermediate, from a dienyne, via domino double enyne ring-closing metathesis, ruthenium alkylidene complex catalyzed 100, 101

648 Keyword Index

→ (–)-Englerin A, key intermediate, from domino cyclization of 1,6-enyn-10-one, gold complex catalyzed 334, 335

→ Enol ethers, acyclic, from alkenyl esters, via carbonyl–alkene metathesis, using an in situ generated titanium–alkylidene complex 143, 144

→ Enol ethers, cyclic, from alkenyl esters, via carbonyl–alkene metathesis, using an in situ generated titanium–alkylidene complex, comparison of reagents 144–146

→ Enol ethers, cyclic, from alkenyl esters, via carbonyl–alkene metathesis, using Tebbe or Petasis reagent, comparison 140, 142

→ Enol ethers, cyclic, from alkenyl esters, via carbonyl–alkene metathesis, using Tebbe reagent 140, 141

Enol ethers, cyclic, reaction with amine-substituted nickel porphyrins, lanthanum catalyzed, via domino condensation/hetero-Diels–Alder process, bis(hydroxyalkyl)-substituted pyrido[2,3-b]-porphyrin synthesis 585, 586

→ Enol ethers, macrocyclic, from macrocyclic alkenyl esters, via carbonyl–alkene metathesis, using in situ generated titanium–alkylidene complex 146

→ Enol ethers, silyl, from reaction of acyloins with lithiated organosilanes, via addition/1,2-Brook rearrangement cascade sequence 357

→ Enol ethers, silyl, from reaction of acylsilanes with α-sulfinyl carbanions, via nucleophilic addition/1,2-Brook rearrangement/elimination cascade sequence 359

→ Enol ethers, silyl, from reaction of alkyl aryl ketones with a (silyl)diazomethane, Lewis acid catalyzed, via domino nucleophilic addition/1,2-aryl shift/1,3-Brook rearrangement sequence 395

→ Enol ethers, silyl, from reaction of α-phenylsulfanyl acylsilanes with organolithium reagents, via nucleophilic addition/base-induced 1,2-Brook rearrangement/elimination cascade sequence 358, 359

→ Enol ethers, silyl, trisubstituted, from acyl(triphenyl)silanes, via domino enolization/1,2-Brook rearrangement sequence, using copper *tert*-butoxide, then cross coupling with alkyl iodides, palladium catalyzed 380, 381

Enols, diazo, silyl protected, coupling with propargylic esters, Lewis acid catalyzed, alkyne-tethered β-keto-α-diazocarbonyl compound synthesis 522

→ Enol thioethers, silyl, from reaction of α-silylsulfanyl ketones with alkyl halides, via enolization/sulfur-to-oxygen 1,4-Brook rearrangement/electrophilic trapping cascade sequence 430, 431

→ Enones, α-aryl-, from cross coupling of propargyl acetates with arylboronic acids, gold complex catalyzed, via domino acyloxy migration/oxidation/reductive elimination sequence 562

Enones, distal cyclopropane containing, radical fragmentation/bicyclization, visible light mediated, lanthanum(III) facilitated, ruthenium complex catalyzed, octahydropentalene synthesis 211, 212

→ Enones, α-fluoro-, from reaction of propargyl acetates with Selectfluor, gold complex catalyzed, via domino rearrangement/allene halogenation sequence 561, 562

→ Enones, α-halo-, from reaction of propargyl acetates with N-halosuccinimides, gold complex catalyzed, via domino rearrangement/allene halogenation sequence 561

Enones, polycyclic, intramolecular peroxidation/cyclization, using oxygen, manganese(III) mediated, (+)-salvadione B intermediate synthesis 176, 177

→ Enones, α-(1-siloxyalkyl)-, from α-silyl carbinols by modified Morita–Baylis–Hillman reaction, using aldehydes and N-heterocyclic carbenes, via deprotonation/1,2-Brook rearrangement/silyl migration/nucleophilic addition/O-silylation cascade sequence 381, 382

Enones, β-silyl-, reaction with aldehydes/tributylphosphine, via Michael addition/1,4-Brook rearrangement/Wittig alkenation cascade sequence, siloxy-substituted 1,3-diene synthesis 420, 421

1,7-Enyne benzoates, cycloisomerization, cationic gold complex catalyzed, functionalized 3-azabicyclo[4.2.0]oct-5-ene synthesis, via intramolecular domino 1,3-ester migration/[2+2] cycloaddition sequence 565, 566

→ Enynediols, chiral, from enantioselective reaction of propargyl-substituted acylsilanes with alkynylsilanes, chiral diarylprolinol catalyzed, zinc promoted, via nucleophilic alkynylation/1,2-Brook rearrangement/ene–allene carbocyclization/electrophilic trapping/oxidation cascade sequence 373, 374

Enynediones, tandem cyclization, DABCO catalyzed, tetrahydroisochromen-8-one synthesis 344

Enyne metathesis catalysts 68, 69

Enyne metathesis versus metallotropic [1,3]-shift 69, 70

Enynes, cross-metathesis reactions, ruthenium alkylidene complex catalyzed, ene first mechanistic support 72, 73

Enynes, cross-metathesis reactions, ruthenium alkylidene complex catalyzed, regioselectivity 76–78

Enynes, cross-metathesis reactions, ruthenium alkylidene complex catalyzed, stereoselectivity 79

Enynes, cross-metathesis reactions, ruthenium alkylidene complex catalyzed, yne first versus ene first mechanisms 71

Enynes, domino cross metathesis with ethene/ring-closing metathesis, ruthenium alkylidene complex catalyzed, (–)-longithorone A key intermediate synthesis 115, 116

Keyword Index

Enynes, domino ring-closing metathesis versus ring-closing metathesis/metallotropic [1,3]-shift, ruthenium alkylidene complex catalyzed 69, 70

Enynes, enantioselective polycyclization, chiral di-gold–bisphosphine complex catalyzed, chiral hexahydrophenanthrene synthesis 22

Enynes, intramolecular metathesis reaction, ruthenium alkylidene complex catalyzed, 1,3-diene synthesis 68

Enynes, metal-catalyzed skeletal reorganization, metal catalyzed, organometallic versus Lewis acid pathways 75

Enynes, metallotropic [1,3]-shift, propargylic alkylidene formation, regioselectivity 80

Enynes, ring-closing metathesis reactions, ruthenium alkylidene complex catalyzed, 1,3-diene synthesis, selectivity in natural product synthesis 86, 87

Enynes, ring-closing metathesis reactions, ruthenium alkylidene complex catalyzed, *exo/endo*-mode selectivity 78, 79

Enynes, ring-closing metathesis reactions, ruthenium alkylidene complex catalyzed, stereoselectivity 79

Enynes, silyl ynol ether containing, ring-closing metathesis, ruthenium alkylidene complex catalyzed, then hydrogen fluoride treatment and hydrogenation, β-eremophilane intermediate synthesis 93, 94

1,5-Enynes, 3-(2-iodopyranyloxy) substituted, reductive tandem radical cyclization, using sodium borohydride, cobaloxime catalyzed, tricyclic cycloalkene synthesis 227, 228

1,5-Enynes, 3-(2-iodopyranyloxy) substituted, tandem radical cyclization, oxygen catalyzed, triethylborane mediated, tricyclic iodocycloalkene synthesis 227, 228

1,5-Enynes, reaction with nitrosobenzene, gold complex catalyzed, via cyclization/[3+2] cycloaddition domino reaction, functionalized tricyclic isoxazolidine synthesis 545

1,5-Enynes, ring-closing metathesis, ruthenium alkylidene complex catalyzed, then hydrogenation and deprotection, (±)-grandisol synthesis 92, 93

1,6-Enynes, alkoxy substituted, vinyl radical tandem radical cyclization, using tributyltin hydride/2,2′-azobisisobutyronitrile, alkoxytetrahydrofuran-substituted cis-hydrindane synthesis 219

1,6-Enynes, cycloisomerization domino reaction, gold complex catalyzed, tricyclic carbocycle synthesis, via 1,5-migration 542–544

1,6-Enynes, domino cyclization, gold complex catalyzed, fused bicyclic 5,7-ring system synthesis 332, 333

1,6-Enynes, ester linked, domino enyne ring-closing metathesis/Diels–Alder dimerization, ruthenium alkylidene complex catalyzed, (±)-differolide synthesis 124

1,6-Enynes, monocyclic, platinum-catalyzed enyne metathesis to give ring-bridged macrocycle, metacycloprodigiosin key intermediate synthesis 129, 130

1,6-Enynes, monocyclic, platinum-catalyzed enyne metathesis to give ring-bridged macrocycle, roseophilin key intermediate synthesis 128, 129

1,6-Enynes, ring-closing metathesis, ruthenium alkylidene complex catalyzed, then deprotection and Diels–Alder reaction with acrylate, valerenic acid key intermediate synthesis 124, 125

1,7-Enynes, reaction with but-3-en-2-ol, via domino cross metathesis/ring-closing metathesis sequence, ruthenium alkylidene complex catalyzed, (+)-panepophenanthrin key intermediate synthesis 114, 115

1,7-Enynes, reaction with a terminal alkene, via domino enyne ring-closing metathesis/cross metathesis sequence, ruthenium alkylidene complex catalyzed, then Diels–Alder cycloaddition with an acetylenedicarboxylate, isofregenedadiol key intermediate synthesis 126, 127

1,7-Enynes, ribose derived, ring-closing metathesis, ruthenium alkylidene complex catalyzed, (+)-pericosine C key intermediate synthesis 95, 96

1,7-Enynes, ring-closing metathesis, ruthenium alkylidene complex catalyzed, (+)-ferruginine key intermediate synthesis 97, 98

1,7-Enynes, ring-closing metathesis, ruthenium alkylidene complex catalyzed, isofagomine key intermediate synthesis 94, 95

1,7-Enynes, ring-closing metathesis, ruthenium alkylidene complex catalyzed, then Diels–Alder cycloaddition with naphtho-1,4-quinones, elimination/aromatization, oxidation, and deprotection, angucyclinone-type natural product synthesis 125, 126

1,7-Enynes, serine derived, ring-closing metathesis, ruthenium alkylidene complex catalyzed, (+)-valienamine key intermediate synthesis 96, 97

1,8-Enynes, reaction with methyl vinyl ketone, via domino ring-closing metathesis/cross metathesis sequence, ruthenium alkylidene complex catalyzed, (+)-8-epi-xanthatin synthesis 113

1,8-Enynes, ring-closing metathesis, ruthenium alkylidene complex catalyzed, then methylation and cross metathesis with methyl vinyl ketone, ruthenium alkylidene complex catalyzed, (–)-11α,13-dihydroxanthatin synthesis 113, 114

1,8-Enynes, ring-closing metathesis, ruthenium alkylidene complex catalyzed, then oxidative cleavage and deprotection, (+)-anatoxin-a synthesis 97, 98

Enynones, reaction with terminal alkynes, rhodium catalyzed, via cyclization/cyclopropenation/ring expansion cascade sequence, 6-arylcyclopenta[b]-furan synthesis 520, 521

1,6-Enyn-10-ones, domino cyclization, gold complex catalyzed, oxatricyclic compound synthesis 332, 334

650 Keyword Index

→ (±)-Epimeloscine, key intermediate, from divinylcyclopropanes, via intramolecular domino radical cyclization, tributyltin radical mediated 191

Epoxides – *see also* Oxiranes

Epoxides, bicyclic, domino epoxide opening/Wagner–Meerwein rearrangement/cyclization process, Lewis acid induced, benzo[d]xanthene synthesis 303, 304

Epoxides, bicyclic, rearrangement/ring contraction, Lewis acid promoted, carbonyl compound synthesis 282

Epoxides, diene, conversion into decalins, ethylaluminum dichloride catalyzed, via cationic polyene cyclization 15, 16

Epoxides, diol, intramolecular epoxide opening/cyclization, base catalyzed, 5-(hydroxymethyl)tetrahydrofuran-3-ol synthesis 250, 251

Epoxides, distal azide substituted, domino intramolecular Schmidt reaction, Lewis acid catalyzed, then reduction, indolizine synthesis 300–302

Epoxides, distal O-protected phenol-substituted, domino cyclization, Lewis acid induced, substituted hexahydroxanthen-2-ol synthesis 298, 299

Epoxides, hydroxy, endo-selective intramolecular opening/cyclization, acid catalyzed, using directing groups, cyclic ether synthesis 248, 249

Epoxides, hydroxy, exo-selective intramolecular opening/cyclization, acid catalyzed, cyclic ether synthesis 246

Epoxides, hydroxy, intramolecular opening/cyclization, using an internal oxygen nucleophile, acid catalyzed, cyclic ether synthesis 245

Epoxides, α-hydroxy, stereoselective tandem rearrangement/reduction/ring contraction, triethylaluminum promoted, chiral 1,3-diol synthesis 282–284

Epoxides, α-hydroxy, tandem rearrangement/reduction/ring contraction, samarium(II) iodide promoted, chiral 1,3-diol synthesis 282

Epoxides, intramolecular opening/cyclization, using an internal oxygen nucleophile, base induced, cyclic ether synthesis 244

Epoxides, polyene, stereoselective domino epoxide opening/[4+2] cycloaddition sequence, Lewis acid induced, polyfunctionalized decahydrofluorene synthesis 305, 306

Epoxides, polyene, tricyclization, Lewis acid induced, pentacyclic triterpene intermediate synthesis 290, 291

Epoxides, reaction with lithiated 2-silylacetonitrile, via ring opening/1,4-Brook rearrangement cascade sequence, then acidic hydrolysis, γ-lactone synthesis 404

→ (+)-Epoxydictymene, key cyclopentenone intermediate, from a functionalized enyne and octacarbonyldicobalt, via intramolecular Pauson–Khand cyclization 602, 603

→ (+)-Equilenin, tetracyclic intermediate, from 1-alkenyl-2-arylcyclobutanols, via tandem semipinacol rearrangement/direct arylation reaction sequence, palladium catalyzed 324, 325

→ β-Eremophilane, key intermediate, from silyl ynol ether containing enyne, via ring-closing metathesis, ruthenium alkylidene complex catalyzed, then hydrogen fluoride treatment and hydrogenation 93, 94

→ (±)-Erythrocarine, from an azadienyne hydrochloride salt, via domino double enyne ring-closing metathesis, ruthenium alkylidene complex catalyzed 109, 110

Esters – *see also* Acetates, Pivalates, etc.

Esters, alkenyl, carbonyl–alkene metathesis, using an in situ generated titanium–alkylidene complex, acyclic enol ether synthesis 143, 144

Esters, alkenyl, carbonyl–alkene metathesis, using an in situ generated titanium–ethylidene complex, bryostatin 1 fragment intermediate synthesis 147

Esters, alkenyl, carbonyl–alkene metathesis, using an in situ generated titanium–alkylidene complex, cyclic enol ether synthesis, comparison of reagents 144, 145

Esters, alkenyl, carbonyl–alkene metathesis, using Tebbe or Petasis reagent, comparison, cyclic enol ether synthesis 140, 142

Esters, alkenyl, carbonyl–alkene metathesis, using Tebbe reagent, cyclic enol ether synthesis 140, 141

Esters, alkenyl, macrocyclic, carbonyl–alkene metathesis, using an in situ generated titanium–ethylidene complex, macrocyclic enol ether synthesis 146

Esters, ω-alkenyl, Wittig/Takai–Utimoto alkenation/alkene metathesis, Schrock molybdenum complex catalyzed, C1 glycal synthesis 137, 138

Esters, α-amino, domino three-component Strecker reaction with cyclohexanone and silyl cyanides, ytterbium catalyzed, N-(cyanoalkyl)amino ester synthesis 583

Esters, α-cyano dienyl, Julia–Snider tandem radical cyclization, using dibenzoyl peroxide, octahydroindene synthesis 218

→ Esters, α-cyano-α-siloxy, chiral, from enantioselective benzoin-type coupling of acylsilanes with cyanoformates, chiral aluminum Lewis acid catalyzed, via nucleophilic addition/1,2-Brook rearrangement cascade sequence 370, 371

→ Esters, cycloalkyl, ring contracted, from cyclic α-halo ketones, via Favorskii rearrangement, base promoted 260

Esters, cyclopropylpropargyl, cyclopentannulation, via intramolecular domino Rautenstrauch rearrangement/ring opening/cyclization sequence, gold/silver catalyzed, then hydrolysis, cyclopent-1-enyl ketone synthesis 566, 567

Esters, cyclopropylpropargyl, intramolecular domino cycloisomerization sequence, gold/silver catalyzed, then hydrolysis, cyclohex-2-enone synthesis 568

→ Esters, 2,3-dihydroxy, from reaction of silyl glyoxylates with ketones, Lewis acid catalyzed, via Oppenauer oxidation/1,2-Brook rearrangement/aldol reaction cascade sequence 369, 370

Esters, α-hydroxy, spirocyclobutyl, retro-benzilic acid rearrangement, base promoted, spirocyclic cyclopentenone synthesis 260, 261

Esters, β-keto, four-component domino condensation process with primary alkylamines and α-diketones, cerium catalyzed, pentasubstituted pyrrole synthesis 591, 592

→ Esters, nitroaldol cyano, chiral, from dialdehydes, via one-pot enantioselective sequential cyanation/nitroaldol process with cyanoformate and then nitromethane, chiral yttrium complex catalyzed 580, 581

→ Esters, β-oxo, bicyclic, from radical polyene oxidative cyclization of polyunsaturated β-oxo esters, manganese mediated 26

Esters, β-oxo, dienyl substituted, double radical carbocyclization, using manganese(III) acetate/copper(II) acetate, bicyclic ketone synthesis 221

Esters, propargyl, domino intramolecular rearrangement reaction, gold complex catalyzed, 3-alkylidenealkane-2,4-dione synthesis 557, 558

Ethanols, 2-alkoxy-1-(cycloprop-1-enyl)-, rhodium-catalyzed rearrangement/cyclization cascade sequence, 4-alkylidenetetrahydropyran-3-ol synthesis 518, 519

→ Ethanols, 1-(1,2-dioxan-3-yl)-, from peroxy radical cyclization of hex-4-enyl hydroperoxide, di-tert-butyl peroxyoxalate initiated 158, 159

→ Ethanols, 2-(octahydro-2,2′-bifuran-5-yl)-, from diepoxides, via exo-selective epoxide-opening cascade, oxocarbenium ion initiated, via photooxidative cleavage 46, 47

→ Ethanols, 1-phenyl-, from reaction of acetyl(dimethyl)phenylsilanes with fluoride ion, via 1,2-aryl shift/1,2-Brook rearrangement cascade sequence, then acidification 386

→ Etheromycin, intermediate, from a hydroxy 1,5,9-triepoxide, via exo-selective domino intramolecular opening/cyclization, acid catalyzed 247, 248

Ethers, bis(aryl), domino aryl-Claisen/[3,3]-sigmatropic rearrangement sequence, europium catalyzed, bis(phenol) synthesis 596

→ Ethers, cyclic, from epoxides, via intramolecular opening/cyclization, using an internal oxygen nucleophile, base induced 244

→ Ethers, cyclic, from epoxy alcohols, via intramolecular epoxide opening/cyclization, base catalyzed 250

→ Ethers, cyclic, from epoxyallenes, via intramolecular cyclization domino reaction, gold/silver catalyzed 554

→ Ethers, cyclic, from hydroxy epoxides, via endo-selective intramolecular opening/cyclization, acid catalyzed, using directing groups 249

→ Ethers, cyclic, from hydroxy epoxides, via intramolecular opening/cyclization, using an internal oxygen nucleophile, acid catalyzed 245

Ethers, enol – see Enol ethers

→ Ethers, macrocyclic, from reaction of epoxyalkynes with N-halosuccinimides, gold complex catalyzed, via double ring-expansion domino reaction 553, 554

Ethers, propargylic, diene substituted, chiral, polyene cyclization, indium catalyzed, chiral polycyclic compound synthesis 17

Ethers, silyl – see Silyl ethers

F

Farnesyl acetate, tandem radical cyclization, using dibenzoyl peroxide/copper benzoate, polysubstituted decalin synthesis 219

→ Fasicularin, spirocyclic intermediate, from 2,3-epoxy silyl ether, via semipinacol rearrangement/ring expansion, titanium(IV) chloride catalyzed 270

Favorskii rearrangement, cyclic α-halo ketones to ring-contracted carboxylic acid derivatives via cyclopropanones, base promoted 286, 287

Favorskii rearrangement, cyclic α-halo ketones to ring-contracted cyclic acids/esters, base promoted 260

→ (+)-Ferruginine, key intermediate, from a 1,7-enyne, via ring-closing metathesis, ruthenium alkylidene complex catalyzed 97, 98

→ (−)-Flueggine A, key intermediate, from a dienyne, via domino double enyne ring-closing metathesis, ruthenium alkylidene complex catalyzed 110, 111

→ Fluorenes, decahydro-, polyfunctionalized, from a polyene epoxide, via stereoselective domino epoxide opening/[4+2] cycloaddition sequence, Lewis acid induced 305, 306

→ Fluorenes, dihydro-, from cycloisomerization of 1,6-diyne esters, cationic gold complex catalyzed, via intramolecular domino 1,2-acyloxy migration/cyclopropenation/Nazarov cyclization sequence 564

→ (−)-Flustramine B, tricyclic intermediate, from a 1-alkenyltryptamine and propenal, via enantioselective domino addition/cyclization, chiral cyclic secondary amine catalyzed 346, 347

→ FR66979 alkaloid, key step, 1,3-Brook rearrangement of a tricyclic hydroxy(silyl)aziridine 400

→ Fredericamycin A, spirocyclic core, from epoxy camphanoate, via tandem rearrangement/reduction/ring contraction, Lewis acid promoted 285, 286

652 Keyword Index

→ Frondosin A, key step, [4+3] cycloaddition of propargyl acetates with 1,3-dienes, cationic gold complex catalyzed, via domino 1,2-acetoxy migration/cyclopropanation/homo-Cope rearrangement sequence 569

→ Fulvenes, (hydroxymethyl)acyl-, from functionalized allenynes, via intramolecular Pauson–Khand cyclization, using hexacarbonylmolybdenum 604

→ Furan-3-amines, disubstituted, from three-component reaction of arylglyoxals with secondary amines and terminal alkynes, gold catalyzed, via cyclization domino reaction 541

Furan-2-carbaldehydes, reaction with secondary amines, via domino condensation/ring opening/ electrocyclization process, dysprosium catalyzed, 4,5-diaminocyclopent-2-enone synthesis 584, 585

Furan-3-carbaldehydes, 2-silyl-, ortho-functionalization, via lithiation/1,4-Brook rearrangement/ electrophilic capture cascade sequence, 2-alkyl-3-(1-siloxyalkyl)furan synthesis 424, 425

→ Furan-3-ols, 5-(hydroxymethyl)tetrahydro-, from diol epoxides, via intramolecular epoxide opening/cyclization, base catalyzed 250, 251

→ Furans, 2-alkyl-3-(1-siloxyalkyl)-, from *ortho*-functionalization of furan-3-carbaldehydes, via lithiation/1,4-Brook rearrangement/electrophilic capture cascade sequence 424, 425

→ Furans, 2-ethoxy-4,5-diphenyl-, from rhodium-catalyzed rearrangement of 2,3-diphenylcycloprop-2-ene-1-carboxylate, via cascade sequence 516, 517

→ Furans, functionalized, from reaction of 2-formylcyclopropane-1,1-dicarboxylates with 1,3-diketones, scandium catalyzed, via domino aldol/ring-opening sequence 592, 593

→ Furans, tetrahydro-, from hydroxy epoxides, via *exo*-selective intramolecular opening/cyclization, acid catalyzed 246

→ Furans, tetrahydro-, substituted, chiral, from tandem radical cyclization of N-alkoxyphthalimides, using tributyltin hydride/2,2′-azobisisobutyronitrile 230, 231

→ Furo[3,4-*d*][1,2]dioxins, tetrahydro-, from 1,6-dienes, via intramolecular cyclization/endoperoxidation, using oxygen, ruthenium complex photocatalyzed 182, 183

→ Furo[2,3-*b*]indoles, tetrahydro-, chiral, from tryptophols and enals, via enantioselective domino addition/cyclization, chiral cyclic secondary amine catalyzed 345, 346

→ Furo[3,4-*b*]pyrans, hexahydro-, from intramolecular radical cyclization cascade of bis(vinyl ethers), triphenyltin radical mediated 188–190

→ (+)-Fusarisetin A, tricyclic cyclopentenone intermediate, from a functionalized enyne and octacarbonyldicobalt, via intramolecular Pauson–Khand cyclization 618, 619

G

→ Glabrescol, from diol pentaepoxides, via *exo*-selective epoxide-opening cascade, Brønsted base promoted 45

→ Glabrescol, intermediate, from a dihydroxy tetraepoxide, via *exo*-selective domino intramolecular opening/cyclization, acid catalyzed 247, 248

→ Gloiosiphone A, dimethyl, key intermediate, from intramolecular cyclization of allylic acetate-diene, palladium/phosphine catalyzed, via oxidative addition/sequential Heck cyclization cascade sequence 494, 495

→ Glycals, C1 substituted, from ω-alkenyl esters, via Wittig/Takai–Utimoto alkenation/alkene metathesis, Schrock molybdenum complex catalyzed 138

→ Glycolates, 2,3-dihydroxy-, from reaction of silyl glyoxylates with magnesium alkoxides, via nucleophilic addition/Oppenauer oxidation/ 1,2-Brook rearrangement/aldol reaction cascade sequence 369

Glyoxals, aryl-, three-component reaction with secondary amines and terminal alkynes, gold catalyzed, disubstituted furan-3-amine synthesis, via cyclization domino reaction 541

Glyoxylates, silyl, allyl esters, reaction with organometallic reagents, via nucleophilic addition/1,2-Brook rearrangement/Ireland–Claisen rearrangement cascade sequence, γ,δ-unsaturated α-siloxy carboxylic acid synthesis 368

Glyoxylates, silyl, reaction with aldehydes, cyanide ion/Lewis acid catalyzed, via cyanide addition/ 1,2-Brook rearrangement/nucleophilic addition/ 1,4-silyl migration/cyanide elimination cascade sequence, benzoin-type product synthesis 377

Glyoxylates, silyl, reaction with ketones, lanthanum(III) cyanide catalyzed, via cyanide addition/ 1,2-Brook rearrangement/nucleophilic addition/ 1,4-silyl migration/cyanide elimination cascade sequence, α-siloxy ketone synthesis 377, 378

Glyoxylates, silyl, reaction with ketones, Lewis acid catalyzed, via Oppenauer oxidation/1,2-Brook rearrangement/aldol reaction cascade sequence, 2,3-dihydroxy ester synthesis 369, 370

Glyoxylates, silyl, reaction with magnesium alkoxides, via nucleophilic addition/Oppenauer oxidation/1,2-Brook rearrangement/aldol reaction cascade sequence, 2,3-dihydroxyglycolate synthesis 368, 369

Glyoxylates, silyl, Reformatsky reagent addition/1,2-Brook rearrangement/ketone trapping/transesterification/lactonization cascade sequence, pentasubstituted γ-butyrolactone synthesis 363

Glyoxylates, silyl, three-component reaction with magnesium acetylides and nitroalkenes, via nucleophilic addition/1,2-Brook rearrangement cascade sequence, tetrasubstituted allene synthesis 368

Keyword Index

653

Glyoxylates, silyl, three-component reaction with Reformatsky reagents and β-lactones, via nucleophilic addition/1,2-Brook rearrangement/ring opening cascade sequence, polysubstituted ketone synthesis 366, 367

→ Gorgonian sesquiterpene, key intermediate, from 3-methylpenta-1,4-diene, via deprotonation/aldehyde addition/1,4-Brook rearrangement/electrophilic capture cascade sequence 442

→ (±)-Grandisol, from a 1,5-enyne, via ring-closing metathesis, ruthenium alkylidene complex catalyzed, then hydrogenation and deprotection 93

→ 6,12-Guaianolides, tricyclic cyclopentenone intermediate, from a functionalized allenyne, via intramolecular Pauson–Khand cyclization, rhodium catalyzed 624, 625

→ Guanacastepene A skeleton, from dienynes, via domino double enyne ring-closing metathesis, ruthenium alkylidene complex catalyzed 99, 100

→ Gymnocin A, tetracyclic ring fragment, from 3-hydroxytetrahydropyran-containing triepoxides, via *endo*-selective epoxide-opening cascade, water promoted 57, 58

H

→ Hamigerans, tetracyclic cyclopentenone intermediate, from a functionalized arylenyne and octacarbonyldicobalt, via intramolecular Pauson–Khand cyclization 609

→ (−)-Harveynone, key intermediate, from an enediyne, via ring-closing metathesis/metallotropic [1,3]-shift domino sequence, ruthenium alkylidene complex catalyzed 120, 121

→ (+)-Harziphilone, from tandem cyclization of an enynedione, DABCO catalyzed 344, 345

Hepta-1,6-diene-3,5-diol, conversion into bicyclic furan lactone, via carbonylative oxacyclization cascade sequence, palladium catalyzed 500, 501

Hept-6-enoates, 2-allyl-3-oxo-, radical polyene oxidative cyclization, manganese mediated, 6-methylene-2-oxobicyclo[3.2.1]octane-1-carboxylate synthesis 26

→ (±)-Herbertene, key intermediate, from an α-hydroxy epoxide, via tandem rearrangement/reduction/ring contraction, samarium(II) iodide promoted 284

Hexanols, 1-(cycloprop-1-enyl)-, rhodium-catalyzed rearrangement/cyclization, 2-alkylidenecyclopentanol synthesis, via cascade sequence 518, 519

Hex-4-enyl hydroperoxide, peroxy radical cyclization, di-tert-butyl peroxyoxalate initiated, 1-(1,2-dioxan-3-yl)ethanol synthesis 158

→ (±)-Hinesol, key intermediate, from spirocyclobutyl α-hydroxy ester, via retro-benzilic acid rearrangement, base promoted 260, 261

→ (−)-Hippodamine, key intermediate, from three-component asymmetric reaction of 1-aminoinden-2-ol, 2-iodo-4-oxobut-2-enoate, and an alkenylstannane, palladium/phosphine catalyzed, via condensation/Stille coupling/electrocyclization cascade sequence 454

→ Hirsutellone B, tricyclic intermediate, from a polyene epoxide, via stereoselective domino epoxide opening/[4+2] cycloaddition sequence, Lewis acid induced 305, 306

→ (+)-Hirsutene, key cyclopentenone intermediate, from a functionalized enyne and octacarbonyldicobalt, via intramolecular Pauson–Khand cyclization 603, 604

→ (±)-Hirsutene, key triquinane intermediate, from a dienylcyclopropane, via domino [(5+2)+1] cycloaddition/aldol reaction sequence, rhodium catalyzed 326, 327

Homoallylic alcohols, distal aryl substituted, reaction with aldehydes, via domino Prins reaction/polyene cyclization sequence, indium(III) bromide promoted, hexahydrobenzo[f]isochromene synthesis 307–309

→ (±)-Horsfiline, precursor, from tandem radical cyclization of azido-N-(2-iodoaryl)acrylamides, using tris(trimethylsilyl)silane/2,2′-azobisisobutyronitrile 223, 224

→ Huperzine Q, bicyclic cyclopentenone intermediate, from a functionalized enyne and octacarbonyldicobalt, via intramolecular Pauson–Khand cyclization 614

→ Huperzine Q, tricyclic intermediate, from conversion of a substituted cyclohex-2-enone into a spirocyclic macrocycle, then carbonyl–alkene metathesis, Schrock molybdenum complex mediated 148, 149

Hydrazines, phenyl-, three-component domino condensation process with aldehydes and acetoacetate, scandium catalyzed, tetrasubstituted pyrazole synthesis 588, 589

→ *cis*-Hydrindane, alkoxytetrahydrofuran substituted, from alkoxy-substituted 1,6-enyne, via vinyl radical tandem radical cyclization, using tributyltin hydride/2,2′-azobisisobutyronitrile 219

Hydroperoxides, polyene, peroxy radical domino cyclization, copper trifluoromethanesulfonate/oxygen induced, substituted 1,2-dioxolane synthesis 161

Hydroperoxides, polyene, peroxy radical domino cyclization, di-tert-butyl peroxyoxalate mediated, endoperoxide synthesis 158–160

Hydroperoxides, polyene, peroxy radical domino cyclization, samarium(II) iodide/oxygen induced, substituted 1,2-dioxane synthesis 162

→ α-Hydroxy acids, ring contracted, from cyclic α-diketones, via benzilic acid rearrangement, base promoted 258

654 Keyword Index

→ 12-Hydroxymonocerin, 12-hydroxy-, key intermediate, from reaction of lithiated 2-silyl-1,3-dithianes with two epoxides, via 1,4-Brook rearrangement cascade sequence 440, 441

I

Indenes, alkynyl-, domino cycloisomerization, transition-metal catalyzed, fused cycloheptadiene synthesis 328, 329
→ Indenes, 1,2-diphenyl-, from rearrangement of 1,2,3-triphenylcyclopropenes, rhodium catalyzed, via cascade sequence 516, 517
→ Indenes, octahydro-, from α-cyano dienyl esters, via Julia–Snider tandem radical cyclization, using dibenzoyl peroxide 218
→ Inden-1-ols, hexahydro-, from enol-substituted bridging cyclohexenes, via metathesis polycyclization cascade, ruthenium carbene complex catalyzed 33, 34
→ Inden-5-ones, hexahydro-, chiral, from polyene cyclization of chiral 4-(pent-3-enyl)cyclohex-2-enones, ethylaluminum dichloride catalyzed, via Wagner–Meerwein shift 21
→ Indeno[5,4-*b*]oxepins, 3-siloxy-8-vinylidenedodecahydro-, from diepoxyalkenes, via *endo*-selective epoxide-opening cascade, silyl trifluoromethanesulfonate initiated, alkene terminated 58, 59
Indoles, 3-(2-aminoethyl)-2-[(methylsulfanyl)vinyl]-, N-protected, reaction with propynal, via asymmetric tandem Diels–Alder/amine cyclization sequence, chiral secondary amine catalyzed, chiral bridged tetracyclic nitrogen heterocycle synthesis 338, 339
→ Indoles, 3,4-fused, from intramolecular dearomatizing [3+2] annulation of 4-(3-arylpropyl)-1-sulfonyl-1,2,3-triazoles, rhodium catalyzed, via cascade sequence 515, 516
→ Indoles, tetracyclic, from indole α-amino esters, via domino three-component Strecker reaction with ketones and silyl cyanides, ytterbium catalyzed 583
→ (−)-Indolizidine, key intermediate, from reaction of lithiated 2-silyl-1,3-dithianes with an epoxide and N-tosylaziridine, via 1,4-Brook rearrangement cascade sequence 442, 443
→ Indolizidine alkaloids, key intermediates, from distal azide substituted epoxides, via domino intramolecular Schmidt reaction, Lewis acid catalyzed, then reduction 301, 302
→ Indolizines, from distal azide substituted epoxides, via domino intramolecular Schmidt reaction, Lewis acid catalyzed, then reduction 300, 301
→ Indolo[2,3-*a*]quinolizines, from three-component domino condensation process of tryptamine with β-keto esters and α,β-unsaturated aldehydes, cerium catalyzed 589, 590

→ (+)-Ingenol, tetracyclic cyclopentenone intermediate, from a functionalized allenyne and carbon monoxide, via intramolecular Pauson–Khand-type cyclization, rhodium catalyzed 626
→ (−)-Irofulven, key intermediate, from dienyne, via domino double enyne ring-closing metathesis, ruthenium alkylidene complex catalyzed, then tetrabutylammonium fluoride treatment 102, 103
→ Isochromen-8-ones, tetrahydro-, from tandem cyclization of enynediones, DABCO catalyzed 344
→ (±)-Isocomene, from tricyclic cyclobutane, via Wagner–Meerwein ring expansion, acid induced 254
→ 4-Isocymobarbatol, precursor, from cationic polyene cyclization of 1,5-dienes, bromonium salt catalyzed 17, 18
→ Isofagomine, key intermediate, from a 1,7-enyne, via ring-closing metathesis, ruthenium alkylidene complex catalyzed 94, 95
→ Isofregenedadiol, key intermediate, from reaction of a 1,7-enyne with a terminal alkene, via domino enyne ring-closing metathesis/cross metathesis sequence, ruthenium alkylidene complex catalyzed, then Diels–Alder cycloaddition with acetylenedicarboxylate 126, 127
Isoprenols, epoxy-, radical polyene cyclization, titanocene catalyzed, 6-hydroxy-2-methylenedecahydronaphthalene synthesis 27
→ Isosteres, fluoroalkene amide type, from reaction of γ,γ-difluoro-α,β-enoylsilanes with camphor-derived sulfonamides and potassium cyanide, via cyanide addition/1,2-Brook rearrangement/fluoride elimination/protonation/nucleophilic addition/cyanide elimination cascade sequence 378
→ Isoxazolidines, tricyclic, functionalized, from reaction of 1,5-enynes with nitrosobenzene, gold complex catalyzed, via cyclization/[3+2] cycloaddition domino reaction 545

J

→ (+)-Jaspine B, key intermediate, from intramolecular cyclization of a bromoallene, palladium catalyzed, via oxidative addition/sequential cyclization cascade sequence 496, 497
→ (+)-Jaspine B, key intermediate, from intramolecular cyclization of a propargylic chloride, palladium catalyzed, via oxidative addition/sequential cyclization cascade sequence 497
→ (−)-Jiadifenin, tricyclic cyclopentenone intermediate, from a functionalized enyne alcohol and octacarbonyldicobalt, via intramolecular Pauson–Khand cyclization 615, 616

→ (−)-Jiadifenin, tricyclic intermediate, from intramolecular cyclization of an allylic cyclic carbonate, palladium/chiral phosphine catalyzed, via oxidative addition/Tsuji–Trost cyclization/lactonization cascade sequence 492, 493

Julia–Snider tandem radical cyclization, α-cyano dienyl esters, using dibenzoyl peroxide, octahydroindene synthesis 218

K

→ (±)-Kelsoene, intermediates, tricyclic, from bicyclic β-sulfonyloxy ketones, via homo-Favorskii rearrangement, base promoted 263

→ (+)-Kempene-2, from a dienyne, via domino double enyne ring-closing metathesis, ruthenium alkylidene complex catalyzed, then deprotection and acetylation 103, 104

→ Ketals, peroxy-, from cyclic 1,3-diketones, via coupling with alkenes and oxygen by C—H radical abstraction, 2,2′-azobisisobutyronitrile initiated 174

Ketones, alkyl aryl, reaction with a (silyl)diazomethane, Lewis acid catalyzed, via domino nucleophilic addition/1,2-aryl shift/1,3-Brook rearrangement sequence, silyl enol ether synthesis 395

→ Ketones, alkyl silyl, from [1-(triorganosiloxy)vinyl]stannanes, via domino tin–lithium exchange/retro-1,2-Brook rearrangement/alkylation sequence 388

→ Ketones, β-amino, from reaction of 2-acylaziridines with organosilyllithium reagents, via addition/1,2-Brook rearrangement/ring-opening cascade sequence 358

→ Ketones, azaspiro, chiral, from (4-aminobut-1-ynyl)cyclobutanols, via asymmetric domino hydroamination/semipinacol rearrangement sequence, chiral silver complex catalyzed 335, 336

→ Ketones, bicyclic, from double radical carbocyclization of dienyl-substituted β-oxo esters, using manganese(III) acetate/copper(II) acetate 221

Ketones, bis-silyl, reaction with thiols, via nucleophilic addition/1,2-Brook rearrangement/electrophilic trapping cascade sequence, [siloxy(sulfanyl)methyl]silane synthesis 376

→ Ketones, cyclopent-1-enyl, from cyclopentannulation of 1-cyclopropylpropargyl esters, via intramolecular domino Rautenstrauch rearrangement/ring opening/cyclization sequence, gold/silver catalyzed, then hydrolysis 566, 567

Ketones, diazo, gem-diethyl, alkyne tethered, carbene metathesis/cyclization/C—H insertion cascade sequence, rhodium catalyzed, bicyclic cyclopentenone synthesis 525, 526

Ketones, 2,3-epoxy, reaction with organosilyllithium reagents, via addition/1,2-Brook rearrangement/ ring opening cascade sequence, β-hydroxy ketone synthesis 357

Ketones, α-halo, cyclic, Favorskii rearrangement, base promoted, ring-contracted cyclic acid/ester synthesis 261

Ketones, α-hydroxy, tricyclic, rearrangement, base catalyzed, cyclopenta[b]benzofuran synthesis 263, 264

→ Ketones, β-hydroxy, fluorinated, from reaction of (trifluoroacetyl)triphenylsilanes with Grignard reagents and aldehydes, Lewis acid catalyzed, via nucleophilic addition/1,2-Brook rearrangement/β-elimination/Mukaiyama aldol addition cascade sequence 361

→ Ketones, β-hydroxy, from reaction of 2,3-epoxy ketones with organosilyllithium reagents, via addition/1,2-Brook rearrangement/ring opening cascade sequence 357

→ Ketones, β-hydroxy, spirocyclic, from 2,3-epoxy alcohols, via semipinacol rearrangement/ring expansion, Lewis acid catalyzed 269–271

→ Ketones, nitro, bicyclic, chiral, from diketo aldehydes, via domino asymmetric intermolecular nitroaldol reaction with nitromethane/ intramolecular nitroaldol reaction/epimerization sequence, praseodymium complex catalyzed 578

Ketones, polyenyl epoxy, tandem radical cyclization, using titanium(III), tricyclic terpenoid synthesis 222

→ Ketones, polysubstituted, from three-component reaction of silyl glyoxylates with Reformatsky reagents and β-lactones, via nucleophilic addition/1,2-Brook rearrangement/ring opening cascade sequence 366, 367

→ Ketones, α-siloxy, from cross-benzoin reaction of acylsilanes with aldehydes, chiral phosphite catalyzed, via nucleophilic addition/1,2-Brook rearrangement/addition/1,4-silyl migration/phosphite cleavage cascade sequence 379, 380

→ Ketones, α-siloxy, from reaction of silyl glyoxylates with ketones, lanthanum(III) cyanide catalyzed, via cyanide addition/1,2-Brook rearrangement/nucleophilic addition/1,4-silyl migration/cyanide elimination cascade sequence 377, 378

Ketones, α-silylsulfanyl, reaction with alkyl halides, via enolization/sulfur-to-oxygen 1,4-Brook rearrangement/electrophilic trapping cascade sequence, silyl enol thioether synthesis 430, 431

Ketones, β-sulfonyloxy, bicyclic, homo-Favorskii rearrangement, base promoted, tricyclic (±)-kelsoene intermediate synthesis 263

→ Ketones, α,β-unsaturated, from rearrangement of 3-methoxy-1,2,3-triphenylcyclopropenes, rhodium catalyzed, via cascade sequence 516, 517

Ketones, γ,δ-unsaturated, reaction with a (silyl)diazomethane, base catalyzed, via nucleophilic addition/1,3-Brook rearrangement/intramolecular 1,3-dipolar cycloaddition cascade sequence, polycyclic pyrazole synthesis 395–398

656 Keyword Index

→Kuehneromycin A, tricyclic intermediate, from a diynol acid, via domino cycloisomerization, gold complex catalyzed 332

→(±)-Kumausallene, bicyclic intermediate, from condensation of 1-vinylcyclopentane-1,2-diol with 2-(benzoyloxy)acetaldehyde, via ring-enlarging annulation, Lewis acid promoted, by sequential cationic cyclization/pinacol rearrangement 279, 280

→(±)-Kumausallene, key intermediate, from intramolecular bicyclization of hepta-1,6-diene-3,5-diol, via carbonylative oxacyclization cascade sequence, palladium catalyzed 500, 501

L

Lactams, alkenyl, carbonyl–alkene metathesis, using in situ generated titanium–ethylidene complex, bicyclic enamide synthesis 146

→Lactams, from addition of silyl(disulfanyl)methanes to bromo isocyanates, via lithiation/nucleophilic addition/1,3-aza-Brook rearrangement/ring closure cascade sequence 400, 401

→Lactams, hydroxy, fused, from α-siloxy epoxy azides, via tandem semipinacol/Schmidt domino process, titanium(IV) chloride promoted 273, 274

Lactones, allenyl/alkenyl substituted, reductive radical cyclization cascade, samarium(II) iodide/water mediated, octahydroazulene-3a,6-diol synthesis 194, 195

→Lactones, bicyclic, from reaction of cyclooctyne with a bis-silyl acetylenedicarboxylate, via 1,3-cycloaddition/retro-1,4-Brook rearrangement cascade sequence 431, 433

Lactones, dialkenyl substituted, alkene/alkene reductive radical cyclization cascade, samarium(II) iodide/water mediated, polysubstituted azulene synthesis 193, 194

→γ-Lactones, from reaction of epoxides with lithiated 2-silylacetonitrile, via ring opening/1,4-Brook rearrangement cascade sequence, then acidic hydrolysis 404

→Lancifodilactone G, tricyclic intermediate, from reaction of a dienynyl bromide with a dienylstannane, palladium catalyzed, via Stille cross coupling/8π-electrocyclization/6π-electrocyclization cascade sequence 456, 457

→Lanosterol, from 2,3-oxidosqualene, biosynthesis, using oxidosqualene cyclase 13

→Laureatin, key intermediate, from domino relay cross metathesis of a 1,6-dien-8-yne with a 2-allyl-3,9-dioxabicyclo[4.2.1]nonane, ruthenium alkylidene complex catalyzed 122, 123

→Lennoxamine, key intermediate, from reaction of an aryl bromide/terminal ynamide with an arylboronic acid, palladium/phosphine catalyzed, via carbopalladation/cross coupling cascade sequence 475

→*ent*-Lepadins, key intermediates, from dienynes, via domino double enyne ring-closing metathesis, ruthenium alkylidene complex catalyzed 106–108

(R)-Limonene, domino hydroperoxidation/cyclization, via thiol–alkene co-oxygenation reaction, using di-tert-butyl peroxyoxalate/benzenethiol/oxygen, endoperoxide synthesis 169, 170

(S)-Limonene, domino hydroperoxidation, using oxygen, cobalt catalyzed, then metathesis with triethylsilane, bicyclic endoperoxide synthesis 164, 165

→(+)-Linoxepin, key intermediate, from three-component Catellani reaction of aryl iodide, alkyl iodide, and acrylate, palladium/phosphine catalyzed, via oxidative addition/carbopalladation/C—H activation/reductive elimination cascade sequence 476, 477

→(+)-Linoxepin, pentacyclic intermediate, from intramolecular cascade cyclization of an aryl bromide/alkyne/allylsilane, palladium/phosphine catalyzed 475, 476

→(−)-Longithorone A, key intermediate, from an enyne, via domino cross metathesis with ethene/ring-closing metathesis, ruthenium alkylidene complex catalyzed 115–117

→Lycoflexine, from a dienyne, via domino double enyne ring-closing metathesis, ruthenium alkylidene complex catalyzed, then hydrogenation, hydroboration, oxidation, and Mannich reaction with formaldehyde 106

→(−)-Lycojapodine A, from oxidation of (+)-alopecuridine, using manganese(IV) oxide 271

→(+)-Lysergic acid, tetracyclic intermediate, from an aryl bromide/allene/sulfonamide, palladium catalyzed, via aminopalladation/cyclization cascade sequence 480, 481

M

Macrocycles, bisenone, intramolecular anionic polycyclization, tetrabutylammonium fluoride promoted, octahydronaphtho[2,1-c]pyran-4,10-dione synthesis, via transannular double Michael addition cascade 38

→(−)-Magellanine, key cyclopentenone intermediates, from functionalized enynes and octacarbonyldicobalt, via two intramolecular Pauson–Khand cyclizations 607, 608

→(−)-Magellanine, tetracyclic intermediate, from a siloxy dienyl acetal, via ring-enlarging cyclopentane annulation, tin(IV) chloride promoted, by Prins cyclization/pinacol rearrangement 280, 281

→(+)-Magellaninone, key cyclopentenone intermediates, from functionalized enynes and octacarbonyldicobalt, via two intramolecular Pauson–Khand cyclizations 607, 608

Malonates, 2,2-bis(2-methylallyl)-, reductive radical polyene cyclization, manganese catalyzed, cyclopentane-1,1-dicarboxylate synthesis 29

Keyword Index

Malonates, 2,2-diallyl-, radical polyene cyclization/azidation, using azodicarboxylates, manganese catalyzed, 3-(hydrazinomethyl)cyclopentane-1,1-dicarboxylate synthesis 28, 29

Manganese(III), tris(acetylacetonato)-, conversion into 1,3-dicarbonyl radicals, then oxidative addition to alkenes/cyclization, endoperoxide synthesis 175, 176

→ Manzamine A core, from reaction of hetaryl iodide with aminoallene, palladium/phosphine catalyzed, via carbopalladation/cyclization cascade sequence 481, 482

→ Marasmene, tricyclic intermediate, from diynediol, via domino cycloisomerization, gold complex catalyzed 331, 332

→ Martinelline, chiral intermediate, from cyclization of a 2-aminobenzaldehyde with an enal, via enantioselective domino aza-Michael/aldol reaction, chiral diarylprolinol silyl ether catalyzed, then oxidation 348

→ Meloscine, analogues, key intermediates, from divinylcyclopropanes, via intramolecular domino radical cyclization, tributyltin radical mediated 191, 192

→ (+)-Merobatzelladine B, key intermediate, from reaction of an alkenyl bromide with an aminoalkene, palladium/phosphine catalyzed, via oxidative addition/complexation/aminopalladation/reductive elimination 463, 464

→ Merrilactone A, tricyclic furan-2-one intermediate, from functionalized ynal and tricarbonyltris(dimethylformamide)molybdenum, via intramolecular hetero-Pauson–Khand cyclization 621, 622

→ Metacycloprodigiosin, key intermediate, from a monocyclic 1,6-enyne, via platinum-catalyzed enyne metathesis to give a ring-bridged macrocycle 129, 130

Methanes, silyl(disulfanyl)-, addition to bromo isocyanates, via lithiation/nucleophilic addition/1,3-aza-Brook rearrangement/ring closure cascade sequence, lactam synthesis 400, 401

→ Methanols, pyrrolo[3,2-f]quinolin-1-yl-, from radical cyclization/trapping of N-allyl-5-iodoquinolin-6-amines, using tris(trimethylsilyl)silane/2,2,6,6-tetramethylpiperidin-1-oxyl, then reduction 233

→ (±)-β-Microbiotene, key intermediate, from tetrasubstituted epoxides, via tandem rearrangement/reduction/ring contraction, Lewis acid promoted 284, 285

→ (±)-Moluccanic acid methyl ester, key intermediate, from reaction of distal aryl-substituted homoallylic alcohols with acetals, via domino Prins reaction/polyene cyclization sequence, indium(III) bromide promoted 310

→ Morphine, precursor, from polysubstituted aryl bromides, via tandem radical cyclization, using tributyltin hydride/2,2′-azobisisobutyronitrile, then reductive amination, using lithium/ammonia 230

→ Morphine, precursor, from radical cyclization cascade of a polyene, tributyltin hydride initiated 31, 32

→ Mycoticins, key intermediate, from reaction of lithiated 2-silyl-1,3-dithianes with epoxide and bis-epoxide, via 1,4-Brook rearrangement cascade sequence 437

N

→ (+)-Nakadomarin A, tricyclic cyclopentenone intermediate, from a functionalized enyne and octacarbonyldicobalt, via intramolecular Pauson–Khand cyclization 611, 612

→ Naphthalene-1,4a-dicarboxylates, 2,5-dioxooctahydro-, from intermolecular anionic polycyclization of 6-oxocyclohex-1-ene-1-carboxylates with 3-hydroxyalka-2,4-dienoate-substituted diketals, cesium carbonate promoted, via intramolecular Diels–Alder cycloaddition/aldol condensation 37

→ Naphthalenes, 1,2-dihydro-, from reaction of 2-alkynylbenzaldehydes with vinyl ethers, bulky cationic gold complex catalyzed, via benzannulation domino reaction 535, 536

→ Naphthalenes, 6-hydroxy-2-methylenedecahydro-, from radical polyene cyclization of epoxyisoprenols, titanocene catalyzed 27

→ Naphthalen-1-ones, 2-(2-oxoalkyl)-3,4-dihydro-, from unsymmetrical silyl bis(enol ethers), via domino oxidative radical rearrangement, cerium(IV) mediated 201–203

→ Naphtho[2,1-c]pyran-4,10-diones, octahydro-, from intramolecular anionic polycyclization of bisenone macrocycles, tetrabutylammonium fluoride promoted, via transannular double Michael addition cascade 38

Naphtho-1,4-quinones, three-component domino condensation process with β-keto esters and arylamines, cerium catalyzed, benzo[g]indole synthesis 589, 590

→ Neostenine, fused tricyclic intermediate, from reaction of a dienyl azide with a cyclic enone, via tandem intermolecular Diels–Alder/intramolecular Schmidt reaction sequence, Lewis acid promoted 342, 343

Nitriles – see Cyanides

→ Nitrogen heterocycles, bridged tetracyclic, chiral, from reaction of N-protected 3-(2-aminoethyl)-2-[(methylsulfanyl)vinyl]indoles with propynal, via asymmetric tandem Diels–Alder/amine cyclization sequence, chiral secondary amine catalyzed 338, 339

→ Nitrogen heterocycles, complex, via dinitrogen extrusion from 1-sulfonyl-1,2,3-triazoles, rhodium catalyzed, then cascade sequence 512, 513

→ Nitrogen heterocycles, fused, from reaction of dienyl azides with enones, via tandem intermolecular Diels–Alder/intramolecular Schmidt reaction sequence, Lewis acid promoted 341, 342

658 Keyword Index

→ Nitrogen heterocycles, tetracyclic, chiral, from enantioselective polyene cyclization of 1-(alk-3-enyl)-5-hydroxypyrrolidin-2-ones, acyliminium ion initiated, chiral substituted thiourea catalyzed 24, 25
→ Nitrogen heterocycles, tricyclic, from N-chloro-N-[2-(cyclohex-2-enyl)ethyl]prop-2-ynamines, via tandem radical cyclization, using tributyltin hydride/2,2′-azobisisobutyronitrile 229, 230
Norbornenes, tricyclic, ring-opening alkene metathesis/carbonyl cyclization, using Tebbe reagent, capnellene key intermediate synthesis 139, 140
→ (−)-Norsecurinine, from a dienyne, via domino double enyne ring-closing metathesis, ruthenium alkylidene complex catalyzed, then allylic bromination and deprotection-induced N-alkylation 110, 111

O

→ (+)-Omaezakianol, precursor, from diepoxides, via *exo*-selective epoxide-opening cascade, Brønsted base promoted 45
→ (+)-α-Onocerin, from domino cyclization of a polyene epoxide, Lewis acid induced 291, 292
→ Ophiobolin A, tricyclic intermediate, from reaction of an enynyl bromide with a dienylstannane, palladium catalyzed, via oxidative addition/carbopalladation/Stille cross coupling/8π-electrocyclization/4π-electrocyclic ring-opening/tautomerization cascade sequence 455, 456
→ (+)-Orientalol F, key intermediate, from domino cyclization of a 1,6-enyn-10-one, gold complex catalyzed 334, 335
2-Oxabicyclo[2.2.2]oct-5-en-3-ones, 8-(2-amino-3-bromopropanoyl)-, radical cyclization/trapping, using triethylborane/allyltributylstannane, then subsequent radical trapping and radical cyclization, (+)-scholarisine A precursor synthesis 233, 234
1-Oxa-2-silacyclopentanes, reaction with allyl halides, copper(I) tert-butoxide promoted, via organolithium addition/1,4-Brook rearrangement/electrophilic trapping cascade sequence, 2-allylbenzyl alcohol synthesis 429, 430
→ Oxathiazepanes, fused, from allyloxyalkyne-tethered sulfamates, via alkyne–nitrene metathesis/oxonium ylide cyclization/2,3-sigmatropic rearrangement cascade sequence, rhodium catalyzed, then reduction 528–531
→ Oxatricyclic compounds, from domino cyclization of 1,6-enyn-10-ones, gold complex catalyzed 332, 334
→ 1,3-Oxazines, 3,6-dihydro-, 3,4,5-trisubstituted, from three-component domino condensation process of acetylenedicarboxylates with amines and formaldehyde, ytterbium catalyzed 590, 591

→ [1,4]Oxazino[4,3-*a*]indoles, chiral, from N-(4-hydroxybut-2-enyl)-2-(3-hydroxyprop-1-ynyl)anilines, via enantioselective intramolecular cyclization domino reaction, chiral gold complex catalyzed 550
→ Oxazolidin-2-imines, from three-component reaction of aldimines with terminal alkynes and tosyl isocyanate, gold/silver catalyzed, via cyclization domino reaction 538
→ Oxepino[3,2-*b*]oxepins, 3-bromodecahydro-, from distal alkene containing diepoxides, via *endo*-selective epoxide-opening cascade, bromonium ion initiated, external nucleophile terminated 53, 54
→ Oxepino[3,2-*b*]pyrano[2,3-*f*]oxepin-2-ones, 8-siloxydodecahydro-, from tert-butyl ester containing triepoxides, via *endo*-selective epoxide-opening cascade, Lewis acid promoted, then silylation 48
Oxiranes – *see also* Epoxides
Oxiranes, 2-phenyl-, homo-Peterson reaction with [tris(silyl)methyl]lithium, via 1,4-Brook rearrangement/cyclization cascade sequence, phenylcyclopropane synthesis 404, 405
Oxiranes, reaction with dihalo(silyl)methyllithium, via lithiation/ring opening/1,4-Brook rearrangement/electrophilic capture cascade sequence, β,β-dihaloalkanol synthesis 416
Oxiranes, reaction with lithiated allyltriphenylsilane, via nucleophilic addition/1,4-Brook rearrangement/electrophilic capture cascade sequence, siloxyalkene synthesis 417, 418
Oxiranes, reaction with lithiated prop-1-ene-1,3-diylbis(triphenylsilane), via nucleophilic addition/1,4-Brook rearrangement/electrophilic capture cascade sequence, alk-4-enol synthesis 417, 418
Oxiranes, vinyl-, [4+1] cycloaddition with lithiated monosilyl dithioacetals, via ring opening/1,4-Brook rearrangement/ring closure cascade sequence, functionalized cyclopentene synthesis 408, 409
Oxiranes, vinyl-, 2-silyl-1,3-dithianyl functionalized, reaction with organometallic reagents, via ring opening/1,4-Brook rearrangement/electrophilic capture cascade sequence, functionalized alkenol synthesis 415, 416
→ Oxocin-4-ones, 7-chlorotetrahydro-, from reaction of epoxycyclopropylalkynes with N-halosuccinimides, gold complex catalyzed, via double ring-expansion domino reaction 552, 553
→ Oxygen heterocycles, tetracyclic 6,6,6,6-fused, from 3-hydroxytetrahydropyran-containing triepoxides, via *endo*-selective epoxide-opening cascade, water promoted 57, 58
→ Oxygen heterocycles, tetracyclic 6,7,7,7-fused, from tert-butyl carbonate containing tetraepoxides, via *endo*-selective epoxide-opening cascade, Lewis acid promoted 49, 50

Keyword Index

→ Oxygen heterocycles, tricyclic 6,6,6-fused, from 5-hydroxy-1,3-dioxane-containing diepoxides, via *endo*-selective epoxide-opening cascade, water promoted 57
→ Oxygen heterocycles, tricyclic 6,7,7-fused, from *tert*-butyl carbonate containing triepoxides, via *endo*-selective epoxide-opening cascade, Lewis acid promoted 48–50
→ Oxygen heterocycles, tricyclic 6,7,7-fused, from distal alkene containing diepoxides, via *endo*-selective epoxide-opening cascade, bromonium ion initiated, internal nucleophile terminated 52, 53
→ Oxygen heterocycles, tricyclic 6,7,7-fused, from *endo*-selective epoxide-opening cascade, oxocarbenium ion initiated, via photooxidative cleavage 54, 55

P

→ Paecilomycine A, tricyclic cyclopentenone intermediate, from a functionalized enyne and octacarbonyldicobalt, via intramolecular Pauson–Khand cyclization 606
→ (–)-Panacene, tricyclic intermediate, from cyclization of a 2-(1-hydroxyallyl)phenol, palladium catalyzed, via carbonylative oxypalladation cascade sequence 502
→ Panaxytriol, key intermediate, from reaction of an enediyne with 1,4-diacetoxybut-2-ene, via domino relay metathesis, double metallotropic [1,3]-shift, and cross metathesis sequence, ruthenium alkylidene complex catalyzed 121, 122
→ (±)-Pancracine, hydroindolone skeleton, from domino aza-Cope rearrangement/Mannich reaction sequence of advanced intermediate, Lewis acid promoted 317, 318
→ (+)-Panepophenanthrin, key intermediate, from reaction of a dienyne but-3-en-2-ol, via domino ring-closing metathesis/cross metathesis sequence, ruthenium alkylidene complex catalyzed 114, 115
→ (+)-Panepophenathrin, key intermediate, from reaction of a 1,7-enyne with but-3-en-2-ol, via domino cross metathesis/ring-closing metathesis sequence, ruthenium alkylidene complex catalyzed 114, 115
→ (+)-Paniculatine, key cyclopentenone intermediates, from functionalized enynes and octacarbonyldicobalt, via two intramolecular Pauson–Khand cyclizations 607, 608
Pauson–Khand reaction, cyclopentenone synthesis from alkenes, alkynes, and octacarbonyldicobalt, overview 601, 602
→ Penostatin B, bicyclic cyclopentenone intermediate, from functionalized dienyne and octacarbonyldicobalt, via intramolecular Pauson–Khand cyclization 617
→ Pentacycles, tetraspiro, from polyene zipper cyclization of pentaenynes, palladium catalyzed, via ene–yne cycloisomerization 35

Pentaenynes, polyene zipper cyclization, palladium catalyzed, tetraspiro pentacycle synthesis, via ene–yne cycloisomerization 35
Pentaepoxides, diol, exo-selective epoxide-opening cascade, Brønsted base promoted, (–)-glabrescol synthesis 45
→ (–)-Pentalenene, tricyclic cyclopentenone intermediate, from functionalized a cyclopropenyne and octacarbonyldicobalt, via intramolecular Pauson–Khand cyclization 610, 611
→ Pentalenes, octahydro-, from distal cyclopropane containing enones, via radical fragmentation/bicyclization, visible light mediated, lanthanum(III) facilitated, ruthenium complex catalyzed 211, 212
→ Pentalenolactone A, methyl ester, bicyclic cyclopentenone intermediate, from a functionalized enyne and octacarbonyldicobalt, via intramolecular Pauson–Khand cyclization 617, 618
→ Pentanols, 1-(2-alkylphenyl)-, from *ortho*-functionalization of 2-silylbenzaldehydes, via nucleophilic attack of an alkyllithium/1,4-Brook rearrangement/electrophilic capture cascade sequence 423, 424
→ (+)-Pericosine C, key intermediate, from ribose-derived 1,7-enyne, via ring-closing metathesis, ruthenium alkylidene complex catalyzed 95, 96
→ Peroxides, silylated, from Mukaiyama hydration/hydroperoxidation of alkenes, using oxygen, metal complex catalyzed, then metathesis with triethylsilane 163, 164
→ (–)-PF-1018, entire core, from reaction of a dienylstannane with a dienyl iodide, palladium/phosphine catalyzed, via Stille coupling/8π-electrocyclization/intramolecular Diels–Alder cycloaddition cascade sequence 453, 454
→ Phenanthrenes, hexahydro-, chiral, from enantioselective polycyclization of enynes, chiral digold–bisphosphine complex catalyzed 22
→ Phenanthrenes, hexahydro-, chiral, from enantioselective polyene cyclization of allylic alcohols, iridium/chiral dioxaphosphepine catalyzed 24
→ Phenanthrenes, 2-iodooctahydro-, chiral, from enantioselective polyene halocyclization of 1,5-dienes, using *N*-iodosuccinimide, chiral phosphoramidite catalyzed 20, 21
→ Phenanthrenes, octahydro-, substituted, from reaction of aryl-substituted dienes with acetals, via intermolecular domino Prins reaction/polyene cyclization sequence, tin(IV) chloride promoted 310–312
→ Phenanthrenes, 2-(phenylsulfanyl)hexahydro-, from polyene cyclization of 1,5-dienes, episulfonium ion initiated 23
→ Phenanthren-2-ols, octahydro-, from aryl-substituted epoxyalkenes, via *endo*-selective epoxide-opening cascade, indium bromide initiated, arene terminated 60–62

660 Keyword Index

→ Phenols, bicyclic, functionalized, from enediynols, via intramolecular cyclization domino
reaction, carbene gold complex catalyzed
552, 553

*Phenols, 2,6-dibromo-, O-silylated, domino double
lithiation/retro-1,2-Brook rearrangement/di-
methylformamide trapping sequence, 2-hydroxy-
3-silylbenzaldehyde synthesis* 389, 390

*Phenols, reaction with cyclic 1,3-dienes, gold/silver
catalyzed, polycyclic benzo[b]furan synthesis, via
cyclization domino reaction* 538, 539

→ Phomoidride B, core structure, from tandem
Diels–Alder/radical cyclization/radical fragmentation sequence of alkynyl-substituted
tricyclic compounds, using triphenyltin hy-
dride/triethylborane then samarium(II) iodide 225, 226

→ Phomoidride B, tetracyclic intermediate, from
intramolecular cyclization of an alkenyl trifluoromethanesulfonate, palladium catalyzed, via carbonylative spiroketalization/
[3,3]-sigmatropic rearrangement cascade sequence 466, 467

*Phthalimides, N-alkoxy-, tandem radical cycliza-
tion, using tributyltin hydride/2,2'-azobisisobut-
yronitrile, chiral substituted tetrahydrofuran
synthesis* 230, 231

→ Physostigmine, pyrrolo[2,3-b]indol-2-one intermediate, from an alkyne carbodiimide and
octacarbonyldicobalt, via intramolecular aza-
Pauson–Khand cyclization 620, 621

→ Phytoalexin, from O-protected 2,4-dihydroxy-
benzoic acid, via conversion into acid chloride and coupling with a substituted phenol,
then Takai–Utimoto alkenation/alkene metathesis, Schrock molybdenum complex catalyzed, and hydrogenation 137

Pinacol rearrangement, vicinal tertiary cyclo-
hexanediols, acid catalyzed, cycloheptanone
synthesis 256

*Pivalates, propargylic, domino intramolecular rear-
rangement reaction, gold complex catalyzed,
2-(pivaloyloxy)-1,3-diene synthesis* 558

→ Podophyllotoxin, precursor, from radical carboxyarylation of aryl thionocarbonates, using
tris(trimethylsilyl)silane/2,2'-azobisisobutyro-
nitrile 222, 223

→ Polycyclic compounds, bridged, from rhodium-
catalyzed rearrangement/cyclization of al-
kyne-substituted diazoacetates, via cascade
sequence 521, 522

→ Polycyclic compounds, chiral, from chiral di-
ene-substituted propargylic ethers, via indi-
um-catalyzed polyene cyclization 17

→ Polycyclic compounds, from tail-to-head polycyclization of polyenes, ethylaluminum dichloride promoted 25, 26

*Poly(cyclopropyls), vinyl substituted, domino ring
opening/peroxidation/cyclization cascade, using
oxygen/light, 2,2'-azobisisobutyronitrile/diphenyl
diselenide catalyzed, polydioxolane synthesis*
172, 173

→ Polydioxolanes, from vinyl-substituted poly(cyclopropyls), via domino ring opening/peroxi-
dation/cyclization cascade, using oxygen/
light, 2,2'-azobisisobutyronitrile/diphenyl diselenide catalyzed 172, 173

*Polyene epoxides, stereoselective domino epoxide
opening/[4+2] cycloaddition sequence, Lewis acid
induced, polyfunctionalized decahydrofluorene
synthesis* 305, 306

*Polyene epoxides, tricyclization, Lewis acid induced,
pentacyclic triterpene intermediate synthesis*
290, 291

*Polyene hydroperoxides, peroxy radical domino cy-
clization, copper trifluoromethanesulfonate/oxy-
gen induced, substituted 1,2-dioxolane synthesis*
161

*Polyene hydroperoxides, peroxy radical domino cy-
clization, di-tert-butyl peroxyoxalate mediated,
endoperoxide synthesis* 158–160

*Polyene hydroperoxides, peroxy radical domino cy-
clization, samarium(II) iodide/oxygen induced,
substituted 1,2-dioxane synthesis* 162

*Polyenes, asymmetric domino oxidative radical cy-
clization, organo-SOMO activated, chiral imida-
zolidin-4-one catalyzed, chiral polycyclic aldehyde
synthesis* 198–200

*Polyenes, radical cyclization cascade, tributyltin hy-
dride initiated, morphine precursor synthesis* 31,
32

*Polyenes, tail-to-head polycyclization, ethylalumi-
num dichloride promoted, polycyclic compound
synthesis* 25, 26

→ Poly(tetrahydrofurans), from hydroxy 1,5-diep-
oxides, via exo-selective domino intramolecular opening/cyclization, acid catalyzed 247

→ Poly(tetrahydropyrans), from hydroxy
1,6,11-triepoxides, via exo-selective domino
intramolecular opening/cyclization, acid catalyzed 247

→ (+)-Preussin, key intermediate, from reaction of
an alkenylcarbamate with bromobenzene,
palladium/phosphine catalyzed, via amino-
palladation/cyclization cascade sequence
478, 479

*Propanals, 3-phenyl-, reaction with [3-(trimethylsi-
lyl)prop-1-en-2-yl]lithium, via nucleophilic addi-
tion/1,4-Brook rearrangement/electrophilic capture cascade sequence, substituted allyl alcohol
synthesis* 418, 419

→ Propan-2-ols, 2-(dodecahydro-2,2':5',2''-terfuran-
5-yl)-, from diepoxides, via exo-selective epoxide-opening cascade, 10-camphorsulfonic
acid promoted 43, 44

Keyword Index

661

Propenes, 1,3-bis(silyl)-, lithiation/1,4-Brook rearrangement/electrophilic capture cascade sequence, silylated alkenol synthesis 417, 418

→ Prop-2-enones, 2-[aryl(siloxy)methyl]-, from sila-Morita–Baylis–Hillman reaction of arylaldehydes with aryl 1-silylvinyl ketones, triarylphosphine catalyzed, via 1,3-Brook rearrangement 394

Prop-2-ynamines, N-chloro-N-[2-(cyclohex-2-enyl)-ethyl]-, tandem radical cyclization, using tributyltin hydride/2,2′-azobisisobutyronitrile, tricyclic nitrogen heterocycle synthesis 229, 230

→ Prostaglandin $F_{2\alpha}$, precursor, from photolytic radical cyclization/trapping of 3-(2-iodoethoxy)cyclopentenes, using tributyltin hydride/2-silyloct-1-en-3-one 232

→ Prostaglandin G_2, methyl ester, from peroxy radical domino cyclization of a polyene hydroperoxide, samarium(II) iodide/oxygen induced 162

→ Pyrano[3,2-*d*][1,3]dioxin-2-ones, 7-hydroxyhexahydro-, from *N,N*-dimethylcarbamate-containing diepoxides, via *endo*-selective epoxide-opening cascade, Lewis acid promoted 47, 48

→ Pyran-3-ols, 4-alkylidenetetrahydro-, from rhodium-catalyzed rearrangement/cyclization of 1-(cycloprop-1-enyl)-2-alkoxyethanols, via cascade sequence 518–520

→ Pyrano[2′,3′:5,6]pyrano[3,2-*d*][1,3]dioxins, from 5-hydroxy-1,3-dioxane-containing diepoxides, via *endo*-selective epoxide-opening cascade, water promoted 57

→ Pyrano[3,2-*b*]pyran-3-ols, octahydro-, from epoxy alcohols, via *endo*-selective intramolecular epoxide opening/cyclization, base catalyzed 251

→ Pyrano[3,2-*b*]pyrans, octahydro-, from bromo-substituted diepoxides, via *endo*-selective epoxide-opening cascade, carbocation initiated by halide abstraction, using silver trifluoromethanesulfonate 55, 56

Pyrano[3,4-c]pyridines, 6-iodo-7-(prop-2-ynyl)-, [4+1] annulation, via tandem radical trapping/cyclization, using phenyl isocyanide/hexamethyldistannane, camptothecin synthesis 236

→ Pyrazoles, 2,3-dihydro-, polyfunctionalized, from three-component reaction of aldehydes with terminal alkynes and hydrazines, gold/silver catalyzed, via domino cyclization reaction 540

→ Pyrazoles, polycyclic, from reaction of γ,δ-unsaturated ketones with (silyl)diazomethane, base catalyzed, via nucleophilic addition/1,3-Brook rearrangement/intramolecular 1,3-dipolar cycloaddition cascade sequence 395–398

→ Pyrazoles, tetrasubstituted, from three-component domino condensation process of phenylhydrazine with aldehydes and acetoacetate, scandium catalyzed 588, 589

→ Pyrazoles, tricyclic, polyfunctionalized, from three-component reaction of 1,2-dialkynylbenzenes with aldehydes and hydrazines, gold/silver catalyzed, via domino cyclization reaction 540

→ Pyridines, 1,4-dihydro-, from three-component domino condensation process of arylamines with β-keto esters and α,β-unsaturated aldehydes, cerium catalyzed 589, 590

→ Pyrido[2,3-*b*]porphyrins, bis(hydroxyalkyl) substituted, from reaction of cyclic enol ethers with amine-substituted nickel porphyrins, lanthanum catalyzed, via domino condensation/hetero-Diels–Alder process 585, 586

→ Pyrrole-3-carbonitriles, 2-amino-4,5-dihydro-, from reaction of N-protected aziridines with malononitrile, scandium catalyzed, via domino ring-opening/intramolecular cyclization sequence 593, 594

→ Pyrroles, 2,3-fused, from dedinitrogenative rearrangement of alkene-tethered 1-sulfonyl-1,2,3-triazoles, rhodium catalyzed, via cascade sequence 514, 515

→ Pyrroles, 3,4-fused, from dedinitrogenative rearrangement of alkyne-tethered 1-sulfonyl-1,2,3-triazoles, rhodium catalyzed, via cascade sequence 513, 514

→ Pyrroles, 3,4-fused, from dedinitrogenative rearrangement of allene-tethered 1-sulfonyl-1,2,3-triazoles, rhodium catalyzed, via cascade sequence 513, 514

→ Pyrroles, pentasubstituted, from four-component domino condensation process of β-keto esters with primary alkylamines and α-diketones, cerium catalyzed 591, 592

→ Pyrrolidines, 2-benzyl-, functionalized, from reaction of aminoalkenes with arylboronic acids, gold complex catalyzed, via intramolecular aminoarylation domino reaction 554

→ Pyrrolidines, 4-(prop-1-en-2-yl)-, from radical retrocycloisomerization of N-(2-methylallyl)-but-2-enamines, cobalt–salen complex catalyzed 30

→ Pyrrolidinones, tricyclic, from bromo(N,N-dipropargylamido), via domino radical cyclization/rearrangement, light mediated, iridium complex photocatalyzed 207–210

Pyrrolidin-2-ones, 1-(alk-3-enyl)-5-hydroxy-, enantioselective polyene cyclization, acyliminium ion initiated, chiral substituted thiourea catalyzed, chiral tetracyclic nitrogen heterocycle synthesis 24, 25

→ Pyrrolo[2,3-*d*]carbazoles, tetrahydro-, chiral, from reaction of N-protected 3-(2-aminoethyl)-2-[(methylsulfanyl)vinyl]indoles with propynal, via asymmetric tandem Diels–Alder/amine cyclization sequence, chiral secondary amine catalyzed 338–340

Keyword Index

→ Pyrrolo[2,3-*b*]indoles, hexahydro-, chiral, from tryptamines and enals, via enantioselective domino addition/cyclization, chiral cyclic secondary amine catalyzed 345, 346

Q

Quasi-Favorskii rearrangement, α-halo cyclic ketones into ring-contracted carboxylic acid derivatives, base promoted 287

Quinolin-6-amines, N-allyl-5-iodo-, radical cyclization/trapping, using tris(trimethylsilyl)silane/2,2,6,6-tetramethylpiperidin-1-oxyl, then reduction, pyrrolo[3,2-f]quinolin-1-ylmethanol synthesis 233

→ Quinoline-3-carbaldehydes, chiral, from cyclization of 2-aminobenzaldehydes with enals, via enantioselective domino aza-Michael/aldol reaction, chiral diarylprolinol silyl ether catalyzed 347, 348

→ Quinolines, 2,4-disubstituted, from three-component domino condensation process of arylamines with aldehydes and terminal alkynes, ytterbium catalyzed 586, 587

→ Quinolines, 4-ferrocenyl-, from three-component domino condensation process of aldehydes with ferrocenylacetylenes and anilines, cerium catalyzed 588

→ Quinoxalines, tricyclic, from reaction of 2-furylcarbinols with benzene-1,2-diamines, lanthanum catalyzed, via domino Piancatelli rearrangement/C—N coupling/Michael addition sequence 597

R

→ Resiniferatoxin, key intermediate, from three-component sequential radical trapping of tricyclic (phenylselanyl)cyclohexanes with cyclopent-2-en-1-ones and allylstannanes, 1,1′-azobis(cyclohexanecarbonitrile) promoted 237

→ Retigeranic acid A, key intermediate, from Favorskii rearrangement of a 2-bromocyclohexanone, using pyrrolidine 287, 288

→ (±)-Rhazinal, key intermediate, from Catellani reaction of an iodopyrrole with 1-bromo-2-nitrobenzene, palladium/phosphine catalyzed, via carbopalladation/C—H activation/Heck coupling cascade sequence 478

→ Ripostatin B, key intermediate, from reaction of lithiated 2-silyl-1,3-dithianes with two epoxides, via 1,4-Brook rearrangement cascade sequence 437, 438

→ Roflamycoin, key intermediate, from reaction of lithiated 2-silyl-1,3-dithianes with two epoxides, via 1,4-Brook rearrangement cascade sequence 433, 434

→ Roseophilin, key intermediate, from monocyclic 1,6-enyne, via platinum-catalyzed enyne metathesis to give ring-bridged macrocycle 128, 129

→ (+)-Rubiginone B2, from a 1,7-enyne, via ring-closing metathesis, ruthenium alkylidene complex catalyzed, then Diels–Alder cycloaddition with a naphtho-1,4-quinone, elimination/aromatization, oxidation, and deprotection 125, 126

→ Rubriflordilactone A, tricyclic intermediate, from a bromoenediyne, via intramolecular electrocyclization cascade sequence, palladium catalyzed 461

S

→ Salicylaldehydes, silyl substituted, precursors, from O-silylated 2,6-dibromophenols, via domino double lithiation/retro-1,2-Brook rearrangement/dimethylformamide trapping sequence 389, 390

→ (+)-Salvadione B, key intermediate, from a polycyclic enone, via intramolecular peroxidation/cyclization, using oxygen, manganese(III) mediated 176, 177

→ (±)-Salviasperanol, tricyclic intermediate, from an alkynylindene, via domino cycloisomerization, transition-metal catalyzed 329, 330

→ Salvinorin A, precursor, from intramolecular anionic polycyclization of bisenone macrocycles, tetrabutylammonium fluoride promoted, via transannular double Michael addition cascade 38

→ Sanguinarine, hexacyclic intermediate, from reaction of an aryl iodide with a bridged azacycle, palladium catalyzed, via carbopalladation/cyclization cascade sequence 482, 483

→ Scalarenedial, key intermediate, from domino cyclization of a polyene epoxide, Lewis acid induced 291, 292

→ Scalarenedial, key intermediate, from reaction of an acylsilane with an α-sulfonyl lithium derivative, via carbonyl addition/1,2-Brook rearrangement/β-sulfonyl elimination cascade sequence 432, 433

→ Scandine, key intermediate, from reaction of a vinylcyclopropane with a nitroalkene, palladium/phosphine catalyzed, via oxidative addition/[3+2] cycloaddition cascade sequence 493, 494

→ Schindilactone A, hexacyclic cyclopentenone intermediate, from a functionalized cyclooctenyne and octacarbonyldicobalt, via intramolecular Pauson–Khand cyclization 612, 613

→ Schindilactone A, tetracyclic intermediate, from intramolecular bicyclization of a bicyclic enediol, via carbonylative oxacyclization cascade sequence, palladium catalyzed 500, 501

→ (+)-Scholarisine A, precursor, from radical cyclization/trapping of 8-(2-amino-3-bromopropanoyl)-2-oxabicyclo[2.2.2]oct-5-en-3-ones, using triethylborane/allyltributylstannane, then subsequent radical trapping and radical cyclization 233, 234

Keyword Index

→ Schweinfurthin A, key synthetic step, tandem *endo*-selective epoxide-opening/electrophilic substitution cascade of O-protected phenol-substituted epoxyalkenes, Lewis acid initiated, protected phenol terminated 63, 64

→ (−)-Scopadulcic acid A, key intermediate, from intramolecular cascade cyclization of a methylenecycloheptene-tethered iodoalkene, palladium/phosphine catalyzed 449, 450

→ (−)-Securinine, from a dienyne, via domino double enyne ring-closing metathesis, ruthenium alkylidene complex catalyzed, then allylic oxidation, allylic bromination, ring closure, and deprotection 108, 109

→ Serratine, 13-deoxy-, key cyclopentenone intermediate, from a functionalized enyne and octacarbonyldicobalt, via intramolecular Pauson–Khand cyclization 606

→ Sesquiterpenes, gorgonian, key intermediate, from 3-methylpenta-1,4-diene, via deprotonation/aldehyde addition/1,4-Brook rearrangement/electrophilic capture cascade sequence 442

→ Sesquiterpenes, microbiotol type, key intermediates, from tetrasubstituted epoxides, via tandem rearrangement/ring contraction, Lewis acid promoted 284, 285

→ Sesquiterpenoids, uncommon, bicyclic cyclopentenone intermediate, from a functionalized allenene and carbon monoxide, via intramolecular Pauson–Khand-type cyclization, rhodium/phosphine catalyzed 625

→ (+)-Sieboldine A, from oxidation of (+)-alopecuridine, using 3-chloroperoxybenzoic acid/mercury(II) oxide 270, 271

Silanes, acetyl(dimethyl)phenyl-, reaction with fluoride ion, via 1,2-aryl shift/1,2-Brook rearrangement cascade sequence, then acidification, 1-phenylethanol synthesis 386

Silanes, acyl-, γ-alkenoyl substituted, [6+2] cyclization with vinyllithium reagents, via nucleophilic addition/1,2-Brook rearrangement/ring closure/anionic oxy-Cope rearrangement cascade sequence, substituted cyclooctenone synthesis 373

Silanes, acyl-, alkenyl substituted, enantioselective reaction with alkynes, chiral imino alcohol catalyzed, zinc promoted, via nucleophilic alkynylation/1,2-Brook rearrangement/ene–allene carbocyclization/electrophilic trapping cascade sequence, chiral 1-alkynylcyclopentanol synthesis 375, 376

Silanes, acyl-, allyl substituted, enantioselective reaction with alkynylsilanes, chiral diarylprolinol catalyzed, zinc promoted, via nucleophilic alkynylation/1,2-Brook rearrangement/ene–allene carbocyclization/electrophilic trapping/oxidation cascade sequence, chiral alkynediol synthesis 373, 374

Silanes, acyl-, β-bromo, reaction with cyanide ion, via nucleophilic addition/1,2-Brook rearrangement/intramolecular S_N2 displacement cascade sequence, 1-cyano-1-siloxycyclopropane synthesis 371, 372

Silanes, acyl-, β-chloro substituted, [3+4] cycloaddition with vinyl ketone enolates, via nucleophilic addition/1,2-Brook rearrangement/cyclization cascade sequence, cycloheptenedione synthesis 363, 364

Silanes, acyl-, cross-benzoin reaction with aldehydes, chiral phosphite catalyzed, via nucleophilic addition/1,2-Brook rearrangement/addition/1,4-silyl migration/phosphite cleavage cascade sequence, α-siloxy ketone synthesis 379, 380

Silanes, acyl-, γ,γ-difluoro-α,β-unsaturated, reaction with camphor-derived sulfonamides and potassium cyanide, via cyanide addition/1,2-Brook rearrangement/fluoride elimination/protonation/nucleophilic addition/cyanide elimination cascade sequence, amide-type fluoroalkene isostere synthesis 378

Silanes, acyl-, enantioselective benzoin-type coupling with cyanoformates, chiral aluminum Lewis acid catalyzed, via nucleophilic addition/1,2-Brook rearrangement cascade sequence, chiral α-cyano-α-siloxy ester synthesis 370, 371

Silanes, acyl-, γ,δ-epoxy-δ-silyl α,β-unsaturated, [3+4] cycloaddition with vinyl ketone enolates, via nucleophilic addition/1,2-Brook rearrangement/ring opening/1,2-Brook rearrangement/intramolecular Michael addition cascade sequence, tetrasubstituted cycloheptenone synthesis 365, 366

Silanes, acyl-, α-phenylsulfanyl, reaction with organolithium reagents, via nucleophilic addition/base-induced 1,2-Brook rearrangement/elimination cascade sequence, silyl enol ether synthesis 358, 359

Silanes, acyl-, propargyl substituted, enantioselective reaction with alkynylsilanes, chiral diarylprolinol catalyzed, zinc promoted, via nucleophilic alkynylation/1,2-Brook rearrangement/ene–allene carbocyclization/electrophilic trapping/oxidation cascade sequence, chiral enynediol synthesis 373, 374

Silanes, acyl-, reaction with enolized N-sulfinyl ketimines, via nucleophilic addition/1,2-Brook rearrangement/ring closure cascade sequence, aminocyclopropanol synthesis 374, 375

Silanes, acyl-, reaction with lithiated epoxysilanes, via nucleophilic addition/1,2-Brook rearrangement/ring opening cascade sequence, allylic alcohol synthesis 360

Silanes, acyl-, reaction with α-sulfinyl carbanions, via nucleophilic addition/1,2-Brook rearrangement/elimination cascade sequence, silyl enol ether synthesis 359

664   Keyword Index

Silanes, acyl-, reaction with trimethyl(trifluoromethyl)silane and α,β-unsaturated carbonyl
compounds, via nucleophilic addition/1,2-Brook
rearrangement/β-elimination/Mukaiyama aldol
addition cascade sequence, fluorinated 1,5-dike-
tone synthesis 360, 361

Silanes, acyl-, sila-Stetter reaction with enones, N-
heterocyclic carbene catalyzed, via nucleophilic
addition/1,2-Brook rearrangement/Michael addi-
tion/carbene cleavage cascade sequence, 1,4-dike-
tone synthesis 378, 379

Silanes, acyl-, β-silyl substituted, [3+4] cycloaddition
with cycloheptenone enolates, via nucleophilic
addition/1,2-Brook rearrangement/cyclization
cascade sequence, then α-hydroxylation and oxidative cleavage, tetrafunctionalized cyclooctene
synthesis 365

Silanes, acyl-, β-silyl substituted, [3+4] cycloaddition
with vinyl ketone enolates, via nucleophilic addi-
tion/1,2-Brook rearrangement/cyclopropanation/
anionic oxy-Cope rearrangement cascade sequence, cycloheptenone synthesis 363, 364

Silanes, acyl-, three-component reaction with imines
and amide-derived enolates, via nucleophilic
addition/1,2-Brook rearrangement/imine addition cascade sequence, γ-amino-β-hydroxy amide
synthesis 367

Silanes, acyl-, α,β-unsaturated, β,β-dichloro substituted, reaction with ketone enolates, via nucleophilic addition/1,2-Brook rearrangement/cycliza-
tion cascade sequence, cyclopentenone synthesis
362, 363

Silanes, acyl-, α,β-unsaturated, β-sulfanyl substituted, reaction with ketone enolates, via nucleophilic addition/1,2-Brook rearrangement/cycliza-
tion cascade sequence, cyclopentanone silyl enol
ether synthesis 362

Silanes, acyl-, β,γ-unsaturated, reaction with phenyllithium, via nucleophilic addition/1,2-Brook
rearrangement/intramolecular Michael addition
cascade sequence, functionalized cyclobutane synthesis 371

Silanes, acyl(triphenyl)-, domino enolization/1,2-
Brook rearrangement sequence, using copper
tert-butoxide, then cross coupling with alkyl iodides, palladium catalyzed, trisubstituted silyl
enol ether synthesis 380, 381

Silanes, alkenynoyl-, enantioselective reaction with
chiral lithium amides and furan-2,5-diones, via
reduction/1,2-Brook rearrangement/Diels–Alder
cycloaddition cascade sequence, chiral substituted
tetrahydrobenzo[c]furan-1,3-dione synthesis 372

→ Silanes, [alkyl(siloxy)methyl]-, from α,α′-bis-silyl
carbinols, via domino deprotonation/1,2-
Brook rearrangement sequence, then coupling with alkyl halides 381

Silanes, epoxy-, α-cyano functionalized, reaction
with bis-electrophiles, via deprotonation/ring
opening/1,2-Brook rearrangement/nucleophilic
addition cascade sequence, 1-cyano-1-(2-siloxyvi-
nyl)cycloalkane synthesis 383–385

Silanes, epoxy-, diene functionalized, reaction with
2-chloroacetamides, via enolization/Michael
addition/ring opening/1,2-Brook rearrangement/
intramolecular alkylation cascade sequence, polyfunctionalized cyclopropane synthesis 385, 386

Silanes, epoxy-, reaction with [(phenylsulfanyl)(trimethylsilyl)methyl]lithium, via nucleophilic addi-
tion/1,3-Brook rearrangement/β-elimination cascade sequence, vinylsilane synthesis 399

Silanes, epoxy-, O-silyl cyanohydrin substituted,
reaction with alkyl halides, via deprotonation/
ring opening/1,2-Brook rearrangement/alkylation
cascade sequence, 2-alkyl-2,4-disiloxybut-3-eneni-
trile synthesis 382, 383

Silanes, epoxy-, vinyl substituted, palladium com-
plexation/1,2-Brook rearrangement cascade sequence, siloxydiene synthesis 387

→ Silanes, [siloxy(sulfanyl)methyl]-, from reaction
of bis-silyl ketones with thiols, via nucleophilic addition/1,2-Brook rearrangement/elec-
trophilic trapping cascade sequence 376

Silanes, (trifluoroacetyl)triphenyl-, reaction with
Grignard reagents and aldehydes, Lewis acid catalyzed, via nucleophilic addition/1,2-Brook rear-
rangement/β-elimination/Mukaiyama aldol
addition cascade sequence, fluorinated β-hydroxy
ketone synthesis 361

→ Silanes, vinyl-, from reaction of epoxysilanes
with [(phenylsulfanyl)(trimethylsilyl)methyl]-
lithium, via nucleophilic addition/1,3-Brook
rearrangement/β-elimination cascade sequence 399

→ Silanes, vinyl-, functionalized, from reaction of
α-(disilylmethyl) enals with alkyllithiums
then electrophiles, via nucleophilic addition/
1,4-Brook rearrangement/electrophilic capture cascade sequence 419, 420

→ Silanes, vinyl-, trisubstituted, from reaction of
bis(silyl) enals with alkyllithiums then alkyl
halides, copper(I) cyanide promoted, via nucleophilic addition/1,4-Brook rearrangement/
alkylation cascade sequence 426

→ (–)-Silvestrol, core, from rearrangement of tricyclic α-hydroxy ketones, base catalyzed 263,
264

Silyl enol ethers – see Enol ethers, silyl; Bis(enol
ethers), silyl

Silyl enol thioethers – see Enol thioethers, silyl

→ Silyl ethers, α,α-dichloro-, from reaction of aldehydes with [(silyl)dichloromethyl]lithium,
via domino nucleophilic addition/1,3-Brook
rearrangement/electrophilic capture sequence 393, 394

Silyl ethers, 2,3-epoxy, semipinacol rearrangement/
ring expansion, titanium(IV) chloride catalyzed,
spirocyclic fasicularin intermediate synthesis 270

Silyl ethers, propargyl, domino α-lithiation/retro-
1,2-Brook rearrangement sequence, then methoxymethyl chloride capture, lithiation, and reaction with electrophiles, tetrasubstituted allene
synthesis 392

Silyl ethers, propargylic, domino α-lithiation/retro-1,2-Brook rearrangement/1,2-silyl migration sequence, then Lewis acid treatment, α,β-bis-silylated enal synthesis 390, 391

→ (±)-Sirenin, key intermediate, from quasi-Favorskii rearrangement of an α-chlorocyclobutanone, using silver nitrate 289

→ SNF4435C/D, from isomerization/cyclization cascade sequence of a tetraene, palladium catalyzed 452, 453

→ SNF4435C/D, from reaction of a dienyl iodide with a dienylstannane, palladium catalyzed, via Stille coupling/8π-electrocyclization/6π-electrocyclization cascade sequence 451

→ Sophoradiol, key intermediate, from domino pentacyclization of a fluoropolyene, trifluoroacetic acid induced 292–294

→ Spiro[cyclopentane-indene]s, from Heck polyene cyclization of iodophenyl-substituted 1,5-dienes, palladium/phosphine catalyzed 32, 33

→ Spiro[cyclopentane-naphthalene]s, from Heck polyene cyclization of vinyl trifluoromethanesulfonate-substituted 1,5-dienes, palladium catalyzed 32, 33

→ Spiro[indene-4,3′-pyran]-2′-ones, decahydro-, from dialdehydes, via radical-anionic cyclization cascade, samarium(II) iodide mediated 196, 197

→ Spirolactones, carbocycle fused, from dialdehydes, via radical-anionic cyclization cascade, samarium(II) iodide mediated 196, 197

→ (–)-Spirotryprostatin B, protected, from intramolecular Heck/Tsuji–Trost cascade cyclization of a diene-aryl iodide-diketopiperidine, palladium/phosphine catalyzed 474, 475

→ Spiro[5.5]undecane-1,7-diols, from 2,3-epoxy alcohols, via reductive rearrangement/ring expansion, aluminum triisopropoxide catalyzed 272, 273

→ Spongistatins, key intermediate, from reaction of lithiated 2-silyl-1,3-dithianes with two epoxides, via 1,4-Brook rearrangement cascade sequence 434, 435

Squalene, 2,3-oxido-, biosynthesis conversion into lanosterol, using oxidosqualene cyclase 13

→ Stachyflin, key intermediate, from a bicyclic epoxide, via enantioselective domino epoxide opening/Wagner–Meerwein rearrangement/cyclization process, Lewis acid induced 304, 305

Stannanes, (α-siloxyalkyl)-, reaction with aldehydes, via domino tin–lithium exchange/retro-1,2-Brook rearrangement/electrophilic trapping sequence, monosilyl-protected diol synthesis 389

Stannanes, [1-(triorganosiloxy)vinyl]-, domino tin–lithium exchange/retro-1,2-Brook rearrangement/alkylation sequence, alkyl silyl ketone synthesis 388

→ (±)-Stemonamine, key intermediate, from α-siloxy epoxy azides, via tandem semipinacol/Schmidt domino process, titanium(IV) chloride promoted, then oxidation, ozonolysis, and aldol reaction 275, 276

→ (–)-Stenine, fused tricyclic intermediate, from reaction of dienyl azides with cyclic enones, via tandem intermolecular Diels–Alder/intramolecular Schmidt reaction sequence, Lewis acid promoted 342, 343

→ (–)-Strychnine, tetracyclic intermediate, from reaction of N-protected 3-(2-aminoethyl)-2-[(methylsulfanyl)vinyl]indoles with propynal, via asymmetric tandem Diels–Alder/amine cyclization sequence, chiral secondary amine catalyzed 340

→ Styrenes, β-alkyl-, from reaction of trisubstituted alkenes with arenecarbaldehydes, via carbonyl–alkene cross metathesis, trityl cation catalyzed 151, 152

Styrenes, α-aryl substituted, [2+2+2] photooxygenation with oxygen, 9,10-dicyanoanthracene sensitized, 3,3,6,6-tetraaryl-1,2-dioxane synthesis 180, 181

Sulfamates, allyloxyalkyne tethered, conversion into fused oxathiazepanes, via alkyne–nitrene metathesis/oxonium ylide cyclization/2,3-sigmatropic rearrangement cascade sequence, rhodium catalyzed, then reduction 528–531

Sulfonates, epoxy, [4+1] cycloaddition with lithiated monosilyl dithioacetals, via ring opening/1,4-Brook rearrangement/ring closure cascade sequence, then acidic hydrolysis, functionalized cyclopentanol synthesis 405–407

T

→ Taxane, framework, from a bromoene-yne-ene, via intramolecular electrocyclization cascade sequence, palladium catalyzed 462, 463

→ Terpenoids, tricyclic, from tandem radical cyclization of polyenyl epoxy ketones, using titanium(III) 222

→ Tetracycline skeleton, from intramolecular cyclization of an allyl acetate-aryl iodide, palladium/phosphine catalyzed, via oxidative addition/Tsuji–Trost cyclization/Heck cyclization cascade sequence 491, 492

Tetraepoxides, tert-butyl carbonate containing, endo-selective epoxide-opening cascade, Lewis acid promoted, tetracyclic 6,7,7,7-fused oxygen heterocycle synthesis 49, 50

Tetraepoxides, dihydroxy, exo-selective domino intramolecular opening/cyclization, acid catalyzed, glabrescol intermediate synthesis 247, 248

Tetraepoxides, exo-selective epoxide-opening cascade, 10-camphorsulfonic acid promoted, 1,4-bis(octahydro-2,2′-bifuran-5-yl)butane-1,4-diol synthesis 43, 44

→(+)-Tetrangomycin, from a 1,7-enyne, via ring-closing metathesis, ruthenium alkylidene complex catalyzed, then Diels–Alder cycloaddition with a naphtho-1,4-quinone, elimination/aromatization, oxidation, and deprotection 125, 126

→Tetraols, from diol diepoxides, via *exo*-selective epoxide-opening cascade, Brønsted base promoted 45

Thionocarbonates, aryl, radical carboxyarylation, using tris(trimethylsilyl)silane/2,2′-azobisisobutyronitrile, podophyllotoxin precursor synthesis 222, 223

Thiophene-3-carbaldehydes, 2-silyl-, annulation, using α,α-disubstituted ester enolates, via nucleophilic attack/1,4-Brook rearrangement/cyclization cascade sequence, 4-siloxycyclopenta[b]thiophen-6-one synthesis 424, 425

→Thunberginol A, precursor, from coupling of an aryl iodide with a ketone enolate, palladium/phosphine catalyzed, then carbonylative cyclization cascade, palladium/phosphine catalyzed 484, 485

→α-Tocopherol, key chiral intermediate, from asymmetric reaction of an alkenylphenol with an acrylate, palladium/chiral bis-oxazoline catalyzed, via oxypalladation/Heck cyclization cascade sequence 498, 499

→α-Tocopherol, key intermediate, from cyclization of a 2-hydroxybenzaldehyde with a hexadec-2-enal, via enantioselective domino aldol/oxa-Michael reaction, chiral diarylprolinol silyl ether catalyzed 349, 350

Tosylates, ω-aryl propargyl, domino C—H functionalization sequence, scandium catalyzed, tricyclic compound synthesis 594, 595

→Totaradiol, precursor, from aryl-substituted epoxyalkenes, via *endo*-selective epoxide-opening cascade, indium bromide initiated, arene terminated 60, 62

1,2,3-Triazoles, 4-(3-arylpropyl)-1-sulfonyl-, intramolecular dearomatizing [3+2] annulation, rhodium catalyzed, 3,4-fused indole synthesis, via cascade sequence 515, 516

1,2,3-Triazoles, 1-sulfonyl-, alkene tethered, dedinitrogenative rearrangement, rhodium catalyzed, 2,3-fused pyrrole synthesis, via cascade sequence 514, 515

1,2,3-Triazoles, 1-sulfonyl-, alkyne tethered, dedinitrogenative rearrangement, rhodium catalyzed, 3,4-fused pyrrole synthesis, via cascade sequence 513, 514

1,2,3-Triazoles, 1-sulfonyl-, allene tethered, dedinitrogenative rearrangement, rhodium catalyzed, 3,4-fused pyrrole synthesis, via cascade sequence 513, 514

1,2,3-Triazoles, 1-sulfonyl-, dinitrogen extrusion, rhodium catalyzed, then complex nitrogen heterocycle synthesis, via cascade sequence 512, 513

→(−)-Tricholomenyn A, from an enediyne, via ring-closing metathesis/metallotropic [1,3]-shift domino sequence, ruthenium alkylidene complex catalyzed, then deprotection and oxidation 120, 121

Tricyclic compounds, alkynyl substituted, tandem Diels–Alder/radical cyclization/radical fragmentation sequence, using triphenyltin hydride/triethylborane then samarium(II) iodide, tetracyclic phomoidride B core structure synthesis 225, 226

→Tricyclic compounds, chiral, from aryl-substituted dienes, via enantioselective domino cyclization sequence, chiral Brønsted acid catalyzed, Lewis acid assisted 318, 319

→Tricyclic compounds, from ω-aryl propargyl tosylates, via domino C—H functionalization sequence, scandium catalyzed 594, 595

→Tricyclic compounds, from diynes, via domino cycloisomerization, gold complex catalyzed 330, 331

→Tricyclic 5,7,3-ring systems, fused, from domino cyclization of dienynes, gold complex catalyzed 332, 333

1,3,6-Trienes, thiol–alkene co-oxygenation reaction, using di-tert-butyl peroxyoxalate/benzenethiol/oxygen, endoperoxide synthesis 170, 171

1,3,7-Trienes, enantioselective polyene cyclization, chiral binaphthol–Lewis acid complex catalyzed, chiral dehydroabietic acid precursor synthesis 20

Triepoxides, tert-butyl carbonate containing, endo-selective epoxide-opening cascade, Lewis acid promoted, 8-hydroxydecahydro[1,3]dioxino[5,4-b]oxepino[2,3-f]oxepin-2-one synthesis 48

Triepoxides, tert-butyl ester containing, endo-selective epoxide-opening cascade, Lewis acid promoted, then silylation, 8-siloxydodecahydro-oxepino[3,2-b]pyrano[2,3-f]oxepin-2-one synthesis 49

Triepoxides, distal alkene containing, endo-selective epoxide-opening cascade, bromonium ion initiated, internal nucleophile terminated, ent-dioxepandehydrothyrsiferol intermediate synthesis 53

Triepoxides, 3-hydroxytetrahydropyran containing, endo-selective epoxide-opening cascade, water promoted, tetracyclic 6,6,6,6-fused oxygen heterocycle synthesis 57, 58

1,5,9-Triepoxides, hydroxy, exo-selective domino intramolecular opening/cyclization, acid catalyzed, etheromycin intermediate synthesis 247, 248

1,6,11-Triepoxides, hydroxy, exo-selective domino intramolecular opening/cyclization, acid catalyzed, poly(tetrahydropyran) synthesis 247

→Trikentrin, didesmethyl-, azatricyclic intermediate, from a bromoene-amide-diyne, via intramolecular electrocyclization cascade sequence, palladium catalyzed 461, 462

→Triquinanes, from dienylcyclopropanes, via domino [(5+2)+1] cycloaddition/aldol reaction sequence, rhodium catalyzed 325, 326

→ Triterpenes, pentacyclic, intermediates, from tricyclization of polyene epoxides, Lewis acid induced 290, 291
→ Triterpenoid natural products, intermediates, from domino pentacyclization of fluoropolyenes, Lewis acid induced 292–294
1,3,5-Triynes, intramolecular reactions with tethered alkenes, ruthenium alkylidene complex catalyzed, initiation of metallotropic [1,3]-shift via relay metathesis 83–85
Tryptamines, reaction with enals, via enantioselective domino addition/cyclization, chiral cyclic secondary amine catalyzed, chiral hexahydropyrrolo[2,3-b]indole synthesis 345, 346
Tryptamines, three-component domino condensation process with β-keto esters and α,β-unsaturated aldehydes, cerium catalyzed, indolo[2,3-a]quinolizine synthesis 589, 590
Tryptophols, reaction with enals, via enantioselective domino addition/cyclization, chiral cyclic secondary amine catalyzed, chiral tetrahydrofuro[2,3-b]indole synthesis 345, 346
→ (−)-Tylophorine, key intermediate, from asymmetric reaction of a carbamate with a bromophenanthrene, palladium chiral phosphoramidite catalyzed, via aminopalladation/cyclization cascade sequence 479, 480

V

→ Valerenic acid, key intermediate, from ring-closing metathesis of a 1,6-enyne, ruthenium alkylidene complex catalyzed, then deprotection and Diels–Alder reaction with acrylate 124, 125
→ (+)-Valienamine, key intermediate, from a serine-derived 1,7-enyne, via ring-closing metathesis, ruthenium alkylidene complex catalyzed 96, 97
→ (+)-Vincadifformine, tetracyclic intermediate, from reaction of N-protected 3-(2-aminoethyl)-2-[(methylsulfanyl)vinyl]indoles with propynal, via asymmetric tandem Diels–Alder/amine cyclization sequence, chiral secondary amine catalyzed 340, 341
→ (+)-Virosaine, key intermediate, from a dienyne, via domino double enyne ring-closing metathesis, ruthenium alkylidene complex catalyzed 110, 111
→ Vitamin D_3, 1α,25-dihydroxy-, from reaction of an alkenyl bromide with an enyne, palladium/phosphine catalyzed, via cross-coupling/electrocyclization cascade sequence 460
→ Vitamin D_3, 1α,25-dihydroxy-, from reaction of an enol trifluoromethanesulfonate with an alkenylpinacol boronate, palladium catalyzed, via Suzuki cross-coupling/electrocyclization cascade sequence 459, 460

W

Wagner–Meerwein ring expansion, polycyclic cyclobutanes, acid induced, polycyclic cyclopentane synthesis 254

X

→ (−)-Xanthatin, 11α,13-dihydro-, from ring-closing metathesis of a 1,8-diyne, ruthenium alkylidene complex catalyzed, then methylation and cross metathesis with methyl vinyl ketone, ruthenium alkylidene complex catalyzed 113, 114
→ (+)-8-*epi*-Xanthatin, from reaction of a 1,8-enyne with methyl vinyl ketone, via domino ring-closing metathesis/cross metathesis sequence, ruthenium alkylidene complex catalyzed 113
→ Xanthenes, hexahydro-, from enantioselective polyene cyclization of 1,5-dienes, Lewis acid–Brønsted acid complex catalyzed 19
→ Xanthenes, hexahydro-, from polyene cyclization of 1,5-dienes, cationic palladium complex mediated 23, 24
→ Xanthenes, 9-substituted, from reaction of arylaldehydes with arenes, scandium catalyzed, via domino intermolecular/intramolecular C—H functionalization sequence 595, 596
→ Xanthen-2-ols, hexahydro-, substituted, from distal O-protected phenol-substituted epoxides, via domino cyclization, Lewis acid induced 298, 299
→ Xanthen-2-ols, hexahydro-, substituted, from O-protected phenol-substituted epoxyalkenes, via *endo*-selective epoxide-opening cascade, Lewis acid initiated, protected phenol terminated 63
→ Xanthen-2-ols, hexahydro-, substituted, from O-protected phenol-substituted epoxyalkenes, via tandem *endo*-selective epoxide-opening/electrophilic substitution cascade, Lewis acid initiated, protected phenol terminated 63, 64
→ (+)-Xestoquinone, pentacyclic intermediate, from carbopalladation cascade reaction of a dienylnaphthol, via aryl trifluoromethanesulfonate formation/asymmetric zipper-Heck reaction/cyclization, palladium/chiral phosphine catalyzed 473, 474

Y

→ Yingzhaosu C, from peroxy radical domino cyclization of an alkene hydroperoxide, samarium iodide/oxygen induced 162, 163
Ynepentaenes, polyene zipper cyclization, palladium catalyzed, tetraspiro pentacycle synthesis, via ene–yne cycloisomerization 35

Author Index

In this index the page number for that part of the text citing the reference number is given first. The number of the reference in the reference section is given in a superscript font following this.

A

Abe, H. 303[79], 304[79], 305[79]
Abe, I. 24[68]
Abe, M. 158[9], 164[9], 165[9], 166[9]
Abe, T. 457[42], 457[43], 457[44], 458[42], 458[43], 470[43]
Abele, S. 10[15], 10[16]
Abelman, M. M. 32[121], 33[121]
Abramo, G. P. 28[90]
Aburano, D. 620[171]
Abu Sohel, S. M. 542[28]
Ackermann, L. 234[99]
Acs, G. 626[216]
Adams, T. E. 263[119], 264[119]
Adeyemi, O. O. 602[31]
Adiwidjaja, G. 249[57], 390[85], 391[85], 404[105], 405[105], 406[105], 407[105], 408[105], 413[115], 414[115]
Adlington, R. M. 114[113], 450[27], 452[27], 452[30], 452[31], 453[31], 469[31]
Adrio, J. 22[59], 620[156]
Agama, K. 236[102]
Ager, D. J. 356[8]
Aggarwal, P. 499[125], 500[125], 503[125]
Aggarwal, V. K. 97[86], 98[86], 138[41], 244[25]
Agnel, G. 449[17]
Agnoli, A. 620[169]
Ahmar, M. 622[180], 622[181], 622[182], 622[183]
Ahn, J. S. 618[140]
Ahrendt, K. A. 611[100]
Aikawa, H. 535[16]
Akai, S. 285[40], 285[41], 286[40], 286[41]
Alabugin, I. V. 217[15], 221[65]
Alameda-Angulo, C. 261[115], 262[115]
Albicker, M. 304[81]
Albini, A. 207[60]
Albrecht, U. 227[77], 228[77]
Albuquerque, E. X. 300[67], 620[163]
Alcaide, B. 552[44]
Alder, R. W. 626[218]
Alfassi, Z. B. 217[31]
Alford, J. S. 514[26], 515[26]
Allison, B. D. 15[37]
Allwein, S. P. 135[9], 135[10], 142[10], 143[9], 143[10], 144[9], 144[10]
Almendros, P. 552[44]
Almy, J. 287[42]
Alonso, C. M. A. 585[14], 586[14]
Alonso, J. M. 552[44]
Alvarado, S. 623[191]
Álvarez de Cienfuegos, L. 218[51]
am Ende, C. W. 228[80], 229[80]
Amijs, C. H. M. 557[65]

An, G. 187[5], 217[3]
An, H. J. 117[116], 117[117], 118[116], 118[117]
An, S. E. 564[91]
Anada, M. 516[28]
Ananth, S. L. 527[43]
Andersen, N. G. 473[78], 474[78], 487[78]
Anderson, B. D. 236[102]
Anderson, E. A. 449[3], 456[40], 457[40], 461[57], 461[58], 462[58], 470[40], 471[57], 471[58], 498[121], 498[122]
Ando, E. 362[25], 363[25]
Ando, K. 516[28]
Ando, M. 624[205]
Ando, S. 277[26]
Andrews, G. C. 15[22]
Andriantsiferana, R. 297[60]
Añorbe, L. 601[4]
Anslyn, E. V. 26[72]
Antonsen, Ø. 389[83]
Antony, S. 236[102]
Antras, F. 622[180], 622[182], 622[183]
Anzovino, M. E. 211[67]
Aoki, K. 613[110], 626[110]
Apeloig, Y. 373[54], 374[54]
Arai, H. 348[159], 350[159]
Arai, Y. 306[83]
Arcadi, A. 321[118]
Archambeau, A. 518[33], 519[33], 520[33], 521[33]
Arigoni, D. 15[20]
Arita, H. 244[31]
Armanetti, L. 218[56]
Aronstam, R. S. 300[67]
Arrhenius, T. S. 258[97]
Arteaga, J. F. 193[25]
Asami, Y. 618[140]
Asano, K. 626[220]
Asao, N. 535[11], 535[12], 535[13], 535[14], 535[15], 535[16], 535[17], 535[18], 535[19]
Ashton, K. 198[41], 198[42]
Astle, C. J. 97[86], 98[86]
Atodiresei, I. 511[5]
Aubé, J. 301[73], 341[144], 341[145], 341[146], 341[147], 342[144], 342[145], 342[147], 343[145], 343[147]
Aubele, D. L. 46[14], 47[14]
Aubert, C. 549[37]
Aue, D. H. 547[36], 548[36], 549[36]
Austin, J. F. 345[155], 346[155], 347[155], 350[155]
Avendaño, C. 589[20], 590[20]
Aver, W. A. 606[45]
Avetta, C. T. 207[59]
Avetta, C. T., Jr. 204[58], 205[58], 206[58], 207[58]

Ay, M. 461[53]
Ayer, W. A. 606[44]
Aylward, J. H. 626[213]

B

Baars, H. 461[57], 471[57]
Baati, R. 356[9]
Baba, M. 626[214]
Babij, N. R. 463[65], 472[65]
Babu, Y. S. 527[43]
Bachi, M. D. 157[3], 169[44], 169[45], 170[44], 170[45], 171[45]
Bäckvall, J.-E. 463[61]
Badine, D. M. 244[25]
Bae, H. J. 564[91]
Baeyer, A. 15[16], 15[17]
Baggio, R. 218[56]
Bagheri, V. 516[30], 517[30]
Bah, J. 151[56], 152[56], 153[56]
Bai, Y. 535[10]
Baik, M.-H. 28[90]
Baillie, T. A. 624[208]
Baine, N. H. 219[60], 219[61]
Bajracharya, G. B. 75[44]
Baker, H. K. 357[12]
Baker, R. T. 67[6], 76[6]
Baker, T. M. 135[21]
Bakshi, K. 522[40]
Baldwin, J. E. 43[4], 114[113], 245[33], 450[27], 452[27], 452[30], 452[31], 453[31], 469[31]
Ball, C. J. 467[75]
Ball, L. T. 555[49], 555[50], 556[58]
Ban, H. 602[23]
Ban, Y. 623[194]
Bandini, M. 551[42], 552[42]
Banert, K. 431[136], 432[136]
Bantia, S. 527[43]
Banwell, M. G. 229[81], 230[81], 250[73]
Bao, J. 593[25]
Barabé, F. 30[116]
Baralle, A. 217[28], 217[29]
Baran, P. S. 1[1], 1[3], 2[3], 8[11], 28[112], 28[113], 569[100], 626[222], 627[222]
Barber, D. M. 481[99]
Barber, L. L., Jr. 621[176]
Bardagí, J. I. 217[9]
Barker, P. J. 170[47], 171[47]
Barker, T. J. 28[110], 28[111], 163[31]
Barluenga, S. 35[135], 36[135]
Barma, D. 356[9]
Barnes-Seeman, D. 1[2], 2[2]
Baro, A. 610[87]
Baroudi, A. 217[28], 217[29]
Barrero, A. F. 27[80], 28[80], 193[25], 218[46]
Barrett, A. G. M. 233[92]
Bartlett, P. D. 158[10], 160[10]

670 Author Index

Bartlett, W. R. 292[53], 293[53]
Barton, D. H. R. 527[47], 527[48]
Barton, V. E. 164[34]
Basabe, P. 246[37], 246[40], 256[93]
Baskar, B. 564[91]
Baskaran, S. 301[74], 302[74], 305[74]
Bastow, K. F. 624[201]
Basu, B. 250[69]
Batey, R. A. 584[13], 585[13], 596[30], 597[30]
Batten, R. J. 158[15]
Battioni, P. 527[50]
Baud, C. 527[52]
Bauer, K. 259[106]
Bausch, C. C. 377[61], 377[62]
Bax, M. N. 467[76], 468[76], 473[76]
Bayer, A. 439[146], 440[146]
Bazan, G. C. 67[6], 69[24], 76[6], 136[24], 136[25], 137[24], 137[25]
Bazdi, B. 218[44]
Beaudry, C. M. 450[24], 450[26], 451[24], 451[29], 469[29]
Becker, J. J. 23[63]
Beckwith, A. L. J. 158[14], 168[43], 169[43], 170[46], 170[47], 171[47]
Beckwith, R. E. J. 512[14]
Bedi, G. 527[50]
Beeler, A. B. 9[13]
Beeson, T. D. 198[41]
Begum, K. 158[9], 164[9], 165[9], 166[9]
Behrmann, T. L. 135[7], 138[7]
Beier, C. 404[105], 405[105], 406[105], 407[105], 408[105]
Beifuss, U. 67[1], 355[5]
Bell, H. P. 491[111], 492[111], 495[111]
Bella, M. 10[14]
Beller, M. 466[71], 466[72], 601[11]
Beller, S. A. 620[166]
Benet-Buchholz, J. 72[38]
Benitez, D. 555[54], 555[55], 556[54]
Bennasar, M. L. 138[42], 138[43], 142[43]
Benneche, T. 389[83]
Bennett, C. S. 434[140], 434[142], 435[140], 435[142]
Bensa, D. 243[1]
Bentley, T. W. 14[9]
Benzing, E. P. 158[10], 160[10]
Bergman, R. G. 75[50]
Bergquist, P. R. 609[76]
Bernal, F. 313[94], 313[95], 313[99], 313[100], 314[94]
Berrien, J.-F. 158[17], 159[17]
Bertelsen, S. 349[160]
Bertrand, M. B. 463[66], 478[92], 479[92], 489[92]
Beutler, J. A. 297[57], 297[58], 298[61]
Bevins, R. 236[102]
Bharadwaj, A. R. 378[65], 379[65]
Bhatia, G. S. 438[145], 439[145]
Bhuniya, D. 126[125], 127[125]
Biller, S. 476[87]
Bilodeau, M. T. 527[51]
Bio, M. M. 466[73], 473[73]

Bischoff, M. 475[85], 476[85], 476[87], 488[85]
Bisseret, P. 135[2]
Blakey, S. B. 528[57], 529[57], 530[57], 530[58], 531[58], 532[57]
Blanchard, N. 135[2]
Blanco, J. G. 624[203]
Blanco-Urgoiti, J. 601[4]
Blechert, S. 106[102], 107[102]
Blond, G. 455[37], 455[38], 462[59], 463[59], 472[59]
Bloodworth, A. J. 158[16]
Bluhm, H. 27[79]
Blumberg, P. M. 626[216]
Bocco, J. L. 624[203]
Boche, G. 444[154]
Bochmann, S. 431[136], 432[136]
Boddy, I. 247[45], 248[45]
Bodurow, C. 464[67], 500[67], 500[128]
Boger, D. L. 28[110], 28[111], 163[31], 218[50], 233[93], 235[93]
Bohlmann, F. 623[191]
Bohn, M. A. 217[27]
Boivin, J. 233[91]
Boldi, A. M. 414[117], 415[117], 433[117], 434[117], 434[138], 434[139], 434[141], 435[138], 435[139], 435[141]
Bolm, C. 358[17], 359[17], 373[54], 374[54], 601[11]
Bom, D. 236[101]
Boñaga, L. V. R. 601[8]
Bookser, B. C. 258[97]
Boone, M. A. 58[35], 59[35]
Bora, U. 175[54]
Borčić, S. 15[39]
Bordwell, F. G. 287[42]
Borkowsky, S. L. 358[16], 359[16]
Boudhar, A. 455[37], 462[59], 463[59], 472[59]
Boukouvalas, J. 162[23], 163[23], 502[137], 505[137]
Bour, C. 455[38], 564[90]
Bowden, N. 68[22]
Bowry, V. 233[90]
Boyce, C. W. 233[93], 235[93]
Boyce, G. R. 368[42], 368[43]
Boyd, M. R. 297[57], 297[58]
Boydell, A. J. 412[114], 413[114]
Boyer, F.-D. 99[90], 99[91], 100[90], 100[91]
Boyle, G. M. 626[213]
Bozell, J. J. 500[127]
Bradley, P. A. 193[29]
Brancour, C. 188[16]
Brandes, B. D. 244[19]
Brandes, B. 349[160]
Brasche, G. 67[5], 187[4], 243[4], 449[1]
Bräse, S. 198[47], 461[56]
Braslau, R. 217[25]
Bräuer, N. 404[105], 405[105], 406[105], 407[105], 408[105], 408[107], 409[109], 410[109], 411[109]
Braun, I. 545[34], 545[35], 546[34], 547[34], 548[34]

Braunton, A. 349[162]
Bravo, F. 3[5], 4[5], 48[15], 48[17], 50[15], 50[17], 249[54]
Brazeau, J.-F. 555[51]
Brecker, L. 444[155]
Breitmaier, E. 13[1]
Brenneman, J. B. 97[87], 98[87], 113[109]
Brenzovich, W. E., Jr. 555[51], 555[54], 556[54], 556[57]
Breslow, R. 219[59], 527[49]
Brigaud, T. 360[20], 360[21], 361[20]
Brinza, I. M. 217[35]
Briones, J. F. 516[29]
Brönnimann, R. 10[16]
Brook, A. G. 355[1], 392[1]
Brook, C. S. 434[138], 434[141], 435[138], 435[141]
Brooks, J. L. 243[2]
Brooner, R. E. M. 22[57]
Brossi, A. 606[45]
Brown, M. J. 277[22], 277[24], 278[22], 280[28], 281[22]
Brown, R. S. 53[22]
Brown-Wensley, K. A. 136[29], 139[29]
Brückner, S. 452[30]
Brucks, A. P. 17[43], 17[44], 17[45]
Brummond, K. M. 316[108], 601[7], 604[37], 622[184], 622[187], 622[188], 624[198], 624[210], 625[198]
Bruno, M. 624[201]
Büchi, G. 261[111]
Buchwald, S. L. 136[29], 139[29], 601[14], 602[15], 619[147], 619[148]
Bulger, P. G. 33[128], 34[131], 217[30], 449[2], 449[7]
Bullock, R. M. 28[86], 28[87]
Bultinck, P. 481[99]
Buñuel, E. 27[80], 28[80], 75[45], 218[46]
Burgess, J. M. 93[77], 98[77], 99[77]
Burghart, A. 201[53], 204[53], 206[53]
Burgstahler, A. W. 15[21]
Buriez, O. 177[63]
Burke, P. J. 608[74]
Burnell, D. J. 610[88]
Burns, B. 449[10], 449[11]
Burns, N. Z. 8[11]
Burton, J. W. 217[18]
Bush, J. B., Jr. 26[73]
Bushey, D. F. 254[89]
Buzas, A. 557[74], 557[79]
Byers, J. A. 56[27], 56[28], 56[31]
Bzowej, E. I. 136[36], 139[36]

C

Cai, S. 535[10]
Cai, S.-J. 618[145]
Calimente, D. 135[6], 135[7], 138[6], 138[7]
Callahan, M. P. 298[63], 299[63], 300[63]
Callens, M. 617[137]
Callier-Dublanchet, A.-C. 188[13]
Caltagirone, C. 620[167]

Author Index

671

Calvo, C. 607[56], 607[58], 607[59]
Cambie, R. C. 609[76]
Campagne, J.-M. 92[76]
Campaña, A. G. 218[51]
Campbell, A. N. 321[122], 322[122], 323[122], 327[122]
Campbell, C. D. 461[58], 462[58], 471[58]
Campbell, M. J. 538[22]
Cane, D. E. 43[2], 243[8], 617[135], 617[138]
Cañeque, T. 549[38]
Canham, S. M. 335[138]
Cannizzo, L. 136[29], 139[29]
Cao, Q. 233[95]
Cao, R. 58[33], 58[35], 59[33], 59[35], 60[33], 294[55]
Cao, S. 297[60]
Carcache, D. A. 615[122], 615[123]
Cárdenas, D. J. 27[80], 28[80], 75[45], 218[46], 564[90]
Cargill, R. L. 254[89]
Cariou, K. 549[39], 550[39]
Carless, H. A. J. 158[15]
Caro-Diaz, J. E. 618[143], 618[144]
Carpenter, J. 464[69], 465[69], 472[69]
Carpenter, N. E. 32[122], 33[122]
Carreira, E. M. 24[64], 24[65], 28[93], 28[101], 28[102], 28[103], 28[104], 28[105], 28[106], 28[107], 28[108], 28[109], 29[93], 166[36]
Carretero, J. C. 22[59], 620[156]
Carter, R. G. 492[112], 495[112], 615[126]
Casarez, A. D. 198[45]
Casey, C. P. 70[27]
Cassady, J. M. 246[36]
Cassayre, J. 188[13], 604[43], 605[43], 606[50]
Castilla-Alcalá, J. A. 193[26]
Castillo, M. 607[56], 607[57], 607[58], 607[59]
Castle, S. L. 217[38]
Castro, J. 603[35], 604[35]
Castro, V. 623[191]
Catellani, M. 476[88]
Cavaleiro, J. A. S. 585[14], 586[14]
Cavallo, L. 72[38]
Cazes, B. 622[180], 622[181], 622[182], 622[183]
Ceccon, J. 332[133], 543[32]
Celmer, W. D. 43[2], 243[8]
Cera, G. 551[42], 552[42]
Cha, J. Y. 197[36]
Chabanis, O. 622[181]
Chan, A. S. C. 269[4], 271[4]
Chan, C. 458[46]
Chan, P. W. H. 142[47], 564[92], 565[92], 565[94], 566[94]
Chan, T. L. 234[97]
Chand, P. 527[43]
Chand, S. 250[73]
Chand, S. S. 598[34]
Chandrasekaran, S. 256[91], 257[91]
Chang, C.-J. 608[69]

Chang, H.-H. 94[83]
Chang, S. 244[20]
Chang, W. K. 615[124]
Chang, Y. 612[107]
Chapman, H. A. 461[58], 462[58], 471[58]
Charpenay, M. 455[37]
Charych, D. 76[57]
Chatani, N. 75[43], 602[16], 619[150], 619[151]
Chatgilialoglu, C. 193[28], 217[5], 217[6], 217[7], 217[8], 217[9], 217[10], 217[11], 217[12], 217[13], 217[14], 217[15], 217[16], 217[17], 217[18], 217[19], 217[20], 217[21], 217[22], 217[23], 217[24], 217[25], 217[26], 217[27], 217[28], 217[29], 217[31], 217[40], 217[41]
Che, C.-M. 527[53]
Chen, C. 314[102], 315[102]
Chen, C.-H. 545[33]
Chen, D. W. C. 259[101]
Chen, D. Y.-K. 313[95], 313[96], 313[97], 313[100], 458[48], 459[48], 471[48]
Chen, G.-F. 614[116]
Chen, H. 622[184]
Chen, J. 500[135], 501[135], 504[135], 616[127], 622[177]
Chen, J.-H. 500[136], 501[136], 505[136], 602[25], 613[111], 613[112], 613[113], 613[114], 613[115]
Chen, J. S. 449[4]
Chen, M. 535[17]
Chen, M. Z. 415[118], 416[118]
Chen, R. 246[39]
Chen, S. 588[18]
Chen, S.-F. 604[38]
Chen, X. 244[12], 334[136], 335[136]
Chen, X. P. 614[117]
Chen, Y. 249[65]
Chen, Y. Z. 272[10]
Chen, Z.-H. 275[15], 275[16], 275[17]
Chen, Z.-M. 275[15], 275[16]
Chen, Z.-X. 500[136], 501[136], 505[136], 613[115]
Cheng, Q. 76[58]
Cheng, S. 24[67]
Cheng, X.-M. 13[2], 33[2], 34[2]
Cheong, J. Y. 564[91]
Chernyak, N. 513[20]
Cheung, A. W.-H. 500[126], 501[126], 503[126]
Chiarucci, M. 551[42], 552[42]
Chidester, C. G. 617[134]
Chinn, M. S. 520[35]
Chino, K. 623[193], 623[195]
Chng, S.-S. 311[88], 312[88]
Cho, E. J. 80[69], 81[69], 82[69], 121[119], 122[119]
Cho, H.-N. 136[25], 137[25]
Cho, Y. S. 615[122], 615[123]
Choi, J. 28[88]
Choi, S. Y. 91[75], 92[75]
Choi, Y. K. 602[28]

Choquesillo-Lazarte, D. 27[81], 222[68]
Choquette, K. A. 193[23]
Choshi, T. 457[43], 458[43], 470[43]
Chou, F. L. 449[19]
Chougnet, A. 349[161], 350[161]
Choy, P. Y. 234[97]
Christensen, S. 624[199]
Christianson, D. W. 24[69]
Christie, S. D. R. 601[9]
Christmann, M. 437[144], 438[144]
Christoffers, J. 610[87]
Chtchemelinine, A. 323[123], 324[123], 327[123]
Chun, S. 10[16]
Chung, d. 464[69], 465[69], 472[69]
Chung, H. S. 91[75], 92[75]
Chung, W. J. 360[22]
Chung, Y. K. 91[75], 92[75], 601[10], 602[21]
Chuprakov, S. 513[19], 513[20], 513[21], 513[23]
Cicio, F. 623[191]
Ciufolini, M. A. 400[97], 400[98]
Claiborne, C. F. 135[4], 140[4], 141[4], 142[4]
Clardy, J. 618[141]
Clark, J. S. 135[14]
Clark, M. C. 358[15]
Claverie, C. K. 75[47], 332[130], 333[130], 334[130]
Clavier, H. 72[38]
Clawson, L. 136[29], 139[29]
Clayton, R. B. 15[27]
Clerc, J. 475[85], 476[85], 476[87], 488[85]
Clift, M. D. 201[55], 202[55], 202[57], 203[55]
Cochrane, N. A. 23[63]
Cockfield, D. M. 611[95]
Cohen, T. 258[99], 375[56], 375[57], 375[58], 376[58]
Cointeaux, L. 158[17], 159[17]
Cole, K. P. 313[101], 314[101]
Cole, T. W. 261[113]
Colsman, M. R. 520[35]
Colyer, J. T. 422[126], 423[126]
Commeiras, L. 114[113]
Conli, L. 620[169]
Conrad, J. C. 198[48]
Cook, J. M. 608[72]
Cook, S. P. 486[107]
Coote, S. C. 193[28], 217[19]
Copéret, C. 449[21]
Coquerel, Y. 243[1]
Corbet, M. 549[40], 550[40]
Cordell, G. A. 300[68], 606[45], 608[71]
Cordonnier, M.-C. A. 456[40], 457[40], 470[40]
Córdova, A. 347[157], 348[157]
Corey, E. J. 1[2], 2[2], 3[4], 13[2], 13[3], 15[13], 15[25], 15[26], 15[28], 15[29], 15[30], 15[32], 15[33], 15[34], 15[38], 16[29], 16[30], 17[30], 17[40], 19[3], 19[48], 20[48], 20[49], 21[3], 21[52],

23[3], 24[3], 26[75], 33[2], 34[2], 44[8], 162[21], 247[46], 248[46], 256[91], 257[91], 290[46], 290[48], 290[49], 291[48], 291[49], 291[50], 291[51], 291[52], 292[50], 292[51], 292[52], 296[49], 432[137], 433[137], 610[85]
Cornwall, R. G. 244[13]
Correa, A. 72[38]
Cossy, J. 475[83], 488[83], 518[33], 519[33], 520[33], 521[33]
Couladouros, E. A. 1[3], 2[3]
Couty, S. 475[83], 488[83]
Covel, J. A. 135[22], 147[22]
Cowley, A. R. 450[27], 452[27]
Cox, J. M. 135[9], 135[10], 135[15], 137[15], 142[10], 143[9], 143[10], 144[9], 144[10], 147[15]
Coxon, J. M. 245[34]
Cozzi, S.-J. 626[213]
Cragg, G. M. 297[58]
Cramer, N. 304[81], 400[98]
Crawley, M. L. 34[133], 491[109]
Creary, X. 177[64]
Crich, D. 217[7]
Crick, P. J. 188[12]
Cristofoli, W. A. 473[78], 474[78], 487[78]
Crone, B. 321[120], 557[66]
Crossley, S. W. M. 28[83], 29[83], 30[83], 30[116], 166[37]
Crowe, W. E. 602[20], 602[33], 602[34], 603[33], 603[34], 619[149]
Csupor, D. 626[215]
Cubillo, M. A. 256[93]
Cuerva, J. M. 27[80], 27[81], 28[80], 218[44], 218[46], 218[51], 222[68]
Cui, L. 556[59], 557[81], 562[88]
Cummins, C. C. 88[71], 88[72]
Cummins, T. J. 135[22], 147[22]
Cumpstey, I. 217[7]
Curini, M. 590[23], 591[23]
Curran, D. P. 6[9], 7[9], 191[18], 191[19], 192[18], 217[37], 236[101], 236[102], 485[104], 486[104], 490[104]
Curtis, R. J. 158[16]
Cusella, P. P. 10[14]
Czakó, B. 300[65], 302[75], 307[85], 316[85]

D

Dai, L.-Z. 598[32]
Dai, M. 500[131], 500[135], 501[135], 504[135]
Dai, M. J. 613[112], 613[113], 613[114]
Dai, W.-M. 249[65]
Dake, G. R. 258[98], 270[5], 270[6], 270[7]
Daly, A. M. 138[41]
Daly, J. W. 300[66], 300[67]
Danheiser, R. L. 256[91], 257[91]
Daniel, M. 217[28], 217[29]
Daniels, D. S. B. 498[121], 498[122]
Daniewski, W. M. 258[99]
Danishefsky, S. J. 99[92], 261[114], 607[52], 615[122], 615[123]

Da Silva, M. 196[32], 196[33], 197[32], 197[33], 197[34]
Datta, S. 550[41], 551[41]
Dauban, P. 527[45], 527[55]
Davies, H. M. L. 512[13], 512[14], 512[15], 512[16], 514[26], 515[26], 516[29], 598[33]
Davies, K. A. 188[17], 189[17], 190[17]
Davies, P. W. 321[115], 535[1]
Davis, A. L. 358[13]
Davis, B. M. 620[164], 620[165]
Davis, K. L. 620[164], 620[165]
Davis, W. M. 88[71]
De, S. R. 259[103]
Debleds, O. 92[76]
de Bruin, B. 28[97]
Decorzant, R. 244[29]
de Frémont, P. 557[73], 557[75]
de Haro, T. 555[48], 559[82], 560[82], 561[87], 562[87]
Dehghani, A. 527[43]
Delgado, G. 625[211]
Delpont, N. 101[94], 332[133], 334[135], 335[135], 543[32]
de Meijere, A. 461[54], 461[55], 461[56]
Demuth, M. 31[118]
Deng, H. 611[103]
Deng, J. 618[142]
Deng, L. 202[56], 244[18], 500[135], 501[135], 504[135]
Deng, L.-J. 602[25], 613[112]
Denmeade, S. R. 624[199]
Depezay, J.-C. 411[110]
DePuy, C. H. 177[61]
Dequirez, G. 527[45]
De Sarno, P. 614[118]
Deschamps, N. M. 612[105], 612[106]
Deslongchamps, P. 37[137], 37[138]
Desphpande, S. S. 620[163]
Dessau, R. M. 26[74]
Devarie-Baez, N. O. 425[129], 426[129]
Devery, J. J., III 202[56], 511[4]
Devin, P. 188[16], 225[74]
de Vries, J. G. 71[32], 72[32], 73[32]
Dexter, D. L. 604[38]
Dhanya, B. P. 598[34]
Dhara, D. 591[24], 592[24]
Dhimane, A.-L. 188[16], 217[32], 549[39], 550[39]
Dickhaut, J. 217[4]
Diedrichs, N. 313[99]
Dieltiens, N. 71[33], 73[33], 74[33]
Díez, D. 246[37], 246[40], 256[93]
Díez-González, S. 557[75]
Di Giuseppe, S. 321[118]
DiMare, M. 69[24], 136[24], 137[24]
Dineen, T. A. 105[100]
Dinér, P. 349[160]
Ding, D. 541[27]
Ding, S. 136[28]
Dintinger, T. 411[112], 412[112]
Dionne, C. 624[199]
Diorazio, L. 258[96]
Dirsch, V. M. 624[204]

Diver, S. T. 71[30], 71[31], 71[34], 72[31], 74[30], 79[67]
Dixon, D. J. 481[97], 481[98], 481[99], 482[97], 489[97], 611[95]
Do, B. 3[5], 4[5], 48[15], 48[16], 48[17], 50[15], 50[16], 50[17], 249[53], 249[54]
Dobler, W. 464[67], 500[67]
Dodd, R. 527[55]
Doedens, R. J. 277[20]
Doherty, A. M. 610[79]
Doherty, G. A. 128[126], 129[126]
Doi, S. 224[72]
Doi, T. 494[115], 496[115]
Do Khac Manh, D. 253[81], 253[82], 254[86]
Dolle, R. E. 244[28]
Dombroski, M. A. 26[76], 27[76], 221[66], 221[67]
Domingo, V. 193[25]
Domínguez, G. 601[4]
Donaldson, S. M. 72[36], 124[36]
Dondi, D. 207[60]
Dong, G.-B. 602[25], 613[114]
Dong, J. 541[27]
Dong, Q.-L. 348[158]
Dong, X. 623[190]
Donohoe, T. J. 258[96]
Dornan, L. M. 233[95]
Dorrity, M. J. 449[13]
Dos Santos, E. N. 511[2]
Dougherty, D. A. 26[72]
Doughty, V. A. 434[139], 434[140], 434[142], 435[139], 435[140], 435[142]
Dounay, A. B. 250[74], 252[74]
Doyle, M. P. 7[10], 512[9], 512[10], 512[11], 512[12], 516[30], 517[30], 520[35], 522[37], 522[38], 522[39], 522[40], 522[41], 522[42], 523[42], 524[42], 525[42], 526[42], 527[42]
Drapela, N. E. 261[116]
Drees, S. 246[42], 248[42]
Dreeßen, S. 409[109], 410[109], 411[109]
Du, J. 211[67], 211[68]
Du, W. 6[9], 7[9], 485[104], 486[104], 490[104]
Duan, X.-H. 496[116], 498[116]
Dubé, P. 557[68]
Du Bois, J. 527[54], 527[56]
Duchamp, D. J. 617[134]
Ducray, R. 400[97], 400[98]
Dudley, G. B. 99[92]
Dudnik, A. S. 557[67]
Duefert, S.-C. 475[85], 476[85], 476[87], 488[85]
Duesler, E. N. 347[156], 348[156]
Duffels, A. 611[97]
Duffey, M. O. 418[122], 419[122]
Duffy, R. 512[10]
Dugan, K. C. 204[58], 205[58], 206[58], 207[58]
Duggan, M. E. 624[209]
Dumas, J. 460[51], 460[52], 471[52]
Duncan, K. K. 28[111], 163[31]
Duplan, V. 35[135], 36[135]

Author Index

Dureault, A. 527[50]
Duschek, A. 335[137]
Dussault, P. 157[4]
Duszenko, M. 617[136]
Dzik, W. I. 28[97]

E

Eade, S. J. 452[30]
Earley, W. G. 316[109]
East, S. P. 626[218]
Eaton, P. E. 261[113]
Echavarren, A. M. 22[59], 75[45], 75[46], 75[47], 101[94], 321[117], 332[130], 332[131], 332[132], 332[133], 332[134], 333[130], 333[131], 333[132], 334[130], 334[131], 334[135], 335[131], 335[135], 337[132], 535[5], 542[29], 542[31], 543[31], 543[32], 544[31], 557[65], 564[90]
Ecker, J. R. 244[18]
Edmonds, D. J. 217[30], 449[2]
Edstrom, E. D. 23[61]
Eduardo, S. 459[49], 460[49], 471[49]
Egger, B. 261[111]
Eisenberg, D. C. 28[91]
Elford, T. G. 261[112]
El-Kattan, Y. 527[43]
Elliott, A. J. 527[43]
Ellison, R. H. 252[78], 252[79], 253[78], 253[79], 254[78]
El Sous, M. 263[119], 264[119]
Enders, D. 345[153], 511[3]
Endsley, K. E. 422[126], 423[126]
Enholm, E. J. 231[85]
Eno-Amooquaye, E. 608[74]
Enoki, N. 602[29]
Enomoto, M. 244[30]
Enomoto, S. 617[128]
Enriz, R. D. 624[206]
Epa, W. R. 498[123], 500[126], 501[126], 503[126]
Epifano, F. 590[23], 591[23]
Eriksson, L. 347[157], 348[157]
Erkkilä, A. 345[152]
Escarcena, R. 246[40]
Eschenmoser, A. 15[18], 15[19], 15[20]
Escudero-Adán, E. C. 72[38]
Espino, C. G. 527[54]
Estes, D. P. 28[89], 29[89]
Estes, L. A. 604[39]
Estieu, K. 259[105]
Estrella, A. 246[37]
Esumi, T. 109[105]
Eustache, J. 135[2]
Evans, D. A. 15[37], 38[139], 136[33], 137[33], 139[33], 527[51]
Evans, M. A. 113[110]

F

Fagnoni, M. 207[60]
Falck, J. R. 356[9]
Fallis, A. G. 217[35]
Fan, C. A. 269[4], 271[4]

Fan, C.-A. 269[1], 275[13], 275[14], 276[13], 282[32]
Fan, J.-Q. 94[83]
Fang, B. 440[148], 441[148]
Fang, L. 330[129], 331[129], 332[129], 337[129]
Fang, L.-C. 619[146]
Fang, M. 244[12]
Fang, X. 58[33], 59[33], 60[33], 294[55]
Fang, Y.-Q. 467[77], 468[77], 473[77]
Faraday, M. 198[37]
Farid, S. 180[72]
Faul, M. M. 527[51]
Favorskii, A. 261[108], 261[109]
Fazakerley, N. J. 197[35]
Fehr, C. 557[62]
Felding, J. 626[222], 627[222]
Feldman, K. S. 171[48], 171[49], 171[50], 171[51], 172[48], 172[49], 172[50], 173[48], 173[49], 173[50]
Feng, H. X. 53[22]
Fenster, M. D. B. 258[98], 270[5], 270[6], 270[7]
Fensterbank, L. 188[16], 217[28], 217[29], 217[32], 225[74], 549[37], 549[39], 550[39], 557[73]
Ferraccioli, R. 476[89]
Ferreira, L. M. 300[70]
Ferrer, C. 75[47]
Fetizon, M. 253[81], 253[82], 254[86]
Fielder, S. 162[22], 163[22]
Fielding, M. R. 163[30], 166[30], 484[102], 485[102], 490[102], 611[99]
Fielenbach, D. 349[162]
Filippini, M.-H. 243[1]
Findley, T. J. K. 197[34]
Finelli, F. G. 198[50]
Finkbeiner, H. 26 73]
Fioravanti, M. 620[169]
Fischer, D. 106[102], 107[102]
Fischer, M.-R. 404[104], 405[104], 406[104]
Fish, P. V. 292[54], 293[54], 294[54]
Fisher, K. D. 622[184]
Flament, J. P. 254[86]
Fleckhaus, A. 217[23]
Fleming, I. 136[30], 136[31], 139[30], 139[31], 217[37], 358[14], 381[14], 404[103], 405[103]
Fletcher, A. J. 467[76], 468[76], 473[76], 601[9]
Fleury, M. 270[7]
Floreancig, P. E. 46[14], 47[14], 54[23], 55[23]
Florence, G. J. 250[74], 252[74]
Flörke, H. 249[57]
Flowers, R. A., II 193[23], 193[27], 193[28], 202[56], 217[19]
Floyd, C. D. 404[103], 405[103]
Flügge, S. 88[73], 89[73]
Flury, A. 624[203]
Fogg, D. E. 511[2]
Fokas, D. 31[120], 32[120], 230[82]

Fokin, V. V. 513[20], 513[21], 513[22], 513[23]
Fong, H. H. S. 612[109]
Foreman, M. I. 601[2]
Forsyth, C. J. 249[67], 250[74], 252[74]
Fox, J. M. 610[89], 611[89]
Fox, M. E. 449[23], 450[23], 468[23]
France, D. J. 335[138], 499[125], 500[125], 503[125]
France, M. B. 136[27]
Franck-Neumann, M. 75[51], 75[52]
Frankowski, K. J. 341[147], 342[147], 343[147]
Franzén, J. 151[56], 152[56], 153[56]
Fraser-Reid, B. 287[43], 288[43]
Frazier, T. L. 422[127], 423[127]
Fréchette, Y. 162[23], 163[23]
Frederick, M. O. 313[97], 313[98], 313[99], 313[100], 313[101], 314[101]
Frelek, J. 579[8]
Friedrich, D. 607[64], 607[65]
Friesner, R. A. 28[90]
Frignani, F. 476[88]
Frohn, M. 244[22]
Frontier, A. J. 243[2]
Fruit, C. 527[46]
Fu, C. 193[22]
Fu, G. C. 137[38], 148[38], 148[52]
Fu, X. 616[127]
Fuess, H. 246[42], 248[42]
Fuji, K. 610[86]
Fujii, N. 480[95], 480[96], 489[96], 496[119], 497[119], 497[120], 498[119], 498[120], 540[25], 540[26], 541[25]
Fujimoto, K.-i. 362[25], 363[25]
Fujimura, O. 137[38], 148[38]
Fujioka, H. 285[40], 285[41], 286[40], 286[41]
Fujioka, K. 617[131]
Fujisawa, M. 362[24]
Fujita, M. 391[86], 392[86]
Fujita, S. 454[34], 470[34]
Fujiwara, K. 55[24], 55[25], 56[24], 56[25], 249[62], 249[63], 249[64]
Fujiwara, M. 383[72], 626[214]
Fujiwara, T. 142[48], 302[76]
Fukatsu, Y. 193[24]
Fukumoto, H. 109[105]
Fukumoto, Y. 602[16], 619[150], 619[151]
Fukuyama, T. 549[39], 550[39]
Fukuyama, Y. 492[112], 495[112], 615[121], 615[126], 621[172], 621[173]
Funakoshi, Y. 515[27], 516[27]
Fung, Y. 170[47], 171[47]
Funk, M. O. 157[7], 157[8], 158[7], 158[8], 160[7]
Funk, R. L. 611[102]
Furey, A. 313[92]
Fürstner, A. 22[54], 22[55], 88[73], 89[73], 128[127], 129[128], 130[128], 321[115], 535[1], 549[40], 550[40], 557[63], 557[70], 611[96], 611[97]

674 Author Index

Furukawa, H. 249[59], 249[61], 249[66]
Furukawa, J. 603[36], 607[54], 607[55]
Furukawa, N. 620[157]
Furukawa, Y. 76[59]
Furusaki, A. 602[29]
Furuta, H. 249[61]
Furuta, T. 256[95]

G

Gabor, B. 22[54], 129[128], 130[128], 611[97]
Gagné, M. R. 23[62], 23[63], 24[62], 321[121], 321[122], 322[121], 322[122], 323[121], 323[122], 327[122], 554[47]
Gagosz, F. 188[13], 557[74], 557[79], 606[50]
Gainotti, G. 620[167]
Galan, B. R. 71[30], 74[30]
Galindo, J. 557[62]
Galliford, C. V. 367[40]
Gallimore, A. R. 244[11]
Galvin, J. M. 244[20]
Gamber, G. G. 612[105]
Gandon, V. 549[39], 550[39]
Gansäuer, A. 27[79], 217[23], 218[51]
Gao, D. 622[188]
Gao, L. 355[7], 356[7], 419[123], 420[123]
Gao, P. 620[158], 622[177]
Gao, S. 227[79]
Gao, S.-H. 282[35], 283[35], 618[145]
Gao, Y. 500[131], 500[132]
Garayalde, D. 555[48], 566[95], 566[96], 567[96], 567[97], 567[98], 568[96], 569[96], 570[98], 571[98], 572[98]
Garber, J. A. O. 198[50]
García, A. 625[211]
Garcia, J. 35[135], 36[135]
Garcia, P. 549[37]
García-Díaz, D. 138[42], 138[43], 142[43]
García-López, M. T. 583[11], 583[12], 584[11]
García-Ocaña, M. 218[45]
García-Ruiz, J. M. 27[81], 222[68]
Gardner, J. M. 626[213]
Gareau, Y. 75[53]
Gaspar, B. 8[12], 9[12], 28[93], 28[101], 28[106], 28[107], 28[108], 28[109], 29[93], 453[33], 454[33], 469[33]
Gasperi, T. 10[14]
Gasteiger, J. 624[200]
Gatzweiler, S. 565[93]
Gaudin, J. 32[125], 35[125]
Gaunt, M. J. 345[151]
Gauthier, J. 622[181]
Gaydou, M. 332[133], 543[32]
Gayen, K. S. 591[24], 592[24]
Gee, A. D. 561[86], 562[86]
Geib, S. J. 191[18], 192[18], 622[184]
Geis, O. 601[12]
Geissler, H. 414[116]
Geissman, T. A. 316[105]
Gelin, C. F. 135[20], 142[20]
Gellman, S. H. 527[49]

Genêt, J.-P. 535[6], 564[6], 564[89]
Genin, E. 564[89]
Genovese, S. 590[23], 591[23]
Genti-Raimondi, S. 624[203]
Geoffroy, P. 75[51], 75[52]
Georgiadis, G. M. 399[93]
Georgian, V. 259[102]
Gericke, K. M. 67[5], 187[4], 243[4], 449[1]
Gerkin, R. M. 282[30]
Gesmundo, N. J. 181[80], 182[80], 183[80]
Gevorgyan, V. 394[89], 513[19], 513[20], 513[25], 514[25], 516[25], 557[67]
Gewert, J. A. 414[116]
Ghannam, A. 387[80], 387[81], 388[80], 388[81]
Ghorai, M. K. 593[26], 593[27], 594[26]
Ghorai, S. 547[36], 548[36], 549[36]
Ghorai, S. K. 259[103]
Ghorbanpour, K. 401[100]
Ghoroku, K. 427[131], 427[132], 428[132]
Ghosh, S. 591[24], 592[24]
Giacobini, E. 614[118]
Gibson, D. H. 177[61]
Gibson, S. E. 601[5], 601[6]
Gibson, V. 412[113], 412[114], 413[113], 413[114]
Giese, B. 217[4]
Giessert, A. J. 71[30], 71[31], 72[31], 74[30], 79[67]
Giessert, R. E. 479[94]
Gil, R. R. 624[203]
Gilchrist, T. L. 512[17]
Giles, M. 258[96]
Gilliom, L. R. 139[45], 140[45]
Gilman, A. 620[161]
Gilmore, D. 157[8], 158[8]
Gilmore, K. 217[15]
Giordano, O. S. 624[206]
Giuffredi, G. T. 561[86], 562[86]
Gnanadesikan, V. 579[5], 579[6], 580[5], 581[5]
Göbel, T. 217[4]
Gockel, B. 461[57], 471[57]
Goddard, J.-P. 217[28], 217[29]
Goddard, R. 549[40], 550[40]
Goddard, W. A., III 555[54], 555[55], 556[54]
Goeke, A. 566[96], 567[96], 567[97], 568[96], 569[96]
Goess, B. C. 93[77], 98[77], 99[77]
Gogoi, P. 459[49], 460[49], 471[49]
Goh, S. S. 461[57], 471[57]
Goldberg, A. F. G. 493[114], 494[114], 495[114]
Golden, J. E. 341[144], 341[147], 342[144], 342[147], 343[147]
Gollnick, K. 179[68], 180[71], 180[74], 181[74]
Gómez-Bengoa, E. 567[97]
Gong, J.-X. 500[136], 501[136], 505[136], 613[115]

González, I. C. 249[67]
González, J. M. 594[28], 595[28]
González, M. J. 520[34], 521[34]
González-Gómez, J. C. 28[104]
González-Vera, J. A. 583[11], 583[12], 584[11]
Goodman, R. S. 620[161]
Gopakumar, G. 549[40], 550[40]
Gordon, A. R. 602[22]
Gorin, D. J. 22[58], 535[2], 542[30], 557[68], 557[69], 557[71]
Goto, T. 28[28]
Gottwald, T. 217[36], 230[36]
Gouverneur, V. 561[86], 562[86]
Govindasamy, M. 313[95], 313[96], 313[100]
Graham, T. H. 198[46]
Graham, T. J. A. 93[77], 98[77], 99[77]
Gränicher, C. 516[31], 516[32], 517[31], 517[32]
Gravestock, M. B. 243[7]
Gravier-Pelletier, C. 411[110], 411[111], 411[112], 412[112]
Gray, E. E. 93[77], 98[77], 99[77]
Greaney, M. F. 626[221]
Greb, M. 246[42], 248[42]
Greenaway, R. L. 461[58], 462[58], 471[58]
Greene, A. E. 603[35], 604[35]
Greene, B. 620[170], 626[170]
Greenmald, B. S. 620[164], 620[165]
Grese, T. A. 279[27], 280[27]
Gress, T. 557[70]
Greszler, S. N. 363[29], 366[39], 367[39], 368[44], 369[47], 370[47]
Griesbeck, A. G. 181[77]
Griffith, A. K. 150[55], 151[55], 152[55]
Grigg, R. 449[10], 449[11], 449[12], 449[13], 475[84]
Grignon, J. 231[84]
Grillet, F. 624[198], 624[210], 625[198]
Grondal, C. 345[153], 511[3]
Grossman, R. B. 602[15]
Groves, J. T. 219[59]
Grubbs, R. H. 4[6], 68[22], 68[23], 69[23], 72[35], 136[26], 136[27], 136[28], 136[29], 136[30], 136[31], 136[33], 136[34], 136[35], 137[33], 137[38], 139[29], 139[30], 139[31], 139[33], 139[34], 139[35], 139[44], 139[45], 139[46], 140[44], 140[45], 140[46], 143[44], 148[38], 148[52]
Gu, P. 273[12], 274[12], 275[12], 275[13], 275[14], 276[12], 276[13]
Gu, Y. 500[126], 501[126], 503[126]
Gu, Z. 478[91], 488[91]
Guerreiro, E. 624[206]
Gui, J. 28[113]
Guiéritte-Voegelein, F. 297[59]
Guiry, P. J. 473[79]
Gunaydin, H. 54[23], 55[23]
Gunjigake, Y. 232[88]
Guo, L.-N. 496[116], 498[116]
Guo, S. 182[81], 183[81]

Author Index

Guo, X. 328[128], 329[128], 330[128], 511[8]
Gupta, R. 17[41]
Gupta, S. K. 512[18]
Guth, O. 611[96], 611[97]
Gutierrez, O. 415[118], 416[118]
Guzmán, J. 624[206]
Guzman-Perez, A. 610[85]
Gymer, G. E. 512[17]

H

Haeuseler, A. 201[54], 204[54], 206[54]
Häffner, T. 552[45], 553[45]
Haganuma, A. 606[51]
Hagishita, S. 617[129], 617[130]
Hagiwara, H. 624[205]
Haïdour, A. 27[80], 28[80], 218[46]
Halcomb, R. L. 244[32]
Halder, S. 535[7]
Hale, K. J. 438[145], 439[145]
Hall, D. G. 261[112]
Hall, N. 67[4]
Haller, F. 444[154]
Halloway, A. 217[2]
Halonen, N. 608[71]
Halpern, J. 28[84], 28[85]
Hamada, M. 402[101], 403[101]
Hamada, Y. 348[159], 350[159]
Hamaker, L. K. 608[72]
Hamilton, K. 53[22]
Hammerschmidt, F. 444[155]
Hammond, G. B. 535[20], 536[20], 537[20], 555[53]
Han, A. 30[117]
Han, J. 597[31]
Han, X. 465[70], 466[70], 472[70]
Han, Y. F. 614[117]
Hanaoka, M. 249[49], 249[50], 249[51], 249[52], 622[186]
Handa, S. 536[21], 537[21]
Hanhinen, P. 608[71]
Hanna, I. 99[90], 99[91], 100[90], 100[91]
Hannen, P. 557[63]
Hansen, E. C. 70[28], 78[28], 78[64], 112[107]
Hansen, J. 512[15]
Hara, H. 608[68], 623[194]
Harada, K. 492[112], 495[112], 615[126]
Haraguchi, H. 373[53]
Harayama, T. 606[46], 606[47], 606[48], 606[49]
Hardcastle, K. I. 3[5], 4[5], 48[15], 48[16], 48[17], 49[19], 50[15], 50[16], 50[17], 50[19], 51[19], 58[33], 58[35], 59[33], 59[35], 60[33], 249[53], 249[54], 294[55]
Hardin, A. R. 328[128], 329[128], 330[128]
Harding, K. E. 289[44], 289[45]
Harmon, R. E. 512[18]
Harms, G. 413[115], 414[115]
Harms, K. 444[154]
Harper, J. K. 618[141]
Harrak, Y. 188[16]
Harring, L. S. 461[53]

Harrison, S. J. 626[221]
Harrison, T. 277[24], 280[28]
Hart, D. J. 217[1]
Hartel, A. M. 357[10], 357[11], 357[12], 358[11], 358[13]
Hartley, R. C. 135[1], 135[3]
Hartshorn, M. P. 245[34]
Hartung, J. 29[115], 30[117], 217[36], 230[36], 246[42], 248[42]
Harvey, J. N. 452[32]
Harvey, M. E. 527[56]
Hasaba, S. 620[153]
Hasegawa, M. 306[83]
Hashida, I. 179[69], 180[69], 181[76]
Hashimoto, A. 365[37]
Hashimoto, S. 516[28]
Hashizume, D. 618[140]
Hashmi, A. S. K. 321[116], 321[119], 535[3], 535[4], 535[19], 545[34], 545[35], 546[34], 547[34], 548[34], 552[45], 553[45], 565[93]
Hasler, C. M. 626[216]
Hasserodt, J. 24[69]
Hatae, N. 457[43], 458[43], 470[43]
Hatakeyama, N. 352[27]
Hatakeyama, S. 109[105]
Hatsui, T. 260[107], 261[107]
Hattori, A. 255[90], 256[90]
Hattori, N. 302[76]
Hauptmann, H. 75[55]
Hawkins, A. 481[97], 481[99], 482[97], 489[97]
Hawkins, B. C. 263[119], 264[119]
Hay, E. B. 191[18], 192[18]
Hayashi, N. 55[24], 55[25], 56[24], 56[25]
Hayashi, Y. 625[212]
Haynes, R. K. 161[20], 180[70]
Heffron, T. P. 51[20], 52[20], 249[58], 251[76], 252[76]
Hegedus, L. S. 28[92]
Heiba, E. I. 26[74]
Heinemann, C. 31[118]
Heinrich, M. R. 217[10]
Helliwell, M. 196[32], 196[33], 197[32], 197[33]
Hellkamp, S. 491[111], 492[111], 495[111]
Helm, M. D. 196[32], 196[33], 197[32], 197[33], 197[34], 197[35]
Helmreich, W. 168[41]
Henbest, H. B. 244[26]
Heravi, M. H. 459[50]
Herr, R. J. 479[94]
Herranz, R. 583[11], 583[12], 584[11]
Herrebout, W. 481[99]
Herrero-Gómez, E. 75[46], 332[134], 564[90]
Herrinton, P. M. 277[22], 277[24], 278[22], 281[22]
Herzon, S. B. 28[114], 166[38]
Hesse, O. 620[160]
Heusser, H. 15[19]
Hewitt, J. F. M. 499[125], 500[125], 503[125]
Hey, D. H. 234[98]

Hibino, S. 457[43], 458[43], 470[43]
Hicks, F. A. 601[14], 619[147], 619[148]
Hiemstra, H. 259[104]
Hierold, J. 193[21], 502[138], 503[138], 505[138]
Higashino, M. 391[86], 392[86]
Higashiya, S. 360[22]
Highton, A. 188[12]
Higuchi, K. 285[40], 285[41], 286[40], 286[41]
Hili, R. 527[44]
Hiller, W. 437[144], 438[144]
Hilt, G. 217[27]
Hinrichs, J. 313[94], 313[99], 314[94]
Hioki, H. 492[112], 495[112], 615[126]
Hirama, M. 135[8], 135[11], 135[12], 142[8], 142[11], 143[8], 143[11], 143[12]
Hirao, H. 564[92], 565[92]
Hirao, T. 235[100]
Hirasawa, A. 540[26]
Hirner, S. 263[119], 264[119]
Hiroi, M. 578[4]
Hirose, T. 623[192]
Hirst, G. C. 277[25], 278[25], 279[25], 280[29], 281[25], 281[29], 607[63]
Hirt, E. 341[145], 342[145], 343[145]
Hite, G. 261[110], 262[110]
Hnawia, E. 297[59]
Ho, S. 136[29], 139[29]
Hofferberth, J. E. 263[118]
Hoffmann, H. M. R. 227[77], 227[78], 228[77], 624[197]
Hoffmann, N. 217[17]
Hofmann, A. 624[200]
Hohmann, J. 626[215]
Hojo, M. 500[129], 500[133]
Holland, H. L. 607[56], 607[58], 607[59]
Holloway, G. 263[119], 264[119]
Holmes, P. E. 249[56]
Holtan, R. C. 358[16], 358[17], 359[16], 359[17]
Holton, O. T. 461[58], 462[58], 471[58]
Honda, M. 359[18]
Honda, T. 108[103], 109[103]
Hong, B. 148[54], 149[54]
Hong, J. 198[41], 198[43]
Hong, Y.-T. 619[152], 620[154]
Hoos, R. 169[45], 170[45], 171[45]
Hopkins, M. H. 277[21], 277[22], 277[23], 277[24], 278[22], 281[22]
Hopkinson, M. N. 561[86], 562[86]
Hori, Y. 75[56], 78[56]
Horikawa, Y. 142[48]
Horiuchi, Y. 579[5], 579[6], 580[5], 581[5]
Horn, L. L. 527[43]
Horneff, T. 513[20]
Horning, B. D. 338[143], 339[143]
Horowitz, R. M. 316[105]
Horvath, T. B. 620[164], 620[165]
Hoshino, O. 608[67], 608[68]
Hoshino, T. 24[66], 614[119]
Hosoi, Y. 76[59]
Hotta, K. 244[12]

676   Author Index

Hou, Z. 540[26]
Houghton, P. J. 608[73], 608[74]
Houk, K. N. 54[23], 55[23], 158[13]
Hours, A. 549[39], 550[39]
Hoveyda, A. H. 69[25], 69[26], 78[25], 78[26]
Howard, B. E. 135[10], 142[10], 143[10], 144[10]
Howard, P. N. 277[25], 278[25], 279[25], 281[25]
Howard, T. R. 136[34], 139[34]
Hoye, T. R. 45[10], 68[15], 68[16], 68[17], 71[15], 71[16], 72[36], 94[79], 94[80], 124[36], 250[75]
Hu, S. 176[60], 177[60]
Hu, W. 511[8]
Hu, X. 4[7], 5[7], 166[39], 167[39]
Hu, X.-D. 282[32]
Hu, X. E. 246[36]
Hua, Z. H. 615[123]
Huang, A. X. 290[49], 291[49], 296[49]
Huang, C.-F. 624[198], 625[198]
Huang, J. 15[34], 619[146]
Huang, J.-M. 615[121], 621[172], 621[173]
Huang, J.-S. 527[53]
Huang, K. 482[100], 483[100], 490[100]
Huang, S.-X. 612[107], 612[108]
Huang, X. 559[82], 560[82], 567[97]
Huang, Y. 389[84], 390[84]
Huang, Z. 135[20], 142[20]
Huber, P. 28[104]
Hudrlik, A. M. 249[56]
Hudrlik, P. F. 249[56]
Hughes, N. L. 233[95]
Hummersone, M. G. 438[145], 439[145]
Humski, K. 15[39]
Hund, D. 483[101], 484[101], 490[101]
Hung, C.-I. 618[144]
Hutchinson, K. D. 277[24], 279[27], 280[27]
Hutchison, T. L. 527[43]
Hwang, C.-K. 248[47], 248[48], 249[47], 249[48], 250[47]
Hwang, F. W. 513[19]
Hwang, I.-C. 564[91]

I

Ibrahem, I. 347[157], 348[157]
Ichimura, A. S. 75[54]
Ichinose, N. 179[66]
Ihara, M. 324[124], 325[124]
Ihee, H. 74[42]
Ihle, A. 431[136], 432[136]
Iijima, Y. 494[115], 496[115]
Iimori, T. 44[9], 247[44]
Iio, K. 285[40], 285[41], 286[40], 286[41]
Ijichi, K. 626[214]
Ikeda, H. 181[79]
Ikeda, M. 237[104]
Ikeda, T. 457[42], 457[43], 458[42], 458[43], 470[43]
Ikeda, Y. 249[49], 249[50], 249[51]

Imahori, T. 94[81], 94[82], 95[82]
Imai, A. 492[112], 495[112], 615[126]
Inagaki, F. 611[104], 612[104]
Inatomi, S. 606[51]
Inauen, R. 10[15]
Ingate, S. T. 601[13]
Ingold, K. U. 233[90]
Inoki, S. 163[26], 163[27]
Inoue, A. 535[17]
Inoue, H. 75[43]
Inoue, M. 135[11], 135[12], 142[11], 143[11], 143[12], 225[73], 226[73], 227[73], 237[103], 238[103], 250[72], 303[78], 304[78]
Inoue, S. 28[96]
Inoue, Y. 586[15]
Inubushi, Y. 606[46], 606[47], 606[48], 606[49]
Inuki, S. 480[95], 480[96], 489[96], 496[119], 497[119], 497[120], 498[119], 498[120]
Ironmonger, A. 481[97], 482[97], 489[97]
Isaac, R. 157[7], 157[8], 158[7], 158[8], 160[7]
Isaacs, J. T. 624[199]
Isaacs, R. C. A. 219[63]
Isayama, S. 163[25], 163[26], 163[27]
Ischay, M. A. 182[81], 183[81], 211[67]
Ishibashi, H. 188[6], 188[7], 188[8], 188[9], 318[112], 318[114], 319[112], 320[114]
Ishibashi, M. 611[91]
Ishida, A. 180[73]
Ishida, R. 602[29]
Ishihara, J. 109[105]
Ishihara, K. 19[47], 20[50], 21[50], 318[111], 318[112], 318[113], 318[114], 319[111], 319[112], 320[111], 320[114]
Ishii, Y. 404[102]
Ishikura, M. 457[41], 457[42], 457[43], 457[44], 458[42], 458[43], 470[43]
Ishita, A. 188[6]
Ishiuchi, K. 614[119]
Ishizaki, M. 608[67], 608[68]
Ismail, M. A. H. 324[124], 325[124]
Isoe, S. 174[52], 175[52]
Isogai, Y. 535[13]
Itami, K. 602[27]
Ito, K. 604[40], 604[41]
Ito, Y. 235[100], 602[27]
Itoh, M. 399[95]
Iwahara, Y. 175[58]
Iwai, T. 45[11], 46[11]
Iwamoto, C. 617[129], 617[130]
Iwamura, H. 75[54]
Iwasaki, K. 28[83], 29[83], 30[83], 166[37]
Iwata, A. 480[96], 489[96]
Iwata, T. 302[76]
Iyer, K. 135[17], 135[23], 136[17], 144[17], 145[17], 146[17], 147[23], 148[17]

J

Jackenkroll, S. 502[138], 503[138], 505[138]
Jacobine, A. M. 164[33]
Jacobs, J. 538[23]
Jacobsen, E. J. 277[19], 277[20], 316[104]
Jacobsen, E. N. 25[70], 244[17], 244[18], 244[19], 244[20], 601[114]
Jacobsen, J. E. 316[109]
Jacobsen, M. F. 452[31], 453[31], 469[31]
Jacquier, Y. 527[52]
Jagadeesh, S. G. 284[37], 285[37]
Jakobi, U. 414[116]
Jakoby, V. 218[51]
Jakubec, P. 481[97], 482[97], 489[97], 611[95]
Jakupovic, J. 623[191]
James, K. J. 313[92], 313[93]
Jamison, T. F. 43[1], 47[1], 49[18], 51[18], 51[20], 52[20], 52[21], 53[18], 53[21], 54[18], 54[21], 56[26], 56[27], 56[28], 56[29], 56[30], 56[31], 57[26], 57[29], 57[30], 57[32], 58[26], 58[32], 245[35], 249[58], 251[76], 252[76], 295[56], 296[56], 602[33], 602[34], 603[33], 603[34]
Jamora, C. 618[144]
Janda, K. D. 24[69]
Jang, H. 198[43]
Jang, H.-Y. 28[100]
Jang, J.-H. 618[140]
Jang, J.-P. 618[140]
Jansone-Popova, S. 522[36], 525[36]
Jasch, H. 217[10]
Jasinski, J. 555[53]
Jatkowitz, S. 620[168]
Jato, J. 297[58]
Jáuregui, E. A. 624[206]
Jautelat, R. 1[1]
Jeanty, M. 345[153], 511[3]
Jee, N. 522[41]
Jeger, O. 15[20]
Jeker, O. F. 24[65]
Jenkins, S. A. 45[10]
Jenson, C. 14[11]
Jeon, K. O. 191[18], 192[18]
Jeon, Y. T. 626[221]
Jeong, N. 601[11], 602[21], 602[28]
Ji, Y. 541[27]
Jia, Y. X. 282[34], 283[34], 284[34], 284[36], 286[34]
Jiang, C. 610[84]
Jiang, H. 555[56]
Jiang, W. 347[156], 348[156]
Jiao, J. 202[56]
Jiao, L. 325[125], 326[125], 327[125], 328[125]
Jiao, Z. 500[132]
Jijy, E. 598[34]
Jiménez-Núñez, E. 75[46], 321[117], 332[130], 332[131], 332[132], 333[130], 333[131], 333[132], 334[130], 334[131],

335[131], 337[132], 535[5], 542[29], 542[31], 543[31], 544[31]
Jin, J. 249[65]
Jing, P. 440[148], 441[148]
Jobst, J. 620[160]
Johannson, M. J. 22[58]
Johansson, C. C. C. 345[151]
Johansson, M. J. 557[69]
Johns, C. A. 620[164], 620[165]
Johns, J. 626[213]
Johnson, H. W. B. 135[10], 135[15], 135[16], 137[15], 137[16], 142[10], 143[10], 144[10], 147[15], 147[16]
Johnson, J. 512[18]
Johnson, J. S. 363[29], 366[39], 367[39], 368[41], 368[42], 368[43], 368[44], 368[45], 369[46], 369[47], 370[47], 370[48], 370[49], 371[49], 377[60], 377[61], 377[62], 377[63], 378[63], 379[66], 380[66]
Johnson, R. A. 244[15]
Johnson, R. P. 114[112]
Johnson, T. 297[57]
Johnson, T. O., Jr. 280[29], 281[29], 607[63]
Johnson, W. S. 14[4], 15[4], 15[22], 24[67], 243[7], 292[53], 292[54], 293[53], 293[54], 294[54], 307[86]
Johnston, J. N. 14[5], 258[100]
Johnstone, C. 602[22]
Jones, A. S. 498[122]
Jones, B. 626[213]
Jones, B. T. 207[59]
Jones, C. M. 198[46]
Jones, S. B. 338[141], 338[142], 339[141], 340[142], 343[142]
Jones, T. 24[69]
Jongbloed, L. 545[35]
Jørgensen, K. A. 349[160], 349[162]
Jørgensen, L. 626[222], 627[222]
Jorgensen, W. L. 14[11]
Josien, H. 236[101]
Jouikov, V. 588[19], 589[19]
Jui, N. T. 198[46], 198[49], 198[50]
Julia, M. 218[54], 218[55], 218[57], 218[58]
Jung, A. 400[99], 401[99]
Jung, S. K. 117[116], 117[117], 118[116], 118[117]
Justicia, J. 27[80], 27[81], 28[80], 218[44], 218[46], 218[51], 222[68]

K

Kabawawa, Y. 256[95]
Kablaoui, N. M. 619[147], 619[148]
Kache, R. 356[9]
Kadota, I. 142[47]
Kadowaki, C. 142[47]
Kahle, K. 476[86]
Kajikawa, S. 175[59]
Kakeya, H. 255[90], 256[90]
Kakimoto, M.-a. 277[18], 316[103]
Kakiuchi, T. 136[37], 137[37]
Kaliappan, K. P. 125[122], 125[123], 125[124], 126[122], 126[123], 126[124]

Kamata, M. 181[78]
Kambara, H. 45[12], 46[12]
Kamigauchi, T. 302[76], 302[77]
Kamimura, A. 232[88]
Kamitamari, M. 244[31]
Kamitani, A. 619[151]
Kamiyama, N. 179[65], 179[66]
Kan, S. B. J. 456[40], 457[40], 470[40]
Kaneda, K. 108[103], 109[103]
Kaneko, C. 602[23]
Kang, B. C. 395[91]
Kang, H.-M. 624[202]
Kang, M. 26[75]
Kang, S.-K. 619[152]
Kapeller, D. C. 444[155]
Karikomi, M. 256[54]
Karisch, A. 467[77], 468[77], 473[77]
Karni, M. 373[54], 374[54]
Kasahara, T. 535[18]
Katagiri, K. 579[6]
Katan, E. 373[54], 374[54]
Kataoka, T. 624[205]
Kataoka, Y. 136[37], 137[37]
Kates, S. A. 26[76], 27[76], 221[66], 221[67]
Kato, K. 163[26], 163[27]
Kato, N. 244[31]
Kato, R. 306[83]
Kato, T. 256[95]
Katoh, S.-i. 237[103], 238[103]
Katoh, T. 303[78], 303[79], 303[80], 304[78], 304[79], 304[80], 305[79], 305[80]
Katsuki, T. 15[31], 43[5], 244[14], 244[16]
Katsumura, S. 454[34], 470[34]
Katsuura, K. 626[214]
Katz, L. 218[54], 218[57], 218[58]
Katz, T. J. 68[13], 68[14], 71[13], 71[14]
Kauffman, F. C. 620[163]
Kawabata, N. 607[54], 607[55]
Kawahata, M. 365[37], 365[38], 366[38], 372[52], 383[71], 383[72], 385[77], 386[77]
Kawakami, J.-i. 596[30], 597[30]
Kawamura, S.-i. 500[129]
Kawamura, Y. 302[76]
Kawanishi, E. 383[69], 383[70]
Kawasaki, N. 611[90]
Keawpradub, N. 608[73], 608[74]
Keay, B. A. 473[78], 474[78], 487[78]
Keck, G. E. 135[22], 147[22], 231[85]
Keimatsu, I. 604[40], 604[41]
Keister, J. B. 71[30], 71[34], 74[30]
Kelly, W. H., IV 360[22]
Kelner, M. J. 604[38], 604[39]
Kent, J. L. 601[7]
Kerekes, A. D. 622[184], 622[187]
Kerr, M. A. 611[94], 611[101]
Kerr, W. J. 602[22]
Khamarui, S. 591[24], 592[24]
Khan, F. A. 188[15]
Khand, I. U. 601[1], 601[2], 601[3]
Kharasch, M. S. 168[42]
Khoo, M. L. 263[119], 264[119]

Kigoshi, H. 613[110], 626[110]
Kikuchi, H. 606[51]
Kikuchi, T. 303[80], 304[80], 305[80]
Kim, B. G. 79[66]
Kim, B. Y. 618[140]
Kim, C. 464[67], 500[67], 500[126], 501[126], 503[126]
Kim, C. H. 117[116], 117[117], 118[116], 118[117]
Kim, C.-Y. 244[12]
Kim, D. 122[120], 123[120]
Kim, D.-S. 355[6], 356[6], 442[150], 442[151], 442[152], 442[153], 443[151], 443[153], 444[151], 444[152]
Kim, H. 122[120], 123[120], 198[44]
Kim, H.-K. 624[202]
Kim, H.-S. 158[9], 164[9], 165[9], 166[9], 181[78]
Kim, I. J. 249[52]
Kim, J. 220[64], 261[116]
Kim, J.-H. 624[202]
Kim, K. H. 74[42]
Kim, K.-J. 619[152]
Kim, M. 75[48], 76[60], 78[60], 78[61], 78[62], 78[63], 80[60], 81[60], 82[60], 83[60], 84[70]
Kim, M. B. 60[37], 62[37]
Kim, S. 122[120], 123[120]
Kim, S.-G. 345[155], 346[155], 347[155], 350[155], 464[69], 465[69], 472[69]
Kim, S.-H. 68[22]
Kim, S.-O. 618[140]
Kim, W.-S. 355[6], 356[6], 423[128], 424[128], 425[129], 426[129], 430[134]
Kim, Y. J. 105[98]
Kimura, T. 399[94]
Kinebuchi, M. 611[104], 612[104]
King, S. M. 28[114], 166[38]
Kingston, D. G. I. 297[60]
Kinnel, R. B. 307[86]
Kinoshita, A. 68[21], 72[21]
Kinoshita, T. 45[11], 46[11]
Kirby, G. C. 608[73]
Kirchen, R. P. 252[77]
Kirsch, S. F. 321[120], 335[137], 557[66]
Kirschning, A. 249[57], 376[59], 404[104], 404[105], 405[104], 405[105], 406[104], 406[105], 407[105], 408[105], 417[120]
Kishimoto, T. 402[101], 403[101]
Kita, Y. 97[88], 97[89], 98[88], 98[89], 285[40], 285[41], 286[40], 286[41]
Kitagaki, S. 285[40], 285[41], 286[40], 286[41], 625[212]
Kitagawa, H. 577[3]
Kitajima, M. 614[120]
Kitamura, T. 73[39], 73[40], 97[88], 97[89], 98[88], 98[89]
Kitson, R. R. A. 624[196]
Kjærsgaard, A. 349[162]
Klahn, P. 335[137]
Klebe, J. 75[49]
Klotz, P. 455[35], 455[36]
Knight, S. D. 316[106]
Knochel, P. 217[38], 218[52], 269[3]

Knowles, R. R. 25[70]
Knox, G. R. 601[1], 601[2]
Ko, H. M. 91[75], 92[75]
Ko, S.-B. 236[101]
Kobayakawa, Y. 306[83]
Kobayashi, J. 88[73], 89[73], 611[90], 611[91], 614[119]
Kobayashi, S. 75[56], 78[56], 193[24], 224[72], 577[3]
Kobayashi, T. 454[34], 470[34], 500[129], 602[18], 622[185]
Kobs, H.-D. 15[36]
Koch, O. 400[99], 401[99]
Koch, W. 444[154]
Kočovský, P. 463[61]
Kodet, J. G. 63[40], 64[40], 298[64], 299[64]
Koehl, W. J., Jr. 26[74]
Koelle, G. B. 620[161]
Koert, U. 246[38], 247[43]
Koftis, T. V. 313[95], 313[96], 313[97], 313[98], 313[99], 313[100], 313[101], 314[101]
Koga, N. 75[54]
Koga, Y. 602[18], 622[179], 622[185]
Kogure, N. 614[120]
Koh, J. H. 23[62], 24[62]
Koh, M. J. 564[92], 565[92]
Kohler, R. E. 15[14]
Koiwai, A. 420[124], 421[124]
Koizumi, T. 362[25], 362[26], 363[25], 363[31]
Kojima, M. 180[73]
Kolesnikova, M. D. 21[51]
Komatsu, K.-I. 181[78]
Konaklieva, M. I. 624[207]
Kondo, T. 602[17]
Kondo, Y. 372[52]
Kone, M. 253[82]
Kong, J. 198[48]
Kong, L.-L. 618[145]
Konishi, K. 617[128]
Konkol, L. C. 204[58], 205[58], 206[58], 207[58]
Konno, A. 181[75], 182[75]
Konno, K. 626[214]
Konstantinovski, L. 169[45], 170[45], 171[45]
Kopping, B. 217[4]
Korb, M. 431[136], 432[136]
Koreeda, M. 263[117]
Korkowski, P. F. 68[15], 71[15]
Korous, A. A. 358[13]
Korp, J. D. 175[55], 176[55], 177[55]
Korshin, E. E. 157[3], 169[44], 169[45], 170[44], 170[45], 171[45]
Kosaka, M. 135[8], 135[11], 142[8], 142[11], 143[8], 143[11]
Koshar, R. J. 621[176]
Kostenko, A. 373[54], 374[54]
Kotha, S. 535[7]
Kotian, P. 527[43]
Kotomori, Y. 383[72]
Kouzuki, S. 302[76], 302[77]
Koyanagi, J. 434[140], 435[140]

Koyanagi, T. 76[59]
Kozaka, T. 608[70]
Kozmin, S. A. 22[56], 72[37], 93[78], 94[78]
Kraebel, C. M. 171[51]
Krafft, M. E. 601[8]
Kraft, S. 70[27]
Kraus, G. A. 220[64]
Krause, H. 557[70]
Krautwald, S. 24[64]
Kravina, A. G. 24[65]
Krinsky, J. L. 567[99]
Krische, M. J. 28[100]
Krishna, P. R. 96[85], 97[85]
Kronja, O. 15[39]
Krüger, K. 567[98], 570[98], 571[98], 572[98]
Kubo, M. 492[112], 495[112], 615[126]
Kubota, T. 88[73], 89[73], 614[119]
Kucera, D. J. 32[122], 32[123], 33[122], 33[123], 449[22]
Kui, L. 360[19]
Kulicke, K. J. 217[4]
Kumabe, R. 175[56], 175[57]
Kumar, A. 586[16], 587[16]
Kumar, V. S. 46[14], 47[14]
Kumari, K. 588[19], 589[19]
Kumazawa, K. 318[113]
Kummer, D. A. 113[109]
Kundu, K. 522[37], 522[38]
Kundu, N. 259[102]
Kunikawa, S. 158[9], 163[24], 164[9], 164[24], 165[9], 165[24], 166[9], 166[24]
Kunimoto, K.-K. 359[18]
Kuninobu, Y. 586[15]
Kure, T. 540[26]
Kurhade, S. E. 126[125], 127[125]
Kuriyama, Y. 181[76]
Kurosawa, K. 175[55], 175[57], 175[58], 175[59], 176[55], 177[55]
Kürti, L. 300[65], 302[75], 307[85], 316[85]
Kuttruff, C. A. 626[222], 627[222]
Kuwahara, S. 244[30]
Kuwajima, I. 626[219], 626[220]
Kwok, S. W. 513[21]
Kwon, B.-M. 624[202]
Kwon, H. 388[82], 389[82]
Kwon, H. K. 91[75], 92[75]
Kwong, F. Y. 234[97]
Kyne, S. H. 217[13]

L

Lackner, A. D. 555[54], 556[54], 556[57]
Lacoske, M. H. 615[125], 618[144]
Lacôte, E. 217[28], 217[29], 217[32]
Laforteza, B. N. 198[48]
Lai, Y.-C. 307[87], 308[87], 309[87], 310[87], 315[87]
Lam, W. W.-L. 577[3]
Lamaty, F. 32[127], 449[14], 449[15]
Lambert, T. H. 150[55], 151[55], 152[55]

Lan, Y.-T. 227[79]
Lang, H. 431[136], 432[136]
Lanter, J. C. 258[100]
Laplaza, C. E. 88[71]
LaPorte, M. G. 191[18], 192[18]
Larionov, O. 88[73], 89[73]
Larraufie, M.-H. 217[28], 217[29]
Latini, D. 201[52]
Lautens, M. 458[45], 458[46], 467[77], 468[77], 473[77], 476[90], 477[90], 486[106], 486[108], 488[90], 491[108]
Lauterbach, T. 332[132], 333[132], 337[132], 542[31], 543[31], 544[31], 564[90], 565[93]
Lauth, N. 617[137]
Lavallée, J. 37[137]
Laver, G. W. 527[43]
Lawrence, A. J. 358[14], 381[14]
Lawrence, J. F. 38[139]
Layton, M. E. 115[114], 115[115], 116[114], 314[102], 315[102]
Lazare, S. 253[81]
Le, T. T. T. 626[213]
Le Bideau, F. 387[79]
Le Brazidec, J.-Y. 135[8], 135[11], 142[8], 142[11], 143[8], 143[11]
Lecher, C. S. 422[127], 423[127]
Lee, B. Y. 602[21]
Lee, C. W. 91[75], 92[75], 136[28]
Lee, D. 70[28], 75[48], 76[60], 78[28], 78[60], 78[61], 78[62], 78[63], 78[64], 78[65], 80[60], 80[69], 81[60], 81[69], 82[60], 82[69], 83[60], 84[70], 90[74], 92[74], 105[98], 105[99], 112[99], 112[107], 114[111], 119[111], 120[118], 121[119], 122[119], 122[120], 123[118], 123[120], 395[92], 396[92], 397[92], 398[92], 440[147]
Lee, D. C. 449[18]
Lee, E. 91[75], 92[75], 117[116], 117[117], 118[116], 118[117]
Lee, E. C. Y. 198[49]
Lee, F.-K. 608[69]
Lee, H. 122[120], 123[120], 624[202]
Lee, H. Y. 219[63]
Lee, H.-Y. 74[42], 79[66]
Lee, J. 100[93], 101[93], 290[48], 291[48]
Lee, J. B. 136[34], 139[34]
Lee, J.-H. 620[154]
Lee, K.-H. 624[201]
Lee, M. S. 243[10]
Lee, S. 535[12], 535[18]
Lee, S. H. 602[21]
Lee, S.-H. 624[202]
Lee, S. J. 554[47]
Lee, Y.-J. 69[25], 78[25]
Lefebvre, O. 360[20], 360[21], 361[20]
Leggans, E. K. 28[111], 163[31]
Le Goffic, F. 218[54], 218[55], 218[58]
Lehmkuhl, H. 15[36]
Lei, C. 612[107]
Lei, J. 419[123], 420[123], 428[133], 429[133]
Lei, X. 114[112], 148[54], 149[54]
Lei, Y. 341[147], 342[147], 343[147]

Author Index

Leighton, J. L. 441[149], 466[73], 473[73]
Lelais, G. 345[150]
Le Merrer, Y. 411[110], 411[111], 411[112], 412[112]
Lemière, G. 549[39], 550[39], 557[73]
Lenarczyk, A. 626[213]
Lennox, A. J. J. 452[32]
Lense, S. 58[35], 59[35]
Lepore, S. D. 394[90], 395[90]
Lerner, R. A. 24[69]
Leseurre, L. 564[89]
Lettan, R. B., II 367[40]
Leung, J. C. T. 230[83], 231[83]
Leung, L. M. H. 412[113], 412[114], 413[113], 413[114]
Levin, J. 316[104]
Levin, M. D. 555[52]
Levine, B. H. 449[20]
Levinson, A. M. 218[52]
Levy, L. M. 491[111], 492[111], 495[111]
Lew, W. 500[126], 501[126], 503[126]
Li, A. 618[142]
Li, B. 307[87], 308[87], 309[87], 310[87], 315[87], 588[18]
Li, C. 217[21], 449[23], 450[23], 468[23], 486[105]
Li, C.-c. 330[129], 331[129], 332[129], 337[129]
Li, C.-C. 110[106], 617[139], 618[139], 619[146]
Li, C.-J. 535[8], 535[9], 538[24], 539[24]
Li, D. 564[92], 565[92]
Li, D. R. 282[33], 282[34], 283[33], 283[34], 284[34], 286[34]
Li, D.-R. 282[35], 283[35]
Li, G. 28[89], 29[89], 30[117], 187[5], 217[3], 478[91], 488[91], 557[80], 558[80], 559[80], 560[80]
Li, G.-G. 623[190]
Li, H. 148[54], 149[54], 347[156], 348[156], 426[130], 430[135], 431[135], 440[148], 441[148]
Li, J. 114[111], 119[111], 120[118], 123[118], 135[3], 378[64], 541[27]
Li, J. Y. 618[141]
Li, L. 360[19], 588[18]
Li, M. 481[98], 481[99]
Li, P.-F. 45[13], 46[13]
Li, R.-T. 612[107], 612[108]
Li, S. 28[97], 94[83]
Li, S.-W. 584[13], 585[13]
Li, X. 282[34], 282[35], 283[34], 283[35], 284[34], 286[34]
Li, Y. 313[95], 313[96], 313[97], 313[100], 500[136], 501[136], 505[136], 613[115]
Li, Y. M. 615[122], 615[123]
Li, Y.-Z. 617[139], 618[139]
Li, Z. 261[115], 262[115], 500[131], 500[132], 598[33], 611[103]
Li, Z.-W. 500[134], 501[134], 504[134]
Liang, B. 613[113], 613[114]
Liang, Y.-M. 496[116], 498[116]
Liao, C.-C. 607[66]
Liao, H.-H. 553[46], 554[46]

Liao, H.-Y. 550[41], 551[41]
Lichtor, P. A. 15[35]
Liébert, C. 335[137]
Liebl, M. 611[97]
Liegault, B. 475[83], 488[83]
Light, M. E. 412[114], 413[114]
Lim, D. S. 360[22]
Lim, Y.-H. 451[28], 469[28]
Lin, G. 617[139], 618 139]
Lin, G.-Y. 550[41], 551[41]
Lin, J. S. 291[51], 292[51], 432[137], 433[137]
Lin, L.-Z. 612[109]
Lin, Q. 434[138], 434[139], 434[141], 435[138], 435[139], 435[141]
Lin, S. 15[29], 15[32], 16[29], 25[70], 99[92], 291[50], 292[50]
Lin, T.-H. 527[43]
Lin, X. 355[7], 356[7], 419[123], 420[123]
Lin, Z. 419[123], 420 123]
Lin, Z.-W. 612[109]
Linclau, B. 412[113], 412[114], 413[113], 413[114]
Linderman, R. J. 387[80], 387[81], 388[80], 388[81]
Ling, T. 313[95], 313[96], 313[97], 313[98], 313[100]
Linghu, X. 369[46], 370[48], 377[62], 379[66], 380[66]
Liou, S.-Y. 449[21]
Lippstreu, J. J. 70[29], 71[29], 74[29], 77[29]
List, B. 345[149]
Lithgow, A. M. 246[37]
Liu, B. 374[55], 375[55]
Liu, D. R. 13[3], 19[5], 21[3], 23[3], 24[3]
Liu, D.-Z. 157[2]
Liu, F. 449[21]
Liu, G. 110[106], 463[60]
Liu, G.-S. 348[158]
Liu, H. 193[23], 395[92], 396[92], 397[92], 398[92], 436[105]
Liu, J. 597[31]
Liu, J.-K. 157[2]
Liu, J.-P. 612[107]
Liu, J.-T. 227[79]
Liu, K. 349[161], 350[161]
Liu, L. 135[7], 138[7], 541[27]
Liu, L.-P. 535[20], 536[20], 537[20]
Liu, Q. 617[139], 618[139]
Liu, R.-S. 542[28], 545[33], 550[41], 551[41], 553[46], 554[46]
Liu, S. 368[43]
Liu, X.-W. 535[10]
Liu, Y. 512[12], 522[40], 522[41]
Liu, Y.-F. 500[136], 501[136], 505[136], 613[115]
Liu, Z. 426[130]
Livinghouse, T. 23[61]
Lizos, D. E. 223[71], 224[71]
Llera, J. M. 287[43], 288[43]
Llewellyn, G. 14[9]
Lloyd-Jones, G. C. 71[32], 72[32], 73[32], 452[32], 555[49], 555[50], 556[58]

Lo, J. C. 28[112], 28[113]
Lobkovsky, E. 618[141]
Lodeiro, S. 21[51]
Loebach, J. L. 244[17]
Loertscher, B. M. 217[38]
Loganathan, V. 475[84]
Loh, T.-P. 60[36], 61[36], 62[36], 307[87], 308[87], 309[87], 310[87], 311[88], 311[89], 311[90], 311[91], 312[88], 313[91], 315[87]
Lohmann, J. J. 75[51]
Lohrenz, J. C. W. 444[154]
Loizidou, E. 313[99]
Loizidou, E. Z. 313[101], 314[101]
Lokare, K. S. 148[53]
Long, R. 619[146]
López, E. 520[34], 521[34]
López, S. 75[46], 332[134], 564[90]
López-Carrillo, V. 557[65]
López Pérez, J. L. 193[25]
Lord, K. E. 15[27]
Loreto, M. A. 376[59]
Lough, A. L. 596[30], 597[30]
Lounasmaa, M. 608[71]
Lourenço, A. M. 300[70]
Lovely, C. J. 609[77], 610[77]
Loyola, L. A. 607[56], 607[57], 607[58], 607[59]
Lu, B.-L. 623[189]
Lu, C.-D. 374[55], 375[55]
Lu, J. 316[108], 355[7], 356[7]
Lu, J.-L. 604[37]
Lu, X. 465[70], 466[70], 472[70]
Lu, Y. 612[107]
Lu, Z. 182[81], 183[81], 211[69], 212[69], 491[110]
Lu, Z.-Y. 618[142]
Luche, J. L. 607[53]
Luiken, S. 376[59]
Lumeras, W. 246[40]
Luo, F. T. 449[19]
Luo, G. 291[51], 292[51], 432[137], 433[137]
Luo, T. 330[129], 331[129], 332[129], 337[129], 500[135], 501[135], 504[135]
Luo, T. P. 613[112]
Lupton, D. W. 193[21], 229[81], 230[81]
Lusztyk, J. 233[90]
Lv, P. 482[100], 483[100], 490[100]
Lv, X. 193[23]
Lynch, V. 28[99], 28[100], 163[28], 166[28], 611[99]
Lyons, S. E. 195[31]

M

Ma, D. 334[136], 335[136]
Ma, H. 440[148], 441[148]
Ma, L. 502[138], 503[138], 505[138]
Ma, S. 193[22], 217[2], 449[21], 491[110]
Ma, X. 28[114], 166[38], 478[91], 488[91]
Ma, X.-Q. 614[116]
Ma, Y. 273[12], 274[12], 275[12], 276[12]
Maaß, C. 475[85], 476[85], 488[85]
McAlpine, D. 626[213]

McBriar, M. D. 434[139], 434[141], 435[139], 435[141]
McCarry, B. E. 243[7]
McCartney, D. 473[79]
McDonald, F. E. 3[5], 4[5], 48[15], 48[16], 48[17], 49[19], 50[15], 50[16], 50[17], 50[19], 51[19], 58[33], 58[34], 58[35], 59[33], 59[34], 59[35], 60[33], 249[53], 249[54], 294[55]
McDonald, I. M. 225[73], 226[73], 227[73]
MacDonald, J. R. 604[38]
McDonald, R. I. 463[60]
McDowell, P. 623[190]
McKerrall, S. J. 626[222], 627[222]
McKervey, M. A. 512[9], 512[11]
McKiernan, G. J. 135[1], 135[3]
MacLean, D. B. 607[56], 607[58], 607[59]
McMahon, T. 313[92], 313[93]
MacMillan, D. W. C. 31[119], 198[41], 198[42], 198[43], 198[44], 198[45], 198[46], 198[48], 198[49], 198[50], 198[51], 199[51], 200[51], 207[63], 338[140], 338[141], 338[142], 338[143], 339[141], 339[143], 340[142], 343[142], 345[150], 345[155], 346[155], 347[155], 350[155], 464[69], 465[69], 472[69]
McMorris, T. C. 604[38], 604[39]
McNally, A. 345[151]
MacNeil, S. 502[137], 505[137]
McPhail, A. T. 159[18], 159[19], 160[18]
MacPherson, D. T. 458[46]
Maddaford, S. P. 473[78], 474[78], 487[78]
Maddess, M. L. 188[16]
Madelaine, C. 177[63]
Madin, A. 316[109]
Madu, C. E. 609[77], 610[77]
Maes, B. U. W. 485[103]
Maestri, G. 217[28], 217[29], 549[38]
Maggio, A. 624[201]
Magnus, P. 28[98], 28[99], 163[28], 163[29], 163[30], 166[28], 166[29], 166[30], 258[96], 611[99]
Mahindaratne, M. P. D. 177[62], 178[62]
Mahy, J.-P. 527[50]
Mai, D. N. 479[93], 489[93]
Maier, M. E. 147[51], 439[146], 440[146]
Maifeld, S. V. 78[61], 78[65]
Maimone, T. J. 4[7], 5[7], 166[39], 167[39], 569[100]
Main, C. A. 135[3]
Mainetti, E. 188[16]
Mainolfi, N. 601[6]
Maio, W. A. 430[134]
Maiti, D. K. 591[24], 592[24]
Maity, P. 394[90], 395[90]
Majander, I. 345[152]
Majetich, G. 176[60], 177[60]
Majumdar, K. C. 449[5]
Majumder, U. 135[13], 135[15], 135[16], 137[15], 137[16], 142[13], 144[13], 147[15], 147[16]

Makino, K. 348[159], 350[159]
Makino, T. 362[24]
Mal, D. 259[103]
Malacria, M. 188[16], 217[28], 217[29], 217[32], 218[49], 225[74], 387[79], 549[37], 549[38], 549[39], 550[39], 557[73]
Male, L. 193[20]
Malerich, J. P. 450[24], 451[24]
Malhotra, D. 535[20], 536[20], 537[20]
Malinowski, J. T. 366[39], 367[39], 368[44]
Malisch, W. 201[53], 204[53], 206[53]
Malone, J. F. 449[13]
Mamane, V. 557[70]
Manabe, S. 219[63]
Mandai, T. 496[117]
Mandal, M. 99[92]
Mangold, G. L. 604[38]
Mankad, N. P. 555[55]
Manna, V. 620[169]
Manning, J. R. 512[16]
Mansuy, D. 527[50]
Mantell, G. J. 168[42]
Mantilli, L. 549[40], 550[40]
Mao, D. 586[17], 587[17], 588[17]
Marco-Contelles, J. 557[60], 601[13]
Marcos, I. S. 246[37], 246[40], 256[93]
Marek, I. 373[54], 374[54], 375[56], 375[57], 375[58], 376[58]
Margue, R. G. 71[32], 72[32], 73[32]
María, A. O. M. 624[206]
Marigo, M. 349[160], 349[162]
Marino, J. P., Jr. 449[23], 450[23], 468[23]
Marion, N. 557[64], 557[73], 557[75]
Marquez, R. 450[27], 452[27]
Marsch, M. 444[154]
Marshall, J. A. 252[78], 252[79], 253[78], 253[79], 254[78], 464[68]
Marshall, J. E. 71[34]
Marti, R. 10[16]
Martin, C. L. 316[107]
Martin, D. G. 617[134]
Martin, S. F. 97[87], 98[87], 113[109], 609[75]
Martin, V. I. 530[58], 531[58]
Martin, V. S. 244[16]
Martín-Lasanta, A. 27[81], 222[68]
Martucci, N. 620[169]
Maruyama, T. 620[170], 626[170]
Masamune, T. 250[70]
Mashuta, M. S. 535[20], 536[20], 537[20]
Mason, I. 247[45], 248[45]
Mastracchio, A. 198[41], 338[142], 340[142], 343[142]
Masu, H. 383[74], 383[76], 384[76], 385[76], 579[6]
Masuda, T. 75[54]
Masullo, C. 620[167]
Masuyama, A. 158[9], 163[24], 164[9], 164[24], 165[9], 165[24], 166[9], 166[24]
Mathé, A. A. 620[164], 620[165]

Mathew, J. 373[54], 374[54]
Maton, W. 411[110], 411[111], 411[112], 412[112]
Matsubara, H. 193[29], 193[30], 194[30], 195[30]
Matsuda, C. 617[128]
Matsuda, I. 404[102]
Matsuda, S. P. T. 21[51]
Matsumoto, T. 254[88], 383[70], 383[76], 384[76], 385[76], 602[29]
Matsunaga, S. 580[9], 581[9]
Matsuo, J.-i. 188[6]
Matsuya, Y. 420[124], 421[124]
Mattes, S. L. 180[72]
Mattson, A. E. 378[65], 379[65]
Mauleón, P. 557[72], 567[99]
Máximo, P. 300[70]
May, J. A. 522[36], 525[36]
Mayer, P. 8[12], 9[12], 453[33], 454[33], 469[33]
Mayer, T. 22[60]
Mayrargue, J. 158[17], 159[17]
Mazzanti, A. 551[42], 552[42]
Mecke, D. 617[136]
Meervelt, L. V. 538[23]
Meerwein, H. 14[6], 15[6], 21[6], 254[84]
Mehta, G. 607[60], 607[62], 610[78]
Meier, G. P. 277[18], 316[109]
Meier, M. 464[67], 500[67]
Meinhardt, D. 136[29], 139[29]
Melhado, A. D. 556[57]
Melnikson, L. T. 277[19]
Mendenhall, G. D. 158[11], 160[11]
Menéndez, J. C. 589[20], 589[21], 589[22], 590[20], 590[21], 590[22]
Meng, F.-K. 500[136], 501[136], 505[136], 613[115]
Meng, X. 270[8], 271[8]
Menggenbateer 535[13], 535[14], 535[17]
Mente, N. R. 63[38], 298[62], 299[62]
Meragelman, T. L. 624[203]
Merfort, I. 624[200]
Mergardt, B. 390[85], 391[85]
Merino, E. 51[20], 52[20], 251[76], 252[76]
Metz, P. 103[97], 104[97]
Meyer, C. 475[83], 488[83], 518[33], 519[33], 520[33], 521[33]
Meyer, F. E. 461[54]
Meyers, A. I. 607[61]
Mi, Y. 291[52], 292[52]
Michael, J. P. 300[68], 300[69], 300[71]
Michel, C. 404[104], 404[105], 405[104], 405[105], 406[104], 406[105], 407[105], 408[105], 408[107]
Michelet, V. 535[6], 564[6], 564[89]
Miege, F. 518[33], 519[33], 520[33], 521[33]
Migliorini, A. 376[59]
Miguel, D. 27[81], 218[51], 222[68]
Mihara, Y. 94[82], 95[82]
Millán, A. 27[81], 222[68]
Millemaggi, A. 624[196]

Miller, J. S. 297[60]
Miller, K. 177[64]
Miller, K. A. 609[75]
Miller, R. E. 332[133], 543[32]
Miller, R. L. 76[60], 78[60], 78[65], 80[60], 81[60], 82[60], 83[60], 120[118], 123[118]
Miller, S. J. 15[35]
Mills, B. M. 452[32]
Min, S.-J. 607[52]
Minagawa, K. 302[76], 302[77]
Minami, I. 496[118]
Minato, D. 420[124], 421[124]
Minato, H. 79[68]
Minger, T. L. 34[130]
Minor, K. P. 277[26]
Minoura, K. 617[128], 617[129], 617[130]
Minozzi, M. 217[34]
Mioskowski, C. 356[9]
Miranda, M. A. 217[8]
Mishra, P. 277[22], 277[24], 278[22], 281[22]
Misra, S. 535[7]
Mistry, N. 158[16]
Mitsudera, H. 232[88]
Mitsudo, T. 602[17]
Mitsuoka, S. 249[61]
Mitura, M. 579[7]
Miura, K. 393[88], 394[88], 416[119]
Miura, T. 515[27], 516[27]
Miyagawa, Y. 606[51]
Miyake, H. 282[31]
Miyakoshi, N. 608[70], 611[104], 612[104], 620[171], 623[192], 625[212]
Miyanohana, Y. 75[43]
Miyashi, T. 181[75], 181[79], 182[75]
Miyashita, M. 626[220]
Miyazaki, F. 474[80]
Miyazaki, K. 135[12], 143[12]
Mizsak, S. 617[134]
Mizuno, K. 179[65], 179[66], 179[69], 180[69], 181[76], 237[104]
Mizutani, H. 108[103], 109[103]
Mlynarski, J. 579[7], 579[8]
Mobley, W. C. 615[125]
Mocci, R. 551[42], 552[42]
Modi, A. 355[4]
Moessner, C. 10[15]
Moga, I. 17[45]
Mohamed, R. K. 221[65]
Mohamud, M. M. 461[53]
Mohr, B. J. 459[50]
Mohr, P. J. 244[32]
Mohs, R. C. 620[164], 620[165]
Molander, G. A. 217[38], 218[52], 269[3], 577[1]
Molawi, K. 101[94], 332[131], 332[132], 333[131], 333[132], 334[131], 334[135], 335[131], 335[135], 337[132], 542[31], 543[31], 544[31]
Møller, J. 624[199]
Molnár, J. 626[215]
Mondal, M. 175[54]
Monerris, M. 138[42], 138[43], 142[43]

Monks, B. M. 486[107]
Montgomery, J. A. 527[43]
Mook, R., Jr. 219[62]
Moon, D. O. 618[14C]
Moonen, K. 71[33], 73[33], 74[33]
Moorthie, V. A. 244[25]
Morales, C. A. 115[114], 115[115], 116[114]
Morales, G. 607[56], 607[57], 607[58], 607[59]
Morales, L. P. 218[45]
Morales-Alcázar, V. M. 193[26]
Morcillo, S. P. 27[81], 222[68]
Morency, L. 22[55]
Morgan, L. R. 527[47]
Mori, M. 68[18], 68[19], 68[20], 68[21], 68[23], 69[23], 72[21], 73[39], 73[40], 97[88], 97[89], 98[88], 98[89], 109[104], 110[104]
Mori, Y. 249[59], 249[60], 249[61], 249[66]
Morimoto, T. 602[16], 619[150], 619[151]
Morimoto, Y. 43[7], 44[7], 45[11], 45[12], 46[11], 46[12], 249[55], 250[71]
Morin, C. 603[35], 604[35]
Moriya, O. 623[193], 623[195]
Morken, J. P. 113[110]
Moro, R. F. 246[40], 256[93]
Morris, M. 626[213]
Morten, C. J. 56[27], 56[29], 56[30], 56[31], 57[29], 57[30]
Morton, D. 512[13]
Moser, W. H. 422[-26], 422[127], 423[126], 423[127], 434[138], 434[139], 434[141], 435[138], 435[139], 435[141]
Moses, J. E. 114[113], 450[27], 452[27], 452[30], 452[31], 453[31], 469[31]
Mota, A. J. 27[81], 222[68]
Mouriño, A. 459[49], 460[49], 471[49]
Mousseau, J. J. 56[29], 57[29]
Movassaghi, M. 102[95], 102[96], 103[95], 103[96]
Moyano, A. 603[35], 604[35]
Muci, A. R. 244[18]
Mukai, C. 249[49], 249[50], 249[51], 249[52], 602[26], 608[70], 611[104], 612[104], 620[26], 620[171], 621[26], 622[186], 623[192], 625[212]
Mukaiyama, T. 28[82], 163[25], 163[26], 163[27]
Mukherjee, S. 105[99], 112[99], 440[147]
Muldoon, M. J. 233[95]
Mullen, C. A. 321[121], 321[122], 322[121], 322[122], 323[121], 323[122], 327[122]
Müller, P. 516[30], 516[31], 516[32], 517[30], 517[31], 517[32], 527[46], 527[52]
Mulzer, J. 106[101], 124[121], 125[121], 128[121]
MuniRaju, C. 95[84], 96[84]
Muñoz, M. P. 22[59], 75[45], 75[46]
Muñoz-Bascón, J. 193[26], 218[45]

Murai, A. 55[24], 55[25], 56[24], 56[25], 249[62], 249[63], 249[64], 250[70]
Murai, K. 237[103], 238[103]
Murai, S. 602[16], 619[150], 619[151]
Murakami, M. 198[40], 515[27], 516[27], 602[27]
Muraoka, O. 188[7]
Murase, N. 434[139], 435[139]
Murata, M. 79[68]
Murata, S. 404[102]
Muratsubaki, M. 76[59]
Murdzek, J. S. 69[24], 136[24], 137[24]
Muroi, M. 618[140]
Murphy, G. K. 511[7]
Murphy, J. A. 217[16], 223[70], 223[71], 224[71], 224[72]
Murphy, M. M. 459[50]
Murphy, P. D. 620[170], 626[170]
Murso, A. 246[42], 248[42]
Muscoplat, C. C. 604[38]
Muzart, J. 449[9], 552[43]
Mynott, R. 22[54], 129[128], 130[128], 611[97]

N

Nadin, A. 135[5], 140[5]
Näf, F. 244[29]
Näf, R. 244[29]
Nagai, K. 624[205]
Nagamani, S. A. 284[37], 284[38], 285[37], 285[38]
Nagamoto, Y. 255[90], 256[90]
Naganuma, S. 244[31]
Nagaoka, H. 226[76]
Nagata, T. 611[92], 611[98]
Nagumo, Y. 135[8], 135[11], 142[8], 142[11], 143[8], 143[11]
Naidu, V. R. 151[56], 152[56], 153[56]
Nakagawa, M. 611[92], 611[93], 611[98]
Nakahata, N. 606[51], 614[119]
Nakai, H. 302[76]
Nakai, M. 617[128]
Nakai, Y. 365[38], 366[38], 383[70]
Nakajima, A. 363[30], 363[31], 364[30]
Nakajima, N. 402[101], 403[101]
Nakajima, T. 359[18]
Nakamura, I. 75[44]
Nakamura, M. 303[78], 304[78]
Nakamura, S. 19[47], 318[111], 319[111], 320[111]
Nakamura, T. 135[21], 626[220]
Nakane, D. 363[33]
Nakanishi, I. 540[26]
Nakatani, M. 303[78], 304[78]
Nakatani, S. 174[52], 175[52]
Nakayama, A. 614[120]
Nakayama, I. 362[28]
Nakayama, K. 434[139], 434[141], 435[139], 435[141]
Nakazaki, A. 193[24]
Nambu, H. 28[93], 28[104], 28[105], 29[93], 516[28]

Author Index

682

Namiki, H. 108[103], 109[103]
Nanni, D. 217[20], 217[34]
Naoe, S. 540[25], 541[25]
Naoki, H. 313[92], 313[93]
Narasaka, K. 198[40], 602[18], 622[179], 622[185]
Narayanam, J. M. R. 207[61], 207[65]
Narayanan, T. K. 300[67]
Narjes, F. 417[120]
Naruse, S. 402[101], 403[101]
Nawaz Khan, F. 535[13]
Negishi, E.-i. 32[126], 32[127], 449[14], 449[15], 449[16], 449[17], 449[21], 461[53]
Neighbors, J. D. 63[38], 63[39], 64[39], 298[61], 298[62], 298[63], 299[62], 299[63], 300[63]
Neiwert, W. A. 3[5], 4[5], 48[15], 48[17], 49[19], 50[15], 50[17], 50[19], 51[19], 249[54]
Nemoto, H. 324[124], 325[124]
Neukom, J. D. 463[66]
Neumaier, J. 147[51]
Neumann, H. 466[71], 466[72]
Nevado, C. 75[45], 75[46], 75[47], 555[48], 557[61], 559[82], 560[82], 561[87], 562[87], 566[95], 566[96], 567[96], 567[97], 567[98], 568[96], 569[96], 570[98], 571[98], 572[98]
Neverov, A. A. 53[22]
Neves, M. G. P. M. S. 585[14], 586[14]
Newcomb, M. 217[6]
Newman, S. G. 486[106]
Ng, S.-S. 49[18], 51[18], 52[21], 53[18], 53[21], 54[18], 54[21], 295[56], 296[56]
Ngo, S. C. 360[22]
Ngoc, D. T. 304[81]
Nguyen, J. D. 218[47]
Nguyen, R.-V. 538[24], 539[24]
Nguyen, S. T. 136[26]
Nguyen, V.-H. 175[57]
Nicewicz, D. A. 181[80], 182[80], 183[80], 368[41], 370[48], 370[49], 371[49]
Nicholas, K. M. 602[32]
Nicholls, B. 244[26]
Nickel, A. 620[170], 626[170]
Nicolaou, K. C. 1[1], 1[3], 2[3], 33[128], 34[131], 135[4], 135[5], 135[20], 135[21], 140[4], 140[5], 141[4], 142[4], 142[20], 198[47], 217[30], 244[28], 248[47], 248[48], 249[47], 249[48], 250[47], 305[82], 306[82], 313[94], 313[95], 313[96], 313[97], 313[98], 313[99], 313[100], 313[101], 314[94], 314[101], 449[2], 449[4], 449[7], 450[25]
Niethe, A. 106[102], 107[102]
Nieto-Oberhuber, C. 75[45], 75[46], 332[130], 332[134], 333[130], 334[130], 564[90]
Nihei, H. 620[157]
Niimi, Y. 608[67], 608[68]
Nijs, A. 373[54], 374[54]
Nilson, M. G. 611[102]

Nishida, A. 611[92], 611[93], 611[98]
Nishikawa, Y. 43[7], 44[7], 45[11], 46[11], 249[55], 250[71]
Nishimura, J. 607[54], 607[55]
Nishino, H. 175[55], 175[56], 175[57], 175[58], 175[59], 176[55], 177[55]
Nissen, P. 624[199]
Niu, Q. 193[23]
Nixon, J. 157[8], 158[8]
Njardarson, J. T. 225[73], 226[73], 227[73]
No, Z. S. 74[42]
Noble, A. R. 557[75]
Noda, K. 75[56], 78[56]
Noda, Y. 449[14]
Noe, M. C. 15[32]
Nogami, T. 535[11], 535[12]
Nojima, M. 158[9], 163[24], 164[9], 164[24], 165[9], 165[24], 166[9], 166[24]
Nolan, S. P. 72[38], 557[64], 557[73], 557[75]
Nomoto, K. 617[128], 617[129]
Nomura, I. 622[186]
Noro, M. 75[54]
Norris, A. 297[60]
Northrup, A. B. 464[69], 465[69], 472[69]
Norton, J. R. 28[88], 28[89], 28[90], 28[91], 29[89], 29[115], 30[117]
Nösel, P. 545[34], 545[35], 546[34], 547[34], 548[34], 565[93]
Nozaki, K. 422[125]
Nozaki, N. 256[94]
Nozari, M. S. 621[174]
Nozoe, S. 603[36]
Nudenberg, W. 168[42]
Nugent, W. A. 27[77], 27[78]
Numa, M. 411[110]
Numata, A. 617[128], 617[129], 617[130]
Nuñez-Zarur, F. 74[41]
Nuss, G. W. 602[30]
Nuss, J. M. 449[20], 459[50]

O

Obara, Y. 614[119]
O'Boyle, B. M. 483[101], 484[101], 490[101]
Obrey, S. J. 67[6], 76[6]
Ochiai, H. 500[129]
Ochifuji, N. 68[18], 68[20]
O'Connor, M. J. 395[92], 396[92], 397[92], 398[92]
O'Connor, S. J. 32[123], 33[123], 449[22]
Odani, A. 602[26], 620[26], 621[26]
Odom, A. L. 88[71], 148[53]
O'Donnell, C. J. 316[109]
Ofuji, K. 313[92], 313[93]
Ogawa, Y. 617[133]
Ogbourne, S. M. 626[213]
Oh, H. 618[140]
Oh, T. 316[109]
Ohashi, Y. 493[113]
Ohfune, Y. 254[88], 391[86], 392[86]
Ohkatsu, Y. 28[96]

Ohnishi, Y. 371[51], 372[51]
Ohno, H. 449[8], 480[95], 480[96], 489[96], 496[8], 496[119], 497[119], 497[120], 498[119], 498[120], 540[25], 541[25]
Ohno, M. 28[96]
Ohshima, T. 138[40], 579[5], 579[6], 580[5], 581[5], 582[10], 583[10]
Ohta, M. 181[78]
Ohtani, M. 606[46], 606[47]
Ohtani, Y. 362[25], 363[25], 363[32], 364[32]
Oishi, S. 480[95], 480[96], 489[96], 496[119], 497[119], 497[120], 498[119], 498[120], 540[25], 540[26], 541[25]
Oishi, T. 79[68], 135[8], 135[11], 142[8], 142[11], 143[8], 143[11]
Ojima, H. 94[81], 94[82], 95[82]
Ojima, I. 244[15]
Oka, S. 28[94], 28[95], 163[32]
Oka, T. 617[130]
Okada, A. 138[40]
Okada, M. 359[18]
Okada, S. 76[57]
Okada, T. 602[17]
Okaguchi, T. 198[40]
Okamoto, N. 385[77], 386[77]
Okamoto, T. 28[94], 28[95], 163[32]
Okamoto, Y. 363[31], 373[53]
Okawara, M. 623[193], 623[195]
Okazaki, M. 277[18]
Oki, M. 606[46], 606[47]
Okita, T. 45[12], 46[12]
Okpaka, O. 602[31]
Okpo, O. S. 602[31]
Okugawa, S. 383[74]
Olah, G. A. 14[7], 26[7]
Old, D. W. 316[109]
Olin, S. S. 219[59]
Oller-López, J. L. 27[80], 28[80], 218[44], 218[46]
Ollivier, C. 217[28], 217[29]
Ollivier, J. 259[105]
Olson, R. E. 358[15]
Oltra, J. E. 27[80], 28[80], 193[26], 218[44], 218[45], 218[46]
O'Malley, S. J. 441[149]
Onak, C. S. 21[51]
Onda, K. 256[94]
O'Neill, P. M. 164[34], 164[35]
Onishi, K. 380[67], 381[67], 399[95]
Ono, K. 611[93]
Ono, M. 250[70]
Onuki, K. 626[220]
Ooi, T. 28[96]
Opel, A. 444[154]
Oppedisano, A. 28[83], 29[83], 30[83], 166[37]
Opperdoes, F. R. 617[137]
Oppolzer, W. 32[125], 35[125], 35[134]
O'Regan, M. 69[24], 136[24], 137[24]
Orellana, A. 323[123], 324[123], 327[123]
Orlović, M. 15[39]
Orru, R. V. A. 449[6], 485[103]

Author Index

Ortiz de Montellano, P. R. 15[25], 290[46]
Osada, H. 618[140]
Oshima, K. 393[88], 394[88], 416[119], 417[121], 418[121]
Oshima, Y. 313[93], 606[51]
Oskam, J. H. 136[25], 137[25]
Otaka, A. 378[64]
Otani, T. 620[153], 620[155], 620[157]
Otsuji, Y. 179[65], 179[66], 179[69], 180[69], 181[76], 237[104]
Ott, K. C. 136[35], 139[35]
Overall, J. E. 620[166]
Overman, L. E. 32[121], 32[122], 32[123], 33[121], 33[122], 33[123], 277[18], 277[19], 277[20], 277[21], 277[22], 277[23], 277[24], 277[25], 277[26], 278[22], 278[25], 279[25], 279[27], 280[27], 280[28], 280[29], 281[22], 281[25], 281[29], 307[84], 316[84], 316[103], 316[104], 316[106], 316[107], 316[109], 317[110], 318[110], 335[138], 449[22], 449[23], 450[23], 468[23], 474[81], 474[82], 487[82], 607[63]
Owczarczyk, Z. 449[15], 461[53]
Owen, D. J. 263[119], 264[119]
Oyamada, K. 365[36]

P

Padial, N. M. 218[45]
Padwa, A. 75[53], 511[6], 512[6], 521[6]
Pairaudeau, G. 316[106]
Pallerla, M. K. 610[89], 611[89]
Palomas, D. 594[28], 595[28]
Pan, C.-M. 28[113]
Panda, G. 595[29], 596[29]
Pandit, P. 591[24], 592[24]
Paolobelli, A. B. 201[52]
Papish, E. T. 28[90]
Paquette, L. A. 256[92], 258[100], 263[118], 607[64], 607[65], 610[79], 610[80], 610[82]
Park, C.-H. 142[47]
Park, L. Y. 136[25], 137[25]
Park, P. K. 441[149]
Park, S. 78[61], 78[62], 120[118], 123[118]
Parker, C. D. 527[43]
Parker, K. A. 31[120], 32[120], 100[93], 101[93], 228[80], 229[80], 230[82], 451[28], 469[28]
Parmar, D. 193[29], 193[30], 194[30], 195[30]
Parr, B. T. 598[33]
Parrish, J. D. 182[81], 183[81]
Parshall, G. W. 136[32], 137[32], 139[32]
Parsons, P. G. 626[213]
Parsons, P. J. 67[7], 76[7], 355[2], 393[2], 461[55]
Parvez, M. 171[48], 172[48], 173[48]
Paschall, C. M. 24[69]
Paterson, I. 247[45], 248[45]
Paton, R. S. 498[122]
Patra, A. 259[103]

Patrick, B. O. 258[98], 270[7]
Patro, B. 223[70]
Pattenden, G. 188[10], 188[11]
Paul, A. 217[27]
Paulmann, U. 179[58]
Pauson, P. L. 601[1], 601[2], 601[3], 622[178]
Pautex, N. 516[30], 517[30]
Pavlakos, I. 193[20]
Paxton, T. J. 67[12]
Payne, A. H. 28[99], 163[28], 166[28]
Payne, G. B. 250[68]
Paz, F. A. A. 585[14], 586[14]
Pearson, W. H. 301[72]
Peet, N. P. 254[89]
Peixoto, P. A. 458[48], 459[48], 471[48]
Pekari, K. 344[148], 345[148], 350[148]
Peláez, R. 193[25]
Pellegrinet, S. C. 617[132]
Pelletier, S. W. 300[66], 608[72]
Pelucchini, C. 590[23], 591[23]
Peng, S. 76[57]
Peng, X.-S. 500[134], 501[134], 504[134]
Peng, Y. 557[81], 562[88]
Pengsuparp, T. 612[109]
Penkett, C. S. 67[7], 76[7], 355[2], 393[2]
Pereira, M. M. A. 300[70]
Pereshivko, O. P. 538[23]
Pereyre, M. 231[84], 388[82], 389[82]
Pérez-Castells, L. 601[4]
Pérez-Galán, P. 332[134], 564[90]
Pérez-Martán, I. 217[26]
Pérez-Prieto, J. 217[8]
Pérez-Serrano, L. 601[4]
Periasamy, M. 602[24]
Pericàs, M. A. 603[35], 604[35]
Perie, J. 617[137]
Perumal, P. T. 589[20], 590[20]
Peshkov, V. A. 538[23]
Pestchanker, M. J. 624[206]
Petasis, N. A. 136[36], 139[36], 450[25]
Peters, N. R. 182[81], 183[81]
Peterson, P. W. 221[65]
Petri, A. F. 439[146], 440[146]
Petrignet, J. 462[59], 463[59], 472[59]
Petrone, D. A. 486[108], 491[108]
Petrovic, G. 313[97], 313[98], 313[99], 313[100], 313[101], 314[101]
Petter, R. C. 624[208]
Pezzuto, J. M. 612[109]
Pfaltz, A. 601[14]
Pfeiffer, M. W. B. 34[129]
Pham, P. V. 198[42]
Phillips, A. J. 34[129], 34[130]
Pierobon, M. 27[79]
Piersanti, G. 102[95], 102[96], 103[95], 103[96]
Pihko, P. M. 313[94], 313[99], 313[100], 314[94], 345[152]
Piizzi, G. 102[95], 102[96], 103[95], 103[96]
Pillow, T. H. 67[12]
Pinard, E. 607[64], 607[65]
Pincock, R. E. 158[10], 160[10]

Pine, S. H. 136[30], 136[31], 136[33], 137[33], 139[30], 139[31], 139[33]
Pink, M. 422[127], 423[127]
Pirrung, M. C. 254[85], 255[85]
Pitram, S. M. 5[8], 6[8], 437[143]
Pitsinos, E. N. 1[3], 2[3]
Plefka, O. 431[136], 432[136]
Pleixats, R. 74[41]
Plotkin, B. J. 624[207]
Plummer, M. S. 292[53], 293[53]
Polborn, K. 181[77]
Pommerville, J. 289[45]
Pommier, Y. 236[102]
Pons, V. 527[45]
Porco, J. A., Jr. 9[13], 114[112]
Portella, C. 360[20], 360[21], 361[20]
Porter, L. A. 604[42]
Porter, N. A. 157[7], 157[8], 158[7], 158[8], 158[12], 159[18], 159[19], 160[7], 160[18]
Posner, G. H. 164[33], 164[35], 169[45], 170[45], 171[45]
Posner, T. 168[40]
Postema, M. H. D. 135[4], 135[5], 135[6], 135[7], 138[6], 138[7], 140[4], 140[5], 141[4], 142[4]
Potashman, M. H. 624[209]
Potnick, J. R. 379[66], 380[66]
Poudel, Y. B. 135[22], 147[22]
Pouliot, M. 502[137], 505[137]
Pouliot, R. 162[23], 163[23]
Powell, D. R. 70[27]
Prakash, B. 598[34]
Prasad, C. V. C. 248[47], 248[48], 249[47], 249[48], 250[47]
Preethanuj, P. 598[34]
Prestwich, G. D. 24[68]
Price, K. 193[29], 193[30], 194[30], 195[30]
Prier, C. K. 207[63]
Probert, M. K. S. 180[70]
Procter, D. J. 193[27], 193[28], 193[29], 193[30], 194[30], 195[30], 195[31], 196[32], 196[33], 197[32], 197[33], 197[34], 197[35], 217[19]
Promo, M. A. 94[80]
Pronin, S. V. 25[71], 26[71], 36[71]
Protasiewicz, J. D. 88[71]
Pu, J.-X. 612[108]
Pulling, M. E. 28[89], 29[89], 29[115]
Pye, P. 258[96]

Q

Qi, J. 9[13]
Qian, W. 313[94], 313[95], 313[96], 313[99], 314[94]
Qian, Y. 7[10], 522[42], 523[42], 524[42], 525[42], 526[42], 527[42]
Qiao, C. 110[106]
Qin, G. 243[10]
Qiu, D. 486[105]
Qiu, S.-X. 612[109]
Qu, J. 45[13], 46[13]
Quan, J. M. 617[139], 618[139]
Qui, W. 3[4], 17[40]

684 Author Index

Quiclet-Sire, B. 188[13], 217[39], 261[115], 262[115]
Quílez del Moral, J. F. 193[25]
Qureshi, Z. 476[90], 477[90], 488[90]

R

Rabe, J. 624[197]
Raccuglia, R. A. 624[201]
Rackelmann, N. 67[8]
Radhakrishnan, K. V. 598[34]
Raducan, M. 75[46], 75[47], 332[132], 333[132], 337[132], 542[31], 543[31], 544[31]
Raghuvanshi, D. S. 588[19], 589[19]
Rainier, J. D. 135[9], 135[10], 135[13], 135[15], 135[16], 135[17], 135[18], 135[19], 135[23], 136[17], 137[15], 137[16], 142[10], 142[13], 143[9], 143[10], 144[9], 144[10], 144[13], 144[17], 144[18], 145[17], 146[17], 146[19], 146[49], 146[50], 147[15], 147[16], 147[18], 147[23], 148[17]
Rajadhyaksha, V. J. 75[50]
RajanBabu, T. V. 27[77], 27[78]
Rajendar, G. 20[49]
Rajesh, T. 602[24]
Rakiel, B. 579[8]
Ramachary, D. B. 345[154]
Ramadhar, T. R. 596[30], 597[30]
Ramharter, J. 106[101], 124[121], 125[121], 128[121]
Rangoni, A. 476[88]
Rankic, D. A. 207[63]
Rao, A. S. 246[41]
Rao, B. V. 95[84], 96[84]
Rao, J. P. 95[84], 96[84]
Rao, V. K. 586[16], 587[16]
Rao, W. 564[92], 565[92], 565[94], 566[94]
Rappé, A. 28[90]
Rappoport, Z. 157[5], 160[5]
Rasamison, V. E. 297[60]
Raschke, T. 476[86]
Ratananukul, P. 449[10]
Ratnikov, M. 512[10], 512[12]
Ratnikov, M. O. 522[39]
Ratovoson, F. 297[60]
Raubo, P. 399[96]
Ravelli, D. 207[60]
Ravikumar, V. 125[123], 125[124], 126[123], 126[124], 126[125], 127[125]
Rawal, V. H. 389[84], 390[84]
Razafitsalama, J. 297[60]
Reddy, D. S. 72[37], 93[78], 94[78], 126[125], 127[125], 341[145], 342[145], 343[145]
Reddy, G. S. 136[32], 137[32], 139[32]
Reddy, M. S. 607[60], 607[62]
Reddy, P. G. 301[74], 302[74], 305[74]
Reddy, P. S. 96[85], 97[85]
Reddy, S. P. 292[53], 293[53]
Reddy, Y. V. 345[154]
Rédei, D. 626[215]
Redert, T. 491[111], 492[111], 495[111]
Reed, C. A. 14[8], 17[8]

Rehberg, G. M. 68[16], 68[17], 71[16]
Reich, H. J. 358[15], 358[16], 358[17], 359[16], 359[17]
Reingruber, R. 198[47]
Reising, J. 201[53], 204[53], 206[53]
Reisman, S. E. 197[36]
Ren, J. 593[25]
Ren, S. K. 269[4], 271[4]
Ren, W.-W. 500[136], 501[136], 505[136], 613[115]
Renaud, P. 187[1], 217[12], 217[31]
Rendler, S. 31[119], 198[51], 199[51], 200[51]
Rennels, R. A. 449[20], 459[50]
Resa, S. 27[81], 222[68]
Reynolds, A. J. 222[69], 223[69]
Reynolds, S. C. 357[11], 358[11]
Reynolds, T. E. 381[68], 382[68]
Rhee, J. Y. 80[69], 81[69], 82[69]
Rhee, Y. H. 564[91]
Rheingold, A. L. 44[9], 247[44]
Ricard, L. 99[90], 100[90]
Richard, J. P. 14[10]
Richardson, R. D. 358[14], 381[14]
Rickards, B. 622[184]
Rickborn, B. 282[30]
Rickett, D. L. 620[163]
Riera, A. 603[35], 604[35]
Ries, M. 400[99], 401[99]
Riguet, E. 217[17]
Rios, R. 347[157], 348[157]
Rise, F. 458[47]
Rishton, G. M. 277[23]
Risley, E. A. 602[30]
Rizzacasa, M. A. 263[119], 264[119]
Robbins, J. 69[24], 136[24], 137[24]
Roberts, B. E. 21[52]
Roberts, S. W. 135[18], 144[18], 147[18]
Robertson, B. D. 357[10]
Robichaud, J. 502[137], 505[137]
Robinson, R. 243[6]
Roca, T. 138[42], 138[43], 142[43]
Roden, B. A. 607[64], 607[65]
Rodini, D. J. 21[53]
Rodríguez, A. M. 624[206]
Rodríguez, C. 564[90]
Rodriguez, J. 243[1]
Rodríguez, J. R. 3[5], 4[5], 48[17], 50[17]
Rodríguez, L. 246[40]
Rodriguez, R. 549[38]
Rodríguez-García, I. 218[45]
Rodríguez-Santiago, L. 74[41]
Rodriquez, M. 313[99]
Roe, A. N. 159[18], 159[19], 160[18]
Rogan, L. 233[95]
Rogers-Evans, M. 97[86], 98[86]
Roh, Y. 28[100]
Rohanna, J. 135[23], 147[23]
Rohanna, J. C. 146[49]
Rohde, J. M. 316[107]
Rohmer, M. 24[68]
Roizen, J. L. 527[56]
Roldan-Molina, E. 218[45]
Romero, D. L. 458[47]

Romero, J. A. C. 577[1]
Rominger, F. 545[34], 545[35], 546[34], 547[34], 548[34], 552[45], 553[45], 565[93]
Rosales, A. 27[80], 28[80], 193[26], 218[45], 218[46]
Rosati, O. 590[23], 591[23]
Rosellón, A. 564[90]
Rosen, M. D. 474[81], 474[82], 487[82]
Ross, A. G. 261[114]
Rosselli, S. 624[201]
Rossi, R. A. 217[9]
Rossiter, L. M. 479[94]
Rouillard, A. 37[138]
Rouse, M. B. 626[221]
Roush, W. R. 105[100]
Rowan, D. D. 162[22], 163[22]
Rowand, D. A. 246[39]
Rowlands, G. J. 187[2], 187[3], 218[42], 218[43]
Roy, S. 250[69]
Rozenberg, H. 169[44], 170[44]
Rubio, E. 594[28], 595[28]
Rudolph, M. 545[34], 545[35], 546[34], 547[34], 548[34], 552[45], 553[45], 565[93]
Rudra, A. 135[22], 147[22]
Rueping, M. 511[5]
Rüffer, T. 431[136], 432[136]
Ruijter, E. 449[6], 485[103]
Rumbo, A. 611[96]
Russell, A. E. 522[37]
Russell, C. A. 555[49], 555[50], 556[58]
Russey, W. E. 15[25], 15[26]
Rutledge, P. S. 609[76]
Ruzicka, L. 15[19], 15[20]
Ruzziconi, R. 201[52]
Rydberg, D. B. 68[15], 71[15]
Ryland, B. L. 233[96]
Rys, D. J. 233[92]
Ryu, D. H. 395[91]
Ryu, I. 224[72]
Ryu, Z. H. 14[9]

S

Saad, J. R. 624[206]
Sabitha, G. 246[41]
Sachinvala, N. D. 250[69]
Sadig, J. E. R. 467[74]
Sadlek, O. 181[77]
Saegusa, T. 235[100]
Safa, K. D. 401[100]
Sahashi, Y. 606[51]
Sahasrabudhe, K. 301[73]
Saito, A. 250[74], 252[74]
Saito, M. 620[157]
Saito, T. 618[140], 620[153], 620[155], 620[157]
Saka, K. 249[63]
Sakaguchi, K. 174[52], 175[52], 391[86], 392[86]
Sakaguchi, T. 454[34], 470[34]
Sakakibara, K. 422[125]
Sakakura, A. 20[50], 21[50], 613[110], 626[110]

Author Index

Sakamoto, Y. 434[142], 435[142]
Sakurai, J. 303[79], 303[80], 304[79], 304[80], 305[79], 305[80]
Sakwa, S. A. 479[94]
Salaün, J. 259[105]
Salem, B. 455[35], 455[36], 455[38], 455[39], 456[39], 470[39]
Salomon, R. G. 250[69]
Salway, A. H. 620[159]
Sammis, G. M. 230[83], 231[83]
Samsel. E. G. 28[86], 28[87]
Samson, K. M. 604[39]
Sanchawala, A. I. 126[125], 127[125]
Sande, T. F. 300[66]
Sandeep, A. 246[41]
Sandham, D. A. 607[61]
Sanière, L. 527[55]
Sankawa, U. 603[36]
Sanner, M. 464[67], 500[67]
Sano, A. 362[26]
Santarsiero, B. D. 139[44], 140[44], 143[64]
Sanz, F. 256[93]
Sarlah, D. 24[64], 33[128], 34[131], 198[47], 305[82], 306[82], 313[99], 449[7]
Sarpong, R. 328[126], 328[127], 328[128], 329[126], 329[127], 329[128], 330[126], 330[127], 330[128], 337[127], 513[24], 514[24]
Sasai, H. 578[4]
Sasaki, K. 217[7]
Sasaki, M. 250[72], 365[35], 365[36], 365[37], 372[52], 383[69], 383[70], 383[71], 383[72], 383[73], 385[77], 386[77]
Satake. M. 313[92], 313[93], 313[97]
Sato, K. 142[47], 535[14]
Sato, T. 24[66], 256[94], 363[31], 500[127]
Sato, Y. 73[39], 73[40], 97[89], 98[89]
Satterfield, A. D. 369[46]
Sattler, W. I. 17[44]
Satyanarayana, G. 284[39], 285[39]
Sautier, B. 195[31]
Savage, G. P. 250[73], 263[119], 264[119]
Sawada, Y. 365[34], 365[35]
Scammells, P. J. 263[119], 264[119]
Scarpino Schietroma, D. M. 10[14]
Schädlich, J. 545[34], 546[34], 547[34], 548[34]
Schafroth, M. A. 24[64]
Schaumann, E. 249[57], 390[85], 391[85], 400[99], 401[99], 404[104], 404[105], 405[104], 405[105], 406[104], 406[105], 407[105], 408[105], 408[107], 408[108], 409[108], 409[109], 410[109], 411[109], 413[115], 414[115], 417[120]
Scheerer, J. R. 38[139]
Scheidt, K. A. 367[40], 378[65], 379[65], 381[68], 382[68]
Schiesser, C. H. 158[14], 217[13]
Schmalz, H.-G. 601[12]
Schmeda-Hirschmann, G. 623[191]

Schmidt, D. R. 441[149]
Schmidt, J. 259[106]
Schmidt, P. 246[42], 248[42]
Schmitt, D. C. 368 45]
Schmittel, M. 201[53], 201[54], 204[53], 204[54], 206[53], 206[54]
Schnatterer, A. 180[71], 180[74], 181[74]
Schneider, L. 304[81]
Scholl, M. 136[28]
Schramm, M. P. 72[37]
Schreiber, J. V. 291[52], 292[52]
Schreiber, S. L. 602[20], 602[33], 602[34], 603[33], 603[34]
Schrock, R. R. 69[24], 69[25], 69[26], 78[25], 78[26], 136[24], 136[25], 137[24], 137[25]
Schubert, M. 103[97], 104[97]
Schubert, U. 24[67]
Schulte, B. 234[99]
Schultz, A. J. 28[100]
Schultz, D. M. 207[64], 463[64]
Schultz, E. E. 513[24], 514[24]
Schultz, G. E. 13[3], 19[3], 21[3], 23[3], 24[3]
Schwab, P. 136[27]
Schweinitz, A. 323[123], 324[123], 327[123]
Schweizer, S. 461[55]
Schwier, T. 557[67]
Scott, A. J. 222[69], 223[69]
Scott, D. A. 28[98], 28[99], 163[28], 163[29], 163[30], 166[28], 166[29], 166[30]
Scott, T. M. 626[213]
Sebren, L. J. 511[4]
Seebach, D. 407[106]
Segi, M. 359[18]
Sei, Y. 579[6]
Seidel, G. 611[96], 611[97]
Seiler, M. P. 290[47], 291[47]
Selander, N. 513[23]
Semmelhack, M. F. 464[67], 498[123], 500[67], 500[126], 500[127], 500[128], 501[126], 503[126]
Seo, J. H. 135[20], 142[20]
Seo, J.-M. 624[202]
Seo, Y. 624[202]
Serba, C. 35[135], 36[135]
Serra, V. I. V. 585[14], 586[14]
Servens, C. 231[84]
Sethofer, S. G. 22[60]
Seto, H. 610[83]
Sévenet, T. 297[59]
Severin, R. 458[48], 459[48], 471[48]
Seya, Y. 535[17]
Sfouggatakis, C. 434[140], 434[142], 435[140], 435[142]
Sha, C.-K. 608[69]
Shair, M. D. 115[114], 115[115], 116[114], 314[102], 315[102]
Shaktikumar, L. 284[39], 285[39]
Shambayati, S. 602[20], 602[33], 602[34], 603[33], 603[34]

Shan, Z.-H. 500[136], 501[136], 505[136], 613[115]
Shanahan, C. S. 7[10], 522[42], 523[42], 524[42], 525[42], 526[42], 527[42]
Shanley, E. S. 157[6], 160[6]
Shao, H. 270[8], 270[9], 271[8], 271[9]
Shapiro, N. D. 557[72]
Sharp, L. A. 188[13], 188[14]
Sharp, M. J. 316[109]
Sharpless, K. B. 15[31], 43[5], 244[14], 244[15]
Shaw, J. T. 60[37], 62[37]
She, X. 440[148], 441[148]
Sheehan, S. M. 314[102], 315[102]
Sheldon, R. A. 67[10]
Shell, A. J. 67[7], 76[7], 355[2], 393[2]
Shen, H.-J. 619[146]
Shen, M. 211[69], 212[69]
Shen, Q. 597[31]
Shen, Z.-L. 307[87], 308[87], 309[87], 310[87], 315[87]
Shenvi, R. A. 15[30], 16[30], 17[30], 25[71], 26[71], 28[83], 29[83], 30[83], 30[116], 36[71], 166[37]
Sher, P. M. 232[86], 232[87], 234[86], 235[87]
Sherburn, M. S. 162[22], 163[22], 217[5], 222[69], 223[69]
Sherry, B. D. 542[30], 557[72]
Sheth, K. A. 94[83]
Shi, H. 330[129], 331[129], 332[129], 337[129]
Shi, L. 282[33], 283[33], 330[129], 331[129], 332[129], 337[129]
Shi, L.-L. 619[146]
Shi, M. 232[89], 598[32], 623[189], 623[190]
Shi, X. 557[71], 561[83]
Shi, Y. 32[124], 35[124], 43[6], 244[13], 244[21], 244[22], 244[23], 244[24], 500[136], 501[136], 505[136], 513[25], 514[25], 516[25], 613[115]
Shibasaki, M. 138[40], 474[80], 577[2], 578[4], 579[5], 579[6], 580[5], 580[9], 581[5], 581[9], 582[10], 583[10]
Shibata, S. 603[36]
Shibata, T. 76[59], 602[19], 622[179]
Shigenaga, A. 378[64]
Shigeta, S. 626[214]
Shim, J. 317[110], 318[110]
Shim, S. C. 179[67]
Shim, S. Y. 395[91]
Shimada, N. 516[28]
Shimizu, I. 493[113], 496[118]
Shimizu, K. 109[104], 110[104]
Shimohigashi, Y. 378[64]
Shimono, Y. 254[87]
Shin, J. 624[202]
Shin, K.-S. 618[140]
Shin, W. K. 117[116], 117[117], 118[116], 118[117]
Shinokubo, H. 393[88], 394[88], 416[119], 417[121], 418[121]
Shiotani, T. 620[153]
Shirahama, H. 254[88]

Shirakami, S. 434[142], 435[142]
Shirakawa, Y. 383[71]
Shiro, M. 363[31], 516[28]
Shishido, K. 617[131]
Shoemaker, R. H. 297[57]
Shoji, M. 135[8], 135[11], 142[8], 142[11], 143[8], 143[11]
Shunatona, H. P. 555[54], 556[54]
Sibi, M. P. 187[1], 217[2], 217[14]
Siedle, B. 624[200]
Siefken, M. W. 621[175]
Siegel, D. S. 102[95], 102[96], 103[95], 103[96]
Sigüeiro, R. 459[49], 460[49], 471[49]
Silke, J. 313[92], 313[93]
Sill, P. C. 622[184]
Silva, A. M. S. 585[14], 586[14]
Simmons, B. 338[141], 338[142], 339[141], 340[142], 343[142]
Simmons, E. M. 328[126], 328[127], 328[128], 329[126], 329[127], 329[128], 330[126], 330[127], 330[128], 337[127]
Simonneau, A. 549[37]
Simonsen, K. B. 1[1], 1[3], 2[3]
Simpkins, N. 193[20]
Simpkins, N. S. 188[12]
Simpson, G. L. 51[20], 52[20], 251[76], 252[76]
Simpson, R. E. 171[48], 171[49], 172[48], 172[49], 173[48], 173[49]
Singh, I. 607[56], 607[58], 607[59]
Singh, J. 624[208]
Singh, K. N. 588[19], 589[19]
Singh, R. 595[29], 596[29]
Sinha, B. 449[5]
Sinz, C. J. 345[155], 346[155], 347[155], 350[155]
Sit, R. K. 513[23]
Sivavec, T. M. 68[13], 68[14], 71[13], 71[14]
Six, Y. 177[63]
Skell, P. S. 75[49]
Skouta, R. 535[8], 535[9]
Skrydstrup, T. 193[27], 193[28], 217[19]
Skytte, D. 624[199]
Slack, R. D. 164[33]
Slater, M. L. 479[94]
Slaughter, L. M. 536[21], 537[21]
Slomp, G. 617[134]
Smirnov, P. 373[54], 374[54]
Smissman, E. E. 261[110], 262[110]
Smith, A. B., III 5[8], 6[8], 355[6], 356[6], 414[117], 415[117], 415[118], 416[118], 418[122], 419[122], 423[128], 424[128], 425[129], 426[129], 430[134], 433[117], 434[117], 434[138], 434[139], 434[140], 434[141], 434[142], 435[138], 435[139], 435[140], 435[141], 435[142], 437[143], 442[150], 442[151], 442[152], 442[153], 443[151], 443[153], 444[151], 444[152]
Smith, C. D. 499[125], 500[125], 503[125]
Smith, D. M. 29[115]
Smith, M. W. 233[94], 234[94], 235[94], 236[94]
Snapper, M. L. 79[66]
Snider, B. B. 21[53], 26[76], 27[76], 175[53], 218[48], 218[56], 221[66], 221[67]
Snieckus, V. 502[137], 505[137]
Snyder, S. A. 17[41], 17[42], 17[43], 17[44], 17[45], 18[42], 18[46], 218[52], 233[94], 234[94], 235[94], 236[94]
Sobukawa, M. 434[139], 434[141], 435[139], 435[141]
Sodeoka, M. 15[28]
Sodupe, M. 74[41]
Sohn, J.-H. 74[42]
Sohng, J.-K. 617[135], 617[138]
Sokol, J. G. 23[63]
Solans-Monfort, X. 74[41]
Söllner, R. 201[53], 204[53], 206[53]
Solorio, C. R. 332[132], 333[132], 337[132], 542[31], 543[31], 544[31]
Somers, P. K. 248[47], 248[48], 249[47], 249[48], 250[47]
Sommer, K. M. 499[124], 503[124]
Son, K.-H. 624[202]
Song, J. 76[58]
Song, J. S. 179[67]
Song, Z. 355[7], 356[7], 360[19], 419[123], 420[123], 426[130], 428[133], 429[133], 430[135], 431[135]
Song, Z.-L. 269[1], 282[32]
Sorensen, E. J. 344[148], 345[148], 350[148]
Sörensen, H. 603[35], 604[35]
Sorensen, T. S. 252[77]
Soriano, E. 557[60]
Sorimachi, M. 602[26], 620[26], 621[26]
Spagnolo, P. 217[20], 217[34]
Spain, M. 193[29], 193[30], 194[30], 195[30]
Spande, T. F. 300[67]
Spanevello, R. A. 617[132]
Spangler, J. E. 514[26], 515[26]
Spataru, T. 30[117]
Spehar, K. 217[36], 230[36]
Spellmeyer, D. C. 158[13]
Spencer, J. B. 244[11]
Spevak, W. 76[57]
Spiegel, D. A. 225[73], 226[73], 227[73]
Spielvogel, D. 10[15]
Spiess, E. 500[127]
Spiteller, G. 157[1]
Sridharan, V. 449[10], 449[11], 449[12], 449[13], 475[84], 589[20], 589[21], 589[22], 590[20], 590[21], 590[22]
Srikrishna, A. 284[37], 284[38], 284[39], 285[37], 285[38], 285[39], 610[78]
St. Laurent, D. 607[64]
St. Laurent, D. R. 607[65]
Staas, D. D. 15[38]
Staben, S. T. 22[58], 557[69]
Stahl, S. S. 233[96], 463[60]
Staley, D. L. 44[9], 247[44]
Stalke, D. 246[42], 248[42], 475[85], 476[85], 488[85]
Stanley, F. 512[18]
Stark, L. M. 344[148], 345[148], 350[148]
Starratt, A. N. 527[48]
Stecker, F. 499[124], 503[124]
Steele, J. C. P. 608[73]
Stein, M. 246[38], 247[43]
Stephenson, C. R. J. 207[61], 207[65], 207[66], 208[66], 209[66], 210[66], 218[47], 511[4]
Stern, C. A. 381[68], 382[68]
Stern, C. L. 204[58], 205[58], 206[58], 207[58]
Stevenazzi, A. 601[5]
Stevens, C. V. 71[33], 73[33], 74[33]
Stevens, E. D. 557[73]
Stevens, R. C. 76[58]
Stevenson, P. 449[10], 449[11]
Steward, K. M. 377[60]
Stewart, C. 511[7]
Stieglitz, J. 15[15]
Still, W. C. 44[9], 247[44]
Stille, J. R. 4[6], 136[29], 139[29], 139[44], 139[46], 140[44], 140[46], 143[44]
Stodulski, M. 579[8]
Stoker, D. A. 188[10], 188[11]
Stoltz, B. M. 493[114], 494[114], 495[114]
Stork, G. 15[21], 219[60], 219[61], 219[62], 219[63], 232[86], 232[87], 234[86], 235[87]
Straten, J. V. 21[53]
Straub, B. F. 70[29], 71[29], 74[29], 77[29]
Straus, D. 136[29], 139[29]
Strickland, J. B. 289[45]
Strobel, G. A. 618[141]
Studer, A. 193[28], 217[5], 217[6], 217[7], 217[8], 217[9], 217[10], 217[11], 217[12], 217[13], 217[14], 217[15], 217[16], 217[17], 217[18], 217[19], 217[20], 217[21], 217[22], 217[23], 217[24], 217[25], 217[26], 217[27], 217[28], 217[29], 234[99]
Stuppner, H. 624[204]
Sturino, C. F. 138[39]
Suárez, E. 217[26]
Suárez-Pantiga, S. 594[28], 595[28]
Subrahmanyam, A. V. 125[122], 126[122]
Sucunza, D. 196[32], 196[33], 197[32], 197[33], 197[34]
Suda, K. 624[205]
Suda, M. 359[18]
Suehiro, K. 602[29]
Suenaga, K. 613[110], 626[110]
Suffert, J. 455[35], 455[36], 455[37], 455[38], 455[39], 456[39], 462[59], 463[59], 470[39], 472[59]
Suga, H. 282[31]
Sugaya, K. 614[118]
Sugihara, T. 602[23]

Sugimoto, K. 420[124], 421[124]
Sugimoto, M. 253[80]
Sugimoto, Y.-i. 249[49], 249[50], 249[51]
Sugiura, M. 577[3]
Sugizaki, K. 620[155]
Suhadolnik, J. C. 250[75]
Suhartono, M. 476[90], 477[90], 488[90]
Suhrbier, A. 626[213]
Sui, X. 478[91], 488[91]
Sukirthalingam, S. 449[10], 449[11], 449[12], 449[13]
Sumi, K. 365[34]
Sun, C. 395[92], 396[92], 397[92], 398[92], 428[133], 429[133], 430[135], 431[135]
Sun, H.-D. 612[107], 612[108], 612[109]
Sun, J. 541[27]
Sun, L. D. 272[11], 273[11]
Sun, T.-W. 500[136], 501[136], 505[136], 613[115]
Sun, X. 360[19], 426[130], 428[133], 429[133], 430[135], 431[135]
Sundén, H. 347[157], 348[157]
Sung, B. K. 602[28]
Surendra, K. 3[4], 17[40], 19[48], 20[48], 20[49]
Surry, D. S. 358[14], 381[14]
Suryavanshi, P. A. 589[21], 589[22], 590[21], 590[22]
Surzur, J. M. 218[57], 218[58]
Susanti, D. 565[94], 566[94]
Suszczyńska, A. 579[8]
Sutherland, K. P. 626[213]
Suyama, T. 620[155], 620[157]
Suzuki, A. 303[78], 304[78]
Suzuki, H. 391[86], 392[86], 604[40], 604[41]
Suzuki, I. 253[80]
Suzuki, N. 602[17]
Suzuki, T. 624[205]
Suzuki, Y. 540[25], 540[26], 541[25], 613[110], 626[110]
Svoboda, I. 246[42], 248[42]
Swallow, W. H. 245[34]
Swann, A. C. 620[166]
Sweany, R. L. 28[84]
Swenson, R. E. 32[125], 35[125]
Swern, D. 157[6], 160[6]
Syntrivanis, L.-D. 551[42], 552[42]
Szillat, H. 22[54], 129[128], 130[128]
Szpilman, A. M. 169[44], 169[45], 170[44], 170[45], 171[45]
Szymoniak, J. 598[34]

T

Tachibana, K. 250[72]
Tacke, R. 386[78]
Tadaki, M. 427[131], 427[132], 428[132]
Tadd, A. C. 484[102], 485[102], 490[102]
Taetle, R. 604[39]
Taguchi, H. 427[131], 427[132], 428[132]
Takagi, K. 602[19]

Takahashi, C. 617[128]
Takahashi, K. 535[11], 535[18]
Takahashi, N. 457[44]
Takahashi, O. 303[80], 304[80], 305[80]
Takahashi, T. 494[115], 496[115]
Takahashi, Y. 88[73], 89[73], 181[75], 181[79], 182[75], 383[69], 608[68], 626[220]
Takahata, H. 94[81], 94[82], 95[82]
Takai, K. 136[37], 137[37], 586[15]
Takai, T. 163[26], 163[27]
Takaishi, M. 250[71]
Takaku, K. 417[121], 418[121]
Takamatsu, K. 250[72]
Takamuku, S. 180[73]
Takamura, H. 142 47
Takaoka, D. 249[63]
Takasaki, M. 494[115], 496[115]
Takase, T. 249[61]
Takasu, K. 255[90], 256[90]
Takatani, M. 606[43], 606[49]
Takayama, H. 614[120]
Takeda, K. 254[87], 362[24], 362[25], 362[26], 362[27], 362[28], 363[25], 363[30], 363[31], 363[32], 363[33], 364[30], 364[32], 365[34], 365[35], 365[36], 365[37], 365[38], 366[38], 371[50], 371[51], 372[51], 372[52], 373[53], 383[69], 383[70], 383[71], 383[72], 383[73], 383[74], 383[75], 383[76], 384[76], 385[76], 385[77], 386[77], 516[28]
Takeda, M. 363[30], 363[31], 363[33], 364[30]
Takeda, T. 142[48], 380[67], 381[67], 399[95], 427[131], 427[132], 428[132]
Takemoto, Y. 255 90], 256[90]
Takeshita, H. 244 31], 260[107], 261[107]
Takeuchi, K. 188[6]
Takeuchi, M. 434[140], 435[140]
Takeuchi, S. 617[133]
Takimoto, M. 109[104], 110[104]
Talukdar, R. 593[27]
Tamai, T. 179[69], 180[69], 181[76]
Tamaru, Y. 500[129], 500[133]
Tan, C. 330[129], 331[129], 332[129], 337[129]
Tan, C.-H. 614[116]
Tan, D. S. 99[92]
Tan, E. H. P. 452[32]
Tanabe, G. 188[7]
Tanaka, K. 198[40], 365[37], 383[75]
Tanaka, M. 621[173]
Tanaka, T. 249[55], 371[50]
Tang, H. 620[170], 626[170]
Tang, J. 586[17], 587[17], 588[17]
Tang, L. 28[88], 28[90]
Tang, L.-F. 45[13], 46[13]
Tang, M. 269[3]
Tang, W. 313[97], 313[98], 313[100], 500[130], 501[130], 504[130]
Tang, X. C. 614[117], 614[118]

Tang, Y. 500[131], 500[135], 501[135], 504[135], 602[25]
Tang, Y.-F. 500[136], 501[136], 505[136], 613[112], 613[115]
Tang, Z.-Y. 17[41]
Tangirala, R. S. 236[102]
Tani, H. 302[76]
Taniguchi, H. 75[56], 78[56]
Taniguchi, T. 188[7], 188[8], 188[9]
Tanino, K. 626[219], 626[220]
Tanoury, G. J. 458[46]
Tansakul, C. 217[25]
Tantillo, D. J. 14[12], 35[136]
Tanuwidjaja, J. 49[18], 51[18], 52[21], 53[18], 53[21], 54[18], 54[21], 295[56], 296[56]
Tarr, J. C. 377[63], 378[63]
Tarrade, A. 527[55]
Tarrant, J. 258[96]
Tarselli, M. A. 554[47]
Tatami, A. 135[12], 143[12]
Tategami, S.-i. 175[55], 176[55], 177[55]
Tateyama, H. 94[82], 95[82]
Taubert, F. 431[136], 432[136]
Taylor, C. N. 201[55], 202[55], 203[55], 204[58], 205[58], 206[58], 207[58]
Taylor, R. J. K. 624[196]
Tebbe, F. N. 136[32], 137[32], 139[32]
Telfer, S. J. 24[67]
Teller, H. 549[40], 550[40]
Tellier, C. 411[112], 412[112]
Terasoma, N. 224[72]
Terfloth, L. 624[200]
Terui, Y. 302[76]
Thadani, A. N. 389[84], 390[84]
Thangadurai, D. T. 564[91]
Theodorakis, E. A. 615[124], 615[125], 618[143], 618[144]
Thiel, W. 549[40], 550[40]
Thoison, O. 297[59]
Thom, S. J. 258[96]
Thoma, G. 217[4]
Thomas, A. 607[62]
Thompson, A. L. 498[121], 498[122]
Thomson, J. J. 198[38]
Thomson, N. M. 188[10]
Thomson, R. J. 201[55], 202[55], 202[57], 203[55], 204[58], 205[58], 206[58], 207[58], 207[59]
Thorand, S. 142[47]
Thornton, A. R. 528[57], 529[57], 530[57], 530[58], 531[58], 532[57]
Thümmler, C. 444[154]
Tian, J. 580[9], 581[9]
Tian, Y. 615[122], 615[123]
Tiecco, M. 234[98]
Tietze, L. F. 67[1], 67[2], 67[5], 67[8], 187[4], 187[5], 217[3], 243[3], 243[4], 243[5], 355[3], 355[4], 355[5], 414[116], 449[1], 475[85], 476[85], 476[86], 476[87], 488[85], 491[111], 492[111], 495[111], 499[124], 502[138], 503[124], 503[138], 505[138], 511[1]
Timokhin, V. I. 217[11], 217[40]

Author Index

Tius, M. A. 392[87]
Tiwari, D. P. 593[26], 593[27], 594[26]
Tkatchouk, E. 555[54], 555[55], 556[54]
Toda, S. 237[104]
Todd, R. 547[36], 548[36], 549[36]
Todres, Z. V. 198[39]
Toganoh, M. 3[5], 4[5], 48[17], 50[17]
Tokan, W. M. 461[55]
Tokeshi, B. K. 392[87]
Tokiwano, T. 249[62], 249[64]
Tokuhisa, K. 626[214]
Tokumaru, K. 181[76]
Tokuyasu, T. 158[9], 163[24], 164[9], 164[24], 165[9], 165[24], 166[9], 166[24]
Tomé, A. C. 585[14], 586[14]
Tomita, T. 97[88], 97[89], 98[88], 98[89]
Tomo, Y. 399[94]
Tong, R. 58[33], 58[34], 58[35], 59[33], 59[34], 59[35], 60[33], 294[55], 430[134]
Tong, X. 486[105]
Tong, Z. 611[103]
Topczewski, J. J. 63[39], 63[40], 64[39], 64[40], 298[63], 298[64], 299[63], 299[64], 300[63]
Tosaki, S.-y. 582[10], 583[10]
Toscano, P. J. 360[22]
Toste, F. D. 22[58], 22[60], 535[2], 535[19], 538[22], 542[30], 555[51], 555[52], 555[54], 555[55], 556[54], 556[57], 557[68], 557[69], 557[71], 557[72], 567[99]
Toteva, M. M. 14[10]
Toullec, P. Y. 535[6], 564[6], 564[89]
Townsend, S. D. 261[114]
Toyata, E. 457[43], 458[43], 470[43]
Toyooka, N. 420[124], 421[124]
Trach, F. 217[4]
Trauner, D. 8[12], 9[12], 450[24], 450[26], 451[24], 451[29], 453[33], 454[33], 469[29], 469[33]
Treitler, D. S. 17[42], 17[43], 17[44], 17[45], 18[42], 18[46]
Tries, F. 408[108], 409[108]
Trifonov, L. S. 606[45]
Trofimov, A. 394[89]
Trost, B. M. 32[124], 34[132], 34[133], 35[124], 67[9], 67[11], 128[126], 129[126], 136[30], 136[31], 139[30], 139[31], 217[37], 259[101], 259[104], 449[18], 458[45], 458[46], 458[47], 460[51], 460[52], 471[52], 483[101], 484[101], 490[101], 491[109], 610[81], 610[84]
Trotter, J. W. 289[44]
Trout, R. E. L. 434[138], 435[138]
Truscott, F. M. 549[38]
Trzoss, L. 615[124], 615[125], 618[143]
Tsai, Y.-C. 545[33]
Tseng, C.-C. 458[48], 459[48], 471[48]
Tsubouchi, A. 380[67], 381[67], 399[95], 427[131], 427[132], 428[132]
Tsuchikawa, H. 454[34], 470[34]
Tsuda, M. 611[90], 611[91]

Tsuji, J. 493[113], 496[117], 496[118]
Tsuji, R. 582[10], 583[10]
Tsujimoto, G. 540[26]
Tsutsui, K. 76[59]
Tu, Y. 193[23], 244[21], 244[22]
Tu, Y. Q. 269[2], 269[4], 271[4], 272[10], 272[11], 273[11], 273[12], 274[12], 275[12], 276[12], 282[33], 282[34], 283[33], 283[34], 284[34], 284[36], 286[34]
Tu, Y.-Q. 269[1], 269[3], 270[8], 270[9], 271[8], 271[9], 275[13], 275[14], 275[15], 275[16], 275[17], 276[13], 282[32], 282[35], 283[35], 335[139], 336[139]
Tucker, J. W. 207[65], 207[66], 208[66], 209[66], 210[66]
Turner, C. I. 222[69], 223[69]
Tzeng, T. 227[79]

U

Ubayama, H. 362[26]
Uchiro, H. 306[83]
Ueba, C. 249[55]
Ueda, K. 285[41], 286[41]
Ueda, M. 224[72]
Uehara, H. 135[8], 135[11], 142[8], 142[11], 143[8], 143[11]
Uemura, D. 626[214]
Ueno, M. 256[94]
Ueno, Y. 623[193], 623[195]
Uenoyama, Y. 224[72]
Uesaka, N. 313[94], 313[95], 313[96], 313[99], 313[100], 314[94]
Ukai, A. 20[50], 21[50]
Ulman, M. 72[35]
Umezawa, T. 135[20], 142[20]
Underwood, B. S. 49[18], 51[18], 53[18], 54[18]
Undheim, K. 389[83]
Ungeheuer, F. 626[222], 627[222]
Unger, R. 375[56], 375[57], 375[58], 376[58]
Uotsu, K. 474[80]
Upadhyay, S. K. 188[15]
Urabe, D. 237[103], 238[103]
Urano, H. 244[30]
Urones, J. G. 246[37], 246[40], 256[93]
Utimoto, K. 136[37], 137[37], 393[88], 394[88], 416[119]
Uto, K. 492[112], 495[112], 615[126]
Utschick, G. 180[74], 181[74]
Uyehara, T. 253[80], 256[94], 256[95]

V

Vaillard, S. E. 217[24], 234[99]
Vaillard, V. A. 217[9]
Valdivia, M. 27[80], 28[80], 218[46]
Valentine, J. C. 49[19], 50[19], 51[19], 58[33], 59[33], 60[33], 294[55]
Valot, G. 35[135], 36[135]
Van der Eycken, E. V. 538[23]
Van de Weghe, P. 135[2]
Van Dyke, A. R. 56[27], 57[32], 58[32]
van Emster, K. 14[6], 15[6], 21[6]

Vanos, C. M. 150[55], 151[55], 152[55]
van Tamelen, E. E. 15[23], 15[24], 15[27], 290[47], 291[47]
Van Vranken, D. L. 34[132]
Vasas, A. 626[215]
Vassilikogiannakis, G. 1[1], 1[3], 2[3]
Vawter, E. J. 449[14], 449[15]
Velluz, A. 244[29]
Verlhac, J.-B. 388[82], 389[82]
Verma, V. A. 67[12]
Viana, G. B. 620[163]
Vicente, R. 520[34], 521[34]
Vidali, V. P. 1[3], 2[3]
Villa, M. 460[52], 471[52]
Villiger, V. 15[16], 15[17]
Vilotijevic, I. 43[1], 47[1], 56[26], 56[27], 57[26], 58[26], 245[35]
Visca, V. 10[14]
Vlaar, T. 449[6], 485[103]
Vo, N. T. 345[151]
Vogt, M. 376[59]
Volchkov, I. 90[74], 92[74]
Volkmann, R. A. 15[22]
Volla, C. M. R. 511[5]
Vollmar, A. M. 624[204]
Von Hoff, D. D. 604[138]
Vonwiller, S. C. 161[20]
von Zezschwitz, P. 461[56]
Vos, T. J. 72[36], 124[36]
Vu, A. T. 619[149]
Vyskocil, S. 313[95], 313[96], 313[97], 313[98], 313[100]

W

Wada, M. 402[101], 403[101]
Wada, Y. 76[59]
Waddell, T. G. 244[27]
Wagner, G. 254[83]
Wagner, H. 246[38], 247[43]
Wagner, R. D. 168[43], 169[43], 170[46]
Wagner, S. 624[200]
Wagstaff, K. E. 252[77]
Wakamatsu, T. 623[194]
Walavalkar, R. 301[72]
Waldecker, B. 502[138], 503[138], 505[138]
Waldmann, H. 67[3]
Walker, M. B. 620[162]
Wallace, D. J. 112[108]
Wallentin, C.-J. 218[47]
Walling, C. 168[41]
Walther, T. 431[136], 432[136]
Wan, H. 622[187]
Wan, K. K. 28[83], 29[83], 30[83], 166[37]
Wan, S. 54[23], 55[23]
Wang, B. 269[2]
Wang, B.-M. 282[32], 282[35], 283[35]
Wang, C. 618[145]
Wang, C.-H. 613[113], 613[114]
Wang, D. 561[83]
Wang, D. Z. 500[132]
Wang, G. 500[131]
Wang, G. C. 38[139]
Wang, G.-X. 617[139], 618[139]

Author Index

Wang, G. Y. 626[214]
Wang, H.-L. 256[92]
Wang, J. 347[156], 348[156]
Wang, J.-J. 260[107], 261[107]
Wang, K.-P. 80[69], 81[69], 82[69]
Wang, L. 586[17], 587[17], 588[17], 597[31]
Wang, L.-C. 28[100]
Wang, P. Z. 272[11], 273[11]
Wang, Q. 244[13], 566[96], 567[96], 568[96], 569[96]
Wang, R. T. 449[19]
Wang, S. 557[76], 557[78], 558[76], 563[78]
Wang, S.-H. 269[3], 270[9], 271[9]
Wang, T.-L. 246[36]
Wang, W. 347[156], 348[156], 555[53], 586[17], 587[17], 588[17], 604[39]
Wang, X. 3[5], 4[5], 28[100], 48[16], 48[17], 50[16], 50[17], 193[23], 249[53]
Wang, Y. 547[36], 548[36], 549[36], 556[59]
Wang, Y.-F. 613[111]
Wang, Z. 162[21], 500[131], 593[25]
Wang, Z.-X. 244[21], 244[22]
Wang, Z.-Y. 612[109]
Ward, S. A. 164[34]
Waring, M. J. 28[98], 28[99], 163[28], 163[29], 166[28], 166[29]
Wartchow, R. 227[77], 228[77]
Waser, J. 28[93], 28[101], 28[102], 28[103], 28[104], 28[105], 28[108], 29[93], 166[36]
Wasilke, J.-C. 67[6], 76[6]
Watanabe, D. 611[90]
Watanabe, H. 496[118]
Watanabe, K. 79[68], 244[12], 303[79], 303[80], 304[79], 304[80], 305[79], 305[80], 613[110], 626[110]
Watanabe, M. 142[48], 623[193], 623[195]
Watanuki, S. 68[18], 68[19], 68[20]
Wataya, Y. 158[9], 164[9], 165[9], 166[9], 181[78]
Waters, W. A. 234[98]
Watts, W. E. 601[1], 601[2]
Wayland, B. B. 28[97]
Webb, M. R. 195[31]
Weber, K. 390[85], 391[85]
Webster, R. 8[12], 9[12], 453[33], 454[33], 469[33]
Wei, H. 110[106]
Wei, X. 3[5], 4[5], 48[17], 50[17]
Wei, Y. 623[189]
Weingarten, M. D. 511[6], 512[6], 521[6]
Weinstabl, H. 106[101], 476[90], 477[90], 486[108], 488[90], 491[108]
Weintritt, H. 128[127]
Welch, J. T. 360[22]
Wellington, K. D. 609[76]
Wells, C. 611[99]
Wendel, G. H. 624[206]
Wender, P. A. 67[12], 612[105], 612[106]

Wendt, K. U. 13[3], 19[3], 21[3], 23[3], 24[3]
Wengryniuk, S. E. 357[11], 358[11]
Wenkert, E. 258[97]
Werness, J. B. 500[130], 501[130], 504[130]
West, F. 219[63]
West, F. G. 511[7]
West, M. C. 358[14], 381[14]
Westersund, M. 608[71]
Westley, J. W. 43[2], 243[8]
White, A. J. P. 246[37]
White, J. D. 261[116]
Whitty, A. 624[208]
Wicha, J. 399[96]
Wickman, H. B. 177[62], 178[62]
Widenhoefer, R. A. 22[57]
Wiemer, D. F. 63[38], 63[39], 63[40], 64[39], 64[40], 298[61], 298[62], 298[63], 298[64], 299[62], 299[63], 299[64], 300[63]
Wiener, J. J. M. 464[69], 465[69], 472[69]
Wierenga, W. 290[47], 291[47]
Wieteck, M. 545[34], 545[35], 546[34], 547[34], 548[34]
Wiley, M. R. 231[85]
Wille, U. 218[53]
Willett, J. D. 15[27]
Williams, D. J. 246[37]
Williams, J. L. 611[101]
Williams, J. P. 607[64], 607[65]
Williams, L. 499[125], 500[125], 503[125]
Williams, R. M. 611[100]
Williams, T. J. 612[106]
Williard, P. G. 617[135]
Willis, M. C. 467[74], 467[75], 467[76], 468[76], 473[76], 484[102], 485[102], 490[102]
Willson, M. 617[137]
Wilmot, I. D. 180[70]
Wilson, J. E. 198[45]
Wilson, S. R. 244[17], 399[93]
Wilson, W. K. 21[51]
Wimalasena, K. 177[62], 178[62]
Wink, D. J. 84[70], 395[92], 396[92], 397[92], 398[92]
Winkler, J. D. 626[221]
Winne, J. M. 188[11]
Winssinger, N. 35[135], 36[135]
Winter, C. A. 602[30]
Winter, P. 437[144], 438[144]
Witham, C. A. 557[72]
Woggon, W.-D. 349[161], 350[161]
Wolf, A. 177[64]
Wolfe, J. P. 463[62], 463[63], 463[64], 463[65], 463[66], 472[65], 478[92], 479[92], 479[93], 489[92], 489[93]
Woltering, T. J. 227[78]
Wong, F. T. 244[12]
Wong, H. N. C. 500[134], 501[134], 504[134]
Wong, J. C. Y. 138[39]
Wong, O. A. 43[6], 244[24]

Wong, Y.-H. 307[87], 308[87], 309[87], 310[87], 315[87]
Woo, S. K. 117[116], 117[117], 118[116], 118[117]
Wood, H. B., Jr. 15[13]
Wood, J. L. 225[73], 226[73], 227[73], 620[170], 626[170]
Woodward, C. C. 367[40]
Worakun, T. 449[10], 449[11]
Worrell, B. T. 513[23]
Wrobleski, A. 301[73]
Wu, B. 282[34], 283[34], 284[34], 286[34]
Wu, G.-z. 32[127], 449[17]
Wu, J. 148[54], 149[54], 249[65]
Wu, L. 361[23]
Wu, N. 500[132], 617[139], 618[139]
Wu, S. 586[17], 587[17], 588[17]
Wu, T. R. 305[82], 306[82]
Wu, W. 555[56]
Wu, X.-F. 466[71], 466[72]
Wu, Y. 234[97]
Wuest, W. M. 423[128], 424[128]
Wulff, J. E. 188[17], 189[17], 190[17]
Wulff, W. 500[127]

X

Xia, G. 465[70], 466[70], 472[70]
Xia, W. J. 282[33], 283[33]
Xian, M. 355[6], 356[6], 425[129], 426[129], 442[150]
Xiang, J. 613[114]
Xiang, K. 335[139], 336[139]
Xiao, P. 430[135], 431[135]
Xiao, Q. 500[136], 501[136], 505[136], 613[115]
Xiao, W.-J. 207[62], 345[155], 346[155], 347[155], 350[155]
Xiao, W.-L. 612[107], 612[108]
Xiao, X.-L. 179[68]
Xie, H. 347[156], 348[156]
Xie, L. 482[100], 483[100], 490[100]
Xie, X. 440[148], 441[148]
Xie, Y. 586[17], 587[17], 588[17]
Xiong, Q. B. 21[51]
Xiong, Z. 44[8], 247[46], 248[46], 290[49], 291[49], 296[49]
Xu, B. 555[53]
Xu, J. 615[124], 615[125], 618[143], 618[144]
Xu, L. 202[56]
Xu, L.-M. 500[136], 501[136], 505[136], 613[111], 613[115]
Xu, P.-F. 620[158]
Xu, S. L. 75[53]
Xu, X. 482[100], 483[100], 490[100], 522[39]
Xu, X.-S. 500[134], 501[134], 504[134]
Xu, Y. 355[7], 356[7]
Xuan, J. 207[62]

Y

Yabe, Y. 28[112], 28[113]
Yadav, J. S. 246[41]
Yaegashi, K. 249[59], 249[66]
Yagi, S. 302[76]

Y

Yamada, A. 624[205]
Yamada, J. 256[95]
Yamada, K. 457[44]
Yamada, M. 602[23]
Yamada, T. 28[82], 163[26], 163[27], 175[55], 175[58], 176[55], 177[55], 617[128]
Yamada, Y. M. A. 313[97], 313[98], 313[101], 314[101], 578[4]
Yamagiwa, N. 580[9], 581[9]
Yamaguchi, K. 359[18], 362[24], 365[37], 365[38], 366[38], 372[52], 383[69], 383[70], 383[71], 383[72], 383[74], 383[76], 384[76], 385[76], 385[77], 386[77], 579[6]
Yamaguchi, M. 602[23]
Yamaguchi, S. 249[52]
Yamaki, Y. 378[64]
Yamamoto, H. 19[47], 318[111], 318[112], 318[113], 318[114], 319[111], 319[112], 320[111], 320[114], 601[14]
Yamamoto, K. 399[94]
Yamamoto, Y. 75[44], 142[47], 253[80], 256[95], 535[11], 535[12], 535[14], 535[16], 535[17], 535[18], 535[19]
Yamanishi, K. 622[186]
Yamaoka, M. 193[24]
Yamashita, S. 135[12], 143[12]
Yamawaki, K. 362[27]
Yan, L. 426[130], 428[133], 429[133]
Yanada, R. 457[42], 457[43], 457[44], 458[42], 458[43], 470[43]
Yanagisawa, S. 226[76]
Yang, C.-S. 615[121], 621[172], 621[173]
Yang, C.-Y. 550[41], 551[41]
Yang, D. X. 29[115]
Yang, L.-B. 612[107]
Yang, L. M. 272[10]
Yang, M. G. 15[37]
Yang, P. 45[13], 46[13]
Yang, X. 611[103]
Yang, Y. 616[127], 622[177]
Yang, Y.-H. 217[14]
Yang, Z. 110[106], 330[129], 331[129], 332[129], 337[129], 500[131], 500[132], 500[135], 500[136], 501[135], 501[136], 504[135], 505[136], 602[25], 613[111], 613[112], 613[113], 613[114], 613[115], 617[139], 618[139], 619[146]
Yanik, M. M. 464[68]
Yao, C.-F. 227[79]
Yao, X. 538[24], 539[24]
Yao, Y.-S. 348[158]
Yao, Z.-J. 348[158]
Yasumoto, T. 313[92], 313[93]
Yasutake, M. 175[57]
Yata, H. 43[7], 44[7]
Yates, C. M. 370[49], 371[49]
Yates, J. B. 231[85]
Ye, L. 555[56]
Ye, Q.-D. 500[136], 501[136], 505[136], 613[115]
Ye, T. 512[9]
Ye, X. 561[83]
Yen, C.-F. 607[66]

Yen, J. R. 328[127], 329[127], 330[127], 337[127]
Yeoman, J. T. S. 197[36]
Yepremyan, A. 547[36], 548[36], 549[36]
Yin, J. 618[145]
Yin, R. 193[23]
Yin, Z. 355[7], 356[7]
Yoder, B. J. 297[60]
Yoder, R. A. 14[5]
Yokoe, H. 617[131]
Yokota, T. 626[214]
Yokoyama, R. 615[121], 621[172]
Yokoyama, S. 232[88]
Yonehara, H. 610[83], 617[133]
Yoo, S.-E. 602[21]
Yoon, H. 486[108], 491[108]
Yoon, T. P. 182[81], 183[81], 207[64], 211[67], 211[68], 211[69], 212[69]
Yoshida, J.-I. 174[52], 175[52]
Yoshida, M. 324[124], 325[124], 617[131]
Yoshida, T. 602[26], 620[26], 620[171], 621[26]
Yoshida, Y. 285[40], 285[41], 286[40], 286[41]
Yoshida, Z.-i. 500[129], 500[133]
Yoshii, E. 254[87], 362[24], 362[25], 362[26], 362[28], 363[25], 363[30], 363[31], 364[30]
Yoshikawa, N. 577[2]
Yoshimitsu, T. 226[75], 226[76]
Yoshimitsu, Y. 496[119], 497[119], 497[120], 498[119], 498[120]
Yoshimoto, J. U. N. 302[76]
Yoshimura, Y. 94[81]
Yoshitomi, Y. 348[159], 350[159]
You, L. 500[136], 501[136], 505[136], 613[115]
You, Z.-J. 613[114]
Young, I. S. 611[94], 611[101]
Yousfi, M. 233[91]
Yu, C.-M. 620[154]
Yu, F. 622[177]
Yu, H.-X. 618[142]
Yu, J. 597[31]
Yu, L.-C. 258[99]
Yu, M. 561[84], 561[85]
Yu, R.-C. 613[111]
Yu, W. 117[116], 117[117], 118[116], 118[117]
Yu, X.-Q. 527[53]
Yu, Z.-X. 325[125], 326[125], 327[125], 328[125]
Yuan, C. 325[125], 326[125], 327[125], 328[125]
Yuan, W. 623[190]
Yudin, A. K. 244[25], 527[44]
Yue, E. W. 135[5], 140[5]
Yue, G.-Z. 617[139], 618[139]
Yun, H. 99[92]
Yun, S. Y. 80[69], 81[69], 82[69], 84[70]
Yusuff, N. 620[170], 626[170]
Yuuya, S. 624[205]

Z

Zabicky, J. 157[5], 160[5]
Zagorski, M. G. 243[10]
Zahler, R. 136[33], 137[33], 139[33]
Zard, S. Z. 188[13], 188[14], 217[22], 217[33], 217[39], 233[91], 261[115], 262[115], 604[43], 605[43], 606[50]
Zask, A. 500[127]
Zavalij, P. 522[40]
Zavalij, P. Y. 522[39]
Zeng, Z. 535[10]
Zeng, R. 193[22]
Zeng, Y. 341[145], 341[146], 341[147], 342[145], 342[147], 343[145], 343[147]
Zhai, H. 611[103], 616[127], 620[158], 622[177]
Zhan, W. 305[82], 306[82]
Zhang, F. 273[12], 274[12], 275[12], 276[12]
Zhang, F. M. 282[33], 283[33]
Zhang, F.-M. 270[8], 270[9], 271[8], 271[9], 275[14], 275[15], 275[16], 275[17]
Zhang, G. 556[59], 557[80], 557[81], 558[80], 559[80], 560[80], 561[84], 561[85], 562[88]
Zhang, H. 191[18], 191[19], 192[18]
Zhang, H.-J. 275[14]
Zhang, J. 15[33], 148[54], 149[54], 422[127], 423[127]
Zhang, J.-R. 244[22]
Zhang, L. 22[56], 263[117], 547[36], 548[36], 549[36], 556[59], 557[76], 557[77], 557[78], 557[80], 557[81], 558[76], 558[80], 559[80], 560[80], 561[84], 561[85], 562[88], 563[78], 586[17], 587[17], 588[17]
Zhang, N. 464[67], 500[67], 500[126], 501[126], 503[126]
Zhang, Q. 9[13], 275[13], 276[13]
Zhang, Q.-W. 275[14], 335[139], 336[139]
Zhang, S.-Y. 275[17], 335[139], 336[139]
Zhang, T. C. 335[139], 336[139]
Zhang, W. 244[17], 244[18], 330[129], 331[129], 332[129], 337[129], 535[17]
Zhang, X.-M. 270[8], 270[9], 271[8], 271[9], 335[139], 336[139]
Zhang, Y. 32[126], 135[19], 135[23], 146[19], 147[23], 202[56], 430[135], 431[135], 449[16], 449[17], 500[135], 501[135], 504[135], 522[41]
Zhang, Y.-D. 602[25], 613[112]
Zhang, Y.-J. 500[134], 501[134], 504[134]
Zhang, Y.-Q. 275[15], 275[16]
Zhao, C. 440[148], 441[148]
Zhao, G.-L. 232[89], 347[157], 348[157]
Zhao, H. 94[79], 588[18]
Zhao, J.-F. 60[36], 61[36], 62[36]
Zhao, M.-Z. 500[136], 501[136], 505[136], 613[115]
Zhao, X. Z. 282[34], 283[34], 284[34], 284[36], 286[34]
Zhao, X.-Z. 282[35], 283[35]

Author Index

Zhao, Y. 69[26], 78[26], 307[87], 308[87], 309[87], 310[87], 315[87]
Zhao, Y.-J. 60[36], 61[36], 62[36], 311[88], 311[89], 311[90], 311[91], 312[88], 313[91]
Zhao, Y.-M. 273[12], 274[12], 275[12], 275[13], 275[14], 276[12], 276[13], 335[139], 336[139]
Zheng, Q.-T. 612[107]
Zhou, J. 135[23], 146[50], 147[23]
Zhou, L. 193[23], 512[10]
Zhou, Q. 334[136], 335[136]
Zhou, X.-G. 527[53]
Zhou, Z. 228[80], 229[80]

Zhu, B. 618[142]
Zhu, D.-Y. 614[116]
Zhu, H. 230[83], 231[83]
Zhu, M. 597[31]
Zhu, R. 478[91], 488[91]
Zhu, S. 76[58], 555[56], 622[177]
Zhu, W. 434[142], 435[142]
Zhu, X. 94[83]
Zhu, X. D. 614[117]
Zhu, Y. 244[13]
Zhu, Y.-Y. 610[88]
Zhuang, L. 434[138], 434[139], 434[141], 435[138], 435[139], 435[141]
Zibinsky, M. 513[22]

Zilch, H. 386[78]
Ziller, J. W. 136[26], 136[27]
Zimmerman, J. 217[2]
Zinngrebe, J. 499[124], 503[124]
Zipkin, R. E. 450[25]
Zou, G. 176[60], 177[60]
Zou, N. 313[99]
Zou, Y. 566[96], 567[96], 568[96], 569[96]
Zu, L. 347[156], 348[156]
Zuccarello, J. L. 554[47]
Zuraw, P. J. 158[12]

Abbreviations

Chemical

Name Used in Text	Abbreviation Used in Tables and on Arrow in Schemes	Abbreviation Used in Experimental Procedures
(*R*)-1-amino-2-(methoxymethyl)pyrrolidine	RAMP	RAMP
(*S*)-1-amino-2-(methoxymethyl)pyrrolidine	SAMP	SAMP
ammonium cerium(IV) nitrate	CAN	CAN
2,2′-azobisisobutyronitrile	AIBN	AIBN
barbituric acid	BBA	BBA
benzyltriethylammonium bromide	TEBAB	TEBAB
benzyltriethylammonium chloride	TEBAC	TEBAC
N,*O*-bis(trimethylsilyl)acetamide	BSA	BSA
9-borabicyclo[3.3.1]nonane	9-BBNH	9-BBNH
borane–methyl sulfide complex	BMS	BMS
N-bromosuccinimide	NBS	NBS
tert-butyldimethylsilyl chloride	TBDMSCl	TBDMSCl
tert-butyl peroxybenzoate	TBPB	*tert*-butyl peroxybenzoate
10-camphorsulfonic acid	CSA	CSA
chlorosulfonyl isocyanate	CSI	chlorosulfonyl isocyanate
3-chloroperoxybenzoic acid	MCPBA	MCPBA
N-chlorosuccinimide	NCS	NCS
chlorotrimethylsilane	TMSCl	TMSCl
1,4-diazabicyclo[2.2.2]octane	DABCO	DABCO
1,5-diazabicyclo[4.3.0]non-5-ene	DBN	DBN
1,8-diazabicyclo[5.4.0]undec-7-ene	DBU	DBU
dibenzoyl peroxide	DBPO	dibenzoyl peroxide
dibenzylideneacetone	dba	dba
di-*tert*-butyl azodicarboxylate	DBAD	di-*tert*-butyl azo-dicarboxylate
di-*tert*-butyl peroxide	DTBP	DTBP
2,3-dichloro-5,6-dicyanobenzo-1,4-quinone	DDQ	DDQ
dichloromethyl methyl ether	DCME	DCME
dicyclohexylcarbodiimide	DCC	DCC
N,*N*-diethylaminosulfur trifluoride	DAST	DAST
diethyl azodicarboxylate	DEAD	DEAD
diethyl tartrate	DET	DET
2,2′-dihydroxy-1,1′-binaphthyllithium aluminum hydride	BINAL-H	BINAL-H
diisobutylaluminum hydride	DIBAL-H	DIBAL-H
diisopropyl tartrate	DIPT	DIPT

Chemical (cont.)

Name Used in Text	Abbreviation Used in Tables and on Arrow in Schemes	Abbreviation Used in Experimental Procedures
1,2-dimethoxyethane	DME	DME
dimethylacetamide	DMA	DMA
dimethyl acetylenedicarboxylate	DMAD	DMAD
2-(dimethylamino)ethanol	Me$_2$N(CH$_2$)$_2$OH	2-(dimethylamino)ethanol
4-(dimethylamino)pyridine	DMAP	DMAP
dimethylformamide	DMF	DMF
dimethyl sulfide	DMS	DMS
dimethyl sulfoxide	DMSO	DMSO
1,3-dimethyl-3,4,5,6-tetrahydro-pyrimidin-2(1H)-one	DMPU	DMPU
ethyl diazoacetate	EDA	EDA
ethylenediaminetetraacetic acid	edta	edta
hexamethylphosphoric triamide	HMPA	HMPA
hexamethylphosphorous triamide	HMPT	HMPT
iodomethane	MeI	MeI
N-iodosuccinimide	NIS	NIS
lithium diisopropylamide	LDA	LDA
lithium hexamethyldisilazanide	LiHMDS	LiHMDS
lithium isopropylcyclohexylamide	LICA	LICA
lithium 2,2,6,6-tetramethylpiperidide	LTMP	LTMP
lutidine	lut	lut
methylaluminum bis(2,6-di-*tert*-butyl-4-methyl-phenoxide)	MAD	MAD
methyl ethyl ketone	MEK	methyl ethyl ketone
methylmaleimide	NMM	NMM
4-methylmorpholine N-oxide	NMO	NMO
1-methylpyrrolidin-2-one	NMP	NMP
methyl vinyl ketone	MVK	methyl vinyl ketone
petroleum ether	PE[a]	petroleum ether
N-phenylmaleimide	NPM	NPM
polyphosphoric acid	PPA	PPA
polyphosphate ester	PPE	polyphosphate ester
potassium hexamethyldisilazanide	KHMDS	KHMDS
pyridine	pyridine[b]	pyridine
pyridinium chlorochromate	PCC	PCC
pyridinium dichromate	PDC	PDC
pyridinium 4-toluenesulfonate	PPTS	PPTS
sodium bis(2-methoxyethoxy)aluminum hydride	Red-Al	Red-Al
tetrabutylammonium bromide	TBAB	TBAB

[a] Used to save space; abbreviation must be defined in a footnote.
[b] py used on arrow in schemes.

Abbreviations

Chemical (cont.)

Name Used in Text	Abbreviation Used in Tables and on Arrow in Schemes	Abbreviation Used in Experimental Procedures
tetrabutylammonium chloride	TBACl	TBACl
tetrabutylammonium fluoride	TBAF	TBAF
tetrabutylammonium iodide	TBAI	TBAI
tetracyanoethene	TCNE	tetracyanoethene
tetrahydrofuran	THF	THF
tetrahydropyran	THP	THP
2,2,6,6-tetramethylpiperidine	TMP	TMP
trimethylamine *N*-oxide	TMANO	trimethylamine *N*-oxide
N,N,N′,N′-tetramethylethylenediamine	TMEDA	TMEDA
tosylmethyl isocyanide	TosMIC	TosMIC
trifluoroacetic acid	TFA	TFA
trifluoroacetic anhydride	TFAA	TFAA
trimethylsilyl cyanide	TMSCN	TMSCN

Ligands

acetylacetonato	acac
2,2′-bipyridyl	bipy
1,2-bis(dimethylphosphino)ethane	DMPE
2,3-bis(diphenylphosphino)bicyclo[2.2.1]hept-5-ene	NORPHOS
2,2′-bis(diphenylphosphino)-1,1′-binaphthyl	BINAP
1,2-bis(diphenylphosphino)ethane	dppe (not diphos)
1,1′-bis(diphenylphosphino)ferrocene	dppf
bis(diphenylphosphino)methane	dppm
1,3-bis(diphenylphosphino)propane	dppp
1,4-bis(diphenylphosphino)butane	dppb
2,3-bis(diphenylphosphino)butane	Chiraphos
bis(salicylidene)ethylenediamine	salen
cyclooctadiene	cod
cyclooctatetraene	cot
cyclooctatriene	cte
η^5-cyclopentadienyl	Cp
dibenzylideneacetone	dba
6,6-dimethylcyclohexadienyl	dmch
2,4-dimethylpentadienyl	dmpd
ethylenediaminetetraacetic acid	edta
isopinocampheyl	Ipc
2,3-*O*-isopropylidene-2,3-dihydroxy-1,4-bis(diphenylphosphino)butane	Diop
norbornadiene (bicyclo[2.2.1]hepta-2,5-diene)	nbd
η^5-pentamethylcyclopentadienyl	Cp*

Radicals

acetyl	Ac
aryl	Ar
benzotriazol-1-yl	Bt
benzoyl	Bz
benzyl	Bn
benzyloxycarbonyl	Cbz
benzyloxymethyl	BOM
9-borabicyclo[3.3.1]nonyl	9-BBN
tert-butoxycarbonyl	Boc
butyl	Bu
sec-butyl	*s*-Bu
tert-butyl	*t*-Bu
tert-butyldimethylsilyl	TBDMS
tert-butyldiphenylsilyl	TBDPS
cyclohexyl	Cy
3,4-dimethoxybenzyl	DMB
ethyl	Et
ferrocenyl	Fc
9-fluorenylmethoxycarbonyl	Fmoc
isobutyl	iBu
mesityl	Mes
mesyl	Ms
4-methoxybenzyl	PMB
(2-methoxyethoxy)methyl	MEM
methoxymethyl	MOM
methyl	Me
4-nitrobenzyl	PNB
phenyl	Ph
phthaloyl	Phth
phthalimido	NPhth
propyl	Pr
isopropyl	iPr
tetrahydropyranyl	THP
tolyl	Tol
tosyl	Ts
triethylsilyl	TES
triflyl, trifluoromethanesulfonyl	Tf
triisopropylsilyl	TIPS
trimethylsilyl	TMS
2-(trimethylsilyl)ethoxymethyl	SEM
trityl [triphenylmethyl]	Tr

General

absolute	abs
anhydrous	anhyd
aqueous	aq
boiling point	bp
catalyst	no abbreviation
catalytic	cat.
chemical shift	δ
circular dichroism	CD
column chromatography	no abbreviation
concentrated	concd
configuration (in tables)	Config
coupling constant	J
day	d
density	d
decomposed	dec
degrees Celsius	°C
diastereomeric ratio	dr
dilute	dil
electron-donating group	EDG
electron-withdrawing group	EWG
electrophile	E^+
enantiomeric excess	ee
enantiomeric ratio	er
equation	eq
equivalent(s)	equiv
flash-vacuum pyrolysis	FVP
gas chromatography	GC
gas chromatography-mass spectrometry	GC/MS
gas–liquid chromatography	GLC
gram	g
highest occupied molecular orbital	HOMO
high-performance liquid chromatography	HPLC
hour(s)	h
infrared	IR
in situ	in situ
in vacuo	in vacuo
lethal dosage, e.g. to 50% of animals tested	LD_{50}
liquid	liq
liter	L
lowest unoccupied molecular orbital	LUMO
mass spectrometry	MS
medium-pressure liquid chromatography	MPLC
melting point	mp
milliliter	mL
millimole(s)	mmol
millimoles per liter	mM
minute(s)	min
mole(s)	mol
nuclear magnetic resonance	NMR
nucleophile	Nu^-
optical purity	op
phase-transfer catalysis	PTC
proton NMR	^{1}H NMR

General (cont.)

quantitative	quant
reference (in tables)	Ref
retention factor (for TLC)	R_f
retention time (chromatography)	t_R
room temperature	rt
saturated	sat.
solution	soln
temperature (in tables)	Temp (°C)
thin layer chromatography	TLC
ultraviolet	UV
volume (literature)	Vol.
via	via
vide infra	*vide infra*
vide supra	*vide supra*
yield (in tables)	Yield (%)

List of All Volumes

Science of Synthesis, Houben–Weyl Methods of Molecular Transformations

Category 1: Organometallics

1   Compounds with Transition Metal—Carbon π-Bonds and Compounds of Groups 10 – 8 (Ni, Pd, Pt, Co, Rh, Ir, Fe, Ru, Os)

2   Compounds of Groups 7–3 (Mn···, Cr···, V···, Ti···, Sc···, La···, Ac···)

3   Compounds of Groups 12 and 11 (Zn, Cd, Hg, Cu, Ag, Au)

4   Compounds of Group 15 (As, Sb, Bi) and Silicon Compounds

5   Compounds of Group 14 (Ge, Sn, Pb)

6   Boron Compounds

7   Compounds of Groups 13 and 2 (Al, Ga, In, Tl, Be ··· Ba)

8a Compounds of Group 1 (Li ··· Cs)

8b Compounds of Group 1 (Li ··· Cs)

Category 2: Hetarenes and Related Ring Systems

9   Fully Unsaturated Small-Ring Heterocycles and Monocyclic Five-Membered Hetarenes with One Heteroatom

10 Fused Five-Membered Hetarenes with One Heteroatom

11 Five-Membered Hetarenes with One Chalcogen and One Additional Heteroatom

12 Five-Membered Hetarenes with Two Nitrogen or Phosphorus Atoms

13 Five-Membered Hetarenes with Three or More Heteroatoms

14 Six-Membered Hetarenes with One Chalcogen

15 Six-Membered Hetarenes with One Nitrogen or Phosphorus Atom

16 Six-Membered Hetarenes with Two Identical Heteroatoms

17 Six-Membered Hetarenes with Two Unlike or More than Two Heteroatoms and Fully Unsaturated Larger-Ring Heterocycles

Category 3: Compounds with Four and Three Carbon—Heteroatom Bonds

18 Four Carbon—Heteroatom Bonds: X—C≡X, X=C=X, X_2C=X, CX_4

19 Three Carbon—Heteroatom Bonds: Nitriles, Isocyanides, and Derivatives

20a Three Carbon—Heteroatom Bonds: Acid Halides; Carboxylic Acids and Acid Salts

20b Three Carbon—Heteroatom Bonds: Esters and Lactones; Peroxy Acids and R(CO)OX Compounds; R(CO)X, X = S, Se, Te

21 Three Carbon—Heteroatom Bonds: Amides and Derivatives; Peptides; Lactams

22 Three Carbon—Heteroatom Bonds: Thio-, Seleno-, and Tellurocarboxylic Acids and Derivatives; Imidic Acids and Derivatives; Ortho Acid Derivatives

23 Three Carbon—Heteroatom Bonds: Ketenes and Derivatives

24 Three Carbon—Heteroatom Bonds: Ketene Acetals and Yne—X Compounds

Category 4: Compounds with Two Carbon—Heteroatom Bonds

25 Aldehydes

26 Ketones

27 Heteroatom Analogues of Aldehydes and Ketones

28 Quinones and Heteroatom Analogues

29 Acetals: Hal/X and O/O, S, Se, Te

30 Acetals: O/N, S/S, S/N, and N/N and Higher Heteroatom Analogues

31a Arene—X (X = Hal, O, S, Se, Te)

31b Arene—X (X = N, P)

32 X—Ene—X (X = F, Cl, Br, I, O, S, Se, Te, N, P), Ene—Hal, and Ene—O Compounds

33 Ene—X Compounds (X = S, Se, Te, N, P)

Category 5: Compounds with One Saturated Carbon—Heteroatom Bond

34 Fluorine

35 Chlorine, Bromine, and Iodine

36 Alcohols

37 Ethers

38 Peroxides

39 Sulfur, Selenium, and Tellurium

40a Amines and Ammonium Salts

40b Amine N-Oxides, Haloamines, Hydroxylamines and Sulfur Analogues, and Hydrazines

41 Nitro, Nitroso, Azo, Azoxy, and Diazonium Compounds, Azides,
Triazenes, and Tetrazenes

42 Organophosphorus Compounds (incl. RO—P and RN—P)

Category 6: Compounds with All-Carbon Functions

43 Polyynes, Arynes, Enynes, and Alkynes

44 Cumulenes and Allenes

45a Monocyclic Arenes, Quasiarenes, and Annulenes

45b Aromatic Ring Assemblies, Polycyclic Aromatic Hydrocarbons, and Conjugated Polyenes

46 1,3-Dienes

47a Alkenes

47b Alkenes

48 Alkanes